Biology of Sharks and Their Relatives

Second Edition

CRC
MARINE BIOLOGY
SERIES

The late Peter L. Lutz, Founding Editor
David H. Evans, Series Editor

PUBLISHED TITLES

Biology of Marine Birds
E. A. Schreiber and Joanna Burger

Biology of the Spotted Seatrout
Stephen A. Bortone

The Biology of Sea Turtles, Volume II
Peter L. Lutz, John A. Musick, and Jeanette Wyneken

Early Stages of Atlantic Fishes: An Identification Guide
for the Western Central North Atlantic
William J. Richards

The Physiology of Fishes, Third Edition
David H. Evans and James B. Claiborne

Biology of the Southern Ocean, Second Edition
George A. Knox

Biology of the Three-Spined Stickleback
Sara Östlund-Nilsson, Ian Mayer, and Felicity Anne Huntingford

Biology and Management of the World Tarpon and Bonefish Fisheries
Jerald S. Ault

Methods in Reproductive Aquaculture: Marine and Freshwater Species
Elsa Cabrita, Vanesa Robles, and Paz Herráez

Sharks and Their Relatives II: Biodiversity, Adaptive Physiology, and Conservation
Jeffrety C. Carrier, John A. Musick, and Michael R. Heithaus

Artificial Reefs in Fisheries Management
Stephen A. Bortone, Frederico Pereira Brandini, Gianna Fabi, and Shinya Otake

Biology of Sharks and Their Relatives, Second Edition
Jeffrey C. Carrier, John A. Musick, and Michael R. Heithaus

Biology of Sharks and Their Relatives

Second Edition

Edited by
Jeffrey C. Carrier
John A. Musick
Michael R. Heithaus

CRC Press
Taylor & Francis Group
Boca Raton   London   New York

CRC Press is an imprint of the
Taylor & Francis Group, an **informa** business

CRC Press
Taylor & Francis Group
6000 Broken Sound Parkway NW, Suite 300
Boca Raton, FL 33487-2742

© 2012 by Taylor & Francis Group, LLC
CRC Press is an imprint of Taylor & Francis Group, an Informa business

No claim to original U.S. Government works

Printed on acid-free paper
Version Date: 20150130

International Standard Book Number-13: 978-1-4398-3924-9 (Hardback)

Library of Congress Cataloging-in-Publication Data

Biology of sharks and their relatives / editors, Jeffrey C. Carrier, John A. Musick, Michael R. Heithaus. -- 2nd ed.
    p. cm. -- (Marine biology ; 13)
  Includes bibliographical references and index.
  ISBN 978-1-4398-3924-9
  1. Chondrichthyes. I. Carrier, Jeffrey C. II. Musick, John A. III. Heithaus, Michael R.

QL638.6.B56 2012
597.3--dc23
                        2011034454

Visit the Taylor & Francis Web site at
http://www.taylorandfrancis.com

and the CRC Press Web site at
http://www.crcpress.com

We, the editors, are profoundly appreciative of the enduring support and patience of our wives through all our endeavors, scientific and otherwise. When not with us in the field or lab as colleagues, photographers, fishers, or taggers, they have provided all aspects of support for our work and have tolerated long and recurring absences while we pursued our fascination with these fishes. To Carol, Bev, and Linda, we thank you for all that you do, for all of the support through the years, and for being such willing partners in our research, our writing, and our careers. This volume is dedicated to you.

Contents

Section I Phylogeny and Zoogeography

Section II Form, Function, and Physiological Processes

Section III Ecology and Life History

Preface

We developed the content and published *Biology of Sharks and Their Relatives* (Volume I) because of a need for an updated, timely reference volume on the biology of sharks, skates, rays, and chimaeras. In the Preface to that first volume, we noted that little in the way of comprehensive summaries of chondrichthyan biology had been done since the volumes of research papers produced by Perry Gilbert in 1963 and 1967. *Sharks and Survival* and *Sharks, Skates, and Rays*, at that time, provided a comprehensive examination of shark research and served as points of departure for future studies. It was over 20 years after the publication of Dr. Gilbert's edited volumes that *Elasmobranchs as Living Resources*, edited by Wes Pratt, Sonny Gruber, and Toru Taniuchi, was published in 1990, followed nearly a decade later by Will Hamlett's extensive review (1999) of the anatomy and fine structure of elasmobranch fishes.

Much has changed in the world of elasmobranch biology since then. When we first considered developing a modern synthesis of the biology of sharks and their relatives, we were forced to look at what major changes have occurred in our world, how those changes have influenced the worldwide status of sharks and their relatives, and how advances in technology and analytical techniques have changed, not only how we approach problem solving and scientific investigations, but how we formulate questions.

At the time of publication of the first edition, we identified at least three major influences that were having a profound influence on more modern approaches to studies of these animals. Perhaps foremost among them is the tremendous interest in sharks and their relatives by the public, perhaps influenced by their main character roles in movies and popular literature. Such an interest has resulted in the development of public displays and public encounter exhibits that cater to our curiosity about these animals. Captive facilities have increased their basic research components in order to develop better ways to maintain these animals in captive environments and to create aquatic "petting zoos" where rays are stroked and fed by hand. Shark and ray dive adventures have proliferated, and public interest continues to help drive basic research into aspects of shark behavior. In some regards, we have seen a shift from blind fascination with shark attacks to a greater interest in the intricacies of their lives. The media fascination never seems to dwindle, however, and stories of predators driven to maniacal attacks on humans still sell papers and television shows, testifying to our more morbid interests in these animals, despite our emerging understanding of the natural behavior of these predators.

A second factor is the significant commercial value of these animals and the resultant worldwide threat to populations that is a result of commercial overexploitation. We have seen areas where populations of sharks and rays have been so reduced that encounters are now almost nonexistent. This forced biologists studying these animals to dramatically increase their focus on studies of life histories. As we developed a better understanding of age, growth, and reproduction, we discovered that these animals, who survived so well for 400 million years, do not possess high rates of natural replenishment. Partially due to this low reproductive rate, the past decade has seen a tremendous increase in conservation and management initiatives around the world that hope to recover depleted populations.

Finally, virtually every area of research associated with these animals has been strongly impacted by the revolutionary growth in technology, and the questions we can now ask are very different than those reported in Gilbert's work not so long ago. A careful reading of the chapters we have presented in this work will show conclusions based on emergent technologies that have revealed some long-hidden secrets of these animals. Modern immunological and genetic techniques, satellite telemetry and archival tagging, modern phylogenetic analysis, GIS, and bomb radiation dating are just a few of the techniques and procedures that have become a part of our investigative lexicon.

We were pleased with the acceptance of the first edition of *Biology of Sharks and Their Relatives* and responded to the call for additional material with what was to become Volume II, *Sharks and Their Relatives II: Ecology, Physiological Adaptations, and Conservation*, a volume that assembled an entirely new collection of authors recognized for their expertise in fields of study that we were unable to cover with the first collection.

In this second edition of Volume I, we have again assembled a respected group of authors and have asked that they update the information presented almost ten years ago with chapters that assess advances in their respective areas of expertise. In some cases, we have eliminated chapters and added new chapters that show emerging fields of interest and technology that will serve to offer new insights into the biology and behavior of these animals. Further, we have enlisted different sets of authors for some chapters in order to present different approaches and different perspectives for some areas covered by other authors in the first edition.

We have continued our practice of soliciting chapters from some of the most eminent chondrichthyan biologists in the field, as well as some of its most promising "rising stars." We hope that these works will not only provide a synopsis of our current understanding of elasmobranchs and how the fields have changed since the first edition was published in 2004 but also show the continuing gaps in our knowledge and help to stimulate further studies.

Acknowledgments

The editors wish to express their deep gratitude to John Sulzycki, senior editor with CRC Press/Taylor & Francis, for his timely advice and encouragement; for his role as mentor, editor, and referee; and for supporting our efforts through three volumes. You have earned your honorary dorsal fin. We are also deeply grateful to Jessica Vakili for her skills in translating our rough manuscripts into a wonderfully attractive and final work. Her attention to detail and quality is unmatched.

This work represents VIMS Contribution No. 3161 from Virginia Institute of Marine Science, Contribution No. 50 from the Shark Bay Ecosystem Research Project (www.sberp.org), and Contribution No. 1205 from Mote Marine Laboratory.

Editors

Jeffrey C. Carrier, PhD, is professor emeritus of biology at Albion College, Michigan, where he was a faculty member from 1979 to 2010. He earned a bachelor of science degree in biology in 1970 from the University of Miami and completed a doctorate in biology from the University of Miami in 1974. While at Albion College, Dr. Carrier received multiple awards for teaching and scholarship and held the A. Merton Chickering and W.W. Diehl Endowed Professorships in Biology. His primary research interests center on various aspects of the physiology and ecology of nurse sharks in the Florida Keys. His most recent work investigated the reproductive biology and mating behaviors of this species in a long-term study from an isolated region of the Florida Keys. Dr. Carrier has been a member of the American Elasmobranch Society, the American Society of Ichthyologists and Herpetologists, Sigma Xi, the Society for Animal Behavior, and the Council on Undergraduate Research. He served as secretary, editor, and president of the American Elasmobranch Society and received multiple distinguished service awards from the society. He holds an appointment as an adjunct research scientist with Mote Marine Laboratory's Center for Shark Research.

John A. (Jack) Musick, PhD, is the Marshall Acuff Professor Emeritus in Marine Science at the Virginia Institute of Marine Science (VIMS), College of William and Mary, where he has served on the faculty since 1967. He earned his bachelor of arts degree in biology from Rutgers University in 1962 and his master's degree and doctorate in biology from Harvard University in 1964 and 1969, respectively. While at VIMS he has successfully mentored 37 masters and 49 doctoral students. Dr. Musick has been awarded the Thomas Ashley Graves Award for Sustained Excellence in Teaching from the College of William and Mary, the Outstanding Faculty Award from the State Council on Higher Education in Virginia, and the Excellence in Fisheries Education Award by the American Fisheries Society. In 2008, Dr. Musick was awarded the Lifetime Achievement Award in Science by the State of Virginia. He has published more than 150 scientific papers and coauthored or edited 16 books focused on the ecology and conservation of sharks, marine fisheries management, and sea turtle ecology. In 1985, he was elected a fellow by the American Association for the Advancement of Science. He has received distinguished service awards from both the American Fisheries Society and the American Elasmobranch Society (AES), for which he has served

as president. In 2009, the AES recognized him as a distinguished fellow. Dr. Musick also has served as president of the Annual Sea Turtle Symposium (now the International Sea Turtle Society) and as a member of the International Union for Conservation of Nature (IUCN) Marine Turtle Specialist Group. Dr. Musick served as co-chair of the IUCN Shark Specialist Group for 9 years and is currently the vice chair for science. Since 1979, Dr. Musick has served on numerous Stock Assessment and Scientific and Statistics committees for the Atlantic States Marine Fisheries Commission (ASMFC), the Mid-Atlantic Fisheries Management Council, the National Marine Fisheries Service, and the Chesapeake Bay Stock Assessment Program. He has chaired the ASMFC Shark Management Technical Committee and ASMFC Summer Flounder Scientific and Statistics Committee. His recent consultancies have included fisheries and environmental assessments for the commercial fishing industry, the Natural Resources Defense Council, Food and Agriculture Organization, the Australian government, and the Marine Stewardship Council.

Michael R. Heithaus, PhD, is an associate professor in the Department of Biological Sciences at Florida International University (FIU) in Miami, where he has been a faculty member since 2003. He received his bachelor of arts degree in biology from Oberlin College, Ohio, in 1995 and his doctorate from Simon Fraser University, Burnaby, British Columbia, in 2001. He was a postdoctoral scientist and staff scientist at the Center for Shark Research and also served as a research fellow at the National Geographic Society's Remote Imaging Department. At FIU, Dr. Heithaus served as the director of the Marine Sciences Program before becoming the founding director of the School of Environment, Arts, and Society. Dr. Heithaus is a behavioral and community ecologist. His main research interests are in understanding the ecological roles of top predators, especially their potential to impact community structure through nonconsumptive effects. His work also explores the factors influencing behavioral decisions, especially of large marine taxa, including marine mammals, sharks and rays, and sea turtles, and the importance of individual variation in behavior in shaping ecological interactions. Dr. Heithaus has been studying the ecological role of tiger sharks and their large-bodied prey in Shark Bay, Western Australia, since 1997 and co-founded the Shark Bay Ecosystem Research Project. He now also has ongoing projects in the coastal Everglades of southwest Florida and the Gulf of Mexico.

Contributors

Allen H. Andrews
National Oceanic and Atmospheric
  Administration
National Marine Fisheries Service
Pacific Islands Fisheries Science Center
Aiea, Hawaii

Neil C. Aschliman
Department of Biology
St. Ambrose University
Davenport, Iowa

Diego Bernal
Department of Biology
University of Massachusetts
  Dartmouth
North Dartmouth, Massachusetts

Joseph J. Bizzarro
School of Aquatic and Fishery Science
University of Washington
Seattle, Washington

Ashby B. Bodine
Department of Animal and Veterinary
  Sciences
Clemson University
Clemson, South Carolina

Elizabeth N. Brooks
National Oceanic and Atmospheric
  Administration
National Marine Fisheries Service
Northeast Fisheries Science Center
Woods Hole, Massachusetts

Gregor M. Cailliet
Moss Landing Marine Laboratories
Moss Landing, California

Janine N. Caira
Department of Ecology and
  Evolutionary Biology
University of Connecticut
Storrs, Connecticut

John K. Carlson
National Oceanic and Atmospheric
  Administration
National Marine Fisheries Service
Southeast Fisheries Science Center
Panama City, Florida

Brandon M. Casper
Department of Biology
University of Maryland
College Park, Maryland

Kerin M. Claeson
Department of Biomedical Sciences
Ohio University Heritage College of
  Osteopathic Medicine
Athens, Ohio

Christina L. Conrath
National Marine Fisheries Service
Alaska Fisheries Science Center
Kodiak Laboratory
Kodiak, Alaska

Enric Cortés
National Oceanic and Atmospheric
  Administration
National Marine Fisheries Service
Southeast Fisheries Science Center
Panama City, Florida

Leo S. Demski
Division of Natural Sciences
New College of Florida
Sarasota, Florida

Dominique A. Didier
Millersville University
Department of Biology
Millersville, Pennsylvania

David A. Ebert
Pacific Shark Research Center
Moss Landing Marine Laboratories
Moss Landing, California

Andrew N. Evans
Marine Science Institute
University of Texas
Port Aransas, Texas

Jayne M. Gardiner
Department of Integrative Biology
University of South Florida
Tampa, Florida
and
Center for Shark Research
Mote Marine Laboratory
Sarasota, Florida

Todd Gedamke
National Oceanic and Atmospheric
  Administration
National Marine Fisheries Service
Southeast Fisheries Science Center
Miami, Florida

James Gelsleichter
Department of Biology
University of North Florida
Jacksonville, Florida

Adrian C. Gleiss
Department of Pure Biosciences
Institute College of Science
Swansea University
Singleton Park, Swansea, United
  Kingdom

Kenneth J. Goldman
Alaska Department of Fish and Game
Division of Commercial Fisheries
Homer, Alaska

Emily Greenfest-Allen
Computational Biology and Informatics
  Laboratory
Penn Center for Bioinformatics
University of Pennsylvania
Philadelphia, Pennsylvania

Eileen D. Grogan
Biology Department
St. Joseph's University
Philadelphia, Pennsylvania

Claire J. Healy
Department of Natural History
Royal Ontario Museum, and
Department of Ecology and
  Evolutionary Biology
University of Toronto
Toronto, Ontario

Edward J. Heist
Fisheries and Illinois Aquaculture
  Center
Department of Zoology
Southern Illinois University
Carbondale, Illinois

Michael R. Heithaus
Department of Biological Sciences
Florida International University
Miami, Florida

Michelle R. Heupel
Australian Institute of Marine Science
and
School of Earth and Environmental
  Sciences
James Cook University
Townsville, Queensland, Australia

Daniel R. Huber
Department of Biology
University of Tampa
Tampa, Florida

Robert E. Hueter
Center for Shark Research
Mote Marine Laboratory
Sarasota, Florida

Kirsten Jensen
Department of Ecology and
  Evolutionary Biology
University of Kansas
Lawrence, Kansas

Jenny M. Kemper
Pacific Shark Research Center
Moss Landing Marine Laboratories
Moss Landing, California

Clemens Lakner
Bioinformatics Research Center
North Carolina State University
Raleigh, North Carolina

George V. Lauder
Organismic and Evolutionary Biology
Harvard University
Cambridge, Massachusetts

Christopher G. Lowe
Department of Biological Sciences
California State University Long Beach
Long Beach, California

Carl A. Luer
Center for Shark Research
Mote Marine Laboratory
Sarasota, Florida

Richard Lund
Section of Vertebrate Paleontology
Carnegie Museum of Natural History
Pittsburgh, Pennsylvania

Anabela M. R. Maia
Department of Biology
Ghent University
Ghent, Belgium

David A. Mann
College of Marine Science
University of South Florida
Tampa, Florida

Karen P. Maruska
Biology Department
Stanford University
Palo Alto, California

John D. McEachran
Department of Wildlife and Fisheries
  Sciences
Texas A&M University
College Station, Texas

Philip J. Motta
Department of Integrative Biology
University of South Florida
Tampa, Florida

John A. Musick
Virginia Institute of Marine Science
Gloucester Point, Virginia

Lisa J. Natanson
National Oceanic and Atmospheric
  Administration
National Marine Fisheries Service
Narragansett Laboratory
Narragansett, Rhode Island

Gavin J. P. Naylor
Department of Biology
College of Charleston
Charleston, South Carolina

Yannis P. Papastamatiou
Florida Museum of Natural History
University of Florida
Gainesville, Florida

Kerri A. M. Rosana
Department of Scientific Computing
Florida State University
Tallahassee, Florida

Colin A. Simpfendorfer
Fishing and Fisheries Research Centre
School of Earth and Environmental
  Sciences
James Cook University
Townsville, Queensland, Australia

Joseph A. Sisneros
Department of Psychology and Biology
University of Washington
Seattle, Washington

Nicolas Straube
Sektion Ichthyologie
Zoologische Staatssammlung München
München, Germany

Jeremy J. Vaudo
Department of Biological Sciences
Florida International University
Miami, Florida

Catherine J. Walsh
Center for Shark Research
Mote Marine Laboratory
Sarasota, Florida

Bradley M. Wetherbee
Department of Biological Sciences
University of Rhode Island
Kingston, Rhode Island

Nicholas M. Whitney
Center for Shark Research
Mote Marine Laboratory
Sarasota, Florida

Cheryl A.D. Wilga
Department of Biological Sciences
University of Rhode Island
Kingston, Rhode Island

Section I

Phylogeny and Zoogeography

Section 1

Phylogeny and Zoogeography

1

The Origin and Relationships of Early Chondrichthyans

Eileen D. Grogan, Richard Lund, and Emily Greenfest-Allen

CONTENTS

1.1 Introduction

Chondrichthyan fishes are probably the most successful of all fishes if success is measured in terms of historical endurance. Indeed, they have survived the mass extinctions of the last 400 million years or so. They are essentially defined by a cartilaginous skeleton that is superficially mineralized by prismatic calcifications (tesserae) and by the modification, within males, of mixopterygia (claspers) for the purpose of internal fertilization. It has been generally accepted that the Class Chondrichthyes is a monophyletic group divisible into two sister taxa, the Elasmobranchii and Holocephali, and that extant chondrichthyans (sharks, skates, rays,

and chimaeras) are derivable from Mesozoic forms. Yet, how the extant forms relate to the distinctly more diverse Paleozoic forms and the relationship of the Chondrichthyes to all other fishes are poorly resolved issues. Furthermore, some paleontologists currently question whether fossils attributed to chondrichthyans support a monophyletic class. The purpose of this chapter is to discuss the evidence for the origin, diversification, and life histories of the early Chondrichthyes; to address trends in their morphological divergence and innovation; and to explore the possible relationships between fossil and modern forms. In a general discussion of relationships, we adopt the classification scheme for shark and shark-like fishes put forth by Compagno (2001), as a consensus of the analyses of

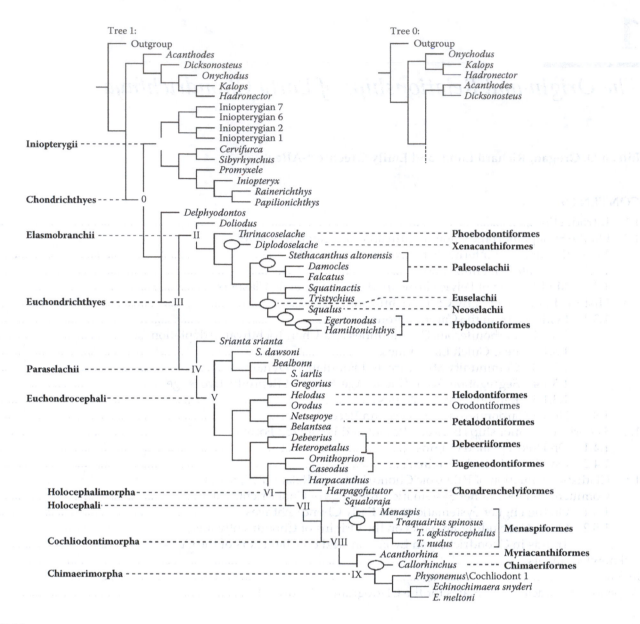

FIGURE 1.1

Cladogram of the matrix Chon55 (120 characters, 53 taxa) produced using the traditional search procedure of TNT (Goloboff et al., 2008) and its tree bisection–reconnection algorithm. The best score is 911, with 5 hits out of 10 random addition sequences retained, resulting in the retention of two trees. Tree 1 differs from tree 0 only as shown in the figure. The characters and states for this matrix are listed in the Appendix. The left and right columns follow the higher systematic usage of Nelson (2006). The term *Euchondrichthyes* is introduced here to designate the node that subsumes the two higher clades of the Class Chondrichthyes.

Compagno (1984), Shirai (1996), and de Carvalho (1996). The classification scheme used to describe the relationships of all Chondrichthyes is that originally developed in Lund and Grogan (1997a,b; 2004a,b) and Grogan and Lund (2000), but subsequently modified given new finds (Grogan and Lund, 2008, 2009, 2011). Details of this scheme have been refined based on discussions of higher chondrichthyan systematics with Dr. Joseph Nelson (University of Alberta), author of *Fishes of the World* (2006).

1.2 On the Synapomorphic Chondrichthyan Characters: Tesserate Mineralization and Internal Fertilization by Male Claspers

Molecular and morphological analyses, including that herein, strongly support a monophyletic Chondrichthyes (Grogan and Lund, 2000, 2008, 2009; Heinicke et al., 2009; Janvier, 1996; Lund and Grogan, 1997a; Maisey, 1984, 2001) (Figure 1.1). Although a variety of characters has been proposed to define this monophyletic group, two synapomorphies are generally accepted to define these fish: tesserate endoskeletal mineralization and pelvic claspers.

1.2.1 Tesserate Mineralization

The tesserate mode of mineralizing endoskeletal tissues peripherally is *the* critical defining character of the group (see Coates and Sequeira, 1998; Lund and Grogan, 1997a; Maisey, 1984, 2000). It is therefore unfortunate that the term *tesserae* has been applied to both the chondrichthyan and placoderm conditions (Applegate, 1967; Denison, 1978; Ørvig, 1951), for they represent two different phenomena. At best, they share an extremely remote relationship derived from common vertebrate patterns of skeletal tissue determination, regulation, and, therefore, development. Chondrichthyan tesserae represent a developmental deviation from the pattern of endoskeletal tissue formation that characterizes primitive gnathostomes. Previous studies and current work in progress address the difficult question of the transition from the primitive gnathostome condition of perichondral bone (Basden et al., 2000; Janvier, 1996) to the chondrichthyan states of perichondral and endochondral mineralized cartilage (see Applegate, 1967; Grogan and Yucha, 1999; Grogan et al., in prep.; Kemp and Westrin, 1979; Ørvig, 1951; Rosenberg, 1998; Yucha, 1998). All data generally support the idea that the endoskeletal mineralization of chondrichthyans represents an autapomorphic condition relative to other gnathostomes. Developmental responses to mechanical and growth parameters (possibly even including regulatory features associated with the pituitary:gonadal axis) led to variants of the common mineralized plan.

"Prismatic" calcification, as used here, refers to the macroscopically visible state of separate, peripheral mineralized units (Ørvig, 1951). Thin sections typically reveal either the more primitive state of spheritic (globular) calcified cartilage or a highly ordered, star-shaped architecture parallel to the cartilage surface and an hourglass microstructure perpendicular to that surface when viewed under a crossed polarizer and analyzer (Applegate, 1967; Ørvig, 1951; Yucha et al., pers. obs.). The latter configuration is due to the mineralized unit having two subunits, Kemp and Westrin's (1979) cap and body components. Subsequent studies further indicated that these tessera subunits are distinct in their origin and the extent of their development, thus offering an explanation for apparent differences in tesserate appearance within and across taxa (Fluharty and Grogan, 1999; Grogan and Yucha, 1999; Grogan et al., in prep.; Rosenberg, 1998; Yucha, 1998). In keeping with these observations and with observations of fossil forms, then, we use "continuous" calcified cartilage,

in the sense of Ørvig (1951), to refer to a modified tesserate condition wherein adjacent tesserae undergo an early ontogenetic fusion and, therefore, do not exhibit the more typical prismatic microstructure.

The primitive gnathostome condition, in contrast, is most likely to be that of endoskeletal elements having a cartilaginous core (with spheritic mineralization) covered by perichondral bone (Janvier, 1996; Ørvig, 1951).

1.2.2 Modification of Pelvic Girdle in Males to Generate Claspers

Chondrichthyan claspers (mixopterygia) are extensions of the endoskeletal axis of the male pelvic fin or girdle that form sperm-conducting structures (copulatory organs) to facilitate internal fertilization of a female. These axial modifications may or may not be accompanied by modifications of the fin radials or modifications of the adjacent squamation. The development of claspers, however, also involves the coordinated development of the musculature necessary to pump sperm and the musculature necessary to maneuver the claspers. A model of clasper development and its morphoclinal transition within Chondrichthyes has been presented in Lund and Grogan (1997a). The alar scales or plates of some skates and iniopterygians (Zangerl and Case, 1973) and the prepelvic tenaculae of chimaeroids and some cochliodonts are accessory sexual structures.

All mature male chondrichthyans display intromittent organs. The Upper Devonian *Cladoselache* has often been claimed to be the exception, yet such arguments ignore the high probability that the recovered forms are strictly female. Zangerl (1981) considered preservational concerns for the lack of evidence for claspers in this instance. There is, however, a plausible alternative. The fossil deposits from which clasper-lacking specimens are derived are shallow, epicontinental, and marine and appear to indicate a paleoenvironment that was like that of a coastal margin/shelf or contiguous bay (Ettensohn and Barron, 1981; Hansen, 1999). This combination of evidence is consistent with the interpretation that evidence of females rather than males may be due to a life history style involving sexual segregation, like that reminiscent of extant elasmobranchs (Klimley, 1987; Springer, 1967). It is also clearly established that other Upper Devonian male elasmobranchs, including the comparatively smaller *Diademodus* from the same deposit as *Cladoselache*, displayed pelvic claspers. We also maintain that the preponderance of evidence supports the view that all mature male chondrichthyans are identifiable by their possession of claspers since all other members of the cladodont group have claspers to identify males. The feature of claspers in male chondrichthyans is plesiomorphous for the group.

The only other Paleozoic group of fishes reported to exhibit claspers are the placoderms, specifically some pyctodonts, phyllolepids, and an arthrodire (Long, 1984, 1997; Long et al., 2008, 2009; Miles and Long, 1977; Ørvig, 1961; Watson, 1938). There is apparently some variation in clasper design across the group; the relatively derived pyctodonts exhibit three denticulated dermal plates supported by a core of cartilages, whereas the less derived phyllolepids have a broad dermal pelvic plate in articulation with a dermal basipterygial-type element that likely supported a distal-most cartilage or series of cartilages (Long et al., 2008, 2009). It remains unresolved whether placoderms primitively possessed claspers because they have not been found in all groups of placoderms to date. What is known of their distribution across these groups does not currently support claspers as a plesiomorphous character for Placodermi. Although placoderm claspers were used for internal fertilization as in chondrichthyans, these structures are not believed to be homologous with those of the Chondrichthyes (Ahlberg et al., 2009; Long et al., 2009).

1.3 Historic Evidence of Early Chondrichthyans

Turner (2004) provided an excellent summary of the record of research on early vertebrate remains (scales and spines) attributed to chondrichthyans and highlighted the dilemmas facing workers using either fossil micro- or macroremains alone. See Turner and Miller (2005) for a less technical discussion of the evidence for scales, spines, and teeth in relation to chondrichthyan origins. In brief, the oldest evidence of scales attributed to chondrichthyans occurs in the Silurian to Ordovician where simple micromeric scales (assumed by some to be primitive for chondrichthyans) as well as complex forms (polyodontode scales) occur. Scales and spines attributed to chondrichthyans range from the Lower Silurian, and more diverse forms of these scales are generally abundant within the Devonian (Cappetta et al., 1993; Goujet, 1976; Karatajute-Talimaa, 1992; Karatajute-Talimaa and Predtechenskyj, 1995; Rodina and Ivanov, 2002; Zhu, 1998). With regard to scales, Karatajute-Talimaa (1992) presented a scenario for the evolution of early chondrichthyans based on putative chondrichthyan morphotypes, noting that the complexity of the polyodontode element shares morphological similarity to that of micromeric scales. This scenario depicts an Ordovician or Silurian origin that led to a maximal adaptive radiation in the Early Devonian that gave rise to ctenacanthid, hybodontid, and protacrodontid forms, lineages that undeniably extend into the Carboniferous

and beyond. Such a progression in the evolution of scale types is quite plausible. Yet, we add the qualifier that more than one of these scale forms can occur simultaneously within the same organism. We have found (1) Upper Silurian *Elegestolepis*-type scales and Devonian *Ctenacanthus*-type scales within the Carboniferous elasmobranch *Falcatus* as cranial and buccopharyngeal denticles, respectively, and (2) that both *Elegestolepis* and Devonian *Protacrodus*-type scales correlate, respectively, with the generalized and specialized cranial scales of the Carboniferous euchondrocephalan *Venustodus argutus* St. John and Worthen 1875 (Bear Gulch specimen, CM 41097). So it is possible that various taxa identified on the basis of microremains (e.g., solely on the basis of a simple vs. complex scale design) may have a real biological origin in one form. Furthermore, there may be an ontogenetic and morphological continuum between thelodont, acanthodian, and chondrichthyan scales and buccopharyngeal denticles (Lund and Grogan, pers. obs.; Rodina, 2002). This could account for Turner's (2004) observation of a high degree of similarity between the buccopharyngeal scale design of sharks and thelodonts. If so, the situation devolves to identifying what is plesiomorphous for vertebrate forms vs. that which is characteristic for chondrichthyans or any other fish group. The complexity of this problem is further accentuated by recent studies of Devonian-age fishes that are argued to have chondrichthyan-like scales but, on the balance, exhibit morphological features that are characteristic of acanthodians (Hanke and Wilson, 2010). Overall, we concur with Turner (2004) that, whenever possible, both macro- and microremains should be collectively used to study the early vertebrates and their relationships. By corollary, we maintain that the nature and affinity of the earliest of these elements will remain subject to skepticism without further developmental and histological studies of these tissues across vertebrates. It appears that we are making progress toward this end. Partially articulated material of Devonian chondrichthyan taxa has emerged and includes information on squamation and/or branchial denticles (Heidtke and Krätschmer, 2001; Miller et al., 2003; Soler-Gijón and Hampe, 2003). We await comparable finds from the preceding geologic time periods.

The earliest evidence of chondrichthyan teeth dates back to the Lower Devonian with the appearance of *Leonodus*, originally described as a diplodont tooth-organ taxon (Mader, 1986). Ginter (2004) and Ginter et al. (2010) described the range of reported tooth forms while introducing order and logic to the distribution and relationships of these remains by considering biostratigraphy. Paleozoic chondrichthyan teeth are organized into three major categories: (1) primitive forms (cladodont, phoebodont, and diplodont forms); (2) euselachian and euchondrocephalan forms (derived forms found in protacrodonts, hybodonts, and euchondrocephalans); and (3) others, those that defy grouping into either the primitive or euselachian categories. The latter include Devonian tooth forms identified as omalodontiform and the radically different petalodont teeth from the Carboniferous and Permian. Assignment to a category is based on characters such as tooth dimensions and number, height, orientation, fusion, and/or ornamentation of cusps, as well as the presence or absence of an articular structure that would have functioned between adjacent teeth in a tooth family. Ginter (initially in his 2004 work, but fully developed in the 2010 collaboration) described his vision for the evolutionary diversification of teeth from an initial diplodont form toward the phoebodont and cladodont forms. He roots the origin of the euselachian and euchondrocephalan tooth forms in the Mid-Devonian in a diplodont antarticlamnid-style. Along the evolutionary trajectory from this point the cladodont and euselachian/euchondrocephalan subsequently diverge, as the latter specialized more for crushing and grinding through the reduction of cusp height and fusion of adjacent cusps, as represented by the protacrodont-style tooth. This overall scheme for the derivation of all chondrichthyan tooth forms from a primitive diplodont condition may or may not contrast with that proposed by Zangerl (1981) depending upon one's perspective. We propose that the scenario of Ginter et al. offers finer resolution than that of Zangerl by clarifying how the diplodont can lead to the triplodont condition and beyond. (We find that the cladodont-style tooth is an autapomorphous condition and so is not a feasible plesiomorphous chondrichthyan state.) What remains to be further assessed in Zangerl's proposal is the timing and means of transformation from a monocuspid to diplodont state and evolution of teeth from denticle-like odontodes. The question is whether or not this transition occurred before the origin of the Chondrichthyes. Evidence of the Mid-Devonian *Pucapampella* suggests that a monocuspid condition existed early in the chondrichthyan lineage if Janvier and Maisey (2010) are correct in their attribution of isolated palatoquadrate and Meckel's cartilages bearing a single series of conical-shaped teeth to this form. The apparently edentulous condition of Mid-Devonian *Gladbachus* also factors into this quandary. Does this represent a primary or secondary condition within Chondrichthyes? Zangerl (1981) referred to Upper Carboniferous iniopterygian teeth and denticles to illustrate his vision of a primitive chondrichthyan dentition. We find evidence of variations on this dental condition in a range of Lower Carboniferous chondrichthyans from the Bear Gulch of Montana that suggest a position basal to the euchondrichthyan (Elasmobranchii + Paraselachii) node, on the basis of endoskeletal, dental, and other characters (Figure 1.1).

Is there evidence of this condition in the earlier chondrichthyan history? As with the debate on scales, the promise for resolution lies in recovery of new, articulated material from the Lower Devonian or earlier.

A more reliable indicator of the presence of chondrichthyans would logically be in the form of elements that exhibit a diagnostic chondrichthyan feature, the tesserate mode of cartilage mineralization. Yet, there is a bias against preservation of such elements in the fossil record given the nature of calcified cartilage in contrast to the more heavily mineralized elements of bone and perichondral bone. Nonetheless, fossils of Devonian age and beyond are found with this style of mineralization, including the earliest partially articulated evidence of a chondrichthyan, *Doliodus problematicus* (Early Devonian, ca. 409 mya) (Miller et al., 2003). Other reports of chondrichthyan-type calcified cartilage (*sensu* Applegate, 1967; Ørvig, 1951) are from the marine Devonian deposits of Bolivia. The frequency with which these calcified cartilage fragments occur suggests that the chondrichthyans were the most abundant of all vertebrates in a marine environment in which agnathan thelodonts, actinopterygians, acanthodians, and placoderms are also indicated (Gagnier et al., 1989; Janvier and Suarez-Riglos, 1986). This site also yielded the first evidence of *Pucapampella* (Janvier and Suarez-Riglos, 1986). Subsequently, other material expanded the evidence of this chondrichthyan form (Anderson et al., 1999; Maisey, 2001; Maisey and Anderson, 2001). This was a remarkable advance for chondrichthyan research not only in presenting the (then) oldest evidence for the lineage but also in the combination of features that the specimens revealed. This included not only some stem gnathostome features but also variation among the specimens. Comparison of the crania from South Africa and Bolivia suggests that distinct cranial morphs with holocephalan vs. selachian affinities were already established by this time (Grogan, pers. obs.).

Since the initial discoveries of the articulated *Doliodus problematicus* and *Pucapampella* crania, the field of vertebrate paleontology has experienced a sea change of sorts, as there has been a tremendous expansion of information not only pertaining to the chondrichthyan lineage but also implicating all of the gnathostome groups. The record of Lower Devonian chondrichthyans currently boasts, in order of oldest to youngest, articulated material of *Doliodus*, *Pucapampella*, *Antarctilamna*, and *Gladbachus* (Heidtke and Krätschmer, 2001). With the finding of pectoral fin spines in *Doliodus problematicus* (Miller et al., 2003) and arguments for the same in *Antarctilamna priscus* (Miller et al., 2003; Wilson et al., 2007), what was once considered to be an acanthodian synapomorphy is now being reassessed as a likely gnathostome synapomorphy because paired fin spines are also reported in placoderms and a basal osteichthyan

(Zhu et al., 1999, 2009). At the same time, *Ptomacanthus*, which was previously considered to be acanthodian, is now hypothesized to be either a basal stem chondrichthyan or the sister group of all gnathostomes (Brazeau, 2009). Comparing the morphological evidence among the new finds reveals that some of these Lower Devonian chondrichthyans (*Pucapampella*, *Doliodus*, *Antarctilamna*) display a neurocranial condition previously reported for all gnathostomes to the exclusion of placoderm and chondrichthyan members, notably an oticooccipital fissure and/or ventral fissure. There are significant differences in the design of these crania, and each presents features that may be considered primitive and derived. Thus, any resolution of the phylogenetic relationships between these chondrichthyans and of the Chondrichthyes to the other gnathosomes will require additional finds. In short, these ancient chondrichthyan finds have proved enriching for all early vertebrate workers by providing greater clues to the picture of the stem gnathostome. They also emphasize the diversity of chondrichthyans present by the Early to Mid-Devonian.

In hindsight, recovery of such data reinforces what was previously known, in general, about the Devonian. The radiations associated with this "Age of Fishes" were supported by an increasingly diverse suite of estuarine, brackish to freshwater, and marine continental margin environments in equatorial Euramerica and in the southern continent of Gondwana. Especially in Euramerica, shallow seas supported extensive reef building by stromatoporoids and corals and so were likely to have favored the retention and diversification of many fishes along or near the continental margins as a consequence of high primary productivity and habitat diversification. In keeping with this, there is significant evidence to document that, by the Middle to Late Devonian, the chondrichthyans were represented by a number of strikingly different forms that inhabited environments ranging from fresh and brackish water to continental margins and oceans (Ivanov and Rodina, 2002a,b). Yet, perhaps because of their propensity for poor holomorphic preservation, they apparently remained relatively scarce compared to the placoderms and actinopterygians.

The freshwater xenacanthids included *Leonodus*, *Aztecodus*, and *Portalodus* (Young, 1982). *Antarctilamna's* designation as a xenacanth now is in flux and may actually represent a more primitive form. Elasmobranchs and euchondrocephalans, including *Diademodus*, *Siamodus*, *Ctenacanthus*, *Plesioselachus*, *Phoebodus*, *Thrinacodus*, *Orodus*, *Protacrodus*, *Stethacanthus*, and hybontids, are reported from marine, estuarine, and coastal lagoonal environments at some point in the Devonian (Anderson et al., 1999; Gagnier et al., 1989; Ginter, 1999; Ivanov and Rodina, 2002a,b; Janvier and Suarez-Riglos, 1986; Lelievre and Derycke, 1998). Chondrichthyan

microremains (teeth, scales) from Germany suggest a wide distributional range for the group at the end of the Devonian (Famennian–Tournasian) but with a progressive partitioning of forms according to an environmental gradient (Ginter, 1999; Ivanov and Rodina, 2002a). Protacrodonts primarily occupied shallow epicontinental seas and the proximal aspect of continental margins, the tooth-taxon *Jalodus* was principally associated with deep marine waters, and the cladodonts reflected a more cosmopolitan distribution as they exhibited more of an ocean-roaming habit. The broad, blunt, durophagous teeth of orodonts, helodonts, and *Psephodus*-like forms, as well as those of *Ageleodus*-like forms (whose teeth are closest to those of the Debeeriidae), are found in what are probable estuarine to freshwater deposits toward the Upper Devonian (Downs and Daeschler, 2001). No evidence currently exists for cochliodont tooth plates in Devonian deposits, possibly indicating that the Holocephali *sensu stricto* had yet to evolve or to diversify. It is true that, if the Holocephali *sensu stricto* had evolved by the Devonian and if they inhabited deep waters, then any fossilized remains of them would be the least likely of all forms to be recovered. Yet, the morphological, chronological, and developmental data all support the view that, at best, paraselachian-type holocephalan ancestors existed during this phase of vertebrate life.

In terms of community structure, arthrodiran placoderms were the apex predators of the Devonian. The known elasmobranchs and protacrodonts were lower-trophic-level forms; the former predominantly bore piercing teeth, whereas the latter possessed lower, blunt-crowned teeth. On the other hand, the various and well-established xenacanths may have vied for the apex predator level with Crossopterygii in freshwater environments.

1.3.1 Evidence from the Carboniferous

1.3.1.1 Carboniferous Communities and Chondrichthyan Adaptations

The Lower Carboniferous witnessed the extinction of the Placodermi, the reduction of the formerly diverse Acanthodii to one or two toothless genera, and slow diversification of freshwater Amphibia. Crossopterygii were limited to very few freshwater and marine species. Coelacanths, however, diversified to an extent that correlated with highly specialized habitat preferences. Small actinopterygians broadly diversified within their primary consumer trophic-level specializations in the marine environment. Chondrichthyans, in contrast, radiated rapidly and expansively in all available aquatic regimes. Marine waters included the stethacanthids, protacrodonts, petalodonts, "helodonts," and a host of other forms known only from teeth (Rodina and

Ivanov, 2002a,b). The freshwater environs, on the other hand, were inhabited by forms such as *Hybodus* and *Helodus*, in addition to various xenacanths (Lund, 1976; Romer, 1952). Carboniferous marine deposits that offer information beyond isolated chondrichthyan remains are those of Glencartholm, Scotland (Moy-Thomas and Dyne, 1938; Traquair, 1888a,b), and Bearsden, Scotland (Coates, 1988). The Lower Carboniferous tooth and spine faunas of Armagh, Ireland (Davis, 1883), and the upper Mississippi Valley of the United States (e.g., Newberry and Worthen, 1866, 1870; St. John and Worthen, 1875, 1883) are either small and limited deposits or organ-taxon deposits. The fish fauna of the penecontemporaneous Upper Mississippian Bear Gulch Limestone includes forms comparable to those uncovered in these deposits but has additional advantages. The biota of the Bear Gulch and the dimensions and conditions of the shallow, tropical, marine bay from which they came are sufficiently detailed by lagerstätte-type preservation to permit what is likely to be the most comprehensive and reliable documentation of community structure and ecology reported for Upper Paleozoic fish to date.

We believe that the community structure of the adjacent epicontinental and open waters was not inconsistent with that of the Bear Gulch to the extent that the Paleozoic Bear Gulch bay was obviously accessible to migratory forms and provided breeding and nursery grounds for those not endemic to the bay. Given this and the continuity between the other Carboniferous deposits noted above (i.e., select genera or even species in common), it is likely that the diversity of the Bear Gulch fauna may be representative of Upper Mississippian marine faunas. Yet, detailed faunal analyses of the Glencartholm and Bearsden deposits would be required to evaluate this possibility further. It is known that the Bear Gulch fauna exhibits a higher diversity and species richness than later Pennsylvanian deposits, which are characterized by both freshwater and marine fishes (Lund and Poplin, 1999; Schultze and Maples, 1992). Yet,

the latter situation may simply reflect the correlation between lower diversity in newly developing ecosystems or in areas of recent disturbance and invasion by generalists or ecological opportunists (e.g., Downs and Daeschler, 2001). By contrast, analyses of the paleoenvironment during the time of the Bear Gulch do not indicate catastrophic or revolutionary change (Grogan and Lund, 2002). The conditions prevailing during the deposition of this deposit suggest periodic disturbances, but not of a magnitude that would dramatically reduce diversity and richness and lead to a permanent shift in community composition. In any event, it is certain that the preserved remains of Bear Gulch chondrichthyans provide a rare view and index of the range of chondrichthyan diversity evident at this early stage in the evolution of the group.

1.3.1.2 Bear Gulch Limestone

The data that follow are based on both published and nonpublished material. Specimen abbreviation codes are as follows: CM, Carnegie Museum of Natural History, Section of Vertebrate Fossils; MV, University of Montana Geological Museum.

1.3.1.2.1 Overview of the Bear Gulch Bay

The extent of this ancient bay is approximately 57 km² in surface exposure area, with a maximum width of slightly over 8 km, length of 16 km, and a maximum depth of 40 meters near its mouth (Figure 1.2). Sampling has occurred at 103 sites across the entire deposit. The geologic and biologic data collected over the years and span of these excavations have revealed distinct macro-habitat zones in the bay (Grogan and Lund, 2002; Lund et al., in prep.). We report below on the community structure, fish diversity, and fish ecomorphology for four major habitat zones to provide an informed view of the chondrichthyans in the context of their natural environment 318 million years ago.

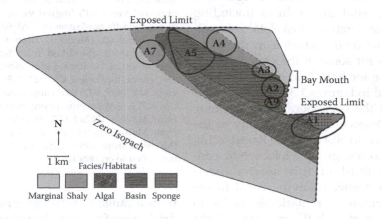

FIGURE 1.2
Surface map of the Bear Gulch Bay (scale in kilometers). Alphanumerics (A1, A2, etc.) refer to habitat sampling areas.

TABLE 1.1

Numbers of Chondrichthyan Specimens
and Species of this Study

Chondrichthyan Clade	Number of Specimens (% of Total Fish Specimens)	Number of Species (% of All Chondrichthyan Species)
Elasmobranchii	417 (40.8%)	21 (25.3%)
Holocephali	238 (23.3%)	19 (22.9%)
Paraselachii (exclusive of Holocephali)	208 (20.3%)	31 (37.3%)
Basal forms	158 (15.5%)	12 (14.5%)
Total	1021	83

Given that the bay was formed and filled in a tectonically active trough, we have the greatest confidence in the data arising from deeper aspects of the deposit compared to the shallower zones, which were subject to greater changes in basin conformation and water depth and conditions over the estimated 1000 years of the bay's existence. The fish fauna consists of chondrichthyans (59% of all species, 20% of all fish specimens), osteichthyans (38% of all species, 79% of all fish specimens), coelacanths (4% of all species, 12% of all fish specimens), and one species each of acanthodian and lamprey (2% and 0.0005% of all fish specimens, respectively). The Class Chondrichthyes (Table 1.1) is represented in this analysis by four groups: the two major clades of the Euchondrichthyes (the Elasmobranchii and Paraselachii), the clade Holocephali (a crown clade within the Paraselachii), and forms that are stem-ward to the Euchondrichthyes (basal forms or their relicts). The data on these forms reveal that the latter assemblage clearly does not fit into the elasmobranch, paraselachian, or holocephalan taxonomic groupings on the basis of newly revealed morphologic features, including variations in a multi-unit neurocranium (in contrast to the continuous vault of all other chondrichthyans) and/ or premandibular endoskeletal elements (Grogan and Lund, 2009, in prep.). Sexual dimorphism (including claspers in males) and tesserate endoskeletal mineralization, however, support their inclusion in the Class Chondrichthyes. Images for some of the chondrichthyans discussed in this chapter, both Bear Gulch and non-Bear Gulch, are provided in Figures 1.3 to 1.5.

Raw data reveal that the elasmobranchs contribute only about 25% of the known species diversity to the Chondrichthyes compared to 60.4% for the Paraselachii. Within the latter, the crown group Holocephali contributes just over one third of this species diversity. Furthermore, those forms resolving basal to the Euchondrichthyes contribute a remarkable 14.5% to overall diversity. As some of both the smaller and the larger Bear Gulch chondrichthyans are also known

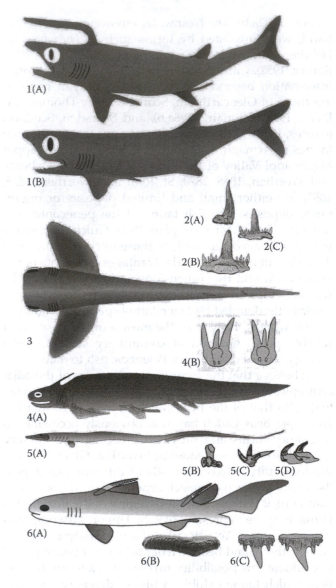

FIGURE 1.3

Examples of fossil Elasmobranchii. 1(A) Restoration of male *Falcatus falcatus*, MV 5385; 1(B) restoration of female *F. falcatus*, MV 5386. 2. General cladodont tooth morphology as represented by *Cladodus springeri*. (Modified from St. John and Worthen, 1875, Plate II.) 2(A) proximal view; 2(B) lingual view; 2(C) labial view. 3. Restoration of *Squatinactis caudispinatus*, CM 62701 (immature specimen). 4(A) *Xenacanthus* sp. (Modified from Carroll, 1988.) 4(B) *Xenacanthus parallelus* teeth. (Modified from Schneider and Zajic, 1994.) 5(A) Restoration of *Thrinacoselache gracia*, CM 62724; 5(B) *Thrinacoselache* tooth in occlusal view, CM 62724; 5(C) *Thrinacoselache* tooth in labial view, MV 7699; 5(D) two successional *T. gracia* teeth in lateral view, CM 62724. 6(A) Reconstruction of the Jurassic *Hybodus* sp. (Modified from Maisey, 1982a.) 6(B) *Hamitonichthys mapesi* tooth. (Modified from Maisey, 1989.) 6(C) Teeth of the Jurassic *Hybodus basanus*. (Modified from Maisey, 1983.) *Note:* Different genera or species are not scaled to one another.

from a range of other geographic locations, this chondrichthyan fauna can be taken as representative of tropical marine conditions at the time.

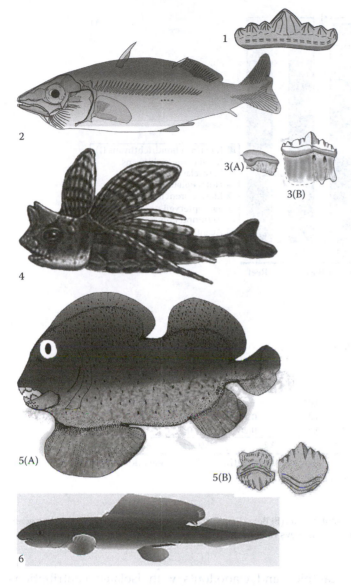

FIGURE 1.4
Examples of euchondrocephalans. 1. *Protacrodus vetustus* tooth. (Modified from Obruchev, 1967.) 2. Composite restoration of *Gregorius rexi*. 3(A) Parasymphysial and 3(B) mandibular teeth of *Srianta srianta*. 4. Iniopterygian species 1. (Courtesy of R. Troll, troll-art.com.) 5(A) Petalodont, restoration of *Belantsea montana*, MV 7698. (Modified from Lund, 1989.) 5(B) Teeth of *Belantsea montana*, MV 7698. 6. Reconstruction of male *Heteropetalus elegantulus*. *Note:* Different genera or species are not scaled to one another.

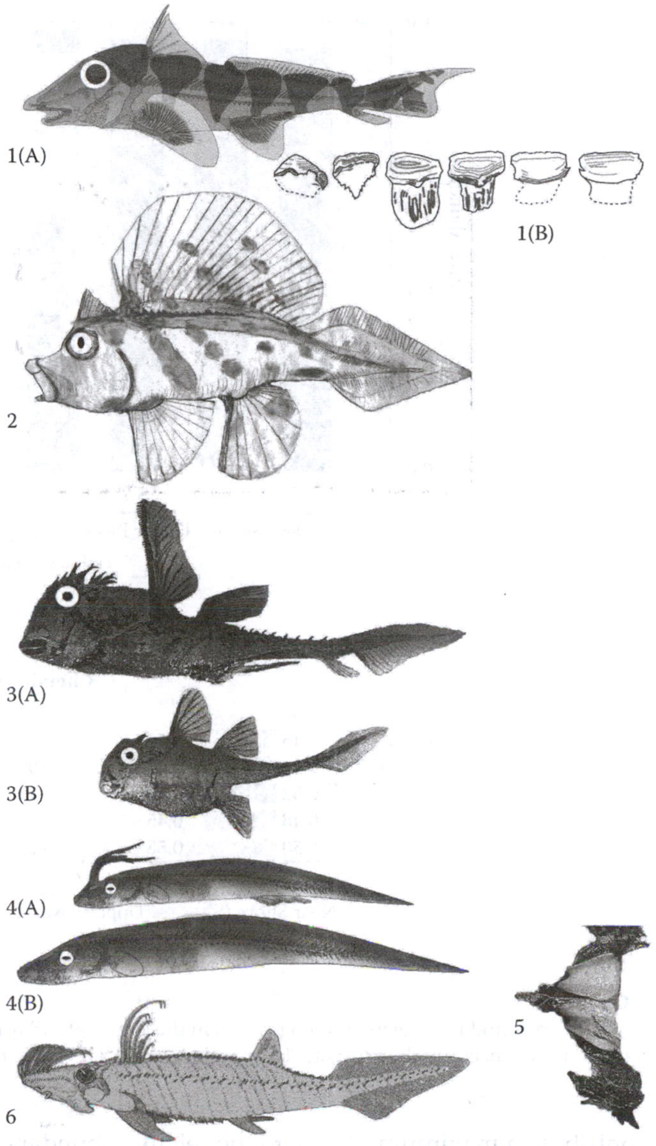

FIGURE 1.5
Further examples of euchondrocephalans. 1(A) Restoration of male *Debeerius ellefseni* with preserved pigment pattern, ROM 41073; 1(B) teeth of *Debeerius ellefseni*, ROM 41073. (Both modified from Grogan and Lund, 2000.) 2. Restoration of undescribed female fish code-named Elweir (CM 41033) with pigment pattern as preserved. 3. Restoration of 3(A) male (CM 30630) and 3(B) female (CM 25588) *Echinchimaera meltoni* depicting the relative size of male to female. 4(A) Reconstruction of male (MV 7700) and 4(B) restoration of female (MV 5370) *Harpagofututor volsellorhinus* depicting the relative male to female size. 5. *Traquairius nudus* (CM 46196) lower jaw and tooth plates in dorsal view. 6. *Traquairius agkistrocephalus* (CM 48662), reconstructed male. *Note:* Different genera or species are not scaled to one another.

Subsequent analysis is based at the genus level, an approach that is required to ensure sufficient numbers in a taxon for statistical evaluation. Specimens not identifiable to genus were not included. Although genus-level analysis may result in underestimating the diversity of a species-rich genus, more inclusive genus- and higher-level groupings of taxa can accurately reflect large-scale biological patterns (for a review, see Roy et al., 1996).

1.3.1.2.2 Relative Abundance of Chondrichthyes and Osteichthyes in the Bear Gulch Bay

The relative abundance of genera in each habitat zone was calculated as absolute abundance standardized to 100% of the fishes in the habitat zone. The Chondrichthyes

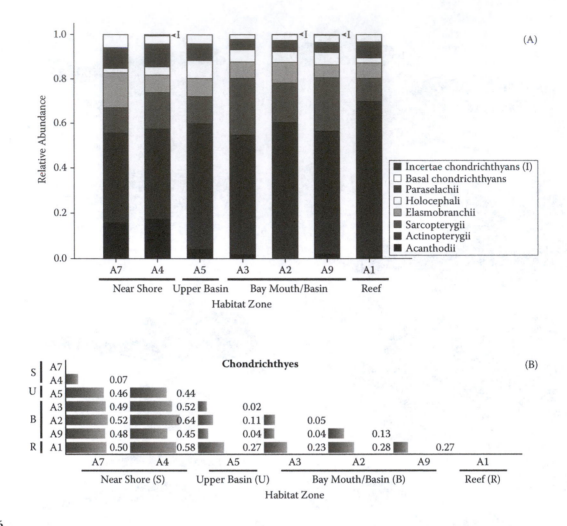

FIGURE 1.6
(A) Relative abundances of gnathostomes fish in the Bear Gulch Bay by habitat zone. (B) Beta diversity of Chondrichthyes, comparing between-habitat occurrences, by the Morisita–Horn index (Horn, 1977). The habitat zone keys refer to the map of Figure 1.2.

contribute a maximum of 33% to the relative abundance of fishes at a nearshore site and a minimum of 19 to 22% in the lower basin/bay mouth and reef areas (Figure 1.6). The balance of the relative abundance of Bear Gulch fishes is principally attributed to the osteichthyans (78 to 80% at the lower basin/mouth areas and reef zone compared to 67 to 74% in the nearshore environs). Within these percentages, the sarcopterygian and actinopterygian fishes contribute 51 to 80% of relative abundance across all zones, of which the numbers are greatest in the reef and lower basin/mouth areas. Of all the major fish groups considered, the actinopterygian fishes present the greatest contribution to relative abundance with a maximum of 69% in the reef and a low of 39% in the nearshore areas.

1.3.1.2.3 Chondrichthyan Distribution and Abundance

The Bear Gulch Elasmobranchii is represented by paleoselachians and euselachians (Figure 1.7A). The Paleoselachii is principally represented by the stetha-

canthids and cladodonts with isolated contributions from other forms, including *Thrinacoselache*, the tooth taxon *Carcharopsis*, and ctenacanth and symmoriid sharks. These elasmobranchs exhibit a significant size range (ca. 150 mm to 3 m), with larger species (including juveniles) representing apex predators. At least one of the smaller species appears to occur in mating aggregations (Lund, 1985a). As a unit, the paleoselachians represent the greatest contribution to elasmobranch abundance in the bay (92 to 100% by area). The stethacanthids are distributed across all areas of the bay and are the predominant abundance component (59 to 96%) within each zone-based community. The cladodonts also occur across all areas but, relative to the stethacanthids, they represent a dramatically smaller contribution (4 to 24%) to each community. *Falcatus* and *Damocles*, the smallest of the stethacanthids, provide the greatest proportion of Stethacanthidae abundance across most zones (30 to 90% for *Falcatus*, 0 to 70% for *Damocles*), but

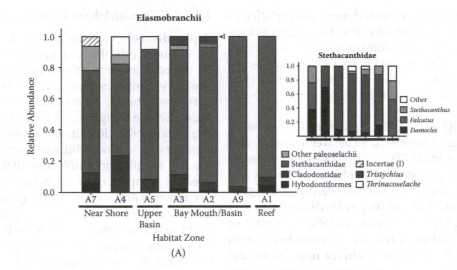

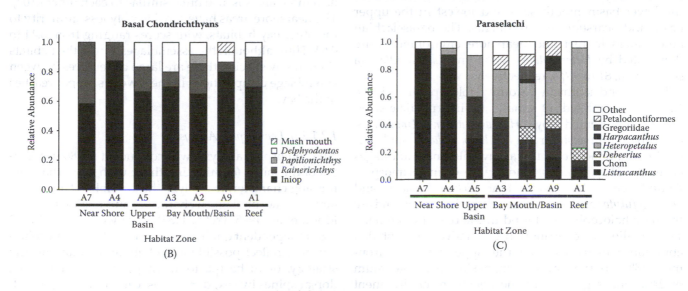

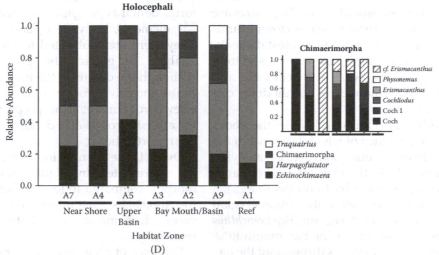

FIGURE 1.7

Relative abundances of chondrichthyan groups by habitat zone. (A) Elasmobranchii and its stethacanthoid component. (B) Basal Chondrichthyes. (C) Non-holocephalan Paraselachii. (D) Holocephali and its chimaerimorph component. The habitat zone keys refer to the map of Figure 1.2.

it is most notable that their abundances shift relative to one another across the habitats. The Euselachii, principally represented by hybodontiforms and *Tristychius*, provide a smaller contribution (0 to 9%) to overall elasmobranchian abundance than the Paleoselachii. They occur in slightly greater numbers in nearshore (6%) and reef (5%) habitats.

Paraselachians (Figure 1.7C), first identified as fishes with morphologies intermediate between sharks and holocephalans (Lund, 1977), tend to the small to intermediate size range for Bear Gulch chondrichthyans and display a range of highly specialized dentitions. In general, they are the most morphologically diverse group of chondrichthyans. They contribute 4 to 5% of the relative abundance of all fishes at the lower basin/bay mouth area, 7% in the reef, and about 9% in the nearshore zone. The diversity of these fishes by genera is greatest near the lower basin/mouth sites and lowest in the upper basin and nearshore environments. The paraselachian abundances in reef and nearshore environments are dominated by *Heteropetalus* (73%) and *Listracanthus* (a scale taxon; 81 to 95%), respectively.

The chondrichthyan group Holocephali (Figure 1.7D) is comprised of small- to intermediate-sized fishes represented by *Harpagofututor*, *Traquairius*, and the Chimaerimorpha (*Echinochimaera* and the Cochliodontidae). Fetal, larval, juvenile, and adult stages occur for *Harpagofututor* and/or the Chimaerimorpha. Abundance data reveal that *Echinochimaera* and *Harpagofututor* are found across all environments and are the only holocephalans found in the reef environment. The cochliodont contribution to relative holocephalan abundance increases from the upper basin environment (8%) to the bay mouth/basin area (maximum of 24%) but is greatest in the nearshore environment (50%). Members of the menaspoid genus *Traquairius* are restricted to the bay mouth/basin area. A closer examination of the cochliodonts reveals the greatest diversity of their genera in the lower basin/bay mouth and total absence in the reef zone. Toward the nearshore and upper basin areas, the abundance of cochliodonts can be attributed to one species.

Basal chondrichthyans (Figure 1.7B) are the chondrichthyans we have evidence of which fall basal to the euchondrichthyan node in our phylogenetic analysis. Of the forms in our analysis, the majority of abundance across all zones is represented by Iniop (58 to 88%), a unit comprised of six diverse undescribed iniopterygian species. Of the remaining iniopterygians, *Papilionichthys* is entirely restricted to one site at the bay mouth (6%), while *Rainerichthys* generally occurs throughout the bay (7 to 42%). The live-bearing *Delphyodontos* extends from the lower basin/bay mouth (7 to 20%) and reef (10%) to the upper basin (17%), with a greater contribution to basal chondrichthyan abundance along the basin axis.

1.3.1.2.4 *Chondrichthyan Diversity in the Bear Gulch Bay*

Alpha-diversity, within-habitat diversity, is measured by richness (numbers of genera in this analysis) and by the contributions of common and rare genera. Beta-diversity compares the similarity in species composition between habitats, as measured by the Morisita–Horn index (where a high index score indicates low similarity) (Horn, 1977). Results for alpha-diversity reveal that numbers of chondrichthyan genera increase toward the bay mouth and emphasize that rare genera are the greatest component of all habitats. Regional trends in occurrences are evident in all higher taxonomic groups. Chondrichthyan beta diversity (Figure 1.6B) indicates that the bay mouth and upper basin sites are most similar to each other in generic composition, ranging from 0.02 to 0.28. The chondrichthyan composition of the nearshore areas is also quite similar to each other (0.07). The nearshore areas have considerably less similarity to the other bay habitats, with scores ranging from 0.44 to 0.64. Thus, although the assemblages across the habitats share many species, there are large differences between assemblage compositions in the lower vs. upper reaches of the bay.

1.3.1.2.5 *Ecomorphic Analysis*

An ecomorphic analysis was conducted for 80 fish genera of the fauna (eliminating those with more than one missing character) to analyze the possible relationships between morphological characters and habitat use (e.g., Motta et al., 1995; Webb, 1984). Taxa were coded for 13 size-independent qualitative features of gross morphology that reflect possible microhabitat adaptations, life strategy, or niche (pectoral fin position, squamation, dorsal spines by sex, dorsal fins, caudal outline, mouth form, dental type, gape size, feeding type, functional jaw elements, body form, and dorsal spine number). They were then subjected to multiple correspondence analysis to extract the patterns of relationships. Although the details of the methodology and results of the all-inclusive ecomorphic analysis are to be reported elsewhere (Lund et al., in prep.), summary information for all fish groups and details of the chondrichthyans are reported here.

Overall, the results indicated that Chondrichthyes are represented by eight unique and diverse ecomorphs, compared to four in the Osteichthyes and one for *Acanthodes*. One other ecomorph, complex maneuverers, had members of both chondrichthyans and osteichthyans.

Evidence of nine distinct chondrichthyan ecomorphs and the abundance evaluation of these ecomorphs across the defined habitats (Figure 1.8) reveal that these early Carboniferous chondrichthyans were finely attuned to their habitats. Pectoral swimmers (E1, illustrated by the

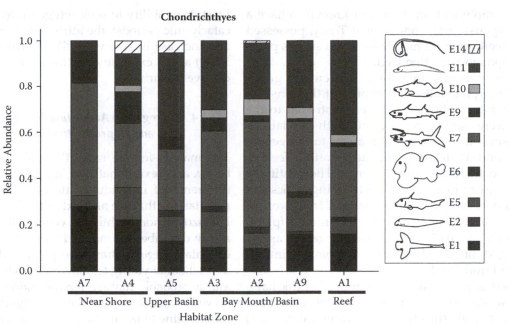

FIGURE 1.8
Relative abundances of the nine chondrichthyan ecomorphs in the Bear Gulch Bay by habitat zone: E1, pectoral swimmers; E2, funnel mouths; E5, demersal and durophagous feeders; E6, complex maneuverers; E7, apex Paleoselachii; E9, Euselachii; E10, spiny-finned heterodont fish; E11, demersal nibblers; and E14, a unique form with a gar-like head and an eel-like body. The habitat zone keys refer to the map of Figure 1.2.

iniopterygians and *Squatinactis*) increased in relative abundance away from the bay mouth. Complex maneuverers (E6, illustrated by *Belantsea* and Elweir) were a small and persistent component at all but the nearshore habitats. Apex predators and smaller Paleoselachii (E7) were the dominant chondrichthyan component of all habitats, while the euselachian ecomorph (E9) increased in relative abundance into the nearshore. Spiny-finned heterodont fish (E10) were a notable component of the bay mouth fauna and were absent from nearshore. Demersal nibblers (E11), principally Holocephali, were a significant faunal component in all areas except the nearshore habitats. The high level of morphological diversity that distinguishes this group from the other gnathostome groups at this time apparently provided the basis upon which the range of ecomorphs identified in the multiple correspondence analysis could be generated.

In sum, the significant changes in chondrichthyan ecomorphic contribution and diversity across the habitats underlie the fine-scale changes in their community structure throughout the bay. The diversity of the Chondrichthyes in this bay probably reflects very fine niche partitioning and microhabitat use, the effects of transient large predators, scarcity of a resource base for many of the highly specialized ecomorphs, and different reproductive strategies than actinopterygians. These analyses of the Bear Gulch Limestone community provide a measure of the fundamental adaptive radiation of the class during the Lower Carboniferous.

1.3.1.3 Community Structure and Population Dynamics

Akin to today's sharks (reviewed by Camhi et al., 1998; Musick et al., 2004), most Bear Gulch and other Carboniferous chondrichthyans appear to have been restricted in their distribution, with the majority confined to continental shelf and slope waters and only a small percentage of the larger taxa being long-distance migrants. The Paleozoic forms reflect a range of habitats and exhibit feeding-based specializations.

The predatory stethacanthids (3 to 3.5 m) represent some of the largest of the Bear Gulch fishes. Their size, the nature of their preserved remains (principally as large but disarticulated remains or of patently juvenile individuals) (Lund, 1985b), and the proximity of these fossils to the deeper aspects of the bay support the interpretation of these as migratory or opportunistic vagrants from adjacent waters. Also, the geographic distribution of stethacanthid sharks ranges as far as current Scotland and Moscow, Russia. Most other paleoselachians (e.g., *Falcatus*, *Damocles*, *Squatinactis*, code-name Tristy) are smaller (~150-mm adults). Because of their size, they were likely to have been persistent inhabitants of, or spent the majority of their time in, the more protective environs of the bay even though they had the ability to extend into the epicontinental sea (Lund, 1985a, 1986a). The highly specialized, eel-shaped form with *Thrinacodus*-style teeth, *Thrinacoselache*, is likely to have ranged from open water to the reef environment.

The only elasmobranch in the fauna known to have a hyostylic suspensorium, code-named Tristy, possessed diminutive teeth in small mobile jaws and so was probably a microphagous suction feeder.

Euchondrocephalans, including the crown group Holocephalimorpha, make up the bulk of the constitutive bay inhabitants, but not all were restricted to the bay. There were various petalodonts with specialized dentitions, several heterodont fish allied to the genera *Chomatodus* and *Venustodus*, and near-bottom feeders such as the cochliodonts and *Debeerius*. The dentitions of this speciose group suggest varied feeding tactics and diets. Fossil data reflect that diets ranged from worms, shrimp, and mollusks to amorphous bituminous (plant) remains that were likely to have been ingested as the fish scavenged or sifted through bottom sediments (Grogan and Lund, 2000).

Basal chondrichthyan taxa range from a benthic habitat (*Delphyodontos*; two predatory iniopterygian taxa, Iniop 1 and 2, with shrimp-filled abdominal cavities) to mid- and upper-water-column swimmers/flyers (*Rainerichthys*, *Papilionichthys*) to the taxon code-named Elweir. This fish had very expansive fins, lacked marginal teeth, bore greatly enlarged labial cartilages, and had an oral rim mechanism resembling that of modern Clupeid teleosts; it was most probably a suction feeder (Grogan, 1993).

Overall, the comparatively smaller chondrichthyans, which also represent the majority of the chondrichthyan forms, display a distribution restricted to the epicontinental sea margins and adjacent shallow bodies of water. Like the smaller shark species of the coastal to inshore environs (Smith et al., 1998), it is probable that these smaller Bear Gulch chondrichthyans matured earlier and were shorter lived compared to the larger, apex predatory stethacanthids. (See Grogan and Lund, 2011, for a greater elaboration of Bear Gulch chondrichthyan reproductive strategies.) By virtue of the range of morphological designs that had become possible earlier in their history and, apparently, by retaining considerable developmental plasticity in cranial and feeding design, the smaller chondrichthyans were able to expand quickly into a variety of habitats and niches and to outnumber non-chondrichthyans in terms of taxonomic diversity (e.g., as specifically demonstrated by the species richness of holocephalimorph forms). Rapid vertebrate diversification such as this can be explained by duplication of body-pattern-determining genes, such as Hox-gene homologues, which permit rapid diversification in form, and by promoting heterochronic manipulation of a common developmental plan through neoteny, progenesis, and/or peramorphosis. Over the subsequent evolution of the group, however, as the majority of surviving forms became increasingly predatory or large (with an oceanic lifestyle or ability to seek refuge in deep waters during cataclysmic periods), the attributes of continued growth and increase in size were likely favored (heterochronically) at the expense of a timely progression to reproductive maturity.

1.3.1.4 Segregation According to Age, Sex, and Reproductive Stage

The male-to-female ratio, the isolation of individuals by sex and sexual maturity, and the sex-associated size difference of individuals are characteristics of extant populations that are also exhibited by some, but not all, Paleozoic Chondrichthyes. Within the Bear Gulch, the newly described (Lund and Grogan, 2004) euchondrocephalan *Gregorius* has been preserved in one instance as a school of young to subadult specimens (both sexes) and a probable mature female. *Falcatus* has been preserved *en masse* and, in the most spectacular case, in a ratio of nine to ten mature males to one immature male and one (supposedly adult) female. The form code-named Orochom has been recovered as a group kill of four to six very immature (neonatal?) individuals. Yet, other forms (e.g., *Debeerius ellefseni*) are principally known from a single sex (males) and show no evidence of group or school-based distribution. Similarly, the holocephalic *Harpagofututor volsellorhinus* is typically found individually, and data across the fossil deposit demonstrate an almost one-to-one ratio of male to female specimens.

Taxa show great variation in size range. Male and female specimens of *Harpagofututor* demonstrate sexual maturity in individuals 110 mm in total length (Grogan and Lund, 1997, 2011), with 155 mm being the longest specimen recovered to date. By contrast, specimens of *Stethacanthus productus* are generally large, with maturity indicated for those attaining 1.5 to 2 m in total length. (The largest *S. "productus"* recovered to date was an estimated 3 to 3.5 m in total length.) Although appreciable evidence is lacking thus far for subadult to adult specimens of smaller size, one partial specimen of *S. altonensis* that shows only the earliest stages of neurocranial mineralization but significant mineralization of vertebral elements was approximately 38 cm. Unfortunately, pelvic information was lacking, due to preservational conditions.

1.3.1.5 Reproductive Strategies

Paleozoic chondrichthyan reproductive strategies have purportedly included oviparity but, more often than not, the precise nature or the ownership of supposed fossil egg cases is difficult to demonstrate. The Devonian egg cases reported by Crookall (1928) are sufficiently different from any known chondrichthyan egg case morphology that it raises the question of whether ownership is just as likely for another coexisting vertebrate with internal fertilization (e.g., placoderms) (cf. Ahlberg et al., 2009; Long et al., 2008, 2009). There is morphological similarity between a chimaeroid-type egg case and material reported from the Devonian Bokkeveld Series of South Africa (Chaloner et al., 1980). Yet, the question remains whether morphological similarity alone correlates with an accurate identification. (In this instance, evidence of Devonian holocephalan fishes is lacking.) *Paleoxyris* fossils (Carboniferous to the Cretaceous) have generally been attributed to hybodonts, as these sharks are in close proximity to finds of *Paleoxyris*, but in most cases there was no evidence for corresponding hybodont remains in the area of the egg capsule recovery. This disparity is plausibly attributed to hybodonts depositing their egg capsules in an environment of different salinity than is typical for the adult (Böttcher, 2010; Fischer et al., 2010).

Our 41 years of excavation across all areas and sections of the Bear Gulch deposit have resulted in the discovery of only *one* possible egg case (Grogan and Lund, in prep.). Microscopic examination suggests similarity to some euselachian egg cases in its general appearance and in details, including a series of respiratory-type apertures. A preliminary CT scan of the fossil indicated structure within the fossilized mass, but a higher resolution examination is necessary for further evaluation. At this stage, the data suggest that this fossil may be an egg case, and, if so, it is more likely to have belonged to a chondrichthyan than a non-chondrichthyan. Given the (1) size of the putative egg case (40 mm in total length, 30 mm minus tendris/horns) and (2) the paucity of other specimens across this well-sampled, fossil-rich deposit, we believe that it probably belonged to a larger, oppoortunistic visitor to the Bear Gulch bay or to a larger resident elasmobranch of the bay. If it is eventually confirmed as an egg case, then the low recovery rate of this material may simply indicate that these egg cases do not fossilize well or that egg cases were not typically deposited in the confines of the Paleozoic Bear Gulch bay. Alternatively, it could be interpreted as an indication that oviparity with an egg case was a rarity at best for the resident chondrichthyan fishes within this bay.

There is evidence of viviparity in Bear Gulch. Direct evidence includes the demonstration of both viviparity and intrauterine feeding in *Delphyodontos* (Lund, 1980), a basal taxon (Figure 1.1). Superfetative viviparity, the first report of such in any fossil fish, is now recognized for *Harpagofututor volsellorhinus* (Grogan and Lund, 2011). Indirect evidence is indicated in a proportionate number of other Bear Gulch euchondrocephalans,

including an aborted cochliodont fetus (MV 6207) and other neonatal-sized cochliodonts and *Echinochimaera* that show evidence of post-embryonic tooth plate development and proportionate total body growth (CM 62713, CM 43164, CM 30625). Tooth plate wear has been observed in one of the latter. Other euchondrocephalans (*Heteropetalus*, *Debeerius*, the orochoms) and another, as yet undescribed, basal form (Elweir) reveal a size range through to sexual maturity. Yet, there is no evidence of a yolk sac in even the smallest of these fossils and no indication of the type of allometric growth that would be expected between embryonic forms and the young or adult states. Given these data, ovoviviparity or viviparity is equally plausible as a reproductive mode for these smaller chondrichthyans but oviparity is not. That viviparity is indicated in a basal Bear Gulch chondrichthyan as well as in a number of holocephalans argues to viviparity as primitive for the Class.

With the acquisition of internal fertilization, oviparity, ovoviviparity, and viviparity all become possible. Of these, oviparity is generally assumed to be the primitive condition. Yet, there is evidence of live birth in two of the early gnathostome groups, the Chondrichthyes and the Placodermi. In the placoderm condition, yolk sac nutrition is implicated in primitive as well as derived taxa; for chondrichthyans there are adelphophagy and superfetation. All are considered relatively derived reproductive modes, yet they are remarkably achieved relatively early in the history of vertebrates (Grogan and Lund, 2011). Obviously, fishes within the Bear Gulch chondrichthyan community were, to their benefit, able to capitalize upon the continuum between oviparity, ovoviparity, and viviparity given prevailing or changing environmental conditions. This undoubtedly contributed to the success of the Chondrichthyes during the Carboniferous, while male display and female choice likely contributed to their morphological diversity (cf. Sander van Doorn et al., 2009).

1.3.2 Upper Carboniferous and Permian Record

The Pennsylvanian (Upper Carboniferous) and Permian chondrichthyan faunas are diminished in diversity above the period boundary but continue the major lineages of the Mississippian adaptive radiation. Edestoid euchondrocephalans extended into the Permian, giving rise to forms such as *Helicoprion* and *Parahelicoprion* (Karpinski, 1899, 1925) and achieving a wide geographic distribution by the latter part of the Early Permian (Chorn, 1978; Nassichuk, 1971). The petalodonts also extended into the Permian with several genera, *Janassa*, "*Janassa*" *korni*, and *Megactenopetalus* (Brandt, 1996; Hansen, 1978; Malzahn, 1968). Their trend was toward a greatly increased size, highly specialized dentition, and an increasingly oceanic distribution. Similarly, the xenacanths continued to flourish in freshwater environments. The hybodonts and helodonts extended into a variety of aquatic environments (Romer, 1952) as the cochliodontimorphs continued as morphologically radical forms such as the Menaspiformes (Lund and Grogan, 1997b; Schaumberg, 1992). Whereas the Carboniferous holocephalimorphs were shallow-water forms that exhibited high morphological diversity, data suggest that a limited number of these lineages later survived the Permo–Triassic extinction, probably by having the ability to extend to deeper waters (see below).

1.4 Theorized Relationships between Recent and Fossilized Forms

1.4.1 On Holocephalan Origins

The chimaeroid/holocephalimorph grouping apparently achieved its greatest diversity during the Carboniferous, and most of the descendant forms appear to have become extinct by the end of the Permian. Thereafter, the holocephalimorphs are represented by the Jurassic *Squaloraja*, *Acanthorhina*, the myriacanthoids, and the chimaeroids. All extant forms are believed to be traced to the last, with *Eomanodon* (Ward and Duffin, 1989) purported to be the oldest. That there are no chimaeroid/holocephalimorph data spanning the hundreds of millions of years from the Permian to the Jurassic has been used to argue that today's chimaeroid fishes are not likely to share a direct ancestry with the Paleozoic forms because it is unlikely for lineages to persist for such an extended period of time (Stahl, 1999). Yet, there is paleontological evidence of chondrichthyan lineages persisting from the Devonian to the Triassic (the xenacanths, ctenacanths) and from the Carboniferous to the Mesozoic (the hybodonts).

Stahl (1999) argued that there are no Paleozoic holocephalans after the Permian, that a decline in the number of fossil finds (which are principally tooth plates) reflects a holocephalan decline after the Carboniferous, and that the Mesozoic forms are of two groups that diverged from some yet to be discovered Permian basal group. We disagree. We find the logic posed in the first part of this argument to be faulty, because a lack of evidence does not equate as evidence of extinction or loss. Rather, it is plausible that some holocephalans survived the Permian by having sought refuge in or having adopted a deeper water lifestyle, as is evidenced by the cochliodonts of the Permian Phosphoria Formation. Any remains of these forms would,

necessarily, have a very low probability of preservation and recovery due to inaccessibility, lower potential of fossilization, and loss due to subductive forces acting on the ocean floor. More importantly, however, and in response to the second half of the argument, the "unusual" or odd morphologies of Jurassic myriacanthids, Squalorajidae, and chimaeroids show such confluence with Carboniferous taxa that it is difficult to dismiss direct developmental links between the Permo–Carboniferous and Mesozoic forms (Lund and Grogan, 1997b, 2004). Repeated cladistic analyses (Figure 1.1) using revised and expanded character matrices consistently associate these forms in a highly stable, robust topology, which traces both modern chimaeroid and the other Mesozoic holocephalan lineages from the Carboniferous Holocephali, a derived group of euchondrocephalans (Lund and Grogan, 1997b). Thus, the Holocephali comprise (Squalorajiformes + [Chondrenchelyiformes + Menaspiformes]) and (cochliodonts + Myriacanthiformes + Chimaeriformes). Of these groups, only the Chondrenchelyiformes and the Menaspiformes appear to not have survived the Permo–Triassic extinction or, inversely, to have direct links across the Permo–Triassic boundary.

1.4.2 On Elasmobranch Origins

Neoselachians (*sensu* de Carvalho, 1996) are monophyletically defined as a common ancestor of all living forms plus all of its descendants. Gaudin (1991), Shirai (1996), and de Carvalho (1996) have provided cladistic models for the relationships between the Recent chondrichthyans (principally elasmobranchs) and Mesozoic forms. (Paleozoic forms were included in the analyses, but the treatment of these select chondrichthyans is so limited as to essentially render them into a Hennegian ladder of distant sister group associations.) The latter analyses resolve neoselachians into two major divisions (de Carvalho's Galeomorphi and Squalea vs. Shirai's Galea and Squalea), with origins in the Jurassic and an appreciable diversification in the Cretaceous. The early neoselachians, like many Paleozoic forms, were principally nearshore predators but offshore predators by the mid-Cretaceous. So, the Jurassic–Cretaceous neoselachian radiation is attributed to an increased availability of basal neopterygians as prey (Thies and Reif, 1985).

Euselachians are neoselachians plus those Paleozoic and Mesozoic forms deemed as the closest allies to neoselachians. Data indicate that cladodonts, the stethacanthids and their allies (*Thrinacoselache*), and the xenacanths are all primitive elasmobranchian members of the chondrichthyan crown group; therefore, these paleoselachians have no direct relationship to any recent form generally referred to as a shark. Ctenacanths and hybodonts, however, are often invoked as likely neoselachian allies. Zangerl and Case (1973) proposed that recent forms, the Paleozoic ctenacanths and the Paleozoic and Mesozoic hybodonts (their phalacanthous sharks), are monophyletic on the basis of dorsal fin spine structure and a tribasal pectoral fin. Maisey (1975) subsequently specified that dorsal fin spine morphology supports a closer relationship between modern elasmobranchs and the ctenacanths rather than hybodonts but later demonstrated paraphyly for the ctenacanth assemblage (and, consequently, the paraphyly of Zangerl's phalacanthous grouping). No significant new ctenacanth evidence has been presented since. As such, then, the Paleozoic ctenacanth information is too incomplete for further cladistic discussion with euselachians at this time. Although ctenacanth spine and scale evidence clearly extends at least to the Devonian, overall the evidence of the group is scanty and generally ranges from slightly informative to uninformative. The Bear Gulch ctenacanths have preserved very poorly and provide little information beyond occurrence; consequently, where and how these forms lived and any qualitative indication of either their numbers or diversity remains essentially elusive (Maisey, 1981, 1982b, 1983).

As for the early hybodont record, microremains are purported to exist as far back as the Devonian (Lelievre and Derycke, 1998), and the Mesozoic *Hybodus* originally appeared to provide a possible link to the neoselachians through *Heterodontus* (reviewed in Maisey, 1982a, 1989). In an attempt to link neoselachians to Paleozoic and Mesozoic hybodonts, Young (1982) proposed *Hybodus*, *Tristychius*, and *Onychoselache* as the sister group to living elasmobranchs. Maisey refined this further, ultimately arguing that hybodonts are a monophyletic sister group to the neoselachians, with the latter grouping comprising modern elasmobranchs plus Mesozoic forms including *Synechodus* and *Paleospinax* (Maisey, 1982a, 1984, 1985, 1987, 1989, 2004). This particular paradigm of relationships was especially strengthened by the discovery of the first appreciably detailed and whole-body evidence of the Upper Carboniferous *Hamiltonichthys* (Maisey, 1989). These specimens permitted the first qualitative morphological comparison of a Paleozoic hybodont with the Mesozoic forms and thus helped to firmly establish the phylogenetic position that is most likely for hybodonts relative to neoselachians (i.e., as the monophyletic sister group). Klug (2010) now provides evidence that a monophyletic Synechodontiformes with a suite of neoselachian characters is a sister group to all living sharks. On the basis of this study, modern sharks have their point of origin in the Late Permian (250 mya).

1.5 Cladistic Evaluation of Paleozoic Chondrichthyan Relationships and Comments on the Higher Systematic Groupings of Chondrichthyans

The matrix (ver. Chon55) of 120 characters and 53 taxa used in this analysis (see the Appendix) is an extensive revision of previously referenced matrixes (e.g., Grogan and Lund, 2008) that incorporates new information and more taxa. The matrix was analyzed with the software program TNT version 1.1 (Goloboff et al., 2008). Every effort has been made to incorporate a broad suite of characters and states into the analysis. The distributions of characters are: 0 to 39, cranial; 40 to 66, branchial and dental; 67 to 85, squamation; and 86 to 119, postcranial. Character states are treated non-additively (unordered). A zero-based hypothetical taxon, one acanthodian, one placoderm, and three osteichthyans serve as outgroups. Two trees result from the analysis (Figure 1.1): one in which the Chondrichthyes are the sister group of all other gnathostomes and another in which the Osteichthyes are basal to a cluster of the placoderm *Dicksonosteus* and the acanthodian *Acanthodes*, which in turn is the sister group to the Chondrichthyes. The Chondrichthyes resolves into two groups basal to the Euchondrichthyes, and the Euchondrichthyes resolves into the Elasmobranchii and the Paraselachii. The Paraselachii consists of a series of orders with autodiastylic suspensoriums plus the holostylic Holocephali.

There has been much discussion about the merits, or potentially destabilizing effects upon trees, of including taxa that have a large percent of missing data (Goloboff et al., 2008; Kearney and Clark, 2003; Santini and Tyler, 2004; Wiens, 1998). Few taxa of Paleozoic chondrichthyans offer multiple or complete specimens. Each of the taxa of this analysis that are missing the most characters contains information that does not introduce significant ambiguities concerning their position on the tree. *Traquairius nudus* (63% of matrix characters), *Ornithoprion hertwigi* (67% of matrix characters), and *Srianta iarlis* (76% of matrix characters) have been included in previous versions of the matrix (Grogan and Lund, 2009; Lund and Grogan, 2004a) and have been relatively consistent in their relationships throughout, although the relative positions of species of *Srianta* and *Traquairius* shift with different analytical procedures. *Doliodus* possesses only 36% of matrix characters; however, the position of *Doliodus* on our trees has remained stable through several matrix revisions as it does display characters that are critical to resolving its position among the lower Elasmobranchii (Grogan and Lund, 2009). The high percentage and critical nature of missing data (only 17% of matrix characters) for the chondrichthyan taxon *Pucapampella* produce significant ambiguities in its position near the base of the tree. For this reason, *Pucapampella* was not included in this analysis. For the same reasons, it is clearly not possible to include scale or tooth taxa in the current type of analysis. The so-called "symmoriid" elasmobranchs consist of composite individuals that lack the necessary data for higher systematic-level analysis (Zangerl, 1981; Zangerl and Case, 1973), but they are anatomically and cladistically indistinguishable from juvenile and female Stethacanthoidei. It is not currently possible to resolve the position of "symmoriids" relative to the Stethacanthoidei without diagnostic information about intact individuals of recognizable life history stages and sexes.

1.5.1 On the Higher Systematic of the Early Chondrichthyes

(See Figure 1.1 and the Appendix.) This analysis produces a well-supported monophyletic Class Chondrichthyes. There are no unambiguous relationships between the Chondrichthyes and any other gnathostome class. Iniopterygii is a monophyletic taxon basal to all other Chondrichthyes, and the families Iniopterygidae and Sibyrhynchidae (Zangerl and Case, 1973) are sustained, although not on the originally suggested characters of Zangerl and Case (1973) or Zangerl (1981, 1997). The Upper Mississippian iniopterygians *Rainerichthys* and *Papilionichthys* (Grogan and Lund, 2009) cluster in the Iniopterygidae. The remainder of the Bear Gulch Limestone Iniopterygii map as a separate suborder. *Delphyodontos dacriformes* is placed as a unique basal taxon of the Chondrichthyes below the Euchondrichthyes. The Euchondrichthyes is a well-supported node that marks the divergence of the Elasmobranchii and Paraselachii, the two principle adaptive radiations of Carboniferous chondrichthyans that lead to the modern crown groups Euselachii and Chimaeriformes. The topology and contents of the Paraselachii have remained consistent through many analyses, with the exception of early coding attempts for the Pennsylvanian Iniopterygii (Lund and Grogan, 2004; Nelson, 2006; Stahl, 1999; Zangerl, 1981).

The putative family Gregoriidae and genus *Srianta* (Lund and Grogan, 2004a) are weakly supported here, reflecting the absence of many data points in this species assemblage. The Orders Petalodontiformes and Orodontiformes include many predominantly tooth taxa; there are few holomorphous specimens and a paucity of phyletically useful information obtainable from these holomorphs. The Order Eugeneodontiformes is supported by abundant cranial, dental, and postcranial data, albeit from composite individuals (Zangerl, 1981). The Holocephalimorpha are a clade supported by many characters and states. Subsumed above this node

are the Holocephali *sensu stricto*, those chondrichthyans with a specific form of neurocranium, holocephalic suspensorium, branchial arrangement, and tooth plates. A number of organ taxa, such as psephodonts, psammodonts, and copodonts, have been traditionally treated as the equivalent of holocephalimorphs because of their close similarity in tooth histology to the cochliodonts (Patterson, 1965). However, the state of their suspensoriums and the morphology of their heads are unknown.

The Cochliodontimorpha are holocephalan fish and isolated tooth plates of common morphology and histology that have been studied by Owen (1867), Newberry and Worthen (1870), St. John and Worthen (1883), Davis (1883), and subsequent scholars. Unfortunately, *Cochliodus* itself is known only by upper and lower dentitions (Davis, 1883; Grogan and Lund, 1997; Owen, 1867). The Chondrenchelyiformes (Lund, 1982) and Squalorajiformes (*Squaloraja*) (Patterson, 1965) are uniquely specialized cochliodontimorph orders. Menaspiformes are a well-supported order. The fishes included in the Chimaerimorpha, Chimaeriformes + *Acanthorhina*, are conservative in head, body, and dental characteristics as far as they are known. The node for the Chimaeriformes has been placed to exclude *Acanthorhina* because the characters supporting its inclusion are principally dorsal fin characters, whereas those characters supporting a more restricted Chimaeriformes involve more wide-ranging morphological transformations.

1.5.2 Other Concluding Remarks on the Origins of Chondrichthyans, Trends in Chondrichthyan Evolution, and on Characters of the Class

The Chondrichthyes are a monophyletic clade and are principally distinguished based on two unique autapomorphous character sets. These are the development of tesserate endoskeletal mineralization and internal fertilization with copulation by means of intromittent organs (claspers and their supporting structures) developed from extensions of the pelvic plate or axis of mature males. The Chondrichthyes share the basic patterns of their scale development with several agnathan scale types, but there is yet no morphological evidence to support the assignment of some Ordovician and Silurian scale types to the Chondrichthyes. Serious questions about the phylogenetic significance of isolated scales render it impossible to use placoid scales as a distinguishing character of the Chondrichthyes. Furthermore, it is important to consider that chondrichthyan scales are not restricted to the placoid form for those with squamation.

As chondrichthyan teeth are absent from the earlier scale-bearing deposits, a toothless condition or one in which teeth are not differentiated from scales may still be considered a viable state for the earliest chondrichthyans. However, patterns of development suggest a progressive transition from individual homodont teeth derived from single odontodes of simple topography to partially or entirely fused organs derived from multiple odontodes with complex topographies. Beyond this, the types and positions of dental organs vary depending on the status of premandibular and mandibular arch elements and the presence of appropriate preoral, palatal, mandibular, and buccopharyngeal inductive fields. The holocephalan dental developmental trend culminates in expansive dental organ fusions, which are identified as tooth plates. A symphysial tooth organ or complex is apparently plesiomorphous for all gnathostomes minus Placodermi, but is absent in Elasmobranchii.

Outgroup comparisons led to the conclusion that the pelvic fin radials plesiomorphously articulated with a horizontally oriented, paired pelvic plate. There is a notable trend to shift the articulation of the radials onto the basipterygium in both Elasmobranchii and Euchondrocephali. In the hybodont Elasmobranchii, a further change in pelvic structure results in the development of a puboischiadic bar that separates the pelvic fins widely across the midline. Holocephalans maintain the primitive and plesiomorphic condition of separate halves to the pelvic girdle. Some Holocephali, however, adopt an alternative mode of pelvic muscular support by the elaboration of an iliac process and the differentiation of a prepelvic tenaculum from a disjunct anterior fin radial. A tall iliac process has also evolved independently in some Petalodontiformes and Iniopterygii.

Neurocrania of basal Chondrichthyes are composed of several units separated by a variety of fissures; the occipital unit is particularly variable among them. Aspects of these conditions approach the neurocranial conditions of the other gnathostome classes. Most Elasmobranchii and Paraselachii demonstrate single-unit neurocrania, although there is still variable occurrence of a separate or partially separate occipital unit in lower taxa (e.g., *Rainerichthyes, Papilionichthyes, Dephyodontos, Pucapampella, Doliodus*). The Elasmobranchii and Paraselachii express divergent trends in the proportions of the neurocranium, the structure, and the support of the visceral skeleton, which reflect on their inhabitation of distinctly different feeding styles and modes. The scheme of chondrichthyan relationships presented here reflects a diversity of the chondrichthyans during the Carboniferous and, possibly, into the Permian, which was at least an order of magnitude greater than that of osteichthyans. Moreover, the nature of the diversity is striking. The morphological diversity of the actinopterygians reflects what are, fundamentally, transitions in *individual* characters (e.g., cranial bone shape), whereas the diversity of the chondrichthyans reflects significant morphological modification and coordination in *suites* of characters. (Thus, Bear Gulch chondrichthyans show a far greater range of ecomorphs compared to the actinopterygians.) What this may say about the evolutionary history, reproductive history, or even the developmental genetics of these distinct vertebrate groupings has yet to be fully resolved; yet, it is clear that the differences between these early vertebrates were already emerging by the end of the Devonian, and that any common evolutionary history was already distant by this time.

For the chondrichthyan fishes to have capitalized on the emerging diversity of ecological and environmental settings of the Devonian and Mississippian would have required, *a priori*, some degree of genetically inherent adaptiveness (e.g., due to the generation of Hox-gene paralogues) that might best be correlated with an earlier (pre-Devonian) radiation event. Alternatively, it would have depended on the environmental/evolutionary selection for those Devonian forms that, by that point in time, were phenotypically expressing the consequences of such duplication. In this context, then, it is conceivable that a Silurian basal radiation would have fundamentally supported the range of chondrichthyan diversity that is identified by the Lower Devonian (and that fueled the subsequent apex of Paleozoic diversity) while also allowing the possibility for retention of gnathostome stem group features in some members. For example, some Lower to Mid-Devonian chondrichthyans display variation in a cranial feature that, until recently, was strictly associated with acanthodians and osteichthyans (Janvier, 1996). An oticooccipital fissure, albeit variable in the extent of its development, has now been confirmed within a range of elasmobranchs: *Orthacanthus* and *Tamiobatis* (Schaeffer, 1981); the Bearsden *Stethacanthus* (Coates and Sequeira, 1998); and *Cladodoides wildungensis* (Maisey, 2005), *Gutturensis* (*Cladodus*) *nielseni* (Sequiera and Coates, 2000), and *Cobelodus* (Maisey, 2000) from the Devonian and Carboniferous. Of the chondrichthyans noted to date, only the Devonian pucapampellids retain the primitive gnathostome cranial fissure generated by the confluence between an oticooccipital fissure and a ventral otic fissure (Maisey, 2001). A midorbital intracranial joint and separate oticooccipital unit have been identified in Upper Mississippian Iniopterygii (Grogan and Lund, 2009). The Mississippian phoebodont *Thrinacoselache* retains a series of occipital zygal elements (Grogan and Lund, 2008). Furthermore, the morphology of articulated endoskeletal, tooth, and scale information from Carboniferous protacrodonts, orodonts, and gregoriids (Lund and Grogan, 2004a), and that of the protacrodontid organ (scale)-species that extends to the Devonian collectively suggest that these are basal forms, not far removed from stem chondrichthyans.

Acknowledgments

The authors are indebted to the Montana families and friends who have made the Bear Gulch fieldwork possible. We thank them and our field crew for their special contributions to this and other Bear Gulch research. We also acknowledge the contributions of the reviewers and thank many colleagues, including Jack Musick, for numerous enthusiastic discussions of both fossil and extant chondrichthyans. Our discussions with, and the contributions of, the late Joe Nelson to chondrichthyan systematics are particularly appreciated by the authors. He will be missed. The Willi Hennig Society has made the software program TNT available on its website.

References

Ahlberg, P., Trinajstic, K., Johanson, S., and Long, J. (2009). Pelvic claspers confirm chondrichthyan-like internal fertilization in arthrodires. *Nature* 460:888–889.

Anderson, M.E., Almond, J.E., Evans, F.J., and Long, J.A. (1999). Devonian (Emsian–Eifelian) fish from the Lower Bokkeveld Group (Ceres Subgroup), South Africa. *J. Afr. Earth Sci.* 29:179–184.

Applegate, S.P. (1967). A survey of shark hard parts. In: Gilbert, P.W., Matthewson, R.F., and Rall, D.P. (Eds.), *Sharks, Skates, and Rays.* The Johns Hopkins University Press, Baltimore, MD, pp. 37–67.

Basden, A.M. and Young, G.C. (2001). A primitive actinopterygian neurocranium from the early Devonian of southeastern Australia. *J. Vertebr. Paleontol.* 21:754–766.

Basden, A.M., Young, G.C., Coates, M.I., and Ritchie, A. (2000). The most primitive osteichthyan braincase? *Nature* 403:185–188.

Böttcher, R. (2010). Description of the shark egg capsule *Palaeoxyris friessi* n. sp. from the Ladinian (Middle Triassic) of SW Germany and discussion of all known egg capsules from the Triassic of the Germanic Basin. *Palaeodiversity* 3:123–139.

Brandt, S. (1996). *Janassa korni* (Weigelt)—Neubeschreibung eines petalodonten Elasmobranchiers aus dem Kupferschiefer und Zechsteinkalk (Perm) von Eisleben (Sachsen-Anhalt). *Paläontol. Z.* 70:505–520.

Brazeau, M.D. (2009). The braincase and jaws of a Devonian "acanthodian" and modern gnathostome origins. *Nature* 457:305–308.

Camhi, M., Fowler, S., Musick, J., Bräutigam, A., and Fordham, S. (1998). *Sharks and Their Relatives: Ecology and Conservation,* Occasional Papers of the IUCN Species Survival Commission, No. 20. International Union for Conservation of Nature, Gland, Switzerland, 39 pp.

Cappetta, H., Duffin, C.J., and Zidek, J. (1993). Chondrichthyes. In: Benton, M.J. (Ed.), *The Fossil Record,* Vol. 2. Chapman & Hall, London, pp. 593–609.

Carroll, R.L. (1988). *Vertebrate Paleontology and Evolution.* Freeman, New York, 648 pp.

Chaloner, W.G., Forey, P.L., Gardiner, B.G., Hill, A.J., and Young, V.T. (1980). Devonian fish and plants from the Bokkeveld Series of South Africa. *Ann. S. Afr. Mus.* 81:127–156.

Chorn, J. (1978). *Helicoprion* (Elasmobranchii, Edestidae) from the Bone Spring formation (Lower Permian of West Texas). *Univ. Kans. Paleontol. Contrib.* 89:2–4.

Coates, M.I. (1988). A New Fauna of Namurian (Upper Carboniferous) Fish from Bearsden, Glasgow, PhD dissertation, University of Newcastle upon Tyne, England.

Coates, M.I. and Sequeira, S.E.K. (1998). The braincase of a primitive shark. *Trans. R. Soc. Edinburgh Earth Sci.* 89:63–85.

Compagno, L.J.V. (1984). *FAO Species Catalogue.* Vol. 4. *Sharks of the World: An Annotated and Illustrated Catalogue of Shark Species Known to Date.* Part 1. *Hexanchiformes to Lamniformes.* United Nations Food and Agriculture Organization, Rome, 250 pp.

Compagno, L.J.V. (2001). *FAO Species Catalogue for Fishery Purposes. Sharks of the World: An Annotated and Illustrated Catalogue of Shark Species Known to Date.* Vol. 2. *Bullhead, Mackerel and Carpet Sharks (Heterodontiformes, Lamniformes, and Orectolobiformes).* United Nations Food and Agriculture Organization, Rome, 269 pp.

Crookall, R. (1928). Paleozoic species of *Vetacapsula* and *Palaeoxyris. Rep. Geol. Surv. Great Britain, Summary of Progress for 1927* 2:87–107.

Daniel, J.F. (1934). *The Elasmobranch Fishes.* University of California Press, Berkeley, 332 pp.

Davis, J.W. (1883). On the fossil fishes of the Carboniferous Limestone Series of Great Britain. *Sci. Trans. R. Dublin Soc.* 2:327–600.

de Carvalho, M.R. (1996). Higher-level elasmobranch phylogeny, basal squaleans, and paraphyly. In: Stiassny, M., Parenti, L.R., and Johnson, G.D. (Eds.), *Interrelationships of Fishes.* Academic Press, New York, pp. 35–62.

Denison, R. (1978). Placodermi. In: Schultze, H.-P. (Ed.), *Handbook of Paleoichthyology,* Vol. 2. Gustav Fischer, Stuttgart, 128 pp.

Downs, J.P. and Daeschler, E.B. (2001). Variation within a large sample of *Ageleodus pectinatus* teeth (Chondrichthyes) from the late Devonian of Pennsylvania, USA. *J. Vertebr. Paleontol.* 21:811–814.

Ettensohn, F.R. and Barron, L.S. (1981). Depositional model for the Devonian–Mississippian black shales of North America: a paleoclimatic–paleogeographic approach. In: Roberts, T.G. (Ed.), *GSA Cincinnati 1981 Field Trip Guidebooks,* Vol. 2. American Geological Institute, Alexandria, VA, pp. 344–357.

Farris, J.S. (1988). *Hennig86,* version 1.5. Distributed by author, Stony Brook, NY.

Fischer, J., Axsmith, B.J., and Ash, S.R. (2010). First unequivocal record of the hybodont shark egg capsule *Palaeoxyris* in the Mesozoic of North America. *N. Jb. Geol. Paläont. Abh.* 255:327–344.

Fluharty, C. and Grogan, E.D. (1999). Chondrichthyan calcified cartilage: chimaeroid cranial tissue with reference to *Squalus.* In: *Abstracts of the 1999 Joint Meeting of the American Elasmobranch Society/American Society of Ichthyologists and Herpetologists,* Pennsylvania State University, University Park, p. 105.

Friedman, M. and Brazeau, M.D. (2010). A reappraisal of the origin and basal radiation of the Osteichthyes. *J. Vertebr. Paleontol.* 30:36–56.

Gagnier, P.Y., Paris, F., Racheboeuf, P., Janvier, P., and Suarez-Riglos, M. (1989). Les vertebres de Bolivie: Sonnées biostratigraphiques et anatomiques complémentaires. *Bull. Inst. Fr. Études Andines* 18:75–93.

Gaudin, T.J. (1991). A re-examination of elasmobranch monophyly and chondrichthyan phylogeny. *N. Jahrb. Geol. Paleontol. Abh.* 182:133–160.

Ginter, M. (1999). Famennian–Tournaisian chondrichthyan microremains from the Eastern Thuringian Slate Mountains. *Abh. Ber. Naturkundemus. Görlitz* 21:25–47.

Ginter, M. (2000). Chondrichthyan biofacies in the Upper Famennian of Western USA. In: *Abstracts of the Ninth Annual International Symposium, Early Vertebrates/Lower Vertebrates*, Flagstaff, AZ, pp. 8–9.

Ginter, M. (2001). Chondrichthyan biofacies in the Late Famennian of Utah and Nevada. *J. Vertebr. Paleontol.* 21:714–729.

Ginter, M. (2004). Devonian sharks and the origin of Xenacanthiformes. In: Arratia, G., Wilson, M.V.H., and Cloutier, R. (Eds.), *Recent Advances in the Origin and Early Radiation of Vertebrates*. Verlag Dr. Friedrich Pfeil, Munich, pp. 473–486.

Ginter, M., Hampe, O., and Duffin, C. (2010). Chondrichthyes IV: Paleozoic Elasmobranchii: Teeth. In: Schultz, H.-P. (Ed.), *Handbook of Paleoichthyology*, Vol. 3D. Verlag Dr. Friedrich Pfeil, Munich, 168 pp.

Goloboff, P.A., Farris, J.S., and Nixon, K.C. (2008). TNT, a free program for phylogenetic analysis. *Cladistics* 24:774–786.

Goujet, D. (1976). Les poissons in les schistes et calcaires éodévoniens de Saint-Céneré (Massif Amouricain, France). *Mem. Soc. Geol. Mineral. Bretagne* 19:313–323.

Grogan, E.D. (1993). The Structure of the Holocephalan Head and the Relationships of the Chondrichthyes, PhD dissertation, College of William and Mary, School of Marine Sciences, Gloucester Point, VA, 240 pp.

Grogan, E.D. and Lund, R. (1997). Soft tissue pigments of the Upper Mississippian chondrenchelyid *Harpagofututor volsellorhinus* (Chondrichthyes, Holocephali) from the Bear Gulch Limestone, Montana, USA. *J. Paleontol.* 71:337–342.

Grogan, E.D. and Lund R. (2000). *Debeerius ellefseni* (fam. nov., gen. nov., spec. nov.), an autodiastylic chondrichthyan from the Mississippian Bear Gulch Limestone of Montana (USA), the relationships of the Chondrichthyes, and comments on Gnathostome evolution. *J. Morphol.* 243:219–245.

Grogan, E.D. and Lund R. (2002). The geological and biological environment of the Bear Gulch Limestone (Mississippian of Montana, USA) and a model for its deposition. *Geodiversitas* 24:295–315.

Grogan, E.D. and Lund R. (2008). A basal elasmobranch, *Thrinacoselache gracia* n. gen. & sp. (Thrinacodontidae, New Family), from the Bear Gulch Limestone, Serpukhovian of Montana, USA. *J. Vertebr. Paleontol.* 28:970–988.

Grogan, E.D. and Lund R. (2009). Two new Iniopterygians (Chondrichthyes) from the Mississippian (Serpukhovian) Bear Gulch Limestone of Montana and a new form of chondrichthyan neurocranium. *Acta Zool. (Stockholm)* 90(Suppl. 1):134–151.

Grogan, E.D. and Lund R. (2011). Superfoetative viviparity in a Carboniferous chondrichthyan and reproduction in early gnathostomes. *Zool. J. Linnean Soc.* 161:587–594.

Grogan, E.D. and Yucha, D.T. (1999). Endoskeletal mineralization in *Squalus*: types, development, and evolutionary implications. In: *Abstracts of the 1999 Joint Meeting of the American Elasmobranch Society/American Society of Ichthyologists and Herpetologists*. Pennsylvania State University, University Park, p. 108.

Hammer, Ø., Harper, D.A.T., and Ryan, P.D. (2001). PAST: paleontological statistics software package for education and data analysis. *Palaeontol. Electronica* 4:1–9 (http://palaeo-electronica.org/2001_1/past/issue1_01.htm).

Hanke, G.F. and Wilson, M.V.H. (2010). The putative stem-group chondrichthyans *Kathemacanthus* and *Seretolepis* from the Lower Devonian MOTH locality, Mackenzie Mountains, Canada. In: Elliott, D.K., Maisey, J.G., Yu, X., and Miao, D. (Eds.), *Morphology, Phylogeny and Paleobiogeography of Fossil Fishes*. Verlag Dr. Friedrich Pfeil, Munich, pp. 159–182.

Hansen, M.C. (1978). A presumed lower dentition and a spine of a Permian petalodontiform chondrichthyan, *Megactenopetalus kaibabanus*. *J. Paleontol.* 52:55–60.

Hansen, M.C. (1999). The geology of Ohio—the Devonian. *Ohio Geol.* 1:1–7.

Heidtke, U. and Krätschmer, K. (2001). *Gladbachus adentatus* nov. gen. et sp., ein primitiver Hai aus dem Oberen Givetium (Oberes Mitteldevon) der Bergisch Gladbach-Paffrath-Mulde (Rheinisches Schiefergebirge). *Mainzer Geowiss. Mitt.* 30:105–122.

Heinicke, M.P., Naylor, G.J.P., and Hedges, S.B. (2009). Cartilaginous fishes (Chondrichthyes). In: Hedges, S.B. and Kumar, S. (Eds.), *The Timetree of Life*. Oxford University Press, Oxford, UK, pp. 320–327.

Horn, H.S. (1977). Measurement of "overlap" in comparative ecological studies. *Am. Nat.* 100(914):419–423.

Ivanov, A.O. and Rodina, O.A. (2002a). Givetian–Famennian phoebodontid zones and their distribution. In: *Proceedings of the Fifth Baltic Stratigraphical Conference*, Vilnius, Lithuania, pp. 66–68.

Ivanov, A.O. and Rodina, O.A. (2002b). Change of chondrichthyan assemblages in the Frasnian/Famennian boundary of Kuznetsk Basin. In: Yushkin, N.P., Tsyganko, V.S., and Männik, P. (Eds.), *Geology of the Devonian System: Proceedings of the International Symposium*, July 9–12, 2002, Syktyvkar, Komi Republic, Russia, pp. 84–87.

Janvier, P. (1996). *Early Vertebrates*. Oxford, New York, 408 pp.

Janvier, P. and Maisey, J.G. (2010). The Devonian vertebrates of South America and their biogeographical relationships. In: Elliott, D.K., Maisey, J.G., Yu, X., and Miao, D. (Eds.), *Morphology, Phylogeny and Paleobiogeography of Fossil Fishes*. Verlag Dr. Friedrich Pfeil, Munich, pp. 431–459.

Janvier, P. and Suarez-Riglos, M. (1986). The Silurian and Devonian vertebrates of Bolivia. *Bull. Inst. Fr. Études Andines* 15:73–114.

Karatajute-Talimaa, V. (1992). The early stages of the dermal skeleton formation in chondrichthyans. In: Mark-Kurik, E. (Ed.), *Fossil Fishes as Living Animals*. Academy of Sciences of Estonia, Tallinn, pp. 223–232.

Karatajute-Talimaa, V. and Predtechenskyj, N. (1995). La repartition des vertébrés dans l'Ordovicien terminal et le Silurien inférieur des paléobassins de la Plateforme sibérienne. *Bull. Mus. Natl. Hist. Nat. C* 17:39–55.

Karpinsky, A. (1899). Über die Reste von Edestiden und die neue Gattung *Helicoprion*. *Verh. Russ.-Kais. Mineral Ges.* 36:361–475.

Karpinsky, A. (1925). Sur une nouvelle troouvaille de restes de *Parahelicoprion* et sur relations de ce genre avec *Campodus*. *Livre Jubil. Soc. Géol. Belg.* 1:127–137.

Kearney, M. and Clark, J.M. (2003). Problems due to missing data in phylogenetic analyses including fossils: a critical review. *J. Vertebr. Paleontol.* 23:263–274.

Kemp, N.E. and Westrin, S.K. (1979). Ultrastructure of calcified cartilage in the endoskeletal tesserae of sharks. *J. Morphol.* 160:75–102.

Klimley, A.P. (1987). The determinants of sexual segregation in the scalloped hammerhead, *Sphyrna lewini*. *Environ. Biol. Fishes* 18:27–40.

Klug, S. (2010). Monophyly, phylogeny and systematic position of the †Synechodontiformes (Chondrichthyes, Neoselachii). *Zool. Scripta* 39:37–49.

Lelievre, H. and Derycke, C. (1998). Microremains of vertebrates near the Devonian–Carboniferous boundary of southern China (Hunan Province) and their biostratigraphical significance. *Rev. Micropaléontol.* 41:297–320.

Long, J.A. (1984). New phyllolepid placoderms from Victoria and the relationships of the group. *J. Linn. Soc. New South Wales* 107:263–308.

Long, J. A. (1997). Ptyctodontid fishes (Vertebrata, Placodermi) from the Late Devonian Gogo Formation, Western Australia, with a revision of the European genus Ctenurella Ørvig, 1960. *Geodiversitas* 19:515–555.

Long, J.A., Trinajstic, K., Young, G.C., and Senden, T. (2008). Live birth in the Devonian period. *Nature* 453:650–652.

Long, J.A., Trinajstic, K., and Johanson, Z. (2009). Devonian arthrodire embryos and the origin of internal fertilization in vertebrates. *Nature* 457:1124–1127.

Lund, R. (1976). General geology and vertebrate biostratigraphy of the Dunkard basin. In: Falke, H. (Ed.), *The Continental Permian in Central, West, and South Europe*, NATO Science Series C. Springer, New York, pp. 225–239.

Lund, R. (1977). New information on the evolution of the Bradyodont Chondrichthyes. *Fieldiana Geol.* 33:521–539.

Lund, R. (1980). Viviparity and intrauterine feeding in a new holocephalan fish from the Lower Carboniferous of Montana. *Science* 209:697–699.

Lund, R. (1985a). The morphology of *Falcatus falcatus* (St. John and Worthen), a Mississippian stethacanthid chondrichthyan from the Bear Gulch Limestone of Montana. *J. Vertebr. Paleontol.* 5:1–19.

Lund, R. (1985b). Stethacanthid elasmobranch remains from the Bear Gulch Limestone (Namurian E2b) of Montana. *Am. Mus. Nat. Hist. Novit.* 2828:124.

Lund, R. (1986a). On *Damocles serratus* nov. gen. et sp. (Elasmobranchii: Cladodontida) from the Upper Mississippian Bear Gulch Limestone of Montana. *J. Vertebr. Paleontol.* 6:12–19.

Lund, R. (1986b). The diversity and relationships of the Holocephali. In: Uyeno, T., Arai, R., Taniuchi, T., and Matsuura, K. (Eds.), *Indo-Pacific Fish Biology: Proceedings of the Second International Conference on Indo-Pacific Fishes*. Ichthyological Society of Japan, Tokyo, pp. 97–116.

Lund, R. (1989). New Petalodonts (Chondrichthyes) from the Upper Mississippian Bear Gulch Limestone (Namurian E2b) of Montana. *J. Vertebr. Paleontol.* 9:350–368.

Lund, R. (1990). Chondrichthyan life history styles as revealed by the 320 million years old Mississippian of Montana. *Environ. Biol. Fishes* 27:1–19.

Lund, R. and Grogan, E.D. (1997a). Relationships of the Chimaeriformes and the basal radiation of the Chondrichthyes. *Rev. Fish Biol. Fisher.* 7:65–123.

Lund, R. and Grogan, E.D. (1997b). Cochliodonts from the Bear Gulch Limestone (Mississippian, Montana, USA) and the evolution of the Holocephali. In: Wolberg, D.L., Stump, E., and Rosenberg, G. (Eds.), *Dinofest International: Proceedings of a Symposium Sponsored by Arizona State University*. Academy of Natural Sciences, Philadelphia, PA, pp. 477–492.

Lund, R. and Grogan, E.D. (2004a). Five new euchondrocephalan Chondrichthyes from the Bear Gulch Limestone (Serpukhovian, Namurian E2b) of Montana and their impact on the Class Chondrichthyes. In: Arratia, G., Cloutier, R., and Wilson, M.V.H. (Eds.), *Recent Advances in the Origin and Early Radiation of Vertebrates*. Verlag Dr. Friedrich Pfeil, Munich, pp. 505–532.

Lund, R. and Grogan, E.D. (2004b). Two tenaculum-bearing Holocephalimorpha (Chondrichthyes) from the Bear Gulch Limestone (Chesterian, Serpukhovian) of Montana, USA, and their impact on the evolution of the Holocephali. In: Arratia, G., Wilson, M.V.H., and Cloutier, R. (Eds.), *Recent Advances in the Origin and Early Radiation of Vertebrates*. Verlag Dr. Friedrich Pfeil, Munich, pp. 171–188.

Lund, R. and Poplin, C. (1999). Fish diversity of the Bear Gulch Limestone, Namurian, Lower Carboniferous of Montana, USA. *Geobios* 32:285–295.

Mader, H. (1986). Schuppen und Zähne von Acanthodiern und Elasmobranchiern aus dem Unter-Devon Spaniens (Pisces). *Göttinger Arb. Geol. Paläont.* 28:1–59.

Maisey, J.G. (1975). The interrelationships of phalacanthous selachians. *Neues Jahrb. Geol. Paläont.* 9:553–567.

Maisey, J.G. (1981). Studies on the Paleozoic Selachian genus *Ctenacanthus* Agassiz, No. 1: Historical review and revised diagnosis of *Ctenacanthus*, with a list of referred taxa. *Am. Mus. Novit.* 2718:1–22.

Maisey, J.G. (1982a). The anatomy and interrelationships of Mesozoic hybodont sharks. *Am. Mus. Novit.* 2724:1–48.

Maisey, J.G. (1982b). Studies on the Paleozoic Selachian genus *Ctenacanthus* Agassiz: No. 2. *Bythiacanthus* St. John and Worthen, *Amelacanthus*, new genus, *Eunemacanthus* St. John and Worthen, *Sphenacanthus* Agassiz, and *Wodnika* Münster. *Am. Mus. Novit.* 2722:1–24.

Maisey, J.G. (1983). Studies on the Paleozoic Selachian genus *Ctenacanthus* Agassiz. No. 3. Nominal species referred to *Ctenacanthus*. *Am. Mus. Novit.* 2774:1–20.

Maisey, J.G. (1984). Chondrichthyan phylogeny: a look at the evidence. *J. Vertebr. Paleontol.* 4:359–371.

Maisey, J.G. (1985). Cranial morphology of the fossil elasmobranch *Synechyodus dubrisiensis*. *Am. Mus. Novit.* 2804:1–28.

Maisey, J.G. (1987). Cranial anatomy of the Lower Jurassic Shark *Hybodus reticulatus* (Chondrichthyes: Elasmobranchii), with comments on hybodontid systematics. *Am. Mus. Novit.* 2878:1–39.

Maisey, J.G. (1989). *Hamiltonichthys mapesi*, g. & sp. nov. (Chondrichthyes; Elasmobranchii), from the Upper Pennsylvanian of Kansas. *Am. Mus. Novit.* 2931:1–42.

Maisey, J.G. (2000). CT-scan reveals new cranial features in Devonian chondrichthyan "*Cladodus*" *wildungensis*. *J. Vertebr. Paleontol.* 21:807–810.

Maisey, J.G. (2001). A primitive chondrichthyan braincase from the Middle Devonian of Bolivia. In: Ahlberg, P.E. (Ed.), *Major Events in Early Vertebrate Evolution: Paleontology, Phylogeny, Genetics and Development.* Taylor & Francis, London, pp. 263–288.

Maisey, J.G. (2004). Endocranial morphology in fossil and recent chondrichthyans. In: Arratia, G., Wilson, M.V.H., and Cloutier, R. (Eds.), *Recent Advances in the Origin and Early Radiation of Vertebrates.* Verlag Dr. Friedrich Pfeil, Munich, pp. 139–170.

Maisey, J.G. (2005). Braincase of the Upper Devonian shark *Cladodoides wildungensis* (Chondrichthyes, Elasmobranchii), with observations on the braincase in early chondrichthyans. *Bull. Am. Mus. Nat. Hist.* 288:1–103.

Maisey, J.G. and Anderson, M.E. (2001). A primitive chondrichthyan braincase from the Early Devonian of South Africa. *J. Vertebr. Paleontol.* 21:702–713.

Maisey, J.G., Miller, R., and Turner, S. (2009). The braincase of the chondrichthyan *Doliodus* from the Lower Devonian Campbellton Formation of New Brunswick, Canada. *Acta Zool. (Stockholm)* 90(Suppl. 1):109–122.

Malzahn, E. (1968). Uber neue funde von *Janassa bituminosa* (Schloth.) in neiderrheinischen Zechstein. *Geol. Jahrb.* 85:67–96.

Miles, R.S. and Young, G.C. (1977). Placoderm interrelationships reconsidered in the light of new ptyctodontids from Gogo, Western Australia. In: Andrews, S.M., Miles, R.S., and Walker, A.D. (Eds.), *Problems in Vertebrate Evolution.* Academic Press, London, pp. 123–198.

Miller, R., Cloutier, R., and Turner, S. (2003). The oldest articulated chondrichthyan from the Early Devonian period. *Nature* 425:501–504.

Motta, P.J., Clifton, K.B., Hernandez, L.P., and Eggold, B.E. (1995). Ecomorphological correlates in ten species of subtropical seagrass fishes: diet and microhabitat utilization. *Environ. Biol. Fish.* 44:37–60.

Moy-Thomas, J.A. and Dyne, M.B. (1938). The actinopterygian fishes from the Lower Carboniferous of Glencartholm, Eskdale, Dumfriesshire. *Trans. R. Soc. Edinburgh* 59:437–480.

Musick, J.A., Harbin, M.M., and Compagno, L.J.V. (2004). Historical zoogeography of the Selachii. In: Carrier, J.C., Musick, J.A., and Heithaus, M.R. (Eds.), *Biology of Sharks and Their Relatives*, CRC Press, Boca Raton, FL, pp. 33–78.

Nassichuk, W.W. (1971). *Helicoprion* and *Physonemus*: Permian vertebrates from the Assistance Formation Canadian Arctic archipelago. *Can. Geol. Surv. Bull.* 192:83–93.

Nelson, J.S. (2006). *Fishes of the World*, 4th ed. Wiley, New York, 624 pp.

Newberry, J.S. and Worthen, A.H. (1866). Descriptions of new genera and species of vertebrates mainly from the sub-Carboniferous limestone and Coal Measures of Illinois. *Geol. Surv. Illinois* 2:9–134.

Newberry, J.S. and Worthen, A.H. (1870). Descriptions of fossil vertebrates. *Geol. Surv. Illinois* 4:347–374.

Nixon, K.C. (1999). Winclada (BETA) ver. 0.9.9, published by the author, Ithaca, NY.

Obruchev, D.V. (1967). Agnatha, Pisces. In: Orlov, Y.A. et al. (Eds.), *Fundamentals of Paleontology*, Vol. 11. Israel Program for Scientific Translations, Jerusalem, 825 pp.

Ørvig, T. (1951). Histologic studies of placoderms and fossil elasmobranchs. I. The endoskeleton, with remarks on the hard tissues of lower vertebrates in general. *Ark. Zool.* 2:321–456.

Ørvig, T. (1961). New finds of acanthodians, arthrodires, crossopterygians, ganoids, and dipnoans in the Upper Middle Devonian Calcareous Flags (Oberer Plattenkalk) of the Bergisch–Paffrath Trough (part 1). *Paläont. Zeit.* 34:295–335.

Owen, R. (1867). On the mandible and mandibular teeth of cochliodonts. *Geol. Mag.* 4:59–63.

Patterson, C. (1965). The phylogeny of the chimaeroids. *Phil. Trans. R. Soc. B Biol. Sci.* 249:101–219.

Reif, W.E. (1978). Bending-resistant enameloid in carnivorous teleosts. *N. Jahrb. Geol. Paläontol. Abh.* 157:173–175.

Reif, W.E. (1979). Structural convergences between enameloid of actinopterygian teeth and of shark teeth. *Scanning Electron Microsc.* 1979(2):546–554.

Rodina, O.A. and Ivanov, A.O. (2002). Chondrichthyans from the Lower Carboniferous of Kuznetsk Basin. In: Yushkin, N.P., Tsyganko, V.S., and Männik, P. (Eds.), *Geology of the Devonian System: Proceedings of the International Symposium*, July 9–12, 2002, Syktyvkar, Komi Republic, Russia, pp. 263–268.

Romer, A.S. (1952). Late Pennsylvanian and Early Permian vertebrates from the Pittsburgh–West Virginia region. *Ann. Carnegie Mus.* 33:47–112.

Rosenberg, L. (1998). A Study of the Mineralized Tissues of Select Fishes of the Bear Gulch Limestone and Recent Fishes, master's thesis, Adelphi University, Garden City, NY.

Roy, K., Jablonski, D., and Valentine, J.W. (1996). Higher taxa in biodiversity studies: patterns from eastern Pacific marine molluscs. *Phil. Trans. R. Soc. B Biol. Sci.* 351:1605–1613.

Sander van Doorn, G., Edelaar, P., and Weissing, F.J. (2009). On the origin of species by natural and sexual selection. *Nature* 326:1704–1707.

Santini, F. and Tyler, J.C. (2004). The importance of even highly incomplete fossil taxa in reconstructing the phylogenetic relationships of the Tetraodontiformes (Acanthomorpha: Pisces). *Integr. Compar. Biol.* 44:349–357.

Schaeffer, B. (1981). The xenacanth shark neurocranium, with comments on elasmobranch monophyly. *Bull. Am. Mus. Natl. Hist.* 169:3–66.

Schaumberg, G. (1992). Neue informationen zu *Menaspis armata* Ewald. *Paläont. Z.* 66:311–329.

Schneider, J.W. and Zajic, J. (1994). Xenacanthiden (Pisces, Chondrichthyes) des mitteleuropäischen Oberkarbon und Perm. Revision der originale zu Goldfuss 1847, Beyrich 1848, Kner 1867, und Fritsch 1879–1890. *Freib. Forschungsh.* 452:101–151.

Schultze, H.-P. and Maples, C. (1992). Comparison of the late Pennsylvanian faunal assemblage of Kinney Brick Company quarry, New Mexico, with other Late Pennsylvanian lägerstatten. *N.M. Bur. Mines Miner. Resour. Bull.* 138:231–235.

Sequiera, E.K. and Coates, M.I. (2000). Reassessment of "*Cladodus*" *neilseni* Traquair, a primitive shark from the lower Carboniferous of East Kilbride, Scotland. *Palaeontology* 43:153–172.

Shirai, S. (1996). Phylogenetic interrelationships of Neoselachians (Chondrichthyes: Euselachii). In: Stiassny, M., Parenti, L.R. and Johnson, G.D. (Eds.), *Interrelationships of Fishes*. Academic Press, New York, pp. 9–34.

Smith, S.E., Au, D.W., and Snow, C. (1998). Intrinsic rebound potentials of 26 species of Pacific sharks. *Mar. Freshwater Res.* 49:663–678.

Soler-Gijón, R. and Hampe, O. (2003). *Leonodus*, a primitive chondrichthyan from the early Devonian. *Ichthyolith Issues Spec. Publ.* 7:47.

Springer, S. (1967). Social organization of shark populations. In: Gilbert, P.W., Mathewson, R.F., and Rall, D.P. (Eds.), *Sharks, Skates and Rays*. The Johns Hopkins University Press, Baltimore, MD, pp. 149–174.

St. John, O.H. and Worthen, A.H. (1875). Descriptions of fossil fishes. *Geol. Surv. Illinois* 6:245–488.

St. John, O.H. and Worthen, A.H. (1883). Descriptions of fossil fishes: a partial revision of the Cochliodonts and Psammodonts. *Geol. Surv. Illinois* 7:55–264.

Stahl, B.J. (1999). Mesozoic holocephalians. In: Arratia, G. and Schultze, H.-P. (Eds.), *Mesozoic Fishes*. Vol. 2. *Systematics and the Fossil Record*. Verlag Dr. Friedrich Pfeil, Munich, pp. 9–19.

Thies, D. and Reif, W.E. (1985). Phylogeny and evolutionary ecology of Mesozoic Neoselachii. *N. Jahrb. Geol. Paläeontol. Abh.* 3:333–361.

Traquair, R.H. (1888a). Notes on Carboniferous Selachii. *Geol. Mag. Ser. 3*, 5:81–86.

Traquair, R.H. (1888b). Further notes on Carboniferous Selachii. *Geol. Mag. Ser. 3* 5:101–104.

Turner, S. (2004). Early vertebrates: analysis from microfossil evidence. In: Arratia, G., Wilson, M.V.H., and Cloutier, R. (Eds.), *Recent Advances in the Origin and Early Radiation of Vertebrates*. Verlag Dr. Friedrich Pfeil, Munich, pp. 67–94.

Turner, S. and Miller, R.F. (2005). New ideas about old sharks. *Am. Sci.* 93:244–252.

Walker, T.I. (1998). Can shark resources be harvested sustainably? A question revisited with a review of shark fisheries. *Mar. Freshwater Res.* 49:553–572.

Ward, D. and Duffin, C.J. (1989). Mesozoic chimaeroids. 1. A new chimaeroid from the Early Jurassic of Gloucestershire, England. *Mesozoic Res.* 2:45–51.

Watson, D.M.S. (1938). On Rhamphodopsis, a ptyctodont from the Middle Old Red Sandstone of Scotland. *Trans. R. Soc. Edinburgh* 59:397–410.

Webb, P.W. (1984). Body form, locomotion and foraging in aquatic vertebrates. *Am. Zool.* 24:107–120.

Wiens, J.J. (1998). Does adding characters with missing data increase or decrease phylogenetic accuracy? *Syst. Biol.* 47:625–640.

Wilson, M.V.H., Hanke, G., and Märss, T. (2007). Paired fins of jawless vertebrates and their homologies across the "agnathan"–ganthostome transition. In: Anderson, J.S. and Sues, H.D. (Eds.), *Major Transitions in Vertebrate Evolution*. Indiana University Press, Bloomington, pp. 122–149.

Young, G.C. (1982). Devonian sharks from south-eastern Australia and Antarctica. *Paleontology* 25:817–850.

Young, G.C. (1986). The relationships of placoderm fishes. *Zool. J. Linn. Soc.* 88:1–57.

Young, G.C. (1997). Ordovician microvertebrate remains from the Amadeus Basin, Central Australia. *J. Vertebr. Paleontol.* 17:1–25.

Yucha, D.T. (1998). A Qualitative Histological Analysis of Calcified Cartilage in *Squalus acanthias*, master's thesis, Saint Joseph's University, Philadelphia, PA, 49 pp.

Zalisko, E.J. and Kardong, K. (1998). *Comparative Vertebrate Anatomy*. McGraw-Hill, New York, 214 pp.

Zangerl, R. (1981). Chondrichthyes I: Paleozoic Elasmobranchii. In: Schultze, H.-P. (Ed.) *Handbook of Paleoichthyology*, Vol. 3A. Gustav Fischer, Stuttgart.

Zangerl, R. (1997). *Cervifurca nasuta* n. gen. et sp., an interesting member of the Iniopterygidae (Subterbranchialia, Chondrichthyes) from the Pennsylvanian of Indiana, USA. *Fieldiana Geol. N.S.* 35:1–24.

Zangerl, R. and Case, G.R. (1973). Iniopterygia, a new order of chondrichthyan fishes from the Pennsylvanian of North America. *Fieldiana Geol. Mem.* 6, 67 pp.

Zhu, M. (1998). Early Silurian sinacanths (Chondrichthyes) from China. *Paleontology* 41:157–171.

Zhu, M., Yu, X., and Janvier, P. (1999). A primitive fossil fish sheds light on the origin of bony fishes. *Nature* 397:607–610.

Zhu, M., Zhao, W.J., Kia, L.T., Lu, J., Qiao, T., and Qu, Q.M. (2009). The oldest articulated osteichthyan reveals mosaic gnathostome characters. *Nature* 458:469–474.

Zidek, J. (1976). Oklahoma Paleoichthyology, Part V: Chondrichthyes. *Oklahoma Geol. Notes* 36:175–192.

Appendix. Characters and States for the Cladogram of Figure 1.1

Key: (Number); Character; State 0; State 1; State 2; State 3; State 4; State 5; State 6

(0); mineralized endoskeletal tissue; perichondral bone; small tabular blocks; tesserate mineralization

(1); male frontal clasper; absent; median; median, extremely elongate; single pair; multiple pairs

(2); extended neurocranial rostrum; absent; present

(3); precerebral fontanelle; absent; large; closed/small opening

(4); supraorbital cartilage; absent; present

(5); ethmoid % of neurocranium; <25%; 25–40%; 40–50%; >50%

(6); orbital % of neurocranium; 35–46%; <35%; >46%

(7); postorbital % of neurocranium; 25–40%; 15–25%; >40%; <15%

(8); neurocranial length:width; ≥1.5:1; ~1:1

(9); otic roof/otic floor; parallel; convex posterodorsally; dorsal overhangs ventral margin

(10); anterior ventral braincase (X-section); narrow v-shaped; platybasic, narrow; platybasic wide

(11); posterior ventral braincase (X-section); stenobasal; narrow shelf; wide shelf

(12); PQ–ethmoid support; none; ethmoid articulation; ethmoid ligament; ethmoid fusion

(13); PQ–basal support; none; basitrabecular (btp) process articulation; btp–postorbital process articulation; basal process/basipterygoid ligament; orbitostyly; fused

(14); PQ–posterior support; none; otic wall; hyomandibular + postorbital process articulation; hyomandibular only; bony palate sutured to osteichthyan hyomandibular; fused

(15); postorbital process; small dorsal protuberance; distinct laterally extended process; stout dorsoventral ridge/postorbital wall; absent

(16); palatoquadrate–postorbital articulation; absent; under postorbital process; on posterior side of postorbital process; posterior to postorbital process; hyomandibular to otic; on antotic process

(17); palatobasal–palatoquadrate support; minor; principal; none

(18); basitrabecular process attitude; in line with post orbit edge/postorbital; flared ventrolateral to postorbital; flared anteroventrolaterally; rudimentary/absent

(19); basitrabecular–palatoquadrate articulation; postorbital; orbital; antorbital; absent

(20); chondrocranial construction (adult); 3, ethmosphenoid otic and occipital; 2, ethmosphenoid and oticooccipital; continuous vault; continuous vault + rhinocapsule

(21); occipital moiety; dorsal occipital + separate ventral vertebral elements; dorsal occipital and ventral basioccipital; separate dorsal, fused ventral elements; single (D+V) separate unit; occipital fused to otic moiety; separate parachordal

(22); oticooccipital fissure; dorsally open; continuous with ventral otic fissure; continuous with ventral occipital fissure; none/fused

(23); ethmosphenoid region; ethmosphenoid fissure; none; intracranial joint

(24); suspensorium; autodiastyly; hyostyly; amphistyly; holostyly; methyostyly

(25); extravisceral cephalic cartilages; premandibular feeding mechanism; labials integrated with mandibular arch; few/reduced labials; none; premaxilla, maxilla, palate

(26); anterior extravisceral cartilages; oral hood; primary biting; otterboard mouth rim; prehensile lip support; lips supplemental to jaws; bones form external jaw arcade; absent

(27); principal skeletal oral margin; premandibular/labial complex; premaxilla, (maxilla), dentary; supra and infragnathals; palatoquadrate and Meckel's; palatoquadrate and Meckel's ossifications

(28); palatoquadrate anteriad; ends at nasal capsule; parallel (parasymphysial) extension; median symphysis

(29); palatoquadrate–otic (pterygoid) process shape; slight/undeveloped; dorsally recurved; posteriorly extended; dorsally expanded

(30); Meckel's–quadrate articulation; postorbital; orbital; preorbital

(31); mandibular symphysis; mobile symphysial cartilage(s); symphysials fused to Meckel's; ligamentous mandibular junction; mandibles fused at symphysis

(32); mandibular mineralization; perichondral bones; spheritic/tabular; random tesserae; solid fibrocartilage; prismatic calcified cartilage

(33); mandible, bony Meckelian cartilage(s); present; absent

(34); upper parasymphysial cartilage(s); multiple; reduced/few; few, anteriorly extended; absent

(35); lower symphysial cartilage(s); multiple; one/two, between mandibles; extended anteriad; absent

(36); lower symphysial cartilage(s) width; narrow; wide; absent

(37); gill openings; membraneous opercular valve; bony opercular valve; separate gill openings

(38); epal hyoid; opercular support; mandibular arch support

(39); hyomandibula–cranial articulation; epihyal, no hyomandibula; rear of lateral commissure (postorbital); posteroventral otic; posterolateral otic; posterodorsal otic

(40); buccopharyngeal/palatal denticles; simple to compound odontodes; absent

(41); branchial basket; subotic to postcranial; subcranial; principally postcranial

(42); Tooth row/jaw length; absent/not apparent (NA); >50%; <40%

(43); Lingual torus/crown length; absent/NA; shorter than crown; as long as/longer than crown

(44); In line cusp relationship; teeth absent; unicuspid; parallel; highly divergent from base; divergent, twisted; cusps suppressed/NA

(45); Relative cusp sizes; teeth absent; unicuspid; tricuspid; laterals largest, small/no central; centrals largest, others low; laterals reduced/suppressed; no cusps

(46); Cusp cross section; teeth absent; rounded; reduced/suppressed; compressed bladelike

(47); Cusp numbers; teeth absent; unicuspid; 2 to 3; about 5 to 7; one main cusp; variable heterodont; cusps suppressed/NA

(48); Crown/root height; teeth absent/NA; equal/subequal; hypsodont; brachydont

(49); Cusp separation; teeth absent; distinctly separate on base; cusps confluent; cusps reduced/suppressed

(50); functional jaw tooth families; teeth absent; 1 to 2 per family; pavement occlusion; teeth and tooth plates

(51); tooth shapes on jaw; teeth absent; homodont; heterodont; teeth and tooth plates; plates alone

(52); lower symphysial family; absent/NA; size as other dentition; prominent; whorl; fused plate; paired whorl; multiple families

(53); lower symphysial whorls; paired only; median + 5–6 pairs; median + 3–4 pairs; median + 2 pairs; median + 1 pair; median only; absent

(54); upper parasymphysial family; absent/NA; size as other dentition; prominent; whorl; fused plate; paired whorl; multiple families

(55); crown base; absent/NA; generalized; lingual heel; lingual, labial ridges; basin and ridges

(56); Crown linguo/labial buttresses; absent/NA; crenellated; buttressed

(57); tooth root; absent/NA; short below crown; long below crown; extended lingual; fused; below lingual edge

(58); root shape; absent/NA; straight below crown; lingual s-shape; lingual shelf; proximodistally arched; fused

(59); orthodentine; absent/NA; present

(60); osteodentine; absent/NA; present

(61); paired, upper dental positions; absent/NA; >3 (teeth); 2 anterior, 1 posterior; 1 anterior, 1 posterior; 1 posterior

(62); paired anterior lowers; absent/NA; >3 (teeth); 2 anterior; 1 anterior; absent

(63); middle, lower condition; absent/NA; tooth; whorl; plate

(64); posterior dental histology; absent/NA; coronal tooth tissues; tritoral dentine; pleromin tritors; other

(65); anterior and middle dental surfaces; absent/NA; coronal tooth tissues; limited tritors; complete coverage; bone, no tritor

(66); anterior dental histology; absent/NA; coronal tooth tissues; tritoral dentine; pleromin tritors; bone, no tritor

(67); scale type; ganoid; placoid; zonal growth; fused placoid; absent; bony acanthodian units

(68); head dermal skeleton; generalized; acanthodian; placoderm; osteichthyan; placoid/chondrichthyan

(69); chondrichthyan head dermal skeleton; NA; placoid; few radial denticulated (Iniopterygian); many radial denticulated (Iniopterygian); bony plates (cochliodont); conical scales; absent

(70); Chondrichthyan biting/masticatory structures; absent/NA; labial denticles; labial plates; labial/mandibular arch plates; mandibular arch teeth/plates

(71); head scale coverage; macromeric; placoid; both generalized and specialized; specialized areas only; absent

(72); head scale modifications; macromeric; placoid; enlarged scales/spines; plates; absent

(73); head lateral line scales; small, simple, oriented; thin rings; bony plates; canal(s) enclosed in plates; absent; in cranial bones

(74); ethmorostral scales (cochliodont); NA; placoid; spikes; enlarged denticles; few plates; absent

(75); supraorbital scales (cochliodont); NA; placoid; enlarged scales/spines; plates; absent

(76); otic scales (cochliodont); NA; placoid; enlarged scales/spines; plates; absent

(77); occipital scales; macromeric; placoid; enlarged scales/spines; compound plates and spines; absent; *Onchus*-type spines

(78); mandibular squamation (cochliodont); NA; placoid; posterior specialization; longitudinal specialization (plates); absent

(79); posterior mandibular scales; macromeric; placoid; small sharp spine; broad spine; buttressed plate; absent

(80); basitrabecular rim scales (cochliodont); NA; placoid; few large denticles; plate; plate and spine; absent

(81); body scales; macromeric; placoid; generalized and specialized areas; only specialized areas; absent; Acanthodian

(82); enlarged paired middorsal body scales (cochliodont); NA; placoid; first dorsal to caudal; between fins; past second dorsal; absent

(83); dorsal fin squamation; complete covering; crest of fin only; upon radials; absent

(84); squamation/sex; monomorphic; sexually dimorphic

(85); lateral line scales of body; small, simple, oriented; rings; canal through scales; none

(86); Shoulder girdle–neurocranial link; exoskeletal plates (DL, MD); exoskeletal bones (PT, SC, AC); scapular cartilage chain; scapula near neurocranium; scapula remote from neurocranium

(87); pectoral fin position; ventrolateral; mid-flank; nape of the neck

(88); pectoral girdle; principally endoskeletal; principally exoskeletal

(89); coracoid length; normal; extended anteriorly; truncated anteriorly

(90); pectoral fin radial support; >50% on anterior basals; >50% on metapterygium; all on metapterygium, axis specialized; >50% on post-metapterygial axis; all borne by girdle

(91); pectoral fin; uniserial; partially biserial; entirely biserial

(92); pectoral fin base; multibasal, eurybasal; unibasal, stenobasal; tribasal

(93); pectoral post-metapterygial axis; absent; 1 to 4 small elements; >4 small elements

(94); pectoral leading edge denticles (iniopterygian); NA; many small; few rows large; single large row

(95); pelvic girdle mineralization; bone; superficial calcified cartilage; solid, three-dimensional; absent

(96); pelvic dorsal process; absent; short; tall; separate dorsal cartilage

(97); pelvic basipterygium; minor or absent; elongate

(98); pelvic radials; majority on girdle; ~50% on basipterygium; majority on basipterygium; all on basal plate artic to girdle

(99); pelvic girdles; separate across midline; close contact across midline; puboischiadic bar in males; puboischiadic bar both sexes

(100); prepelvic tenaculum; absent; anterior edge of pelvic fin; posterior edge of girdle; anterior edge of girdle; tenacular hooks on fin

(101); fertilization; without claspers in males; pelvic claspers in males

(102); anal fin and/or anal plate; absent; present

(103); dorsal fin numbers; two fins; one fin

(104); dorsal spines; anterior fin only; two spines; no spines; posterior dorsal alone

(105); anterior dorsal fin; large; small flap; absent; rod

(106); anterior dorsal fin and/or spine support; basal plate/radials; synarcuum; on head; shoulder girdle; absent

(107); anterior dorsal spine; deeply fixed; long, mobile; superficial insertion; absent

(108); anterior dorsal presence; found in both sexes; absent in both sexes; sexually dimorphic

(109); anterior dorsal fin/spine development; at birth; at puberty; does not develop

(110); anterior dorsal spine shape; posteriorly directed, narrow; triangular; forwardly curved; absent

(111); anterior dorsal spine enameloid, dentine; present; absent; no spine

(112); posterior dorsal fin; short; elongate; absent

(113); posterior dorsal fin base; all radials; triangular basal plate and radials; vertical basal plate; few supraneurals; absent

(114); posterior dorsal spine; absent; superficial insertion; deeply fixed

(115); caudal fin; heterocercal; extended heterocercal; homocercal; diphycercal

(116); caudal endoskeleton; serial hypochordal; epichordal component; fusions/expansions; homocercal; abbreviate heterocercal

(117); notochordal calcification; uncalcified; chordacentra; complete centra

(118); vertebral arcual mineralization; uncalcified; regionalized; entire column

(119); ribs; absent; present

2

Elasmobranch Phylogeny: A Mitochondrial Estimate Based on 595 Species

Gavin J.P. Naylor, Janine N. Caira, Kirsten Jensen, Kerri A.M. Rosana, Nicolas Straube, and Clemens Lakner

CONTENTS

2.1 Introduction

2.1.1 Background

Interest in elasmobranch biodiversity and taxonomy has grown in recent years, catalyzed primarily by four influences: (1) the large number of new species that have been described over the past 30 years (e.g., Last and Stevens, 2009); (2) the recognition that many species of elasmobranchs, several of which have not yet been formally described, may be threatened with extinction from fishing pressures and habitat destruction (Stevens et al., 2000); (3) the growing interest in DNA "barcoding" as a tool to augment taxonomic description (e.g., Ward et al., 2007); and (4) an emerging recognition of the important role that elasmobranchs play as top predators in marine ecosystems (Heithaus et al., 2008).

Increasingly, elasmobranch workers across a wide range of fields of science, both pure and applied, are recognizing the importance of an accurate species-level taxonomy. Fisheries scientists are ever more keenly aware of the need for accurate species-level assessments of catches to manage fisheries effectively. Ecologists have become more careful to ensure that the animals to which life history attributes are ascribed constitute distinct species rather than assemblages of closely related congeners with potentially different ecological roles and life history attributes. Finally, conservation biologists are beginning to recognize how critically important it is to have an accurate understanding of species compositions based on careful taxonomy to prioritize and manage units of biodiversity for conservation (Griffiths et al., 2010; Iglésias et al., 2009; White and Kyne, 2010).

2.1.2 Motivation

While interest in the taxonomy of elasmobranchs is probably at an all-time high, efforts to understand their phylogenetic interrelationships have lagged behind (Thomson and Shaffer, 2010). Contributions to our understanding of elasmobranch phylogeny have thus far been restricted to studies focusing on the interrelationships of particular groups, including *Arctoraja* (Spies et al., 2011), Batoidea (Aschliman et al., in press; McEachran and Aschliman, 2004; Rocco et al., 2007), *Carcharhinus* (Dosay-Akbulut, 2008), Carcharhinidae (Naylor, 1992), Carcharhiniformes (Compagno, 1988), Dasyatidae (Sezaki et al., 1999), *Dasyatis* (Rosenberger, 2001), Etmopteridae (Shirai and Nakaya, 1990; Straube et al., 2010), Lamnidae (Dosay-Akbulut, 2007; Martin, 1997), Laminiformes (Martin and Naylor, 1997; Naylor et al., 1997; Shimada, 2005), Myliobatiformes (de Carvalho et al., 2004; Dunn et al., 2003; Gonzalez-Isáis and Dominguez, 2004; Lovejoy, 1996; Nishida, 1990), Orectolobidae (Corrigan and Beheregaray, 2009), Rajiformes (McEachran and Dunn, 1998; McEachran and Miyake, 1990; Turan, 2008), Scyliorhinidae (Human et al., 2006; Iglésias et al., 2005), Sphyrnidae (Cavalcanti, 2007; Lim et al., 2010), Selachii (Vélez-Zuazo and Agnarsson, 2011), *Squatina* (Stelbrink et al., 2009), and Triakidae (López et al., 2006), or studies focusing on the relationships among major lineages using a few carefully chosen exemplars for multiple lineages (Compagno, 1977; de Carvalho, 1996; Douady et al., 2003; Heinicke et al., 2009; Maisey, 1984a,b; Maisey et al., 2004; Mallatt and Winchell, 2007; Naylor et al., 2005; Shirai, 1992, 1996; Winchell et al., 2004). To our knowledge, no phylogenetic studies of elasmobranchs have incorporated dense taxon sampling at the species level across the entire breadth of elasmobranch diversity. This is, in part, because the most immediate concerns have centered on documenting the extant diversity as quickly as possible, before fishing pressure and habitat destruction drive it to extinction. However, it is also a result of the fact that obtaining samples from the broad spectrum of taxa required for a comprehensive phylogenetic analysis is particularly challenging. Nonetheless, a phylogenetic perspective provides a context for understanding the historical forces that have shaped extant biodiversity. This information can be helpful to conservation efforts and effective fisheries management because degree of relatedness can often be a good predictor of life history attributes and sensitivity to environmental change.

2.1.3 Barcodes, GenBank, and Phylogenetic Estimation

We believe that the lag in interest in generating a comprehensive phylogeny for elasmobranchs is unlikely to last long. As CO1 barcode sequences pour into GenBank for a diversity of elasmobranchs (e.g., Holmes et al., 2009; Mariguela et al., 2009; Quattro et al., 2006; Richards et al., 2009; Serra-Pereira et al., 2011; Smith et al., 2008; Spies et al., 2006; Straube et al., 2010, 2011; Toffoli et al., 2008; Ward and Holmes, 2007; Ward et al., 2005, 2007, 2008, 2009; Wong et al., 2009; Wynen et al., 2009; Zemlak et al., 2009), driven in part by the various Barcode of Life initiatives, it will only be a short time before enterprising efforts are made to estimate phylogenetic trees from these barcode sequences. We anticipate that when this happens, there will be a profusion of trees forwarded in the literature that suggest conflicting phylogenetic relationships. If past experience is any guide, trees derived this way will contain a high proportion of accurate and credible relationships interspersed with a few erroneous groupings. Unfortunately, it will be difficult to tell which relationships are erroneous, as the misleading inferences are likely to vary from study to study, depending on the taxon-sampling scheme and individual specimens used. More insidiously, it is also likely that congruent misleading inferences will surface in different studies, unwittingly leading to strong confidence in an erroneous consensus topology. This can occur, for example, in instances of model misspecification exacerbated by missing data (Lemmon et al., 2009) and uneven taxon sampling.

Barcode sequences downloaded from GenBank are especially prone to yielding misleading estimates of phylogeny, in large part because the 650-bp CO1 barcode fragment has become the *de facto* standard for molecular identification of species. It is now routine to remove tissue samples from specimens in the field and send them off to sequencing centers for "barcoding." Unfortunately, some of the specimens from which tissue samples are derived are misidentified when collected, and, because there is no expertly curated reference dataset against which to compare sequences, many are added to GenBank with their original incorrectly assigned identities (Bridge et al., 2003; Vilgalys, 2003; Wesche et al., 2004).

Notwithstanding the potential problems with misidentification, barcode sequences are not well suited to phylogenetic analysis from the outset. Being relatively short (approximately 650 bp in most vertebrate taxa), they do not provide a large number of characters upon which to base phylogenetic inferences; also, being relatively fast evolving, they are generally not useful for estimating relationships among deeply divergent taxa. There are, of course, other sources of error that can lead to incorrect phylogenetic inferences such as labeling errors, sequencing errors, dissonance between gene trees and the species trees that contain them, sampling errors due to stochasticity of the evolutionary process, model violation, and exacerbations of model violation caused by sparse taxon sampling, and missing data. These problems are not unique to CO1 barcode data, however, and can affect any molecular phylogenetic study.

2.1.4 The Current Study

We take the position that accurate and reliable estimates of phylogeny are best achieved through analysis of congruence among a carefully selected suite of independent single-copy markers whose patterns of evolutionary change can be accommodated with simple i.i.d. (independently and identically distributed) models. However, as an interim measure, our goal for the current chapter is to provide a phylogenetic analysis of a densely taxon-sampled, mitochondrial, protein-coding gene. In an effort to minimize the types of errors referred to above, we have avoided using any sequences derived from GenBank. All of our sequences have been generated *de novo* from samples taken from specimens collected by the authors or identified by taxonomic experts. Our primary motivation for this contribution is to provide a baseline against which future phylogenetic studies can be contrasted and evaluated, particularly those based on datasets compiled from GenBank submissions. As noted above, these potentially include sequences from misidentified specimens or sequences of questionable provenance, and a substantial fraction of missing data that can often yield peculiar results. This is exemplified by the recent paper of Vélez-Zuazo and Agnarsson (2011), whose inferences were based on a dataset with 85% missing entries, a Bayesian analysis that had not fully converged, and a taxon sampling scheme that included several misidentified specimens (evidenced by their untenable placements on the tree presented). The inferred relationships we present doubtlessly depict several relationships that are incorrect; however, we anticipate that any errors are likely to be the consequence of model violation of one type or another or differential fixation of ancestral polymorphism in descendant lineages, rather than those associated with questionable specimen provenance or misidentification, although we are certainly not immune to these type of problems, either.

2.2 Methods

2.2.1 Taxon Sampling

We recently completed a survey of sequence variation in elasmobranchs using the NADH2 mitochondrial gene (Naylor et al., in press). That study was based on an analysis of sequences derived from a total of 4283 specimens of elasmobranchs representing 574 (of approximately 1200 described) species in 157 (of 193 described) genera in 56 (of 57 described) elasmobranch families. Its primary goal was to better understand the taxonomy and species boundaries among elasmobranchs from a genetic perspective based on mitochondrial sequence variation. In that study, we were careful to point out that the summary tree presented, being based on "p" distances and a neighbor-joining cluster analysis, was not to be interpreted as a phylogeny, although many of the clusters among closely related forms likely do reflect phylogenetic groupings.

In contrast, the current study, explicitly sets out to conduct a model-based Bayesian phylogenetic analysis of a representative subset consisting of 585 of those 4283 NADH2 sequences from the Naylor et al. (in press) study. This subset includes single representatives of 570 species as well as two to seven replicates each of four problematic potential species complexes (i.e., *Rhinoptera steindachneri*, *Bathyraja kincaidii/interupta*, *Amblyraja hyperborea/badia/jensenae*, and *Dipturus batis/oxyrinchus*) included in the Naylor et al. (in press) analysis. These sequences have been augmented here to include NADH2 data from an additional two genera (*Miroscyllium* and *Trigonognathus*) and 21 species, one of which (*Etmopterus viator*) (Straube et al., 2011b) was replicated from widely separated parts of its range. The identities assigned to these specimens in the current analysis are given in Table 2.1, along with names assigned in the previous analyses of Straube et al. (2010, 2011a). Exemplars representing each species were selected on the basis of the availability of photographic or voucher material associated with the specimen from which the sequence was derived. In total, 69% of the specimens used in this study are represented by images in our online database (http://tapewormdb.uconn.edu/index.php/hosts/specimen_search/elasmobranch) or their identifications have been verified by taxonomic experts. Of these specimens, 35% have been deposited in museums; several are types. The elasmobranch sample in the current study represents approximately 50% (i.e., 595) of all known species, 83% (i.e., 159) of all genera, and 98% (i.e., 56) of all families; these numbers represent a relatively even spread across sharks and rays. Locality data, voucher information, and GenBank accession numbers are provided in Naylor et al. (in press) for the 585 sequences taken from that study and in Table 2.1 for the specimens new to this study. In addition, four chimaeroid species, representing two genera and two families, were used to represent the outgroup. These are *Hydrolagus bemisi*, *Hydrolagus colliei*, *Hydrolagus novaezealandiae*, and *Rhinochimaera pacifica*.

2.2.2 Sequence Generation

Although most of the sequences for this study were taken directly from those used in the Naylor et al. (in press) study, sequences for 22 specimens were generated

TABLE 2.1

Voucher Information for the 22 Specimens Added in This Study

Order and Family	Species	Database ID	GenBank ID	Museum Voucher No.	Locality	Tissue Sample No.
Carcharhiniformes						
Carcharhinidae	*Carcharhinus leiodon*[a]	GN5013	JQ400110	BMNH 2010.2.8.1 (jaws only)	Kuwait City, Kuwait, Persian (Arabian) Gulf	BW-A6072
Scyliorhinidae	*Bythaelurus canescens*	GN7459	JQ400111	ZSM-33566	Chile, Pacific Ocean	ZSM-P-CH_0290
Squaliformes						
Centrophoridae	*Centrophorus acus*	GN7425	JQ400112	Photo voucher	Suruga Bay, Japan, Pacific Ocean	ZSM-P-CH_0076
Centrophoridae	*Deania profundorum*	GN7456	JQ400113	OCA-P-20061202.3C; photo voucher	Okinawa, Japan, Pacific Ocean	ZSM-P-CH_0257
Echinorhinidae	*Echinorhinus* sp. 1[b]	GN7438	JQ400114	No voucher specimen	Oman, Indian Ocean	ZSM-P-CH_0149
Etmopteridae	*Centroscyllium ritteri*	GN7428	JQ400115	No voucher specimen	Suruga Bay, Japan, Pacific Ocean	ZSM-P-CH_0082
Etmopteridae	*Centroscyllium nigrum*	GN7443	JQ400116	No voucher specimen	Chile, Pacific Ocean	ZSM-P-CH_0210
Etmopteridae	*Centroscyllium granulatum*[c]	GN7445	JQ400117	No voucher specimen	Chile, Pacific Ocean	ZSM-P-CH_0212
Etmopteridae	*Etmopterus* sp. B[d]	GN7398	JQ400118	No voucher specimen	Norfolk Ridge, Tasman Sea, Pacific Ocean	ZSM-P-CH_0017
Etmopteridae	*Etmopterus granulosus*[c]	GN7399	JQ400119	NMV A25150-016	Norfolk Ridge, Tasman Sea, Pacific Ocean	ZSM-P-CH_0022
Etmopteridae	*Etmopterus sentosus*[c]	GN7402	JQ400120	SAIAB 82362	Mozambique, Indian Ocean	ZSM-P-CH_0040
Etmopteridae	*Etmopterus* sp. 1	GN7406	JQ400121	No voucher specimen	South Africa, Atlantic Ocean	ZSM-P-CH_0045
Etmopteridae	*Etmopterus* sp. 2[e]	GN7409	JQ400122	No voucher specimen	South Africa, Atlantic Ocean	ZSM-P-CH_0050
Etmopteridae	*Etmopterus viator*[f]	GN7412	JQ400123	NMNZ P.42742	Chatham Rise, New Zealand, Pacific Ocean	ZSM-P-CH_0053
Etmopteridae	*Etmopterus viator*	GN7415	JQ400124	MNHN 20071666	Kerguel Plateau, Indian Ocean	ZSM-P-CH_0059
Etmopteridae	*Etmopterus schultzi*[c]	GN7418	JQ400125	Photo voucher	USA, Gulf of Mexico	ZSM-P-CH_0065
Etmopteridae	*Etmopterus polli*	GN7420	JQ400126	No voucher specimen	Angola Basin, Western Africa, Atlantic Ocean	ZSM-P-CH_0070
Etmopteridae	*Etmopterus brachyurus*	GN7423	JQ400127	Photo voucher	Suruga Bay, Japan, Pacific Ocean	ZSM-P-CH_0074
Etmopteridae	*Etmopterus unicolor*[c]	GN7434	JQ400128	Photo voucher	Suruga Bay, Japan, Pacific Ocean	ZSM-P-CH_0097
Etmopteridae	*Miroscyllium sheikoi*[c]	GN7440	JQ400129	No voucher specimen	Tashi Fish market, Illan, Taiwan, Pacific Ocean	ZSM-P-CH_0151
Etmopteridae	*Trigonognathus kabeyai*[c]	GN7431	JQ400130	HMD 2003-18	Japan Pacific Ocean	ZSM-P-CH_0093
Rajiformes						
Rajidae	*Dipturus trachyderma*	GN7449	JQ400131	Photo voucher	Huinay Fjord, Chile, Pacific Ocean	ZSM-P-CH_0246

Note: Specimen information for the remaining 585 specimens can be found in Naylor et al. (in press).

Abbreviations: BMNH, Natural History Museum (formerly British Museum [Natural History]), London, UK; HMD, Hekinan Seaside Aquarium, Hekinan City, Aichi Prefecture, Japan; MNHN, Muséum National d'Histoire Naturelle, Paris, France; NMNZ, Museum of New Zealand, Te Papa Tongarewa, Wellington, New Zealand; NMV, Museum Victoria (formerly National Museum of Victoria), Melbourne, Victoria, Australia; OCA, Okinawa Churaumi Aquarium, Okinawa, Japan; SAIAB, South African Institute for Aquatic Biodiversity, Grahamstown; ZSM, Zoologische Staatssammlung München, Munich, Germany (tissue collection).

[a]   Specimen included in Moore et al. (2011).
[b]   Specimen included in Straube et al. (2010) (as *Echinorhinus brucus*).
[c]   Specimen included in Straube et al. (2010).
[d]   *Sensu* Last and Stevens (1994).
[e]   Specimen included in Straube et al. (2011a) (as *Etmopterus baxteri*).
[f]   Specimen included in Straube et al. (2010, 2011a) (as *Etmopterus* cf. *granulosus*).

de novo using the same primer sets and amplification conditions described in Naylor et al. (in press). All sequences used in the current study have been entered in GenBank. In an effort to minimize problems associated with missing data (see Lemmon et al., 2009), only specimens for which close to the full sequence complement of NADH2 (see below) was available were included. The proportion of missing data in the final full matrix was less than 0.25%.

2.2.3 Sequence Alignment

Electropherogram trace files were assessed for quality and base assignments were made using the software package Phred (Ewing et al., 1998), and fragments were subsequently assembled using Phrap (Ewing et al., 1998). A script was written (by CL) to translate the assembled nucleotide sequences to amino acids, subject the amino acid sequences to alignment using ClustalW (Thompson et al., 1994), and to back translate the aligned amino acids to their original nucleotide sequences. The final alignment across all 607 elasmobranch sequences was 1044 bp long.

2.2.4 Phylogenetic Analysis

2.2.4.1 Model Choice

There is a trade-off between the number of model parameters used to estimate a quantity of interest and the statistical power underlying the estimate. The Akaike Information Criterion (AIC) (Akaike, 1973, 1983), or its Bayesian equivalent, provides a general framework to assess the trade-off between the accuracy gained by adding parameters and the attendant loss in statistical power across both nested and non-nested models. For molecular phylogenetic datasets with few taxa, or datasets with patterns of low complexity, AIC will favor models with fewer parameters. Datasets with more complex patterns often require a larger number of parameters, which, in turn, compromises their statistical power. When estimating phylogenetic trees from parameter-rich models, it is important that the dataset to which the model is applied can meaningfully inform the additional parameters of the model. In general, analyses of large, taxon-rich datasets that sample the evolutionary process in an even and balanced way are more likely to benefit from parameter-rich models. It should be pointed out, however, that complex models often yield a better fit to datasets with sparse or biased sampling than do simpler models, but they do not necessarily guarantee a better tree. The critical issue is that parameterizations should be tailored to capture the salient aspects of the process rather than to simply account for variance.

One of the more commonly used parameter-rich models in phylogenetic analyses is the GTR+I+Γ model. If the six substitution rates among the different nucleotides are constrained to sum to one, the substitution rate component of the model has ten free parameters, the values of which must be estimated: five parameters for the relative substitution rates, three for the base frequencies, and two to capture patterns of rate variation over sites (the proportion of invariant sites and the shape parameter of the distribution) (Gu et al., 1995; Waddell and Penny, 1996; Yang, 1994). Nevertheless, this model (GTR+I+Γ) does not accommodate the fact that patterns of nucleotide change are generally different among codon positions due to the architectural constraints of the genetic code (differences in patterns of change among the three codon positions due to redundancy of the code at third positions and hydrophobicity constraints at second positions). We used AIC and its small-sample-size corrected counterpart, AICc (Hurvich, and Tsai, 1989) to determine if modeling each codon position separately or pooling the information across codon positions was warranted for the assembled NADH2 dataset. Although results from AIC indicated that a separate GTR+I+Γ model for each codon position was warranted, the (approximate) AICc measure indicated that the model pooled over codon positions had the better score (Table 2.2).

Both models yielded very similar, although not identical, tree topologies. We take this to indicate that the hierarchical signal in the dataset is not highly sensitive to model choice, at least not between the two models we tested. We speculate that this is probably due to the dense and evenly balanced taxon sampling scheme used (595 distinct elasmobranch species sampled across the diversity of the class). Given the AICc scores, the tree presented here was generated using the GTR+I+Γ model pooled across codon positions. Instances in which the topologies of the trees resulting from the two models differed with respect to monophyly or placement of taxa are indicated in the relevant sections below.

2.2.4.2 Bayesian Analysis

The Bayesian phylogenetic analysis was performed with the parallel implementation of MrBayes (Altekar et al., 2004; Huelsenbeck and Ronquist, 2001; Ronquist and

TABLE 2.2

Model Comparison (K = Number of Free Parameters Including Branch Lengths)

Model	K	AIC	AICc
GTR+I+Γ (pooled across codons)	1227	289248.7	272870.9
GTR+I+Γ (each codon position)	3681	288611.8	278336.2

Huelsenbeck, 2003), version 3.2 (http://sourceforge.net/projects/mrbayes/). For each model, four independent analyses were run, each with one cold chain and three heated chains, using the default heating and chain-swapping parameter settings. The proposal mechanism autotuning feature was used for substitution-model parameters, and chains were sampled every 500 generations. Topological convergence was assessed by comparing the standard deviations of split frequencies (SDSF) between the tree samples of the runs. For each of the other parameters, the potential scale reduction factor (PSRF) (Gelman and Rubin, 1992) is also given.

2.3 Results

After discarding the first 25% of samples as burn-in, comparing the samples of the cold chains strongly indicated that all of the independent runs had converged to the same stationary distribution. The MrBayes runs that estimated parameters separately for each codon position were stopped after 28,771,000 generations to yield a largely resolved tree. The average (maximum) SDSF across runs at the time the analysis was stopped was <0.0076 (maximum, <0.06). The average (maximum) PSRF for the substitution model parameters was 1.0006 (maximum, 1.003), and for the branch length parameters it was 1.001 (maximum, 1.067). The runs for which the model parameters were shared across codon positions were stopped after 22,813,500 generations at an average (maximum) SDSF of 0.0085 (maximum, 0.126). The average (maximum) PSRF for the substitution model parameters was 1.0015 (maximum, 1.004), and for the branch length parameters it was 1.001 (maximum, 1.017).

The fact that the NADH2 dataset yielded a tree that was well resolved (see Figures 2.1 to 2.11) was unexpected. NADH2 was selected for the original Naylor et al. (in press) study to distinguish among closely related species because it ranks as one of the fastest evolving protein-coding genes in the mitochondrial genome. The decision to subject a representative subset of these sequences to phylogenetic analysis was expected to yield a tree poorly resolved at its base with a few well-resolved subsets toward the tips. The tree resulting from the analysis was not only resolved at multiple levels of divergence but was also largely consistent with existing elasmobranch taxonomy and classification. We speculate that much of the unanticipated phylogenetic signal in this dataset may be due to the dense and balanced taxon sampling scheme employed. We discuss the topology obtained from the analysis in light of current taxonomy and classification below.

2.3.1 Assessing Monophyly of Previously Recognized Groups

2.3.1.1 Monophyly of Genera

Of the 159 genera sampled, 43 are known to be monotypic. An additional 30 are represented by only a single species in this analysis and thus are not amenable to tests of monophyly here. Assessments of the monophyly of the 86 genera for which two or more species were included in the study are addressed below. Our results suggest that 54 of these 86 are monophyletic. Most of these are supported by robust posterior probabilities. Caution is advised in interpretation because we do not regard robust posterior probabilities as definitive evidence of the monophyly of a group because they can vary tremendously across models and taxon-sampling schemes. The remaining 32 genera are inferred not to be monophyletic in this study. They include several in which the monophyly is compromised by the inclusion of either a single monotypic genus or a closely related subset of taxa or by the exclusion of a single species (see Section 2.3.1.1.1), as well as a number of more problematic cases in which genera are rendered non-monophyletic through the inclusion of species collectively assigned to several different genera or in which multiple species ostensibly assigned to the same genus fall in disparate parts of the tree (see Section 2.3.1.1.2).

2.3.1.1.1 Simple Cases of Generic Non-Monophyly

Simple cases in which monophyly of a genus is compromised by the inclusion of a single monotypic genus or monophyletic subset of taxa in the clades resulting from the analysis are treated below in alphabetical order:

- *Aetomylaeus* (Figure 2.11)—Our analysis included all four species of *Aetomylaeus* and one of the two described species of *Pteromylaeus*. *Pteromylaeus bovinus* is deeply nested within the otherwise monophyletic genus *Aetomylaeus*. This result questions the wisdom of recognizing *Pteromylaeus* as a distinct genus.

- *Aetoplatea* (Figure 2.11)—*Aetoplatea zonura* groups deeply among the eight species of *Gymnura* included here. The second species of *Aetoplatea* recognized by Compagno (2005), *A. tentaculata*, was not included in our sampling. The current analysis supports the reassignment of *A. zonura* to *Gymnura* following Naylor et al. (in press), as suggested, for example, by Smith et al. (2009) and Jacobson and Bennet (2009).

- *Centrophorus* (Figure 2.7)—The monophyly of the genus *Centrophorus* is potentially compromised by its inclusion of *Deania* species that group in a polytomy with two lineages of

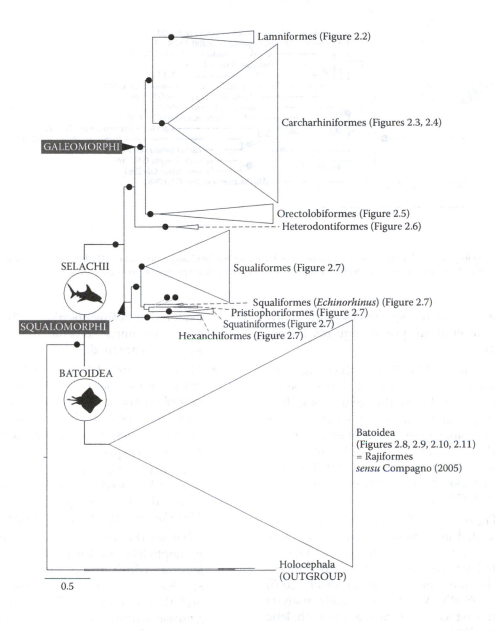

FIGURE 2.1

Summary of the phylogenetic relationships among elasmobranch orders based on NADH2 sequence data (1044 bp) of 595 elasmobranch species inferred from a Bayesian analysis using a separate GTR+I+Γ model pooled over codon positions. Black dots indicate posterior probabilities of >95%.

Centrophorus. We have included 12 of the 14 recognized species of *Centrophorus* and all but one of the species of *Deania*. Interestingly, a similar nesting of *Deania* within *Centrophorus* was obtained by Straube et al. (2010), although with a different suite of molecular markers. This polytomy clearly requires additional investigation.

- *Chiloscyllium* (Figure 2.5)—The analysis included six of the eight known species of *Chiloscyllium* and one of the nine known species of *Hemiscyllium*. Although poorly supported, the species of *Hemiscyllium* grouped among the species of *Chiloscyllium*; however, the phylogenetic placement of the specimen of *Hemiscyllium*

ocellatum was found to be model dependent, and this grouping may merely reflect the close relationships between these two genera.

- *Dipturus* (Figure 2.9)—Our analysis suggests that *Dipturus* is monophyletic only if the species currently placed in *Zearaja* are included. This is a relatively robust result given how deeply the *Zearaja* species, which include three of the four known members of this genus, are nested among *Dipturus* species; the position of *Spiniraja whitleyi* further potentially compromises the monophyly of *Dipturus*. In addition, our results support Compagno's (2005) suggestion that the generic placement of *Dipturus linteus*, which

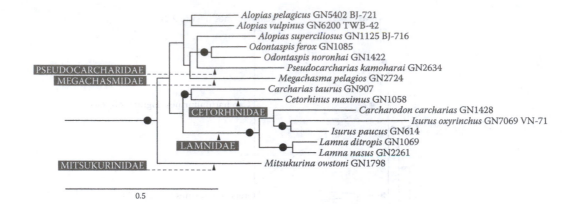

FIGURE 2.2

Hypothesis of the phylogenetic relationships of Lamniformes based on NADH2 sequence data (1044 bp) inferred from a Bayesian analysis using a separate GTR+I+Γ model pooled over codon positions. Black dots indicate posterior probabilities of >95%.

was represented by five specimens in the analysis of Naylor et al. (in press), requires further investigation.

- *Etmopterus* (Figure 2.7)—The analysis included approximately 50% of the greater than 40 species currently described from this genus, as well as the monotypic *Miroscyllium*. *Miroscyllium sheikoi* was found to fall squarely within an otherwise monophyletic *Etmopterus*. This placement was also seen in the multi-gene study of the phylogenetic relationships among etmopterids by Straube et al. (2010).

- *Galeus* (Figure 2.3)—The monophyly of this genus is called into question by the placement of *Galeus sauteri* in a clade outside of that containing its four congeners. This result is inconsistent with the work of Iglésias et al. (2005) based on 16S rDNA sequences; their analysis suggested that *Galeus* forms a monophyletic group including *G. sauteri*.

- *Halaelurus* (Figure 2.3)—Our results suggest that the monophyly of *Halaelurus* is compromised by the inclusion of *Haploblepharus* among *Halaelurus* species. The South African species *Haploblepharus edwardsii* grouped with the two South African species of *Halaelurus* (*H. lineatus* and *H. natalensis*) to the exclusion of the Australian and Southeast Asian representatives of *Halaelurus* (*H. sellus* and *H. buergeri*). The relationships among these two genera warrant further exploration with additional taxon sampling and markers.

- *Hexanchus* (Figure 2.7)—The monotypic *Heptranchias* clusters as sister to *Hexanchus griseus*, potentially compromising the monophyly of the genus *Hexanchus*; however, the topology presented is weakly supported. Further

exploration of this question with nuclear markers is recommended before any taxonomic reassignments are made.

- *Mobula* (Figure 2.11)—The analysis included five of the nine species of *Mobula* and one of the two described species of *Manta*. The latter species was deeply nested within the otherwise monophyletic genus *Mobula* in the current analysis.

- *Mustelus* (Figure 2.3)—The genus *Mustelus* is monophyletic only if the clade containing *Scylliogaleus quecketti* and *Triakis megalopterus* is included. The grouping of these two taxa within *Mustelus* was also found by López et al. (2006).

- *Okamejei* (Figure 2.9)—The genus *Okamejei* is monophyletic but for the exclusion of *Okamejei jensenae*, which groups outside the three other species of *Okamejei* with a cluster of *Rostroraja* and *Raja* and species. The membership of *O. jensenae* within the genus *Okamejei* should be reassessed.

- *Pristiophorus* (Figure 2.7)—The analysis included two of the eight known species of *Pristiophorus*, as well as the monotypic *Pliotrema*. In the presented analysis, *Pliotrema warreni* grouped as the sister of *Pristiophorus japonicus*, rendering the genus *Pristiophorus* non-monophyletic. This result was model dependent. When a GTR+I+Γ model was run for each codon separately, *Pliotrema* was inferred to be the sister to a monophyletic *Pristiophorus*.

- *Sphyrna* (Figure 2.4)—Our study included all but one of the described species of *Sphyrna*, as well as *Eusphyra blochii* (*S. media* was not included). The current analysis implies that recognition of the monotypic genus *Eusphyra* may be unwarranted; this result is consistent

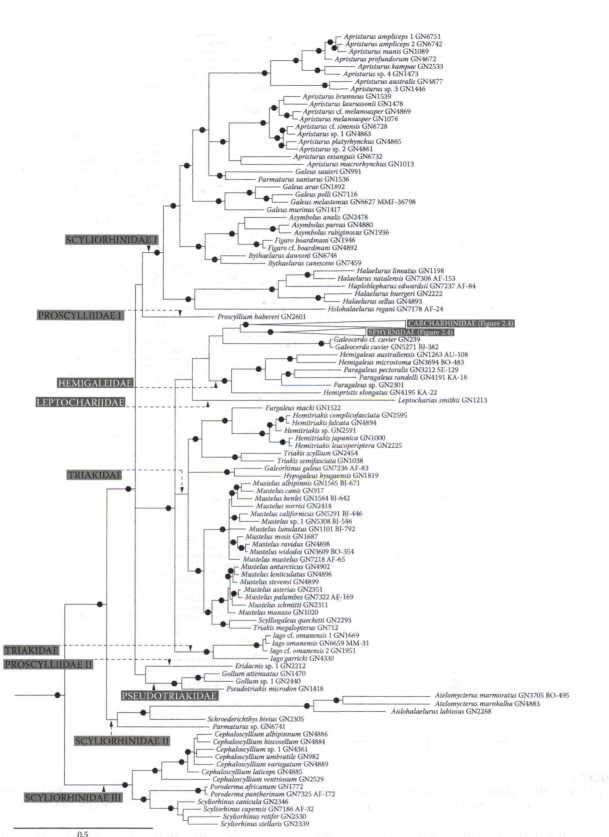

FIGURE 2.3

Hypothesis of the phylogenetic relationships of Carcharhiniformes based on NADH2 sequence data (1044 bp) inferred from a Bayesian analysis using a separate GTR+I+Γ model pooled over codon positions. Carcharhinid and sphyrnid relationships are shown in detail in Figure 2.4. Paraphyletic families are indicated with black text in family label boxes. Black dots indicate posterior probabilities of >95%.

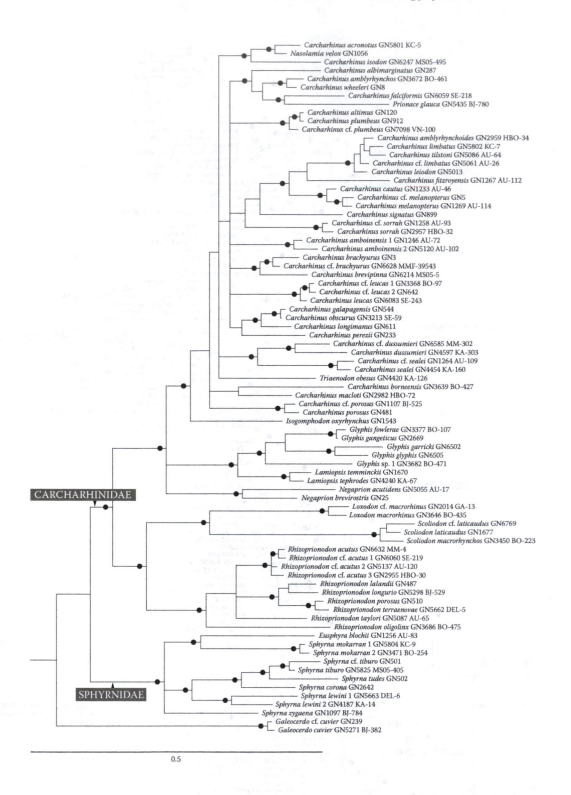

FIGURE 2.4

Hypothesis of the phylogenetic relationships of Carcharhinidae and Sphyrnidae based on NADH2 sequence data (1044 bp) inferred from a Bayesian analysis using a separate GTR+I+Γ model pooled over codon positions. Black dots indicate posterior probabilities of >95%.

with Compagno (1988). Nonetheless, a recent multi-gene study (Lim et al., 2010) focusing on the relationships among hammerhead sharks revealed variation in inferred relationships across genes. Lim et al. (2010) concluded that

the signal among all of the genes they analyzed was most consistent with a basal placement of the monotypic *Eusphyra*, which is consistent with the current taxonomy and different from our findings with NADH2.

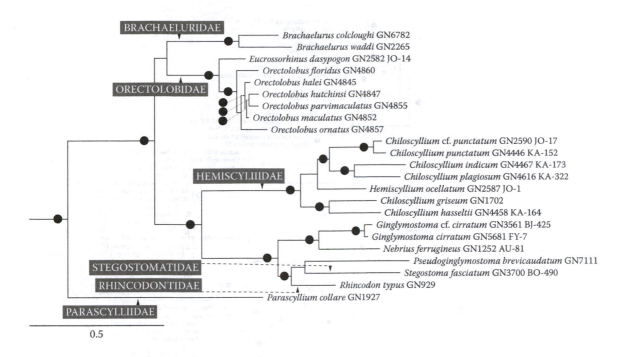

FIGURE 2.5

Hypothesis of the phylogenetic relationships of Orectolobiformes based on NADH2 sequence data (1044 bp) inferred from a Bayesian analysis using a separate GTR+I+Γ model pooled over codon positions. Black dots indicate posterior probabilities of >95%.

- *Squaliolus* (Figure 2.7)—The analysis included both described species of *Squaliolus* as well as the monotypic *Euprotomicrus*. The latter taxon grouped as sister to *S. aliae* with strong support. The justification for recognizing *Euprotomicrus* as a distinct genus warrants closer scrutiny and further analysis with additional markers.

- *Squalus* (Figure 2.7)—The 17 species of *Squalus* included in the analysis comprise a monophyletic group only if the genus also includes the two (of a total of three) species of *Cirrhigaleus*. *Squalus acanthias* and *S. suckleyi* were found to group with the two *Cirrhigaleus* species, albeit with relatively weak support and a short branch, as sister to the remaining *Squalus* species. If the close relationship between *Cirrhigaleus* and these two species of *Squalus* is borne out with further data, consideration should be given to

expanding *Squalus* to include the three known species of *Cirrhigaleus*, especially given that *S. acanthias* is the type species of the genus.

- *Taeniura* (Figure 2.11)—*Taeniura grabata* groups with *Taeniurops meyeni*, well away from the two other putative forms of *Taeniura lymma*. Our analysis supports the transfer of *Taeniura grabata* to *Taeniurops*.

- *Triakis* (Figure 2.3)—Although *Triakis scyllium* groups with *Triakis semifasciata*, their congener, *Triakis megalopterus*, groups with *Scylliogaleus quecketti*. This same result was seen in the multi-gene analysis conducted by López et al. (2006). We suggest that the generic assignment of *T. megalopterus* be re-examined.

- *Urobatis* (Figure 2.11)—Our analysis included three species of *Urobatis* from the Gulf of California and one from the Caribbean Sea.

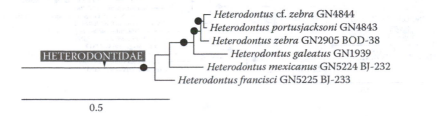

FIGURE 2.6

Hypothesis of the phylogenetic relationships of Heterodontiformes based on NADH2 sequence data (1044 bp) inferred from a Bayesian analysis using a separate GTR+I+Γ model pooled over codon positions. Black dots indicate posterior probabilities of >95%.

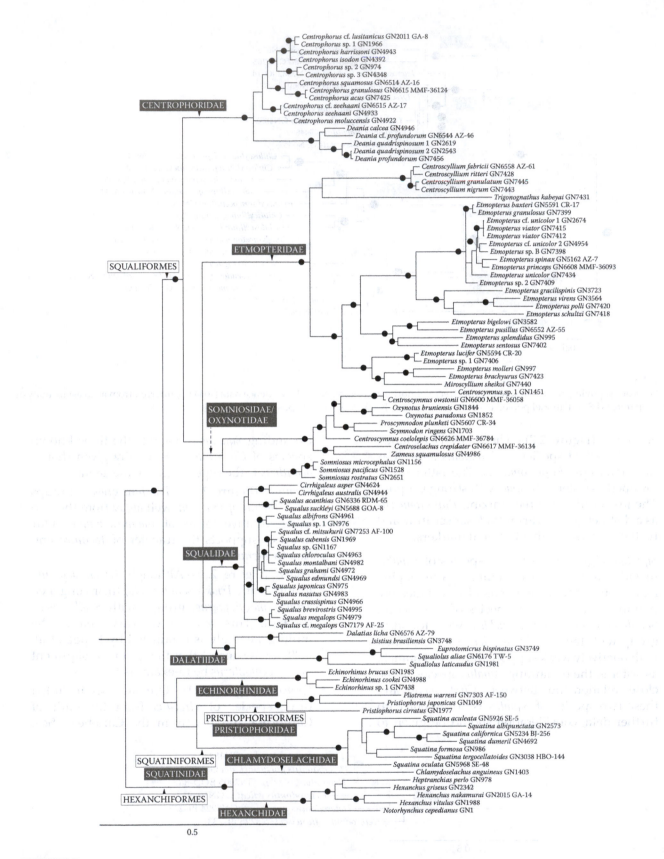

FIGURE 2.7

Hypothesis of the phylogenetic relationships of Squaliformes, Pristiophoriformes, Squatiniformes, and Hexanchiformes based on NADH2 sequence data (1044 bp) inferred from a Bayesian analysis using a separate GTR+I+Γ model pooled over codon positions. Black dots indicate posterior probabilities of >95%.

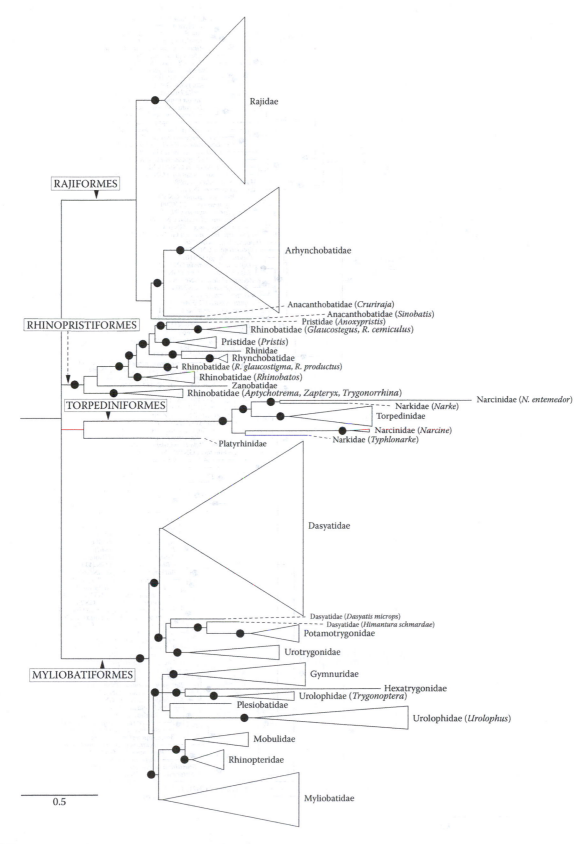

FIGURE 2.8

Summary of the phylogenetic relationships among the families and orders of Batoidea (i.e., Rajiformes *sensu* Compagno, 2005) based on NADH2 sequence data (1044 bp) inferred from a Bayesian analysis using a separate GTR+I+Γ model pooled over codon positions. Details of relationships for individual families are shown in Figures 2.9, 2.10, and 2.11. Black dots indicate posterior probabilities of >95%.

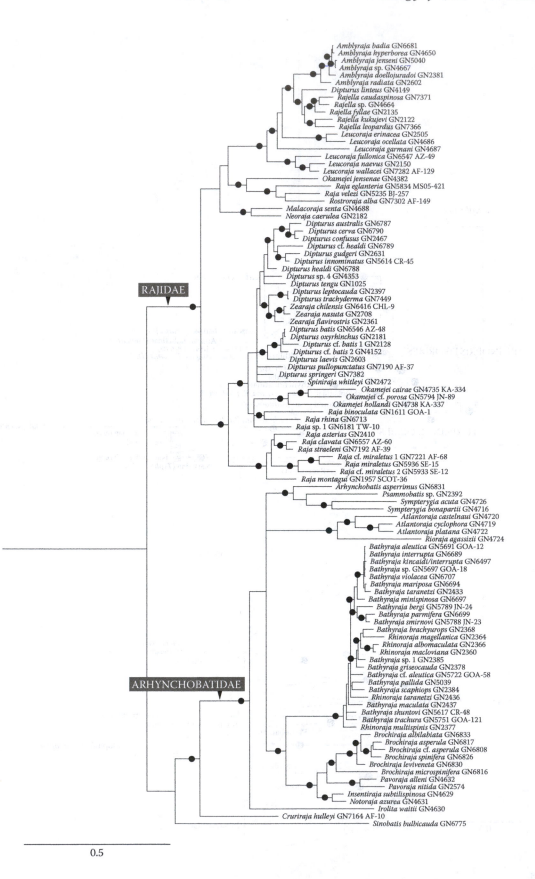

FIGURE 2.9

Hypothesis of the phylogenetic relationships of Rajiformes *sensu stricto* based on NADH2 sequence data (1044 bp) inferred from a Bayesian analysis using a separate GTR+I+Γ model pooled over codon positions. Black dots indicate posterior probabilities of >95%.

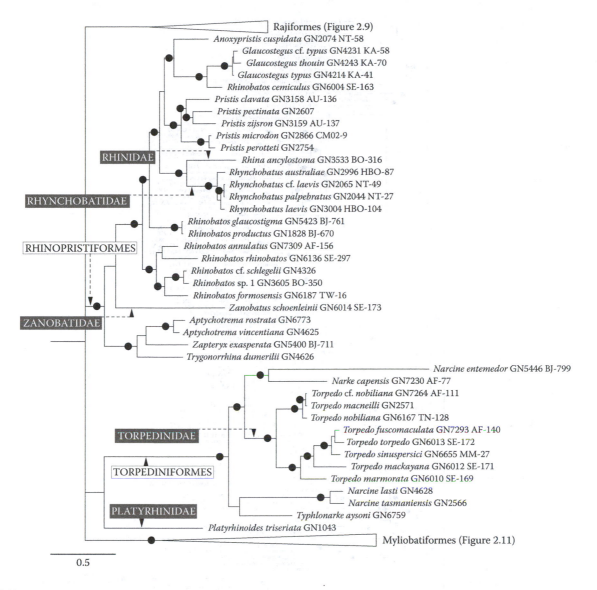

FIGURE 2.10
Hypothesis of the phylogenetic relationships of Rajiformes, Rhinopristiformes, Torpediniformes, *Platyrhinoides*, and Myliobatiformes based on NADH2 sequence data (1044 bp) inferred from a Bayesian analysis using a separate GTR+I+Γ model pooled over codon positions. Myliobatiform relationships are shown in detail in Figure 2.11. Black dots indicate posterior probabilities of >95%.

Our current results suggest that the genus is monophyletic only if either the two species of *Urotrygon* included in the analysis are also considered members of the genus or *Urobatis jamaicensis* is transferred to *Urotrygon*.

2.3.1.1.2 Complex Cases of Generic Non-Monophyly

The monophyly of the following genera is more problematic. These genera were rendered non-monophyletic in the trees resulting from our analysis, either because species collectively assigned to two or more genera are intermingled within a single clade or because multiple species ostensibly assigned to the same genus were found to fall in disparate parts of the tree. These genera are treated below in alphabetical order:

- *Alopias* (Figure 2.2)—Results of the current analysis suggest that the genus *Alopias* is non-monophyletic, although with a low posterior probability. The lack of monophyly for the genus, while highly surprising and seemingly unlikely, is consistent with analyses based on whole mitochondrial genomes (Ferrara, unpublished master's thesis) and also with preliminary analyses based on nuclear markers (GJPN, unpublished). Nonetheless, previous analyses of cytochrome *b* (Martin and Naylor, 1997) and cytochrome *b*/NADH2 (Naylor et al., 1997) data indicate that the genus *Alopias* is monophyletic, as did the morphological analysis of Shimada (2005). It is important to note that the

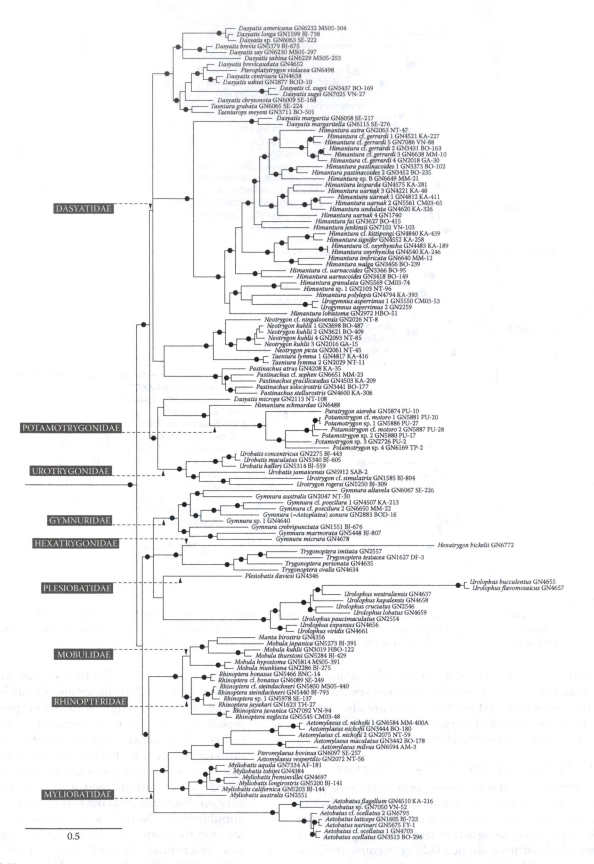

FIGURE 2.11

Hypothesis of the phylogenetic relationships of Myliobatiformes based on NADH2 sequence data (1044 bp) inferred from a Bayesian analysis using a separate GTR+I+Γ model pooled over codon positions. Black dots indicate posterior probabilities of >95%.

Lamniformes, as a group, seem to be especially recalcitrant to phylogenetic analysis. Inferred topologies for this group yield trees with long pendant edges separated by short internodal branches for most taxa, and the topology fluctuates from gene to gene. The order is clearly old and spans considerable morphological diversity, but it contains relatively few species. It is comprised of seven families, four of which are monotypic. Given these patterns of diversification, it is not surprising that conclusive branching patterns are difficult to estimate and that inferences vary across genes and models.

- *Apristurus* (Figure 2.3)—The monophyly of *Apristurus* is potentially compromised by the placement of one of the five species of *Galeus* (*G. sauteri*) and one of the two species of *Parmaturus* (*P. xaniurus*) included in this study as part of a polytomy with the *Apristurus* species. Assuming our specimen identifications are correct, these results call into question not only the monophyly of *Apristurus* but also the monophyly of both *Galeus* and *Parmaturus* as currently circumscribed. Interestingly, a molecular analysis based on 16S rDNA sequences by Iglésias et al. (2005) yielded a result in which *G. sauteri* formed a monophyletic group with the other *Galeus* species in their study. Clearly, further work is needed before any conclusions can be drawn.

- *Bathyraja* and *Rhinoraja* (Figure 2.9)—These two genera are among the most problematic in the current study. All five species of *Rhinoraja* included here were found to be interspersed among the 20 species of *Bathyraja*. This result is consistent with the analysis of Spies et al. (2011), in which the species of *Rhinoraja* nested among the *Bathyraja* species. These results suggest that these two genera are unlikely to be reciprocally monophyletic; however, as noted in Naylor et al. (in press), several of the species in these genera are difficult to identify. Clarification of relationships will require additional samples and careful specimen identity validation.

- *Carcharhinus* (Figure 2.4)—The monophyly of *Carcharhinus* is challenged by the inclusion of the three monotypic genera: *Nasolamia*, *Prionace*, and *Triaenodon*. The relationships among members of the genus *Carcharhinus* and their immediate close relatives are highly unstable, even when based on whole mitochondrial genome analyses (GJPN, unpublished). Accordingly, we do not recommend any taxonomic reassignment until a comprehensive study is undertaken

that includes sampling of geographic variants within species across multiple nuclear markers. That said, although the relationships among the different species of *Carcharhinus* are collectively unclear, a consistently close relationship is seen between *Nasolamia* and *Carcharhinus acronotus* across datasets. These relationships were suggested previously by Compagno (1984, 1988), based on morphological data. Similarly, the blue shark, *Prionace glauca*, is almost always inferred to be deeply nested within the genus *Carcharhinus*, and most often allied with the silky shark, *Carcharhinus falciformis*, as is the case in the current analysis. The nesting of *P. glauca* among *Carcharhinus* species has been observed in previous studies (e.g., Compagno, 1988; Dosay-Akbulut, 2008; Naylor, 1992), but in each case with different specific affinities than recovered here. In contrast, the phylogenetic placement of the whitetip reef shark, *Triaenodon obesus*, was unresolved with respect to species in the genus *Carcharhinus*.

- *Centroscymnus* (Figure 2.7)—Although *Centroscymnus owstoni* and *Centroscymnus* sp. 1 (*sensu* Naylor et al., in press) cluster together, *Centroscymnus coelolepis* was placed at the base of a clade including these two species, as well as *Oxynotus*, *Proscymnodon*, and *Scymnodon*. This result is independently supported by nuclear gene data (GJPN, unpublished).

- *Dasyatis* (Figure 2.11)—Our analysis raises several issues with the current composition of *Dasyatis*. First, it seems likely that *Dasyatis microps* does not belong in the genus because it grouped well away from its 14 putative congeners, most closely with *Himantura schmardae* and the potamotrygonids. Second, the species from Senegal (*D. margarita* and *D. margaritella*) appear to be more closely allied with *Himantura* species than with the majority of the *Dasyatis* species; this result is generally consistent with Rosenberger (2001). Third, the monotypic *Pteroplatytrygon* groups among the main group of 14 species of *Dasyatis*; some consideration should be given to whether a unique genus designation is appropriate for *P. violacea*. Perhaps the biggest issue raised by our analysis is that *Taeniura grabata* and *Taeniurops meyeni* group among *Dasyatis* species. This arrangement has been suggested previously by Lovejoy (1996) based on morphological data. It is interesting that the relationships among *Dasyatis* species seen here generally differed from those seen by Rosenberger (2001).

- *Himantura* (Figure 2.11)—There are two issues associated with the monophyly of *Himantura*. First, as noted above, the North American species *H. schmardae* groups well outside of the other *Himantura* species, as the sister to the South American freshwater stingrays as observed by Lovejoy (1996). Second, the two putative *Urogymnus* species (*sensu* Naylor et al., in press) included here fall among the Indo–Pacific *Himantura* species in this analysis. This result suggests that *Himantura* is monophyletic only if it includes the *Urogymnus* species. Interestingly, the squamation pattern at the base of the tail in both *Himantura granulata* and *Urogymnus* are similar, supporting the placement seen in the current analysis (Last, pers. comm.).

- *Leucoraja* (Figure 2.9)—The analysis included 6 of the 14 recognized species of *Leucoraja*. The genus overall was conspicuously paraphyletic in that three of the included species grouped in a strongly supported clade with *Rajella* and *Amblyraja* species, away from the three other *Leucoraja* species.

- *Narcine* (Figure 2.10)—*Narcine lasti* and *Narcine tasmaniensis* group together in this analysis, but the third included species, *Narcine entemedor*, grouped with *Narke capensis*. This result is supported with a strong posterior probability and appears robust across models for this dataset, suggesting that the generic placements among the various species of *Narcine* and *Narke* warrant further exploration with a denser taxon sampling across the 25 currently recognized species of *Narcine*.

- *Parmaturus* (Figure 2.3)—The two species identified as *Parmaturus* (*P. xaniurus* and *Paramaturus* sp.) appear in different parts of the tree. The specimen identified as *P. xaniurus* falls out as the sister taxon to *Galeus sauteri* (see above). This clade, in turn, appears as the sister group to the genus *Apristurus*. The specimen identified as *Parmaturus* sp. is sister to a clade containing *Atelomycterus*, *Aulohalaelurus*, and *Schroederichthys*. Given these patterns, it is possible, indeed likely, that the specimen identified as *Parmaturus* sp. is an as of yet undescribed genus, rather than a different species of *Parmaturus*.

- *Raja* (Figure 2.9)—Our analysis validates much of the comparative anatomical work conducted over the last decade on the substructure within the Rajidae. Many of the proposed new genera are supported as monophyletic in the current molecular analysis. Several issues remain, however. Species recognized in the genus *Raja* by Compango (2005) appear in three different places on the tree. The clade most appropriately considered to represent *Raja*, because it contains the three variants of the type species for the genus (i.e., *Raja miraletus*), also includes *R. clavata*, *R. straeleni*, *R. asterias*, and *R. montagui*; however, other species typically assigned to *Raja* group outside this clade: *Raja binoculata*, *R. rhina*, and *Raja* sp. 1 group with the *Dipturus*, *Zearaja*, *Spiniraja*, and most of the *Okamejei* species. These are assigned to the "new genus 1" by McEachran and Dunn (1998); our results support the recognition of these as a distinct genus. Finally, *Raja eglanteria* and *R. velezi* group together with *Rostroraja alba* and *Okamejei jensenae*. Both *R. eglanteria* and *R. velezi* were assigned to the "new genus 2" by McEachran and Dunn (1998). The results presented herein support the recognition of this grouping as members of *Rostroraja*.

- *Rhinobatos* (Figure 2.10)—Species currently placed in *Rhinobatos* appear at three different points on the tree, suggesting that the genus is not monophyletic as currently configured. The West African *R. cemiculus* is strongly allied with the three included species of *Glaucostegus*. The two species of *Rhinobatos* from the Gulf of California (*R. productus* and *R. glaucostigma*) group robustly together as the sister taxon to a clade consisting of *Rhynchobatus*, *Pristis*, *Anoxypristis*, and *Glaucostegus*, as well as *R. cemiculus*. The remaining five species, which include the type, *R. rhinobatos*, comprise a clade that is sister to the previous clade.

The following observations regarding the monophyly of the higher level groups (i.e., orders and families) are made in the context of the classification of elasmobranchs presented in Compagno (2005).

2.3.1.2 Monophyly of Families

Our analysis included representation of 56 of the 57 families recognized by Compagno (2005), with only the Hypnidae missing from consideration. Ten of these families are monotypic: Cetorhinidae, Hexatrygonidae, Leptochariidae, Megachasmidae, Mitsukurinidae, Plesiobatidae, Pseudocarchariidae, Rhincodontidae, Rhinidae, and Stegostomatidae. An additional ten families are monogeneric: Alopiidae, Chlamydoselachidae, Echinorhinidae, Heterodontidae, Oxynotidae, Rhinopteridae, Rhynchobatidae, Squatinidae, Torpedinidae, and Zanobatidae. Finally, three families (Narcinidae, Parascylliidae, Platyrhinidae), although not monogeneric, were represented by species belonging to only a single genus in our sampling scheme, providing little to explore with respect to family-level monophyly. The monophyly of each of the 33 families that were represented by two or more genera in our analysis is addressed below.

Our results support the monophyly of 17 of the 33 elasmobranch families for which this issue can be addressed here; these are treated below in alphabetical order. In each case, the relative representation for the sample is given. The monophyletic families are as follows: Arhynchobatidae (43 species representing 10 of 12 genera) (Figures 2.8 and 2.9); Centrophoridae (17 species representing both genera) (Figure 2.7); Dalatiidae (5 species representing 4 of 7 genera) (Figure 2.7); Etmopteridae (29 species representing 4 of 5 genera) (Figure 2.7); Gymnuridae (9 species representing both genera) (Figures 2.8 and 2.11); Hemiscylliidae (7 species representing both genera) (Figure 2.5); Hexanchidae (5 species representing all 3 genera) (Figure 2.7); Lamnidae (all 5 species in all 3 genera represented) (Figure 2.2); Myliobatidae (20 species representing all 4 genera) (Figures 2.8 and 2.11); Orectolobidae (7 species representing 2 of 3 genera) (Figure 2.5); Potamotrygonidae (7 species in 2 of 4 recognized genera, including that of Ishihara and Taniuchi, 1995) (Figures 2.8 and 2.11); Pristiophoridae (3 species representing both genera) (Figure 2.7); Pseudotriakidae (3 species in 2 of 3 genera, including an undescribed species of *Gollum*) (Figure 2.3); Rajidae (60 species representing 11 of 17 families) (Figures 2.8 and 2.9); Sphyrnidae (7 of 8 species in both genera) (Figures 2.3 and 2.4); Squalidae (19 species representing both genera) (Figure 2.7); and Urotrygonidae (6 species representing both genera) (Figures 2.8 and 2.11).

Our results call into question the monophyly of 13 families as currently circumscribed (e.g., Compagno 2005). These are treated below in alphabetical order.

- Anacanthobatidae (Figure 2.9)—Two of the three genera in this family, each represented by a single species, were included in the analysis. Results for the current study fail to provide support for the monophyly of this family. Whereas *Cruriraja hulleyi* grouped as the sister to the Arhynchobatidae, *Sinobatis bulbicauda* grouped as the sister to that larger group. It is possible, however, that the groupings observed here are the consequence of model misspecification associated with long branch taxa. Nonetheless, a similar pattern was seen in the recent work on batoid phylogeny by Aschliman et al. (in press). In their morphological analysis, however, McEachran and Dunn (1998) found the two genera of anacanthobatids to group together, but in a clade comprised essentially of the Rajidae as recovered here.

- Carcharhinidae (Figure 2.4)—All 12 genera of carcharhinids were represented in the analysis. This family was well supported as monophyletic, with the exception of a clade comprising the two putative species of *Galeocerdo* (*sensu* Naylor et al., in press). The *Galeocerdo* clade was grouped as the sister to a clade comprising the Carcharhinidae and Sphyrnidae. Although this node was poorly supported in the current analysis, similar results were obtained by López et al. (2006). Furthermore, independent nuclear gene evidence (GJPN, unpublished) and its unique reproductive biology suggest that *Galeocerdo* does not belong in the Carcharhinidae.

- Dasyatidae (Figure 2.11)—Our analysis included representation of all eight dasyatid genera. The analysis yielded a clade consisting of 61 of the 63 species included in this study. The two exceptions were *Dasyatis microps* and *Himantura schmardae*, both of which grouped outside of the dasyatids, along with the potamotrygonids.

- Ginglymostomatidae (Figure 2.5)—The monophyly of this family was not supported. Our analysis included representation of all four species in the three genera recognized in this family by Compagno (2005). Although both *Ginglymostoma* species (*sensu* Naylor et al., in press) grouped with the monotypic *Nebrius*, the monotypic *Pseudoginglymostoma* grouped in a clade with the monotypic Stegostomatidae and Rhincodontidae.

- Narkidae (Figure 2.10)—The two species representing two of the five genera included in the analysis grouped well away from one another. *Narke capensis* grouped with *Narcine entemedor*, while *Typhlonarke aysoni* grouped, albeit with relatively low support, with the two other species of *Narcine* included in the analysis. These results, although preliminary, suggest that the monophyly of the families of Torpediformes requires further study.

- Odontaspidae (Figure 2.2)—All three species in both genera were included. The analysis suggests that the family is not monophyletic, an observation consistent with previous analyses (Human et al., 2006; Martin and Naylor, 1997; Naylor et al., 1997). Although the two species of *Odontaspis* grouped together, they clustered well away from *Carcharias taurus*, which grouped, with strong support, as the sister to the basking shark.

- Pristidae (Figure 2.10)—The analysis included six of the seven species in both genera. Our results suggest that the family may not be monophyletic as currently configured. Curiously, although the six species of *Pristis* comprise a robust clade, they do so to the exclusion of the monotypic *Anoxypristis cuspidatus*. The latter species grouped in a clade comprised of three species of *Glaucostegus*, and one of the seven included species of *Rhinobatos* (i.e., *R. cemiculus* from Senegal). These results are surprising and at odds with the work of Faria (unpublished doctoral dissertation), who found the Pristidae to be strongly monophyletic but who notably did not include any specimens of *Rhinobatos* in his study. The NADH2 results are also at odds with an analysis based on combined nuclear and mitochondrial genes (Aschliman, pers. comm.). Clearly, these relationships require further work before firm conclusions can be drawn.

- Proscylliidae (Figure 2.3)—The two included species, representing two of the three genera, did not form a monophyletic group in the current analysis. *Proscyllium habereri* (Proscylliidae I) was placed at the base of one of the three clades of scyliorhinids (Scyliorhinidae I), whereas the specimen identified as *Eridacnis* sp. 1 (Proscyliidae II) grouped as sister to the clade consisting of the Pseudotriakidae; however, the placement of neither species is strongly supported. In contrast, independent data from the more slowly evolving RAG1 gene places these taxa together, thus rendering the Proscyllidae a monophyletic group (GJPN, unpublished).

- Rhinobatidae (Figure 2.10)—Four of the five genera (including the newly resurrected *Glaucostegus*) were represented by a total of 15 species in the analysis. Our results suggest that this group is monophyletic only if it also includes the Pristidae, Rhinidae, Rhynchobatidae, and Zanobatidae; however, the inclusion of *Zanobatus* in the group was found to be model dependent for the current dataset.

- Scyliorhinidae (Figure 2.3)—The analysis included 55 species representing 15 of the 17 families of catsharks. Our results indicate that the family is not monophyletic; rather, there exist three distinct, paraphyletic lineages, two of which are relatively well supported. The first, and largest, group (Scyliorhinidae I) consists of *Apristurus*, *Galeus*, *Asymbolus*, *Figaro*, *Bythaelurus*, *Halaelurus*, *Haploblepharus*, and *Holohalaelurus* species, and *Parmaturus xaniurus*; the second group (Scyliorhinidae II) consists of *Atelomycterus*, *Aulohalaelurus*, and *Schroederichthys* and *Parmaturus* species; and

the third group (Scyliorhinidae III), consists of *Cephaloscyllium*, *Poroderma*, and *Scyliorhinus* species. The primary issue is that the Proscylliidae, Carcharhinidae, Sphyrnidae, Hemigaleidae, Leptochariidae, Triakidae, and Pseudotriakidae are interspersed among these three catshark clades, thereby compromising their collective monophyly. The non-monophyly of the Scyliorhinidae has been previously documented by Iglésias et al. (2005) and Human et al. (2006), although in both cases based on a much less representative sample of taxa.

- Somniosidae (Figure 2.7)—The ten species in this family included in the current study represent six of the seven genera. Our analysis suggests that the family is monophyletic only if the Oxynotidae are included. This result appears to be robust; the two species of *Oxynotus* included grouped deeply within the Somniosidae. These results corroborate the similar placement for *Oxynotus* seen in the recent work by Straube et al. (2010) using a different suite of molecular markers.

- Triakidae (Figure 2.3)—A total of 34 species, representing eight of the nine triakid genera, were included in this study. Analysis indicates that all but *Iago* form a monophyletic group, albeit with weak support. The four putative species of *Iago* grouped as part of a larger polytomy that also contained the Hemigaleidae, Carcharhinidae, and Sphyrnidae, as well as several monotypic families and problematic groups treated above. These results are consistent with the patterns observed by López et al. (2006).

- Urolophidae (Figure 2.11)—The current analysis, based on 13 species representing both genera of this family, failed to support the monophyly of this family. Not only did the species in its two constituent genera not group with one another, but also the clade containing the *Trygonoptera* species was placed as the sister group to *Hexatrygon bickelii*, and the clade of *Urolophus* species grouped as sister to *Plesiobatis daviesi*, albeit with relatively weak support. Further work with nuclear markers will be required before any firm conclusions can be drawn.

2.3.1.3 Monophyly of Orders

All but one of the nine orders of elasmobranchs recognized by Compagno (2005) were strongly supported as monophyletic (Figure 2.1). The exception was the Squaliformes, the monophyly of which was compromised by the placement of the three species of

Echinorhinus (family Echinorhinidae) as the sister taxon to a clade consisting of the Pristiophoriformes plus the Squatiniformes (Figures 2.1 and 2.7). Although support for these groupings was weak, generally similar results were obtained based on the nuclear gene RAG1 by Maisey et al. (2004). These findings are provocative and warrant further exploration with multiple nuclear genes.

Although Compagno (2005) recognized only the single batoid order Rajiformes, our results indicate that the subdivision of the batoids into several orders is warranted. McEachran and Aschliman (2004) and Nelson (2006) recognized the four batoid orders Torpediniformes, Pristiformes, Rajiformes, and Myliobatiformes; Compagno (1999) recognized the Rhiniformes and Rhinobatiformes in addition to these four orders. The constituencies of these orders differed somewhat among the classification schemes presented by these authors; our results support the following blend of their scenarios. Regardless of the circumscription of its families, all three sets of authors recognized the electric rays (Torpediniformes); our results strongly support the monophyly of this order (Figure 2.8). Our analysis included one of the three known species of platyrhinids. This species grouped as the sister taxon to the electric rays. This is an interesting result in light of the fact that the Platyrhinidae was considered to belong among the Rhinobatiformes by Compagno (1999) and among the Myliobatiformes by McEachran and Aschliman (2004) and Nelson (2006). Similarly, regardless of family-level organization, all three sets of authors recognized an order comprised of the skates (i.e., Rajiformes *sensu stricto*); our results provide support for the monophyly of this order (Figure 2.8). Family-level organization aside, the concept of the Myliobatiformes was also remarkably consistent among these authors. At a minimum, the order was considered to include the Dasyatidae, Gymnuridae, Hexatrygonidae, Mobulidae, Myliobatidae, Potamotrygonidae, Rhinopteridae, Urolophidae, and the Urotrygonidae in all three studies. Differences among these studies stem in part from the fact that the Plesiobatidae was included in the order by Compagno (1999) and Nelson (2006) but was not represented in the analysis of McEachran and Aschliman (2004); also, whereas McEachran and Aschliman (2004) and Nelson (2006) considered the Platyrhinidae and Zanobatidae to belong to the Myliobatiformes, Compagno (1999) considered these families to belong to the Rhinobatiformes. Our results provide robust support for the Myliobatiformes as circumscribed by Compagno (1999).

The greatest differences between our results and the classification schemes of these previous authors center around the guitarfish (Rhinobatidae), wedgefish (Rhynchobatidae), sharkrays (Rhinidae), and sawfish (Pristidae). In all three previous studies, the Pristidae was placed in its own order, the Pristiformes. Compagno (1999) also recognized the order Rhiniformes for the monotypic Rhinidae. McEachran and Aschliman (2004) and Nelson (2006) placed the guitarfish, wedgefish, and sharkrays in the Rajiformes along with the skates. But, Compagno (1999) placed the guitarfish and wedgefish in their own order, the Rhinobatiformes (along with the Platyrhinidae and Zanobatidae). Our results suggest a slightly different scenario; there may be some merit to the recognition of an order consisting of the Pristidae, Rhinidae, Rhynchobatidae, Rhinobatidae, and Zanobatidae. The results of the recent analysis of Aschliman et al. (in press) are generally consistent with the latter grouping, except that the group did not include *Zapteryx*, *Trygonorrhina*, or *Zanobatus* species. To our knowledge, no name for the group comprised of the above five families of batoids currently exists. We propose that the name Rhinopristiformes be considered. With respect to the ordinal placement of the Platyrhinidae, our analysis supports the scheme of Compagno (2005) in recognizing this family in its own higher taxonomic category, perhaps at the ordinal level.

2.3.1.4 Interrelationships among Orders

Our results do not support the Hypnosqualea concept of Shirai (1992, 1996), endorsed by de Carvalho (1996); rather, the reciprocal monophyly of sharks (i.e., Selachii) and of the batoids (i.e., Batoidea) was recovered. This result is consistent with the findings of many previous authors working with a diversity of data types (e.g., Arnason et al., 2001; Douady et al., 2003; Human et al., 2006; Maisey et al., 2004; Naylor et al., 2005; Schwartz and Maddock, 2002; Winchell et al., 2004).

Within the Selachii, our results support division of the group into two major subgroups (Figure 2.1). There is support for a subgroup consisting of the four orders Carcharhiniformes, Lamniformes, Orectolobiformes, and Heterodontiformes (Maisey, 1984b). This subgroup is the Galea of Shirai (1992, 1996) and the Galeomorphi of Compagno (2001), reiterated in Nelson (2006); we have followed Compagno (2001) here in referring to this clade as the Galeomorphi. Our results also support the hypothesis of the interrelationships among these four orders obtained by Shirai (1992) and Naylor et al. (2005) in suggesting that the Heterodontiformes (Figure 2.6) are the immediate sister to a group consisting of the Orectolobiformes and the Lamniformes + Carcharhiniformes. These were not, however, the relationships obtained by Douady et al. (2003), Winchell et al. (2004), or Vélez-Zuazo and Agnarsson (2011), each of whom recovered relationships among these orders that differed from our result and also from each other.

The second major subgroup of Selachii recovered in our analysis, albeit with less support, is a clade consisting of the orders Squaliformes, Squatiniformes,

Pristiophoriformes, and Hexanchiformes. This group is the Squalomorphii of Nelson (2006) and is consistent with the concept of the orbitostylic sharks first forwarded by Maisey (1980), based on the type of articulation between the jaw and the brain case. Although the proposed interrelationships among these orders have differed among studies (e.g., de Carvalho, 1996; Shirai, 1996), our results support a clade consisting of the Squaliformes, Pristiophoriformes, and Squatiniformes, to the exclusion of the Hexanchiformes. However, the interrelationships among the Echinorhinidae, Pristiophoriformes, and Squatiniformes remain unclear given the weak support in the current dataset.

We are hesitant to make direct comparisons with the results of Vélez-Zuazo and Agnarsson's (2011) analysis of 229 species of sharks. In a number of respects their results are incongruent with current generic and familial level taxonomy, but they are also incongruent in many respects with ours. As noted by these authors, their analysis suffers from a substantial amount of missing data. Furthermore, the critical step of specimen identity verification was not taken because their data came directly from GenBank. As a consequence, it is unclear which differences to attribute to which issue.

The proposed relationships among the four major lineages of batoids recognized above (i.e., Rajiformes, Torpediniformes [+ Platyrhinidae], Rhinopristiformes, and Myliobatiformes) (Figure 2.8) have also differed among various studies. McEachran et al. (1996) and Rocco et al. (2007) found the Torpediniformes to be basal to a group consisting of the Myliobatiformes and essentially the Rhinopristiformes + Rajiformes *sensu stricto*. Alternatively, Shirai (1996) considered the Rhinopristiformes to be basal to a group consisting of the Torpediniformes and the Rajiformes *sensu stricto* + Myliobatiformes. The current results contribute little to our understanding of these interordinal relationships, for these relationships are unresolved in our analysis.

2.4 Discussion

We have presented a phylogenetic tree depicting inferred relationships among most of the major groups of elasmobranchs. Because the dataset is itself a subset of an even more comprehensive sample of over 4200 specimens that included replicates of most species (Naylor et al., in press), we feel that the results from this survey should provide a good baseline against which to compare subsequent studies. Nonetheless, it is important to remember that the results we present were derived from a single, fast-evolving mitochondrial gene and thus should be regarded as only the first tentative step

toward understanding phylogenetic relationships. Our inference will, at best, constitute a single locus gene tree that may or may not reflect species-level relationships at different levels in the hierarchy due to population-level processes differentially affecting patterns of lineage coalescence. At worst, it could be an inaccurate gene tree, depending on how well the model used for inference captures the dynamics of the sequence evolution among the sequences used.

The current dataset was certainly not expected to yield an accurate phylogenetic signal because NADH2 is one of the fastest evolving protein-coding genes in the mitochondrial genome and is thus especially questionable for assessing relationships among deeply diverged groups. However, most of the deep-level relationships retrieved are surprisingly concordant with the current classification. We speculate that this is due to the relative absence of missing data and the density and evenness of the taxon-sampling scheme used. Lakner et al. (2011) have shown explicitly that model inadequacies can be ameliorated by balanced and judicious taxon sampling. Interestingly, convergence occurred much more quickly for our 607 specimen NADH2 dataset than it did for the 229 shark species sequence dataset of Vélez-Zuazo and Agnarsson (2011) which we ran again in an effort to estimate the phylogenetic signal in that dataset (because the published analysis was based on an analysis that had not converged) (Agnarsson, pers. comm.). It would appear that the complexity and shape of the posterior distribution for the larger NADH2 dataset were much less difficult to approximate than was the case for the smaller Vélez-Zuazo and Agnarsson (2011) dataset. We speculate that this may be due to the differences in the amount of missing data in the two datasets—Vélez-Zuazo and Agnarsson (2011) dataset, 85% missing data; current NADH2 dataset, <0.25% missing data.

Although model inadequacies may sometimes be ameliorated by balanced and dense taxon sampling schemes (*sensu* Lakner et al., 2011), such strategies can only be deployed in instances in which taxa actually exist to break up long branches that have accrued extensive evolutionary change. For some groups, the required taxa are simply not available. The lamniform sharks, for example, comprise a sparse collection of highly divergent long branch taxa. Thus, sampling all of the available extant forms in this order does not ameliorate the difficulties associated with estimating an accurate phylogeny. In such instances, it is necessary to use a battery of slowly evolving independent markers in conjunction with carefully parameterized inference models to obtain accurate results, and even under such circumstances accuracy is never guaranteed. These cautions notwithstanding, the taxon sampling in the current study is reasonably good, at least to the extent that the diversity recognized by the current taxonomy is represented.

Finally, it is important to remember that there will likely be a few groups whose phylogenetic relationships will be recalcitrant to analysis even under the best-case scenario (i.e., complete taxon sampling and examination of multiple independent markers). This is because certain patterns of diversification and extinction tend not to leave unambiguous hierarchical traces in DNA sequences or morphological character state distributions. Rapid diversification of multiple lineages over a short period of time followed by subsequent lineage pruning due to extinction can often lead to situations in which ancestral polymorphism in slowly evolving markers is fixed in such a way as to be phylogenetically misleading. Such scenarios can lead to the situation where fast evolving markers are "saturated" and rendered uninformative because they have been overprinted with substitutions while slowly evolving markers leave a clear but misleading pattern of allele fixation.

2.4.1 Closing Remarks

In a number of instances, the results of our analysis call into question the monophyly of recognized genera and families, and even orders, as currently circumscribed. Some of the groupings suggested by the current analysis are clearly questionable as they conflict both with morphological data and assessments based on other genes. Most conspicuously, these include, but are not limited to, the non-monophyly of the following families: Pristidae, Alopiidae, Narkidae, Proscylliidae, and Urolophidae, as well as the non-monophyly of the following genera: *Halaelurus*, *Hexanchus*, and *Urobatis*, and the placement of *Eusphyra* within *Sphyrna*. In such cases, we have described the taxonomic implications revealed by our results, but these discussions are presented solely to inform future taxonomic work. They should not be interpreted as formal taxonomic actions. Instead, we hope they serve to encourage further exploration based on additional nuclear genes and a more thorough sampling of the taxa that were underrepresented in the current study, as well as the morphological work required to fully assess the taxonomic implications of our results.

Acknowledgments

We thank Neil Aschliman for comments on the manuscript and for comparing our results with those he has obtained based on whole mitochondrial genome and nuclear marker analysis. In addition, Joanna Cielocha, Maria Pickering, and Veronica Bueno provided helpful comments on the manuscript. This work was supported in part with funds from NSF grants DEB 8708121, DEB 9707145, DEB 0089533, DEB 0415486, DEB 1036500, DEB 9300796, DEB 9521943, DEB 0118882, DEB 0103640, DEB 0542846, DEB 0818696 DEB 0542941, and DEB 0818823.

References

Akaike, H. (1973). Information theory as an extension of the maximum likelihood principle. In: Petrov, B.N. and Csaki, F. (Eds.), *Second International Symposium on Information Theory*, Akademiai Kiado, Budapest, Hungary, pp. 267–281.

Akaike, H. (1983). Information measures and model selection. *Int. Stat. Inst.* 44:277–291.

Altekar, G., Dwarkadas, S., Huelsenbeck, J.P., and Ronquist, F. (2004). Parallel Metropolis-coupled Markov chain Monte Carlo for Bayesian phylogenetic inference. *Bioinformatics* 20:407–415.

Arnason, U., Gullberg, A., and Janke, A. (2001). Molecular phylogenetics of gnathostomous (jawed) fishes: old bones, new cartilage. *Zool. Scr.* 30:249–255.

Aschliman, N.C., Nishida, M., Miya, M., Inoue, J.G., Rosana, K.M., and Naylor, G.J.P. (in press). Body plan convergence in the evolution of skates and rays (Chondrichthyes: Batoidea). *Mol. Phylogenet. Evol.*

Bridge, P.D., Spooner, B.M., Roberts, P.J., and Panchal, G. (2003). On the unreliability of published DNA sequences. *New Phytol.* 160:43–48.

Cavalcanti, M. (2007). A phylogenetic supertree of the hammerhead sharks (Carcharhiniformes: Sphyrnidae). *Zool. Stud.* 46(1):6–11.

Compagno, L.J.V. (1977). Phyletic relationships of living sharks and rays. *Am. Zool.* 17:303–322.

Compagno, L.J.V. (1984). *FAO Species Catalogue.* Vol. 4. *Sharks of the World: An Annotated and Illustrated Catalogue of Shark Species Known to Date.* Part 2. *Carcharhiniformes.* U.N. Food and Agriculture Organization, Rome, pp. 251–655.

Compagno, L.J.V. (1988). *Sharks of the Order Carcharhiniformes.* Princeton University Press, Princeton, NJ.

Compagno, L.J.V. (1999). Checklist of living elasmobranchs. In: Hamlett, W.C. (Ed.), *Sharks, Skates, and Rays: The Biology of Elasmobranch Fishes.* The Johns Hopkins University Press, Baltimore, MD, pp. 471–498.

Compagno, L.J.V. (2001). *FAO Species Catalogue for Fishery Purposes. Sharks of the World: An Annotated and Illustrated Catalogue of Shark Species Known to Date.* Vol. 2. *Bullhead, Mackerel and Carpet Sharks (Heterodontiformes, Lamniformes, and Orectolobiformes).* United Nations Food and Agriculture Organization, Rome, 269 pp.

Compagno, L.J.V. (2005). Global checklist of living chondrichthyan fishes. In: Fowler, S.L., Cavanagh, R.D., Camhi, M. et al. (Eds.), *Sharks, Rays and Chimaeras: The Status of Chondrichthyan Fishes.* International Union for the Conservation of Nature, Gland, Switzerland, pp. 401–423.

Corrigan, S. and Beheregaray, L.B. (2009). A recent shark radiation: molecular phylogeny, biogeography and speciation of wobbegong sharks (family: Orectolobidae). *Mol. Phylogenet. Evol.* 52(1):205–216.

de Carvalho, M.R. (1996). Higher-level elasmobranch phylogeny, basal squaleans, and paraphyly. In: Stiassny, M.L.J., Parenti, L.R., and Johnson, G.D. (Eds.), *Interrelationships of Fishes*. Academic Press, San Diego, CA, pp. 35–62.

de Carvalho, M.R., Maisey, J.G., and Grande, L. (2004). Freshwater stingrays of the Green River formation of Wyoming (early Eocene), with the description of a new genus and species and an analysis of its phylogenetic relationships (Chondrichthyes: Myliobatiformes) *Bull. Am. Mus. Nat. Hist.* 284:1–136.

Dosay-Akbulut M. (2007) What is the relationship within the family Lamnidae? *Turk. J. Biol.* 31:109–113.

Dosay-Akbulut, M. (2008). The phylogenetic relationship within the genus *Carcharhinus*. *C. R. Biol.* 331(7):500–509.

Douady, C.J., Dosay, M., Shivji, M.S., and Stanhope, M.J. (2003). Molecular phylogenetic evidence refutes the hypothesis of Batoidea (rays and skates) as derived sharks. *Mol. Phylogenet. Evol.* 26:215–221.

Dunn, K.A., McEachran, J.D., and Honeycutt, R.L. (2003). Molecular phylogenetics of myliobatiform fishes (Chondrichthyes: Myliobatiformes), with comments on the effects of missing data on parsimony and likelihood. *Mol. Phylogenet. Evol.* 27:259–270.

Ewing, B., Hillier, L., Wendl, M.C., and Green, P. (1998). Base-calling of automated sequencer traces using Phred. I. Accuracy assessment. *Genome Res.* 8(3):175–185.

Gelman, A. and Rubin, D.B. (1992). Inference from iterative simulation using multiple sequences. *Stat. Sci.* 7:457–511.

González-Isáis, M. and Domínguez, H.M.M. (2004). Comparative anatomy of the superfamily Myliobatoidea (Chondrichthyes) with some comments on phylogeny. *J. Morphol.* 262:517–535.

Griffiths, A.M., Sims, D.W., Cotterell, S.P., El Nagar, A., Ellis, J.R., Lynghammar, A., McHugh, M., Neat, F.C., Pade, N.G., Queiroz, N., Serra-Pereira, B., Rapp, T., Wearmouth, V.J., and Genner, M.J. (2010). Molecular markers reveal spatially segregated cryptic species in a critically endangered fish, the common skate (*Dipturus batis*). *Proc. R. Soc. B Biol. Sci.* 277:1497–1503.

Gu, X., Fu, Y.X., and Li, W.H. (1995). Maximum likelihood estimation of the heterogeneity of substitution rate among nucleotide sites. *Mol. Biol. Evol.* 12:546–557.

Heinicke, M.P., Naylor, J.P., and Hedges, S.B. (2009). Cartilaginous fishes (Chondrichthyes). In: Hedges, S.B., and Kumar, S. (Eds.), *The Timetree of Life*. Oxford University Press, New York, pp. 320–327.

Heithaus, M.R., Frid, A., Wirsing, A.J., and Worm, B. (2008). Predicting ecological consequences of marine top predator declines. *Tr. Ecol. Evol.* 23:202–210.

Holmes, B.H., Steinke, D., and Ward, R.D. (2009). Identification of shark and ray fins using DNA barcoding. *Fish. Res.* 95(2–3):280–288.

Huelsenbeck, J.P. and Ronquist, F. (2001). MRBAYES: Bayesian inference of phylogeny. *Bioinformatics* 17:754–755.

Human, B.A., Owen, E.P., Compagno, L.J.V., and Harley, E.H. (2006). Testing morphologically based phylogenetic theories within the cartilaginous fishes with molecular data, with special reference to the catshark family (Chondrichthyes; Scyliorhinidae) and the interrelationships within them. *Mol. Phylogenet. Evol.* 39:384–391.

Hurvich, C.M. and Tsai, C.L. (1989). Regression and time series model selection in small samples. *Biometrika* 76:297–307.

Iglésias, S.P., Lecointre, G., and Sellos, D.Y. (2005). Extensive paraphylies within sharks of the order Carcharhiniformes inferred from nuclear mitochondrial genes. *Mol. Phylogenet. Evol.* 34:569–583.

Iglésias, S.P., Toulhoat, L., and Sellos, D.Y. (2009). Taxonomic confusion and market mislabeling of threatened skates: important consequences for their conservation status. *Aquat. Conserv. Mar. Freshwat. Ecosyst.* 20(3):319–333.

Jacobson, I.P. and Bennett, M.B. (2009). A taxonomic review of the Australian butterfly ray *Gymnura australis* (Ramsay & Ogilby, 1886) and other members of the family Gymnuridae (Order Rajiformes) from the Indo-West Pacific. *Zootaxa* 2228:1–28.

Lakner, C., Holder, M.T., Goldman, N., and Naylor, G.J.P. (2011). What's in a likelihood? Simple models of protein evolution and the contribution of structurally viable reconstructions to the likelihood. *Syst. Biol.* 60:161–174.

Last, P.R. and Stevens, J.D. (1994). *Sharks and Rays of Australia*. CSIRO Publishing, Melbourne.

Last, P.R. and Stevens, J.D. (2009). *Sharks and Rays of Australia*, 2nd ed. CSIRO Publishing, Melbourne.

Lemmon, A.R., Brown, J.M., Stanger-Hall, K., and Lemmon, E.M. (2009). The effect of missing data on phylogenetic estimates obtained by maximum-likelihood and Bayesian inference. *Syst. Biol.* 58:130–145.

Lim, D.D., Motta, P., Mara, K., and Martin, A.P. (2010). Phylogeny of hammerhead sharks (Family Sphyrnidae) inferred from mitochondrial and nuclear genes. *Mol. Phylogenet. Evol.* 55(2):572–579.

López, J.A., Ryburn, J.A., Fedrigo, O., and Naylor, G.J.P. (2006). Phylogeny of sharks of the family Triakidae (Carcharhiniformes) and its implications for the evolution of carcharhiniform placental viviparity. *Mol. Phylogenet. Evol.* 52:50–60.

Lovejoy, N.R. (1996). Systematics of Myliobatoid elasmobranchs: with emphasis on the phylogeny and historical biogeography of neotropical freshwater stingrays (Potamotrygonidae: Rajiformes). *Zool. J. Linn. Soc.* 117:207–257.

Maisey, J.G. (1980). An evaluation of jaw suspension in sharks. *Am. Mus. Nov.* 2706:1–17.

Maisey, J.G. (1984a). Chondrichthyan phylogeny: a look at the evidence. *J. Vertebr. Paleontol.* 4:359–371.

Maisey, J.G. (1984b). Higher elasmobranch phylogeny and biostratigraphy. *Zool. J. Linn. Soc.* 82:33–54.

Maisey, J.G., Naylor, G.J.P., and Ward, D.J. (2004). Mesozoic elasmobranchs, neoselachian phylogeny and the rise of modern elasmobranch diversity. In: Arratia, G. and Tintori, A. (Eds.), *Mesozoic Fishes 3: Systematics, Paleoenvironments, and Biodiversity*. Verlag Dr. Friedrich Pfeil, Munich, pp. 17–56.

Mallatt, J. and Winchell, C.J. (2007). Ribosomal RNA genes and deuterostome phylogeny revisited: more cyclostomes, elasmobranchs, reptiles, and a brittle star. *Mol. Phylogenet. Evol.* 43:1005–1022.

Mariguela, T.C., De-Franco, B., Almeida, T.V.V., Mendonça, F.F., Gadig, O.B.F., Foresti, F., and Oliveira, C. (2009). Identification of guitarfish species *Rhinobatos percellens*,

R. horkelli, and *Zapteryx brevirostris* (Chondrichthyes) using mitochondrial genes and RFLP technique. *Conserv. Genet. Resour.* 1:393–396.

Martin, A.P. (1997). Systematics of the Lamnidae and the origination time of *Carcharodon carcharias* inferred from the comparative analysis of mitochondrial DNA sequences. In: Klimley, P. and Ainsley, D. (Eds.), *The Biology of the White Shark*. Academic Press, New York, pp. 49–54.

Martin, A.P. and Naylor, G.J.P. (1997). Independent origins of filter-feeding in megamouth and basking sharks (Order Lamniformes) inferred from phylogenetic analysis of cytochrome *b* gene sequences. In: Yano, K., Morrissey, J.F., Yabumoto, Y., and Nakaya, K. (Eds.), *Biology of the Megamouth Shark*. Tokai University Press, Tokyo, pp. 39–50.

McEachran, J.D. and Aschliman, N. (2004). Phylogeny of Batoidea. In: Carrier, J.C., Musick, J.A., and Heithaus, M.R. (Eds.), *Biology of Sharks and Their Relatives*. CRC Press, Boca Raton, FL, pp. 79–113.

McEachran, J.D. and Dunn, K.A. (1998). Phylogenetic analysis of skates, a morphologically conservative clade of elasmobranchs (Chondrichthyes: Rajidae). *Copeia* 1998(2):271–290.

McEachran, J.D. and Miyake, T. (1990). Phylogenetic interrelationships of skates: a working hypothesis (Chondrichthyes: Rajoidei). In: Pratt, H.L., Gruber, S.H., and Taniuchi, T. (Eds.), *Elasmobranchs as Living Resources: Advances in the Biology, Ecology, Systematics, and the Status of the Fisheries*, NOAA Technical Report NMFS 90. National Oceanic and Atmospheric Administration, Washington, D.C., pp. 285–304.

McEachran, J.D., Dunn, K.A., and Miyake, T. (1996). Interrelationships of the batoid fishes (Chondrichthyes: Batoidea). In: Stiassny, M.J., Parenti, L.R., and Johnson, G.D. (Eds.), *Interrelationships of Fishes*. Academic Press, London, pp. 63–82.

Moore, A., White, W., Ward, B., Naylor, G., and Peirce, R. (2011). Rediscovery and redescription of the smoothtooth blacktip shark, *Carcharhinus leiodon* (Carcharinidae), from Kuwait, with notes on its possible conservation status. *Mar. Freshwat. Res.* 62:528–539.

Naylor, G.J.P. (1992). The phylogenetic relationships among requiem and hammerhead sharks: inferring phylogeny when thousands of equally most parsimonious trees result. *Cladistics* 8:295–318.

Naylor, G.J.P., Martin, A.P., Matisson, E.G., and Brown, W.M. (1997). The inter-relationships of lamniform sharks: testing hypotheses with sequence data. In: Kocher, T.D. and Stepien, C.A. (Eds.), *Molecular Systematics of Fishes*. Academic Press, San Diego, pp. 195–218.

Naylor, G.J.P., Ryburn, J.A., Fedrigo, O., and López, J.A. (2005). Phylogenetic relationships among the major lineages of modern elasmobranchs. In: Hamlett, W.C. and Jamieson, B.G.M. (Eds.), *Reproductive Biology and Phylogeny*, Vol. 3. Science Publishers, EnWeld, NH, pp. 1–25.

Naylor, G.J.P., Caira, J.N., Jensen, K., Rosana, K.A.M., White, W.T., and Last, P.R. (in press). A sequence based approach to the identification of shark and ray species and its implications for global elasmobranch diversity and parasitology. *Bull. Am. Mus. Nat. Hist.*

Nelson, J.S. (2006). *Fishes of the World*. John Wiley & Sons, New York.

Nishida, K. (1990). Phylogeny of the suborder Myliobatidoidei. *Mem. Fac. Fish. Hokkaido Univ.* 37:1–108.

Quattro, J.M., Stoner, D.S., Driggers, W.B., Anderson, C.A., Priede, K.A., Hoppmann, E.C., Campbell, N.H., Duncan, K.M., and Grady, J.M. (2006). Genetic evidence of cryptic speciation within hammerhead sharks (Genus *Sphyrna*). *Mar. Biol.* 148:1143–1155.

Richards, V.P., Henning, M., Witzell, W., and Shivji, M.S. (2009). Species delineation and evolutionary history of the globally distributed spotted eagle ray (*Aetobatus narinari*). *J. Hered.* 100:273–283.

Rocco, L., Liguori, I., Costagliola, D., Morescalchi, M.A., Tinti, F., and Stingo, V. (2007). Molecular and karyological aspects of Batoidea (Chondrichthyes, Elasmobranchi) phylogeny. *Gene* 389:80–86.

Ronquist, F. and Huelsenbeck, J.P. (2003). MRBAYES 3: Bayesian phylogenetic inference under mixed models. *Bioinformatics* 19:1572–1574.

Rosenberger, L.J. (2001). Phylogenetic relationships within the stingray genus *Dasyatis* (Chondrichthyes: Dasyatidae). *Copeia* 2001:615–627.

Schwartz, F.J. and Maddock, M.B. (2002). Cytogenetics of the elasmobranchs: genome evolution and phylogenetic implications. *Mar. Freshwat. Res.* 53:491–502.

Serra-Pereira, B., Moura, T., Griffiths, A.M., Serrano Gordo, L., and Figueiredo, I. (2010). Molecular barcoding of skates (Chondrichthyes: Rajidae) from the southern Northeast Atlantic. *Zool. Scrip.* 40:76–84.

Sezaki, K., Begum, R.A., Wongrat, P., Srivastava, M.P., SriKantha, S., Kikuchi, K., Ishihara, H., Tanaka, S., Taniuchi, T., and Watabe, S. (1999). Molecular phylogeny of Asian freshwater and marine stingrays based on the DNA nucleotide and deduced amino acid sequences of the cytochrome *b* gene. *Fish. Sci. (Tokyo)* 65:563–570.

Shimada, K. (2005). Phylogeny of lamniform sharks (Chondrichthyes: Elasmobranchii) and the contribution of dental characters to lamniform systematics. *Paleontol. Res.* 9:55–72.

Shirai, S. (1992). *Squalean Phylogeny: A New Framework of "Squaloid" Sharks and Related Taxa*. Hokkaido University Press, Sapporo.

Shirai, S. (1996). Phylogenetic interrelationships of neoselachians (Chondrichthyes: Euselachii). In: Stiassny, M.L.J., Parenti, L.R., and Johnson, G.D. (Eds.), *Interrelationships of Fishes*. Academic Press, San Diego, pp. 9–34.

Shirai, S. and Nakaya, K. (1990). Interrelationships of the Etmopterinae (Chondrichthyes, Squaliformes). In: Pratt, H.L., Gruber, S.H., and Taniuchi, T. (Eds.), *Elasmobranchs as Living Resources: Advances in the Biology, Ecology, Systematics, and the Status of the Fisheries*, NOAA Technical Report NMFS 90. National Oceanic and Atmospheric Administration, Washington, D.C., pp. 347–356.

Smith, P.J., Steinke, D., Mcveagh, S.M., Stewart, A.L., Struthers, C.D., and Roberts, C.D. (2008). Molecular analysis of Southern Ocean skates (*Bathyraja*) reveals a new species of Antarctic skate. *J. Fish Biol.* 73:1170–1182.

Smith, W.D., Bizzarro, J.J., Richards, V.P., Nielsen, J., Márquez-Flarías, F., and Shivji, M.S. (2009). Morphometric convergence and molecular divergence: the taxonomic status and evolutionary history of *Gymnura crebripunctata* and *Gymnura marmorata* in the eastern Pacific Ocean. *J. Fish Biol.* 75:761–783.

Spies, I.B., Gaichas, S., Stevenson, D.E., Orr, J.W., and Canino, M.F. (2006). DNA-based identification of Alaska skates (*Amblyraja*, *Bathyraja* and *Raja*: Rajidae) using cytochrome *c* oxidase subunit I (COI) variation. *J. Fish Biol.* 69:283–292.

Spies, I.B., Stevenson, D.E., Orr, J.W., and Hoff, G.R. (2011). Molecular systematics of the skate subgenus *Arctoraja* (*Bathyraja*: Rajidae) and support for an undescribed species, the leopard skate, with comments on the phylogenetics of *Bathyraja*. *Ichthyol. Res.* 58:77–83.

Stelbrink, B., von Rintelen, T., Cliff, G., and Kriwet, J. (2010). Molecular systematics and global phylogeography of angel sharks (genus *Squatina*). *Mol. Phylogenet. Evol.* 54:395–404.

Stevens, J.D., Bonfil, R., Dulvy, N.K., and Walker, P.A. (2000). The effects of fishing on sharks, rays, and chimaeras (chondrichthyans), and the implications for marine ecosystems. *ICES J. Mar. Sci.* 57:476–494.

Straube, N., Iglésias, S.P., Sellos, D.Y., Kriwet, J., and Schliewen, U.K. (2010). Molecular phylogeny and node time estimation of bioluminescent lantern sharks (Elasmobranchii: Etmopteridae). *Mol. Phylogenet. Evol.* 56(3):905–917.

Straube, N., Kriwet, J., and Schliewen, U.K. (2011a). Cryptic diversity and species assignment of large lantern sharks of the *Etmopterus spinax* clade from the Southern Hemisphere (Squaliformes, Etmopteridae). *Zool. Scrip.* 40:61–75.

Straube, N., Duhamel, G., Gasco, N., Kriwet, J., and Schliewen U.K. (2011b). Description of a new deep-sea lantern shark *Etmopterus viator* sp. nov. (Squaliformes: Etmopteridae) from the Southern Hemisphere. In: Duhamel, G. and Welsford, D., Eds., *The Kerguelen Plateau: Marine Ecosystem and Fisheries*, Société Française d'Ichtyologie, Paris, pp. 135–148.

Thompson, J.D., Higgins, D.G., and Gibson, T.J. (1994). ClustalW: improving the sensitivity of progressive multiple sequence alignment through sequence weighting, position-specific gap penalties and weight matrix choice. *Nucl. Acids Res.* 22(22):4673–4680.

Thomson, R.C. and Shaffer, H.B. (2010). Rapid progress on the vertebrate tree of life. *BMC Biol.* 8:19.

Toffoli, D., Hrbek, T., Araújo, M.L.G., Almeida, M.P., Charvet-Almeida, P., and Farias, I.P. (2008). A test of the utility of DNA barcoding in the radiation of the freshwater stingray genus *Potamotrygon* (Potamotrygonidae, Myliobatiformes). *Genet. Mol. Biol.* 31(1, Suppl.):324–336.

Turan, C. (2008). Molecular systematic analyses of Mediterranean skates (Rajiformes). *Turk. J. Zool.* 32:437–442.

Vélez-Zuazo, X. and Agnarsson, I. (2011). Shark tales: a molecular species-level phylogeny of sharks (Selachimorpha, Chondrichthyes). *Mol. Phylogenet. Evol.* 58:207–217.

Vilgalys, R. (2003). Taxonomic misidentifications in public DNA databases. *New Phytol.* 160:4–5.

Waddell, P. and Penny, D. (1996). Evolutionary trees of apes and humans from DNA sequences. In: Lock, A.J. and Peters, C.R. (Eds.), *Handbook of Symbolic Evolution*, Clarendon Press, Oxford, pp. 53–73.

Ward, R. and Holmes, B. (2007). An analysis of nucleotide and amino acid variability in the barcode region of cytochrome *c* oxidase I (cox1) in fishes. *Mol. Ecol. Notes* 7(6):899–907.

Ward, R.D., Zemlak, T.S., Innes, B.H., Last, P.R., and Hebert, P.D.N. (2005). DNA barcoding Australia's fish species. *Phil. Trans. R. Soc. Lond. B* 360(1462):1847–1857.

Ward, R.D., Holmes, B.A., Zemlak, T.S., and Smith, P.J. (2007). Part 12: DNA barcoding discriminates spurdogs of the genus *Squalus*. In: Last, P.R., White, W.T., and Pogonoski, J.J. (Eds.), *Descriptions of New Dogfishes of the Genus Squalus (Squaloidea: Squalidae)*, CSIRO Marine and Atmospheric Research Paper 14. Commonwealth Scientific and Industrial Research Organisation, Canberra, Australia, pp. 114–130.

Ward, R.D., Bronwyn, H.H., White, W.T., and Last, P.R. (2008). DNA barcoding Australasian Chondrichthyans: results and potential uses in conservation. *Mar. Freshwat. Res.* 59:57–71.

Ward, R.D., Hanner, R., and Hebert, P.D.N. (2009). The campaign to DNA barcode all fishes, FISH-BOL. *J. Fish Biol.* 74(2):329–356.

Wesche, P.L., Gaffney, D.J., and Keightley, P.D. (2004). DNA sequence error rates in GenBank records estimated using the mouse genome as a reference. *DNA Seq.* 15:362–364.

White, W.T. and Kyne, P.M. (2010). The status of chondrichthyan conservation in the Indo-Australasian region. *J. Fish Biol.* 76:2090–2117.

Winchell, C.J., Martin, A.P., and Mallatt, J. (2004). Phylogeny of elasmobranchs based on LSU and SSU ribosomal RNA genes. *Mol. Phylogenet. Evol.* 31:214–224.

Wong, E.H.K., Shivji, M.S., and Hanner, R.H. (2009). Identifying sharks with DNA barcodes: assessing the utility of a nucleotide diagnostic approach. *Mol. Ecol. Res.* 9:243–256.

Wynen, L., Larson, H., Thorburn, D., Peverell, S., Morgan, D., Field, I., and Gibb, K. (2009). Mitochondrial DNA supports the identification of two endangered river sharks (*Glyphis glyphis* and *Glyphis garricki*) across northern Australia. *Mar. Freshwat. Res.* 60:554–562.

Yang, Z. (1994). Maximum likelihood phylogenetic estimation from DNA sequences with variable rates over sites: approximate methods. *J. Mol. Evol.* 39:306–314.

Zemlak, T.S., Ward, R.D., Connell, A.D., Holmes, B.H., and Hebert, P.D.N. (2009). DNA barcoding reveals overlooked marine fishes. *Mol. Ecol. Res.* 9(Suppl. s1):237–242.

3

Phylogeny of Batoidea

Neil C. Aschliman, Kerin M. Claeson, and John D. McEachran

CONTENTS

3.1 Introduction

Chondrichthyan fishes (sharks, rays, and ratfishes) comprise one of only two surviving lineages of jawed vertebrates, or gnathostomes (Carroll, 1988). Their sister group, Osteichthyes, includes ray-finned fishes and tetrapods and encompasses the vast majority of vertebrate species. Despite the important outgroup perspective that chondrichthyans lend to higher level studies of osteichthyans, little attention has yet been paid to chondrichthyan interrelationships; that is, although these fishes are employed as proxies for ancestral jawed vertebrates, there is no well-supported framework for interpreting the polarity of character state changes within the group. This is particularly the case for skates, rays, and their allies (Batoidea, hereafter "batoids"), which comprise the majority of chondrichthyan species diversity (≈630 of ≈1170 species) and morphological disparity, including departures from a shark-like ancestral body

plan (Compagno, 1999, 2005). The spectrum of batoid body plans rivals that of many other vertebrate groups and includes such disparate forms as sawfishes with elongate rostral saws, 7-meter-wide planktivorous mantas, and benthic, saucer-shaped torpedo rays capable of generating powerful electric discharges from modified branchial muscles (Davy, 1829; Stiassny et al., 2004).

The relatively few attempts to construct an in-group phylogeny for batoids can be roughly classified as: (1) higher level phylogenies using morphological (neontological) and/or fossil data; (2) finer scale phylogenies focused on interrelationships within an order or family, also based on morphological data; and (3) molecular analyses, usually addressing higher level phylogeny. Early work in elasmobranch (sharks and rays, excluding ratfishes) systematics dates to the mid-19th century (reviewed briefly by Naylor et al., 2005), but modern attempts are generally considered (e.g., McEachran et al., 1996) to have begun with Compagno (1973). Here, we review this more recent (1973 to current) body of work.

3.2 Review of Previous Morphological Phylogenies

3.2.1 Higher Level Morphological Phylogenies

Modern morphology-based attempts to resolve the phylogeny of batoids began with Compagno (1973). This effort to define the major groups and interrelationships of extant elasmobranchs described a large number of characters that were used by other investigators for the next three decades. Compagno's classification scheme and "highly provisional" phylogeny are phenetic, based on similarities and differences rather than shared derived characters. He established three superorders of living sharks and one superorder of rays (Batoidea). Batoidea was further subdivided into four orders: Rajiformes (including suborders Rhinobatoidei [guitarfishes] and Rajoidea [skates]), Pristiformes (sawfishes), Torpediniformes (electric rays), and Myliobatiformes (stingrays). He identified Pristiformes as sister to all other extant batoids, with Torpediniformes sister to Rajiformes + Myliobatiformes.

Compagno's tree (1973, Fig. 5) is not a cladogram but rather an attempt to depict his largely unresolved phenetic assessment of batoid interrelationships. He suggested that guitarfishes are the "central or axial" group of batoids and are probably ancestral to all other forms, grading from Jurassic guitarfish-like batoids (hereafter Jurassic "guitarfishes") to modern taxa. However, he then identified sawfishes as sister to all other extant batoids because of skeletal similarities to those Jurassic "guitarfishes" (†*Spathobatis*, †*Belemnobatis*), including a short synarcual (series of fused cervical vertebrae or vertebral elements), short and unsegmented propterygia, and a wide gap between the propterygia and antorbital cartilages. He did not state which of the numerous electric rays he examined exhibits a long synarcual, and later authors (e.g., McEachran et al., 1996) rejected this assertion. Electric rays were allied to guitarfishes by their well-developed tail and dorsal fins, as well as by the trough-shaped rostrum and nasal capsule arrangement of the Narcinidae (numbfishes). Compagno noted that electric rays do not closely resemble any extant group of guitarfishes (but see Garman, 1913). Stingrays were suggested to be derived guitarfishes due to the similarity of the narrow scapula/suprascapular joint of these groups relative to the joint's wide articulation in skates.

Compagno (1977) followed this phenetic work with a cladistic analysis. He expanded his dataset to include additional jaw suspension and neurocranial characters and attempted to define interrelationships using synapomorphies. This study described six shared derived characters uniting batoids, many of which were heavily modified by later authors and are discussed below.

Compagno's cladogram (Compagno, 1977, Fig. 15) identified electric rays rather than sawfishes as sister to all other extant batoids, with sawfishes sister to a trichotomy of stingrays, skates, and guitarfishes. Two characters influenced the rootward shift of electric rays. First, in one group of electric rays (Narkidae), the ceratohyals are very well developed and attach by a strong ligament to the hyomandibula. This condition is unique among batoids and resembles the primitive condition exhibited by fossil (and extant) sharks. Second, Compagno (1977) considered the number of hypobranchials in the gill arches. Jurassic "guitarfishes" exhibit the outgroup condition of four pairs of hypobranchials, some extant guitarfishes have three pairs, and some electric rays have two pairs articulating with the basibranchial copula in a shark-like fashion. Sawfish hypobranchials are fused into a single median plate. Although Miyake and McEachran (1991) challenged the specific numbers of hypobranchials described by Compagno, they agreed that the electric ray *Torpedo* exhibits the "generalized or typical chondrichthyan ventral gill arch structure" (discussed further below). Compagno (1977) retained sawfishes as sister to all batoids other than electric rays because of their very short synarcual and the failure of their propterygia to reach the antorbitals.

Heemstra and Smith (1980) proposed a dissenting batoid phylogeny in their description of a six-gilled stingray family, Hexatrygonidae. They recovered a phylogeny (Heemstra and Smith, 1980, Fig. 14) resembling Compagno's (1973) phenetic hypothesis in which electric rays and sawfishes are successive sister groups to all other extant batoids. Characters contributing to the sister relationship of sawfishes to all other batoids included the following shark-like states: (12) basal angle on ventral surface of neurocranium present, (13) propterygia not reaching head, and (14) pectoral and pelvic fins aplesodic, with well-developed ceratotrichia (Heemstra and Smith, 1980).

Maisey (1984) addressed the disagreement between Compagno (1977) and Heemstra and Smith (1980) by mining additional morphological characters from the literature and reexamining key Jurassic "guitarfishes." He listed ten synapomorphies for Batoidea, although character 24, "last hypobranchial articulates or is fused with the shoulder girdle," is erroneous, as the last ceratobranchial articulates with the scapula. Maisey noted that the hyoid arch of †*Spathobatis* and †*Belemnobatis* is "probably more specialized" than that of extant electric rays and agreed with Compagno's (1977) assessment of the relatively derived condition of the sawfish hypobranchial skeleton. He tentatively concluded that electric rays are sister to all other extant batoids and that the contact between the antorbitals and propterygia in this clade is convergent with the condition in derived batoids other than sawfishes. Maisey strongly criticized the crownward portions of Heemstra and Smith's phylogeny, noting that there was no synapomorphy defining their Rajiformes (skates, guitarfishes) clade and that the relationship of this group to stingrays was defined by a symplesiomorphy. Guitarfishes could then have represented either a paraphyletic assemblage or a clade of unknown affinity to other derived batoids.

Nishida (1990) broadly surveyed batoids in order to maximize the number of outgroups for his stingray phylogeny. His thorough descriptions and illustrations of characters were valuable resources for later studies of batoid interrelationships. Nishida's work was the first to code a number of guitarfish genera as individual operational taxonomic units (OTUs). His phylogeny (Nishida, 1990, Fig. 4), like previous trees, recovered skates as sister to stingrays and suggested that the highly depressed, circular or lozenge-shaped pectoral discs of these two groups are homologous. This skate and stingray clade was nested within a largely unresolved but paraphyletic assemblage of guitarfishes. Working rootward, successive sister groups to this polytomy were electric rays, the shark-like guitarfish genera *Rhynchobatus* and *Rhina*, and finally sawfishes.

The most salient element of Nishida's tree may be his removal of *Rhynchobatus* and *Rhina* from the larger, poorly defined guitarfish group described by previous authors. Nishida identified three characters for which *Rhynchobatus* and *Rhina* exhibit the shark-like, or plesiomorphic, state relative to most other batoids: (28) pectoral propterygia and radials failing to reach the nasal capsules, (29) pectoral fin posterior corner not reaching level of pelvic fin, and (30) caudal fin with pronounced dorsal and ventral lobes. Sawfishes share the plesiomorphic states with these two genera and sharks. Characters 28 and 29 may be developmentally correlated, redundantly measuring the extent of the pectoral fin expansion characteristic of most batoids. Nishida's topology required him to assert that an elongated synarcual fused to (or articulating with) (Claeson, 2010; Regan, 1906) the suprascapulae, exhibited by *Rhynchobatus* and *Rhina*, is convergent with that exhibited by all other guitarfishes, skates, and stingrays. He found this scenario more likely than a reversal to a short synarcual in electric rays. McEachran et al. (1996) reanalyzed Nishida's character matrix using the contemporary phylogenetics software package PAUP v3.1.1 (Swofford, 1993), but they did not recover his published phylogeny. The strict consensus of 14 resulting trees was largely unresolved, with sawfishes sister to all other batoids.

Nishida's (1990) work included brief surveys of a number of different batoid skeletal and muscle systems. It was closely followed by a suite of papers that investigated several of these systems in greater detail and taxonomic breadth. Two of these focal systems were the ventral gill arch skeleton (Miyake and McEachran, 1991) and rostral cartilages (Miyake et al., 1992b). These components of Miyake's (1988) doctoral dissertation on the interrelationships of the stingray family Urolophidae were novel in that they incorporated ontogenetic information. Batoid ontogeny is difficult to study and remains poorly understood because these animals are typically large and mature slowly, and all groups except skates retain their young during development (Dulvy and Reynolds, 1997).

Miyake and McEachran (1991) demonstrated that not only is the ventral gill arch skeleton highly derived within each major batoid group, but it also exhibits many presumably convergent losses or arrangements of elements. That is, a number of synapomorphies unite taxa within the major groups (e.g., all skates), but few to no phylogenetically informative characters unite two or more of these higher taxa. The authors noted, for example, that, although it is "highly unlikely" that stingrays and sawfishes are sister taxa, in both groups several hypobranchials coalesce into a medial plate through which pass the afferent branches of the ventral aorta. Similarly, the basihyal is lost apparently independently in electric rays, the stingray genus *Urotrygon*, and the pelagic stingrays. The broad taxonomic sampling of this study also demonstrated that some characters that were previously considered diagnostic for a clade are more broadly distributed than expected. The medial fusion of the pseudohyal and ceratobranchials 1 and 2 was thought to be exclusive to the freshwater stingray genera *Potamotrygon* and *Plesiotrygon* (Rosa et al., 1987), but is actually scattered throughout several other genera.

Of particular importance to higher level batoid systematics, the authors reevaluated previous claims (e.g., Compagno, 1977) that the electric ray branchial skeleton approximates the ancestral chondrichthyan condition. Compagno (1977) interpreted this pattern as evidence that electric rays are sister to all other extant batoids (introduced above). Miyake and McEachran (1991) demonstrated that only *Torpedo* exhibits this generalized pattern, whereas its presumed sister genus *Hypnos* and the more distantly related electric rays have a very different condition in which the hypobranchials are arranged parallel to the longitudinal axis of the body and are fused to various degrees. Even if the state in *Torpedo* were nearly identical to the ancestral condition, it is equally parsimonious to interpret the observed patterns as independent gains of the derived condition by *Hypnos* and the other clade of electric rays, or that the common ancestor of all electric rays exhibited the derived state and *Torpedo* reverted to the "ancestral" state. The authors preferred the first interpretation and speculated that the developmental processes behind the similar longitudinal arrangements of the hyomandibulae differ between groups, but they offered no supporting data.

Shirai conducted comprehensive morphological surveys of squalean sharks and batoids (1992b) and of all extant elasmobranchs (1996). Although batoids were represented by only one operational taxonomic unit in Shirai's landmark study (1992b), his subsequent analysis included more genera in order to resolve batoid interrelationships. These studies identified a number of new myological synapomorphies for batoids, as well as skeletal and muscle characters allying batoids with sawsharks and angel sharks. Shirai (1992b) was the first to describe in a cladistic framework the hypothesis

that sawsharks, which resemble sawfishes, are sister to batoids. Some authors (e.g., Compagno, 1977) previously suggested that sawsharks are intermediate forms between other squalean sharks and batoids, but this was proposed as only one of several untested hypotheses.

Shirai's phylogeny (1996, Fig. 2) (Figure 3.1A) resembled that of Nishida (1990), again describing sawfishes as sister to all other extant batoids and indicating that guitarfishes are an artificial assemblage. *Rhynchobatus* and *Rhina* were recovered as sister to all remaining batoids, placed crownward to sawfishes by one unambiguous

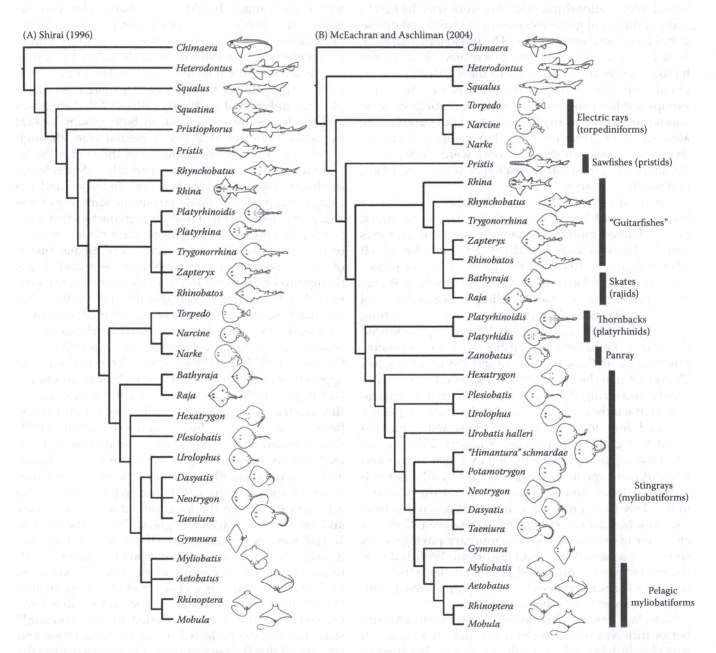

FIGURE 3.1

Frequently cited morphological phylogenies of Batoidea. Trees are modified to better reflect taxa included in the present study. (A) From Shirai, S., in *Interrelationships of Fishes*, Stiassny, M.J. et al., Eds., Academic Press, London, 1996, pp. 9–34. (B) From McEachran, J.D. and Aschliman, N., in *Biology of Sharks and Their Relatives*, Carrier, J.C. et al., Eds., CRC Press, Boca Raton, FL, 2004, pp. 79–113.

(68, plesodic pelvic fins) and two ambiguous character states. Electric rays were allied to a stingray and skate clade by a misdescribed but appropriately coded character (92, nasoral groove present). The skate and stingray sister relationship was defined by two (a third character is erroneously repeated) ambiguous and two unambiguous changes (80, dorsal fin reduced or absent; 83, first dorsal radials plesodic) that are all potentially correlated with a gross reduction in tail development.

McEachran et al. (1996) built on the phylogenetic analyses of Maisey (1984) and Shirai (1992b) using the extensive anatomical data generated by a number of recent descriptive works. Because guitarfishes, despite being the presumed "central or axial" (Compagno, 1973) group of batoids, were largely overlooked by previous studies, they included all but one genus in their new analysis. Their matrix incorporated new characters from external morphology, the lateral line, chondroskeleton, and cephalic and branchial musculature for 28 batoids. The outgroups comprised two squalean sharks that Shirai (1992b, 1996) suggested were successive sister groups to batoids: the sawshark, *Pristiophorus*, and the angel shark, *Squatina*. These two genera are distinct among squalean sharks in exhibiting batoid-like characters, so the presumption of (*Pristiophorus*, Batoidea) monophyly strongly influenced the polarity of character states in this analysis.

This study identified eight synapomorphies for Batoidea that remain the currently accepted diagnosis: (1) upper eyelid absent, (2) palatoquadrate does not articulate with cranium, (3) pseudohyal present, (4) last ceratobranchial articulates with scapulocoracoid, (5) synarcual present, (6) suprascapulae articulate with vertebral column (but see Section 3.4), (7) antorbital cartilage joins propterygium and nasal capsule, and (8) levator and depressor rostri present (with the levator rostri secondarily lost in stingrays). Eleven synapomorphies united electric rays, including the presence of large electric organs and absence of a cephalic lateral line on the ventrum. Eight were diagnostic for stingrays, including the presence of a serrated stinging spine, second (thoracolumbar) synarcual, and a synovial joint between the scapula and first (cervical) synarcual.

McEachran et al. (1996, Fig. 12) recovered electric rays as sister to all other extant batoids, with sawfishes sister to the remaining taxa. This was due to the exclusion of two characters used in previous studies that placed sawfishes sister to all other batoids and also to the addition of three new characters in which electric rays exhibit the outgroup (sawshark-like) condition. The first excluded character was the degree of forward extension of the pectoral propterygium. The authors suggested that the state in electric rays is convergent with that of skates and/or stingrays and also may be correlated with other characters such as the anterolaterally expanded antorbital cartilages and the presence of massive electric organs in the branchial region. This assumption of convergence is likely inappropriate and is one that phylogenetic analysis should help resolve. Nishida's (1990) separate characters for the aplesodic condition (ceratotrichia present) in the pectoral and the pelvic fins were combined (effectively excluding a second character) because of probable covariation in fin development. In all batoids, the pectoral fin state is identical with that of the pelvic fin.

The three new characters for which electric rays exhibit the outgroup condition were (31) jugal arch absent (also absent in stingrays), (47) scapulocoracoid condyles not horizontal, and (57) ethmoideo-parethmoidalis absent. Character 47 may only apply to the electric ray family Narkidae, the only batoids to retain a well-developed ceratohyal (Compagno, 1977). The condylar arrangement in other electric rays does not reflect the diagonal pattern of sawsharks, but rather an irregular arrangement in which the metacondyle is level with or dorsal to the mesocondyle. McEachran et al. (1996) noted that previous authors (Compagno, 1977; Miyake, 1988) depicted narcinid electric rays with the horizontal condylar orientation of other batoids. Despite the above selections favoring the position of electric rays as sister to all other extant batoids, a Bremer decay index (BDI) of 1 was recovered for this node.

Crownward in the tree, stingrays were included in a largely unresolved assemblage of guitarfishes and skates. One novel aspect of this topology was the placement of the guitarfish genus *Zanobatus* as sister to stingrays, with the thornbacks (*Platyrhina*, *Platyrhinoidis*) sister to this larger clade. Two characters united *Zanobatus* and stingrays: (51) distal section of propterygium extends posterior to procondyle, and (54) puboischiadic bar narrow and arched. The thornbacks were allied to this clade by only one unambiguous character: (65) coracohyoideus diagonal to body axis. This hypothesis was revisited in 2004 by the most recent large-scale morphological phylogeny of Batoidea.

McEachran and Aschliman (2004) expanded the matrix of McEachran et al. (1996) to include 35 batoids (32/72 genera), 4 outgroups, and 82 characters. Because of a growing body of molecular evidence casting doubt on the affinity of batoids to sawsharks and angel sharks (see Section 3.5.3), this study used a different and taxonomically broader set of outgroup taxa that included a ratfish. Major character additions in this iteration included tooth root vascularization patterns, squamation, and clasper (intromittent organ) structure. These new character suites helped resolve the tree toward the tips but were largely uninformative for resolving the interrelationships of major groups of batoids.

The primary difference between the resulting topology (Figure 3.1B) (McEachran and Aschliman, 2004, Fig. 3.9) and that of McEachran et al. (1996) was the resolution of the guitarfishes–skates polytomy into a clade with *Rhynchobatus* and *Rhina* as successive sister groups to a trichotomy of skates, *Trygonorrhina*, and *Rhinobatos* + *Zapteryx*. The tree is misprinted in early runs of *Biology of Sharks and Their Relatives* and suggests that skates and *Trygonorrhina* are sister taxa; this was only proposed in the tree generated by successive approximations reweighting (McEachran and Aschliman, 2004, Fig. 3.10). These five guitarfish genera and skates were united by one unambiguous synapomorphy: (25) rostral appendices present. The paired cartilaginous elements on the anterior aspect of the rostrum in thornbacks are of unknown homology to true rostral appendices (coded as missing data for these taxa). Two unambiguous characters supported the trichotomy that included skates: (53) mesocondyle of scapulocoracoid closer to procondyle than to metacondyle, and (58) direct articulation of several pectoral radials with the scapulocoracoid posterior to the mesocondyle. The authors failed to note that character 53 is indeed an unambiguous character in the matrix. Characters 53 and 58 are likely correlated, in that expansion of the scapulocoracoid between the mesopterygium and metapterygium appears to be a precondition for the articulation there of free radials with the scapulocoracoid.

The tenuous position of thornbacks as sister to *Zanobatus* + stingrays first proposed by McEachran et al. (1996) was strengthened by a change in character: (57) proximal section of propterygium extends behind procondyle. The state in thornbacks was changed from absent to present, as in *Zanobatus* and stingrays. This change was not explicitly mentioned in the text, and the condition in *Platyrhinoidis* (NCA, pers. obs.) and *Platyrhina* (Garman, 1913, Plate 66) appears to be less well developed than in even some skates (Garman, 1913, Plates 68 and 69).

The McEachran and Aschliman (2004) topology also corroborated the monophyly of the butterfly ray, *Gymnura*, with the pelagic stingrays (e.g., *Mobula*), which are phenetically similar but for which few shared derived characters have been described. In this study, only one unambiguous character state united these taxa: (32) postorbital process very broad and shelflike, and located in the orbital region. This process is broad and shelflike in all stingrays. McEachran and Aschliman (2004) did not acknowledge that it extends into the orbital region in additional stingray taxa, such as *Urotrygon microphthalmum* and *Himantura bleekeri* (Nishida, 1990, Figs. 12B and 14B). A more appropriate scheme would code all stingrays as state "2," omitting the orbital qualification. Two ambiguous states that probably redundantly measure the condition of the mesopterygium in *Gymnura*, pelagic stingrays, and the guitarfish, *Zanobatus*, are (53) relative location of mesocondyle on scapulocoracoid, and (59) mesopterygium fragmented or absent. These taxa are coded as an unexplained state "3" for character 53, which probably implies that the mesocondyle is

effectively absent. A proper arrangement obviating the correlation between the above characters would code these taxa as missing data for character 53, as did the authors for other absent structures.

McEachran and Aschliman (2004) suggested that the dorsoventrally flattened, circular or lozenge-shaped disc of skates and stingrays is convergent, being derived from two different groups of more shark-like guitarfishes. The authors further speculated that morphological differences between rhinobatid guitarfishes (sister to skates) and thornbacks (sister to *Zanobatus* + stingrays) might have constrained the way in which skates and stingrays achieved their depressed disc shapes (see Section 3.6). Finally, McEachran and Aschliman (2004) attempted to summarize potential sources of discordance between theirs and previous phylogenies, particularly in the relative branching orders of electric rays and sawfishes. The authors suggested that disagreements may be due to the high degree of specialization in these two taxa and to the scarcity of synapomorphies uniting major batoid groups. The pectoral girdle of electric rays is highly modified to accommodate their voluminous electric organs, and the neurocranium and anterior vertebral column of sawfishes are subject to the mechanical constraints of manipulating the saw-like rostrum. Both of these taxa exhibit different sets of plesiomorphic characters not exhibited by other batoids. Sawfishes exhibit a basal cranial angle as seen in squalean sharks, as well as shark-like aplesodic paired fins. Narkid electric rays exhibit a diagonal arrangement of the scapular condyles and a ceratohyal–hyomandibular connection.

Shortly following McEachran and Aschliman (2004), two papers incorporating fossils into morphological datasets were published in a collected volume on Mesozoic fishes. The first of these, by Kriwet (2004), reexamined articulated skeletons, isolated teeth, and rostral spines of the fossil sclerorhynchid "sawfishes." Sclerorhynchids, known only from the Cretaceous, appear in some ways intermediate between sawfishes and sawsharks, Shirai's putative sister group to batoids. Kriwet's phylogeny (Kriwet, 2004, Fig. 3) indicated that sclerorhynchids are sister to all other batoids, with sawfishes sister to all extant forms. The interrelationships of extant batoids were left almost entirely unresolved. This study includes a number of important errors in character definitions and coding; specific ones are highlighted below.

Bootstrap (BS) values were poor for both the node placing sclerorhynchids sister to extant batoids (BS = 53) and the node separating sawfishes from the other extant forms (BS = 51). Both of these nodes were supported only by homoplasious character states. Kriwet favored unparsimonious character state transformations: Although the synarcual is short in the earliest known articulated batoids (†*Spathobatis*, †*Belemnobatis*) and in the two taxa most frequently identified as sister to all other extant batoids (electric rays and sawfishes), Kriwet proposed that the ancestral state is a *long* synarcual, exhibited by sclerorhynchids and derived batoids such as stingrays (Kriwet, 2004, p. 66). The topology suggests that, among batoids included in this study, it is equally parsimonious to interpret the ancestral state as either a long or a short synarcual.

The three successive sister groups to the other extant batoids in Kriwet's topology—sawfishes, sclerorhynchids, and sawsharks—each exhibit an elongated rostrum bearing enlarged placoid scales on the lateral margins, giving the extant groups their common names. Kriwet acknowledged that, although these spiny rostra are superficially similar across groups and one may be tempted to presume that they are homologous, their fundamentally different architectures, as well as the discovery of small-spined sclerorhynchid forms, suggests that these structures were independently derived in each of the three groups. That is, evidence of substantial convergence in functionally important structures is apparent in batoids, even when interpreted against a phylogeny that would suggest otherwise.

A second study of higher level batoid interrelationships in the same volume attempted to determine the phylogenetic positions of a number of fossil guitarfishes. Brito and Dutheil (2004) scored eight well-preserved Late Cretaceous guitarfishes for 28 characters from previous studies. Thirteen additional taxa covering all major non-stingray batoid groups, a sawshark, and an angel shark were included. The resulting phylogeny (Brito and Dutheil, 2004, Fig. 3) again identified sawfishes as sister to all other extant batoids, followed by *Rhynchobatus* (plus two fossil taxa), with electric rays sister to a largely unresolved group of remaining batoids. Support indices were not generated, and the structure of the tree is supported mostly by ambiguous characters, some of which are probably correlated. For example, separate characters were used to describe the aplesodic/plesodic condition of the pectoral and pelvic fins (Nishida, 1990).

3.2.2 Intraordinal Morphological Phylogenies

Finer scale investigations of batoid interrelationships are more common than are higher level phylogenies. Often accompanying descriptions of new taxa, these studies draw upon neontological and fossil morphological characters. They usually include members of only one order or family, although the taxonomic arrangement of batoids is inconsistent and largely depends on whether the investigator interprets batoids as the sister group to sharks or as derived squalean sharks sister to sawsharks. The former group of workers recognizes several distinct batoid orders (e.g., four by McEachran and Aschliman, 2004), while the latter subsumes all batoids into the single

order Rajiformes (e.g., Compagno, 2005). As sharks are arranged into eight orders in the most frequently cited, current classification (Compagno, 2005), the Rajiformes scheme is inherently unbalanced in that it encompasses the majority of chondrichthyan morphological disparity and species diversity into a single order.

Finer scale (where appropriate, hereafter "intraordinal" *sensu* McEachran and Aschliman, 2004) phylogenies often imply that the patterns of convergence and specialization obscuring the deep interrelationships of batoids have similarly affected some smaller groups. A comprehensive treatment of this large body of work is well beyond the scope of this review, but examples from the two largest batoid groups, both apparently influenced by convergence, will be summarized: Rajidae (skates) and Myliobatiformes (stingrays and pelagic rays).

Rajidae comprises nearly half of all batoid species (283/631) and a quarter of all chondrichthyans (283/1168) (Compagno, 2005). It is also the most thoroughly studied by systematists, although very few analyses have included many supraspecific taxa. McEachran and Miyake (1990) cited 100 systematics studies on regional skate faunae, after which the interrelationships of most subgenera and genera remained obscure. Of several previous attempts to establish a skate phylogeny, most suffered from sparse, region-limited sampling or failed to distinguish between derived and primitive character states. McEachran and Miyake (1990) combined new and previously reported characters from the neurocranium, hypobranchial skeleton, scapulocoracoid, and particularly the claspers for all of the 29 recognized (at the time) supraspecific skate taxa. A strict consensus (McEachran and Miyake, 1990, Fig. 1) of their guitarfish-rooted phylogenies described two major assemblages. Group I was united by three synapomorphies: (1) basihyal cartilage with laterally directed projections, (2) claspers with spoon-shaped ventral terminal cartilage, and (3) dorsal and ventral terminal cartilages arranged in parallel. Group II was less well defined by derived characters, but members exhibited: (20) greatly expandable clasper glans, and (21) clasper glans with component shield. The interrelationships of the four subgroups within Group II remained unresolved. The authors described numerous and widespread homoplasious gains and reversals in skates: a total of 36 in the most parsimonious topologies. Reduction of the rostral cartilage, which is stout in guitarfishes, appears to have occurred independently many times among skates.

McEachran and Dunn (1998) revisited this problem, expanding the matrix of McEachran and Miyake (1990) to include 28 skate genera (including two "assemblages"), three outgroups, and 55 characters. The resulting phylogeny (McEachran and Dunn, 1998, Fig. 1) was still in many respects poorly resolved but corroborated much of McEachran and Miyake's hypothesis. The authors formalized the previous study's Group I as subfamily Arhynchobatinae and Group II as Rajinae using a mostly novel suite of diagnostic characters. Arhynchobatinae had two unambiguous synapomorphies: (1) basihyal cartilage with lateral projections, and (2) clasper glans with component projection. Three ambiguous characters united Rajinae: (1) scapulocoracoids lack anterior bridge, (2) claspers are distally expanded, and (3) clasper glans has component rhipidion (erectile tissue). Finally, the authors used the phylogeny to identify at least two independent reductions of the rostral cartilage in each subfamily. In Arhynchobatinae, reduction is indicated to begin with segmentation and shortening at the base, near the neurocranium. Conversely, in Rajinae the distal elements are progressively reduced instead. McEachran and Dunn (1998) suggested that the various states of reduction resemble different stages in the well-characterized ontogeny of skates, describing these reductions as paedomorphic. The authors concluded that the limited morphological disparity of this species-rich group, the prevalence of homoplasious character states, and the failure of skates to explore new niches (as have the morphologically similar stingrays) suggest that the evolution of the skate body plan is constrained. Several different lineages of skates appear to have adapted independently to a deep-sea habitat, with each acquiring a reduced rostrum for grubbing in soft substrates as well as enlarged nasal rosettes.

The prevalence among skates of convergent gains or reversals in several character suites—the "mosaic evolution" (McEachran and Miyake, 1990) and character state cycling described above—impedes not only the recovery of their higher interrelationships but also the diagnosis of smaller units such as genera. Before McEachran and Dunn's (1998) work, most skates were subsumed into an enormous, polyphyletic genus, *Raja*. McEachran and Dunn elevated nine subgenera of *Raja* to generic status and recognized the need for two more describing the North Pacific and amphi-American *Raja* assemblages. A number of skate genera remain difficult to diagnose; for example, *Okamejei* and *Raja* differ only in exhibiting or lacking dark pigmentation around the ampullar pores on the ventral disc, as well as a few subtle differences in clasper morphology (McEachran and Dunn, 1998).

Myliobatiformes (stingrays and pelagic rays) is the second most speciose clade of batoids. These fishes also exhibit apparent patterns of convergence that obscure their interrelationships. Despite the greater morphological disparity exhibited by this group relative to skates, it has been less often subjected to phylogenetic analysis. Two areas of inquiry have driven much of the research on this group: identifying the sister group to the South American freshwater stingrays (Potamotrygonidae), which necessitates addressing the questionable monophyly of several stingray genera in the family Dasyatidae

(Lovejoy, 1996; Rosenberger, 2001b), and whether or not the demersal butterfly rays (Gymnuridae) are sister to the pelagic rays, such as eagle and manta rays (González-Isáis and Domínguez, 2004).

In the course of his extensive work on Neotropical biogeography, Lovejoy (1996) broadly surveyed myliobatiforms in an attempt to identify the closest extant relatives of Potamotrygonidae. He compiled a matrix of 39 new and previously employed characters from the lateral line, chondroskeleton (but not claspers), muscles, physiology, and embryology for 18 taxa. Lovejoy identified numerous characters diagnosing different myliobatiform groups, several of which corroborated hypotheses of a contemporary study by McEachran et al. (1996). He found that the presence of angular cartilages in the ligament spanning the hyomandibula and lower jaw is synapomorphic for Potamotrygonidae and the two amphi-American species of the dasyatid genus *Himantura*. This conflicted with a previous parasite-based hypothesis (Brooks et al., 1981) that allied Potamotrygonidae with the stingaree genus *Urobatis* and implied that the river rays invaded South America from the north, around the Caribbean, rather than potentially from the Pacific, where most extant *Urobatis* are distributed.

Lovejoy's work indicated that both *Himantura* and the similar stingray genus *Dasyatis* are non-monophyletic (Lovejoy, 1996, Fig. 15). He noted that this was long suspected by other authors because species were previously assigned to these genera based on a loose suggestion by Garman (1913) that *Dasyatis* exhibits cutaneous tail finfolds while *Himantura* does not. Lovejoy identified two distinct groups of *Dasyatis* but was unable to further resolve them. Butterfly rays and pelagic rays were united by only one character in this study, despite their apparent similarities in body plan and mode of swimming: (27) mesopterygium composed of several fragments that all articulate with the scapulocoracoid. Some pelagic rays lack an evident mesopterygium, and this structure is also absent in the guitarfish, *Zanobatus* (discussed above).

Rosenberger (2001b) followed Lovejoy in addressing the non-monophyly of *Dasyatis* and *Himantura* and the relationship of butterfly rays to other myliobatiforms. Her matrix included 32 morphological characters similar to those employed by Lovejoy (1996). She coded 14 of (at the time) 35 *Dasyatis* species, including all of the Neotropical taxa as well as other genera that Lovejoy (1996) suggested may interdigitate with *Dasyatis*. Her phylogeny (Rosenberger, 2001b, Fig. 7) corroborated the non-monophyly of both *Dasyatis* and *Himantura* and identified no characters that could be reinterpreted as synapomorphic for either genus under a constrained, reciprocally monophyletic topology. Rosenberger's tree nested the butterfly ray, *Gymnura*, well within derived dasyatids. This conflicted with most previous and subsequent (McEachran and Aschliman, 2004) hypotheses and is based entirely on homoplasious character states. Rosenberger noted that most of her characters were homoplasious—the tree had a low confidence index (CI) value of 0.345—and she concluded that the morphology of the disc and tail is prone to convergence in dasyatids.

González-Isáis and Domínguez (2004) investigated the interrelationships of butterfly rays and pelagic rays using a set of 76 morphological characters, many of which were previously employed by Nishida (1990), McEachran et al. (1996), and others. Ten characters supported the sister relationship of butterfly rays and pelagic rays (together Myliobatoidea), but on closer inspection a number of these are problematic and reflect poor outgroup sampling. Examining these is made difficult by apparent errors in the order of characters in the reported matrix.

Two characters (discussed above) uniting Myliobatoidea are dubious and occur in unselected outgroup taxa: (2) first postorbital process located in orbital region, and (4) first postorbital process developed. Two characters probably redundantly measure the laterally expanded disc of these fishes: (9) interorbital width larger than 45% of cranium length, and (70) disc more than 1.5 times wider than long. A third character correlated with this lateral expansion masks differences in the mode of locomotion between butterfly and pelagic rays: (76) flapping locomotion. Character 37 is not explained in the text, is apparently miscoded in the reported matrix, and ignores outgroups that exhibit the "shared" state: rounded synarcual articular surface (see Lovejoy, 1996, Fig. 8; Nishida, 1990, Fig. 38). Character 41, number and size of synarcual ventral condyles, is incorrectly presented as a synapomorphy. Another character again ignores potential outgroup taxa that exhibit the state: (46) pelvic girdle with lateral prepelvic processes reduced or absent (see Lovejoy, 1996, Fig. 11; Nishida, 1990, Fig. 36). A new character describing the lateral expansion of the disc, subsuming characters 9, 70, and possibly 13, would appear to be an independent and potentially valid character uniting butterfly and pelagic rays.

The authors also recovered an unusual topology (González-Isáis and Domínguez, 2004, Fig. 11) in which cownose rays (Rhinopteridae) are sister to bat and eagle rays (Myliobatidae) rather than to devil rays (Mobulidae), in disagreement with other studies (Lovejoy, 1996; McEachran and Aschliman, 2004; McEachran et al., 1996). González-Isáis and Domínguez (2004) identified five characters uniting Rhinopteridae and Myliobatidae. Two are probably correlated with the specialized crushing mode of feeding exhibited by these taxa, while mobulids are instead highly adapted for filter feeding: (61) well-developed mandibular complex muscles, and (66) pavement-like arrangement of teeth. Rhinopteridae and Myliobatidae share two other muscle characters, while a third instead supports (Rhinopteridae, Mobulidae): (55)

superficial transverse muscle present. Supporting the hypothesis of González-Isáis and Domínguez (2004), a final character—(67) the number of turns of the intestinal spiral valve—is again potentially functionally correlated with mode of feeding.

McEachran and Aschliman (2004) presented two synapomorphies for (Rhinopteridae, Mobulidae) that are less obviously correlated with feeding mode: (26) dorsolateral components of nasal capsule form a pair of projections that support the cephalic lobes or cephalic fins, and (29) preorbital process absent (convergent with the electric ray, *Temera*). A third character unites Myliobatidae, and potentially Gymnuridae, to the exclusion of the others: (60) some pectoral fin radials expanded distally.

Despite several exhaustive attempts, the sister relationship of demersal butterfly rays and pelagic rays has yet to be convincingly demonstrated. A number of fundamental differences between the groups have been described, however, suggesting that this is perhaps another example of convergence. The highly derived nature of taxa such as mobulids has confounded the confident reconstruction of their interrelationships with other stingrays, as between higher groups of batoids. These patterns of morphological convergence and autapomorphism occur at both high and low taxonomic levels across Batoidea. This has led some workers to investigate batoid interrelationships using molecular sequence data (see Section 3.5.3).

3.3 Analyses

The following character complexes were surveyed for characters that exhibited variation that was thought to reflect phylogenetic relationships: (1) external morphological structures; (2) squamation; (3) tooth root vascularization patterns; (4) lateral line patterns; (5) neurocranium and branchial skeleton; (6) synarcual and pectoral girdle skeleton; (7) pelvic girdle, claspers, and caudal skeleton; and (8) cephalic and branchial musculature. Specimens were cleared and stained, dissected, radiographed, or CT scanned to observe internal structures. Anatomical terminology follows Miyake (1988). Characters were also interpreted from Regan (1906), Garman (1913), Melouk (1948), Ishiyama (1958), Nelson (1969), Hulley (1970, 1972), Stehmann (1970), Compagno (1973, 1977), Capapé and Desoutter (1979), Reif (1979), Chu and Wen (1980), Heemstra and Smith (1980), Compagno and Roberts (1982), Rosa (1985), Rosa et al. (1987), Miyake (1988), Nishida (1990), Miyake and McEachran (1991), Miyake et al. (1992a,b), Shirai (1992a,b), Lovejoy (1996), McEachran and Konstantinou (1996), McEachran et al.

(1996), Herman et al. (1994, 1995, 1996, 1997, 1998, 1999), Rosenberger (2001b), Carvalho (2004), Carvalho et al. (2004), González-Isáis and Domínguez (2004), Claeson (2008, 2010), and Claeson and Hilger (2011). In nearly all cases, observations based on the literature were verified with independent observations.

Representatives of 60 of the approximately 80 genera of batoids and 4 outgroup taxa were examined (Appendix 3.1). For species-rich genera (e.g., *Torpedo, Narcine, Rhinobatos, Bathyraja, Raja, Urolophus, Urotrygon, Dasyatis, Himantura, Potamotrygon, Gymnura, Myliobatis, Rhinoptera, Mobula*), several to many species were examined but only one to three were included in the data matrix because character states were usually consistent for a majority or for all of the species in the genus. In cases where character states varied within a genus, additional species were included in the matrix to represent the character state variability. Outgroups included a holocephalan (Chimaeridae), as well as relatively underived galeomorph (Heterodontidae) and squalomorph sharks (Chlamydoselachidae, Hexanchidae) (Didier, 1995; Maisey et al., 2004). Most of the characters used are binary; those with multiple character states were unordered to reduce subjectivity.

The data matrix (Appendix 3.2) includes 4 outgroup taxa, 36 batoid taxa, and 89 characters. Characters and character states are described in Appendix 3.3. Maximum parsimony (MP) analyses were performed in PAUP* v.4.0b10 (Swofford, 2003) using the heuristic search option with random sequence addition (100 replicates), tree bisection–reconnection (TBR) branch swapping, and retained trees limit (MaxTrees) set to 5000. Tree robustness was assessed by: (1) calculating Bremer decay indices (Bremer, 1994) with TreeRot v.3 (Sorenson and Franzosa, 2007), and (2) performing a bootstrap (BS) analysis (1000 replicates) using the same heuristic search options in PAUP*. Finally, a successive approximations approach (Carpenter, 1994; Farris, 1969) was employed by reweighting characters based on their rescaled consistency (RC) index and performing a new heuristic search.

3.4 Matrix Revisions

3.4.1 Synarcual

Three new characters and two modified characters from McEachran and Aschliman (2004) are described following recent studies that comprehensively examined the batoid synarcual (Claeson, 2008, 2010; Claeson and Hilger, 2011):

- Character 6, modified character, suprascapulae: fused medially (6,1) in all batoids.

- Character 7, new character, synarcual lip (odontoid process): 0 = absent, 1 = present. In all examined batoids except *Typhlonarke*, the ventral rim projects slightly, forming a lip or an odontoid process (7,1) (Claeson, 2008; Domínguez and González-Isáis, 2007; González-Isáis and Domínguez, 2004; Melouk, 1948).

- Character 50, modified character, suprascapular–vertebral association: 0 = free of vertebral column, 1 = articulates with vertebral column, 2 = fused medially to synarcual, forming pectoral arch, 3 = fused medially and laterally to synarcual. The suprascapulae were previously considered to be tightly connected to the synarcual in all batoids except torpediniforms (electric rays) (Compagno, 1973; Miyake, 1988) and pristids (McEachran et al., 1996). The suprascapulae of torpediniforms are connected medially and situated dorsal to the vertebral column but are completely free of the synarcual (50,0) (Figure 3.2) and resting closely above the vertebral column or widely separated from it by musculature (Claeson, 2010; Regan, 1906). The state in *Pristis* (sawfish) is unknown (50,?). Among guitarfishes (e.g., *Rhinobatos, Zapteryx, Trygonorrhina*), there is no direct connection between the suprascapulae and synarcual; instead, the suprascapulae articulate with the neural arches of the vertebrae that are directly posterior to the synarcual cartilages (50,1) (Figure 3.3). This articulation is tight in mature specimens but was never observed as a medial fusion (Claeson, 2010; KMC, pers. obs.). In rajids (skates), the suprascapulae are fused to the median crest of the synarcual, forming the pectoral arch (50,2) (Figure 3.4) (Claeson, 2008; Garman, 1913). In myliobatiforms (stingrays), the suprascapulae are fused to the median crest of the synarcual and are continuous with the posterior margin of the lateral stay, creating a well-defined bridge (50,3) (Lovejoy, 1996; Miyake, 1988).

- Character 51, new character, orientation of lateral stay processes: 0 = posteriorly directed, 1 = laterally directed, 2 = dorsally directed. Pristid synarcuals exhibit wide lateral stays with gracile, posteriorly directed processes (51,0) (Garman, 1913, Plate 64). Torpediniforms exhibit laterally directed processes (51,1), except for *Electrolux* and *Typhlonarke*, as these taxa lack lateral stays (51,?). The lateral stays of guitarfishes, rajids, and myliobatiforms exhibit U-shaped, dorsally directed processes (51,2). The lateral stays of myliobatiforms are much smaller than those of other batoid taxa (Claeson, 2008, 2010; Garman, 1913; KMC, pers. obs.).

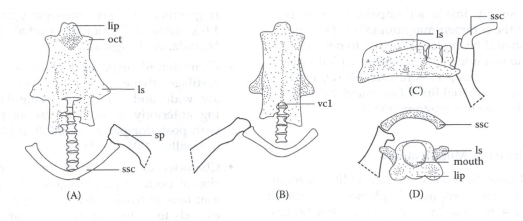

FIGURE 3.2
Synarcual cartilages of *Narke dipterygia* (ZMB 33991). (A) Dorsal view, (B) ventral view, (C) lateral view, (D) anterior view. *Abbreviations:* lip, synarcual lip/odontoid process; ls, lateral stay; mc, median crest; oct, occipital cotyle; sp, scapular process; ssc, suprascapular cartilage; vc1, first free vertebral centrum.

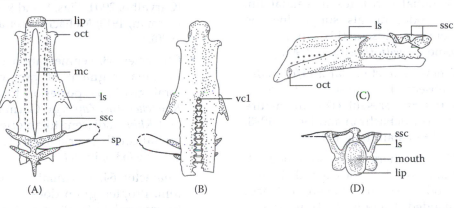

FIGURE 3.3
Synarcual cartilages of *Rhinobatos blochii* (ZMB 12452). (A) Dorsal view, (B) ventral view, (C) lateral view, (D) anterior view. *Abbreviations:* lip, synarcual lip/odontoid process; ls, lateral stay; mc, median crest; oct, occipital cotyle; sp, scapular process; ssc, suprascapular cartilage; vc1, first free vertebral centrum.

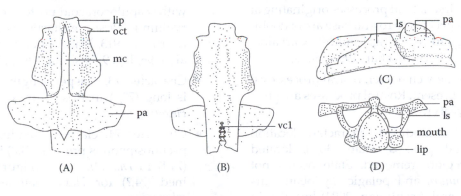

FIGURE 3.4
Synarcual cartilages of *Bathyraja parmifera* (TCWC 9609.02). (A) Dorsal view, (B) ventral view, (C) lateral view, (D) anterior view. *Abbreviations:* lip, synarcual lip/odontoid process; ls, lateral stay; mc, median crest; oct, occipital cotyle; pa, pectoral arch (suprascapular cartilage fused to synarcual); sp, scapular process; ssc, suprascapular cartilage; vc1, first free vertebral centrum.

• Character 52, new character, vertebral–occipital articulation: 0 = synarcual lip fitted into notch in basicranium; 1 = synarcual lip rests in foramen magnum. In all examined torpediniforms except *Torpedo*, the synarcual lip rests within an unpaired notch beneath the foramen magnum (52,0); this concavity is paired in *Torpedo*. In all examined guitarfishes and myliobatiforms, the lip fits snugly inside an unpaired concavity beneath the foramen magnum (52,0). This condition should be explored further to determine if it is homologous with the basioccipital fovea present in sharks (Claeson, 2010; Shirai, 1992b, 1996). The synarcual lip of *Pristis* and *Raja* rests inside the foramen magnum (52,1).

3.4.2 Miscellaneous Modifications

A number of miscellaneous character additions, modifications, and corrections were implemented, some of which were discussed in the above review (see Section 3.2.1) of McEachran and Aschliman (2004). The following are supported by new observations or the literature:

- Character 13, radial cartilages in caudal fin. Rudimentary caudal radials supporting the ventral tail fold are present (13,0) in *Taeniura lymma* and some *Dasyatis* (Nishida, 1990).

- Character 17, new character, alar and/or malar thorns: 0 = absent, 1 = present. Alar and/or malar thorns are present (17,1) in mature males of all rajids (McEachran and Dunn, 1998; McEachran and Konstantinou, 1996).

- Character 27, new character, rostral node: 0 = expanded laterally, 1 = not expanded laterally. Rostral node is spatula shaped, flattened, and not expanded laterally (27,1) in *Pristis*, *Rhynchobatus*, *Rhina*, and rajids.

- Character 28, rostral appendices. Rostral appendices are absent (28,0) in platyrhinids (thornbacks); the paired lateral processes originating at the ventral aspect of the rostral base are of doubtful homology to rostral appendices (Carvalho, 2004, Fig. 8; McEachran et al., 1996, Fig. 5).

- Character 29, new character, rostral processes: 0 = absent, 1 = present. Rostral processes are present (29,1) in platyrhinids.

- Character 36, modified character, location of postorbital process; state (36,3), located in orbital region, removed. State occurs not only in *Gymnura* and pelagic myliobatiforms (McEachran and Aschliman, 2004) but also in unselected taxa including *Urotrygon micropthalmum* and *Himantura bleekeri* (Nishida, 1990, Figs. 12B and 14B).

- Character 43, labial cartilages. Labial cartilages are present (43,0) in *Rhinobatos* (Garman, 1913).

- Character 59, location of mesocondyle. Data are missing (59,?) for *Zanobatus*, *Gymnura*, and pelagic myliobatiforms, which exhibit fragmented or absent mesopterygia (Garman, 1913, Plates 73–75; McEachran et al., 1996, Fig. 9; Nishida, 1990, Figs. 31 and 32).

- Character 61, expanded character, antorbital cartilage shape and orientation. Cartilages are wide and somewhat triangular, projecting anteriorly at least as far as posteriorly, with posterior processes (61,2) in platyrhinids (Carvalho, 2004, Fig. 8).

- Character 62, new character, anterior extension of pectoral propterygium: 0 = propterygium fails to reach anterior margin of disc, 1 = extends to or to near anterior margin of disc. Pectoral propterygia extend to near the anterior disc margin or snout tip (62,1) in platyrhinids, *Zanobatus*, *Bathyraja*, and myliobatiforms (Carvalho, 2004, Figs. 8 and 9; Compagno, 1977; Garman, 1913; McEachran et al., 1996; Nishida, 1990).

- Character 63, segmentation of propterygium. Proximal segment of propterygium of pectoral girdle is posterior to mouth (63,0) in *Rhynchobatus*, *Zapteryx* (Garman, 1913, Plate 65), and *Rhina*; adjacent to anterior margin of antorbital cartilage or anterior to margin of nasal capsule (63,3) in *Aetobatus*.

- Character 64, proximal section of propterygium. Propterygium does not extend posterior to procondyle (64,0) in platyrhinids (Garman, 1913, Plate 66).

- Character 65, articulation of pectoral fin radials. One or more radials directly articulate with scapulocoracoid posterior to the mesopterygium in *Pristis*, guitarfishes, and rajids (65,1) (Garman, 1913, Plates 64 and 65; McEachran et al., 1996, Fig. 9; Nishida, 1990, Fig. 32).

- Character 73, clasper length. Clasper length is long (73,1) in *Pristis*; unexamined (73,?) for *Hexatrygon*.

- Character 74, clasper pseudosiphon. Clasper pseudosiphon is present (74,0) in *Pristis*; absent (74,1) in *Dasyatis brevis* (Figure 3.5); unexamined (74,?) for *Hexatrygon* and "*Himantura*" *schmardae*.

- Character 75, dorsal marginal clasper cartilage. Medial flange is absent (75,0) in *Trygonorrhina*, *Neotrygon* (Figure 3.6A,B), and *Taeniura lymma* (Figure 3.6C,D).

- Character 76, dorsal terminal clasper cartilage. Smooth margin (76,0) found in *Trygonorrhina*; rough or crenate margin (76,1) in *Neotrygon* and *Taeniura lymma*.

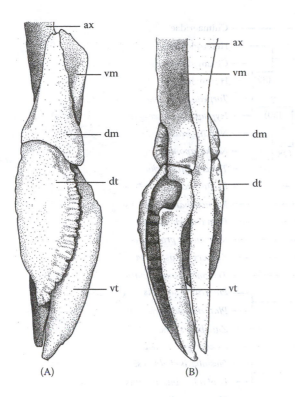

FIGURE 3.5

Clasper cartilages of *Dasyatis brevis* (TCWC 12099.01) in (A) dorsal and (B) ventral view with ventral covering piece removed. *Abbreviations:* ax, axial; dm, dorsal marginal cartilage; dt, dorsal terminal cartilage; vm, ventral marginal cartilage; vt, ventral terminal cartilage.

- Character 77, cartilage forming component claw. It is present (77,0) in *Trygonorrhina*; absent in *Taeniura lymma*.
- Character 78, ventral terminal cartilage shape. It is simple (78,0) in *Trygonorrhina*; forms complex flange (78,2) in *Taeniura lymma*.
- Character 79, ventral terminal cartilage attachment. Ventral cartilage is attached over length to axial cartilage (79,0) in *Trygonorrhina*; free of axial cartilage (79,1) in *Neotrygon*, *Taeniura lymma*, and *Himantura signifer*.

3.5 Results and Discussion

3.5.1 Phylogenies and Robustness

The heuristic search based on unweighted characters yielded two equally parsimonious trees of 201 steps, with a consistency index (CI) of 0.6318 and retention index (RI) of 0.9008. The modest CI indicates that a number of characters experienced homoplasious changes on the tree, while the high RI suggests that a number of these homoplasious changes are also phylogenetically informative. The strict consensus of the two most parsimonious trees (Figure 3.7) closely resembles

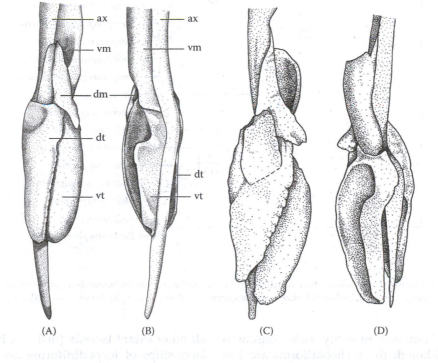

FIGURE 3.6

Clasper cartilages of (A, B) *Neotrygon kuhlii* (TCWC 12770.01) and (C, D) *Taeniura lymma* (TCWC 12772.01). (A) and (C) are dorsal views and (B) and (D) are ventral views with the ventral covering piece removed. *Abbreviations:* ax, axial; dm, dorsal marginal cartilage; dt, dorsal terminal cartilage; vm, ventral marginal cartilage; vt, ventral terminal cartilage.

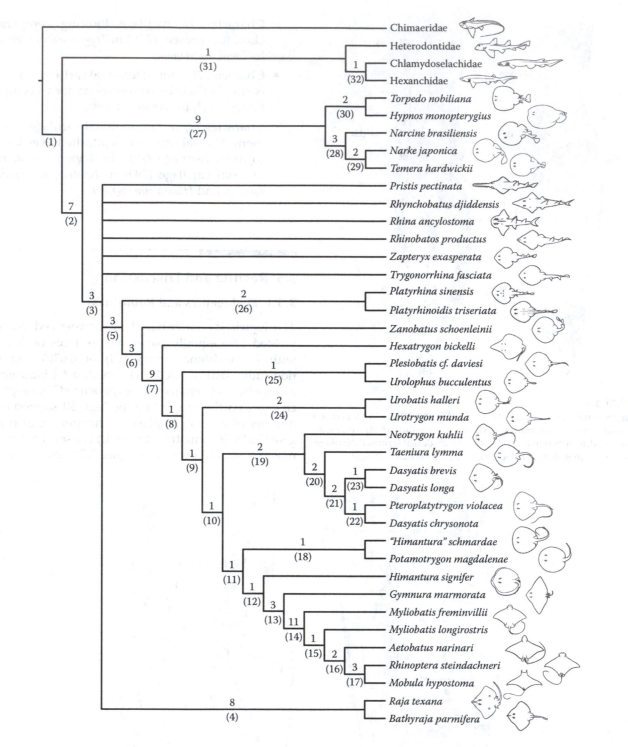

FIGURE 3.7

Strict consensus of the two most parsimonious trees resulting from maximum parsimony analysis, characters unweighted. Numbers above internal branches are Bremer decay indices; labels below branches correspond to apomorphy lists in Appendix 3.4.

the 1000-replicate bootstrap majority rule consensus tree (Figure 3.8), although the myliobatiforms are less resolved in the latter. Batoids are strongly supported as monophyletic (BDI = 7, BS = 98), with good support for a sister relationship between torpediniforms and

all other extant batoids (BDI = 3, BS = 85). The interrelationships of torpediniforms are well resolved, with (*Torpedo, Hypnos*) indicated to be sister to *Narcine* and *Narke* + *Temera* (BDI = 3, BS = 94). The strict consensus left the branching orders of *Pristis, Rhynchobatus, Rhina,*

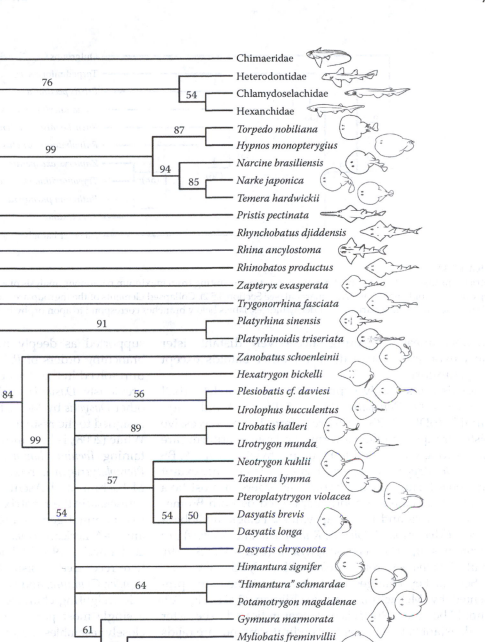

FIGURE 3.8

Majority rule consensus of the 1000-replicate bootstrap analysis. Numbers above internal branches are bootstrap values.

Rhinobatos, Zapteryx, Trygonorrhina, rajids, and a clade containing all remaining batoids unresolved. Except for the novel inclusion of *Pristis* in that polytomy, this aspect of the topology resembles that of McEachran et al. (1996).

Among these taxa, the lack of resolution in the strict consensus is due to highly discordant topologies between the two summarized trees. One (Figure 3.9)

of the two most parsimonious trees closely resembles the successive approximations tree (Figure 3.10), in which *Pristis* is sister to a clade including guitarfishes and rajids and *Trygonorrhina* is indicated to be sister to *Zapteryx*. In the other most parsimonious tree (not shown), *Zapteryx, Trygonorrhina,* rajids, *Rhinobatos, Rhynchobatus, Rhina,* and *Pristis* are recovered as

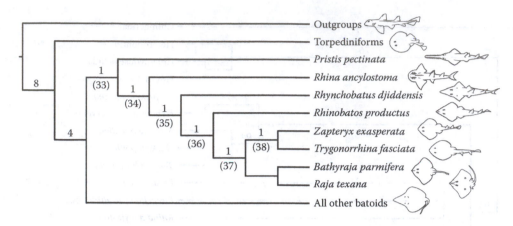

FIGURE 3.9
Strict consensus of the four most parsimonious trees resulting from maximum parsimony analysis of a modified matrix assuming nonhomology of the batoid and chimaerid synarcual (see Section 3.5.2). Collapsed elements of the ingroup are identical to those in Figure 3.7. Numbers above internal branches are Bremer decay indices; labels below branches correspond to apomorphy lists in Appendix 3.4.

distinct lineages forming successively distant sister taxa (working rootward) to all other batoids except torpediniforms.

Rajids are strongly supported as monophyletic (BDI = 8, BS = 99). *Zanobatus* (BDI = 3, BS = 84) and platyrhinids (BDI = 3, BS = 82) are inferred to be successive sister groups to myliobatiforms, and myliobatiforms are also strongly indicated to be monophyletic (BDI = 9, BS = 99). *Hexatrygon* is recovered as sister to all other extant myliobatiforms, but the branching orders of most taxa within the clade are poorly supported by both Bremer decay indices and bootstrap values. Pelagic myliobatiforms (*Myliobatis*, *Aetobatus*, *Rhinoptera*, and *Mobula*) are strongly supported as monophyletic (BDI = 11, BS = 99), with *Rhinoptera* sister to *Mobula* (BDI = 3, BS = 83).

Several key changes from the unweighted tree presented by McEachran and Aschliman (2004, Fig. 3.9) should be noted. *Pristis* is no longer indicated to be sister to all extant batoids except torpediniforms, nor are rajids

supported as deeply nested within guitarfishes; the branching orders of *Pristis*, rajids, and guitarfishes are unresolved in the strict consensus tree. *Neotrygon kuhlii*, previously *Dasyatis kuhlii* but noted as distinct from other *Dasyatis* by McEachran and Aschliman (2004) and assigned to the resurrected genus *Neotrygon* by Last and White (2008), is here indicated to be sister to a clade containing *Taeniura lymma*, *Dasyatis*, and *Pteroplatytrygon*. *Pteroplatytrygon* is recovered as nested within *Dasyatis*; McEachran and Aschliman (2004) did not include *D. chrysonota* in their matrix, leaving the relationship of these two nominal genera ambiguous. *Potamotrygon* and the amphi-American "*Himantura*" *schmardae* (see Aschliman, 2011; Lovejoy, 1996; McEachran and Aschliman, 2004) are here recovered as sister to a clade including *Himantura signifer*, *Gymnura*, and pelagic myliobatiforms.

Reweighting characters by the RC index resulted in a single most parsimonious tree (Figure 3.10), which closely resembles the successive approximations (by the

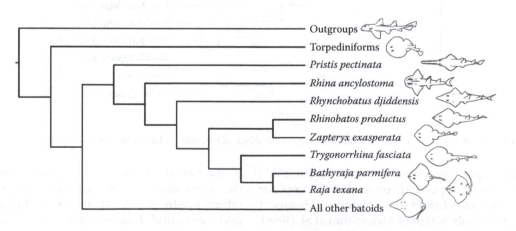

FIGURE 3.10
Single most parsimonious tree resulting from maximum parsimony analysis after character reweighting based on the rescaled consistency (RC) index.

RI) tree of McEachran and Aschliman (2004). *Pristis* is recovered as sister to a clade including guitarfishes and rajids, with rajids nested deeply within the clade and sister to *Trygonorrhina*. The successive approximations tree of McEachran and Aschliman (2004) instead placed *Pristis* one node rootward in the tree, sister to all other extant batoids except torpediniforms.

3.5.2 Character Evolution

The character additions and modifications to McEachran and Aschliman (2004) suggest an updated diagnosis for Batoidea. Batoids lack an upper eyelid, with the cornea directly attached to the skin around the eyes (1,1); exhibit a palatoquadrate that does not articulate with the neurocranium (2,1); exhibit a pseudohyal (3,1); exhibit a last (posteriormost) ceratobranchial that articulates with the scapulocoracoid (4,1); exhibit a synarcual comprised of fused cervical vertebrae or vertebral elements (5,1), but see discussion of chimaerid synarcual below; exhibit a suprascapula that is fused medially to its antimere (6,1); exhibit a lip on the anteroventral aspect of the synarcual (7,1); exhibit an antorbital cartilage that directly or indirectly articulates with the propterygium and nasal capsule (8,1); and exhibit levator and depressor rostri muscles (8,1), but the depressor rostri is secondarily lost in myliobatiforms.

Recent investigations (Claeson, 2008, 2010; Claeson and Hilger, 2011) of the synarcual and anterior vertebral column have improved the understanding of variation in this structure and its relationship to the suprascapulae among batoid taxa. Earlier studies (e.g., McEachran and Aschliman, 2004; McEachran et al., 1996) were overly simplistic in categorizing the synarcual as either short or long, depending on whether or not it extended to and articulated or fused with the suprascapulae. This overlooked a suite of discrete character states (character 50) describing various degrees of articulation or fusion of the suprascapulae to the vertebral column. Notably, rajids do not exhibit the same state as do guitarfishes and myliobatiforms.

The lateral stay processes (character 51) are U-shaped and dorsally directed in all batoids except torpediniforms and pristids, but this character is of questionable phylogenetic utility. Torpediniforms are unique in exhibiting a laterally directed process (when present), while *Pristis* exhibits a gracile, posteriorly directed process, as do chimaerids. The condition in *Pristis* resembles that of *Rhinobatos*, but with the posterodorsal apex of the process in *Rhinobatos* elaborated into an elongated, posteriorly directed process in *Pristis* (Garman, 1913, Plates 55.3 and 55.4). Given the tree topology and the highly derived nature of the chimaerid bauplan, the chimaerid synarcual is most likely convergent with the batoid synarcual. The chimaerid synarcual is objectively coded as present

(5,1) and with posteriorly directed lateral stay processes (51,0) in the present study. Maximum parsimony analysis of a modified matrix in which the chimaerid state was changed to "?" for these characters resulted in an ingroup topology (Figure 3.9) identical to one of the two most parsimonious trees in the original analysis and similar to the successive approximations tree (Figure 3.10) except for the placement of *Trygonorrhina*.

A large number of characters are indicated to be synapomorphic for individual clades; for example, the monophyly of torpediniforms is supported by a BDI of 9 and 12 character changes with CI ≥ 0.5. However, despite the addition of new and modified characters, relatively few phylogenetically informative characters unite two or more major batoid groups (e.g., platyrhinids with myliobatiforms). Several of these relationships have slightly stronger support (BDI increases of 1 to 2) in the present study than in the analyses of McEachran and Aschliman (2004).

Torpediniforms are strongly supported as monophyletic based on a large number of character states. Torpediniforms lack a cephalic lateral line (20,1); exhibit a ventrolaterally expanded nasal capsule (31,1); lack a supraorbital crest (34,1), postorbital process (36,1), and basihyal (48,2); exhibit a long and posteriorly displaced scapular process (56,1); exhibit an antorbital cartilage that is anteriorly expanded and fan or antler like (61,1); lack a cartilage forming the clasper component claw (77,1); exhibit an intermandibularis, which originates on the hyomandibula and inserts on the posterior margin of Meckel's cartilage (82,2); exhibit a divided spiracularis, with one bundle entering the dorsal oral membrane (85,1); exhibit branchial electric organs (86,1); and lack a coracohyoideus (89,1). The sister relationship of torpediniforms to all other extant batoids (that is, exclusion from remaining batoids) is also supported by a number of characters with CI ≥ 0.5. Torpediniforms exhibit a laterally expanded rostral node (27,0), lack a jugal arch (39,0), lack direct articulation of the suprascapula with the vertebral column (50,0), exhibit a laterally directed lateral stay process (51,1), exhibit scapulocoracoid condyles that are not arranged horizontally (58,0), and lack an ethmoideo-parethmoidalis (81,0). Torpediniforms exhibit the outgroup condition for each of these except character 51. It has been noted (see Section 3.2.1) that torpediniforms and pristids are described by non-overlapping sets of characters for which they exhibit the outgroup condition. The present study increased the number of characters that suggest that torpediniforms are sister to all other batoids, and also reduced the number that suggest that pristids may occupy this position (e.g., character 50).

The interrelationships of the torpediniform genera are unchanged from McEachran and Aschliman (2004) and are corroborated by subsequent studies on the clade

(Carvalho, 2010; Claeson, 2010). *Torpedo* and *Hypnos* are united to the exclusion of other torpediniforms based on the longitudinally expanded medial section of the hyomandibula that spans the oticooccipital region of the neurocranium (44,1). *Narcine, Narke,* and *Temera* exhibit a component claw cartilage embedded in integument (77,2), a ligamentous sling at the symphysis of Meckel's cartilage supporting several muscles (83,1), and a coracobranchialis consisting of a single component (87,1). *Narke* and *Temera* exhibit a fully developed ceratohyal (49,0), which along with other narkids (*sensu* Compagno, 2005) is unique among batoids, as well as a coracohyomandibularis with separate origins on the fascia (88,1).

The interrelationships of *Pristis*, guitarfishes, and rajids are unresolved in the strict consensus and bootstrap analyses but are recovered as a clade in the modified search assuming the non-homology of the chimaerid and batoid synarcual (Figure 3.9) and in the successive approximations tree (Figure 3.10). In the modified search, this potential clade exhibits a rostral node that is not expanded laterally (27,1) and pectoral radials that articulate directly with the scapulocoracoid posterior to the mesopterygium (65,1). *Zanobatus* and myliobatiforms exhibit an extremely reduced (*Zanobatus*) or absent rostrum, so the state is unknown for these taxa (27,?). Character 65 is also relatively ambiguous for *Pristis* in comparison to the pronounced state in other members of this potential clade. These taxa exhibit elongated claspers (73,1), as do chimaerids and *Platyrhina* (CI = 0.333). In the modified search, guitarfishes and rajids are united to the exclusion of *Pristis* by exhibiting tooth roots with large pulp cavities (18,0), with implied reversals or independent gains of small pulp cavities in *Rhinobatos* and *Zapteryx*. Guitarfishes and rajids also exhibit a rostral appendix, a thin sheet of cartilage on each side of the developing rostrum that is carried forward with the anterior growth of the rostral cartilage (28,1). The rostral appendix is indicated to be an unambiguous synapomorphy for this group. No character state changes with CI ≥ 0.5 unite *Rhynchobatus* and crownward taxa. In *Rhinobatos, Zapteryx, Trygonorrhina,* and rajids, the scapulocoracoid is elongated between the mesocondyle and metacondyle (59,1; unambiguous synapomorphy), and the antorbital cartilage directly joins the propterygium and nasal capsule (60,1). No character state changes with CI ≥ 0.5 support the sister relationship between *Zapteryx* and *Trygonorrhina*.

Rajids are again strongly supported as monophyletic (McEachran and Aschliman, 2004; McEachran and Dunn, 1998; McEachran and Miyake, 1990; McEachran et al., 1996). Rajids exhibit alar and/or malar thorns in mature males (17,1); osteodentine in the roots of large teeth only (19,1); a pectoral arch formed by fusion of the suprascapula to the median crest of the synarcual (50,2); a synarcual lip resting inside the foramen magnum (52,1); a rodlike compound pelvic radial articulating with single radial segments in series (71,2); distinctly separated pelvic girdle condyles for the compound radial and basipterygium, with several radials directly articulating with the pelvic girdle between the two condyles (72,1); a ventral terminal clasper cartilage forming the component shield (77,3); and an accessory terminal 1 cartilage that forms the component sentinel or component projection (78,1). The present study did not include a character describing the mode of reproduction, but rajids are unique among batoids in exhibiting oviparity (McEachran and Miyake, 1990).

Platyrhinids exhibit a pair of lateral processes originating at the ventral aspect of the rostral cartilage (29,1), a wide and somewhat triangular antorbital cartilage with a posterior process (61,2), and a pair of triangular, posteriorly directed processes on the puboischiadic bar (69,1). All three of these states are unique to platyrhinids. The sister relationship between platyrhinids and (*Zanobatus,* myliobatiforms) first proposed by McEachran et al. (1996) and reinforced by McEachran and Aschliman (2004) is again recovered. The BDI for this node increased by one from McEachran and Aschliman (2004), despite the correction of the inaccurate coding scheme for character (64) that was previously indicated to be synapomorphic for these taxa (see Section 3.4.2) (McEachran and Aschliman, 2004, character 57). Several character states with CI ≥ 0.5 support this clade. Platyrhinids, *Zanobatus,* and myliobatiforms exhibit an incomplete or absent rostral cartilage (26,1–2), a pectoral propterygium that extends to the anterior disc margin (62,1), pectoral radials that all articulate with pterygials (65,0), and a coracohyoideus running diagonally from first two gill slits to the posteromedial aspect of the basihyal or first basibranchial (89,3). Of these, only character 89 is indicated to be an unambiguous synapomorphy for these taxa.

Zanobatus is again indicated to be sister to myliobatiforms (McEachran and Aschliman, 2004; McEachran et al., 1996). These taxa lack a rostral cartilage (reduced to a filament in *Zanobatus;* 26,2) and exhibit a proximal section of the propterygium that extends posterior to the procondyle (64,1), a narrow and arched puboischiadic bar without distinct lateral prepelvic processes (70,1), and a clasper ventral terminal cartilage folded ventrally along its long axis to form a convex flange (78,2). Myliobatiforms (*sensu stricto,* excluding *Zanobatus*) are strongly supported as monophyletic. Myliobatiforms exhibit an inferred secondary loss of the levator and depressor rostri muscles (9,0), a well-developed anterior nasal lobe forming a nasal curtain (11,2–3), serrated tail stings (14,1), an infraorbital loop of the suborbital and infraorbital lateral line canals (21,1), a scapular loop of the trunk lateral line (25,1), and a broad and shelflike postorbital process (36,2); lack a jugal arch (39,0); exhibit a suprascapula fused medially and laterally to the synarcual (50,3), a ball and socket articulation between the scapular process and synarcual (53,1), and a second (thoracolumbar) synarcual (54,1); lack ribs (55,1); and exhibit a component claw cartilage embedded in integument (77,2), a spiracularis split into lateral and medial bundles, with the medial bundle inserting onto the posterior surface of Meckel's cartilage and the lateral bundle inserting onto the dorsal edge of the hyomandibula (85,2), and a coracohyomandibularis with separate origins on the anterior portion of the gill arch region and on the pericardial membrane (88,2).

Branching orders among the major lineages of myliobatiforms are weakly supported. Myliobatiforms to the exclusion of *Hexatrygon* exhibit a nasal curtain that extends to or just anterior to the mouth (11,3) and ventrolaterally expanded nasal capsules (31,1). A poorly supported (*Plesiobatis*, *Urolophus*) clade exhibits a basihyal and first hypobranchial that are both present and unsegmented (48,0). Crownward myliobatiforms except *Gymnura* exhibit some degree of loss or segmentation of these elements. *Urobatis* and *Urotrygon* exhibit a spiracular tentacle during development (12,1) and branched lateral tubules of the subpleural loop of the lateral line (23,1), which are both indicated to be unambiguous synapomorphies for this clade.

Among *Neotrygon*, *Taeniura lymma*, and *Dasyatis* (including *Pteroplatytrygon*), phylogenetic signal is weak and mostly based on tooth morphology, lateral line canals, and clasper structure. McEachran and Aschliman (2004) did not examine claspers from *T. lymma*, and some character states for *Neotrygon* were miscoded. *Dasyatis* (except *D. chrysonota*), *Pteroplatytrygon*, *Neotrygon*, and *T. lymma* exhibit claspers with a crenate margin on the dorsal terminal cartilage (76,1) (Figures 3.5 and 3.6). Osteodentine is widespread in tooth roots (19,2) of *Dasyatis*, *Pteroplatytrygon*, and *Taeniura lymma*, but not *Neotrygon*. *Dasyatis brevis* and *D. longa* exhibit a subpleural loop of the hyomandibular lateral line canal forming a lateral hook (22,1); the state is unknown for *D. chrysonota* (22,?).

"*Himantura*" *schmardae* (western Atlantic) and *Potamotrygon* exhibit two small cartilages in the hyomandibular–Meckelian ligament (46,2). Support for the sister relationship between this clade and remaining batoids (Indo–West Pacific *Himantura*, *Gymnura*, pelagic myliobatiforms) is weak, based on the pattern of division of the spiracularis in *Himantura* (*sensu lato*) and *Potamotrygon* (85,3). The spiracularis is undivided (85,0) or of unknown state (85,?) in *Gymnura* and pelagic myliobatiforms.

Gymnura and pelagic myliobatiforms exhibit a fragmented or absent mesopterygium (66,1), some pectoral fin radials that are expanded distally (67,1), and a cartilage forming the component claw that is embedded in integument (77,2). The branching orders of the two

Myliobatis and *Aetobatus* lack support, being based in part on characters for which several taxa have missing or unique states. However, *Rhinoptera* and *Mobula* are well supported as sister taxa in that they exhibit a paired and discontinuous cephalic lobe (10,3) and dorsolateral components of the nasal capsule (30,1), lack a preorbital process (33,1), and exhibit pectoral fin radials that are not expanded distally (67,0).

3.5.3 Comparison with Molecular Phylogenies

To date there have been few attempts to use molecular data to test hypotheses proposed by the extensive morphological surveys of batoids. The exceptions are limited in taxonomic scope and in the diversity and length of markers used. Most molecular studies including batoids have, to date, been focused either on assessing the premise that batoids are derived squalean sharks (the "hypnosqualean hypothesis" of Shirai, 1992, 1996) or on recovering the phylogeny of sharks or more inclusive clades, such as Vertebrata (e.g., Arnason et al., 2001; Rasmussen and Arnason, 1999). Notable studies evaluating the monophyly of extant sharks to the exclusion of batoids or using batoids as outgroups in shark phylogenies include Dunn and Morrissey (1995), Kitamura et al. (1996), Douady et al. (2003), Maisey et al. (2004), Winchell et al. (2004), Naylor et al. (2005), Human et al. (2006), and Vélez-Zuazo and Agnarsson (2010). The topologies generated by early molecular studies, while based on few taxa and sequence data, described novel phylogenetic hypotheses at odds with morphology. Many of these preliminary hypotheses would later be corroborated by studies including broad taxonomic sampling, multiple independent molecular markers, and careful assessment of the robustness of the phylogenetic signal in the data (e.g., Aschliman, 2011; Maisey et al., 2004; Naylor et al., 2005).

Kitamura et al. (1996) sequenced a fragment of cytochrome *b* for four sharks and four batoids (*Pristis*, *Rhinobatos*, and two myliobatiforms). Maximum likelihood (ML) and maximum parsimony (MP) analyses of the translated amino acid sequences (Kitamura et al., 1996, Fig. 3) recovered a novel topology in which myliobatiforms were sister to *Pristis* + *Rhinobatos*, which had not been proposed by morphological studies. Dunn et al. (2003) conducted the first molecular study specifically addressing batoid interrelationships, albeit with a focus on myliobatiforms. The authors sequenced 1528 bp of mitochondrial (mt) DNA (12S, four tRNAs, fragments of ND1 and ND2) for ten taxa spanning the morphological breadth of myliobatiforms. *Pristis*, *Raja*, *Rhinobatos*, and *Torpedo* were included as outgroups, marking the first time that each of these taxa was included in the same molecular phylogeny. Maximum likelihood analysis (Dunn et al., 2003, Fig. 3) identified *Pristis* + *Rhinobatos* as sister to all other batoids, with *Raja* + *Torpedo* sister to myliobatiforms. The interrelationships of myliobatiforms were poorly resolved but generally congruent with previous morphological hypotheses. One exception was the rootward shift of *Gymnura*, implying that its general body plan is convergent with pelagic myliobatiforms (see Section 3.2.2). However, a second ML analysis that included partial cytochrome *b* sequences (Dunn et al., 2003, Fig. 5) recovered a sister relationship between *Gymnura* and pelagic myliobatiforms (BS ≤ 50) and moved skates to a trichotomy at the base of the tree. The authors noted widespread changes in the rate of molecular evolution among lineages, assuming that their topology is accurate.

Maisey et al. (2004) provided the most comprehensive treatment of the discordance between morphological and molecular elasmobranch phylogenies. They sequenced 34 sharks and four batoids (*Pristis*, *Raja*, and two myliobatiforms) for RAG1 (nuclear) and ND2 (mt) genes and examined both morphological and molecular datasets for potential sources of systematic error. Maximum parsimony analysis of RAG1 identified batoids as sister to sharks, but the ML tree proposed a unique topology allying batoids with galeoid sharks (support values not provided) (Maisey et al., 2004, Fig. 9). *Raja* was placed sister to all other batoids, with *Pristis* sister to myliobatiforms. Among-lineage rate variation in RAG1 appeared to be the only major potential source of error, but the authors demonstrated that the marker was not saturated and ML should be largely insensitive to long branch attraction. ND2 corroborated the sister relationship of batoids and sharks (Maisey et al., 2004, Fig. 8B). The authors suggested that many of Shirai's (1992, 1996) morphological synapomorphies for batoids and batoid-like sharks may be convergent, influenced by constraints imposed by a benthic niche. The authors identified several other synapomorphies that are probably not related to the benthic mode of life and conceded that morphology presents a self-consistent signal nesting batoids within squalean sharks.

Naylor et al. (2005) followed this study by analyzing RAG1 and three mitochondrial genes (ND2, ND4, cytochrome *b*) for the four batoids sequenced by Maisey et al. (2004) and for 13 sharks. They used a suite of methods to identify potential sources of error in their data, including tests for partition homogeneity, saturation, stationarity in base composition, and codon usage bias. The prevailing signal through their numerous analyses indicated that batoids are sister to sharks (Naylor et al., 2005, Fig. 1.5). Among the four batoids, the pattern of interrelationships was identical to that of Maisey et al. (2004).

Human et al. (2006) sequenced one to three of the mitochondrial genes used by Naylor et al. (2005) for *Pristis*, *Rhinobatos*, and *Amblyraja*, as well as five myliobatiforms that were used as outgroups for a study emphasizing the

interrelationships of catsharks (Scyliorhinidae). Most of the included batoids were sequenced for cytochrome *b* only. The authors suggest (Human et al., 2006, Table 2) that sequence data were generated for several additional taxa, including *Narke*, but these taxa were apparently excluded from phylogenetic analysis. Bayesian inference on the cytochrome *b* dataset allied *Pristis* with *Rhinobatos* and indicated that myliobatiforms are monophyletic (Human et al., 2006, Fig. 2).

Aschliman (2011) constructed a resolved and time-calibrated batoid phylogeny using the protein-coding complement of the mitochondrial genome (mtGenome), independent nuclear markers (RAG1 and SCFD2), and fossils. Sequence data were generated for 37 batoid species representing 22 of 23 families (Compagno, 2005), with five sharks and a chimaerid used as outgroups. Aschliman (2011) evaluated data partitioning schemes, potential biases in the sequence data, and the relative informativeness of each genetic marker and fossil. The resulting phylogeny is largely congruent with morphology crownward in the tree, but the branching orders of major batoid groups are mostly novel or corroborate earlier molecular studies. While Bayesian analyses generated resolved and well-supported trees, ML analyses and a strict consensus of the phylogenies resulting from analyses of individual markers (RAG1, SCFD2, mtGenomes) were less so. This robust but more conservative phylogeny will here be compared and contrasted with morphology.

The phylograms from Aschliman's (2011) analyses indicated that several lineages have long internal branches with species-rich crowns. The primary lineages of batoids are indicated to have arisen relatively rapidly, with short internodes between branching events. This pattern might have been anticipated given the scarcity of phylogenetically informative characters uniting higher batoid taxa (e.g., platyrhinids with myliobatiforms) relative to the large number of synapomorphies for some major taxa (e.g., rajids). That is, there is little time to accumulate shared derived characters on a short internode, while many such changes are expected to arise on the long internal branches characterizing major batoid taxa. This pattern indicative of rapid radiation or lineage pruning by extinction is repeated among major myliobatiform lineages as well (Aschliman, 2011).

Aschliman's (2011) more conservative topology includes a four-way polytomy at the base of the tree between torpediniforms, rajids, platyrhinids, and a clade containing all other extant batoids. The placement of torpediniforms at or near the base of the tree is predicted by morphology; however, recent morphological studies suggest that rajids and platyrhinids occupy derived positions in the tree (distant from the root), with rajids nested within guitarfishes (McEachran and Aschliman, 2004; McEachran et al., 1996). The unweighted MP and bootstrap analyses (Figures 3.7 and 3.8) in the present study more closely resemble Aschliman's (2011) tree in that rajids are included in the large polytomy sister to torpediniforms; however, the modified matrix and successive approximations trees (Figures 3.9 and 3.10) place rajids sister to *Trygonorrhina* + *Zapteryx* or to *Trygonorrhina*, respectively. Both the molecular and morphological signals determining the placement of rajids appear to be robust and are difficult to reconcile. As discussed above (see Section 3.5.2), rajids may be allied with guitarfishes based on similarities in the rostral node, rostral appendices, and scapulocoracoid. The highly derived and often autapomorphic nature of rajids relative to other batoids, such as the unique arrangement of the claspers (McEachran and Miyake, 1990), complicates morphology-based attempts to ally them to other major taxa.

The rootward placement of platyrhinids in the molecular tree (Aschliman, 2011) is likewise unexpected under morphology (McEachran and Aschliman, 2004; McEachran et al., 1996). The sister relationship between platyrhinids and *Zanobatus* + myliobatiforms has moderate support in the present study, with the only unambiguous synapomorphy uniting these taxa describing the orientation of the coracohyoideus. Aschliman (2011) described strong support for the molecular placements of both rajids and platyrhinids in analyses of the combined dataset and of each individual marker, with no apparent systematic biases in the data. For example, platyrhinids were separated from *Zanobatus* + myliobatiforms by several intervening nodes with Bayesian posterior probabilities of 1.00 and ML bootstrap values of 95 to 100.

Aschliman's (2011) conservative topology arranged remaining batoids into a trichotomy, between (1) *Trygonorrhina* + *Zapteryx*, (2) *Pristis* + other guitarfishes, and (3) *Zanobatus* + myliobatiforms. In the present study, the modified matrix approach (plus one of the two most parsimonious trees in the unweighted analysis) indicated that *Trygonorrhina* is sister to *Zapteryx* (Figure 3.9), although this clade was unsupported by any character changes with CI ≥ 0.5. *Pristis* was very strongly supported as a member of a clade including the remaining guitarfishes in all of Aschliman's (2011) analyses. The present study is the first based on morphology to approximate this molecular hypothesis. This is at odds with previous studies suggesting that pristids are sister to all other extant batoids or to all except torpediniforms (see Section 3.2.1). It likely reflects more precise characterization of pristid morphology, as well as outgroup choices and characters that are not based in the increasingly doubted premise that batoids are derived squalean sharks (e.g., Aschliman, 2011; Maisey et al., 2004; Naylor et al., 2005; Vélez-Zuazo and Agnarsson, 2010). Also within this potential clade, Aschliman's (2011)

combined dataset and all markers except for RAG1 very strongly indicated that *Rhina* and *Rhynchobatus* are sister taxa. The interrelationships of these two genera have been difficult to resolve under morphology (Nelson, 2006).

Both morphology (McEachran and Aschliman, 2004; McEachran et al., 1996; present study) and molecules (Aschliman, 2011) indicate that *Zanobatus* is sister to myliobatiforms. The interrelationships of major myliobatiform taxa are largely unresolved by both morphology and molecules. Internodes at the base of the clade are extremely short and would permit little time for the accumulation of both anatomical and molecular synapomorphies (Aschliman, 2011). *Hexatrygon* is weakly supported as sister to all other myliobatiforms by both morphology and molecules.

Both types of data support the monophyly of *Dasyatis*, *Neotrygon*, and *Taeniura lymma*, with particularly strong support from molecules. *Pteroplatytrygon* is nested within *Dasyatis* by morphology (present study) and molecules (subsequent analyses in Aschliman, 2011). Pending corroboration by examining more species of *Dasyatis*, it should be reassigned as *D. violacea*. Amphi-American "*Himantura*" and potamotrygonids are very strongly supported as sister taxa by molecules and are united by exhibiting two cartilages in the hyomandibular–Meckelian ligament. Molecules provide moderate support for urotrygonids as sister to this clade (Aschliman, 2011). Molecular data are largely uninformative regarding the relationship of *Gymnura* to pelagic myliobatiforms, which has weak to modest support from morphology. *Rhinoptera* and mobulids are very strongly supported as sister taxa by molecules, and strongly so by morphology (Aschliman, 2011; present study).

3.5.4 Provisional Classification of Batoids

A definitive classification of batoids is not possible at this time because interrelationships have not been completely resolved, some of the resolved nodes are not robustly supported, and in some cases molecular data provide a strong and self-consistent signal at odds with morphology-based phylogenies. The following partially annotated classification is presented as a working hypothesis. It recognizes that subsuming all batoids into a single order, Rajiformes *sensu* Compagno (2005), is inherently unbalanced and based on the increasingly doubted hypnosqualean hypothesis (see Section 3.5.3). Rajiformes *sensu* Compagno (2005) encompasses most extant chondrichthyan morphological disparity and species diversity into a single order, in comparison to the eight orders into which extant sharks are currently arranged. If a Linnean system of classification is to be preserved for these taxa, it should more evenly reflect the evolutionary distinctness and morphological disparity of extant chondricththyan lineages. This provisional scheme is similar to that of McEachran and Aschliman (2004) and Nelson (2006). Some families *sensu* Compagno (2005) are here assigned as subfamilies following a preference to unite morphologically similar taxa containing relatively few genera.

Class Chondrichthyes
Subclass Neoselachii
Cohort Batoidea
Order Torpediniformes
Family Torpedinidae Bonaparte, 1838
Subfamily Torpedininae Bonaparte, 1838; 1 genus: *Torpedo* Duméril, 1806
Subfamily Hypninae Gill, 1862; 1 genus: *Hypnos* Duméril, 1852

Torpedo and *Hypnos* are indicated to be sister taxa with relatively high bootstrap support. Because Torpedinidae and Hypnidae *sensu* Compagno (2005) each comprise a single genus, they are here considered subfamilies in agreement with Nelson (2006).

Family Narcinidae Gill, 1862
Subfamily Narcininae Gill, 1862; 4 genera: *Benthobatis* Alcock, 1898; *Diplobatis* Bigelow and Schroeder, 1948; *Discopyge* Heckel, 1846; *Narcine* Henle, 1834
Subfamily Narkinae Fowler, 1934; 6 genera: *Crassinarke* Takagi, 1951 (genus questionable; see Compagno and Heemstra, 2007; Nelson, 2006); *Electrolux* Compagno and Heemstra, 2007; *Heteronarce* Regan, 1921; *Narke* Kaup, 1826; *Temera* Gray, 1831; *Typhlonarke* Waite, 1909

Narcine and *Narke* + *Temera* form a clade with very high bootstrap support; this clade is consistent with phylogenetic studies focused on torpediniforms (e.g., Carvalho, 2010). Because Narcinidae and Narkidae *sensu* Compagno (2005) comprise a total of about 37 species in nine or ten genera, they are here considered subfamilies following Nelson (2006).

Order Rajiformes
Family Pristidae Bonaparte, 1838; 2 genera: *Anoxypristis* White and Moy-Thomas, 1941; *Pristis* Latham, 1794

Pristids are here included in Rajiformes based on morphological evidence for the monophyly of this larger clade (present study), which although weak is corroborated by extensive molecular evidence suggesting that pristids are nested within guitarfishes (Aschliman, 2011).

Incertae sedis Rhina Bloch and Schneider, 1801
Incertae sedis Rhynchobatus Müller and Henle, 1837

These genera should be considered *incertae sedis* until they can be examined in greater detail. Molecular evidence suggests that they are likely sister taxa (Aschliman, 2011). Skeletal structures were largely unavailable for the present study.

Family Rhinobatidae Müller and Henle, 1837; 5 genera: *Aptychotrema* Norman, 1926; *Glaucostegus* Forsskål, 1775; *Rhinobatos* Link, 1790; *Trygonorrhina* Müller and Henle, 1837; *Zapteryx* Jordan and Gilbert, 1880

Placing these genera in the same family is highly provisional because the relationships within the inferred clade comprising *Rhinobatos*, *Trygonorrhina*, *Zapteryx*, and rajids are not fully resolved, and the clade was only recovered (with weak support) under the modified matrix or successive approximations approaches. *Aptychotrema* was not examined. Molecular evidence suggests that this nominal family may be paraphyletic to the exclusion of pristids, *Rhina*, and/or *Rhynchobatus*, or, alternatively, to the inclusion of *Trygonorrhina* + *Zapteryx*. This latter pair of genera may comprise a distinct clade of guitarfishes (Aschliman, 2011). Nelson (2006) considers *Glaucostegus* to be a synonym of *Rhinobatos*, but Last and Stevens (2009) and molecular evidence (Aschliman, 2011) suggest that it is distinct.

Family Rajidae Bonaparte, 1831
Subfamily Rajinae; 19 genera, of which 17 are named: *Amblyraja* Malm, 1877; *Anacanthobatis* von Bonde and Swart, 1923; *Breviraja* Bigelow and Schroeder, 1948; *Cruriraja* Bigelow and Schroeder, 1948; *Dactylobatus* Bean and Weed, 1909; *Dipturus* Rafinesque, 1810; *Fenestraja* McEachran and Compagno, 1982; *Gurgesiella* de Buen, 1959; *Leucoraja* Malm, 1877; *Malacoraja* Stehmann, 1970; *Neoraja* McEachran and Compagno, 1982; *Okamejei* Ishiyama, 1959; *Raja* Linnaeus, 1758; *Rajella* Stehmann, 1970; *Rostroraja* Hulley, 1972; *Sinobatis* Hulley, 1973; *Zearaja* Müller and Henle, 1841; undescribed Genus *A* Assemblage of McEachran and Dunn (1998); undescribed Genus *B* Assemblage of McEachran and Dunn (1998)
Subfamily Arhynchobatinae; 12 genera: *Atlantoraja* Menni, 1972; *Arhynchobatis* Waite, 1909; *Bathyraja* Ishiyama, 1958; *Brochiraja* Last and McEachran, 2006; *Irolita* Whitley,

1931; *Notoraja* Ishiyama, 1958; *Pavoraja* Whitley, 1939; *Psammobatis* Günther, 1870; *Pseudoraja* Bigelow and Schroeder, 1954; *Rhinoraja* Ishiyama, 1952; *Rioraja* Whitley, 1939; *Sympterygia* Müller and Henle, 1837

Morphological and molecular data are strongly at odds regarding the relationship of rajids to other extant batoids. Recent morphological phylogenies (e.g., McEachran and Aschliman, 2004; McEachran et al., 1996) suggest that rajids are nested deeply within a larger clade of guitarfishes (herein Rajiformes). The results of the present study are more ambiguous, having failed to resolve the placement of rajids in the unweighted analysis of the unmodified matrix. Molecular data indicate that rajids are a distinct lineage of batoids not closely related to extant guitarfishes and may be sister to all other extant batoids (Aschliman, 2011). Pending more detailed investigations of the discordance in signal between morphology and molecules, rajids are here provisionally assigned to Rajiformes. Compagno (1999) divided Rajidae into three families: Rajidae, Arhynchobatidae, and Anacanthobatidae Hulley, 1972. Anacanthobatidae is comprised of *Anacanthobatis*, *Cruriraja*, and *Sinobatis*, genera that are nested in Rajidae according to McEachran and Dunn (1998). *Sinobatis* was recently elevated from a subgenus of *Anacanthobatis* (Last and Séret, 2008). Elevating these three genera to familial status would make Rajidae paraphyletic. The monophyly of Anacanthobatidae *sensu* Compagno (1999) is itself in question. The three constituent genera are unique in exhibiting particularly elongated crurae, or leg-like anterior lobes of the pelvic fins, but *Cruriraja* is otherwise morphologically distinct from the others, and molecular data suggest that the family is polyphyletic (Aschliman, 2011). Rajidae, as conceived herein, is a large clade with nearly 300 described species that are morphologically very similar. It does not seem practical to divide rajids into multiple families.

Incertae sedis (not Rajiformes) *Platyrhina* Müller and Henle, 1838; *Platyrhinoidis* Garman, 1881

McEachran and Aschliman (2004) and Nelson (2006) included platyrhinids in Myliobatiformes, sister to *Zanobatus*, stingrays, and pelagic rays (= Myliobatoidei). This relationship is again recovered with moderate support and one unambiguous synapomorphy (89,3; orientation of coracohyoideus) in the present study. However, molecular data strongly indicate that platyrhinids are not sister to *Zanobatus* + Myliobatoidei but are instead distantly related to them with several intervening lineages and are possibly sister to torpediniforms (Aschliman, 2011). Reconciling these observations will require further study (see Section 3.6).

Order Myliobatiformes

Suborder Zanobatoidei

Family Zanobatidae; 1 genus: *Zanobatus* Garman, 1913

Suborder Myliobatoidei

The superfamilies of McEachran and Aschliman (2004) and Nelson (2006) are here excluded. This is a conservative measure due to uncertainty in the interrelationships of many families within Myliobatoidei under both morphology (McEachran and Aschliman, 2004; present study) and molecules (Aschliman, 2011).

Family Hexatrygonidae Heemstra and Smith, 1980; 1 genus: *Hexatrygon* Heemstra and Smith, 1980

Family Urolophidae Müller and Henle, 1841; 3 genera: *Plesiobatis* Nishida, 1990; *Trygonoptera* Müller and Henle, 1841; *Urolophus* Müller and Henle, 1837

Family Urotrygonidae McEachran et al., 1996; 2 genera: *Urobatis* Garman, 1913; *Urotrygon* Gill, 1864

Family Dasyatidae Jordan, 1888; 6 genera: *Dasyatis* Rafinesque, 1810; *Neotrygon* Castelnau, 1872; *Pastinachus* Rüppell, 1829; *Pteroplatytrygon* Fowler, 1910; *Taeniura* Müller and Henle, 1837; *Urogymnus* Müller and Henle, 1837

The composition of Dasyatidae is provisional because the node is weakly supported and most of the species have not been surveyed. *Pteroplatytrygon* is indicated to be nested within *Dasyatis* by morphology (present study) and molecules (Aschliman, 2011) but is here kept distinct until it can be formally reassigned. Molecular data suggest that *Pastinachus* and *Urogymnus* may be sister taxa and in turn sister to a clade containing the other dasyatid genera (Aschliman, 2011); these two genera are included in Dasyatidae following Nelson (2006). Molecular data (Aschliman, 2011) suggest that *Taeniura* is polyphyletic; "*Taeniurops*" appears to be reentering usage for at least *T. meyeni* but awaits formal redescription.

Family Potamotrygonidae Garman, 1913; 5 genera: *Heliotrygon* Carvalho and Lovejoy, 2011; amphi-American *Himantura* non-Müller and Henle, 1837; *Paratrygon* Duméril, 1865; *Plesiotrygon* Rosa, Castello and Thorson, 1987; *Potamotrgyon* Garman, 1877

Incertae sedis Indo–West Pacific *Himantura* Müller and Henle, 1837

Support for the sister relationship between Indo–West Pacific *Himantura* and (*Gymnura*, pelagic myliobatiforms) is weak, and few species of *Himantura* were surveyed in the present study.

> Family Gymnuridae Fowler, 1934; 1 genus: *Gymnura* Kuhl in van Hasselt, 1823
>
> Family Myliobatidae Bonaparte, 1816; 7 genera: *Aetobatus* Blainville, 1816; *Aetomylaeus* Garman, 1908; *Manta* Bancroft, 1828; *Mobula* Rafinesque, 1810; *Myliobatis* Cuvier, 1817; *Pteromylaeus* Garman, 1913; *Rhinoptera* Kuhl in Cuvier, 1829

3.6 Summary and Conclusions

Incorporating new characters from the synarcual, as well as a number of other additions, deletions, and modifications (see Section 3.4), resulted in several key changes from McEachran and Aschliman's (2004) hypothesis. *Pristis* is no longer indicated to be sister to all extant batoids except torpediniforms and may be a member of a clade including guitarfishes and rajids. There is less support for the position of rajids as deeply nested within guitarfishes, although this relationship remains supported by several potential synapomorphies. A sister relationship between *Trygonorrhina* and *Zapteryx* is weakly supported. *Pteroplatytrygon* is indicated to be nested within *Dasyatis*. As in McEachran and Aschliman (2004), torpediniforms are recovered as sister to all other extant batoids.

The evolutionary implications of the morphological phylogeny in the present study are largely consistent with those of McEachran and Aschliman (2004), albeit with added uncertainty regarding the position of rajids in the tree. Rajids and myliobatiforms are indicated to have independently achieved a strongly depressed, lozenge-shaped or circular disc generally supported to the snout tip by pectoral fin radials, and with reduced development of the tail. Other batoids retain a shark-like habitus with the tail and caudal and dorsal fins well developed and in most cases a pectoral disc that is no more than moderately expanded. Morphology (present study) suggests that rajids derived this depressed disc shape from their common ancestor with *Trygonorrhina* (and possibly *Zapteryx*), while molecules (Aschliman, 2011) suggest that it was derived from their common ancestor with a probably extinct, as yet unidentified taxon. Myliobatiforms, including *Zanobatus*, are inferred to have achieved the disc from a common ancestor either with platyrhinids (morphology) or with a clade of guitarfishes (molecules).

The phylogenies of McEachran and Aschliman (2004) and the present study suggest that morphological differences between rhinobatids (*Rhinobatos*, *Zapteryx*, and *Trygonorrhina*) and platyrhinids (*Platyrhina* and *Platyrhinoidis*) might have constrained the manner in which rajids and myliobatiforms achieved their highly depressed discs and might likewise have affected their respective modes of locomotion. Presumably, the trend for anteroposterior expansion of the scapulocoracoid evident in the (rhinobatids, rajids) clade and in the (platyrhinids, myliobatiforms) clade is related to undulatory–oscillatory modes of pectoral fin locomotion (Rosenberger, 2001a). The two clades appear to have achieved the expansion by alternative means. Rhinobatids and rajids have predominantly expanded the scapulocoracoid toward the posterior, between the mesocondyle and the metacondyle. These taxa exhibit one or more pectoral radials that articulate directly with the shoulder girdle posterior to the mesopterygium, and this condition may be the result of the posterior expansion of the scapulocoracoid without compensatory expansion of the mesopterygium.

Conversely, the platyrhinids + myliobatiforms clade appears to have expanded the scapulocoracoid to the anterior, between the procondyle and the mesocondyle. In this clade, the propterygium extends distinctly behind the procondyle in *Zanobatus*, and in other myliobatiforms additionally forms a synovial-like joint with the scapulocoracoid posterior to the procondyle (Howes, 1890; McEachran et al., 1996). It is possible that these trends established in the ancestors of rajids and of myliobatiforms constrained both their present-day anatomical structures and locomotor abilities. Expansion of the scapulocoracoid between the procondyle and the mesocondyle, followed by posterior expansion of the proximal section of the propterygium into a socketlike process and the development of a synovial joint between the scapular process and the synarcual, might have enabled the pelagic myliobatiforms to achieve an oscillatory mode of swimming.

Rosenberger and Westneat (1999) and Rosenberger (2001a) demonstrated that dasyatids (*Dasyatis*, *Pteroplatytrygon*, and *Taeniura*) increase swimming velocity by increasing fin-beat frequency and wave speed, whereas rajids (*Raja*) decrease wave number and increase wave speed. Rosenberger (2001a) concluded that this difference between dasyatids and rajids may be due to independent derivations of pectoral fin locomotion in the two clades. Both rajids and myliobatiforms have a full range of pectoral fin undulation, but only the pelagic myliobatiforms achieved pectoral fin oscillation. Differences between the two clades in linear expansion of the scapulocoracoid (toward the anterior in myliobatiforms vs. toward the posterior in rajids) might have allowed myliobatiforms to achieve oscillation and expansion into a pelagic niche while imposing a constraint against the same in rajids. It should be noted that a pelagic lifestyle evolved twice in myliobatiforms (separately in *Pteroplatytrygon violacea* and in myliobatids) (Rosenberger, 2001a).

Phylogenies estimated from morphological (McEachran and Aschliman, 2004; present study) and recent molecular (Aschliman, 2011) data are largely congruent toward the tips and in allying certain higher taxa. The datasets indicate that *Zanobatus* and myliobatiforms are sister taxa, and molecules strongly support a close relationship between pristids and at least some guitarfishes. In some cases, morphological signal is relatively ambiguous while molecular data strongly support a relationship— for example, that *Rhina* and *Rhynchobatus* are sister taxa. There are at least two elements of the phylogeny for which morphology and molecules currently appear to be difficult to reconcile, each self-consistent and with evidently strong support. The first is the position of rajids as nested within guitarfishes (morphology, several potential synapomorphies; see Section 3.5.2) or as non-monophyletic with guitarfishes (molecules, Bayesian posterior probabilities, 1.00; ML bootstrap values, 95 to 100) and possibly sister to all other extant batoids. Second is the position of platyrhinids as sister to *Zanobatus* + myliobatiforms, suggested by morphology (one unambiguous synapomorphy), or as distantly related to that clade as indicated by molecules (Bayesian posterior probabilities, 1.00; ML bootstrap values, 95 to 100). Attempting to identify potential sources of disagreement between these datasets and to reconcile them will require a thorough investigation of the underlying drivers of each: for morphology, homology assessment through ontogenetic data and interpretation against newly described fossil taxa; for molecules, identifying and controlling for potential sources of systematic error and corroboration by additional independent markers. The short internodes apparent in Aschliman's (2011) phylogenies between major batoid lineages and again among myliobatiforms suggest that few morphological synapomorphies can be predicted to have arisen during these brief periods. Further developing character sets describing structures expected to evolve quickly, such as claspers, may be a profitable means by which to better resolve relatively recent patterns of diversification.

Acknowledgments

Curators of the following museums generously provided access to specimens used in the present study: American Museum of Natural History (AMNH); Academy of Natural Sciences in Philadelphia (ANSP); Bernice P. Bishop Museum (BPBM); Biozentrum Grindel und Zoologisches Museum (ZMH); British Museum of Natural History (BMNH); California Academy of Sciences (CAS); Field Museum of Natural History (FMNH); JLB Smith Institute of Ichthyology (RUSI); Kyoto University, Department of Fisheries, Faculty of Agriculture (FAKU); Los Angeles County Museum (LACM); Museum of Comparative Zoology (MCZ); Muséum national d'Histoire naturelle (MNMH); Museum für Naturkunde (ZMB); National Museum of Natural History, Smithsonian Institution (USNM); Scripps Institution of Oceanography (SIO); Senkenberg Museum (SMF); Texas Cooperative Wildlife Collection (TCWC); and the sea-going crew at the Ecosystems Surveys Branch (ESB) of NOAA. Data used in the study were collected under grants from the National Science Foundation to JDM (DEB82-04661 and BSR87-00292). Travel funds to visit the MCZ were provided by the Ernst Mayr Grant Fund to JDM and Tsutomu Miyake. Travel funds to visit the AMNH, ANSP, FMNH, and European collections were provided by the Banks Fellowship for Paleontology, the Predoctoral Fellowship from the Society of Vertebrate Paleontology, and the University of Texas Continuing Fellowship to KMC. Dave Ebert, Dean Grubbs, Austin Mast, Gavin Naylor, and Scott Steppan provided helpful suggestions during the preparation of this manuscript.

References

Arnason, U., Gullberg, A., and Janke, A. (2001). Molecular phylogenetics of gnathostomous (jawed) fishes: old bones, new cartilage. *Zool. Scr.* 30:249–255.

Aschliman, N. (2011). The Batoid Tree of Life: Recovering the Patterns and Timing of the Evolution of Skates, Rays and Allies (Chondrichthyes: Batoidea), PhD dissertation, Florida State University, Tallahassee.

Bremer, K. (1994). Branch support and tree stability. *Cladistics* 10:295–304.

Brito, P.M. and Dutheil, D.B. (2004). A preliminary systematic analysis of Cretaceous guitarfishes from Lebanon. In: Arriata, G. and Tintori, A. (Eds.), *Mesozoic Fishes 3*. Verlag Dr. Friedrich Pfeil, Munich, pp. 101–109.

Brooks, D.R., Thorson, T.B., and Mayes, M.A. (1981). Freshwater stingrays (Potamotrygonidae) and their helminth parasites: testing hypotheses of evolution and coevolution. In: Funk, V.A. and Brooks, D.R. (Eds.), *Advances in Cladistics*. New York Botanical Garden, New York, pp. 147–175.

Capapé, C. and Desoutter, M. (1979). Etude morphologique des pterygopodes de *Torpedo* (*Torpedo*) *marmorata* Risso, 1810 (Pisces, Torpedinidae). *Neth. J. Zool.* 29:443–449.

Carpenter, J.M. (1994). Successive weighting, reliability and evidence. *Cladistics* 10:215–220.

Carroll, R.L. (1988). *Vertebrate Paleontology and Evolution*. W. H. Freeman, New York, pp. 62–83.

Carvalho, M.R. (2004). A Late Cretaceous thornback ray from southern Italy, with a phylogenetic reappraisal of the Platyrhinidae (Chondrichthyes: Batoidea). In: Arriata, G. and Tintori, A. (Eds.), *Mesozoic Fishes 3*. Verlag Dr. Friedrich Pfeil, Munich, pp. 75–100.

Carvalho, M.R. (2010). Morphology and phylogenetic relationships of the giant electric ray from the Eocene of Monte Bolca, Italy (Chondrichthyes: Torpediniformes). In: Elliot, D.K., Maisey, J., Yu, X., and Miao, D. (Eds.), *Morphology, Phylogeny and Paleobiogeography of Fossil Fishes: Honoring Meemann Chang*. Verlag Dr. Friedrich Pfeil, Munich, 448 pp.

Carvalho, M.R., Maisey, J.G., and Grande, L. (2004). Freshwater stingrays of the Green River Formation of Wyoming (Early Eocene), with the descriptions of a new genus and species and an analysis of its phylogenetic relationships (Chondrichthyes: Myliobatiformes). *Bull. Am. Mus. Nat. Hist.* 284:1–136.

Chu, Y.T. and Wen, M.C. (1980). *A Study of the Lateral-Line Canals System and That of Lorenzini Ampullae and Tubules of Elasmobranchiate Fishes of China*, Monograph of Fishes of China 2. Shanghai Science and Technology Press, Shanghai, People's Republic of China.

Claeson, K.M. (2008). Variation of the synarcual in the California ray, *Raja inornata* (Elasmobranchii: Rajidae). *Acta Geol. Pol.* 58:121–126.

Claeson, K.M. (2010). Trends in Evolutionary Morphology: A Case Study in the Relationships of Angel Sharks and Batoid Fishes, PhD dissertation, The University of Texas, Austin.

Claeson, K.M. and Hilger, A. (2011). Morphology of the anterior vertebral region in elasmobranchs: special focus, Squatiniformes. *Foss Rec.* 14(2):129–140.

Compagno, L.J.V. (1973). Interrelationships of living elasmobranchs. In: Greenwood, P.H., Miles, R.S., and Patterson, C. (Eds.), *Interrelationships of Fishes*. Academic Press, London, pp. 15–61.

Compagno, L.J.V. (1977). Phyletic relationships of living sharks and rays. *Am. Zool.* 17:303–322.

Compagno, L.J.V. (1999). Systematics and body form. In: Hamlett, W.C. (Ed.), *Sharks, Skates, and Rays: The Biology of Elasmobranch Fish*. The Johns Hopkins University Press, Baltimore, MD, pp. 1–42.

Compagno, L.J.V. (2005). Checklist of living chondrichthyan fishes. In: Fowler, S.L., Cavanagh, R.D., Camhi, M., Burgess, G.H., Caillet, G.M., Fordham, S.V., Simpfendorfer, C.A., and Musick, J.A. (Eds.), *Sharks, Rays, and Chimaeras: The Status of the Chondrichthyan Fishes*, Status Survey. International Union for the Conservation of Nature, Gland, Switzerland, pp. 401–423.

Compagno, L.J.V. and Heemstra, P.C. (2007). *Electrolux addisoni*, a new genus and species of electric ray from the east coast of South Africa (Rajiformes: Torpedinoidei: Narkidae), with a review of torpedinoid taxonomy. *Smithiana Bull.* 7:15–49.

Compagno, L.J.V. and Roberts, T.R. (1982). Freshwater stingrays (Dasyatidae) of southeast Asia, with a description of a new species of *Himantura* and reports of unidentified species. *Environ. Biol. Fish.* 7:321–339.

Compagno, L.J.V., Ebert, D.A., and Smale, M.J. (1989). *Guide to the Sharks and Rays of Southern Africa*. New Holland Publishers, London.

Davy, H. (1829). An account of some experiments on the Torpedo. *Phil. Trans. R. Soc. Lond. B* 119:15–18.

Didier, D.A. (1995). Phylogenetic systematics of extant chimaeroid fishes (Holocephali, Chimaeroidei). *Am. Mus. Novit.* 3119:1–86.

Domínguez, H.M.M. and González-Isáis, M. (2007). Contribution to the knowledge of anatomy of species of genus *Mobula* Rafinesque 1810 (Chondricthyes [sic]: Mobulinae). *Anat. Rec.* 290:920–931.

Douady, C.J., Dosay, M., Shivji, M.S., and Stanhope, M.J. (2003). Molecular phylogenetic evidence refuting the hypothesis of Batoidea (rays and skates) as derived sharks. *Mol. Phylogenet. Evol.* 26:215–221.

Dulvy, N.K. and Reynolds, J.D. (1997). Evolutionary transitions among egg-laying, live-bearing and maternal inputs in sharks and rays. *Proc. R. Soc. Lond. B Biol. Sci.* 264:1309–1315.

Dunn, K.A. and Morrissey, J.F. (1995). Molecular phylogeny of elasmobranchs. *Copeia* 1995(3):526–531.

Dunn, K.A., McEachran, J.D., and Honeycutt, R.L. (2003). Molecular phylogenetics of myliobatiform fishes (Chondrichthyes: Myliobatiformes), with comments on the effects of missing data on parsimony and likelihood. *Mol. Phylogenet. Evol.* 27:259–270.

Farris, J.S. (1969). A successive approximation approach to character weighting. *Syst. Zool.* 18:374–385.

Fechhelm, J.D. and McEachran, J.D. (1984). A revision of the electric ray genus *Diplobatis* with notes on the interrelationships of Narcinidae (Chondrichthyes, Torpediniformes). *Bull. Fla. State Mus. Biol. Sci.* 29:173–209.

Garman, S. (1913). The Plagiostoma (sharks, skates, and rays). *Mem. Mus. Comp. Zoöl. Harvard College* 36:1–528.

González-Isáis, M. and Domínguez, H.M.M. (2004). Comparative anatomy of the superfamily Myliobatoidea (Chondrichthyes) with some comments on phylogeny. *J. Morphol.* 262:517–535.

Heemstra, P.C. and Smith, M.M. (1980). Hexatrygonidae, a new family of stingrays (Myliobatiformes: Batoidea) from South Africa, with comments on the classification of batoid fishes. *Ichthyol. Bull. J.L.B. Smith Inst. Ichthyol.* 43:1–17.

Herman, J., Hovestadt-Euler, M., Hovestadt, D.C., and Stehmann, M. (1994). Contributions to the study of the comparative morphology of teeth and other relevant ichthyodorulites in living supra-specific taxa of Chondrichthyan fishes. Part B. Batomorphi No. 1a: Order Rajiformes–Suborder Rajoidei–Family Rajidae–Genera and Subgenera: *Anacanthobatis* (*Schroederobatis*), *Anacanthobatis* (*Springeria*), *Breviraja, Dactyobatus, Gurgesiella* (*Gurgesiella*), *Gurgesiella* (*Fenstraja*), *Malacoraja, Neoraja* and *Pavoraja. Bull. Inst. R. Sci. Nat. Belg. Biol.* 64:165–207.

Herman, J., Hovestadt-Euler, M., Hovestadt, D.C., and Stehmann, M. (1995). Contributions to the study of the comparative morphology of teeth and other relevant ichthyodorulites in living supra-specific taxa of Chondrichthyan fishes. Part B. Batomorphi No. 1b: Order

Rajiformes–Suborder Rajoidei–Family Rajidae–Genera and Subgenera: *Bathyraja* (with a deep-water, shallow water and transitional morphotypes), *Psammobatis, Raja (Amblyraja), Raja (Dipturus), Raja (Leucoraja), Raja (Raja), Raja (Rajella)* (with two morphotypes), *Raja (Rioraja), Raja (Rostroraja), Raja lintea,* and *Sympterygia. Bull. Inst. R. Sci. Nat. Belg. Biol.* 65:237–307.

Herman, J., Hovestadt-Euler, M., Hovestadt, D.C., and Stehmann, M. (1996). Contributions to the study of the comparative morphology of teeth and other relevant ichthyodorulites in living supra-specific taxa of Chondrichthyan fishes. Part B. Batomorphi No. 1c: Order Rajiformes–Suborder Rajoidei–Family Rajidae–Genera and Subgenera: *Arhynchobatis, Bathyraja richardsoni*-type, *Cruriraja, Irolita, Notoraja, Pavoraja (Insentiraja), Pavoraja (Pavoraja), Pseudoraja, Raja (Atlantoraja), Raja (Okamejei)* and *Rhinoraja. Bull. Inst. R. Sci. Nat. Belg. Biol.* 66:179–236.

Herman, J., Hovestadt-Euler, M., Hovestadt, D.C., and Stehmann, M. (1997). Contributions to the study of the comparative morphology of teeth and other relevant ichthyodorulites in living supra-specific taxa of Chondrichthyan fishes. Part B. Batomorphi No. 2: Order Rajiformes–Suborder Pristiodei–Family Prisidae–Genera: *Anoxypristis* and *Pristis.* No. 3: Suborder Rajoidei–Superfamily Rhinobatoidea–Families Rhinidae–Genera: *Rhina* and *Rhynchobatus,* and Rhinobatidae–Genera: *Aptychotrema, Platyrhina, Platyrhinoidis, Rhinobatos, Trygonorrhyna, Zanobatus* and *Zapteryx. Bull. Inst. R. Sci. Nat. Belg. Biol.* 67:107–162.

Herman, J., Hovestadt-Euler, M., Hovestadt, D.C., and Stehmann, M. (1998). Contributions to the study of the comparative morphology of teeth and other relevant ichthyodorulites in living supra-specific taxa of Chondrichthyan fishes. Part B. Batomorphi No. 4a: Order Rajiformes–Suborder Mylioabtoidei–Superfamily Dasyatoidea–Family Dasyatidae–Subfamily Dasyatinae–Genera: *Amphotistius, Dasyatis, Himantura, Pastinachus, Pteroplatytrygon, Taeniura, Urogymnus* and *Urolophoides* (incl. Supraspecific taxa of uncertain status and validity), Superfamily Myliobatoidea–Family Gymnuridae–Genera: *Aetoplatea* and *Gymnura,* Superfamily Plesiobatoidea–Family Hexatrygonidae–Genus *Hexatrygon. Bull. Inst. R. Sci. Nat. Belg. Biol.* 68:145–197.

Herman, J., Hovestadt-Euler, M., Hovestadt, D.C., and Stehmann, M. (1999). Contributions to the study of the comparative morphology of teeth and other relevant ichthyodorulites in living supra-specific taxa of Chondrichthyan fishes. Part B: Batomorphi No. 4b. Order Rajiformes–Suborder Myliobatoidei–Superfamily Dastyatoidea–Family Dasyatidae–Subfamily Dasyatinae–Genera: *Taeniura, Urogymnus* and *Urolophoides*–Subfamily Potamotrygoninae–Genera: *Disceus, Plesiotrygon* and *Potamotrygon* (incl. Supraspecific taxa of uncertain status and validity), Family Urolophidae–Genera: *Trygonoptera, Urolophus* and *Urotrygon*–Superfamily Myliobatidea–Family Gymnuridae–Genus: *Aetoplatea. Bull. Inst. R. Sci. Nat. Belg. Biol.* 69:161–200.

Holst, R.J. and Bone, Q. (1993). On bipedalism in skates and rays. *Phil. Trans. R. Soc. Lond. B* 339:105–108.

Howes, G.B. (1890). Observations on the pectoral fin-structure of the living batoid fishes and of the extinct genus *Squaloraja,* with especial reference to the affinities of the same. *Proc. Zool. Soc. Lond.* 675–688.

Hulley, P.A. (1970). An investigation of the Rajidae of the west and south coasts of southern Africa. *Ann. S. Afr. Mus.* 55:151–220.

Hulley, P.A. (1972). The origin, interrelationships and distribution of southern African Rajidae (Chondrichthyes, Batoidei). *Ann. S. Afr. Mus.* 60:1–103.

Human, B.A., Owen, E.P., Compagno, L.J.V., and Harley, E.H. (2006). Testing morphologically based phylogenetic theories within the cartilaginous fishes with molecular data, with special reference to the catshark family (Chondrichthyes; Scyliorhinidae) and the interrelationships within them. *Mol. Phylogenet. Evol.* 39:384–391.

Ishiyama, R. (1958). Studies on the rajid fishes (Rajidae) found in the waters around Japan. *J. Shimonoseki Coll. Fish.* 7:1–394.

Kitamura, T., Takemura, A., Watabe, S., Taniuchi, T., and Shimizu, M. (1996). Molecular phylogeny of the sharks and rays of superorder Squalea based on mitochondrial cytochrome *b* gene. *Fish. Sci.* 62(3):340–343.

Kriwet, J. (2004). The systematic position of the Cretaceous sclerorhynchid sawfishes (Elasmobranchii, Pristiorajea). In: Arriata, G. and Tintori, A. (Eds.), *Mesozoic Fishes 3.* Verlag Dr. Friedrich Pfeil, Munich, pp. 57–73.

Last, P.R. and Séret, B. (2008). Three new legskates of the genus *Sinobatis* (Rajoidei: Anacanthobatidae) from the Indo–West Pacific. *Zootaxa* 1671:33–58.

Last, P.R. and Stevens, J.D. (1994). *Sharks and Rays of Australia.* Commonwealth Scientific and Industrial Research Organisation, Canberra, Australia.

Last, P.R. and Stevens, J.D. (2009). *Sharks and Rays of Australia.* Commonwealth Scientific and Industrial Research Organisation, Canberra, Australia, 640 pp.

Last, P.R. and White, W.T. (2008). Resurrection of the genus *Neotrygon* Castelnau (Myliobatoidei: Dasyatidae) with the description of *Neotrygon picta* sp. nov., a new species from northern Australia. In: Last, P.R., White, W.T., and Pogonoski, J.J. (Eds.), *Descriptions of New Australian Chondrichthyans.* CSIRO Marine and Atmospheric Research, Hobart, Tasmania, pp. 315–325.

Leviton, A.E., Gibbs, Jr., R.H., Heal, E., and Dawson, C.E. (1985). Standards in herpetology and ichthyology. Part 1. Standard symbolic codes for institutional resource collections in herpetology and ichthyology. *Copeia* 1985:802–832.

Lovejoy, N.R. (1996). Systematics of myliobatoid elasmobranchs: with emphasis on the phylogeny and historical biogeography of neotropical freshwater stingrays (Potamotrygonidae: Rajiformes). *Zool. J. Linn. Soc. Lond.* 117:207–257.

Lucifora, L.O. and Vassallo, A.I. (2002). Walking in skates (Chondrichthyes, Rajidae): anatomy, behavior and analogies to tetrapod locomotion. *Biol. J. Linn. Soc.* 77:35–41.

Maisey, J.G. (1984). Higher elasmobranch phylogeny and biostratigraphy. *Zool. J. Linn. Soc. Lond.* 82:33–54.

Maisey, J.G., Naylor, G.J.P., and Ward, D.J. (2004). Mesozoic elasmobranchs, neoselachian phylogeny and the rise of modern elasmobranch diversity. In: Arriata, G. and Tintori, A. (Eds.), *Mesozoic Fishes 3*. Verlag Dr. Friedrich Pfeil, Munich, pp. 17–56.

McEachran, J.D. and Aschliman, N. (2004). Phylogeny of Batoidea. In: Carrier, J.C., Musick, J.A., and Heithaus, M.R. (Eds.), *Biology of Sharks and Their Relatives*, CRC Press, Boca Raton, FL, pp. 79–113.

McEachran, J.D. and Dunn, K.A. (1998). Phylogenetic analysis of skates, a morphologically conservative clade of elasmobranchs (Chondrichthyes: Rajidae). *Copeia* 1998:271–290.

McEachran, J.D. and Konstantinou, H. (1996). Survey of the variation in alar and malar thorns in skates: phylogenetic implications (Chondrichthyes: Rajoidei). *J. Morphol.* 228:165–178.

McEachran, J.D. and Miyake, T. (1990). Phylogenetic interrelationships of skates: a working hypothesis (Chondrichthyes: Rajoidei). In: Pratt, H.L., Gruber, S.H., and Taniuchi, T. (Eds.), *Elasmobranchs as Living Resources: Advances in the Biology, Ecology, Systematics, and the Status of the Fisheries*, NOAA Technical Report NMFS 90. National Oceanic and Atmospheric Administration, Washington, D.C., pp. 285–304.

McEachran, J.D., Dunn, K.A., and Miyake, T. (1996). Interrelationships of the batoid fishes (Chondrichthyes: Batoidea). In: Stiassny, M.J., Parenti, L.R., and Johnson, G.D. (Eds.), *Interrelationships of Fishes*. Academic Press, London, pp. 63–82.

Melouk, M.A. (1948). On the relation between the vertebral column and the occipital region of the chondrocranium in the Selachii and its phylogenetics significances. *Publ. Mar. Biol. Stat. Al-Ghardaqa, Egypt, Red Sea, Egypt* 6:45–51.

Miyake, T. (1988). The Systematics of the Stingray Genus *Urotrygon* with Comments on the Interrelationships within Urolophidae (Chondrichthyes: Myliobatiformes), PhD dissertation, Texas A&M University, College Station.

Miyake, T. and McEachran, J.D. (1991). The morphology and evolution of the ventral gill arch skeleton in batoid fishes (Chondrichthyes: Batoidea). *Zool. J. Linn. Soc. Lond.* 102:75–100.

Miyake, T., McEachran, J.D., and Hall, B.K. (1992a). Edgeworth's legacy of cranial development with an analysis of muscles in the ventral gill arch region of batoid fishes (Chondrichthyes: Batoidea). *J. Morphol.* 212:213–256.

Miyake, T., McEachran, J.D., Walton, P.J., and Hall, B.K. (1992b). Development and morphology of rostral cartilages in batoid fishes (Chondrichthyes: Batoidea), with comments on homology within vertebrates. *Biol. J. Linn. Soc.* 46:259–298.

Naylor, G.J.P., Ryburn, J.A., Fedrigo, O., and Lopez, A. (2005). Phylogenetic relationships among the major lineages of modern elasmobranchs. In: Hamlett, W.C. and Jamieson, B.G.M. (Eds.), *Reproductive Biology and Phylogeny of Chondrichthyes: Sharks, Batoids, and Chimaeras*, Vol. 3. Science Publishers, Enfield, NH, pp. 1–25.

Nelson, G.J. (1969). Gill arches and the phylogeny of fishes, with notes on the classification of vertebrates. *Bull. Am. Mus. Nat. Hist.* 141:477–552.

Nelson, J.S. (2006). *Fishes of the World*, 4th ed. John Wiley & Sons, Hoboken, NJ, 601 pp.

Nishida, K. (1990). Phylogeny of the suborder Myliobatoidei. *Mem. Fac. Fish. Hokkaido Univ.* 37:1–108.

Rasmussen, A. and Arnason, U. (1999). Molecular studies suggest that cartilaginous fishes have a terminal position in the piscine tree. *Proc. Natl. Acad. Sci. USA* 96:2177–2182.

Regan, C.T. (1906). A classification of the selachian fishes. *Proc. Zool. Soc. Lond.* 1906:722–758.

Reif, W.-E. (1979). Morphogenesis and histology of large scales of batoids (Elasmobranchii). *Paläontol. Z.* 53:26–37.

Rosa, R.S. (1985). A Systematic Revision of the South American Freshwater Stingrays (Chondrichthyes: Potamotrygonidae), PhD dissertation, College of William and Mary, Williamsburg, VA.

Rosa, R.S., Castello, H.P., and Thorson, T.B. (1987). *Plesiotrygon iwamae*, a new genus and species of neotropical freshwater stingray (Chondrichthyes: Potamotrygonidae). *Copeia* 1987:447–458.

Rosenberger, L.J. (2001a). Pectoral fin locomotion in batoid fishes: undulation versus oscillation. *J. Exp. Biol.* 204:379–394.

Rosenberger, L.J. (2001b). Phylogenetic relationships within the stingray genus *Dasyatis* (Chondrichthyes: Dasyatidae). *Copeia* 2001:615–627.

Rosenberger, L.J. and Westneat, M.W. (1999). Functional morphology of undulatory pectoral fin locomotion in the stingray *Taeniura lymma* (Chondrichthyes: Dasyatidae). *J. Exp Biol.* 202:3523–3539.

Shirai, S. (1992a). Phylogenetic relationships of the angel sharks, with comments on elasmobranch phylogeny (Chondrichthyes, Squatinidae). *Copeia* 1992:505–518.

Shirai, S. (1992b). *Squalean Phylogeny: A New Framework of Squaloid Sharks and Related Taxa*. Hokkaido University Press, Sapporo, pp. 1–138.

Shirai, S. (1996). Phylogenetic interrelationships of neoselachians (Chondrichthyes: Euselachii). In: Stiassny, M.J., Parenti, L.R., and Johnson, G.D. (Eds.), *Interrelationships of Fishes*. Academic Press, London, pp. 9–34.

Smith, J.L.B. (1964). Fishes collected by Dr. T. Mortenson off the coast of South Africa in 1929, with an account of the genus *Cruriraja* Bigelow and Schroeder, 1954, in South Africa. *Vidensk. Meddel. Dansk Natur. Foren.* 126:283–300.

Sorenson, M.D. and Franzosa, E.A. (2007). *TreeRot, Version 3*. Boston University, Boston, MA.

Stehmann, M. (1970). Vergleichende morphologische und anatomische Untersuchungen zur Neuordnung der Systematik der nordostatlantischen Rajidae (Chondrichthyes, Batoidei). *Arch. Wiss.* 21:73–164.

Stiassny, M.L.J., Wiley, E.O., Johnson, G.D., and de Carvalho, M.R. (2004). Gnathostome fishes. In: Cracraft, J. and Donoghue, M.J. (Eds.), *Assembling the Tree of Life*. Oxford University Press, Oxford, pp. 410–429.

Swofford, D.L. (1993). *PAUP: Phylogenetic Analysis Using Parsimony, Version 3.1.1*. Smithsonian Institution, Washington, D.C.

Swofford, D.L. (2003). *PAUP*: Phylogenetic Analysis Using Parsimony (*and Other Methods), Version 4.0b10*. Sinauer Associates, Sunderland, MA.

Vélez-Zuazo, X. and Agnarsson, I. (2010). Shark tales: a molecular species-level phylogeny of sharks (Selachimorpha, Chondrichthyes). *Mol. Phylogenet. Evol.* 58(2):207–217.

Winchell, C.J., Martin, A.P., and Mallatt, J. (2004). Phylogeny of elasmobranchs based on LSU and SSU ribosomal RNA genes. *Mol. Phylogen. Evol.* 31:214–224.

Appendix 3.1. Specimens Examined

Where possible, institutional abbreviations follow Leviton et al. (1985) and are as listed at http://www.asih.org/codons.pdf.

- **Chimaeridae:** *Hydrolagus alberti* (TCWC 10940.01); *H. collei* (AMNH 58243, AMNH 58248)

- **Heterodontidae:** *Heterodontus francisci* (TCWC 3284.01 [private collection of A. Summers AL21, AL22, P1999]); *H. mexicanus* (TCWC 7581.01)

- **Hexanchidae:** *Heptranchias perlo* (TCWC 8534.01)

- **Torpedininae:** *Torpedo californica* (MCZ 43); *T. marmorata* (MCZ 42); *T. nobiliana* (ESB 200608 23 001, ESB 200707 310 001, ESB 200707 322 001, TCWC uncataloged); *T. torpedo* (MCZ Glass 143–167, MCZ Glass 156–178, MCZ Glass 203–257, ZMB Glass 5, ZMB Glass 14a, ZMB Glass 382–407F, ZMB Glass 408–459F, ZMB Glass 460–529F, ZMB Glass 501–28or, ZMB Glass 529–49or, ZMB Glass 1295–32F, ZMB Glass 1333–70F, ZMB Glass 1371–03F, ZMB 33933); *T. tremens* (TCWC 12124.01); *Torpedo* sp. (ZMB 28/782A–C)

- **Hypninae:** *Hypnos monopterygius* (MCZ 38602); *H. "subnigrum"* (MCZ S985, ZMB 33928, ZMH 10427)

- **Narkinae:** *Heteronarce mollis* (ZMH 113459); *Narke dipterygia* (ZMB 33911); *N. japonica* (MCZ 1339); *Typhlonarke aysoni* (FAKU 46477, FAKU 47178); *Typhlonarke tarakea* (ZMH 119562); *Typhlonarke* sp. (ANSP 120293); *Temera hardwickii* (BMNH 1887.4.16.14, BMNH 1984.1.18.6)

- **Narcininae:** *Benthobatis marcida* (MCZ 41171, TCWC 442.01, TCWC 1903.01, ZMH 119660, ZMH 119661, ZMH 119863); *Diplobatis pictus* (MCZ 40377, TCWC 1900.01, TCWC 1909.01, TCWC 5291.01, ZMH 123096); *Discopyge tschudii* (CJU uncataloged, FAKU 105040, FAKU 105043, ZMH 104818); *Narcine bancroftii* (TCWC 2923.01, TCWC 6808.01, TCWC 12125.01); *N. brasiliensis* (AMNH 90769, AMNH 92321.a, AMNH 218276, TNHC 18512 A–C, ZMB 11889); *N. rierai* (ZMH 113381); *N. tasmaniensis* (AMNH 95343); *Narcine* sp. (ZMB 33929, ZMB 33930)

- **Pristidae:** *Pristis pectinata* (FMNH 1939, MCZ 36960); *Pristis pristis* (BMNH 1872.10.18:142)

- *Incertae sedis Rhina ancylostoma* (TCWC uncataloged)

- *Incertae sedis Rhynchobatus djiddensis* (MCZ 806)

- **Rhinobatidae:** *Glaucostegus granulatus* (SMF 27180); *G. typus* (AMNH 98724, SMF 8120); *Rhinobatos blochii* (ZMB 12452 Upper–Lower); *R. lentiginosus* (MCZ 51799, MCZ 57799, MCZ 153663, TCWC 2191.02); *R. percellens* (MCZ 40025); *R. planiceps* (TCWC uncataloged); *R. productus* (CAS 65978); *Trygonorrhina fasciata* (MCZ 982S); *Zapteryx exasperata* (MCZ 833S, SMF 26135a–b, SMF 30674, TCWC 7581.01); *Z. xyster* (TCWC 10846.01)

- **Rajidae:** *Amblyraja hyperborea* (TCWC 3846.01); *A. radiata* (MCZ S870, MCZ 4049, MCZ 98243 sm, MCZ 98243 md, MCZ 98243 lg, TCWC 2722.02); *Anacanthobatis americanus* (TCWC 2802.01); *A. marmoratus* (ZMH 25465); *Arhynchobatis asperrimus* (MCZ 40268); *Atlantoraja castelnaui* (TCWC uncataloged); *Bathyraja maculata* (TCWC 12040.07); *B. multispinis* (BMNH 1936.8.26.95); *B. parmifera* (TCWC 6385.01); *B. taranetzi* (RN 15060, RN 15181, UMPM 46194, ZIN 46193, ZIN 46194); *Breviraja claramaculata* (TCWC 2728.02); *Breviraja* sp. (UF 213899); *Cruriraja hulleyi* ("*parcomaculata*" *sensu* Smith 1964) (TCWC 3093.03, ZMH 122862); *C. rugosa* (UF 29861); *Dactylobatus clarkii* (TCWC 2703.01); *Dipturus batis* (TCWC 2819.05, ZMB slides 68–75, ZMB slides 76–89, ZMB slides 99–134); *D. olseni* (TCWC 6839.29); *Fenestraja plutonia* (TCWC 6964.01); *Gurgesiella atlantica* (TCWC 3364.01); *Irolita waitei* (WAM P702); *Leucoraja circularis* (MNHN 1334); *L. erinacea* (AMNH 225755SW, TCWC 5260.01); *L. naevus* (SMF 13317, SMF 17513, SMF 17514 a–b); *Malacoraja senta* (MCZ 34226, MCZ 98252 sm, MCZ 98252 lg, MCZ 162432, TCWC 4179.01); *Neoraja caerulea* (ISH 720/74, ZMH uncataloged); *Notoraja ochroderma* (CSIRO H248501); *Okamejei acutispina* (MCZ 40330); *O. kenojei* (MCZ S834, MCZ S1240, MCZ 40331, MCZ 40332, ZMB 15512), *Pavoraja alleni* (FSFRL EB-070); *Psammobatis extenta* (TCWC 3488.01); *Pseudoraja fischeri* (TCWC uncataloged); *Raja eglanteria* (TCWC 839.01); *R. inornata* (FMNH 2754 A–G), *R. miraletus* (TCWC 6454.01); *R. velezi* (SMF 30672), *R. whitleyi* (AMNH 92321 a–b), *Rajella bigelowi* (TCWC 2811.01, UF 227051); *R. fuliginea* (TCWC 2701.01); *Rhinoraja longicauda* (HUMZ 34923, HUMZ 49128); *Rioraja agassizi* (FSFRL EM-101); *Rostroraja alba* (SMF 604, TCWC 3093.04;

Sympterygia bonapartii (BMNH 1879.5.14.415, ZMH 106632); *S. brevicaudata* (TCWC 5445.01); *S. lima* (ZMH 10281, ZMH 10284, ZMH 10285)

- *Incertae sedis Platyrhina sinensis* (CAS 15919); *Platyrhinoidis triseriata* (CAS 31248)

- **Zanobatidae:** *Zanobatus schoenleinii* (USNM 222120, TCWC uncataloged)

- **Urolophidae:** *Plesiobatis daviesi* (RUSI 7861, BPBM 24578, RUSI 7861, TCWC uncataloged); *Urolophus bucculentus* (FSFRL EC-361)

- **Urotrygonidae:** *Urobatis concentricus* (LACM 31771-2; TCWC 7563.07); *U. halleri* (FMNH 42601, TCWC 7586.05); *U. jamaicensis* (TCWC 815.01); *Urotrygon aspidura* (CAS 51834, CAS 51835-13); *U. asterias* (LACM 7013-4); *U. chilensis* (FMNH 62371, FMNH 93737, LACM 7013, USNM 29542); *U. microphthalmum* (USNM 222692); *U. munda* (USNM 220612-4); *U. rogersi* (LACM W50-51-12); *U. venezuelae* (USNM 121966, TCWC 7054.02)

- **Dasyatidae:** *Dasyatis americana* (TCWC 2749.01, TCWC 5820.01); *D. brevis* (TCWC 12099.01); *D. longa* (TCWC 12102.01); *D. margarita* (TCWC 7273.01); *D. sabina* (TCWC 2790.01, TCWC 5824.01); *D. say* (FMNH 40223, TCWC 2791.01); *Neotrygon kuhlii* (TCWC 12770.01); *Pteroplatytrygon violacea* (TCWC 10251.01); *Taeniura lymma* (TCWC 5278.01, TCWC 12772.01, ZMB 4657, ZMB 5718)

- **Potamotrygonidae:** *"Himantura" pacifica* (TCWC uncataloged); *"Himantura" schmardae* (TCWC uncataloged); *Potamotrygon constellata* (MCA 2955); *P. hystrix* (ZMB 16863); *P. magdalenae* (TCWC uncataloged); *Potamotrygon* sp. (ZMB 33206)

- *Incertae sedis Himantura walga* (ZMB 21716)

- **Gymnuridae:** *Gymnura marmorata* (TCWC uncataloged); *G. micrura* (FMNH 89990, TCWC 642.08, UF 26491)

- **Myliobatidae:** *Aetobatus narinari* (MCZ 1400, TCWC 12107.01); *Mobula hypostoma* (MCZ 36406); *Myliobatis californicus* (MCZ 395, TCWC 12105.01); *M. freminvillii* (TCWC uncataloged); *M. goodei* (MCZ S638, MCZ 1343, TCWC 3699.01); *M. longirostris* (TCWC 12106.01); *Rhinoptera bonasus* (TCWC 4423.01); *R. steindachneri* (TCWC uncataloged)

Appendix 3.2. Data Matrix

See table on page 87.

Appendix 3.3. Character Descriptions

Characters Supporting Batoid Monophyly

- **Character 1.** Upper eyelid: 0 = present, 1 = absent

- **Character 2.** Palatoquadrate: 0 = articulates with neurocranium, 1 = does not articulate with neurocranium

- **Character 3.** Pseudohyal: 0 = absent, 1 = present

- **Character 4.** Last ceratobranchial: 0 = free of scapulocoracoid, 1 = articulates with scapulocoracoid

- **Character 5.** Synarcual: 0 = absent, 1 = present

- **Character 6.** Suprascapulae: 0 = not fused medially, 1 = fused medially

- **Character 7.** Synarcual lip (odontoid process): 0 = absent, 1 = present

- **Character 8.** Antorbital cartilage: 0 = free of propterygium, 1 = articulates with propterygium and nasal capsule

- **Character 9.** Levator and depressor rostri muscles: 0 = absent, 1 = present

External Morphological Structures

- **Character 10.** Cephalic lobes: 0 = absent, 1–3 = present. The pelagic stingrays (*Myliobatis, Aetobatus, Rhinoptera,* and *Mobula*) possess cephalic lobes anterior to the neurocranium supported by the pectoral girdle (McEachran et al., 1996). *Myliobatis* exhibits a single lobe that is continuous with the pectoral fin (10,1). *Aetobatus* exhibits a single lobe that is discontinuous with the pectoral fin (10,2). *Rhinoptera* and *Mobula* exhibit paired discontinuous lobes (10,3).

- **Character 11.** Anterior nasal lobe: 0 = poorly developed, 1–3 = well developed. Lobe is moderately expanded medially to cover most of the medial half of the naris and extends medially onto the internarial space in *Zapteryx, Platyrhina, Platyrhinoidis,* and *Zanobatus* (11,1) (McEachran et al., 1996, Fig. 1). In *Hexatrygon,* the anterior lobe extends medially to join its antimere and forms a nasal curtain that falls short of the mouth (11,2). Torpediniforms, *Trygonorrhina,* rajids, and myliobatiforms (except *Hexatrygon*) exhibit a nasal curtain that extends to or just anterior to the mouth (11,3) (McEachran et al., 1996, Fig. 1). In *Plesiobatis daviesi,* from South Africa, the nasal curtain falls short of the mouth according

Appendix 3.2. Data Matrix

	1	11	21	31	41	51	61	71	81
Chimaeridae	0000100000	0010200??0	0000000000	0201200000	0000000000	0200110000	0000000000	0012000000	000000000
Heterodontidae	0000000000	0000000??0	0000000000	0100000000	0000000000	??00000000	0000000000	0007000000	000000000
Chlamydoselachidae	0000000000	0000000??0	0000000000	0000000000	0000000000	??00000000	0000000000	000?????0	000000000
Hexanchidae	0000000000	0000000??0	0000000000	0000000000	0000000000	??00000000	0000000000	000?????0	000000000
Torpedo nobiliana	1111111110	3000200??1	0000000000	1101217?00	0011000210	1000010000	1000000100	1000001070	020011001
Hypnos monopterygius	1111111110	3000200??1	0000000000	1101217?00	0011000210	1000010700	1000000100	100??????0	020011001
Narcine brasiliensis	1111111110	3000200??1	0000000000	1101217?00	0000000210	1000010000	1000000100	1000102000	021011101
Narke japonica	1111111110	3000200??1	0000000000	1101217?00	0000000200	1000010000	1000000100	1000102070	021011111
Temera hartwickii	1111111110	3000200??1	0000000000	1111717?00	0000000200	1000010000	1000000100	1000102070	021011111
Pristis pectinata	1111111110	0000000200	0000001000	0000000010	0010000011	0100000100	0000100000	2010?????0	110000002
Rhynchobatus djiddensis	1111111110	0000010000	0001001100	0100000010	0000000011	2200000100	0000100101	2017?????0	110000000
Rhina ancylostoma	1111111110	0000010000	0000001100	0100000010	0000000011	2200000100	0000100107	2017?????0	110000000
Rhinobatos productus	1111111110	0000010200	0001001100	0100000010	0000000011	2000000111	0010100100	1010000000	110000000
Zapteryx exasperata	1111111110	1000010200	0002001100	0100000010	0010000011	2000000111	0000100100	1010000000	110000000
Trygonorrhina fasciata	1111111110	3000010000	0002001100	0100000010	0020000011	2000000111	0010100100	1010000000	110000000
Platyrhina sinensis	1111111110	1000010200	0002010010	0100000010	0000000011	2000000101	2110000110	1010000000	110000003
Platyrhinoidis triseriata	1111111110	1000010200	0002010010	0100000010	0020000011	2000000101	2110000110	1000100000	110000003
Zanobatus schoenleinii	1111111110	1000010200	0002022000	0100000010	0020110011	2000000171	0121010101	1000100200	110000003
Bathyraja parmifera	1111111110	3000111010	0000001100	0100000010	0010000012	2100000111	0100100100	2111003110	110000000
Raja texana	1111111110	3000111010	0001001100	0100000010	0010000012	2100000111	0000100100	2111003110	110000000
Hexatrygon bickelli	1111111100	2001200100	???22??00	0100020000	0010000013	2011100121	0121000101	10?????0	11012007?
Plesiobatis cf. daviesi	1111111100	3001100??0	1000127000	1100021000	0010100013	2011100121	0111000101	1000102200	110020023
Urolophus bucculentus	1111111100	3001210100	1000127000	1100021000	0010101013	2011100121	0111000101	1000102200	110020023
Urobatis halleri	1111111100	3101210100	1010127000	1100020000	0010100113	2011101121	0111000101	1000102200	110020023
Urotrygon munda	1111111100	3101210100	1010127000	1100020000	0010100213	2011101121	0111000101	1000102200	110020023
Pteroplatytrygon violacea	1111111100	3011210320	1000127000	1100021000	0010100113	2011101121	0131000101	1001011211	110020023
Dasyatis brevis	1111111100	3001210320	1100127000	1100020000	0010100113	2011101121	0131000101	1001011211	110020023
Dasyatis chrysonota	1111111100	3011210320	1200127000	1100020000	0010100113	2011101121	0171000101	1001001211	110020023
Dasyatis longa	1111111100	3001210320	2100127000	1100020000	0010100113	2011101121	0131000101	1001011211	110020023
Neotrygon kuhlii	1111111100	3001210100	1000127000	1100020000	0010100113	2011101121	0121000101	1000011211	110020023
Taeniura lymma	1111111100	3001210320	1000127000	1100020000	0010100113	2011101121	0121000101	1000011211	110020023
Himantura signifer	1111111100	3011210100	2100127000	1100020000	0010100113	2011101121	0131000101	1000101211	110030023
"Himantura" schmardae	1111111100	3011210??0	1000127000	1100020000	0010120113	2011101121	0121000101	102?????1	110030023
Potamotrygon magdalenae	1111111100	3011210100	3000127000	1100020000	0010120113	2011100121	0121000103	1000101201	110030023
Gymnura marmorata	1111111100	3011200100	1100127000	1100020000	0010000013	2011100171	0131011101	1000102201	110020023
Myliobatis freminvillii	1111111101	3011210320	1200127000	1100120100	1110101313	2011101171	0131011102	1000102201	110100024
Myliobatis longirostris	1111111101	3011210320	1200127000	1100120101	1110101213	2011101171	0131011102	1000102201	110100024
Aetobatus narinari	1111111102	3011200320	1200127001	1100121101	1110101313	2011101171	0131011102	1000102201	110170024
Rhinoptera steindachneri	1111111103	3011200320	1200127001	1110121101	1110101313	2011101171	0171010102	1000102201	110140024
Mobula hypostoma	1111111103	3011200??0	1200127001	1110121101	0110001313	2011101171	0171010102	1001?????1	110170024

to Compagno et al. (1989) and Nishida (1990); thus, *P.* cf. *daviesi*, which is from Hawaii, may represent another taxon.

- **Character 12.** Spiracular tentacle: 0 = absent, 1 = present. *Urobatis* and *Urotrygon* are unique in exhibiting a tentacle on the inner margin of the spiracle during their later embryonic stages (12,1) (McEachran et al., 1996).

- **Character 13.** Radial cartilages in caudal fin: 0 = present, 1 = absent. *Pteroplatytrygon*, some *Dasyatis*, *Himantura*, *Potamotrygon*, *Gymnura*, *Myliobatis*, *Aetobatus*, *Rhinoptera*, and *Mobula* lack caudal fins and caudal fin radials (13,1) (McEachran et al., 1996; Nishida, 1990).

Squamation

- **Character 14.** Serrated tail stings: 0 = absent, 1 = present. All but one myliobatiform genus and most species possess serrated stings (14,1) (McEachran et al., 1996). *Urogymnus* appears to have secondarily lost a serrated spine, as have several species within *Gymnura*, *Aetomylaeus*, and *Mobula*.

- **Character 15.** Placoid scales: 0 = uniformly present, 1–2 = limited to absent. Placoid scales are uniformly present in Galeomomorphii and Squalomorphii and are very limited in holocephalans (15,0). Rajids, with very few exceptions, are sparsely to densely covered with placoid scales on the dorsal surface only (15,1). Some genera including *Atlantoraja*, *Rioraja*, *Irolita*, *Anacanthobatis*, *Dipturus*, *Okamejei*, *Raja*, *Rostroraja*, the North Pacific Assemblage, and the amphi-American Assemblage are largely free of denticles, but this state is considered derived within Rajidae (McEachran and Dunn, 1998) and only *Anacanthobatis*, *Sinobatis*, and *Irolita* are totally free of denticles (except for alar and/or malar thorns in males). Torpediniforms and myliobatiforms (except *Plesiobatis*) are largely to totally free of denticles over the entire body surface (15,2).

- **Character 16.** Enlarged placoid scales: 0 = absent, 1 = present. According to Reif (1979), enlarged placoid scales are a derived character state of Rhinobatidae, Rajidae, and Dasyatidae. They also occur in *Rhynchobatus*, *Rhina*, *Platyrhina*, *Platyrhinoidis*, and *Zanobatus* (16,1).

- **Character 17.** Alar and malar thorns: 0 = absent, 1 = present. Alar and/or malar thorns are present in mature males of all rajids (17,1) (McEachran and Dunn, 1998; McEachran and Konstantinou, 1996).

Tooth Root Vascularization and Structure

- **Character 18.** Pulp cavities in tooth roots: 0 = large, 1–3 = elongated to absent. *Rhynchobatus*, *Rhina*, *Trygonorrhina*, and rajids have tooth roots with large pulp cavities (18,0) (Herman et al., 1994, 1995, 1996, 1997). *Hexatrygon*, *Urolophus*, *Urotrygon*, *Neotrygon*, *Himantura*, *Potamotrygon*, and *Gymnura* have broad and elongated pulp cavities in tooth roots (18,1) (Herman et al., 1997, 1998). *Pristis*, *Rhinobatos*, *Zapteryx*, *Platyrhina*, *Platyrhinoidis*, and *Zanobatus* have tooth roots with small pulp cavities (18,2), and *Dasyatis*, *Taeniura lymma*, *Myliobatis*, *Aetobatus*, and *Rhinoptera* (Herman et al., 1998, 1999) have tooth roots lacking pulp cavities (18,3).

- **Character 19.** Osteodentine: 0 = absent, 1–2 = present to widespread. Osteodentine is present in the roots of large teeth only in rajids (Herman et al., 1994, 1995, 1996), and this state is thought to be derived (19,1) and different from widespread occurrence of osteodentine in tooth roots (19,2) of *Dasyatis*, *Pteroplatytrygon*, *Taeniura lymma*, *Myliobatis*, *Aetobatus*, and *Rhinoptera* (Herman et al., 1998, 1999).

Lateral Line Canals

- **Character 20.** Cephalic lateral line canal on ventral surface: 0 = present, 1 = absent. Cephalic lateral line is present on the ventral side of the body in outgroups and all batoids except torpediniforms (20,1) (McEachran et al., 1996).

- **Character 21.** Infraorbital loop of suborbital and infraorbital canals: 0 = absent, 1–3 = present. Infraorbital loop is unique to myliobatiforms, with an unknown state in *Hexatrygon* (McEachran et al., 1996). In *Plesiobatis*, *Urolophus*, *Urobatis*, *Urotrygon*, *Pteroplatytrygon*, *Dasyatis brevis*, *Neotrygon kuhlii*, *Taeniura lymma*, and amphi-America "*Himantura*" it forms a simple posterolaterally directed loop (21,1) (Lovejoy, 1996, Figs. 3a,b; Rosenberger, 2001b). In *D. longa* and Indo–West Pacific *Himantura* it forms a complex reticular pattern or a number of loops (21,2) (Lovejoy, 1996, Fig. 3c,d; Rosenberger, 2001b). In *Potamotrygon*, the loop is directed to the anterior (21,3) (Lovejoy, 1996, Fig. 4a).

- **Character 22.** Subpleural loop of the hyomandibular canal: 0 = broadly rounded, 1–2 = not broadly rounded. Loop forms lateral hook in *Dasyatis*, Indo–West Pacific *Himantura*, and *Gymnura* (22,1) (McEachran et al., 1996, Fig. 2a; Rosenberger, 2001b). In the pelagic stingrays

(*Myliobatis*, *Aetobatus*, *Rhinoptera*, and *Mobula*), the lateral aspects of the subpleural loop are nearly parallel (22,2) (McEachran et al., 1996, Fig. 2b).

- **Character 23.** Lateral tubules of subpleural loop: 0 = unbranched, 1 = branched. In *Urobatis* and *Urotrygon*, subpleural loop exhibits dichotomously branched lateral tubules (23,1) (McEachran et al., 1996, Fig. 4).

- **Character 24.** Abdominal canal on coracoid bar: 0 = absent, 1–2 = present. Cephalic lateral line forms abdominal canal on coracoid bar in *Rhynchobatus*, *Rhinobatos*, *Trygonorrhina*, *Zapteryx*, *Platyrhina*, *Platyrhinoidis*, *Zanobatus*, and *Raja*. In *Rhynchobatus*, *Rhinobatos*, and *Raja* the canal is in a groove (24,1) (McEachran et al., 1996, Fig. 3). In *Trygonorrhina*, *Zapteryx*, *Platyrhina*, *Platyrhinoidis*, and *Zanobatus*, canals are represented by pores (24,2).

- **Character 25.** Scapular loops of scapular canals: 0 = absent, 1 = present. The trunk lateral line forms scapular loop dorsally over the pectoral girdle in myliobatiforms, with an unknown state in *Hexatrygon* (25,1) (McEachran et al., 1996, Fig. 4).

Neurocranium and Branchial Skeletal Structures

- **Character 26.** Rostral cartilage: 0 = complete, 1–2 = incomplete or absent. The rostral cartilage fails to reach the tip of the snout in *Platyrhina* and *Platyrhinoidis* (26,1) (Carvalho, 2004, Figs. 8 and 9; McEachran et al., 1996, Fig. 5). In *Zanobatus* and myliobatiforms, the rostral cartilage is either vestigial or completely lacking (26,2) (McEachran et al., 1996, Fig. 6).

- **Character 27.** Rostral node: 0 = expanded laterally, 1 = not expanded laterally. The rostral node is spatulate and flattened but not expanded laterally in *Pristis*, *Rhynchobatus*, *Rhina*, and rajids (27,1) (Garman, 1913).

- **Character 28.** Rostral appendices: 0 = absent, 1 = present. Rostral appendices are present in *Rhynchobatus*, *Rhina*, *Rhinobatos*, *Zapteryx*, *Trygonorrhina*, and rajids (28,1) (McEachran et al., 1996, Fig. 7).

- **Character 29.** Rostral processes: 0 = absent, 1 = present. Paired lateral processes originating at the ventral aspect of the rostral cartilage are present in *Platyrhina* and *Platyrhinoidis* (29,1) (Carvalho, 2004, Figs. 8 and 9; McEachran et al., 1996, Fig. 5).

- **Character 30.** Dorsolateral components of nasal capsule: 0 = absent, 1 = present. In *Rhinoptera* and *Mobula*, the dorsolateral components of the nasal capsule form a pair of projections that support the cephalic lobes or cephalic fins (30,1) (McEachran et al., 1996).

- **Character 31.** Nasal capsules: 0 = laterally expanded, 1 = ventrolaterally expanded. Nasal capsules are ventrolaterally expanded in torpediniforms and myliobatiforms, except for *Hexatrygon* (31,1) (McEachran et al., 1996).

- **Character 32.** Basal angle of neurocranium: 0 = present, 1 = absent. Basal angle on the ventral surface of the neurocranium is absent in all batoids (32,1) except *Pristis* (32,0) and present in outgroups Chlamydoselachidae and Hexanchidae, but absent in Heterodontidae (32,0) and unknown for Chimaeridae (32,?) (Compagno, 1977; Shirai, 1992b).

- **Character 33.** Preorbital process: 0 = present, 1 = absent. Preorbital process is absent in *Temera*, *Rhinoptera*, and *Mobula* (33,1) (McEachran et al., 1996).

- **Character 34.** Supraorbital crest: 0 = present, 1 = absent. Supraorbital crest is absent in torpediniforms (34,1) (McEachran et al., 1996). Chimaerids also lack the supraorbital crest (Didier, 1995).

- **Character 35.** Anterior preorbital foramen: 0 = dorsally located, 1 = anteriorly located. Anterior preorbital foramen opens on the anterior aspect of the nasal capsule in pelagic myliobatiforms (35,1) (McEachran et al., 1996). The state in torpediniforms is unknown, possibly because they lack a supraorbital crest (35,?).

- **Character 36.** Postorbital process: 0 = narrow; in otic region, 1–2 = absent or broad. Postorbital process is absent in torpediniforms (36,1). In myliobatiforms, the postorbital process is very broad and shelflike (36,2) (McEachran et al., 1996; Nishida, 1990, Figs. 10–17).

- **Character 37.** Postorbital process: 0 = separated from triangular process, 1 = fused with triangular process. Postorbital process is distally fused with the triangular process of the supraorbital crest with the groove between the processes represented by a foramen in *Plesiobatis*, *Urolophus*, *Pteroplatytrygon*, *Aetobatus*, *Rhinoptera*, and *Mobula* (37,1) (Carvalho et al., 2004; McEachran et al., 1996; Nishida, 1990, Fig. 17).

- **Character 38.** Postorbital process: 0 = projects laterally, 1 = projects ventrolaterally. Lateral margin of postorbital process is prolonged and projects ventrolaterally to form a cylindrical protuberance in *Myliobatis, Aetobatus, Rhinoptera,* and *Mobula* (38,1) (McEachran et al., 1996).

- **Character 39.** Jugal arch: 0 = absent, 1 = present. Hyomandibular facet and posterior section of otic capsule are joined by an arch in *Pristis, Rhynchobatus, Rhina, Rhinobatos, Trygonorrhina, Zapteryx, Platyrhina, Platyrhinoidis, Zanobatus,* and rajids (39,1) (McEachran et al., 1996, Fig. 7).

- **Character 40.** Antimeres of upper and lower jaws: 0 = separate, 1 = fused. Antimeres of upper and lower jaws are fused in *Aetobatus, Rhinoptera, Mobula,* and some species of *Myliobatis* (40,1) (McEachran et al., 1996; Nishida, 1990, Fig. 21).

- **Character 41.** Meckel's cartilage: 0 = not expanded medially, 1 = expanded medially. Meckel's cartilage is expanded and thickened near the symphysis in *Myliobatis, Aetobatus,* and *Rhinoptera* (41,1) (McEachran et al., 1996; Nishida, 1990, Figs. 20 and 21).

- **Character 42.** Winglike processes on Meckel's cartilage: 0 = absent, 1 = present. Meckel's cartilage has posteriorly expanded, winglike process in *Myliobatis, Aetobatus, Rhinoptera,* and *Mobula* (42,1) (McEachran et al., 1996; Nishida, 1990, Figs. 20 and 21).

- **Character 43.** Labial cartilages: 0 = present, 1 = absent. Labial cartilages are present (43,0) in the majority of squalomorphs and galeomorphs, except in *Lamna, Pseudocarcharias,* carcharhinoids (Shirai, 1992b), and chimaerids (Didier, 1995). Labial cartilages are absent in *Torpedo, Hypnos, Pristis, Zapteryx,* rajids, and myliobatiforms (43,1) (Compagno, 1977; Nishida, 1990; Shirai, 1992b).

- **Character 44.** Medial section of hyomandibula: 0 = narrow, 1 = expanded. Medial section of the hyomandibula is longitudinally expanded and spans the entire length of the oticooccipital region of the neurocranium in *Torpedo* and *Hypnos* (44,1) (McEachran et al., 1996).

- **Character 45.** Hyomandibular–Meckelian ligament: 0 = absent, 1 = present. Distal tip of the hyomandibula and Meckel's cartilage are joined by a long ligament (hyomandibular–Meckelian ligament, but called tendon in McEachran et al., 1996) in *Zanobatus, Plesiobatis, Urolophus, Urobatis, Urotrygon,* dasyatids, *Himantura,* potamotrygonids, *Myliobatis, Aetobatus,* and *Rhinoptera* (45,1) (Lovejoy, 1996, Fig. 6; McEachran et al., 1996, Fig. 8; Nishida, 1990, Figs. 20 and 21).

- **Character 46.** Ligamentous cartilage(s): 0 = absent, 1–2 = present. A broad and triangular cartilage is embedded in the posterior section of the hyomandibular–Meckelian ligament

in *Zanobatus* (46,1) (McEachran et al., 1996, Fig. 8b). Two small cartilages lie in parallel in the ligament in *"Himantura" schmardae* (western Atlantic species) and *Potamotrygon* (46,2) (Lovejoy, 1996, Fig. 6; McEachran et al., 1996; Nishida, 1990, Fig. 24).

- **Character 47.** Small cartilage associated with hyomandibular–Meckelian ligament: 0 = absent, 1 = present. Small cartilage or cartilages, free of ligament, are located between the hyomandibula and Meckel's cartilage in *Urolophus, Myliobatis, Aetobatus, Rhinoptera,* and *Mobula* (47,1) (Garman, 1913, Plates 73–75; Lovejoy, 1996, Fig. 6; McEachran et al., 1996).

- **Character 48.** Basihyal and first hypobranchial: 0 = both present and unsegmented, 1–3 = segmented or absent. The basihyal is located between the paired first hypobranchial cartilages in most neoselachians (Nelson, 1969; Shirai, 1992b). In *Urobatis,* dasyatids, *Himantura,* and potamotrygonids, the basihyal is segmented (48,1) (Lovejoy, 1996, Fig. 7; McEachran et al., 1996; Miyake and McEachran, 1991, Fig. 8; Nishida, 1990, Fig. 27). In torpediniforms and *Urotrygon,* the basihyal is absent (48,2) (McEachran et al., 1996; Nishida, 1990, Fig. 27d; Shirai, 1992b, Plate 32b, Fig. 7). In *Aetobatus, Rhinoptera,* and *Mobula,* the basihyal and first hypobranchial cartilages are absent (48,3) (McEachran and Miyake, 1991, Fig. 8; McEachran et al., 1996; Nishida, 1990, Fig. 28). *Myliobatis* lacks the basihyal and either has or lacks the first hypobranchial cartilage (48,2 or 48,3, respectively).

- **Character 49.** Ceratohyal: 0 = fully developed, 1 = reduced or absent. Ceratohyal articulates with the basihyal and hyomandibula in most neoselachians (Nelson, 1969; Shirai, 1992b). It is partially or totally replaced by the pseudohyal in all batoids (49,1) except for *Narke* and *Temera* (49,0) (Lovejoy, 1996, Fig. 7; McEachran et al., 1996; Miyake and McEachran, 1991, Fig. 6).

Synarcual and Pectoral Girdle Skeletal Structures

- **Character 50.** Suprascapulae: 0 = free of vertebral column, 1 = articulates with vertebral column, 2 = fused medially to synarcual (= pectoral arch), 3 = fused medially and laterally to synarcual. Suprascapulae of torpediniforms are connected medially and situated dorsal to the vertebral column but are completely free of the synarcual (50,0) and resting closely above the vertebral column or widely separated from it by musculature (Claeson, 2010; Regan, 1906, p. 754;

KMC, pers. obs.). The state in *Pristis* is unknown (50,?). Among guitarfishes, suprascapulae articulate with the neural arches of the vertebrae that are directly posterior to the synarcual (50,1). In rajids, the suprascapulae are fused to the median crest of the synarcual, forming the pectoral arch (50,2) (Claeson, 2008; Garman, 1913). In myliobatiforms, the suprascapulae are fused to the median crest of the synarcual and are continuous with the posterior margin of the lateral stay, creating a well-defined bridge (50,3) (Lovejoy, 1996; Miyake, 1988).

- **Character 51.** Orientation of lateral stay processes: 0 = posteriorly directed, 1 = laterally directed, 2 = dorsally directed. Torpediniforms exhibit laterally directed lateral stay processes (51,1), except for *Electrolux* and *Typhlonarke*, as these taxa lack lateral stays (51,?). Pristid synarcuals have wide lateral stays with gracile, posteriorly directed processes (51,0) (Garman, 1913, Plate 64). The lateral stays of guitarfishes, rajids and myliobatiforms have U-shaped, dorsally directed processes (51,2) (Claeson, 2008, 2010).

- **Character 52.** Vertebral–occipital articulation: 0 = synarcual lip fitted into notch in basicranium, 1 = synarcual lip rests in foramen magnum. In torpediniforms, the synarcual lip rests within an unpaired notch beneath the foramen magnum (52,0); this concavity is paired in *Torpedo*. In all examined guitarfishes and myliobatiforms, the lip fits snugly inside an unpaired concavity beneath the foramen magnum (52,0). The synarcual lip of *Pristis* and *Raja* rests inside the foramen magnum (52,1) (Claeson, 2008; Claeson and Hilger, 2011; Garman, 1913).

- **Character 53.** Ball and socket articulation between scapular process and synarcual: 0 = absent, 1 = present. Suprascapular process of the shoulder girdle forms a ball-and-socket articulation with the synarcual in myliobatiforms (53,1) (McEachran et al., 1996).

- **Character 54.** Second synarcual: 0 = absent, 1 = present. Second (thoracolumbar) synarcual, generally separated from the first synarcual by several free vertebral centra, is exhibited by myliobatiforms (54,1) (McEachran et al., 1996; Nishida, 1990, Fig. 36).

- **Character 55.** Ribs: 0 = present, 1 = absent. Ribs are absent in myliobatiforms (55,1) (McEachran et al., 1996).

- **Character 56.** Scapular process: 0 = short, 1 = long. Scapular process of the shoulder girdle is long and posteriorly displaced in torpediniforms (56,1) (McEachran et al., 1996).

- **Character 57.** Scapular process: 0 = without fossa, 1 = with fossa. Scapular process of the shoulder girdle exhibits a fossa or foramen in *Urobatis*, *Urotrygon*, dasyatids, *Myliobatis*, *Aetobatus*, *Rhinoptera*, and *Mobula* (57,1) (Lovejoy, 1996, Fig. 9; McEachran et al., 1996; Nishida, 1990, Figs. 30 and 31).

- **Character 58.** Scapulocoracoid condyles: 0 = not horizontal, 1 = horizontal. Lateral aspect of the scapulocoracoid has three horizontally arranged condyles that articulate with the propterygium, mesopterygium, and metapterygium in all batoids (58,1) except torpediniforms; however, the character state is unknown for *Hypnos* (58,?) (McEachran et al., 1996).

- **Character 59.** Mesocondyle: 0 = equidistant, 1–2 = closer to procondyle or to metacondyle. Scapulocoracoid is elongated between the mesocondyle and the metacondyle in *Rhinobatos*, *Zapteryx*, *Trygonorrhina*, and rajids (59,1) (McEachran et al., 1996, Fig. 9; Nishida, 1990, Fig. 32). In myliobatiforms, the scapulocoracoid is elongated between the procondyle and the mesocondyle (59,2) (Lovejoy, 1996, Fig. 9; McEachran et al., 1996, Fig. 9; Nishida, 1990, Figs. 30 to 32). *Zanobatus*, *Gymnura*, and the pelagic myliobatiforms exhibit fragmented or absent mesopterygia, so the state cannot be confidently determined (59,?).

- **Character 60.** Antorbital cartilage: 0 = indirectly joins propterygium to nasal capsule, 1 = directly joins cartilages. Antorbital cartilage directly joins the propterygium of the shoulder girdle to the nasal capsule in all batoids (60,1) except for torpediniforms, *Pristis*, *Rhynchobatus*, and *Rhina* (McEachran et al., 1996).

- **Character 61.** Antorbital cartilage: 0 = not anteriorly expanded, 1 = anteriorly expanded and fan- or antlerlike, 2 = wide and somewhat triangular, with posterior processes. Antorbital cartilage is anteriorly expanded and fan- or antlerlike in torpediniforms (61,1) (McEachran et al., 1996; Miyake et al., 1992b, Fig. 16). Antorbital cartilage of *Platyrhina* and *Platyrhinoidis* is wide and somewhat triangular, projecting anteriorly at least as far as posteriorly, with posterior processes (61,2) (Carvalho, 2004, Figs. 8 and 9).

- **Character 62.** Anterior extension of propterygium: 0 = pectoral propterygia fail to reach anterior margin of disc, 1 = extend to near the anterior margin of the disc. Pectoral propterygia extend to the anterior disc margin or snout tip in *Platyrhina*, *Platyrhinoidis*, *Zanobatus*,

Bathyraja, and myliobatiforms (62,1) (Carvalho, 2004, Figs. 8 and 9; Compagno, 1977; Garman, 1913; McEachran et al., 1996; Nishida, 1990).

- **Character 63.** Segmentation of propterygium: 0 = posterior to mouth, 1–3 = anterior to mouth to anterior to nasal capsule. Proximal segment of propterygium of pectoral girdle is between mouth and antorbital cartilage in *Rhinobatos, Trygonorrhina, Platyrhina, Platyrhinoidis, Plesiobatis, Urolophus, Urobatis,* and *Urotrygon* (63,1). In *Zanobatus, Hexatrygon, Neotrygon, Taeniura lymma, "Himantura" schmardae,* and *Potamotrygon,* the first segment is adjacent to the nasal capsule (63,2) (Lovejoy, 1996, Fig. 10; Rosenberger, 2001b, Fig. 2). In *Pteroplatytrygon, Dasyatis brevis, D. longa, Himantura signifer, Gymnura, Myliobatis,* and *Aetobatus,* the first segment is adjacent to anterior margin of antorbital cartilage or anterior to margin of nasal capsule (63,3) (Garman, 1913, Plate 73; Lovejoy, 1996, Fig. 10; Rosenberger, 2001b, Fig. 2).

- **Character 64.** Proximal section of propterygium: 0 = does not extend posterior to procondyle, 1 = extends behind procondyle. Proximal section of propterygium of shoulder girdle extends behind procondyle and articulates with scapulocoracoid between pro- and mesocondyles in *Zanobatus* and myliobatiforms (64,1) (Carvalho et al., 2004; Lovejoy, 1996, Fig. 10; McEachran et al., 1996, Fig. 9; Nishida, 1990, Figs. 30 and 31; Rosenberger, 2001b, Fig. 2).

- **Character 65.** Pectoral fin radials: 0 = all articulate with pterygials or directly with scapulocoracoid between propterygium and mesopterygium, 1 = one or more radials articulate with scapulocoracoid posterior to mesopterygium. Some pectoral fin radials articulate directly with scapulocoracoid posterior to mesopterygium in *Pristis,* guitarfishes, and rajids (65,1) (Garman, 1913, Plates 64 and 65; McEachran et al., 1996, Fig. 9; Nishida, 1990, Fig. 32).

- **Character 66.** Mesopterygium: 0 = present and single, 1 = fragmented or absent. Mesopterygium is fragmented or absent in *Zanobatus, Gymnura, Myliobatis, Aetobatus, Rhinoptera,* and *Mobula* (66,1) (McEachran et al., 1996, Fig. 9; Nishida, 1990, Figs. 31 and 32).

- **Character 67.** Pectoral fin radials: 0 = not expanded distally, 1 = some pectoral fin radials expanded distally. Some fin radials supported by the propterygium are expanded distally and articulate with the surface of adjacent radials in *Gymnura, Myliobatis,* and *Aetobatus* (67,1) (McEachran et al., 1996; Nishida, 1990, Fig. 34).

- **Character 68.** Paired fin rays: 0 = aplesodic, 1 = plesodic. Pectoral and pelvic fins are plesodic, radials extend to margin of fins, and ceratotrichia are reduced or absent in all batoids except *Pristis* (68,1) (McEachran et al., 1996).

Pelvic Girdle, Claspers, and Caudal Skeletal Structures

- **Character 69.** Puboischiadic bar: 0 = without postpelvic processes, 1 = with postpelvic processes. Puboischiadic bar of *Platyrhina* and *Platyrhinoidis* exhibits a pair of triangular, posteriorly directed processes (69,1) (McEachran et al., 1996; Nishida, 1990, Fig. 36q).

- **Character 70.** Puboischiadic bar: 0 = platelike, 1–3 = narrow and moderately to greatly arched. Puboischiadic bar of the pelvic girdle is narrow and moderately to strongly arched without distinct lateral prepelvic processes in *Rhynchobatus, Zanobatus, Hexatrygon, Plesiobatis, Urolophus, Urobatis, Urotrygon,* dasyatids, *Himantura,* and *Gymnura* (70,1) (Heemstra and Smith, 1980, Fig. 12; Hulley, 1970, Fig. 1; Lovejoy, 1996, Fig. 11; McEachran et al., 1996, Fig. 10; Nishida, 1990, Fig. 36; Rosenberger, 2001b, Fig. 6). In *Myliobatis, Aetobatus,* and *Mobula,* the puboischiadic bar is narrow and strongly arched, with a triangular medial prepelvic process (70,2) (Hulley, 1972, Fig. 1; Lovejoy, 1996, Fig. 11; McEachran et al., 1996; Nishida, 1990, Fig. 36). In *Potamotrygon,* the puboischiadic bar is narrow and moderately arched, with a barlike medial prepelvic process (70,3) (Lovejoy, 1996, Fig. 11; McEachran et al., 1996; Nishida, 1990, Fig. 36).

- **Character 71.** First pelvic radial: 0 = bandlike, 1–2 = variable. The first pelvic radial is thickened, variously shaped, and variably associated with distal radial segments. In torpediniforms, *Rhinobatos, Zapteryx, Trygonorrhina, Platyrhina, Platyrhinoidis, Zanobatus,* and myliobatiforms, the compound radial is bandlike and slightly expanded distally, articulating with several radial segments in parallel fashion (71,1). In rajids, the compound radial is rodlike and articulates with single radial segments in serial fashion (71,2) (Holst and Bone, 1993, Fig. 1; Lucifora and Vassallo, 2002, Fig. 2).

- **Character 72.** Pelvic girdle condyles: 0 = close together, 1 = separated. Pelvic girdle condyles for the compound radial and the basipterygium are distinctly separated and several radials articulate directly with the pelvic girdle between the two condyles in rajids (72,1).

- **Character 73.** Clasper length: 0 = short, 1 = long. Clasper is elongated and slender in Chimaeridae, *Pristis*, *Rhynchobatus*, *Rhina*, *Rhinobatos*, *Zapteryx*, *Trygonorrhina*, *Platyrhina*, and rajids (73,1) (Didier, 1995; Ishiyama, 1958; Last and Stevens, 1994). Claspers of the other outgroups and batoids are relatively short and usually rather stout (73,0) (Capapé and Desoutter, 1979, Fig. 1; Compagno and Roberts, 1982, Fig. 10; Fechhelm and McEachran, 1984, Fig. 14; Nishida, 1990, Figs. 59 and 60).

- **Character 74.** Pseudosiphon: 0 = present, 1 = absent. Clasper component pseudosiphon, a blind cavity situated on the ventromedial aspect of the clasper and formed in part by the medial margin of the ventral covering piece cartilage, is absent in rajids, *Dasyatis*, *Pteroplatytrygon*, and *Mobula* (74,1) (Hulley, 1972, Fig. 12; Nishida, 1990, Fig. 59).

- **Character 75.** Dorsal marginal clasper cartilage: 0 = lacks medial flange, 1 = possesses medial flange. The dorsal marginal clasper cartilage possesses a medial flange that extends most of the length of the cartilage in *Narcine*, *Narke*, *Temera*, *Zanobatus*, *Plesiobatis*, *Urolophus*, *Urobatis*, *Urotrygon*, *Himantura signifer*, *Potamotrygon*, *Gymnura*, *Myliobatis*, *Aetobatus*, and *Rhinoptera* (75,1) (Compagno and Roberts, 1982, Fig. 10; Hulley, 1972, Fig. 46; Nishida, 1990, Fig. 60).

- **Character 76.** Dorsal terminal cartilage: 0 = smooth margin, 1 = crenate margin. Dorsal terminal clasper cartilage has a crenate lateral margin in *Pteroplatytrygon*, some *Dasyatis*, *Neotrygon*, and *Taeniura lymma* (76,1).

- **Character 77.** Cartilage forming component claw: 0 = present, 1–3 = absent, not visible externally, or forms component shield. Cartilage is absent in *Torpedo*, dasyatids, *Himantura*, and *Potamotrygon* (77,1) (Capapé and Desoutter, 1979, Fig. 2; Nishida, 1990, Fig. 60; Rosa et al., 1987). In *Narcine*, *Narke*, *Temera*, *Urolophus*, *Urobatis*, *Urotrygon*, *Gymnura*, *Myliobatis*, *Aetobatus*, and *Rhinoptera*, cartilage is embedded in integument and is not visible externally (77,2) (Nishida, 1990, Fig. 60, as small cartilage 1). In rajids, the ventral terminal clasper cartilage lines the inner ventral margin of the clasper glans and often forms the component shield (77,3) (Hulley, 1972; Ishiyama, 1958; Stehmann, 1970; McEachran and Miyake, 1990).

- **Character 78.** Ventral terminal cartilage (accessory terminal 1 cartilage in rajids): 0 = simple, 1–2 = forming component sentinel or projection, or complex. Ventral terminal clasper cartilage is free distally and forms component sentinel or is fused with ventral marginal cartilage and forms component projection in rajids (78,1) (Hulley, 1972; Ishiyama, 1958; McEachran and Miyake, 1990; Stehmann, 1970). In *Zanobatus*, *Plesiobatis*, *Urolophus*, *Urobatis*, *Urotrygon*, dasyatids, *Potamotrygon*, *Gymnura*, *Myliobatis*, *Aetobatus*, and *Rhinoptera*, the ventral terminal cartilage is folded ventrally along its long axis to form a convex flange (78,2) (Compagno and Roberts, 1982, Fig. 10, as ventral terminal cartilage; Hulley, 1972, Fig. 46, as accessory terminal 1 cartilage).

- **Character 79.** Ventral terminal cartilage (accessory terminal 1 cartilage in rajids): 0 = attached over length to axial cartilage, 1 = free of axial. Ventral terminal clasper cartilage is free of axial cartilage in rajids, all examined dasyatids, and *Himantura* (79,1) (Hulley, 1972; Ishiyama, 1958; McEachran and Miyake, 1990; Stehmann, 1970).

- **Character 80.** Caudal vertebrae: 0 = diplospondylous, 1 = fused. Caudal vertebrae distal to serrated tail sting are fused into a tube in dasyatids, potamotryognids, *Gymnura*, and pelagic myliobatiforms (80,1) (Lovejoy, 1996, Fig. 12; McEachran et al., 1996).

Cephalic and Branchial Musculature

- **Character 81.** Ethmoideo-parethmoidalis: 0 = absent, 1 = present. The cranial muscle ethmoideo-parethmoidalis is present in all batoids (81,1) except for torpediniforms (81,0) (McEachran et al., 1996; Nishida, 1990, Figs. 43, 45, 46).

- **Character 82.** Intermandibularis: 0 = present, 1–2 = absent or modified. The mandibular plate muscle intermandibularis is present in sharks but absent in batoids (82,1) except for torpediniforms (McEachran et al., 1996). In torpediniforms, the intermandibularis is a narrow band of muscle that originates on the hyomandibula and inserts on the posterior margin of Meckel's cartilage (82,2) (McEachran et al., 1996).

- **Character 83.** Ligamentous sling on Meckel's cartilage: 0 = absent, 1 = present. In *Narcine*, *Narke*, and *Temera*, a ligamentous sling at the symphysis of Meckel's cartilage supports the intermandibularis, coracomandibularis, and depressor mandibularis muscles (83,1) (McEachran et al., 1996; Miyake et al., 1992a).

- **Character 84.** Depressor mandibularis: 0 = present, 1 = absent. The depressor mandibularis is either absent or does not exist as an independent muscle in the pelagic myliobatiforms (84,1) (McEachran et al., 1996).

- **Character 85.** Spiracularis: 0 = undivided, 1–4 = divided in various ways. In torpediniforms, the mandibular plate muscle spiracularis is divided and one bundle enters the dorsal oral membrane underlying the neurocranium (85,1) (McEachran et al., 1996; Miyake et al., 1992a). In *Plesiobatis, Urolophus, Urobatis, Urotrygon, Pteroplatytrygon, Dasyatis,* and *Neotrygon,* the spiracularis splits into lateral and medial bundles, with the medial bundle inserting onto the posterior surface of Meckel's cartilage and the lateral bundle inserting onto the dorsal edge of the hyomandibula (85,2) (McEachran et al., 1996; Miyake et al., 1992a). In *Taeniura lymma, Himantura,* and potamotrygonids, the muscle extends beyond the hyomandibula and Meckel's cartilage (85,3) (Lovejoy, 1996, Fig. 13b; McEachran et al., 1996; Miyake et al., 1992a). The spiracularis muscle is subdivided proximally and inserts separately onto the palatoquadrate and the hyomandibula in *Rhinoptera* (85,4) (McEachran et al., 1996).

- **Character 86.** Branchial electric organs: 0 = absent, 1 = present. Electric organs derived from branchial muscles are present in torpediniforms (86,1).

- **Character 87.** Coracobranchialis: 0 = consists of three to five components, 1 = single component. The coracobranchialis of the branchial muscle plate consists of a single component in *Narcine, Narke,* and *Temera* (87,1) (McEachran et al., 1996; Miyake et al., 1992a).

- **Character 88.** Coracohyomandibularis: 0 = single origin, 1–2 = separate origins. The coracohyomandibularis of the hypobranchial muscle plate has separate origins on the fascia supporting the insertion of the coracoarcualis and on the pericardial membrane in *Narke* and *Temera* (88,1) (McEachran et al., 1996). In the myliobatiforms, the muscle has separate origins on the anterior portion of the ventral gill arch region and on the pericardial membrane (88,2) (McEachran et al., 1996).

- **Character 89.** Coracohyoideus: 0 = parallel to body axis, 1–4 = absent, parallel to body axis or short, or diagonal to body axis. The coracohyoideus of the hypobranchial muscle plate is absent in torpediniforms (89,1) (McEachran et al., 1996; Miyake et al., 1992a). In *Pristis,* the muscle runs parallel to the body axis and is very short (89,2) (McEachran et al., 1996; Miyake et al., 1992a). In *Platyrhina, Platyrhinoidis, Zanobatus,* and benthic myliobatiforms (*Plesiobatis, Urolophus, Urobatis, Urotrygon, Pteroplatytrygon, Dasyatis, Neotrygon,*

Taeniura, Himantura, Potamotrygon, and *Gymnura*), the muscle runs diagonally from the wall of the first two gill slits to the posteromedial aspect of the basihyal or first basibranchial (89,3) (McEachran et al., 1996, Fig. 11b; Nishida, 1990, Fig. 53a). In pelagic myliobatiforms, each muscle fuses with its antimere at a raphe near its insertion on the first hypobranchial (89,4) (McEachran et al., 1996; Nishida, 1990, Fig. 53b).

Appendix 3.4. Apomorphies Supporting Phylogeny

Following are apomorphies supporting the phylogenies in Figures 3.7 and 3.9. Character changes with a confidence index (CI) $\geq$ 0.500 are reported for each internal branch as labeled in Figures 3.7 and 3.9; $(x > y)$ denotes a change from state x to state y.

Node 1　34 (1 > 0, CI = 0.500); 51 (0 > 1, 0.667); 55 (1 > 0, 0.500); 56 (1 > 0, 0.500)

Node 2　1 (0 > 1, 1.000); 2 (0 > 1, 1.000); 3 (0 > 1, 1.000); 4 (0 > 1, 1.000); 5 (0 > 1, 1.000); 6 (0 > 1, 1.000); 7 (0 > 1, 1.000); 8 (0 > 1, 1.000); 9 (0 > 1, 0.500); 49 (0 > 1, 0.500); 68 (0 > 1, 0.500); 71 (0 > 1, 1.000); 82 (0 > 1, 1.000)

Node 3　27 (0 > 1, 0.500); 39 (0 > 1, 0.500); 50 (0 > 1, 1.000); 51 (1 > 2, 0.667); 58 (0 > 1, 1.000); 81 (0 > 1, 1.000)

Node 4　17 (0 > 1, 1.000); 19 (0 > 1, 0.667); 50 (1 > 2, 1.000); 52 (0 > 1, 0.500); 71 (1 > 2, 1.000); 72 (0 > 1, 1.000); 77 (0 > 3, 0.500); 78 (0 > 1, 1.000)

Node 5　26 (0 > 1, 1.000); 27 (1 > 0, 0.500); 62 (0 > 1, 0.500); 65 (1 > 0, 0.500); 89 (0 > 3, 1.000)

Node 6　26 (1 > 2, 1.000); 64 (0 > 1, 1.000); 70 (0 > 1, 0.750); 78 (0 > 2, 1.000)

Node 7　9 (1 > 0, 0.500); 11 (1 > 2, 0.500); 14 (0 > 1, 1.000); 21 (0 > 1, 0.750); 25 (0 > 1, 1.000); 36 (0 > 2, 1.000); 39 (1 > 0, 0.500); 50 (1 > 3, 1.000); 53 (0 > 1, 1.000); 54 (0 > 1, 1.000); 55 (0 > 1, 0.500); 77 (0 > 2, 0.500); 85 (0 > 2, 0.667); 88 (0 > 2, 1.000)

Node 8　11 (2 > 3, 0.500); 31 (0 > 1, 0.500)

Node 9　48 (0 > 1, 0.500)

Node 10　77 (2 > 1, 0.500); 80 (0 > 1, 1.000)

Node 11　85 (2 > 3, 0.667)

Node 12　22 (0 > 1, 0.667)

Node 13　48 (1 > 0, 0.500); 66 (0 > 1, 0.500); 67 (0 > 1, 0.500); 77 (1 > 2, 0.500); 85 (3 > 0, 0.667)

Node 14 10 (0 > 1, 1.000); 19 (0 > 2, 0.667); 22 (1 > 2, 0.667); 35 (0 > 1, 1.000); 38 (0 > 1, 1.000); 41 (0 > 1, 0.500); 42 (0 > 1, 1.000); 47 (0 > 1, 0.500); 48 (0 > 3, 0.500); 70 (1 > 2, 0.750); 84 (0 > 1, 0.500); 89 (3 > 4, 1.000)

Node 15 40 (0 > 1, 1.000)

Node 16 10 (1 > 2, 1.000); 85 (0 > 4, 0.667)

Node 17 10 (2 > 3, 1.000); 30 (0 > 1, 1.000); 33 (0 > 1, 0.500); 67 (1 > 0, 0.500)

Node 18 46 (0 > 2, 1.000)

Node 19 76 (0 > 1, 0.500)

Node 20 19 (0 > 2, 0.667)

Node 21 none with CI > 0.500

Node 22 none with CI > 0.500

Node 23 22 (0 > 1, 0.667)

Node 24 12 (0 > 1, 1.000); 23 (0 > 1, 1.000)

Node 25 none with CI > 0.500

Node 26 29 (0 > 1, 1.000); 61 (0 > 2, 1.000); 69 (0 > 1, 1.000)

Node 27 20 (0 > 1, 1.000); 31 (0 > 1, 0.500); 34 (0 > 1, 0.500); 36 (0 > 1, 1.000); 48 (0 > 2, 0.500); 56 (0 > 1, 0.500); 61 (0 > 1, 1.000); 77 (0 > 1, 0.500); 82 (1 > 2, 0.500); 85 (0 > 1, 0.667); 86 (0 > 1, 1.000), 89 (0 > 1, 1.000)

Node 28 77 (1 > 2, 0.500); 83 (0 > 1, 1.000); 87 (0 > 1, 1.000)

Node 29 49 (1 > 0, 0.500); 88 (0 > 1, 1.000)

Node 30 44 (0 > 1, 1.000)

Node 31 5 (1 > 0, 1.000)

Node 32 32 (1 > 0, 0.500)

Node 33 27 (0 > 1, 1.000); 65 (0 > 1, 1.000)

Node 34 18 (2 > 0, 0.500); 28 (0 > 1, 1.000)

Node 35 none with CI > 0.500

Node 36 59 (0 > 1, 1.000); 60 (0<1, 0.500)

Node 37 11 (0 > 3, 0.500)

Node 38 none with CI > 0.500

4

Phylogeny, Biology, and Classification of Extant Holocephalans

Dominique A. Didier, Jenny M. Kemper, and David A. Ebert

CONTENTS

4.1 Introduction

The extant holocephalans belong to the Subclass Euchondrocephali, a group of chondrichthyan fishes defined by the presence of a primitive autodiastylic suspensorium with ethmoid region shifted forward. Closely related to the holocephalans, and also grouped within the Euchondrocephali, are the extinct iniopterygians as well as a group of enigmatic euchondrocephalan forms known collectively as the Paraselachii (Grogan and Lund, 2004; Grogan et al., 1999; Lund and Grogan, 1997). The holocephalans are characterized by a unique form of holostyly derived from the primitive autodiastylic condition in which the upper jaw fuses to the neurocranium, the ethmoid region is extended anteriorly, and the branchial arches are shifted anteriorly to lie below the cranium (Didier, 1995; Grogan et al., 1999). In addition, the holocephalans exhibit a trend toward reduced number of tooth elements and the formation of nonreplaceable, hypermineralized tooth plates (Didier, 1995; Grogan and Lund, 2004; Lund and Grogan, 1997; Maisey, 1986). The living holocephalans and their closest fossil relatives belong to the Order Chimaeriformes, which includes three families—Callorhinchidae, Rhinochimaeridae, and Chimaeridae—each of which is distinguished by a unique snout morphology (Figure 4.1). The living holocephalans as a group are commonly

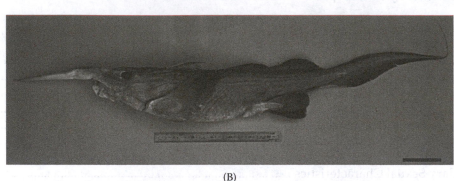

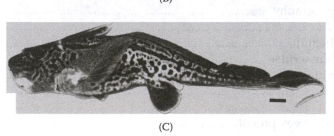

FIGURE 4.1

Representative species from each of the three families of holocephalans. (A) *Callorhinchus milii* of Family Callorhinchidae; note the plow-shaped snout and heterocercal tail. Scale = 2 cm. (Courtesy of the American Museum of Natural History.) (B) *Rhinochimaera pacifica*, a long-snouted chimaera of Family Rhinochimaeridae (CBM-ZF6140). Scale = 10 cm. (Courtesy of the Ichthyological Society of Japan.) (C) *Chimaera panthera*, one of several new species in Family Chimaeridae (NSMT P32122). Scale = 5 cm. (Courtesy of the Ichthyological Society of Japan.)

referred to as the chimaeroid fishes, or chimaeras; however; this general category for all living chimaeriform fishes is not to be confused with the chimaerid fishes, which refers specifically to members of Family Chimaeridae (also known as "ratfishes"). A current classification and a list of all valid species of chimaeroids are presented in Table 4.1.

Most early work on the morphology and relationships of chimaeroid fishes focused primarily on studies of fossil forms (summarized in Didier, 1995). Most notable was the work of Dean, who studied both fossil and living forms (Dean, 1903, 1904a–c, 1906, 1912). In terms of taxonomic work, Linnaeus (1758) described the first two species, *Callorhinchus callorynchus* and *Chimaera monstrosa*, but it was not until Bory de Saint-Vincent (1823), nearly 65 years later, that the next chimaera species was named (*Callorhinchus milii*). Throughout the remainder of the 19th century, only another six species were described, bringing the total number to nine by the turn of the century. A large number of chimaeroid species

were described in the early part of the, 20th century (Collett, 1904; Fowler, 1910; Garman, 1908; Gilbert, 1905; Holt and Byrne, 1909; Jordan and Snyder, 1900, 1904; Tanaka, 1905), doubling the number of known species to 18. Garman (1911) provided the first taxonomic review of the chimaeroid fishes that later served as the basis of Fowler's review published in 1941, and in 1953 Bigelow and Schroeder published a taxonomic summary of chimaeroid fishes of the northwestern Atlantic. In the 80-year time span from 1920 until 2000, 12 additional new species were described (Bigelow and Schroeder, 1951; Bullis and Carpenter, 1966; Compagno et al., 1990; de Buen, 1959; Didier, 1998; Didier and Stehmann, 1996; Gilchrist, 1922; Hardy and Stehmann, 1990; Howell-Rivero, 1936; Karrer, 1972; Schnakenbeck, 1929; Whitley, 1939), bringing the total number to 30 at the close of the 20th century. The first decade of the 21st century has seen another resurgence in chimaeroid taxonomy, with deep-water exploration, as well as emerging deep-water fisheries, resulting in the discovery of many new species (e.g., Barnett et al., 2006; Compagno et al., 1990; Didier, 2008; Didier and Séret, 2002; Didier et al., 2008; Ebert, 2003; James et al., 2009; Kemper et al., 2010a,b; Last and Stevens, 2009; Paulin et al., 1989; Quaranta et al., 2006). In the last decade, between 2002 and 2011, 17 new species have been described, and today there are 47 described species of extant holocephalans (Table 4.1).

4.2 General Morphological Features

4.2.1 External Features

The chimaeroid fishes are characterized by long, tapering bodies and large heads. Adults range in size from small-bodied slender fishes averaging around 60 cm in total length including the whip-like tail (e.g., *Hydrolagus mirabilis*) to massive fishes exceeding 1 m in length with large bulky heads and bodies (e.g., *Chimaera lignaria*, *Hydrolagus affinis*). The skin is completely scaleless in adults; it can be quite fragile in some species and more elastic in others. The terms *deciduous* or *nondeciduous* are used to describe skin condition based on whether or not the skin easily flakes off in patches or remains intact. Hatchlings and small juveniles have tiny denticles embedded in the skin along the dorsal surface of the trunk and head. These denticles are arranged in a horseshoe shape atop the head anterior to the first dorsal fin and in two rows between the first and second dorsal fins and between the second dorsal and upper caudal fins. These denticles are lost at the time of hatching or very shortly thereafter.

The gill arches are concentrated underneath the neurocranium and covered by a fleshy operculum supported by cartilaginous rays extending off of the hyoid arch. The operculum in holocephalans evolved in conjunction with gill arch appendage reduction that is genetically controlled by expression of Sonic hedgehog (Gillis et al., 2011). Respiratory water passes from the mouth cavity through the several interbranchial spaces into a small common parabranchial chamber, bounded laterally by the operculum. Unlike in sharks and rays, this chamber has only a single exit: a small opercular opening located anterior to the pectoral fin base on each side of the body. Current research on the ventilation anatomy and mechanics indicates that the ventilation process in chimaeras is as divergent as the anatomy, characterized by ventilatory pressures that are one to two orders of magnitude smaller than those observed in other fishes (M. Dean, Max Planck Institute of Colloids and Interfaces, pers. comm.). Adults lack a spiracle, although it is present in embryos (Didier et al., 1998). The ventrally positioned mouth is small, connected to the nostrils by deep grooves. A complex arrangement of labial cartilages supports the labial folds, forming deep naso-oral grooves through which water flows via buccopharyngeal pumping into the nostrils (Didier, 1995; Lisney, 2010). The incisor-like anterior tooth plates and large nostrils give the appearance of a rabbitlike mouth, and the common name "rabbitfish" is often applied to members of Family Chimaeridae.

Most species possess large elliptical eyes, although small eyes distinguish a few species. The eyes of chimaeroids possess a well-developed reflective tapetum, which gives them their characteristic green eyeshine. Variation in tapetal structure suggests interspecific differences that may relate to visual sensitivity of species living at different depths (Lisney, 2010). In addition to anatomical differences in eye structure, it appears that larger eyes tend to evolve in species that are more reliant on vision, and the differences in eye size of chimaeroids may be correlated with depth range and habitat (Lisney, 2010; Lisney and Didier, unpublished data).

4.2.1.1 Lateral Line Canals

The first comparative analysis of lateral line canals in chondrichthyans included descriptions of the morphology in several species of chimaeroids (Garman, 1888). Other early studies focused on innervation of the canals (Cole 1896a,b; Cole and Dakin, 1906) and histology (Reese, 1910). Reese (1910) was the first to note the morphological differences between the undilated and dilated canals, which he distinguished as "type 1" and "type 2" canals, respectively. More recently, several studies have focused on lateral line canal and neuromast morphology and evolution in primitive fishes, including a variety of elasmobranchiomorphs, but not chimaeroids specifically (e.g., Maruska and Tricas, 1998; Northcutt, 1989; Peach and Rouse, 2000; Webb and Northcutt, 1997).

With the exception of the callorhynchids, which have enclosed, pored, lateral line canals that sit in the dermis, the living chimaeroid fishes have a unique lateral line system on the head and trunk that is composed of a series of open grooves supported by open, C-shaped, cartilaginous rings. Members of the Family Chimaeridae are further distinguished by a modification of the grooves on the head in which the anterior portions that extend onto the snout are widened and have enlarged dilations between small series of cartilaginous rings. The morphology of the calcified rings is significant at higher taxonomic levels, but the number of cartilaginous rings between dilations varies between three and seven for all species examined and does not seem to be a useful character for species identification. Preliminary evidence suggests that the number of lateral line canal dilations in chimaerids is correlated with size and may be significant for generic distinctions, but

TABLE 4.1

Classification and Geographic Range of Species of Extant Holocephali

Class Chondrichthyes
  Subclass Euchondrocephali
    Order Chimaeriformes
      Family Callorhinchidae
        Genus *Callorhinchus*
          Callorhinchus callorhynchus (Linnaeus, 1758)
            American elephantfish; South America; Argentina, Chile, Peru
          Callorhinchus capensis (Dumeril, 1865)
            Cape elephantfish or St. Joseph; South Africa
          Callorhinchus milii (Bory de St. Vincent, 1823)
            Elephantfish; Southwestern Pacific: Southern New Zealand and Australia
      Family Rhinochimaeridae
        Genus *Harriotta*
          Harriotta haeckeli (Karrer, 1972)
            Small-spine chimaera; Northeastern Atlantic, Southwestern Pacific: New Zealand
          Harriotta raleighana (Goode and Bean, 1895)
            Narrow-nose chimaera; Atlantic and Pacific Ocean/circumglobal
        Genus *Neoharriotta*
          Neoharriotta carri (Bullis and Carpenter, 1966)
            Caribbean chimaera; Western Central Atlantic: Caribbean Sea
          Neoharriotta pinnata (Schnakenbeck, 1929)
            Sickle-fin chimaera; Eastern Atlantic off West Africa
          Neoharriotta pumila (Didier and Stehmann, 1996)
            Dwarf chimaera; Northern Indian Ocean
        Genus *Rhinochimaera*
          Rhinochimaera africana (Compagno, Stehmann, and Ebert, 1990)
            Paddle-nose chimaera; Southeast Atlantic, Indian Ocean, Western North Pacific: Taiwan, Japan, southern Africa;
             Southeastern Pacific: Peru
          Rhinochimaera atlantica (Holt and Byrne, 1909)
            Atlantic longnose chimaera; Atlantic Ocean
          Rhinochimaera pacifica (Mitsukuri, 1895)
            Pacific longnose chimaera; Pacific Ocean
      Family Chimaeridae
        Genus *Chimaera*
          Chimaera argiloba (Last, White, and Pogonoski, 2008)
            White-fin chimaera; Eastern Indian Ocean: Western Australia
          Chimaera bahamaensis (Kemper, Ebert, Didier, and Compagno, 2010)
            Bahamas ghostshark; Western North Atlantic: Bahamas Islands
          Chimaera cubana (Howell-Rivero, 1936)
            Cuban chimaera; Western Central Atlantic: Caribbean Sea
          Chimaera fulva (Didier, Last, and White, 2008)
            Southern chimaera; Eastern Indian and Southwestern Pacific: southern Australia, Tasmania
          Chimaera jordani (Tanaka, 1905)
            Jordan's chimaera; Western North Pacific: Japan, possibly western Indian Ocean
          Chimaera lignaria (Didier, 2002)
            Carpenter's chimaera; Southwestern Pacific: Tasmania, Australia, New Zealand
          Chimaera macrospina (Didier, Last, and White, 2008)
            Long-spine chimaera; Eastern Indian and Southwestern Pacific: Australia
          Chimaera monstrosa (Linnaeus, 1758)
            Rabbitfish; Northern Atlantic, Mediterranean
          Chimaera notafricana (Kemper, Ebert, Compagno, and Didier, 2010)
            Cape chimaera; Southeastern Atlantic: Southern Africa
          Chimaera obscura (Didier, Last, and White, 2008)
            Short-spine chimaera; Southwestern Pacific: Australia
          Chimaera opalescens (Luchetti, Iglésias, and Sellos, 2011)
            Opal chimaera; Eastern North Atlantic: British Isles, France, Greenland
          Chimaera owstoni (Tanaka, 1905)
            Owston's chimaera; Western North Pacific: Japan

TABLE 4.1 (continued)

Classification and Geographic Range of Species of Extant Holocephali

---

Chimaera panthera (Didier, 1998)
  Leopard chimaera; Southwestern Pacific: New Zealand
Chimaera phantasma (Jordan and Snyder, 1900)
  Silver chimaera; Western North Pacific
Genus *Hydrolagus*
Hydrolagus affinis (Brito Capello, 1867)
  Small-eyed chimaera; North and South Atlantic
Hydrolagus africanus (Gilchrist, 1922)
  African rabbitfish; Southeastern Atlantic: Southern Africa
Hydrolagus alberti (Bigelow and Schroeder, 1951)
  Gulf chimaera; Western Central Atlantic: Gulf of Mexico, Caribbean Sea
Hydrolagus alphus (Quaranta, Didier, Long, and Ebert, 2006)
  White-spot ghostshark; Southeastern Pacific: Galapagos Islands
Hydrolagus barbouri (Garman, 1908)
  Nine-spot chimaera; Western North Pacific: Japan
Hydrolagus bemisi (Didier, 2002)
  Pale ghostshark; Southwestern Pacific: New Zealand
Hydrolagus colliei (Lay and Bennett, 1839)
  White-spotted chimaera; Eastern North Pacific; Gulf of Alaska to Gulf of California
Hydrolagus homonycteris (Didier, 2008)
  Black ghostshark; Southwestern Pacific: Tasmania, Australia, New Zealand
Hydrolagus lemures (Whitley, 1939)
  Black-fin ghostshark; Eastern Indian, Western Central and Southwestern Pacific: Australia
Hydrolagus lusitanicus (Moura, Figueiredo, Bordalo-Machado, Almeida, and Gordo, 2005)
  Portuguese chimaera; Eastern North Atlantic: Portugal
Hydrolagus macrophthalmus (de Buen, 1959)
  Big-eye chimaera; Eastern Central and Southern Pacific: Mexico to Chile
Hydrolagus marmoratus (Didier, 2008)
  Marbeled ghostshark; Southwestern Pacific: Eastern Australia
Hydrolagus matallanasi (Soto and Vooren, 2004)
  Striped chimaera; Southwest Atlantic: southern Brazil
Hydrolagus mccoskeri (Barnett, Didier, Long, and Ebert, 2006)
  Galapagos ghostshark; Southeastern Pacific: Galapagos Islands
Hydrolagus melanophasma (James, Ebert, Long, and Didier, 2009)
  Eastern Pacific black ghostshark; Eastern Pacific: Southern California to Peru
Hydrolagus mirabilis (Collett, 1904)
  Large-eyed rabbitfish; Northern Atlantic, Gulf of Mexico
Hydrolagus mitsukurii (Dean, 1904)
  Mitsukuri's chimaera; Western North Pacific: Japan to the Philippines
Hydrolagus novaezealandiae (Fowler, 1910)
  New Zealand chimaera; Southwestern Pacific: New Zealand
Hydrolagus ogilbyi (Waite, 1898)
  Ogilby's ghostshark; Southeastern Australia
Hydrolagus pallidus (Hardy and Stehmann, 1990)
  Pale chimaera; Eastern North Atlantic
Hydrolagus purpurescens (Gilbert, 1905)
  Purple chimaera; Central and Western North Pacific: Hawaii, Japan
Hydrolagus trolli (Didier and Seret, 2002)
  Pointy-nosed blue chimaera; Southwest and Western Pacific: New Zealand, New Caledonia, possibly Hawaii and Eastern Pacific

---

the number of canal dilations has not proved useful for species determination (Wilmot et al., 2001). The pattern of lateral line canals on the head, in particular the branching pattern of canals below the eye, can be useful for distinguishing species; however, due to intraspecific variation that may occur, this feature is most useful in combination with other characters (Didier, 1995, 1998; Didier and Nakaya, 1999; Didier and Séret, 2002; Didier and Stehmann, 1996). The terminology shown in Figure 4.2 follows a historical morphological approach to canal

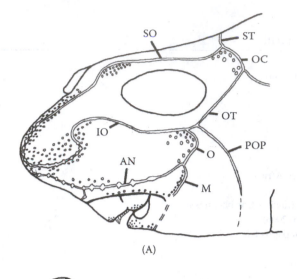

(A)

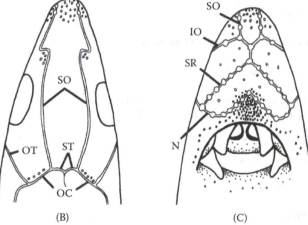

(B)   (C)

FIGURE 4.2
General pattern of lateral line canals on the head and snout of chimaeroid fishes as shown in a representative chimaerid fish: (A) lateral view, (B) dorsal view, (C) ventral view. Canals generally follow the same basic pattern in callorhinchids and rhinochimaerids although the canal positions will vary slightly due to elongation of snouts in members of these two families. *Abbreviations:* AN, angular; IO, infraorbital; M, mandibular; N, nasal; O, oral; OC, occipital; OT, otic; POP, preopercular; SO, suborbital; SR, subrostral; ST, supratemporal. (Courtesy of the American Museum of Natural History.)

designation and is based on Garman (1888) with terminological modifications from Didier (1995, 1998) and Compagno et al. (1990). A different terminology adopted by Fields et al. (1993) was based on a study of the nervous innervation of the canals in *Hydrolagus colliei*. Although terminology based on nervous innervation is more informative in terms of understanding homology, it cannot at present be adapted to represent the variation of canal branching patterns observed among all species of chimaeroids.

Studies on the functional capabilities of the mechanoreceptive lateral line in chimaeroid fishes have not been conducted; however, research on lateral line function in

other chondrichthyan species indicates that open canals and canals with dilations have different functional capabilities than enclosed canals, suggesting that open and dilated canals may confer a sensory advantage in low light conditions (Lisney, 2010). Additionally, the spatial distribution of the lateral line canals on expanded or elongate snouts, such as in the rhinochimaerids, also may serve to enhance sensory input (Lisney, 2010). Adjacent to the lateral line canals of the head are clusters of ampullary pores that have most recently been described in detail for *Hydrolagus colliei* (Fields et al., 1993). Morphological, behavioral, and neurophysiological studies of *H. colliei* confirm that these ampullary structures respond to electric fields and are homologous to the ampullae of Lorenzini in elasmobranchs (Fields and Lange, 1980). The distribution and number of ampullary pores vary among species of chimaeroids and therefore have not proved to be of taxonomic significance. Research on a variety of chondrichthyans has shown that differences in density and size of ampullary pores may relate to functional and behavioral differences (Lisney, 2010).

4.2.1.2 Fins and Fin Spines

All chimaeroids possess two dorsal fins, a caudal fin, and paired pectoral and pelvic fins, all with delicate fin webs supported by cartilaginous rays (ceratotrichia). The first dorsal fin is erectile, triangular in shape, and preceded by a long, stout spine. The dorsal fin spine is triangular in cross section with a keel-like ridge along the anterior surface and two rows of serrations present along the distal posterior edge of the upper half of the spine. The lower half of the spine remains attached to the first dorsal fin. As individuals mature, the anterior keel becomes worn away, so the spine in large adults is often smooth or possesses only a very narrow keel, whereas juveniles and subadults will generally have a much more prominent keel. Likewise, the serrations along the distal posterior edge of the spine become worn with age; therefore, juveniles will have obvious serrations in two rows on the posterior surface of the spine while the serrations in adults are not as prominent and may be partly or mostly worn away. The fin spine of *Hydrolagus colliei* has been found to be mildly venomous (Halstead and Bunker, 1952). The fin spines of other species of chimaeroids, such as *Callorhinchus milii*, are known to inflict painful wounds that result in several days of swelling and redness (DD, pers. obs.), and it is likely that the spines of most species of chimaeroids are venomous.

The second dorsal fin in most species is elongate and in some species may have a central indentation that nearly separates the second dorsal fin into two parts. All callorhinchids and *Neoharriotta* species possess a prominent anal fin with an internal skeletal structure. The anal fin in chimaerids is small and lacks internal skeletal support. The caudal fin in most species is leptocercal with upper and lower caudal lobes of nearly equal size, although in some species of rhinochimaerids the caudal fin appears externally heterocercal. The callorhinchids are the exception in having a heterocercal tail. All species possess a distal caudal filament, which is usually in the form of an elongate whip but may be quite short or absent in the adults of some species. The pectoral and pelvic fins are nearly uniform in shape for all species of chimaeroids. The large, broad triangular pectoral fins look and function like wings, propelling the fish underwater by a flapping motion. The much smaller pelvic fins are usually squared or rounded along their distal edge. In a few species, the shape of the pelvic fins is sexually dimorphic.

4.2.2 Skeleton

The skeleton is completely cartilaginous; however, like their elasmobranch relatives, chimaeroids possess calcified tissues in the dentition, denticles, and fin spines. Several unique skeletal features characterize the Holocephali; the most distinctive is the holostylic jaw in which the palatoquadrate is completely fused to the neurocranium. Holostyly in chimaeroids is derived from an ancestral autodiastylic state (Grogan et al., 1999; Lund and Grogan, 1997). Related to the holostylic jaw, and perhaps the most unusual skeletal feature, is the complete nonsuspensory hyoid arch. Articulating with the hyoid arch is the opercular cartilage and hyoid rays, which support the fleshy operculum. In addition to the hyoid arch are five regular gill arches, which are concentrated beneath the neurocranium.

Several unique features characterize the neurocranium of chimaeroids. Anterior to the orbits and dorsal to the nasal capsules is the ethmoid canal, which is an enclosed passage for nerves and blood vessels to the anterior-most region of the snout. The snout is supported by three rostral cartilages arranged as a single dorsal cartilage and paired ventral cartilages. An interorbital septum separates the orbits and is formed by a sheet of dense connective tissue rather than cartilage. Articulating with the occipital region of the neurocranium is the synarcual formed by the fusion of the first ten vertebral segments. The fin spine and first dorsal fin articulate with the synarcual. True vertebral centra are absent from the vertebral column of chimaeroids. Within the notochordal sheath are calcified rings, which are not segmentally organized but appear to increase in number and density as the fish mature. Callorhynchids lack these notochordal rings. Other features of the skeletal anatomy of chimaeroids are summarized in Didier (1995).

4.2.3 Tooth Plates

Holocephalans are characterized by the possession of ever-growing, nonreplaceable hypermineralized tooth plates. All chimaeroids have six tooth plates in three pairs, a single pair in the lower jaw and two pairs in the upper jaw. The lower mandibular tooth plates are characterized by a large symphysial tritor, which is sculpted into a prominent point at the symphysial edge, and together the mandibular tooth plates form a distinct double-pointed beak at the symphysis. Incisor-like vomerine tooth plates are located at the anterior edge of the upper jaw and occlude with the mandibular tooth plates. Together the mandibular and vomerine tooth plates form a beaklike bite. Posterior to the vomerine tooth plates are the palatine tooth plates that lie flat on the roof of the mouth and occlude with the tongue and posterior edges of the mandibular tooth plates.

Because they fossilize well, the tooth plates are among the only fossil remains that are known to exist and have long been central to evolutionary studies of holocephalans (e.g., Bendix-Almgreen, 1968; Dean, 1906, 1909; Lund, 1977, 1986, 1988; Moy-Thomas, 1939; Ørvig, 1967, 1985; Patterson, 1965; Zangerl, 1981; see Stahl, 1999, for a review). Of particular interest are the mineralized tissues of the tooth plates. The bulk of the tooth plate is comprised of a matrix of trabecular dentine (Peyer, 1968; "osteodentine" of Ørvig, 1967) surrounding hypermineralized tritors that are composed of a tissue that has been identified by various workers as tubular dentine (Moy-Thomas, 1939), pleromin (Ørvig, 1967, 1985), and orthotrabeculine (Zangerl et al., 1993). On the oral surface of the tooth plate, the tritors exhibit two distinct morphologies. Hypermineralized rods (Didier, 1995) are usually located at or near the edge of the tooth plate and appear as beads on a string ("pearlstrings" of Bargmann, 1933), while hypermineralized pads (Didier, 1995) are single large tritors located at or near the center of the tooth plate.

Details of the orientation, development, and growth of the tooth plates are important for understanding the evolution of the holocephalan dentition. Schauinsland (1903) observed that each tooth plate developed from a single primordium; therefore, it was interpreted that tooth plates probably did not evolve from separate tooth primordia like tooth families. A comparison of the development of the tooth plates of lungfishes and chimaeroids supported this hypothesis (Kemp, 1984). However, new embryological studies have shown that chimaeroid tooth plates exhibit a compound structure with individual tooth plates formed from multiple growth regions, suggesting that tooth plates may represent the fusion of members of a tooth family (Didier et al., 1994). Further support that tooth plates are derived from an ancestral chondrichthyan dentition is based on a reinterpretation of the growth and orientation of chimaeroid tooth plates and a new, more informative nomenclature for tooth-plate surfaces that indicates that the chimaeroid dentition is lyodont (growing in a lingual to labial direction) and is similar to that of other chondrichthyans (Patterson, 1992).

4.2.4 Secondary Sexual Characteristics

Males and females are sexually dimorphic, and males possess several secondary sexual structures, including a frontal tenaculum, paired prepelvic tenacula, and paired pelvic claspers. Juvenile males lack the frontal tenaculum but have tiny developing pelvic claspers and small slitlike pouches on the ventral surface of the trunk. Development of the frontal tenaculum and growth of the prepelvic and pelvic claspers occur as sexual maturity is reached. Most early studies of secondary sexual characteristics focused on morphology and histology of the urogenital system, and these are summarized in the more recent comprehensive morphological work of Stanley (1963).

The frontal tenaculum is a small clublike structure with a bulbous tip armed with numerous sharp denticles located on top of the head just anterior to the eyes. This structure, unique to chimaeroid fishes, has long been assumed to play a role in mating, and only recently has it been observed that males use the frontal tenaculum to grasp the pectoral fin of the female during copulation (D. Powell, Monterey Bay Aquarium, pers. comm.). The frontal tenaculum varies among species (Didier, 1995) and may be a useful character for species identification when considered in combination with other characters (e.g., Didier and Séret, 2002).

Paired prepelvic tenaculae are in the form of flat, spatulate blades with a row of prominent denticles along the medial edge. The prepelvic tenaculae articulate with the anterior edge of the pelvic girdle and are housed in pouches on the ventral side of the trunk. As they emerge from their pouches, the tenaculae flex anteriorly and aid in anchoring the male to the ventral side of the female. The number of denticles on the medial edge of prepelvic tenaculae ranges from five to seven in every adult male specimen examined. Denticles are always located in a single line along the medial edge, with the largest spines most proximal and distal spines the smallest. The only variation on this pattern occurs in *Hydrolagus africana*, and the presence of additional denticles on the prepelvic tenaculae appears to be a diagnostic character for this species.

The pelvic claspers extend from the medial edge of the pelvic fins and serve to transport sperm to the oviducts of the female. A comparative morphological study of the secondary sexual characteristics of elasmobranchs included descriptions of the morphology of pelvic claspers in several species of chimaeroids (Leigh-Sharpe, 1922, 1926). Pelvic claspers are phylogenetically useful

characters (Didier, 1995); however, for species identification, clasper characters are useful only when considered with other characters that are found in both males and females (e.g., Didier, 1998, 2002; Didier and Séret, 2002). Within the Family Chimaeridae, males have pelvic claspers that are described as either bifurcate or trifurcate, and this feature has been used as diagnostic characters at the genus level (Fowler, 1941; Garman, 1911). The taxonomic value of the bifurcate vs. trifurcate condition needs to be reexamined. Pelvic claspers of chimaerids consist of an internal cartilaginous support that divides into two branches, each with a fleshy denticulate lobe at its distal end (Figure 4.3). In some species, a separate fleshy lobe, continuous with the fleshy tissue of the lateral branch, encircles the base of the medial branch. This fleshy lobe usually lacks internal cartilaginous support, but in some species a thin strip of cartilage that does not appear to originate from or articulate with the internal skeleton of the pelvic clasper may support this third fleshy lobe. In most individuals, the fleshy lobe is closely associated with the medial branch of the clasper, but in some it can be separated by the clasper groove, thus appearing as a third branch. The clasper groove is unrelated to the actual bifurcation of the clasper itself and runs the entire length of the clasper. Whether claspers are interpreted as bifurcate vs. trifurcate may depend on how visible or separated the third fleshy lobe appears. A detailed morphological comparison of the pelvic claspers in chimaerids would help clarify the morphological distinction between bifurcate and trifurcate claspers.

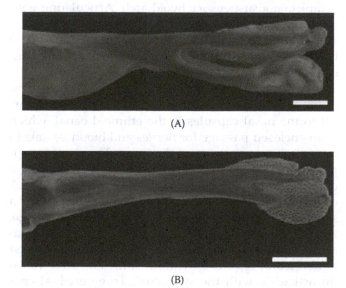

(A)

(B)

FIGURE 4.3
(See color insert.) Cleared and stained pelvic claspers of representative specimens of *Chimaera* (A) and *Hydrolagus* (B) showing bifurcate internal skeletal morphology. Morphology of pelvic claspers, particularly the point at which the internal skeleton divides, is useful for species identification when used in combination with other morphological characters. Scale = 1 cm.

4.3 Classification and Zoogeography

4.3.1 Classification

Although it has long been accepted that the holocephalans belong within a monophyletic Chondrichthyes (Didier, 1995; Maisey, 1986), their relationships have historically been poorly understood. This is primarily due to the fact that attempts to resolve relationships among holocephalans have relied on scant fossil evidence (Lund, 1986; Patterson, 1965). Recent work on a diversity of exceptionally preserved chondrichthyan forms from the Bear Gulch of Montana has shed new light on the diversity of Paleozoic Chondrichthyes, as well as providing new evidence regarding the relationships among holocephalan fishes (Grogan and Lund, 2000, 2004; Lund and Grogan, 1997). Holocephalans (= Holocephali or Holocephalimorpha) are currently classified within the Subclass Euchondrocephali, along with iniopterygians and paraselachians (Grogan et al., 1999; Lund and Grogan, 1997). This new evidence seems to suggest that the holocephalans are not a monophyletic grouping but rather an assemblage of forms sharing a suite of cranial modifications and dental morphologies not shared by other euchondrocephalans. The extant holocephalans, Order Chimaeriformes, fall within the Holocephalimorpha and include the three families of extant forms as well as their closest fossil ancestors, such as the Echinochimaeroidei, Squalorajiformes, Menaspiformes, and Echinochimaeroidei (Nelson, 2006).

4.3.1.1 *Callorhinchidae*

(See Figure 4.1A.) A prominent plow-shaped snout, extending forward from the front of the head, characterizes the callorhinchid fishes, also commonly known as the plow-nosed chimaeras or elephantfishes. A stiff cartilaginous rod supports the dorsal surface of the snout, and at the distal end is a fleshy ovoid or leaf-shaped flap of tissue. Within this monogeneric family, the three recognized species all occur in the Southern Hemisphere: *Callorhinchus milii* from New Zealand and Australia, *C. capensis* from southern Africa, and *C. callorynchus* from southern South America (Didier, 1995; Nelson, 2006). In addition to the unique snout morphology, callorhinchids differ from other chimaeroids in their more torpedo-like body shape, heterocercal tail, and a large skeletally supported anal fin. The callorhinchids are the most primitive living chimaeroids based on interpretation of a variety of characters of the tooth plates, skeleton, and musculature (Didier, 1995).

Morphologically the three species are nearly indistinguishable. All callorhinchids are silvery in color and black along the dorsal midline, with saddle-like bands on the dorsal side of the head and along the dorsal surface of the trunk and sometimes with dark blotches along the sides of the trunk, as well. Unlike other chimaeroids, the callorhinchids have lateral line canals that are enclosed, visible on the body surface as narrow canals underneath the dermis, rather than open grooves. The eye is small. The second dorsal fin is not elongate; it is usually nearly equal to the length of the pectoral to pelvic space and very tall anteriorly, sloping posteriorly to a low, evenly tall fin with the height of the anterior portion about five times that of the posterior portion. Males possess simple scrolled pelvic claspers lacking fleshy lobes and denticulations (Figure 4.4A). The frontal tenaculum is flat and not deeply curved, with very short denticles on the distal bulb. There appear to be some distinctions among the frontal tenacula of males of the three species, and this may prove a useful identifying feature, although useful only for distinguishing males of the species. Prepelvic tenacula are complex cartilaginous structures consisting of a flat cartilaginous blade, the fleshy portion adorned with flat multicuspid denticles, and a small cartilaginous tubelike structure lacking denticles. No large denticles are present along the medial edge of the blade. Rudimentary prepelvic pouches are present in females, visible on the ventral surface anterior to the pelvic girdle. Female callorhynchids produce large egg capsules, averaging 20 cm in length and 9 cm in width (Figure 4.5A). The single egg is encased in a central spindle-shaped cavity, and extending around the lateral edge of the central spindle is a flexible ridged flange that gives the egg capsule an overall ovoid shape. The dorsal side of the egg capsule is convex, and the ventral side is concave.

In the absence of reliable morphological characters for distinguishing the three species, the only means at present for species identification is by geographic location. Color pattern is highly variable and may be of limited usefulness for distinguishing species (in the key provided in the Appendix, note some suggested color variations that might be helpful for identification). Many authors have suggested that perhaps the traditional three-species concept of Callorhinchidae is incorrect and that they might all be one wide-ranging species (Bigelow and Schroeder, 1953; Krefft, 1990; Norman, 1937); however, differences in the shape of the egg capsules and some variation in the morphology of the frontal tenacula of males may support the validity of these species.

Not only does the Family Callorhinchidae have a long and confusing taxonomic history, but identification among the three species is also difficult. Gronovius (1756, 1763), in various works, was the first to describe callorhinchids, whereas the first available description of a callorhinchid was *Chimaera callorynchus* (Linnaeus, 1758). Lacépède (1798) recognized this species as separate from other chimaeras and placed it in the genus

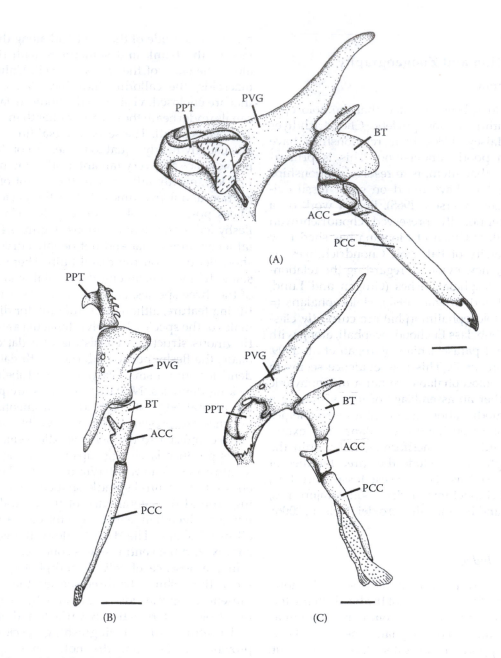

FIGURE 4.4

Skeletal morphology of the left side of the pelvic girdle showing the anatomy of pelvic claspers and prepelvic tenaculae in the three families of chimaeroids: (A) Callorhinchidae, (B) Rhinochimaeridae, and (C) Chimaeridae. Scale = 2 cm. *Abbreviations:* ACC, anterior clasper cartilage; BT, basipterygium; PCC, posterior clasper cartilage; PPT, prepelvic tenaculum; PVG, pelvic girdle. (Courtesy of the American Museum of Natural History.)

Callorhinchus. Much later, Garman (1901) placed all chimaeroids distinguished by a flexible plow-shaped snout and tubular lateral line canals in the Family Callorhynchidae. The family name Callorhynchidae was based on the type species, *Callorhinchus callorhynchus*, and is historically the most common family spelling used. The family name should correctly be Callorhinchidae (ICZN Code Article 32.3, 32.4) (Didier, 1995; Fowler, 1941; Paxton et al., 1989). The first spelling of the genus was *Callorynchus* (Gronovius 1754); however, that work was pre-Linnaean, and confusion continues

to this day regarding the correct spelling of the genus name. Recent research indicates that Lacépède (1798) is the first valid work containing the genus name, and the currently accepted spelling is *Callorhinchus* (Didier, 1995; Eschmeyer, 1998).

4.3.1.2 Rhinochimaeridae

(See Figure 4.1B.) Family Rhinochimaeridae includes all chimaeroids possessing a long, tapering fleshy snout extending anterior to the head. Rhinochimaerids are

FIGURE 4.5
(See color insert.) Egg capsules from representative species of the three families of chimaeroid fishes. (A) *Callorhinchus milii*, shown in dorsal view (left) and ventral view (right). Scale = 3 cm. (B) *Rhinochimaera atlantica*, preserved specimen, shown in dorsal view (left) and ventral view (right). Scale = 3 cm. (C) *Hydrolagus colliei* shown in dorsolateral view (left) and ventral view (right). Scale = 1 cm.

commonly referred to as the longnose chimaeras or spookfish. These fishes generally inhabit deep waters and are usually found at depths around 1000 m to more than 2000 m. This is a small family with only eight species in three genera. In general, rhinochimaerids are medium- to large-bodied fishes with an elongate spear-like snout and a somewhat compressed, elongate body tapering to a narrow tail with an elongate distal filament. In some species, the tail appears externally heterocercal with a dorsal caudal fin lobe that is very narrow and a much deeper ventral caudal fin web. In all species, the color is usually grayish or brownish, often lighter or white ventrally, without any distinct color pattern. Hatchlings and very small juveniles tend to be much paler in color but dark in the region of the opercular flap, with very dark brown or black fins. The snout is also disproportionately long in juveniles when compared to adults of the same species. Adult males possess slender, rod-like pelvic claspers with small, fleshy denticulate tips (Figure 4.4B). Females produce an ovoid egg capsule, similar in shape to that of callorhynchids, in which a fanlike lateral web surrounds a hollow central spindle-shaped chamber; however, the lateral flange is usually much narrower than that of callorhinchid egg capsules, and the central spindle is longer and somewhat

indented at either or both ends (Figure 4.5B). As a result of their deepwater habitat, species of rhinochimaerids have been poorly studied, and almost nothing is known of their biology and reproduction.

Morphologically, there appear to be two distinct lineages of rhinochimaerids: the Rhinochimaerinae, which includes the genus *Rhinochimaera*, and the Harriottinae, comprised of the remaining two genera, *Harriotta* and *Neoharriotta* (Didier, 1995). Within the Harriottinae, the genus *Neoharriotta* is distinguished from *Harriotta* by the possession of a distinct, separate anal fin (Bigelow and Schroeder, 1950). Gill (1893) first named the Harriottinae, and Dean (1904c) supported the hypotheses that distinctions among rhinochimaerids warranted separation into two groups. Species of *Rhinochimaera* are distinguished from the harriottines by several morphological features, including tooth plates in the form of smooth shearing blades rather than raised hypermineralized tritors on the surface, tubercles on the dorsal caudal fin, a dorsoventrally compressed neurocranium, and the presence of the retractor mesioventralis pectoralis muscle (Didier, 1995). The presence of tubercles on the dorsal caudal fin in *Neoharriotta pinnata* was noted by Bigelow and Schroeder (1950); however, this characteristic has not been observed in subsequent examination of this species (DD, pers. obs.). The harriottine lineage suggested by Didier (1995) is not supported by any morphological synapomorphies, and in fact a recent genetic analysis strongly supports a monophyletic Rhinochimaeridae (Inoue et al., 2010).

4.3.1.3 Chimaeridae

(See Figure 4.1C.) Commonly known as shortnose chimaeras, ratfishes, or ghost sharks, the chimaerid fishes are characterized by a conical fleshy snout that is bluntly pointed at the tip. Members of this family are distinguished from other chimaeroids by lateral line canals on the snout that are expanded with wide dilations. Species of chimaerids have somewhat compressed, elongate bodies tapering to a whiplike tail with an elongate filament. Most species are a uniform brown, gray, or black, but some species also exhibit color patterns with spots and stripes. In all species, the eyes are large, usually a bright green in fresh specimens. Body size can be quite variable, with some species remaining small and almost dwarflike at maturity (e.g., *Hydrolagus mirabilis*) while other species attain massive sizes with large bulky heads and bodies and mature at over 1 m in length (e.g., *Chimaera lignaria*). All males have divided claspers with fleshy denticulate tips (Figure 4.4C; discussed above in Section 4.2.4). Females produce egg capsules that are slender and spindle shaped without broad lateral flanges (Figure 4.5C). Dean (1912) noted differences among the egg capsules of chimaerids, and it is likely that egg capsules are species specific; however, egg capsules for most species have not been identified. The chimaerids are widespread geographically, with species known from every ocean region with the exception of the far northern and southern polar regions. They are known to occur at depths ranging from nearshore surface waters to deeper than 2000 m.

Family Chimaeridae is the most speciose, with 36 recognized species in two separate genera. There are currently 14 recognized species of *Chimaera* and 22 species of *Hydrolagus*, and it is likely that several additional species still remain to be described in this family. Morphologically, the two genera are remarkably similar, the only difference being the presence of an anal fin separated from the ventral caudal fin by a notch in *Chimaera* and a continuous ventral caudal fin without a notch separating the anal fin in *Hydrolagus*. Species of chimaerids are difficult to distinguish because of morphological similarity and are best identified by combinations of characters such as body color, lateral line canal pattern, fin shape, and the relative size or shape of the eyes, snout, or fin spine. The anal fin may not be sufficient for separation of the two genera. Indeed, this single character does not hold up well when considered in combination with other characters (Hardy

and Stehmann, 1990; DD, pers. obs.). A revision of the family, perhaps incorporating molecular techniques, will be needed to fully resolve taxonomic relationships among the chimaerids (see discussion in the Appendix).

4.3.2 Zoogeography

The chimaeroids are marine fishes inhabiting all of the world's oceans with the exception of Antarctic waters. Table 4.1 identifies the approximate geographic region where each species most commonly occurs. More than half of all known species occur in high-latitude seas (Ebert and Winton, 2010). The Western Indo–Pacific region appears to exhibit the highest diversity of chimaeroids, followed by the North Atlantic (Kyne and Simpfendorfer, 2010). The genera *Chimaera* and *Hydrolagus* (Family Chimaeridae) are the most diverse, with 14 and 22 species, respectively, currently recognized. Many of these species exhibit a high degree of endemism (e.g., *H. alphus, H. mccoskeri*). Some species are known from a relatively restricted range vertically and horizontally (e.g., *H. barbouri, Chimaera panthera*), whereas other chimaeroid species seem to be widespread (e.g., *C. monstrosa, H. colliei*). Members of the Family Rhinochimaeridae tend to have a broad but widely scattered distribution, in some instances throughout an entire ocean basin (e.g., *Harriotta raleighana, Rhinochimaera pacifica*), while members of the Family Callorhinchidae are restricted to the Southern Hemisphere.

Most chimaeroids are deepwater dwellers of the shelf and slope off continental landmasses, oceanic islands, seamounts, and underwater ridges, generally occurring at depths of around 500 m and deeper. A few species inhabit shallower coastal waters, most notably *Hydrolagus colliei* off the west coast of the United States and all three species in the Family Callorhinchidae. *H. colliei* tends to occupy sandy or muddy bottoms at depths ranging from the surface to 971 m (Ebert, 2003). *Callorhinchus callorhynchus* in the southern Atlantic occurs at a maximum depth of 116 m (Di Giácomo, 1992). The discovery of *C. callorhynchus* in the Beagle Channel provides strong evidence for a continuous widespread geographic range that extends from the southern Atlantic off Uruguay to the southeastern Pacific off of Chile (López et al., 2000). New collection records continue to expand the known range of many species of chimaeroid fishes, and additional zoogeographic studies may indicate that species currently known from restricted ranges are in fact more widespread (e.g., González-Acosta et al., 2010); however, some range extensions, such as the report of a single specimen of the northeastern Atlantic *Hydrolagus pallidus* from the southeastern Pacific (Andrade and Pequeño, 2006), should be considered with caution.

Very little information exists on the ecology and behavior of chimaeroid fishes due to the deepwater habitat of these fishes. Most information has been obtained through observations of the few commercially fished species that occur in nearshore waters: *Hydrolagus colliei*, *Callorhinchus milii*, *C. callorhynchus*, and *C. capensis*. These species appear to be locally migratory and exhibit seasonal inshore migration for breeding and spawning (Gorman, 1963; Lopez et al., 2000). Studies of *H. colliei* provide evidence that this species aggregates by both sex and size. Males and females form separate groups, with the sex ratio becoming more skewed to females with depth (Barnett, 2008). Juveniles tend to aggregate in deeper waters, while larger fish exhibit seasonal migrations to shallower waters, which is more noticeable with increasing latitude (Barnett, 2008; Mathews, 1975; Quinn et al., 1980). Other species of chimaeroids may exhibit similar aggregation and migration patterns (Di Giácomo, 1992).

4.4 Life Histories

4.4.1 Reproduction and Development

Chimaeroids, like their shark relatives, have internal fertilization in which males, equipped with pelvic claspers, transfer sperm directly into the female reproductive tract. All chimaeroids are oviparous. Females produce large, yolky eggs, pale yellow in color and similar in size to those of elasmobranchs. Fertilization occurs in the upper end of the reproductive tract, and eggs pass through the oviducal gland. Eggs are encased individually in a tough, leathery egg capsule within the oviduct (Dean, 1906; Didier, 1995; Didier et al., 1998). Egg capsules remain attached to the oviducal gland within the oviduct for several days (Dean, 1903, 1906; Sathyanesan, 1966) and are deposited directly on the seafloor. The shape of the egg capsules is characteristic for each family (Didier, 1995). Dean (1912) noted differences in the morphology of egg capsules among species of *Chimaera* and identified primitive and derived character states; he defined this gradual variation of egg capsule morphology as "orthogenesis," and these early studies indicate that egg capsule characters may be of phylogenetic significance. Egg capsules also may be taxonomically significant at the species level; however, it is difficult to reliably associate egg capsules with a species because the ranges of many species overlap, and the egg capsules are not usually collected in association with the females that laid them.

The egg is supported by a very fragile vitelline membrane and thick jellylike material that fills the inside of the egg capsule. As the embryo matures, the vitelline membrane will toughen and the jelly material breaks down. At the anterior, or blunt, end of the central spindle portion of the egg capsule is a raised seam that is tightly sealed when first laid, but as the embryo develops and the egg capsule wears with age the seal gradually softens and opens slightly. At the time of hatching, the egg capsule breaks open along this seam to release the fully developed embryo. Additional slits at the posterior end of the spindle also gradually open to facilitate the flow of water through the egg capsule for gas exchange and removal of waste products (Dean, 1903, 1904a, 1906; Didier, 1995).

Spawning generally occurs on flat, muddy or sandy substrates, but spawned egg capsules have also been observed on pebbly bottoms and in beds of seaweed. Females spawn two egg capsules simultaneously, one from each oviduct, which are deposited onto the ocean floor (Dean, 1906; Didier, 1995). Several pairs of eggs are laid each season, but the exact number is unknown. Captive *Hydrolagus colliei* have been observed to lay a pair of eggs every 7 to 10 days (Didier et al., 1998; Sathyanesan, 1966; K. Wong, pers. obs.), and other species of chimaeroids probably spawn at a similar rate, with females laying a pair of eggs every fortnight for several months (Gorman, 1963). It is likely that all females store sperm, as evidenced by a recent study of *Callorhinchus milii* (Smith et al., 2001). A single collection of embryos in the field will contain embryos of all stages; therefore, it is likely that the spawning season lasts several months, perhaps up to 6 months. Gravid females have been found in summer and winter months, but the number of mature ova are generally more abundant in late summer and early fall (Barnett et al., 2009a; Sathyanesan, 1966), which coincides with the maximum observed egg case deposition (Dean, 1906). Barnett et al. (2009a) found *H. colliei* to have approximately 6 to 8 months of parturition per year. The gestation period is suspected to take from 9 to 12 months in *H. colliei* and 6 to 12 months in *C. milii* before the fully developed embryo hatches from its egg capsule (Dean, 1903, 1906; Didier et al., 1998; Gorman, 1963). Undoubtedly, temperature plays a role in determining developmental rates and timing. The only estimate of annual fecundity was determined by Barnett et al. (2009a) at 19.5 to 28.9 by using captive egg deposition rates and calculated length of parturition in *H. colliei*.

Embryological development has been observed and described for only 2 of the 47 recognized species of chimaeroids. The earliest descriptive embryological studies were of *Callorhinchus milii* (Schauinsland, 1903) and *Hydrolagus colliei* (Dean, 1903, 1906). These studies were based on only a few embryos and lacked many critical early developmental stages. More recently, Didier et al. (1998) described a complete, post-neurula, developmental series of *C. milii*. Details of reproduction and spawning are based primarily on studies of *C. milii* and *H. colliei*, and it is assumed that reproductive biology for other species of chimaeroids is similar. During development, the embryo relies completely on the large yolk for nourishment. As the embryo develops, the yolk sac takes on a characteristic bulged shape with lateral extensions that come up around the developing head region (Figure 4.6). This sculpting of the yolk sac during development appears to be unique to chimaeroids among vertebrate animals.

4.4.2 Age and Growth

Methods for determining age and growth rates in chimaeroid fishes have been based on studies of *Hydrolagus colliei*, *Callorhinchus milii*, *C. capensis*, and *Chimaera monstrosa*. Johnson and Horton (1972) tested a variety of morphological measurements for *H. colliei*, including eye-lens weights, vertebral radii, basal sections of the dorsal spine and left pectoral fin, body-length frequencies, and tooth-plate ridges, none of which provided accurate results. Because tooth plates are continually growing and are known to change morphology as the fish grows, it is unlikely that fish age can be determined on the basis of tooth-plate morphology (Bigelow and Schroeder, 1950; Didier et al., 1994; Garman, 1904). The most common method is examination of the banding patterns in the dorsal fin spine, based on the assumption of annual ring deposition. Using this method, Sullivan (1978) estimated that *C. milii* males mature at about 3 years and 50 cm fork length (FL) and females at 4.5 years and 70 cm FL, with the oldest male and female at 4 and 6 years, respectively. In a more recent study of growth rates of *C. milii*, based on length–frequency and tag–recapture data, Francis (1997) found that growth rates varied among populations. For example, males collected in the 1960s matured at 4+ years and males collected in the 1980s matured at 2+ to 3+ years. This variation in growth rates may be related to biomass of the population. Freer and Griffiths (1993a) estimated age at maturity in *C. capensis* to be 3.3 years for males (43.5 cm FL) and 4.2 years for females (49.6 cm FL), and suggested that total spine length was related to the age of the individual. Moura et al. (2004) estimated the ages of male and female *C. monstrosa* at 0 to 15 years and 0 to 17 years, respectively. Males showed a slower growth rate than females, and females reached a larger maximum length. Other studies have also found growth patterns to differ between sexes (Francis, 1997; Freer and Griffiths, 1993a). Another study of *C. monstrosa* estimated ages from 3 to 30 years in males and 4 to 26 years in females (Calis et al., 2005). Although several studies have employed the use of age determination via dorsal fin spine banding patterns, none has validated their results. Barnett et al. (2009b) found the use of the dorsal fin spine in chimaeroids as an age estimation structure unreliable. They

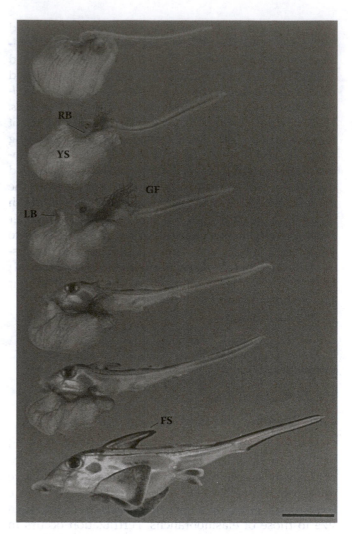

FIGURE 4.6
An embryonic series of *Callorhinchus milii* collected from the Marlborough Sounds, New Zealand, using SCUBA. Embryos in this series range in size from 53 mm TL (top) to 122 mm TL (bottom). The lateral extensions (LB) of the yolk sac (YS) are clearly visible. Also shown are external gill filaments (GF) and the rostral bulb (RB). The developing fin spine (FS) does not appear until late in development. Scale = 2 cm. (Courtesy of the American Museum of Natural History.)

found that band counts differed in sections taken at differing distances from the spine tip which could provide inaccurate age estimates depending on the area sectioned. Barnett et al. (2009b) also looked at vertebrae and neural arches, oxytetracycline (OTC) injection, and mineral density gradients as other methods for assessing ages. They found that, because the vertebrae and neural arches are poorly calcified, they are likely poor structures for age determination; mineral density gradients were not present in the dorsal fin spine and thus cannot show growth zones; and OTC was not incorporated internally and thus cannot be used as a means of validating age. Using assessments of maturity status of reproductive organs, Barnett et al. (2009a) determined the size at 50% maturity for *H. colliei* to be 202.8 mm snout to vent length (SVL) for females and 157.2 mm SVL for males, indicating that females mature at a larger size. Further research is needed to address these issues, especially in determining alternative techniques for age determination and validation, including further investigation into the use of the dorsal fin spine.

4.4.3 Diet and Feeding Ecology

Chimaeras are a unique group in relation to other chondrichthyans in that the stomach is absent (Kobegenova, 1993). Instead, their digestive tract consists of the esophagus region, which leads to a digestive tube, similar to a spiral valve, then to the rectum and cloaca (Di Giácomo and Perier, 1996). Few studies have been conducted on chimaeroid fishes in relation to their diet and feeding ecology, and they have mainly been constrained to a few species (e.g., *Chimaera monstrosa*, *Hydrolagus mirabilis*, *Harriotta raleighana*). In general, chimaeras seem to be opportunistic benthic feeders that prey upon benthic invertebrates and small fishes (Gorman, 1963; Johnson and Horton, 1972; Quinn et al., 1980; Stehmann and Bürkel, 1984). Chimaeroids have hypermineralized tooth plates that are used to crush hard-bodied prey such as crustaceans, mollusks, and echinoderms, including hydrothermal vent mussels (Graham, 1939; Macpherson and Roel, 1987; Marques and Porteiro, 2000); however, soft-bodied prey (e.g., salps, tunicates, jellyfish, polychaetes, amphipods) are also consumed (Graham, 1956).

The diet of chimaeroids has been found to vary with geographic region within a species. The greatest number of dietary studies, for example, has been conducted on *Chimaera monstrosa* from different geographic regions, including Rockall Trough (Mauchline and Gordon, 1983), western Mediterranean Sea (Macpherson, 1980), southern Portuguese continental slope (Moura et al., 2005a), Adriatic Sea (Ungaro, 1996), and Skagerrak (Bergstad et al., 2003). Mauchline and Gordon (1983) found that *C. monstrosa* consumed predominantly anemones and their tubes. Macpherson (1980) found that *C. monstrosa* consumed ophiuroids, benthic crustaceans, and polychaetes. Moura et al. (2005a) found decapod crustaceans and amphipods to be the most frequent prey of *C. monstrosa*. Ungaro (1996) found over half of *C. monstrosa* digestive tracts to be empty, and the important prey groups of those with contents were crustaceans and bivalves. Bergstad et al. (2003) found *C. monstrosa* of all size classes to consume benthic prey (i.e., polychaetes and bivalves), and larger fish to also consume large decapod crustaceans. Another example is *Hydrolagus mirabilis* in the northeast Atlantic that had a diet of polychaetes, small benthic crustaceans, and spatangoids (Mauchline

and Gordon, 1983). The diet of *H. mirabilis* in the northwest Atlantic is dominated by echinoderms and scyphozoans on the Grand Bank, caprellids and priapulids in the Flemish Cap, and scyphozoans, roundnose grenadier, and tunicates in the Flemish Pass (Gonzales et al., 2007). On the other hand, the diet of *Harriotta raleighana* from New Zealand was dominated by polychaetes and mollusks (Dunn et al., 2010), similar to the findings reported for *H. raleighana* in the north Atlantic (Gonzalez et al., 2007; Mauchline and Gordon, 1983).

Diet shifts have been observed ontogenetically in many chondrichthyans and are also present in chimaeroids. In general, there is a trend of smaller fish consuming more soft-bodied prey, including amphipods and polychaetes, and larger fish consuming hard-shelled prey such as large decapod crustaceans. Large *Chimaera monstrosa* typically consumed a greater proportion of anemones, spatangoids (Mauchline and Gordon, 1983), and decapod crustaceans (Macpherson, 1980; Mauchline and Gordon, 1983; Moura et al., 2005a). Juvenile *C. monstrosa* typically consumed a greater proportion of polychaetes and amphipods (Macpherson, 1980; Mauchline and Gordon, 1983; Moura et al., 2005a). Moura et al. (2005a) concluded that smaller *C. monstrosa* had a generalist feeding strategy while larger fish, which consumed mostly decapods, had a more specialized feeding strategy. This trend has also been observed in *Hydrolagus mirabilis* (Gonzalez et al., 2007; Mauchline and Gordon, 1983). Diet differences have also been observed between sexes (Di Giácomo and Perier, 1996).

Dietary overlap within and among species was examined in three chimaeroid species—*Harriotta raleighana*, *Hydrolagus bemisi*, and *Hydrolagus novaezealandiae*—from the Chatham Rise, New Zealand (Dunn et al., 2010). The diet of *H. bemisi* in cool, deep water had a greater similarity to *H. raleighana* in similar habitat than to other *H. bemisi* inhabiting warm, shallower water, suggesting that these species are flexible in their diets and likely respond to local variation in prey availability. Also, Dunn et al. (2010) found that the best predictors of dietary variability were the subtropical front in *H. raleighana* and *H. novaezealandiae* and bottom temperature in *H. bemisi*, thus demonstrating the importance of environmental variation and probably fluctuation in prey availability in determining diet.

The presence of sediment mixed with prey in digestive tracts suggests that most chimaeras are benthic feeders, consuming prey located on the seafloor (González et al., 2007); however, the presence of squid beaks in *Harriotta raleighana* indicates that they may be capable of consuming pelagic organisms as well (Mauchline and Gordon, 1983). Scavenging is also likely; it seems that most chimaeroids are opportunistic feeders, and several species have been captured on baited hooks, including *Hydrolagus affinis* (Forster, 1964), *Hydrolagus pallidus* (Marques and Porteiro, 2000), *Chimaera mon-strosa* (Clarke et al., 2005), and *Hydrolagus melanophasma* (James et al., 2009).

Little is known about predation on chimaeroid fishes, but it is likely that sharks are their primary predators. Adult *Callorhinchus milii* are consumed by the New Zealand carpet shark, *Cephaloscyllium isabellum*, and by school sharks, *Galeorhinus galeus* (Gorman, 1963; Didier pers. obs.). *Callorhinchus capensis* is preyed upon in southern African waters by bluntnose sixgill shark (*Hexanchus griseus*), sevengill shark (*Notorynchus cepedianus*), *G. galeus*, and great white sharks (*Carcharodon carcharias*) (Ebert, 1991, 1994; unpubl. data). *Hydrolagus colliei* appears to be an important prey item for several shark species, including *H. griseus*, *N. cepedianus*, *G. galeus*, and North Pacific spiny dogfish (*Squalus suckleyi*) (Ebert, 2003; unpubl. data). It also is consumed by several large teleost species, including giant sea bass (*Stereolepis gigas*), lingcod (*Ophiodon elongatus*), rockfish (*Sebastes* spp.), Pacific halibut (*Hippoglossus stenolepis*), and marine mammals, including pinnipeds and cetaceans (Ebert, 2003). Two occurrences of cannibalism were found in *H. colliei*: an egg capsule and a caudal fin found in the stomach of a conspecific (Johnson and Horton, 1972).

4.4.4 Habitat Association

Relatively little is known about the habitat associations of chimaeroids, mainly due to their deep-sea habitat. Manned submersibles and remotely operated vehicles (ROVs), however, have allowed for observations and collections to be made in deep-sea habitats. Two Galapagos Islands species, *Hydrolagus alphus* and *H. mccoskeri*, were observed to be associated with areas of high rocky relief containing volcanic boulders, cobbles, and pebbles, interspersed with patches of sand and silt (Barnett et al., 2006; Quaranta et al., 2006). Rocky areas contained few to many benthic invertebrates, including stony corals, sponges, crinoids, hydroids, gorgonians, bryozoans, holothurans, and ophiuroids (Quaranta et al., 2006). *Hydrolagus* cf. *trolli*, seen by ROVs at the Davidson Seamount, off central California, has been observed swimming over rocky substrate with high vertical relief (Figure 4.7). These observations are in contrast to *Harriotta raleighana* (Figure 4.8) or *Hydrolagus melanophasma* (Figure 4.9) (James et al., 2009) that typically occur over soft-bottom habitats or cobble patches with minimal vertical relief. In all cases, regardless of habitat type, these specimens were observed within a few meters of the seafloor. *Hydrolagus affinis* and *H. pallidus* have been observed and collected at the Lucky Strike hydrothermal vent site from the mid-Atlantic Ridge (Marques and Porteiro, 2000). Aggregations of juveniles on the soft bottom habitat of the shelf break near Cordell Bank, off northern California, along with

FIGURE 4.7
Hydrolagus cf. *trolli* shown swimming over high rocky relief substrate on the Davidson Seamount off the central California coast at a depth of 1624 m. (Photograph courtesy of Monterey Bay Aquarium Research Institute.)

FIGURE 4.8
Harriotta raleighana shown swimming within a few meters of a soft-bottom substrate in the Gulf of California at a depth of 1554 m. (Photograph courtesy of Monterey Bay Aquarium Research Institute.)

FIGURE 4.9
Hydrolagus melanophasma shown swimming over low-relief cobble substrate in the Gulf of California coast at a depth of 1664 m. (Photograph courtesy of Monterey Bay Aquarium Research Institute.)

collections of hundreds of *Hydrolagus colliei* egg cases found in sand troughs in the vicinity of Cordell Bank, indicate that this habitat may be a nursery area for *H. colliei* (Barnett, 2008).

4.5 Fisheries and Conservation

4.5.1 Fisheries

Chimaeroids are both targeted in commercial fisheries and taken as either retained bycatch or discarded at sea. With many fisheries moving to the deep sea, chimaeras are at increasing risk of exploitation. Little information exists on chimaera life histories and behaviors, but the slow growth, late maturation, and low fecundity typical of chondrichthyans are likely for chimaeroids; thus, chimaeras may be vulnerable to fishing pressure. Catches of chimaeras are rarely reported, hampering the ability to gain data that could provide essential information about these fishes as well as data for conservation efforts. *Chimaera monstrosa*, however, is caught by deepwater trawl fisheries in the northeast Atlantic—landed as bycatch or discarded—and constitutes 13 to 15% of discarded bycatch in deepwater trawls off Ireland (Calis et al., 2005).

Historically, targeted fisheries have existed for a few species of chimaeroids, primarily those nearshore species of *Callorhinchus*. *Callorhinchus milii* has been a targeted species in New Zealand dating back to as early as 1914 (Francis, 1998). From the 1950s to the 1970s, there was a steady increase in *C. milii* landings, with a mean annual landing of 1075 t (Francis, 1998). High demand for *C. milii* was due mainly to its export for fillets as well as livers for oil (Francis, 1998). Most catch was taken during spring and summer months when *C. milii* migrate inshore for mating and spawning (Francis, 1998). After 1971, landings began to decline, and in 1986 *C. milii* was considered to be severely overfished; the stock since then has appeared to rebuild due to total allowable catches (TACs) being implemented (Francis, 1998). Today *C. milii* is typically caught as bycatch in other targeted fisheries (Francis, 1998).

Callorhinchus capensis is caught through a directed gillnet fishery off South Africa and by trawl and line fishing (Freer and Griffiths, 1993b). Annual catches are stable at 700 to 900 t, and fishing is regulated by the number of gillnet permits allowed (Freer and Griffiths, 1993b). *Callorhinchus callorhynchus* is caught commercially by trawl fisheries and recreationally by line fishing. There seems to be a "boom or bust" cycle in landings, with increased catch one year and sharp declines the next, which could indicate the potential for overfishing (Di Giácomo and Perier, 2005). Other species of chimaeroids that are fished, but for which little catch information exists, include *Chimaera phantasma* from Taiwanese waters and *Neoharriotta pinnata* off India, with large numbers of individuals showing up in markets (Figure 4.10).

FIGURE 4.10
A collection of *Neoharriotta pinnata* collected as part of a bycatch fishery in India. (Photograph courtesy of K.V. Akhilesh.)

4.5.2 Conservation

Of the 47 described species of chimaeras, under the International Union for Conservation of Nature (IUCN) Red List of Threatened Species, 24 chimaeroids are data deficient, 15 are of least concern, 3 are near threatened, and 5 have not been evaluated. Because little information exists on the biology and ecology of chimaeroids and several new species have been described in recent years, the predominance of data-deficient species or those not evaluated is not surprising. Chimaeroid research is a high priority, especially given that many species have rather restricted distributions and fishing pressures may increase (see Section 4.5.1 on Fisheries).

The three near threatened species are *Chimaera monstrosa, Hydrolagus mirabilis*, and *Hydrolagus ogilbyi*. *C. monstrosa* was evaluated as near threatened due to its late age at maturity and that it represents a rather large portion of both landed and discarded bycatch in deepwater trawl fisheries, implying a high rate of mortality. *H. mirabilis* was assessed as near threatened because its locale and depth distribution put it at a higher risk to deepwater fisheries. *H. ogilbyi* has a rather restricted range, and surveys have documented huge declines in catch rates because of fishing pressures; thus, this species was given a near threatened status.

4.6 Summary and Conclusions

The holocephalan fishes have historically been a poorly studied group, and in particular very little research, aside from anatomical studies, has been conducted on the living chimaeroids through the latter part of the, 20th century. The last decade has seen a resurgence of interest in chimaeroid fishes and significant research advances. New discoveries have resulted in the description of 17 new species as well as new understanding of the distributions of chimaeroids worldwide. Emerging fisheries have resulted in increased attention to the fishery potential of chimaeroids as well as the management and conservation potential for chimaeroids. Despite the relatively low diversity in this group of fishes, there is great potential as well as need for increased research. In addition, the use of chimaeroid fishes in the field of evolutionary development is shedding light on major patterns of vertebrate evolution. Whole genome studies of *Callorhinchus milii* have shed new light on the importance of chimaeroids as representing a model genome for evolutionary studies (Inoue et al., 2010; Venkatesh et al., 2007). Molecular studies, including a recent comparative analysis of Hox gene clusters in primitive vertebrates that included the elephant fish (*C. milli*), indicates that chimaeroid fishes hold a pivotal role in understanding the genomics behind major events in vertebrate evolution such as the evolution of jaws and appendages (Gillis et al., 2011; Venkatesh et al., 2007).

Acknowledgments

DD is grateful to the many curators and collection managers who provided specimens and museum support for this research, especially T. Abe; N. Feinberg (AMNH); M. McGrouther and D. Hoese (AMS), S. Schaefer and B. Saul (ANSP); N. Merrett (BMNH); D. Catania (CAS); P. Last, J. Stevens, A. Graham, and G. Yearsley (CSIRO, Department of Marine Research); Mary Ann Rodgers (FMNH); K. Nakaya and students (HUMZ); M. Stehmann (ISH); K. Hartel (MCZ); P. Pruvost, X. Gregorio, and B. Séret (MNHN); C. Roberts and A. Stewart (NMNZ); K. Matsuura and G. Shinohara (NSMT); L. Compagno and M. Van der Merwe (SAM); T. Pietsch (UW); G. Burgess (UF); and L. Knapp, J. Finan, and S. Jewett (USNM). My thanks to students who have participated in aspects of chimaeroid research: L. Rosenberger, D. VanBuskirk, E. LeClair, B. Marquardt, A. Wilmot, T. Meckley, and J. Moyer. Funded by grants from the National Science Foundation (NSF DEB-9510735 and NSF DEB-0097541) and the National Geographic Society (5414-95). Additional support was received from the American Philosophical Society, JSPS, Jessup Fellowships (to E. LeClair and L. Rosenberger), a fellowship to the MNHN, the New Zealand Foundation for Scientific Research and Technology, Biosystematics of New Zealand EEZ Fishes Project (contract MNZ603), and a Millersville University faculty research grant.

DAE and JMK would like to thank M. Bougaardt (South African Museum); J. Williams (USNM); H.J. Walker (SIO); D. Long (Oakland Museum of California); L. Barnett (University of California–Davis); and K. James and K. Quaranta (MLML). Photographs for Figures 4.7, 4.8, and 4.9 were kindly provided by the Monterey Bay Aquarium Research Institute and for Figure 4.10 by K.V. Akhilesh. Funding support for DAE and JMK was provided by NOAA/NMFS to the National Shark Research Consortium and Pacific Shark Research Center and the David and Lucile Packard Foundation.

References

Andrade, I. and Pequeño, G. (2006). Primer registro de Hydrolagus *pallidus* Hardy & Stehmann, 1990 (Chondrichthyes: Chimaeridae) en el Océano Pacífico, con comentarios sobre los holocéfalos de Chile. *Rev. Biol. Mar. Ocean.* 41:111–115.

Bargmann, W. (1933). Die Zahnplatten von *Chimaera monstrosa*. *Z. Zell. Mikr. Anat.* 19:537–561.

Barnett, L.A.K. (2008). Life History, Abundance, and Distribution of the Spotted Ratfish, *Hydrolagus colliei*, master's thesis, California State University, Long Beach.

Barnett, L.A.K., Didier, D.A., Long, D.J., and Ebert, D.A. (2006). *Hydrolagus mccoskeri* sp. nov., a new species of chimaeroid fish from the Galapagos Islands (Holocephali: Chimaeriformes: Chimaeridae). *Zootaxa* 1328:27–38.

Barnett, L.A.K., Earley, R.L., Ebert, D.A., and Cailliet, G.M. (2009a). Maturity, fecundity, and reproductive cycle of the spotted ratfish, *Hydrolagus colliei* (Lay & Bennett, 1839). *Mar. Biol.* 156:301–316.

Barnett, L.A.K., Ebert, D.A., and Cailliet, G.M. (2009b). Assessment of the dorsal fin spine for chimaeroid (Holocephali: Chimaeriformes) age estimation. *J. Fish Biol.* 75:1258–1270.

Bendix-Almgreen, S.A. (1968). The bradyodont elasmobranchs and their affinities; a discussion. In: Ørvig, T. (Ed.), *Current Problems of Lower Vertebrate Phylogeny: Proceedings of the Fourth Nobel Symposium*. Amlqvist and Wiksell, Stockholm, pp. 153–170.

Bergstad, O.A., Wik, Å.D., and Hildre, Ø. (2003). Predator–prey relationships and food sources of the Skagerrak deep-water fish assemblage. *J. Northw. Atl. Fish. Sci.* 31:165–180.

Bigelow, H.B. and Schroeder, W.C. (1950). New and little known cartilaginous fishes from the Atlantic. *Bull. Mus. Comp. Zool.* 103:35–408.

Bigelow, H.B. and Schroeder, W.C. (1951). Three new skates and a new chimaeroid fish from the Gulf of Mexico. *J. Wash. Acad. Sci.* 41:383–392.

Bigelow, H.B. and Schroeder, W.C. (1953). Chimaeroids. In: *Fishes of the Western North Atlantic*, Vol. 1, Part 2. Sears Foundation for Marine Research, Yale University, New Haven, pp. 539–541.

Bory de Saint-Vincent, J.B. (1823). *Dictionnaire Classique D'Histoire Naturelle*, Vol. 3. Paris.

Bullis, Jr., H.R. and Carpenter, J.C. (1966). *Neoharriotta carri*, a new species of Rhinochimaeridae from the southern Caribbean Sea. *Copeia* 1966:443–450.

Calis, E., Jackson, E.H., Nolan, C.P., and Jeal, F. (2005). Preliminary age and growth estimates of the rabbitfish, *Chimaera monstrosa*, with implications for future resource management. *J. Northw. Atl. Fish. Sci.* 35:15–26.

Capello, B. (1867). Descripção de dois peixes novos proveniéntes dos mares de Portugal. *J. Sci. Math Phys. Nat. Lisboa* 1:314–317.

Clarke, M.W., Borges, L., and Officier, R.A. (2005). Comparisons of trawl and longline catches of deepwater elasmobranchs west and north of Ireland. *J. Northw. Atl. Fish. Sci.* 35:429–442.

Cole, F.J. (1896a). On the cranial nerves of *Chimaera monstrosa* (Linn. 1754); with a discussion of the lateral line system, and of the morphology of the corda tympani. *Trans. R. Soc. Edinburgh* 38:631–680.

Cole, F.J. (1896b). On the sensory and ampullary canals of *Chimaera*. *Anat. Anz.* 12:172–182.

Cole, F.J. and Dakin, W.J. (1906). Further observations on the cranial nerves of *Chimaera*. *Anat. Anz.* 28:595–599.

Collett, R. (1904). Diagnoses of four hitherto undescribed fishes from the depths south of the Faroe Islands. *Forhand. Vidensk. Christ.* 9:5.

Compagno, L.J.V., Stehmann, M., and Ebert, D.A. (1990). *Rhinochimaera africana*, a new longnose chimaera from southern Africa, with comments on the systematics and distribution of the genus *Rhinochimaera* Garman, 1901 (Chondrichthyes, Chimaeriformes, Rhinochimaeridae). *S. Afr. J. Mar. Sci.* 9:201–222.

de Buen, F. (1959). Notas preliminares sobre la fauna marina preabismal de Chile, con descripción de una familia de rayas, dos géneros y siete especies nuevos. *Bol. Mus. Nac. Hist. Nat.* 27:171–201.

Dean, B. (1903). An outline of the development of a chimaeroid. *Biol. Bull.* 4:270–286.

Dean, B. (1904a). The egg cases of chimaeroid fishes. *Am. Nat.* 38:486–487.

Dean, B. (1904b). Two Japanese species, *C. phantasma* Jordan and Snyder and *C. Mitsukurii* N.S., and their egg cases. *J. Coll. Sci. Imp. Univ. Tokyo* 19:5–9; pl. 1.

Dean, B. (1904c). Notes on the long-snouted chimaeroid of Japan, *Rhinochimaera* (*Harriotta*) *pacifica*. *J. Coll. Sci. Imp. Univ. Tokyo* 19:1–20.

Dean, B. (1906). *Chimaeroid Fishes and Their Development*, Publ. No. 32. Carnegie Institute, Washington, D.C.

Dean, B. (1909). Studies on fossil fishes (sharks, chimaeroids, and arthrodires). *Mem. Am. Mus. Nat. Hist.* 9:209–287.

Dean, B. (1912). Orthogenesis in the egg capsules of *Chimaera*. *Bull. Am. Mus. Nat. Hist.* 31:35–40.

Di Giácomo, E.E. (1992). Distribucion de la poblacion del pez gallo (*Callorhynchus callorhynchus*) en el Golfo San Matias, Argentina. *Frente Marit.* 12:113–118.

Di Giácomo, E.E. and Perier, M.R. (1996). Feeding habits of cockfish, *Callorhincus callorhynchus* (Holocephali: Callorhynchidae), in Patagonian waters (Argentina). *Mar. Freshw. Res.* 47:801–808.

Di Giácomo, E.E. and Perier, M.R. (2005). The fishing of *Callorhinchus callorhynchus* in the SW Atlantic Ocean. In: *Condrictios del golfo San Matias: biologia, ecologia y explotacion pesquera*, Technical Report. National University of Comahue, Neuquén, Argentina.

Didier, D.A. (1995). Phylogenetic systematics of extant chimaeroid fishes (Holocephali, Chimaeroidei). *Am. Mus. Nat. Hist. Novit.* 3119:1–86.

Didier, D.A. (1998). The leopard *Chimaera*, a new species of chimaeroid fish from New Zealand (Holocephali, Chimaeriformes, Chimaeridae). *Ichthyol. Res.* 45:281–289.

Didier, D.A. (2002). Two new species of chimaeroid fishes from the southwestern Pacific Ocean (Holocephali, Chimaeridae). *Ichthyol. Res.* 49:299–306.

Didier, D.A. (2008). Two new species of the genus *Hydrolagus* Gill (Holocephali: Chimaeridae) from Australia. In: Last, P.R., White, W.T., and Pogonoski, J.J. (Eds.), *Descriptions of New Australian Chondrichthyans*, Research Paper No. 022. CSIRO Marine and Atmospheric Research, Hobart, Tasmania, pp. 349–356.

Didier, D.A. (2006) Chimaeriformes. In: Carpenter, K.E. (Ed.), *The Living Marine Resources of the Eastern Central Atlantic. FAO Species Identification Guide for Fishery Purposes*. Food and Agricultural Organization of the United Nations, Rome.

Didier, D.A. and Nakaya, K. (1999). Redescription of *Rhinochimaera pacifica* (Mitsukuri) and first record of *R. africana* Compagno, Stehmann & Ebert from Japan (Chimaeriformes: Rhinochimaeridae). *Ichthyol. Res.* 46:139–152.

Didier, D.A. and Séret, B. (2002). Chimaeroid fishes of New Caledonia with description of a new species of *Hydrolagus* (Chondrichthyes, Holocephali). *Cybium* 26:225–233.

Didier, D.A. and Stehmann, M. (1996). *Neoharriotta pumila*, a new species of longnose chimaera from the northwestern Indian Ocean (Pisces, Holocephali, Rhinochimaeridae). *Copeia* 1996:955–965.

Didier, D.A., Stahl, B.J., and Zangerl, R. (1994). Compound tooth plates of chimaeroid fishes (Holocephali: Chimaeroidei). *J. Morphol.* 222:7389.

Didier, D.A., LeClair, E.E., and VanBuskirk, D.R. (1998). Embryonic staging and external features of development of the chimaeroid fish *Callorhinchus milii* (Holocephali, Callorhinchidae). *J. Morphol.* 236:25–47.

Didier, D.A., Last, P.R., and White, W.T. (2008). Three new species of the genus *Chimaera* Linnaeus (Chimaeriformes: Chimaeridae) from Australia. In: Last, P.R., White, W.T., and Pogonoski, J.J. (Eds.), *Descriptions of New Australian Chondrichthyans*, Research Paper No. 022. CSIRO Marine and Atmospheric Research, Hobart, Tasmania, pp. 327–339.

Duméril, A. (1865). *Histoire Naturelle des Poissons ou Ichthyologie Genéralé.* Vol. 1. *Elasmobranches, Plagiostomes et Holocéphales.* Paris.

Dunn, M.R., Griggs, L., Forman, J., and Horn, P. (2010). Feeding habits and niche separation among the deep-sea chimaeroid fishes *Harriotta raleighana*, *Hydrolagus bemisi* and *Hydrolagus novaezealandiae*. *Mar. Ecol. Prog. Ser.* 407:209–225.

Ebert, D.A. (1991). Diet of the sevengill shark *Notorynchus cepedianus* in the temperate coastal waters of southern Africa. *S. Afr. J. Mar. Sci.* 11:565–572.

Ebert, D.A. (1994). Diet of the sixgill shark *Hexanchus griseus* off southern Africa. *S. Afr. J. Mar. Sci.* 14:213–218.

Ebert, D.A. (2003). *The Sharks, Rays and Chimaeras of California.* University of California Press, Berkeley, p. 284.

Ebert, D.A. and Winton, M.V. (2010). Chondrichthyans of high latitude seas. In: Carrier, J., Musick, J.A., and Heithaus, M. (Eds.), *Sharks and Their Relatives II: Biodiversity, Adaptive Physiology, and Conservation.* CRC Press, Boca Raton, FL, pp. 115–158.

Eschmeyer, B. (1998). *Catalog of Fishes*, Vols. 1–3. California Academy of Sciences, San Francisco.

Fields, R.D. and Lange, G.D. (1980). Electroreception in the ratfish (*Hydrolagus colliei*). *Science* 207:57–548.

Fields, R.D., Bullock, T.H., and Lange, G.D. (1993). Ampullary sense organs, peripheral, central and behavioral electroreception in chimeras (*Hydrolagus*, Holocephali, Chondrichthyes). *Brain Behav. Evol.* 41:269–289.

Forster, G.R. (1964). Line-fishing on the Continental Shelf. *J. Mar. Biol. Assoc. U.K.* 44:277–284.

Fowler, H.W. (1910). Notes on chimaeroid and ganoid fishes. *Proc. Acad. Nat. Sci. Philadelphia* 62:603–612.

Fowler, H.W. (1941). The fishes of the groups Elasmobranchii, Holocephali, Isospondyli and Ostarophysi obtained by the U.S. Bureau of Fisheries steamer "Albatross" in 1907 to 1910, chiefly in the Philippine Islands and adjacent seas. *Bull. U.S. Nat. Mus.* 100:486–510.

Francis, M.P. (1997). Spatial and temporal variation in the growth rate of elephantfish (*Callorhinchus milii*). *N.Z. J. Mar. Freshw. Res.* 31:9–23.

Francis, M.P. (1998). New Zealand shark fisheries: development, size and management. *Mar. Freshw. Res.* 49:579–591.

Freer, D.W.L. and Griffiths, C.L. (1993a). Estimation of age and growth in the St. Joseph *Callorhinchus capensis* (Dumeril). *S. Afr. J. Mar. Sci.* 13:75–81.

Freer, D.W.L. and Griffiths, C.L. (1993b). The fishery for, and general biology of, the St. Joseph *Callorhinchus capensis* (Dumeril) off south-western Cape, South Africa. *S. Afr. J. Mar. Sci.* 13:63–74.

Garman, S. (1888). On the lateral canal system of the Selachia and Holocephala. *Bull. Mus. Comp. Zool.* 17:57–120.

Garman S. (1901). Genera and families of the chimaeroids. *Proc. N. Engl. Zool. Club* 2:75–77.

Garman, S. (1904). The chismopnea especially *Rhinochimaera* and its allies. *Bull. Mus. Comp. Zool.* 41:245–272.

Garman, S. (1908). New plagiostomia and chismopnea. *Bull. Mus. Comp. Zool.* 51:251–256.

Garman, S. (1911). The chismopnea (chimaeroids). *Mem. Mus. Comp. Zool.* 15:81–101.

Gilbert, C.H. (1905). Deep sea fishes of the Hawaiian Islands. *Bull. U.S. Fish Comm.* 23:575–713.

Gilchrist, J.D.F. (1922). Deep-sea fishes procured by the S.S. "Pickle" (Part 1). *Rep. Fish. Mar. Biol. Surv. Union S. Afr. Fish.* 2:41–79.

Gill, T. (1893). Families and subfamilies of fishes. *Nat. Acad. Sci. Mem.* 6:127–138.

Gillis, J.A., Rawlinson, K.A., Bell, J., Lyon, W.S., Baker, C.V.H., and Shubin, N.H. (2011). Holocephalan embryos provide evidence for gill arch appendage reduction and opercular evolution in cartilaginous fishes. *Proc. Nat. Acad. Sci. USA* 108:1507–1512.

González, C., Teruel, J., López, E., and Paz, X. (2007). Feeding habits and biological features of deep-sea species of the northwest Atlantic: large-eyed rabbitfish (*Hydrolagus mirabilis*), narrownose chimaera (*Harriotta raleighana*) and black dogfish (*Centroscyllium fabricii*). Northwest Atlantic Fisheries Organization Scientific Council Meeting, September, Serial No. N5423, SCR Doc. 07/63, 9 pp.

González-Acosta, A.F., Castro-Aguirre, J.L., Didier, D.A., Vélez-Marín, R., and Burnes-Romo, L.A. (2010). Occurrence of *Hydrolagus macrophthalmus* (Chondrichthyes: Holocephali: Chimaeridae) in the northeastern Pacific. *Rev. Mex. Biodiv.* 81:197–201.

Goode, G.B. and Bean, T.H. (1895). *Oceanic Ichthyology.* U.S. National Museum, Washington, D.C.

Gorman, T.B.S. (1963). *Biological and Economic Aspects of the Elephant Fish* Callorhynchus milii *Bory, in Pegasus Bay and the Canterbury Bight*, Fisheries Technical Report No. 8. New Zealand Ministry of Agriculture and Fisheries, Wellington, pp. 1–54.

Graham, D.H. (1939). Food of the fishes of Otago Harbour and adjacent sea. *Trans. R. Soc. N.Z.* 68:421–436.

Graham, D.H. (1956). *A Treasury of New Zealand Fishes*. Reed, Wellington.

Grogan, E.D. and Lund, R. (2000). *Debeerius ellefseni* (fan. nov., gen. nov., spec. nov.), an autodiastylic chondrichthyan from the Mississippian Bear Gulch Limestone of Montana (USA), the relationships of the Chondrichthyes, and comments on Gnathostome evolution. *J. Morphol.* 243:219–245.

Grogan, E.D. and Lund, R. (2004). The origin and relationships of early Chondrichthyes. In: Carrier, J.C., Musick, J.A., and Heithaus, M.R. (Eds.), *The Biology of Sharks and Their Relatives*, CRC Press, Boca Raton, FL, pp. 3–31.

Grogan, E.D., Lund, R., and Didier, D. (1999). A description of the chimaerid jaw and its phylogenetic origins. *J. Morphol.* 239:45–59.

Gronovius, L.T. (1756). *Museum Ichthyologicum. Vol. II. Sistens piscium indegenorum quorumdam exoticorum, qui in Museo Laurentii Theodori Gronovii adservantur, descriptiones ordine systematico. Accedunt nonnullorum exoticorum piscium icones aeri incisae.* Theodorum Haak, Leiden, The Netherlands.

Gronovius, L.T. (1763). *Zoophylacii Gronoviani Fasciculus Primus Exhibens Animalia Quadrupeda, Amphibia atque Pisces, quae in Museo suo Adservat, Rite Examinavit, Systematice Disposuit, Descripsit atque Iconibus Illustravit Laur. Theod. Gronovius, J.U.D.* Theodorum Haak, Leiden, The Netherlands.

Halstead, B.W. and Bunker, N.C. (1952). The venom apparatus of the ratfish, *Hydrolagus colliei*. *Copeia* 1952:128–138.

Hardy, G.S. and Stehmann, M. (1990). A new deep-water ghost shark, *Hydrolagus pallidus* n. sp. (Holocephali, Chimaeridae), from the eastern North Atlantic, and redescription of *Hydrolagus affinis* (Brito Capello, 1867). *Arch. Fischereiwiss* 40:229–248.

Holt, E.W.L. and Byrne, L.W. (1909). Preliminary note on some fishes from the Irish Atlantic Slope. *Ann. Mag. Nat. Hist.* 8:279–280.

Howell-Rivero, L. (1936). Some new, rare and little-known fishes from Cuba. *Proc. Boston Soc. Nat. Hist.* 41:41–76.

Inoue, J.G., Masaki, M., Lam, K., Tay, B.-H., Danks, J.A., Bell, J., Walker, T.I., and Venkatesh, B. (2010). Evolutionary origin and phylogeny of the modern holocephalans (Chondrichthyes: Chimaeriformes): a mitogenomic perspective. *Mol. Biol. Evol.* 27:2576–2586.

James, K.C., Ebert, D.A., Long, D.J., and Didier, D.A. (2009). A new species of chimaera, *Hydrolagus melanophasma* sp. nov. (Chondricthyes: Chimaeriformes: Chimaeridae), from the eastern North Pacific. *Zootaxa* 2218:59–68.

Johnson, A.G. and Horton, H.F. (1972). Length–weight relationship, food habits, parasites, and sex and age determination of the ratfish, *Hydrolagus colliei* (Lay and Bennett). *Fish. Bull. (NOAA)* 70:421–429.

Jordan, D.S. and Snyder, J.O. (1900). A list of fishes collected in Japan by Keinosuke Otaki, and by the United States Fish Commission steamer "Albatross," with descriptions of fourteen new species. *Proc. U.S. Nat. Mus.* 23:335–380.

Jordan, D.S. and Snyder, J.O. (1904). On the species of white chimaera from Japan. *Proc. U.S. Nat. Mus.* 27:223–226.

Karrer, C. (1972). Die Gattung *Harriotta* Goode and Bean, 1895 (Chondrichthyes, Chimaeriformes, Rhinochimaeridae). *Mitt. Zool. Mus. (Berlin)* 48:203–221.

Kemp, A. (1984). A comparison of the developing dentition of *Neoceratodus forsteri* and *Callorhynchus milii*. *Proc. Linn. Soc. NSW* 107:245–262.

Kemper, J.M., Ebert, D.A., Compagno, L.J.V., and Didier, D.A. (2010a). *Chimaera notafricana* sp. nov. (Chondrichthyes: Chimaeriformes: Chimaeridae), a new species of chimaera from southern Africa. *Zootaxa* 2532:55–63.

Kemper, J.M., Ebert, D.A., Didier, D.A., and Compagno, L.J.V. (2010b). Description of a new species of chimaerid, *Chimaera bahamaensis* sp. nov., from the Bahamas (Holocephali: Chimaeridae). *Bull. Mar. Sci.* 86(3):649–659.

Kobegenova, S.S. (1993). Morphology and morphogenesis of the digestive tract of some cartilaginous fishes (Chondrichthyes). *J. Ichthyol.* 33:111–126.

Krefft, G. (1990). Callorhynchidae. In: Quèro, J.C. et al. (Eds.), *Check-list of the Fishes of the Eastern Tropical Atlantic*, Vol. 1. UNESCO, Lisbon.

Kyne, P.M. and Simpfendorfer, C.A. (2010). Deepwater chondrichthyans. In: Carrier, J., Musick, J.A., and Heithaus, M. (Eds.), *Sharks and Their Relatives II: Biodiversity, Adaptive Physiology, and Conservation*. CRC Press, Boca Raton, FL, pp. 37–113.

Lacépède, B.G.E. (1798). *Histoire Naturelle des Poissons*, Vol. 1. Paris.

Last, P.R. and Stevens, J.D. (2009). *Sharks and Rays of Australia*. Harvard University Press, Cambridge, MA, p. 644.

Last, P.R., White, W.T., and Pogonoski, J.J. (2008). *Chimaera argiloba* sp. nov., a new species of chimaerid (Chimaeriformes: Chimaeridae) from northwestern Australia. In: Last, P.R., White, W.T., and Pogonoski, J.J. (Eds.), *Descriptions of New Australian Chondrichthyans*. CSIRO Marine and Atmospheric Research, Hobart, Tasmania, pp. 341–348.

Lay, G.T. and Bennett, E.T. (1839). Fishes. In: Richardson, J. et al. (Eds.), *The Zoology of Captain Beechey's Voyage*. Henry G. Bohn, London, pp. 71–75.

Leigh-Sharpe, W.H. (1922). The comparative morphology of the secondary sexual characters of elasmobranch fishes, memoirs IV and V. *J. Morphol.* 36:199–243.

Leigh-Sharpe, W.H. (1926). The comparative morphology of the secondary sexual characters of elasmobranch fishes, memoir X. *J. Morphol.* 42:335–348.

Linnaeus, C. (1758). *Systema Naturae, sive regna tria naturae systematice proposita per secundum classes, ordines, genera, et species cum characteribus, differentiis, synonymis, locis*, 10th ed. L. Salvi, Holmiae, Sweden, 824 pp.

Lisney, T.J. (2010). A review of the sensory biology of chimaeroid fishes (Chondrichthyes; Holocephali). *Rev. Fish Biol. Fisher.* 20:571–590.

López, H.L., San Román, N.A., and Di Giácomo, E.E. (2000). On the South Atlantic distribution of *Callorhinchus callorhynchus* (Holocephali: Callorhynchidae). *J. Appl. Ichthyol.* 16:39.

Lund, R. (1977). New information on the evolution of the bradyodont Chondrichthyes. *Field Geol.* 33:521–539.

Lund, R. (1986). The diversity and relationships of the Holocephali. In: Uyeno, T., Arai, R., Taniuchi, T., and Matsuura, K. (Eds.), *Indo-Pacific Fish Biology: Proceedings of the Second International Conference on Indo–Pacific Fishes.* Ichthyological Society of Japan, Tokyo, pp. 97–106.

Lund, R. (1988). New Mississippian Holocephali (Chondrichthyes) and the evolution of the Holocephali. *Mem. Mus. Nat. Hist. Nat. (Paris) C* 53:195–205.

Lund, R. and Grogan, E.D. (1997). Relationships of the Chimaeriformes and the basal radiation of the Chondrichthyes. *Rev. Fish Biol. Fisher.* 7:65–123.

Macpherson, E. (1980). Food and feeding of *Chimaera monstrosa*, Linnaeus, 1758, in the western Mediterranean. *ICES J. Mar. Sci.* 39:26–29.

Macpherson, E. and Roel, B.A. (1987). Trophic relationships in the demersal fish community off Namibia. *S. Afr. J. Mar. Sci.* 5:585–596.

Maisey, J.G. (1986). Heads and tails: a chordate phylogeny. *Cladistics* 2:201–256.

Marques, A. and Porteiro, F. (2000). Hydrothermal vent mussel *Bathymodiolus* sp. (Mollusca: Mytilidae): diet item of *Hydrolagus affinis* (Pisces: Chimaeridae). *Copeia* 2000:806–807.

Maruska, K.F. and Tricas, T.C. (1998). Morphology of the mechanosensory lateral line system in the Atlantic stingray, *Dasyatis sabina*: the mechanotactile hypothesis. *J. Morphol.* 238:1–22.

Mathews, C.P. (1975). Note on the ecology of the ratfish, *Hydrolagus colliei*, in the Gulf of California. *Calif. Fish Game* 61:47–53.

Mauchline, J. and Gordon, J.D.M. (1983). Diets of the sharks and chimaeroids of the Rockall Trough, northeastern Atlantic Ocean. *Mar. Biol.* 75:269–278.

Mitsukuri, K. (1895). On a new species of the chimaeroid group *Harriotta*. *Zool. Mag. Tokyo* 7:97–98.

Moura, T., Figueiredo, I., Bordalo-Machado, P., and Gordo, L.S. (2004). Growth pattern and reproductive strategy of the holocephalan *Chimaera monstrosa* along the Portuguese continental slope. *J. Mar. Biol. Assoc. U.K.* 84:801–804.

Moura, T., Figueiredo, I., Bordalo-Machado, P., and Gordo, L.S. (2005a). Feeding habits of *Chimaera monstrosa* L. (Chimaeridae) in relation to its ontogenetic development on the southern Portuguese continental slope. *Mar. Biol. Res.* 1:118–126.

Moura, T., Figueiredo, I., Bordalo-Machado, P., Almeida, C., and Gordo, L.S. (2005b). A new deep-water chimaerid species, *Hydrolagus lusitanicus* n. sp., from off mainland Portugal with a proposal of a new identification key for the genus *Hydrolagus* (Holocephali: Chimaeridae) in the north-east Atlantic. *J. Fish Biol.* 67:742–751.

Moy-Thomas, J.A. (1939). The early evolution and relationships of the elasmobranchs. *Biol. Rev.* 14:1–26.

Nelson, J.S. (2006). *Fishes of the World*, 4th ed. John Wiley & Sons, New York.

Norman, J.R. (1937). Coast fishes. II. The Patagonian region. *Disc. Rep.* 16:1–150.

Northcutt, G.R. (1989). The phylogenetic distribution and innervation of craniate mechanoreceptive lateral lines. In: Coombs, S. et al. (Eds.), *The Mechanosensory Lateral Line Neurobiology and Evolution.* Springer, New York.

Ørvig, T. (1967). Phylogeny of tooth tissues: evolution of some calcified tissues in early vertebrates. In: Miles, A.E.W. (Ed.), *Structural and Chemical Organization of Teeth.* Academic Press, London, pp. 45–110.

Ørvig, T. (1985). Histologic studies of ostracoderms, placoderms and fossil elasmobranchs 5. Ptyctodontid tooth plates and their bearing on holocephalans ancestry: the condition of chimaerids. *Zool. Sci.* 14:55–79.

Patterson, C. (1965). The phylogeny of the chimaeroids. *Phil. Trans. R. Soc. Lond. B Biol. Sci.* 249:101–219.

Patterson, C. (1992). Interpretation of the toothplates of chimaeroid fishes. *Zool. J. Linn. Soc.* 106:33–61.

Paulin, C., Stewart, A., Roberts, C., and McMillan, P. (1989). *New Zealand Fish: A Complete Guide*, Miscellaneous Series 19. National Museum of New Zealand, Wellington.

Paxton, J.R., Hoese, D.F., Allen, G.R., and Hanley, J.E. (Eds.). (1989). *Zoological Catalogue of Australia.* Vol. 7. *Pisces.* Australian Government Publishing Service, Canberra.

Peach, M.B. and Rouse, G.W. (2000). The morphology of the pit organs and lateral line canal neuromasts of *Mustelus antarcticus* (Chondrichthyes: Triakidae). *J. Mar. Biol. Assoc. U.K.* 80:155–162.

Peyer, B. (1968). *Comparative Odontology.* University of Chicago Press, Chicago.

Quaranta, K.L., Didier, D.A., Long, D.J., and Ebert, D.A. (2006). A new species of chimaeroid, *Hydrolagus alphus* sp. nov. (Chimaeriformes: Chimaeridae), from the Galapagos Islands. *Zootaxa* 1377:33–45.

Quinn, T.P., Miller, B.S., and Wingert, R.C. (1980). Depth distribution and seasonal and diel movements of ratfish, *Hydrolagus colliei*, in Puget Sound. *Wash. Fish. Bull.* 78:816–821.

Reese, A.M. (1910). The lateral line system of *Chimaera colliei*. *J. Exp. Zool.* 9:349–371.

Sathyanesan, A.G. (1966). Egg-laying of the chimaeroid fish *Hydrolagus colliei*. *Copeia* 1966:132–134.

Schauinsland, H. (1903). Beiträge zur Entwicklungsgeschichte und Anatomie der Wirbeltiere. I. *Sphenodon, Callorhynchus, Chamaeleo. Zoologica* 16:5–32, 58–89.

Schnakenbeck, W.V. (1929). Über einige Meeresfische aus Südwestafrika. *Mitt. Zool. Staats Zool. Mus. Hamburg* 44:38–45.

Smith, R.M., Day, R.W., Walker, T.I., and Hamlett, W.C. (2001). Microscopic organization and sperm storage in the oviducal gland of the elephant fish, *Callorhynchus milii*, at different stages of maturity [abstract]. 6th Indo–Pacific Fish Conference, May 20–25, Durban, South Africa.

Soto, J.M.R. and Vooren, C.M. (2004). *Hydrolagus matallanasi* sp. nov. (Holocephali, Chimaeridae), a new species of rabbitfish from southern Brazil. *Zootaxa* 687:1–10.

Stahl, B.J. (1999). Chondrichthyes III: Holocephali. In: Schultze, H.-P. (Ed.), *Handbook of Paleoichthyology*, Vol. 4. Verlag Dr. Friedrich Pfeil, Munich.

Stanley, H.P. (1963). Urogenital morphology in the chimaeroid fish *Hydrolagus colliei* (Lay and Bennett). *J. Morphol.* 112:97–127.

Stehmann, M. and Bürkel, D.L. (1984). Chimaeridae. In: Whitehead, P.J.P., Bauchot, M.L., Hureau, J.C., Nielsen, J., and Tortonese, E. (Eds.), *Fishes of the Northeastern Atlantic and the Mediterranean*, Vol. 1. UNESCO, Paris, pp. 212–215.

Sullivan, K.J. (1978). Age and growth of the elephant fish *Callorhinchus milii* (Elasmobranchii: Callorhynchidae). *N.Z. J. Mar. Freshw. Res.* 11:745–753.

Tanaka, S. (1905). On two new species of *Chimaera*. *J. Coll. Sci. Imp. Univ. Tokyo* 20:1–14.

Ungaro, N. (1996). Ruolo dei crostacei decapodi reptanti nello spettro trofico di *Chimaera monstrosa* L. nell'Adriatico sud-occidentale. *Biol. Mar. Medit.* 3:582–585.

Venkatesh, B., Kirkness, E.F., Loh, Y.-H., Halpern, A.L., Lee, A.P., Johnson, J., Dandona, N., Viswanathan, L.D., Tay, A., Venter, J.C., Strausberg, R.L., and Brenner, S. (2007). Survey sequencing and comparative analysis of the elephant shark (*Callorhinchus milii*) genome. *PLoS Biol.* 5:e101.

Waite, E.R. (1898). Sea fisheries report upon trawling operations off the coast of New South Wales between the Manning River and Jervis Bay, carried on by H.M.C.S. "Thetis," *Sci. Rep. Fish.* 41–42.

Webb, J.F. and Northcutt, G.R. (1997). Morphology and distribution of pit organs and canal neuromasts in non-teleost bony fishes. *Brain Behav. Evol.* 50:139–151.

Whitley, G.P. (1939). Taxonomic notes on sharks and rays. *Aust. Zool.* 9:227–262.

Wilmot, A., Didier, D.A., and Webb, J. (2001). Morphology of the lateral line canals of the head in chimaerid fishes (Family Chimaeridae) [abstract]. *Am. Zool.* 40:1261.

Zangerl, R. (1981). *Handbook of Paleoichthyology*. Vol. 3A. *Chondrichthyes I. Paleozoic Elasmobranchii*. Gustav Fischer Verlag, New York.

Zangerl, R., Winter, H.P., and Hansen, M.C. (1993). Comparative microscopic dental anatomy in the Petalodontida (Chondrichthyes, Elasmobranchii). *Field Geol. Ser.* 26:1–43.

Appendix. Provisional Key to Species

This key is based on study of more than 1000 specimens, almost exclusively from museum collections. Live specimens may vary from preserved specimens, particularly in coloration and overall body shape, and subtle variations in color must be taken into consideration when using this key to identify fresh specimens. This is intended as a provisional key only and is designed for identification of adult or near-mature specimens of all valid species of chimaeroids. Known but undescribed species are not included. Very small juveniles and hatchlings may not be identified using this key because they can vary considerably from adults in body proportions and coloration. It is for this reason that small juveniles and hatchings have not been positively identified for most species.

Species of *Hydrolagus* are particularly troublesome to identify, and some species can only be positively identified on the basis of color; for example, couplet 25 distinguishes *H. ogilbyi* and *H. lemures* solely on the basis of subtle differences in body color, which may not be sufficient for accurate identification. The description of *Hydrolagus lemures* was based on two small juvenile specimens, and it is possible that this species is actually a color variant of *H. ogilbyi*. Likewise, several *Hydrolagus* species are distinguished by light and dark coloration in couplet 27. In some cases, species with similar coloration and geographic range can also be separated by morphological features; for example, *H. affinis* (step 29a) and *H. pallidus* (step 30a) are also reliably distinguished by the proportion of the eye-to-head length, a feature described in a recent FAO guide (Didier, 2006). The only apparent distinguishing feature to separate *H. pallidus* and *H. lusitanica* seems to be pectoral fin proportion based on a comparison of only two specimens of *H. pallidus*. Because pectoral fin proportions vary ontogenetically, additional data would be desirable to confirm that *H. lusitanicus* is indeed different from *H. pallidus* and not a color variant. By far the greatest challenge is the separation of a complex of small-bodied species, all of which are pale brown with slender tapering bodies (steps 14a and 26b). Separation of these species may often depend on geographic location. Morphological distinction of *Chimaera obscura* and *C. macrospina* (step 16) is particularly difficult; however, differences in the CO1 gene support the distinction of these two species. Identification of *H. alberti*, *H. bemisi*, and *H. mitsukurii* (step 33) is also difficult. Although *H. mirabilis* can be distinguished from *H. alberti* by the indentation of the second dorsal fin (step 31), this character may not always be sufficient for identification of these very similar species, which also overlap in geographic range. Separation of these two species may require comparison of the head length to the distance from the junction of the common branch of the oral and preopercular canal to the junction of the trunk lateral line canal. The canal distance is usually longer in *H. mirabilis* and is <2.5 times the head length and generally >2.5 times the head length in *H. alberti*.

1a.   Plow-shaped snout extending forward from the head; body silvery with black saddlelike bands across
        the dorsal surface and dark blotches on the head and trunk; heterocercal tail; large anal fin precedes
        caudal; pelvic claspers in males unbranched, tubelike, lacking a fleshy denticulate tip.................................. 2

1b.   Elongate, spear-shaped snout extending forward from the head; body color an even brown, without
        distinct markings; pelvic claspers unbranched, slender rods with denticulate bulbous tip 3

1c.   Blunt fleshy snout, slightly pointed at the tip; body tapering to whip-like tail; lateral line canals on the
        snout expanded with large dilations; males with branched pelvic claspers bearing fleshy denticulate
        lobes at the tips 10

2a.   Locality South America; dark spots on trunk along lateral line canal may be fused to form large blotches
        usually numbering less than six .... *Callorhinchus callorhynchus*

2b.   Locality South Africa; trunk may be pale with few, usually about three, dark spots or blotches.................... *Callorhinchus capensis*

2c.   Locality New Zealand, Australia; spots on trunk along and above the lateral line canal often numbering
        six or greater; spots usually rounded, not fused into large blotches... *Callorhinchus milii*

3a.   Tooth plates with raised hypermineralized tritors on the surface... 4

3b.   Tooth plates smooth, lacking raised hypermineralized tritors on the surface.. 8

4a.   Separate anal fin located anterior to the ventral lobe of the caudal fin.. 5

4b.   Anal fin absent.. 7

5a.   Pelvic fins rounded along distal margin; second dorsal fin uniform in height; oral and preopercular
        lateral line canals separated by a large space .... *Neoharriotta pinnata*

5b.   Pelvic fins with straight posterior margin; second dorsal fin not uniform in height, sloping posteriorly;
        oral and preopercular lateral line canals separated by a narrow space... 6

6a.   Anal fin originates at, or anterior to, insertion of second dorsal fin; snout evenly slender along its length...... *Neoharriotta carri*

6b.   Anal fin originates posterior to insertion of second dorsal fin; snout wide at base, tapering to a slender
        distal tip... *Neoharriotta pumila*

7a.   Eye is large; dorsal fin spine equal to or longer than height of first dorsal fin............................. *Harriotta raleighana*

7b.   Eye is small; dorsal fin spine significantly shorter than height of first dorsal fin .... *Harriotta haeckeli*

8a.   Body color an even dark brown, snout broad and paddle-shaped; eye is small; junction of supraorbital
        and infraorbital canals on ventral side of snout closer to the tip of the snout than to the nasal canal .... *Rhinochimaera africana*

8b.   Body color a pale brownish gray with dark fins; snout narrow and conical shaped; junction of
        supraorbital and infraorbital canals on ventral side of snout nearly equidistant between the tip of the
        snout and the nasal canal 9

9a.   Locality Pacific Ocean; number of denticulations on upper lobe of caudal fin usually 41 to 68.................... *Rhinochimaera pacifica*

9b.   Locality Atlantic Ocean; number of denticulations on upper lobe of caudal fin usually, 19 to 33.................... *Rhinochimaera atlantica*

10a.   Anal fin present, separated from the ventral caudal fin by a notch 11

10b.   No anal fin present; ventral caudal fin a continuous ridge along base of tail.. 20

11a.   Body color gray, brown or silvery with distinct mottled pattern 12

11b.   Body evenly colored, pale silvery, gray, tan, brown, black, gray-blue, or lavender................................... 14

12a.   Body color gray or tan with chocolate brown reticulations and spots; posterior margin of first dorsal fin
        white; pelvic fins with rounded distal margin .... *Chimaera panthera*

12b.   Pelvic fins with straight or squared distal margin 13

13a.   Body color silvery, mottled with brown spots, unpaired fin edges black; pectoral fins broad, reaching to
        origin of pelvic fin base when depressed, rarely extending to posterior edge of pelvic fin base; males with
        long pelvic claspers, divided for the distal half of length, total length of claspers >20% of body length;
        preopercular and oral canals share a small common branch .... *Chimaera monstrosa*

13b.   Body color dark brown and marbled with small pale spots; pectoral fins elongate and slender, reaching
        posterior edge of pelvic fin base or beyond when depressed; males with pelvic claspers divided for the
        distal one third of length, total length of claspers <20% of body length; preopercular and oral canals do
        not share a common branch.. *Chimaera owstoni*

14a.   Body color an uniform dark brown or black.. 15

14b.   Body color pale, silvery, gray, tan, gray-blue, or lavender.. 17

15a.   Trunk lateral line canal nearly straight along length 16

15b. Trunk lateral line canal with slight undulations along length; large body; body color caramel brown; locality northwestern Atlantic (Bahamas) .... *Chimaera bahamaensis*

16a. Pectoral fin weakly convex anteriorly, convex posteriorly; preopercular and oral lateral line canals share a common branch, or not sharing; body color dark brown to black; locality New South Wales .... *Chimaera obscura*

16b. Pectoral fin almost straight anteriorly, convex posteriorly; preopercular and oral canals share a common branch; body color chocolate brown; locality Australia .... *Chimaera macrospina*

16c. Pectoral fin when depressed extends to origin of pelvic fin; male pelvic claspers divided for distal one-third of their length; preopercular and oral canals share a common branch; body color blackish-brown with dark bluish streaking; locality southern Africa... *Chimaera notafricana*

16d. Body color even brown; locality Japan .... *Chimaera jordani*

17a. Preopercular and oral lateral line canals share a common branch from the infraorbital canal; trunk lateral line canal without sinuous undulations along its length.. 18

17b. Preopercular and oral lateral line canals branch separately from the infraorbital canal; trunk lateral line canal with tight sinuous undulations or broad undulations anterior to pelvic fin 19

18a. Body color silvery pink to pale brown; pectoral fins when depressed extend slightly posterior to pelvic fin origin; pelvic fins large, paddle-shaped; male pelvic claspers divided for less than half of their length .... *Chimaera fulva*

18b. Body color gray-blue or lavender; pectoral fins when depressed extend to pelvic fin origin or beyond; pelvic fins large and rounded; adults massive in size; male pelvic claspers divided for distal one-third of their length.. *Chimaera lignaria*

18c. Body color iridescent beige, tan, or bronze; pectoral fins do not extend to pelvic fin origin when depressed; male pelvic claspers divided for distal one-third of their length... *Chimaera opalescens*

19a. Body color silvery grayish; pectoral fins when depressed extend well posterior to pelvic fin insertion; pelvic fins broadly triangular; male pelvic claspers extend posterior to distal edges of pelvic fins, divided for distal one half to two thirds length .... *Chimaera argiloba*

19b. Body color silvery white below, darker above, dorsal and anal fin edges blackish; faint dark longitudinal stripes along the lateral line canal and trunk; pectoral fins when depressed extend to one half length of pelvic fin... *Chimaera phantasma*

19c. Body color silver gray, unpaired fin edges black; trunk lateral line canal with broad undulations, if present, anterior to pelvic fin; pectoral fins when depressed extend halfway along length of pelvic fin; male pelvic claspers divided for distal three fifths length... *Chimaera cubana*

20a. Distinct pattern of white spots and or blotches on the head and trunk.. 21

20b. Body color even, lacking pattern of spots or stripes... 24

21a. Oral and preopercular canals share a common branch from the infraorbital canal... 22

21b. Oral and preopercular canals branch separately from the infraorbital canal 23

22a. Body color medium brown with numerous narrow, circular and elongate white blotches .... *Hydrolagus mccoskeri*

22b. Body color dark brown with one white spot on side above pectoral fin; second dorsal fin dark anterior and posteriorly with middle indented and white.. *Hydrolagus alphus*

22c. Body color dark brown or black with about nine large white spots; pelvic fins rounded................................ *Hydrolagus barbouri*

22d. Body with brown reticulations and spots on head and trunk... *Hydrolagus matallansi*

23a. Body color brown or reddish brown with small white spots on head and trunk .... *Hydrolagus colliei*

23b. Body color dark brown or gray, sometimes almost black, with white spots fusing to elongate blotches on the head and trunk.. *Hydrolagus novaezealandiae*

23c. Body color pale brown with prominent, but faint, dark reticulations on trunk... *Hydrolagus marmoratus*

24a. Trunk lateral line canal with regular, small sinuous undulations along its length 25

24b. Trunk lateral line canal lacking sinuous undulations.. 26

25a. Body color usually a uniform pale cream, brown, or tan, sometimes paler ventrally, fins darker with distal margins black; a pale indistinct brownish stripe may be visible along the trunk................................... *Hydrolagus lemures*

25b. Body color pale, white, silvery, or tan, lighter ventrally, snout sometimes yellowish, fins dark, usually charcoal to black in color.. *Hydrolagus ogilbyi*

26a. Large-bodied fish, adults sometimes massive; body color dark black, purplish, blue, or gray........................ 27

26b. Small-bodied, slender fish, some adults almost dwarf-like; body color pale, brown, tan, or silvery/gray 31

27a. Body color dark brown, black, or purplish 28

27b.   Body color pale, blue, gray, or light brown.. 30
28a.   Second dorsal fin about equal in height throughout its length.. 29
28b.   Second dorsal fin depressed in center with greatest height at the anterior and posterior portions; long
         slender body with pointed snout and large eye; body color an even dark black; locality Chile and Peru ...... *Hydrolagus macrophthalmus*

29a.   Massive body with large blunt head; body color a dark black or purplish-black; locality western and
         eastern North Atlantic; eye diameter >5 times in head length... *Hydrolagus affinis*
29b.   Large body with blunt snout; body color dark black; locality southern California and Baja California .... *Hydrolagus melanophasma*
29c.   Medium body with a gently point snout and large eyes; body color an even black or blackish-brown;
         adults usually <1 m in length; locality southeastern Australia.. *Hydrolagus homonycteris*
29d.   Large head and body; body color an even purplish or purple-black; adults very large................................ *Hydrolagus purpurescens*

30a.   Massive size with large blunt head; color pale gray or bluish; locality eastern North Atlantic; eye
         diameter <5 times in head length.. *Hydrolagus pallidus*
30b.   Large bodied with large blunt head; color pale brown or rose; locality eastern North Atlantic; pectoral
         fin length 1.9 to 2.3 times width.. *Hydrolagus lusitanicus*
30c.   Large bodied with distinctly pointed snout; color pale blue or blue-gray with dark black margin around
         orbit and along trunk lateral line canal... *Hydrolagus trolli*

31a.   Second dorsal fin deeply indented in the center, nearly dividing the fin into two parts; eye is large; body
         color a pale brown or gray-brown with dark fins; body shape with a tendency toward concentrated
         mass in the trunk, tapering rapidly to a long slender tail with long caudal filament........................... *Hydrolagus mirabilis*
31b.   Second dorsal fin straight along distal margin, or only slightly indented in the center 32

32a.   Second dorsal fin only slightly indented in the center; long, curved fin spine usually equal to or
         sometimes exceeding height of the first dorsal fin; males with lateral patch of denticles on the prepelvic
         tenaculae; body color pale brown or tan .... *Hydrolagus africana*
32b.   Second dorsal fin evenly tall along its length, not indented in center; prepelvic tenaculae possess only a
         single medial row of denticles.. 33

33a.   Dorsal fin spine usually exceeds height of first dorsal fin and reaches beyond origin of second dorsal fin
         when depressed; color pale gray-brown with dark almost black fins; locality Japan .... *Hydrolagus mitsukurii*
33b.   Dorsal fin spine usually equal to height of first dorsal fin, just reaches origin of second dorsal fin when
         depressed; color pale silvery/gray in life, pale brown or tan in fixative, pale white or cream ventrally;
         locality New Zealand .... *Hydrolagus bemisi*
33c.   Dorsal fin spine usually nearly equal to height of first dorsal fin; body color an even brown with dark
         fins; locality Gulf of Mexico and Caribbean... *Hydrolagus alberti*

Section II

Form, Function, and Physiological Processes

5

Biomechanics of Locomotion in Sharks, Rays, and Chimaeras

Anabela M.R. Maia, Cheryl A.D. Wilga, and George V. Lauder

CONTENTS

5.1 Introduction

The body form of sharks is notable for the distinctive heterocercal tail with external morphological asymmetry present in most taxa and the ventrolateral winglike pectoral fins extending laterally from the body (Figure 5.1) that give the appearance of powerful yet effortless locomotion. In contrast, expansion of the pectoral fins coupled with a dorsoventrally flattened body in rays and skates resulted in modification of locomotor mode from trunk based to pectoral based, while the chimaera body shape is similar to that of actinopterygian fishes in terms of lateral compression. These features are distinct from the variety of body forms present in actinopterygian fishes (Lauder, 2000) and have long been of interest to researchers wishing to understand the functional design of sharks (Aleev, 1969; Garman, 1913; Grove and Newell, 1936; Harris, 1936; Magnan, 1929; Thomson, 1971).

5.1.1 Approaches to Studying Locomotion in Chondrichthyans

Historically, many attempts have been made to understand the function of the median and paired fins in sharks and rays, and these studies have included work with models (Affleck, 1950; Harris, 1936; Simons, 1970), experiments on fins removed from the body (Aleev, 1969; Alexander, 1965; Daniel, 1922; Harris, 1936), and quantification of body form and basic physical modeling (Thomson, 1976; Thomson and Simanek, 1977). More recently, direct quantification of fin movement using videography has allowed a better understanding of fin conformation and movement (Ferry and Lauder, 1996; Fish and Shannahan, 2000; Flammang, 2010; Wilga and Lauder, 2000), although such studies have to date been limited to relatively few species. Obtaining high-resolution, three-dimensional (3D) data on patterns of shark fin motion is a difficult task, and these studies have

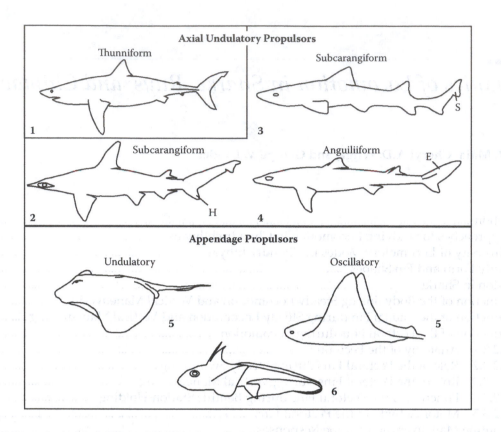

FIGURE 5.1
Propulsion mechanisms in chondrichthyans. Numbers indicate body groups (see text). E, epicaudal lobe; H, hypochordal lobe; S, subterminal lobe. (Based on Webb, 1984; Webb and Blake, 1985.)

been confined to a highly controlled laboratory environment where sharks swim in a recirculating flow tank. Although locomotion of sharks and rays under these conditions does not allow the range of behaviors seen in the wild, the ability to obtain data from precisely controlled horizontal swimming as well as specific maneuvering behaviors has been vital to both testing classical hypotheses of fin function and to the discovery of new aspects of locomotory mechanics. A key general lesson learned from recent experimental kinematic and hydrodynamic analyses of shark locomotion is the value of understanding the 3D pattern of fin movement and the requirement for experimental laboratory studies that permit detailed analyses of fin kinematics and hydrodynamics.

Two new laboratory-based approaches in recent years have been particularly fruitful in clarifying the biomechanics of shark locomotion. Chief among these has been the use of two- and three-camera high-speed video systems to quantify patterns of fin motion in 3D (e.g., Ferry and Lauder, 1996; Standen and Lauder, 2005; Wilga and Lauder, 2000). Two-dimensional (2D) analyses are subject to very large errors when motion occurs in 3D, and the orientation of a planar surface element in 3D can be opposite to the angle appearing in a single 2D view; an example of this phenomenon relevant to the study of shark tails is given in Lauder (2000). The use

of two or more simultaneous high-speed video cameras permits determination of the x, y, and z locations of individual points and hence the 3D orientation of fin and body surface elements and distortion to be extracted from the images (Lauder and Madden, 2008). Three-dimensional kinematic analysis has been identified as the new challenge in fish locomotion (Tytell et al., 2008).

The second new approach to studying shark locomotor biomechanics has been the application of flow visualization techniques from the field of fluid mechanics. Briefly, the technique of particle image velocimetry (PIV) (Krothapalli and Lourenco, 1997; Willert and Gharib, 1991) allows direct visualization of water flow around the fins of swimming sharks and quantification of the resulting body and fin wake (e.g., Lauder and Drucker, 2002; Lauder et al., 2003; Wilga and Lauder, 2002). We now have the ability to understand the hydrodynamic significance of different fin and body shapes and to measure forces exerted on the water as a result of fin motion (Lauder and Drucker, 2002). This represents a real advance over more qualitative previous approaches, such as injection of dye to gain an impression of how the fins of fishes function.

Additional techniques that have provided new avenues for research in fish locomotion and are being applied to chondrichthyan locomotion are computational fluid dynamics (CFD) (Tytell et al., 2010) and material property testing on cartilaginous locomotor structures (Porter and Long, 2010; Porter et al., 2006, 2007; Schaefer and Summers, 2005). Finally, more traditional experimental techniques such as electromyography to quantify the timing of muscle activation, in combination with newer techniques such as sonomicrometry (Donley and Shadwick, 2003; Donley et al., 2005), are revealing new aspects of shark muscle function during locomotion.

5.1.2 Diversity of Locomotory Modes in Chondrichthyans

Sharks, rays, and chimaeras have had a long evolutionary history leading to the locomotor modes observed in extant forms (Carroll, 1988). Chondrichthyans have a remarkable diversity of body forms and locomotor modes for a group containing so few species (Figure 5.1). All sharks swim using continuous lateral undulations of the axial skeleton; however, angel sharks, which are dorsoventrally depressed, may supplement axial propulsion with undulations of their enlarged pectoral fins. Four modes of axial undulatory propulsion have been described, based on decreasing proportion of the body that is undulated during locomotion, which form a continuum from anguilliform to thunniform (Donley and Shadwick, 2003; Webb and Blake, 1985; Webb and Keyes, 1982). In anguilliform swimmers, the entire trunk and tail participate in lateral undulations where more than one wave is present. This mode is characteristic of many elongate sharks such as orectolobiforms, *Chlamydoselachus*, and more benthic carcharhiniform sharks such as scyliorhinids. More pelagic sharks, such as squaliforms, most carcharhiniforms, and some lamniforms, are carangiform swimmers (Breder, 1926; Donley and Shadwick, 2003; Gray, 1968; Lindsey, 1978), whose undulations are mostly confined to the posterior half of the body with less than one wave present. The amplitude of body motion increases markedly over the posterior half of the body (Donley and Shadwick, 2003; Webb and Keyes, 1982). Only the tail and caudal peduncle undulate in thunniform swimmers, which is a distinguishing feature of lamniform sharks, most of which are high-speed cruisers (Donley et al., 2005).

Most batoids (skates and rays) have short, stiff head and trunk regions with slender tails and reduced dorsal fins; therefore, they must swim by moving the pectoral fins. Two modes of appendage propulsion are exhibited by batoids: undulatory and oscillatory (Figure 5.1) (Webb, 1984). Similar to axial swimmers, undulatory appendage propulsors swim by passing undulatory waves down the pectoral fin from anterior to posterior (Daniel, 1922). Most batoids are undulatory appendage propulsors; however, some myliobatiforms, such as eagle and manta rays, swim by flapping their

pectoral fins up and down in a mode known as oscillatory appendage propulsion (Rosenberger, 2001). In addition, batoids can augment thrust by punting off the substrate with the pelvic fins (Koester and Spirito, 1999; Macesic and Kajiura, 2010). Holocephalans are appendage propulsors and utilize a combination of flapping and undulation of the pectoral fins for propulsion and maneuvering, much like many teleost fishes (Combes and Daniel, 2001; Foster and Higham, 2010).

5.1.3 Body Form and Fin Shapes

Most species of sharks have a fusiform-shaped body that varies from elongate in species such as bamboo sharks to the more familiar torpedo shape of white sharks; however, angel sharks and wobbegong sharks are dorsoventrally depressed. There is great variability in the morphology of the paired and unpaired fins. Four general body forms have been described for sharks that encompass this variation (Thomson and Simanek, 1977), with two additional body forms that include batoids and holocephalans.

Sharks with body type 1 (Figure 5.1) have a conical head; a large, deep body; large pectoral fins; a narrow caudal peduncle with lateral keels; and a high-aspect-ratio tail (high heterocercal angle) that is externally symmetrical. These are typically fast-swimming pelagic sharks such as *Carcharodon*, *Isurus*, and *Lamna*. As is typical of most high-speed cruisers, these sharks have reduced pelvic, second dorsal, and anal fins, which act to increase streamlining and reduce drag; however, *Cetorhinus* and *Rhincodon*, which are slow-moving filter feeders, also fit into this category. In these sharks, the externally symmetrical tail presumably results in more efficient slow cruising speeds in large-bodied pelagic sharks, aligns the mouth with the center of mass and the center of thrust from the tail, and probably increases feeding efficiency.

Sharks with body type 2 (Figure 5.1) have a more flattened ventral head and body surface, a less deep body, large pectoral fins, and a lower heterocercal tail angle, and they lack keels. These are more generalized, continental swimmers such as *Alopias*, *Carcharias*, *Carcharhinus*, *Galeocerdo*, *Negaprion*, *Prionace*, *Sphyrna*, *Mustelus*, and *Triakis*. *Alopias* is similar to these sharks despite the elongate pectoral and caudal fins. Similarly, hammerheads, with the exception of the cephalofoil, also fit into this category. These sharks probably have the greatest range of swimming speeds. They also retain moderately sized pelvic, second dorsal, and anal fins and therefore remain highly maneuverable over their swimming range.

Sharks with body type 3 (Figure 5.1) have relatively large heads, blunt snouts, more anterior pelvic fins, more posterior first dorsal fins, and a low heterocercal tail angle with a small to absent hypochordal lobe and a large subterminal lobe. These sharks are slow-swimming epibenthic, benthic, and demersal sharks such as *Scyliorhinus*, *Ginglymostoma*, *Chiloscyllium*, *Galeus*, *Apristurus*, *Pseudotriakis*, and Hexanchiformes. Pristiophoriforms and pristiforms may fit best into this category. Although the body morphology of hexanchiform sharks is most similar to these, they have only one dorsal fin that is positioned more posterior on the body than the pelvic fins.

Body type 4 (Figure 5.1) is united by only a few characteristics and encompasses a variety of body shapes. These sharks lack an anal fin and have a large epicaudal lobe. Only squalean or dogfish sharks are represented in this category. Most of these species are deep-sea sharks and have slightly higher pectoral fin insertions (i.e., *Squalus*, *Isistius*, *Centroscymnus*, *Centroscyllium*, *Dalatius*, *Echinorhinus*, *Etmopterus*, and *Somniosus*). *Squalus* also frequent continental waters and have higher aspect tails similar to those in type 2.

A fifth body type (Figure 5.1) can be described based on dorsoventral flattening of the body, enlarged pectoral fins, and a reduction in the caudal half of the body. This type would include batoids, except for pristiforms and guitarfishes. These chondrichthyans are largely benthic but also include the pelagic myliobatiform rays. Rajiforms and myliobatiforms locomote by undulating the pectoral fins, whereas torpediniforms undulate the tail and rhinobatiforms undulate both the pectoral fins and tail.

Holocephalans or chimaeras represent the sixth body type. They resemble teleosts in that they are laterally compressed and undulate the pectoral fins rather than the axial body in steady horizontal swimming. Tail morphology ranges from long and tapering (leptocercal) to distinctly heterocercal.

5.2 Locomotion in Sharks

5.2.1 Function of the Body during Steady Locomotion and Vertical Maneuvering

The anatomy of the various components of shark fin and body musculature and skeleton has previously been reviewed (Bone, 1999; Compagno, 1999; Kemp, 1999; Liem and Summers, 1999) and is not covered again here, where our focus is the biomechanics of fin and body locomotion. It is worth noting, however, that there are very few detailed studies of the musculature and connective tissue within fins and little knowledge of how myotomal musculature is modified at the caudal peduncle (Gemballa et al., 2006; Reif and Weishampel,

1986; Wilga and Lauder, 2001). Such studies will be particularly valuable for understanding how muscular forces are transmitted to paired and median fins.

One of the most important factors in shark locomotion is the orientation of the body, because this is the primary means by which the overall force balance (considered in detail below) is achieved during swimming and maneuvering. When sharks are induced to swim horizontally so that the path of any point on the body is at all times parallel to the x (horizontal) axis with effectively no vertical (y) motion, the body is tilted up at a positive angle of attack to oncoming flow (Figure 5.2). This positive body angle occurs even though sharks are swimming steadily and not maneuvering and are maintaining their vertical position in the water. This positive body angle ranges from 11° to 4° in *Triakis* and *Chiloscyllium*, respectively, at slow swimming speeds of 0.5 l/s. The angle of body attack varies with speed, decreasing to near zero at 2 l/s swimming speed (Figure 5.2). During vertical maneuvering in the water column, the angle of the body is altered as well (Figure 5.3). When leopard sharks rise so that all body points show increasing values along the y-axis, the body is tilted to a mean angle of 22° into the flow. During sinking in the water, the body is oriented at a negative angle of attack averaging –11° in *Triakis* (Figure 5.3). These changes in body orientation undoubtedly reflect changes in lift forces necessary either to maintain body position given the negative buoyancy of most sharks or to effect vertical maneuvers.

The locomotor kinematics of the body in sharks at a variety of speeds has been studied by Webb and Keyes (1982). Recent studies have presented electromyographic recordings of body musculature to correlate activation patterns of red myotomal fibers with muscle strain patterns and body movement (Donley and Shadwick, 2003; Donley et al., 2005). Red muscle fibers in the body myotomes of *Triakis* are activated to produce the body wave at a consistent relative time all along the length of the body (Donley and Shadwick, 2003). The onset of muscle activation always occurred as the red fibers were lengthening, and these fibers were deactivated consistently during muscle shortening. The authors concluded that the red muscle fibers along the entire length of the body produce positive power and hence contribute to locomotor thrust generation, in contrast to some previous hypotheses suggesting that locomotion in fishes is powered by anterior body muscles alone. Strain in the white axial musculature, which is indicative of force transmission, was measured in mako sharks, *Isurus oxyrinchus* (thunniform swimmers), and showed that there is a decoupling of red muscle activity and local axial bending (Donley et al., 2005). The presence of well-developed hypaxial lateral tendons that differ markedly from those in teleost fishes lends support to this hypothesis (Donley et al., 2005). Recent studies on musculotendinous anatomy revealed

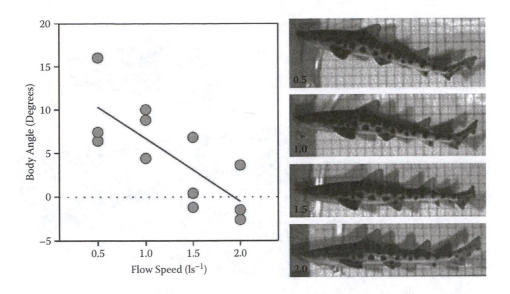

FIGURE 5.2

Plot of body angle vs. flow speed to show the decreasing angle of the body with increasing speed. Each symbol represents the mean of five body angle measurements (equally spaced in time) for five tail beats for four individuals. Images show body position at the corresponding flow speeds in *l*/s, where *l* is total body length (flow direction is left to right). At all speeds, sharks are holding both horizontal and vertical position in the flow and not rising or sinking in the water column. Body angle was calculated using a line drawn along the ventral body surface from the pectoral fin base to the pelvic fin base and the horizontal (parallel to the flow). A linear regression ($y = 15.1 - 7.4x$, adjusted $r^2 = 0.43$, $P < 0.001$) was significant and gives the best fit to the data. (From Wilga, C.D. and Lauder, G.V., *J. Exp. Biol.*, 203, 2261–2278, 2000. With permission.)

significant implications for force transmission in thunniform sharks (Gemballa et al., 2006). This study compared red muscle and tendon changes in subcarangiform to thunniform swimmers. The subcarangiform species have myosepta with one main anterior-pointing cone and two posterior-pointing ones. Within each myoseptum the cones are connected by longitudinal tendons, hypaxial and epaxial lateral tendons, and myorhabdoid

tendons, while connection to the skin and vertebral axis is made through epineural and epipleural tendons with a mediolateral orientation. The lateral tendons do not extend more than 0.075 total length (TL) of the shark, and the red muscles insert in the mid-region of these lateral tendons. The thunniform swimmer (mako), however, has a very different condition, thought to have evolved as a result of the demands of this locomotor mode. The

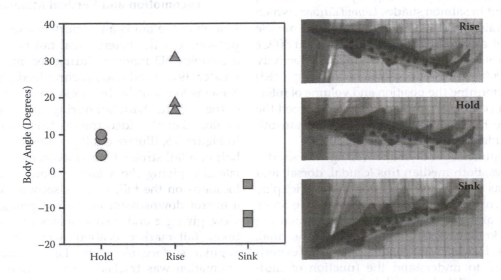

FIGURE 5.3

Plot of body angle vs. behavior during locomotion at 1.0 *l*/s. Circles indicate holding behavior, triangles show rising behavior, and squares reflect sinking behavior. Body angle was calculated as in Figure 5.2. Each point represents the mean of five sequences for each of four individuals. To the right are representative images showing body position during rising, holding, and sinking behaviors. Body angle is significantly different among the three behaviors (ANOVA, $P = 0.0001$). (From Wilga, C.D. and Lauder, G.V., *J. Exp. Biol.*, 203, 2261–2278, 2000. With permission.)

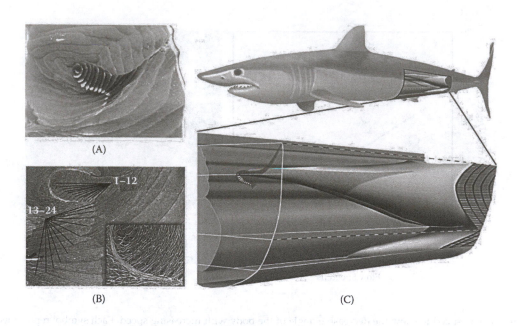

FIGURE 5.4

Muscle and tendon architecture in a thunniform swimmer, *Isurus oxyrinchus*. (A, B) Transverse sections through main anterior cone and adjacent hypaxial musculature, lateral to the left. (A) Fresh specimen illustrating the deep position of red muscles within the white muscles. (B) Histological section at 0.54 with 24 hypaxial lateral tendons visible (1 to 12 within red muscles, and 13 to 24 within white muscles). Dorso- and ventromedially, the red muscles are separated from the white muscles by a sheath of connective tissue. (C) Three-dimensional reconstruction of a posterior myoseptum. Notice the sections of hypaxial lateral tendons within the red muscle and the correspondence with the sections shown in (A) and (B). (From Gemballa, S. et al., *J. Morphol.*, 267, 477–493, 2006. With permission.)

red muscle is internalized and surrounded by a lubricating connective tissue sheath, and it inserts onto the anterior hypaxial lateral tendon, which increases caudally, spanning as much as 0.19 TL. In addition, the medio-lateral fibers are not organized into tendons as in sub-carangiform species (Figure 5.4) (Gemballa et al., 2006). Additional specializations for high-speed swimming have been found in salmon sharks, *Lamna ditropis*, which inhabit cold waters and have internalized red muscle that function at elevated temperatures (20°C and 30°C); thus, this species is closer to mammals in muscle activity (Bernal et al., 2005). Magnetic resonance imaging (MRI) was used to determine the position and volume of internalized red muscle in salmon sharks and confirmed the position of the hypaxial lateral tendons that transmit force to the caudal peduncle (Perry et al., 2007).

During propulsion and maneuvering in sharks, skates, and rays, both median fins (caudal, dorsal, and anal) as well as paired fins (pectoral and pelvic) play an important role. In this chapter, however, we focus on the caudal and pectoral fins, as virtually nothing quantitative is known about the function of dorsal, anal, and pelvic fins. Harris (1936) conducted specific experiments designed to understand the function of multiple fins using model sharks placed in an unnatural body position in a wind tunnel. The first dorsal fin in white sharks, *Carcharodon carcharias*, has been hypothesized to function as a dynamic stabilizer during steady swimming based on dermal fiber arrangement, which

may allow internal hydrostatic pressure to increase (Lingham-Soliar, 2005). The role of the dorsal, anal, and pelvic fins during locomotion in elasmobranchs is a key area for future research on locomotor mechanics.

5.2.2 Function of the Caudal Fin during Steady Locomotion and Vertical Maneuvering

Motion of the tail is an important aspect of shark propulsion, and the heterocercal tail of sharks moves in a complex 3D manner during locomotion. Ferry and Lauder (1996) used two synchronized high-speed video cameras to quantify the motion of triangular segments of the leopard shark tail during steady horizontal locomotion. Sample video frames from that study, shown in Figure 5.5, illustrate tail position at six times during half of a tail stroke. One video camera viewed the tail laterally, giving the x and y coordinates of identified locations on the tail, while a second camera aimed at a mirror downstream of the tail provided a posterior view, giving z and y coordinates for those same locations. Tail marker locations were connected into triangular surface elements (Figure 5.6A,B), and their orientation was tracked through time. This approach is discussed in more detail by Lauder (2000). Analysis of surface element movement through time showed that for the majority of the tail beat cycle the caudal fin surface was inclined at an angle greater than 90° to the horizontal (Figure 5.6), suggesting that the downwash

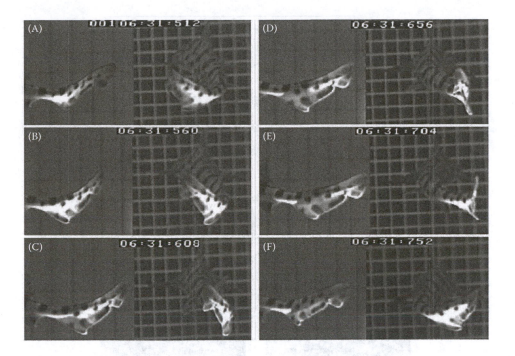

FIGURE 5.5
Composite video sequence of the tail beating from the leftmost extreme (A), crossing the midline of the beat (B, C, and D), and beating to the rightmost extreme or maximum lateral excursion (reached in E and F). In (F), the tail has started its beat back to the left. Times for each image are shown at the top, with the last three digits indicating elapsed time in milliseconds. Each panel contains images from two separate high-speed video cameras, composited into a split-screen view. (From Ferry, L.A. and Lauder, G.V., *J. Exp. Biol.*, 199, 2253–2268, 1996. With permission.)

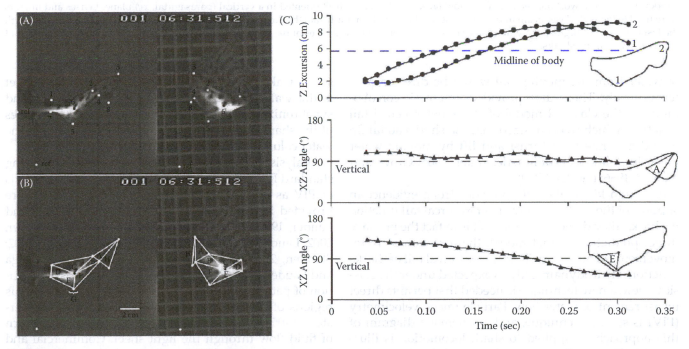

FIGURE 5.6
Images of the tail of a representative leopard shark, *Triakis semifasicata*, swimming in the flow tank. Landmarks (1–8) are shown in (A) with both lateral and posterior views and in (B) with the points joined to form the triangles (A–H) for analysis. Points marked "ref" were digitized as reference points. Both views were identically scaled using the grid in the lateral view (1 box = 2 cm); the smaller grid visible in the posterior view is the upstream baffle reflected in the mirror toward which the shark is swimming. (C) Heterocercal tail kinematics in a representative leopard shark swimming steadily at 1.2 *l*/s; z-dimension excursions (upper panel) of two points on the tail and the three-dimensional angles of two tail triangles with the *xz* plane. Note that for most of the tail beat, the orientation of these two triangular elements is greater than 90°, indicating that the tail is moving in accordance with the classical model of heterocercal tail function. (From Ferry, L.A. and Lauder, G.V., *J. Exp. Biol.*, 199, 2253–2268, 1996; Lauder, G.V., *Am. Zool.*, 40, 101–122, 2000. With permission.)

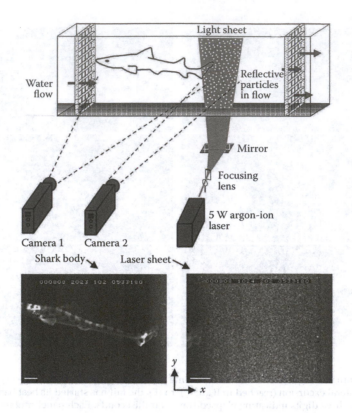

FIGURE 5.7
Schematic diagram of the working section of the flow tank illustrating the defocusing digital particle image velocimetry (DDPIV) system. Sharks swam in the working section of the flow tank with the laser sheet oriented in a vertical (parasagittal, xy) plane. Lenses and mirrors were used to focus the laser beam into a thin light sheet directed vertically into the flow tank. The shark is shown with the tail cutting through the laser sheet. Two high-speed video cameras recorded synchronous images of the body (camera 1) and particles in the wake (camera 2) of the freely swimming sharks.

of water from the moving tail would be directed posteroventrally. These data provided kinematic corroboration of the classical model of shark heterocercal tail function, which hypothesized that the shark caudal fin would generate both thrust and lift by moving water posteriorly and ventrally (Alexander, 1965; Grove and Newell, 1936; Lauder, 2000).

Although kinematic data provide strong evidence in support of the classical view of heterocercal tail function in sharks, they do not address what is in fact the primary direct prediction of that model: the direction of water movement. To determine if the heterocercal tail of sharks functions hydrodynamically as expected under the classical view, a new technique is needed that permits direct measurement of water flow. Particle image velocimetry (PIV) is such a technique, and a schematic diagram of this approach as applied to shark locomotion is illustrated in Figure 5.7. Sharks swim in a recirculating flow tank, which has been seeded with small (12-μm mean diameter) reflective hollow glass beads. A 5- to 10-W laser is focused into a light sheet 1 to 2 mm thick and 10 to 15 cm wide, and this beam is aimed into the flow tank using focusing lenses and mirrors. Sharks are induced to swim with the tail at the upstream edge of the light sheet

so the wake of the shark passes through the light sheet as this wake is carried downstream. Generally, a second synchronized high-speed video camera takes images of the shark body so orientation and movements in the water column can be quantified.

Analysis of wake flow video images proceeds using standard PIV processing techniques, and further details of PIV as applied to problems in fish locomotion are provided in a number of recent papers (Drucker and Lauder, 1999, 2005; Lauder, 2000; Lauder and Drucker, 2002; Lauder et al., 2002, 2003; Nauen and Lauder, 2002; Standen, 2010; Standen and Lauder, 2005, 2007; Wilga and Lauder, 1999, 2000, 2001, 2002). Briefly, cross-correlation of patterns of pixel intensity between homologous regions of images separated in time is used to generate a matrix of velocity vectors, which reflect the pattern of fluid flow through the light sheet. Commercial and freeware versions of PIV analysis software are available and used widely (Raffel et al., 2007; Stamhuis, 2006). Sample PIV data are presented in Figure 5.8. From these matrices of velocity vectors the orientation of fluid accelerated by the tail can be quantified and any rotational movement measured as fluid vorticity. Recent research on fish caudal fin function has shown that the caudal

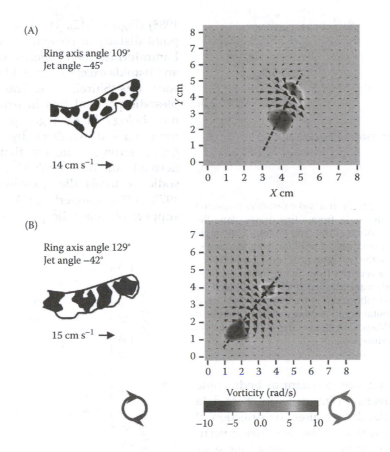

FIGURE 5.8

Defocusing digital particle image velocimetry (DDPIV) analysis of the wake of the tail of representative (A) *Triakis semifasciata* and (B) *Chiloscyllium punctatum* sharks during steady horizontal locomotion at 1.0 *l*/s. On the left is a tracing depicting the position of the tail relative to the shed vortex ring visible in this vertical section of the wake. The plot to the right shows fluid vorticity with the matrix of black velocity vectors representing the results of DPIV calculations based on particle displacements superimposed on top. A strong jet, indicated by the larger velocity vectors, passes between two counterrotating vortices representing a slice through the vortex ring shed from the tail at the end of each beat. The black dashed line represents the ring axis angle. *Note:* Light gray color indicates no fluid rotation, dark gray color reflects clockwise fluid rotation, and medium gray color indicates counterclockwise fluid rotation. To assist in visualizing jet flow, a mean horizontal flow of $u = 19$ and $u = 24$ cm/s was subtracted from each vector for *T. semifasciata* and *C. punctatum*, respectively. (From Wilga, C.D. and Lauder, G. V., *J. Exp. Biol.*, 205, 2365–2374, 2002. With permission.)

fin of fishes sheds momentum in the form of vortex loops as the wake rolls up into discrete torus-shaped rings with a central high-velocity jet flow (Drucker and Lauder, 1999; Lauder and Drucker, 2002). By quantifying the morphology of these wake vortex rings, we can determine the direction of force application to the water by the heterocercal tail by measuring the direction of the central vortex ring momentum jet. In addition, the absolute force exerted on the water by the tail can be calculated by measuring the strength and shape of the vortex rings (Dickinson, 1996; Drucker and Lauder, 1999; Lauder and Drucker, 2002).

Using the two-camera arrangement illustrated in Figure 5.7, Wilga and Lauder (2002) studied the hydrodynamics of the tail of leopard sharks during both steady horizontal locomotion and vertical maneuvering. They measured the orientation of the body relative to the horizontal, the path of motion of the body through

the water, and the orientation and hydrodynamic characteristics of the vortex rings shed by the tail (Figure 5.9). Representative data from that study are shown in Figure 5.8, which illustrates the pattern of water velocity and vortex ring orientation resulting from one tail beat in two species of sharks. Tail vortex rings are inclined significantly to the vertical and are tilted posterodorsally. The central high-velocity water jet through the center of each vortex ring is oriented posteroventrally at an angle between 40° and 45° below the horizontal. These data provide unequivocal support for the classical model of heterocercal tail function in sharks by demonstrating that the tail accelerates water posteroventrally and that there must necessarily be a corresponding reaction force with dorsal (lift) and anterior (thrust) components.

Analysis of the changing orientation of tail vortex rings as sharks maneuver vertically in the water demonstrates that the relationship between vortex ring

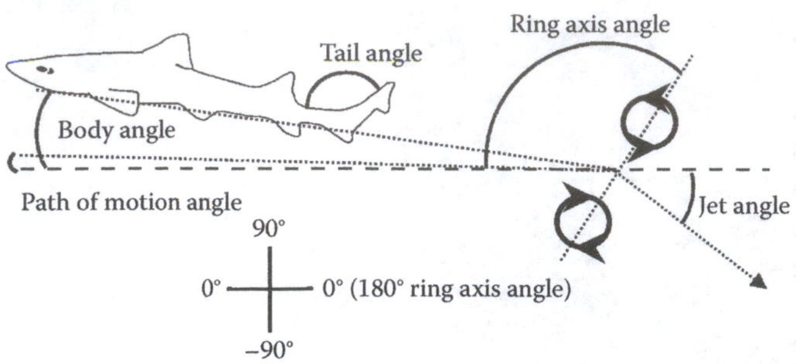

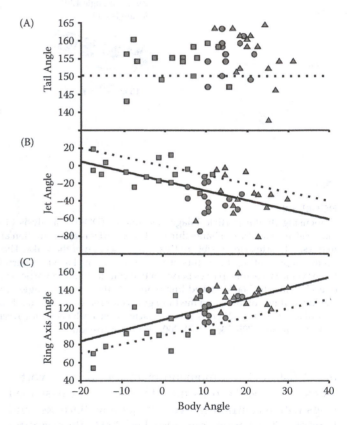

FIGURE 5.9

Schematic summary illustrating body and wake variables measured relative to the horizontal: body angle, from a line drawn along the ventral body surface; path of motion of the center of mass; tail angle between the caudal peduncle and dorsal tail lobe; ring axis angle, from a line extending between the two centers of vorticity; and mean vortex jet angle. Angle measurements from the variables of interest (dotted lines) to the horizontal (dashed line) are indicated by the curved solid lines. Angles above the horizontal are considered positive and below the horizontal negative. Ring axis angle was measured from 0° to 180° (From Wilga, C.D. and Lauder, G. V., *J. Exp. Biol.*, 205, 2365–2374, 2002. With permission.)

angle and body angle remains constant as body angle changes during maneuvering (Figure 5.10). These data show that leopard sharks do not alter the direction of force application to the water by the tail during vertical maneuvering, in contrast to previous data from sturgeon that demonstrated the ability to actively alter tail vortex wake orientation as they maneuver (Liao and Lauder, 2000).

A newly described intrinsic radialis tail muscle may function to stiffen the fin to change tail conformation (Flammang, 2010). The radialis muscle extends ventral to the axial myomeres and is composed of red fibers angled dorsoposteriorly. A similar arrangement exists in all sharks examined, with slight changes in angel sharks and rays and absence in skates and chimaeras. Muscle activity in spiny dogfish at slow speed follows an anterior to posterior pattern prior to activation of red axial muscle in the caudal myomeres (Figure 5.11). In contrast, at higher speed, only the anterior portion of the radialis muscle shows activity (Flammang, 2010).

5.2.3 Function of the Pectoral Fins during Locomotion

5.2.3.1 Anatomy of the Pectoral Fins

There are two distinct types of pectoral fins in sharks based on skeletal morphology. In aplesodic fins, the cartilaginous radials are blunt and extend up to 50% into the fin with the distal web supported only by ceratotrichia. In contrast, plesodic fins have radials that extend more than 50% into the fin to stiffen it and supplement the support of the ceratotrichia (Compagno,

1988) (Figure 5.12). The last row of radials tapers to a point distally in plesodic fins. Plesodic fins appear in Lamniformes, hemigaleids, carcharhinids, sphyrnids, and batoids except for pristids; other groups have aplesodic fins (Shirai, 1996). The restricted distribution of plesodic pectoral fins in extant sharks, the different morphology in each group, and their occurrence in more derived members (by other characters) of each group strongly suggest that plesodic pectorals are derived and have evolved independently from aplesodic pectorals (Bendix-Almgreen, 1975; Compagno, 1973, 1988; Zangerl, 1973). The decreased skeletal support of aplesodic pectoral fins over plesodic fins

FIGURE 5.10

Plot of body angles. (A) Tail angle, (B) jet angle, and (C) ring axis angle in leopard sharks, *Triakis semifasciata*, while swimming at 1.0 *l*/s. Solid lines indicate a significant linear regression, and the dotted line represents the predicted relationship. The lack of significance of the tail vs. body angle regression ($P = 0.731$, $r^2 = 0.003$) indicates that the sharks are not altering tail angle as body angle changes but instead are maintaining a constant angular relationship regardless of locomotor behavior. Jet angle decreases with increasing body angle ($P < 0.001$, $r^2 = 0.312$, $y = -17 - 1.087x$) at the same rate as the predicted parallel relationship, indicating that the vortex jet is generated at a constant angle to the body regardless of body position. Ring axis angle increases with body angle at the same rate as the predicted perpendicular relationship ($P < 0.001$, $r^2 = 0.401$, $y = 107 + 1.280x$). Circles, triangles, and squares represent holds, rises, and sinks, respectively. (From Wilga, C.D. and Lauder, G. V., *J. Exp. Biol.*, 205, 2365–2374, 2002. With permission.)

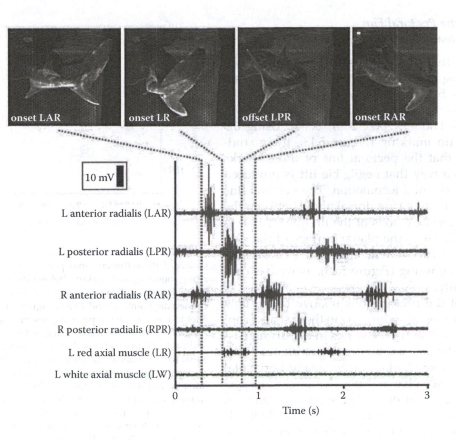

FIGURE 5.11
Tail kinematics and electromyographic recordings of tail muscles of a spiny dogfish swimming steadily at 0.5 *l*/s. Note the anterior to posterior activation of the radialis muscle. (From Flammang, B.E., *J. Morphol.*, 271, 340–352, 2010. With permission.)

allows greater freedom of motion in the distal web of the fin and may function to increase maneuverability. *Chiloscyllium* (Orectolobiformes) frequently "walk" on the substrate using both the pectoral and pelvic fins (Pridmore, 1995) in a manner similar to that of salamanders. They can bend the pectoral fins such that an acute angle is formed ventrally when rising on the substrate, and angles up to 165° are formed dorsally when station-holding on the substrate. *Chiloscyllium* are even able to walk backward using both sets of paired fins (AMRM and CDW, pers. obs.). In contrast, the increased skeletal support of plesodic fins stiffens and streamlines the distal web, which reduces drag. Furthermore, the extent of muscle insertion into the pectoral fin appears to correlate with the extent of radial support into the fin and thus pectoral fin type. In sharks with aplesodic fins, the pectoral fin muscles insert as far as the third (and last) row of radial pterygiophores, well into the fin. In contrast, those sharks with plesodic fins have muscles that insert only as far as the second row (of three) of radials.

Streamlined rigid bodies are characteristic of fishes that are specialized for cruising and sprinting, whereas flexible bodies are characteristic of fishes that are specialized for accelerating or maneuvering (Webb, 1985, 1988). Applying this analogy to shark pectoral fins, it may be

that plesodic fins are specialized for cruising (fast-swimming pelagic sharks) and aplesodic fins are specialized for accelerating or maneuvering (slow-cruising pelagic and benthic sharks).

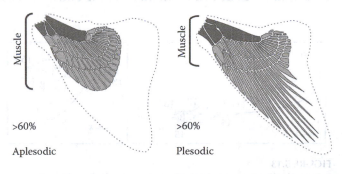

FIGURE 5.12
(Left) Skeletal structure of the pectoral fins in aplesodic sharks, such as leopard, bamboo, and dogfish (Wilga and Lauder, 2001); (right) plesodic sharks, such as lemon, blacktip, and hammerhead (redrawn from Compagno, 1988). The left pectoral fin for each species is shown in dorsal view. Dark gray elements are propterygium, mesopterygium, and metapterygium from anterior to posterior; light gray elements are radials. The dotted line delimits the extent of ceratotrichia into the fin web. Muscle insertion extends to the end of the third row of radials in aplesodic sharks and to the end of the second row or middle of the third row of radials in plesodic sharks.

5.2.3.2 Role of the Pectoral Fins during Steady Swimming

The function of the pectoral fins during steady horizontal swimming and vertical maneuvering (rising and sinking) has been tested experimentally in *Triakis semifasciata*, *Chiloscyllium plagiosum*, and *Squalus acanthias* (Wilga and Lauder, 2000, 2001, 2004). Using 3D kinematics and fin marking (Figure 5.13), these studies have shown that the pectoral fins of these sharks are held in such a way that negligible lift is produced during steady horizontal locomotion. The pectoral fins are cambered with an obtuse dorsal angle between the anterior and posterior regions of the fin (mean, 190° to 191°) (Figure 5.14). Thus, the planar surface of the pectoral fin is held concave downward relative to the flow during steady swimming (Figure 5.15), as well as concave mediolaterally.

The posture of the pectoral fins relative to the flow during steady horizontal swimming in these sharks contrasts markedly to those of the wings in a cruising passenger aircraft. The anterior and posterior planes of the pectoral fins in these sharks during steady horizontal swimming are at negative and positive angles, respectively, to the direction of flow (Figure 5.15). When both planes are considered together, the chord angle is –4° to –5° to the flow. Conversely, the wings of most cruising passenger aircraft have a positive attack angle to the direction of oncoming air, which generates positive lift.

The planar surface of the pectoral fins of these sharks is held at a negative dihedral angle (fin angle relative to the horizontal) from –6° (*Chiloscyllium plagiosum*) to –23° (*Triakis semifasciata*) during steady horizontal swimming (Figure 5.16). The pectoral fins are destabilizing in this position (Simons, 1994; Smith, 1992; Wilga and Lauder, 2000) and promote rolling motions of the body, such as

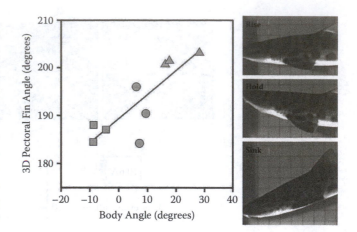

FIGURE 5.14
Graph of three-dimensional pectoral fin angle vs. body angle for rising, holding, and sinking behaviors at 1.0 *l*/s in leopard sharks. Symbols are as in Figure 5.3. Body angle was calculated using the line connecting points 12 and 13 (see Figure 5.11) and the horizontal (parallel to the flow). Each point represents the mean of five sequences for each of four individuals. Images to the right show sample head and pectoral fin positions during each behavior. Pectoral fin angles equal to 180° indicate that the two fin triangles (see Figure 5.11) are coplanar; angles less than 180° indicate that the fin surface is concave dorsally; and angles greater than 180° indicate that the fin surface is concave ventrally. The three-dimensional internal pectoral fin angle is significantly different among the three behaviors (ANOVA, P = 0.0001). The least-squares regression line is significant (slope, 0.41; adjusted r^2 = 0.39; $P < 0.001$). (From Wilga, C.D. and Lauder, G.V., *J. Exp. Biol.*, 203, 2261–2278, 2000. With permission.)

those made while maneuvering in the water column. For example, in a roll, the fin with the greatest angle to the horizontal meets the flow at a greater angle of attack, resulting in a greater force (F_x) directed into the roll, while the angle of attack of the more horizontally oriented fin is reduced by the same amount. This is in direct contrast to previous studies suggesting that the pectoral fins of sharks are oriented to prevent rolling, as in the keel of a ship (Harris, 1936, 1953). Wings that are tilted at a positive angle with respect to the horizontal have a positive dihedral angle, as in passenger aircraft, and are self-stabilizing in that they resist rolling motions of the fuselage (Figure 5.16) (Simons, 1994; Smith, 1992). When a passenger aircraft rolls, the more horizontally oriented wing generates a greater lift force than the inclined wing (Simons, 1994; Smith, 1992). In this way, a corrective restoring moment arises from the more horizontal wing, which opposes the roll, and the aircraft is returned to the normal cruising position. Interestingly, the negative dihedral wings of fighter aircraft, which are manufactured for maneuverability, function similarly to shark pectoral fins.

The flow of water in the wake of the pectoral fins during locomotion in these three species was quantified using PIV to estimate fluid vorticity and the forces exerted by the fin on the fluid (see Drucker and Lauder,

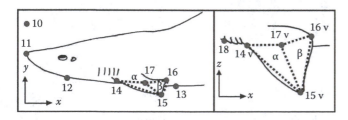

FIGURE 5.13
Schematic diagram of a shark illustrating the digitized points on the body and pectoral fin. Lateral view of the head and pectoral fin (left) and ventral view of pectoral fin region (right). Note that the reference axes differ for lateral (*x,y*) and ventral (*x,z*) views. Data from both views were recorded simultaneously. Points 14 to 16 are the same points in lateral and ventral views, and points 17 and 17v represent the same location on the dorsal and ventral fin surfaces. These three-dimensional coordinate data were used to calculate a three-dimensional planar angle between the anterior and posterior fin planes (α and β), as shown in B. (From Wilga, C.D. and Lauder, G.V., *J. Exp. Biol.*, 203, 2261–2278, 2000. With permission.)

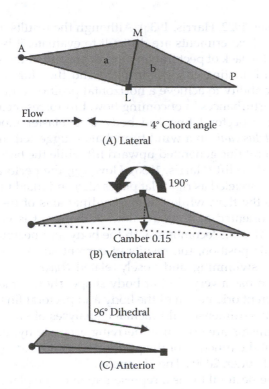

FIGURE 5.15

Orientation of the two pectoral fin planes (*a* and *b*) in three-dimensional space during pelagic holding in bamboo sharks, *Chiloscyllium plagiosum* (leopard and dogfish sharks show similar conformations). Panels show (A) lateral, (B) ventrolateral, and (C) posterior views of the fin planes. Points defining the fin triangles correspond to the following digitized locations in Figure 5.11: A, anterior, point 14, black circle; L, point 15, black square; P, posterior, point 16; M, medial, point 17. Chord angle to the flow is given in the lateral view, camber and internal fin angles between planes *a* and *b* are given in the ventrolateral view, and the dihedral angle is shown in the posterior view. (Note that in the posterior view the angles are given as acute to the *xy* plane.) (From Wilga, C.D. and Lauder, G.V., *J. Morphol.*, 249, 195–209, 2001. With permission.)

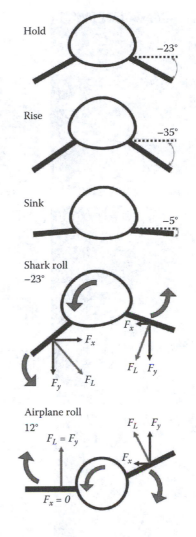

FIGURE 5.16

Schematic diagram of the dihedral orientation of the pectoral fins in a shark during holding, rising, and sinking behaviors. Forces during a roll are illustrated below for the pectoral fins of a shark and the wings of an airplane. The body and fin are represented as a cross-section at the level of plane α of the pectoral fin (see Figure 5.11). Thin, gray, double-headed arrows represent the dihedral angle between the plane α (dotted line) and pectoral fin. Thick arrows show the direction of movement of the body and fins or wing during a roll. Note that positive dihedrals (such as those used in aircraft design) are self-stabilizing, while fins oriented at a negative dihedral angle, as in sharks, are destabilizing in roll and tend to amplify roll forces. F_x, horizontal force; F_y, vertical force; F_L, resultant force. (From Wilga, C.D. and Lauder, G.V., *J. Exp. Biol.*, 203, 2261–2278, 2000. With permission.)

1999; Wilga and Lauder, 2000). These results further corroborate the conclusion from the 3D kinematic data that the pectoral fins generate negligible lift during steady horizontal swimming. There was virtually no vorticity or downwash detected in the wake of the pectoral fins during steady horizontal swimming, which shows that little or no lift is being produced by the fins (Figure 5.17). According to Kelvin's law, vortices shed from the pectoral fin must be equivalent in magnitude but opposite in direction to the theoretical bound circulation around the fin (Dickinson, 1996; Kundu, 1990); therefore, the circulation of the shed vortex can be used to estimate the force on the fin. Mean downstream vertical fluid impulse calculated in the wake of the pectoral fins during steady horizontal swimming was not significantly different from zero. This indicates that the sharks are holding their pectoral fins in such a way that the flow speed and pressure are equivalent on the dorsal and ventral surfaces of the fin. Furthermore, if the pectoral

fins were generating lift to counteract moments generated by the heterocercal tail, there would necessarily be a downwash behind the wing to satisfy Kelvin's law. The lack of an observable and quantifiable downwash indicates clearly that, during holding behavior, pectoral fins generate negligible lift.

These results showing that the pectoral fins of these sharks do not generate lift during steady forward swimming stand in stark contrast to previous findings on sharks with bound or amputated fins (Aleev, 1969;

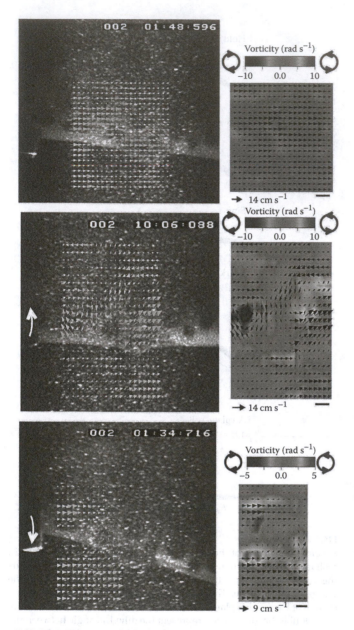

FIGURE 5.17

DPIV data from leopard shark pectoral fins during (top) holding vertical position, (middle) sinking, and (bottom) rising behaviors at 1.0 *l*/s (patterns for bamboo and dogfish sharks are similar). The video image (on the left) is a single image of a shark with the left pectoral fin located just anterior to the laser light sheet. Note that the ventral body margin is faintly visible through the light sheet. The plot on the right shows fluid vorticity with velocity vectors with conventions as in Figure 5.8. Note that the fin in the holding position is held in a horizontal position, and that the vorticity plot shows effectively no fluid rotation. Hence, the pectoral fins in this position do not generate lift forces. During sinking, note that there is a clockwise vortex (dark gray region of rotating fluid to the right) that resulted from the upward fin flip (curved white arrow) to initiate the sinking event. During rising, note that the fin has flipped ventrally (curved white arrow) to initiate the rising event and that a counterclockwise vortex (medium gray region of rotating fluid to the right) has been shed from the fin. To assist in visualizing the flow pattern, a mean horizontal flow of *U* = 33 cm/s was subtracted from each vector. (Adapted from Wilga, C.D. and Lauder, G.V., *J. Morphol.*, 249, 195–209, 2001.)

Daniel, 1922; Harris, 1936). Although the results of such radical experiments are difficult to evaluate, it is likely that the lack of pectoral fin motion prevented the sharks from initiating changes in pitch and therefore limited their ability to achieve a horizontal position and adjust to perturbances in oncoming flow. Lift forces measured on the pectoral fins and body of a plaster model of *Mustelus canis* in a wind tunnel also suggested that the pectoral fins generated upward lift while the body generated no lift (Harris, 1936). However, the pectoral fins were modeled as rigid flat plates (2D) and tilted upward 8° to the flow, while the longitudinal axis of the body was oriented at 0° to the flow. Although it is possible that *M. canis* locomotes with the body and pectoral fins in this position, the results of current studies on live, freely swimming, and closely related *Triakis semifasciata*, which has a very similar body shape, show a radically different orientation of the body and pectoral fins.

Three-dimensional kinematic analyses of swimming organisms are crucial to deriving accurate hypotheses about the function of the pectoral fins and body (Wilga and Lauder, 2000). The 2D angle of the anterior margin of the pectoral fin as a representation of the planar surface of the pectoral fin in sharks is extremely misleading. Although the pectoral fin appears to be oriented at a positive angle to the flow in lateral view, 3D kinematics reveals that the fin is actually concave downward with a negative dihedral. When viewed laterally, this negative-dihedral, concave-downward orientation of the pectoral fin creates a perspective that suggests a positive angle of attack when the angle is, in fact, negative.

5.2.3.3 Role of the Pectoral Fins during Vertical Maneuvering

Triakis semifasciata, *Chiloscyllium plagiosum*, and *Squalus acanthias* actively adjust the angle of their pectoral fins to maneuver vertically in the water column (Wilga and Lauder, 2000, 2001, 2004). Rising in the water column is initiated when the posterior plane of the fin is flipped downward to produce mean obtuse dorsal fin angles around 200°, while the leading edge of the fin is rotated upward relative to the flow. This downward flipping of the posterior plane of the fin increases the chord angle to +14, and as a result the shark rises in the water. In contrast, to sink in the water the posterior plane of the pectoral fin is flipped upward relative to the anterior plane, which produces a mean obtuse dorsal fin angle of 185°. At the same time, the leading edge of the fin is rotated downward relative to the flow such that the chord angle is decreased to –22°, and the shark sinks in the water.

The dihedral angle of shark pectoral fins changes significantly during vertical maneuvering in the water column (Figure 5.16). The dihedral angle increases to –35° during rising and decreases to –5° during sinking. This may be due to a need for greater stability during sinking behavior because the heterocercal tail generates a lift force that tends to drive the head ventrally. Holding the pectoral fins at a low dihedral angle results in greater stability during sinking compared to rising. The greater negative dihedral angle increases maneuverability and allows rapid changes in body orientation during rising.

These angular adjustments of the pectoral fins are used to maneuver vertically in the water column and generate negative and positive lift forces, which then initiate changes in the angle of the body relative to the flow. As the posterior plane of the pectoral fin is flipped down to ascend, a counterclockwise vortex, indicating upward lift force generation, is produced and shed from the trailing edge of the fin and pushes the head and anterior body upward (Figure 5.17). This vortex is readily visible in the wake as it rolls off the fin and is carried downstream. The opposite flow pattern occurs when sharks initiate a sinking maneuver in the water column. A clockwise vortex, indicating downward lift force generation, is visualized in the wake of the pectoral fin as a result of the dorsal fin flip and pulls the head and anterior body of the shark downward (Figure 5.17).

Lift forces produced by altering the planar surface of the pectoral fin to rise and sink appear to be a mechanism to reorient the position of the head and anterior body for maneuvering. Changing the orientation of the head will alter the force balance on the body as a result of interaction with the oncoming flow and will induce a change in vertical forces that will move the shark up or down in the water column. Forces generated by the pectoral fins are significantly greater in magnitude during sinking than during rising. This may be due to the necessity of reorienting the body through a greater angular change to sink from the positive body tilt adopted during steady swimming. A shark must reposition the body from a positive body tilt of 8° (mean holding angle) down through the horizontal to a negative body tilt of –11° (mean sinking angle), a change of 19°. In contrast, to rise a shark simply increases the positive tilt of the body by 14° (mean rise – hold difference), which should require less force given that the oncoming flow will assist the change from a slightly tilted steady horizontal swimming position to a more inclined rising body position.

5.2.3.4 Function of the Pectoral Fins during Benthic Station-Holding

Chiloscyllium plagiosum have a benthic lifestyle and spend much of their time resting on the substrate on and around coral reefs where current flows can be strong. To maintain position on the substrate during significant current flow, these sharks shift their body posture to reduce drag (Wilga and Lauder, 2001). The

sharks reorient the longitudinal axis of the body to the flow with the head pointing upstream during current flow, but they do not orient when current flow is negligible or absent. Body angle steadily decreases from 4° at 0 l/s to 0.6° at 1.0 l/s as they flatten their body against the substrate with increasing flow speed. This reduces drag in higher current flows, thereby promoting station-holding. This behavior is advantageous in fusiform benthic fishes that experience a relatively high flow regime, such as streams where salmon parr are hatched (Arnold and Webb, 1991) and inshore coral reefs where bamboo sharks dwell (Compagno, 1984).

Chiloscyllium plagiosum also reorient the pectoral fins to generate negative lift, increase friction, and oppose downstream drag during station-holding in current flow (Wilga and Lauder, 2001). They hold the pectoral fins in a concave upward orientation, similar to that in sinking, which decreases from a mean planar angle of 174° at 0 l/s to a mean of 165° at 1.0 l/s. At the same time, the chord angle steadily decreases from a mean of 2.7° at 0 l/s to a mean of –3.9° at 1.0 l/s. Flattening the body against the substrate lowers the anterior edge of the fin, whereas elevating the posterior edge of the fin to decrease the planar angle significantly decreases the chord angle (Figure 5.18). In this orientation, water flow is deflected up and over the fin and produces a clockwise vortex that is shed from the fin tip. The clockwise vortex produces significant negative lift (mean –0.084 N) directed toward the substrate that is eight times greater than that generated during sinking. As the clockwise vortex shed from the fin rotates just behind the fin, flow recirculates upstream and pushes against the posterior surface of the fin, which opposes downstream drag. These movements generate negative lift that is directed toward the substrate and acts to increase total downward force and friction force, thereby promoting station-holding as predicted by previous studies (Arnold and Web, 1991; Webb and Gerstner, 1996), as well as a novel mechanism leading to vortex shedding that opposes downstream drag to further aid benthic station-holding (Wilga and Lauder, 2001).

5.2.3.5 Motor Activity in the Pectoral Fins

Movement of the posterior plane of the pectoral fin during sinking and rising is actively controlled by *Triakis semifasciata*. At the beginning of a rise, the pectoral fin depressors (ventral fin muscles, adductors) are active to depress the posterior portion of the pectoral fin (Figure 5.19). Small bursts of activity in the lateral hypaxialis, protractor, and levator muscles are sometimes present during rising, probably to stabilize pectoral fin position. In contrast, the pectoral fin levators (dorsal fin muscles, adductors), as well as the cucullaris and ventral hypaxialis, are strongly active during elevation of the posterior

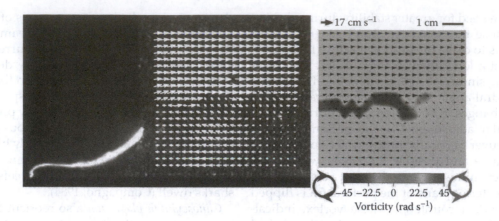

FIGURE 5.18

DPIV data from the pectoral fins of a representative bamboo shark, *Chiloscyllium plagiosum*, while station-holding on the substrate. The video image on the left shows a shark with the left pectoral fin located in the anterior end of the laser light sheet; other conventions are as in Figures 5.8 and 5.15. Note that the fin is held at a negative chord angle to the flow. A clockwise vortex (negative vorticity) was produced in the wake of the pectoral fins, which continued to rotate just behind the fin for several seconds until it was carried downstream by the flow (as seen here), after which a new vortex forms in the wake of the fin. (Adapted from Wilga, C.D. and Lauder, G.V., *J. Morphol.*, 249, 195–209, 2001.)

portion of the fin at the beginning of sinking behavior. Virtually no motor activity is present in the pectoral fin muscles while holding position at 0.5 and 1.0 l/s, indicating that the pectoral fins are not actively held in any particular position during steady horizontal locomotion. At higher flow speeds (1.5 l/s), however, recruitment of epaxial and hypaxial muscles occurs with slight activity in the pectoral fin muscles that may function to maintain stability.

Epaxial or hypaxial muscles are recruited to elevate or depress the head and anterior body during rising or sinking, respectively. At the initiation of rising behavior, simultaneously with the head pitching upward, a strong burst of activity occurs in the cranial epaxialis, while it is virtually silent during holding and sinking. Similarly, a strong burst of activity occurs in the ventral hypaxialis

during the initiation of sinking behavior, again with virtually no activity during holding and rising. This shows that the head is actively elevated or depressed to rise or sink, respectively, and that conformational changes in the anterior body assist the forces generated by the pectoral fins to accomplish vertical maneuvers. Finally, antagonistic pectoral fin muscles become active as rising or sinking slows or during braking (i.e., the levators are active as rising stops and the depressors are active as sinking stops).

5.2.4 Routine Maneuvers and Escape Responses

Less well studied than steady swimming, routine maneuvers and escape responses have recently become the focus of several shark locomotion studies. Foraging turn kinematics have been analyzed in juveniles of three species: *Sphyrna tiburo*, *Sphyrna lewini*, and *Carcharhinus plumbeus* (Kajiura et al., 2003). Scalloped hammerhead sharks, *Sphyrna lewini*, are more maneuverable than sandbar sharks, *Carcharhinus plumbeus*, based on variables such as turning radius, velocity, and banking (Kajiura et al., 2003). Hammerheads do not roll the body during turns, thus rejecting the hypothesis that the cephalofoil functions as a steering wing. The cephalofoil might still have hydrodynamic functions by providing stability during maneuvers (Kajiura et al., 2003). Further investigation with larger individuals and flow visualization techniques would clarify cephalofoil function. Compared to sandbar sharks, hammerhead sharks have greater lateral flexure. This may be due to a smaller second moment of area in hammerhead sharks, which is related to cross-sectional shape of vertebrae, rather than vertebral count (Kajiura et al., 2003).

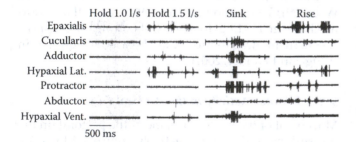

FIGURE 5.19

Electromyographic data from selected pectoral fin and body muscles during locomotion in *Triakis semifasciata* at 1.0 l/s for four behaviors: holding position at 1.0 and 1.5 l/s and sinking and rising at 1.0 l/s. Note the near absence of fin muscle activity while holding position at 1.0 l/s and recruitment of body and fin muscles at 1.5 l/s. The hypaxialis was implanted in both lateral (mid-lateral dorsal and posterior to pectoral fin base) and ventral (posterior to coracoid bar) positions. All panels are from the same individual. Scale bar represents 500 ms.

Body curvature has been assessed in shark species during routine maneuvers to determine which features of axial morphology are good predictors of maneuverability (Porter et al., 2009). The species studied were *Triakis semifasciata, Heterodontus francisci, Chiloscyllium plagiosum, Chiloscyllium punctatum,* and *Hemiscyllium ocellatum.* The best predictor of body curvature is the second moment of area of the vertebral centrum, followed by length and transverse height of vertebral centra. Body total length, fineness ratio, and width also appear to influence maneuverability (Porter et al., 2009).

Another important behavior in terms of selective pressure is escape behavior, which enables individuals to elude predators. Escape behaviors in sharks have been poorly studied, and only one study has been published to date (Domenici et al., 2004). Spiny dogfish perform C-start escape responses, which are characterized by an initial bend of the body into a "C" shape in stage I (Domenici et al., 2004). This initial conformation allows the body to accelerate in stage II when the fish straightens by thrusting the tail back to start moving away from the stimulus (Domenici and Blake, 1997). Spiny dogfish appear to have two types of escape response resulting in a bimodal distribution in duration, velocity, and acceleration (Domenici et al., 2004). Fast and slow escape responses have maximum turning rates of 766 and 1023 deg. s^{-1} and 434 and 593 deg. s^{-1}, respectively (Figure 5.20). It appears that spiny dogfish are capable of modulating the escape response based on some perceived stimulus or have two neural circuits for escape responses. Compared to bony fishes, escape responses in spiny dogfish are relatively slow; however, turning rate and turning radius are comparable (Domenici et al., 2004).

Traditionally routine and escape maneuvers have been analyzed using 2D approaches; however, for sharks that can quickly swim in any direction, except backward (Wilga and Lauder, 2000, 2001), 3D analyses and fluid dynamics studies would enable relations of body type and maneuverability to be made with escape response behavior. For example, juvenile spiny dogfish perform vertically oriented escape responses, whereas hatchling skates move horizontally (AMRM, pers. obs.).

5.2.5 Synthesis

The data presented above on pectoral and caudal fin function and body orientation in the shark species studied permit construction of a new model of the overall force balance during swimming (Figure 5.21). It is useful to discuss separately the vertical force balance and the rotational (torque) balance. During steady horizontal locomotion, when sharks are holding vertical position, body weight is balanced by lift forces generated

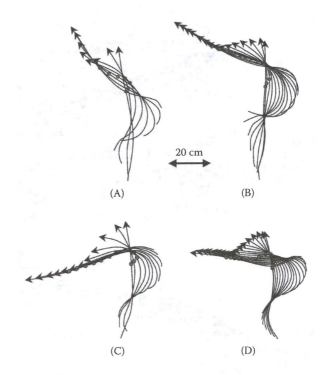

FIGURE 5.20
Midline kinematics of spiny dogfish during escape responses. Center of mass in represented in gray circles, and the head is indicated by an arrow. Consecutive lines are 40 ms apart after the onset of escape. Traces (A) and (C) are representative of fast responses; (B) and (D) represent slow responses. Note the distance covered by the center of mass in the same time for fast and slow responses (From Domenici, P. et al., *J. Exp. Biol.*, 207, 2339–2349, 2004. With permission.)

by the heterocercal tail and ventral body surface. The ventral surface generates lift both anterior and posterior to the center of body mass by virtue of its positive angle of attack to the oncoming water. Sharks adjust their body angle to modulate the total lift force produced by the body and can thus compensate for changes in body weight over both short and longer time frames.

Rotational balance is achieved by balancing the moments of forces around the center of mass. It has not been generally appreciated that the ventral body surface generates both positive and negative torques corresponding to the location of the ventral surface anterior and posterior to the center of mass. Water impacting the ventral body surface posterior to the center of mass will generate a counterclockwise torque of the same sign as that generated by the heterocercal tail. In contrast, water impacting the ventral body anterior to the center of mass will generate a clockwise torque, which is opposite in sign to that generated by the ventral body and tail posterior to the center of mass. Experimental data show that shark pectoral fins do not generate lift or torque during steady horizontal locomotion (Wilga and Lauder, 2000, 2001) as a result of their orientation relative to the flow. This stands in contrast to the textbook

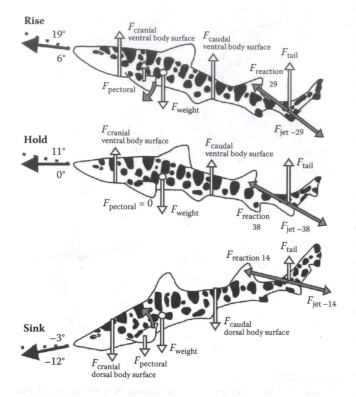

FIGURE 5.21
Schematic diagram of a force balance on swimming sharks during holding position, rising, and sinking behaviors (also representative of bamboo sharks, *Chiloscyllium punctatum*, and spiny dogfish, *Squalus acanthias*). The white circle represents the center of mass and vectors indicate forces *F* exerted by the fish on the fluid. Lift forces are generated by the ventral body surface, both anterior and posterior to the center of mass. The jet produced by the beating of the tail maintains a constant angle relative to body angle and path angle and results in an anterodorsally directed reaction force oriented dorsal to the center of mass during all three behaviors, supporting the classical model. Tail vortex jet angles are predicted means. (From Wilga, C.D. and Lauder, G.V., *J. Exp. Biol.*, 205, 2365–2374, 2002. With permission.)

depiction of shark locomotion in which the pectoral fins play a central role in controlling body position during horizontal locomotion. In our view, experimental kinematic and hydrodynamic data obtained over the last 10 years on benthic and benthopelagic species demonstrate that control of body orientation is the key to modulating lift and torques during horizontal swimming, and the pectoral fins are not used for balancing forces during horizontal swimming.

During maneuvering, however, the pectoral fins do play a key role in generating both positive and negative lift forces and hence torques about the center of mass (Figure 5.21). To rise in the water, sharks rapidly move the trailing pectoral fin edge ventrally, and a large vortex is shed, generating a corresponding lift force. This force has a clockwise rotational moment about the center of mass pitching the body up, thus increasing the angle of the body and hence the overall lift force. As a result, sharks move vertically in the water even while maintaining horizontal position via increased thrust produced by the body and caudal fin.

To stop this vertical motion or to maneuver down (sink) in the water, the trailing pectoral fin edge is rapidly elevated and sheds a large vortex, which produces a large negative lift force (Figure 5.21). This generates a counterclockwise torque about the center of mass, pitching the body down, exposing the dorsal surface to incident flow, and producing a net sinking motion. Pectoral fins thus modulate body pitch.

Overall, the force balance on swimming sharks is maintained and adjusted by small alterations in body angle and this in turn is achieved by elevation and depression of the pectoral fins. Pectoral fins thus play a critical role in shark locomotion by controlling body position and facilitating maneuvering, but they do not function to balance tail lift forces during steady horizontal locomotion.

5.3 Locomotion in Skates and Rays

Most batoids either undulate or oscillate the pectoral fins to move through the water (Figure 5.22). Basal batoids, such as guitarfishes, sawfishes, and electric rays, locomote by undulating their relatively thick tails similar to those of laterally undulating sharks (Rosenberger, 2001). Interestingly, *Rhinobatos lentiginosus*, which has a sharklike trunk and tail like all guitarfishes, also adopts a positive body angle to the flow during steady horizontal swimming (Rosenberger, 2001). Sawfishes and most electric rays are strict axial undulators and use only the tail for locomotion, whereas guitarfishes and some electric rays may supplement axial locomotion with undulations of the pectoral fin (Rosenberger, 2001). Most rays use strict pectoral fin locomotion; however, some rays, such as *Rhinoptera* and *Gymnura*, fly through the water by oscillating the pectoral fins in broad up and down strokes in a manner that would provide vertical lift similar to that of aerial bird flight (Rosenberger, 2001). Although skates undulate the pectoral fins to swim when in the water column, they have enlarged muscular appendages on the pelvic fins that are modified for walking or "punting" off the substrate (Koester and Spirito, 1999) in a novel locomotor mechanism. Although they lack the modified pelvic appendages of skates, some rays with similar habitats and prey also use punting locomotion (Macesic and Kajiura, 2010). Punting kinematics were similar across one skate and three ray species (*Raja eglanteria*, *Narcine brasiliensis*, *Urobatis jamaicensis*, and *Dasyatis sabina*), with protraction of the

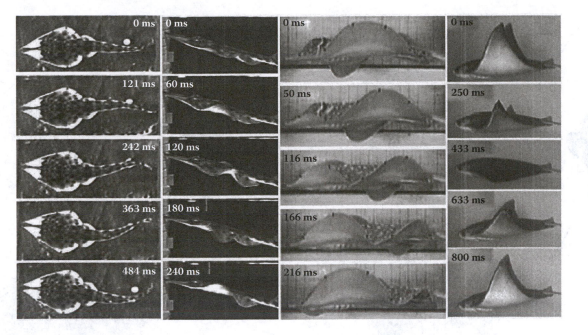

FIGURE 5.22
Successive dorsal video images of Atlantic guitarfish (*Rhinobatos lentiginosus*, left) and lateral video images of *R. lentiginosus* (second from left), blue-spotted stingray (*Taeniuralymma*, second from right), and cownose rays (*Rhinoptera bonasus*, right) swimming in a flow tank. Like sharks, *Rhinobatos lentiginosus* swims primarily with its thick shark-like tail. (From Rosenberger, L.J. and Westneat, M.W., *J. Exp. Biol.*, 202, 3523–3539, 1999; Rosenberger, L.J., *J. Exp. Biol.*, 204, 379–394, 2001. With permission.)

anterior edge of the pelvic fins followed by contact with the substrate and then retraction to push off (Macesic and Kajiura, 2010). *Raja eglanteria* and *N. brasiliensis* are true punters in which the pelvic fins are used to punt while pectorals are held horizontally. In contrast, *U. jamaicensis* and *D. sabina* augment punting by undulations of the pectoral fins, although this does not increase performance as measured by the distance traveled during a punting cycle. The musculature of true punters is highly specialized with robust crura, mobile distal joints, and specialized propterygium levators, depressors, and protractors originating from lateral processes on the pelvic girdle (Figure 5.23) (Macesic and Kajiura, 2010). In contrast, augmented punters have only one levator and depressor muscle controlling the protopterygium with a reduced pelvic girdle that limits movements (Macesic and Kajiura, 2010).

Some rays are able to vary the mechanics of the pectoral fins during locomotion (Rosenberger, 2001). There appears to be a trade-off between the amplitude of undulatory waves and fin beat frequency: Those that have higher wave amplitudes have fewer waves and *vice versa* (Rosenberger, 2001). This phenomenon appears to be correlated with lifestyle. Fully benthic rays and skates that are mostly sedentary, such as *Dasyatis sabina* and *D. say*, have low-amplitude waves with high fin beat frequencies, permitting high maneuverability at low speeds, which is more suited for swimming slowly along the substrate to locate food items (Rosenberger,

2001). Fully pelagic rays are able to take advantage of the 3D environment of the water column and oscillate the pectoral fins using high-amplitude waves and low fin beat frequencies (Rosenberger, 2001). Rays and skates that have both benthic and pelagic lifestyles, such as *Raja* sp. and *Dasyatis americana*, are typically more active and have intermediate values of amplitude and frequency (Rosenberger, 2001).

Oscillatory appendage propulsors that feed on benthic mollusks and crustaceans, such as cownose and butterfly rays, do not extend the fins below the ventral body axis during swimming, presumably so they can use the lateral line canals to detect prey and also to avoid contact with the substrate (Rosenberger, 2001). In contrast, oscillatory appendage propulsors that feed in the water column (i.e., filter feeders such as manta and mobulid rays) extend the pectoral fins equally above and below the body axis during swimming (Rosenberger, 2001). Some batoids are capable of modifying the swimming mechanism dependent on habitat; *Gymnura* undulates the pectoral fins when swimming along a substrate and oscillates them when swimming in the water column (Rosenberger, 2001). Undulatory mechanisms are efficient at slow speeds, offer reduced body and fin drag, and are highly maneuverable (Blake, 1983a,b; Lighthill and Blake, 1990; Rosenberger, 2001; Walker and Westneat, 2000). In contrast, oscillatory mechanisms are efficient at fast cruising and generate greater lift but are less well suited for maneuvering

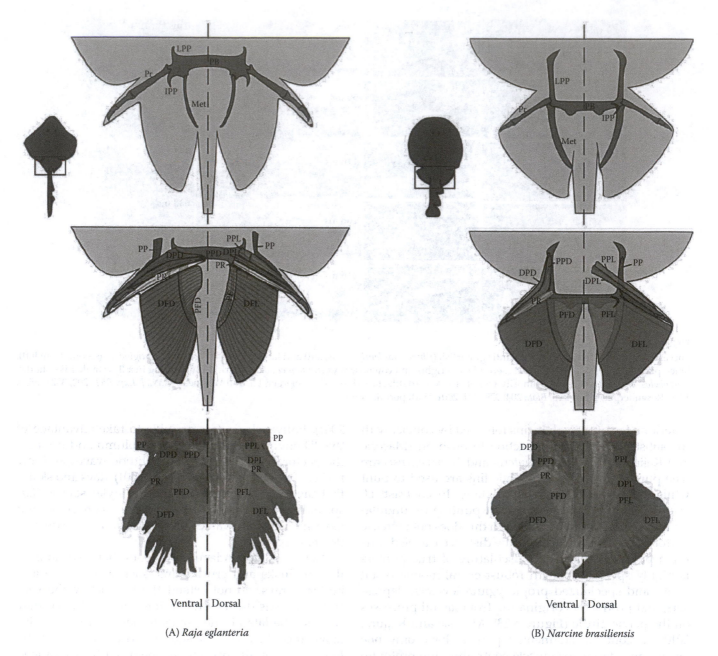

FIGURE 5.23
Schematic representation (top) and photographs (bottom) of pelvic fin skeletal elements and musculature of two benthic true punters: (A) *Raja eglanteria*, and (B) *Narcine brasiliensis*. Skeletal elements are shown in the top and middle illustrations: puboischiac bar (PB), iliac pelvic process (IPL), lateral pelvic process (LPP), metapterygium (Met), and propterygium (Pr). Pelvic musculature is shown at the middle and bottom illustrations: proximal fin depressor (PDF), distal fin depressor (DFP), distal propterygium depressor (DPD), proximal propterygium depressor (PPD), proximal fin levator (PFL), distal fin levator (DFL), proximal propterygium levator (PPL), and distal propterygium levator (DPL). The propterygium retractors (PR) and protractors (PP) were found on both dorsal and ventral sides (PP is occluded from view in the dorsal photograph of *N. brasiliensis*). Note the specializations in propterygium depressors and levators. (From Macesic, L.J. and Kajiura, S.M., *J. Morphol.*, 271, 1219–1228, 2010. With permission.)

(Blake, 1983b; Cheng and Zhaung, 1991; Chopra, 1974; Rosenberger, 2001). A recent study suggested that sensory coverage area is inversely related to the proportion of the wing used for propulsion in batoids (Jordan, 2008). However, *Myliobatis californica*, which uses oscillatory propulsion and thus is expected to have a smaller

sensory area, has extensions of the lateral line and electrosensory systems along the anterior edge of the pectoral fins (Jordan, 2008).

Different strategies are employed to increase swimming speed in various batoid species (Rosenberger, 2001). Most *Dasyatis* species increase fin beat frequency, wave

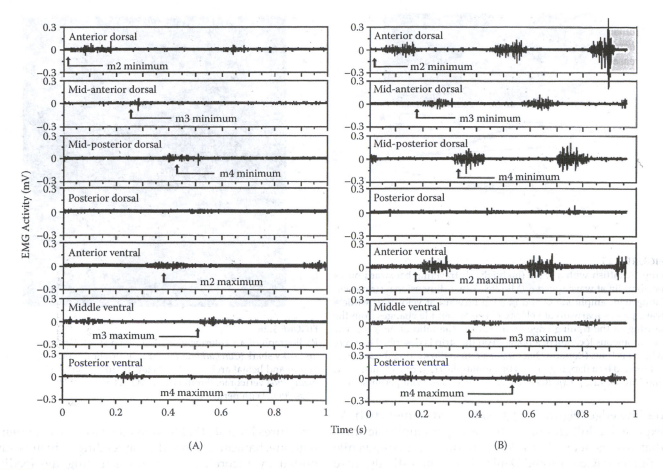

FIGURE 5.24
Electromyographic (EMG) data illustrating the muscle activity for the pectoral fin undulation of blue-spotted stingrays, *Taeniura lymma*, at a low speed of 1.2 disk length/s (A) and at a higher speed of 3.0 disk length/s (B). The electrode recordings are taken from the following muscles: anterior dorsal, mid-anterior dorsal, mid-posterior dorsal, posterior dorsal, anterior ventral, middle ventral, and posterior ventral. The arrows below the EMG activity indicate the point during the fin-beat cycle at which the anterior, middle, and posterior fin markers are at their maximum (peak upstroke) and minimum (peak downstroke) excursion. (From Rosenberger, L.J. and Westneat, M.W., *J. Exp. Biol.*, 202, 3523–3539, 1999; Rosenberger, L.J., *J. Exp. Biol.*, 204, 379–394, 2001. With permission.)

speed, and stride length to increase swimming speed while amplitude is held constant; however, *Taeniura lymma* and *D. americana* increase fin beat frequency and wave speed but decrease wave number while holding amplitude constant to increase speed (Rosenberger, 2001; Rosenberger and Westneat, 1999). Similarly, *Raja eglanteria* increases wave speed and decreases wave number to swim faster (Rosenberger, 2001; Rosenberger and Westneat, 1999). Oscillatory propulsors, such as *Rhinoptera* and *Gymnura*, increase wave speed in addition to fin-tip velocity to increase swimming speed (Rosenberger, 2001; Rosenberger and Westneat, 1999). Interestingly, *Gymnura* pauses between each fin beat at high flow speeds, similar to the burst and glide flight mechanisms of aerial birds (Rosenberger, 2001; Rosenberger and Westneat, 1999).

As expected, the dorsal and ventral fin muscles are alternately active during undulation of the pectoral fin from anterior to posterior (Figure 5.24) (Rosenberger and Westneat, 1999). The intensity of muscle contraction is increased to swim faster in *Taeniura lymma*, and the ventral muscles are also active longer than the respective dorsal muscles, indicating that the downstroke is the major power-producing stroke (Rosenberger and Westneat, 1999). Chondrichthyans are negatively buoyant; thus, lift must be generated to counter the weight of the fish as well as for locomotion. Interburst duration is decreased in *T. lymma* at higher swimming speeds with the fin muscles firing closer together (Rosenberger and Westneat, 1999).

5.4 Locomotion in Holocephalans

Chimaeras have large flexible pectoral fins that have been described as both undulatory and oscillatory. The leading edge of the pectoral fin is flapped, which then passes an undulatory wave down the pectoral fin to the

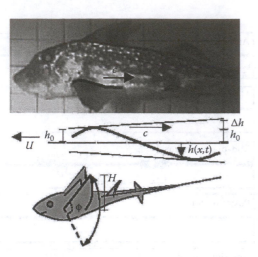

FIGURE 5.25
(Top) A ratfish with a wave (highlighted) traveling backward on its pectoral fin at wave speed c. (Middle) A two-dimensional strip oscillating with amplitude h_0 and moving forward at velocity U while a wave passes rearward at velocity c. The amplitude changes from the leading to the trailing edge by a factor ε, the ratio of Δh to h_0. The instantaneous location of a point (x) on the strip is described by $h(x,t)$, where t is time. (Bottom) Diagram of a ratfish illustrating the angle (φ) subtended by a flapping fin and tip amplitude (H). (From Combes, S.A. and Daniel, T.L., *J. Exp. Biol.*, 204, 2073–2085, 2001. With permission.)

trailing edge (Figure 5.25) (Combes and Daniel, 2001). As expected, adult chimaeras had a larger amplitude wave that was generated at a lower frequency than juvenile chimaeras (Combes and Daniel, 2001). Interestingly, there is no net chordwise bend in the pectoral fin, which averages a 0° angle of attack to the flow over a stroke cycle (Combes and Daniel, 2001). Potential flow models, based on kinematic and morphological variables measured on the chimaeras, for realistic flexible fins and theoretical stiff fins emphasize the importance of considering flexion in models of animal locomotion; significantly higher values for thrust were calculated when the fin was assumed to be stiff rather than flexible as in reality (Combes and Daniel, 2001). As predicted by the high degree of motion in the pectoral fins of white-spotted ratfish, *Hydrolagus colliei*, muscle mass corrected for body mass and the proportion of muscle inserting into the fin is greater in ratfish than spiny dogfish (Foster and Higham, 2010).

5.5 Material Properties of Chondrichthyan Locomotor Structures

Recent studies on the material properties of skeletal elements in elasmobranchs has focused on understanding the biomechanics of cartilaginous structures during locomotion (Dean and Summers, 2006; Dean et al., 2009; Porter and Long, 2010; Porter et al., 2006, 2007; Schaeffer and Summers, 2005). Material property studies of head

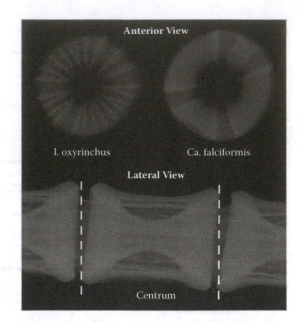

FIGURE 5.26
Radiograph of an anterior view of a mako shark (*Isurus oxyrinchus*) and silky shark (*Carcharhinus falciformis*) vertebral centra with excised neural and hemal arches. Lateral view of gulper shark (*Centrophorus granulosus*) vertebrae. (From Porter, M.E. et al., *J. Exp. Biol.*, 209, 2920–2928, 2006. With permission.)

structures indicate that tendon and cartilage experience high mechanical stresses during feeding, which is supported by theoretical studies calculating the feeding forces generated (Summers and Koob, 2002; Summers et al., 2003).

Evolution of the vertebral column and myomeres allowed the strong compressions needed to undulate the body through a dense medium such as water (Liem et al., 2001). The vertebral column is composed of multiple vertebrae, each with a central body and dorsal neural and ventral hemal arches. Figure 5.26 presents radiographs of vertebral centra. Unlike cartilage in the remaining skeletal structures of the body in living elasmobranchs that are tessellated, vertebral centra are formed by areolar calcification (Dean and Summers, 2006). Material testing revealed that vertebral centra have stiffness and strength of the same order of magnitude as mammalian trabecular bone (Porter et al., 2006). Swimming speed appears to be a good predictor of vertebral cartilage stiffness and strength in elasmobranchs. In addition, chemical composition of vertebral centra is correlated with collagen content and material stiffness and strength, except for proteoglycan content (Figure 5.27) (Porter et al., 2006). In a separate study, vertebrae were tested under compressive loads with neural arches attached or removed; failure was seen in the vertebral centra but not in the neural arches. Thus, vertebral centra are likely the main load-bearing structures of the vertebral column in sharks during routine swimming (Porter and Long, 2010).

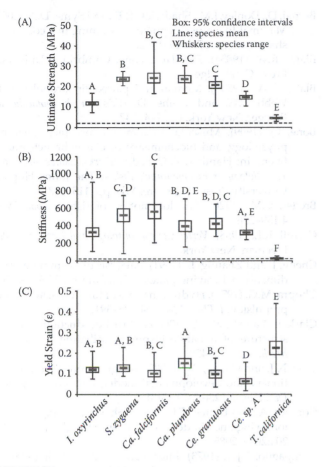

FIGURE 5.27

Material properties of mineralized cartilage in shark vertebral centra. (A) Ultimate strength (MPa) of vertebral cartilages from seven elasmobranch species showing significant differences ($F_{6,151} = 182.8$, $P < 0.001$). The broken horizontal line represents the lower limits of trabecular bone. Letters above the box and whisker plot denote significant differences between species. (B) Material stiffness was significantly different among the species ($F_{6,151} = 54.4$, $P < 0.001$). *Torpedo californica*, the only batoid, was less stiff than all shark species ($P < 0.001$). The horizontal line shows the lower limits of stiffness for trabecular bone. (C) Yield strain was significantly different among the species ($F_{6,147} = 27.6$, $P < 0.001$). *T. californica* had the greatest yield strain of all species ($P < 0.007$). (From Porter, M.E. et al., *J. Exp. Biol.*, 209, 2920–2928, 2006. With permission.)

Locomotor mode is also a good predictor of batoid wing skeletal morphology, indicating a role of flexural stiffness (Schaeffer and Summers, 2005). Undulatory swimmers such as *Dasyatis* have chain-like catenated calcification patterns with staggered joints in the radials; oscillatory swimmers have radials completely covered with mineralization in a crustal calcification pattern. Some oscillatory species also had cross-bracing between adjacent radials that is thought to increase fin stiffness. Theoretical stiffness of fin morphology agreed with observed stiffness (Schaeffer and Summers, 2005). These studies provide further evidence of the role of material properties in determining the function and evolution of skeletal elements and locomotor modes in Chondrichthyans.

5.6 Future Directions

The diversity of shark species for which we have even basic functional data on locomotor mechanics is extremely limited. Most papers to date have focused on leopard (*Triakis*), spiny dogfish (*Squalus*), and bamboo (*Chiloscyllium*) sharks swimming under controlled laboratory conditions. A high priority for future studies of locomotion in sharks, skates, and rays is to expand the diversity of taxa studied, especially for analyses of shark mechanics. The data obtained by Rosenberger (2001) on batoid locomotion are exemplary for their broadly comparative character, but studies like this are rare, perhaps necessarily so when detailed functional data must be obtained for a variety of behaviors.

Experimental studies of kinematics and hydrodynamics would benefit from increased spatial and temporal resolution so that a more detailed picture could be obtained of patterns of fin deformation and the resulting hydrodynamic wake, especially during unsteady maneuvering behaviors. New high-resolution, high-speed digital video systems will permit a new level of understanding of fin function and its impact on locomotor performance. Such increased resolution may also permit further observations of boundary layer flows in relation to surface denticle patterns to follow up on the observation by Anderson et al. (2001) that the boundary layer of *Mustelus* swimming at 0.5 *l*/s did not separate and remained attached along the length of the body. Defocusing digital particle image velocimetry (DDPIV) and stereoscopic PIV can be used to compute flow in three dimensions and offer a new avenue for comprehensive studies of fluid dynamics in the wake of swimming elasmobranchs (Gordon et al., 2002; Lauder, 2010; Raffel et al., 2007). These studies can be coupled with computational fluid dynamics (CFD) (Lauder, 2010) to investigate limitations and compare different locomotor modes, similar to what has been done for actinopterygian fishes (Tytell et al., 2010). Biorobotics and biomimetics studies that are emerging use elasmobranchs as models and have focused mainly on batoids (Clark and Smits, 2006; Gao et al., 2009; Xu et al., 2007; Yang et al., 2009; Zhou and Low, 2010). Biorobotics offers a way to test scenarios that are not possible with living animals, such as altering Reynolds and Strouhal numbers, aspect ratios, and material properties (Clark and Smits, 2006; Lauder, 2010). Biomimetics has great potential to advance the field through the demand for detailed kinematics, material properties, and activation patterns of elasmobranch locomotion for application purposes.

More studies on the mechanical properties of elasmobranch connective tissue elements and of the role these play in transmitting forces to hydrodynamic fin control surfaces are needed. This is a key area in which *in vitro* studies of material properties and *in vivo* analyses of how elasmobranch connective tissues function can greatly enhance our understanding of elasmobranch locomotor mechanics.

Finally, new information on locomotor structures such as fin placement and diversity might come from advances in developmental and molecular biology methods. Shark models have been used to study the evolution of paired and median fins (Cole and Currie, 2007) as well as skeletogenesis and the early origins of bone (Eames et al., 2007). To the extent that equipment and elasmobranch behavior permits, it would be extremely valuable to have quantitative 3D field data over the natural locomotor behavioral repertoire to answer such questions as what are routine swimming speeds, what are typical vertical and lateral maneuvering velocities, and what is the natural range of body angles observed during diverse locomotor behaviors? Advances in technology, such as accelerometers, can be used in shark research and can provide great insight to natural routine locomotor behaviors (Sims, 2010). Such data would serve as a link between experimental laboratory studies of shark biomechanics and locomotor performance in nature.

Acknowledgments

Support for preparation of this paper was provided by MCTES/FCT/SFRH/BD/36852/2007 to AMRM, by NSF grant DBI 97-07846 to CDW, and grants ONR N00014-09-1-0352, ONR N00014-03-1-0897, and NSF EFRI-0938043 to GVL.

References

Affleck, R.J. (1950). Some points in the function, development, and evolution of the tail in fishes. *Proc. Zool. Soc. Lond.* 120:349–368.

Aleev, Y.G. (1969). *Function and Gross Morphology in Fish,* trans. from the Russian by M. Raveh. Keter Press, Jerusalem.

Alexander, R.M. (1965). The lift produced by the heterocercal tails of Selachii. *J. Exp. Biol.* 43:131–138.

Anderson, E.J., McGillis, W., and Grosenbaugh, M.A. (2001). The boundary layer of swimming fish. *J. Exp. Biol.* 204:81–102.

Arnold, G.P. and Webb, P.W. (1991). The role of the pectoral fins in station-holding of Atlantic salmon parr (*Salmo salar* L). *J. Exp. Biol.* 156:625–629.

Bendix-Almgreen, S.E. (1975). The paired fins and shoulder girdle in *Cladoselache*, their morphology and phyletic significance. *Colloq. Int. C.N.R.S. Paris* 218:111–123.

Bernal, D., Donley, J.M., Shadwick, R.E., and Syme, D.A. (2005). Mammal-like muscles power swimming in a cold-water shark. *Nature* 437:1349–1352.

Blake, R.W. (1983a). *Fish Locomotion.* Cambridge University Press, Cambridge.

Blake, R.W. (1983b). Median and paired fin propulsion. In: Webb, P.W. and Weihs, D. (Eds.), *Fish Biomechanics.* Praeger, New York, pp. 214–247.

Bone, Q. (1999). Muscular system: microscopical anatomy, physiology, and biochemistry of elasmobranch muscle fibers. In: Hamlett, W.C. (Ed.), *Sharks, Skates, and Rays: The Biology of Elasmobranch Fishes.* The Johns Hopkins University Press, Baltimore, MD, pp. 115–143.

Breder, C.M. (1926). The locomotion of fishes. *Zool. (N.Y.)* 4:159–256.

Carroll, R.L. (1988). *Vertebrate Paleontology and Evolution.* WH Freeman, New York.

Cheng, J. and Zhaung, L. (1991). Analysis of swimming three-dimensional waving plates. *J. Fluid Mech.* 232:341–355.

Chopra, M.G. (1974). Hydrodynamics of lunate-tail swimming propulsion. *J. Fluid Mech.* 64:375–391.

Clark, R.P. and Smits, A.J. (2006). Thrust production and wake structure of a batoid-inspired oscillating fin. *J. Fluid Mech.* 562: 415–429.

Cole, N.J. and Currie, P.D. (2007). Insights from sharks: evolutionary and developmental models of fin development. *Dev. Dynam.* 236:2421–2431.

Combes, A. and Daniel, T.L. (2001). Shape, flapping and flexion: wing and fin design for forward flight. *J. Exp. Biol.* 204:2073–2085.

Compagno, L.J.V. (1973). Interrelationships of living elasmobranchs. In: Greenwood, P.H., Miles, R.S., and Patterson, C. (Eds.), Interrelationships of fishes, *Zool. J. Linn. Soc.* 53(Suppl. 1):15–61.

Compagno, L.J.V. (1984). *Sharks of the World.* United Nations Development Program, Rome.

Compagno, L.J.V. (1988). *Sharks of the Order Carcharhiniformes.* Princeton University Press, Princeton, NJ.

Compagno, L.J.V. (1999). Endoskeleton. In: Hamlett, W.C. (Ed.), *Sharks, Skates, and Rays: The Biology of Elasmobranch Fishes.* The Johns Hopkins University Press, Baltimore, MD, pp. 69–92.

Daniel, J.F. (1922). *The Elasmobranch Fishes.* University of California Press, Berkeley.

Dean, M.N. and Summers, A.P. (2006). Mineralized cartilage in the skeleton of chondrichthyan fishes. *Zoology* 109:164–168.

Dean, M.N., Mull, C.G., Gorb, S.N., and Summers, A.P. (2009). Ontogeny of the tessellated skeleton: insight from the skeletal growth of the round stingray *Urobatis halleri*. *J. Anat.* 215:227–239.

Dickinson, M.H. (1996). Unsteady mechanisms of force generation in aquatic and aerial locomotion. *Am. Zool.* 36:537–554.

Domenici, P. and Blake, R.W. (1997). Fish fast-start kinematics and performance. *J. Exp. Biol.* 200:1165–1178.

Domenici, P., Standen, E.M., and Levine, R.P. (2004). Escape manoeuvres in the spiny dogfish (*Squalus acanthias*). *J. Exp. Biol.* 207:2339–2349.

Donley, J. and Shadwick, R. (2003). Steady swimming muscle dynamics in the leopard shark *Triakis semifasciata. J. Exp. Biol.* 206:1117–1126.

Donley, J.M., Shadwick, R.E., Sepulveda, C.A., Konstantinidis, P., and Gemballa, S. (2005). Patterns of red muscle strain/activation and body kinematics during steady swimming in a lamnid shark, the shortfin mako (*Isurus oxyrinchus*). *J. Exp. Biol.* 208:2377–2387.

Drucker, E.G. and Lauder, G.V. (1999). Locomotor forces on a swimming fish: three-dimensional vortex wake dynamics quantified using digital particle image velocimetry. *J. Exp. Biol.* 202:2393–2412.

Drucker, E.G. and Lauder, G.V. (2005). Locomotor function of the dorsal fin in rainbow trout: kinematic patterns and hydrodynamic forces. *J. Exp. Biol.* 208:4479–4494.

Eames, B.F., Allen, N., Young, J., Kaplan, A., Helms, J.A., and Schneider, R.A. (2007). Skeletogenesis in the swell shark *Cephaloscyllium ventriosum. J. Anat.* 210:542–554.

Ferry, L.A. and Lauder, G.V. (1996). Heterocercal tail function in leopard sharks: a three-dimensional kinematic analysis of two models. *J. Exp. Biol.* 199:2253–2268.

Fish, F.E. and Shannahan, L.D. (2000). The role of the pectoral fins in body trim of sharks. *J. Fish Biol.* 56:1062–1073.

Flammang, B.E. (2010). Functional morphology of the radialis muscle in shark tails. *J. Morphol.* 271:340–352.

Foster, K.L. and Higham, T.E. (2010). How to build a pectoral fin: functional morphology and steady swimming kinematics of the spotted ratfish (*Hydrolagus colliei*). *Can. J. Zool.* 88:774–780.

Gao, J., Bi, S., Li, J., and Liu, C. (2009). Design and experiments of robot fish propelled by pectoral fins. *ROBIO* 2009:445–450.

Garman, S. (1913). The Plagiostoma (sharks, skates, and rays). *Mem. Mus. Comp. Zool. Harvard Coll.* 36.

Gemballa, S., Konstantinidis, P., Donley, J.M., Sepulveda, C., and Shadwick, R.E. (2006). Evolution of high-performance swimming in sharks: transformations of the musculotendinous system from subcarangiform to thunniform swimmers. *J. Morphol.* 267:477–493.

Gordon, M.S., Hove, J.R., and Bartol, I.K. (2002). Dynamics and energetics of animal swimming and flying: introduction. *Integr. Comp. Biol.* 42:960–963.

Gray, J. (1968). *Animal Locomotion.* Weidenfeld & Nicolson, London.

Grove, A.J. and Newell, G.E. (1936). A mechanical investigation into the effectual action of the caudal fin of some aquatic chordates. *Ann. Mag. Nat. Hist.* 17:280–290.

Harris, J.E. (1936). The role of the fins in the equilibrium of the swimming fish. I. Wind tunnel tests on a model of *Mustelus canis* (Mitchell). *J. Exp. Biol.* 13:476–493.

Harris, J.E. (1953). Fin patterns and mode of life in fishes. In: Marshall, S.M. and Orr, A.P. (Eds.), *Essays in Marine Biology.* Oliver & Boyd, Edinburgh, pp. 17–28.

Jordan, L.K. (2008). Comparative morphology of stingray lateral line canal and electrosensory systems. *J. Morphol.* 269:1325–1339.

Kajiura, S.M., Forni, J.B., and Summers, A.P. (2003). Maneuvering in juvenile carcharhinid and sphyrnid sharks: the role of the hammerhead shark cephalofoil. *Zoology* 106:19–28.

Kemp, N.E. (1999). Integumentary system and teeth. In: Hamlett, W.C. (Ed.), *Sharks, Skates, and Rays: The Biology of Elasmobranch Fishes.* The Johns Hopkins University Press, Baltimore, MD, pp. 43–68.

Koester, D.M. and Spirito, C.P. (1999). Pelvic fin locomotion in the skate, *Leucoraja erinacea. Am. Zool.* 39:55A.

Krothapalli, A. and Lourenco, L. (1997). Visualization of velocity and vorticity fields. In: Nakayama, Y. and Tanida, Y. (Eds.), *Atlas of Visualization,* Vol. 3. CRC Press, Boca Raton, FL, pp. 69–82.

Kundu, P. (1990). *Fluid Mechanics.* Academic Press, San Diego.

Lauder, G.V. (2000). Function of the caudal fin during locomotion in fishes: kinematics, flow visualization, and evolutionary patterns. *Am. Zool.* 40:101–122.

Lauder, G.V. (2010). Swimming hydrodynamics: ten questions and the technical approaches needed to resolve them. In: Taylor, G.K., Triantafyllou, M.S., and Tropea, C. (Eds.), *Animal Locomotion.* Springer-Verlag, Berlin, pp. 3–15.

Lauder, G.V. and Drucker E. (2002). Forces, fishes, and fluids: hydrodynamic mechanisms of aquatic locomotion. *News Physiol. Sci.* 17:235–240.

Lauder, G.V. and Madden, P.G.A. (2008). Advances in comparative physiology from high-speed imaging of animal and fluid motion. *Annu. Rev. Physiol.* 70:143–163.

Lauder, G.V., Nauen, J., and Drucker, E.G. (2002). Experimental hydrodynamics and evolution: function of median fins in ray-finned fishes. *Integr. Comp. Biol.* 42:1009–1017.

Lauder, G.V., Drucker, E.G., Nauen, J., and Wilga, C.D. (2003). Experimental hydrodynamics and evolution: caudal fin locomotion in fishes. In: Bels, V., Gasc, J.P., and Casinos, A. (Eds.), *Vertebrate Biomechanics and Evolution,* Bios Scientific Publishers, Oxford, pp. 117–135.

Liao, J. and Lauder, G.V. (2000). Function of the heterocercal tail in white sturgeon: flow visualization during steady swimming and vertical maneuvering. *J. Exp. Biol.* 203:3585–3594.

Liem, K.F. and Summers, A.P. (1999). Muscular system. In: Hamlett, W.C. (Ed.), *Sharks, Skates, and Rays: The Biology of Elasmobranch Fishes,* The Johns Hopkins University Press, Baltimore, MD, pp. 93–114.

Liem, K.F., Bemis, W.E., Walker, Jr., W.F., and Grande, L. (2001). *Functional Anatomy of the Vertebrates: An Evolutionary Perspective,* 3rd ed. Harcourt, New York.

Lighthill, J. and Blake, R. (1990). Biofluid dynamics of balistiform and gymnotiform locomotion. Part 1. Biological background and analysis by elongated-body theory. *J. Fluid. Mech.* 212:183–207.

Lindsey, C.C. (1978). Form, function, and locomotory habits in fish. In: Hoar, W.S. and Randall, D.J. (Eds.), *Fish Physiology.* Vol. 7. *Locomotion.* Academic Press, New York, pp. 1–100.

Lingham-Soliar, T. (2005). Dorsal fin in the white shark, *Carcharodon carcharias*: a dynamic stabilizer for fast swimming. *J. Morphol.* 263:1–11.

Macesic, L.J. and Kajiura, S.M. (2010). Comparative punting kinematics and pelvic fin musculature of benthic batoids. *J. Morphol.* 271:1219–1228.

Magnan, A. (1929). Les charactéristiques géométriqes et physiques des poisons. *Ann. Sci. Nat. Zool.* 10:1–132.

Nauen, J.C. and Lauder, G.V. (2002). Quantification of the wake of rainbow trout (*Oncorhynchus mykiss*) using three-dimensional stereoscopic digital particle image velocimetry. *J. Exp. Biol.* 205:3271–3279.

Perry, C.N., Cartamil, D.P., Bernal, D. et al. (2007). Quantification of red myotomal muscle volume and geometry in the shortfin mako shark (*Isurus oxyrinchus*) and the salmon shark (*Lamna ditropis*) using T1-weighted magnetic resonance imaging. *J. Morphol.* 268:284–292.

Porter, M.E. and Long, Jr., J.H. (2010). Vertebrae in compression: mechanical behavior of arches and centra in the gray smooth-hound shark (*Mustelus californicus*). *J. Morphol.* 271:366–375.

Porter, M.E., Beltrán, J.L., Koob, T.J., and Summers, A.P. (2006). Material properties and biochemical composition of mineralized vertebral cartilage in seven elasmobranch species (Chondrichthyes). *J. Exp. Biol.* 209:2920–2928.

Porter, M.E., Koob, T.J., and Summers, A.P. (2007). The contribution of mineral to the material properties of vertebral cartilage from the smooth-hound shark *Mustelus californicus*. *J. Exp. Biol.* 210:3319–3327.

Porter, M.E., Roque, C.M., and Long, Jr., J.H. (2009). Turning maneuvers in sharks: predicting body curvature from axial morphology. *J. Morphol.* 270:954–965.

Pridmore, P.A. (1995). Submerged walking in the epaulette shark *Hemiscyllium ocellatum* (Hemiscyllidae) and its implications for locomotion in rhipidistian fishes and early tetrapods. *Zoology* 98:278–297.

Raffel, M., Willert, C.E., Wereley, S.T., and Kompenhans, J. (2007). *Particle Image Velocimetry: A Practical Guide*, 2nd ed. Springer, Berlin.

Reif, W.E. and Weishampel, D.B. (1986). Anatomy and mechanics of the lunate tail in lamnid sharks. *Zool. Jahrb. Anat.* 114:221–234.

Rosenberger, L. (2001). Pectoral fin locomotion in batoid fishes: undulation *versus* oscillation. *J. Exp. Biol.* 204:379–394.

Rosenberger, L.J. and Westneat, M.W. (1999). Functional morphology of undulatory pectoral fin locomotion in the stingray *Taeniura lymma* (Chondrichthyes: Dasyatidae). *J. Exp. Biol.* 202:3523–3539.

Schaeffer, J.T. and Summers, A.P. (2005). Batoid wing skeletal structure: novel morphologies, mechanical implications, and phylogenetic patterns. *J. Morphol.* 264:298–313.

Shirai, S. (1996). Phylogenetic interrelationships of neoselachians (Chondrichthyes: Euselachii). In: Stiassny, M., Parenti, L., and Johnson, G.D. (Eds.), *Interrelationships of Fishes*. Academic Press, San Diego, pp. 9–34.

Simons, J.R. (1970). The direction of the thrust produced by the heterocercal tails of two dissimilar elasmobranchs: the Port Jackson shark, *Heterodontus portusjacksoni* (Meyer), and the piked dogfish, *Squalus megalops* (Macleay). *J. Exp. Biol.* 52:95–107.

Simons, M. (1994). *Model Aircraft Aerodynamics*. Argus Books, Herts, U.K.

Sims, D.W. (2010). Tracking and analysis techniques for understanding free-ranging shark movements and behavior. In: Carrier, J.C., Musick, J.A., and Heithaus, M.R. (Eds.), *Sharks and Their Relatives II*. CRC Press, Boca Raton, FL, pp. 351–392.

Smith, H.C. (1992). *The Illustrated Guide to Aerodynamics*. TAB Books, New York.

Stamhuis, E.J. (2006). Basics and principles of particle image velocimetry (PIV) for mapping biogenic and biologically relevant flows. *Aquat. Ecol.* 40:463–479.

Standen, E.M. (2010). Muscle activity and hydrodynamic function of pelvic fins in trout (*Oncorhynchus mykiss*). *J. Exp. Biol.* 213:831–841.

Standen, E.M. and Lauder, G.V. (2005). Dorsal and anal fin function in bluegill sunfish *Lepomis macrochirus*: three-dimensional kinematics during propulsion and maneuvering. *J. Exp. Biol.* 208:2753–2763.

Standen, E.M. and Lauder, G.V. (2007). Hydrodynamic function of dorsal and anal fins in brook trout (*Salvelinus fontinalis*). *J. Exp. Biol.* 210:340–356.

Summers, A.P. and Koob, T.J. (2002). The evolution of tendon: morphology and material properties. *Comp. Biochem. Phys. A* 133:1159–1170.

Summers, A.P., Koob-Emunds, M.M., Kajiura, S.M., and Koob, T.J. (2003). A novel fibrocartilaginous tendon from an elasmobranch fish (*Rhinoptera bonasus*). *Cell Tissue Res.* 312:221–227.

Thomson, K.S. (1971). The adaptation and evolution of early fishes. *Q. Rev. Biol.* 46:139–166.

Thomson, K.S. (1976). On the heterocercal tail in sharks. *Paleobiology* 2:19–38.

Thomson, K.S. and Simanek DE. (1977). Body form and locomotion in sharks. *Am. Zool.* 17:343–354.

Tytell, E.D., Standen, E.M., and Lauder, G.V. (2008). Escaping flatland: three dimensional kinematics and hydrodynamics of median fins in fishes. *J. Exp. Biol.* 211:187–195.

Tytell, E.D., Borazjani, I., Sotiropoulos, F. et al. (2010). Disentangling the functional roles of morphology and motion in the swimming of fish. *Int. Comp. Biol.* 50:1140–1154.

Walker, J.A. and Westneat, M.W. (2000). Mechanical performance of aquatic rowing and flying. *Proc. R. Soc. Lond. B* 267:1875–1881.

Webb, P.W. (1984). Form and function in fish swimming. *Sci. Am.* 251:72–82.

Webb, P.W. (1988). Simple physical principles and vertebrate aquatic locomotion. *Am. Zool.* 28:709–725.

Webb, P.W. and Blake, R.W. (1985). Swimming. In: Hildebrand, M. et al. (Eds.), *Functional Vertebrate Morphology*. Harvard University Press, Cambridge, MA, pp. 110–128.

Webb, P.W. and Gerstner, C.L. (1996). Station-holding by the mottled sculpin, *Cottus bairdi* (Teleostei: Cottidae) and other fishes. *Copeia* 1996:488–493.

Webb, P.W. and Keyes, R.S. (1982). Swimming kinematics of sharks. *Fish Bull.* 80:803–812.

Wilga, C.D. and Lauder, G.V. (1999). Locomotion in sturgeon: function of the pectoral fins. *J. Exp. Biol.* 202:2413–2432.

Wilga, C.D. and Lauder GV. (2000). Three-dimensional kinematics and wake structure of the pectoral fins during locomotion in leopard sharks *Triakis semifasciata*. *J. Exp. Biol.* 203:2261–2278.

Wilga, C.D. and Lauder, G.V. (2001). Functional morphology of the pectoral fins in bamboo sharks, *Chiloscyllium plagiosum*: benthic versus pelagic station holding. *J. Morphol.* 249:195–209.

Wilga, C.D. and Lauder, G.V. (2002). Function of the heterocercal tail in sharks: quantitative wake dynamics during steady horizontal swimming and vertical maneuvering. *J. Exp. Biol.* 205:2365–2374.

Wilga, C.D. and Lauder, G.V. (2004). Biomechanics of locomotion in sharks, rays and chimeras. In: Carrier, J.C., Musick, J., and Heithaus, M. (Eds.), *Biology of Sharks and Their Relatives*. CRC Press: Boca Raton, FL, pp. 139–164.

Willert, C.E. and Gharib, M. (1991). Digital particle image velocimetry. *Exp. Fluids* 10:181–193.

Xu, Y., Zong, G., Bi, S., and Gao, J. (2007). Initial development of a flapping propelled unmanned underwater vehicle (UUV). *ROBIO* 2007:514–529.

Yang, S., Qiu, J., and Han, X. (2009). Kinematics modeling and experiments of pectoral oscillation propulsion robotic fish. *J. Bionic Eng.* 6:174–179.

Zangerl, R. (1973). *Interrelationships of Early Chondrichthyans*. Academic Press, London.

Zhou, C. and Low, K. (2010). Better endurance and load capacity: an improved design of Manta Ray Robot (RoMan-II). *J. Bionic Eng.* 7(Suppl.):S137–S144.

6

Prey Capture Behavior and Feeding Mechanics of Elasmobranchs

Philip J. Motta and Daniel R. Huber

CONTENTS

6.1 Introduction

Perhaps the most remarkable thing about the elasmobranch feeding mechanism is its functional diversity despite its morphological simplicity. Compared to the teleost skull, which has approximately 63 bones (excluding the branchiostegal, circumorbital, and branchial bones), the feeding apparatus of a shark is composed of just 10 cartilaginous elements: the chondrocranium, paired palatoquadrate and Meckel's cartilages, hyomandibulae, ceratohyal, and a basihyal. Furthermore, the elasmobranchs lack pharyngeal jaws and the ability to further process food by this secondary set of decoupled jaws as do bony fishes. Despite this, sharks, skates, and rays display a diversity of feeding mechanisms and behaviors that, although they do not match those of the

bony fishes, is truly remarkable, especially considering there are only approximately 1100+ species of elasmobranchs compared to about 24,000 species of teleost fishes (Compagno, 2001; Compagno et al., 2005; Nelson, 1994). The elasmobranchs capture prey by methods as diverse as ram, biting, suction, and filter feeding, and they feed on prey ranging from plankton to marine mammals and giant squid (Cherel and Duhamel, 2004; Frazzetta, 1994; Moss, 1972; Motta and Wilga, 2001; Motta et al., 2010). Understanding the elasmobranch feeding mechanism will shed light on how this functional versatility is achieved and whether or not it parallels that of the bony fishes.

Understanding the feeding mechanism of elasmobranchs is also important to biologists from an evolutionary perspective. The chondrichthyan fishes represent a basal group of jawed fishes that share a common ancestor with bony fishes (Carroll, 1988; Long, 1995; Schaeffer and Williams, 1977); therefore, they provide insight into the evolution of lower vertebrate feeding mechanisms. Studies on chondrichthyan fishes have provided an understanding of the evolution of the jaw depression mechanism in aquatic vertebrates (Wilga et al., 2000) and the evolution and function of jaw suspension systems in vertebrates (Grogan and Lund, 2000; Grogan et al., 1999; Wilga, 2002). Studies on elasmobranch teeth also provide insight into the evolution of dermal teeth and armor and the patterns of tooth replacement in vertebrates (Reif, 1978, 1980; Reif et al., 1978).

Despite a tremendous increase in the knowledge of bony fish feeding mechanisms in the last three decades (Lauder, 1985; Liem, 1978; Westneat, 2004), there have been fewer studies on elasmobranchs and even fewer on batoids (Bray and Hixon, 1978; Dean and Motta, 2004a,b; Marion, 1905; Summers, 2000) than on sharks (Moss, 1972; Nobiling, 1977; Shirai and Nakaya, 1992; Wilga, 2008). Numerous embryological and anatomical studies on the head of sharks in the previous century or early part of this century (reviewed in Motta and Wilga, 1995, 1999) were influential in our understanding of the evolution and development of the skull and branchial arches; however, following some earlier anatomical studies (Moss, 1972, 1977b; Springer, 1961), there have been relatively fewer studies that incorporate cineradiography, high-speed photography, electromyography, and biomechanical modeling of the feeding apparatus (Ferrara et al., 2011; Ferry-Graham, 1998a,b; Huber et al., 2005; Motta et al., 1997; Wilga and Motta, 1998a,b, 2000; Wilga et al., 2007; Wu, 1994).

The goal of this chapter is to provide a review of the feeding anatomy, behavior, and biomechanics of extant elasmobranchs with an emphasis on the structure and function of the feeding apparatus. To place prey capture and mechanics in a more meaningful framework it is necessary to outline how elasmobranchs approach their prey; consequently, prey approach behavior is briefly discussed. Feeding behavior is considered to be pre-capture behaviors (e.g., stalking, ambushing), whereas prey capture refers to the process beginning with opening of the mouth as the fish approaches the prey and usually ends with the prey grasped between the jaws. Because so little is known of postcapture manipulation or processing, this topic is covered only briefly. During manipulation, the prey is reduced in size by cutting or crushing, often combined with head shaking, and then it is transported from the buccal cavity through the pharyngeal cavity into the esophagus. Similarly, because so little is known of batoid feeding mechanisms, sharks are emphasized more than skates or rays. In some instances, food is used to refer to pieces of whole items offered to an animal under experimental conditions, whereas prey refers to dietary items captured during natural feeding.

The review does not cover feeding ecology and diet (see Cortés, 1999; Wetherbee et al., Chapter 8 of this volume), although diet is occasionally referred to when discussing feeding behaviors and mechanisms.

6.2 Ethology of Predation

6.2.1 Predatory Behaviors

Sharks, skates, and rays must first approach their prey before they can capture it. When the prey is within grasp of the predator, the capture event is usually very rapid as compared to the approach, and at this point either the prey may be held within the grasp of the teeth or it may be transported directly through the mouth to the entrance of the esophagus. If the prey is grasped by the teeth, one or a series of manipulation/processing bites can reduce the prey in size prior to the final transport event. In this manner, we speak of capture bites, manipulation/processing bites, and hydraulic transport, the last of which invariably involves suction of the water with the entrained food (Motta and Wilga, 2001). The mechanics of swallowing—that is, getting the food into and through the esophagus—is still unresolved.

Because of the inherent difficulty of studying elasmobranchs in their natural environment, predatory behavior is generally poorly understood, especially as compared to that of bony fishes. Large or pelagic sharks are perhaps the least understood, but their foraging patterns are being revealed due to the advent of telemetry studies (Domeier and Nasby-Lucas, 2008; Holland et al., 1999; Klimley et al., 2001) and the attachment of small animal-borne video cameras and other biosensors (e.g., accelerometers) to free-swimming sharks (Heithaus et al., 2001, 2002a; Sims, 2010; Heithaus and Vaudo, Chapter 17 of this volume). A great deal of what we know of predatory behavior is from anecdotal or one-of-a-kind observations (Pratt et al., 1982; Strong, 1990), telemetry studies (Klimley et al., 2001), behavioral studies of shallow-water benthic elasmobranchs (Fouts and Nelson, 1999; Strong, 1989), and laboratory studies (Lowry et al., 2007; Sasko et al., 2006) or is inferred from morphology (Compagno, 1990; Myrberg, 1991). Surprisingly, the more accessible batoids are vastly understudied as compared to sharks (Belbenoit and Bauer, 1972; Lowe et al., 1994).

How sharks and rays approach and hunt their prey is perhaps the least understood aspect of their feeding biology. Most elasmobranchs are probably very opportunistic in what they prey on and how they acquire their prey (see Chapter 17 of this volume). When hunting by speculation, the fish searches an area that it expects to have prey or it follows another organism expecting that animal to flush prey out by its presence (Curio, 1976). *Dasyatis* rays will position themselves at regions of higher tidal water movement, such as near beach promontories, waiting for prey organisms to be swept by. Large aggregations of rays may be found at these locations during periods of swift tidal movement (Motta, pers. obs.). Large tiger sharks, *Galeocerdo cuvier*, occur most frequently in Shark Bay, Western Australia, during the season that dugongs, an important prey item for this size class of sharks, are present (Heithaus, 2001; Wirsing et al., 2007). Tiger sharks aggregate at the northwestern Hawaiian Islands during June and July, coinciding with the summer fledging period of blackfooted and Laysan albatross birds, upon which they prey (Lowe et al., 2003), and white sharks, *Carcharodon carcharias*, aggregate at Seal Island, South Africa, to prey on Cape fur seals (Martin et al., 2005). Each March and April, whale sharks, *Rhincodon typus*, aggregate on the continental shelf of the central western Australian coast, particularly at Ningaloo Reef, in response to coral spawning events that occur each year (Gunn et al., 1999; Taylor and Pearce, 1999). Perhaps the largest aggregation of whale sharks in the world occurs off the Yucatan Peninsula, where they feed on rich plankton blooms in relatively shallow water (Hueter et al., 2008; Motta et al., 2010). White sharks, *C. carcharias*, spend a lot of time patrolling near seal colonies off the South Farallon Islands and Año Nuevo Island, California. Most of the shark's movement is back and forth parallel and near to the shoreline as it intercepts seals and sea lions that are departing from and returning to their shore-based rookeries. In some cases, the sharks pass within 2 m of the shore. Prey capture, however, is infrequent compared to the time spent patrolling (Klimley et al., 2001).

Ambushing involves the predator trying to conceal or advertise (aggressive mimicry) its presence while lying in wait for the prey (Curio, 1976). By partially burying themselves in the soft substrate, Pacific angel sharks, *Squatina californica*, ambush demersal fishes. These sharks appear to actively select ambush sites within localized areas adjacent to reefs (Fouts, 1995; Fouts and Nelson, 1999). Pacific electric rays, *Torpedo californica*, either ambush their prey from the bottom or use a search-and-attack behavior from the water column. During the day, the rays ambush their prey of mostly fishes by burying themselves in sand and jumping over the prey. After swimming over the prey, the rays cup their pectoral fins around the prey while electrically discharging. They then pivot over the stunned prey so as to swallow it head first. At night, the rays are seen swimming or hovering in the water column 1 to 2 m above the substratum. The rays then lunge forward over the prey, cup their pectorals over the prey while discharging, and either pin the prey to the bottom or, using frontal somersaults and peristaltic-like movements of the disk, move

the prey closer to the mouth for swallowing (Bray and Hixon, 1978; Lowe, 1991; Lowe et al., 1994). Similar stereotyped prey capture behavior has also been described for the electric rays *T. marmorata*, *T. ocellata*, and *T. nobiliana* (Belbenoit and Bauer, 1972; Michaelson et al., 1979; Wilson, 1953; reviewed in Belbenoit, 1986).

Ambushing behavior of rays and sharks has been observed at the inshore spawning grounds of chokka squid (*Loligo vulgaris reynaudii*) off South Africa. Diamond rays, *Gymnura natalensis*, camouflage themselves in the substrate and then lunge out toward female squid as they try to spawn on the bottom. Large numbers of sharks and rays aggregate at these spawning grounds. In addition to pyjama catsharks, *Poroderma africanum*, and leopard catsharks, *P. pantherinum*, ambushing the spawning squid from the rocky reef substrate, the rays and sharks also chase down the squid to capture them or simply bite off the attached egg masses from the substrate (Smale et al., 1995, 2001).

In contrast to ambushing, the stalking predator approaches the prey while concealed and then makes a sudden assault (Curio, 1976). White sharks, *Carcharodon carcharias*, will stalk prey downstream in oceanic or tidal currents (Pyle et al., 1996), and they stalk Cape fur seals primarily within 2 hours of sunrise when light levels are low (Martin et al., 2005). Sevengill sharks, *Notorynchus cepedianus*, capture elusive prey using a stealthy underwater approach with very little body movement and only slight undulatory motions of the caudal fin. They move within striking distance and make a quick dash at the prey, which can include fur seals (Ebert, 1991). Using animal-borne video technology, Heithaus et al. (2002a) observed tiger sharks, *Galeocerdo cuvier*, stalking their benthic prey from above, in some cases getting as close as 2 m from large teleost fishes before the shark was detected.

Other elasmobranchs may lure prey to them. Luminescent tissue on the upper jaw of the megamouth shark, *Megachasma pelagios*, might attract euphausid shrimp and other prey into its mouth (Compagno, 1990). The white tips on the pectoral fins of oceanic whitetip sharks, *Carcharhinus longimanus*, might act as visual lures to aid in the capture of its rapid moving prey (Myrberg, 1991), and bioluminescence in the cookie-cutter shark, *Isistius brasiliensis*, might serve to lure pelagic predators from which it gouges chunks of flesh (Jones, 1971; Papastamatiou et al., 2010; Widder, 1998).

Most elasmobranchs will scavenge food when given the opportunity. Sevengill sharks, *Notorynchus cepedianus*, will feed on marine mammals, including whale and dolphin carcasses, bait left on fishing hooks, and even human remains (Ebert, 1991). Tiger sharks, *Galeocerdo cuvier*, are notorious opportunistic feeders; in addition to their regular diet, they will scavenge food ranging from dead dugongs to human refuse (Heithaus,

2001; Lowe et al., 1996; Randall, 1992; Smale and Cliff, 1998). Blue sharks, *Prionace glauca*, will similarly scavenge human refuse and dead or injured birds, although they have been observed to stalk resting birds and perhaps scavenge mesopelagic cephalopods (Henderson et al., 2001; Markaida and Sosa-Nishizaki, 2010; Stevens, 1973, cited in Henderson et al., 2001). Sleeper sharks, *Somniosus microcephalus*, will scavenge fishery offal and carrion, including fur seals (Cherel and Duhamel, 2004). White sharks, *Carcharodon carcharias*, often scavenge whale carcasses (Casey and Pratt, 1985; Curtis et al., 2006; Dicken, 2008; Dudley et al., 2000; Long and Jones, 1996; McCosker, 1985; Pratt et al., 1982). Large gray reef sharks, *Carcharhinus amblyrhynchos*, at Enewetak Island, Marshall Islands, follow carangid jacks as both scavengers and predators (Au, 1991), and velvet belly lanternsharks, *Etmopterus spinax*, undergo a dietary shift, with larger individuals incorporating scavenging of fish and cephalopods (Neiva et al., 2006).

Although many species of sharks forage solitarily, in some cases aggregations of sharks will come together to feed. Blacktip reef sharks, *Carcharhinus melanopterus*, and lemon sharks, *Negaprion brevirostris*, were observed to apparently herd schools of fish against the shoreline and then feed on them (Eibl-Eibesfeldt and Hass, 1959; Morrissey, 1991), and oceanic whitetip sharks, *Carcharhinus longimanus*, were observed to herd squid at night (Strasburg, 1958). Thresher sharks (*Alopias*) are reported to apparently work in groups to capture fish, using their long caudal fins to herd and stun fish (Aalbers et al., 2010; Budker, 1971; Castro, 1996; Coles, 1915; Compagno, 1984). Sevengill sharks, *Notorynchus cepedianus*, will circle a seal and prevent its escape. The circle is tightened, and eventually one shark initiates the attack that stimulates the others to begin feeding (Ebert, 1991). Although some authors have considered these behaviors cooperative, they could simply reflect aggregations of animals at a prey item and not cooperative foraging (Motta and Wilga, 2001; see Chapter 17 of this volume for a definition of cooperative foraging). So-called feeding frenzies of sharks appear to be nothing more than highly motivated feeding events involving generally many individuals. The sharks have been described as attacking prey or food items indiscriminately, moving at an accelerated speed, and disregarding any injuries they may receive in the attack. Injured or hooked sharks are often attacked and consumed by the other sharks. These feeding bouts, which can involve as few as six sharks to hundreds of sharks, can end as abruptly as they begin (Gilbert, 1962; Hobson, 1963; Nelson, 1969; Springer, 1967; Vorenberg, 1962). The feeding ecology, behavior, and diet of epipelagic, deepwater, and tropical marine elasmobranchs are also reviewed by Kyne and Simpfendorfer (2010), Stevens (2010), and White and Sommerville (2010).

6.2.2 Feeding Location and Prey Capture

Sharks approach their prey on the surface, in midwater, or on the bottom. One of the older misconceptions was that sharks must roll on their side to take prey in front of them because of their subterminal mouth (Budker, 1971). In fact, the mouths of modern sharks do not preclude them from feeding on prey in front of or above them, and sharks will approach surface or underwater food with a direct head-on approach or will roll on their side to bite at the food (Budker, 1971; Motta, pers. obs.). White sharks, *Carcharodon carcharias*, will approach in their normal orientation, roll on their side, or roll completely over so their ventral side is up when they feed on underwater bait or a floating whale carcass (Dicken, 2008; Pratt et al., 1982; Tricas and McCosker, 1984). During surface feeding, *C. carcharias* may bite such prey as elephant seals and then retreat until the prey lapses into shock or bleeds to death. The shark then returns to feed on the prey (McCosker, 1985; Tricas and McCosker, 1984). Tricas and McCosker referred to this as the "bite and spit" strategy. Klimley (1994) and Klimley et al. (1996) proposed, however, that white sharks hold the pinniped prey tightly in their mouth and drag it below the surface, often removing a bite from the prey in the process. The prey may be released underwater, after which it floats or swims to the surface and dies by exsanguination. Meanwhile, the shark follows the prey to the surface to begin feeding after it dies. Martin et al. (2005) also found no support for the "bite and spit" behavior in white shark predation on Cape fur seals off South Africa. The sharks often performed a subsurface carry whereby the shark carried a dead or incapacitated seal underwater before feeding on it. After capturing a seal or decoy in its jaws, the shark often repurchased or repositioned the prey by lifting its snout, removing the upper teeth with the lower teeth remaining in the prey, then quickly protruding the upper jaw and bringing the teeth in contact with the prey again. This would bring the longitudinal axis of the food item in line with the longitudinal axis of the shark. These sharks displayed a variety of surface attacks on seals including subsurface attacks that launched the shark partially or completely out of the water in a vertical to near vertical orientation and inverted breaches with the body in an inverted position.

Blue sharks, *Prionace glauca*, approach schools of squid on the surface with an underwater approach or a surface charge. Small anchovies are captured from a normal swimming posture, but when capturing larger whole mackerel from behind blue sharks may roll on their side (Tricas, 1979). Large schools of oceanic whitetip sharks, *Carcharhinus longimanus*, have been observed swimming erratically in a sinuous course on the surface with their mouths wide open. These sharks made no attempt to snap up the small tuna through which they were swimming;

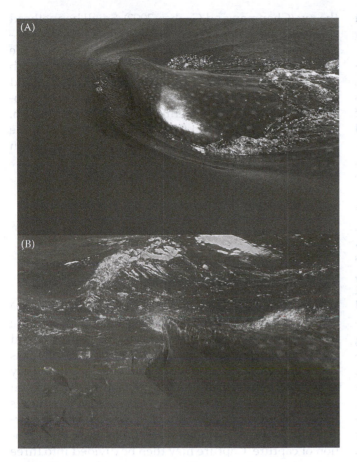

FIGURE 6.1

(A) Surface and (B) subsurface view of two surface ram filter feeding whale sharks, *Rhincodon typus* (size between 5 and 8 m TL). (From Motta, P.J. et al., *Zoology*, 113, 199–212, 2010. With permission.)

rather, they appeared to simply wait for the fish to swim or leap into their mouths (Bullis, 1961). Surface-feeding blacknose sharks (*Carcharhinus acronotus*), oceanic whitetip sharks (*C. longimanus*), white sharks (*C. carcharias*), and Caribbean reef sharks (*Carcharhinus perezi*) may raise the head just prior to prey capture (Bullis, 1961; Frazzetta and Prange, 1987; Motta and Wilga, 2001; Tricas and McCosker, 1984). This might place the open mouth in line with food as the shark approaches (Frazzetta and Prange, 1987). Whale sharks, *Rhincodon typus*, will make regular dives through the water column foraging for food. They will also swim slowly (average 1.1 m/s) at or near the surface with their body at an angle and the top of the head clear of the surface while feeding (Figure 6.1) (Gunn et al., 1999; Motta et al., 2010).

Rays and skates will also feed off the bottom. The ventral mouth of Pacific electric rays, *Torpedo californica*, does not preclude them from foraging in the water column in addition to sitting on the bottom. After stunning the prey, which can result in breaking of the vertebral column, they manipulate the prey toward the mouth with the pectoral fins or force the stunned prey to the

substrate (Bray and Hixon, 1978; Lowe et al., 1994). The thorny skate, *Raja radiata*, is primarily a benthivorous feeder as a juvenile or adolescent, but benthopelagic food items, including fishes, become important to larger individuals (Skjaeraasen and Bergstad, 2000). Dietary items indicate that *Dasyatis say* and *D. centroura* in Delaware Bay frequently feed off the bottom on free-swimming organisms (Hess, 1961). The midwater-swimming cownose ray, *Rhinoptera bonasus*, will descend to the bottom and excavate benthic invertebrates, including bivalves (Sasko et al., 2006). The lesser electric ray, *Narcine bancroftii*, captures buried benthic prey by protruding its jaws up to 100% of its head length beneath the substrate and sucking the prey into its mouth (Dean and Motta, 2004b). Mobulid rays, including *Manta birostris* and *Mobula tarapacana*, filter feed both at the surface and in midwater, extending their cephalic wings to funnel prey and water through the mouth. Upon encountering a patch of prey, they will often swim in a circular formation or somersault while filter feeding to stay within the patch (Notarbartolo di Sciara and Hillyer, 1989; Motta, pers. obs.).

Some sharks will also take prey buried within the substrate or capture prey on the bottom. Leopard sharks, *Triakis semifasciata*, can apparently suck worms out of their burrows in addition to biting pieces off their benthic prey (Compagno, 1984; Talent, 1976). The epaulette shark, *Hemiscyllium ocellatum*, and the whitespotted bamboo shark, *Chiloscyllium plagiosum*, occasionally thrust their heads into the sediment up to the level of the first gill slit, apparently using suction to capture their benthic prey of worms and crabs. They then winnow the prey from the sand in the buccopharyngeal cavity and eject the sand through the first gill slit (Heupel and Bennett, 1998; Wilga, pers. obs.). Skates and rays primarily feed in or on the bottom by biting pieces of sessile invertebrates or excavating buried prey, although they will feed in the water column (Abd El-Aziz, 1986; Ajayi, 1982; Babel, 1967; Ebert et al., 1991; Edwards, 1980; Goitein et al., 1998; Gray et al., 1997; Hess, 1961; Hines et al., 1997; Holden and Tucker, 1974; Howard et al., 1977; Lucifora et al., 2000; Muto et al., 2001; Orth, 1975; Rudloe, 1989; Sasko et al., 2006; Sherman et al., 1983; Skjaeraasen and Bergstad, 2000; Stokes and Holland, 1992; Thrush et al., 1991; Valadez-Gonzalez et al., 2001; VanBlaricom, 1976). Rays dig up prey by pectoral "wing-flapping," or they hydraulically mine the prey by jetting water through the mouth (Gregory et al., 1979; Howard et al., 1977; Muto et al., 2001; Sasko et al., 2006; VanBlaricom, 1976). As discussed above, the cownose ray, *Rhinoptera bonasus*, uses a combination of wing flapping and water jetting to expose prey in the wild (Sasko et al., 2006; Schwartz, 1967, 1989); however, in the laboratory, the rays rest on the substrate on the tips of their pectoral fins and use repeated jaw opening and closing movements at 2.4 to

2.9 cycles per second to generate water flow in and out of the buccal cavity. The ventrally directed jet of water resuspends the sand and bivalve food, resulting in the effective separation of food and sand so the rays can capture the food. The large subrostral lobes are depressed, forming a chamber around the food item that it encloses laterally and partially anteriorly; the two lobes have been observed to move independently and push food toward the mouth (Sasko, 2000; Sasko et al., 2006).

Large-scale destruction of eelgrass, *Zostera marina*, beds in the Chesapeake Bay has been attributed to the excavation behavior of *Rhinoptera bonasus* (Orth, 1975). Excavation of benthic prey by rhythmic flapping of the rostrum and pectoral fins is common in other rays (Babel, 1967; Hines et al., 1997; Howard et al., 1977; Thrush et al., 1991; VanBlaricom, 1976). Southern stingrays, *Dasyatis americana*, excavate lancelets, *Branchiostoma floridae*, from the sandy substrate, and the presence in the gut of only medium- and large-sized prey led Stokes and Holland (1992) to speculate that the rays are winnowing out the sand and smaller lancelets while retaining the larger ones. Winnowing prey from ingested sediment is perhaps common in rays. The lesser electric ray, *Narcine bancroftii*, which specializes in wormlike prey including polychaete worms and anguilliform fishes, uses suction to capture the prey along with some sediment and ejects the latter out of the mouth, spiracle, or gill slits (Dean and Motta, 2004b; Funicelli, 1975; Rudloe, 1989). Similarly, during food processing, *R. bonasus* can separate prey from sand, flushing the sand out of the mouth and gill slits. This ray can also strip unwanted parts of the food item (e.g., mussel shell, skin and vertebral column of fish, shrimp shell) from the edible parts and eject the unwanted pieces. Larger pieces are ejected from the mouth, and smaller particles such as sand exit through the gill slits (Sasko et al., 2006).

Bottom-feeding horn sharks, *Heterodontus francisci*, use suction and biting to remove benthic invertebrates such as anemone tentacles, polychaetes, and urchins. They remove their prey with a "pecking-like" motion, often while they are raised on their pectoral fins (Edmonds et al., 2001; Strong, 1989). Gray reef sharks, *Carcharhinus amblyrhynchos*, in Hawaii primarily feed near the bottom on reef-associated teleosts and supplement their diet with invertebrates (Wetherbee et al., 1997). Rays are often taken by sharks, particularly hammerhead sharks (Budker, 1971; Gudger, 1907). Great hammerhead sharks, *Sphyrna mokarran*, have been observed to use their head to deliver powerful blows and to restrain rays on the substrate prior to biting pieces off the ray (Strong, 1990). They also exhibit a "pin and pivot" behavior during which the shark forcibly presses the ray against the substrate with the ventral surface of the cephalofoil and then, with a twisting motion of the body, pivots its head while remaining atop the ray as it engulfs part or all of

the ray (Chapman and Gruber, 2002). Small bonnethead sharks, *Sphyrna tiburo*, capture their food by depressing the mandible considerably as they swim over the food, catching the food either within the mouth or with the anterior mandibular teeth (Wilga, 1997; Wilga and Motta, 2000). Perhaps the strangest means of prey processing occurs when juvenile lesser spotted dogfish, *Scyliorhinus canicula*, use their dermal denticles on the tail to anchor food items so bite-sized pieces can be torn away by the jaws (Southall and Sims, 2003).

6.3 Feeding Mechanism

6.3.1 Mechanics of Prey Capture

When the shark, skate, or ray is within striking distance of its prey it begins the capture sequence. Prey capture is generally very rapid compared to the approach and typically lasts from about 100 to 400 ms. Capture begins when the mouth starts to open and lasts until the prey is grasped between the teeth or the jaws are closed on the prey (Motta et al., 2002). In some cases, the mouth is briefly closed just prior to opening, and under those circumstances this closing may be said to mark the initiation of capture. Capture may then be divided into three or four phases for heuristic purposes, although they are all continuous and rapid. If the slightly agape mouth is closed prior to mouth opening, this is termed the *preparatory phase* and is more common in suction-feeding bony fishes than elasmobranchs (Ajemian and Sanford, 2007; Lauder, 1985). An expansive phase follows during which there might be cranial (head) elevation accompanied by depression of the lower jaw. The branchial apparatus may also be expanded and the paired labial cartilages that lie at the edges of the mouth extended during this phase. The compressive phase begins at peak gape, and as the lower jaw is elevated the upper jaw (palatoquadrate cartilage) might be protruded toward the lower jaw. Cranial depression also occurs during this phase in many sharks, although surface-feeding *Carcharodon carcharias* can keep the cranium elevated until the recovery phase. At the end of the compressive phase, either the prey is grasped between the teeth or the food is already well within the buccal cavity. The recovery phase is marked by retraction of the upper jaw and the recovery of the other elements (hyomandibula, ceratohyal, basihyal, and branchial arches) back to their original resting positions (Figure 6.2) (Ajemian and Sanford, 2007; Frazzetta, 1994; Frazzetta and Prange, 1987; Matott et al., 2005; Moss, 1972, 1977b; Motta and Wilga, 2001; Motta et al., 1997, 2008; Tricas and McCosker, 1984; Wilga and Sanford, 2008).

Sharks and batoids capture their prey in a variety of ways. Ram feeding is perhaps the most common prey capture method in sharks, especially in carcharhinid and lamnid sharks. During ram capture, the shark swims over the relatively stationary prey and engulfs it whole or seizes it in its jaws. The food is then moved from the mouth through the pharyngeal cavity into the esophagus by hydraulic suction. Bonnethead sharks, *Sphyrna tiburo*, ram feed benthic food by depressing the mandible and scooping the food up as they swim over it (Wilga and Motta, 2000). White sharks, *Carcharodon carcharias*, primarily ram capture their food, sometimes approaching the food at such great speeds that they leave the water when feeding on surface-dwelling prey (Klimley, 1994; Klimley et al., 1996; Martin et al., 2005; Tricas, 1985; Tricas and McCosker, 1984).

Inertial suction feeding, or simply suction feeding, involves a decrease in the pressure of the buccopharyngeal chamber such that the prey or food is pulled into the mouth. There is a functional continuum from pure ram to pure inertial suction, and fishes can, and often do, use a combination of both (Norton and Brainerd, 1993; Wilga and Motta, 1998a). Caribbean reef sharks, *Carcharhinus perezi*, taking pieces of food will primarily over-swim the food item by ram but also employ some suction as witnessed by the food being sucked into the mouth rapidly when it is very close to the approaching shark. Sixgill sharks, *Hexanchus griseus*, will also position themselves close to bait, sitting on the bottom and sucking it into their mouth (Motta, pers. obs.).

Sharks specialized for suction prey capture, such as the nurse shark, *Ginglymostoma cirratum*, and the white-spotted bamboo shark, *Chiloscyllium plagiosum*, exhibit a suite of kinematic and morphological characters, including a relatively small mouth (generally less than one-third head length) as compared to ram-feeding sharks, small teeth, a mouth laterally enclosed by large labial cartilages, hypertrophied abductor muscles, and rapid buccal expansion (Ajemian and Sanford, 2007; Lowry and Motta, 2008; Lowry et al., 2007; Matott et al., 2005; Moss, 1965, 1977b; Motta and Wilga, 1999; Motta et al., 2002, 2008; Nauwelaerts et al., 2008). Suction feeding appears to be the predominant prey capture behavior in some clades, including the orectolobiforms and batoids. Specialization for suction feeding apparently evolved independently in conjunction with a benthic lifestyle, and these suction specialists feed on both elusive and non-elusive prey that live in or on the substrate, are attached to it, or are associated with the bottom (Ajemian and Sanford, 2007; Belbenoit, 1986; Clark and Nelson, 1997; Edmonds et al., 2001; Ferry-Graham, 1998b; Fouts, 1995; Fouts and Nelson, 1999; Heupel and Bennett, 1998; Moss, 1977b; Motta et al., 2002; Robinson and Motta, 2002; Tanaka, 1973; Wilga, 1997; Wilga and Motta, 1998a,b; Wu, 1994). Suction feeding near the

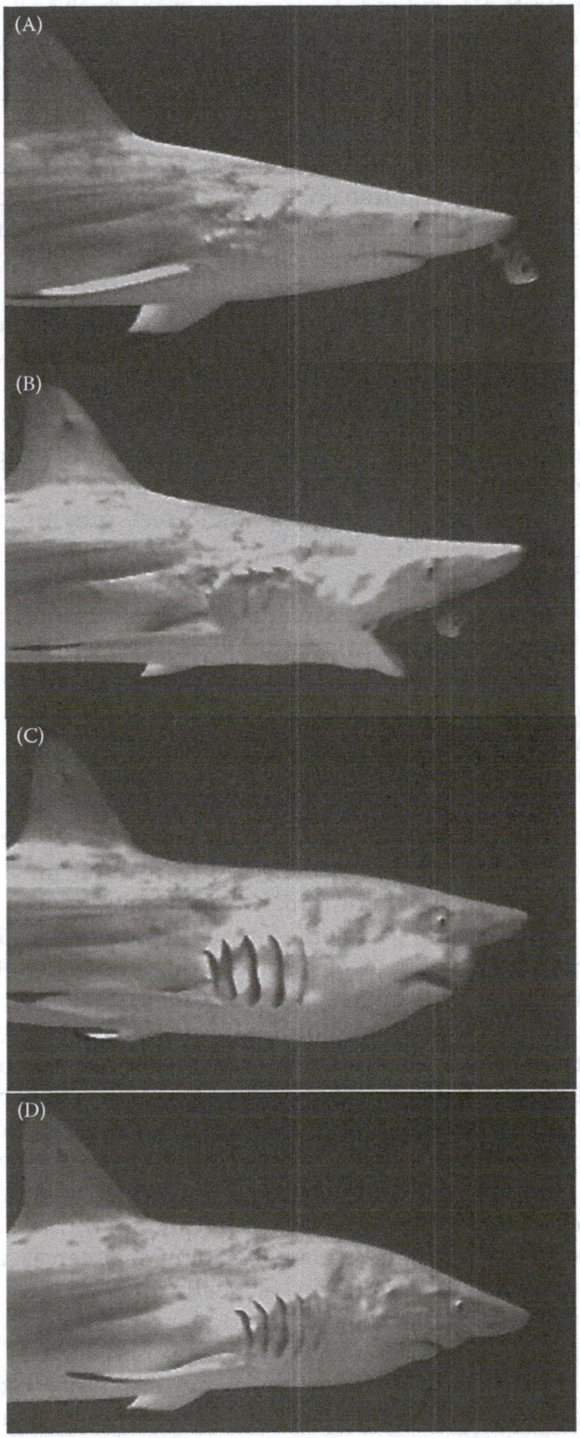

FIGURE 6.2

Food capture sequence of the blacktip shark, *Carcharhinus limbatus*. (A) Start of prey capture before mandible depression or cranial elevation. (B) The expansive phase characterized by mandible depression and head elevation. The slight bulge below the lower jaw is the basihyal being depressed. (C) During the compressive phase, the upper jaw is protruded (note bulge of upper jaw) and the mandible elevated as the prey is engulfed. The pharyngohyoid apparatus is depressed as the food is being transported posteriorly. (D) During the recovery phase, the upper jaw is retracted (partially retracted here), and the hyoid is mostly elevated.

substrate also extends the distance over which suction is effective; consequently, a predator can be effective further from the prey (Nauwelaerts et al., 2007). The prevalence of suction capture in batoids (Belbenoit and Bauer, 1972; Collins et al., 2007; Dean and Motta, 2004a,b; Sasko et al., 2006; Wilga and Motta, 1998b) might be related to the fact that fish often comprise a significant portion of the diet in many rays and skates, particularly in larger individuals (Abd El-Aziz, 1986; Ajayi, 1982; Babel, 1967; Barbini et al., 2010; Belbenoit and Bauer, 1972; Bray and Hixon, 1978; Ebert et al., 1991; Edwards, 1980; Funicelli, 1975; Holden and Tucker, 1974; Lucifora et al., 2000; Muto et al., 2001; Skjaeraasen and Bergstad, 2000; Smale and Cowley, 1992). Rapid suction combined with jaw protrusion might be an effective way to catch such elusive prey. Suction feeding might also be better suited for feeding off the bottom, allowing batoids to pick prey from the substrate.

Biting, which may accompany ram feeding, may also occur when an elasmobranch approaches its prey or food, ceases swimming, and simply bites the prey or pieces off the prey. The cookie-cutter shark, *Isistius brasiliensis*, shows a unique biting behavior in which it employs its modified pharyngeal muscles, upper jaw, and hyoid and branchial arches to suck onto its prey of pelagic fishes or marine mammals. Forming a seal with its fleshy lips, it then sinks its hooklike upper teeth and sawlike modified lower teeth into the prey and twists about its longitudinal axis to gouge out a plug of flesh, leaving a craterlike wound (Compagno, 1984; Jones, 1971; LeBoeuf et al., 1987; Papastamatiou et al., 2010; Shirai and Nakaya, 1992). The related kitefin shark, *Dalatias licha*, has dentition similar to that of the cookie-cutter shark and apparently feeds in the same manner (Clark and Kristof, 1990), as does the Greenland shark, *Somniosus microcephalus*. The latter apparently slowly stalks unsuspecting seals at breathing holes in the ice. Its slow movements and cryptic coloration may facilitate an element of surprise. Skomal and Benz (pers. obs.) observed Greenland sharks grasping seal carcasses in their jaws while oriented vertically in the water column. The sharks slowly rolled their bodies left and right allowing the band of closely opposed and elevated lower jaw teeth to carve out large hunks of flesh. In addition to ingesting whole sea turtles, Mediterranean white sharks, *Carcharodon carcharias*, and tiger sharks, *Galeocerdo cuvier*, in Shark Bay, Western Australia, often bite off pieces of the turtle, including limbs, often resulting in the turtle surviving (Fergusson et al., 2000; Heithaus et al., 2002b).

Continuous ram filter feeding, such as in the basking shark, *Cetorhinus maximus*, occurs when the shark continuously swims forward with the mouth open. In this manner, these sharks will actively seek and locate zooplankton patches on the surface. Basking sharks forage for longer periods in patches with high zooplankton density, and these high-density patches produce the most prolonged area-restricted searching during which the sharks follow convoluted swimming paths to stay within the plankton patches (Sims and Merrett, 1997; Sims and Quayle, 1998; Sims et al., 1997). The whale shark, *Rhincodon typus*, can employ intermittent suction filter feeding, generating suction with aperiodic pulses (Clark and Nelson, 1997; Diamond, 1985; Martin and Naylor, 1997; Sanderson and Wassersug, 1993; Taylor et al., 1983). Whale sharks can also use continuous ram filter feeding or hang vertically in the water column. In the latter case, they will suck prey into the mouth or rise vertically out of the water and sink back under water, thus creating an inflow of water and prey into their open mouths (Budker, 1971; Colman, 1997; Gudger, 1941a,b; Motta et al., 2010; Springer, 1967). The feeding behavior of the megamouth shark, *Megachasma pelagios*, has been inferred from its morphology. Although its feeding was speculated to employ ram (Taylor et al., 1983) or suction (Compagno, 1990), a more recent anatomical study proposed engulfment feeding similar to that of rorqual and humpback whales. During this behavior, the shark employs a low-velocity, high-volume suction during which the prey and water are drawn into the mouth. The continuous ram swimming then distends the elastic buccopharyngeal cavity. The engulfed prey and water are then driven across the gill rakers when the mouth is closed and the buccopharyngeal cavity compressed (Figure 6.3) (Nakaya et al., 2008).

6.3.2 Evolution of the Feeding Mechanism

The stem gnathostomes and early chondrichthyans had a jaw apparatus quite unlike modern sharks. In these, the upper jaw was braced against the braincase at multiple locations. This type of jaw suspension, termed *autodiastyly*, was possibly the ancestral type for the Chondrichthyes. Autodiastyly is characterized by a nonsuspensory hyoid arch that articulated with the palatoquadrate, with the hyoid arch being similar in morphology to the branchial arches. The palatoquadrate had ethmoidal and orbital articulations with the cranium (Figure 6.4) (Grogan and Lund, 2000; Lund and Grogan, 1997; reviewed in Wilga, 2002, and Wilga et al., 2007). The earliest sharks, the cladoselachians, had a large and almost terminal mouth with multicuspid teeth, relatively small labial cartilages, and a long palatoquadrate and Meckel's cartilage. The upper jaw of these sharks had an ethmoidal and a large postorbital articulation between the upper jaw and the cranium, as well as a hyomandibula that supposedly contributed little to jaw support. This type of jaw support is termed *amphistylic*. The body and caudal fin of these sharks were similar to modern fast-swimming pelagic sharks

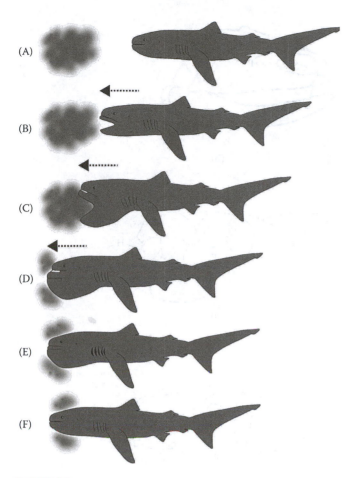

FIGURE 6.3
Engulfment feeding of the megamouth shark *Megachasma pelagios*. (From Nakaya, K. et al., *J. Fish Biol.*, 73, 17–34, 2008. With permission.)

(Figure 6.5). Their teeth were apparently suited for seizing and tearing prey rather than shearing or sawing, and it is speculated that they captured prey by biting, although suction may have played a role in prey capture (Carroll, 1988; Lund and Grogan, 1997; Moy-Thomas and Miles, 1971; Schaeffer, 1967; Wilga, 2002, 2005; Wilga et al., 2007; but see Maisey, 2008, for an alternative view). The xenacanthids that followed also had an amphistylic jaw suspension, a grasping dentition, long jaws, and a large gape suggesting a biting or ram-feeding mechanism (Carroll, 1988; Wilga, 2002; but see Maisey, 2008). The ctenacanthid sharks that followed likewise had an amphistylic jaw suspension, but they gave rise to the neoselachians, which include all modern sharks, skates, and rays (Figure 6.4).

During the evolution of modern elasmobranchs there was a general trend that involved shortening of the jaws and increased kinesis of the jaw suspension, facilitating upper jaw protrusion. Modern sharks have a subterminal mouth, shorter jaws, more movable hyomandibula that suspends the jaws, more protrusible upper jaw with a smaller otic process, and a dentition suited for sawing

and shearing (but see Section 6.5). In the modern galean sharks, the ethmoidal articulation between the ethmoid process of the palatoquadrate and the ethmoid region of the cranium is the only anterior connection to the cranium, and it is joined by an ethmopalatine ligament, but the hyomandibula is thought to contribute more to jaw support than the anterior ligaments (Figure 6.6). This type of jaw suspension is termed *hyostylic* (Figure 6.4) (Carroll, 1988; Schaeffer, 1967; reviewed in Wilga, 2002, 2005). Some groups of squalomorph sharks, including *Chlamydoselachus*, Squaliformes, and Hexanchiformes, have an orbitostylic jaw suspension in which the orbital process articulates with the orbital wall while the hyomandibula contributes significantly to support of the jaws (Maisey, 1980; Wilga, 2002, 2005). Hexanchiform sharks, however, retained the postorbital articulation while acquiring the orbitostylic articulation and therefore possess two jaw suspension types, orbitostyly and amphistyly (Wilga, 2002; Wilga et al., 2007). The batoids have a euhyostylic jaw suspension, which is perhaps the most kinetic jaw system. This type has no cranial–palatoquadrate articulation, the hyomandibula is the sole means of support for the jaws, and the hyoid arch is "broken up," with the hyomandibula losing its connection to the ceratohyal (Compagno, 1999; Miyake and McEachran, 1991; Wilga, 2002, 2005). The hyostylic and euhyostylic jaw suspension plays a key role in the functioning of the elasmobranch feeding mechanism; therefore, from a biting ancestor there were multiple forays into biting, suction, and filter-feeding behaviors (Figure 6.6) (Wilga et al., 2007).

6.3.3 Functional Morphology of the Feeding Mechanism

6.3.3.1 Sharks

Despite numerous studies on the anatomy of the head and cranium (e.g., Allis, 1923; Compagno, 1988; Daniel, 1915, 1934; Edgeworth, 1935; Frazzetta, 1994; Gadow, 1888; Gohar and Mazhar, 1964; Goodey, 1910; Goto, 2001; Lightoller, 1939; Luther, 1909; Marinelli and Strenger, 1959; Moss, 1972, 1977b; Motta and Wilga, 1995, 1999; Nobiling, 1977; Shirai and Okamura, 1992; Waller and Baranes, 1991; Wu, 1994), the functional morphology of the feeding mechanism is only understood for some representative species and is perhaps best understood for the carcharhiniform, squaliform, and orectolobiform sharks.

The feeding mechanism is perhaps best known in the spiny dogfish, *Squalus acanthias*, and the lemon shark, *Negaprion brevirostris*. As previously discussed, squaloids have an orbitostylic jaw suspension in which the hyomandibula suspends the jaws from the cranium, and the palatoquadrate articulates with the orbital

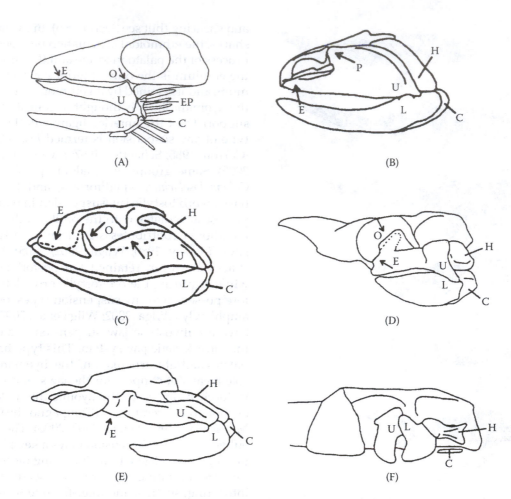

FIGURE 6.4

Left lateral views of select gnathostomes showing articulations involved with the jaw suspension. (A) Autodiastylic ancestor; (B) *Pleuracanthus*, Xenacanthida; (C) *Chlamydoselachus*, Chlamydoselachida; (D) *Squalus*, Squaliformes; (E) *Sphyrna*, Carcharhiniformes; (F) *Rhinobatos*, Batoidea. *Abbreviations:* C, ceratohyals; E, ethmoidal articulation; EP, epihyal; H, hyomandibula; O, orbital articulation; L, lower jaw; P, postorbital articulation; U, upper jaw. (From Wilga, C.D., *Biol. J. Linn. Soc.*, 75, 483–502, 2002. With permission.)

wall of the cranium by a relatively long orbital process (Figure 6.7) (Maisey, 1980; Marinelli and Strenger, 1959; Wilga and Motta, 1998a). The lemon shark has a hyostylic suspension in which the jaws are suspended from a more posteroventrally oriented hyomandibula, in contrast to the more laterally directed hyomandibula of the dogfish (Figure 6.8). The orbital process of the lemon shark is bound somewhat more loosely to the cranium by the elastic ethmopalatine ligament. The distal hyomandibula is braced against the mandibular knob of the mandible, and the ceratohyal is ligamentously bound to the distal hyomandibula and the mandible (Moss, 1965, 1972, 1977b; Motta and Wilga, 1995). In both species, the hyomandibula is ligamentously bound to the ceratohyal and in turn to the ventral basihyal, which rests somewhat dorsal to the mandibular symphysis.

Electromyographic analyses reveal that during jaw opening a relatively conservative series of events occurs in both species. Similar to the expansive phase described for teleost fishes (Lauder, 1985; Liem, 1978), the cranium is elevated by contraction of the epaxialis muscle, although cranial elevation need not occur (Motta et al., 1991, 1997; Wilga and Motta, 1998a) (Figures 6.9 and 6.10). Almost simultaneously, the mandible is depressed, primarily by the action of the coracomandibularis muscle, and the basihyal–ceratohyal apparatus begins to depress due to contraction of the coracoarcualis and coracohyoideus muscles. The branchial apparatus is depressed by the action of the coracobranchiales muscles. In the dogfish, in particular, the labial cartilages are extended as the mandible is depressed and laterally occlude the mouth (Motta et al., 1991, 1997; Wilga and Motta, 1998a). The compressive phase begins at peak gape as the mouth is maximally open, which is followed by the beginning of upper jaw protrusion and elevation of the mandible. Jaw adduction in both species is accomplished by contraction of the quadratomandibularis muscle. Various

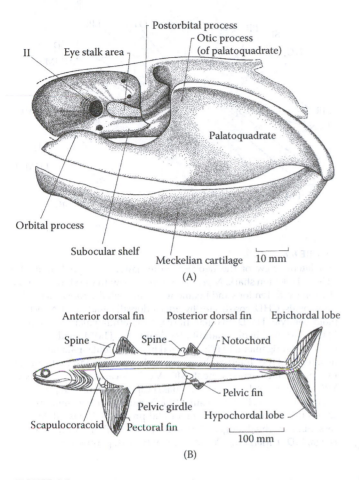

FIGURE 6.5

(A) Restoration of the neurocranium of *Cladodus*, and (B) restoration of *Cladoselache fyleri*. The cladodont palatoquadrate in panel A had a large otic process that is not well represented in the whole animal reconstruction shown in (B). The narrow suborbital ramus also extends anteriorly to the rostrum. (From Moy-Thomas, J.A. and Miles, R.S., *Paleozoic Fishes*, Chapman & Hall, London, 1971. With permission.)

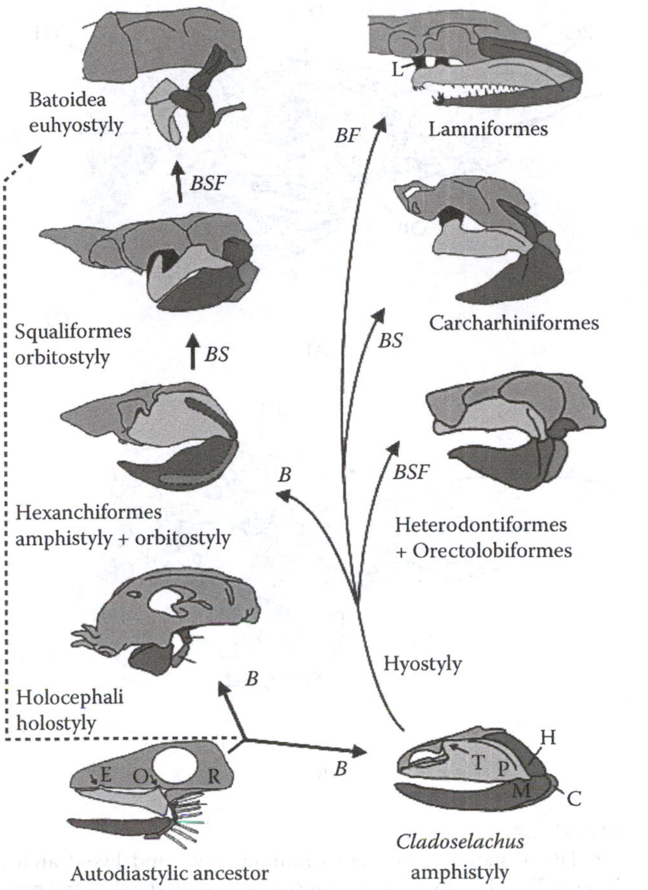

FIGURE 6.6

(See color insert.) Morphoclinal transition predicted for the evolution of jaw suspension and feeding modes in chondrichthyans. Biting appears to be the basal feeding mode. Other feeding modes evolved with hyostyly, orbitostyly, and euhyostyly. *Abbreviations:* C, ceratohyals; E, ethmoidal articulation; H, hyomandibula; L, palatobasal articulation; M, lower jaw; O, orbital articulation; P, palatoquadrate; R, cranium; T, postorbital articulation; B, bite feeding; F, filter feeding; S, suction feeding. (From Wilga, C.D. et al., *Integr. Comp. Biol.*, 47, 55–69, 2007. With permission.)

combinations of the preorbitalis and levator palatoquadrati muscles that are particular to each taxon protrude the upper jaw. In squaliform sharks, such as in *Squalus acanthias*, the preorbitalis muscle (homologous to the ventral preorbitalis in carcharhiniform sharks; see Compagno, 1988, and Moss, 1972) is horizontally directed, representing the ancestral condition (Wilga, 2005). The contraction of the preorbitalis produces an anteriorly directed force near the posterior region of the jaw (Figure 6.11). This forces the orbital process of the upper jaw to slide ventrally along the orbital wall and the ethmopalatine groove to protrude the upper jaw. As the upper jaw is protruding, the orbital process slides ventrally within the sleevelike ethmopalatine ligament until the ligament becomes taut, at which time the upper jaw protrusion is complete. As the upper jaw protrudes, the entire jaw moves anteroventrally while the hyomandibula passively follows. The distal end of the hyomandibula is pulled ventrally and only

slightly anteriorly. In species with laterally or anteriorly directed hyomandibulae (e.g., *Chiloscyllium plagiosum*, *Squalus acanthias*), the distal ends of the hyomandibulae are adducted during jaw opening. In contrast, in sharks with posteriorly directed hyomandibulae such as *Isurus oxyrinchus*, *Carcharhinus plumbeus*, and *Negaprion brevirostris*, the distal ends of the hyomandibulae swing outward, forward, and downward, resulting in lateral expansion of the hyoid and increasing the gape width (Wilga, 2008). Because the action of an adductor muscle is to bring two elements closer together, contraction of the quadratomandibularis not only elevates the lower jaw but may also pull the upper jaw away from the cranium toward the lower jaw. In this way, the quadratomandibularis may assist the preorbitalis in protruding the upper jaw (Wilga and Motta, 1998a).

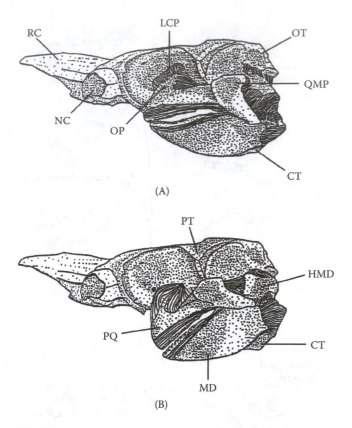

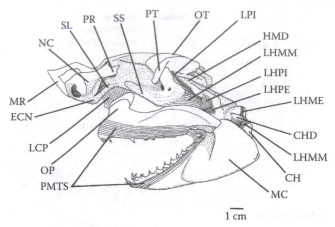

FIGURE 6.8

Left lateral view of the neurocranium, jaws, and hyoid arch of a 122-cm TL lemon shark, *Negaprion brevirostris*, with the skin and muscles removed. Tendons and ligaments are indicated. *Abbreviations:* CH, ceratohyals; CHD, constrictor hyoideus dorsalis tendon; ECN, ectethmoid condyle; HMD, hyomandibula; LCP, ethmopalatine ligament; LHME, external hyoid–mandibular ligament; LHMM, medial hyoid–mandibular ligament; LHPE, external hyomandibula–palatoquadrate ligament; LHPI, internal hyomandibula–palatoquadrate ligament; LPI, postspiracularis ligament; MC, Meckel's cartilage or lower jaw; MR, medial rostral cartilage; NC, nasal capsule; OP, orbital process of palatoquadrate; OT, otic capsule; PMTS, palatoquadrate–mandibular connective tissue sheath; PR, preorbital process; PT, postorbital process; SL, suborbital ledge; SS, suborbital shelf. (From Motta, P.J. and Wilga, C.D., *J. Morphol.*, 226, 309–329, 1995. With permission.)

FIGURE 6.7

Left lateral view of the neurocranium, jaws, and hyoid arch of a 74.5-cm TL spiny dogfish, *Squalus acanthias*, with the skin and muscles removed. (A) At resting position, and (B) at peak upper jaw protrusion. *Abbreviations:* CT, ceratohyals; HMD, hyomandibula; LCP, ethmopalatine ligament; MD, mandible or lower jaw; NC, nasal capsule; OP, orbital process of palatoquadrate; OT, otic capsule of cranium; PQ, palatoquadrate cartilage or upper jaw; PT, postorbital process; QMP, quadratomandibularis process of palatoquadrate; RC, rostral cartilage. (From Wilga, C.D. and Motta, P.J., *J. Exp. Biol.*, 201, 1345–1358, 1998. With permission.)

The mechanism of upper jaw protrusion in carcharhiniform sharks differs slightly from that in squaliform sharks. The carcharhiniform mechanism has been proposed in several studies (Frazzetta, 1994; Frazzetta and Prange, 1987; Luther, 1909; Moss, 1972) and has largely been supported in functional studies of feeding in *Negaprion brevirostris*, and the bonnethead shark, *Sphyrna tiburo* (Motta et al., 1997; Wilga, 1997; Wilga and Motta, 2000). Carcharhiniform sharks have a derived condition in which the levator palatoquadrati muscle is oriented more anteroposteriorly instead of dorsoventrally as in dogfish (Figure 6.9) (Compagno, 1988; Moss, 1972; Nakaya, 1975). In this orientation, the levator palatoquadrati muscle can assist the dorsal and ventral preorbitalis muscles (carcharhiniform sharks have two divisions of the preorbitalis muscle) in protruding the upper jaw (Figure 6.10). The dorsal division of the preorbitalis pulls the palatoquadrate ventrally as

the ventral division of the preorbitalis and the levator palatoquadrati muscles pull it anterodorsally. Similar to the dogfish, the orbital process of the palatoquadrate is forced to glide on the ethmopalatine groove, and the resultant reaction force drives the upper jaw anteriorly and ventrally to protrude it. As the upper jaw is protruded, the ropelike ethmopalatine ligament unfolds (it is folded in the resting position) until it becomes taut, halting upper jaw protrusion. As the upper jaw protrudes, the jaws and the distal end of the hyomandibula also swing anteroventrally but to a greater extent than in the spiny dogfish, and the distal ceratohyal and basihyal complex pivots posteroventrally (Motta and Wilga, 1995; Motta et al., 1997; Wilga, 2005; Wilga and Motta, 2000). Contraction of the quadratomandibularis muscle might also assist upper jaw protrusion as described above (Moss, 1965). Lamnid sharks such as *Carcharodon carcharias* have shifted the insertion of the preorbitalis and levator hyomandibularis muscles so they have more forceful and controlled movement of the upper jaw, and white sharks are known to protrude and retract the upper jaw several times during a gape cycle (Tricas and McCosker, 1984; Wilga, 2005). Numerous functions for protrusion have been proposed, including more efficient biting of the prey, gouging of the upper jaw into large prey, reorientation of the teeth, simultaneous

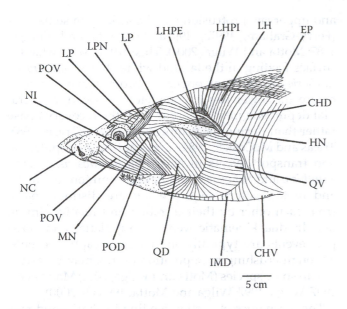

FIGURE 6.9

Left lateral view of the head and muscles of a 229-cm TL lemon shark, *Negaprion brevirostris*, with the skin removed and muscle fiber direction indicated. Myosepta of the epaxialis muscle (W-shape) are indicated in addition to the muscle fiber direction. The chondrocranial–palatoquadrate connective tissue sheath is removed. *Abbreviations:* CHD, constrictor hyoideus dorsalis; CHV, constrictor hyoideus ventralis; EP, epaxialis; HN, hyomandibular nerve; IMD, intermandibularis; LH, levator hyomandibularis; LHPE, external hyomandibula–palatoquadrate ligament; LHPI, internal hyomandibula–palatoquadrate ligament; LP, levator palatoquadrati; LPN, levator palpebrae nictitantis; MN, mandibular branch of trigeminal nerve; NC, nasal capsule; NI, nictitating membrane; POD, dorsal preorbitalis; POV, ventral preorbitalis; QD, quadratomandibularis dorsal; QV, quadratomandibularis ventral. (From Motta, P.J. and Wilga, C.D., *J. Morphol.*, 226, 309–329, 1995. With permission.)

closing of the upper and lower jaws, and reducing the time to close the jaws on the prey, to name a few. Only one function, reducing the time to close the jaws, has been experimentally verified (reviewed in Motta, 2004; Wilga, 2005).

Peak hyoid depression occurs in the latter half of the compressive phase. In *Squalus acanthias*, *Negaprion brevirostris*, and *Sphyrna tiburo* the mandible meets the maximally protruded upper jaw either with the food grasped between the teeth or after the food has been engulfed and passes through the buccal cavity. Finally, the recovery phase occurs as the palatoquadrate is retracted into its cranial seat. In the dogfish, the dorsoventrally oriented levator palatoquadrati assists in its retraction, whereas in the carcharhinids the elastic ethmopalatine ligament assists. It is not known if the ethmopalatine ligament of squaloids is elastic. In both species, however, the levator hyomandibularis retracts the hyomandibula, helping to elevate the entire jaw apparatus (Motta et al., 1997; Wilga and Motta, 1998a, 2000; see Wilga, 2005, for a discussion of the ethmopalatine ligament and jaw support in lamniform sharks).

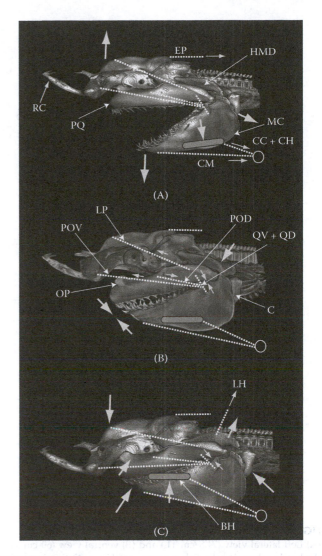

FIGURE 6.10

A reconstruction of chondrocranial, mandibular, and hyoid arch kinetics during feeding in the blacktip shark *Carcharhinus limbatus*. (A) Expansive phase, characterized by depression of the mandible, elevation of the cranium, hyoid depression, and movement of the hyomandibula such that its anterior end is depressed along with the basihyal and its posterior end pivots about its joint with the distal hyomandibula. (B) Compressive phase, characterized by elevation of the mandible to the protruded upper jaw and anteroventral movement of the distal hyomandibula. (C) Recovery phase, characterized by hyomandibular and palatoquadrate retraction, elevation of the basihyal, and depression of the cranium. The cranial elements were manipulated into the approximate anatomical positions for the CT scan; however, the hyomandibula and basihyal were most likely not depressed to the extent possible during a feeding event. Consequently, the anterior end of the ceratohyal is not fully depressed. The branchial arches are not manipulated and their movement is not indicated. Dotted white lines indicate approximate muscle origins and insertions, large arrows indicate the movement of specific elements, and small arrows indicate direction of muscle contraction. *Abbreviations:* BH, basihyal; C, ceratohyals; CC, coracoarcualis; CH, coracohyoideus; CM, coracomandibularis; EP, epaxialis; HMD, hyomandibula; LH, levator hyomandibularis; LP, levator palatoquadrati; MC, Meckel's cartilage or lower jaw; OP, orbital process of the palatoquadrate; POD, dorsal preorbitalis; POV, ventral preorbitalis; PQ, palatoquadrate cartilage or upper jaw; QD, quadratomandibularis dorsal; QV, quadratomandibularis ventral; RC, rostral cartilage.

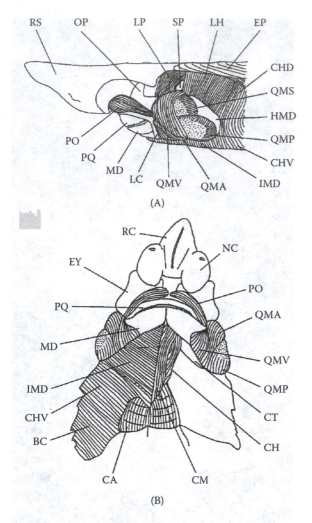

FIGURE 6.11

(A) Left lateral view (74.5-cm TL) and (B) ventral view (60-cm TL) of the head of the spiny dogfish, *Squalus acanthias*, with the skin and eye removed and muscle fiber direction indicated. Skin over the rostrum and cranium is left intact. Myosepta only of the epaxialis muscle are indicated. Raphe overlying quadratomandibularis is indicated by stippling. Anterior and posterior margins of the interhyoideus (deep to intermandibularis) are indicated by dotted lines. *Abbreviations:* BC, branchial constrictors; CA, coracoarcualis; CH, coracohyoideus; CHD, constrictor hyoideus dorsalis; CHV, constrictor hyoideus ventralis; CM, coracomandibularis; CT, ceratohyals; EP, epaxialis; EY, eye; HMD, hyomandibula; IMD, intermandibularis; LC, labial cartilages; LH, levator hyomandibularis; LP, levator palatoquadrati; MD, mandible or lower jaw; NC, nasal capsule; OP, orbital process of palatoquadrate; PO, preorbitalis; PQ, palatoquadrate or upper jaw; QMA, quadratomandibularis anterior; QMS, quadratomandibularis superficial; QMP, quadratomandibularis posterior; QMV, quadratomandibularis ventral; RC, rostral cartilage; RS, rostrum; SP, spiracularis. (From Wilga, C.D. and Motta, P.J., *J. Exp. Biol.*, 201, 1345–1358, 1998. With permission.)

The kinematic sequence described above is similar to that reported for carcharhiniform sharks such as the blacknose (*Carcharhinus acronotus*), blacktip (*C. limbatus*), swell (*Cephaloscyllium ventriosum*), and Caribbean reef (*Carcharhinus perezi*) sharks, although cranial elevation

and upper jaw protrusion may be lacking in some bites (Ferry-Graham, 1997a, 1998a; Frazzetta and Prange, 1987; Motta and Wilga, 2001). This differs somewhat for surface feeding in the lamnid white shark, *Carcharodon carcharias*, in that peak upper jaw protrusion occurs well before the lower jaw is completely elevated, and cranial depression may not occur until the recovery phase rather than during the compressive phase (Tricas, 1985; Tricas and McCosker, 1984). Prey capture, manipulation, and transport events in *Negaprion brevirostris*, *Squalus acanthias*, and *Sphyrna tiburo* have a common kinematic and motor pattern sequence but are distinguishable from each other by their duration and relative timing of individual kinematic events. Manipulation and transport events are typically shorter than capture events, although crushing manipulation events may be extensive in some species (Motta and Wilga, 2001; Motta et al., 1997; Wilga, 1997; Wilga and Motta, 1998a,b, 2000).

The mechanics of suction feeding in sharks and rays is understood from kinematic and electromyographic analyses (Clark and Nelson, 1997; Edmonds et al., 2001; Ferry-Graham, 1997b, 1998b; Lowry and Motta, 2007, 2008; Lowry et al., 2007; Matott et al., 2005; Motta, pers. obs.; Motta et al., 2002, 2008; Nauwelaerts et al., 2008; Robinson and Motta, 2002; Wilga and Sanford, 2008; Wu, 1994). A variety of extant and primarily benthic elasmobranchs use inertial suction to some degree as their dominant feeding method: whitespotted bamboo shark (*Chiloscyllium plagiosum*) (Lowry and Motta, 2007, 2008; Nauwelaerts et al., 2008; Wilga and Sanford, 2008; Wilga et al., 2007); spiny dogfish (*Squalus acanthias*) (Wilga and Motta, 1998a; Wilga et al., 2007); leopard shark (*Triakis semifasciata*) (Ferry-Graham, 1998b; Lowry and Motta, 2008; Russo, 1975; Talent, 1976); wobbegong (*Orectolobus maculatus*), nurse shark (*Ginglymostoma cirratum*), whale shark (*Rhincodon typus*), and zebra shark (*Stegostoma fasciatum*) (Clark and Nelson, 1997; Matott et al., 2002, 2005, 2008, 2010; Motta, pers. obs.; Robinson and Motta, 2002; Wu, 1994); horn shark (*Heterodontus francisci*) (Edmonds et al., 2001; Strong, 1989); chain catshark (*Scliorhinus retifer*) (Ajemian and Sanford, 2007); little skates (*Leucoraja erinacea*) (Wilga et al., 2007); guitarfish (*Rhinobatos lentiginosus*) (Wilga and Motta, 1998b); cownose ray (*Rhinoptera bonasus*) (Sasko et al., 2006); lesser electric ray (*Narcine bancroftii*) (Dean and Motta, 2004a,b); spotted torpedo ray (*Torpedo marmorata*) (Belbenoit, 1986; Belbenoit and Bauer, 1972; Michaelson et al., 1979; Wilson, 1953); and perhaps the angel shark (*Squatina californica*) (Fouts and Nelson, 1999). Inertial suction-feeding elasmobranchs are found in at least eight families, often nested within clades that contain ram and compensatory suction feeders, indicating that specialization for inertial suction feeding has most likely evolved independently in several elasmobranch lineages (Motta and Wilga, 2001; Motta et al., 2002; Wilga et al., 2007).

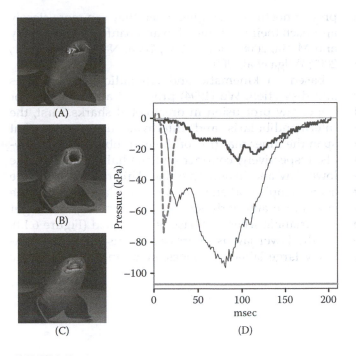

FIGURE 6.12
(See color insert.) Suction food capture in an 85-cm TL nurse shark, *Ginglymostoma cirratum*. (A) Mandible depression, which averages 26 ms, is occurring during the expansive phase. (B) Peak gape, which occurs at 32 ms, is visible with the food entering the mouth (36 ms). (C) Upper jaw protrusion is visible as the white band inside the mouth during the compressive phase. Total bite time averages 92 ms. (D) Three representative buccal pressure profiles from different semicaptive adult *Ginglymostoma cirratum* demonstrating the variability in suction performance during feeding. Some captures are extremely rapid with large subambient pressures (green dashed line), others can approach –1 atmosphere but be more prolonged (blue solid line), whereas others may generate little subambient pressure (red dash–dot line). The lower gray line indicates –1 atmosphere pressure at the average depth of the probe (~0.5 m). (From Motta, P.J. et al., *J. Morphol.*, 269, 1041–1055, 2008. With permission.)

Ginglymostoma cirratum (Ginglymostomatidae), *Triakis semifasciata* (Triakidae), and *Heterodontus francisci* (Heterodontidae) appear to exhibit an abbreviated kinematic sequence in which cranial elevation is reduced or lacking during many capture bites. In contrast, carcharhiniform and lamniform sharks usually consume relatively large prey with their ventrally located mouth, and as such they elevate the cranium and depress the mandible to open the mouth as wide as possible and direct the gape more anteriorly toward the prey. In contrast, *G. cirratum*, *T. semifasciata*, *H. francisci*, and perhaps most suction-feeding sharks primarily capture relatively small prey with a mouth that is almost terminal when maximally open (e.g., *G. cirratum*) or a mouth that is protruded anteroventrally to capture prey below them (e.g., *Squalus acanthias*). Consequently, lifting of the cranium during prey capture may not always be necessary (Lowry and Motta, 2007; Matott et al., 2005; Motta et al., 2002, 2008). In these suction-feeding sharks, the labial cartilages

protrude anteriorly as the lower jaw is depressed to effectively form a lateral enclosure of the mouth (Figure 6.12). This not only directs the suction anteriorly but may also prevent the food from escaping from the sides of the mouth (Edmonds et al., 2001; Ferry-Graham, 1997b, 1998b; Motta and Wilga, 1999; Wilga and Motta, 1998a).

In the suction-feeding whitespotted bamboo shark, *Chiloscyllium plagiosum*, the progression of the anterior to posterior expansion of the buccal, hyoid, and pharyngeal cavities is accompanied by sequential onset of subambient pressures in these cavities as the prey is drawn into the mouth. The increased velocity of hyoid area expansion is primarily responsible for generating peak subambient pressure in the buccal and hyoid regions (Wilga and Sanford, 2008). In the nurse shark, *Ginglymostoma cirratum*, peak subambient pressure is not related to shark size but instead to the rate of buccopharyngeal expansion (Motta et al., 2008). Buccopharyngeal expansion in these suction-feeding elasmobranchs appears to be mostly due to ventral expansion of the hyoid cavity, not lateral expansion, with the orientation of the hyomandibular cartilages making the difference (Wilga, 2010). Concomitant with an allometric increase in the relative contribution of suction over ontogeny in *C. plagiosum*, the hyoid muscles that expand the buccopharyngeal chamber hypertrophy (Lowry and Motta, 2007).

Not surprisingly, bite duration, from the beginning of mandible depression to retraction of the jaws to their resting position, is generally shorter for the suction-feeding sharks (*Chiloscyllium plagiosum*, 69 ms; *Ginglymostoma cirratum*, 100 ms; *Heterodontus francisci*, 113 to 148 ms; *Triakis semifasciata*, 150 to 180 ms) than for ram-feeding sharks (*Sphyrna tiburo*, 302 ms; *Negaprion brevirostris*, 309 ms; *Carcharhinus perezi*, 383 ms; *Cephaloscyllium ventriosum*, 367 to 419 ms; *Carcharodon carcharias*, 405 ms). Bite duration is 200 ms for suction-feeding and 280 ms for ram-feeding sequences in the dogfish. Time to maximum gape from mouth opening is similarly much faster in suction-feeding sharks (*Orectolobus maculatus*, 30 ms; *G. cirratum*, 32 ms; *H. francisci*, 47 to 64 ms) compared to the ram-feeding sharks (*N. brevirostris*, 81 ms; *C. perezi*, 120 ms; *S. tiburo*, 162 ms) (Edmonds et al., 2001; Ferry-Graham, 1997a; Lowry and Motta, 2008; Motta et al., 1997, 2002; Tricas, 1985; Tricas and McCosker, 1984; Wilga and Motta, 1998a,b; Wu, 1994). Faster buccopharyngeal expansion for suction-feeding sharks is expected because the dominant force that suction-feeding fishes exert on their prey is the fluid pressure gradient experienced by the prey (Wainwright and Day, 2007). The forces exerted on the prey can be elevated by increasing the rate of expansion or by reducing the size of the mouth aperture (Carroll et al., 2004; Lauder, 1980; Muller et al., 1982; Sanford and Wainwright, 2002; Svanback et al., 2002; Van Wassenbergh et al., 2006; Wainwright and Day, 2007; Wainwright et al., 2001a,b). Suction pressure in specialized suction-feeding sharks can be large. *Chiloscyllium plagiosum* can generate subambient pressures as low as –99 kPa, and *Ginglymostoma cirratum* as low as –110 kPa. Nurse sharks may use a spit–suck manipulation to dismember larger prey (Matott et al., 2005; Motta and Wilga, 1999; Motta et al., 2002, 2008; Robinson and Motta, 2002; Tanaka, 1973; Wilga and Sanford, 2008). Despite large suction pressures, the parcel of water and consequently the prey that is effectively sucked into the mouth only extend a few centimeters in front of the mouth, although feeding just above the substrate can extend this effective distance approximately 2.5 times. As a result, these suction-feeding elasmobranchs need to closely approach their prey to capture them or to thrust their heads into crevices. This functional limitation may lead to stalking or ambushing of

prey or nocturnal foraging when they can more closely approach their prey (Ajemian and Sanford, 2007; Lowry and Motta, 2008; Motta et al., 2008; Nauwelaerts et al., 2007; Wilga et al., 2007).

Based on kinematic and cineradiograhic analysis and dissection, Wu (1994) proposed a mechanism for upper jaw protrusion in orectolobid sharks. First, the intermandibularis and interhyoideus muscles that span the inner margins of the mandible and ceratohyals, respectively, contract and medially compress the lower jaw and hyomandibulae. This results in a more acute symphyseal angle of the lower jaw such that the jaws move anteriorly similar to the change in height of a triangle when the base is shortened (Figure 6.13). As the lower jaw is depressed it pushes on the relatively large labial cartilages, swinging them laterally

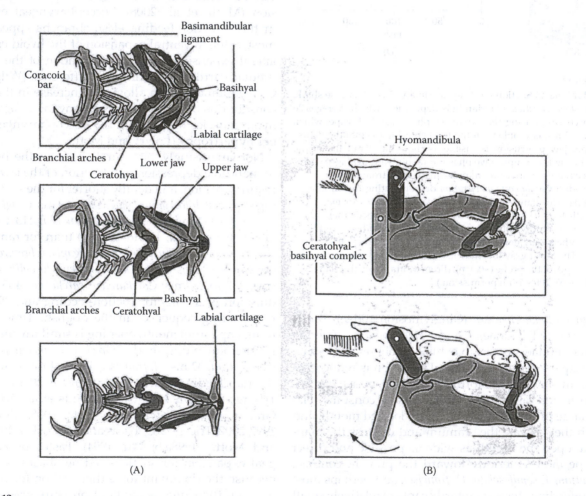

FIGURE 6.13
Feeding mechanism of *Orectolobus maculatus*. (A) Ventral view of the head, branchial arch, and pectoral girdle skeleton during jaw protrusion. In the top figure, the shark is shown with its mouth closed. In the center figure, the jaws are partly protruded, showing retraction of the basihyal, lateral compression of the jaw joints, and anterolateral swing of the labial cartilages. In the bottom figure, the jaws are completely protruded, showing the continued compression of the jaw joints and the branchial arches. The labial cartilages reach their maximum arc. (B) Schematic of the ceratohyal–hyomandibular mechanism of jaw protrusion. In the upper figure, the ceratohyal and the hyomandibula are represented as two links of a kinematic chain. In the lower figure, as the ceratohyal rotates around the posterior process of the lower jaw, the dorsal end pushes against the hyomandibula. The hyomandibula rotates forward against the mandibular knob and pushes the lower jaw forward. (From Wu, E.H., *J. Morphol.*, 222, 175–190, 1994. With permission.)

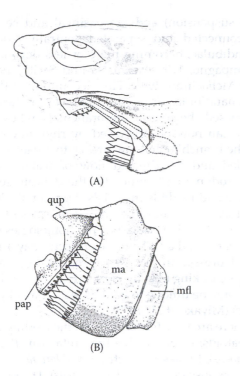

FIGURE 6.14

(A) Lateral view of the mouth of *Isistius brasiliensis* with the upper jaw protruded. Labial cartilages are indicated by broken line. (B) Lateral view of the upper and lower jaw showing hinge on the upper jaw. *Abbreviations:* Ma, mandibula or lower jaw; mfl, mandibular flap (a flexible, weakly chondrified plate at its posteroventral edge); pap, palatine process of palatoquadrate; qup, quadrate plate of palatoquadrate. (From Shirai, S. and Nakaya, K., *Zool. Sci.*, 9, 811–821, 1992. With permission.)

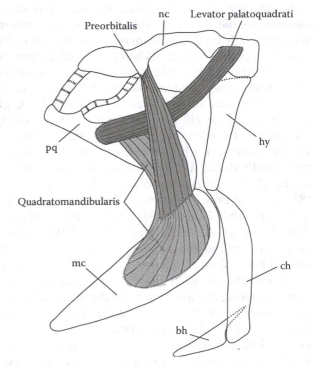

FIGURE 6.15

(See color insert.) Diagram of the skeleton, muscles, and ligaments associated with jaw movements in the megamouth shark, *Megachasma pelagios*. *Abbreviations:* bh, basihyal; ch, ceratohyals; hy, hyomandibula; mc, Meckel's cartilage; nc, neurocranium; pq, palatoquatrate. (From Nakaya, K. et al., *J. Fish Biol.*, 73, 17–34, 2008. With permission.)

and anteriorly and moving the oral aperture forward to form a round mouth opening. In addition, Wu proposed that the ceratohyals rotate around a process on the lower jaw, pushing the hyomandibulae anteroventrally, which in turn pushes the jaw articulation ventrally and anteriorly to protrude the jaws (Figure 6.13). This putative mechanism awaits electromyographic confirmation.

The cookie-cutter shark, *Isistius brasiliensis*, employs a unique behavior and mechanism to gouge out pieces of its prey. It anchors itself to the prey with its hooklike upper teeth and sinks its large sawlike lower teeth into the prey as it apparently sucks onto its prey, forming a seal with its fleshy lips. Twisting about its longitudinal axis, it gouges out a piece of flesh, leaving a craterlike wound (Compagno, 1984; Jones, 1971; LeBoeuf et al., 1987; Shirai and Nakaya, 1992). The upper jaw of this small shark is reduced in size and composed of two pieces: an anterior section that can pivot dorsally and a posterior section. The lower jaw is relatively large and robust (Figure 6.14). Presumably the upper jaw pivots at this juncture when the shark has gripped its prey with its upper jaw, allowing the shark to pivot dorsally about this joint and sink its large lower jaw teeth into the prey.

The adductor mandibulae and preorbitalis muscles are modified, apparently to facilitate the gouging function of the lower jaw (Shirai and Nakaya, 1992).

The megamouth shark, *Megachasma pelagios*, is apparently a slow, weak swimmer that filter feeds on small, deepwater prey such as euphausid shrimp. This shark has a large terminal mouth, no labial cartilage, densely packed papillose gill rakers, and relatively small gill openings, and the upper jaw is very protrusible. Anatomical investigation suggests that bioluminescent tissue in its mouth likely attracts prey. The shark has long palatorostral and ethmopalatine ligaments and long hyomandibular and ceratohyal cartilages. Together with a stretchy skin, these features give it a very kinetic jaw mechanism that is capable of extreme jaw protrusion and lateral expansion and depression. The large gape creates a low-velocity, high-volume suction that pulls the prey into the mouth. As the shark swims forward, the water and prey are forced into the distended buccopharyngeal cavity by ram; when the cavity is fully distended, the mouth is closed and the water is forced over the gill rakers, filtering out the prey (Figures 6.3 and 6.15). Thus, this shark uses a combination of suction, ram, and engulfment, similar to that of balenopterid whales (Nakaya et al., 2008).

The basking shark, *Cetorhinus maximus*, has slender jaws that are hardly protrusible. The jaws of *C. maximus* swing ventrally on the cranium and spread apart to form a circular hooplike mouth. When the mouth is open, two rows of bristle-like gill rakers stretch across each gill slit with an inter-raker distance of about 0.8 mm. The rakers do not greatly impede water flow through the gills and out the large gill openings but catch microscopic crustaceans. The filtering apparatus of the basking shark is better suited for a higher rate of water flow than the megamouth shark, and the former is better suited for sustained, powerful swimming, which may average 0.85 m/s as it ram filter feeds (Clark and Nelson, 1997; Compagno, 1990; Gudger, 1941a,b; Matthews and Parker, 1950; Taylor et al., 1983; Sims, 2000, 2008). Seasonal change in feeding morphology occurs in *C. maximus*. The gill raker sieve is apparently shed sporadically and nonsynchronously each year during late autumn or winter, a period during which the sharks were believed not to feed; however, some basking sharks have been caught with gill rakers in autumn and winter, and it is now evident that basking sharks can continue to feed at plankton densities much lower than previously thought possible (Francis and Duffy, 2002; Parker and Boeseman, 1954; Sims, 1999, 2008; Sims et al., 1997).

The whale shark, *Rhinocodon typus*, employs either a ram or pulsatile suction-filtering mechanism during which the shark may swim at (0.3 to 1.5 m/s) or below (0.2 to 0.5 m/s) the surface, or they may slow down and even cease swimming, assuming a horizontal or nearly vertical position to suction food into the mouth (Clark and Nelson, 1997; Heyman et al., 2001; Motta et al., 2010; Nelson and Eckert, 2007; Taylor, 2007). The diet of the whale shark is primarily composed of plankton, which they filter through 20 unique filtering pads that occlude the pharyngeal openings. A reticulated mesh lies on the proximal surface of the pads; the openings average 1.2 mm in diameter (Figure 6.16). These sharks can feed on fish eggs that are smaller in diameter than the pore openings of the pads (Heyman et al., 2001; Hoffmayer et al., 2007). This may be due to cross-flow filtration whereby the pads lie at an acute angle to the incoming water and plankton. If this mechanism does occur, the plankton would accumulate at the posterior end of the buccopharyngeal chamber while the water would exit through the pads. Such a mechanism would reduce clogging of the pads and concentrate the food into a bolus for swallowing (Motta et al., 2010).

6.3.3.2 Batoids

The feeding mechanics of batoids differs from that of sharks in cranial anatomy and function. The hyoid arch of batoids is modified in that the hyomandibula is the only major support for the jaws (euhyostylic

jaw suspension) and the basihyal and ceratohyal are disconnected and separated ventrally from the hyomandibulae, becoming more or less degenerate or lost (Compagno, 1999; Heemstra and Smith, 1980; Miyake and McEachran, 1991). This decoupling of the jaws and hyomandibulae from the branchial arches may have increased the role of the branchial arches in feeding. Prey can now be processed by rhythmic contractions of the branchial and jaw arches to create a highly controlled and coordinated flow of water, essentially a "hydrodynamic tongue" for the delicate separation of edible and inedible materials (Dean et al., 2005, 2007b). Furthermore, while the cranial muscles of batoids are generally similar to sharks, the homologies of some are unclear (e.g., the "X" muscle of electric rays), the muscles are depressed in form (e.g., preorbitalis), some muscles may be lacking (e.g., intermandibularis), and some muscles may be unique to batoids (e.g., coracohyomandibularis) (Miyake et al., 1992).

There are very few studies on the feeding mechanism of batoids; three involve the guitarfish (*Rhinobatos lentiginosus*), the lesser electric ray (*Narcine bancroftii*), and the cownose ray (*Rhinoptera bonasus*). The guitarfish captures its food by suction. The suction captures, manipulation bites, and suction transport of the food through the buccal cavity are all similar in the relative sequence of kinematic and motor activity but differ in the absolute muscle activation time, the presence or absence of muscle activity, and in the duration of muscle activity (Figure 6.17). A preparatory phase, which is often present prior to food capture, is marked by activity of the levator palatoquadrati muscle as the upper jaw is being retracted. The expansive phase is characterized by mouth opening, during which posteroventral depression of the lower jaw is initiated by the coracomandibularis. Midway through the expansive phase, the hyomandibula is depressed ventrally by the coracohyomandibularis and occasionally by the depressor hyomandibularis, which expands the orobranchial cavity. Movement of the food toward the mouth occurs during the activity of the hyomandibular depressors. The compressive phase begins with elevation of the lower jaw and the beginning of upper jaw protrusion. Maximum upper jaw protrusion is attained just prior to complete closure of the jaws. The compressive phase is represented by motor activity in the jaw adductors. Protrusion appears to be a coordinated effort of the quadratomandibularis and preorbitalis. The quadratomandibularis not only elevates the lower jaw but also protrudes the upper jaw by pulling the upper jaw ventrally toward the lower jaw. As the preorbitalis pulls the jaws anteroventrally, the upper jaw is protruded and the lower jaw is elevated by the quadratomandibularis until the jaws are closed. In the final recovery phase, the head and jaws are returned to their resting position. The upper jaw is retracted by the levator palatoquadrati

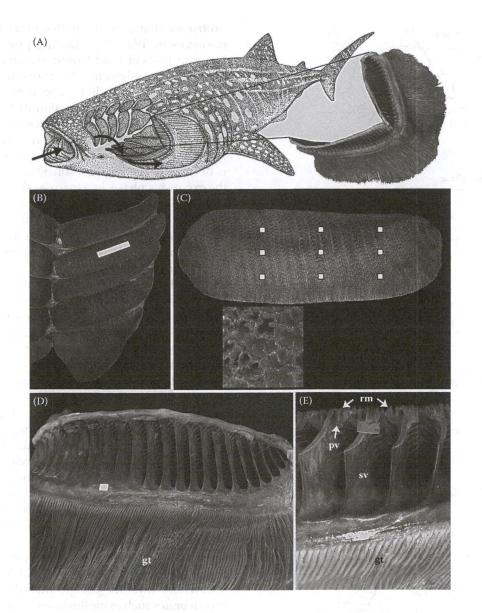

FIGURE 6.16

(See color insert.) (A) Schematic representation of a surface ram filter feeding whale shark, *Rhincodon typus*, showing the approximate position of the filtering pads and the direction of water flow through them. Inset shows a lateral view of the vanes deep to the filtering mesh, as well as the primary gill filaments on the first branchial arch over which the water flows. (B) Gross morphology of the whale shark filtering pads. Dorsal view of the lower filtering pads of a shark of approximately 622-cm TL. The fifth most posterior lower pad at the bottom is triangular in shape, and the lateral side of the pads is to the left. The lateral raphe between the lower and upper pads is visible toward the left. All other soft tissue has been removed. White ruler is 15 cm. (C) The upper second filtering pad of a shark of approximately 593-cm TL. Because it is an upper pad, lateral is to the left and posterior toward the top. Upper pads are not as falcate on their medial margin as the lower pads. The 1-cm squares indicate areas sampled to measure mesh diameter, and the inset is a representative 1-cm square area showing the irregularly shaped holes of the reticulated mesh. (D) External view of the first upper left pad of 622-cm TL shark with lateral margin toward the left. Note that the secondary vanes direct water laterally into the parabranchial chamber and over the gill tissue (gt) before it exits the pharyngeal slit (not shown). White square is 1 cm. (E) Close-up of a section through the third left lower filtering pad of 622-cm TL shark showing the reticulated mesh (rm), primary vanes (pv), secondary vanes (sv), and gill tissue (gt). Water flow is through the mesh, between the primary and secondary vanes, and over the gill tissue. White square is 1 cm. (From Motta, P.J. et al., *Zoology*, 113, 199–212, 2010. With permission.)

and the hyomandibula is retracted by the levator hyomandibularis. Hyomandibular elevation also elevates the jaws because the mandible is attached to the hyomandibula. The cranium is finally elevated to its resting position by the epaxialis and the levator rostri (Wilga and Motta, 1998b).

The lesser electric ray, *Narcine bancroftii*, has a remarkably protrusible and versatile mouth that it uses to probe beneath the substrate and suction feed on benthic invertebrates such as polychaete worms. Based on high-speed videographic analysis and anatomical dissection, Dean and Motta (2004a,b) proposed a novel mechanism for jaw

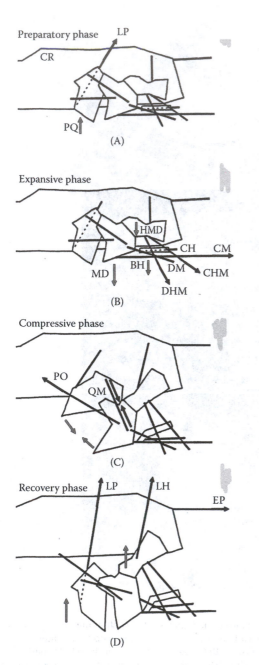

FIGURE 6.17
Schematic diagram of the functional components involved in jaw protrusion and jaw retraction during suction capture in *Rhinobatos lentiginosus*. (A) Upper jaw retraction during the preparatory phase; (B) lower jaw and hyomandibular depression during the expansive phase; (C) upper jaw protrusion and lower jaw elevation during the compressive phase; (D) hyomandibular, upper jaw, and lower jaw retraction during the recovery phase. Solid black lines represent muscles, and dark gray arrows indicate their direction of travel. Open elements represent skeletal elements, and their direction of movement is indicated by light gray arrows. *Abbreviations:* BH, basihyal; CH, coracohyoideus; CHM, coracohyomandibularis; CM, coracomandibularis; CR, cranium; DHM, depressor hyomandibularis; DM, depressor mandibularis; EP, epaxialis; HMD, hyomandibula; LH, levator hyomandibularis; LP, levator palatoquadrati; MD, mandible or lower jaw; PO, medial preorbitalis; PQ, palatoquadrate or upper jaw; QM, anterior quadratomandibularis. (From Wilga, C.D. and Motta, P.J., *J. Exp. Biol.*, 201, 3167–3184, 1998. With permission.)

protrusion that is similar to that proposed for *Orectolobus maculatus* by Wu (1994). During protrusion, which can be up to 100% of head length, the stout hyomandibulae are moved medioventrally, transmitting that motion to the attached mandible. The hyomandibular motion results in a medial compression of the entire jaw complex, shortening the distance between the right and left posterior corners of the jaws and forming a more acute symphyseal angle. As the angle between the mandibles is decreased, the jaws are forced anteroventrally during the expansive phase, in a manner similar to a scissor jack (Figure 6.18). The euhyostylic jaw suspension permits a degree of ventral protrusion that is impossible in the orectolobid sharks. The food item and sand are consequently sucked into the buccal region before maximum protrusion is reached. Suction pressures of ≤31 kPa can be generated in this manner. Food processing, when present, involves repeated, often asymmetrical, protrusion of the jaws, while sand is expelled from the spiracles, gills, and mouth. A pronounced difference between this mechanism and that of sharks is the degree of asymmetrical control during protrusion, which may be due to the highly subdivided and duplicated cranial musculature of batoids, the jaws being suspended only by the hyomandibula, the lack of an ethmopalatine ligament, and a flexible jaw symphysis (Dean and Motta, 2004a,b; Dean et al., 2007b; Gerry et al., 2008). In fact, bilateral implantation of electromyographc leads in four species of elasmobranchs revealed the greatest asymmetry of muscle firing in the batoid *Leucoraja erinacea* compared to three other shark species (Gerry et al., 2008). It should be noted, however, that the morphological restrictions that permit the unique protrusion mechanism of *N. bancroftii*, including the coupled jaws and narrow gape, most likely constrain its dietary breadth (Dean and Motta, 2004a,b).

Cownose rays are pelagic rays that feed on benthic invertebrates such as mollusks and crustaceans (Collins et al., 2007; Nelson, 1994; Orth, 1975; Schwartz, 1989; Smith, 1980). Food is captured by suction in a conservative series of expansive, compressive, and recovery phases similar to that of other elasmobranchs, then crushed between the platelike teeth (Figure 6.19). Prey is excavated by repeated opening and closing of the jaws to fluidize the surrounding sand and prey. The food is then surrounded laterally by the mobile cephalic lobes, which are anterior extensions of the pectoral fins, and sucked into the mouth. The cephalic lobes, which are also covered with electroreceptive ampullae (Mulvany, pers. comm.), may be used to herd elusive prey into the range of the mouth, as well. During capture, the spiracle, mouth, and gill movements are timed such that water enters only the mouth (Sasko et al., 2006). In mobulid rays, the cephalic lobes are paired and presumably used to direct water flow and plankton into the terminal mouth of these filter feeding rays (Sasko et al., 2006).

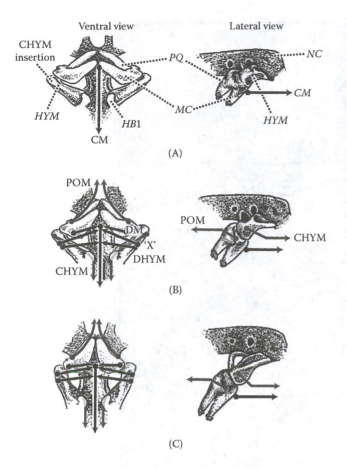

FIGURE 6.18

Proposed muscular basis of jaw protrusion for *Narcine bancroftii* based on muscle morphology and manipulations of fresh specimens. The protrusion mechanism is illustrated in ventral (left column) and left lateral (right column) view, with anterior up and to the left, respectively. Only contracting muscles are labeled (dark arrows), and dashed arrows denote skeletal structures (note that the coracohyomandibularis tendon runs beneath the first hypobranchial). The lower jaw depresses slowly (A), before protrusion onset (B) and peak protrusion (C). Lower jaw depression is effected by the coracomandibularis (CM), followed by medial compression of the jaws and hyomandibulae through coracohyomandibularis (CHYM), depressor hyomandibularis (DHYM), depressor mandibularis (DM), and 'X' ('X') muscle contraction. The halves of the upper jaw may also be adducted and extended by the preorbitalis medialis (POM). *Abbreviations:* HB1, first hypobranchial; HYM, hyomandibula; MC, Meckel's cartilage; NC, neurocranium; PQ, palatoquadrate. (From Dean, M. and Motta, P.J., *J. Morphol.*, 262, 462–483, 2004. With permission.)

6.4 Feeding Biomechanics

6.4.1 Skeletal Materials and Mechanics

Chondrichthyans are unique in that their skeletons consist of a combination of mineralized and unmineralized cartilage. Nearly all other vertebrates (agnathans withstanding) possess unmineralized cartilage as the developmental precursor to soon-to-be bony skeletal elements and at the articular surfaces of load-bearing

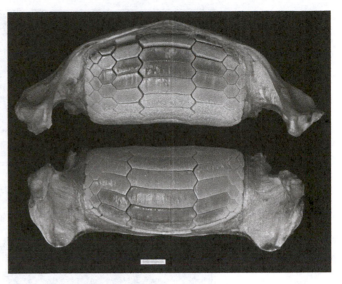

FIGURE 6.19

Posterior view of the upper and lower tooth plates of a cownose ray *Rhinoptera bonasus* (disk width ~60 cm) showing the imbricated tooth plates. The jaws are separated to better show the tooth plates. Scale bar is 1 cm.

joints, as well as mineralized cartilage as an ephemeral tissue that exists solely at the cartilage–bone interface of developing skeletal elements. Not only is chondrichthyan cartilage a composite of mineralized and unmineralized forms, but the latter is also a composite of solid and liquid phases consisting of chondrocytes in a gelatinous extracellular matrix (ECM) permeated with collagen fibers (Dean and Summers, 2006; Liem et al., 2001). Mineralized chondrichthyan cartilage is quite stiff due to the presence of calcium phosphate hydroxyapatite, whereas unmineralized chondrichthyan cartilage is remarkably elastic due to high concentrations of proteoglycans within the ECM. Negatively charged glycosaminoglycan side chains (e.g., chondroitin sulfate and keratin sulfate) on the proteoglycans attract water, which infiltrates the ECM and causes the tissue to swell. This swelling is resisted by the collagen network and mineralized cortex, thereby creating turgor. Water is forced out of the ECM during compression only to rush back into the matrix once the force is removed, due to polar attractions with the glycosaminoglycan side chains. The tissue is brought back to its original state provided that the loading has not resulted in plastic deformation (Carter and Wong, 2003; Liem et al., 2001).

The structural arrangement of the mineralized and unmineralized cartilage in skeletal elements of elasmobranch feeding mechanisms consists of a cortical mesh of mineralized tiles known as tesserae that surround a soft core of unmineralized tissue (Figure 6.20) (Kemp and Westrin, 1979; Orvig, 1951; Summers, 2000). Mineralization in holocephalans also consists of tesserae, but these are arranged in lamellar sheets that pass

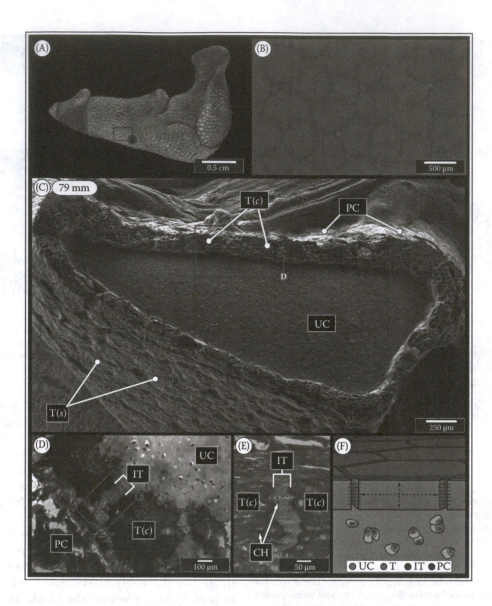

FIGURE 6.20

(See color insert.) Anatomy of the tessellated lower jaw skeleton of an elasmobranch; lettered and inset red boxes reference other panels in the figure (e.g., the box in panel A references panel B). Elasmobranch skeletal elements are tiled superficially with abutting mineralized blocks called tesserae (T) (panel A: microCT scan, left lateral view; panel B: cleared and stained tissue), overlain by a fibrous perichondrium (PC) and surmounting a monolithic core of uncalcified cartilage (UC) (panel C: cryoSEM cross-section). Tesserae can be seen in cross-section [T(c)] at the top of the image and in surface view [T(s)], covered by perichondrium, at the bottom. At higher magnifications (panels D and E: hematoxylin and eosin stained cross-sections), the margins of tesserae are less regular, and vital chondrocytes (CH) can be seen in mineralized lacunae in tesserae and extending into the intertesseral fibrous joints (IT). The tessellated skeleton can therefore be thought of simply as unmineralized cartilage wrapped in a composite fibro-mineral bark (panel F: schematic cross-section). (From Dean, M.N. et al., *J. Anat.*, 215, 227–239, 2009. With permission.)

through skeletal elements, the mechanical properties and homology of which are less well understood (Lund and Grogan, 1997; Rosenberg, 1998). Individual tesserae exhibit diverse forms, but all are roughly polygonal; those of the round stingray, *Urobatis halleri*, are hexagonal in shape and rectangular in cross-section (Dean et al., 2009a; Kemp and Westrin, 1979). Tesserae consist of an outer crystalline layer of prismatic calcification, which is connected to the perichondrium via Sharpey's fibers, and an inner layer of globular calcification consisting of

fused spherules of hydroxyapatite (Dean and Summers, 2006; Kemp and Westrin, 1979; Summers, 2000); globular calcification may also appear within the ECM (Dean and Summers, 2006).

Tesserae are linked together via collagenous ligaments and contain live chondrocytes in lacunae which are connected via canaliculi that presumably facilitate intercellular communication and nutrient diffusion (Dean et al., 2010; Kemp and Westrin, 1979; Moss, 1977a; Orvig, 1951; Rosenberg, 1998; Summers, 2000).

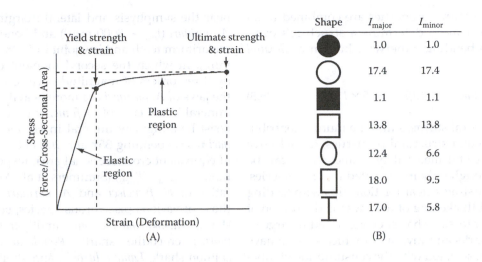

FIGURE 6.21
(A) Representative stress–strain curve indicating the elastic and plastic regions of material behavior. The slope of the elastic region is the stiffness (Young's modulus) of the material. The transition point from the elastic to plastic regions represents the yield strength and strain, after which the material has undergone permanent deformation. Ultimate strength and strain occur at the point of material failure. The area under the stress–strain curve represents the quantity of work required to cause material failure (work to fracture). (Adapted from Liem, K.F. et al., *Functional Anatomy of the Vertebrates: An Evolutionary Perspective*, 3rd ed., Brooks Cole, Belmont, CA, 2001.) (B) Second moments of area along the major and minor axes of representative geometric cross-sections. Wall thickness of hollow sections is constant, and the major axis is 1.5× the length of the minor axis in asymmetrical sections. (Adapted from Wainwright, S.A. et al., *Mechanical Design in Organisms*, Princeton University Press, Princeton, NJ, 1976.)

Tesserae develop embryonically as globular calcification surrounds "strings" of chondrocytes aligned parallel to the perichondral surface of the jaw (Dean et al., 2009a; Summers, 2000). The initial deposition of mineral occurs in association with collagen fibrils and appears to be related to alkaline phosphatase activity (Eames et al., 2007). Tesserae thicken by continuing to engulf chondrocytes at the endochondral surface in globular mineral while the calcification at the perichondral surface attains a more organized crystalline structure; tesserae widen by continued accretion of mineral at the surfaces of the intertesseral joints. Through this process, the tesserae of *U. halleri* were observed to increase in size by two to three times over ontogeny (Dean et al., 2009a).

Chondrichthyan mineralization differs from that of bony vertebrates in several key ways. Chondrocytes at the mineralizing front do not hypertrophy or form columns perpendicular to the perichondral surface, and mineral in chondrichthyans occurs in association with both Type I and II collagen, whereas vertebrate bone contains exclusively Type I collagen (Currey, 2008; Dean et al., 2009a; Eames et al., 2007; Kittiphattanabawon et al., 2010; Rama and Chandrakasan, 1984). Unlike vertebrate bone, chondrichthyan cartilage is incapable of remodeling (Ashhurst, 2004; Clement, 1992). The tessellated design of the skeleton therefore appears to be a means of facilitating growth in a system that does not reabsorb calcium. If the mineralized cortex of an elasmobranch jaw was solid it would not be able to accommodate growth of the ECM without remodeling.

Tesselation allows for increases in ECM volume as the individual tesserae grow via mineral deposition on all surfaces (Dean et al., 2009a).

The composite nature of the mineralized and unmineralized portions of skeletal elements in the elasmobranch feeding mechanism yields emergent mechanical properties that would otherwise not be found in a single-phase structure. The mineralized cortex confers stiffness, while the unmineralized core confers damping, two properties that are generally at odds in isolated materials (Dean et al., 2009b,c; Meyers et al., 2008). Both the type (material properties) and distribution (structural properties) of materials within skeletal elements determine their mechanical performance, which is inherently linked to aspects of ecological performance (e.g., jaw performance and prey capture). Material properties are reflected by the stiffness (*E*, Young's modulus) of those materials, whereas structural properties are reflected by their second moment of area (*I*), the contribution of a structure's cross-sectional shape to its resistance to bending. Young's modulus is the slope of the elastic region of a material's stress–strain curve, and the second moment of area is the distribution of material about the neutral axis of a skeletal element in the direction of loading (Figure 6.21):

$$E = \frac{\text{Stress}}{\text{Strain}} \qquad (6.1)$$

$$I = \int y^2 \, dA \qquad (6.2)$$

Material and structural properties are combined in an elegantly simple way to determine a structure's overall resistance to bending, otherwise known as *flexural stiffness*:

$$\text{Flexural stiffness} = E \times I \qquad (6.3)$$

By parsing flexural stiffness into the mutual contributions of material and structural properties, the selective pressures on skeletal material type and shape can be delineated. Although only investigated in a few species, jaw shape demonstrates clear relationships with feeding ecology. Cortical thickening of the jaws via deposition of multiple layers of tesserae has been identified in large or durophagous sharks and rays, all of which tend to have high bite forces associated with the consumption of functionally difficult prey (Dingerkus et al., 1991; Summers, 2000; Summers et al., 1998). Cortical thickening directly impacts the second moment of area, which is greatest in durophagous species such as the spotted eagle ray, *Aetobatus narinari*, and horn shark, *Heterodontus francisci* (Figure 6.22). The second moment of area is highest

near the symphysis and lateral margins of the jaws in *A. narinari* ($I_{max} \sim 8000$ mm^4) and beneath the posterior molariform teeth and jaw joint in *H. francisci* ($I_{max} \sim 2000$ mm^4), in which the second moment of area increases by three orders of magnitude over ontogeny. Although the jaws of *A. narinari* are more heavily mineralized, the mineral in the jaws of *H. francisci* is better positioned to resist bending; jaw mineral in *H. francisci* and *A. narinari* resists bending 35× and 20× better than a solid rod of equivalent cross-sectional area, respectively (moment ratio = $I_{specimen}/I_{circle}$) (Summers et al., 2004). The moment ratios of *H. francisci* and *A. narinari* are higher than those of various piscivorous species, such as the goblin shark, *Mitsukurina owstoni*; sandtiger shark, *Carcharias taurus*; crocodile shark, *Pseudocarcharias kamoharai*; salmon shark, *Lamna ditropis*; and shortfin mako shark, *Isurus oxyrinchus* ($I_{specimen}/I_{circle}$, 5 to 18) (Goo et al., 2010). Feeding ecology aside, all of the species investigated thus far exhibit similar changes in the second moment of area along the lengths of the upper and lower jaws, have peaks in the second moment of area beneath the jaw joints and anterior biting surfaces of the jaws, and their

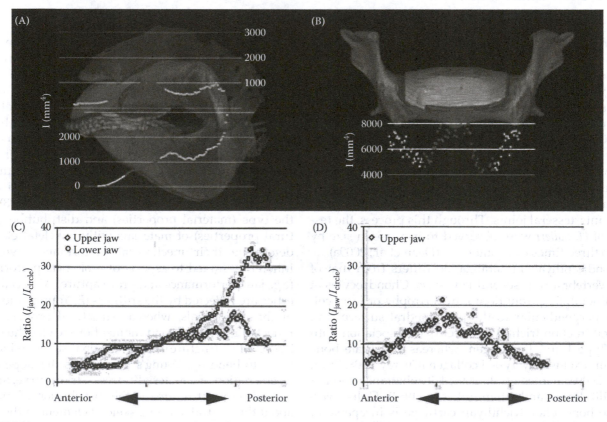

FIGURE 6.22

(See color insert.) (A) Second moment of area of the cross-section of the upper and lower jaws of the horn shark, *Heterodontus francisci*, and (B) upper jaw of the spotted eagle ray, *Aetobatus narinari*. The *x*-axis position of each point on the graph corresponds to the position of the section through jaws in the background. Points corresponding to sections with teeth involved in crushing hard prey are in red. (C) Moment ratio (ratio of the second moment of area of the jaw cross-section to the second moment of area of a circle with the same cross-sectional area) plotted vs. position along the jaw for the upper and lower jaws of *H. francisci*. (D) Moment ratio vs. position along the jaw for the upper jaw of *A. narinari*. (Adapted from Summers, A.P. et al., *J. Morphol.*, 260, 1–12, 2004. With permission.)

high moment ratios are correlated with orthogonal orientation between the major axis of the jaws and occlusal surface of the teeth (Goo et al., 2010; Summers et al., 2004).

Although cortical thickening is the principal means of stiffening the jaws of cartilaginous fish, trabecular reinforcement is also found in myliobatid stingrays and the lesser electric ray, *Narcine bancroftii*. Trabeculae, a means of optimizing weight and strength found throughout animal skeletons, are mineralized struts that pass through the lumen of a skeletal element, connecting and transmitting stress between its cortical layers. These struts are orthogonal to the tooth plates in the jaws of myliobatid stingrays so as to resist jaw flexion during hard prey consumption (Summers, 2000). Trabeculae in *N. bancroftii* are found near the jaw joints and symphyses and in the hyomandibulae near their joints with the cranium. Those oriented transverse to applied loads resist jaw flexion as in the myliobatids, whereas those in the parasymphyseal regions of the jaws form cross-braces parallel to the occlusal surface that resist buckling during ballistic protrusion of the jaws into the sediment in search of prey (Figure 6.23) (Dean et al., 2006). Although these reinforcements certainly play a role in maintaining the functional integrity of the feeding mechanism, their presence prior to birth in the cownose ray, *Rhinoptera bonasus*, and in the planktivorous manta ray, *Manta birostris*, suggests that they are phylogenetic in origin and not functionally induced (Summers, 2000).

As with structural properties, the material properties of chondrichthyan skeletons have seldom been examined. The dentine and enameloid in elasmobranch teeth range in stiffness from 22.49 to 28.44 GPa and 68.88 to 72.61 GPa, respectively (Table 6.1) (Whitenack et al., 2010). Mineralized jaw cartilage from the round stingray, *Urobatis halleri*, has a stiffness of 4.05 GPa, which is at the low end of the range of stiffness for vertebrate bone (Table 6.1) (Currey, 1987; Currey and Butler, 1975; Erickson et al., 2002; Hudson et al., 2004; Kemp et al., 2005; Rath et al., 1999; Rho et al., 2001; Wroe et al., 2008). The stiffness of unmineralized jaw cartilage ranges from 0.029 to 0.056 GPa and is generally greater than that of mammalian articular cartilage but exhibits little correlation with feeding ecology in durophagous and piscivorous species (Table 6.1) (Jagnandan and Huber, 2010; Laasanen et al., 2003; Summers and Long, 2006; Tanne et al., 1991). Selective pressure for mechanical adaptation in unmineralized jaw cartilage may be weak due to the principal role of mineralized cartilage in withstanding stress (Ferrara et al., 2011). Mineral content is the single largest determinant of stiffness in calcified connective tissues; as little as a 16% increase in mineral content can cause a 95% increase in the stiffness (Porter et al., 2006). Regardless, variation in the properties of unmineralized jaw cartilage may be associated with water and collagen content (Porter et al., 2006).

Structural analysis of whole shark jaws (including mineralized and unmineralized portions) has identified correlations among their mechanical performance, feeding ecology, and chondrichthyan evolution. Fahle and Thomason (2008) found that jaw viscoelasticity in the lesser spotted dogfish, *Scyliorhinus canicula*, decreases over ontogeny, which may contribute to an ontogenetic dietary shift toward harder and larger prey in adults. Using finite element analysis (FEA), Ferrara et al. (2011) determined that the mechanical performance of the jaws varies with gape angle such that species consuming large prey must have jaws better able to withstand stress. Finally, Wroe et al. (2008) used FEA to simulate the mechanical performance of the jaws of the white shark, *Carcharodon carcharias*, using realistic cartilaginous jaw models and hypothetical bony jaw models to explore the mechanical consequences of the loss of bone in the chondrichthyan skeleton. As expected, bony jaws exhibited higher stress and lower strain, but bite force was only 4.4% lower for cartilaginous jaws, suggesting that the adoption of a more compliant skeletal system has not compromised the biting performance of cartilaginous fishes (Wroe et al., 2008).

6.4.2 Musculoskeletal Lever Mechanics

Assuming that the jaws are functioning as rigid levers, the mechanical determinants of feeding performance are the forces produced by the cranial muscles and the leverage with which they act (a.k.a. mechanical advantage). Force generation (F_I, force input) is a function of muscular structure and geometry, whereas mechanical advantage (MA) determines the proportion of that force that is transmitted either to the fluid medium during jaw abduction or to prey items during jaw adduction (F_O, force output):

$$F_O = F_I \times MA \qquad (6.4)$$

Although expansive phase force generation plays a pivotal role in suction feeding performance (see Section 6.3.3.1), the output force of the mandibular lever system during jaw adduction represents none other than bite force. In recent years, bite force has become an increasingly relied upon measure of vertebrate feeding performance because it affects prey capture energetics and dietary diversity within species, as well as the partitioning of dietary and reproductive resources among species. High bite forces improve prey capture efficiency by reducing prey handling time and enabling the consumption of relatively large prey, thereby improving the net energy return per feeding event (Herrel et al., 2001b; van der Meij et al., 2004; Verwaijen et al., 2002). High bite forces are also frequently associated with reduced dietary diversity, niche specialization, and ontogenetic dietary shifts because high-performance

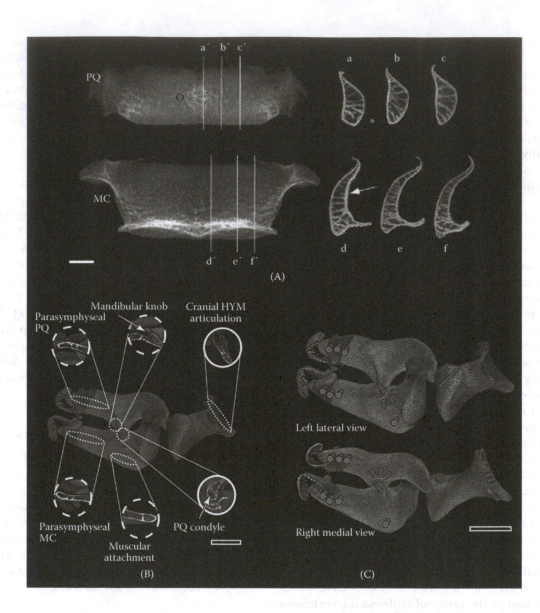

FIGURE 6.23

(A) Radiograph of the palatoquadrate (PQ) and Meckel's cartilage (MC) of an adult cownose ray, *Rhinoptera bonasus*, with the tooth plates removed (scale bar, 1 cm). The lower (MC) and upper (PQ) jaws are shown in anteroposterior and dorsoventral view, respectively, and the crushing area of the tooth plates would have been near the bottom of both images. Parasagittal sections are shown on the right (anterior on left, dorsal on top), with each section taken in approximately the plane indicated by the corresponding section line and trabecular reinforcement evident in all sections. An area of high tooth wear on the upper jaw is indicated by an asterisk, and an area of particularly thick prismatic cartilage on the lower jaw is indicated by an arrow. Hollow trabeculae can be seen end-on in some areas of the whole jaw radiographs, as indicated by the black circle on the upper jaw. (From Summers, A.P., *J. Morphol.*, 243, 113–126, 2000. With permission.) (B) Trabeculation in the jaws and hyomandibula of the lesser electric ray *Narcine bancroftii* (scale bar, 1 cm). Expanded white circles illustrate trabeculation within a given area with text indicating relevant regional landmarks; solid circles present elements in lateral view (in the same orientation as the larger image of the jaws and hyomandibula), while dashed circles provide a dorsal view. *Abbreviations:* HYM, hyomandibula; MC, Meckel's cartilage; PQ, palatoquadrate. (C) The orientation of groups of trabeculae shown schematically in left lateral (top) and right lingual (bottom) views, with trabeculae represented as columns (scale bar, 1 cm). (From Dean, M.N. et al., *J. Morphol.*, 267, 1137–1146, 2006. With permission.)

biting facilitates access to functionally difficult (large or hard) prey resources. This may reduce interspecific competition, because the lower absolute performance capacities of sympatric species preclude them from consuming such prey (Anderson et al., 2008; Christiansen and Wroe, 2007; Clifton and Motta, 1998; Hernandez and Motta, 1997; Herrel et al., 2002, 2004; Huber et al., 2006, 2009; Kolmann and Huber, 2009; Wainwright, 1988; Wyckmans et al., 2007). Although the relationship between bite force and fitness has not been directly quantified, Lappin and Husak (2005) and Husak et al. (2006) have demonstrated that bite force in male lizards is correlated with territory size and access to female conspecifics such that potential reproductive output

TABLE 6.1

Material Properties of Connective Tissues in Elasmobranchs and Other Vertebrates

Major Group	Species	Tissue	Young's Modulus (GPa)[a]	Ultimate Strength (MPa)[a]	Hardness (GPa)[a]	Source
Petromyzontiformes	*Petromyzon marinus*	Cartilage (annular)	0.001	—	—	Courtland et al. (2003)
Elasmobranchimorphii	*Sphyrna tiburo*	Dentine (orthodentine)	22.490	—	0.970	Whitenack et al. (2010)
Elasmobranchimorphii	*Carcharias taurus*	Dentine (osteodentine)	28.440	—	1.210	Whitenack et al. (2010)
Elasmobranchimorphii	*Carcharodon carcharias*	Dentine (osteodentine)	—	—	0.250	Chen et al. (2008)
Elasmobranchimorphii	*Carcharias taurus*	Enameloid	72.610	—	3.200	Whitenack et al. (2010)
Elasmobranchimorphii	*Carcharodon carcharias*	Enameloid	—	—	1.500	Chen et al. (2008)
Elasmobranchimorphii	*Sphyrna tiburo*	Enameloid	68.880	—	3.530	Whitenack et al. (2010)
Elasmobranchimorphii	*Urobatis halleri*	Cartilage (jaw, mineralized)	4.050	—	—	Wroe et al. (2008)
Elasmobranchimorphii	*Carcharhinus limbatus*	Cartilage (jaw, unmineralized)	0.051	—	—	Huber (unpublished data)
Elasmobranchimorphii	*Heterodontus francisci*	Cartilage (jaw, unmineralized)	0.056	—	—	Jagnandan and Huber (2010)
Elasmobranchimorphii	*Negaprion brevirostris*	Cartilage (jaw, unmineralized)	0.043	—	—	Jagnandan and Huber (2010)
Elasmobranchimorphii	*Rhinoptera bonasus*	Cartilage (jaw, unmineralized)	0.029	41.000	—	Summers and Long (2006)
Elasmobranchimorphii	*Carcharhinus falciformis*	Cartilage (vertebrae)	0.560	24.300	—	Porter et al. (2006)
Elasmobranchimorphii	*Carcharhinus plumbeus*	Cartilage (vertebrae)	0.400	23.700	—	Porter et al. (2006)
Elasmobranchimorphii	*Centrophorus granulosus*	Cartilage (vertebrae)	0.430	20.800	—	Porter et al. (2006)
Elasmobranchimorphii	*Isurus oxyrinchus*	Cartilage (vertebrae)	0.330	11.900	—	Porter et al. (2006)
Elasmobranchimorphii	*Sphyrna zygaena*	Cartilage (vertebrae)	0.520	23.800	—	Porter et al. (2006)
Elasmobranchimorphii	*Torpedo californica*	Cartilage (vertebrae)	0.025	4.600	—	Porter et al. (2006)
Osteichthyes	*Clupea harengus*	Bone (rib)	3.5–19.0	155.000	—	Rho et al. (2001)
Osteichthyes	*Lepidosiren paradoxa*	Dentine	—	—	0.430	Currey and Abeysekera (2003)
Osteichthyes	*Lepidosiren paradoxa*	Petrodentine	—	—	2.490	Currey and Abeysekera (2003)
Amphibia	*Cryptobranchus alleganiensis*	Bone (femur)	22.300	—	—	Erickson et al. (2002)
Amphibia	*Cyclorana alboguttata*	Bone (tibiofibula)	8.800	—	—	Hudson et al. (2004)
Reptilia	*Crocodylus sp.*	Bone (frontal)	5.600	—	—	Currey (1987)
Reptilia	*Varanus xanthematicus*	Bone (femur)	22.800	—	—	Erickson et al. (2002)
Aves	*Gallus gallus*	Bone (tibia)	0.600	—	—	Rath et al. (1999)
Aves	*Phoenicopterus ruber*	Bone (tarsometatarsus)	27.800	—	—	Currey (1987)
Mammalia	*Homo sapiens*	Bone (femur)	123.406	—	—	Currey and Butler (1975)
Mammalia	*Canis lupus familiaris* (pit bull)	Bone (humerus)	3.200	—	—	Kemp et al. (2005)
Mammalia	*Bos taurus*	Cartilage (articular, tibia)	0.0001	—	—	Laasanen et al. (2003)
Mammalia	*Canis lupus familiaris*	Cartilage (articular, mandible)	0.092	—	—	Tanne et al. (1991)
Mammalia	*Loxodonta africana*	Dentine	7.700	—	0.430	Currey (1998); Currey and Abeysekera (2003)
Mammalia	*Homo sapiens*	Dentine	19.890	—	0.920	Waters (1980); Mahoney et al. (2000)
Mammalia	*Bos taurus*	Enamel	73.000	—	3.000	Currey (1998); Currey and Abeysekera (2003)
Mammalia	*Homo sapiens*	Enamel	87.500	—	3.900	Habelitz et al. (2001)

[a] Non-elasmobranch tissues with more than one value per major group represent minimum and maximum known values for that major group

(based on female fecundity and probability of insemination) is greater for individuals with higher bite forces. An inquiry of this type has not been conducted in cartilaginous fishes.

6.4.2.1 Muscle Force and Leverage

The force produced by a muscle (P_O, maximum tetanic tension; F_I of Equation 6.4) is determined by the specific tension (T_S) and geometric arrangement of its fibers, the latter of which is approximated by its cross-sectional area (*CSA*):

$$P_O = T_S \times CSA \qquad (6.5)$$

Specific tension is the maximum stress (force per unit area) that a muscle fiber can generate. Although likely an oversimplification (Hernandez and Morgan, 2009), for the purposes of biomechanical modeling cartilaginous fishes are considered to have either "white" or "red" jaw muscles. White muscles contain glycolytic fibers that produce greater power (18.3 W/kg) and stress (28.9 N/cm^2) than their red counterparts (6.6 W/kg and 14.2 N/cm^2), although the latter are oxidative and dramatically better at resisting fatigue (Curtin et al., 2010; Lou et al., 2002). Current estimates of the specific tension of chondrichthyan muscle are based on axial myomeres and may underestimate the physiological capacity of jaw adductors. "Superfast" isoforms of masticatory myosin capable of producing greater stress than locomotory muscles have been identified in the jaw adductors of numerous vertebrates, including the blacktip shark, *Carcharhinus limbatus* (Hoh, 2002; Qin et al., 2002).

Masticatory myosin aside, there is preponderance of white muscle tissue in the jaw adductors of cartilaginous fishes with limited exception. The deepest subdivisions of the jaw adductors of carcharhinid and orectolobiform sharks and holocephalan ratfishes have red fibers, which are believed to be associated with rhythmic respiratory movements of the jaw apparatus and stabilization of the jaw joint, respectively (Huber et al., 2008; Motta, pers. obs.). The only group possessing jaw adductors dominated by red muscle tissue are the myliobatid stingrays such as the cownose ray, *Rhinoptera bonasus*, in which the fatigue resistance of this oxidative tissue likely aids in the rhythmic crushing of epibenthic fauna (Gonzalez-Isais, 2003; Peterson et al., 2001; Smith and Merriner, 1985). However, all other parameters held constant, a chondrichthyan with "white" jaw adductors can generate static bite forces approximately two times greater than one with "red" jaw adductors, due to differences in specific tension.

Muscles in which the fibers are arranged parallel to the mechanical line of action are the norm among cartilaginous fish. The cross-sectional areas of these parallel-fibered muscles can be estimated rather simply from digital images of muscle sections taken perpendicular to the principal fiber direction through the center of mass. The prevalence of parallel fibered muscles in cartilaginous fish is believed to be due to the low pull-out strength of the tessellated cartilaginous skeleton, which necessitates the presence of aponeurotic surface insertions for cranial muscles. These broad insertions decrease the applied stress by spreading muscular force over a considerable skeletal area (Liem and Summers, 1999; Summers et al., 2003). Tendinous point insertions, such as those of pinnate muscles in which fibers insert onto a central tendon at acute angles, are likely to cause considerable point stresses and local instability along the tessellated skeleton. Not surprisingly, pinnate muscles are fairly rare among cartilaginous fishes; nonetheless, the few known examples generally occur in concert with skeletal structures that ameliorate the stresses created by tendinous point insertions.

Estimating the force produced by a pinnate muscle requires a more elaborate calculation of cross-sectional area due to the angular insertion of muscle fibers onto a central tendon. In such cases, it is necessary to determine "physiological" cross-sectional area, which estimates the portion of the muscle making a mechanically relevant input to a musculoskeletal system by accounting for the fact that some of the force generated by muscle fibers not parallel to the central line of action of the muscle will be lost during contraction. Physiological cross-sectional area is calculated as:

$$PCSA = \frac{\text{Muscle mass}}{\text{Muscle density}} \times \cos\Theta \times \frac{1}{\text{Fiber length}} \qquad (6.6)$$

in which Θ is the average angle of pinnation from the central tendon of the muscle and the density of fish muscle is 1.05 g/cm^3 (Powell et al., 1984; Wainwright, 1988). Although this loss of force may sound problematic, pinnate muscles actually generate greater forces than comparably sized parallel muscles because pinnate architecture allows greater packing of muscle fibers per unit volume. In fact, pinnate fiber architecture is typical of the primary jaw adductors of most bony fishes (Gans and Gaunt, 1991; Liem et al., 2001; Winterbottom, 1974). Parallel-fibered muscles are not without their merit, however. The fibers of parallel muscles typically have more sarcomeres arranged in series, facilitating larger, more rapid contractions, which is advantageous for consumers of elusive prey (Liem et al., 2001).

Given the force-generating advantages of pinnate muscles, it is no surprise that some cartilaginous fishes have found ways to circumvent the problem of pinnate muscle-point stress on a cartilaginous skeleton. Pinnate muscle architecture has been found in the feeding mechanisms of carcharhinid and orectolobiform sharks, myliobatid stingrays, and holocephalan ratfishes. Subdivisions of the dorsal quadratomandibularis muscle in the lemon shark, *Negaprion brevirostris*, and blacktip shark, *Carcharhinus limbatus*, gradually shift to pinnate fiber architecture during early ontogeny; however, the central tendons of these subdivisions insert onto the mid-lateral raphe of the quadratomandibularis, avoiding interaction with the tessellated skeleton (Huber et al., 2006; Motta and Wilga, 1995). The mid-lateral raphe is a connective tissue sheath onto which fibers from the dorsal and ventral divisions of the quadratomandibularis merge.

There are numerous tendons throughout the feeding musculature of the nurse shark, *Ginglymostoma cirratum*, and cownose ray, *Rhinoptera bonasus*, both of which possess heavily mineralized jaws that presumably preclude the problems associated with tendinous point insertions (Motta and Wilga, 1999; Summers, 2000). The adductor complex of *R. bonasus* is particularly unique in that the major subdivision (adductor mandibulae major) is a multipinnate muscle that originates from its antimere beneath the lower jaw and extends parallel to the occlusal surface until wrapping around the corner of the lower jaw and inserting onto the upper jaw via a stout tendon. This tendon-wrapping redirects the adductor force perpendicular to the occlusal plane, thereby yielding bite force from a muscle not inherently positioned to do so (Figure 6.24). Fibrocartilaginous pads found where the muscle–tendon complex wraps around the corners of the jaw are believed to be mechanical responses to compressive and shear loading, analogous to the "wrap around" tendons of mammals such as the bovine deep digital flexor tendon (Summers, 2000; Summers et al., 2003). The presence of red muscle fibers and pinnate architecture in the jaw adductors of *R. bonasus* appears to be a way of optimizing fatigue resistance and force production in its feeding mechanism. The jaws of holocephalan ratfishes are barely (if at all) mineralized, necessitating an altogether different solution to the problem of pinnate muscle-point stress on a cartilaginous skeleton. The adductor mandibulae anterior of these fish is a bipinnate muscle that inserts onto the lower jaw via a tendinous sling that twists as it wraps beneath the mandible (Figure 6.25). Fiber pinnation angle increases over ontogeny, serving to maintain isometric force production despite hypoallometric growth of cranial volume, while the twisted tendon topology equalizes strain throughout the muscle such that all of its fibers operate at equal positions on their length–tension curves regardless of the size of the gape (Dean et al., 2007a; Didier, 1995; Huber et al., 2008).

As one might expect, the forces produced by the cranial muscles of sharks exhibit significant correlations with feeding mode and ecology. Suction feeders tend to have hypertrophied abductors of the oropharyngeal

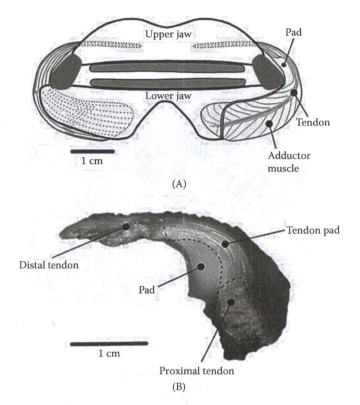

FIGURE 6.24
(A) Ventral view of the jaws of the cownose ray, *Rhinoptera bonasus*, dissected to show the position of the adductor mandibularis medialis muscle and the associated tendon. The right side of the ventral margin of the lower jaw (Meckel's cartilage) has been cut away and the muscle sectioned along its long axis to reveal the central tendon. The upper and lower jaws are connected by strong, short ligaments shown in dark gray. (B) Photograph of a section of the tendon showing the fibrocartilaginous pad and the linear fibers of the tendon lateral to it. The orientation of the tendon is approximately the same as in the right side of panel A. (From Summers, A.P. et al., *Cell Tissue Res.*, 312, 221–227, 2003. With permission.)

cavity, whereas those taxa relying on some variant of a biting mechanism tend to have hypertrophied jaw adductors. Heterodontiform and orectolobiform sharks exhibit remarkable suction and biting performance, having hypertrophied virtually everything in their heads (Edmonds et al., 2001; Huber et al., 2005; Motta and Wilga, 1999; Motta et al., 2008; Nobiling, 1977; Ramsay and Wilga, 2007; Wilga and Sanford, 2008). Interestingly, Habegger et al. (in review) found jaw adductor size to be a significant predictor of bite force in a phylogenetically informed analysis of chondrichthyan feeding biomechanics, suggesting that species having higher than expected bite forces for their body size have achieved this through convergent evolution of hypertrophied jaw adductors. Nonetheless, size is not the only thing that matters; muscle position can play a significant role in determining function (Maas et al., 2004). Ancestrally, the preorbitalis muscle of sharks originated posterior to the nasal capsule and inserted onto the mid-lateral

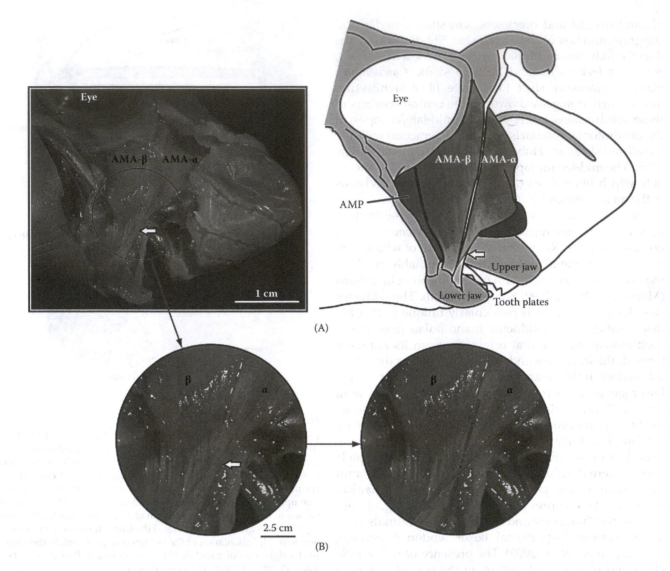

FIGURE 6.25

(See color insert.) Right lateral view of the cranial musculature of the spotted ratfish, *Hydrolagus colliei*. (A) The schematic on the right illustrates the musculature labeled on the left. (B) The tendon (circled in panel A) has been magnified to show the twisted portion. Although all three adductors insert on the lower jaw, only the anterior adductor (AMA-α) exhibits a pronounced twist in its tendon (its approximate middle indicated by a white arrow in A and B) where the anterior face (red arrow) inserts more posteriorly than the posterior face (blue arrow). *Abbreviations:* AMA-α, anterior subdivision of the adductor mandibulae anterior; AMA-β, posterior subdivision of the adductor mandibulae anterior; AMP, adductor mandibulae posterior. (From Dean, M.N. et al., *J. Exp. Biol.* 210, 3395–3406, 2007. With permission.)

raphe of the quadratomandibularis muscle. Horizontal orientation enabled the preorbitalis to actuate upper jaw protrusion, which facilitates rapid jaw closure and gouging of prey and is further augmented in carcharhinid sharks by the derived motor pattern and horizontal orientation of the levator palatoquadrati (Frazzetta, 1994; Motta et al., 1997; Wilga, 2005; Wilga et al., 2001). However, heterodontiform and orectolobiform sharks possess a derived vertical preorbitalis which inserts onto the lower jaw anterior to the quadratomandibularis, giving this muscle higher leverage over jaw adduction. Consequently, species such as the horn shark, *Heterodontus francisci*, and whitespotted bamboo shark,

Chiloscyllium plagiosum, have among the highest leverage jaws of all cartilaginous fishes (Table 6.2) (Habegger et al., in review; Huber, 2006; Huber et al., 2005, 2008; Wilga, 2005; Wilga et al., 2001).

Positioning muscles perpendicular to their associated skeletal element is a key way to optimize force transmission in lever systems, as illustrated by the preorbitalis muscles of heterodontiform and orectolobiform sharks. Recent work by Ferrara et al. (2011) has demonstrated a novel role of the mid-lateral raphe in maintaining the orthogonal arrangement of the musculoskeletal elements in shark feeding mechanisms as well. Through computational modeling of jaw

TABLE 6.2

Morphometrics, Biomechanical Parameters of the Feeding Mechanism, and Dietary Categorizations of Various Cartilaginous Fishes

Scientific Name	TL (cm)	Mass (g)	PBL (cm)	HW (cm)	HH (cm)	JL (cm)	AM CSA (cm²)	Ant MA	Post MA	Ant BF (N)	Post BF (N)	% Cephalopod	% Decapod	% Fish	% Mollusk	Trophic Level
Carcharhinus acronotus[a]	103	—	—	—	—	7.3	7.2	0.33	1.18	67	270.1	0	1.2	98.2	0.6	4.2
Carcharhinus leucas[b]	285	192,976	47.4	41.1	40.8	29.0	112.0	0.37	1.10	2128	5914	0.5	2.6	52.3	0.2	4.3
Carcharhinus limbatus[c]	152	9833	34.0	26.0	17.0	15.6	30.0	0.42	1.33	423	1083	4.1	2.1	88.9	0.1	4.2
Carcharodon carcharias[d]	250	240,000	42.2	45.0	17.9	24.1	78.4	0.40	1.36	1602	3131	3.6	1.8	35.5	0.4	4.5
Chiloscyllium plagiosum[e]	71	1219	10.2	7.5	4.9	3.9	4.0	0.44	0.83	93	168	2.3	33.5	30.3	7.8	3.7
Chimaera monstrosa[f]	41	310	5.0	3.8	5.2	2.2	1.3	0.69	1.13	34	61	0.0	34.4	0.0	17.2	3.2
Etmopterus spinax[g]	38	191	6.3	3.9	2.8	2.7	0.7	0.57	0.95	21	28	19.7	22.7	33.3	0.0	3.8
Eusphyra blochii[a]	132	—	—	—	—	7.8	7.1	0.26	0.93	52.1	171.8	4	12.8	82.9	0	4.1
Galeus melastomus[g]	71	742	9.9	6.1	3.5	1.8	0.7	0.33	0.84	12	30	3.8	32.3	32.9	0.0	3.7
Heptranchias perlo[e]	85	1614	11.9	7.0	8.0	9.0	13.2	0.34	0.95	245	845	40.0	13.3	40.0	0.0	4.2
Heterodontus francisci[h]	63	1616	8.7	9.5	7.5	7.5	5.4	0.55	1.21	117	318	0.1	27.0	0.0	71.3	3.2
Hydrolagus colliei[i]	46	515	5.4	4.0	6.7	0.9	1.9	0.68	1.80	89	175	20.0	17.2	0.1	34.4	3.2
Negaprion brevirostris[e]	61	1219	10.3	7.0	6.0	5.6	4.0	0.36	0.98	79	220	0.0	4.3	92.9	0.0	4.2
Rhizoprionodon terraenovae[a]	88	—	—	—	—	6.7	4.6	0.3	1.18	38.6	157.7	1.8	31.6	66.4	0.1	4
Sphyrna lewini[a]	257	—	—	—	—	14.6	23.4	0.24	0.76	207.4	623.1	15.5	22	61.9	0.1	4.1
Sphyrna mokarran[j]	286	—	—	—	—	20.4	68.3	0.26	0.84	642.2	1839.4	3.3	11.2	43.5	0.0	4.3
Sphyrna tiburo[k]	88	2920	14	8.6	5.2	5.8	3.5	0.22	0.84	18.2	71.1	2.2	71.5	1.6	0.0	3.2
Sphyrna tudes[a]	93	—	—	—	—	6.0	5.1	0.24	0.88	38.4	139	0	83.3	16.7	0	3.6
Sphyrna zygaena[a]	263	—	—	—	—	16.3	32.1	0.12	1.01	288.5	1210	68.9	0.4	29.8	0	4.2
Squalus acanthias[l]	55	673	9.9	3.2	7.4	3.6	1.6	0.20	0.50	12	30	5.2	3.5	41.6	0.4	3.9

Note: Individuals represent the highest mass-specific bite force among the adults sampled for each species, other than the lemon shark, for which no adult data are available.

Abbreviations: AM CSA, adductor mandibulae cross-sectional area; Ant BF, anterior bite force; Ant MA, anterior mechanical advantage; HH, head height; HW, head width; JL, jaw length; PBL, prebranchial length; Post BF, posterior bite force; Post MA, posterior mechanical advantage; TL, total length.

[a] Cortés (1999); Mara (2010).
[b] Habegger et al. (in review); Cortés (1999).
[c] Cortés (1999); Huber et al. (2006).
[d] Cortés (1999); Wroe et al. (2008).
[e] Cortés (1999); Huber (2006).
[f] Claes, Huber, and Mallefet (unpublished data); MacPherson (1980); Mauchline and Gordon (1983).
[g] Claes, Huber, and Mallefet (unpublished data); Cortés (1999).
[h] Cortés (1999); Kolmann and Huber (2009).
[i] Mauchline and Gordon (1983); Huber et al. (2008); Dunn et al. (2010).
[j] Cortés (1999); Huber et al. (2009); Mara (2010).
[k] Cortés (1999); Mara (2010); Mara et al. (2010).
[l] Cortés (1999); Huber and Motta (2004).

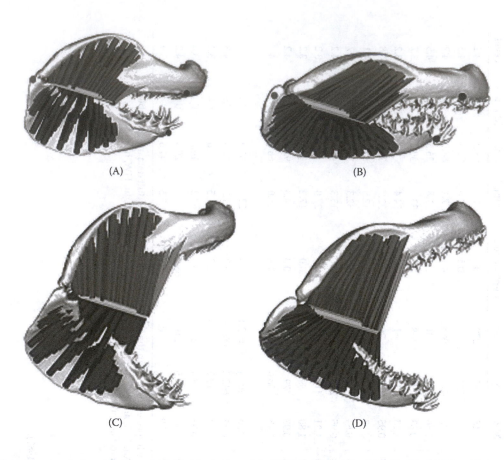

(A)   (B)

(C)   (D)

FIGURE 6.26
(See color insert.) Arrangement of muscle fibers in finite element models (FEMs) of the jaws of (A, C) white sharks, *Carcharodon carcharias*, and (B, D) sandtiger sharks, *Carcharias taurus*, at 15° (A, B) and 55° (C, D) gape angles. Each jaw adductor muscle group inserts on the mid-lateral raphe (yellow) and is represented by a series of trusses that are used to approximate muscle forces and insertion angles of muscle fibers. In both species, the angle of muscle trusses becomes more orthogonal at 55° due to their insertion on the MLR. Truss colors correspond to the following muscle groups: blue, dorsal quadratomandibularis (QMD); orange, medial division of dorsal quadratomandibularis (sandtiger only); green, preorbitalis; red, ventral quadratomandibularis. (From Ferrara, T.L. et al., *J. Biomech.*, 44(3), 430–435, 2011. With permission.)

mechanics in the white shark, *Carcharodon carcharias*, and sandtiger shark, *Carcharias taurus*, Ferrara et al. (2011) found that bite force increases as gape increases because the angle between the adductor muscle fibers and the lower jaw becomes increasingly orthogonal due to their insertion on the mid-lateral raphe (Figure 6.26). This is in stark contrast to models of mammalian jaw mechanics, in which bite force decreases significantly at wider gapes due to increasingly acute adductor insertion angles (Bourke et al., 2008; Dumont and Herrel, 2003). The muscle arrangement characteristic of shark jaw adductors therefore creates the potential to maintain jaw leverage and generate high bite forces across a much wider range of gape angles than is observed in mammalian predators.

Jaw leverage is most commonly investigated through analyses of mechanical advantage (*MA*), the ratio of a jaw's in-lever distance (L_I) to its out-lever distance (L_O):

$$MA = \frac{L_I}{L_O} \qquad (6.7)$$

The in-lever is the distance from the jaw joint to the point of insertion of a particular muscle, whereas the out-lever is the distance from the jaw joint to a relevant bite point such as the anteriormost tooth of the functional row (Figure 6.27). The ratio of these distances determines the proportion of the force applied *to* the lever system (i.e., muscle force) that will be transmitted *by* the lever system (i.e., bite force). Though conceptually simple, mechanical advantage is difficult to determine in cartilaginous fishes because the aponeurotic attachments of most jaw adductors do not have a clear insertion points. Muscle insertions must be approximated from the intersection of the muscle's mechanical line of action with the jaw, which can be determined by following the principal fiber direction through the muscle's center of mass. The presence of multiple adductor divisions then requires calculation of a resultant in-lever based on a weighted average of the individual in-levers and their respective forces. Mechanical advantage can subsequently be calculated as the ratio of the weighted in-lever to a relevant out-lever (Huber et al., 2005).

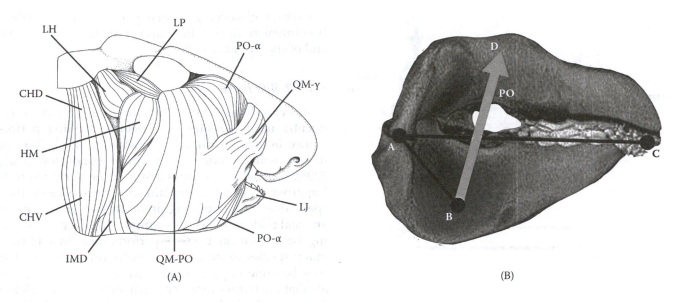

FIGURE 6.27

(A) Right lateral view of the cranial musculature of the horn shark, *Heterodontus francisci*. *Abbreviations:* CHD, dorsal hyoid constrictor; CHV, ventral hyoid constrictor; HM, hyomandibulo–mandibularis; IMD, intermandibularis; LH, levator hyomandibularis; LJ, lower jaw; LP, levator palatoquadrati; QM–PO, quadratomandibularis–preorbitalis complex; PO-α, preorbitalis-α; QM-γ, quadratomandibularis-γ; (B) Right lateral view of the jaws of *H. francisci* indicating measurements used in calculating mechanical advantage: A–B, resolved in-lever for jaw adduction; A–C, out-lever; B–D, adductive muscle force vector; PO, maximum tetanic tension. (From Kolmann, M.A. and Huber, D.R., *Zoology*, 112, 351–361, 2009. With permission.)

Methodological complications aside, mechanical advantage exhibits a strong ecomorphological signal among cartilaginous fishes. Those species that consume a significant amount of functionally difficult (large or hard) prey typically have mechanical advantage ratios ≥ 0.40, while those specializing on soft-bodied prey typically have mechanical advantages ratios ≤ 0.40 (Table 6.2). Four of the top five mechanical advantage ratios among cartilaginous fish are durophagous taxa, with the holocephalans *Chimaera monstrosa* and *Hydrolagus colliei* ranking highest (0.69 and 0.68, respectively). Despite the lack of mechanical advantage data for batoids, Summers (2000) proposed that asymmetrical jaw adductor contraction and fused jaw symphyses in the durophagous cownose ray, *Rhinoptera bonasus*, create a "nutcracker" mechanism capable of amplifying the force produced by the adductor musculature (mechanical advantage > 1.0) (Figure 6.28). Blue crabs, *Callinectes sapidus*, constitute 72% of the diet of the bonnethead shark, *Sphyrna tiburo*, yet it has a rather low mechanical advantage (0.22), suggesting that chemical digestion may play a greater role than mechanical processing in its durophagous ecology (Cortes, 1999; Mara, 2010; Mara et al., 2010). In fact, hammerhead sharks in general have low-leverage feeding mechanisms (mechanical advantage, 0.12 to 0.26) (Table 6.2), which is consistent with other species in which diet is dominated by elusive prey such as teleosts, cephalopods, and other elasmobranchs (Mara, 2010). Although force transmission is of obvious importance for prey capture, low-leverage jaws can

be advantageous to consumers of elusive prey because force and velocity are inversely proportional in mechanical lever systems (De Schepper et al., 2008; Wainwright et al., 2000; Westneat, 1994). Low-leverage jaws reach higher angular velocities, which can be augmented by increasing adductor muscle mass at any mechanical advantage, thereby aiding in the capture of elusive prey (Van Wassenbergh et al., 2005; Wainwright and Shaw, 1999). Nonetheless, very-high-leverage jaws appear to be a luxury that only predators of sessile benthic epifauna can afford.

Ecomorphological variation among jaw leverage, feeding modality, and prey type has been well documented in teleost fishes as well (Hernandez and Motta, 1997; Turingan et al., 1995; Wainwright et al., 2000, 2004; Westneat, 2004). Although durophagous specialists such as the scarid parrotfishes have mechanical advantage ratios approaching 1.0, those teleosts with mechanical advantage ratios greater than 0.34 are considered to have "high-leverage" jaws (Wainwright et al., 2004; Westneat, 2004). Yet, the average mechanical advantage ratio of the cartilaginous fishes that have been investigated is 0.42, begging the question of whether or not cartilaginous fishes inherently have higher leverage feeding mechanisms. The adductor mandibulae complex develops from cranial somitomeres in association with the first visceral arch in cartilaginous and bony fishes (Liem et al., 2001). Although the palatoquadrate and Meckel's cartilages of the first visceral arch develop into the upper and lower jaws

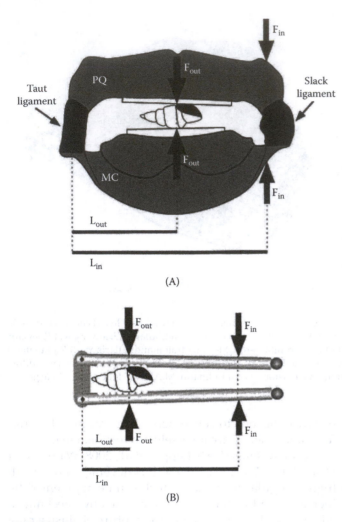

FIGURE 6.28

(A) Ventral view of the jaws of the cownose ray, *Rhinoptera bonasus*, illustrating the "nutcracker" model of the bite force amplification proposed by Summers (2000). A gastropod is shown between the tooth plates, acted upon by a second-class lever powered by the jaw adductors. The ligaments between the upper and lower jaws, shown in black, serve as the fulcrum of the lever system. (B) A nutcracker of similar proportions showing the principles of the second-order lever system. (From Summers, A.P., *J. Morphol.*, 243, 113–126, 2000. With permission.)

of chondrichthyans, these elements are incorporated into the jaw joint and suspensorium of teleosts, and the upper and lower jaws are subsequently derived from dermal bones (de Beer, 1932; Liem et al., 2001). Consequently, the adductor mandibulae complex develops posterior to the teleost lower jaw and inserts via tendons that extend past the jaw joint, whereas this muscle complex develops in direct association with the chondrichthyan lower jaw. Specific adaptations for teleostean jaw leverage notwithstanding, the chondrichthyan adductor mandibulae complex is located relatively more anteriorly and is inherently in a position of higher leverage. This may ultimately be of great significance, as Friedman (2009) has demonstrated that

fish with high-leverage jaws experienced lower extinction intensity during the mass extinction event at the end of the Cretaceous Period.

6.4.2.2 Bite Force

As previously mentioned, bite force has become a frequently used measure of vertebrate feeding performance because it directly impacts the acquisition of dietary and reproductive resources (Anderson et al., 2008). Cartilaginous fishes are a particularly interesting group in which to examine bite force because they span seven orders of magnitude in size, representing an ideal clade in which to examine the effects of changing body size on feeding performance. In addition, many species consume functionally difficult prey that may be quite large or composed of structural materials that are harder than the cartilaginous endoskeleton (Currey, 1980; Huber et al., 2009; Kolmann and Huber, 2009; Summers and Long, 2006; Wainwright et al., 1976; Wroe et al., 2008). Whether these feats of prey capture are a byproduct of large body size or are the result of size-specific selection for enhanced feeding performance (allometric growth) is largely unknown (Huber et al., 2009).

The bite forces of apex predators such as the white shark (*Carcharodon carcharias*), bull shark (*Carcharhinus leucas*), and great hammerhead (*Sphyrna mokarran*) rank among the highest of all extant vertebrates (Table 6.2), and the extinct *Carcharodon megalodon* may have had the highest bite force of any predator in vertebrate evolutionary history (Habegger et al., in review; Huber et al., 2009; Mara, 2010; Wroe et al., 2008). Recent phylogenetically informed analyses of numerous species spanning five orders of magnitude in size have determined that these tremendous bite forces can be attributed to large body size, not size-specific adaptations for high-performance feeding; bite force scales isometrically with body mass among species (Habegger et al., in review; Huber et al., 2009). It is likely that the high absolute bite forces generated by these larger shark species allow them to overcome constraints set by prey durability, thereby eliminating the selective pressure for size-specific adaptation (e.g., positive allometry of jaw adductor force or leverage) at large sizes. Huber et al. (2009) identified a possible transition point for this change in selective pressure at approximately 100 N of bite force, above which the slope of the relationship between bite force and body mass among species decreases dramatically. In other words, bite force increases with size more rapidly among smaller shark species perhaps because their lower bite forces represent a greater constraint with respect to the physical properties of their prey. Interestingly, a recent analysis of the cutting performance of teeth from 14 species of extant and extinct sharks found the highest force required to penetrate a range of prey items to be 114 N (Whitenack and Motta, 2010).

Perhaps the most intriguing examples of feeding performance among cartilaginous fishes are those that consume hard prey, as the physical properties of such prey items represent a significant ecological constraint and the prey may be even harder than their own skeletons. Most durophagous chondrichthyans are small benthic predators that likely experience selective pressure for high-performance biting due to the physical demands of their trophic niche. As one might expect, durophagous species such as the spotted ratfish, *Hydrolagus colliei*, and horn shark, *Heterodontus francisci*, have the highest mass-specific bite forces of all species, and durophagy is significantly correlated with high mass-specific bite forces over evolutionary history (Habegger et al., in review; Huber et al., 2008, 2009; Mara et al., 2010). Surprisingly, the piscivorous sharpnose sevengill shark, *Heptranchias perlo*, has a very high mass-specific bite force, which appears to be a function of disproportionately large jaw adductors occupying deep fossae in the upper and lower jaws, characteristic of the ancestral cranial morphotype for sharks (Compagno, 1977; Huber, 2006). The bonnethead shark, *Sphyrna tiburo*, has one of the lowest mass-specific bite forces of all species, again suggesting that its durophagous ecology can be attributed to behavioral handling of prey as well as digestive enzymes, and not the pulverizing of prey (Habegger et al., in review; Huber et al., 2008, 2009; Mara, 2010; Mara et al., 2010). As with mechanical advantage, hammerhead sharks generally have low size-specific bite forces, although this low performance does not appear to be a consequence of the evolution and expansion of the cephalofoil (Mara, 2010). Although bite force has seldom been examined in batoids, at 60-cm disk width the cownose ray, *Rhinoptera bonasus*, can generate bite forces of at least 200 N, which is 20× greater than the forces at which their coquina clam prey (*Donax* sp.) begin to fracture, but considerably less than the force needed to crush large oysters and clams (Fisher et al., 2011; Maschner, 2000; Sasko, 2000; Sasko et al., 2006).

Phylogenetic analyses have identified jaw adductor cross-sectional area, mechanical advantage, and widening of the head as the biomechanical correlates of high mass-specific biting performance over chondrichthyan evolutionary history (Habegger et al., in review). This corroborates previous studies that have posited that a suite of morphological characters has convergently evolved among durophagous cartilaginous fishes, including hypertrophied jaw adductors, high-leverage jaws, and a pavement-like dentition (whether of individual teeth as in elasmobranchs or tooth plates as in holocephalans), with well-mineralized jaws and fused mandibular symphyses present among some taxa as well (Huber et al., 2005, 2008; Summers, 2000; Summers

et al., 2004). Although a pavement-like dentition is common among durophagous chondrichthyans, it must be noted that the interaction between biting performance and tooth morphology is little understood. Sustained adductor contraction, force amplification through asymmetrical biting, and cyclical loading of prey items have been identified as behavioral correlates of durophagy in these species as well (Huber et al., 2005; Summers, 2000; Wilga and Motta, 2000). Phylogenetic analyses have also indicated that evolution toward increased bite force has gone hand in hand with the evolution of wider heads among sharks (Habegger et al., in review; Huber et al., 2009). These results corroborate findings on the blacktip shark, *Carcharhinus limbatus*; horn shark, *Heterodontus francisci*; spotted ratfish, *Hydrolagus colliei*; and other vertebrate lineages in which head width is an excellent predictor of biting performance as it best approximates the cross-sectional area of the jaw adductors (Herrel et al., 1999, 2001a, 2002, 2004, 2005; Huber et al., 2006, 2008; Kolmann and Huber, 2009; Verwaijen et al., 2002).

6.4.2.3 Scaling of Feeding Biomechanics

Organismal performance changes over ontogeny as the musculoskeletal systems underlying animal behavior change in relative size and shape. As performance is largely a determinant of ecology, ontogenetic changes in the former can influence the latter. Thus, a thorough understanding of organismal ecology requires knowledge of how the functional integrity of morphological systems is maintained or enhanced during growth (Kolmann and Huber, 2009; Schmidt-Nielsen, 1984). Inquiries of this type are made through analyses of scaling patterns, which indicate the rate of change of morphological and performance measures with respect to body size. The null hypothesis of scaling analyses is one of geometric similarity, in which the chosen dependent parameter grows isometrically (in direct proportion) relative to body size, whereas positive or negative allometry indicates relatively faster or slower growth, respectively. Evidence of allometric growth is generally thought to indicate selective pressure for deviation from geometric similarity (Herrel and Gibb, 2006). For example, positive allometry of feeding performance is associated with ontogenetic dietary shifts and niche partitioning because enhanced performance enables the consumption of functionally difficult prey that other species, or younger members of the same species, cannot consume (Aguirre et al., 2003; Hernandez and Motta, 1997; Herrel and O'Reilly, 2006). Alternatively, positive allometry of feeding performance during early life history stages allows juveniles to rapidly reach adult performance levels, after which selective pressure for allometry maybe relaxed (Habegger et al., in review). Scaling relationships are examined by comparing the observed slope of a relationship to the hypothetical isometric slope, which is the ratio of the exponents of the dependent and independent variables in the analysis. The isometric slope for a comparison of bite force against total length is 2.0 because bite force is a function of the cross-sectional area of the jaw muscles (x^2) and total length is a linear function (x^1). Given this logic, the isometric conditions for biomechanical parameters such as bite force and muscle force, in-levers and out-levers, and mechanical advantage are 2, 1, and 0 respectively.

Positive allometry of bite force has now been identified in horn sharks, *Heterodontus francisci*; blacktip sharks, *Carcharhinus limbatus*; juvenile bull sharks, *Carcharhinus leucas*; and spotted ratfish, *Hydrolagus colliei*. This is consistent with intraspecific findings from other vertebrates (Binder and Van Valkenburgh, 2000; Erickson et al., 2003; Habegger et al., in review; Hernandez and Motta, 1997; Herrel and Gibb, 2006; Herrel et al., 1999; Huber et al., 2006, 2008; Kolmann and Huber, 2009; Meyers et al., 2002). The biomechanical determinants of these scaling patterns were positive allometry of jaw adductor cross-sectional area in the three shark species and positive allometry of mechanical advantage at the anterior or posterior bite point in *H. francisci*, *C. limbatus*, *H. colliei*, and juvenile *C. leucas* (Habegger et al., in review; Huber et al., 2006, 2008; Kolmann and Huber, 2009). Conversely, isometry of bite force has been identified within adult *C. leucas* and from interspecific analyses of shark species spanning five orders of magnitude of size (Habegger et al., in review; Huber et al., 2009). These findings contrast with interspecific analyses of other vertebrate clades that have identified positive allometry of bite force among bats and turtles (Aguirre et al., 2002; Herrel et al., 2002), but negative allometry within carnivoran mammals (Christiansen and Wroe, 2007).

The collective results of scaling analyses within and among shark species suggest that positive allometry of bite force in small species or during the early ontogeny of large species plays a key role in determining adult performance and the ability to capture functionally difficult prey, and it may confer a competitive advantage over isometric ontogenetic trajectories, providing access to *relatively* competitor-free trophic niches earlier in life (Figure 6.29). For example, consumption of hard or large prey increases ontogenetically in *Heterodontus francisci*, *Carcharhinus limbatus*, and *Carcharhinus leucas*,

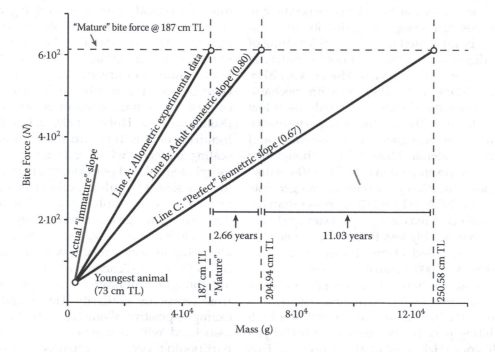

FIGURE 6.29

Simulation of bite force (*N*) with respect to body mass (g) in the bull shark, *Carcharhinus leucas*, illustrating the selective advantage gained due to the period of positive allometry early in its life history. Line A is drawn between this study's smallest animal (73 cm TL) and the smallest examined "mature" animal (187 cm TL). This is compared to lines B and C—bearing the experimentally calculated adult slope (0.80) and a perfectly isometric slope (0.67), respectively—to determine the masses (vertical dashed lines) at which animals on these curves would reach line A's adult bite force (horizontal dashed line). These masses are converted to units of time based on Branstetter and Stiles' (1987) equations for bull shark growth. The simulation indicates that the early growth rate of bull shark bite force allows it to reach mature bite forces many years before hypothetical isometric members of the same cohort; this performance and time gain could therefore have been a major driver in the evolution of positive allometry in shark species. Furthermore, the slope of line A is an underestimate of the allometric slope for all sampled juvenile *C. leucas* (short gray line on the far left of the graph), so this effect could be even more pronounced over the course of the life history of *C. leucas*. (From Habegger, M.L. et al., *Zoology*, in review. With permission of Elsevier.)

and positive allometry of bite force allows *H. francisci* to begin consuming purple sea urchins, *Strongylocentrotus purpuratus*, during its second year of life, whereas an isometric ontogenetic trajectory would have delayed access to this resource for at least another year (Bethea et al., 2004; Cliff and Dudley, 1991; Kolmann and Huber, 2009; Strong, 1989). Thus, selective pressure for size-related performance adaptations may be expected in small species or the young of large species, whereas extremely large sharks likely experience reduced pressure because the sheer magnitude of their bite forces is great enough to overcome the physical constraints of any prey item. The discrepancy between interspecific analyses of sharks and other vertebrate groups may be due to the fact that sharks simply reach larger sizes and have larger absolute bite forces than other groups of vertebrates.

6.4.3 Biomechanics of Chondrichthyan Jaw Suspension Mechanisms

Jaw suspension has historically been regarded as a major determinant of jaw kinesis in chondrichthyan feeding mechanisms (Huxley, 1876; Moss, 1977b; Wilga, 2002, 2008). Although Wilga (2002) found little correlation between the diversity of extant jaw suspension mechanisms and hyomandibular morphology with upper jaw protrusion ability, it can be expected in a macroevolutionary sense that the transition from Paleozoic forms to neoselachians has brought about increased movement of the jaws relative to the cranium, and that this undoubtedly plays a key role in the predatory ability of modern elasmobranchs (Carroll, 1988; Dean and Motta, 2004b; Frazzetta and Prange, 1987; Maisey, 2008; Motta and Wilga, 2001; Moy-Thomas and Miles, 1971; Schaeffer, 1967; Tricas and McCosker, 1984; Wilga et al., 2001).

While the postorbital articulation of extinct taxa such as *Cobelodus*, *Orthacanthus*, and *Pucapampella* is believed to have severely restricted palatoquadrate protrusion (Maisey, 2008), that of the hexanchid sharks readily disengages facilitating limited dorsoventral protrusion (Compagno, 1988; Wilga, 2002), and the remaining 99.6% of modern elasmobranchs lack a postorbital articulation altogether. Spanning the continuum of jaw suspensions, it is clear that euhyostylic batoids are capable of far greater jaw protrusion than amphistylic hexanchids, although the hyostylic and orbitostylic sharks show no conclusive pattern (Figure 6.6) (Dean and Motta, 2004b; Wilga, 2002). Among those species that have been experimentally investigated, the length of the ethmopalatine ligament, or its absence in the case of euhyostylic batoids, appears to be the primary determinant of jaw protrusion distance (Wilga, 2002).

Despite the ambiguity between suspension type and jaw mobility, hyomandibular morphology and behavior have been linked with other aspects of elasmobranch feeding. During prey capture, depression of the basihyal cartilage pulls the ceratohyal cartilages posteroventrally, causing rotation of the hyomandibular cartilages about the cranium. The short, laterally directed hyomandibulae of the whitespotted bamboo shark, *Chiloscyllium plagiosum*, and spiny dogfish, *Squalus acanthias*, as well as the anteriorly directed hyomandibulae of the little skate, *Leucoraja erinacea*, are depressed ventrally and compressed medially during this sequence. Although hyomandibular adduction temporarily delays peak suction pressure generation, the combined effect of ventral and medial movement of the hyomandibulae is expansion of the oropharyngeal cavity and hydraulic transport of water and prey into the mouth (Wilga, 2008; Wilga and Sanford, 2008). The posteriorly directed hyomandibulae of the shortfin mako shark, *Isurus oxyrinchus*, and sandbar shark, *Carcharhinus plumbeus*, swing ventrolaterally during feeding, causing expansion of the hyoid arch and an increase in the area of the mouth opening (Wilga, 2008). Although it is difficult to determine which hyomandibular orientation is plesiomorphic among modern elasmobranchs, it appears that the laterally directed hyomandibulae of heterodontiform, orectolobiform, and squaliform sharks facilitate suction feeding, whereas the posteriorly directed hyomandibulae of carcharhiniform, lamniform, and hexanchiform sharks facilitate a large gape for the biting of large prey (Wilga, 2008).

The general kinematic pattern of the hyoid arch experimentally verified by Wilga (2008) suggests that the hyomandibulae are acting in tension, which is reassuring considering that they have been referred to as "suspensory" elements for well over a century. Huber (2006) verified this role via biomechanical modeling of jaw suspensions in hyostylic species (e.g., horn shark, *Heterodontus francisci*; whitespotted bamboo shark, *Chiloscyllium plagiosum*; lemon shark, *Negaprion brevirostris*) and an amphistylic species (e.g., sharpnose sevengill shark, *Heptranchias perlo*). The hyomandibular and anterior craniopalatine articulations (ethmoidal/orbital) of *H. francisci*, *C. plagiosum*, and *H. perlo* are loaded in tension and compression, respectively, the magnitude of which is proportional to the force generated by the preorbitalis muscle (other adductor muscles act between the jaws with no net effect on suspensorial elements). In these species, the preorbitalis attaches in front of the anterior craniopalatine articulation, which remains intact throughout the gape cycle. Contraction of the preorbitalis compresses the upper jaw into the articulation and generates a torque about this point that rotates the posterior region of the upper jaw anteroventrally, pulling the hyomandibular cartilages in tension. Conversely, *N. brevirostris* is capable of protruding the upper jaw far enough to disengage its ethmoidal articulation (Motta and Wilga, 1995). In the absence of anterior craniopalatine contact, contraction of the preorbitalis and levator palatoquadrati causes upward translation of the jaws and compression of the hyomandibulae (Huber, 2006). Thus, it appears that the hyomandibulae are tensile elements among species in which the anterior cranio-palatine articulation remains intact during feeding, and they act as compressive elements among species in which the anterior craniopalatine articulation either disengages (some carcharhinid and lamniform sharks) (Motta and Wilga, 1995; Wilga, 2005) or is absent alto-gether (batoids) (Huber, 2006). The euhyostylic batoids possess several novel jaw and hyomandibular muscles that appear to pivot the jaws and hyomandibulae about the cranium, allowing for high-precision, asymmetri-cal movements of the feeding mechanism (Dean and Motta, 2004a; Dean et al., 2005; McEachran et al., 1996; Wilga and Motta, 1998b). Furthermore, trabecular rein-forcement in the jaws and hyomandibulae of the lesser electric ray, *Narcine bancroftii*, resist buckling, as the jaws are ballistically protruded into the sediment in search of prey (Dean et al., 2006). Although a suspensorial model-ing analysis has not been performed for the euhyostylic mechanism, these findings corroborate the role of the hyomandibulae as compressive elements in batoids.

These suspensorial mechanics are suggestive of a broad evolutionary pattern in elasmobranch feed-ing mechanisms. The amphistylic jaw suspension of *Heptranchias perlo* restricts jaw protrusion, thereby reducing selective pressure for well-developed preor-bitalis muscles; consequently, the hyomandibulae and craniopalatine articulations of *H. perlo* experience negli-gible loading (Huber, 2006). Although not homologous, the amphistylic jaw suspensions of modern hexanchid sharks and archaeostylic Paleozoic species are mechani-cally analogous (Maisey, 2008). The postorbital articula-tion of these Paleozoic forms precluded jaw protrusion (Maisey, 2008), so it can be assumed that these sharks also lacked well-developed preorbitalis muscles. Given that the force generated by the preorbitalis is the pri-mary determinant of suspensorial loading (Huber, 2006), neither of these groups is therefore likely to have experienced selective pressure for structural modifi-cations to their long, thin, poorly calcified, posteriorly directed hyomandibular cartilages. Thus, the hyoid arch retained the appearance of the postmandibular visceral arch from which it was derived (Maisey, 1980; Mallat, 1996; Zangerl and Williams, 1975). When the postor-bital articulation was lost in neoselachians via reduc-tion of the otic and postorbital processes of the upper jaw and cranium, respectively (Carroll, 1988; Maisey and de Carvalho, 1997; Schaeffer, 1967), architectural changes to the preorbitalis (enlargement, subdivision, reorientation) would have facilitated enhanced force production and jaw kinesis. Enhanced force produc-tion by the preorbitalis may then have provided the mechanical impetus for structural modifications to the

hyomandibular cartilages and the evolution of hyostyly in neoselachians. During this process, the hyoman-dibular cartilages became shorter, thicker, and rotated anteriorly into a more orthogonal position relative to the cranium, and they developed deep articular facets against the cranium, facilitating directionally specific motion (Cappetta, 1987; Schaeffer, 1967; Wilga, 2002). The increase in load-bearing ability of the hyomandibular cartilages and enhanced jaw kinesis associated with the evolution of hyostyly appear to have increased the func-tional versatility of the feeding mechanism, resulting in the evolution of ram, suction, biting, and filter-feeding mechanisms in modern elasmobranchs (Moss, 1977b).

Although the holocephalans have a much simpler jaw suspension mechanism, it would be remiss to neglect them in this discussion because they are one of the more curious groups of chondrichthyans. The upper jaw of holocephalans fuses to the nasal, trabecular, and parachordal cartilages of the cranium early in develop-ment, resulting in an akinetic holostylic jaw suspension (Grogan et al., 1999; Wilga, 2002). The fused upper jaw is located directly below the vaulted ethmoidal region of the cranium; along with hypermineralized tooth plates, these are considered adaptations for durophagy in holo-cephalans (Didier, 1995; Grogan and Lund, 2004). Many holocephalans regularly consume hard prey despite the fact that their feeding mechanisms are poorly if at all mineralized. Because hard prey can generate large bite reaction forces and unmineralized cartilage has poor compressive stiffness, the unique cranial morphotype of holocephalans is believed to stabilize the feeding mechanism against dorsoventral flexion, thus repre-senting a wholly different strategy for cranial stability than is found among elasmobranchs (Huber et al., 2008; Wroe et al., 2008).

6.5 Tooth Form and Function

6.5.1 Arrangement and Terminology

Elasmobranch teeth are arranged in rows on the pal-atoquadrate and Meckel's cartilage, such as in most sharks and many rays, or they form large pavement-like tooth plates for crushing prey, as in many batoids. Elasmobranch teeth are polyphyodont, meaning that they develop in rows similar to the teeth of bony fishes and are replaced at a regular interval. A tooth in the functional position at the edge of the jaw and its replacement teeth constitute a tooth row (file, fam-ily). The number of tooth rows/families varies from 1 per jaw in some rays to more than 300 in the whale shark; in most sharks, there are 20 to 30 tooth rows. A tooth series refers to a line of teeth along the jaw that is parallel to the jaw axis and includes teeth from all rows (Compagno, 1984; James, 1953; Reif, 1976, 1984). The rate of replacement is species specific; is affected by age, diet, seasonal changes, and water temperature; and may vary between the upper and lower jaw (Moss, 1967). Most species only replace a few teeth at a time, although the cookie-cutter shark, *Isistius brasiliensis*, differs in that its relatively large lower triangular teeth are shed together as a complete set (Strasburg, 1963). Replacement rates, as measured by the rate of movement of a tooth from the row lingual to the functional row to that of the functional row, vary from 9 to 12 days in the leopard shark, *Triakis semifasciata* (Reif et al., 1978); from 9 to 28 for the nurse shark, *Ginglymostoma cirratum*, in the summer and from 51 to 70 days in the winter (Luer et al., 1990; Reif et al., 1978); from 8 to 10 days for the lemon shark, *Negaprion brevirostris* (Moss, 1967); about 4 weeks for *Heterodontus* (Reif, 1976); and 5 weeks for *Scyliorhinus canicula* (Botella et al., 2009; Märkel and Laubier, 1969). Primitive chondrichthyan fishes, such as the Early Devonian *Leonodus carlsi*, are believed to have an extremely slow dental replacement rate (Botella et al., 2009). The teeth of myliobatid rays are arranged as a central file of thick, flattened, usually hexagonal teeth that are fused together and three lateral files of smaller teeth on each side. Other myliobatid rays, such as the spotted eagle ray, *Aetobatus narinari*, have only a central file of fused teeth on the upper and lower jaws, in which replacement teeth move toward the occlusal plane where they fuse and become functional (Figure 6.30). *Myliobatis* has three to ten rows of mature, unworn teeth behind the functional rows, and as they are replaced these teeth eventually pass aborally and are lost. *Aetobatus narinari* has an unusual condition in which the lower jaw teeth move anteriorly out of the crushing zone and remain attached to the tooth plate to form a spade-like appendage used to dig up prey items (Bigelow and Schroeder, 1953; Cappetta, 1986a,b; Summers, 2000; A. Collins, pers. comm.).

Within a jaw, homodont teeth are all the same shape and show no abrupt change in size. This is rare in recent and fossil sharks, but apparently exists in *Rhincodon* and *Cetorhinus*. Monognathic heterodonty refers to a significant change in size and shape of the teeth in different parts of the same jaw (upper or lower) and is common in recent and fossil sharks (Applegate, 1965; Compagno, 1988). Horn sharks (Heterodontidae) and bonnethead sharks (Sphyrnidae) both have anterior cuspidate teeth for grasping and posterior molariform crushing teeth (Figure 6.31) (Budker, 1971; Compagno, 1984; Nobiling, 1977; Peyer, 1968; Reif, 1976; Smith, 1942; Taylor, 1972). Carcharhinid sharks have dignathic heterodonty, with more cuspidate lower jaw teeth lacking serrations and more blade-like, serrated teeth in the upper jaw (Bigelow

and Schroeder, 1948; Compagno, 1984, 1988). Sexual heterodonty occurs in many elasmobranchs, and in many cases the teeth of adult males differ in shape from those of females and immature males. The dimorphism is often confined to the anterior teeth, and in the carcharhinoids it is mostly confined to species less than one meter in length. Sexual heterodonty in sharks and particularly rays appears to be related to courtship, during which the male holds onto a female with his mouth, rather than to feeding (Cappetta, 1986b; Compagno, 1970, 1988; Ellis and Shackley, 1995; Feduccia and Slaughter, 1974; Herman et al., 1995; Kajiura and Tricas, 1996; McCourt and Kerstitch, 1980; McEachran, 1977; Nordell, 1994; Smale and Cowley, 1992; Springer, 1967).

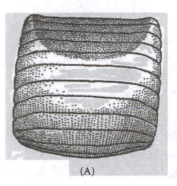

(A)

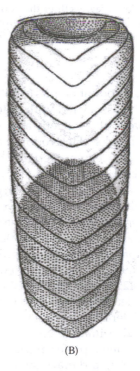

(B)

FIGURE 6.30

(A) Upper and (B) lower tooth plate of *Aetobatus narinari*. In the lower plate, the front tooth is lowermost. (From Bigelow, H.B. and Schroeder, W.C., *Mem. Sears Found. Mar. Res.*, 1(2), 1–588, 1953. Courtesy of the Peabody Museum of Natural History, Yale University.)

FIGURE 6.31
Dorsal view of the lower jaw teeth of the horn shark, *Heterodontus francisci*, showing the grasping teeth in the front of the jaw and the molariform or grinding teeth behind. Rostral tip of the jaw is at the top of the picture.

Ontogenetic heterodonty refers to ontogenetic changes in dentition, often associated with ontogenetic changes in diet. The shape of the teeth and number of tooth cusps in horn sharks (Heterodontidae) change with ontogeny. Rear replacement teeth gradually lose cusps, broaden at the base, and flatten along the crown. The more anterior recurved teeth have larger central cusps and fewer overall cusps with age. Juvenile Port Jackson shark, *Heterodontus portusjacksoni*, have more pointed teeth and apparently take more soft-bodied prey than the adults (Compagno, 1984; McLaughlin and O'Gower, 1971; Nobiling, 1977; Peyer, 1968; Reif, 1976; Shimada, 2002b; Smith, 1942; Taylor, 1972). White sharks less than 1.5 m TL (total length) have relatively long and narrow teeth with lateral cusplets (Hubbell, 1996). Smaller white sharks feed primarily on fish, while larger animals with broader teeth prefer marine mammals (Tricas and McCosker, 1984), a dietary switch that is reflected in the isotopic signature of their vertebrae (Estrada et al., 2006). Lamniform sharks have an embryonic peg-like dentition before parturition, and at about 30 to 60 cm TL they transition into the adult lamnoid type of

dentition just before or after birth. The early stage of the adult dentition often possesses bluntly pointed crowns without distinct cutting edges, serrations, and lateral cusplets of the adult teeth. This is perhaps to prevent the developing embryos, which are often consuming eggs and embryos *in utero*, from damaging the mother's uterus (Shimada, 2002b).

6.5.2 Evolutionary and Functional Patterns

It is suggested that the earliest sharks for which there are no fossil teeth, just denticles (placoid scales), were microphagous filter feeders. Presumably with a selection for larger teeth there was a concomitant change to a macrophagous diet (Williams, 2001). Many of the early Paleozoic sharks, including the cladodont, xenacanthid, hybodont, and ctenacanthid lineages, had a dentition apparently suited for piercing, holding, and slashing. Most of the Early Devonian and Carboniferous sharks have a tooth pattern often referred to as "cladodont" in form (Figure 6.32). These grasping teeth have a broad base with a single major cusp and smaller lateral cusps and apparently slow replacement. In *Xenacanthus*, the lateral cusps are enlarged, and the central cusp is reduced. Hybodont and ctenacanthid sharks in general also had a tooth morphology, composed of two of more elongated cusps, that appears suited for piercing and holding prey. Even within these early lineages, as in modern forms, there were repeated evolutionary forays into a benthic lifestyle and development of crushing, pavement-like teeth (Cappetta, 1987; Carroll, 1988; Hotton, 1952; Moy-Thomas and Miles, 1971; Schaeffer, 1967; Williams, 2001; Zangerl, 1981).

The tooth microstructure of these ancestral lineages was characterized by a single crystallite enameloid monolayer with random crystallite orientation (Gillis and Donoghue, 2007). The lack of microstructural diversity in these teeth is believed to have limited the functional diversification of feeding mechanisms in non-neoselachian elasmobranchs because microstructural diversity is related to the mechanical integrity of the teeth (Preuschoft et al., 1974). Neoselachian elasmobranchs other than batoids possess teeth with a triple-layered enameloid structure consisting of a layer of single crystallite enameloid and layers of parallel-fibered enameloid and tangle-fibered enameloid; parallel-fibered enameloid is believed to resist crack propagation and confer tensile strength, whereas tangle-fibered enameloid confers compressive strength (Gillis and Donoghue, 2007; Preuschoft et al., 1974). Triple-layered enameloid has been identified in basal members of the neoselachian crown groups Galea and Squalea (Gillis and Donoghue, 2007; Reif, 1977). Though lacking this extent of microstructural diversification, certain highly predatory Paleozoic species

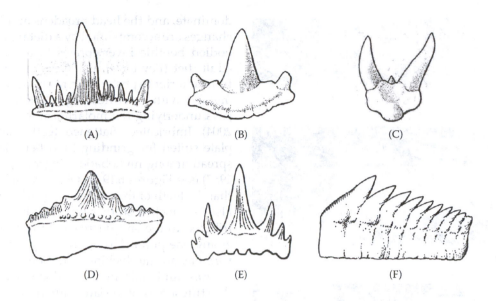

FIGURE 6.32
Ancestral shark tooth types: (A, B) acrodont teeth of *Cladodus* sp.; (C) diplodus teeth of *Xenacanthus* sp.; (D, E) hybodont type teeth; (F) tooth from extant *Hexanchus griseus*. (Parts A to E from Schaeffer, B., in *Sharks, Skates and Rays*, Gilbert, P.W. et al., Eds., The Johns Hopkins University Press, Baltimore, MD, 1967, pp. 3–35. With permission. Part F from Bigelow, H.B. and Schroeder, W.C., *Mem. Sears Found. Mar. Res.*, 1(1), 1–576, 1948. Courtesy of the Peabody Museum of Natural History, Yale University.)

exhibited some specialization. The Carboniferous shark *Carcharopsis prototypus* possessed increased density of enameloid crystallites in the single crystallite enameloid layer of its tooth serrations, as has also been found in fossil neoselachians (Andreev, 2010; Duffin and Cuny, 2008). The general lack of tooth serrations in Paleozoic species has been attributed to the rarity of such microstructural specialization and subsequent inability of the teeth to handle stresses associated with a sawing dentition (Duffin and Cuny, 2008).

Perhaps predicated on the evolution of microstructural diversification, the teeth of extant neoselachians display a considerable diversity of forms that are often ascribed functional roles (e.g., seizing/grasping, tearing, cutting, crushing, grinding) (Cappetta, 1986b, 1987). Teeth that apparently seize prey prior to swallowing are generally small, with multiple rows of lateral cusplets. These may be found on benthic-associated sharks and rays such as in the Orectolobiformes (e.g., *Ginglymostoma cirratum*) and male dasyatid rays (Figure 6.33). Some teeth appear suited for seizing and tearing, as they are long and pointed with narrow cusps. The dagger-like anterior teeth of the upper jaw in the sandtiger shark, *Carcharias taurus*, have a pronounced inward inclination and are thought to puncture and retain struggling prey after it has been grasped by the outwardly inclined anterior teeth of the lower jaw (Lucifora et al., 2001). The shortfin mako shark, *Isurus oxyrinchus*, has similar teeth anteriorly, with more triangular cutting teeth found toward the back of the jaw. The teeth of hexanchoids (*Hexanchus, Heptranchias, Notorhynchus, Chlamydoselachus*) can range from sawlike in *Hexanchus*

to three-pronged and grasping in *Chlamydoselachus* (Figure 6.32) (Cappetta, 1987; Carroll, 1988; Daniel, 1934; Pfeil, 1983). Many squaloid sharks, including *Etmopterus*, have a multicuspid grasping upper dentition and blade-like lower cutting teeth. Sharks with blade-like cutting teeth tend to have one fully erect functional row forming an almost continuous blade in which the bases of the teeth may interlock (e.g., *Dalatias, Etmopterus*) (Figure 6.33) or have edentulous spaces, such as are found in many lamnids (e.g., bigeye thresher, *Alopias superciliosus*; *Carcharodon carcharias*) (Shimada, 2002a; Shirai and Nakaya, 1990). In the tiger shark, *Galeocerdo cuvier*, the anterior and posterior margins of the teeth have coarse serrations and are markedly asymmetrical, with a distinct notch on the distal edge of the crown (Figure 6.33) (Bigelow and Schroeder, 1948; Cappetta, 1987; Williams, 2001). The more curved side of these teeth might serve to slice through tissue as they are dragged across a prey item, while the notch on the other side encounters and concentrates stress in more durable tissues such as collagen, cartilage, and bone (Figure 6.34). Witzell (1987) attributed the ability of *G. cuvier* to bite through whole large chelonid sea turtles to a suite of morphological and behavioral characters, including a single row of cusped, serrated teeth on a broad-based, heavily mineralized jaw that can be extensively protruded (Moss, 1965, 1972), and head shaking, which drags the teeth across the prey. Indigestible pieces of shell are regurgitated by stomach eversion, which has also been noted in other sharks and rays (Bell and Nichols, 1921; Brunnschweiler et al., 2005, 2011; Budker, 1971; Randall, 1992; Sims et al., 2000; Witzell, 1987).

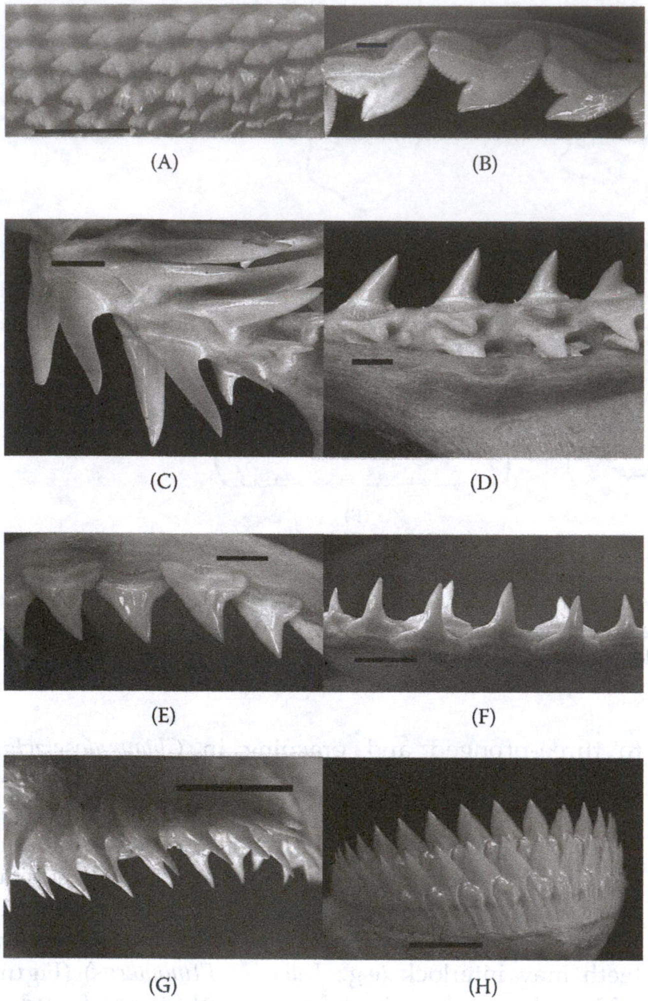

FIGURE 6.33

Modern tooth types: (A) lingual teeth of the nurse shark, *Ginglymostoma cirratum*; (B) upper lateral teeth of the tiger shark, *Galeocerdo cuvier*; (C) upper anterior teeth of shortfin mako shark, *Isurus oxyrinchus*; (D) lower lateral teeth of *I. oxyrinchus*; (E) upper anterior and lateral teeth of sandbar shark, *Carcharhinus plumbeus*; (F) lower anterior and lateral of *C. plumbeus*; (G) upper anterior and lateral teeth of the kitefin shark, *Dalatias licha*; (H) lower teeth of *D. licha*. Scale bar is 1 cm in all cases.

Durophagous dentitions have evolved numerous times among chondrichthyans. *Mustelus* has a crushing-type dentition, in which the teeth are low and have cutting edges with bluntly rounded apices (Bigelow and Schroeder, 1948; Cappetta, 1987). The crushing rear teeth of *Heterodontus* are closely opposed to each other such that the load on any one tooth is distributed to adjacent teeth in the same row (Nobiling, 1977). Ontogenetic differences in dentition occur in the Port Jackson shark, *Heterodontus portusjacksoni*, coinciding with a dietary shift. Juvenile sharks have sharp, cuspidate anterior teeth and the head is shorter and narrower. As they mature, the posterior molariform teeth begin to

dominate, and the head broadens and lengthens. These changes are accompanied by a dietary change from soft-bodied benthic invertebrates to a more durophagous adult diet (Powter et al., 2010). Jaw stiffness increases from anterior to posterior in the horn shark, *Heterodontus francisci*, with the stiffest regions of the upper and lower jaws underlying the molariform teeth (Summers et al., 2004). Imbricated, flattened teeth that form a dental plate suited for grinding hard benthic prey are widespread among myliobatid stingrays (Cappetta, 1986a,b, 1987) (see Figures 6.19 and 6.30). Maschner (2000) found that the teeth of the cownose ray, *Rhinoptera bonasus*, are also interlocked so that point loads are effectively distributed to the jaw, decreasing the stress concentration at any one point. The spotted eagle ray, *Aetobatus narinari*, has an interlocking dentition similar to that of *R. bonasus*, underlain by a fused jaw symphysis which is the stiffest part of the jaw. Although maximum stiffness occurs in different regions of the jaws of *H. francisci* and *A. narinari*, this parameter coincides with the location of molariform teeth in both species (Summers et al., 2004).

In contrast to these rigidly interlocking dentitions, numerous species possess teeth that are quite kinetic. Frazzetta (1988, 1994) has proposed that the relatively loose fibrous connection of shark teeth to the jaw cartilage allows the teeth to conform to irregularities in soft tissue and guide around solid obstructions such as bone. For example, the front teeth of the white shark, *Carcharodon carcharias*, are angled inward, perhaps making them more effective at gouging chunks of flesh, grasping prey items, or preventing prey escape from the mouth. During mouth closure, the crown angle of the anterior teeth initially increases by 8.7° and then decreases by 15.7° as the jaw is adducted through an arc of 35° or more. Although the mechanism is not clear, this is believed to facilitate a plucking action during feeding (Powlik, 1995). In many orectolobiform sharks, such as the whitespotted bamboo shark, *Chiloscyllium plagiosum*, the teeth have an inward inclination in their resting state but are capable of passively rotating about their connection to the jaw to form an imbricated crushing surface when hard prey is contacted (Ramsay and Wilga, 2007). Finally, individual teeth can have diverse functions even when they are firmly attached to the jaw, provided that the jaws themselves are capable of dynamic behaviors. The lesser electric ray, *Narcine bancroftii*, ballistically protrudes its jaws into the sediment in search of benthic prey, during which the halves of its jaws are adducted medially. Medial rotation of the jaw halves is accompanied by medial rotation of the teeth into positions that augment both normal and frictional forces as prey are sucked into the oral cavity. These forces hold prey in place as the buccal cavity is flushed of sediment prior to swallowing (Dean et al., 2008).

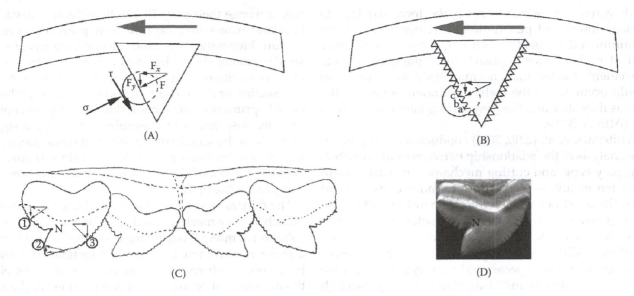

FIGURE 6.34
Proposed cutting mechanism for nonserrated, serrated, and notched shark teeth. (A) When tooth is drawn across an object as indicated by the large gray arrow, the object (denoted as a circular shape) impacts the leading edge of the beveled tooth (triangle). As the nonserrated tooth edge cuts into the object, a force normal to the object (F) can be resolved into a force in the x plane (F_x) and a force in the y plane (F_y). These forces result in a stress normal to the tooth (σ) and a shear stress (τ) that result in the object being deflected toward the tip of the tooth as the tooth edge cuts into the object. The sharp leading edge of the tooth results in stress concentration that helps cut into the object (σ = Force/Area; the sharp leading edge has a very small area in contact with the object at any time). (B) As a serrated tooth is similarly drawn across an object, the object is similarly deflected toward the tooth tip, but the very small area at the tip of each serration further increases the stress, resulting in even greater penetration into the object. For example, when serration (a) encounters a region of the object, it results in stress concentration, resulting in penetration of the tooth margin into the object; similarly, serrations (b) and (c) encounter additional uncut material as the tooth is driven toward (F_x) and across (F_y) the object; in this manner, serrations result in localized regions of high stress that facilitate cutting through the object. These serrations can be linearly arranged, as they are on most fish teeth, and need not be laterally staggered as they are on a carpenter's wood saw. The latter serrations may reduce the entrapment of cut material from among the serrations. (C) Tiger shark, *Galeocerdo cuvier*, teeth are arranged and shaped in the indicated manner about the palatoquadrate symphysis. Different faces of these teeth may serve different functions. On the notched surface of these teeth, objects encountered at positions (1) or (2) are driven toward the notch (N), which is extremely narrow and thin, consequently increasing the stress in this region. This action serves to cut the material in a manner similar to a notched paper cutter or scissors. If the tooth is moving in the other direction, the object (3) is driven toward the tooth tip and cut in the manner explained above. As the shark swings its jaws from side to side while biting down on a prey item, the different faces of the teeth, which are arranged in a mirror image on the opposite jaw, cut through the prey by both of these methods. Tougher material, such as ligaments, tendons, and bundles of collagen fibers, may be cut more easily on the notched side of the tooth. (D) Fourth upper lateral tooth of *G. cuvier* with the notch (N) indicated.

Despite this extensive attribution of function to form, there are almost no quantitative functional studies of tooth use in cartilaginous fishes, and the mechanics of piercing and cutting are poorly understood (Anderson, 2009; Atkins, 2006; Atkins and Xu, 2005; Cappetta, 1986b, 1987; Whitenack and Motta, 2010). Carcharhinid lower jaw teeth may be used to grasp prey during mandibular elevation, after which the serrated, triangular upper jaw teeth descend and saw through the prey, often facilitated by rapid head shaking (Frazzetta, 1988, 1994; Frazzetta and Prange, 1987; Moss, 1972, 1977b; Motta et al., 1997; Smale et al., 1995; Springer, 1961). Some squaloid sharks have blade-like teeth in both jaws with large, laterally directed cusps that cut through prey during lateral head shaking (Compagno, 1984; Wilga and Motta, 1998a), and upper jaw protrusion in both groups may expose the serrated or blade-like upper teeth, facilitating their unobstructed lateral movement through the prey (Motta and Wilga, 2001). These examples illustrate several key points regarding the fracture of tough, extensible tissues. Bladed edges are clearly better at cutting than blunted edges, but the presence of angled or notched blades dramatically increases cutting efficiency relative to a straight blade. In particular, notched blades can decrease the work to fracture of compliant tissues by up to 60% because notching traps the substrate at the cutting surface, thereby concentrating stress and facilitating material rupture (Abler, 1992; Anderson, 2009; Anderson and LaBarbera, 2008). These effects are apparent at multiple scales, ranging from the single large notch in the teeth of the tiger shark, *Galeocerdo cuvier*, to the serrations (small notches) on the teeth of numerous other species, and can be augmented by behaviors such as lateral head shaking. Drawing a blade across a substrate reduces the force required to initiate downward penetration. This penetration force is minimized for blades with larger radii of curvature and at higher "slice-to-push" ratios (tangential blade movement greater than normal blade movement), as would occur when the teeth are drawn rapidly to the side as they sink into the prey during lateral head shaking (Atkins, 2006).

Whitenack et al. (2010, 2011) conducted a comprehensive analysis of the relationship between tooth morphology, prey type, and cutting mechanics in sharks. Teeth from ten extant species and aluminum casts of teeth from three extinct species were punctured and drawn through five prey items of varying thickness and toughness (three teleosts, one elasmobranch, one crustacean). Significant differences in puncture and draw forces were found among species and prey types, with some species incapable of initiating tissue damage through either puncture or draw on particular prey items. For example, teeth with distally inclined cusps were less effective at initiating puncture, although there was little correlation between puncture performance and prey type, whereas teeth from the knifetooth dogfish, *Scymnodon ringens*, were incapable of cutting via draw on any prey type. Teeth from extinct species (*Cladodus*, *Xenacanthus*, *Hybodus*) performed comparably to those of extant species in puncture but were less effective in draw, and broad triangular teeth were generally less efficient at cutting (require greater force to penetrate).

Nonetheless, no clear relationships between cutting performance and phylogeny, tooth shape, or the presence of serrations were found, suggesting that the functional morphotypes to which shark teeth have long been categorized (e.g., tearing, cutting, cutting–clutching) are not supported by their cutting performance. Furthermore, finite element analysis revealed that shark teeth are structurally strong, and the majority of the teeth have stress patterns consistent with a well-designed cantilever beam. Notches, such as those of the tiger shark, *Galeocerdo cuvier*, result in stress concentration and may serve as a weak point, but they are functionally important for cutting prey during lateral head shaking. It is proposed that frequent tooth replacement in sharks is driven by tooth wear, not tooth failure.

6.6 Summary, Conclusions, and Future Directions

The feeding mechanism of sharks and their relatives displays remarkable diversity, especially when considering the simplicity of this system. Elasmobranchs will hunt by speculation, ambush, stalk, or lure and will scavenge their prey. Elasmobranchs are versatile in how and where they capture their prey; however, the capture kinematics and motor action patterns are very similar among sharks but differ somewhat from those of batoids. Sharks capture their prey by ram, biting, and suction or some combination of the three, whereas batoids primarily use suction-dominated prey capture. The jaw suspension in elasmobranchs plays a significant role in the kinesis of the jaws and consequently the feeding mechanisms. Batoids have a unique suspensory apparatus, the most kinetic jaws, and a highly versatile and protrusible mouth.

The mineralized layer of skeletal elements in elasmobranch feeding mechanisms is load bearing, and the distribution of mineral within these skeletal elements plays a major role in their mechanical function. Although it is unknown whether the material properties of either the mineralized or unmineralized portions of the feeding mechanism vary with feeding ecology, it is clear that the tessellated cartilaginous skeleton is an elegant solution to the need to maintain mechanical integrity while permitting growth in a skeletal system that is incapable of remodeling. This skeletal design represents the highly economical use of structural material (i.e., limited mineral ideally located) and facilitates high-performance feeding where intuition would suggest that a system predominantly composed of pliant, unmineralized cartilage would have compromised performance. Nonetheless, cartilaginous fishes are capable of some of the highest bite forces of any animal, and some species are capable of processing prey items that are harder than their own skeletons. The causative factors of this high-performance feeding—muscle cross-sectional area and mechanical advantage—both demonstrate strong ecomorphological signatures, and recent evidence suggests the correlated evolution of these traits in taxa with above-average feeding performance for their body size. Furthermore, smaller species and juveniles of larger species exhibit positive allometry of bite force due to the benefits of attaining high-performance feeding early in their life history, whereas selection for size-specific performance apparently decreases in large species due to their high absolute bite forces.

Elasmobranch teeth are diverse in structure and arrangement, ranging from the flattened pads of mollusk-crushing batoids to the villiform-like teeth of plankton-feeding whale sharks. Throughout evolution there have been multiple forays into different tooth forms, with a general trend toward increased complexity of the enameloid microstructure in modern sharks. Shark teeth have been ascribed functional roles, but only recently have quantitative analyses revealed that teeth from extinct species perform comparably to those of extant species in puncture but are less effective in draw, and broad triangular teeth are generally less efficient at cutting. There appears to be no clear relationships among cutting performance and phylogeny, tooth shape, or the presence of serrations. Regardless, shark teeth are structurally strong, and it is proposed that frequent tooth replacement in sharks is driven by tooth wear, not tooth failure.

Despite great advances in our understanding of the feeding biology of elasmobranchs, major gaps in our knowledge still remain. Even though there have been consistent and excellent studies on the anatomy of the feeding apparatus and diet, the ethology of predation is less studied, most likely because of the inherent difficulties of *in situ* studies of such large, mobile predators. Surprisingly, rays and skates would seem relatively easy to study but are less understood and investigated than sharks. In particular, studies on batoid feeding mechanics are lacking. We are only beginning to understand how prey capture behavior differs within and among species and the link between feeding behavior and the morphology of the feeding apparatus. Feeding mechanics, particularly those of jaw protrusion, have only been investigated in a handful of taxa representing a few families. The putative functions of jaw protrusion are still elusive, as discussed in the first edition of this chapter (Motta, 2004). Studies of tooth performance and the biomechanical properties of shark biomaterials have only just begun, and this remains an area ripe for quantitative analyses. Finally, perhaps the most challenging task lies in our understanding of the evolution of feeding types in the elasmobranchs, a task we can only accomplish with a thorough understanding and comparative analyses of extant forms.

Acknowledgments

We gratefully acknowledge the editors for the invitation to contribute to this book. Many people and institutions contributed to the research conducted in our laboratories, and their contributions are acknowledged. We thank the anonymous reviewers for providing their usual insightful feedback. The work would not be possible without the generous donations of specimens, facilities, and support by the University of South Florida, Mote Marine Laboratory, and The University of Tampa. During the course of all experiments referred to here by the authors, the animals were treated according to the University of South Florida, The University of Tampa, and Mote Marine Laboratory Institutional Animal Care and Use Committee guidelines. Portions of the research reported were supported by grants from the National Science Foundation to PJM and Robert E. Hueter (DEB 9117371, IBN 9807863, IOS 0640133). This chapter is dedicated to our parents and families for believing in and supporting us throughout our endeavors. They are our source of inspiration.

References

Aalbers, S.A., Vernal, D., and Sepulveda, C.A. (2010). The functional role of the caudal fin in the feeding ecology of the common thresher shark *Alopias vulpinus*. *J. Fish Biol.* 76:1863–1868.

Abd El-Aziz, S.H. (1986). Food and feeding habits of *Raja* species (Batoidei) in the Mediterranean waters of Alexandria. *Bull. Inst. Oceanogr. Fish. (Arab. Repub. Egypt)* 12:265–276.

Abler, W.L. (1992). The serrated teeth of tyrannosaurid dinosaurs and biting structures in other animals. *Paleobiology* 18:161–183.

Aguirre, L.F., Herrel, A., Van Damme, R., and Matthysen, E. (2002). Ecomorphological analysis of trophic niche partitioning in a tropical savanna bat community. *Proc. R. Soc. Lond. B Biol. Sci.* 269:1271–1278.

Aguirre, L.F., Herrel, A., Van Damme, R., and Matthysen, E. (2003). The implications of food hardness for diet in bats. *Func. Ecol.* 17:201–212.

Ajayi, T.O. (1982). Food and feeding habits of *Raja* species (Batoidei) in Carmarthen Bay, Bristol Channel. *J. Mar. Biol. Assoc. U.K.* 62:215–223.

Ajemian, M.J. and Sanford, C.P. (2007). Food capture kinematics in the deep-water chain catshark *Scliorhinus retifer*. *J. Mar. Biol. Assoc. U.K.* 87:1277–1286.

Allis, E.P.J. (1923). The cranial anatomy of *Chlamydoselachus anguineus*. *Acta Zool.* 4:123–221.

Anderson, P.S.L. (2009). The effects of trapping and blade angle of notched dentitions on fracture of biological tissues. *J. Exp. Biol.* 212:3627–3632.

Anderson, P.S.L. and LaBarbera, M. (2008). Functional consequences of tooth design: effects of blade shape on energetics of cutting. *J. Exp. Biol.* 211:3619–3626.

Anderson, R.A., McBrayer, L.D., and Herrel, A. (2008). Bite force in vertebrates: opportunities and caveats for use of a nonpareil whole-animal performance measure. *Biol. J. Linn. Soc. Lond.* 93:709–720.

Andreev, P.S. (2010). Enameloid microstructure of the serrated cutting edges in certain fossil carcharhiniform and lamniform sharks. *Microsc. Res. Tech.* 73:704–713.

Applegate, S.P. (1965). Tooth terminology and variation in sharks with special reference to the sand shark, *Carcharias taurus* Rafinesque. *Contrib. Sci. Mus. Nat. Hist. Los Angeles County* 86:1–18.

Ashhurst, D.E. (2004). The cartilaginous skeleton of an elasmobranch fish does not heal. *Matrix Biol.* 23:15–22.

Atkins, A.G. and Xu, X. (2005). Slicing of soft flexible solids with industrial application. *Int. J. Mech. Sci.* 47:479–492.

Atkins, T. (2006). Optimum blade configurations for the cutting of soft solids. *Eng. Fracture Mech.* 73:2523–2531.

Au, D.W. (1991). Polyspecific nature of tuna schools: shark, dolphin, and seabird associates. *U.S. Fish. Bull.* 89:343–354.

Babel, J.S. (1967). Reproduction, life history, and ecology of the round stingray, _Urolophus halleri_ Cooper. _U.S. Fish. Bull._ 137:76–104.

Barbini, S.A., Scenna, L.B., Figueroa, D.E., Cousseau, M.B., and Díaz de Astarloa, J.M. (2010). Feeding habits of the Magellan skate: effects of sex, maturity stage, and body size on diet. _Hydrobiologia_ 641:275–286.

Belbenoit, P. (1986). Fine analysis of predatory and defensive motor events in _Torpedo marmorata_ (Pisces). _J. Exp. Biol._ 121:197–226.

Belbenoit, P. and Bauer, R. (1972). Video recordings of prey capture behaviour and associated electric organ discharge of _Torpedo marmorata_ (Chondrichthyes). _Mar. Biol._ 17:93–99.

Bell, J.C. and Nichols, J.T. (1921). Notes on the food of Carolina sharks. _Copeia_ 1921:17–20.

Bethea, D.M., Buckel, J.A., and Carlson, J.K. (2004). Foraging ecology of the early life stages of four sympatric shark species. _Mar. Ecol. Prog. Ser._ 268:245–264.

Bigelow, H.B. and Schroeder, W.C. (1948). Fishes of the Western North Atlantic: lancelets, cyclostomes, sharks. _Mem. Sears Found. Mar. Res._ 1(1):1–576.

Bigelow, H.B. and Schroeder, W.C. (1953). Fishes of the Western North Atlantic: sawfishes, guitarfishes, skates, and rays. _Mem. Sears Found. Mar. Res._ 1(2):1–588.

Binder, W.J. and Van Valkenburgh, B.V. (2000). Development of bite strength and feeding behavior in juvenile spotted hyenas (_Crocuta crocuta_). _J. Zool._ 252:273–283.

Botella, H., Valenzuela-Ríos, J.I., and Martínez-Pérez, C. (2009). Tooth replacement rates in early chondrichthyans: a qualitative approach. _Lethaia_ 42(3):365–376.

Bourke, J., Wroe, S., Moreno, K., McHenry, C., and Clausen, P. (2008). Effects of gape and tooth position on bite force and skull stress in the dingo (_Canis lupus dingo_) using a 3-dimensional finite element approach. _PLoS ONE_ 3:e2200.

Branstetter, S. and Stiles, R. (1987). Age and growth of the bull shark, _Carcharhinus leucas_, from the northern Gulf of Mexico. _Environ. Biol. Fish._ 20(3):169–181.

Bray, R.N. and Hixon, M.A. (1978). Night-shocker: predatory behavior of the Pacific electric ray (_Torpedo californica_). _Science_ 200:333–334.

Brunnschweiler, J.M., Andrews, P.L.R., Southall, E.J., Pickering, M., and Sims, D.W. (2005). Rapid voluntary stomach eversion in a free-living shark. _J. Mar. Biol. Assoc. U.K._ 85:1141–1144.

Brunnschweiler, J.M., Nielsen, F., and Motta, P. (2011). _In situ_ observation of stomach eversion in a line-caught shortfin mako (_Isurus oxyrinchus_). _Fish Res._ 109:212–216.

Budker, P. (1971). _The Life of Sharks_. Columbia University Press, New York.

Bullis, H.R. (1961). Observations on the feeding behavior of white-tip sharks on schooling fishes. _Ecology_ 42:194–195.

Cappetta, H. (1986a). Myliobatidae nouveaux (Neoselachii, Batomorphii) de l'Ypresien des Ouled Abdoun, Maroc. _Geol. Palaeontol._ 20:185–207.

Cappetta, H. (1986b). Types dentaires adaptatifs chez les selaciens actuels et post-paleozoiques. _Palaeovertebrata_ 16:57–76.

Cappetta, H. (1987). Chondrichthyes II: Mesozoic and Cenozoic Elasmobranchii. In: Schultze, H.P. (Ed.), _Handbook of Paleoichthyology_, Vol. 3B. Gustav Fischer Verlag, Stuttgart, 193 pp.

Carroll, A.M., Wainwright, P.C., Huskey, S.H., Collar, D.C., and Turingan, R.G. (2004). Morphology predicts suction feeding performance in centrarchid fishes. _J. Exp. Biol._ 207:3873–3881.

Carroll, R.L. (1988). _Vertebrate Paleontology and Evolution_. W.H. Freeman, New York.

Carter, D.R. and Wong, M. (2003). Modelling cartilage mechanobiology. _Phil. Trans. R. Soc. Lond. B. Biol. Sci._ 358:1461–1471.

Casey, J.G. and Pratt, Jr., H.L. (1985). Distribution of the white shark, _Carcharodon carcharias_, in the western North Atlantic. _Mem. South. Calif. Acad. Sci._ 9:2–14.

Castro, J.I. (1996). _The Sharks of North American Waters_. Texas A&M University Press, College Station.

Chapman, D.D. and Gruber, S.H. (2002). A further observation of batoid prey handling by the great hammerhead shark, _Sphyrna mokarran_, upon a spotted eagle ray, _Aetobatus narinari_. _Bull. Mar. Sci._ 70:947–952.

Chen, P., Lin, A.Y.M., Lin, Y.S., Seki, Y., Stokes, A.G., Peyras, J., Olevsky, E.A., Meyers, M.A., and McKittrick, J. (2008). Structure and mechanical properties of selected biological materials. _J. Mech. Behav. Biomed. Mater._ 1:208–226.

Cherel, Y. and Duhamel, G. (2004). Antarctic jaws: cephalopod prey of sharks in Kerguelen waters. _Deep-Sea Res. I_ 50:17–31.

Christiansen, P. and Wroe, S. (2007). Bite forces and evolutionary adaptations to feeding ecology in carnivores. _Ecology_ 88:347–358.

Clark, E. and Kristof, E. (1990). Deep sea elasmobranchs observed from submersibles in Grand Cayman, Bermuda and Bahamas. In: Pratt, Jr., H.L., Gruber, S.H., and Taniuchi, T. (Eds.), _Elasmobranchs as Living Resources: Advances in the Biology, Ecology, Systematics, and the Status of the Fisheries_, NOAA Tech. Rep. NMFS 90. National Oceanic and Atmospheric Administration, Washington, D.C., pp. 269–284.

Clark, E. and Nelson, D.R. (1997). Young whale sharks, _Rhincodon typus_, feeding on a copepod bloom near La Paz, Mexico. _Environ. Biol. Fish._ 50:63–73.

Clement, J.G. (1992). Re-examination of the fine structure of endoskeletal mineralization in Chondrichthyans: implications for growth, ageing, and calcium homeostasis. _Aust. J. Mar. Freshwat. Res._ 43:157–181.

Cliff, G. and Dudley, S.F.J. (1991). Sharks caught in the protective gill nets off Natal, South Africa. 4. The bull shark _Carcharhinus leucas_ Valenciennes. _S. Afr. J. Mar. Sci._ 10:253–270.

Clifton, K.B. and Motta, P.J. (1998). Feeding morphology, diet, and ecomorphological relationships among five Caribbean labrids (Teleostei, Labridae). _Copeia_ 1998:953–966.

Coles, R.J. (1915). Notes on the sharks and rays of Cape Lookout, NC. _Proc. Biol. Soc. Wash._ 28:89–94.

Collins, A.B., Heupel, M.R., Hueter, R.E., and Motta, P.J. (2007). Hard prey specialists or opportunistic generalists? An examination of the diet of the Atlantic cownose ray _Rhinoptera bonasus_. _Mar. Freshwat. Res._ 58:135–144.

Colman, J.G. (1997). A review of the biology and ecology of the whale shark. *J. Fish Biol.* 51:1219–1234.

Compagno, L.J.V. (1970). Systematics of the genus *Hemitriakis* (Selachii: Carcharhinidae) and related genera. *Proc. Calif. Acad. Sci. Ser. 4* 38:63–98.

Compagno, L.J.V. (1977). Phyletic relationships of living sharks and rays. *Am. Zool.* 17:303–322.

Compagno, L.J.V. (1984). *FAO Species Catalogue. Vol. 4. Sharks of the World: An Annotated and Illustrated Catalogue of Shark Species Known to Date. Part 1. Hexanchiformes to Lamniformes.* United Nations Food and Agriculture Organization, Rome, 250 pp.

Compagno, L.J.V. (1988). *Sharks of the Order Carcharhiniformes.* Princeton University Press, Princeton, NJ.

Compagno, L.J.V. (1990). Relationships of the megamouth shark, *Megachasma pelagios* (Lamniformes: Megachasmidae), with comments on its feeding habits. In: Pratt, Jr., H.L., Gruber, S.H., and Taniuchi, T. (Eds.), *Elasmobranchs as Living Resources: Advances in the Biology, Ecology, Systematics, and the Status of the Fisheries,* NOAA Tech. Rep. NMFS 90. National Oceanic and Atmospheric Administration, Washington, D.C., pp. 357–380.

Compagno, L.J.V. (1999). Endoskeleton in sharks, skates, and rays. In: Hamlett, W.C. (Ed.), *The Biology of Elasmobranch Fishes.* The Johns Hopkins University Press, Baltimore, MD, pp. 69–92.

Compagno, L.J.V. (2001). *FAO Species Catalogue for Fishery Purposes. Sharks of the World: An Annotated and Illustrated Catalogue of Shark Species Known to Date.* Vol. 2. *Bullhead, Mackerel and Carpet Sharks (Heterodontiformes, Lamniformes, and Orectolobiformes).* United Nations Food and Agriculture Organization, Rome, 269 pp.

Compagno, L.J.V., Dando, M., and Fowler, S. (2005). *Sharks of the World.* Princeton University Press, Princeton, NJ.

Cortés, E. (1999). Standardized diet compositions and trophic levels of sharks. *ICES J. Mar. Sci.* 56:707–717.

Courtland, H., Wright, G.M., Root, R.G., and DeMont, M.E. (2003). Comparative equilibrium mechanical properties of bovine and lamprey cartilaginous tissues. *J. Exp. Biol.* 206:1397–1408.

Curio, E. (1976). *The Ethology of Predation.* Springer–Verlag, Berlin.

Currey, J. (2008). Collagen and the mechanical properties of bone and calcified cartilage. In: Fratzl, P. (Ed.), *Collagen: Structure and Mechanics.* Springer, New York, pp. 397–420.

Currey, J.D. (1980). Mechanical properties of mollusc shell. In: Vincent, J.F.V. and Currey, J.D. (Eds.), *The Mechanical Properties of Biological Materials.* Press Syndicate of the University of Cambridge, Cambridge, U.K., pp. 75–98.

Currey, J.D. (1987). The evolution of the mechanical properties of amniote bone. *J. Biomech.* 20:1035–1044.

Currey, J.D. (1998). Mechanical properties of vertebrate hard tissues. *Proc. Inst. Mech. Eng. Part H* 212:399–411.

Currey, J.D. and Abeysekera, R.M. (2003). The microhardness and fracture surface of the petrodentine of *Lepidosiren* (Dipnoi), and of other mineralized tissues. *Arch. Oral Biol.* 48:439–470.

Currey, J.D. and Butler, G. (1975). The mechanical properties of bone tissue in children. *J. Bone Joint Surg.* 57:810–814.

Curtin, N.A., Lou, F., and Woledge, R.C. (2010). Sustained performance by red and white muscle fibers from the dogfish *Scyliorhinus canicula. J. Exp. Biol.* 213:1921–1929.

Curtis, T.H., Kelly, J.T., Menard, K.L., Laroche, R.K., Jones, R.E., and Klimley, A.P. (2006). Observations on the behavior of white sharks scavenging from a whale carcass at Point Reyes, California. *Calif. Fish Game* 92:113–124.

Daniel, J.F. (1915). The anatomy of *Heterodontus francisci.* II. The endoskeleton. *J. Morphol.* 26:447–493.

Daniel, J.F. (1934). *The Elasmobranch Fishes.* University of California Press, Berkeley.

De Schepper, N., Van Wassenburgh, S., and Adriaens, D. (2008). Morphology of the jaw system in trichiurids: trade-offs between mouth closing and biting performance. *Zool. J. Linn. Soc.* 152:717–736.

Dean, M.N. and Motta, P.J. (2004a). Anatomy and functional morphology of the feeding apparatus of the lesser electric ray, *Narcine basiliensis* (Elasmobranchii: Batoidea). *J. Morphol.* 262:462–483.

Dean, M.N. and Motta, P.J. (2004b). Feeding behavior and kinematics of the lesser electric ray, *Narcine brasiliensis* (Elasmobranchii: Batoidea). *Zoology* 107:171–189.

Dean, M.N. and Summers, A.P. (2006). Mineralized cartilage in the skeleton of chondrichthyan fishes. *Zoology* 109:164–168.

Dean, M.N., Wilga, C.D., and Summers, A.P. (2005). Eating without hands or tongue: specialization, elaboration and the evolution of prey processing mechanisms in cartilaginous fishes. *Biol. Lett.* 1:357–361.

Dean, M.N., Huber, D.R., and Nance, H.A. (2006). Functional morphology of jaw trabeculation in the lesser electric ray *Narcine brasiliensis,* with comments on the evolution of structural support in the Batoidea. *J. Morphol.* 267:1137–1146.

Dean, M.N., Azizi, E., and Summers, A.P. (2007a). Uniform strain in broad muscles: active and passive effects of the twisted tendon of the ratfish, *Hydrolagus colliei. J. Exp. Biol.* 210:3395–3406.

Dean, M.N., Bizzarro, J.J., and Summers, A.P. (2007b). The evolution of cranial design, diet, and feeding mechanisms in batoids fishes. *Integr. Comp. Biol.* 47:70–81.

Dean, M.N., Ramsay, J.B., and Schaefer, J.T. (2008). Tooth reorientation affects tooth function during prey processing and tooth ontogeny in the lesser electric ray, *Narcine brasiliensis. Zoology* 111:123–134.

Dean, M.N., Mull, C.G., Gorb, S.N., and Summers, A.P. (2009a). Ontogeny of the tesselated skeleton: insight from the skeletal growth of the round stingray *Urobatis halleri. J. Anat.* 215:227–239.

Dean, M.N., Youssefpour, H., Earthman, J.C., Gorb, S.N., and Summers, A.P. (2009b). Micro-mechanics and material properties of the tesselated skeleton of cartilaginous fishes. *Integr. Comp. Biol.* 49:e45.

Dean, M.N., Swanson, B.O., and Summers, A.P. (2009c). Biomaterials: properties, variation and evolution. *Integr. Comp. Biol.* 49:15–20.

Dean, M.N., Socha, J.J., Hall, B.K., and Summers, A.P. (2010). Canaliculi in the tesselated skeleton of cartilaginous fishes. *J. Appl. Ichthyol.* 26:263–267.

DeBeer, G.R. (1932). *Vertebrate Zoology.* Macmillan, New York.

Diamond, J.M. (1985). Filter-feeding on a grand scale. *Nature* 316:679–680.

Dicken, M.L. (2008). First observations of young of the year and juvenile white sharks (*Carcharodon carcharias*) scavenging from a whale carcass. *Mar. Freshwat. Res.* 59:596–602.

Didier, D.A. (1995). Phylogenetic systematics of extant chimaeroid fishes (Holocephali, Chimaeroidei). *Am. Mus. Novit.* 3119:1–86.

Dingerkus, G., Seret, B., and Guilbert, E. (1991). Multiple prismatic calcium phosphate layers in the jaws of present-day sharks (Chondrichthyes; Selachii). *Experientia* 47:38–40.

Domeier, M.L. and Nasby-Lucas, N. (2008). Migration patterns of white sharks *Carcharodon carcharias* tagged at Guadalupe Island, Mexico, and identification of an eastern Pacific shared offshore foraging area. *Mar. Ecol. Prog. Ser.* 370:221–237.

Dudley, S.F.J., Anderson-Reade, M.D., Thompson, G.S., and McMullen, P.B. (2000). Concurrent scavenging of a whale carcass by great white sharks, *Carcharodon carcharias*, and tiger sharks, *Galeocerdo cuvier*. *Fish. Bull.* 98:646–649.

Duffin, C.J. and Cuny, G. (2008). *Carcharopsis prototypus* and the adaptations of single crystallite enameloid in cutting dentitions. *Acta Geol. Polon.* 58:181–184.

Dumont, E.R. and Herrel, A. (2003). The effect of gape angle and bite point on bite force in bats. *J. Exp. Biol.* 206:2117–2123.

Dunn, M.R., Griggs, L., Forman, J., and Horn, P. (2010). Feeding habits and niche separation among the deep-sea chimaeroid fishes *Harriotta raleighana*, *Hydrolagus bemisi*, and *Hydrolagus novaezealandiae*. *Mar. Ecol. Prog. Ser.* 407:209–225.

Eames, B.F., Allen, N., Young, J., Kaplan, A., Helms, J.A., and Schneider, R.A. (2007). Skeletogenesis in the swell shark *Cephaloscyllium ventriosum*. *J. Anat.* 210:542–554.

Ebert, D.A. (1991). Observations on the predatory behaviour of the sevengill shark *Notorynchus cepedianus*. *S. Afr. J. Mar. Sci.* 11:455–465.

Ebert, D.A., Cowley, P.D., and Compagno, L.J.V. (1991). A preliminary investigation of the feeding ecology of skates (Batoidea: Rajidae) off the West coast of southern Africa. *S. Afr. J. Mar. Sci.* 10:71–81.

Edgeworth, F.H. (1935). *Cranial Muscles of Vertebrates*. Cambridge University Press, Cambridge, U.K.

Edmonds, M.A., Motta, P.J., and Hueter, R.E. (2001). Food capture kinematics of the suction feeding horn shark *Heterodontus francisci*. *Environ. Biol. Fish.* 62:415–427.

Edwards, R.R.C. (1980). Aspects of the population dynamics and ecology of the white spotted stingray, *Urolophus paucimaculatus* Dixon, in Port Phillip Bay, Victoria. *Aust. J. Mar. Freshwat. Res.* 31:459–467.

Eibl-Eibesfeldt, I. and Hass, H. (1959). Erfahrungen mit Haien. *Z. Tierpsychol.* 16:733–746.

Ellis, J.R. and Shackley, S.E. (1995). Ontogenetic changes and sexual dimorphism in the head, mouth and teeth of the lesser spotted dogfish. *J. Fish Biol.* 47:155–164.

Erickson, G.M., Cavanese III, J.C., and Keaveny, T.M. (2002). Evolution of the biomechanical material properties of the femur. *Anat. Rec.* 268:115–124.

Erickson, G.M., Lappin, A.K., and Vliet, K.A. (2003). The ontogeny of bite-performance in American alligator (*Alligator mississippiensis*). *J. Zool.* 260:317–327.

Estrada, J.A., Rice, A.N., Natanson, L.J., and Skomal, G.B. (2006). Use of isotopic analysis of vertebrae in reconstructing ontogenetic feeding ecology in white sharks. *Ecology* 87:829–834.

Fahle, S.R. and Thomason, J.C. (2008). Measurement of jaw viscoelasticity in newborn and adult lesser spotted dogfish *Scyliorhinus canicula* (L., 1758). *J. Fish Biol.* 72:1553–1557.

Feduccia, A. and Slaughter, B.H. (1974). Sexual dimorphism in skates (Rajidae) and its possible role in differential niche utilization. *Evolution* 28:164–168.

Fergusson, I.K., Compagno, L.J.V., and Marks, M.A. (2000). Predation by white sharks *Carcharodon carcharias* (Chondrichthyes: Lamnidae) upon chelonians, with new records from the Mediterranean Sea and a first record of the ocean sunfish *Mola mola* (Osteichthyes: Molidae) as stomach contents. *Environ. Biol. Fish.* 58:447–453.

Ferrara, T.L., Clausen, P., Huber, D.R., McHenry, C.R., Peddemors, V., and Wroe, S. (2011). Mechanics of biting in great white and sandtiger sharks. *J. Biomech.* 44(3):430–435.

Ferry-Graham, L.A. (1997a). Effects of prey size and elusivity on prey capture kinematics in leopard sharks, *Triakis semifasciata*. *Am. Zool.* 37:82A.

Ferry-Graham, L.A. (1997b). Feeding kinematics of juvenile swellsharks, *Cephaloscyllium ventriosum*. *J. Exp. Biol.* 200:1255–1269.

Ferry-Graham, L.A. (1998a). Feeding kinematics of hatchling swellsharks, *Cephaloscyllium ventriosum* (Scyliorhinidae): the importance of predator size. *Mar. Biol.* 131:703–718.

Ferry-Graham, L.A. (1998b). Effects of prey size and mobility on prey-capture kinematics in leopard sharks, *Triakis semifasciata*. *J. Exp. Biol.* 201:2433–2444.

Fisher, R.A., Call, G.C., and Grubbs, R.D. (2011). Cownose ray (*Rhinoptera bonasus*) predation relative to bivalve ontogeny. *J. Shellfish Res.* 30(1):187–196.

Fouts, W.R. (1995). The Feeding Behavior and Associated Ambush Site Characteristics of the Pacific Angel Shark, *Squatina californica*, at Santa Catalina Island, California, doctoral thesis, California State University.

Fouts, W.R. and Nelson, D.R. (1999). Prey capture by the Pacific angel shark, *Squatina californica*: visually mediated strikes and ambush-site characteristics. *Copeia* 1999:304–312.

Francis, M.P. and Duffy, C. (2002). Distribution, seasonal abundance and bycatch of basking sharks (*Cetorhinus maximus*) in New Zealand, with observations on their winter habitat. *Mar. Biol.* 140:831–842.

Frazzetta, T.H. (1988). The mechanics of cutting and the form of shark teeth (Chondrichthyes, Elasmobranchii). *Zoomorphology* 108:93–107.

Frazzetta, T.H. (1994). Feeding mechanisms in sharks and other elasmobranchs. *Adv. Comp. Environ. Physiol.* 18:31–57.

Frazzetta, T.H. and Prange, C.D. (1987). Movements of cephalic components during feeding in some requiem sharks (Carcharhiniformes: Carcharhinidae). *Copeia* 1987:979–993.

Friedman, M. (2009). Ecomorphological selectivity among marine teleost fishes during the end-Cretaceous extinction. *Proc. Natl. Acad. Sci. USA* 106:5218–5223.

Funicelli, N.A. (1975). Taxonomy, Feeding, Limiting Factors and Sex Ratios of *Dasyatis sabina, Dasyatis americana, Dasyatis sayi,* and *Narcine brasiliensis,* doctoral dissertation, University of Southern Mississippi, Hattiesburg.

Gadow, H. (1888). On the modifications of the first and second visceral arches, with special reference to the homologies of the auditory ossicles. *Phil. Trans. R. Soc. Lond. B. Biol. Sci.* 179:451–485.

Gans, C. and Gaunt, A.S. (1991). Muscle architecture in relation to function. *J. Biomech.* 24:53–65.

Gerry, S.P., Ramsay, J.B., Dean, M.N., and Wilga, C.D. (2008). Evolution of asynchronous motor activity in paired muscles: effects of ecology, morphology, and phylogeny. *Integr. Comp. Biol.* 48:272–282.

Gilbert, P.W. (1962). The behavior of sharks. *Sci. Am.* 207:60–68.

Gillis, J.A. and Donoghue, C.J. (2007). The homology and phylogeny of chondrichthyan tooth enameloid. *J. Morphol.* 268:33–49.

Gohar, H.A.F. and Mazhar, F.M. (1964). The internal anatomy of Selachii from the north western Red Sea. *Publ. Mar. Biol. Stn. Ghardaqa, Red Sea* 13:145–240.

Goitein, R.F., Torres, S., and Signorini, C.E. (1998). Morphological aspects related to feeding of two marine skates *Narcine brasiliensis* and *Rhinobatos horkelli* Muller and Henle. *Acta Sci.* 20:165–169.

Gonzalez-Isais, M. (2003). Anatomical comparison of the cephalic musculature of some members of the superfamily Myliobatoidea (Chondrichthyes): implications for evolutionary understanding. *Anat. Rec.* 271A:259–272.

Goo, B.Y., Dean, M.N., Huber, D.R., and Summers, A.P. (2010). Jaw morphology and structure in lamniform sharks. *Integr. Comp. Biol.* 50:e234.

Goodey, T. (1910). A contribution to the skeletal anatomy of the frilled shark, *Chlamydoselachus anguineus* (Gar.). *Proc. Zool. Soc. Lond.* 2:540–571.

Goto, T. (2001). Comparative anatomy, phylogeny, and cladistic classification of the order Orectolobiformes (Chondrichthyes, Elasmobranchii). *Mem. Grad. Sch. Fish. Sci. Hokkaido Univ.* 48:1–100.

Gray, A.E., Mulligan, T.J., and Hannah, R.W. (1997). Food habits, occurrence, and population structure of the bat ray, *Myliobatis californica,* in Humboldt Bay, California. *Environ. Biol. Fish.* 49:227–238.

Gregory, M.R., Balance, P.F., Gibson, G.W., and Ayling, A.M. (1979). On how some rays (Elasmobranchia) excavate feeding depressions by jetting water. *J. Sed. Petrol.* 49:1125–1130.

Grogan, E.D. and Lund, R. (2000). *Debeerius ellefseni* (fam. nov., gen. nov., spec. nov.), an autodiastylic chondrichthyan from the Mississippian Bear Gulch Limestone of Montana (USA), the relationships of the Chondrichthyes, and comments on gnathostome evolution. *J. Morphol.* 243:219–245.

Grogan, E.D. and Lund, R. (2004). The origin and relationships of early Chondrichthyes. In: Carrier, J.C., Musick, J.A., and Heithaus, M.R. (Eds.), *Biology of Sharks and Their Relatives.* CRC Press, Boca Raton, FL, pp. 3–31.

Grogan, E.D., Lund, R., and Didier, D. (1999). Description of the chimaerid jaw and its phylogenetic origins. *J. Morphol.* 239:45–59.

Gudger, E.W. (1907). A note on the hammerhead shark (*Sphyrna zygaena*) and its food. *Science* 25:1005.

Gudger, E.W. (1941a). The food and feeding habits of the whale shark *Rhineodon typus. J. Elisha Mitchell Sci. Soc.* 57:57–72.

Gudger, E.W. (1941b). The feeding organs of the whale shark, *Rhineodon typus. J. Morphol.* 68:81–99.

Gunn, J.S., Stevens, J.D., Davis, T.L.O., and Norman, B.M. (1999). Observations on the short-term movements and behaviour of whale sharks (*Rhincodon typus*) at Ningaloo Reef, Western Australia. *Mar. Biol.* 135:553–559.

Habegger, M.L., Motta, P.J., Huber, D.R., and Dean, M.N. (In review). Feeding biomechanics in bull sharks (*Carcharhinus leucas*) during ontogeny. *Zoology.*

Habelitz, S., Marshall, S.J., Marshall, Jr., G.W., and Balooch, M. (2001). Mechanical properties of human dental enamel on the nanometre scale. *Arch. Oral Biol.* 46:173–83.

Heemstra, P.C. and Smith, M.M. (1980). Hexatrygonidae, a new family of stingrays (Myliobatiformes: Batoidae) from South Africa, with comments on the classification of batoid fishes. *Ichthyol. Bull. J.L.B. Smith Inst. Ichthyol.* 43:1–17.

Heithaus, M.R. (2001). The biology of tiger sharks, *Galeocerdo cuvier,* in Shark Bay, Western Australia: sex ratio, size distribution, diet, and seasonal changes in catch rates. *Environ. Biol. Fish.* 61:25–36.

Heithaus, M.R., Marshall, G.J., Buhleier, B., and Dill, L.M. (2001). Employing CritterCam to study habitat use and behavior of large sharks. *Mar. Ecol. Prog. Ser.* 209:307–310.

Heithaus, M.R., Dill, L.M., Marshall, G.J., and Buhleier, B. (2002a). Habitat use and foraging behavior of tiger sharks (*Galeocerdo cuvier*) in a seagrass ecosystem. *Mar. Biol.* 140:237–248.

Heithaus, M.R., Frid, A., and Dill, L.M. (2002b). Shark-inflicted injury frequencies, escape ability, and habitat use of green and loggerhead turtles. *Mar. Biol.* 140:229–236.

Henderson, A.C., Flannery, K., and Dunne, J. (2001). Observations on the biology and ecology of the blue shark in the North-East Atlantic. *J. Fish Biol.* 58:1347–1358.

Herman, J., Hovestadt-Euler, M., Hovestadt, D.C., and Stehmann, M. (1995). Contributions to the study of the comparative morphology of teeth and other relevant ichthyodorulites in living supra-specific taxa of Chondrichthyan fishes. *Biologie* 65:237–307.

Hernandez, L.P. and Morgan, R.J. (2009). Size and distribution of muscle fiber types within chondrichthyan muscles. *Integr. Comp. Biol.* 49:E242.

Hernandez, L.P. and Motta, P.J. (1997). Trophic consequences of differential performance: ontogeny of oral jaw-crushing performance in the sheepshead, *Archosargus probatocephalus* (Teleostei, Sparidae). *J. Zool.* 243:737–756.

Herrel, A. and Gibb, A.C. (2006). Ontogeny of performance in vertebrates. *Physiol. Biochem. Zool.* 79:1–6.

Herrel, A. and O'Reilly, J.C. (2006). Ontogenetic scaling of bite force in lizards and turtles. *Physiol. Biochem. Zool.* 79:31–42.

Herrel, A., Spithoven, L., Van Damme, R., and De Vree, F. (1999). Sexual dimorphism of head size in *Gallotia galloti*: testing the niche divergence hypothesis by functional analyses. *Func. Ecol.* 13:289–297.

Herrel, A., Grauw, E., and Lemos-Espinal, J.A. (2001a). Head shape and bite performance in xenosaurid lizards. *J. Exp. Zool.* 290:101–107.

Herrel, A., Van Damme, R., Vanhooydonck, B., and De Vree, F. (2001b). The implications of bite force for diet in two species of lacertid lizards. *Can. J. Zool.* 79:662–670.

Herrel, A., O'Reilly, J.C., and Richmond, A.M. (2002). Evolution of bite performance in turtles. *J. Evol. Biol.* 15:1083–1094.

Herrel, A., Vanhooydonck, B., and Van Damme, R. (2004). Omnivory in lacertid lizards: adaptive evolution or constraint. *J. Evol. Biol.* 17:974–984.

Herrel, A., Podos, J., Huber, S.K., and Hendry, A.P. (2005). Evolution of bite force in Darwin's finches: a key role for head width. *J. Evol. Biol.* 18:669–675.

Hess, P.W. (1961). Food habits of two dasyatid rays in Delaware Bay. *Copeia* 1961:239–241.

Heupel, M.R. and Bennett, M.B. (1998). Observations on the diet and feeding habits of the epaulette shark, *Hemiscyllium ocellatum*, on Heron Island Reef, Great Barrier Reef. *Aust. Mar. Freshwat. Res.* 49:753–756.

Heyman, W.D., Graham, R.T., Kjerfve, B., and Johannes, R.E. (2001). Whale sharks *Rhincodon typus* aggregate to feed on fish spawn in Belize. *Mar. Ecol. Prog. Ser.* 215:275–282.

Hines, A.H., Whitlatch, R.B., Thrush, S.F., Hewitt, J.E., Cummings, V.J., Dayton, P.K., and Legendre, P. (1997). Nonlinear foraging response of a large marine predator to benthic prey: eagle ray pits and bivalves in a New Zealand sandflat. *J. Exp. Mar. Biol. Ecol.* 216:191–210.

Hobson, E.S. (1963). Feeding behavior in three species of sharks. *Pac. Sci.* 17:171–194.

Hoffmayer, E.R., Franks, J.S., Driggers III, W.B., Oswald, K.J., and Quattro, J.M. (2007). Observations of a feeding aggregation of whale sharks, *Rhincodon typus*, in the north central Gulf of Mexico. *Gulf Caribb. Res.* 19:1–5.

Hoh, J.F.Y. (2002). 'Superfast' or masticatory myosin and the evolution of jaw-closing muscles of vertebrates. *J. Exp. Biol.* 205:2203–2210.

Holden, M.J. and Tucker, R.N. (1974). The food of *Raja clavata* Linnaeus 1758, *Raja montagui* Fowler 1910, *Raja naevus* Muller and Henle 1841 and *Raja brachyura* Lafont 1873 in British waters. *J. Cons. Int. Explor. Mer.* 35:189–193.

Holland, K.N., Wetherbee, B.M., Lowe, C.G., and Meyer, C.G. (1999). Movements of tiger sharks (*Galeocerdo cuvier*) in coastal Hawaiian waters. *Mar. Biol.* 134:665–673.

Hotton III, N. (1952). Jaws and teeth of American xenacanth sharks. *J. Paleontol.* 26:489–500.

Howard, J.D., Mayou, T.V., and Heard, R.W. (1977). Biogenic sedimentary structures formed by rays. *J. Sed. Petrol.* 47:339–346.

Hubbell, G. (1996). Using tooth structure to determine the evolutionary history of the white shark. In: Klimley, A.P. and Ainley, D.G. (Eds.), *Great White Sharks: The Biology of Carcharodon carcharias*. Academic Press, New York, pp. 9–18.

Huber, D.R. (2006). Cranial Biomechanics and Feeding Performance in Sharks, doctoral dissertation, University of South Florida, Tampa.

Huber, D.R. and Motta, P.J. (2004). Comparative analysis of methods for determining bite force in the spiny dogfish *Squalus acanthias*. *J. Exp. Zool.* 301A:26–37.

Huber, D.R., Eason, T.G., Hueter, R.E., and Motta, P.J. (2005). Analysis of the bite force and mechanical design of the feeding mechanism of the durophagous horn shark *Heterodontus francisci*. *J. Exp. Biol.* 208:3553–3571.

Huber, D.R., Weggelaar, C.L., and Motta, P.J. (2006). Scaling of bite force in the blacktip shark *Carcharhinus limbatus*. *Zoology* 109:109–119.

Huber, D.R., Dean, M.N., and Summers, A.P. (2008). Hard prey, soft jaws, and the ontogeny of feeding biomechanics in the spotted ratfish *Hydrolagus colliei*. *J. Roy. Soc. Interface* 5:941–952.

Huber, D.R., Claes, J.M., Mallefet, J., and Herrel, A. (2009). Is extreme bite performance associated with extreme morphologies in sharks? *Physiol. Biochem. Zool.* 82:20–28.

Hudson, N.J., Bennett, M.B., and Franklin, C.E. (2004). Effect of aestivation on long bone mechanical properties in the green-striped burrowing frog, *Cyclorana alboguttata*. *J. Exp. Biol.* 207:475–482.

Hueter, R.E., Tyminski, J.P., and de la Parra, R. (2008). The geographic movements of whale sharks tagged with pop-up archival tags off Quintana Roo, Mexico. In: *Proceedings of the Second International Whale Shark Conference*, Holbox, Quintana Roo, Mexico, July 15–20.

Husak, J.F., Lappin, A.K., Fox, S.F., and Lemos-Espinal, J.A. (2006). Bite-force performance predicts dominance in male venerable collared lizards (*Crotaphytus antiquus*). *Copeia* 2006:301–306.

Huxley, T.H. (1876). Contributions to morphology. Ichthyopsida No 1. On *Ceratodus forsteri*, with observations on the classification of fishes. *Proc. Zool. Soc. Lond.* 1876:24–59.

Jagnandan, K. and Huber, D. (2010). Structural and material properties of the jaws of the lemon shark *Negaprion brevirostris* and horn shark *Heterodontus francisci*. *Fla. Sci.* 73:38.

James, W.W. (1953). The succession of teeth in elasmobranchs. *Proc. Zool. Soc. Lond.* 123:419–475.

Jones, E.C. (1971). *Isitius brasiliensis*, a squaloid shark, the probable of crater wounds on fishes and cetaceans. *U.S. Fish. Bull.* 69:791–798.

Kajiura, S.M. and Tricas, T.C. (1996). Seasonal dynamics of dental sexual dimorphism in the Atlantic stingray, *Dasyatis sabina*. *J. Exp. Biol.* 199:2297–2306.

Kemp, N.S. and Westrin, S.K. (1979). Ultrastructure of calcified cartilage in the endoskeletal tesserae of sharks. *J. Morphol.* 160:75–102.

Kemp, T.J., Bachus, K.N., Nairn, J.A., and Carrier, D.R. (2005). Functional trade-offs in the limb bones of dogs selected for running versus fighting. *J. Exp. Biol.* 208:3475–3482.

Kittiphattanabawon, P., Benjakul, S., Visessanguan, W., and Shahidi, F. (2010). Isolation and characterization of collagen from the cartilages of brownbanded bamboo shark (*Chiloscyllium punctatum*) and blacktip shark (*Carcharhinus limbatus*). *LWT Food Sci. Technol.* 43:792–800.

Klimley, P.A. (1994). The predatory behavior of the white shark. *Am. Sci.* 82:122–133.

Klimley, P.A., Pyle, P., and Anderson, S.D. (1996). The behavior of white sharks and their pinniped prey during predatory attacks. In: Klimley, A.P. and Ainley, D.G. (Eds.), *Great White Sharks: The Biology of Carcharodon carcharias*. Academic Press, New York, pp. 175–191.

Klimley, P.A., Leboeuf, B.J., Cantara, K.M., Richert, J.E., Davis, S.F., Van Sommeran, S., and Kelly, J.T. (2001). The hunting strategy of white sharks (*Carcharodon carcharias*) near a seal colony. *Mar. Biol.* 138:617–636.

Kolmann, M.A. and Huber, D.R. (2009). Scaling of feeding biomechanics in the horn shark *Heterodontus francisci*: ontogenetic constraints on durophagy. *Zoology* 112:351–361.

Kyne, P.M. and Simpfendorfer, C. (2010). Deepwater chondrichthyans. In: Carrier, J.C. et al. (Eds.), *Sharks and Their Relatives II: Biodiversity, Adaptive Physiology, and Conservation*. CRC Press, Boca Raton, FL, pp. 37–113.

Laasanen, M.S., Saarakkala, S., and Toyras, J. (2003). Ultrasound indentation of bovine knee articular cartilage *in situ*. *J. Biomech.* 36:1259–1267.

Lappin, A.K. and Husak, J.F. (2005). Weapon performance, not size, determines mating success and potential reproductive output in the collared lizard (*Crotaphytus collaris*). *Am. Nat.* 166:426–436.

Lauder, G.V. (1980). The suction feeding mechanism in sunfishes (*Lepomis*): an experimental analysis. *J. Exp. Biol.* 88:49–72.

Lauder, G.V. (1985). Aquatic feeding in lower vertebrates. In: Hildebrand, M., Bramble, D.M., Liem, K.F., and Wake, D.B. (Eds.), *Functional Vertebrate Morphology*. Belknap Press, Cambridge, MA, pp. 210–229.

LeBoeuf, B.J., McCosker, J.E., and Hewitt, J. (1987). Crater wounds on northern elephant seals: the cookiecutter shark strikes again. *U.S. Fish. Bull.* 85:387–392.

Liem, K.F. (1978). Modulatory multiplicity in the functional repertoire of the feeding mechanisms in cichlid fishes. *J. Morphol.* 158:323–360.

Liem, K.F. (1993). Ecomorphology of the teleostean skull. In: Hanken, J. and Hall, B.K. (Eds.), *The Skull: Functional and Evolutionary Mechanisms*, Vol. 3. University of Chicago Press, Chicago, pp. 422–452.

Liem, K.F. and Summers, A.P. (1999). Muscular system: gross anatomy and functional morphology of muscles. In: Hamlett, W.C. (Ed.), *Sharks, Skates, and Rays: The Biology of Elasmobranch Fishes*. The Johns Hopkins Press, Baltimore, MD, pp. 93–114.

Liem, K.F., Bemis, W.E., Walker, Jr., W.F., and Grande, L. (2001). *Functional Anatomy of the Vertebrates: An Evolutionary Perspective*. Harcourt, New York.

Lightoller, G.H.S. (1939). Probable homologues: a study of the comparative anatomy of the mandibular and hyoid arches and their musculature. Part I. Comparative morphology. *Trans. Zool. Soc. Lond.* 24:349–444.

Long, D.J. and Jones, R.E. (1996). White shark predation and scavenging on cetaceans in the eastern North Pacific Ocean. In: Klimley, A.P. and Ainley, D.G. (Eds.), *Great White Sharks: The Biology of Carcharodon carcharias*. Academic Press, New York, pp. 293–307.

Long, J.A. (1995). *The Rise of Fishes*. The Johns Hopkins University Press, Baltimore, MD.

Lou, F., Curtin, N.A., and Woledge, R.C. (2002). Isometric and isovelocity contractile performance of red muscle fibers from the dogfish *Scyliorhinus canicula*. *J. Exp. Biol.* 205:1585–1595.

Lowe, C.G. (1991). The *in situ* Feeding Behavior and Associated Electric Organ Discharge of the Pacific Electric Ray, *Torpedo californica*, master's thesis, The California State University, Long Beach.

Lowe, C.G., Bray, R.N., and Nelson, D.R. (1994). Feeding and associated electrical behavior of the Pacific electric ray *Torpedo californica* in the field. *Mar. Biol.* 120:161–169.

Lowe, C.G., Wetherbee, B.M., Crow, G.L., and Tester, A.L. (1996). Ontogenetic dietary shifts and feeding behavior of the tiger shark, *Galeocerdo cuvier*, in Hawaiian waters. *Environ. Biol. Fish.* 47:203–211.

Lowe, C.G., Wetherbee, B.M., Holland, K.N., and Meyer, C.G. (2003). Movement patterns of tiger and Galapagos sharks around French Frigate Shoals, Hawaii. In *Abstracts of the American Society of Ichthyologists and Herpetologists Joint Meeting*, Manaus, Brazil, June 26–July 1.

Lowry, D. and Motta, P.J. (2007). Ontogeny of feeding behavior and cranial morphology in the whitespotted bamboo shark *Chiloscyllium plagiosum*. *Mar. Biol.* 151:2013–2023.

Lowry, D. and Motta, P.J. (2008). Relative importance of growth and behavior to elasmobranch suction-feeding performance over early ontogeny. *J. Roy. Soc. Interface* 5:641–652.

Lowry, D., Motta, P.J., and Hueter, R.E. (2007). The ontogeny of feeding behavior and cranial morphology in the leopard shark *Triakis semifasciata* (Girard 1854): a longitudinal perspective. *J. Exp. Mar. Biol. Ecol.* 341:153–167.

Lucifora, L.O., Valero, J.L., Bremec, C.S., and Lasta, M.L. (2000). Feeding habits and prey selection by the skate *Dipterus chilensis* (Elasmobranchii: Rijidae) from the South-Western Atlantic. *J. Mar. Biol. Assoc. U.K.* 80:953–954.

Lucifora, L.O., Menni, R.C., and Escalante, A.H. (2001). Analysis of dental insertion angles in the sand tiger shark, *Carcharias taurus* (Chondrichthyes: Lamniformes). *Cybium* 25:23–31.

Luer, C.A., Blum, P.C., and Gilbert, P.W. (1990). Rate of tooth replacement in the nurse shark, *Ginglymostoma cirratum*. *Copeia* 1990:182–191.

Lund, R. and Grogan, E.D. (1997). Relationships of the Chimaeriformes and the basal radiation of the Chondrichthyes. *Rev. Fish Biol. Fisher.* 7:65–123.

Luther, A. (1909). Untersuchungen über die vom n. trigeminus innervierte Muskulatur der Selachier (Haie und Rochen) unter Berücksichtigung ihrer Beziehungen zu benachbarten Organen. *Acta Soc. Sci. Fenn.* 36:1–176.

Maas, H., Baan, G.C., and Huijing, P.A. (2004). Muscle force is determined by muscle position relative position: isolated effects. *J. Biomech.* 37:99–110.

MacPherson, E. (1980). Food and feeding of *Chimaera monstrosa*, Linnaeus, 1758, in the western Mediterranean. *ICES J. Mar. Sci.* 39:26–29.

Mahoney, E., Holt, A., Swain, M., and Kikpatrick, N. (2000). The hardness and modulus of elasticity of primary molar teeth: an ultra-micro-indentation study. *J. Dent.* 28:589–594.

Maisey, J.G. (1980). An evaluation of jaw suspension in sharks. *Am. Mus. Novit.* 2706:1–17.

Maisey, J.G. (2008). The postorbital palatoquadrate articulation in elasmobranchs. *J. Morphol.* 269:1022–1040.

Maisey, J.G. and de Carvalho, M.R. (1997). A new look at old sharks. *Nature* 385:779–780.

Mallat, J. (1996). Ventilation and the origin of jawed vertebrates: a new mouth. *Zool. J. Linn. Soc.* 117:329–404.

Mara, K.R. (2010). Evolution of the Hammerhead Cephalofoil: Shape Change, Space Utilization, and Feeding Biomechanics in Hammerhead Sharks (Sphyrnidae), doctoral dissertation, University of South Florida, Tampa.

Mara, K.R., Motta, P.J., and Huber, D.R. (2010). Bite force and performance in the durophagous bonnethead shark *Sphyrna tiburo. J. Exp. Zool.* 313A:95–105.

Marinelli, W. and Strenger, A. (1959). *Vergleichende Anatomie und Morphologie der Wirbeltiere.* III. *Lieferung (Squalus acanthias).* Franz Deuticke, Vienna.

Marion, G.E. (1905). Mandibular and pharyngeal muscles of acanthias and raia. *Tufts Coll. Stud.* 2:1–34.

Markaida, U. and Sosa-Nishizaki, O. (2010). Food and feeding habits of the blue shark *Prionace glauca* caught off Ensenada, Baja California, Mexico, with a review on its feeding. *J. Mar. Biol. Assoc. U.K.* 90:977–994.

Märkel, V.K. and Laubier, L. (1969). Zum Zahnerzatz bei Elasmobranchiern. *Zool. Beiträge* 15:41–44.

Martin, R.A. and Naylor, G.J.P. (1997). Independent origins of filter-feeding in megamouth and basking sharks (order Lamniformes) inferred from phylogenetic analysis of cytochrome *b* gene sequences. In: Yano, K., Morrissey, J.F., Yabumoto, Y., and Nakaya, K. (Eds.), *Biology of Megamouth Shark.* Tokai University Press, Tokyo, pp. 39–50.

Martin, R.A., Hammerschlag, N., Collier, R.S., and Fallows, C. (2005). Predatory behavior of white sharks (*Carcharodon carcharias*) at Seal Island, South Africa. *J. Mar. Biol. Assoc. U.K.* 85:1121–1135.

Maschner, Jr., R.P. (2000). Studies of the Tooth Strength of the Atlantic Cow-Nose Ray, *Rhinoptera bonasus,* master's thesis, California State Polytechnic University, Pomona.

Matott, M.P., Motta, P.J., and Hueter, R.E. (2005). Modulation in feeding kinematics and motor pattern of the nurse shark *Ginglymostoma cirratum. Environ. Biol. Fish.* 74:163–174.

Matthews, L.H. and Parker, H.W. (1950). Notes on the anatomy and biology of the basking shark *Cetorhinus maximus* (Gunner). *Proc. Zool. Soc. Lond.* 120:535–576.

Mauchline, J. and Gordon, J.D.M. (1983). Diets of the sharks and chimaeroids of the Rockall Trough, northeastern Atlantic Ocean. *Mar. Biol.* 75:269–278.

McCosker, J.E. (1985). White shark attack behavior: observations of and speculations about predator and prey strategies. *Mem. South. Calif. Acad. Sci.* 9:123–135.

McCourt, R.M. and Kerstitch, A.N. (1980). Mating behavior and sexual dimorphism in dentition in the stingray *Urolophus concentricus* from the Gulf of California. *Copeia* 1980:900–901.

McEachran, J.D. (1977). Reply to "sexual dimorphism in skates (Rajidae)." *Evolution* 31:218–220.

McEachran, J.D., Dunn, K.A., and Miyake, T. (1996). Interrelationships of the batoid fishes (Chondrichthyes: Batoidea). In: Johnson, G.D. (Ed.), *Interrelationships of Fishes.* Academic Press, New York, pp. 63–84.

McLaughlin, R.H. and O'Gower, A.K. (1971). Life history and underwater studies of a heterodont shark. *Ecol. Monogr.* 41:271–289.

Meyers, J.J., Herrel, A., and Birch, J. (2002). Scaling of morphology, bite force and feeding kinematics in an iguanian and scleroglossan lizard. In: Aerts, P., D'Aout, K., Herrel, A., and Van Damme, R. (Eds.), *Topics in Functional and Ecological Vertebrate Morphology.* Shaker Publishing, The Netherlands, pp. 47–62.

Meyers, M.A., Chen, P., Lin, A.Y., and Seki, Y. (2008). Biological materials: structure and mechanical properties. *Prog. Mater. Sci.* 53:1–206.

Michaelson, D.M., Sternberg, D., and Fishelson, L. (1979). Observations on feeding, growth and electric discharge of newborn *Torpedo ocellata* (Chondrichthyes, Batoidei). *J. Fish Biol.* 15:159–163.

Miyake, T. and McEachran, J.D. (1991). The morphology and evolution of the ventral gill arch skeleton in batoid fishes (Chondrichthyes: Batoidea). *Zool. J. Linn. Soc.* 102:75–100.

Miyake, T., McEachran, J.D., and Hall, B.K. (1992). Edgeworth's legacy of cranial muscle development with an analysis of muscles in the ventral gill arch region of batoid fishes (Chondrichthyes: Batoidea). *J. Morphol.* 212:213–256.

Morrissey, J.F. (1991). Home range of juvenile lemon sharks. In: Gruber, S.H. (Ed.), *Discovering Sharks.* American Littoral Society, Highlands, NJ, pp. 85–86.

Moss, M.L. (1977a). Skeletal tissues in sharks. *Am. Zool.* 17:335–342.

Moss, S.A. (1965). The Feeding Mechanisms of Three Sharks: *Galeocerdo cuvieri* (Peron & LeSueur), *Negaprion brevirostris* (Poey), and *Ginglymostoma cirratum* (Bonnaterre), doctoral dissertation, Cornell University, Ithaca, NY.

Moss, S.A. (1967). Tooth replacement in the lemon shark, *Negaprion brevirostris.* In: Gilbert, P.W., Mathewson, R.F., and Rall, D.P. (Eds.), *Sharks, Skates and Rays.* The Johns Hopkins University Press, Baltimore, MD, pp. 319–329.

Moss, S.A. (1972). The feeding mechanism of sharks of the family Carcharhinidae. *J. Zool. Lond.* 167:423–436.

Moss, S.A. (1977b). Feeding mechanisms in sharks. *Am. Zool.* 17:355–364.

Motta, P.J. (2004). Prey capture behavior and feeding mechanics of elasmobranchs. In: Carrier, J.C., Musick, J.A., and Heithaus, M.R. (Eds.), *Biology of Sharks and Their Relatives.* CRC Press, Boca Raton, FL, pp. 165–202.

Motta, P.J. and Wilga, C.D. (1995). Anatomy of the feeding apparatus of the lemon shark, *Negaprion brevirostris. J. Morphol.* 226:309–329.

Motta, P.J. and Wilga, C.D. (1999). Anatomy of the feeding apparatus of the nurse shark, *Ginglymostoma cirratum. J. Morphol.* 241:1–29.

Motta, P.J. and Wilga, C.D. (2001). Advances in the study of feeding behaviors, mechanisms, and mechanics of sharks. *Environ. Biol. Fish.* 60:131–156.

Motta, P.J., Hueter, R.E., and Tricas, T.C. (1991). An electromyographic analysis of the biting mechanism of the lemon shark, *Negaprion brevirostris*: functional and evolutionary implications. *J. Morphol.* 201:55–69.

Motta, P.J., Hueter, R.E., Tricas, T.C., and Summers, A.P. (1997). Feeding mechanism and functional morphology of the jaws of the lemon shark, *Negaprion brevirostris* (Chondrichthyes, Carcharhinidae). *J. Exp. Biol.* 200:2765–2780.

Motta, P.J., Hueter, R.E., Tricas, T.C., and Summers, A.P. (2002). Kinematic analysis of suction feeding in the nurse shark, *Ginglymostoma cirratum* (Orectolobiformes, Ginglymostomatidae). *Copeia* 2002:24–38.

Motta, P.J., Hueter, R.E., Tricas, T.C., Summers, A.P., Huber, D.R., Lowry, D., Mara, K.R., Matott, M.P., Whitenack, L.B., and Wintzer, A.P. (2008). Functional morphology of the feeding apparatus, feeding constraints, and suction performance in the nurse shark *Ginglymostoma cirratum*. *J. Morphol.* 269:1041–1055.

Motta, P.J., Maslanka, M., Hueter, R.E., Davis, R.L., de la Parra, R., Mulvany, S.L., Habegger, M.L., Strother, J.A., Mara, K.R., Gardiner, J.M., Tyminski, J.P., and Zeigler, L.D. (2010). Feeding anatomy, filter-feeding rate, and diet of whale sharks *Rhincodon typus* during surface ram filter feeding off the Yucatan Peninsula, Mexico. *Zoology* 113:199–212.

Moy-Thomas, J.A. and Miles, R.S. (1971). *Paleozoic Fishes*. Chapman & Hall, London.

Muller, M., Osse, J.W.M., and Verhagen, J.H.G. (1982). A quantitative hydrodynamic model of suction feeding in fish. *J. Theor. Biol.* 95:49–79.

Muto, E.Y., Soares, L.S.H., and Goitein, R. (2001). Food resource utilization of the skates *Rioraja agassizii* (Muller and Henle, 1841) and *Psammobatis extenta* (Garman, 1913) on the continental shelf off Ubatuba, south-eastern Brazil. *Rev. Brasil Biol.* 61:217–238.

Myrberg, Jr., A.A. (1991). Distinctive markings of sharks: ethological considerations of visual function. *J. Exp. Zool.* 5:156–166.

Nakaya, K. (1975). Taxonomy, comparative anatomy and phylogeny of Japanese catsharks, Scyliorhinidae. *Mem. Fac. Fish. Hokkaido Univ.* 23:1–94.

Nakaya, K., Matsumoto, R., and Suda, K. (2008). Feeding strategy of the megamouth shark *Megachasma pelagios* (Lamniformes: Megachasmidae). *J. Fish Biol.* 73:17–34.

Nauwelaerts, S., Wilga, C., Sanford, C., and Lauder, G. (2007). Hydrodynamics of pray capture in sharks: effects of substrate. *J. Roy. Soc. Interface* 4:341–345.

Nauwelaerts, S., Wilga, C.D., Lauder, G.V., and Sanford, C.P. (2008). Fluid dynamics of feeding behavior in white-spotted bamboo sharks. *J. Exp. Biol.* 211:3095–3102.

Neiva, J., Coelho, R., and Erzini, K. (2006). Feeding habits of the velvet belly lanternshark *Etmopterus spinax* (Chondrichthyes: Etmopteridae) off the Algarve, southern Portugal. *J. Mar. Biol. Assoc. U.K.* 86:835–841.

Nelson, D.R. (1969). The silent savages. *Oceans* 1:8–22.

Nelson, J.D. and Eckert, S.A. (2007). Foraging ecology of whale sharks (*Rhincodon typus*) within Bahia de Los Angeles, Baja California Norte, Mexico. *Fish. Res.* 84:47–64.

Nelson, J.S. (1994). *Fishes of the World*. John Wiley & Sons, New York.

Nobiling, G. (1977). Die Biomechanik des Kieferapparates beim Stierkopfhai (*Heterodontus portusjacksoni* = *Heterodontus philippi*). *Adv. Anat. Embryol. Cell Biol.* 52:1–52.

Nordell, S.E. (1994). Observations of the mating behavior and dentition of the round stingray, *Urolophus halleri*. *Environ. Biol. Fish.* 39:219–229.

Norton, S.F. and Brainerd, E.L. (1993). Convergence in the feeding mechanics of ecomorphologically similar species in the Centrarchidae and Cichlidae. *J. Exp. Biol.* 176:11–29.

Notarbartolo di Sciara, G. and Hillyer, E.V. (1989). Mobulid rays off Eastern Venezuela. *Copeia* 1989:607–614.

Orth, R.J. (1975). Destruction of eelgrass, *Zostera marina*, by the cownose ray, *Rhinoptera bonasus*, in the Chesapeake Bay. *Chesapeake Sci.* 16:205–208.

Orvig, T. (1951). Histologic studies of placoderm and fossil elasmobranchs. I. The endoskeleton, with remarks on the hard tissues of lower vertebrates in general. *Arkiv. Zool.* 2:321–454.

Papastamatiou, Y.P., Wetherbee, B.M., O'Sullivan, J., Goodmanlowe, G.D., and Lowe, C.G. (2010). Foraging ecology of cookiecutter sharks (*Isistius brasiliensis*) on pelagic fishes in Hawaii, inferred from prey bite wounds. *Environ. Biol. Fish.* 88(4):361–368.

Parker, H.W. and Boeseman, M. (1954). The basking shark (*Cetorhinus maximus*) in winter. *Proc. Zool. Soc. Lond.* 124:185–194.

Peterson, C.H., Fodrie, F.J., Summerson, H.C., and Powers, S.P. (2001). Site-specific and density-dependent extinction of prey by schooling rays: generation of a population sink in top-quality habitat for bay scallops. *Oecologia* 129:349–356.

Peyer, B. (1968). *Comparative Odontology*. University of Chicago Press, Chicago.

Pfeil, F.H. (1983). Zahmorphologische Untersuchungen an rezenten und fossilen Haien der Ordnungen Chlamydoselachiformes und Echinorhiniformes. *Palaeoichthyologica* 1:1–135.

Porter, M.E., Beltran, J.L., Koob, T.J., and Summers, A.P. (2006). Material properties and biochemical composition of mineralized vertebral cartilage in seven elasmobranch species (Chondrichthyes). *J. Exp. Biol.* 209:2920–2928.

Powell, P.L., Roy, R.R., Kanim, P., Bello, M.A., and Edgerton, V.R. (1984). Predictability of skeletal muscle tension from architectural determinations in guinea pigs. *J. Appl. Physiol.* 57:1715–1721.

Powlik, J.J. (1995). On the geometry and mechanics of tooth position in the white shark, *Carcharodon carcharias*. *J. Morphol.* 226:277–288.

Powter, D.M., Gladstone, W., and Platell, M. (2010). The influence of sex and maturity on the diet, mouth morphology and dentition of the Port Jackson shark, *Heterodontus portusjacksoni*. *Mar. Freshwat. Res.* 61:74–85.

Pratt, Jr., H.L., Casey, J.G., and Conklin, R.B. (1982). Observations on large white shark, *Carcharodon, carcharias*, off Long Island, New York. *U.S. Fish. Bull.* 80:153–156.

Preuschoft, H., Reif, W.E., and Muller, W.H. (1974). Funktionsanpassungen in form und struktur an haifischzahnen. *Z. Anat. Entwickl.-Gesch.* 143:315–344.

Pyle, P., Klimley, A.P., Anderson, S.D., and Henderson, R.P. (1996). Environmental factors affecting the occurrence and behavior of white sharks at the Farrallon Islands, California. In: Klimley, A.P. and Ainley, D.G. (Eds.), *Great White Sharks: The Biology of Carcharodon carcharias.* Academic Press, New York, pp. 281–291.

Qin, H., Hsu, M.K.H., Morris, B.J., and Hoh, J.F.Y. (2002). A distinct subclass of mammalian striated myosins: structure and molecular evolution of "superfast" or masticatory myosin heavy chain. *J. Mol. Evol.* 55:544–552.

Rama, S. and Chandrakasan, G. (1984). Distribution of different molecular species of collagen in the vertebral cartilage of shark (*Carcharhinus acutus*). *Connect. Tissue Res.* 12:111–118.

Ramsay, J.B. and Wilga, C.D. (2007). Morphology and mechanics of the teeth and jaws of white-spotted bamboo sharks (*Chiloscyllium plagiosum*). *J. Morphol.* 268:664–682.

Randall, J.E. (1992). Review of the biology of the tiger shark (*Galeocerdo cuvier*). *Aust. J. Mar. Freshwat. Res.* 43:21–31.

Rath, N.C., Balog, J.M., Huff, W.E., Huff, G.R., Kulkarni, G.B., and Tierce, J.F. (1999). Comparative differences in the composition and biomechanical properties of tibiae of seven- and seventy-two-week-old male and female broiler breeder chickens. *Poult. Sci.* 78:1232–1239.

Reif, W.E. (1976). Morphogenesis, pattern formation, and function of the dentition of *Heterodontus* (Selachii). *Zoomorphology* 83:1–46.

Reif, W.E. (1977). Tooth enameloid as a taxonomic criterion. Part 1. A new euselachian shark from the Rhaetic–Liassic boundary. *Neues Jahrbuch Geol. Paläontol. Monatsh.* 1977(9):565–576.

Reif, W.E. (1978). Shark dentitions: morphogenetic processes and evolution. *Geol. Paleontol. Abh.* 157:107–115.

Reif, W.E. (1980). Development of dentition and dermal skeleton in embryonic *Scyliorhinus canicula*. *J. Morphol.* 166:275–288.

Reif, W.E. (1984). Pattern regulation in shark dentitions. In: Malacinski, G.M. and Bryant, S.V. (Eds.), *Pattern Formation: A Primer in Developmental Biology.* Macmillan, New York, pp. 603–621.

Reif, W.E., McGill, D., and Motta, P. (1978). Tooth replacement rates of the sharks *Triakis semifasciata* and *Ginglymostoma cirratum*. *Zoll. Jahrb. Anat. Bd.* 99:151–156.

Rho, J.Y., Mishra, S.R., Chung, K., Bai, J., and Pharr, G.M. (2001). Relationship between ultrastructure and the nanoindentation properties of intramuscular herring bones. *Ann. Biomed. Eng.* 29:1082–1088.

Robinson, M.P. and Motta, P.J. (2002). Patterns of growth and the effects of scale on the feeding kinematics of the nurse shark (*Ginglymostoma cirratum*). *J. Zool.* 256:449–462.

Rosenberg, L.R. (1998). A Comparison of the Mineralized Endoskeletal Tissues of Several Recent and Fossil Chondrichthyans from the Bear Gulch Limestone of Montana, master's thesis, Adelphi University, Long Island, NY.

Rudloe, A. (1989). Captive maintenance of the lesser electric ray, with observations of feeding behavior. *Prog. Fish-Cult.* 51:37–41.

Russo, R.A. (1975). Observations on the food habits of leopard sharks (*Triakis semifasciata*) and brown smooth-hounds (*Mustelus henlei*). *Calif. Fish Game* 61:95–103.

Sanderson, S.L. and Wassersug, R. (1993). Convergent and alternative designs for vertebrate suspension feeding. In: Hanken, J. and Hall, B.K. (Eds.), *The Skull*, Vol. 3. University of Chicago Press, Chicago, pp. 37–112.

Sanford, C.P. and Wainwright, P.C. (2002). Use of sonomicrometry demonstrates the link between prey capture kinematics and suction pressure in largemouth bass. *J. Exp. Biol.* 205:3445–3457.

Sasko, D.E. (2000). The Prey Capture Behavior of the Atlantic Cownose Ray, *Rhinoptera bonasus*, master's thesis, University of South Florida, Tampa.

Sasko, D.E., Dean, M.N., Motta, P.J., and Hueter, R.E. (2006). Prey capture behavior and kinematics of the Atlantic cownose ray, *Rhinoptera bonasus*. *Zoology* 109:171–181.

Schaeffer, B. (1967). Comments on elasmobranch evolution. In: Gilbert, P.W., Matthewson, R.F., and Rall, D.P. (Eds.), *Sharks, Skates and Rays.* The Johns Hopkins University Press, Baltimore, MD, pp. 3–35.

Schaeffer, B. and Williams, M. (1977). Relationship of fossil and living elasmobranchs. *Am. Zool.* 17:293–302.

Schmidt-Nielson, K. (1984). *Scaling: Why Is Animal Size So Important?* Cambridge University Press, Cambridge, U.K.

Schwartz, F.J. (1967). Embryology and feeding behavior of the Atlantic cownose ray *Rhinoptera bonasus*, presented at the Seventh Meeting of the Association of Island Marine Laboratories of the Caribbean, August 24–26, 1966, Barbados, West Indies.

Schwartz, F.J. (1989). Feeding behavior of the cownose ray, *Rhinoptera bonasus* (family Myliobatidae). *Assoc. Southeast Biol. Bull.* 36:66.

Sherman, K.M., Reidenauer, J.A., Thistle, D., and Meeter, D. (1983). Role of a natural disturbance in an assemblage of marine free-living nematodes. *Mar. Ecol. Prog. Ser.* 11:23–30.

Shimada, K. (2002a). Dental homologies in lamniform sharks (Chondrichthyes: Elasmobranchii). *J. Morphol.* 251:38–72.

Shimada, K. (2002b). Teeth of embryos in lamniform sharks (Chondrichthyes: Elasmobranchii). *Environ. Biol. Fish.* 63:309–319.

Shirai, S. and Nakaya, K. (1990). Interrelationships of the Etmopterinae (Chondrichthyes, Squaliformes). In: Pratt, Jr., H.L., Gruber, S.H., and Taniuchi, T. (Eds.), *Elasmobranchs as Living Resources: Advances in the Biology, Ecology, Systematics, and the Status of the Fisheries*, NOAA Tech. Rep. NMFS 90. National Oceanic and Atmospheric Administration, Washington, D.C., pp. 347–356.

Shirai, S. and Nakaya, K. (1992). Functional morphology of feeding apparatus of the cookie-cutter shark, *Isistius brasiliensis* (Elasmobranchii, Dalatiinae). *Zool. Sci.* 9:811–821.

Shirai, S. and Okamura, O. (1992). Anatomy of *Trigonognathus kabeyai*, with comments on feeding mechanism and phylogenetic relationships (Elasmobranchii, Squalidae). *Jpn. J. Icthyol.* 39:139–150.

Sims, D.W. (1999). Threshold foraging behaviour of basking sharks on zooplankton: life on an energetic knife-edge? *Proc. R. Soc. Lond. B Biol. Sci.* 266:1437–1443.

Sims, D.W. (2000). Filter-feeding and cruising swimming speeds of basking sharks compared with optimal models: they filter-feed slower than predicted for their size. *J. Exp. Mar. Biol. Ecol.* 249:65–76.

Sims, D.W. (2008). Sieving a living: a review of the biology, ecology and conservation status of the plankton-feeding basking shark *Cetorhinus maximus*. *Adv. Mar. Biol.* 54:171–220.

Sims, D.W. (2010). Tracking and analysis techniques for understanding free-ranging shark movements and behavior. In: Carrier, J.C., Musick, J.A., and Heithaus, M.R. (Eds.), *Sharks and Their Relatives II: Biodiversity, Adaptive Physiology, and Conservation*. CRC Press, Boca Raton, FL, pp. 351–392.

Sims, D.W. and Merrett, D.A. (1997). Determination of zooplankton characteristics in the presence of surface feeding basking sharks *Cetorhinus maximus*. *Mar. Ecol. Prog. Ser.* 158:297–302.

Sims, D.W. and Quayle, V.A. (1998). Selective foraging behaviour of basking sharks on zooplankton in a small-scale front. *Nature* 393:460–464.

Sims, D.W., Fox, A.M., and Merrett, D.A. (1997). Basking shark occurrence off South-West England in relation to zooplankton abundance. *J. Fish Biol.* 51:436–440.

Sims, D.W., Andrews, P.L.R., and Young, J.Z. (2000). Stomach rinsing in rays. *Nature* 404:566.

Skjaeraasen, J.E. and Bergstad, O.A. (2000). Distribution and feeding ecology of *Raja radiata* in the northeastern North Sea and Skagerrak (Norwegian Deep). *ICES J. Mar. Sci.* 57:1249–1260.

Smale, M.J. and Cliff, G. (1998). Cephalopods in the diets of four shark species (*Galeocerdo cuvier*, *Sphyrna lewini*, *S. zygaena* and *S. mokarran*) from KwaZulu-Natal, South Africa. *S. Afr. J. Mar. Sci.* 20:241–253.

Smale, M.J. and Cowley, P.D. (1992). The feeding ecology of skates (Batoidea: Rajidae) off the Cape South coast, South Africa. *S. Afr. J. Mar. Sci.* 12:823–834.

Smale, M.J., Sauer, W.H.H., and Hanlon, R.T. (1995). Attempted ambush predation on spawning squids *Loligo vulgaris reynaudii* by benthic pyjama sharks, *Poroderma africanum*, off South Africa. *J. Mar. Biol. Assoc. U.K.* 75:739–742.

Smale, M.J., Sauer, W.H.H., and Roberts, M.J. (2001). Behavioural interactions of predators and spawning chokka squid off South Africa: towards quantification. *Mar. Biol.* 139:1095–1105.

Smith, B.G. (1942). The heterodontid sharks: their natural history, and the external development of *Heterodontus japonicus* based on notes and drawings by Bashford Dean. In: Gudger, E.W. (Ed.), *The Bashford Dean Memorial Volume: Archaic Fishes*. American Museum of Natural History, New York, pp. 647–784.

Smith, J.W. (1980). The Life History of the Cownose Ray, *Rhinoptera bonasus* (Mitchell 1815), in Lower Chesapeake Bay, with Notes on the Management of the Species, master's thesis, College of William and Mary, Williamsburg, VA.

Smith, J.W. and Merriner, J.V. (1985). Food habits and feeding behavior of the cownose ray, *Rhinoptera bonasus*, in lower Chesapeake Bay. *Estuaries* 8:305–310.

Southall, E.J. and Sims, D.W. (2003). Shark skin: a function in feeding. *Proc. R. Soc. Lond. B Biol. Sci.* 270(Suppl.):47–49.

Springer, S. (1961). Dynamics of the feeding mechanism of large galeoid sharks. *Am. Zool.* 1:183–185.

Springer, S. (1967). Social organization of shark populations. In: Gilbert, P.W., Mattewson, R.F., and Rall, D.P. (Eds.), *Sharks, Skates and Rays*. The Johns Hopkins Press, Baltimore, MD, pp. 149–174.

Stevens, J.D. (1976). The Ecology of the Blue Shark (*Prionace glauca* L.) in British Waters, doctoral dissertation, University of London.

Stevens, J.D. (2010). Epipelagic oceanic elasmobranchs. In: Carrier, J.C., Musick, J.A., and Heithaus, M.R. (Eds.), *Sharks and Their Relatives II: Biodiversity, Adaptive Physiology, and Conservation*. CRC Press, Boca Raton, FL, pp. 3–35.

Stokes, M.D. and Holland, N.D. (1992). Southern stingray (*Dasyatis americana*) feeding on lancelets (*Branchiostoma floridae*). *J. Fish Biol.* 41:1043–1044.

Strasburg, D.W. (1958). Distribution, abundance, and habits of pelagic sharks in the central Pacific Ocean. *U.S. Fish. Bull.* 58:335–361.

Strasburg, D.W. (1963). The diet and dentition of *Isistius brasiliensis*, with remarks on tooth replacement in other sharks. *Copeia* 1963:33–40.

Strong, Jr., W.R. (1989). Behavioral Ecology of Horn Sharks, *Heterodontus francisci*, at Santa Catalina Island, California, with Emphasis on Patterns of Space Utilization, master's thesis, The California State University, Long Beach.

Strong, Jr., W.R. (1990). Hammerhead shark predation on stingrays: an observation of prey handling by *Sphyrna mokarran*. *Copeia* 1990:836–840.

Summers, A.P. (2000). Stiffening the stingray skeleton: an investigation of durophagy in myliobatid stingrays (Chondrichthyes, Batoidea, Myliobatidae). *J. Morphol.* 243:113–126.

Summers, A.P. and Long, Jr., J.H. (2006). Skin and bones, sinew and gristle: the mechanical behavior of fish skeletal tissues. In: Shadwick, R.E. and Lauder, G.V. (Eds.), *Fish Biomechanics*. Elsevier, San Diego, pp. 141–178.

Summers, A.P., Koob, T.J., and Brainerd, E.L. (1998). Stingray jaws strut their stuff. *Nature* 395:450–451.

Summers, A.P., Koob-Emunds, M.M., Kajiura, S.M., and Koob, T.J. (2003). A novel fibrocartilaginous tendon from an elasmobranch fish (*Rhinoptera bonasus*). *Cell Tissue Res.* 312:221–227.

Summers, A.P., Ketcham, R., and Rowe, T. (2004). Structure and function of the horn shark (*Heterodontus francisci*) cranium through ontogeny: the development of a hard prey specialist. *J. Morphol.* 260:1–12.

Svanback, R., Wainwright, P.C., and Ferry-Graham, L.A. (2002). Linking cranial kinematics, buccal pressure, and suction feeding performance in largemouth bass. *Physiol. Biochem. Zool.* 75:532–543.

Talent, L.G. (1976). Food habits of the leopard shark, *Triakis semifasciata*, in Elkhorn Slough, Monterey Bay, California. *Calif. Fish Game* 62:286–298.

Tanaka, S.K. (1973). Suction feeding by the nurse shark. *Copeia* 1973:606–608.

Tanne, K., Tanaka, E., and Sakuda, M. (1991). The elastic modulus of the temporomandibular joint disc from adult dogs. *J. Dent. Res.* 70:1545–1548.

Taylor, J.G. (2007). Ram filter-feeding and nocturnal feeding of whale sharks (*Rhincodon typus*) at Ningaloo Reef, Western Australia. *Fish. Res.* 84:65–70.

Taylor, J.G. and Pearce, A.F. (1999). Ningaloo reef currents: implications for coral spawn dispersal, zooplankton and whale shark abundance. *J. Roy. Soc. West. Aust.* 82:57–65.

Taylor, L.R. (1972). A Revision of the Sharks of the Family Heterodontidae (Heterodontiformes, Selachii), doctoral dissertation, University of California, San Diego.

Taylor, L.R., Compagno, L.J.V., and Struhsaker, P.J. (1983). Megamouth, a new species, genus, and family of lamnoid shark (*Megachasma pelagios*, family Megachasmidae) from the Hawaiian Islands. *Proc. Calif. Acad. Sci.* 43:87–110.

Thrush, S.F., Pridmore, R.D., Hewitt, J.E., and Cummings, V.J. (1991). Impact of ray feeding disturbances on sandflat macrobenthos: do communities dominated by polychaetes or shellfish respond differently? *Mar. Ecol. Prog. Ser.* 69:245–252.

Tricas, T.C. (1979). Relationships of the blue shark, *Prionace glauca*, and its prey species near Santa Catalina Island, California. *U.S. Fish. Bull.* 77:175–182.

Tricas, T.C. (1985). Feeding ethology of the white shark, *Carcharodon carcharias*. *Mem. South. Calif. Acad. Sci.* 9:81–91.

Tricas, T.C. and McCosker, J.E. (1984). Predatory behavior of the white shark (*Carcharodon carcharias*) with notes on its biology. *Proc. Calif. Acad. Sci.* 43:221–238.

Turingan, R.G., Wainwright, P.C., and Hensley, D.A. (1995). Interpopulation variation in prey use and feeding biomechanics in Caribbean triggerfishes. *Oecologia* 102:296–304.

Valadez-Gonzalez, C., Anguilar-Palomino, B., and Hernandez-Vazquez, S. (2001). Feeding habits of the round stingray *Urobatis halleri* (Cooper, 1863) (Chonrichthyes: Urolophidae) from the continental shelf of Jalisco and Colima, Mexico. *Cien. Mar.* 27:91–104.

van der Meij, M.A.A., Griekspoor, M., and Bout, R.G. (2004). The effect of seed hardness on husking time in finches. *Anim. Biol.* 54:195–205.

Van Wassenbergh, S., Aerts, P., Adriaens, D., and Herrel, A. (2005). A dynamic model of mouth closing movements in clariid fishes: the role of enlarged adductors. *J. Theor. Biol.* 234:49–65.

Van Wassenbergh, S., Aerts, P., and Herrel, A. (2006). Hydrodynamic modeling of aquatic suction performance and intra-oral pressures: limitation for comparative studies. *J. Roy. Soc. Interface* 3:507–514.

VanBlaricom, G.R. (1976). *Preliminary Observations on Interactions between Two Bottom-Feeding Rays and a Community of Potential Prey in a Sublittoral Sand Habitat in Southern California*. National Oceanic and Atmospheric Administration, Astoria, OR, pp. 153–162.

Verwaijen, D., Van Damme, R., and Herrel, A. (2002). Relationships between head size, bite force, prey handling efficiency and diet in two sympatric lacertid lizards. *Funct. Ecol.* 16:842–850.

Vorenberg, M.M. (1962). Cannibalistic tendencies of lemon and bull sharks. *Copeia* 1962:455–456.

Wainwright, P.C. (1988). Morphology and ecology: functional basis of feeding constraints in Caribbean labrid fishes. *Ecology* 69:635–645.

Wainwright, P.C. and Day, S.W. (2007). The forces exerted by aquatic suction feeders on their prey. *J. Roy. Soc. Interface* 4:553–560.

Wainwright, P.C. and Shaw SS. (1999). Morphological basis of kinematic diversity in feeding sunfishes. *J. Exp. Biol.* 202:3101–3110.

Wainwright, P.C., Westneat, M.W., and Bellwood, D.R. (2000). Linking feeding behavior and jaw mechanics in fishes. In: Domenici, P. and Blake, R.W. (Ed.), *Biomechanics in Animal Behavior*. BIOS Scientific, Oxford, pp. 207–221.

Wainwright, P.C., Ferry-Graham, L.A., Waltzek, T.B., Carroll, A.M., Hulsey, C.D., and Grubich, J.R. (2001a). Evaluating the use of ram and suction during prey capture by cichlid fishes. *J. Exp. Biol.* 204:3039–3051.

Wainwright, P.C., Ferry-Graham, L.A., Waltzek, T.B., Hulsey, C.D., Carroll, A.M., and Svanback, R. (2001b). Evaluating suction feeding performance in fishes. *Am. Zool.* 41:1617–1617.

Wainwright, P.C., Bellwood, D.R., Westneat, M.W., Grubich, J.R., and Hoey, A.S. (2004). A functional morphospace for the skull of labrid fishes: patterns of diversity in a complex biomechanical system. *Biol. J. Linn. Soc. Lond.* 82:1–25.

Wainwright, S.A., Biggs, W.D., Currey, J.D., and Gosline, J.M. (1976). *Mechanical Design in Organisms*. Princeton University Press, Princeton, NJ.

Waller, G.N.H. and Baranes, A. (1991). Chondrocranium morphology of northern Red Sea triakid sharks and relationships to feeding habits. *J. Fish Biol.* 38:715–730.

Waters, N.E. (1980). Some mechanical and physical properties of teeth. In: Vincent, J.F.V. and Currey, J.D. (Eds.), *The Mechanical Properties of Biological Materials*. Cambridge University Press, Cambridge, U.K., pp. 99–135.

Westneat, M.W. (1994). Transmission of force and velocity in the feeding mechanisms of labrid fishes (Teleostei, Perciformes). *Zoomorphology* 114:103–118.

Westneat, M.W. (2004). Evolution of levers and linkages in the feeding mechanisms of fishes. *Integr. Comp. Biol.* 44:378–389.

Wesneat, M.W. (2006). Skull biomechanics and suction feeding in fishes. In: Shadwick, R.E. and Lauder, G.V. (Eds.), *Fish Biomechanics*. Elsevier, Amsterdam, pp. 29–68.

Wetherbee, B.M., Crow, G.L., and Lowe, C.G. (1997). Distribution, reproduction, and diet of the gray reef shark *Carcharhinus amblyrhynchos* in Hawaii. *Mar. Ecol. Prog. Ser.* 151:181–189.

White, W.T. and Sommerville, E. (2010). Elasmobranchs of tropical marine ecosystems. In: Carrier, J.C., Musick, J.A., and Heithaus, M.R. (Eds.), *Sharks and Their Relatives II: Biodiversity, Adaptive Physiology, and Conservation*. CRC Press, Boca Raton, FL, pp. 159–239.

Whitenack, L.B. and Motta, P.J. (2010). Performance of shark teeth during puncture and draw: implications for the mechanics of cutting. *Biol. J. Linn. Soc. Lond.* 100:271–286.

Whitenack, L.B., Simkins, Jr., D.C., Motta, P.J., Hirai, M., and Kumar, A. (2010). Young's modulus and hardness of shark tooth materials. *Arch. Oral Biol.* 55:203–209.

Whitenack, L.B., Simkins, Jr., D.C., and Motta, P.J. (2011). Biology meets engineering: the structural mechanics of fossil and extant shark teeth. *J. Morphol.* 272:169–179.

Widder, E.A. (1998). A predatory use of counterillumination by the squaloid shark, *Isistius brasiliensis. Environ. Biol. Fish.* 53:267–273.

Wilga, C.D. (1997). Evolution of Feeding Mechanisms in Elasmobranchs: A Functional Morphological Approach, doctoral dissertation, University of South Florida, Tampa.

Wilga, C.D. (2002). A functional analysis of jaw suspension in elasmobranchs. *Biol. J. Linn. Soc.* 75:483–502.

Wilga, C.D. (2005). Morphology and evolution of the jaw suspension in lamniform sharks. *J. Morphol.* 265:102–119.

Wilga, C.D. (2008). Evolutionary divergence in the feeding mechanism of fishes. *Acta Geol. Polon.* 58:113–120.

Wilga, C.D. (2010). Hyoid and pharyngeal arch function during ventilation and feeding in elasmobranchs: conservation and modification in function. *J. Appl. Ichthyol.* 26:162–166.

Wilga, C.D. and Motta, P.J. (1998a). Conservation and variation in the feeding mechanism of the spiny dogfish *Squalus acanthias. J. Exp. Biol.* 201:1345–1358.

Wilga, C.D. and Motta, P.J. (1998b). Feeding mechanism of the Atlantic guitarfish *Rhinobatus lentiginosus*: modulation of kinematic and motor activity? *J. Exp. Biol.* 201:3167–3183.

Wilga, C.D. and Motta, P.J. (2000). Durophagy in sharks: feeding mechanics of the hammerhead *Sphyrna tiburo. J. Exp. Biol.* 203:2781–2796.

Wilga, C.D. and Sanford, C.P. (2008). Suction generation in white-spotted bamboo sharks *Chiloscyllium plagiosum. J. Exp. Biol.* 211:3128–3138.

Wilga, C.D., Wainwright, P.C., and Motta, P.J. (2000). Evolution of jaw depression mechanics in aquatic vertebrates: insights from Chondrichthyes. *Biol. J. Linn. Soc.* 71:165–185.

Wilga, C.D., Hueter, R.E., Wainwright, P.C., and Motta, P.J. (2001). Evolution of upper jaw protrusion mechanisms in elasmobranchs. *Am. Zool.* 41:1248–1257.

Wilga, C.D., Motta, P.J., and Sanford, C.P. (2007). Evolution and ecology of feeding in elasmobranchs. *Integr. Comp. Biol.* 47:55–69.

Williams, M. (2001). Tooth retention in cladodont sharks: with a comparison between primitive grasping and swallowing, and modern cutting and gouging feeding mechanisms. *J. Vertebr. Paleontol.* 21:214–226.

Wilson, D.P. (1953). Notes from the Plymouth Aquarium II. *J. Mar. Biol. Assoc. U.K.* 32:199–208.

Winterbottom, R. (1974). A descriptive synonymy of the striated muscles of the Teleostei. *Proc. Acad. Nat. Sci. Phil.* 125:225–317.

Wirsing, A.J., Heithaus, M.R., and Dill, L.M. (2007). Can measures of prey availability improve our ability to predict the abundance of large marine predators? *Oecologia* 153:563–568.

Witzell, W.N. (1987). Selective predation on large cheloniid sea turtles by tiger sharks (*Galeocerdo cuvier*). *Jpn. J. Herpetol.* 12:22–29.

Wroe, S., Huber, D.R., Lowry, M., McHenry, C., Moreno, K., Clausen, P., Ferrara, T., Cunningham, E., Dean, M.N., and Summers, A.P. (2008). Three-dimensional computer analysis of white shark jaw mechanics: how hard can a great white bite? *J. Zool.* 276:336–342.

Wu, E.H. (1994). Kinematic analysis of jaw protrusion in orectolobiform sharks: a new mechanism for jaw protrusion in elasmobranchs. *J. Morphol.* 222:175–190.

Wyckmans, M., Van Wassenburgh, S., Adriaens, D., Van Damme, R., and Herrel, A. (2007). Size-related changes in cranial morphology affect diet in the catfish *Clariallabes longicauda. Biol. J. Linn. Soc. Lond.* 92:323–334.

Zangerl, R. (1981). *Chondrichthyes I: Paleozoic Elasmobranchii.* Gustav Fischer Verlag, New York.

Zangerl, R. and Williams, M.W. (1975). New evidence on the nature of the jaw suspension in Paleozoic anacanthous sharks. *Paleobiology* 18:333–341.

7

Energetics, Metabolism, and Endothermy in Sharks and Rays

Diego Bernal, John K. Carlson, Kenneth J. Goldman, and Christopher G. Lowe

CONTENTS

7.1 Energetics

During the last several decades, studies on the aerobic metabolism of elasmobranchs have relied on the use of indirect calorimetry centered on small, generally sedentary species that acclimate well to the tight enclosures of a respirometer chamber and those that are easily maintained in captivity for long periods of time. By contrast, work on larger and obligate ram ventilating species has lagged far behind and typically features a paucity of measurements usually conducted over short durations of time on a few selected species. More recently, field-based empirical models have been used to estimate the metabolic requirements of species too large to work with in the laboratory. The original view of the energetic

demands in sharks and rays was that they typically possessed lower metabolic rates relative to similar sized teleosts; however, this view is beginning to change as new data emerge on the metabolism of several large, actively swimming species. Nonetheless, several general important factors need to be considered when attempting to compare the energetic demands between elasmobranchs and teleosts and among elasmobranchs.

7.1.1 Size

Most studies have access to sharks and rays that are of similar sizes, and the lack of a large span of body mass precludes investigation of mass-specific scaling effects; however, mass-specific estimates over a relatively wide range of sizes have been possible for several species with small maximum sizes. For example, Parsons (1990) determined routine metabolic rates for bonnetheads, *Sphyrna tiburo*, ranging in body mass (M) from 0.9 to 4.7 kg and found that the mass-specific oxygen consumption rate (VO_2, in mg O_2 kg^{-1} hr^{-1}) scaled with $M^{0.87}$ and was described by $VO_2 = 68.9 + (177.8M)$. Work on lesser spotted dogfish, *Scyliorhinus canicula*, reported a mass-specific VO_2 relationship of $VO_2 = 0.104M^{0.855}$ (Sims, 1996). Generally, larger sharks have a lower mass-specific metabolic rate than smaller sharks, and this relationship is similar for rays. For example, Neer et al. (2006) estimated almost a sixfold decrease in VO_2 for cownose rays, *Rhinoptera bonasus*, ranging in body mass from 2.2 to 8.3 kg (332.8 and 55.9 mg O_2 kg^{-1} hr^{-1}, respectively). Interspecifically, Sims (2000) reviewed metabolic data for seven species of sharks ranging from 0.35 to 3.5 kg and developed a mass-specific metabolic rate relationship in which the overall routine oxygen consumption rate increased with $M^{0.86}$, a very similar mass-specific exponent found for bonnethead (Parsons, 1990) and lesser spotted dogfish (Sims, 1996). By comparison, Clarke and Johnston (1999) determined that the mass-specific scaling exponent for 69 teleost species was $M^{0.80}$, which suggests that even when comparing across diverse taxa with a body mass differing by several orders of magnitude, the physical laws that govern gas diffusion during oxygen uptake, transport, and delivery appear to result in a fairly well-conserved mass-specific scaling of VO_2 in fish. Nonetheless, it remains important to increase our understanding of how body mass may affect the metabolic rates of sharks and rays that can range in body mass by up to three or four orders of magnitude and have a wide range of levels of swimming activity. Caution should be taken, however, when applying mass-scaling corrections to compare the metabolic rates between species that are widely different in size, such as a 0.1-kg swell shark, *Cephaloscyllium ventriosum*, and a 10-kg shortfin mako shark, *Isurus oxyrinchus* (Figure 7.1).

7.1.2 Temperature

Most studies use a limited range of ambient temperatures and hence exclude the incorporation of the thermal effects on metabolism. Ambient temperature is a key variable and plays a major role in controlling the metabolic rate of ectothermic fishes. In general, the metabolic rate of ectothermic elasmobranchs is directly correlated with temperature, and, for example, metabolic rates will increase between two and three times for every 10°C elevation in ambient temperature (see Section 7.1.6). Thus, temperature will play a major role in the energetic demands of sharks and rays that undergo diurnal or seasonal changes in thermal habitat as a result of their horizontal (i.e., geographic) or vertical (i.e., depth) movement patterns. On the other hand, several species

of sharks are capable of regional endothermy, and these unique physiological specializations may result in a different thermal effect on metabolic rate (see Section 7.3). It is important to consider the validity of any correction for the thermal effects on metabolic rate between species that have widely different temperature preferences. For example, the adjustment of the metabolic rate of a coldwater shark (which normally inhabits 5°C) to match that of a tropical species (which normally inhabits 25°C) should take into consideration the potential presence of coldwater adaptations, temperature tolerance, and the thermal limits of each species (Clarke, 1991) (Figure 7.1).

7.1.3 Acclimation

The logistical problems of housing captive elasmobranchs for extended periods of time under controlled laboratory conditions (e.g., temperatures, light levels) have resulted in the majority of work focusing on small, relatively inactive and docile elasmobranchs that readily adapt to captivity. For these reasons, there are few studies on larger and obligate ram ventilating species, which can only be studied for short periods of time (hours, days, or weeks) in a captive setting. This lack of proper acclimation may lead to an inaccurate estimation of their energetic requirements, because it is not possible to measure the metabolic rates of sharks that have been fully acclimated to the experimental apparatus (Clarke, 1991). This scenario has been readily observed in recent studies on tunas in which specimens that were acclimated to both the laboratory and respirometer for longer periods of time had significantly lower metabolic rates than those measured in previous studies with shorter acclimation times (e.g., Blank et al., 2007; Dewar et al., 1994). Still not much is known about longer term issues around acclimatization in elasmobranchs.

7.1.4 Experimental Apparatus

The mechanical design by which metabolic rates are measured in fishes may also affect their energetic demands (Steffensen, 1989). Lowe (2001), for example, indicated that juvenile scalloped hammerhead sharks, *Sphyrna lewini*, exhibited poorer swimming performance and entrainment when forced to swim in a flume than while swimming freely in a pond, particularly at lower swimming velocities. Therefore, the actual respirometer chamber may impede optimal swimming performance in sharks and rays, and the current measured values of swimming energetics may include data under non-steady and turbulent conditions that do not truly represent the noncaptive (i.e., free-swimming) energetic demands. Despite the presence of potential inter- and intraspecific differences that may or may not be attributed to size effects, thermal effects, or the experimental techniques used, our body of knowledge on the energetics of sharks and rays has grown considerably during the last 30 years.

7.1.5 Metabolic Rates

In elasmobranchs, VO_2 is generally regarded as the standard in determination of aerobic metabolism in a post-absorptive state (i.e., metabolic rate excluding energy devoted to digestion and assimilation). Although some work on sharks has attempted to correlate heart rate, muscle temperature, and food consumption, or an estimate thereof, to metabolic rate (Sims, 2000; Stillwell and Kohler, 1982; reviewed in Carlson et al., 2004), quantifying a reduction in the dissolved oxygen in the water as a function of time during which the animal respires (i.e., respirometry) is the most common method used. Carlson et al. (2004) gave an overview of the types of respirometers used to measure VO_2 in elasmobranchs, the problems and benefits of each type, and details on how these systems range from simple closed respirometers to more complex swimming chambers or flumes. We do not repeat this review here.

Although it is very difficult to compare metabolic rates among species because of differences in experimental technique, size of animals used, and water temperature (Figure 7.1; see Sections 7.1.1 to 7.1.4), it is still useful to qualitatively examine trends in metabolism in an attempt to provide an overview of energetic requirements of elasmobranchs. The most common metabolic estimates used to compare the energetic demands of resting or swimming elasmobranchs are standard metabolic rate (SMR), the metabolic rate of a fish at rest (Brill, 1987); routine metabolic rate (RMR) (Fry, 1971); and maximum metabolic rate (MMR), the maximum measured metabolic rate (Korsmeyer and Dewar, 2001).

7.1.5.1 Sharks

As reported in Carlson et al. (2004), SMRs vary greatly among species (Table 7.1, Figure 7.1). In general, species that are obligate ram ventilators and swim continuously have the highest measures of metabolism. Even when correcting for temperature differences (i.e., $Q_{10} = 2$), lower estimates of VO_2 are generally found for less active species (Figure 7.1). For example, a temperature-corrected

TABLE 7.1

Summary of Standard Metabolic Rates (VO_2) for a Variety of Elasmobranch Species

Species	Temperature (°C)	Average Mass (kg)	N	Methods[a]	VO_2 (mg O_2 kg^{-1} hr^{-1})	Ref.
Isurus oxyrinchus	18–20	3.9	1	Swimming closed	240[b]	Graham et al. (1990)
Carcharhinus acronotus	28	0.5	10	Circular closed	239[b]	Carlson et al. (1999)
Sphyrna lewini	26	0.5–0.9	17	Swimming closed	189[b]	Lowe (2002)
Sphyrna tiburo	28	1.0	8	Flow-through	168[b]	Carlson and Parsons (2003)
Negaprion brevirostris	25	1.6	7	Annular closed	153[b]	Scharold and Gruber (1991)
Isurus oxyrinchus	16–21	4.4–9.5	—	Swimming closed	124[b]	Sepulveda et al. (2007a)
Carcharhinus plumbeus	24	1.0	—	Annular closed	120	Dowd et al. (2006)
Ginglymostoma cirratum	23	1.3–4.0	5	Flow-through	106	Fournier (1996)
Negaprion brevirostris	22	0.8–1.3	13	Annular closed	95	Bushnell et al. (1989)
Scyliorhinus stellaris	25	2.5	12	Circular flow-through	92	Piiper et al. (1977)
Triakis semifasciata	14–18	2.2–5.8	5	Swimming closed	91.7[b]	Scharold et al. (1989)
Chiloscyllium plagiosum	24.5	0.19	13	Circular closed	91.2	Tullis and Baillie (2005)
Cephaloscyllium ventriosum	16	0.1–0.2	4	Circular closed	44.3	Ferry-Graham and Gibb (2001)
Scyliorhinus canicula	15	1.0	33	Circular closed	38.2	Sims (1996)
Squalus acanthias	10	2.0	6	Circular closed	32.4	Brett and Blackburn (1978)
Squalus suckleyi	10	2.2–4.3	—	Flow-through	31.0	Hanson and Johansen (1970)
Rhinoptera bonasus	22–25	2.2	19	Flow-through	332.7	Neer et al. (2006)
Myliobatus californica	14	5.0	6	Circular flow-through	50	Hopkins and Cech (1994)
Rhinobatus annulatus	15	1.0	10	Circular flow-through	61	DuPreez et al. (1988)
Dasyatis americana	20	0.3	6	Flow-through	164	Fournier (1996)
Raja erinacea	10	0.5	6	Circular flow-through	20	Hove and Moss (1997)
Myliobatus aquila	10	1.1–2.1	5	Flow-through	44.4	DuPreez et al. (1988)
Dasayatis violacea	20	10.7	9	Circular flow-through	39.1	Ezcurra (2001)

Source: Adapted and updated from Carlson, J.K. et al., in *Biology of Sharks and Their Relatives,* Carrier, J.C. et al., Eds., CRC Press, Boca Raton, FL, 2004, pp. 203–224.

[a] Methods indicate the type of respirometer used to measure metabolic rate.

[b] Values of standard metabolic rate were estimated through extrapolation to zero velocity.

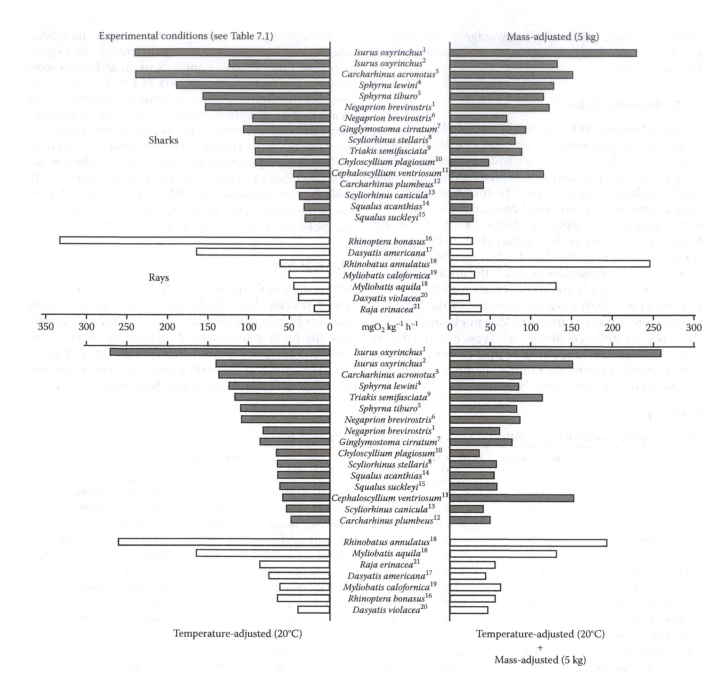

FIGURE 7.1
Standard metabolic rate (SMR) in sharks and rays. Data taken from Table 7.1. SMR values adjusted to 20°C (using a $Q_{10} = 2$), 5-kg body mass (using $M^{0.80}$), and both 20°C and 5-kg body mass. *Sources:* [1]Graham et al. (1990); [2]Sepulveda et al. (2007a); [3]Carlson et al. (1999); [4]Lowe (2002); [5]Carlson and Parsons (2003); [6]Scharold and Gruber (1991); [7]Fournier (1996); [8]Piiper et al. (1977); [9]Scharold et al. (1989); [10]Tullis and Baillie (2005); [11]Ferry-Graham and Gibb (2001); [12]Dowd et al. (2006); [13]Sims (1996); [14]Brett and Blackburn (1978); [15]Hanson and Johansen (1970); [16]Neer et al. (2006); [17]Fournier (1996); [18]DuPreez et al. (1988); [19]Hopkins and Cech (1994); [20]Ezcura (2001); [21]Hove and Moss (1997).

comparison (i.e., at 25°C) between the relatively sedentary lesser spotted dogfish and the more active scalloped hammerhead reveals that the SMR of the former is less than 40% of that of the latter (i.e., 76.4 vs. 189 mg O_2 kg^{-1} hr^{-1}) (Lowe, 2002; Sims, 1996). Nonetheless, the SMRs for sharks appear to encompass a continuum similar to that of teleosts with a comparable level of swimming activity. The mostly benthic dogfish (Family Squalidae) and catsharks (Family Scyliorhinidae) have metabolic rates that are analogous to some coldwater teleosts such as cod, *Gadus morhua* (78.2 mg O_2 kg^{-1} hr^{-1} at 15°) (Schurmann and Steffensen, 1997), while comparably sized largemouth bass, *Micropterus salmoides*, a more active teleost, has SMR values (110 to 190 mg O_2

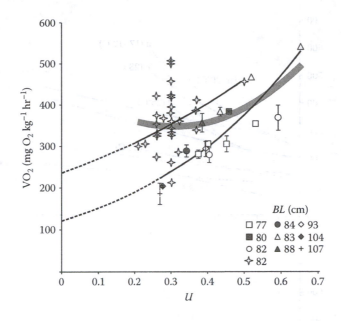

FIGURE 7.2
Oxygen consumption (VO_2) as a function of swimming velocity (U; in BL s^{-1}) for juvenile shortfin mako sharks (*Isurus oxyrinchus*) between 16 and 20°C. Regressions show the untransformed relationship between log(VO_2) and U. Upper line is for the 82-cm FL mako (✧) from Graham et al. (1990): log(VO_2) = 2.3716 + 0.58U. Lower line is for all specimens, except the 82-cm FL mako (✧): log(VO_2) = 2.0937 + 0.97U. Hatched lines indicate extrapolation to zero velocity (i.e., an estimate of SMR). Shaded curve shows the polynomial regression for all data combined (VO_2 = 0.275U^2 − 14.06U + 488). Values are individual data or mean ± SE. (Adapted from Graham, J.B. et al., *J. Exp. Biol.*, 151, 175–192, 1990; Sepulveda, C.A. et al., *Mar. Biol.*, 152, 1087–1094, 2007.)

kg^{-1} hr^{-1} at 25 to 28°C) (Beamish, 1970) that are similar to those of more active sharks (i.e., bonnethead and sandbar shark, *Carcharhinus plumbeus*) (Carlson and Parsons, 2003; Dowd et al., 2006). Therefore, it should be expected that the most actively swimming pelagic shark species such as lamnids (e.g., shortfin mako) would possess SMRs analogous to those of the highly active teleosts (e.g., tunas), which have remarkable similarities in morphology and physiology (Bernal et al., 2001a). Work on a swimming mako by Graham et al. (1990) showed that the SMR was ~240 mg O_2 kg^{-1} hr^{-1} at ~16°C, a value comparable to that of similarly sized yellowfin tuna, *Thunnus albacares* (253 to 257 mg O_2 kg^{-1} hr^{-1} at 25°C) (Dewar and Graham, 1994). A more recent study on swimming makos by Sepulveda et al. (2007a) reported a SMR of 124 mg O_2 kg^{-1} hr^{-1} at 18°C that, although more than twofold lower than previously reported, matches more recent estimates of SMR in Pacific bluefin tuna, *Thunnus orientalis*, and yellowfin tuna (120 and 91 mg O_2 kg^{-1} hr^{-1} at 20°C, respectively) (Blank et al., 2007), suggesting that the physiological capabilities of mako sharks are comparable to those of tunas.

It has been suggested that comparing SMRs between species that are obligate ram ventilators and those that have the capability to stop swimming and adequately ventilate their gills via buccal pumping could lead to erroneous results (Carlson et al., 2004; Dowd et al., 2006; Sepulveda et al., 2007a). In general, for species that never stop swimming, SMRs can be determined either

by extrapolation to zero velocity based on the oxygen consumption–swimming speed relationship or by measuring SMR on immobilized fish. Although validation of SMRs on spinally blocked sharks indicates that this can be an appropriate technique (Carlson and Parsons, 2003; Dowd et al., 2006), the zero-velocity extrapolation method could potentially lead to an overestimated SMR if the swimming speed and VO_2 functions were elevated or if the regression slope was affected by inefficient swimming at low swimming speeds (Brett, 1964), a likely scenario found by Sepulveda et al. (2007a) for shortfin makos (Figure 7.2).

Several reviews have extrapolated the swimming speed–VO_2 relationship data from the Graham et al. (1990) study on a single shortfin mako and determined a SMR of ~240 mg O_2 kg^{-1} hr^{-1} (Bernal et al., 2001a; Carlson et al., 2004; Dickson and Graham, 2004), which is more than double that determined by Sepulveda et al. (2007a) using the swimming speed–VO_2 relationship from nine shortfin makos (Figure 7.2). Although the methods used to exercise the mako sharks were similar, these authors suggested that the differences in SMRs may have arisen from the initial makos swimming at or below the minimum velocity required to maintain hydrostatic equilibrium. This will increase metabolism at the lowest test velocities and lead to a reduced slope of the regression and an increased intercept of the swimming speed–VO_2 relationship at zero velocity (Figure 7.2). Sepulveda et al. (2007a) concluded that, if the extrapolated swimming

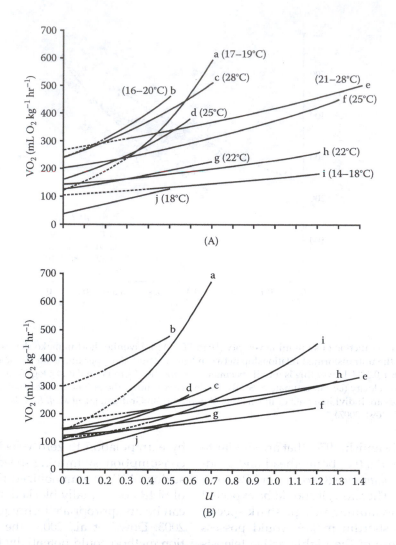

FIGURE 7.3

Oxygen consumption (VO_2) as a function of relative swim speed, U (BL s^{-1}) for sharks. (A) Data at experimental temperatures, and (B) data corrected to 20°C (using $Q_{10} = 2$). Hatched lines indicate an extrapolation to zero swim speed. Labels: a, *Isurus oxyrinchus* ($n = 9$)[1]; b, *I. oxyrinchus* ($n = 1$)[2]; c, *Carcharhinus acronotus* ($n = 8$)[3]; d, *Negaprion brevirostris* ($n = 7$)[4]; e, *Sphyrna lewini* ($n = 17$)[5]; f, combined data for *N. brevirostris* ($n = 1$)[2] and ($n = 7$)[4]; g, *N. brevirostris* ($n = 13$)[6]; h, combined data for *N. brevirostris* ($n = 1$)[2] and ($n = 13$)[6]; i, *Triakis semifasciata* ($n = 5$)[7]; j, *Carcharhinus plumbeus* ($n = 16$)[8] (calculated at 5 kg). *Sources:* [1]Sepulveda et al. (2007a); [2]Graham et al. (1990); [3]Carlson et al. (1999); [4]Scharold and Gruber (1991); [5]Lowe (2001); [6]Bushnell et al. (1989); [7]Scharold et al. (1989); [8]Dowd et al. (2006).

speed–VO_2 relationship is to be used to estimate SMRs, then the VO_2 should only be collected at the minimum speeds the shark swims where there are no apparent changes in swimming angle and there are no erratic side-to-side movements during the experimental trials.

Carlson et al. (2004) noted that for most sharks the average slopes of the power–performance curves (i.e., swim speed vs. VO_2) ranged between 0.27 and 0.36 and were similar among ectothermic species, which all share comparable morphological adaptations for drag reduction. Since that review, only two additional studies have examined the relationship of swimming and VO_2. A study on juvenile sandbar sharks by Dowd et al. (2006) reported that the slope of the power–performance curve was about 0.38, a value that is within the range of other carcharhinid

sharks that possess similar body morphology (Figure 7.3). Sepulveda et al. (2007a), however, found the slope of the swimming speed–VO_2 relationship for juvenile shortfin makos to be 0.92 (i.e., more than double that of sandbar sharks) (Figure 7.3). This higher cost of propulsion for the shortfin mako is surprising given that the swimming performance of this lamnid is hypothesized to approach that of tunas (Bernal et al., 2001a; Gemballa et al., 2006), in which the slope of the power–performance curves is around 0.33 (Blank et al., 2007). Unlike tunas, however, makos lack a swim bladder, have large pectoral fins that do not retract into grooves along the body (i.e., increasing frictional drag), and may have a more inefficient mechanism of force transmission between the swimming muscles and the caudal fin. Taken together, all of these marked differences may result in makos having an elevated cost of swimming relative to tunas. Moreover, the finding that the slope of the power–performance curves was also higher for makos when compared to other sharks suggests that makos may not have evolved the capacity to have a lower cost of transport but rather the capacity for a higher maximum metabolic rate and a large aerobic scope. Recent work by Ezcurra et al. (2012a) suggests that the other lamnids may also have a potential higher cost of locomotion and an elevated metabolic rate; for example, the routine VO_2 of juvenile (22.6 to 36.2 kg) white sharks, *Carcharodon carcharias*, being transported in a large 11,356-L tank (15 to 18°C) was 246 ± 13 mg O_2 kg^{-1} hr^{-1}. When corrected for body mass (mass-specific scaling for lamnids, $458.5 \times M^{0.79}$) (Ezcurra et al., 2012a), this would yield a value similar to that of the RMR in mako sharks (Sepulveda et al., 2007a).

The logistical and technical difficulties surrounding the current methods used to estimate the MMRs in sharks most likely lead to underestimating their total aerobic capacity by not being able to truly measure their MMR and instead measuring values that are limited by experimental conditions (e.g., maximum swim tunnel water velocities, adverse behavioral modification due to confined swimming). In addition, these methods undoubtedly lead to MMR values that not only account for the swimming-related energy expenditures (the target value) but also reflect the simultaneous occurrence of other important aerobic processes that occur during the experimental trial (i.e., recovery from stress, repayment of oxygen debt after anaerobic activity, digestion and assimilation) (Blank et al., 2007; Steffensen, 1989). Nonetheless, these values offer an estimate of the total aerobic capacity of the shark, at least under experimental conditions. In general, sharks that are more active have higher MMRs when compared to sharks that are more sedentary. These differences remain even when adjusting for any temperature effects ($Q_{10} = 2.0$), with the MMRs of active species being from 1.5 to 2.3 times greater than those of more sedentary ones (Figure 7.3).

A comparison at 25°C of a 2.0-kg spiny dogfish, *Squalus acanthias*, and 1.6-kg lemon shark, *Negaprion brevirostris*, shows that the former consumed a maximum of 250 mg O_2 kg^{-1} hr^{-1} (Brett and Blackburn, 1978), compared to 620 mg O_2 kg^{-1} hr^{-1} for the latter (Graham et al., 1990). In addition, at 26°C, scalloped hammerhead sharks swimming at 1.0 body length per second (*BL* s^{-1}) consumed up to 500 mg O_2 kg^{-1} hr^{-1} (Lowe, 2001), while the less active leopard shark swimming at a comparable 0.9 *BL* s^{-1} had a MMR of 334 mg O_2 kg^{-1} hr^{-1} (Scharold et al., 1989). More recent respirometry data show that the MMR in juvenile makos (MMR = 541 mg O_2 kg^{-1} hr^{-1}, $n = 9$, 4.5 to 9.5 kg at 18°C) exceeded that reported for all other comparably sized elasmobranchs (Carlson et al., 2004; Graham et al., 1990; Sepulveda et al., 2007a), with mako

MMRs being similar to those estimated for some tuna species (Blank et al., 2007). The observed MMR for the makos is probably the result of their specialized cardiovascular and swimming muscle physiology (e.g., large gill surface area, relatively larger heart mass, increased muscle capillary density, high hemoglobin and myoglobin concentrations, which allow for an elevated rate of oxygen uptake and delivery) (Bernal et al. 2001a, 2003a; Emery, 1985; Wegner et al., 2010).

While SMR provides a basis for comparing physiological capabilities for basal activities, the question remains whether it is appropriate to use this value for fish that must swim continuously to maintain hydrostatic equilibrium and ventilate their gills and therefore never stop moving. In this case, a more biologically relevant value than either SMR or MMR is the *aerobic scope*, defined as the difference between MMR and SMR (Priede, 1985). This new estimate represents the potential capacity a shark may have to handle multiple simultaneous aerobic demands (e.g., continuous swimming, recovering from oxygen debt, somatic growth, digestion and assimilation) (Brill, 1996; Bushnell et al., 1989; Korsmeyer et al., 1996; Lowe, 2001; Priede, 1985). For typical carcharhind sharks, the metabolic cost of swimming (i.e., aerobic scope × SMR^{-1}) ranges from about 1.4 to 1.8 times the SMR. The swimming cost ratio is 1.4 for scalloped hammerhead sharks (Lowe, 2001), 1.5 for lemon sharks (Brett and Blackburn, 1978), 1.6 for both bonnethead and sandbar sharks (Dowd et al., 2006; Parsons, 1990), and 1.7 for blacknose sharks (Carlson et al., 1999). Not surprisingly, shortfin mako data show that they have the highest costs of swimming (2.7) (Sepulveda et al., 2007a) (Figure 7.3).

The heightened aerobic scope in continuously swimming pelagic predators most probably reflects their physiological ecology, which revolves around their large-scale seasonal migrations and the need to capture fast-moving prey and its physiological consequences (e.g., rapid recovery from burst activity and rapid digestion), and potentially high rates of somatic and gonadal growth (Brill, 1996; Graham and Dickson, 2004; Korsmeyer et al., 1996; Lowe, 2001; Natanson et al., 2006; reviewed by Bushnell and Jones, 1994). Ezcurra et al. (2012b) estimated growth, daily ration, and a simplified energy budget for young-of-the-year white sharks held in captivity. They found that the daily ration for four young-of-the-year white sharks held in Monterey Bay Aquarium and fed a high-caloric diet peaked at 3.5% body mass per day, yielding a mean growth rate of 71.6 ± 8.2 kg yr^{-1}. Based on these captive growth rates and VO_2 measurements, they estimated that young-of-the-year white sharks fed a high-caloric diet expended $46 \pm 2.9\%$ of their consumed energy on metabolic costs. This is 35% higher than estimates of metabolic costs for juvenile scalloped hammerhead sharks (metabolic costs = ~30%) kept in captivity and fed high-caloric diets (Lowe, 2002).

7.1.5.2 Rays

In general, myliobatoids have similar autecologies in that they are specialized for active swimming (McEachran, 1990), and experiments indicate that SMRs are similar among batoid species. For example, at 20°C, the SMRs of cownose rays, *Rhinoptera bonasus* (0.5 to 4.8 kg), and similarly sized pelagic stingrays, *Dasyatis violacea* (5 kg), were very similar (104.3 and 101.7 mg O_2 kg^{-1} hr^{-1}, respectively) (Ezcurra, 2001; Neer et al., 2006). At colder temperatures, the SMRs of both bull rays, *Myliobatis aquila* (77.0 mg O_2 kg^{-1} hr^{-1} at 10°C, 5 kg), and bat rays, *Myliobatis californica* (90 mg O_2 kg^{-1} hr^{-1} at 16°C, 4.3 to 6.8 kg), were also similar (DuPreez et al., 1988; Hopkins and Cech, 1994, respectively) (Figure 7.1).

7.1.6 Temperature Effects

In general, temperature has a profound and positive effect on metabolic rate (Figures 7.1 and 7.3). The increase in oxygen consumption rate caused by a 10°C increase in temperature (Q_{10}) (Schmidt-Nielsen, 1983) typically falls between 2 and 3 for elasmobranchs; however, variability does occur in Q_{10}, primarily related to the acclimation procedure of the experimental animals. For elasmobranchs exposed to rapid temperature changes, Q_{10} values are generally higher. For bat rays (8 to 26°C) (Hopkins and Cech, 1994), leopard sharks (*Triakis semifasciata*) (10 to 26°C) (Miklos et al., 2003), and sandbar sharks (18 to 28°C) (Dowd et al., 2006), Q_{10} values were 3.0, 2.5, and 2.9, respectively. In seasonally acclimated elasmobranchs (19 to 28°C), Q_{10} values were 2.3 in both cownose rays (Neer et al., 2006) and bonnetheads (Carlson and Parsons, 1999), whereas juvenile scalloped hammerhead sharks had a Q_{10} of 1.3 (21 to 29°C) (Lowe, 2001). For animals that were acclimated in the laboratory for longer periods of time (e.g., weeks), Q_{10} values were 1.9 for bull ray and 2.3 for guitarfish, *Rhinobatus annulatus* (10 to 25°C) (DuPreez et al., 1988). Although it is unclear why ectothermic elasmobranchs vary so much in their degrees of metabolic temperature sensitivity, this aspect of their physiological ecology may greatly influence their behavior and use of differing thermal environments (see Section 7.2). Although no data exist on how temperature affects swimming VO_2 in lamnid sharks, any prediction of the thermal effects on metabolic rate will be inherently complicated by their ability to retain metabolic heat and alter whole-body heat balance (see Section 7.3). Lamnids are known to undergo seasonal migrations to higher latitudes and exhibit rapid and repeated diurnal sojourns to deeper (i.e., colder) waters (Domeier and Nasby-Lucas, 2008; Jorgensen et al., 2009; Weng et al., 2005). Despite the fact that lamnids frequent colder waters, their ability to warm swimming muscles will alter contractile function and, to a certain degree, may lead to a decreased thermal

effect on swimming metabolism, as has been shown for some tunas (Carey and Teal, 1966; Dewar et al., 1994; Dizon and Brill, 1979).

7.1.7 Muscle Metabolic Biochemistry

In addition to the direct measurement of VO_2, the capacity for aerobic metabolism can be assessed through the quantification of key tissue-specific biochemical indices (Dickson et al., 1993). Because continuous swimming is powered by red muscle (RM), the aerobic potential of this tissue plays a major role in whole-body metabolism; thus, it is possible to use the metabolic biochemical capacity of the RM as a proxy for the aerobic swimming potential in sharks. Most studies have generally focused on the activity of the enzyme citrate synthase (CS), which catalyzes the first step of the Krebs citric acid cycle and correlates with tissue mitochondrial density. Work on elasmobranchs has shown that when the RM metabolic enzyme activities are compared at the same temperature (e.g., 20°C), there is no marked difference among species, suggesting that the capacity for ATP production in elasmobanchs is similar regardless of the level of swimming activity. However, because lamnid sharks are capable of endothermy and the RM is warm (see Section 7.3), their enzyme activities at *in vivo* temperatures are higher than those of ectothermic sharks at 20°C. This thermal effect may significantly increase the RM CS activity of, for example, mako and salmon sharks by 48% and 123%, respectively, when the warmer *in vivo* temperatures are considered, relative to what it would be at ambient temperature (Bernal et al., 2003b). Nonetheless, the potential benefit of an increased aerobic capacity resulting from endothermy requires an increased supply of both O_2 and aerobic fuels to the RM, and lamnids have cardiorespiratory specializations that increase the uptake of O_2 at the gills and its delivery to the RM (Bernal et al., 2001a, 2003a; Wegner et al., 2010).

7.1.8 Metabolic Rates in the Field

Controlled laboratory studies provide a general basis of elasmobranch metabolism, but the question still remains as to whether those estimates determined in the laboratory are analogous to metabolic rates in the field. Similarly, the large size and high mobility of many elasmobranchs make controlled laboratory studies extremely difficult and field estimates the only practical approach.

Advances in telemetry continue to permit researchers to gather physiological data from captive and free-swimming elasmobranchs (for thorough reviews, see Lowe and Bray, 2006; Lowe and Goldman, 2001), but few studies have bridged the gap between laboratory- and field-based estimates. Sundström and Gruber (1998) used a speed-sensing transmitter on juvenile lemon sharks in the field and correlated that data with a VO_2 relationship obtained in the laboratory to estimate field metabolic rates. Parsons and Carlson (1998) used speed-sensing acoustic transmitters to quantify *in situ* swimming speeds of bonnetheads and correlated those with VO_2 measured in the laboratory under different oxygen concentrations. Using a custom-made tail-beat transmitter, Lowe et al. (1998) calibrated tail-beat frequency in relation to speed and oxygen consumption which allowed for estimates of field metabolic rates by tracking free-ranging sharks with these transmitters. As noted by Lowe (2002), however, using tail-beat frequency alone as a measure of activity may be too simplistic because it does not represent acceleration and deceleration and cannot account for any alteration of tail-beat amplitude and thus reduces accuracy of field metabolic rate estimates.

A new technique to determine the locomotor activity of organisms uses accelerometers (Tanaka et al., 2001; Wilson and McMahon, 2006; Yoda et al., 2001). These sensors measure the cyclic changes in lateral position of the body that can vary with activity level and behavior. Recently, overall dynamic body acceleration has shown promise as a proxy of energy expenditure in vertebrates (Wilson and McMahon, 2006) due to the connection between acceleration and work (Gleiss et al., 2010), with some studies showing that overall dynamic body acceleration is closely correlated with oxygen consumption in a number of taxa (Halsey et al., 2009; Wilson and McMahon, 2006), including sharks (Gleiss et al., 2010). Although integrating respirometry and accelerometry technology has the capability to further bridge the gap between laboratory- and field-based metabolic measurements, it is very important that care be taken to control for the costs of carrying accelerometry data-logging packages or transmitters on smaller animals. Lowe (2002) measured the energetic costs of juvenile hammerhead sharks carrying tail-beat transmitters and found that instrumented sharks had costs of transport 25 to 35% higher than those without transmitters, a similar scenario observed between instrumented and control (non-instrumented) leopard sharks (Scharold et al., 1989).

7.1.9 Anaerobic Metabolism

Unlike aerobic metabolism, which is powered by the RM and can be estimated using swimming VO_2 measurements, there are no simple *in vivo* laboratory-based techniques to determine the capacity for anaerobic metabolism in swimming sharks. Anaerobic metabolism, which is powered by white muscles (WM), which comprise the majority of myotomal muscle in elasmobranchs, is the major metabolic pathway used during burst swimming, and there are currently no swim tunnels that can subject sharks to controlled and repeated

bouts of burst swimming. Thus, most data on the *in vivo* burst swimming capacities in sharks come from field-based observations; for example, telemetry data on mako sharks show that during a rapid ascent they are capable of short-duration bursts swimming at speed in excess of 12 ms^{-1} (C. Sepulveda, pers. comm.). There are also numerous observations of blacktip (*Carcharhinus limbatus*), spinner sharks (*Carcharhinus brevipinna*), common thresher (*Alopias vulpinus*), and white sharks leaping and spinning on their body axes above the water surface, which requires considerable exertion to propel themselves out of the water (Castro, 1996; J. Carlson, pers. obs.).

A closer look at the WM metabolic capacities does reveal some differences among sharks, however. In general, elasmobranchs with similar levels of swimming activity have comparable levels of WM anaerobic metabolism (Dickson et al., 1993); for example, several key biochemical metabolic indices in WM showed that both lactate dehydrogenase (an index of anaerobic capacity) and citrate synthase (an index of the capacity to recover from anaerobic activity) were low in benthic skates and rays (Bernal et al., 2003b; Dickson et al., 1993). By contrast, the greatest capacity for anaerobic metabolism was observed for shortfin mako shark, which have significantly greater WM citrate synthase and lactate dehydrogenase levels and proton-buffering capacities than ectothermic sharks (Bernal et al., 2003b; Dickson et al., 1993). Shortfin mako sharks also have higher WM activities of creatine phosphokinase (an index of adenosine triphosphate production rate during burst swimming) than active ectothermic sharks and teleosts, which allows for redox balance to be retained during anaerobiosis (Bernal et al., 2001a, 2003b; Dickson, 1996). Although not measured to date, it is believed that mako sharks, in a manner similar to tunas, are also able to return blood and muscle lactate levels to pre-exercise more quickly than other sharks (Arthur et al., 1992). This is a consequence of the apparent capacity that the lamnid cardiorespiratory system may have to deliver oxygen and metabolic substrates at rates far above those needed at routine activity levels (Brill and Bushnell, 1991), which taken together appear to be a direct result of the selective pressures in the pelagic environment where food resources are aggregated but widely scattered and where no refuge exists for animals to hide and recover from a bout of strenuous activity (Brill, 1996; Dickson, 1995).

There are, however, some interesting cardiorespiratory differences between young mako and white sharks. Although all lamnids are thought to be obligate ram ventilators, their apparent ability to uptake and deliver oxygen to the working tissues during periods of inactivity (no swimming) vary markedly. Anecdotal observations on mako sharks suggest that, when the sharks are not swimming, they can only withstand brief periods (seconds or minutes) of inactivity before going hypoxic. This limited hypoxia tolerance may be related to the mako's decreased capacity for buccal pumping (lower ventilation volumes) and the potential reliance on myotomal muscle contraction (body bending) to facilitate venous return to the heart and gills. By contrast, white sharks seem to be able to withstand greater periods of inactivity without undergoing any apparent hypoxia (Weng et al., 2012; C.G. Lowe, pers. obs.). It remains unclear how white sharks are able to increase their hypoxia tolerance and withstand these conditions, yet there is growing evidence that this species routinely dives into the oxygen minimum zone (less than 1.5 mL O_2 L^{-1}) for prolonged periods of time while still being able to maintain adequate muscle and tissue function (Nasby-Lucas et al., 2009; Weng et al., 2012).

7.2 Behavioral Thermoregulation of Ectothermic Elasmobranchs

The evolution of acoustic and satellite telemetry technology has also significantly increased the number of studies on movements and habitat use of elasmobranchs (Lowe and Bray, 2006; Lowe and Goldman, 2001). An increasing number of these studies have found evidence of behavioral thermoregulation in ectothermic sharks and rays, which occurs when animals selectively move between thermal environments to achieve some potential energetic benefit. It has been hypothesized that these energetic benefits may influence growth, digestion and assimilation, and reproduction (e.g., Akerman et al., 2000; Campos et al., 2009; Di Santo and Bennett, 2011; Espinoza et al., 2011; Jirik, 2009; Klimley et al., 2005; Sims et al., 2006; Vaudo and Lowe, 2006; Weng et al., 2007). Although it is easy to demonstrate that animals use some thermal environments more than others, it can be very challenging to quantify the actual energetic benefits related to their unique and dynamic thermal preferences.

7.2.1 Foraging and Digestion

One behavioral thermoregulatory benefit for elasmobranchs may come from foraging in warm waters but returning to cooler waters to rest, particularly for species that are highly temperature sensitive (i.e., high Q_{10}). This scenario (hunting in warm water and resting in cool water) is evident in, for example, California bat rays, *Myliobatis californica*, which forage in shallow warmer water during high tide but move to deeper, cooler water during low tides (Matern et al., 2000). In

addition, bat rays have been shown to have a very high metabolic Q_{10} of 6.8 (Hopkins and Cech, 1994; Matern et al., 2000), which may translate into significant energetic savings when resting in cooler water after a foraging event. By contrast, the sympatric leopard sharks, which also move into the warmer tidal flats during high tide, have a significantly lower metabolic Q_{10} of 2.5 (Miklos et al., 2003); thus, this behavior will not result in a similar degree of energetic savings when compared to bat rays.

Work on male *Scyliorhinus canicula* also found foraging-related diel movement patterns into shallower, warmer waters at night with daytime retreats back to cooler, deeper waters (Sims et al., 2006). Even though the thermal effects on metabolic rate in this species are rather typical for ectothermic fishes (Q_{10} = 2.16) (Davenport and Sayer, 1993), a bioenergetic model for this species estimated that this feeding-related diel movement pattern could result in a 4% reduction in overall energetic costs (Sims et al., 2006). It is thus not surprising that similar cost-saving diel movement patterns were observed in laboratory shuttle box experiments on this species (Sims et al., 2006).

It has been hypothesized that by moving to different temperature conditions some elasmobranchs may ultimately alter the rates of digestion, thereby allowing them to maximize their energetic uptake while decreasing their energetic expenditures, such as for small-spotted catsharks (Sims et al., 1996) and bat rays (Miklos et al., 2003). Wallman and Bennett (2006) used laboratory experiments to demonstrate that the eurythermal Atlantic stingray, *Dasyatis sabina*, moved to cooler environments after feeding but returned to warmer environments when foraging. Di Santo and Bennett (2011) also used laboratory experiments to compare rates of digestion between *D. sabina* and the stenothermal whitespotted bamboo shark, *Chiloscyllium plagiosum*. They found that *D. sabina* would digest food more slowly when exposed to cooler conditions, whereas there was no significant change in rate of passage in *C. plagiosum* over the temperature range; therefore, it was concluded that *D. sabina* would benefit more by moving between different thermal environments to maximize their energy uptake depending on their prandial state. These laboratory experiments provide evidence correlating the level of metabolism in elasmobranchs to behaviorally mediated thermal preferences and movement patterns; such evidence can be used to gain knowledge on lesser well known species for which little is understood of their metabolic demands or temperature sensitivities. Endothermy may contribute to rapid digestion of prey in lamnid sharks because the stomach is located not only centrally to the body core but also directly above the suprahepatic rete and below the lateral cutaneous rete (see Section 7.3) (Carey et al., 1981; Goldman, 1997).

7.2.2 Reproduction

Another possible explanation for behavioral thermo-regulation can be seen in species that show sexual segregation and where one gender, particularly mature females, tend to aggregate in warmer environments. It is hypothesized that these mature female elasmobranchs may receive some energetic benefit that enhances embryo development which may lead to shorter gestation periods when exposed to warmer environments during pregnancy (Economakis and Lobel, 1998; Hight and Lowe, 2007; Jirik, 2009). Economakis and Lobel (1998) documented that, after the mating season, gray reef sharks, *Carcharhinus amblyrhynchos*, showed sexual segregation, with numerous mature females aggregating in warm shallow lagoons, and they hypothesized that females stayed in this warmer habitat for a reproductive benefit.

Hight and Lowe (2007) followed the movement patterns and monitored internal body temperatures of aggregating mature female leopard sharks during their breeding season (summer to fall) and found that females moved into the warmest environments throughout the day but dispersed to cooler waters to forage at night. Moreover, during the daytime periods, the core body temperature of the female leopard sharks increased and even remained elevated until late evening. In a similar manner, pregnant round stingrays, *Urobatis halleri*, were found to aggregate in shallow warm embayments in southern California (Mull et al., 2008) and remained in these areas until late summer or early fall, after which time they were observed to leave just prior to parturition in October (Jirik, 2009). (Pregnancy and embryo development over the season were characterized for aggregating females using non-invasive field ultrasonography.) By contrast, no males and few juvenile females were observed in these warm areas over the course of the season, suggesting that pregnant females may be using these warmer areas to increase embryo development under the warmer thermal conditions. In addition, evidence of an increased rate of embryo development in warmer waters has been documented in captive pregnant Atlantic stingrays. A series of thermal preference tests by Wallman and Bennett (2006) found that pregnant Atlantic stingrays selected warmer conditions more than non-pregnant females and that by being in water 1°C warmer could reduce gestation by 2 weeks. Lamnids may also achieve shorter gestation periods, not through behavioral means but by keeping their viscera (and hence reproductive system) warm (Goldman, 2002). This may be particularly true for porbeagle, *Lamna nasus*, and salmon sharks, *Lamna ditropis*, which possess a kidney (or renal) rete not found in the other lamnids.

It is important to consider that for several coastal temperate ectothermic elasmobranchs potential access to stratified thermal habitats may be important during embryo development, and the recent anthropogenic-related changes on coastal habitats (e.g., freshwater influx limitations, degradation of estuarine habitats, thermal effluent from once-through cooling systems) may enhance or degrade the quality of the habitat, thereby altering the potential growth rates or reproductive output of some species (Hoisington and Lowe, 2005; Vaudo and Lowe, 2006).

7.3 Endothermy

In most fishes, the body temperature closely matches that of ambient water temperature because all metabolically produced heat is rapidly lost to the water either by convective transfer via the blood at the gills or by thermal conduction across the body surface (Brill et al., 1994); however, several fishes have evolved the capacity to maintain their body core and other regions of the body at a warmer temperature relative to the water they are swimming in (Bernal et al., 2001a; Block and Finnerty, 1994; Carey and Teal, 1966, 1969a,b; Carey et al., 1971). In sharks, this form of endothermy has been documented in the aerobic swimming muscles (in all lamnids and one alopiid), in the eyes and brains (in all lamnids and suspected in one alopiid), and viscera (in all lamnids and suspected in one alopiid) (Anderson and Goldman, 2001; Bernal and Sepulveda, 2005; Bernal et al., 2001a; Carey and Teal, 1969a; Carey et al., 1971, 1985; Fudge and Stevens, 1996; Goldman et al., 2004; Patterson et al., 2011). The presence of endothermy in these sharks does not appear to be the result of specialized thermogenic tissues but rather the ability to retain the metabolic heat generated by the continuous activity of the aerobic swimming muscles during sustained locomotion as well as by digestion and assimilation (Block and Carey, 1985; Carey and Teal, 1969a; Carey et al., 1981, 1985; Wolf et al., 1988). Retia also occur in the head region of myliobatid rays (Alexander, 1995, 1996); however, no temperature data are available for those species.

7.3.1 Myotomal Muscle Endothermy

In sharks, the locomotor musculature is comprised primarily of red (RM) and white (WM) myotomal muscle fiber types, which not only are morphologically different but are also spatially segregated in the body (Bone, 1988; Johnston, 1981; Rome et al., 1988). The aerobic RM fibers are myoglobin rich and are used

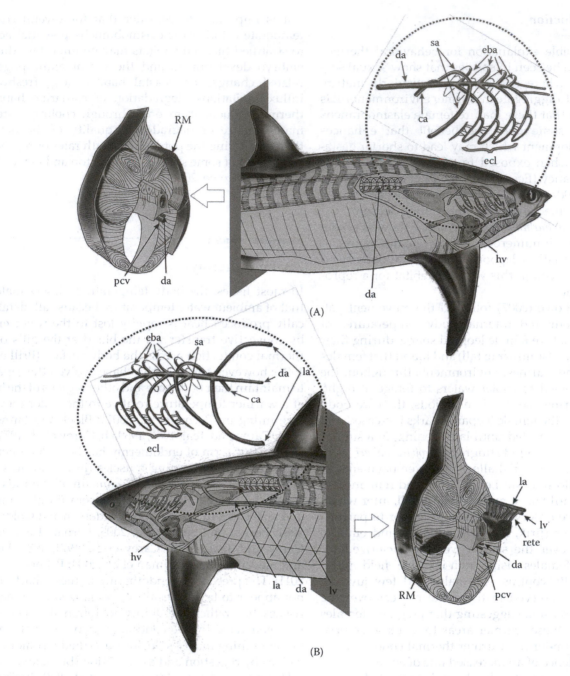

FIGURE 7.4

Representative vascular layout of the anterior region of ectothermic and endothermic sharks. (A) Side view of the ectothermic bigeye thresher shark (*Alopias superciliosus*), and (B) the endothermic common thresher shark (*A. vulpinus*) showing the presence or absence of lateral vasculature. Insets show a detail of the arterial vasculature for each shark and a representative cross-sectional slice detailing the position of the red muscle (RM) and the major systemic vessels. *Abbreviations:* ca, celiac artery; da, dorsal aorta; eba, efferent branchial arteries; ecl, efferent collecting loop; hv, heart ventricle; la, lateral artery; lv, lateral vein; pcv, post-cardinal vein; rete, heat-exchanging rete; sa, subclavian arteries. (Adapted from Patterson, J. et al., *J. Morphol.*, 272(11), 1353–1364, 2011.)

during continuous swimming, whereas the anaerobic WM fibers are myoglobin deficient and used during short-duration burst swimming (e.g., catching prey or predator avoidance) (Bone, 1988; Johnston, 1981). The RM in most sharks is located directly beneath the skin (i.e., laterally) along the length of the body; however, in lamnid sharks and the common thresher shark, the RM is located closer to the vertebral column (i.e., medially) and is predominantly distributed more anteriorly along the body (Bernal et al., 2001a; Carey and Teal, 1969a; Carey et al., 1971, 1985; Sepulveda et al., 2005) (Figures 7.4 and 7.6).

The most common systemic vascular layout in sharks is associated with non-endothermic (i.e., ectothermic) species (e.g., spiny dogfish) with a lateral RM position and is generally described as a central circulation, where the two major systemic vessels (i.e., dorsal aorta and post-cardinal vein) run ventral to the vertebral column and give rise to afferent arteries that radiate to, and efferent veins that return blood from, the myotomal muscles (reviewed by Patterson et al., 2011). The second vascular layout (i.e., lateral circulation pattern) in sharks is present in species with a medial RM position that are capable of RM endothermy, where two large vessels branch from either the efferent branchials or the dorsal aorta to form lateral arteries (one on each side of the body) that run directly beneath the skin along the length of the body. Although these sharks may also have a dorsal aorta, it is generally reduced in size (or may even be absent), as, in a manner similar to tunas, they rely mainly on the lateral arteries for systemic blood supply (Figure 7.4) (Carey and Teal, 1969a,b; Carey et al., 1971; Kishinouye, 1923; reviewed by Bernal et al., 2001a). The major systemic venous return in these sharks is also through enlarged lateral veins that run subcutaneously (very close to the lateral arteries) *en route* back to the heart. The lateral vessel arrangement in sharks with lateral circulation varies from a single artery and vein (i.e., shortfin mako, white shark, salmon shark, common thresher) to an artery and two veins (i.e., porbeagle shark; see Burne, 1923). In addition, the origin of the lateral arteries also differs between lamnids (from the dorsal root of the fourth efferent branchial arch with vascular connections to all arches) (Burne, 1923; Carey and Teal, 1969a) and the common thresher shark (i.e., arising from the dorsal aorta) (Patterson et al., 2011) (Figure 7.4). Nonetheless, in all sharks with a medial RM position, the arterial flow to the myotomal musculature from the lateral arteries is through smaller, thin-walled arteries that branch inward toward the RM, while venous return from the RM is through small thin-walled veins that run outward until joining the lateral veins. This unique blood flow to and from the medially positioned RM forms a network of juxtaposed vessels (retia) that do not allow for the diffusion of dissolved gases but readily allow the transfer of heat (Carey and Gibson, 1983; Carey and Lawson, 1973). This vascular anatomy effectively acts as a countercurrent heat-exchanging system that allows for thermal transfer between the cool arterial blood entering the RM and the warm venous blood leaving the RM (Carey, 1973; Carey and Teal, 1969a,b; Carey et al., 1971) and thus provides the basis for RM endothermy (Bernal et al., 2001a; Brill et al., 1994; Carey, 1973; Graham, 1983).

Endothermic sharks do show some species-specific differences in both the number of vessel rows comprising the lateral retia and whether they form contiguous blocks of blood vessels or are separated by WM fibers into smaller vascular bands. The common thresher sharks appear to only have one or two lateral rete arterial rows, while among lamnids makos have the lowest number (e.g., longfin mako, *Isurus paucus*, has 4 to 6; shortfin mako has ~20), followed by the white shark (20 to 30), porbeagle (42 to 46), and salmon shark (60 to 69) (Bone and Chubb, 1983; Carey et al., 1985). The retia in the white shark, porbeagle, and salmon shark form vascular bands (2 to 10+ vessels) separated from one another by WM fibers, while in the shortfin mako all the vessels in the lateral retia form a dense band extending from the lateral vessels to the RM, without intervening WM fibers (Carey et al., 1985). By contrast, the longfin mako and the common thresher shark have a very simple artery–vein–artery arrangement (Bone and Chubb, 1983; Carey et al., 1985). Taken together, retia size and complexity appear to provide different degrees of heat retention in sharks, with those having the largest and most intricate retia being able to maintain the highest relative RM temperatures and penetrate the coldest waters (Figure 7.5A).

Historically, the degree to which sharks can elevate RM temperatures was typically measured by inserting a temperature probe (thermocouples or thermistors) into freshly caught fish; the resulting temperatures were plotted in relation to the sea surface temperature (SST) to indicate the thermal excess (T_X = tissue temperature – SST) (Anderson and Goldman, 2001; Bernal and Sepulveda, 2005; Bernal et al., 2001a; Carey and Teal, 1969a; Carey et al., 1971, 1985; Goldman et al., 2004; Patterson et al., 2011). In general, the RM T_X of ectothermic sharks (i.e., lateral RM position) is small or may even be negative (i.e., RM temperature below SST) (Figure 7.5A). By contrast, sharks capable of RM endothermy have a large T_X (i.e., T_X of up to 17.5°C), and species having a broad latitudinal or depth distribution show that T_X is largest in cold waters and smallest in warmer waters (Figure 7.5) (Bernal et al., 2001a, 2009; Carey et al., 1971, 1985). In addition, the RM T_X from probed endothermic sharks ranges from 4 to 12°C (up to 17.5°C in salmon sharks) above SST, but stressed and moribund specimens generally have a lower or even negligible T_X relative to live free-swimming sharks (Anderson and Goldman, 2001; Block and Carey, 1985; Carey and Teal, 1969a; Carey et al. 1985; Smith and Rhodes, 1983; D. Bernal and K.J. Goldman, pers. obs.) (Figure 7.5).

How warm do sharks have to be to be considered endothermic? A review of the thermal data available for endothermic and ectothermic fishes by Dickson (1994) proposed that a temperature elevation of at least 2.7°C above ambient be the benchmark for determining whether a species is capable of RM endothermy. Based on this criterion, the *in vivo* temperature measurements

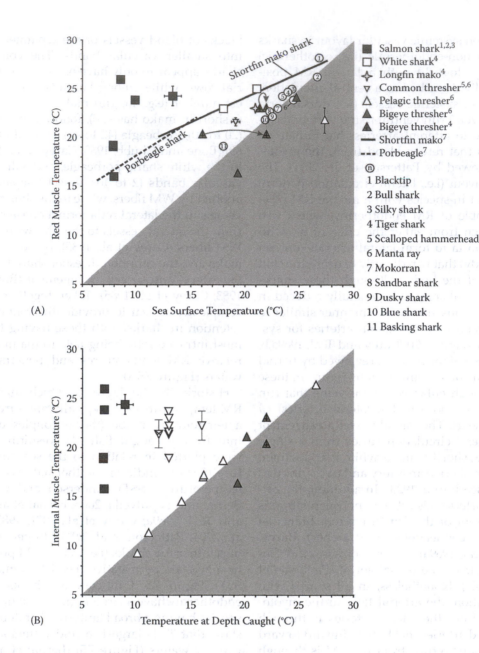

FIGURE 7.5
Muscle temperature elevation in elasmobranchs as a function of (A) sea surface temperature (SST) and (B) temperature at the depth of capture. Temperature measurement in part A are for red muscle (T_{RM}) and in part B for deep internal muscle. Shaded areas indicate the line of equality (i.e., muscle temperature = SST or temperature at depth caught). Data are shown as mean ± SE for common thresher[5,6] (*Alopias vulpinus*, $n = 24$), pelagic thresher[6] (*A. pelagicus*, $n = 7$), and salmon shark[1] (*Lamna ditropis*, $n = 31$). Individual data are shown for all others. Linear regressions in part A (i.e., T_{RM} vs. SST) are shown for shortfin mako shark[7] (*Isurus oxyrinchus*, $n = 38$, $T_{RM} = 13.8 + 0.51$ SST), and porbeagle shark[7] (*Lamna nasus*, $n = 13$, $T_{RM} = 10.8 + 0.72$ SST). Data for the internal muscle temperature of two bigeye thresher sharks in part A are likely shifted to the right (arrows) due to the presence of an inverted thermocline during capture.[4] *Sources:* [1]Anderson and Goldman (2001); [2]Bernal et al. (2001a); [3]Smith and Rhodes (1983); [4]Carey et al. (1971); [5]Bernal and Sepulveda (2005); [6]Patterson et al. (2011); [7]Carey and Teal (1969a); all others are taken from Carey et al. (1971).

of all lamnid sharks and the common thresher shark indicate their capacity for RM endothermy. By contrast, T_X data for other sharks suggest they are not capable of RM endothermy (Figure 7.5). Recent work on pelagic (*Alopias pelagicus*) and bigeye (*Alopias superciliosus*) threshers found that all myotomal muscles (i.e., RM and WM) were colder than SST (i.e., negative T_X) (Figure

7.5A) and showed a marked decrease in temperature from the exterior (lateral RM) toward the WM near the vertebrae (Figure 7.6); this temperature closely matched the ambient temperature at the depth of capture (Figure 7.5B). By contrast, the thermal data collected for lamnids and the common thresher shark indicated that the warmest *in vivo* RM temperature measurements were

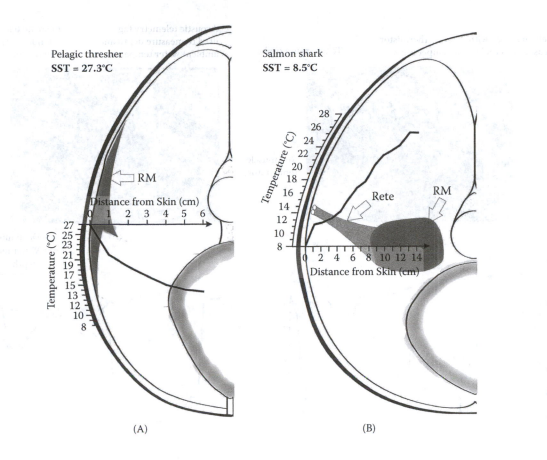

FIGURE 7.6

Thermal profiles inwards from the skin in (A) pelagic thresher (ectotherm) and (B) salmon shark (endotherm). The thermal data are super-imposed on a half-transverse section taken at mid-body showing the position of the red muscle (RM) and heat-exchanging rete (only in part B). SST, sea surface temperature. Temperature probes were inserted along the *x*-axis following the arrow. (From Patterson, J. et al., *J. Morphol.*, 272(11), 1353–1364, 2011. With permission.)

all greater than the SST; the transverse thermal gradient reflected the coolest temperatures proximal to the skin (body periphery), while the warmest measurements were in the vicinity of the medial RM (i.e., near the vertebrae) (Figure 7.6). Further, when compared to the temperatures at the depth of capture, lamnids and the common thresher RM showed a pronounced T_X (Figure 7.5B) (Bernal and Sepulveda, 2005; Patterson et al., 2011).

Previous work on bigeye thresher sharks by Carey et al. (1971) reported that two specimens had a positive T_X, but that study also mentioned that these specimens were captured in waters where a marked thermal inversion was present (SST = 12.7°C, 30 m = 22°C) and that it was not possible to determine the precise depth at which the sharks were swimming prior to capture. Therefore, it is possible that, if those bigeye thresher sharks were swimming within the cold inversion layer prior to capture, their muscle temperatures would be warmer than the SST. The different result from bigeye threshers (Figure 7.5A) illustrates the inherent types of problems with relying solely on SST as a benchmark for determining the presence of RM endothermy and make

it clear that future work not only must consider the thermal stratification of the water column but should also establish the depth at which the sharks were swimming prior to capture.

In general, water temperature in the upper 500 m of the ocean changes with depth (i.e., temperature decreases with increasing depth), and unless a fish swims at a continuous depth for a prolonged period of time it will be subject to changes in water temperatures that will inevitably alter heat balance (i.e., RM T_X). Thus, the ideal benchmark for determining the capacity for endothermy should be the degree to which tissue temperature is elevated relative to the ambient temperature at which the fish was swimming prior to capture. Accordingly, it becomes apparent that the RM temperature of ectothermic sharks closely matches that of the water at depth, and, by contrast, the RM temperatures for lamnids and the common thresher are not only consistently warmer than the SST (Figure 7.5A) but are both markedly warmer than ambient (Figure 7.5B) and the adjacent WM (Figure 7.6). However, the full extent to which the sharks are capable of RM endothermy still

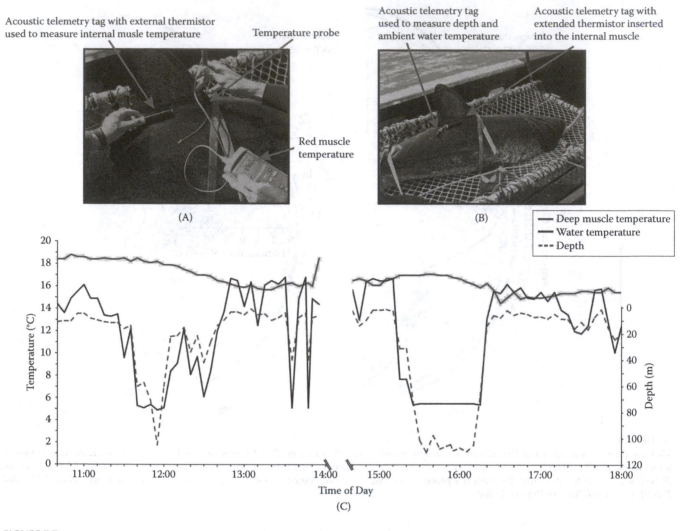

FIGURE 7.7

(See color insert.) (A) Close-up of the dorsal fin area of a 140-kg salmon shark (*Lamna ditropis*) showing the acoustic telemetry tag with an external thermistor and a real-time reading of red muscle temperature (i.e., 26.0°C) using a temperature probe. (B) Final tag placement with the extended thermistor inserted ~15 cm into the internal muscle. (Photographs courtesy of Kenneth J. Goldman.) (C) An example of vertical movement patterns of a salmon shark (shown in part B) in the Gulf of Alaska showing 9 hours of depth, ambient temperature, and internal muscle temperature recorded during an ~15.5-hour acoustic telemetry track (no data available between 14:00 and 14:40). Notice the degree to which the internal muscle temperature remains elevated relative to that of ambient water temperature, particularly during the dive between 15:00 and 16:30. (Data from K.J. Goldman, unpublished.)

remains unresolved, as it requires simultaneous measurements of (1) RM temperature, (2) ambient temperature at which the shark was swimming, (3) the duration of time spent at depth, and (4) an index of the level of swimming activity (i.e., metabolic heat production).

Because RM endothermy relies on the retention of metabolically produced heat during sustained, aerobic swimming, the swimming activity level of a shark prior to sampling can influence the degree to which the shark elevates its RM temperature. Therefore, any reduction in swimming activity due to an interaction with fishing gear that leads to exhaustion could act to reduce RM T_X when the shark is sampled boatside (Bernal and Sepulveda, 2005; Carey et al., 1985; Goldman et al.,

2004). For this reason, thermal studies on sharks routinely select for specimens that were actively swimming when landed; however, a better assessment of the capacity for RM endothermy in sharks comes from acoustic telemetry determinations of tissue temperatures in free-swimming sharks (Figure 7.7) and from laboratory-based work on sharks swimming in a water tunnel. In combination, these thermal data show that lamnids have the capacity to alter thermal balance in response to changes in ambient temperature and offer evidence of physiological thermoregulation (Bernal et al., 2001b; Carey and Lawson, 1973). Specifically, when makos were exposed to changes in the ambient water temperature while swimming at a constant speed, the magnitude and direction of changes in T_X indicated their ability to modulate rates of heat retention and heat loss (presumably by altering blood-flow rate and retial heat-transfer efficiency) (Block and Carey, 1985; Brill et al., 1994; Carey and Teal, 1969a,b; Carey et al., 1982; Dewar et al., 1994; Graham, 1983).

7.3.2 RM Endothermy and Movement Patterns

Several hypotheses have been proposed on the selective advantages of RM endothermy in fishes (Block and Finnerty, 1994; Carey et al., 1985; Dickson and Graham, 2004). One hypothesis that has received much attention and support is that of thermal niche expansion, in which the increased thermal capacity of fishes with RM endothermy may allow them to exploit the additional food resources found in cooler waters at both a greater depth and at higher latitudes (Bernal et al., 2001a). Although this scenario applies to lamnids, in which the capacity for RM endothermy appears to be linked to latitudinal distribution and thermal tolerance, it does not appear to explain the different thermal distributions of the alopiids (see below) nor does it help explain how several pelagic carcharhinids (e.g., blue shark, *Prionace glauca*) (Carey and Scharold, 1990) have the capacity to undergo prolonged dives into cooler waters.

Recent studies documenting the movement patterns and distribution of all three thresher species show that, although there may be overlap in their distributions, the ectothermic bigeye thresher routinely dives to greater depths, the endothermic common thresher enters higher latitudes, and the ectothermic pelagic thresher remains most of the time in tropical waters (Cartamil et al., 2010, 2011; Heberer et al., 2010; Liu et al., 1999; Nakano et al., 2003; Weng and Block, 2004; D. Bernal, J. Martinez, and G. Skomal, unpublished). In addition, tracking and tagging data have shown that the bigeye thresher has the greatest thermal tolerance and is able to penetrate cold ambient temperatures (6 to 10°C) for extended periods of time (6 to 8 hours) (Nakano et al., 2003; Weng and Block, 2004). This ability of the bigeye thresher to routinely penetrate cold temperatures for prolonged periods of time will inevitably lead to a low RM T_X. Recent findings have shown that in lamnid and common thresher sharks a decrease in temperature has a dramatic detrimental effect on the RM if it cools slightly below its *in vivo* operating temperature (Bernal et al., 2005; Donley et al., 2007; J. Donley, C. Sepulveda, and D. Bernal, unpublished), so the fact that bigeye threshers are able to tolerate such cold temperatures for an extended period of time during their dives is perplexing. If, for example, the RM temperature of salmon sharks, which commonly inhabit water cooler than 10°C and as cold as 2°C (Weng et al. 2005), falls below 20°C then this tissue stops producing positive work (Bernal

et al., 2005). In addition, a similar, but less pronounced, muscle performance deterioration has been documented for the RM of mako sharks if cooled below 15°C (Donley et al., 2007), even though this species repeatedly dives below the thermocline to water temperatures cooler than 13°C (Holts and Bedford, 1993; Sepulveda et al., 2004). By contrast, these thermal effects are not as prominent for other sharks that are not capable of RM endothermy (e.g., leopard shark) in which muscle function is still possible, albeit much slower (i.e., lower cycle frequencies) at cooler temperatures (below 15°C) (Donley et al., 2007). Thus, unlike the bigeye thresher, regional RM endothermy enables both lamnids and the common thresher to maintain their RM temperatures within a narrow range even when subject to cool ambient conditions and, therefore, may decrease any thermal-induced loss of muscle function when in deeper (colder) waters. The fact still remains that the bigeye thresher, a species that lacks RM endothermy, routinely experiences cold temperatures and somehow maintains adequate muscle performance over a much greater thermal range than the other two thresher species.

Other pelagic sharks, however, that are not capable of RM endothermy share the vertical (i.e., temperature–depth) and horizontal (latitude) distribution of lamnids and the common thresher shark. For example, blue sharks inhabit similar water temperatures and spend extensive periods below the thermocline (Carey and Scharold, 1990). Upon closer examination of the vertical movement data for blue sharks, lamnids, and the common thresher, there appears to be a small difference in the lower limit of water temperature that these sharks routinely penetrate, but a striking difference becomes apparent in both the frequency and duration at which these species undergo their vertical oscillations. Acoustic telemetry data for a blue shark showed that an approximately 150-minute incursion from the relatively warm surface waters (26°C) down to depths below the thermocline (9°C) resulted in a decrease in deep WM (i.e., body core) temperature from about 21 to 14°C (Carey and Scharold, 1990). If the shark remained at this depth for an additional 300 minutes, its WM (i.e., body core) temperature would ultimately decrease until reaching thermal equilibrium with ambient (Bernal et al., 2009). By contrast, if a mako shark underwent the exact same vertical dive pattern, its physiological ability to alter rates of heat gain and heat loss (Bernal et al., 2001b) would provide it with an overall warmer RM operating temperature throughout the majority of vertical excursions (Bernal et al., 2009). This outcome becomes even more pronounced if these sharks undergo repeated vertical movements with brief periods of basking at the surface (Holts and Bedford, 1993; Sepulveda et al., 2004) (Figure 7.8). In this scenario, the body temperature of the blue sharks would decline progressively with each descent, while the mako

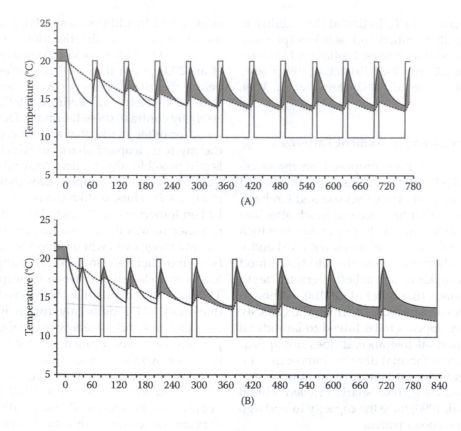

FIGURE 7.8

Body temperature changes in shortfin mako (—) and blue sharks (---) subjected to a series of modeled vertical movement patterns (i.e., rapid changes in ambient temperature, gray line) following (A) a series of identical dives spending 60 minutes at depth in cool (10°C) waters followed by 10 minutes at the surface in warmer (20°C) waters, and (B) a series of dives spending progressively longer durations of time at depth in cool (10°C) waters followed by 10 minutes at the surface in warmer (20°C) waters. Shaded areas in parts (A) and (B) indicate periods of time when the body temperature of the mako shark remained above that of the blue shark. Note that in both modeled scenarios the blue shark body temperature continues to cool with each progressive dive. Thermal rate coefficients (i.e., *k* values) were taken from Bernal et al. (2001a).

shark would maintain a more stable and warmer RM operating temperature (Bernal et al. 2009). Thus, RM endothermy may, therefore, not provide an overall larger tolerance to colder surface water, but rather it may provide the ability to make frequent sojourns into cooler water (Neill et al., 1976). On the other hand, fishes with RM endothermy may also be able to inhabit very cold, highly productive, subpolar waters (e.g., 2 to 10°C) for prolonged periods of time (i.e., numerous months), while maintaining an almost constant RM temperature 20°C or more above ambient (Goldman et al., 2004) as long as there is an ample supply of metabolic heat (i.e., aerobic swimming) to maintain RM endothermy (Figure 7.7).

7.3.3 Eye and Brain Endothermy

Whereas the eye and brain temperatures of most sharks are in thermal equilibrium with ambient water, lamnid sharks are able to elevate the temperatures of these organs through the strategic placement of heat-exchanging retia and other modifications of the vascular supply to these tissues (Anderson and Goldman, 2001; Bernal et al., 2001a; Block and Carey, 1985; Wolf et al., 1988). Unlike specialized teleosts (i.e., billfish, swordfish, tuna, and opah) (Block, 1986, 1987; Carey, 1982; Runcie et al., 2009; Sepulveda et al., 2007b) where vascular and ocular muscle specializations have been reported as the main brain-heat-producing mechanisms, lamnid sharks have a specialized vessel that transports warm venous blood from the RM to the brain and eyes. This unique vein is embedded deep within the RM and proceeds toward the head, where it joins the myelonal vein prior to entering a vascular plexus in the meningeal membrane that covers the brain. Thus, warm blood arriving from the RM drains through the posterior cerebral veins into a large sinus within the orbital cavity (i.e., the orbital sinus) (Wolf et al. 1988) and effectively elevates the temperature of the brain from 3°C above SST in mako (Block and Carey, 1985) to as much as 9.5°C above mean SST in salmon sharks (Anderson and Goldman, 2001; Bernal et al., unpublished).

The general arterial blood supply to the eyes and brain in sharks, including lamnids, is through the efferent hyoidean and pseudobranchial arteries, which deliver oxygenated blood that is at thermal equilibrium with ambient (due to their origin at the gills). However, in lamnids these arteries coil extensively and run anteriomedially and enter the orbital sinus (which is filled with warm venous blood arriving from the RM vein), in which the hyoidean artery branches into many smaller vessels to form a rete (Alexander, 1998; Block and Carey, 1985). In all lamnids, the pseudobranchial artery coils profusely within the sinus, and both its diameter and wall thickness decrease significantly, forming a true pseudobranchial rete in the salmon and porbeagle sharks but not in the mako and white shark (Alexander, 1998; Tubbesing and Block, 2000). Nonetheless, the arteries exiting the orbital sinus perfuse the eye and extraocular muscles with warmed arterial blood that in mako elevates the temperature of the eye 2.8°C above SST (Alexander, 1998; Block and Carey, 1985) and in salmon shark as much as 12.9°C above SST (Anderson and Goldman, 2001; D. Bernal and J. Graham, unpublished). In addition, the eyes receive warm blood from a tributary of the cerebral arteries, which send warmed blood to the brain after passing through the orbital sinus (Alexander, 1998; Block and Carey, 1985). Although the bigeye thresher shark has also been suspected of having cranial endothermy (Block and Finnerty, 1994; Carey, 1982; Weng and Block, 2004), no temperature data are available. Additionally, two species of myliobatoid rays possess cranial retia (Alexander, 1996); however, no temperature measurements have been obtained from these species, so their body temperatures and thermoregulatory abilities (if any) remain unknown.

Some workers have suggested that, in lamnid sharks, the extraocular eye muscles may play a role in producing metabolic heat that aids in brain and eye endothermy (Alexander, 1998; Wolf et al., 1988). Indeed, relative to other sharks, lamnids have more than twice the relative extraocular eye muscle mass (comprising 50 to 60% of the total eye weight), and the extraocular eye muscles are a darker red color, suggesting high levels of aerobic metabolism (Alexander, 1998; Block and Carey, 1985; Wolf et al., 1988). Recent morphological and histological examinations of the six extraocular muscles of the shortfin mako shark and other lamnids indicate that they lack the structural specializations for thermogenesis found in the specialized ocular muscle heater tissues of certain teleosts (Block, 1986, 1987; Carey, 1982; Runcie et al. 2009; Sepulveda et al., 2007b; Dickson, pers. comm.). However, preliminary evidence found that all six extraocular muscles are larger as a percentage of total eye mass in *Isurus oxyrinchus* than in *Prionace glauca*, and the specific activity of CS in the medial rectus extraocular muscles of the shortfin mako is significantly higher than that of the ectothermic blue shark (K. Dickson, pers. comm.). Thus, it is possible that contraction of all six extraocular muscles generates heat for

cranial endothermy in *I. oxyrinchus*, with muscle mass contributing more than CS activity to interspecific differences in heat production capacity.

In fishes, warming of the brain and eye region has been shown to enhance physiological processes such as synaptic transmission, postsynaptic integration, conduction, and, in the eye, temporal resolution (Friedlander et al., 1976; Fritsches et al., 2005; Konishi and Hickman, 1964; Montgomery and Macdonald, 1990; van den Burg et al., 2005). Recent work on swordfish (*Xiphias gladius*) shows that the flicker fusion frequency of the eye is extremely temperature sensitive (thermal coefficient, Q_{10}, of 5.1) and that warming the retina significantly improves temporal resolution (Fritsches et al., 2005). Thus, warming the retina likely enhances the swordfish's ability to detect and capture fast-moving prey at low temperatures and in dimly lit waters (Block, 1986; Fritsches et al., 2005). The convergence of cranial endothermy among billfishes, tunas, and lamnids sharks suggests a strong selection for this trait among pelagic predators that need to conserve sensory and integrative functions while in the cooler and darker deep water (Alexander, 1998; Block, 1991; Block and Carey, 1985; Linthicum and Carey, 1972; Sepulveda et al., 2007b; Wolf et al. 1988).

7.3.4 Visceral Endothermy and Homeothermy

The capacity to elevate and maintain visceral temperatures above ambient is present in all lamnid sharks and has been suspected in the common thresher shark (Bernal et al., 2001a; Carey et al., 1985; Fudge and Stevens, 1996; Goldman, 1997; Goldman et al., 2004; Sepulveda et al., 2004); however, the structural specializations in lamnids and the common thresher differ significantly. In general, blood delivery to the viscera of sharks is through the coelaic artery, but in lamnids the visceral circulation relies mainly on the delivery of arterial blood via greatly enlarged pericardial arteries that arise from the ventral region of the third and fourth efferent branchial arteries (Burne, 1923; Carey et al., 1981). These arteries extend posteriorly and branch repeatedly to form the suprahepatic rete, which is completely enclosed within a venous sinus. Before exiting this venous sinus, the arterial vessels of the rete coalesce to form a large vessel that sends warm blood to the viscera. Thus, unlike in ecothermic sharks, the principal flow of blood to and from the viscera (stomach, liver, spiral valve) in lamnids is through the suprahepatic rete. It has been suggested that visceral thermal balance may be altered by changing blood flow to evade the suprahepatic rete (Carey et al., 1981) by: (1) delivering arterial blood via the dorsal aorta to the relatively reduced celiac, spermatic, and lineogastric arteries, which flow into the viscera; or (2) bypassing the venous return through the suprahepatic rete into the sinus via a large central channel that bypasses the suprahepatic rete and empties directly into the sinus venosus (Burne, 1923; Carey et al., 1981; K. Goldman and D. Bernal, pers. obs.). Carey et al. (1981) described the presence of smooth and circular muscle within the walls of this venous vessel and suggested that this passage may be opened or closed in order to regulate (up to 20% of) returning blood flow through or around the rete. The blood delivery to the viscera of thresher sharks appears to be considerably different from lamnids based on a very brief description of visceral retia from a common thresher shark (Eschricht and Müller, 1835, in Fudge and Stevens, 1996). Eschricht and Müller (1835) described a number of retia, including a single rete along the portal vein to several others associated with the stomach wall and spiral valve. Although these retia appear to be different from those observed in tunas and porbeagle sharks (Burne, 1923), there is still a need to undergo a detailed description of visceral retia in sharks so the differences between alopiid and lamnid sharks can be clearly addressed.

Although lamnids and the common thresher differ in the location and complexity of the retia that enable visceral heat conservation, both groups appear to utilize heat produced from digestion (catabolism) to maintain elevated gut temperatures. For example, probed lamnid visceral temperatures (stomach, liver, spiral valve) range from 4 to 14°C above mean SST (Anderson and Goldman, 2001; D. Bernal and C. Sepulveda, unpublished), and the spiral valve, which digests and assimilates food arriving from the stomach and has a large size and surface area, is among the warmest organs or tissues and thus may be a main source of visceral heat production (Anderson and Goldman, 2001; D. Bernal, unpublished). In salmon and porbeagle sharks, the spiral valve and surrounding area may be assisted in staying warm due to proximity to the kidney or renal rete (Burne, 1923; K.J. Goldman and D. Bernal, pers. obs.); however, the role that this and other organs (e.g., the liver) play in lamnid visceral heat production is unknown. Temperature measurements show that the T_X within the renal rete ranges from 8 to 11.4°C (Anderson and Goldman, 2001; D. Bernal and J. Graham, unpublished), suggesting a highly effective heat-conserving function. Unfortunately, there are no visceral temperature data for any thresher shark species, leaving a gap in our knowledge of that group's ability to elevate and potentially thermoregulate body core temperature.

Whereas white sharks possess a slightly higher absolute mean body temperature than other lamnids, the maximum relative stomach temperature elevation over ambient is 8°C for shortfin mako sharks (Carey et al., 1981), 14.3°C for white sharks (Goldman, 1997), and 21.2°C for salmon sharks (Goldman et al., 2004). Stomach temperature is a good indicator and proxy for body core temperature due to its central location in the viscera and

its proximity to the suprahepatic and lateral cutaneous retia (Goldman, 1997; Goldman et al., 2004). Thermal data obtained via acoustic telemetry from mako, white, and salmon sharks show that stomach temperature remains elevated over ambient, is uniform within a very narrow range, and appears to be independent of changes in ambient temperature (Carey et al., 1981; Goldman, 1997; Goldman et al., 2004; McCosker, 1987; Sepulveda et al., 2004) (Figure 7.9). All stomach and body core temperature data from adult lamnid sharks obtained to date support the homeothermy hypothesis of Lowe and Goldman (2001). Adult lamnids appear to essentially function as homeotherms, in a way analogous to mammals, through a combination of thermal inertia and physiological thermoregulation. Additionally, the presence of elevated visceral temperatures has been considered to be a potential mechanism for enhancing the rate of digestion and assimilation (Carey, 1981; Goldman, 1997; Stevens and McLeese, 1984) and may also be a significant contributor to the warming of the body core, which may allow these fishes to penetrate and inhabit cool waters (reviewed by Dickson and Graham, 2004).

7.3.5 Endothermy and Blood-Oxygen Binding

In all sharks, the temperature of the blood leaving the gills and entering the systemic circulation is in thermal equilibrium with ambient conditions; however, in lamnids, as this cool blood approaches the RM it passes through the retia, where it is rapidly warmed by heat transfer from the venous blood returning from the warm RM. Efficient countercurrent heat exchange within the rete minimizes heat loss from the RM; however, changes in the temperature of the blood have the potential of altering the partial pressure of oxygen (PO_2) and affecting diffusion (Carey et al., 1971, 1985). For example, in sharks the warming of arterial (oxygen-rich) blood is expected to lower the affinity of hemoglobin (Hb) for oxygen, driving bound O_2 off Hb and into the plasma and increasing arterial blood PO_2 (reviewed by Bernal et al., 2009). Thus, because the rete arteries in lamnids and the common thresher shark are in close proximity to the oxygen-poor veins, a diffusion gradient would form and an arterial to venous short circuit for oxygen diffusion could potentially result in decreased O_2 delivery to the RM. Moreover, an acute change in the temperature of the blood modifies Hb quaternary structure (Rossi-Fanelli and Antonini, 1960), and, in addition to O_2 binding, this could also affect CO_2 transport and acid–base regulation (Nikinmaa and Salama, 1998). It is possible that in lamnids the presence of elevated hematocrit, hemoglobin, and myoglobin (equivalent to those of birds and mammals) (Emery, 1985) may play a role in buffering against the potential detrimental effects of a decreased blood-oxygen affinity by warming of the blood.

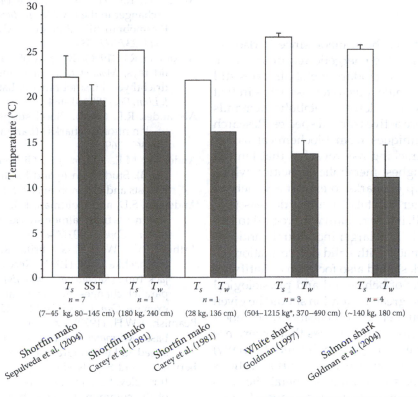

FIGURE 7.9
Stomach (T_s) and water temperature obtained by acoustic telemetry in three species of free-swimming lamnid sharks. Juvenile shortfin mako (male and female) tracks from Sepulveda et al. (2004) ranged from 6.8 to 45.4 consecutive hours, whereas tracks from Carey et al. (1981) were separated by size with the juvenile (136-cm) mako tracked for just over 5 hours and the adult (240 cm) mako tracked continuously for over 4 days. White shark (male and female) tracks ranged from 5.3 to 7.3 non-consecutive hours (data were accumulated over 7- to 10-day periods) on individuals estimated to be between 370 and 490 cm (Goldman, 1997). Salmon shark tracks ranged from 3.8 to 20.7 consecutive hours on females (Goldman et al., 2004). Mean ambient water temperatures are from the swimming depth (or mean swimming depth, T_w) of all shark studies except Sepulveda et al. (2004), which used sea surface temperature (SST). Data for this figure were summarized from text and figures in the above cited papers.

Recent work on blue and mako sharks subjected their blood to rapid temperature changes under PO_2 and PCO_2 conditions that mimicked blood passing through the rete. For blue sharks, the arterial blood-oxygen equilibrium curves showed the expected thermally induced right-shift (O_2 affinity decrease), while mako shark blood-oxygen equilibrium curves were hardly affected by this rapid warming (D. Bernal, J. Graham, and J. Cech, unpublished). This apparent temperature insensitivity decreases the potential for trans-retial oxygen diffusion and ensures the efficient delivery of oxygen to the aerobic RM and other tissues. These results for whole mako blood are similar to findings for both tunas and other lamnids in which neither crystallized Hb nor whole blood preparations showed expected loss of oxygen affinity with an increase in temperature (Andersen et al., 1973; Brill and Bushnell, 1991; Cech et al., 1994; Larsen et al., 2004; Rossi-Fanelli and Antonini, 1960; Sharp, 1975). Thus, the presence of a low thermal effect on oxygen binding in some tunas

and lamnids reflects their convergence for endothermy and the need to prevent premature oxygen dissociation within the rete.

Future studies of the blood-oxygen binding properties in the tropically distributed longfin mako, coldwater salmon, and porbeagle sharks, as well as the three thresher species, may demonstrate the importance of changes in ambient temperature and RM T_X on blood-oxygen affinity. For example, the longfin mako and pelagic thresher are distributed in tropical environments and, to some extent, appear to have either a small RM T_X or none at all (Carey et al., 1985) (Figure 7.5). Therefore, thermal sensitivity in the blood-oxygen binding properties of these species is predicted. In contrast, salmon, porbeagle, and common thresher sharks penetrate cold waters, and in some species a large RM T_X (Figure 7.5) is necessary to maintain proper RM function (Bernal et al., 2005). Thus, for those species an adequate delivery of O_2 to the warm RM may require thermally insensitive blood-oxygen binding properties.

7.4 Summary

Considerable progress has been made since Carlson et al. (2004) on elucidating the energetic requirements of elasmobranchs. Trends in standard metabolic rates still follow the continuum analogous to teleost fishes in that cooler water species have lower metabolic demands with respective to more active tropical species. Research using a suite of techniques from biochemical assays to archival satellite tracking has revealed that lamnid sharks possess the highest metabolic capacities, which may enable this group of sharks to exploit a variety of niches from deepwater habitats to subarctic seas (e.g., Weng et al., 2005). Still, issues remain with regard to laboratory testing, especially of larger individuals, and correlation of these estimates with valid determinations in the field. Further work should also focus on quantifying the energetic tradeoffs of behavioral and physiological thermoregulation, and greater attention should be given to studying the thermal effects on elasmobranch energetics, especially in ectothermic species that experience large changes in temperature (seasonally and diurnally) and those capable of endothermy. In addition, little is known about the effects of scaling on metabolic rates in elasmobranchs, which span a wide range of body mass and include some of the largest fishes in the ocean; future work should combine laboratory and field experiments and incorporate energetic modeling. There also is a need to quantify how endothermy may vary across size and age classes in lamnid and alopiid sharks and the degree to which free-swimming sharks can remain warm and have the capacity to thermoregulate.

Acknowledgments

We thank J. Valdez, T. Tazo, S. Adams, and T. Reposado for their unwavering logistical support. We are indebted to Jeff Carrier for being so patient and flexible with our multiple deadlines. Most importantly, we thank our wives, children, and pets for being so understanding about the time we had to dedicate to this piece of work. This work is dedicated to J.B Graham (grandpa). For some of us, not enough can be said to describe the significant impact he has had on our work and in our lives.

References

Akerman, J.T., Kondratieff, M.C., Matern, S.A., and Cech, J.J. (2000). Tidal influence on spatial dynamics of leopard sharks, *Triakis semifasciata*, in Tomales Bay, California. *Environ. Biol. Fish.* 58:33–43.

Alexander, R.L. (1995). Evidence of a counter–current heat exchanger in the ray *Mobula tarapacana* (Chondrichthyes: Elasmobranchii: Batiodea: Myliobatiformes). *J. Zool. Lond.* 237:377–384.

Alexander, R.L. (1996). Evidence of brain-warming in the mobulid rays, *Mobula tarapacana* and *Manta birostris* (Chondrichthyes: Elasmobranchii: Batiodea: Myliobatiformes). *J. Linn. Soc.* 188:151–164.

Alexander, R.L. (1998). Blood supply to the eyes and brain of lamniform sharks (Lamniformes). *J. Zool. Lond.* 245:363–369.

Andersen, M.E., Olson, J.S., Gibson, Q.H., and Carey, F.G. (1973). Studies on ligand binding to hemoglobins from teleosts and elasmobranchs. *J. Biol. Chem.* 248:331–341.

Anderson, S.D. and Goldman, K.J. (2001). Temperature measurements from salmon sharks, *Lamna ditropis*, in Alaskan waters. *Copeia* 2001:794–796.

Arthur, P.G., West, T.G., Brill, R.W., Schulte, P.M., and Hochachka, P.W. (1992). Recovery metabolism of skipjack tuna (*Katsuwonus pelamis*) white muscle rapid and parallel changes in lactate and phosphocreatine after exercise. *Can. J. Zool.* 70:1230–1239.

Beamish, F.W.H. (1970). Oxygen consumption of largemouth bass, *Micropterus salmoides*, in relation to swimming speed and temperature. *Can. J. Zool.* 48:1221–1228.

Bernal, D. and Sepulveda, C. (2005). Evidence for temperature elevation in the aerobic swimming musculature of the common thresher shark, *Alopias vulpinus*. *Copeia* 1:146–151.

Bernal, D., Dickson, K.A., Shadwick, R.E., and Graham, J.B. (2001a). Review: analysis of the evolutionary convergence for high performance swimming in lamnid sharks and tunas. *Comp. Biochem. Physiol.* 129A:695–726.

Bernal, D., Sepulveda, C., and Graham, J.B. (2001b). Water-tunnel studies of heat balance in swimming mako sharks. *J. Exp. Biol.* 204:4043–4054.

Bernal, D., Sepulveda, C., Mathieu-Costello, O., and Graham, J.B. (2003a). Comparative studies of high performance swimming in sharks. I. Red muscle morphometrics, vascularization and ultrastructure. *J. Exp. Biol.* 206:2831–2843.

Bernal, D., Smith, D., Lopez, G., Weitz, D., Grimminger, T., Dickson, K., and Graham, J.B. (2003b). Comparative studies of high performance swimming in sharks. II. Metabolic biochemistry of locomotor and myocardial muscle in endothermic and ectothermic sharks. *J. Exp. Biol.* 206:2845–2857.

Bernal, D., Donley, J.M., Shadwick, R.E., and Syme, D.A. (2005). Mammal-like muscles power swimming in a cold-water shark. *Nature* 437:1349–1352.

Bernal, D., Sepulveda, C., Musyl, M., and Brill, R. (2009). The eco-physiology of swimming and movement patterns of tunas, billfishes, and large pelagic sharks. In: Domenici, P. and Kapoor, D. (Eds.), *Fish Locomotion: An Etho-Ecological Perspective*. Science Publishers, Enfield, NH, pp. 436–483.

Blank, J.M., Farwell, C.J., Morrissette, J.M., Schallert, R.J., and Block, B.A. (2007). Influence of swimming speed on metabolic rates of juvenile Pacific bluefin tuna and yellowfin tuna. *Physiol. Biochem. Zool.* 80:167–177.

Block, B.A. (1986). Structure of the brain and eye heater tissue in marlins, sailfish, and spearfishes. *J. Morphol.* 190:169–189.

Block, B.A. (1987). Billfish brain and eye heater: a new look at nonshivering heat production. *News Physiol. Sci.* 2:208–213.

Block, B.A. (1991). Endothermy in fish: thermogenesis ecology and evolution. In: P.W. Hochachka and Mommsen, T. (Eds.), *Biochemistry and Molecular Biology of Fishes*, Vol. 1. Elsevier, Amsterdam, pp. 269–311.

Block, B.A. and Carey, F.G. (1985). Warm brain and eye temperatures in sharks. *J. Comp. Physiol. B* 156:229–236.

Block, B.A. and Finnerty, J.R. (1994). Endothermy in fishes: a phylogenetic analysis of constraints, predispositions, and selection pressures. *Environ. Biol. Fish.* 40:283–302.

Bone, Q. (1988). Muscles and locomotion. In: Shuttleworth, T.J. (Ed.), *The Physiology of Elasmobranch Fishes*. Springer-Verlag, New York, pp. 99–141.

Bone, Q. and Chubb, A.D. (1983). The retial system of the locomotor muscles in the thresher shark. *J. Mar. Biol. Assoc. U.K.* 63:239–242.

Brett, J.R. (1964). The respiratory metabolism and swimming performance of young sockeye salmon. *J. Fish. Res. Bd. Can.* 21:1183–1225.

Brett, J.R. and Blackburn, J.M. (1978). Metabolic rate and energy expenditure of the spiny dogfish, *Squalus acanthias*. *J. Fish. Res. Bd. Can.* 35:816–821.

Brill, R.W. (1987). On the standard metabolic rates of tropical tunas, including the effect of body size and acute temperature change. *Fish. Bull.* 85:25–35.

Brill, R.W. (1996). Selective advantages conferred by the high performance physiology of tunas, billfishes, and dolphinfish. *Comp. Biochem. Physiol.* 113A:3–15.

Brill, R.W. and Bushnell, P.G. (1991). Effects of open and closed-system temperature changes on blood oxygen dissociation curves of skipjack tuna (*Katsuwonus pelamis*) and yellowfin tuna (*Thunnus albacares*). *Can. J. Zool.* 69:1814–1821.

Brill, R.W., Dewar, H., and Graham, J.B. (1994). Basic concepts relevant to heat transfer in fishes, and their use in measuring the physiological thermoregulatory abilities of tunas. *Environ. Biol. Fish.* 40:109–124.

Burne, R.H. (1923). Some peculiarities of the blood vascular system of the porbeagle shark, *Lamna cornubica*. *Phil. Trans. R. Soc. Lond. B* 212:209–257.

Bushnell, P.G. and Jones, D.R. (1994). Cardiovascular and respiratory physiology of tuna: adaptations for support of exceptionally high metabolic rates. *Environ. Biol. Fish.* 40:303–318.

Bushnell, P.G., Lutz, P.L., and Gruber, S.H. (1989). The metabolic rate of an active, tropical, elasmobranch, the lemon shark (*Negaprion brevirostris*). *Exp. Biol.* 48:279–283.

Campos, B.R., Fish, M.A., Jones, G., Riley, R.W., Allen, P.J., Klimley, P.A., Cech, J.J., and Kelly, J.Y. (2009). Movements of brown smoothhounds, *Mustelus henlei*, in Tomales Bay, California. *Environ. Biol. Fish.* 85:3–13.

Carey, F.G. (1973). Fishes with warm bodies. *Sci. Am.* 228(2): 36–44.

Carey, F.G. (1982). Warm fish. In: Taylor, C.R., Johansen, K., and Bolis, L. (Eds.), *A Companion to Animal Physiology*, Fifth International Conference on Comparative Physiology, Sandbjerg, Denmark. Cambridge University Press, Cambridge, U.K., pp. 216–234.

Carey, F.G. and Gibson, Q.H. (1983). Heat and oxygen exchange in the *rete mirabile* of the bluefin tuna (*Thunnus thynnus*). *Comp. Biochem. Physiol.* 74:333–342.

Carey, F.G. and Lawson, K.D. (1973). Temperature regulation in free swimming bluefin tuna. *Comp. Biochem. Physiol.* 44:375–392.

Carey, F.G. and Scharold, J.V. (1990). Movements of blue sharks *Prionace glauca* in depth and course. *Mar. Biol.* 106:329–342.

Carey, F.G. and Teal, J.M. (1966). Heat conservation in tuna fish muscle. *Proc. Natl. Acad. Sci. U.S.A.* 56:1464–1469.

Carey, F.G. and Teal, J.M. (1969a). Mako and porbeagle: warm-bodied sharks. *Comp. Biochem. Physiol.* 28:199–204.

Carey, F.G. and Teal, J.M. (1969b). Regulation of body temperature by the bluefin tuna. *Comp. Biochem. Physiol.* 28:205–213.

Carey, F.G., Teal, J.M., Kanwisher, J.W., Lawson, K.D., and Beckett, J.S. (1971). Warm-bodied fish. *Am. Zool.* 11:137–143.

Carey, F.G., Teal, J.M., and Kanwisher, J.W. (1981). The visceral temperatures of mackerel sharks (Lamnidae). *Physiol. Zool.* 54:334–344.

Carey, F.G., Kanwisher, W., Brazier, O., Gabrielson, G., Casey, J.G., and Pratt, Jr., H.L. (1982). Temperature and activity of a white shark, *Carcharodon carcharias*. *Copeia* 1982:254–260.

Carey, F.G., Casey, J.G., Pratt, H.L., Urquhart, D., and McCosker, J.E. (1985). Temperature, heat production, and heat exchange in lamnid sharks. *S. Calif. Acad. Sci. Mem.* 9: 92–108.

Carlson, J.K. and Parsons, G.R. (1999). Seasonal differences in routine oxygen consumption rates of the bonnethead shark. *J. Fish Biol.* 55:876–879.

Carlson, J.K. and Parsons, G.R. (2003). Respiratory and hematological responses of the bonnethead shark, *Sphyrna tiburo*, to acute changes in dissolved oxygen. *J. Exp. Mar. Biol. Ecol.* 294:15–26.

Carlson, J.K., Palmer, C.P., and Parsons, G.R. (1999). Oxygen consumption rate and swimming efficiency of the blacknose shark, *Carcharhinus acronotus*. *Copeia* 1999:34–39.

Carlson, J.K., Goldman, K.J., and Lowe, C.G. (2004). Metabolism, energetic demand, and endothermy. In: Carrier, J.C., Musick, J.A., and Heithaus, M.R. (Eds.), *Biology of Sharks and Their Relatives*. CRC Press, Boca Raton, FL, pp. 203–224.

Cartamil, D., Wegner, N.C., Aalbers, S., Sepulveda, C.A., Baquero, A., and Graham, J.B. (2010). Diel movement patterns and habitat preferences of the common thresher shark *Alopias vulpinus* in the Southern California Bight. *J. Mar. Fresh. Res.* 61:596–604.

Cartamil, D., Sepulveda, C.A., Wegner, N.C., Aalbers, S.A., and Graham, J.B. (2011). Archival tagging of subadult and adult common thresher sharks (*Alopias vulpinus*) off the coast of Southern California. *Mar. Biol.* 158:935–944.

Castro, J.I. (1996). Biology of the blacktip shark, *Carcharhinus limbatus*, off the southeastern United States. *Bull. Mar. Sci.* 59:508–522.

Cech, Jr., J.J., Laurs, R.M., and Graham, J.B. (1984). Temperature-induced changes in blood gas equilibria in the albacore (*Thunnus alalunga*), a warm bodied tuna. *J. Exp. Biol.* 109:21–34.

Clarke, A. (1991). What is cold adaptation and how should we measure it? *Am. Zool.* 31:81–92.

Clarke, A. and Johnston, N.M. (1999). Scaling of metabolic rate with body mass and temperature in teleost fish. *J. Anim. Ecol.* 68:893–905.

Davenport, J. and Sayer, M.D.J. (1993). Physiological determinants of distribution in fishes. *J. Fish Biol.* 43:121–145.

Dewar, H. and Graham, J.B. (1996). Studies of tropical tuna swimming performance in a large water tunnel. I. Energetics. *J. Exp. Biol.* 192:13–31.

Dewar, H., Graham, J.B., and Brill, R.W. (1994). Studies of tropical tuna swimming performance in a large water tunnel. II. Thermoregulation. *J. Exp. Biol.* 192:33–44.

Di Santo, V. and Bennett, W.A. (2011). Is post-feeding thermotaxis advantageous in elasmobranch fishes? *J. Fish Biol.* 78:195–207.

Dickson, K.A. (1994). Tunas as small as 207 mm fork length can elevate muscle temperatures significantly above ambient water temperature. *J. Exp. Biol.* 190:79–93.

Dickson, K.A. (1995). Unique adaptations of the metabolic biochemistry of tunas and billfishes for life in the pelagic environment. *Environ. Biol. Fish.* 42:65–97.

Dickson, K.A. (1996). Locomotor muscle of high-performance fishes: what do comparisons of tunas with ectothermic sister taxa reveal? *Comp. Biochem. Physiol.* 113(1):39–49.

Dickson, K.A. and Graham, J.B. (2004). Evolution and consequences of endothermy in fishes. *Physiol. Biochem. Zool.* 77:998–1018.

Dickson, K.A., Gregorio, M.O., Gruber, S.J., Loefler, K.L., Tran, M., and Terrell, C. (1993). Biochemical indices of aerobic and anaerobic capacity in muscle tissues of California elasmobranch fishes differing in typical activity level. *Mar Biol.* 117:185–193.

Dizon, A.E. and Brill, R.W. (1979). Thermoregulation in tunas. *Am. Zool.* 19:249–265.

Domeier, M. and Nasby-Lucas, N. (2008). Migration patterns of white sharks *Carcharodon carcharias* tagged at Guadalupe Island, Mexico, and identification of an eastern Pacific shared offshore foraging area. *Mar. Ecol. Prog. Ser.* 370:221–237.

Donley, J.M., Shadwick, R.E., Sepulveda, C.A., and Syme, D.A. (2007). Thermal dependence of contractile properties of the aerobic locomotor muscle in the leopard shark and shortfin mako shark. *J. Exp. Biol.* 210:1194–1203.

Dowd, W.W., Brill, R.W., Bushnell, P.G., and Musick, J.A. (2006). Estimating consumption rates of juvenile sandbar sharks (*Carcharhinus plumbeus*) in Chesapeake Bay, Virginia, using a bioenergetics model. *Fish. Bull.* 104:332–342.

DuPreez, H.H., McLauchlan, A., and Marias, J.F.K. (1988). Oxygen consumption of two nearshore elasmobranchs, *Rhinobatus annulatus* (Muller and Henle, 1841) and *Myliobatis aquila* (Linnaeus, 1758). *Comp. Biochem. Physiol.* 89A:283–294.

Economakis, A.E. and Lobel, P.S. (1998). Aggregation behaviour of the grey reef shark, *Carcharhinus amblyrhynchos*, at Johnston Atoll, Central Pacific Ocean. *Environ. Biol. Fish.* 51:129–139.

Emery, S.H. (1985). Hematology and cardiac morphology in the great white shark, *Carcharodon carcharias. S. Calif. Acad. Sci. Mem.* 9:73–80.

Eschricht, D.F. and Müller, J. (1835). Nachtrag zu der Abhandlung der Herren Eschricht und Müller über die Wundernetze an der Leber des Thunfisches. Über die Wundernetze am Darmkanal des *Squalus vulpes* L., Alopecias vulpes Nob. Physikal. Abhandl. d. K.Wissensch, Berlin, pp. 325–328 (after Fudge and Stevens, 1996).

Espinoza, M., Farrugia, T.J., and Lowe, C.G. (2011). Habitat use, movements and site fidelity of the gray smoothhound shark (*Mustelus californicus* Gill 1863) in a newly restored southern California estuary. *J. Exp. Mar. Biol. Ecol.* 401:63–74.

Ezcurra, J.M. (2001). The Mass-Specific Routine Metabolic Rate of Captive Pelagic Stingrays, *Dasyatis violacea*, with Comments on Energetics, master's thesis, The California State University, Moss Landing Marine Laboratory, Stanislaus, 64 pp.

Ezcurra, J.M., Lowe, C.G., Mollet, M.F., Ferry-Graham, L.A., and O'Sullivan, J.B. (2012a). Routine metabolic rate of young-of-the-year white sharks (*Carcharodon carcharias*) at the Monterey Bay Aquarium. In: Domeier, M. (Ed.), *Global Perspectives on the Biology and Life History of the Great White Shark.* Taylor & Francis, New York.

Ezcurra, J.M., Lowe, C.G., Mollet, M.F., Ferry-Graham, L.A., and O'Sullivan, J.B. (2012b). Captive feeding and growth of young-of-the-year white sharks (*Carcharodon carcharias*) at the Monterey Bay Aquarium. In: Domeier, M. (Ed.), *Global Perspectives on the Biology and Life History of the Great White Shark.* Taylor & Francis, New York.

Ferry-Graham, L.A. and Gibb, A.C. (2001). Comparison of fasting and postfeeding metabolic rates in a sedentary shark, *Cephaloscyllium ventriosum. Copeia* 2001:1108–1113.

Fournier, R.W. (1996). The Metabolic Rates of Two Species of Benthic Elasmobranchs, Nurse Sharks and Southern Stingrays, masters thesis, Hofstra University, Hempstead, NY, 29 pp.

Friedlander, M.J., Kotchabhakdi, N., and Prosser, C.L. (1976). Effects of cold and heat on behavior and cerebellar function in goldfish. *J. Comp. Physiol.* 112:19–45.

Fritsches, K.A., Brill, R.W., and Warrant, E.J. (2005). Warm eyes provide superior vision in swordfishes. *Curr. Biol.* 15:55–58.

Fry, F.E. (1971). The effect of environmental factors on the physiology of fish. In: Hoar, W.S. and Randall, D.J. (Eds.), *Fish Physiology*, Vol. VI. Academic Press, New York, pp. 1–98.

Fudge, D.S. and Stevens, E.D. (1996). The visceral retia mirabilia of tuna and sharks: an annotated translation and discussion of the Eschricht & Müller 1835 paper and related papers. *Guelph Ichthyol. Rev.* 4:1–92.

Gemballa, S., Konstantinidis, P., Donley, J.M., Sepulveda, C.A., and Shadwick, R.E. (2006). Evolution of high-performance swimming in sharks: transformations of the musculotendinous system from subcarangiform to thunniform swimmers. *J. Morphol.* 267:477–493.

Gleiss, A.C., Dale, J.J., Holland, J.N., and Wilson, R.P. (2010). Accelerating estimates of activity-specific metabolic rate in fish: testing the applicability of acceleration data-loggers. *J. Exp. Mar. Biol. Ecol.* 385:85–91.

Goldman, K.J. (1997). Regulation of body temperature in the white shark, *Carcharodon carcharias. J. Comp. Physiol. B* 167:423–429.

Goldman, K.J. (2002). Aspects of Age, Growth, Demographics and Thermal Biology of Two Lamniform Shark Species, doctoral dissertation, College of William and Mary, School of Marine Science, Virginia Institute of Marine Science, 220 pp.

Goldman, K.J., Anderson, S.D., Latour, S.J., and Musick, J.A. (2004). Homeothermy in adult salmon sharks, *Lamna ditropis. Environ. Biol. Fish.* 71:403–411.

Graham, J.B. (1983). Heat transfer. In: Webb, P.W. and Weihs, D. (Eds.), *Fish Biomechanics*. Praeger, New York, pp. 248–278.

Graham, J.B, Dewar, H., Lai, N.C., Lowell, W.R., and Arce, S.M. (1990). Aspects of shark swimming performance determined using a large water tunnel. *J. Exp. Biol.* 151:175–192.

Hanson, D. and Johansen, K. (1970). Relationships of gill ventilation and perfusion in Pacific dogfish, *Squalus suckleyi. J. Fish. Res. Bd. Can.* 27:551–564.

Halsey, L.G., Green, J.A., Wilson, R.P., and Frappell, P.B. (2009). Accelerometry to estimate energy expenditure during activity: best practice with data loggers. *Physiol. Biochem. Zool.* 82:396–404.

Heberer, C., Aalbers, S.A., Bernal, D., Kohin, S., DiFioree, B., and Sepulveda, C.A. (2010). Insights into catch-and-release survivorship and stress-induced blood biochemistry of common thresher sharks (*Alopias vulpinus*) captured in the southern California recreational fishery. *Fish. Res.* 106:495–500.

Hight, B.V. and Lowe, C.G. (2007). Behavioral thermoregulation in adult female leopard sharks, *Triakis semifasciata*, in nearshore embayments. *J. Exp. Mar. Biol. Ecol.* 352:114–128.

Hoisington, G. and Lowe, C.G. (2005). Abundance and distribution of the round stingray, *Urobatis halleri*, near a heated effluent outfall. *Mar. Environ. Res.* 60:437–453.

Holts, D.B. and Bedford, D.W. (1993). Horizontal and vertical movements of the shortfin mako, *Isurus oxyrinchus*, in the southern California Bight. *Aust. J. Mar. Fresh. Res.* 44:45–60.

Hopkins, T.E. and Cech, Jr., J.J. (1994). Effect of temperature on oxygen consumption of the bat ray, *Myliobatis californica* (Chondrichthyes, Myliobatidae). *Copeia* 1994:529–532.

Hove, J.R. and Moss, S.A. (1997). Effect of MS-222 on response to light and rate of metabolism of the little skate *Raja erinacea. Mar. Biol.* 128:579–583.

Jirik, K. (2009). Influence of Temperature on the Habitat Use and Movement Patterns of Round Stingrays in a Southern California Estuary, master's thesis, The California State University, Long Beach, 85 pp.

Johnston, I.A. (1981). Structure and function of fish muscles. *Symp. Zool. Soc. Lond.* 48:71–113.

Jorgensen, S.J., Reeb, C.A., Chapple, T.K., Anderson, S., Perle, C., Van Sommeran, S.R., Fritz-Cope, C., Brown, A.C., Klimley, A.P., and Block, B.A. (2009). Philopatry and migration of Pacific white sharks. *Proc. R. Soc. B Biol. Sci.* 277:679–688.

Kishinouye, K. (1923). Contributions to the comparative study of the so-called scombroid fishes. *Immunol. Rev.* 8(3):293–475.

Klimley, A.P., Kelly, J.T., and Kihslinger, R.L. (2005). Directional and non-directional movements of bat rays, *Myliobatis californica*, in Tomales Bay, California. *Environ. Biol. Fish.* 74:79–88.

Kohler, N.E., Casey, C.G., and Turner, P.A. (1995). Length–weight relationships for 13 species of sharks from the western North Atlantic. *Fish. Bull.* 93:412–418.

Konishi, J. and Hickman, C.P. (1964). Temperature acclimation in the central nervous system of rainbow trout (*Salmo gairdnerii*). *Comp. Biochem. Physiol.* 13:433–442.

Korsmeyer, K.E. and Dewar, H. (2001). Tuna metabolism and energetics. In: Block, B.A. and Stevens, E.D. (Eds.), *Tunas: Physiology, Ecology and Evolution*. Academic Press, San Diego, CA, pp. 35–78.

Korsmeyer, K.E., Dewar, H., Lai, N.C., and Graham, J.B. (1996). The aerobic capacity of tunas: adaptation for multiple metabolic demands. *Comp. Biochem. Physiol.* 113A:17–24.

Larsen, C., Malte, H., and Weber, R.E. (2004). ATP-induced reverse temperature effect in isohemoglobins from the endothermic porbeagle shark, *Lamna nasus. J. Biol. Chem.* 278:30741–30747.

Linthicum, D.S. and Carey, F.G. (1972). Regulation of brain and eye temperatures by the bluefin tuna. *Comp. Biochem. Physiol.* 43:425–433.

Liu, K.M., Chen, C.T., Liao, T.H., and Joung, S.J. (1999). Age, growth, and reproduction of the pelagic thresher, *Alopias pelagicus*, in the northwestern Pacific. *Copeia* 1999:68–74.

Lowe, C.G. (2001). Metabolic rates of juvenile scalloped hammerhead sharks (*Sphyrna lewini*). *Mar. Biol.* 139:447–453.

Lowe, C.G. (2002). Bioenergetics of free-ranging juvenile scalloped hammerhead sharks (*Sphyrna lewini*) in Kaneohe Bay, Hawaii. *J. Exp. Mar. Biol. Ecol.* 278(2):139–154.

Lowe, C.G. and Bray, R.N. (2006). Fish movement and activity patterns. In: Allen, L.G., Horn, M.H., and Pondella, D.J. (Eds.), *The Ecology of California Marine Fishes*. University of California Press, Berkeley, pp. 524–553.

Lowe, C.G. and Goldman, K.J. (2001). Thermal and bioenergetics of elasmobranchs: bridging the gap. *Environ. Biol. Fish.* 60:251–266.

Lowe, C.G., Holland, K.N., and Wolcott, T.G. (1998). A new acoustic tailbeat transmitter for fishes. *Fish. Res.* 36:275–283.

Matern, S.A., Cech, Jr., J.J., and Hopkins, T.E. (2000). Diel movements of bat rays, *Myliobatis californica*, in Tomales Bay, California: evidence for behavioral thermoregulation? *Environ. Biol. Fish.* 58:173–182.

McCosker, J.E. (1987). The white shark, *Carcharodon carcharias*, has a warm stomach. *Copeia* 1987:195–197.

McEachran, J.D. (1990). Diversity of rays: why are there so many species? *Chondros* 2:1–6.

Miklos, P., Katzman, S.M., and Cech, Jr., J.J. (2003). Effects of temperature on oxygen consumption of the leopard shark, *Triakis semifasciata. Environ. Biol. Fish.* 66:15–18.

Montgomery, J.C. and Macdonald, J.A. (1990). Effects of temperature on nervous system: implications for behavioral performance. *Am. J. Physiol. Reg. Integr. Comp. Physiol.* 259:191–196.

Mull, C.G., Lowe, C.G., and Young, K.A. (2008). Photoperiod and water temperature regulation of seasonal reproduction in male round stingrays (*Urobatis halleri*). *J. Comp. Biochem. Physiol. A* 151:717–725.

Mull, C.G., Lowe, C.G., and Young, K.A. (2010). Seasonal reproduction of female round stingrays (*Urobatis halleri*): steroid hormone profiles and assessing reproductive state. *Gen. Comp. Endocrinol.* 166:379–387.

Nakano, H., Matsunaga, H., and Hiroaki, M. (2003). Acoustic tracking of bigeye thresher shark, *Alopias superciliosus*, in the eastern Pacific Ocean. *Mar. Ecol. Prog. Ser.* 265:255–261.

Nasby-Lucas, N., Dewar, H., Lam, C.H., Goldman, K.J., and Domeier, M.L. (2009). White shark offshore habitat: a behavioral and environmental characterization of the eastern pacific shared offshore foraging area. *PLoS ONE* 4(12):e8163.

Natanson, L.J., Kohler, N.E., Ardizzone, D., Cailliet, G.M., Wintner, S.P., and Mollet, H.F. (2006). Validated age and growth estimates for the shortfin mako, *Isurus oxyrinchus*, in the North Atlantic Ocean. *Environ. Biol. Fishes* 77:367–383.

Neer, J.A., Carlson, J.K., and Thompson, B.A. (2006). Standard oxygen consumption of seasonally acclimatized cownose rays, *Rhinoptera bonasus* (Mitchill 1815), in the northern Gulf of Mexico. *Fish Physiol. Biochem.* 32:67–71.

Neill, W.H., Randolph, K.C., Chang, K.C., and Dizon, A.E. (1976). Magnitude and ecological implications of thermal inertia in skipjack tuna, *Katsuwonus pelamis* (Linnaeus). *Environ. Biol. Fish.* 1:61–80.

Nikinmaa, M. and Salama, A. (1998). Oxygen transport in fish. In: Perry, S.F. and Tufts, B.I. (Eds.), *Fish Physiology*, Vol. 17. Academic Press, San Diego, CA, pp. 141–184.

Parsons, G.R. (1990). Metabolism and swimming efficiency of the bonnethead shark, *Sphyrna tiburo*. *Mar. Biol.* 104:363–367.

Parsons, G.R. and Carlson, J.K. (1998). Physiological and behavioral responses to hypoxia in the bonnethead shark, *Sphyrna tiburo*: routine swimming and respiratory regulation. *Fish Physiol. Biochem.* 19:189–196.

Patterson, J., Sepulveda, C., and Bernal, D. (2011). The vascular morphology and *in vivo* muscle temperatures of thresher sharks (Alopiidae). *J. Morphol.* 272(11):1353–1364.

Piiper, J., Meyer, H., Worth, H., and Willmer, H. (1977). Respiration and circulation during swimming activity in the dogfish, *Scyliorhinus stellaris*. *Respir. Physiol.* 30:221–239.

Priede, I.G. (1985). Metabolic scope in fishes. In: Tytler, P. and Calow, P. (Eds.), *Fish Energetics: New Perspectives*. The Johns Hopkins University Press, Baltimore, MD, pp. 33–64.

Rome, L.C., Funke, R.P., Alexander, R.M., Lutz, G., Aldridge, H., Scott, F., and Freadman, M. (1988). Why animals have different muscle fibre types. *Nature* 335:824–827.

Rossi-Fanelli, A. and Antonini, E. (1960). Oxygen equilibrium of hemoglobin from *Thunnus thynnus*. *Nature* 186:895–896.

Runcie, R.M., Heidi, D., Hawn, D.R., Frank, L.R., and Dickson, K.A. (2009). Evidence for cranial endothermy in the opah (*Lampris guttatus*) *J. Exp. Biol.* 212:461–470.

Scharold, J. and Gruber, S.H. (1991). Telemetered heart rate as a measure of metabolic rate in the lemon shark, *Negaprion brevirostris*. *Copeia* 1991:942–953.

Scharold, J., Lai, N.C., Lowell, W.R., and Graham, J.B. (1989). Metabolic rate, heart rate, and tailbeat frequency during sustained swimming in the leopard shark *Triakis semifasciata*. *Exp. Biol.* 48:223–230.

Schmidt-Nielsen, K. (1983). *Animal Physiology: Adaptation and Environment*. Cambridge University Press, New York, 619 pp.

Schurmann, H. and Steffensen, J.F. (1997). Effects of temperature, hypoxia and activity on the metabolism of Atlantic cod, *Gadus morhua*. *J. Fish Biol.* 50:1166–118.

Sepulveda, C.A., Kohin, S., Chan, C., Vetter, R., and Graham, J.B. (2004). Movement patterns, depth preferences, and stomach temperatures of free-swimming juvenile mako sharks, *Isurus oxyrinchus*, in the Southern California Bight. *Mar. Biol.* 145:191–199.

Sepulveda, C.A., Wegner, N.C., Bernal, D., and Graham, J.B. (2005). The red muscle morphology of the thresher sharks (family Alopiidae). *J. Exp. Biol.* 208:4255–4261.

Sepulveda, C.A., Graham, J.B., and Bernal, D. (2007a). Aerobic metabolic rates of swimming juvenile mako sharks, *Isurus oxyrinchus*. *Mar. Biol.* 152:1087–1094.

Sepulveda, C.A., Dickson, K.A., Frank, L.R., and Graham, J.B. (2007b). Cranial endothermy and a putative brain heater in the most basal tuna species, *Allothunnus fallai*. *J. Fish Biol.* 70:1720–1733.

Sharp, D. (1975). A comparison of the O_2 dissociation properties of some scombrid hemoglobins. *Comp. Biochem. Physiol.* 51A:683–691.

Sims, D.W. (1996). The effect of body size on the metabolic rate of the lesser spotted dogfish. *J. Fish Biol.* 48:542–544.

Sims, D.W. (2000). Can threshold foraging responses of basking sharks be used to estimate their metabolic rate? *Mar. Ecol. Prog. Ser.* 200:289–296.

Sims, D.W., Wearmouth, V.J., Southall, E.J., Hill, J.M., Moore, P., Rawlinson, K., Hutchingson, N., Budd, G.C., Righton, D., Metcalfe, J., Nash, J.P., and Morritt, D. (2006). Hunt warm, rest cool: bioenergetic strategy underlying diel vertical migration of a benthic shark. *J. Anim. Ecol.* 75:176–190.

Smith, R.L. and Rhodes, D. (1983). Body temperature of the salmon shark, *Lamna ditropis*. *J. Mar. Biol. Assoc. U.K.* 63:243–244.

Steffensen, J.F. (1989). Some errors in respirometry of aquatic breathers: how to avoid and correct for them. *Fish Physiol. Biochem.* 6:49–59.

Stevens, E.D. and McLeese, J.M. (1984). Why bluefin tuna *Thunnus thynnus* have warm tummies: temperature effect on trypsin and chymotrypsin. *Am. J. Physiol.* 246:R487–R494.

Stillwell, C.E. and Kohler, N.E. (1982). Food, feeding habits, and estimates of daily ration of the shortfin mako (*Isurus oxyrinchus*) in the North Atlantic. *Can. J. Fish. Aquat. Sci.* 39:407–414.

Sundström, L.F. and Gruber, S.H. (1998). Using speed-sensing transmitters to construct a bioenergetics model for subadult lemon sharks, *Negaprion brevirostris* (Poey), in the field. *Hydrobiologia* 372:241–247.

Tanaka, H., Takagi, Y., and Naito, Y. (2001). Swimming speeds and buoyancy compensation of migrating adult chum salmon *Oncorhynchus keta* revealed by speed/depth/acceleration data logger. *J. Exp. Biol.* 204:3895–3904.

Tubbesing, V.A. and Block, B.A. (2000). Orbital rete and red muscle vein anatomy indicate a high degree of endothermy in the brain and eye of the salmon shark. *Acta Zool.* 81:49–56.

Tullis, A. and Baillie, M. (2005). The metabolic and biochemical responses of tropical whitespotted bamboo shark *Chiloscyllium plagiosum* to alterations in environmental temperature. *J. Fish Biol.* 67:950–968.

van den Burg, E.H., Peeters, R.R., Verhoye, M., Meek, J., Flik, G., and Van der Linden, A. (2005). Brain responses to ambient temperature fluctuations in fish: reduction of blood volume and initiation of a whole-body stress response. *J. Neurophysiol.* 93:2849–2855.

Vaudo, J.J. and Lowe, C.G. (2006). Movement patterns of round stingrays (*Urobatis halleri*) near a thermal outfall. *J. Fish Biol.* 68:1756–1766.

Wallman, H.L. and Bennett, W.A. (2006). Effects of parturition and feeding on thermal preference of Atlantic stingray, *Dasyatis sabina* (Lesueur). *Environ. Biol. Fish.* 75:259–267.

Wegner, N.C., Sepulveda, C.A., Olsen, K.R., Hyndman, K.A., and Graham, J.B. (2010). Functional morphology of the gills of the shortfin mako, *Isurus oxyrinchus*, a lamnid shark. *J. Morphol.* 271:937–948.

Weng, K.C. and Block, B.A. (2004). Diel vertical migration of the bigeye thresher shark (*Alopias superciliosus*), a species possessing orbital *retia mirabilia*. *Fish. Bull.* 102:221–229.

Weng, K.C., Castilho, P.C., Morrissette, J.M., Landeira-Fernandez, A.M., Holts, D.B., Schallert, R.J., Goldman, K.J., and Block, B.A. (2005). Satellite tagging and cardiac physiology reveal niche expansion in salmon sharks. *Science* 310:104–106.

Weng, K.C., O'Sullivan, J., Lowe, C.G., Winkler, C., Dewar, H., and Block, B. (2007). Movements, behavior and habitat preferences of juvenile white sharks in the eastern Pacific as revealed by electronic tags. *Mar. Ecol. Prog. Ser.* 338:211–224.

Weng, K.C., O'Sullivan, J., Lowe, C.G., Winkler, C., Sippel, T., and Hamilton, R. (2012). Back to the wild: release of juvenile white sharks from the Monterey Bay Aquarium. In: Domeier, M. (Ed.), *Global Perspectives on the Biology and Life History of the Great White Shark*. Taylor & Francis, New York.

Wilson, R.P. and McMahon, C.R. (2006). Measuring devices on wild animals: what constitutes acceptable practice? *Front. Ecol. Environ.* 4:147–154.

Wolf, N.G., Swift, P.R., and Carey, F.G. (1988). Swimming muscle helps warm the brain of lamnid sharks. *J. Comp. Physiol. B* 157:709–715.

Yoda, K., Naito, Y., Sato, K., Takahashi, A., Nishikawa, J., Ropert-Coudert, Y., Kurita, M., and Le Maho, Y. (2001). A new technique for monitoring the behaviour of free-ranging Adelie penguins. *J. Exp. Biol.* 204:685–690.

8

Food Consumption and Feeding Habits

Bradley M. Wetherbee, Enric Cortés, and Joseph J. Bizzarro

CONTENTS

8.1 Introduction

Among the most common statements made in studies of elasmobranch ecology is something to the effect that "sharks play an important role in marine ecosystems." Although there have been few conclusive demonstrations of this role, it is widely recognized that sharks and batoids are major participants in the transfer of energy from lower to upper trophic levels within marine ecosystems (see Heithaus et al., 2010, for a review). However, our understanding of the dynamics of prey consumption and processing of food in elasmobranchs remains rudimentary. To fully comprehend energy flow through elasmobranchs in marine communities it is necessary to know not only what they eat but also the rates at which they ingest, digest, and process energy and nutrients contained in the prey consumed. As with other areas of elasmobranch biology, investigations on the dynamics of feeding and processing food lag behind similar studies on other marine fishes and vertebrates despite the large amount of interest in sharks as predators. By far the most common elasmobranch feeding studies are

simple descriptions of stomach contents for a particular species in a particular location. Rate of consumption, feeding patterns, and the fate of food once ingested, as well as other aspects of feeding ecology and digestive physiology, have been examined for very few species of elasmobranchs.

The spiral valve-type intestine present in elasmobranchs is often viewed as a primitive design, and there has been speculation that food is processed differently as it passes through the digestive systems of elasmobranchs compared with most teleost fishes. The different digestive morphology present in elasmobranchs might be expected to influence the time required for passage of food through the alimentary canal, the efficiency of energy and nutrient absorption, the rate of consumption, the amount of energy available for growth and other needs, and ultimately the amount of energy passing through trophic levels occupied by elasmobranchs.

In this chapter, we review information on patterns of food consumption and processing of food in the digestive tracts of elasmobranchs. In general terms, we examine food consumption from several perspectives: what is eaten, feeding patterns, and how much is eaten. Our discussion includes dietary overlap and dietary breadth among species of elasmobranchs, as well as presumptions that have been made about food partitioning in these species. Second, we review the current state of knowledge concerning the processing of food once it is ingested, including rates of digestion and evacuation of food from the stomachs and entire intestinal tracts of elasmobranchs. Absorption, assimilation, and conversion of ingested food into new tissue are also discussed. For most topics, we include methodological considerations relevant for experimental design and interpretation of results for past or future elasmobranch feeding studies. We conclude by offering some recommendations for future work.

8.2 Diet

The feeding biology of elasmobranchs has been investigated to understand the natural history of a particular species, the role of elasmobranchs in marine ecosystems, the impact of elasmobranch predation on economically valuable or endangered prey, and various other questions. For these reasons, researchers have attempted to describe the diets of elasmobranchs, ranging from the stomach contents of a single shark to detailed examinations of the quantity of each prey item, feeding periodicity, feeding frequency, and trophic level at which the animals feed.

8.2.1 Quantification of Diet

Many early descriptions of the diets of different elasmobranch species were simply lists of prey items recovered from their stomachs (Breeder, 1921; Clark and von Schmidt, 1965; Coles, 1919; Dahlberg and Heard, 1969; Gudger, 1949; Randall, 1967). Other studies have quantified prey types found in stomachs using indices: the number of stomachs with a specific prey (frequency of occurrence, F), the total number of a specific prey found in stomachs (N), or the total weight (W) or volume (V) of a specific prey item (e.g., Matallanas, 1982; Snelson et al., 1984; Stevens, 1973; Stillwell and Kohler, 1982). Each of these terms has shortcomings for accurately expressing the amount of various prey that constitute the diet of a consumer (Bowen, 1996; Liao et al., 2001; Mumtaz Tirasin and Jorgensen, 1999). For example, the expression of stomach contents with counts may give the impression that a specific prey item that occurs very frequently in stomachs represents one of the most important prey items; however, if these prey are small, they may represent only a small proportion of the total food consumed. Similarly, if diet is expressed in terms

of weight or volume, consumption of a single large prey item would imply that this prey is a major component of the diet, when in fact very few individuals may have consumed it. To overcome such limitations, diet has often been reported in terms of a composite index, such as the index of relative importance (IRI) (Cortés, 1997, 1999):

$$IRI = \%F(\%W + \%N) \tag{8.1}$$

Compound expressions of diet provide less biased estimates of the contribution of various prey in the diet of a consumer, but their use remains controversial (Cortés, 1998; Hansson, 1998). Nonetheless, Cortés (1997) suggested that the presentation of stomach contents of sharks in terms of %IRI would both provide estimates of the diet that were intuitive and that would allow more direct comparison among studies.

Reliance on stomach contents to quantify diet of an animal also has limits; for example, rate of digestion of prey items in the stomach may vary with size and type of prey, and therefore items that are digested slowly may be overrepresented in stomachs examined. Capture technique may also influence contents in stomachs. Stomach contents of sharks captured at depth may be regurgitated, or differentially regurgitated, as the sharks are brought to the surface. Similar presumptions have been made in a number of studies where sharks were captured using gillnets. Sharks captured in gillnets are also presumed to represent a more unbiased representation of the diet of sharks, as sharks captured on baited lines might consist of a larger proportion of hungry sharks with less food in their stomachs. Ideally, prey can be identified to species and a list of prey species recovered from stomachs of sharks or rays interpreted as the "diet" of the species; however, diet varies with season, age, or size and location, and the term "diet" must be viewed in relative terms as a snapshot of what the species in question consumed shortly before capture.

Ecological energetics is a common framework for consideration of the fate of food consumed by animals, relating consumption to life activities through a common unit of measure for heat (the calorie) or work (the joule) (Brafield and Llewellyn, 1982; Kleiber, 1975). Expression of diet in energetic terms would refer to the amount of energy that each item ingested contributes toward the total amount of energy consumed by an animal and available for energetic demands. The first law of thermodynamics (conservation of energy) necessitates that all energy consumed by an animal be balanced by energy used (for growth, metabolism, or reproduction) and energy lost (in feces and urine) (Kleiber, 1975); therefore, quantification of diet in energetic terms (the amount of energy contributed by each prey type) would provide a method for expressing diet in standardized and biologically meaningful terms. Difficulties of such an approach include determination of initial size of each prey item consumed and energy content of each prey type (Scharf et al., 1998). An additional consideration far beyond simply quantifying stomach contents would be the inclusion of the energetic costs of capturing various types of prey. Although such analyses would be challenging given the technology currently available, a general understanding of the amount of energy expenditure required to capture specific prey would provide insight into net energy gains resulting from capture and consumption of particular prey types and be useful for evaluation of optimal foraging and comparison of life history characteristics. This approach has yet to be applied to elasmobranchs.

8.2.2 Broad Dietary Groups

As strict carnivores, elasmobranchs consume a limited array of prey in comparison to carnivorous, herbivorous, and omnivorous teleosts; nevertheless, a wide range of prey is consumed by elasmobranchs, ranging from very small plankton to very large whales. Plankton or small crustaceans are consumed by large, filter-feeding elasmobranch species, including manta rays (*Manta birostris*) and basking (*Cetorhinus maximus*), whale (*Rhincodon typus*), and megamouth (*Megachasma pelagios*) sharks (Compagno, 1990; Gudger, 1941; Hallacher, 1977; Sims and Merrett, 1997; Sims and Quayle, 1998). The diet of most species of sharks includes teleosts, and for many species the percentage of stomachs containing teleosts exceeds 90%, particularly for sharks in the genus *Carcharhinus* (Bass et al., 1973; Castro, 1993; Cliff and Dudley, 1992; Dudley and Cliff, 1993; Salini et al., 1992; Stevens and McLoughlin, 1991; Stevens and Wiley, 1986), closely related sharpnose (*Rhizoprionodon*) and hammerhead (*Sphyrna*) species (Simpfendorfer and Milward, 1993; Stevens and Lyle, 1989; Stevens and McLoughlin, 1991), as well as mackerel sharks (Lamnidae) (Gauld, 1989; Stillwell and Kohler, 1982). Elasmobranchs are common prey of many sharks and may form a large portion of the diet of some large carcharhinids (Cliff and Dudley, 1991a; Dudley and Cliff, 1993; Gelsleichter et al., 1999; Wetherbee et al., 1996), hammerheads (Cliff, 1995; Stevens and Lyle, 1989), sixgill (*Hexanchus griseus*) and sevengill (*Notorynchus cepedianus*) sharks (Ebert, 1991, 1994), and white (*Carcharodon carcharias*) and tiger (*Galeocerdo cuvier*) sharks (Cliff et al., 1989; Gudger, 1932; Lowe et al., 1996).

Cephalopods are also common prey items. Many pelagic sharks feed on squid (Backus et al., 1956; Kohler, 1987; Smale, 1991; Stillwell and Casey, 1976), and demersal sharks often feed on octopus (Baba et al., 1987; Carrassón et al., 1992; Castro et al., 1988; Ebert, 1994; Ebert

et al., 1992; Kubota et al., 1991; Mauchline and Gordon, 1983; McElroy et al., 2006; Relini Orsi and Wurtz, 1977; Stevens and McLoughlin, 1991; Waller and Baranes, 1994). Small, benthic catsharks (Scyliorhinidae), smoothhounds (Triakidae), and horn sharks (Heterodontidae) frequently prey upon mollusks (Gelsleichter et al., 1999; Lyle, 1983; Menni, 1985; Segura-Zarzosa et al., 1997; Talent, 1976), and crustaceans form a large portion of the diet of a number of bottom-feeding carcharhinid species (Lyle, 1987; Medved et al., 1985; Salini et al., 1992, 1994; Simpfendorfer and Milward, 1993; Stevens and McLoughlin, 1991), hammerheads (Bush, 2002; Castro, 1989; Cortés et al., 1996), sharpnose (Devadoss, 1989; Gelsleichter et al., 1999; Gómez and Bashirulah, 1984), smoothhounds (King and Clark, 1984; Rountree and Able, 1996; Smale and Compagno, 1997; Talent, 1982; Taniuchi et al., 1983; Vianna and de Amorim, 1995), and catsharks (Cross, 1988; Ebert et al., 1996; Ford, 1921; Heupel and Bennett, 1998; Lyle, 1983; Macpherson, 1980).

Large sharks occasionally consume vertebrates other than fish. Birds have been found in the stomachs of bull sharks (*Carcharhinus leucas*) (Tuma, 1976) and tiger sharks (Carlson et al., 2002; Dodrill and Gilmore, 1978; Heithaus, 2001a; Saunders and Clark, 1962) and may compose a large part of the diet of tiger sharks (Bass et al., 1973; Lowe et al., 1996; Simpfendorfer, 1992; Stevens, 1984) and white sharks (Randall et al., 1988). Reptiles (turtles and snakes) are occasionally eaten by carcharhinid sharks (Cliff and Dudley, 1991a; Heatwole et al., 1974; Lyle, 1987; Lyle and Timms, 1987; Tuma, 1976) and white sharks (Fergusson et al., 2000; Long, 1996) and are common in the stomachs of tiger sharks (Heithaus, 2001a; Lowe et al., 1996; Simpfendorfer, 1992; Stevens and McLaughlin, 1991; Witzell, 1987). Marine mammals are frequently preyed upon by large sharks such as white and tiger sharks (Bell and Nichols, 1921; Cliff et al., 1989; Corkeron et al., 1987; Dudley et al., 2000; Heithaus, 2001a; LeBoeuf et al., 1982; Lowe et al., 1996; Stevens, 1984) and have been found in the stomachs of carcharhinid sharks (Bass et al., 1973; Cliff and Dudley, 1991a; Wetherbee et al., 1996) and of sleeper sharks (*Somniosus*) (Scofield, 1920) and sixgill and sevengill sharks (Hexanchidae) (Ebert, 1991, 1994). The unusual tooth and jaw morphology of cookie-cutter sharks (*Isistius brasiliensis* and presumably *I. plutodon*) enables these sharks to maintain a predominantly parasitic lifestyle by removing plugs of flesh from large vertebrates (tunas, billfish, dolphins, and whales) and from squid (Jahn and Haedrich, 1988; Jones, 1971; Muñoz-Chapuli et al., 1988; Papastamatiou et al., 2010b; Shirai and Nakaya, 1992; Strasburg, 1963). Readers are referred to Cortés (1999) for a summary of the standardized diet compositions of 149 shark species.

The prey spectrum of batoids differs from that of sharks, largely as a function of differences in size and morphology, behavior, and habitat. Sea birds and marine mammals are not viable prey items for batoids; instead, some marine mammals consume batoids (Green et al., 1989; Visser, 1999; Walsh et al., 1988) and the egg cases of skates (Bor and Santos, 2003). Ingestion of elasmobranchs is rare among batoids, largely due to their relatively small size and inability (with the exception of pristids) to dismember vertebrate prey; however, elasmobranchs are supplemental dietary items for some large skates (Rajiformes: Rajidae) (Capapé, 1977a; Dolgov, 2005; Gordon and Duncan, 1989) and have been reported as prey items in other families such as Gymnuridae (Daiber and Booth, 1960), Torpedinidae (Abdel-Aziz, 1994), and Rhinobatidae (Bizzarro, 2005; Marshall et al., 2007). Although batoids are incapable of consuming most large vertebrates, their diets can be extremely diverse and often contain a wide variety of invertebrate prey. In addition to ingesting entire organisms, batoids can selectively remove and consume parts of invertebrates, such as bivalve siphons (Arrighetti et al., 2005; Babel, 1967; Bizzarro, 2005) and ophiuroid disks (Turner et al., 1982).

Most batoids tend to be generalists, with diets that are largely indicative of local prey availability (Ebert and Bizzarro, 2007; Link, 2004). Correspondingly, batoids can respond to episodic prey aggregations by switching their diets or gorging. Examples include *Gymnura natalensis* and *Dasyatis brevicaudata* ingesting squid on their spawning grounds (Smale et al., 2001), *Bathyraja kincaidii* opportunistically foraging on euphausiid swarms concentrated in submarine canyon heads (Rinewalt et al., 2007), and *Raja stellulata* feeding on newly recruited rockfishes (Bizzarro, unpublished). In addition to trophic generality within a certain prey spectrum, some batoids exhibit remarkable dietary plasticity between dissimilar food resources. The cownose ray, *Rhinoptera bonasus*, for example, was historically thought to be a durophagous feeder of hard-bodied prey based on stomach content information (Bigelow and Schroeder, 1953; Blaylock, 1992; Smith and Merriner, 1985) and jaw morphology (Summers, 2000); however, Collins et al. (2007) demonstrated suction feeding on polychaetes and crustaceans. These results suggest that, instead of being a hard prey specialist, *R. bonasus* may be an opportunistic generalist, modifying feeding behavior to consume locally available prey (Collins et al., 2007).

Not all batoids are generalists, however; for example, the diets of *Mobula japanica* and *M. thurstoni* appear to consist largely of a single euphausiid species, *Nyctiphanes simplex* (Notarbartolo di Sciara, 1988; Sampson et al., 2010). In addition, some prey groups (e.g., aquatic insects) represent unique but important prey resources for batoid species (e.g., Potamotrygonidae) (Almeida et al., 2010; Lasso et al., 1996; Shibuya et al., 2009; Silvia and Uieda, 2007). Most batoid species are intimately tied to the benthos and may excavate infaunal food resources

and seek refuge below the seafloor. This exposes them to prey resources not typically available to many sharks. Furthermore, batoids may rely more heavily on nonvisual senses (e.g., mechanoreception, electroreception) to locate prey, and many batoid species forage primarily at night to gain a predatory advantage. These considerations all influence the potential prey spectrum of batoid fishes.

More diet studies with far greater sample sizes have been devoted to skates compared to other batoid families (Ebert and Bizzarro, 2007; Bizzarro, unpublished). Part of this emphasis on skates is due to their considerable taxonomic diversity, as >40% of all extant batiods are rajids (Ebert and Compagno, 2007). Skates are typically abundant in temperate, offshore waters, where they are common incidental catch in commercial groundfish fisheries. For this reason, skate diets are often monitored by government agencies, a practice that can result in large sample sizes (Chuchukalo and Napazakov, 2002; Glubokov and Orlov, 2000; Link and Almeida, 2000). By contrast, diet information of other batoids has often been obtained opportunistically from a small number of specimens (e.g., Daiber and Booth, 1960; Ebert et al., 2002; Yáñez-Arancibia and Amezcua-Linares, 1979).

Crustaceans are the most important prey taxon of batoids, accounting for ≥50% of prey in approximately half of all species studied (Ebert and Bizzarro, 2007; Bizzarro, unpublished). Crustacean-based diets are especially prevalent in skates (Rajidae) (Abdel-Aziz, 1986; Ajayi, 1982; Braccini and Perez, 2005; Capapé, 1975b; Muto et al., 2001), guitarfishes (Rhinobatidae) (Dowton-Hoffman, 2006; Marshall et al., 2007; Rossouw, 1983; Valadez-González et al., 2006), whiptail stingrays (Dasyatidae) (Bizzarro, 2005; Fahmi et al., in press; Ismen, 2003; Raje, 2007), round stingrays (Urotrygonidae) (Almeida et al., 2000; Valadez-González et al., 2001, 2006), and stingarees (Urolophidae) (Edwards, 1980; Marshall et al., 2008; Platell et al., 1998; Treloar and Laurenson, 2004). Decapods are the primary crustacean prey of most batoids and, in aggregate, other crustacean are of relatively minor importance; however, stomatopods may compose a large portion of the diet of nearshore, tropical species (Navia et al., 2007; Raje, 2007; Valadez-González et al., 2006), and amphipods (Brickle et al., 2003; Shibuya et al., 2009; Treloar and Laurenson, 2004; Yang, 2003) and euphausiids (Almeida et al., 2000; Ebert et al., 1991; Platell et al., 1998; Rinewalt et al., 2007) are common in the diets of relatively small species. Consumption of euphausiids is not limited to small species, however, as some large devil rays (Mobulidae) consume euphausiids almost exclusively (Notarbartolo di Sciara, 1988; Sampson et al., 2010). Whereas most crustacean prey, including euphausiids, is consumed when in proximity to the benthos, mobulids may target euphausiid swarms in pelagic or demersal waters.

Teleost fishes are commonly consumed by batoids, and often dominate diets. Three batoid families are primarily piscivorous: torpedo rays (Torpedinidae) (Abdel-Aziz, 1994; Capapé, 1979), butterfly rays (Gymnuridae) (Bizzarro, 2005; Jacobsen et al., 2009), and sawfishes (Pristidae) (Peverell, 2005; Thorburn, 2006). By contrast, fishes are of minor or trivial importance in the diets of round rays (Urotrygonidae) (Almeida et al., 2000; Valadez-González et al., 2006), stingarees (Urolophidae) (Platell et al., 1998; Treloar and Laurenson, 2004), and cownose rays (Rhinopteridae) (James, 1962; Smith and Merriner, 1985). Fishes are absent or extremely rare in the diets of most mobulids and myliobatids but contribute substantially to the diets of a few of these species, including *Manta birostris* (Anderson and Hafiz, 1989; Homma et al., 1999), *Mobula mobular* (Capapé, 1986; Celona, 2004), *Aetomylaeus nichofii* (Capapé and Desoutter, 1979), and *Pteromylaeus bovinus* (Capapé, 1977b). Because of size and gape limitations, only small fishes or early life stages of larger fishes are accessible to most batoids. Fish that are >50 cm TL (total length) are rarely reported and are generally limited to predatory species with relatively wide mouths that attain large maximum sizes (e.g., some rajid, torpedinid, and gymnurid species). One exception to this general trend is observed in the narcinids, which have small, circumscribed mouths. This characteristic limits the width of teleost prey but not the length, and eels of >70 cm TL have been reported in the diet of the giant electric ray, *Narcine entemedor* (Bizzarro, 2005). Another exception may be the sawfishes (Pristidae). Pristids have rather small mouths but reach large sizes (>7.0 m) (Compagno et al., 1989; Last and Stevens, 1994). Although reported to prey primarily on small, demersal fishes (Peverell, 2005; Thorburn, 2006), pristids may consume larger fishes simply as a result of their large size and tendency to sever prey prior to ingestion. Although most studies report predation on benthic and demersal fishes, benthopelagic feeding appears to be increasingly common with ontogeny in some species (e.g., Boyle, 2010; Capapé, 1976; Koen-Alonso et al., 2001). In addition, *Pteroplatytrygon violacea* (Veras et al., 2009) and some mobulids (Celona, 2004; Homma et al., 1999) consume pelagic fishes in association with their typical habitats.

Infaunal organisms contribute substantially to the diets of stingrays (Myliobatiformes) of the following families: Urolophidae (Marshall et al., 2008; Platell et al., 1998), Urotrygonidae (Babel, 1967; Beebe and Tee-Van, 1941; Yáñez-Arancibia and Amezcua-Linares, 1979), Dasyatidae (Ebert and Cowley, 2003; Euzen, 1987; Homma and Ishihara, 1994), Myliobatidae (Gudger, 1914; Talent, 1982; Yamaguchi et al., 2005), and Rhinopteridae (James, 1962; Smith and Merriner, 1985). Bivalves and polychaetes are most commonly consumed, but burrowing crustaceans (e.g., Amphipoda, Thalassinidea) and fishes (e.g., Anguilliformes) are also representative prey

items. Bivalves are typically crushed by myliobatids and rhinopterids, but relatively small or thin-shelled bivalves may be swallowed whole by some species, such as *Raja rhina* (Robinson et al., 2007) and *Dasyatis dipterura* (Bizzarro, 2005), and larger bivalves may be removed from the shell prior to ingestion (*Rhinobatos productus*) (Talent, 1982). Stingrays typically access infauna by the formation of feeding pits (Babel, 1967; Gregory et al., 1979), resulting in considerable disturbance to the benthos. The combination of predation and disturbance by stingrays has been demonstrated to regulate abundance and possibly composition of infaunal communities (Cross and Curran, 2000, 2004; Thrush et al., 1994).

Cephalopods are not primary prey items for batoids but may represent sporadic or supplemental dietary components. Squid contribute substantially to the diets of some rajids (Brickle et al., 2003; Chuchukalo and Napazakov, 2002; Orlov, 1998) and are minor prey items for other taxa, such as dasyatids (Capapé, 1978; Devadoss, 1978; Hess, 1959). In addition, some species (e.g., *Gymnura natalensis*, *Dasyatis brevicaudata*) may opportunistically ingest squid when they occur in dense aggregations, such as spawning events, when major scavenging opportunities arise (Smale et al., 2001). Cuttlefishes have been reported in the diets of *Pteromylaeus bovinus* (Capapé, 1977b), *Dasyatis tortonesei* (Capapé, 1978), and *Rhinobatos rhinobatos* (Capapé and Zaouali, 1979). Octopi are mainly consumed by rajids but are usually minor prey items (Morato et al., 2003; Robinson et al., 2007); however, octopi (and to a lesser degree, cuttlefish and squid) constitute a substantial portion of the diet of *Raja stellulata* off central California (Bizzarro, unpublished). Unlike most skates, this species inhabits rocky substrates of the inner continental shelf; therefore, its greater reliance on cephalopods is probably due to their increased availability in this habitat.

8.2.3 Diet Shifts

Adequate representation of the diet of a species of elasmobranch is complicated by differences in diet that occur within populations among individuals of different sizes and geographical locations, as well as during different seasons. Ontogenetic change in feeding habits is an almost universal phenomenon in fishes; thus, its occurrence in elasmobranchs is not surprising considering that, as many species of sharks and rays increase in size, there also are changes in habitat occupied, movement patterns, swimming speed, size of jaws, teeth and stomachs, energy requirements, experience with prey, vulnerability to predation, and other factors that result in variable exposure to prey or improved ability of larger elasmobranchs to capture different prey items (Ebert and Bizzarro, 2007; Graeber, 1974; Koen-Alonso et al., 2001; Lowe et al., 1996; Marshall et al., 2008; McElroy et al., 2006; Stillwell and Kohler, 1982; Weihs et al., 1981). Although diet shifts are more often reported qualitatively rather than based on rigorous statistical analysis, there are many reports of a shift from a diet of invertebrates to a diet that is more varied or that includes more teleosts (Bizzarro et al., 2007; Capapé, 1974, 1975b; Capapé and Zaouali, 1976; Ellis and Musick, 2006; García de la Rosa and Sánchez, 1997; Jakobsdóttir, 2001; Jones and Geen, 1977; Kao, 2000; Mauchline and Gordon, 1983; Olsen, 1954; Platell et al., 1998; Rinewalt et al., 2007; Robinson et al., 2007; Smale and Cowley, 1992; Smale and Goosen, 1999; Stillwell and Kohler, 1993; Talent, 1976).

Within rajids, dietary shifts from amphipods and small shrimps to larger decapods and polychaetes (and sometimes small fishes) are common for relatively small species, whereas further shifts to fishes often occur in larger species (Ajayi, 1982; Braccini et al., 2005; Brickle et al., 2003; Muto et al., 2001; Treloar et al., 2007; Yeon et al., 1999). Multiple studies have documented increased consumption of elasmobranchs (Cliff and Dudley, 1991a; Cortés and Gruber, 1990; Lowe et al., 1996; Matallanas, 1982; Simpfendorfer et al., 2001a,b; Smale, 1991) and marine mammals (Ebert, 1994; Tricas and McCosker, 1984) with increasing size of shark. A number of studies, however, have found no ontogenetic dietary changes (Avsar, 2001; Clarke et al., 1996; Cliff and Dudley, 1991b; Cortés et al., 1996; Jakobsdóttir, 2001; Kohler, 1987; Marshall et al., 2008; Matallanas et al., 1993; Segura-Zarzosa et al., 1997), or shifts to larger individuals of the same taxa (Lucifora, 2000; Treloar et al., 2007).

There are also examples of geographical differences in the diets of several wide-ranging species of sharks. For example, the diets of spiny dogfish (*Squalus acanthias*) and blue (*Prionace glauca*), sandbar (*Carcharhinus plumbeus*), blacktip (*C. limbatus*), and bull sharks all differed among locations in the Atlantic, Pacific, and Indian Oceans (Cliff and Dudley, 1991a; Cliff et al., 1988; Dudley and Cliff, 1993; Gubanov and Grigoryev, 1975; Gudger, 1948, 1949; Harvey, 1989; Holden, 1966; Jones and Geen, 1977; Kondyurin and Myagkov, 1982; Lowe et al., 1996; McElroy et al., 2006; Medved, 1985; Rae, 1967; Sarangadhar, 1983; Snelson et al., 1984; Stevens, 1973; Stevens et al., 1982; Tricas, 1979; Tuma, 1976; Wass, 1971). Variation of diet among locations is exemplified by the tiger shark, which has a diet that differs substantially among areas sampled worldwide (DeCrosta et al., 1984; Lowe et al., 1996; Simpfendorfer, 1992; Simpfendorfer et al., 2001a). Diet may differ within a species even between locations that are relatively close, as has been found for sandbar sharks (Ellis and Musick, 2006; Lawler, 1976; Medved et al., 1985; Stillwell and Kohler, 1993), lemon sharks (*Negaprion brevirostris*) (Cortés and Gruber, 1990; Schmidt, 1986; Springer, 1950), and the starspotted smoothhound (*Mustelus manazo*) (Yamaguchi and Taniuchi, 2000). Habitat type and water depth have

also been found to influence diet composition (Cortés et al., 1996; Kohler, 1987; Smale and Compagno, 1997; Stillwell and Kohler, 1982, 1993; Webber and Cech, 1998). Several authors have reported differences in the diet between sexes of sharks (Bonham, 1954; Hanchet, 1991; Matallanas, 1982; Simpfendorfer et al., 2001a; Stillwell and Kohler, 1993), which may be related to sexual segregation within species and different sizes attained by males and females. In all, findings of geographical differences in the diets of sharks are not surprising considering the diversity of prey in different regions and the apparent plasticity of feeding behaviors among sharks (see Chapter 17 of this volume for a more complete discussion).

Variation in feeding of sharks is further demonstrated by seasonal differences in diet that have been reported within species (Allen and Cliff, 2000; Capapé, 1974; Cortés et al., 1996; Dudley and Cliff, 1993; Horie and Tanaka, 2000; Jones and Geen, 1977; Kohler, 1987; Lyle, 1983; McElroy et al., 2006; Nagasawa, 1998; Olsen, 1984; Platell et al., 1998; Talent, 1976; Tricas, 1979; Waller and Baranes, 1994). Seasonal differences in diet presumably reflect seasonal migration of sharks or of their prey. Matallanas (1982), for example, reported seasonal shifts in the most important teleosts in the diet of kitefin sharks (*Dalatias licha*), and Stillwell and Kohler (1982) described seasonal shifts between consumption of fish and cephalopods by the mako shark (*Isurus oxyrinchus*). There is also evidence of a diet shift in leopard sharks (*Triakis semifasciata*) sampled at a single location during two periods 25 years apart, which may be indicative of community changes (Kao, 2000).

8.2.4 Feeding Relationships

There have been relatively few investigations comparing diets of sympatric species of elasmobranchs. In several studies, standard ecological indices of similarity were used to calculate dietary overlap among elasmobranch species, among elasmobranchs and teleosts caught in the same location, or among different size classes of a single species. Such comparisons represent initial attempts to characterize niche partitioning and competition among elasmobranchs and co-occurring teleosts. Ecological indices of dietary breadth or diversity have also been calculated for several species of elasmobranchs to examine the degree of feeding specialization.

Evidence indicates that both food partitioning and competition for food resources are likely to occur in marine communities where elasmobranchs occur. Dietary overlap among sympatric species of elasmobranchs has been characterized—qualitatively or using quantitative indices—as low (Baba et al., 1987; Carrassón et al., 1992; Macpherson, 1981; Orlov, 1998), moderate (Orlov, 1998; Relini Orsi and Wurtz, 1977;

Smale and Compagno, 1997), substantial (Ellis et al., 1996; Macpherson, 1980), high (Bethea et al., 2004; Platell et al., 1998; Salini et al., 1990), or variable, depending on the species compared (Euzen, 1987; Macpherson, 1981). Varying degrees of diet overlap have also been described for co-occurring elasmobranchs and teleosts (Ali et al., 1993; Blaber and Bulman, 1987; Clarke et al., 1996) or for elasmobranchs and marine mammals (Clarke et al., 1996; Heithaus, 2001b). At the intraspecific or intrapopulation level, increased dietary overlap is most often encountered between pairs of consecutive size classes (Bethea et al., 2004; Cortés et al., 1996; García de la Rosa and Sánchez, 1997; Kao, 2000; Platell et al., 1998; Koen-Alonso et al., 2002; Simpfendorfer et al., 2001a; Wetherbee et al., 1996, 1997) or between similar size classes of elasmobranchs and teleosts (Platell et al., 1998). Food overlap also tends to be high between adjacent geographic locations (Simpfendorfer et al., 2001a; Yamaguchi and Taniuchi, 2000). There have recently been several studies comparing diets of sympatric species of sharks and rays indicating niche partitioning in terms of diet (Papastamatiou et al., 2006; Vaudo and Heithaus, 2011; White et al., 2004). Papastamatiou et al. (2006) found that sandbar (*Carcharhinus plumbeus*) and gray reef (*C. amblyrhynchos*) sharks had an allopatric distribution; sandbar sharks were common in the main Hawaiian Islands but rare in the Northwestern Hawaiian Islands, and gray reef sharks were rare in the main Hawaiian Islands but common in the Northwestern Hawaiian Islands. The two species exhibited a relatively high dietary overlap, but in areas of sympatry the level of dietary overlap was lower. Sample sizes in overlapping areas were small, but these findings were suggestive of competition and character displacement.

Diets of elasmobranchs vary from highly specialized to very generalized. Specialized diets include those of elasmobranchs that consume zooplankton, crustaceans, and cephalopods as discussed in an earlier section. In contrast, several top predators, such as bull and tiger sharks, have very generalized diets. Varying degrees of specialization have been reported in studies that calculated true measures of diversity (Ali et al., 1993; Blaber and Bulman, 1987; Carrassón et al., 1992; Clarke et al., 1989; Cortés et al., 1996; Ellis et al., 1996; Macpherson, 1981; Simpfendorfer et al., 2001a) or that reported only the total number of different prey types or contained qualitative statements about dietary diversity (Capapé and Zaouali, 1976; Chatwin and Forrester, 1953; Gelsleichter et al., 1999; Segura-Zarzosa et al., 1997; Smale and Compagno, 1997). Dietary breadth tends to increase with size or age in some cases (Cortés and Gruber, 1990; Lowe et al., 1996; Talent, 1976; Wetherbee et al., 1996, 1997) and decrease in others (Platell et al., 1998; Simpfendorfer et al., 2001a; Smale and Compagno, 1997; Yamaguchi and Taniuchi, 2000). A number of studies

have addressed the question of generalist versus specialist feeding at the population level in sharks through the application of various diversity indices (Simpson's Index, Shannon–Weiner Index) based on measures of proportionality (Blaber and Bulman, 1987; Macpherson, 1981; Wetherbee et al., 1996, 1997) or based on stable isotope methods (Matich et al., 2010, 2011; Vaudo and Heithaus, 2011). Few studies have applied measures such as between-individual component (BIC) and within-individual component (WIC) variance to calculate total niche width as an index of degree of dietary specialization (see Bolnick et al., 2002), but recent studies of bull and tiger sharks have employed this approach to investigate dietary diversity at the individual and population level (Matich et al., 2011).

Because of the widespread occurrence of ontogenetic, geographical, and seasonal changes in feeding habits discussed above, very few studies on the diet of sharks have been extensive enough to provide a comprehensive description of the diet for a species. Additionally, the diversity of prey found in stomachs generally increases with the number of stomachs sampled. The issue of sample sufficiency has been addressed by using cumulative prey curves to determine whether a sufficient number of stomachs have been examined to describe precisely the diet of the species in question (see Cortés, 1997; Ferry and Cailliet, 1996; and references therein); however, sufficient sample size for other aspects of feeding such as comparisons among groups may not be indicated by a cumulative prey curve. Clearly, there is ample opportunity for improving our understanding of aspects of the feeding ecology of elasmobranchs at the organism, population, community, and ecosystem level through additional and more focused research.

8.2.5 Feeding Patterns

Understanding a consumer's feeding patterns requires more than knowledge of the prey items that make up its diet. The dynamics of the feeding process must be accounted for, and thus to understand the ecological interaction between predator and prey we must have knowledge of the amount of food ingested and the feeding frequency of the predator. Analysis of stomach contents allows inference of feeding patterns through reconstruction of meal sizes, ingestion times, feeding duration, and feeding frequency. The frequency of occurrence of empty stomachs; the number, weight, and stage of digestion of food items; and knowledge on the gastric evacuation dynamics of each food item all give insight into the feeding pattern of a predator.

The occurrence of high proportions of empty stomachs in shark diet studies and in commercial fisheries operations is common (Wetherbee et al., 1990). Use of longlines to capture sharks may attract more animals with empty stomachs, but this is less likely when using passive gear such as gillnets or active gear such as trawls. The frequent occurrence of empty stomachs, combined with the observation that there are often few food items—many of them in advanced stages of digestion—in shark stomachs, such as in the juvenile sandbar shark (Medved et al., 1985) and the juvenile lemon shark (Cortés and Gruber, 1990), lends support to the notion that many sharks are intermittent rather than continuous feeders, because otherwise one would expect to regularly find multiple food items at different stages of digestion and few empty stomachs. Demersal carnivores that feed on invertebrate prey, such as many skates and rays (Bradley, 1996), and filter-feeding zooplanktivorous sharks are obvious exceptions to this pattern (Baduini, 1995; Sims and Quayle, 1998). They feed more continuously, their stomachs often contain a large number of prey, and empty stomachs occur at a lower rate.

Feeding frequency can be estimated from the total time required to complete gastric evacuation and the proportion of empty stomachs in a sample (Diana, 1979). Based on this method, Jones and Geen (1977) estimated that mature spiny dogfish would feed only every 10 to 16 days after completely filling their stomachs, whereas Medved et al. (1985), Cortés and Gruber (1990), and Bush and Holland (2002) estimated a feeding frequency of 95 hr, 33 to 47 hr, and 10 to 11 hr for juvenile sandbar, lemon, and scalloped hammerhead (*Sphyrna lewini*) sharks, respectively.

Gastric evacuation experiments (see Section 8.3.2) allow development of qualitative scales describing the various stages of digestion of food items. These qualitative scales can then be used to calculate the difference between the least and most advanced stages of digestion of food items found in stomachs of field-sampled animals and infer feeding duration. Medved et al. (1985) and Cortés and Gruber (1990) used this approach to obtain estimates of feeding duration for juvenile sandbar (7 to 9 hr) and lemon (11 hr) sharks. The occurrence of food items in different stages of digestion in stomachs of juvenile lemon and sandbar sharks caught at the same time also indicated that feeding in these two species was asynchronous; that is, there was no preferred feeding time for all individuals of a population, a pattern believed to be prevalent in most shark species. Conversely, Kao (2000) reported some evidence for feeding synchronicity in the leopard shark off the central California coast. Results from Medved et al. (1985) and Cortés and Gruber (1990) for juvenile sandbar and lemon sharks, respectively, did not reveal increased food consumption at night or during a particular tidal phase; however, these studies did not estimate meal ingestion times (see below).

Cortés (1997) reviewed the numerous methodological issues that can affect the interpretation of diel feeding chronology in fishes and elasmobranchs. In addition to the effect of passive vs. active sampling gear, experimental design, and statistical analysis of results, he cautioned against using the weight of stomach contents alone to assess diel feeding continuity or discontinuity and to interpret diel feeding chronology. To estimate preferred feeding times it is also necessary to reconstruct meal ingestion times using qualitative stage-of-digestion scales. Longval et al. (1982) found a cyclical feeding pattern in juvenile lemon sharks in captivity, with peak consumption followed by several days of reduced food intake. The evidence for sharks, as exemplified by work on juvenile lemon sharks, supports the concept of a cyclical pattern of feeding motivation observed in many vertebrates, whereby relatively short feeding bouts would be followed by longer periods of reduced predatory activity until the return of appetite, which in the lesser spotted dogfish (*Scyliorhinus canicula*) was found to be inversely correlated with gastric evacuation rate (Sims et al., 1996).

8.2.6 Trophic Levels

It is commonly accepted that sharks are upper trophic level predators in many marine communities; however, until recently, virtually no quantitative estimates of trophic levels existed for sharks. Cortés (1999) calculated standardized diet compositions and estimated trophic level (TL) for 149 shark species belonging to 23 families using published TLs of prey categories, largely based on the Ecopath II model (Christensen and Pauly, 1992). He concluded that sharks as a group are tertiary consumers (TL > 4) that occupy trophic positions similar to those of marine mammals and higher than those of seabirds. Measurement of stable isotopes of nitrogen and carbon in tissues of marine consumers is an alternative approach to estimating TLs based on stomach contents. A number of studies have used stable isotope ratios to estimate TLs of sharks (Borrell et al., 2011; Domi et al., 2005; Estrada et al., 2003, 2006; Fisk et al., 2002; Kerr et al., 2006; Ostrom et al., 1993). Concentrations of organochlorine contaminants and trace metals have also been used in conjunction with isotope ratios to trace feeding relationships, but these multiple technique methods did not agree well and raised questions about the use of stable isotopes as a single indicator of ecological relationships (Domi et al., 2005; Fisk et al., 2002). Fisk et al. (2002) attributed the lower TLs obtained through stable isotope ($\delta^{15}N$) analysis compared to that from contaminant analysis to urea retention in elasmobranch tissues for osmoregulation, which could result in lower levels of $\delta^{15}N$

and thus underestimate TL. However, Logan and Lutcavage (2010) investigated the effects of urea on stable isotopes of captive elasmobranchs and concluded that urea content did not affect stable isotope ratios. Because N fractionation can vary with ecosystem, trophic level, and environmental conditions, the use of stable isotope methods for determination of absolute trophic levels still includes drawbacks and should be viewed with caution (Caut et al., 2009).

Stable isotope ratios have also been used as a mechanism to identify source of food for sharks and rays—for example, differentiating between pelagic and inshore sources or on a more fine scale such as sea grass or estuaries (Borrell et al., 2011; Domi et al., 2005; Kerr et al., 2006; Matich et al., 2010, 2011; Papastamatiou et al., 2010a; Vaudo and Heithaus, 2011). Several studies have attempted to use stable isotope ratio analysis to identify seasonal or ontogenetic shifts in the diets of sharks, or even for identification of specific prey consumed by sharks (Estrada et al., 2006; Kerr et al., 2006; MacNeil et al., 2005; Matich et al., 2010). Although stable isotope analysis provides information on the long-term feeding history of an animal not available from stomach content analysis, the few studies where sharks were fed known diets while held in captivity have generated results that cast suspicion on the reliability of stable isotope ratio analysis as a single indicator of the various ecological relationships examined using this methodology (Hussey et al., 2010; Logan and Lutcavage, 2010). Consequently, the majority of stable isotope ratio studies have been conducted in conjunction with stomach content analysis or they at least reference published accounts of the diet of the species in question to evaluate the effectiveness of stable isotope methods for accurate identification of prey consumed (Domi et al., 2005; Kerr et al., 2006; Logan and Lutcavage, 2010). Although useful for examining broad-scale questions of relative trophic position and ultimate sources of primary productivity, stable isotope analysis alone provides limited information that includes a number of caveats. Given current disputes about carbon and nitrogen turnover in aquatic systems and flow within animals (Auerswald et al., 2010), stable isotope ratio methods complement, but do not substitute for, stomach content analysis as a means of answering a number of questions about the feeding history of elasmobranchs.

Ebert and Bizzarro (2007) calculated trophic levels for 60 skate (Rajiformes: Rajidae) species, using prey categories expanded from those of Cortés (1999). Trophic level estimates ranged from 3.48 (*Rajella caudaspinosa*) to 4.22 (*Zearaja chilensis*), with a mean of 3.80. Among five genera, *Bathyraja* was found to feed at a statistically higher TL than *Rajella*, but no additional differences were found. A positive relationship was evident between TL and skate total length, a condition that also has been reported among predatory sharks (Cortés, 1999). Ordinal comparisons of TL values showed that skates had lower TLs than all but two shark orders (Heterodontiformes, Orectolobiformes) (Cortés, 1999). In comparisons between skates and sympatric shark families, skate TLs were significantly lower than those of Squalidae but not significantly different from those of Scyliorhinidae, Squatinidae, and Triakidae (Cortés, 1999; Ebert and Bizzarro, 2007). Other than the Ebert and Bizzarro (2007) meta-analysis, TL calculations for skates are rather limited. The six Australian skate species examined by Treloar et al. (2007) were evenly divided between secondary and tertiary consumers based on stomach content data. *Bathyraja brachyurops* was determined to be a tertiary consumer with a consistent TL throughout ontogeny (Belleggia et al., 2008). Mabragaña and Giberto (2007) found spatial differences in TLs of *Psammobatis* spp. in the southwest Atlantic. Stable isotope analysis to identify specific prey in diets has only been performed for a single species of skate or ray, the smallnose fanskate (*Sympterygia bonapartii*). Although TL was not estimated, δ^{15}N results suggested high individual variability in the relative dietary proportion of clams and shrimp (Penchaszadeh et al., 2006).

Synthesized TL information is not yet available for other batoid families; however, preliminary analysis indicates that members of the largely piscivorous families Torpedinidae, Gymnuridae, and Pristidae feed at the highest TLs (Bizzarro, unpublished). In a situation analogous to that of sharks (Cortés, 1999), the largest batoids (family Mobulidae) are mainly planktivorous and occupy the lowest TLs (Bizzarro, unpublished). Species-specific TL information has been estimated for a limited number of rays. Using stomach content data, Bizzarro (2005) estimated TLs ranging from 3.36 (*Dasyatis dipterura*) to 4.24 (*Gymnura marmorata*) in a ray assemblage off the west coast of Mexico. When Vaudo and Heithaus (2011) examined a ray assemblage in Western Australia using stable isotope methods, they concluded that the rays were exclusively secondary consumers with high individual variability. By contrast, devil rays, *Mobula thurstoni* and *M. japanica*, showed low individual variability, and TLs indicated specialized feeding on the euphausiid *Nyctiphanes simplex* (Sampson et al., 2010). Nitrogen signatures for the common eagle ray, *Myliobatis aquila*, also were similar in two Australian estuaries (Svennson et al., 2007). Woodland et al. (2011) found slightly higher TL estimates from stomach content analysis (3.53) of *Myliobatis freminvillii* when compared to those derived from stable isotope analysis (3.32).

8.3 Food Consumption

Feeding ecology is an important aspect of the life-history strategy of a species that can be adequately expressed through determination of food consumption rates. Daily rates of food consumption are in turn dependent on gastric evacuation rates. Measurement of daily rates of food consumption and digestion rates require regular collection of stomach contents of fish caught in the wild and fish held in captivity in the laboratory or field. This poses a particularly difficult problem for those studying elasmobranchs and sharks in particular, because of the difficulty of captive studies and the logistical requirements of extended field sampling. Additionally, rates of consumption in teleost fishes may vary depending upon a myriad of intrinsic (e.g., age, feeding history, reproductive status) and extrinsic factors (e.g., geographical location, habitat type, water temperature, prey availability). The scarcity of information on food consumption rates of elasmobranchs is thus not surprising.

8.3.1 Daily Ration

Daily ration is the mean amount of food consumed on a daily basis by individuals of a population and is generally expressed as a proportion of mean body weight (%BW day^{-1}). Although an individual does not ingest the same amount of food every day and may not even feed daily, daily ration allows standardization of rates of consumption and provides a measure for comparative studies (Wetherbee et al., 1990). There are two basic approaches for estimating daily ration: (1) *in situ* (field-derived) methods, which require knowledge of the amount of food found in stomachs of fish sampled in the wild and of the gastric evacuation dynamics of the ingested foodstuffs; and (2) bioenergetic models, which estimate food consumption based on the other components of the bioenergetic equation (growth, metabolism, excretion, and egestion).

With field-based methods, daily ration cannot be estimated by simply examining stomach contents because the amount of food found in stomachs is a function of both ingestion and digestion rates (Wetherbee et al., 1990). Cortés (1997) reported that there has been very little investigation of the applicability to elasmobranchs of the most common models used to estimate daily ration in teleosts. *In situ* methods for estimation of daily ration applied to elasmobranch studies include models proposed by Diana (1979), Eggers (1979), Elliott and Persson (1978), Pennington (1985), and Olson and Mullen (1986). Cortés (1997) concluded that the Diana and Olson–Mullen methods were a better application for intermittent feeders, such as most sharks, and that

these models were also based on less restrictive assumptions and required comparatively less demanding sampling regimens. Given the absence of error analyses of the estimates of daily ration in elasmobranch studies, Cortés (1997) advocated the use of resampling techniques, such as bootstrapping, or Monte Carlo simulation to enable statistical testing of differences between estimates obtained through different models and generally to provide a picture of the variability associated with those point estimates.

Laboratory approaches to estimating daily ration are based on a bioenergetic or energy budget equation (Winberg, 1960), which relates consumption (C) to growth (G), metabolism (M), excretion (urine, U), and egestion (feces, F):

$$C = G + M + U + F \qquad (8.2)$$

The daily energy required for growth (J day^{-1}) can be derived from laboratory or field estimates of growth (g day^{-1}) multiplied by the energy equivalent of shark tissue (J g^{-1}), which to date has only been determined for juvenile lemon sharks (5.41 kJ g^{-1} wet weight) (Cortés and Gruber, 1994) and scalloped hammerhead pups (6.07 kJ g^{-1}) (Lowe, 2002). The daily energy required for total metabolic expenditures (J day^{-1}) can be obtained from average daily metabolic rate (for example), expressed as mg O$_2$ kg shark^{-1} day^{-1}, multiplied by a standard oxycalorific value of 3.25 cal mL O$_2^{-1}$ (Elliott and Davidson, 1975) or 13.59 J mL O$_2^{-1}$, with adjustments for shark mass (kg). Energy loss in non-assimilated food (urine and feces) has only been measured in the lemon shark (Wetherbee and Gruber, 1993a), where it represented approximately 27% of the total ingested energy. This proportion of energy corresponding to $F + U$ can be substituted into the bioenergetic equation by multiplying $G + M$ by a factor of 1.37 (to account for energy losses). The final step is to divide the energy value of food consumed (J g^{-1}) into $1.37(G + M)$, and express the result as a percentage of body weight. Cortés and Gruber (1990) used a variation of this bioenergetic approach to estimate daily ration for juvenile lemon sharks; that is, they used a laboratory-derived feeding rate–growth rate curve (also known as G–R curve) to estimate daily ration in the wild as the food intake level that corresponded to field-observed growth.

Table 8.1 summarizes studies of food consumption rates in elasmobranchs, including the shape of the model that best described the rate of gastric evacuation, total gastric evacuation time, estimates of daily ration, and gross conversion efficiency. Feeding rates of elasmobranchs—at least on a body weight basis—are considerably lower than those of many teleosts (Brett and Groves, 1979), even with the inclusion of sharks fed to satiation in captivity, and they rarely surpass 3% BW per day (Table 8.1). In addition, consumption rates of adults may decrease by an order of magnitude with respect to those of young sharks, as found for captive sevengill sharks (*Notorynchus cepedianus*) fed to satiation (Van Dykhuizen and Mollet, 1992) (Table 8.1) and in bioenergetic estimates for the bonnethead (*Sphyrna tiburo*) (Bethea et al., 2007).

8.3.2 Gastric Evacuation

Estimation of daily ration through *in situ* methods requires knowledge of gastric evacuation rates. As in many areas of elasmobranch research, the ability to conduct controlled field or laboratory experiments is severely impaired by the difficulty of maintaining large individuals, which has resulted in experiments conducted on small species or juvenile stages of larger species (Cortés, 1997). Cortés (1997) pointed out that there is still considerable debate about the adequacy of the most common mathematical models (linear, exponential, square root, surface area) used to describe gastric evacuation in fishes and that no single model can be used to represent the dynamics of different species consuming different prey under different environmental conditions in all cases. The physiological rationale for the various models of gastric evacuation and the statistical adequacy of the criteria used to select the best model of evacuation have been extensively reviewed elsewhere (see references in Cortés, 1997). Cortés (1997) advocated the use of multiple measures of statistical fit along with formal residual analysis and an examination of residual plots before selecting a model but pointed out that even with thorough analyses results may still be inconclusive. A sensible approach for estimating daily ration through *in situ* methods is therefore to evaluate the effects of various evacuation models.

In addition to the well-known accelerating effect of temperature (Brett and Groves, 1979), meal size and food type also seem to affect the gastric evacuation dynamics of elasmobranchs. Larger meal sizes generally take longer to digest and evacuate (Bush and Holland, 2002; Sims et al., 1996). In general, it appears that small, more friable, and easily digestible items are evacuated more quickly than larger items with lower surface-to-volume ratios (Cortés and Gruber, 1992; Medved, 1985; Nelson and Ross, 1995; Schurdak and Gruber, 1989). Surface area models provided the best fit to gastric evacuation data for the lesser spotted dogfish, especially when the meal included more than one prey item (Macpherson et al., 1989). Most species of elasmobranchs consume different types of prey, which in turn may be evacuated from the stomach at different rates and thus greatly influence estimates of daily ration based on gastric evacuation rate. Medved (1985), for example, found that time required for evacuation of crab and teleost prey from

the stomachs of sandbar sharks could differ by as much as 20 hr. In general, quantification of the effects of food type, number and digestibility of prey, and meal size on gastric evacuation dynamics of elasmobranchs would clearly improve the accuracy of estimates of daily ration and overall rates of consumption.

The sequence of digestion and gastric evacuation of foodstuffs in elasmobranchs has not been fully elucidated. An initial lag phase before the start of gastric evacuation into the intestine, attributed to the time required for gastric juices and enzymatic reactions to take effect, was reported for the sandbar shark (Medved, 1985); however, this delay in the onset of digestion may have resulted from handling and force feeding of experimental animals (Wetherbee et al., 1990). In fishes, initial chemical digestion is generally attributed to pepsin, an acid protease (Holmgren and Nilsson, 1999). Plots of the change in energy content of the ingested meal with time suggested that tissues with higher energy, such as muscle, were evacuated before lower energy tissues, such as exoskeleton, during the earlier stages of gastric evacuation in gray smoothhound sharks (*Mustelus californicus*) (San Filippo, 1995). In contrast, Schurdak and Gruber (1989) reported that carbohydrates were evacuated from stomachs of lemon sharks prior to evacuation of proteins. For a detailed description of the anatomy and physiology of the digestive system of elasmobranchs, readers are referred to Holmgren and Nilsson (1999) and Cortés et al. (2008).

Although research for skates and rays is extremely scarce, emptying of food from the stomachs of elasmobranchs takes considerably longer than in teleosts. With very few exceptions, it takes a minimum of one to (often) several days to completely evacuate a meal from an elasmobranch stomach (Table 8.1). Presumably, lamnid sharks, such as the white shark, and other species capable of elevating stomach temperature above ambient water temperature through countercurrent mechanisms (McCosker, 1987; see also Chapter 7 of this volume) could have rapid rates of digestion, but no gastric evacuation measurements have been made to date on such heterothermic species.

8.4 Excretion and Egestion

A portion of food that is consumed by elasmobranchs is not absorbed by the digestive tract and is egested as feces. Additionally, a portion of the food that is absorbed by intestinal cells is not available for the energetic demands of the animal and is excreted as nitrogenous waste in urine and gill effluent.

TABLE 8.1

Summary of Gastric Evacuation, Daily Ration, and Food Conversion Efficiency Estimates for Elasmobranchs

Species	Stage	GE Curve	TGET (hr)	Daily Ration (%BW day^{-1})	K_1 (%)	Ref.
Carcharhinus acronotus	Juvenile, adult	—	—	0.87–1.56[a] (28)	—	Carlson and Parsons (1998)
Carcharhinus dussumieri	Juvenile (10)	—	—	2.9[b] (26–30)	—	Salini et al. (1999)
Carcharhinus leucas	Pup (6)	—	—	0.50[c] (24)	—	Schmid et al. (1990)
	Pup (5)	—	—	—	5–12[c] (23–25)	Schmid and Murru (1994)
Carcharhinus melanopterus	Juvenile (20)	—	—	0.3–0.8[c] (22–28)	20[c] (22–28)	Taylor and Wisner (1989)
Carcharhinus plumbeus	Juvenile	Gompertz	81–104 (22—26; 17)	0.9–1.3[d] (25; 414)	14.1 (25)	Medved (1985), Medved et al. (1988)
	NR (3)	—	>48 (NR)	—	—	Wass (1973)
	Pup	—	—	1.43[a] (18.5)	—	Stillwell and Kohler (1993)
	Juvenile, adult	—	—	0.86[a] (18.5)	—	Stillwell and Kohler (1993)
	Adult (6)	—	—	0.47[c] (24)	—	Schmid et al. (1990)
Carcharhinus tilstoni	Juvenile (4)	—	—	3.44[b] (26–30)	—	Salini et al. (1999)
Negaprion acutidens	Juvenile (4)	—	—	3.35[b] (26–30)	—	Salini et al. (1999)
Negaprion brevirostris	Juvenile	Linear	28–41 (20–29; 48)	1.5–2.1[d] (23–32; 86)	9.4–13.1 (32) [–64–25][e] (25; 80)	Cortés and Gruber (1990, 1992, 1994)
	Juvenile	Exponential	24 (25; 20)	—	—	Schurdak and Gruber (1989)
	Juvenile	—	—	2.7[b] (25; 6)	[22.4][e] (25; 3)	Gruber (1984); Longval et al. (1982)
Prionace glauca	Juvenile (1), adult (1)	—	—	0.5–1.4[b] (21–29)	—	Clark (1963)
	NR	—	>24 (14–16; 3)	—	—	Tricas (1979)
	Adult	Exponential[f]	164 (19; 2)	0.40–0.65[d] (17; 54)	17.1 (17)	Kohler (1987)
Sphyrna lewini	Juvenile	Multiple[g]	>5–29 (21–29; 64)	2.12–3.54 (22–28; 451)	—	Bush and Holland (2002)
Sphyrna lewini	Juvenile	—	—	—	2.9–9.4[a] (26)	Lowe (2002)
Sphyrna lewini	Juvenile	—	—	—	[–36–34][e] (24–29; 54)	Duncan (2006)
Sphyrna tiburo	All	Logistic	>50 (20–30; 46)	2.16–4.34[d] (20–30; 53)	—	Tyminski et al. (1999)
	All	—	—	0.37–5.46[a] (NR)	—	Bethea et al. (2007)
Triakis semifasciata	All	Linear	28—32 (13–18; 30)	0.85–2.20 (NR; 138)	—	Kao (2000)
Schroederichthys chilensis	NR	Exponential	74 (16; 18)	—	—	Aedo and Arancibia (2001)
Scyliorhinus canicula	All	Surface area[h]	50–>70 (14; 237)	—	—	Macpherson et al. (1989)
	Adult	Exponential	>200 (15; 20)	—	—	Sims et al. (1996)

Species	Stage (n)	GE curve	TGET	Gastric evacuation rate	K_1	Reference
Isurus oxyrinchus	Adult	—	36—48[i]	2.2–3.0[a] (19)	—	Stillwell and Kohler (1982)
Carcharias taurus	Adult (13)	—	—	0.27[c] (24)	—	Schmid et al. (1990)
Ginglymostoma cirratum	Adult (6)	—	—	0.31[c] (24)	—	Schmid et al. (1990)
Notorynchus cepedianus	Pup	—	—	2[c] (12–14)	—	Van Dykhuizen and Mollet (1992)
	Juvenile			0.6[c] (12–14)	25–40	
	Adult			0.2[c] (12–14)	10–15	
Squalus acanthias	Juvenile, adult	—	124 (10;75)	1.3[b] (10;5)	6.1–10.7[j]	Jones and Geen (1977)
				0.4[k] (10)		Holden (1966)
	Adult		—	1.5–2.0[c] (10)		Brett and Blackburn (1978)
	Adult		>48 (15)			Van Slyke and White (1911)
	All			2.60[l] (NR; 3396)		Tanasichuk et al. (1991)
Dasyatis sabina	All	Exponential[f]	—	—		Bradley (1996)
Gymnura altavela	Adult (2)	—	—	2.52 (27–33; 48)	10.8[c] (23)	Henningsen (1996)
Raja erinacea	All	Multiple[g]	12–52[m] (10 and 16; 28)	—	—	Nelson and Ross (1995)
Callorhynchus callorhynchus	All	—	>24 (13, 113)	1.36 (11.5–13, 181)	—	Di Giácomo et al. (1994)

Note: GE curve is the mathematical model that best describes gastric evacuation; TGET is total gastric evacuation time; K_1 is gross conversion efficiency (annual production divided by annual consumption estimates); NR is not reported. Single values in parentheses denote temperature range in degrees Celsius, except for the Stage column, where they indicate sample size; a second value indicates sample size.

a Bioenergetic estimate(s) only.
b Captive sharks fed experimental meal to satiation.
c Food consumed by captive sharks in display aquarium.
d Includes both in situ and bioenergetic estimates.
e Derived in laboratory or aquarium experiments where sharks were fed at varying ration sizes and growth recorded.
f Assumed functional relationship.
g Different models provided the best fit depending on temperature, food type, or meal size.
h Gastric evacuation of small prey items was adequately described by exponential model.
i Assumed values.
j 6.1% is for age 1 dogfish, 10.7% is for age 0 dogfish.
k "Working" bioenergetic estimate.
l Estimated from mean stomach fullness indices.
m Depending on temperature and food type.
n A holocephalan.

8.4.1 Excretion

Energetic losses in gill effluent and urine have not been measured in elasmobranchs but have been presumed to be similar in scale to losses (about 7% of the energy budget) estimated for teleost fishes (Brett and Groves, 1979). Quantification of energy losses through the gills and kidneys of elasmobranchs is problematic due to the large quantity of water involved in housing elasmobranchs, as well as retention of nitrogenous wastes in the form of urea and trimethylamine oxide in blood and tissues for osmoregulatory purposes (Perlman and Goldstein, 1988; Wood, 1993; see Chapter 11 of this volume).

8.4.2 Egestion

Elasmobranchs have a spiral valve intestine, which functions to increase surface area for digestion and absorption of food but which also conserves space in the body cavity for a large liver and development of large embryos (Moss, 1984). The digestive capability of the spiral valve intestine has been investigated in only one species of elasmobranch, the lemon shark (Wetherbee and Gruber, 1993a). These authors used an indirect method of measurement incorporating an inert, naturally occurring marker (acid-insoluble ash) into food (Wetherbee and Gruber, 1993b). In this study, lemon sharks were capable of absorbing energy and nutrients in food with an average efficiency close to 80%, which is similar to many carnivorous teleosts; however, the time required for a meal to be completely eliminated from the digestive tract of lemon sharks was prolonged (70 to 100 hr) in comparison to most teleosts (Wetherbee and Gruber, 1990; Wetherbee et al., 1987). Other studies have reported that food remains in the digestive tract of elasmobranchs for long periods of time (up to 18 days) in comparison to most teleosts (Sims et al., 1996; Wetherbee et al., 1990). The protracted periods of time required for complete food passage, in addition to difficulties involved with maintaining sharks in captivity and the labor-intensive methods required for fecal collection, present major obstacles for studies on digestive efficiency of sharks (Wetherbee and Gruber, 1990).

Prolonged passage of food through digestive tracts of elasmobranchs may be required for spiral valve intestines to accomplish digestion and absorption of food at levels comparable with those of teleosts. There have been several studies on enzymatic digestion in the stomachs of elasmobranchs, but few studies on pancreatic and brush border enzymes that function to break down macromolecules to smaller subunits for absorption across the intestinal epithelium (Caira and Jolitz, 1989; Fänge and Grove, 1979; Papastamatiou, 2003; Sullivan, 1907; Van Slyke and White, 1911). Although the relationship between prolonged food passage time and limitation of enzymatic digestion in elasmobranchs is unknown, it is apparent that prolonged food passage is related to a low rate of consumption in sharks, which in turn limits growth and reproductive rates. Although low rates of food consumption may provide evolutionary advantages for elasmobranch populations, the associated low growth and reproductive rates are life history characteristics that contribute to the vulnerability of the majority of elasmobranch populations to overfishing.

8.5 Production

Production, or growth in body mass, can be measured through laboratory experiments, field mark-recapture methods, or indirectly through size at age relationships. Relative rates of production (expressed as percent body weight) of most teleost species are considerably higher than those of elasmobranchs (Wetherbee et al., 1990), with many teleosts doubling their body weight in less than a week after birth (Brett and Groves, 1979). Relative growth rates in length and mass are much higher for immature than mature individuals in most elasmobranch species (see Chapter 14 of this volume), especially during the first year of life. Branstetter (1990) estimated first-year growth in body length for several shark species, with values ranging from 16 to 100% per year. Wetherbee et al. (1990) reported values of first-year growth in mass of 33, 79, and 138% for the spiny dogfish, sandbar shark, and lemon shark, respectively. In relative terms, small coastal and pelagic species tend to grow at a faster rate than their large coastal counterparts, probably reflecting differences in the risk of predation faced by juveniles. As very few estimates of food consumption are available, it is unclear whether differences in production are a result of different levels of food consumption or differences in energy partitioning.

Growth efficiency measures have been calculated for few elasmobranchs. The efficiency of food conversion to somatic growth, or gross conversion efficiency (K_1), is important ecologically because it measures the proportion of ingested food that will be available to the next trophic level (Warren and Davis, 1967). K_1 values reported for elasmobranchs range from about 3 to 40% (Table 8.1). Van Dykhuizen and Mollet (1992) reported that K_1 values (which they referred to as cumulative total efficiency) decreased with increasing age, from 25 to 40% at age 1 to 3 years to 10 to 15% at age 5 to 6 years in aquarium-fed sevengill sharks. Most K_1 values for elasmobranchs (Table 8.1) are comparable to values reported for teleosts (10 to 25%) (Brett and Groves, 1979), indicating that elasmobranchs are generally capable of converting energy to growth as efficiently as teleosts.

The rate of production and K_1 are functions of the rate of food consumption. Two studies have examined this relationship in elasmobranchs. For both lemon (Cortés and Gruber, 1994) and scalloped hammerhead (Duncan, 2006) sharks, the relationship between production rate and feeding rate was best described by a von Bertalanffy growth-like equation of the form:

$$G_r = G_{max}\left(1 - e^{-k(R-R_m)}\right) \qquad (8.3)$$

where G_r is growth rate, G_{max} is maximum growth rate, k is the rate of change in growth rate with feeding rate, R is feeding rate, and R_m is the maintenance ration (no growth). Cortés and Gruber (1994) reported very similar values of $R_m = 1.06\%$ wet BW day^{-1} and G_s (loss in weight due to starvation) $= 1.11\%$ BW day^{-1}. Duncan (2006) estimated $R_m = 3.4\%$ wet BW day^{-1} and also reported almost identical values of R_m and G_s. Cortés (1991) also estimated a value for R_{opt}, the optimal ration (Pandian, 1982), of 2.15 BW day^{-1} for a 2-kg lemon shark in its first year of life, by drawing a tangent from the origin of coordinates in the G–R curve to the point in the curve with the steepest slope. Cortés and Gruber (1994) found values of K_1 ranging from -64% to 25% and that K_1 slowed, but continued to increase, at ration levels above maintenance. This finding did not support those from several studies with teleosts where a dome-shaped curve was found (Paloheimo and Dickie, 1966), and K_1 rapidly decreased after reaching a peak at an optimum feeding rate. Duncan (2006) estimated values of K_1 ranging from -36% to 34%, that maximum food conversion efficiency occurred at a feeding rate of 5.1% wet BW day^{-1}, but that at the highest feeding rate K_1 decreased to 28%, thus conforming more with the dome-shaped curve hypothesis. The efficiency of conversion of absorbed food to growth, or net conversion efficiency (K_2), has not been measured for any elasmobranchs, except for an estimate of 33% provided by Gruber (1984) for juvenile lemon sharks.

8.6 Conclusions

The major prey item consumed by elasmobranchs is teleost fishes, whereas crustaceans are the most important prey of batoids. However, there are numerous exceptions to this generalization. Accurate descriptions of the diets of elasmobranchs are complicated by the plasticity of their feeding habits, which regularly result in ontogenetic and spatiotemporal shifts. Based on determinations for a limited number of species, sharks appear to exhibit short feeding bouts followed by longer periods of digestion. The food consumption dynamics of elasmobranchs may ultimately be governed by a morphological peculiarity of this group of predators, a spiral valve intestine. This digestive morphology likely dictates slower rates of gastrointestinal emptying and consequent lower food consumption rates, lower production rates, and generally slower food dynamics for elasmobranchs compared to teleosts. From our limited knowledge, however, it appears that elasmobranchs are capable of absorbing food and converting it to growth with efficiencies comparable to those of teleosts. Improved understanding of consumption and channeling of energy within elasmobranchs will advance understanding of the role of sharks and their relatives in marine ecosystems. Clearly, much remains to be learned about food consumption and feeding habits of elasmobranchs. Because of the difficulty of conducting controlled experiments with large, adult individuals of many elasmobranch species, we advocate a pragmatic approach to advance our knowledge of the feeding ecology of this group.

References

Abdel-Aziz, S.H. (1986). Food and feeding habits of *Raja* species (Batoidei) in the Mediterranean waters of Alexandria. *Bull. Inst. Oceanogr. Fish.* 12:165–276.

Abdel-Aziz, S.H. (1994). Observations on the biology of the common torpedo (*Torpedo torpedo*, Linnaeus, 1758) and marbled electric ray (*Torpedo marmorata*, Risso, 1810) from Egyptian Mediterranean waters. *Aust. J. Mar. Freshw. Res.* 45:693–704.

Aedo, G. and Arancibia, H. (2001). Gastric evacuation of the redspotted catshark under laboratory conditions. *J. Fish Biol.* 58:1454–1457.

Ajayi, T.O. (1982). Food and feeding habits of *Raja* species (Batoidei) in Carmarthen Bay, Bristol Channel. *J. Mar. Biol. Assoc. U.K.* 62:215–223.

Al-Habsi, S.H., Sweeting, C.J., Polunin, N.V.C., and Graham, N.A.J. (2008). δ^{15}N and δ^{13}C elucidation of size-structured food webs in a western Arabian Sea demersal trawl assemblage. *Mar. Ecol. Prog. Ser.* 353:55–63.

Ali, T.S., Mohamed, A.R.M., and Hussain, N.A. (1993). Trophic interrelationships of the demersal fish assemblage in the Northwest Arabian Gulf, Iraq. *Asian Fish. Sci.* 6:255–264.

Allen, B.R. and Cliff, G. (2000). Sharks caught in the protective gill nets of Kwazulu-Natal, South Africa. 9. The spinner shark (*Carcharhinus brevipinna*) (Müller and Henle). *S. Afr. J. Mar. Sci.* 22:199–215.

Almeida, Z.S., Nunes, J.S., and Costa, C.L. (2000). Presencia de *Urotrygon microphthalmum* (Elasmobranchii: Urolophidae) en aguas bajas de Maranháo (Brasil) y notas sobre su biología. *Bol. Invest. Mar. Cost.* 29:67–72.

Almeida, M.P., Lins, P.M.O., Charvet-Almeida, P., and Barthem, R.B. (2010). Diet of the freshwater stingray *Potamotrygon motoro* (Chondrichthyes: Potamotrygonidae) on Marajó Island (Pará, Brazil). *Braz. J. Biol.* 70:155–162.

Anderson, C. and Hafiz, A. (1989). *Common Reef Fishes of the Maldives*, Part 2. Novelty Press, Maldives.

Arrighetti, F., Livore, J.P., and Penchaszadeh, P.E. (2005). Siphon nipping of the bivalve *Amiantis purpurata* by the electric ray *Discopyge tschudii* in Mar del Plata, Argentina. *J. Mar. Biol. Assoc. U.K.* 85:1151–1154.

Auerswald, K., Wittmer, M.H.O.M., Zazzo, A., Schaufele, R., and Schnyder, H. (2010). Biases in the analysis of stable isotope discrimination in food webs. *J. Appl. Ecol.* 47:936–941.

Avsar, D. (2001). Age, growth, reproduction and feeding of the spurdog (*Squalus acanthias* Linnaeus, 1758) in the southeastern Black Sea. *Estuar. Coast. Shelf Sci.* 52:269–278.

Baba, O., Taniuchi, T., and Nose, Y. (1987). Depth distribution and food habits of three species of small squaloid sharks off Choshi. *Nippon Suisan Gakkaishi* 53:417–424.

Babel, J.S. (1967). Reproduction, life history and ecology of the round stingray *Urolophus halleri* Cooper. *Calif. Fish Game Comm. Fish. Bull.* 137.

Backus, R.H., Springer, S., and Arnold, E.L. (1956). A contribution to the natural history of the white-tip shark, *Pterolamiops longimanus* (Poey). *Deep Sea Res.* 3:178–188.

Baduini, C.L. (1995). Feeding Ecology of the Basking Shark (*Cetorhinus maximus*) Relative to Distribution and Abundance of Prey, master's thesis, Moss Landing Marine Laboratories, San Jose State University.

Barry, J.P., Yoklavich, M.M., Cailliet, G.M. et al. (1996). Trophic ecology of the dominant fishes in Elkhorn Slough, California, 1974–1980. *Estuaries* 19:115–138.

Bass, A.J., D'Aubrey, J.D., and Kistnasamy, N. (1973). Sharks of the east coast of southern Africa. I. The genus *Carcharhinus* (Carcharhinidae). *S. Afr. Ass. Mar. Biol. Res. Oceanogr. Res. Inst. Invest. Rep.* 33:1–168.

Beebe, W. and Tee-Van, J. (1941). Fishes of the tropical eastern Pacific (from Cedros Island, Lower California, South to the Galápagos Islands and northern Peru). Part 3. Rays, mantas, and chimaeras. *Zoologica* 26:245–280.

Bell, J.C. and Nichols, J.T. (1921). Notes on the food of Carolina sharks. *Copeia* 1921:17–20.

Belleggia, M., Mabragaña, E., Figueroa, D.E., Scenna, L.B., Barbini, S.A., and de Astarloa, J.M.D. (2008). Food habits of the broad nose skate, *Bathyraja brachyurops* (Chondrichthyes: Rajidae), in the southwest Atlantic. *Sci. Mar. (Barc.)* 72:701–710.

Bethea, D., Buckel, J.A., and Carlson, J.K. (2004). Foraging ecology of the early life stages of four sympatric shark species. *Mar. Ecol. Prog. Ser.* 268:245–264.

Bethea, D., Hale, L., Carlson, J.K., Cortés, E., Manire, C.A., and Gelsleichter, J. (2007). Geographic and ontogenetic variation in the diet and daily ration of the bonnethead shark, *Sphyrna tiburo*, from the eastern Gulf of Mexico. *Mar. Biol.* 152:1009–1020.

Bigelow, H.B. and Schroeder, W.C. (1953). *Fishes of the Western North Atlantic*. Part 2. *Sawfishes, Guitarfishes, Skates, Rays and Chimaeroids*. Sears Foundation for Marine Research. Yale University, New Haven, CT.

Bizzarro, J.J. (2005). Fishery Biology and Feeding Ecology of Rays in Bahía Almejas, Mexico, master's thesis, Moss Landing Marine Laboratories, The California State University, San Francisco.

Blaber, S.J.M. and Bulman, C.M. (1987). Diets of fishes of the upper continental slope of eastern Tasmania: content, calorific values, dietary overlap and trophic relationships. *Mar. Biol.* 95:345–356.

Blaylock, R.A. (1992). Distribution, Abundance, and Behavior of the Cownose Ray, *Rhinoptera bonasus* (Mitchill 1815), in Lower Chesapeake Bay, doctoral dissertation, Virginia Institute of Marine Science, College of William and Mary, Williamsburg, VA.

Bolnick, D.I., Yang, L.H., Fordyce, J.A., Davis, J.M., and Svanback, R. (2002). Measuring individual-level resource specialization. *Ecology* 83:2936–2941.

Bonham, K. (1954). Food of the spiny dogfish *Squalus acanthias*. *Wash. Dept. Fish., Fish. Res. Pap.* 1:25–36.

Bor, P.H.F. and Santos, M.B. (2003). Findings of elasmobranch egg cases in the stomachs of sperm whales and other marine organisms. *J. Mar. Biol. Assoc. U.K.* 83:1351–1353.

Borrell, A., Cardona, L., Ramanathan, K.P., and Aguilar, A. (2011). Trophic ecology of elasmobranchs caught off Gujarat, India, as inferred from stable isotopes. *ICES J. Mar. Sci.* 68:547–554.

Bowen, S. (1996). Quantitative description of the diet. In: Murphy, B.R. and Willis, D.W. (Eds.), *Fisheries Techniques*. American Fisheries Society, Bethesda, MD, pp. 513–532.

Boyle, M.D. (2010). Trophic Interactions of *Bathyraja trachura* and Sympatric Fishes, master's thesis, Moss Landing Marine Laboratories, The California State University, San Francisco.

Braccini, J.M. and Perez, J.E. (2005). Feeding habits of the sandskate *Psammobatis extenta* (Garman, 1913): sources of variation in dietary composition. *Mar. Freshw. Res.* 56:395–403.

Bradley, J.L. (1996). Prey Energy Content and Selection, Habitat Use and Daily Ration of the Atlantic Stingray, *Dasyatis sabina*, master's thesis, Florida Institute of Technology, Melbourne.

Brafield, A.E. and Llewellyn, M.J. (1982). *Animal Energetics*. Blackie & Sons, Glasgow.

Branstetter, S. (1990). Early life-history implications of selected carcharhinoid and lamnoid sharks of the northwest Atlantic. In: Pratt, Jr., H.L., Gruber, S.H., and Taniuchi, T. (Eds.), *Elasmobranchs as Living Resources: Advances in the Biology, Ecology, Systematics, and the Status of the Fisheries*, NOAA Tech. Rep. NMFS 90. U.S. Department of Commerce, Washington, D.C., pp. 17–28.

Breeder, Jr., C.M. (1921). The food of *Mustelus canis* (Mitchell) in mid-summer. *Copeia* 101:85–86.

Brett, J.R. and Blackburn, J.M. (1978). Metabolic rate and energy expenditure of the spiny dogfish, *Squalus acanthias*. *J. Fish. Res. Board Can.* 35:816–821.

Brett, J.R. and Groves, T.D.D. (1979). Physiological energetic. In: Hoar, W.S., Randall, D.J., and Brett, J.R. (Eds.), *Fish Physiology*, Vol. 8. Academic Press, New York, pp. 279–352.

Brickle, P., Lapitkhovsky, V., and Pompert, J. (2003). Ontogenetic changes in the feeding habits and dietary overlap between three abundant rajid species on the Falkland Island's shelf. *J. Mar. Biol. Assoc. U.K.* 83:1119–1125.

Bush, A.C. (2002). The Feeding Ecology of Juvenile Scalloped Hammerhead Sharks (*Sphyrna lewini*) in Kane'ohe Bay, O'ahu, Hawai'i, doctoral dissertation, University of Hawaii.

Bush, A.C. and Holland, K.N. (2002). Food limitation in a nursery area: estimates of daily ration in juvenile scalloped hammerheads, *Sphyrna lewini* (Griffith and Smith, 1834) in Kane'ohe Bay, O'ahu, Hawai'i. *J. Exp. Mar. Biol. Ecol* 278:157–178.

Caira, J.N. and Jolitz, E.C. (1989). Gut pH in the nurse shark, *Ginglymostoma cirratum* (Bonnaterre). *Copeia* 1989:192–194.

Capapé, C. (1974). Contribution à la biologie des Scyliorhinidae des côtes tunisiennes. II. *Scyliorhinus canicula* (Linné, 1758): régime alimentaire. *Ann. Inst. Michel Pacha* 7:13–29.

Capapé, C. (1975a). Contribution à la biologie des Scyliorhinidae des côtes tunisiennes. IV. *Scyliorhinus stellaris* (Linné, 1758): régime alimentaire. *Arch. Inst. Pasteur Tunis* 52:383–394.

Capapé, C. (1975b). Contribution a la biologie des Rajidae des cotes tunisiennes. 4. *Raja clavata* (Linne, 1758): regime alimentaire. *Ann. Inst. Michel Pacha* 8:16–32.

Capapé, C. (1976). Étude du régime alimentaire de l'Aigle de mer, *Myliobatis aquila* (L., 1758) des côtes tunisiennes. *Cons. Int. Explor. Mer* 37:29–35.

Capapé, C. (1977a). Contribution a la biologie des Rajidae des cotes tunisiennes. 12. *Raja alba* (Lacepede, 1803): regime alimentaire. *Arch. Inst. Pasteur Tunis* 54:85–95.

Capapé, C. (1977b). Study of the feeding behavior of *Ptermomylaeus bovinus* (Geoffroy Saint-Hilarie, 1817) (Pisces, Myliobatidae) off the Tunisian coast. *J. du Conseil.* 37:214–220.

Capapé, C. (1978). Contribution à la biologie des Dasyatidae des côtes tunisiennes. IV. *Dasyatis tortonesei* (Capapé, 1975): régime alimentaire. *Arch. Inst. Pasteur Tunis* 55:359–370.

Capapé, C. (1979). La Torpille marbreé, *Torpedo marmorata* Risso, 1810 (Pisces, Rajiformes) des côtes tunisiennes: nouvelles données sur l'écologie et la biolgie de la reproduction de l'espèce, avec une comparaison entre les populations Méditerranéenne et Atlantiques. *Ann. Sci. Nat. Zool.* 1:79–97.

Capapé, C. (1986). Données générales sur le régime alitmentaire des Gymnuridae et des Mobulidae (Pisces: Selachii). *Arch. Inst. Pasteur Tunis* 63:241–246.

Capapé, C. and Desoutter, M. (1979). Nouvelle description de *Aetomylaeus nichofi* (Bloch et Schneider, 1801) (Pisces, Myliobatidae). Premières observations biologiques. *Cah. Indo-Pac.* 1:305–322.

Capapé, C. and Zaouali, J. (1976). Contribution à la biologie des Scyliorhinidae des côtes tunisiennes. V. *Galeus melastomus* (Rafinesque, 1810): régime alimentaire. *Arch. Inst. Pasteur Tunis* 53:281–292.

Capapé, C. and Zaouali, J. (1979). Etude du regime alimentaire de deux selaciens communs dans le golfe de Gabes (Tunisie): *Rhinobatos rhinobatos* (Linne, 1758) et *Rhinobatos cemiculus* (Geoffroy Saint-Hilaire, 1817). *Arch. Inst. Pasteur Tunis* 56:285–306.

Carlson, J.K. and Parsons, G.R. (1998). Estimates of daily ration and a bioenergetic model for the blacknose shark, *Carcharhinus acronotus* [abstract]. In: *Proceedings of the American Society of Ichthyologists and Herpetologists 78th Annual Meeting*, July 16–22, University of Guelph, Guelph, Ontario, Canada.

Carlson, J.K. et al. (2002). An observation of juvenile tiger sharks feeding on clapper rails off the southeastern coast of the United States. *Southeast. Nat.* 1:307–310.

Carrassón, M., Stefanescu, C., and Cartes, J.E. (1992). Diets and bathymetric distributions of two bathyal sharks of the Catalan deep sea (western Mediterranean). *Mar. Ecol. Prog. Ser.* 82:21–30.

Castro, J.I. (1989). The biology of the golden hammerhead, *Sphyrna tudes*, off Trinidad. *Environ. Biol. Fish.* 24:3–11.

Castro, J.I. (1993). The biology of the finetooth shark, *Carcharhinus isodon. Environ. Biol. Fish.* 36:219–239.

Castro, J.I., Bubucis, P.M., and Overstrom, N.A. (1988). The reproductive biology of the chain dogfish, *Scyliorhinus retifer. Copeia* 1988:740–746.

Caut, S., Angulo, E., and Courchamp, F. (2009). Variation in discrimination factors (Δ15N and Δ13C): the effect of diet isotopic values and applications for diet reconstruction. *J. Appl. Ecol.* 46:443–453.

Celona, A. (2004). Caught and observed giant devil rays *Modula mobular* (Bonnaterre, 1788) in the Strait of Messina. *Ann. Ser. Hist. Nat.* 14: 11–18.

Chatwin, B.M. and Forrester, C.F. (1953). Feeding habits of dogfish *Squalus suckleyi* (Girard). *Fish Res. Bd. Can. Prog. Rep. Pac. Coast Sta.* 95:35–38.

Christensen, V. and Pauly, D. (1992). The Ecopath II: a software for balancing steady-state models and calculating network characteristics. *Ecol. Model.* 61:169–185.

Chuchukalo, V.I. and Napazakov, V.V. (2002). Feeding and trophic status of abundant skate species (Rajidae) of the western Bering Sea. *Izvestiya TINRO* 130:422–428.

Clark, E. (1963). The maintenance of sharks in captivity, with a report on their instrumental conditioning. In: Gilbert, P.W. (Ed.), *Sharks and Survival*. Heath, Boston, pp. 115–149.

Clark, E. and von Schmidt, K. (1965). Sharks of the central Gulf coast of Florida. *Bull. Mar. Sci.* 15:13–83.

Clarke, M.R., King, K.J., and McMillan, P.J. (1989). The food and feeding relationships of black oreo, *Allocyttus niger*, smooth oreo, *Pseudocyttus maculatus*, and eight other fish species from the continental slope of the southwest Chatham Rise, New Zealand. *J. Fish Biol.* 35:465–484.

Clarke, M.R., Clarke, D.C., Martins, H.R., and da Silva, H.M. (1996). The diet of the blue shark (*Prionace glauca* L.) in Azorean waters. *Arquipélago Life Mar. Sci. (Ponta Delgada)* 14A:41–56.

Cliff, G. (1995). Sharks caught in the protective gill nets off Kwazulu-Natal, South Africa. 8. The great hammerhead shark (*Sphyrna mokarran*) (Ruppell). *S. Afr. J. Mar. Sci.* 15:105–114.

Cliff, G. and Dudley, S.F.J. (1991a). Sharks caught in the protective gill nets of Natal, South Africa. 4. The bull shark (*Carcharhinus leucas*) (Valenciennes). *S. Afr. J. Mar. Sci.* 10:253–270.

Cliff, G. and Dudley, S.F.J. (1991b). Sharks caught in the protective gill nets of Natal, South Africa. 5. The Java shark (*Carcharhinus amboinensis*) (Muller and Henle). *S. Afr. J. Mar. Sci.* 11:443–453.

Cliff, G. and Dudley, S.F.J. (1992). Sharks caught in the protective gill nets of Natal, South Africa. 6. The copper shark (*Carcharhinus brachyurus*) (Gunther). *S. Afr. J. Mar. Sci.* 12:663–674.

Cliff, G., Dudley, S.F.J., and Davis, B. (1988). Sharks caught in the protective gill nets of Natal, South Africa. 1. The sandbar shark (*Carcharhinus plumbeus*) (Nardo). *S. Afr. J. Mar. Sci.* 7:255–265.

Cliff, G., Dudley, S.F.J., and Davis, B. (1989). Sharks caught in the protective gill nets of Natal, South Africa. 2. The great white shark (*Carcharodon carcharias*) (Linnaeus). *S. Afr. J. Mar. Sci.* 8:131–144.

Coles, R.J. (1919). The large sharks off Cape Lookout, North Carolina. The white sharks or maneater, tiger shark and hammerhead. *Copeia* 69:34–43.

Collins, A.B., Heupel, M.R., Hueter, R.E., and Motta, P.J. (2007). Hard prey specialists or opportunistic generalists? An examination of the diet of the cownose ray, *Rhinoptera bonasus*. *Mar. Freshw. Res.* 58:135–144.

Compagno, L.J.V. (1990). Relationships of the megamouth shark, *Megachasma pelagios* (Lamniformes: Megachasmidae), with comments on its feeding habits. In: Pratt, Jr., H.L., Gruber, S.H., and Taniuchi, T. (Eds.), *Elasmobranchs as Living Resources: Advances in the Biology, Ecology, Systematics, and the Status of the Fisheries*, NOAA Tech. Rep. NMFS 90. U.S. Department of Commerce, Washington, D.C., pp. 357–379.

Compagno, L.J.V., Ebert, D.A., and Smale, M.J. (1989). *Guide to the Sharks and Rays of Southern Africa*. New Holland, London.

Corkeron, P.J., Morris, R.J., and Bryden, M.M. (1987). Interactions between bottlenose dolphins and sharks in Moreton Bay, Queensland. *Aquat. Mammals* 13:109–113.

Cortés, E. (1991). Alimentación en el Tiburón Galano, *Negaprion brevirostris* (Poey): Dieta, Hábitos Alimentarios, Digestión, Consumo y Crecimiento, doctoral dissertation, University of Barcelona.

Cortés, E. (1997). A critical review of methods of studying fish feeding based on analysis of stomach contents: application to elasmobranch fishes. *Can. J. Fish. Aquat. Sci.* 54:726–738.

Cortés, E. (1998). Methods of studying fish feeding: reply. *Can. J. Fish. Aquat. Sci.* 55:2708.

Cortés, E. (1999). Standardized diet compositions and trophic levels of sharks. *ICES J. Mar. Sci.* 56:707–717.

Cortés, E. and Gruber, S.H. (1990). Diet, feeding habits, and estimates of daily ration of young lemon sharks, *Negaprion brevirostris* (Poey). *Copeia* 1990:204–218.

Cortés, E. and Gruber, S.H. (1992). Gastric evacuation in the young lemon shark, *Negaprion brevirostris*, under field conditions. *Environ. Biol. Fish.* 35:205–212.

Cortés, E. and Gruber, S.H. (1994). Effect of ration size on growth and gross conversion efficiency of young lemon sharks, *Negaprion brevirostris*. *J. Fish Biol.* 44:331–341.

Cortés, E., Manire, C.A., and Hueter, R.E. (1996). Diet, feeding habits, and diel feeding chronology of the bonnethead shark, *Sphyrna tiburo*, in southwest Florida. *Bull. Mar. Sci.* 58:353–367.

Cortés, E., Papastamatiou, Y., Wetherbee, B.M., Carlson, J.K., and Ferry-Graham, L. (2008). An overview of the feeding ecology and physiology of elasmobranch fishes. In: Cyrino, J.E.P., Bureau, D.P., and Kapoor, B.G. (Eds.), *Feeding and Digestive Functions in Fishes*. Science Publishers, Enfield, NH, pp. 393–443.

Cross, J.N. (1988). Aspects of the biology of two scyliorhinid sharks, *Apristurus brunneus* and *Parmaturus xaniurus*, from the upper continental slope off southern California. *Fish. Bull.* 86:691–702.

Cross, R.E. and Curran, M.C. (2000). Effects of feeding pit formation by rays on an intertidal meiobenthic community. *Estuar. Coast. Shelf Sci.* 51:293–298.

Cross, R.E. and Curran, M.C. (2004). Recovery of meiofauna in intertidal feeding pits created by rays. *Southeast. Nat.* 3:219–230.

Dahlberg, M.D. and Heard, R.W. (1969). Observations on elasmobranchs from Georgia. *Q. J. Fla. Acad. Sci.* 32:21–25.

Daiber, F.C. and Booth, R.A. (1960). Notes on the biology of the butterfly rays, *Gymnura altavela* and *Gymnura micrura*. *Copeia* 1960:137–139.

DeCrosta, M.A., Taylor, Jr., L.R., and Parrish, J.D. (1984). Age determination, growth, and energetics of three species of carcharhinid sharks in Hawaii. In: Grigg, R.W. and Tanouse, K.Y. (Eds.), *Proceedings of the Second Symposium on Resource Investigations in the Northwest Hawaiian Islands*, Vol. 2, May 25–27, 1983, University of Hawaii, Honolulu, Sea Grant Misc. Rep. UNIHI-SEAGRANT-MR-84-01, pp. 75–95.

Devadoss, P. (1978). On the food of rays, *Dasyatis uarnak* (Forskal), *D. alcockii* (Annandale), and *D. sephen* (Forskal). *Indian J. Fish.* 25:9–13.

Devadoss, P. (1989). Observations on the length–weight relationship and food and feeding habits of spade nose shark, *Scoliodon laticaudus* Muller and Henle. *Indian J. Fish.* 36:169–174.

Di Giácomo, E., Parma, A.M., and Orensanz, J.M. (1994). Food consumption by the cock fish, *Callorhynchus callorhynchus* (Holocephali: Callorhynchidae), from Patagonia (Argentina). *Environ. Biol. Fish.* 40:199–211.

Diana, J.S. (1979). The feeding pattern and daily ration of a top carnivore, the northern pike, *Esox lucius*. *Can. J. Zool.* 57:977–991.

Dodrill, J.W. and Gilmore, G.R. (1978). Land birds in the stomachs of tiger sharks *Galeocerdo cuvieri* (Peron and Leseur). *Auk* 95:585–586.

Dolgov, A. (2005). Feeding and food consumption by the Barents Sea skates. *J. Northw. Atl. Fish. Sci.* 35:496–503.

Domi, N., Bouquegneau, J.M., and Das, K. (2005). Feeding ecology of five commercial shark species of the Celtic Sea through stable isotope and trace metal analysis. *Mar. Environ. Res.* 60:551–569.

Dowton-Hoffman, C.A. (2006). Biología del pez guitarra *Rhinobatos productus* (Ayres, 1856), en Baja California Sur, México, doctoral dissertation, Centro Interdisciplinario de Ciencias Marinas.

Dudley, S.F.J. and Cliff, G. (1993). Sharks caught in the protective gill nets of Natal, South Africa. 7. The blacktip shark (*Carcharhinus limbatus*) (Valenciennes). *S. Afr. J. Mar. Sci.* 13:237–254.

Dudley, S.F.J., Anderson-Reade, M.D., Thompson, G.S., and McMullen, P.B. (2000). Concurrent scavenging off a whale carcass by great white sharks, *Carcharodon carcharias*, and tiger sharks, *Galeocerdo cuvier*. *Fish. Bull.* 98:646–649.

Duncan, K. (2006). Estimation of daily energetic requirements in young scalloped hammerhead sharks, *Sphyrna lewini*. *Environ. Biol. Fish.* 76:139–149.

Ebert, D.A. (1991). Diet of the sevengill shark *Notorynchus cepedianus* in the temperate coastal waters of southern Africa. *S. Afr. J. Mar. Sci.* 11:565–572.

Ebert, D.A. (1994). Diet of the sixgill shark *Hexanchus griseus* off southern Africa. *S. Afr. J. Mar. Sci.* 14:213–218.

Ebert, D.A. and Bizzarro, J.J,. (2007). Standardized diet compositions and trophic levels of skates (Chondrichthyes: Rajiformes: Rajoidei). *Environ. Biol. Fish.* 80:221–237.

Ebert D.A. and Compagno, L.J.V. (2007). Biodiversity and systematics of skates (Chondrichthyes: Rajiformes: Rajoidei). *Environ. Biol. Fish.* 80:111–124.

Ebert, D.A. and Cowley, P.D. (2003). Diet, feeding behavior and habitat utilization of the blue stingray *Dasyatis chrysonota* (Smith, 1828) in South African waters. *Mar. Freshw. Res.* 54: 957–965.

Ebert, D.A., Cowley, P.D., and Compagno, L.J.V. (1991). A preliminary investigation of the feeding ecology of skates (Batoidea: Rajidae) off the west coast of southern Africa. *S. Afr. J. Mar. Sci.* 10:71–81.

Ebert, D.A., Compagno, L.J.V., and Cowley, P.D. (1992). A preliminary investigation of the feeding ecology of squaloid sharks off the west coast of southern Africa. *S. Afr. J. Mar. Sci.* 12:601–609.

Ebert, D.A., Cowley, P.D., and Compagno, L.J.V. (1996). A preliminary investigation of the feeding ecology of catsharks (Scyliorhinidae) off the west coast of southern Africa. *S. Afr. J. Mar. Sci.* 17:233–240.

Ebert, D.A., Cowley, P.D., and Compagno, L.J.V. (2002). First records of the longnose spiny dogfish *Squalus blainvillei* (Squalidae) and the deep–water stingray *Plesiobatis daviesi* (Urolophidae) from South African waters. *S. Afr. J. Mar. Sci.* 24:355–357.

Edwards, R.R.C. (1980). Aspects of the population dynamics and ecology of the white spotted stingaree, *Urolophus paucimaculatus* Dixon, in Port Phillip Bay, Victoria. *Aust. J. Mar. Freshw. Res.* 31:459–467.

Eggers, D.M. (1979). Comments on some recent methods for estimating food consumption by fish. *J. Fish. Res. Bd. Can.* 36:1018–1019.

Elliott, J.M. and Davidson, W. (1975). Energy equivalents of oxygen consumption in animal energetics. *Oecologia* 19:195–201.

Elliott, J.M. and Persson, L. (1978). The estimation of daily rates of food consumption for fish. *J. Anim. Ecol.* 47:977–991.

Ellis, J.K. and Musick, J.A. (2006). Ontogenetic changes in the diet of the sandbar shark, *Carcharhinus plumbeus*, in lower Chesapeake Bay and Virginia (USA) coastal waters. *Environ. Biol. Fish.* 76:167–176.

Ellis, J.R., Pawson, M.G., and Shackley, S.E. (1996). The comparative feeding ecology of six species of shark and four species of ray (Elasmobranchii) in the north-east Atlantic. *J. Mar. Biol. Assoc. U.K.* 76:89–106.

Estrada, J.A., Rice, A.N., Lutcavage, M.E., and Skomal, G.B. (2003). Predicting trophic position in sharks of the northwest Atlantic Ocean using stable isotope analysis. *J. Mar. Biol. Assoc. U.K.* 83:1347–1350.

Estrada, J.A., Rice, A.N., Natanson, L.J., and Skomal, G.B. (2006). Use of isotopic analysis of vertebrae in reconstructing ontogenetic feeding ecology in white sharks. *Ecology* 87:829–834.

Euzen, O. (1987). Food habits and diet comparison of some fish of Kuwait. *Kuwait Bull. Mar. Sci.* 1987:65–85.

Fahmi, S.J., Blaber, M., Adrim, M., and Tibbetts, I.R. (In press). Diet overlap and trophic shifts in four sympatric whiprays (*Himantura* spp.) from the Java Sea, Indonesia. *J. Fish Biol.*

Fänge, R. and Grove, D. (1979). Digestion. In: Hoar, W.S., Randall, D.J., and Brett, J.R. (Eds.), *Fish Physiology*, Vol. 8. Academic Press, New York, pp. 161–260.

Fergusson, I.K., Compagno, L.J.V., and Marks, M.A. (2000). Predation by white sharks, *Carcharodon carcharias* (Chondrichthyes: Lamnidae), upon chelonians, with new records from the Mediterranean Sea and a first record of the ocean sunfish *Mola mola* (Osteichthyes: Molidae) as stomach contents. *Environ. Biol. Fish.* 58:447–453.

Ferry, L.A. and Cailliet, G.M. (1996). Sample size and data analysis: are we characterizing and comparing diet properly? In: MacKinlay, D. and Shearer, K. (Eds.), *Feeding Ecology and Nutrition in Fish, International Congress of the Biology of Fishes.* American Fisheries Society, Bethesda, MD, pp. 71–80.

Fisk, A.T., Tittlemier, S.A., Pranschke, J.L., and Norstrom, R.J. (2002). Using anthropogenic contaminants and stable isotopes to assess the feeding ecology of Greenland sharks. *Ecology* 83:2162–2172.

Ford, E. (1921). A contribution to our knowledge of the life histories of the dogfishes landed at Plymouth. *J. Mar. Biol. Assoc. U.K.* 12:468–505.

García de la Rosa, S.B. and Sánchez, F. (1997). Alimentación de *Squalus acanthias* y predación sobre *Merluccius hubbsi* en el Mar Argentino entre 34°47–47°S. *Rev. Invest. Des. Pesq.* 11:119–133.

Gauld, J.A. (1989). *Records of Porbeagles Landed in Scotland, with Observations on the Biology, Distribution and Exploitation of the Species*, Scottish Fish Research Rep. 45. Department of Agriculture and Fisheries for Scotland, Aberdeen.

Gelsleichter, J., Musick, J.A., and Nichols, S. (1999). Food habits of the smooth dogfish, *Mustelus canis*, dusky shark, *Carcharhinus obscurus*, Atlantic sharpnose shark, *Rhizoprionodon terraenovae*, and the sand tiger, *Carcharias taurus*, from the northwest Atlantic Ocean. *Environ. Biol. Fish.* 54:205–217.

Glubokov, A.I. and Orlov, A.M. (2000). Some morphophysiological indices and feeding peculiarities of the Aleutian skate, *Bathyraja aleutica*, from the western Bering Sea. *Russian Fed. Res. Inst. Fish. Oceanogr.* 1:126–149.

Gómez, F.E. and Bashirulah, A.K.M. (1984). Relación longitud-peso y hábitos alimenticios de *Rhizoprionodon porosus* Poey 1861 (Fam. Carcharhinidae) en el oriente de Venezuela. *Bol. Inst. Oceanogr. Univ. Oriente, Venezuela* 23:49–54.

Gordon, J.D.M. and Duncan, J.A.R. (1989). A note on the distribution and diet of deep-water rays (Rajidae) in an area of the Rockall Trough. *J. Mar. Biol. Assoc. U.K.* 69:655–658.

Graeber, R.C. (1974). Food intake patterns in captive juvenile lemon sharks, *Negaprion brevirostris. Copeia* 1974:554–556.

Green, K., Burton, H.R., and Williams, R. (1989). The diet of Antarctic fur seals *Arctocephalus gazella* (Peters) during the breeding season at Heard Island. *Antarctic Sci.* 1:317–324.

Gregory, M., Balance, P.F., Gibson, G.W., and Ayling, A.M. (1979). On how some rays (Elasmobranchia) excavate feeding depressions by jetting water. *J. Sed. Pet.* 49:1125–1130.

Gruber, S.H. (1984). Bioenergetics of the captive and free-ranging lemon shark (*Negaprion brevirostris*). *Proc. Am. Assoc. Parks Aquar.* 1984:341–373.

Gubanov, Y.P. and Grigoryev, V.N. (1975). Observations on the distribution and biology of the blue shark *Prionace glauca* (Carcharhinidae) of the Indian Ocean. *J. Ichthyol.* 15:37–43.

Gudger, E.W. (1914). History of the spotted eagle ray, *Aetobatus narinari*, together with a study of its external structures. *Carnegie Inst. Wash. Publ.* 183:241–332.

Gudger, E.W. (1932). Cannibalism among the sharks and rays. *Sci. Monthly* 34:403–419.

Gudger, E.W. (1941). The food and feeding habits of the whale shark, *Rhineodon typus*. *J. Elisha Mitchell Sci. Soc.* July:57–72.

Gudger, E.W. (1948). The tiger shark, *Galeocerdo tigrinus*, on the North Carolina coast and its food and feeding habits there. *J. Elisha Mitchell Sci. Soc.* 64:221–233.

Gudger, E.W. (1949). Natural history notes on tiger sharks, *Galeocerdo tigrinus*, caught at Key West, Florida, with emphasis on food and feeding habits. *Copeia* 1949:39–47.

Hallacher, L.E. (1977). On the feeding behavior of the basking shark, *Cetorhinus maximus*. *Environ. Biol. Fish.* 2:297–298.

Hanchet, S. (1991). Diet of spiny dogfish, *Squalus acanthias* Linnaeus, on the east coast, South Island, New Zealand. *J. Fish Biol.* 39:313–323.

Hansson, S. (1998). Methods of studying fish feeding: a comment. *Can. J. Fish. Aquat. Sci.* 55:2706–2707.

Harvey, J.T. (1989). Food habits, seasonal abundance, size, and sex of the blue shark, *Prionace glauca*, in Monterey Bay, California. *Calif. Fish Game* 75:33–44.

Heatwole, J., Heatwole, E., and Johnson, C.R. (1974). Shark predation on sea snakes. *Copeia* 1974:780–781.

Heithaus, M.R. (2001a). The biology of tiger sharks, *Galeocerdo cuvier*, in Shark Bay, Western Australia: sex ratio, size distribution, diet, and seasonal changes in catch rates. *Environ. Biol. Fish.* 61:25–36.

Heithaus, M.R. (2001b). Predator–prey and competitive interactions between sharks (order Selachii) and dolphins (suborder Odontoceti): a review. *J. Zool. Lond.* 253:53–68.

Heithaus, M.R., Frid, A., Vaudo, J., Worm, B., and Wirsing, A.J. (2010). Unraveling the ecological importance of elasmobranchs. In: Carrier, J.C., Heithaus, M.R., and Musick, J.A. (Eds.), *Sharks and Their Relatives. II. Biodiversity, Adaptive Physiology, and Conservation.* CRC Press, Boca Raton, FL, pp. 612–637.

Henningsen, A.D. (1996). Captive husbandry and bioenergetics of the spiny butterfly ray, *Gymnura altavela* (Linnaeus). *Zoo Biol.* 15:135–142.

Hess, P.W. (1959). The Biology of Two Sting Rays, *Dasyatis centroura* Mitchill 1815 and *Dasyatis say* Lesueur 1817, in Delaware Bay, master's thesis, University of Delaware, Newark.

Heupel, M.R. and Bennett, M.B. (1998). Observations on the diet and feeding habits of the epaulette shark, *Hemiscyllium ocellatum* (Bonnaterre), on Heron Island Reef, Great Barrier Reef, Australia. *Mar. Freshw. Res.* 49:753–756.

Holden, M.J. (1966). The food of the spurdog, *Squalus acanthias* (L.). *J. Cons. Perm Int. Explor. Mer.* 30:255–266.

Holmgren, S. and Nilsson, S. (1999). Digestive system. In: Hamlett, W.C. (Ed.), *Sharks, Skates, and Rays: The Biology of Elasmobranch Fishes.* The Johns Hopkins University Press, Baltimore, MD, pp. 144–173.

Homma, K. and Ishihara, H. (1994). Food habits of six species of rays occurring at Pohnpei (Ponape) Island (E. Caroline Islands), FSM. *Chondros* 5:4–8.

Homma, K., Maruyama, T., Itoh, T., Ishihara, H., and Uchida, S. (1999). Biology of the manta ray, *Manta birostris* Walbaum, in the Indo-Pacific. In: *Proceedings of the 5th Indo-Pacific Fisheries Conference*, November 3–8, 1997, Nouméa, New Caledonia, pp. 209–216.

Horie, T. and Tanaka, S. (2000). Reproduction and food habits of two species of sawtail sharks, *Galeus eastmani* and *G. nipponensis*, in Suruga Bay, Japan. *Fish. Sci.* 66:812–825.

Hussey, N.E., Brush, J., McCarthy, I.D., and Fisk, A.T. (2010). $\delta^{15}N$ and $\delta^{13}C$ diet-tissue discrimination factors for large sharks under semi-controlled conditions. *Comp. Biochem. Physiol. A* 155:445–553.

Ismen, A. (2003). Age, growth, reproduction, and food of common stingray (*Dasyatis pastinaca* L., 1758) in Iskenderun Bay, the eastern Mediterranean. *Fish. Res.* 60:169–176.

Jacobson, I.P., Johnson, J.W., and Bennett, M.B. (2009). Diet and reproduction in the Australian butterfly ray, *Gymnura australis*, from northern and northeastern Australia. *J. Fish Biol.* 75:2475–2489.

Jahn, A.E. and Haedrich, R.L. (1988). Notes on the pelagic squaloid shark *Isistius brasiliensis*. *Biol. Oceanogr.* 5:297–309.

Jakobsdóttir, K.B. (2001). Biological aspects of two deep-water squalid sharks: *Centroscyllium fabricii* (Reinhardt, 1825) and *Etmopterus princeps* (Collett, 1904) in Icelandic waters. *Fish. Res.* 51:247–265.

James, P.S.B.R. (1962). Observations on shoals of the Javanese Cownose ray *Rhinoptera javanica* Müller & Henle from the Gulf of Mannar, with additional notes on the species. *J. Mar. Biol. Assoc. India* 4:217–224.

Jones, B.C. and Geen, G.H. (1977). Food and feeding of spiny dogfish (*Squalus acanthias*) in British Columbia waters. *J. Fish. Res. Board Can.* 34:2067–2078.

Jones, E.C. (1971). *Isistius brasiliensis*, a squaloid shark, the probable cause of crater wounds on fishes and cetaceans. *Fish. Bull.* 69:791–798.

Kao, J.S. (2000). Diet, Daily Ration and Gastric Evacuation of the Leopard Shark (*Triakis semifasciata*), master's thesis, The California State University, Hayward.

Kerr, L.A., Andrews, A.H., Cailliet, G.M., Brown, T.A., and Coale, K.H. (2006). Investigations of $\Delta^{14}C$, $\delta^{13}C$, and $\delta^{15}N$ in vertebrae of white shark (*Carcharodon carcharias*) from the eastern North Pacific Ocean. *Environ. Biol. Fish.* 77:337–353.

King, K.J. and Clark, M.R. (1984). The food of rig (*Mustelus lenticulatus*) and the relationship of feeding to reproduction and condition in Golden Bay. *N.Z. J. Freshw. Res.* 18:29–42.

Kleiber, M. (1975). *The Fire of Life: An Introduction to Animal Energetics.* Krieger, Huntington, NY.

Koen-Alonso, M., Crespo, E.A., and Garcia, N.A. (2001). Food habits of *Dipturus chilensis* (Pisces: Rajidae) off Patagonia, Argentina. *ICES J. Mar. Sci.* 58:288–297.

Koen-Alonso, M., Crespo, E.A., García, N.A., Pedraza, S.N., Mariotti, A.P., and Mora, N.J. (2002). Fishery and ontogenetic driven changes in the diet of the spiny dogfish, *Squalus acanthias*, in Patagonian waters, Argentina. *Environ. Biol. Fish.* 63:193–202.

Kohler, N.E. (1987). Aspects of the Feeding Ecology of the Blue Shark, *Prionace glauca*, in the Western North Atlantic, doctoral dissertation, University of Rhode Island, Kingston.

Kondyurin, V.V. and Myagkov, N.A. (1982). Morphological characteristics of two species of spiny dogfish, *Squalus acanthias* and *Squalus fernandinus* (Squalidae, Elasmobranchii), from the southeastern Atlantic. *J. Ichthyol.* 22:41–51.

Kubota, T., Shiobara, Y., and Kubodera, T. (1991). Food habits of the frilled shark, *Chlamydoselachus anguineus*, collected from Suruga Bay, Central Japan. *Nippon Suisan Gakkaishi* 57:15–20.

Lasso, C.A., Rial, A., and Lasso-Alcala, O. (1996). Notes of the biology of the freshwater stingrays *Paratrygon aiereba* (Muller & Henle, 1841) and *Potamotrygon orbignyi* (Castelnau, 1855) (Chondrichthyes: Potamotrygonidae) in the Venezuelan llanos. *Aqua (Graffignana)* 2:39–50.

Last, P.R. and Stevens, J.D. (1994). *Sharks and Rays of Australia*. CSIRO Division of Fisheries, East Melbourne, Australia.

Lawler, E.F. (1976). The Biology of the Sandbar Shark, *Carcharhinus plumbeus* (Nardo, 1827), in the Lower Chesapeake Bay and Adjacent Waters, master's thesis, College of William and Mary, Williamsburg, VA.

LeBoeuf, B.L., Riedman, M., and Keyes, R.S. (1982). White shark predation on pinnipeds in California coastal waters. *Fish. Bull.* 80:891–895.

Liao, H., Pierce, C.L., and Larscheid, J.G. (2001). Empirical assessment of indices of prey importance in the diets of predacious fish. *Trans. Am. Fish. Soc.* 130:583–591.

Link, J.S. (2004). Using fish stomachs as samplers of the benthos: integrating long-term and broad scales. *Mar. Ecol. Prog. Ser.* 269:265–275.

Link, J.S. and Almeida, F.P. (2000). *An Overview and History of the Food Web Dynamics Program of the Northeast Fisheries Science Center, Woods Hole, Massachusetts*, National Oceanic and Atmospheric Administration Tech. Memo. NMFS-NE-159, U.S. Department of Commerce, Springfield, VA.

Logan, J.M. and Lutcavage, M.E. (2010). Stable isotope dynamics in elasmobranch fishes. *Hydrobiologia* 644:231–244.

Long, D.J. (1996). Records of white shark-bitten leatherback sea turtles along the California coast. In: Klimley, A.P. and Ainley, D.J. (Eds.), *Great White Sharks: The Biology of Carcharodon carcharias*. Academic Press, New York, pp. 317–319.

Longval, M.J., Warner, R.M., and Gruber, S.H. (1982). Cyclical patterns of food intake in the lemon shark *Negaprion brevirostris* under controlled conditions. *Fla. Sci.* 45:25–33.

Lowe, C.G. (2002). Bioenergetics of free-ranging juvenile scalloped hammerhead sharks (*Sphyrna lewini*) in Kane'ohe Bay, O'ahu, HI. *J. Exp. Mar. Biol. Ecol.* 278:141–156.

Lowe, C.G., Wetherbee, B.M., Crow, G.C., and Tester, A.L. (1996). Ontogenetic dietary shifts and feeding behavior of the tiger shark, *Galeocerdo cuvier*, in Hawaiian waters. *Environ. Biol. Fish.* 47:203–211.

Lucifora, L.O., Valero, J.L., Bremec, C.S., and Lasta, M.L. (2000). Feeding habits and prey selection by the skate *Dipturus chilensis* (Elasmobranchii: Rajidae) from the south-western Atlantic. *J. Mar. Biol. Assoc. U.K.* 80:953–954.

Lyle, J.M. (1983). Food and feeding habits of the lesser spotted dogfish, *Scyliorhinus canicula* (L.), in Isle of Man waters. *J. Fish Biol.* 23:725–737.

Lyle, J.M. (1987). Observations on the biology of *Carcharhinus cautus* (Whitley), *C. melanopterus* (Quoy and Gaimard) and *C. fitzroyensis* (Whitley) from northern Australia. *Aust. J. Mar. Freshw. Res.* 38:701–710.

Lyle, J.M. and Timms, G.J. (1987). Predation on aquatic snakes by sharks from northern Australia. *Copeia* 1987:802–803.

Mabragaña, E. and Giberto, D.A. (2007). Feeding ecology and abundance of two sympatric skates, the shortfin sand skate *Psammobatis normani* McEachran, and the smallthorn sand skate *P. rudis* Gunther (Chondrichthyes, Rajidae), in the southwest Atlantic. *ICES J. Mar. Sci.* 64:1017–1027.

MacNeil, M.A., Skomal, G.B., and Fisk, A.T. (2005). Stable isotopes from multiple tissues reveal diet switching in sharks. *Mar. Ecol. Prog. Ser.* 302:199–206.

Macpherson, E. (1980). Régime alimentaire de *Galeus melastomus* Rafinesque, 1810, *Etmopterus spinax* (L., 1758) et *Scymnorhinus licha* (Bonnaterre, 1788) en Méditerranée occidentale. *Vie Milieu* 30:139–148.

Macpherson, E. (1981). Resource partitioning in a Mediterranean demersal fish community. *Mar. Ecol. Prog. Ser.* 4:183–193.

Macpherson, E., Lleonart, J., and Sánchez, P. (1989). Gastric emptying in *Scyliorhinus canicula* (L.): a comparison of surface-dependent and non-surface dependent models. *J. Fish Biol.* 35:37–48.

Marshall, L.J., White, W.T., and Potter, L.C. (2007). Reproductive biology and diet of the southern fiddler ray, *Trygonorrhina fasciata* (Batoidea: Rhinobatidae), an important trawl bycatch species. *Mar. Freshw. Res.* 58:104–115.

Marshall, A.D., Kyne, P.M., and Bennett, M.B. (2008). Comparing the diet of two sympatric urolophid elasmobranchs (*Trygonoptera testacea* Muller and Henle and *Urolophus kapalensis* Yearsley and Last): evidence of ontogenetic shifts and possible resource partitioning. *J. Fish Biol.* 72: 883–898.

Matallanas, J. (1982). Feeding habits of *Scymnorhinus licha* in Catalan waters. *J. Fish Biol.* 20:155–163.

Matallanas, J., Carrasón, M., and Casadevall, M. (1993). Observations on the feeding habits of the narrow mouthed cat shark, *Schroederichthys bivius* (Chondrichthyes, Scyliorhinidae), in the Beagle Channel. *Cybium* 17:55–61.

Matich, P., Heithaus, M.R., and Layman, C.A. (2010). Size-based variation in intertissue comparisons of stable carbon and nitrogen isotopic signatures of bull sharks (*Carcharhinus leucas*) and tiger sharks (*Galeocerdo cuvier*). *Can. J. Fish. Aquat. Sci.* 67:877–885.

Matich, P., Heithaus, M.R., and Layman, C.A. (2011). Contrasting patterns of individual specialization and trophic coupling in two marine apex predators. *J. Animal Ecol.* 80:294–305.

Mauchline, J. and Gordon, J.D.M. (1983). Diets of the sharks and chimaeroids of the Rockall Trough, northeastern Atlantic Ocean. *Mar. Biol.* 75:269–278.

McCosker, J.E. (1987). The white shark, *Carcharodon carcharias*, has a warm stomach. *Copeia* 1987:195–197.

McElroy, W.D., Wetherbee, B.M., Mostello, C.S., Lowe, C.G., Crow, G.L., and Wass, R. (2006). Food habits and ontogenetic changes in the diet of the sandbar shark (*Carcharhinus plumbeus*) in Hawaii. *Environ. Biol. Fish.* 76:81–92.

Medved, R.J. (1985). Gastric evacuation in the sandbar shark, *Carcharhinus plumbeus*. *J. Fish Biol.* 26:239–253.

Medved, R.J., Stillwell, C.E., and Casey, J.G. (1985). Stomach contents of young sandbar sharks, *Carcharhinus plumbeus*, in Chincoteague Bay, Virginia. *Fish. Bull.* 83:395–402.

Medved, R.J., Stillwell, C.E., and Casey, J.G. (1988). The rate of food consumption of young sandbar sharks (*Carcharhinus plumbeus*) in Chincoteague Bay, Virginia. *Copeia* 1988:956–963.

Menni, R.C. (1985). Distribución y biología de *Squalus acanthias*, *Mustelus schmitti* y *Galeorhinus vitaminicus* en el mar Argentino en agosto–septiembre de 1978 (Chondrichthyes). *Rev. Mus. La Plata n.s.* 13:151–182.

Morato, T., Solà, E., and Grós, M.P. (2003). Diets of thornback ray (*Raja clavata*) and tope shark (*Galeorhinus galeus*) in the bottom longline fishery of the Azores, northeastern Atlantic. *Fish. Bull.* 101:590–602.

Moss, S.A. (1984). *Sharks: An Introduction for the Amateur Naturalist*. Prentice-Hall, Englewood Cliffs, NJ.

Mumtaz Tirasin, E. and Jorgensen, T. (1999). An evaluation of the precision of diet description. *Mar. Ecol. Prog. Ser.* 182:243–252.

Muñoz-Chapuli, R., Salgado, J.C.R., and de La Serna, J.M. (1988). Biogeography of *Isistius brasiliensis* in the northeastern Atlantic, inferred from crater wounds on swordfish (*Xiphas gladius*). *J. Mar. Biol. Assoc. U.K.* 68:315–321.

Muto, E.Y., Soares, L.S.H., and Goitein, R. (2001). Food resource utilization of the skates *Rioraja agassizii* (Mueller and Henle, 1841) and *Psammobatis extenta* (Garman, 1913) on the continental shelf off Ubatuba, south-eastern Brazil. *Braz. J. Biol.* 61:217–238.

Nagasawa, K. (1998). Predation by salmon sharks (*Lamna ditropis*) on Pacific salmon (*Oncorhynchus* spp.) in the North Pacific Ocean. *N. Pac Anadr. Fish Comm. Bull.* 1:419–433.

Navia, A.F., Mejia-Falla, P.A., and Giraldo, A. (2007). Feeding ecology of elasmobranch fishes in coastal waters of the Columbian Eastern Tropical Pacific. *BMC Ecol.* 7:8.

Nelson, G.A. and Ross, M.R. (1995). Gastric evacuation in little skate. *J. Fish Biol.* 46:977–986.

Notarbartolo di Sciara, G. (1988). Natural history of the rays of the genus *Mobula* in the Gulf of California. *Fish. Bull.* 86:45–66.

Olsen, A.M. (1954). The biology, migration and growth rate of the school shark, *Galeorhinus australis* (Mcleay) (Carcharhinidae), in south-eastern Australian waters. *Aust. J. Mar. Freshw. Res.* 5:353–410.

Olsen, A.M. (1984). *Synopsis of Biological Data on the School Shark Galeorhinus australis*, FAO Fisheries Synopsis 139. Fisheries and Aquaculture Department, Food and Agriculture Organization, Rome.

Olson, R.J. and Mullen, A.J. (1986). Recent developments for making gastric evacuation and daily ration determinations. *Environ. Biol. Fish.* 16:183–191.

Orlov, A.M. (1998). On feeding of mass species of deep-sea skates (*Bathyraja* spp., Rajidae) from the Pacific waters of the northern Kurils and southeastern Kamchatka. *J. Ichthyol.* 38:635–644.

Ostrom, P.H., Lien, J., and Macko, S.A. (1993). Evaluation of the diet of Sowerby's beaked whale, *Mesoplodon bidens*, based on isotopic comparisons among northwest Atlantic cetaceans. *Can. J. Zool.* 71:858–861.

Paloheimo, J. and Dickie, L.M. (1966). Food and growth of fishes. III. Relations among food, body size and growth efficiency. *J. Fish. Res. Bd. Can.* 23:1209–1248.

Pandian, T.J. (1982). Contributions to the bioenergetics of a tropical fish. In: Cailliet, G.M. and Simenstad, C.A. (Eds.), *Gutshop '81: Fish Food Habits Studies*. Washington Sea Grant Program, Seattle, WA, pp. 124–131.

Papastamatiou, Y.P. (2003). Gastric pH Changes Associated with Feeding in Leopard Sharks (*Triakis semifasciata*): Can pH Be Used to Study the Foraging Ecology of Sharks? master's thesis, The California State University, Long Beach.

Papastamatiou, Y.P., Wetherbee, B.M., Crow, G.L., and Lowe, C.G. (2006). Dietary and spatial overlap of four large coastal sharks in Hawaiian waters: evidence of niche separation. *Mar. Ecol. Prog. Ser.* 320:239–251.

Papastamatiou, Y.P., Friedlander, A.M., Caselle, J.E., and Lowe, C.G. (2010a). Long-term movement patterns and trophic ecology of blacktip reef sharks (*Carcharhinus melanopterus*) at Palmyra Atoll. *J. Exp. Mar. Biol. Ecol.* 386:94–102.

Papastamatiou, Y.P., Wetherbee, B.M., O'Sullivan, J., Goodmanlowe, G.D., and Lowe, C.G. (2010b). Foraging ecology of cookie cutter sharks (*Isistius brasiliensis*) on pelagic fishes from Hawaii, inferred from prey bite wounds. *Environ. Biol. Fish.* 88:361–368.

Penchaszadeh, P.E., Arrighetti, F., Cledón, M., Livore, J.P., Botto, F., and Iribarne, O.O. (2006). Bivalve contribution to shallow sandy bottom food web off Mar del Plata (Argentina): inference from stomach contents and stable isotope analysis. *J. Shellfish Res.* 25:52–54.

Pennington, M. (1985). Estimating the average food consumption by fish in the field from stomach contents data. *Dana* 5:81–86.

Perlman, D.F. and Goldstein, L. (1988). Nitrogen metabolism. In: Shuttleworth, P.J. (Ed.), *Physiology of Elasmobranch Fishes*. Springer-Verlag, Berlin, pp. 253–276.

Peverell, S.C. (2005). Sawfish (Pristidae) of the Gulf of Carpentaria, Queensland, Australia, master's thesis, James Cook University, Townsville.

Platell, M.E., Potter, I.C., and Clarke, K.R. (1998). Resource partitioning by four species of elasmobranchs (Batoidea: Urolophidae) in coastal waters of temperate Australia. *Mar. Biol.* 131:719–734.

Rae, B.B. (1967). The food of the dogfish, *Squalus acanthias* L. *Mar. Res.* 4:1–19.

Raje, S.G. (2007). Some aspects of the biology of *Himantura bleekeri* (Blyth) and *Amphotistius imbricatus* (Schneider) from Mumbai. *Indian J. Fish.* 54:235–238.

Randall, B.M., Randall, R.M., and Compagno, L.J.V. (1988). Injuries to jackass penguins (*Spheniscus demersus*): evidence for shark involvement. *J. Zool.* 214:589–600.

Randall, J.E. (1967). Food habits of reef fishes of the West Indies. *Stud. Trop. Oceanogr.* 5:665–847.

Relini Orsi, L. and Wurtz, M. (1977). Patterns and overlap in the feeding of two selachians of bathyal fishing grounds in the Ligurian Sea. *Rapp. Comm. Int. Mer Médit.* 24:89–94.

Rinewalt, C.S., Ebert, D.A., and Cailliet, G.M. (2007). The feeding habits of the sandpaper skate, *Bathyraja kincaidii* (Garman, 1908) in central California: seasonal variation in diet linked to oceanographic conditions. *Environ. Biol. Fish.* 80:147–163.

Robinson, C.S., Cailliet, G.M., and Ebert, D.A. (2007). Food habits of the longnose skate, *Raja rhina* (Jordan and Gilbert, 1880), in central California waters. *Environ. Biol. Fish.* 80:165–179.

Rossouw, G.J. (1983). The Biology of the Sand Shark, *Rhinobatos annulatus*, in Algoa Bay with Notes on Other Elasmobranchs, doctoral dissertation, University of Port Elizabeth, Cape Province, South Africa.

Rountree, R.A. and Able, K.W. (1996). Seasonal abundance, growth, and foraging habits of juvenile smooth dogfish, *Mustelus canis*, in a New Jersey estuary. *Fish. Bull.* 94:522–534.

Salini, J.P., Blaber, S.J.M, and Brewer, D.T. (1990). Diets of piscivorous fishes in a tropical Australian estuary, with special reference to predation on penaeid prawn. *Mar. Biol.* 105:363–374.

Salini, J.P., Blaber, S.J.M., and Brewer, D.T. (1992). Diets of sharks from estuaries and adjacent waters of the northeastern Gulf of Carpentaria, Australia. *Aust. J. Mar. Freshw. Res.* 43:87–96.

Salini, J.P., Blaber, S.J.M., and Brewer, D.T. (1994). Diets of trawled predatory fish of the Gulf of Carpentaria, Australia, with particular reference to predation on prawns. *Aust. J. Mar. Freshw. Res.* 45:397–411.

Salini, J.P., Tonks, M., Blaber, S.J.M., and Ross, J. (1999). Feeding of captive, tropical carcharhinid sharks from the Embley River estuary, northern Australia. *Mar. Ecol. Prog. Ser.* 184:309–314.

Sampson, L., Galván-Magaña, F., de Silva-Dávila, R., Aguíñiga-Garca, S., and O'Sullivan, J.B. (2010). Diet and trophic position of the devil rays *Mobula thurstonia* and *Mobula japanica* as inferred from stable isotope analysis. *J. Mar. Biol. Assoc. U.K.* 90:969–976.

San Filippo, R.A. (1995). Diet, Gastric Evacuation and Estimates of Daily Ration of the Gray Smoothhound, *Mustelus californicus*, master's thesis, San Jose State University.

Sarangadhar, P.N. (1983). Tiger shark (*Galeocerdo tigrinus*) Muller and Henle. Feeding and breeding habits. *J. Bombay Nat. Hist. Soc.* 44:101–110.

Saunders, G.B. and Clark, E. (1962). Yellow-billed cuckoo in stomach of tiger shark. *Auk* 79:118.

Scharf, F.S., Yetter, R.M., Summers, A.P., and Juanes, F. (1998). Enhancing diet analysis of piscivorous fishes in the Northwest Atlantic through identification and reconstruction of original prey sizes from ingested remains. *Fish. Bull.* 96:575–588.

Schmid, T.H. and Murru, F.L. (1994). Bioenergetics of the bull shark, *Carcharhinus leucas*, maintained in captivity. *Zoo Biol.* 13:177–185.

Schmid, T.H., Murru, F.L., and McDonald, F. (1990). Feeding habits and growth rates of bull (*Carcharhinus leucas* (Valenciennes)), sandbar (*Carcharhinus plumbeus* (Nardo)), sandtiger (*Eugomphodus taurus* (Rafinesque)) and nurse (*Ginglymostoma cirratum* (Bonnaterre)) sharks maintained in captivity. *J. Aquar. Aquat. Sci.* 5:100–105.

Schmidt, T.W. (1986). Food of young juvenile lemon sharks, *Negaprion brevirostris* (Poey), near Sandy Key, western Florida Bay. *Fla. Sci.* 49:7–10.

Schurdak, M.E. and Gruber, S.H. (1989). Gastric evacuation of the lemon shark *Negaprion brevirostris* (Poey) under controlled conditions. *Exp. Biol.* 48:77–82.

Schwartz, F.J. (1996). Biology of the clearnose skate, *Raja eglanteria*, from North Carolina. *Fla. Sci.* 59:82–95.

Scofield, N.B. (1920). Sleeper shark captured. *Calif. Fish Game* 6:80.

Segura-Zarzosa, J.C., Abitia-Cárdenas, L.A., and Galván-Magaña, F. (1997). Observations on the feeding habits of the shark *Heterodontus francisci* Girard 1854 (Chondrichthyes: Heterodontidae), in San Ignacio Lagoon, Baja California Sur, México. *Cien. Mar* 23:111–128.

Shibuya, A., Araújo, M.L.G., and Zuanon, J.A.S. (2009). Analysis of stomach contents of freshwater stingrays (Elasmobranchii, Potamotrygonidae) from the middle Negro River, Amazonas, Brazil. *Pan-Am. J. Aquat. Sci.* 4:466–475.

Shirai, S. and Nakaya, K. (1992). Functional morphology of feeding apparatus of the cookie-cutter shark, *Isistius brasiliensis* (Elasmobranchii, Dalatiinae). *Zool. Sci.* 9:811–821.

Silva, T.B. and Uieda, V.S. (2007). Preliminary data on the feeding habits of the freshwater stingrays *Potamotrygon falkneri* and *Potamotrygon motoro* (Potamotrygonidae) from the Upper Parana River basin, Brazil. *Biota Neotrop.* 7:221–226.

Simpfendorfer, C.A. (1992). Biology of tiger sharks (*Galeocerdo cuvier*) caught by the Queensland shark meshing program off Townsville, Australia. *Aust. J. Mar. Freshw. Res.* 43:33–43.

Simpfendorfer, C.A. and Milward, N.E. (1993). Utilisation of a tropical bay as a nursery area by sharks of the families Carcharhinidae and Sphyrnidae. *Environ. Biol. Fish.* 37:337–345.

Simpfendorfer, C.A., Goodreid, A.B., and McAuley, R.B. (2001a). Size, sex and geographic variation in the diet of the tiger shark, *Galeocerdo cuvier*, from Western Australian waters. *Environ. Biol. Fish.* 61:37–46.

Simpfendorfer, C.A., Goodreid, A.B., and McAuley, R.B. (2001b). Diet of three commercially important shark species from Western Australian waters. *Mar. Freshw. Res.* 52:975–985.

Sims, D.W. and Merrett, D.A. (1997). Determination of zooplankton characteristics in the presence of surface feeding basking sharks *Cetorhinus maximus*. *Mar. Ecol. Prog. Ser.* 158:297–302.

Sims, D.W. and Quayle, V.A. (1998). Selective foraging behaviour of basking sharks on zooplankton in a small-scale front. *Nature* 393:460–464.

Sims, D.W., Davies, S.J., and Bone, Q. (1996). Gastric emptying rate and return of appetite in lesser spotted dogfish, *Scyliorhinus canicula* (Chondrichthyes: Elasmobranchii). *J. Mar. Biol. Assoc. U.K.* 76:479–491.

Smale, M.J. (1991). Occurrence and feeding of three shark species, *Carcharhinus brachyurus, C. obscurus* and *Sphyrna zygaena*, on the Eastern Cape coast of South Africa. *S. Afr. J. Mar. Sci.* 11:31–42.

Smale, M.J. and Compagno, L.J.V. (1997). Life history and diet of two southern African smoothhound sharks, *Mustelus mustelus* (Linnaeus, 1758) and *Mustelus palumbes* (Smith, 1957) (Pisces: Triakidae). *S. Afr. J. Mar. Sci.* 18:229–248.

Smale, M.J. and Cowley, P.D. (1992). The feeding ecology of skates (Batoidea: Rajidae) off the Cape south coast, South Africa. *S. Afr. J. Mar. Sci.* 12:823–834.

Smale, M.J. and Goosen, A.J.J. (1999). Reproduction and feeding of spotted gully shark, *Triakis megalopterus*, off the Eastern Cape, South Africa. *Fish. Bull.* 97:987–998.

Smale, M.J., Sauer, W.H., and Roberts, M.J. (2001). Behavioural interactions of predators and spawning chokka squid off South Africa: towards quantification. *Mar. Biol.* 139:1095–1105.

Smith, J.W. and Merriner, J.V. (1985). Food habits and feeding behavior of the cownose ray, *Rhinoptera bonasus*, in lower Chesapeake Bay. *Estuaries* 8:305–310.

Snelson, Jr., F.F., Mulligan, T.J., and Williams, S.E. (1984). Food habits, occurrence, and population structure of the bull shark, *Carcharhinus leucas*, in Florida coastal lagoons. *Bull. Mar. Sci.* 34:71–80.

Springer, S. (1950). Natural history notes on the lemon shark *Negaprion brevirostris. Tex. J. Sci.* 2:349–359.

Stevens, J.D. (1973). Stomach contents of the blue shark (*Prionace glauca* L.) off southwest England. *J. Mar. Biol. Assoc. U.K.* 53:357–361.

Stevens, J.D. (1984). Biological observations on sharks caught by sport fishermen off New South Wales. *Aust. J. Mar. Freshw. Res.* 35:573–590.

Stevens, J.D. and Lyle, J.M. (1989). Biology of three hammerhead sharks (*Eusphyra blochii, Sphyrna mokarran*, and *S. lewini*), from northern Australia. *Aust. J. Mar. Freshw. Res.* 40:129–146.

Stevens, J.D. and McLoughlin, K.J. (1991). Distribution, size, and sex composition, reproductive biology and diet of sharks from northern Australia. *Aust. J. Mar. Freshw. Res.* 40:129–146.

Stevens, J.D. and Wiley, P.D. (1986). Biology of two commercially important carcharhinid sharks from northern Australia. *Aust. J. Mar. Freshw. Res.* 37:671–688.

Stevens, J.D., Davis, T.L.O., and Church, A.G. (1982). NT shark gillnetting survey shows potential for Australian fishermen. *Aust. Fish.* 41:39–43.

Stillwell, C.E. and Casey, J.G. (1976). Observations on the bigeye thresher shark, *Alopias superciliosus*, in the western North Atlantic. *Fish. Bull.* 74:221–225.

Stillwell, C.E. and Kohler, N.E. (1982). Food, feeding habits, and estimates of daily ration of the shortfin mako (*Isurus oxyrinchus*) in the Northwest Atlantic. *Can. J. Fish. Aquat. Sci.* 39:407–414.

Stillwell, C.E. and Kohler, N.E. (1993). Food habits of the sandbar shark *Carcharhinus plumbeus* off the U.S. northeast coast, with estimates of daily ration. *Fish. Bull.* 91:138–150.

Strasburg, D.W. (1963). The diet and dentition of *Isistius brasiliensis*, with remarks on tooth replacement in other sharks. *Copeia* 1963:33–40.

Sullivan, M.S. (1907). The physiology of the digestive tract of elasmobranchs. *U.S. Fish Wildl. Serv. Fish. Bull.* 27:3–27.

Summers, A.P. (2000). Stiffening the stingray skeleton: an investigation of durophagy in myliobatid stingrays (Chondrichthyes, Batoidea, Myliobatidae). *J. Morphol.* 243:113–126.

Svennson, C.J., Hyndes, G.A., and Lavery, P.S. (2007). Food web analysis in two permanently open temperate estuaries: consequences of saltmarsh loss? *Mar. Environ. Res.* 62:286–304.

Talent, L.G. (1976). Food habits of the leopard shark, *Triakis semifasciata*, in Elkhorn Slough, Monterey Bay, California. *Calif. Fish Game* 62:286–298.

Talent, L.G. (1982). Food habits of the gray smoothhound, *Mustelus californicus*, the brown-smoothhound, *Mustelus henlei*, the shovelnose guitarfish, *Rhinobatos productus*, and the bat ray, *Myliobatus californica*, in Elkhorn Slough, California. *Calif. Fish Game* 68:224–234.

Tanasichuk, R.W., Ware, D.M., Shaw, W., and McFarlane, G.A. (1991). Variations in diet, daily ration, and feeding periodicity of Pacific hake (*Merluccius productus*) and spiny dogfish (*Squalus acanthias*) off the lower west coast of Vancouver Island. *Can. J. Fish. Aquat. Sci.* 48:2118–2128.

Taniuchi, T., Kuroda, N., and Nose, Y. (1983). Age, growth, reproduction, and food habits of the star-spotted dogfish, *Mustelus manazo*, collected from Choshi. *Bull. Jpn. Soc. Sci. Fish.* 49:1325–1334.

Taylor, L. and Wisner, M. (1989). Growth rates of captive blacktip reef sharks (*Carcharhinus melanopterus*). *Bull. Inst. Océanogr. Monaco* 5:211–217.

Thorburn, J.C. (2006). Biology, Ecology, and Trophic Interactions of Elasmobranchs and Other Fishes in Riverine Waters of Northern Australia, doctoral dissertation, Murdoch University, Perth, Western Australia.

Thrush, S.F., Pridmore, R.D., Hewitt, J.E., and Cummings, V.J. (1994). The importance of predators on a sandflat: interplay between seasonal changes in prey density and predator effects. *Mar. Ecol. Prog. Ser.* 107:211–222.

Treloar, M.A. and Laurenson, L.J.B. (2004). Preliminary observations on the reproduction, growth, and diet of *Urolophus cruciatus* (Lacepede) and *Urolophus expansus*, McCulloch (Urolophidae) in southeastern Australia. *Proc. R. Soc. Victoria* 116:183–190.

Treloar, M.A., Laurenson, L.J.B., and Stevens, J.D. (2007). Dietary comparisons of six skate species (Rajidae) in south-eastern Australian waters. *Environ. Biol. Fish.* 80:181–196.

Tricas, T.C. (1979). Relationships of the blue shark, *Prionace glauca*, and its prey species near Santa Catalina Island, California. *Fish. Bull.* 77:175–182.

Tricas, T.C. and McCosker, J.E. (1984). Predatory behavior of the white shark (*Carcharodon carcharias*), with notes on its biology. *Proc. Calif. Acad. Sci.* 43:221–238.

Tuma, R.E. (1976). An investigation of the feeding habits of the bull shark, *Carcharhinus leucas*, in the lake Nicaragua–Rio San Juan system. In: Thorson, T.B. (Ed.), *Investigations of the Ichthyofauna of Nicaraguan Lakes*, School of Life Sciences, University of Nebraska, Lincoln, pp. 533–538.

Turner, R.L., Heatwole, D.W., and Stancyk, S.E. (1982). Ophiuroid discs in stingray stomachs: Evasive autonomy or partial consumption of prey? In: Lawrence, J.M. (Ed.), *Echinoderms: Proceedings of the International Conference, Tampa Bay*. Balkema, Rotterdam, pp. 331–335.

Tyminski, J.P., Cortés, E., Manire, C.A., and Hueter, R.E. (1999). Gastric evacuation and estimates of daily ration in the bonnethead shark, *Sphyrna tiburo* [abstract]. American Society of Ichthyologists and Herpetologists, 79th Annual Meeting, June 24–30, Pennsylvania State University, State College.

Valadez-González, C.M., Aguilar-Palomino, B., and Hernandez-Vazquez, S. (2001). Feeding habits of the round stingray *Urobatis halleri* (Cooper, 1863) (Chondrichthyes: Urolophidae) from the continental shelf of Jalisco and Colima, Mexico. *Cienc. Mar* 27:91–104.

Valadez-González, C.M., Saucedo-Lozano, A.R., and Raymundo-Huizar, A.R. (2006). Trophic aspects of the benthic rays of Jalisco and Colima, Mexico. In: Jimenez-Quiroz, M.C. and Espino-Barr, M.E. (Eds.), *Los Recursos Pesqueros y Acuicolas de Jalisco Colima y Michoacan*. INP, SAGARPA: Mexico, pp. 235–249.

Van Dykhuizen, G. and Mollet, H.F. (1992). Growth, age estimation and feeding of captive sevengill sharks, *Notorynchus cepedianus*, at the Monterey Bay aquarium. *Aust. J. Mar. Freshw. Res.* 43:297–318.

Van Slyke, D.D. and White, G.F. (1911). Digestion of protein in the stomach and in intestine of the dogfish. *J. Biol. Chem.* 9:209–217.

Vaudo, J.J. and Heithaus, M.R. (2011). Dietary niche overlap in a near shore elasmobranch mesopredator community. *Mar. Ecol. Prog. Ser.* 425:247–260.

Veras, D.P., Júnior, T.V., Hazin, F.H.V., Less, R.P., Travassos, R.E., Tolotti, M.T., and Barbosa, R.M. (2009). Stomach contents of the pelagic stingray (*Pteroplatytrygon violacea*) (Elasmobranchii: Dasyatidae) from the tropical Atlantic. *Braz. J. Oceanogr.* 57:339–343.

Vianna, M. and de Amorim, A.F. (1995). Feeding habits of the shark *Mustelus canis* (Mitchill, 1815), caught in southern Brazil. VII. Reunião do grupo de trabalho sobre pesca e pesquisa de tubarões e raias no Brasil [abstract]. Fundação Universidade do Rio Grande, Rio Grande, Brasil, 20–24 November.

Visser, I. (1999). Benthic foraging on stingrays by killer whales (*Orcinus orca*) in New Zealand waters. *Mar. Mam. Sci.* 15:220–227.

Waller, G.N.H. and Baranes, A. (1994). Food of *Iago omanensis*, a deep water shark from the northern Red Sea. *J. Fish Biol.* 45:37–45.

Walsh, M.T., Beusse, D., Bossart, G.D., Young, W.G., Odell, D.K., and Patton, G.W. (1988). Ray encounters as a mortality factor in Atlantic bottlenose dolphins (*Tursiops truncatus*). *Mar. Mam. Sci.* 4:154–162.

Warren, C.E. and Davis, G.E. (1967). Laboratory studies on the feeding, bioenergetics, and growth of fish. In: Gerking, S.D. (Ed.), *The Biological Basis of Freshwater Fish Production*. Blackwell Scientific, London, pp. 175–214.

Wass, R.C. (1971). A Comparative Study of the Life History, Distribution and Ecology of the Sandbar Shark and the Gray Reef Shark in Hawaii, doctoral dissertation, University of Hawaii, Honolulu.

Wass, R.C. (1973). Size, growth, and reproduction of the sandbar shark, *Carcharhinus milberti*, in Hawaii. *Pac. Sci.* 27:305–318.

Webber, J.D. and Cech, Jr., J.J. (1998). Nondestructive diet analysis of the leopard shark from two sites in Tomales Bay, California. *Calif. Fish Game* 84:18–24.

Weihs, D., Keyes, R.S., and Stalls, D.M. (1981). Voluntary swimming speeds of two species of large carcharhinid sharks. *Copeia* 1981:219–222.

Wetherbee, B.M. and Gruber, S.H. (1990). The effect of ration level on food retention time in juvenile lemon sharks, *Negaprion brevirostris*. *Environ. Biol. Fish.* 29:59–65.

Wetherbee, B.M. and Gruber, S.H. (1993a). Absorption efficiency of the lemon shark *Negaprion brevirostris* at varying rates of energy intake. *Copeia* 1993:416–425.

Wetherbee, B.M. and Gruber, S.H. (1993b). Use of acid-insoluble ash as a marker in absorption efficiency studies with the lemon shark. *Prog. Fish-Cult.* 55:270–274.

Wetherbee, B.M., Gruber, S.H., and Ramsey, A.L. (1987). X-radiographic observations of food passage through digestive tracts of lemon sharks. *Trans. Am. Fish. Soc.* 116:763–767.

Wetherbee, B.M., Gruber, S.H., and Cortés, E. (1990). Diet, feeding habits, digestion, and consumption in sharks, with special reference to the lemon shark, *Negaprion brevirostris*. In: Pratt, Jr., H.L., Gruber, S.H., and Taniuchi, T. (Eds.), *Elasmobranchs as Living Resources: Advances in the Biology, Ecology, Systematics, and the Status of the Fisheries*, NOAA Tech. Rep. NMFS 90. U.S. Department of Commerce, Washington, D.C., pp. 29–47.

Wetherbee, B.M., Lowe, C.G., and Crow, G.L. (1996). Biology of the Galapagos shark, *Carcharhinus galapagensis*, in Hawai'i. *Environ. Biol. Fish.* 45:299–310.

Wetherbee, B.M., Crow, G.L., and Lowe, C.G. (1997). Distribution, reproduction, and diet of the gray reef shark *Carcharhinus amblyrhychos* in Hawaii. *Mar. Ecol. Prog. Ser.* 151:181–189.

White, W.T., Platell, M.E., and Potter, I.C. (2004). Comparisons between the diets of four abundant species of elasmobranchs in a subtropical embayment: implications for resource partitioning. *Mar. Biol.* 144:439–448.

Winberg, G.G. (1960). Rate of metabolism and food requirements of fishes. *Fish. Res. Board Can. Trans. Ser.* 194:1–202.

Witzell, W.N. (1987). Selective predation on large cheloniid sea turtles by tiger sharks (*Galeocerdo cuvier*). *Jpn. J. Herpetol.* 12:22–29.

Wood, C.M. (1993). Ammonia and urea metabolism and excretion. In: Evans, D.H. (Ed.), *The Physiology of Fishes*. CRC Press, Boca Raton, FL, pp. 379–425.

Woodland, R.J., Secor, D.H., and Wedge, M.E. (2011). Trophic resource overlap between small elasmobranchs and sympatric teleosts in Mid-Atlantic Bight nearshore habitats. *Estuar. Coast.* 34:391–404.

Yamaguchi, A. and Taniuchi, T. (2000). Food variations and ontogenetic dietary shift of the starspotted-dogfish, *Mustelus manazo*, at five locations in Japan and Taiwan. *Fish. Sci.* 66:1039–1048.

Yamaguchi, A., Kawahara, I., and Ito, S. (2005). Occurrence, growth, and food of longheaded eagle ray, *Aetobatus flagellum*, in Ariake Sound, Kyushu, Japan. *Environ. Biol. Fish.* 74: 229–238.

Yáñez-Arancibia, A. and Amezcua-Linares, F. (1979). Ecologia de *Urolophus jamaicensis* (Cuvier) en laguna de terminos un sistema estuarino del sur del golf de Mexico (Pisces: Urolophidae). *Anales del Centro de Ciencias del Mar y Limnologia*, Universidad Nacional Autónoma de México 6:123–136.

Yang, M.S. (2003). *Food Habits of the Important Groundfishes in the Aleutian Islands in 1994 and 1997*, AFSC Proc. Rep. 2003-07. Alaska Fisheries Science Center, National Marine Fisheries Service, Seattle, WA.

Yeon, J., Hong, S.H., Cha, H.K., and Kim, S.T. (1999). Feeding habits of *Raja pulchra* in the Yellow Sea. *Bull. Natl. Fish. Res. Dev. Inst. Korea* 57:1–11.

9

Integrative Multisensor Tagging: Emerging Techniques to Link Elasmobranch Behavior, Physiology, and Ecology

Nicholas M. Whitney, Yannis P. Papastamatiou, and Adrian C. Gleiss

CONTENTS

9.1 Introduction

Over the last several decades, advances in telemetry have greatly expanded our ability to track where sharks go and when, but are limited in their ability to provide insights into specific behaviors or reasons for using particular habitats. Until recently, the study of wild elasmobranch movements and habitat use has largely been limited to quantifying movements via acoustic or satellite telemetry (see reviews by Nelson, 1990; Sims, 2010; Sundstrom et al., 2001); however, despite advances in these techniques, results of movement studies often remain disconnected from the behavior and physiology of the animal being tracked. Because of this, researchers have often inferred behavior from horizontal or vertical movements without the ability to test these assumptions empirically. Although brief, direct observations of elasmobranch behaviors are possible in a few cases, we generally lack data on their daily activities even when we may have weeks or months of information about their movements. Due to technological limitations many important questions about elasmobranch behavior have gone unaddressed: When are animals most active? When do they rest? How often do they feed? How often do they mate? What is the energetic cost of different behaviors, and how does this determine where and how they spend their time?

The need to quantify the behavior, physiology, and energetic currencies that underlie elasmobranch movements and ecology is now beginning to be met by a new generation of devices (e.g., Cooke et al., 2004a; Muramoto et al., 2004; Ropert-Coudert and Wilson, 2005; Wilson et al., 2008) that allow the integrated sampling of multiple physical and physiological parameters. Although efforts are underway to incorporate many of these devices into transmitters, most are currently available only as "data loggers," which collect large amounts of information and store it to memory that is only accessible when the tag is retrieved. These tags present a wealth of opportunities for the study of elasmobranchs but also a new set of logistical challenges that must be overcome for them to be used successfully and appropriately. Our goal in this chapter is to identify applications of these devices to the study of elasmobranchs, review some of the biological questions that they have been used to address thus far, and assess both the challenges and potential application of these devices for future elasmobranch research. Because accelerometers in particular can be used to address a variety of the biological questions and their use in marine research is rapidly expanding, we give special attention to the application of these devices in some sections. By doing so, we hope to provide a helpful introduction for those interested in using these devices as well as guidance in their application.

9.2 Measuring Fine-Scale Behaviors

Despite the ability to understand the position of elasmobranchs at a variety of spatial scales using traditional telemetry techniques, our understanding of actual behaviors in any quantitative sense is lacking. Although videocamera loggers (animal-borne video and environmental data collection systems, or AVEDs) have provided unprecedented direct observations of shark behavior *in situ* (Heithaus et al., 2001, 2002; Marshall et al., 2007), their relatively large size and brief record duration limit their utility. Recent advances in video compression and battery technology are resulting in

smaller units with increased recording times, but these have yet to be deployed widely on sharks and are not commercially available (Moll et al., 2007). In this section we focus primarily on *acceleration data loggers* (ADLs), which can provide precise data on specific behaviors that have been inferred from telemetry or other proxy measures in the past.

Although hard-wired acceleration sensors were first applied to the study of fish movement decades ago (DuBois and Ogilvy, 1978; DuBois et al., 1976), their use to study free-living marine organisms has been a more recent development (Tanaka et al., 2001; Wilson et al., 2006; Yoda et al., 2001). Within a short time, these sensors have become available from numerous manufacturers and are beginning to become widely used, primarily on air-breathing vertebrates (Sato et al., 2007; Shepard et al., 2009b; Wilson et al., 2010), but increasingly on gill-breathers, as well (Gleiss et al., 2009a, 2010b; Kawabe et al., 2004; Tanaka et al., 2001; Watanabe & Sato, 2008; Whitney et al., 2007, 2010). Accelerometers can be easily attached to the dorsal fin of sharks (see Figure 1 in Gleiss et al., 2009b), placed into the stomach (Whitney et al., 2007), or implanted into the body cavity (Clark et al., 2010), and they provide a measure of locomotory activity and body posture that can be used for a diverse range of applications (Shepard et al., 2008; Wilson et al., 2006).

Accelerometers measure static acceleration, which is acceleration due to the Earth's gravity (9.8 m/s^2 or $1 \text{ } g$), as well as dynamic acceleration, which is produced by movement of the device or the animal to which it is attached. For those not familiar with their use, it is common to think of velocity or the forward (surge) acceleration of an automobile when first learning about ADLs. In fact, the acceleration typically used in animal studies is either postural or the repetitive oscillations in acceleration produced by an animal's gait (tail-beat for sharks) (Shepard et al., 2008). In fishes, high-sampling-rate ($\geq 5 \text{ Hz}$) ADLs are thus able to quantify fine-scale swimming behavior by recording the frequency and acceleration amplitude of every stroke of the tail or fin (Gleiss et al., 2009a). In order to monitor all phases of the tail-beat acceleration wave and be able to quantify it with time-series analyses, sampling frequency must be high—usually at least double the rate of the highest frequency of movement expected from the animal (the Nyquist frequency); however, approximately 10 samples per tail-beat are desirable (Ropert-Coudert and Wilson, 2005). Thus, a higher sampling frequency would be required for a bonnethead, *Sphyrna tiburo*, than a whale shark, *Rhincodon typus*, to accurately quantify the kinematics of each tail-beat movement. Because sampling resolution is so high and because they record both the animal's movements (dynamic acceleration) and the animal's posture (static acceleration), accelerometers present a way for us to indirectly "observe" animal behavior in the wild by visualizing physical activities from their corresponding acceleration signature. This makes ADLs applicable to a broad range of issues in elasmobranch behavioral ecology and allows derivation of energy expenditures (see Section 9.4).

9.2.1 Swimming and Resting

One of the most fundamental changes in behavior for any animal is the change from a state of rest to motion. Although obligate ram ventilating sharks are in a perpetual state of swimming, many elasmobranchs are capable of resting on the seafloor, and our most commonly used tools (acoustic and satellite telemetry) for studying behavioral ecology are incapable of differentiating swimming from rest. Although acoustic tracking data can be used to infer activity rhythms (e.g., Cartamil et al., 2010; Gruber et al., 1988), tracking fixes provide only a crude measure of the animal's true movement (Cooke et al., 2001; Gruber et al., 1988) and can be misleading in some cases (Payne et al., 2010), making it difficult to detect a significant difference between time periods. The first use of accelerometry on sharks was an attempt to address this problem by attaching ADLs to whitetip reef sharks, *Triaenodon obesus* (Whitney et al., 2007). Using a simple dart-and-tether attachment or gastric implantation, the authors were able to clearly differentiate between swimming and resting in animals held in a semi-natural lagoon. Using relatively low sampling rates (1 Hz), animals could be monitored continuously for up to 17 days at a time and were shown to be primarily nocturnal, despite being regularly fed to satiation for years in captivity and thus having no need to forage (Whitney et al., 2007). Similar general measures of activity can also be obtained with high-resolution depth loggers, which will indicate oscillations in pressure when the animal is swimming but not during periods of rest. Such devices have been used for this purpose in both captive sharks (Jacoby et al., 2010) and field studies on sharks (Fitzpatrick et al., in press; Wearmouth and Sims, 2009). Acceleration transmitters present an additional method of quantifying general activity level (e.g., Murchie et al., 2011; O'Toole et al., 2010) but have limitations associated with the processing required to summarize and transmit data (see Section 9.5.1).

To quantify swimming activity in more detail, ADLs need to be securely attached to one of the fins (usually first or second dorsal) so recorded data are reflective of animal posture and movement. Gleiss et al. (2009a) first used this technique to quantify four separate swimming behaviors in lemon sharks, *Negaprion brevirostris*, maintained in a semi-natural lagoon. These included resting, initiation of swimming, steady swimming (and its associated intensity), and fast-start swimming.

Whitney et al. (2010) applied ADLs to the second dorsal fin of free-living nurse sharks, *Ginglymostoma cirratum*, and showed how various behaviors could be identified by from their acceleration wave characteristics (i.e., frequency and amplitude) (Figure 9.1). The ability to differentiate between types of swimming behavior, combined with body orientation information, is the key to interpreting a number of other behaviors in elasmobranchs and represents a means to "observe" the animals by recreating their physical movements from their acceleration characteristics.

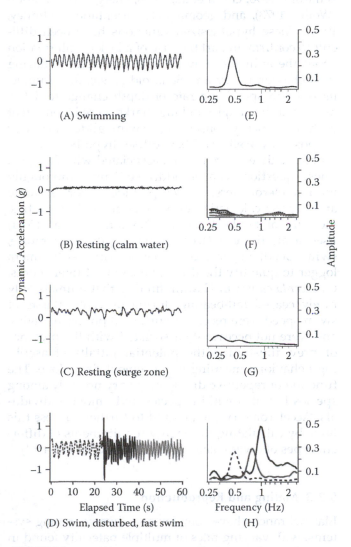

FIGURE 9.1

Dynamic acceleration from the swaying (lateral) axis of a nurse shark showing signatures associated with (A) swimming; (B) resting; (C) incidental fin movements while resting in shallow, surge zone; and (D) the sudden increase in tail-beat frequency (TBF) and signal amplitude typical of a disturbance. (E–H) Acceleration spectra produced from each behavior as identified by *k*-means cluster analysis. Line characteristics in (D) correspond to those of the spectra in (H). All examples shown were confirmed through direct observation from the surface or by divers. (From Whitney, N.M. et al., *Endangered Species Res.*, 10, 71–82, 2010. With permission.)

9.2.2 Diving

Vertical movements of sharks have been measured for decades (e.g., Carey et al., 1981, 1982; Sciarrotta and Nelson, 1977) using pressure sensors, and they can be determined with higher resolution than horizontal movements. Fine-scale vertical movement data have thus been a commonly used proxy for behavior, especially in pelagic species, for which location in the water column can vary dramatically. Many pelagic sharks show repetitive diving behavior that has been attributed to functions including foraging (Carey and Scharold, 1990; Gunn et al., 1999), energy conservation (Weihs, 1973), and geomagnetic orientation (Klimley, 1993). These hypothesized functions have been difficult to confirm, in part because of a lack of information about the animal's body position and activity during the behavior. Carey and Scharold (1990), for example, used swim speed and rate of depth change to determine that dive angles in blue sharks were inconsistent with the energy-conserving "swim–glide" behavior proposed by Weihs (1973), and they hypothesized that repetitive dives were instead associated with foraging. These questions can be addressed more thoroughly with accelerometers, as they provide a more accurate and precise calculation of body angle and also a true measure of tail-beat activity. Nakamura et al. (2011) used a multisensor (acceleration, depth, temperature, swim speed) logger in combination with a still camera logger to quantify the diving behavior of tiger sharks, *Galeocerdo cuvier*, in Hawaii, finding that animals only rarely ceased tail-beating during descents. Bursts of swim speed were recorded during all phases of diving and were not necessarily associated with the presence of prey, illustrating the potential pitfalls of ascribing behavioral meaning to movement data alone. The function of repetitive diving may vary not only among species but also within species and among individuals. Accelerometry can be used to further address this issue by calculating the energy requirements of different types of dives (Gleiss et al., 2010b; see Section 9.4).

9.2.3 Mating and Reproduction

Elasmobranchs have complex and diverse mating systems, with varying rates of multiple paternity found in several species (Daly-Engel et al., 2007; Portnoy et al., 2007) and the evolutionary implications of polyandry unclear (Daly-Engel et al., 2010; DiBattista et al., 2008). Only a few studies have been able to analyze video evidence of multiple mating events in wild sharks (Carrier et al., 1994; Pratt and Carrier, 2001; Whitney et al., 2004), and these indicate that mating behavior invariably involves a suite of postures and movements that are very different from typical non-mating behaviors.

These characteristics make mating behavior a good candidate for study using accelerometry. Whitney et al. (2010) used triaxial acceleration data loggers to identify courtship and mating behavior in female nurse sharks in the Florida Keys, a site that allowed validation of acceleration signatures from direct observation. The static component of acceleration was used to determine shark body angle, and the dynamic component was used to measure tail-beat activity (Figure 9.2). Sharks were shown to mate primarily during daylight hours, with individual mating events lasting from a few seconds to over 20 minutes (Whitney et al., 2010). This work illustrates the potential of accelerometry to quantify the fine-scale timing and location of courtship and mating in other shark species, particularly those for which mating seasonality is already known from bite scars on females or other morphological indicators (Pratt and Carrier, 2001). Because acceleration also provides a proxy for energy expenditure (Gleiss et al., 2010a; Wilson et al., 2006), the same method used to identify mating behavior can be used to test hypotheses regarding the relative energy requirements of each sex during the mating season (Springer, 1967), which may thereby shed light on molecular-based hypotheses regarding sexual conflict and its relation to polyandry (Daly-Engel et al., 2010). The recent development of transmitter–receiver hybrid tags can be used to determine if individuals are associating with tagged conspecifics and over what time scales (e.g., Guttridge et al., 2010; Holland et al., 2009). When used in conjunction with ADLs, these devices provide additional potential to shed light on the nature of shark social structure and mating system dynamics.

Fine details on the nature and timing of parturition in elasmobranchs are also scarce and are known primarily from captive observations (e.g., Schaller, 2004; Uchida et al., 1990). Whereas many species are known to use specific pupping and nursery areas (Heupel et al., 2007), the pupping grounds of some species (e.g., *Sphyrna mokarran*, *Rhincodon typus*) are still unknown, as is the timing of pupping events within a litter. Some oviparous species exhibit unique egg-laying behavior (Castro et al., 1988) that could be detected using acceleration or compass sensors, but live birth in other species could be quantified using an intra-mandibular angle sensor (IMASEN) (Leibsch et al., 2007) (see Section 9.3.5). IMASEN loggers fixed to the pelvic fins or cloaca (Wilson et al., 2004) of pregnant females should be able to detect pupping events and, combined with some form of a geo-location tag, could provide the location of pupping grounds. Although currently only available as data loggers, IMASEN sensor data could be summarized in a concise format with onboard processing methods already used by most satellite tag manufacturers, thus allowing transmission via Argos satellites (see Section 9.6).

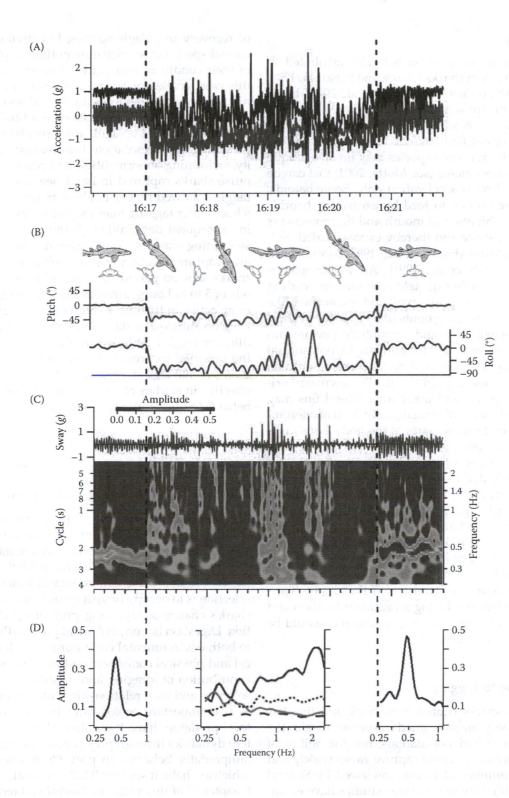

FIGURE 9.2

(See color insert.) A mating event between a logger-equipped female nurse shark and two males as represented by (A) the raw triaxial acceleration data; (B) the static component of acceleration (converted to degrees) from the surge (pitch) and sway (roll) axes showing the female moving into a head-down pitch of as much as 67° while rolling to 90° and 45° to her left and right, respectively; (C) a spectrogram of dynamic acceleration from the swaying (lateral) axis; and (D) the behavioral spectra of the event showing typical swimming (TBF = 0.4 Hz) before the event, sporadic movements across a broad range of frequencies during the event, and faster (TBF = ~0.5 Hz) swimming with high signal amplitude after the event. Vertical dashed lines indicate the start and end of the mating event respectively. Timing and details of this event were confirmed through direct observation from the surface (i.e., researcher in a kayak). (From Whitney, N.M. et al., *Endangered Species Res.*, 10, 71–82, 2010. With permission.)

9.2.4 Foraging

Foraging is the most common behavior attributed to movement patterns in sharks (Carey and Scharold, 1990; Gunn et al., 1999; reviewed by Sims et al., 2008), but it is difficult to confirm without additional measures (see Section 9.3). Although successful bouts of feeding cannot usually be quantified from acceleration data alone, foraging behavior for some species may involve unique postures or fin movements (see Motta, 2004) that can be clearly identified from acceleration data. Some batoids, for example, are known to feed by excavating benthic prey using a combination of mouth and fin movements to stir up the substrate and thereby expose buried mollusks or crustaceans (Hines et al., 1997; Motta, 2004; Sasko, 2000; Thrush et al., 1991). An accelerometer-equipped spotted eagle ray, *Aetobatus narinari*, feeding on clam meat in captivity exhibited excavation-like movements during feeding bouts that consisted of short bursts of high-frequency and -amplitude pectoral fin undulations that could be differentiated from normal swimming using wavelet analysis of acceleration data (Figure 9.3) (Whitney, unpublished). The downward orientation of the pelvic and posterior pectoral fins may also be good indicators of foraging activity if accelerometers are attached to these parts of the body. For sharks feeding on benthic prey, foraging attempts may be indicated by similar tail-beat irregularities combined with a body angle oriented toward the seafloor (cf. Figure 5 in Whitney et al., 2010). However, for many species of shark, foraging attempts may consist of relatively subtle changes in tail-beat characteristics or bursts of speed that are impossible to differentiate from normal variance in swimming activity or disturbances from predators or agonistic interactions. The potential for quantifying foraging behavior from accelerometry is thus highly variable and dependent on the tag attachment location and feeding mode of the species in question and should be validated with other measures.

9.2.5 Response to Tagging

Live release of sharks captured by hook and line has been a commonly implemented response to aid the recovery of overfished populations, but the ability of sharks to survive the stress of capture varies widely and depends on a number of factors (reviewed by Skomal and Bernal, 2010). Whereas various studies have examined the physiological effects of capture stress on blood biochemistry (e.g., Hoffmayer and Parsons, 2001; Manire et al., 2001; Skomal and Chase, 2002), very few have been able to correlate those indices with post-release behavior and survivorship via telemetry (Campana et al., 2009; Moyes et al., 2006; Skomal, 2006). Sublethal effects on post-release diving behavior can be a good measure of recovery in pelagic species, but diving behavior in coastal species is restricted by bottom depth, regardless of their condition, and is therefore less informative. In this case, accelerometry represents an invaluable technique for quantifying movements and body posture (i.e., tail-beat frequency, acceleration amplitude, rolling, or inability to remain upright) at subsecond intervals, thus providing high-resolution information about mortality, swimming abnormalities, and recovery time. Adult nurse sharks captured in hand nets and restrained for tagging showed higher tail-beat frequencies in the first 6 hours after tagging than they did over the same hours in subsequent days, although their percent time spent swimming was not significantly different immediately after capture (Whitney et al., unpublished). Whitetip reef sharks showed periods of constant swimming for periods of 5 to 14 hours immediately after tagging and were more active in the first 18 hours after tagging than they were on subsequent days, indicating a response to handling or tagging (Whitney et al., 2007). ADLs thus have the potential to provide higher resolution information than satellite tags at a fraction of the cost, and they are effective in studies of coastal animals for which diving behavior alone cannot be used as a proxy for recovery.

9.3 Measuring Feeding and Digestion

The development of mechanistic home-range models, where shark movements can be predicted based on environmental conditions and habitat, requires a detailed understanding of shark habitat selection and the factors that drive it. One logical function of habitat selection is to occupy or visit areas that will increase the shark's chance of capturing prey or optimizing digestion. Digestion is a physiological process that is sensitive to both environmental conditions as well as the chemical and physical composition of prey. Knowledge of the contribution of foraging and digestion to shark habitat selection and their relationship with movement patterns is also important ecologically because upper trophic level predators have the potential to influence ecosystem dynamics through prey consumption and inducing antipredator behavior in prey ("risk effects"), both of which are habitat specific (Heithaus et al., 2008, 2010; see Chapter 17 of this volume). Traditional techniques such as stomach content analysis and stable isotopes have provided much insight into shark diet (as well as ontogenetic, geographical, and seasonal dietary shifts), and when used in conjunction with gastric evacuation rates they can estimate field consumption rates and feeding frequency (Cortés et al., 2008; Wetherbee and Cortés, 2004; see Chapter 8 of this volume). These techniques,

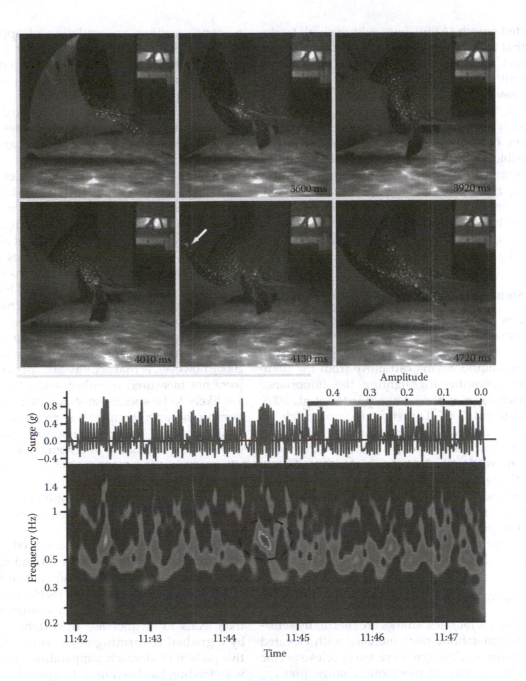

FIGURE 9.3
(See color insert.) Frames from high-speed (125 frames per second) digital video showing an accelerometer-equipped (white arrow) spotted eagle ray feeding on clam meat in captivity and exhibiting fin undulations typical of excavation-type foraging behavior observed in the wild (Whitney, unpublished). Gray lines are raw acceleration data from the surge (anteroposterior) axis with pectoral finbeat frequency and acceleration signal amplitude depicted in the spectrograph. Dashed circle indicates the portion of the spectrograph representing the behavior in the frames above, which is easily distinguished from the adjacent data produced by typical swimming movements.

however, suffer from several setbacks, including: (1) low spatial and temporal resolution of feeding activity, and (2) generally measuring only foraging at the population level. Recent studies have shown that marine predators can show individual specialization in diet and potentially also foraging tactics such as preferred feeding times and daily ration (see Chapter 17 of this volume), which will essentially be lost when using stomach content techniques (Estes et al., 2003; Matich et al., 2011; Woo et al., 2008). The ability to quantify foraging and periods of digestion *in situ* in free-ranging animals will result in major advances in our understanding of elasmobranch ecology.

Traditional telemetry techniques and associated movement metrics such as tortuosity and first passage time (see Sims, 2010) have been used to identify area-restricted search (ARS) zones within shark tracks, assuming that these are synonymous with feeding, but these indirect methods may not always be accurate. Indeed, recent studies show that animals demonstrating ARS behavior may not necessarily be feeding (e.g., Bestley et al., 2008, 2010). Data loggers that measure swim speed or acceleration can be used to infer foraging events, but without more direct measurements it is impossible to differentiate unsuccessful foraging attempts from successful prey capture, or even other non-foraging related behaviors (e.g., predator avoidance). Below, we review the transmitters and data loggers that can be used to study these aspects of shark foraging ecology, in addition to various aspects of their physiological ecology.

9.3.1 The Stomach

Many species of shark consume their prey whole, thus the stomach is the first site of active digestion. Furthermore, the shape of the pyloric junction is such that only semi-liquid chyme can move from the stomach into the intestine, highlighting the importance of the stomach to prey breakdown (Cortés et al., 2008; Papastamatiou et al., 2007). Digestion is completed chemically (by the secretion of gastric acid and enzymes) and sometimes mechanically through gastric contractions, which also move digested material into the intestine (Papastamatiou et al., 2007). From a bio-logging standpoint, the stomach is an ideal location to place sensors because many tags can actually be fed to larger sharks, and natural regurgitation will void the sensors where they can be later recovered (Papastamatiou and Lowe, 2004, 2005).

9.3.1.1 Gastric Temperature

As ectothermic vertebrates, sharks are thermally sensitive, and their metabolic rates increase with elevated body temperatures, although there are species-specific differences in the rate of metabolic change (the Q_{10} value) (Carlson et al., 2004). Stomach temperatures are specifically important because gastric secretion rates, enzyme kinematics, and gastric evacuation rates are all temperature dependent. Generally, gastric evacuation rates increase as the stomach gets warmer (Cortés et al., 2008; Wetherbee and Cortés, 2004), and alterations in the rate at which prey leaves the shark stomach can lead to changes in behavior via appetite. The lesser spotted dogfish, *Scyliorhinus canicula*, demonstrates an inverse relationship between the evacuation of stomach contents and return of appetite, and it was hypothesized that mechanoreceptors in the stomach wall respond to distention and initiate a graded nervous

response leading to regulation of appetite (Sims et al., 1996). Simply put, the faster that food leaves the stomach, the faster the animal can feed again, which can be advantageous in areas where prey distribution may be unpredictable. A caveat to this proposal is that theoretical models predict that digestive efficiency will decrease as gut transit time decreases because prey items will be exposed to gastric acids and enzymes for shorter periods (Hume, 2005).

Based on movement patterns and prey distribution it was suggested that both bat rays, *Myliobatis californica*, and lesser spotted dogfish move into warmer water to hunt, but then move to cooler water to digest and thereby improve digestive efficiency (Matern et al., 2000; Sims et al., 2006). The high water temperatures increase metabolic rate, allowing the elasmobranchs to forage more efficiently. When they move into colder water, metabolic and gastric evacuation rates decrease and digestive efficiency should increase. A computer simulation predicted that energy expenditure may be reduced by 4% for a shark using this behavioral strategy (Sims et al., 2006); however, actual activity and associated behaviors were not measured in either study. In addition, there are likely to be species-specific differences in the relationship between temperature and digestive efficiency, which may not always fit a simple "hunt warm, rest cool" strategy (DiSanto and Bennet, 2011). Future studies should use stomach temperature transmitters or data loggers to quantify the range of stomach temperatures experienced in the wild, in conjunction with laboratory studies that measure absorption efficiency at a variety of water temperatures.

When endothermic predators feed, seawater and cold prey items enter the stomach and cause a precipitous decrease in stomach temperature (Wilson et al., 1992). Specific dynamic action (SDA), the energetic cost of ingesting, digesting, and assimilating prey items, then leads to an increase in metabolic rate followed by a gradual rewarming of the stomach. This distinctive pattern of stomach temperature change associated with feeding has been used to quantify *in situ* foraging events in wild seabirds, marine mammals, and hetero- or homeothermic fishes (e.g., Austin et al., 2006; Bestley et al., 2008; Wilson et al., 1992). The time taken for the stomach to rewarm prey items can also be used to estimate meal size, although absolute meal size has proven more difficult to measure in the field (Gunn et al., 2001; Wilson et al., 1992). Stomach temperature cannot be used to estimate feeding in ectothermic elasmobranchs because they will have body temperatures very similar to that of seawater and of their prey; however, this technique may be applicable to a number of homeothermic sharks. Gastric temperature transmitters were used to actively track and study the thermal physiology of

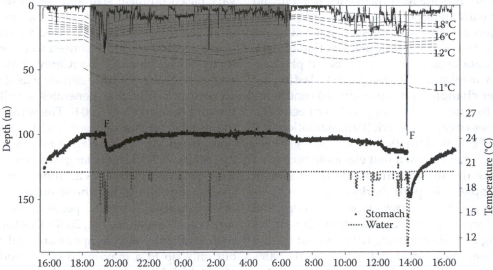

FIGURE 9.4

Continuous measurements of stomach temperature and swimming depth in a juvenile shortfin mako shark in the Southern California Bight, obtained using an acoustic depth/temperature transmitter. Upper trace is swimming depth (with water temperature contours); lower trace is stomach temperature. "F" indicates feeding events; gray area is nighttime. (From Sepulveda, C.A. et al., *Mar. Biol.*, 145(1), 191–199, 2004. With permission.)

salmon sharks, *Lamna ditropis*, in Alaska (Goldman et al., 2004). Sharks elevated and maintained their body temperatures by up to 16°C above ambient seawater during the tracking period (Goldman et al., 2004). At one point during a track, a precipitous decline in stomach temperature occurred, suggesting that the shark had fed. Juvenile shortfin mako sharks, *Isurus oxyrinchus*, were actively tracked in the southern California bight using temperature and depth transmitters placed in the stomach (Sepulveda et al., 2004). Distinctive changes in stomach temperature associated with feeding were observed during several tracks, which the authors verified by catching the tracked shark and examining the stomach contents (Figure 9.4). The majority of feeding events were observed during daytime periods. Finally, a passive monitoring Vemco Radio-Acoustic Positioning (VRAP) system (a passive acoustic monitoring system that can provide real-time active tracks as long as the shark remains within the detection area) was used to measure stomach temperatures in free-ranging white sharks, *Carcharodon carcharias*, off a seal rookery in northern California (Klimley et al., 2001). Although sharks were tracked for up to 78 days, no changes in stomach temperature associated with feeding were observed. This may support the hypothesis that white sharks feed infrequently, although the low resolution of stomach temperature measurements may have made it difficult to distinguish feeding events (Klimley et al., 2001). Although telemetry transmitters provide useful

information over short time periods, deployment of archival temperature loggers will be of most use for studying feeding behavior of warm-bodied sharks. In all of the above studies, the researchers deployed the transmitters by concealing them in bait and feeding them to the animals. Loggers will remain in the stomach for periods of a few days up to several weeks, particularly with the attachment of corrosible hooks (e.g., Goldman et al., 2004; McCosker, 1987; Papastamatiou and Lowe, 2004, 2005). Obviously, the problem becomes how to recover loggers so data can be retrieved, but several options are available (see Section 9.5.2). It may also be possible to surgically implant data loggers within the viscera for long-term retention, as has been done with tunas, although this will require recapturing the animal (Bestley et al., 2008).

9.3.1.2 Gastric pH

The stomachs of carnivorous vertebrates secrete concentrated hydrochloric acid, which aids in the chemical breakdown of prey hard parts, sterilization of the stomach mucosa, and conversion of the inactive zymogen pepsinogen into the protease enzyme pepsin (Cortés et al., 2008; Holmgren and Holmberg, 2005, Papastamatiou et al., 2007). Gastric enzymes are highly sensitive to the pH in the stomach, with most having pH optima in the range of 1.5 to 2.5 (see Cortés et al., 2008); thus, changes in gastric pH can give an indication of the state of digestion and also indicate feeding events. Remote and continuous measurements of gastric pH are possible using data loggers coupled to a pressure-insensitive pH electrode with a free-diffusion liquid junction (Peters, 1997a,b). These data loggers are able to record pH measurements with minimal drift for periods lasting up to 16 days under changing pressure (depth) conditions (Peters, 1997a). To date, pH data loggers have been used with three species of captive shark (Papastamatiou and Lowe, 2004, 2005; Papastamatiou et al., 2007). These studies and others have shown that elasmobranchs can differ in the way the stomach responds to periods of fasting. For example, in leopard sharks (*Triakis semifasciata*), blacktip reef sharks (*Carcharhinus melanopterus*), and spiny dogfish (*Squalus acanthias*), the stomach secretes acid even during periods of fasting (Papastamatiou and Lowe, 2004; Papastamatiou et al., 2007; Wood et al., 2007). In contrast, nurse sharks, *Ginglymostoma cirratum*, stop secreting acid during certain periods of fasting, with the pH going as high as 8.1 (Papastamatiou and Lowe, 2005). The reason for these differences are unclear but may be related to species-specific differences in activity, feeding frequency, diet, or enzyme kinematics (Papastamatiou and Lowe, 2004, 2005; Papastamatiou et al., 2007). The characteristic changes in gastric pH with feeding can also be used as a proxy for feeding in wild sharks (Figure 9.5). Generally, pH increases rapidly after feeding as the prey items and ingested seawater dilute the small amount of acid present in the empty stomach. Stomach distention and the release of secretagogues (hormones produced by the stomach) (Cortés et al., 2008; Holmgren and Holmberg, 2005) cause an increase in acid secretion rates and the subsequent re-acidification of stomach contents. The time taken for pH to decrease to baseline levels is related to meal size, and pH changes may be used to estimate how much the shark has eaten (Papastamatiou and Lowe, 2004; Papastamatiou et al., 2008). Measuring pH in the field will require either attaching acoustic transmitters to pH data loggers to facilitate their recovery or the use of an acoustic pH transmitter so pH can be measured in real time (Papastamatiou et al., 2008). Thus far, such transmitters have been deployed in captive blacktip reef sharks for periods of up to 12 days, and feeding events were clearly distinguishable during that period (Papastamatiou et al., 2008).

9.3.1.3 Gastric Motility

The stomach is a muscle that is responsible for mixing prey items with digestive fluids, mechanical digestion (although this may be of minor importance in elasmobranchs; see Papastamatiou et al., 2007), and evacuation of digested prey items into the intestine. Gastric motility can also be indicative of the state of digestion and will vary based on meal size and composition, abiotic conditions, stress levels, and other factors (Papastamatiou et al., 2007; Peters, 2004). A motility transmitter using a lever system sensor was used to record gastric motility in captive bluefin tuna (Carey et al., 1984). Recently, a more sensitive motility sensor was designed, using a piezoelectric film that generates a voltage every time the film flexes (Peters, 2004). The sensor measures both the number of contractions and the speed of contractions, providing a cumulative measure of gastric activity (Peters, 2004). Data from gastric motility data loggers were obtained in free-swimming captive blacktip reef sharks, suggesting that the strength of gastric contractions varied with meal type, size, and water temperature (Papastamatiou et al., 2007). Contractions increased with ambient water temperature and with meal size, but only up to a maximum of 1% body weight (BW), after which larger meals actually caused lower levels of contractions. Furthermore, immediately after feeding, sharks showed a lag in gastric contractions, which lasted 7 to 12 hours. These post-feeding lags are common in vertebrates and are known as gastric accommodation, which allows for the stomach to expand during feeding (Cortés et al., 2008; Holmgren and Holmberg, 2005; Papastamatiou et al., 2007). These collective measurements enabled predictions of the behavior of sharks in the field; in particular, optimal meal size will be approximately 1% BW and optimal foraging time will be during the early-morning hours when water temperatures are lower (Papastamatiou et al., 2007). Under these tropical conditions, the period of gastric accommodation coincides with increasing water temperatures and enhanced motility. Field measurements were obtained by attaching small acoustic transmitters to motility data loggers and deploying them in blacktip reef sharks at Palmyra Atoll, in the central Pacific Ocean. Sharks were actively tracked until the loggers were regurgitated (from 2 to 21 days), after which they were recovered using a handheld underwater receiver to locate the transmitters. Preliminary data showed that there were diel changes in motility, with periods of increased gastric motility during the early-evening hours just as water temperatures were starting to decline (Figure 9.6). Overall levels of stomach activity appeared to be greater during the night, which may suggest that more active digestion was occurring during this period.

9.3.2 Jaw Movements

The first stage of prey acquisition and handling involves manipulation and processing of prey using the jaws and teeth. As such, quantifying jaw movements may be used *in situ* to determine when feeding is occurring. Jaw movements can be detected by using an IMASEN logger to monitor the distance between a strong magnet

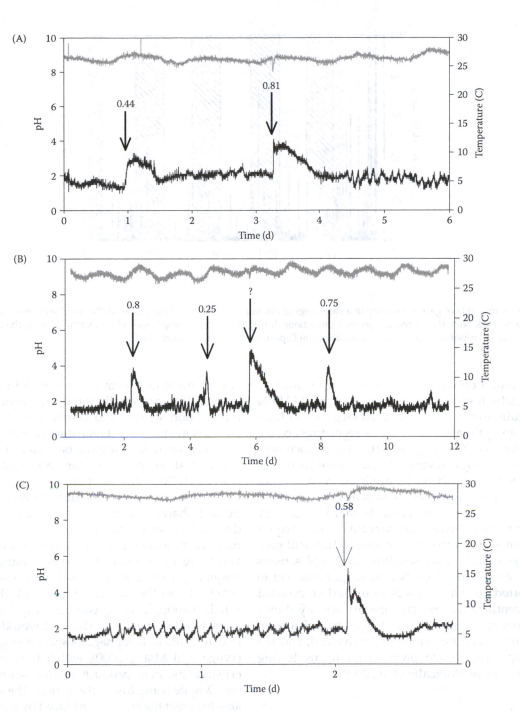

FIGURE 9.5
Continuous measurements of gastric pH (lower black line) and temperature (upper gray line) in three free-swimming blacktip reef sharks. Arrow indicates time of feeding, and meal size is given as %BW. Data were obtained using an acoustic pH/temperature transmitter. Data are from Papastamatiou et al. (2008).

attached to one jaw and a Hall magnetic sensor on the other (Wilson et al., 2002). IMASEN loggers have been used to study the feeding behavior of marine birds, mammals, and recently fish (Liebsch et al., 2007; Metcalfe et al., 2009; Naito, 2007; Sims, 2010; Wilson et al., 2002). Separating jaw movements associated with feeding from those associated with other behaviors, such as breathing ventilations and yawning, would require calibration of the sensors under laboratory conditions. Further validation in the field may be possible by combining IMASEN sensors with videorecorders (see below) to correlate observed foraging with jaw motion signatures. Using recordings to differentiate meal size and prey types will be difficult due to differences in jaw

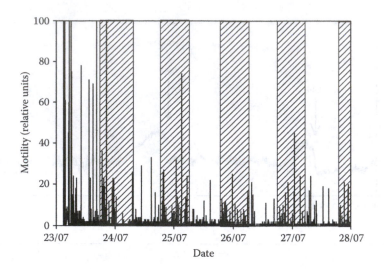

FIGURE 9.6
Continuous measurements of gastric motility in a free-ranging blacktip reef shark, at Palmyra Atoll, Pacific Ocean. Shaded rectangles represent nighttime periods. Note the increased gastric contractions during the early evening hours. High motility during the first day of deployment is likely a stress response from tagging. Data are from Papastamatiou et al. (unpublished).

motion signatures between animals feeding on dead vs. live prey (Liebsch et al., 2007). Also, current IMASEN sensors require extensive memory, so loggers will have to be set so an optimal period of data recording occurs. Despite Nelson (1990) suggesting the use of such sensors in shark ecology several decades ago, there has been very little research on its application to elasmobranch fishes. For a few species of shark, jaw opening and the sequence of jaw movements that lead to prey consumption are relatively conservative (see Chapter 6), which may suggest that prey consumption will generate unique signals from sensitive IMASEN sensors (e.g., Motta, 2004). An IMASEN data logger attached to a lesser spotted dogfish over a 24-hour period recorded breathing ventilations and changes in activity (Sims, 2010). A current version of this logger has also been used to identify feeding events in captive cod, and the technique appears to hold promise for future testing with elasmobranchs (Metcalfe et al., 2009).

9.3.3 Sound

The marine environment is noisy; therefore, the acoustic environment could be a source of additional data about an animal's surroundings and what it is doing. Sharks generally have a good sense of hearing that is most sensitive to low-frequency sounds between 40 and 800 Hz (Myrberg, 2001). Field studies have shown that multiple species of shark are attracted to low-frequency sounds (such as those produced by struggling fish), and laboratory experiments indicate that some species can orient to within 10° of a sound source (Nelson, 1967; Nelson and Gruber, 1963). Measurements of the acoustic

environment may provide insight into shark foraging by detecting potentially interesting acoustic sources of information sounds. Bioacoustic data loggers are simply sensors that record and store sound; they have been used primarily to study the behavior of sound-producing animals such as cetaceans and seals (e.g., Aguilar Soto et al., 2008; Insley et al., 2004). A bioacoustic data logger was surgically implanted into a captive blacktip reef shark, and sounds were recorded over several days (Meyer et al., 2007). The sounds produced by tail movements allowed tail-beat frequency to be measured, and there were relatively distinct changes in tail-beat frequency associated with feeding. The sensor was also able to detect the sounds of the shark thrashing while it fed. Although the sensor has only been tested with a piscivorous species of shark, it would be interesting to determine if durophagous species (e.g., bonnetheads) (Wilga and Motta, 2000), which feed on hard-shelled crustaceans, may produce unique sounds when they eat. Despite being inside the animal, the sensor was also able to detect the sounds produced by passing boats and reef fish vocalizations. Other potential uses of these sensors could be determining if sharks are associating with particular sounds, such as those produced by teleosts during spawning aggregations, quantitative field studies of the role of sound on attraction and orientation in sharks, and determining the effects of anthropogenic sound sources, such as boat traffic, on shark behavior. Although bioacoustic sensors by themselves are unlikely to be a suitable proxy for foraging, when combined with other sensors, they may provide additional insight into the behaviors associated with prey capture and when these occur.

9.3.4 Visual Images

Videocameras (AVEDs) and still cameras have been attached to a few species of shark (e.g., Heithaus et al., 2001, 2002; Nakamura et al., 2011). Videocameras attached to tiger sharks in a shallow bay have been able to detect feeding events and foraging locations over several hours (Heithaus et al., 2002), and a digital camera was able to obtain photographs of a putative foraging event by a tiger shark moving through deeper water (Nakamura et al., 2011). In a shallow Australian bay, tiger sharks appeared to attack prey that were unaware of the shark's presence, while avoiding more vigilant prey, suggesting that they are stealth foragers (Heithaus et al., 2002). In the deeper waters of Hawaii, however, tiger sharks were observed pursuing schools of fish that appeared to move large vertical distances to escape capture (Nakamura et al., 2011). Visual data can also provide insight into other aspects of shark behavior, such as social interactions and competitive interactions with other top predators (Parrish et al., 2008). The challenge with visual digital media is that data can only be recorded over short time periods (maximum a few days) due to the large memory requirements. Furthermore, the technique is only useful in locations with clear water and during daytime hours, where feeding events can be easily visually identified. Finally, the subterminal location of the mouth of most shark species makes it difficult to observe actual feeding via a camera mounted on the dorsal fin; however, recent advances have greatly increased the data-storage capabilities of video logging sensors (eventually recordings can be made over several months) and have allowed them to be combined with other sensors such as accelerometers and magnetometers (e.g., Marshall et al., 2007; Moll et al., 2007). In addition, the use of visual images will allow the calibration and interpretation of signals from other sensors such as accelerometers or bioacoustic probes, which may be impossible under captive conditions (e.g., for large sharks). This also highlights the importance of combining multiple sensors to produce the ultimate "ecological" tag that can paint the most accurate picture of what a shark is doing.

9.4 Measuring Energy Expenditure

Energetics is a central theme in ecology and physiology and provides a critical currency that operates across levels of organization (Brown et al., 2004). A thorough understanding of field energetics can therefore contribute to a deeper understanding of organismal biology of many species (Butler et al., 2004; Nagy et al., 1984),

including sharks and their relatives (Lowe, 2002; Lowe & Goldman, 2001; Sundstrom & Gruber, 1998). Whereas great strides have been made in measuring metabolism of a range of species of shark in the laboratory (Carlson & Parsons, 2001; Dowd et al., 2006; Lowe, 2002; Nixon & Gruber, 1988; see Chapter 7 of this volume), few studies have managed to bridge the gap between laboratory and field studies, mainly due to our inability to observe and quantify fine-scale activity from free-swimming sharks (Lowe & Goldman, 2001). In vertebrate biology as a whole, relatively few approaches are available for estimating metabolic rate in the field (Butler et al., 2004), with most common approaches relating to the remote measurement of heart rate and the use of doubly labeled water. Both of these techniques have been used extensively in homeotherms but present problems for use in fishes. Heart rate (f_H) suffers from high variability in the f_H/MO_2 relationship in fishes (Thorarensen et al., 1996; but see Clark et al., 2010), and particularly sharks (Scharold & Gruber, 1991), and the doubly labeled water technique suffers from large water turnover in fishes, rendering it unusable (Butler et al., 2004). Thus, for fish (and elasmobranchs), remotely measuring activity has been the predominant method for alluding to energetics in a field setting (Cooke et al., 2004a,b). In teleosts, electromyography (EMG) telemetry has been used widely to quantify activity. Here, electromagnetic potential is measured in muscle tissue and relayed via radio telemetry (Cooke et al., 2004a) or acoustic telemetry (Dewar et al., 1999). These sensors have a wide range of applications, from migration studies to the impact of catch and release angling (Cooke et al., 2004b), but they have not been used on sharks due to the complicated surgery necessary to accurately place electrodes into the muscle. The other recently developed technique uses the differential pressure measured near the caudal peduncle of swimming fish (Webber et al., 2001). A study on captive cod demonstrated how both tail-beat frequency and amplitude could be deduced from such data, that oxygen consumption was related to the integrated signal (i.e., a combination of frequency and amplitude), and that a similar approach could also apply to sharks (Voegeli et al., 2001).

It is clear that, despite a number of techniques being commercially available to measure locomotory activity in fish, few have been used in the field. Lowe et al. (1998, 2002) used a custom-made tail-beat transmitter utilizing a pivoting vein coupled to a reed switch to remotely measure tail-beat frequency (TBF) in juvenile scalloped hammerheads, *Sphyrna lewini*. They calibrated TBF in relation to speed and oxygen consumption in a Brett-style respirometer, which subsequently permitted estimates of field metabolic rate (FMR) by tracking free-ranging sharks equipped with these transmitters. Sundstrom and Gruber (1998) used a speed-sensing transmitter and, with allometric assumptions, used a power–performance curve obtained from juvenile lemon sharks to estimate FMR of sub-adult animals. In both of these cases, researchers used custom-built devices that have subsequently not been made commercially available. ADLs now present a widely available and superior alternative to earlier methods.

9.4.1 Movement, Acceleration, and Energy Expenditure

Consider a shark swimming with caudal propulsion. The axial muscle contracts rhythmically, which sends bending waves down the body of the fish, creating the characteristic tail-beat. In its most elemental form, a tail-beat can be characterized by its frequency and amplitude (Bainbridge, 1958). Any segment of the tail therefore experiences cyclical changes in lateral position and associated changes in velocity as it moves from side to side, which is represented in the trace of the lateral acceleration as repetitive peaks (e.g., Figure 9.1A). From these data, locomotory activity can be quantified with a number of different approaches (Watanabe and Sato, 2008; Whitney et al., 2007, 2010), but in the past several years dynamic body acceleration has shown the most promise for studies of energy expenditure *in situ* (Gleiss et al., 2010a; Halsey et al., 2008; Wilson et al., 2006).

Acceleration appeals intuitively as a measure of locomotory activity, being essentially a measure of motion. By definition, energy is consumed when performing mechanical work, and the rate at which mechanical work is performed is power, which can be seen as analogous to oxygen consumption or energy expenditure (i.e., it represents a fraction of metabolic power due to movement, excluding standard metabolism). Through Newton's laws of motion, we can deduce that acceleration is intrinsically linked to power, and if muscle performs mechanical work (i.e., it shortens by contracting) it produces acceleration (Gleiss et al., 2010b). Thus, in theory, acceleration should provide a good proxy for the power requirements of swimming.

Overall dynamic body acceleration (ODBA), or partial dynamic body acceleration (PDBA) when acceleration is only measured along one or two axes, is simply the sum of the absolute value of dynamic acceleration from all acceleration axes being used (Wilson et al., 2006). It is a reliable proxy of energy expenditure in a range of vertebrates from birds to reptiles, mammals, and elasmobranchs (Halsey et al., 2011). It also has been successfully used on various taxa in the field (Gleiss et al., 2010a; Shepard et al., 2009b; Wilson et al., 2010). Importantly, the relationship between ODBA and energy expenditure has been linear in all vertebrates examined thus far (Halsey et al., 2008, 2011), which makes its use as a

proxy for movement-related energy expenditure superior to tail-beat frequency, which is not linearly related to oxygen consumption (e.g., Graham et al., 1990).

9.4.2 Signal Processing: The Need for High-Resolution Data

As described above, raw acceleration data consist of both static (due to the Earth's gravitational pull) and dynamic (due to animal movement) components (see Section 9.2). Whereas the static component is useful for calculating body angle, pitch, and roll, it must be removed in order to estimate locomotory activity and energy use (Halsey et al., 2009). The manner in which this is done is crucial to the proper calculation of ODBA and requires high-resolution data to be done accurately. Typically, the static component is removed by being identified and then subtracted from the raw acceleration data, leaving only the dynamic component. Identification of the static component has been done using low-pass filters based on fast Fourier transforms (Tanaka et al., 2001; Watanabe & Sato, 2008; Whitney et al., 2010) or running means performed over various window sizes (Halsey et al., 2011; Shepard et al., 2009a; Wilson et al., 2006). Although a comprehensive study is needed to determine which method is best, both of these techniques require assessment and calibration for each species and size-range to which they are applied. Selection of an inappropriate window can lead to severe miscalculations of dynamic acceleration and ODBA (Figure 9.7). Shepard et al. (2009a) described a method to determine optimal smoothing window size for a range of species and concluded that data should be smoothed over approximately 2 to 4 seconds. However, the performance of any window varies with the species studied due to variance of locomotory activity and limb-stroke frequency with size (Sato et al., 2007), so no single filtering window will be appropriate in all cases.

This need for appropriate, species-specific filtration to separate static and dynamic acceleration has important implications for the use of devices that only store a single mean value of dynamic acceleration in order to optimize memory (Clark et al., 2010) or enable transmission of data via acoustic telemetry (Murchie et al., 2011; O'Toole et al., 2010; Payne et al., 2010). These summary data do not allow accurate estimates of error inherent in the derived value for activity because appropriate testing of the filtering window cannot be performed (Figure 9.7). Depending on species and attachment location, dynamic acceleration values can be small in swimming sharks (Gleiss et al., 2009a, 2010a); therefore, any oversmoothing can seriously bias data because changes in static acceleration may contribute substantially to the derived estimate (Halsey et al., 2011). Conversely, undersmoothing

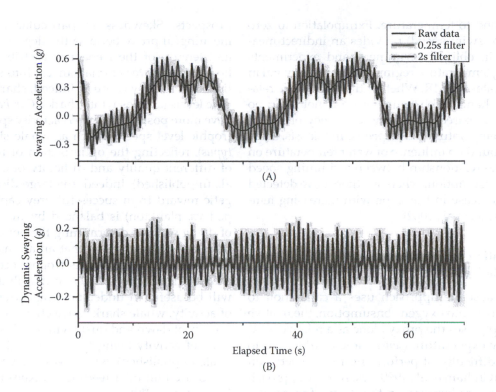

FIGURE 9.7

Swaying (lateral) acceleration trace of a captive black-tip shark (*Carcharinus limbatus*). (A) Filtering the static acceleration from raw acceleration requires constructing a filter with the correct smoothing window (or passband when using FFT filters). Using a window of 0.25 seconds results in undersmoothing (A) and underestimation of dynamic acceleration (B), whereas a filter of 2 seconds provides a better estimate of static acceleration (A) and thus larger amplitudes in the swaying acceleration. Note the severe underestimate in locomotory effort when data are smoothed over 0.25 seconds compared to 2 seconds.

will prevent comparability of the data to other studies. Only the logging of continuous acceleration data permits users to identify and quantify these errors.

9.4.3 Application to Shark Energetics: The Absolute Approach

The application of any proxy to energetics has two distinct lines of approach: the absolute (or quantitative) and the relative (or qualitative) approach (Green, 2010). In order to determine absolute energy expenditure it is necessary to obtain a predictive relationship between the variable being estimated (in this case, oxygen consumption) and the predictor (acceleration) being used to estimate it. This is usually achieved by placing accelerometer-equipped animals in respirometers and monitoring their oxygen consumption in tandem with their movement (Gleiss et al., 2010a; Halsey et al., 2008; Lowe, 2002). The aim here is to calibrate a wide range of activity levels that will be representative of animal movement in the field.

Only one study so far has applied this approach to sharks using acceleration as a proxy. Gleiss et al. (2010a) found a linear relationship between oxygen consumption and acceleration in juvenile scalloped hammerheads,

and the relationship had high predictive power. Future studies using modern respirometry equipment coupled with ADLs present an excellent opportunity to quantify energy expenditure of sharks in the wild, with the size of individuals to be studied limited only by the size of available respirometers (see Section 9.4.4 for application to larger species).

Among the challenges of using acceleration as a proxy is the fact that the energy budget of an organism is not entirely driven by movement-related metabolism, and the largest component of the energy budget of most fish is the maintenance of homeostasis while at rest (i.e., standard metabolic rate, or SMR) (Jobling, 1994). Although SMR can be predicted by allometry (Clarke and Johnston, 1999; Sims, 1996), estimates based on size alone are not accurate enough to create an appropriate energy budget without a species-specific equation (Killen et al., 2010). Furthermore, temperature plays a key role in governing the metabolic rate of ectotherms (Clarke and Johnston, 1999; Gillooly et al., 2001), and sharks are no exception (Carlson and Parsons, 1999). To date, only a few studies have quantified the effect of temperature on the metabolic rate of sharks (Carlson and Parsons, 1999); therefore, the appropriate use of acceleration to predict energy expenditure will require good estimates of how SMR is likely to respond to temperature. Extrapolation to zero acceleration (or swim speed) provides an indirect measure of SMR in obligate swimmers, and experiments under varying temperature regimes will prove pivotal in accurate estimates of FMR. Whether the slope of the relationship will change with temperature is not clear; no studies have compared swimming efficiency of sharks at a range of temperatures and speeds. In teleosts, some studies have found no influence of water temperature on the slope of the relationship between swimming speed and oxygen consumption, whereas others have detected a significant increase in the slope with increasing temperature (Dickson et al., 2002).

9.4.4 Application to Shark Energetics: The Relative Approach

Whereas the absolute approach uses a calibration to convert the proxy into oxygen consumption, the relative approach simply uses the proxy alone as a relative measure of energy expenditure (*sensu* Green, 2010). Due to the logistical difficulty of performing respirometry on larger sharks (Graham et al., 1990), the relative approach will most likely predominate in the study of these species. Given the impressive ability to distinguish behavior from acceleration data (Section 9.2), we are now well equipped to compare the energetic costs of different behaviors and relate these to ecological optima. Gleiss et al. (2010b), for example, were able to show that whale sharks, due to their negative buoyancy, have to invest less locomotory power into descents compared to ascents and that the steeper the sharks ascend the higher their energetic demand (Figure 9.8). Moreover, using the empirical relationship of dynamic body acceleration and pitch derived for these animals, they estimated the relative instantaneous power required for sharks to perform dives of all possible combinations of ascent and descent pitch angles in relation to their horizontal and vertical cost of transport. They found that continuous yo-yo dives were shallow and conformed to predictions made by Weihs (1973) as a means to reduce the horizontal cost of transport. Isolated V-shaped dives, on the other hand, conformed to a strategy minimizing the cost of vertical transport, suggesting that these dives primarily served the purpose of searching the water column.

The ability to compare datasets obtained from different species is desirable; however, this would require stringent standardization of data-logger attachment and analysis, and even this may not suffice when comparing data from different species where the relationships between oxygen consumption and acceleration may differ. Comparative analyses thus require an alternative approach. Characterizing the shape of the frequency distribution of ODBA with scale-invariant indices, such as skewness and kurtosis, offers some

prospects. Skewness in particular is a ecologically meaningful proxy because the degree of positive skew (tail length) of the frequency distribution will reflect how much power animals invest into strenuous activities, such as prey capture. Indeed, sharks of higher trophic levels (such as white sharks, *Carcharodon carcharias*) have more positively skewed activity spectra than lower trophic level species (such as whale sharks, *Rhincodon typus*), reflecting the optimization of foraging on prey of different quality and difficulty of capture (Gleiss et al., unpublished). Indeed, the large dichotomy in energetic reward from successful prey capture (e.g., pinniped vs. plankton) is balanced by the energetic outlay of doing so, thus determining the net energy gain. It is therefore not surprising that an animal's activity budget is directly linked to its trophic ecology. Moreover, characterizing the frequency distributions of activity will be useful in understanding intraspecific patterns of activity; whale sharks have characteristic changes in activity at dawn and dusk, with the frequency distributions of activity being typical of higher skew (Wright et al., unpublished), which was shown to be indicative of surface ram filter feeding (cf. Nelson & Eckert, 2007; Motta et al., 2010).

9.5 Logistical Challenges, Solutions, and Future Directions

Although the use of data-logging tags for wildlife research has been going on for decades, only within the past few years have integrative, high-resolution, multisensor data loggers become commercially available (Ropert-Coudert et al., 2009). Application of these devices to elasmobranch research has also been a recent development, and their use presents nearly as many challenges as it does opportunities. However, the value of the data provided and their application to a broad range of questions and taxa has led to a concerted and increasingly successful effort to overcome these challenges. We conclude by examining some of the most common limitations to the use of these devices in elasmobranch research and how they might be overcome.

9.5.1 Device Limitations

The high sampling rates (often 30 Hz or greater over multiple channels) associated with many of these devices (particularly ADLs) make it impossible to relay their raw datasets via typical acoustic or satellite transmission due to the very limited bandwidth of these systems. Data must therefore be logged to tag memory, which means that record durations have typically been

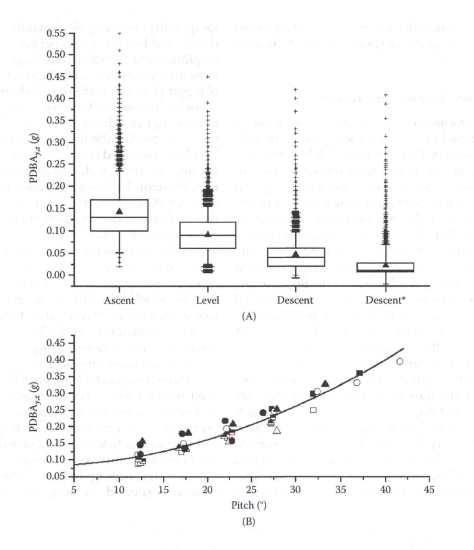

FIGURE 9.8

The impact of movement trajectory on the power requirements of swimming for whale sharks as represented by $PDBA_{y,x}$ (see text). (A) The box plots show power requirements using acceleration for the three main behaviors. The boxes represent the first and third quartiles; the triangle, the mean; and error bars, the standard deviation of instantaneous $PDBA_{y,x}$. Ascents were characterized by the largest mean dynamic body accelerations, whereas descents had the lowest. (B) Mean instantaneous $PDBA_{y,x}$ for all nine whale sharks for a range of ascent pitch angles. Data were placed into 5° bins to visualize the trend of required power in relation to animals' pitch. *Noise from accelerometers was subtracted from all periods with no visible locomotory activity. (From Gleiss, A.C. et al., *Funct. Ecol.*, 25(3), 595–607, 2010. With permission.)

brief. Although advances in digital memory capability promise to lengthen record durations to periods of several months, this problem also can be mitigated using currently available programming options that allow loggers to record intermittently and go into a sleep state between sampling intervals. In the near future, onboard logger processing may allow periods of repetitive behavior such as steady-state swimming to be summarized by their relevant wave characteristics (i.e., frequency, amplitude, and duration) to save memory. For resting species, the device could go into a memory- and power-saving mode if a certain minimum threshold of acceleration was not exceeded or be triggered to record by actions of the animal (the initiation of a dive for instance). These options are being explored by multiple manufacturers and raise the possibility of data reduction, summary, and potentially Argos transmission at some point in the future.

One device, already available, records acceleration intermittently for short periods (several seconds) and then transmits a summarized value of acceleration representing an average of multiple axes over the entire sampling interval (Murchie et al., 2010; O'Toole et al., 2010). Although this provides data for several weeks (as long as the animal stays within range of a receiver), averaging data over time and axes provides an extremely coarse measure of activity. Still, these tags represent a good first step to overcoming the temporal limitations of ADLs and are useful as indicators of mortality as well as long-term changes in activity levels (Murchie et al., 2010; O'Toole et al., 2010) and may be used for broad assessments of energy expenditure (Payne et al., 2011; but see Section 9.4.2).

9.5.2 Attachment, Release, and Recovery

Because the data recorded must be recovered, attachment location and method must be considered more carefully for these new devices than for typical telemetry tags. Gastric data loggers and transmitters must be fed to animals or manually placed in the stomach in a way that maximizes their retention but still allows for their regurgitation at the end of the experiment (Papastamatiou and Lowe, 2004, 2005). Depending on the question being addressed, accelerometers must be positioned not only to allow precise monitoring of fin or body movement but also to allow the tag to release from the fin for recovery. In most cases, this requires the device to be secured at both ends so any movement of the tag is reflective of the part of the body to which it is attached (e.g., Figure 1 in Whitney et al., 2010). Attachment location depends on the species and research question being addressed but must be carefully considered. For many shark species, the first dorsal is the only one large enough to carry the tag package, and this is sufficient for detecting tail-beat activity and animal body angle (Gleiss et al., 2009a; Nakamura et al., 2011). The second dorsal fin is likely to reveal more subtle changes in tail-beat amplitude, and this may aid in differentiating behaviors from acceleration data (Whitney et al., 2010). Either location is feasible

for quantifying energetic expenditure, but more posterior-located loggers will experience greater acceleration amplitude and therefore have larger ODBA values than more anterior-placed units (Figure 9.9). Standardization of logger placement within a study is thus paramount.

Logger release and recovery may be the most significant logistical hurdle that must be overcome to use these devices. Gastric loggers (after being regurgitated) have been relocated on the seafloor using acoustic pingers and an underwater hydrophone (Papastamatiou, unpublished). This acoustic recovery method has been used for ADLs as well (Whitney et al., 2010), but the use of floats and very high frequency (VHF) telemetry is growing more common. In this case, release and recovery are achieved by encasing the ADL and a VHF transmitter in a small float and attaching it with an electronic or galvanic timed release that allows the float to detach from the shark and ascend to the surface within hours or days after tagging (e.g., Gleiss et al., 2009b; Nakamura et al., 2011). Detection and relocation are then achieved using a VHF receiver, which provides detection ranges orders of magnitude greater than those for acoustic telemetry. This method has been used with an 84% recovery rate over distances up to 16 km in Caribbean reef sharks (*Carcharhinus perezi*) over periods of 1 to 3 days and distances up to 27 km in whale sharks (Gleiss, unpublished), as well as recovery rates of 86% in nurse sharks over periods of 1 to 5 days and distances up to 7 km (Whitney, unpublished). Animals expected to range widely can be actively

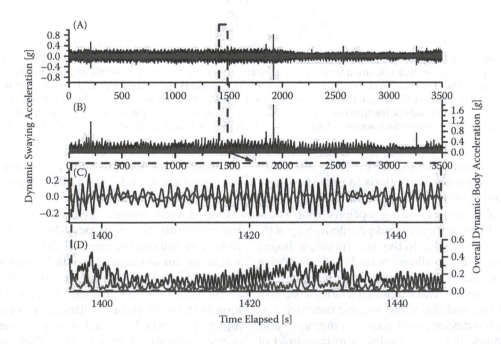

FIGURE 9.9
Swaying (lateral) dynamic acceleration (A, C) and overall dynamic body acceleration (B, D) traces of a captive blacktip shark (*Carcharinus limbatus*) tagged with two accelerometers, one located on the first dorsal fin (gray line) and the other attached to the second dorsal fin (black line). Note the overall larger variability present in the sway (C) and the larger ODBA (D) recorded with loggers attached to the second dorsal fin.

tracked (via acoustic telemetry) to improve the likelihood of logger recovery (Nakamura et al., 2011), or small Argos tags could be incorporated into the float package to get researchers close enough for VHF detection. Thus, data recovery is certainly more challenging than for satellite-transmitting tags but is still feasible for many elasmobranch species; it can also be highly cost effective because most loggers can be reused multiple times as long as they are recovered. There are also other technologies that could increase the likelihood of recovery, such as release systems that could be triggered with an acoustic signal from a vessel (Hammill et al., 1999). These would allow for tagging animals that range widely but return to a location at a predictable time (e.g., white sharks) (Jorgensen et al., 2010).

9.5.3 Data Analysis Limitations

Once the logger is recovered, the high sampling rates and large datasets (often several million data points per day) provide additional challenges for analysis. Although infrequent transmission of summarized data may be feasible for some research questions, the most valuable uses of these tags are currently only accessible by analyzing large amounts of high-resolution data.

These datasets often require specialized software more commonly used in acoustic wave analysis, as well as the use of time-series techniques such as fast Fourier transforms, power spectral analysis, and wavelet analysis (e.g., Nakamura et al., 2011; Whitney et al., 2010). One promising technique for behavioral analysis of large acceleration datasets is the use of wavelet analysis combined with k-means clustering (Sakamoto et al., 2009). This method uses wavelet analysis to estimate frequency and amplitude values for an entire dataset on a per-second basis. It then uses k-means clustering to group periods of similar frequency and amplitude into behavior spectra, each with their own frequency and amplitude characteristics. Because each second of data is assigned a behavior spectrum, the software is able to generate an ethogram showing which behaviors were exhibited at which times and then automatically calculate the percent time spent on each behavior (Figure 9.10). This technique is helpful not only for quantifying known behaviors but also for identifying new behaviors that have not been previously observed or detected. Additional techniques for quantifying and visualizing body orientation data from static acceleration are also in development (Grundy et al., 2009) and are likely to become commercially available in the near future.

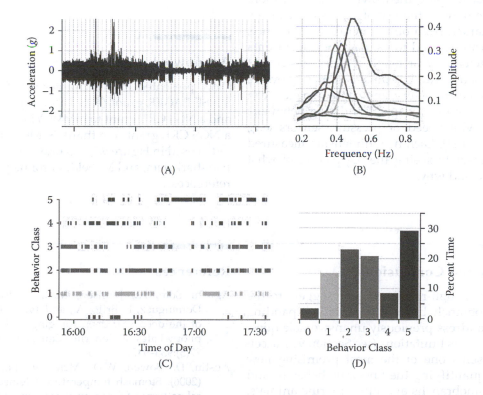

FIGURE 9.10
(See color insert.) Nurse shark behaviors classified from acceleration data using wavelet analysis and k-means clustering using Ethographer software (Japanese Society of Biologging Science; Sakamoto et al., 2009). A 100-minute period of swimming and resting is shown as raw acceleration data from the sway (lateral) axis (A). Behaviors were grouped into six clusters based on their tail-beat frequency and amplitude (B) and used to generate an ethogram (C) showing the animal's behavior on a per-second basis. The percent time spent on each behavior is automatically calculated from the ethogram and presented in a frequency histogram (D).

9.5.4 Future Directions

As the solutions and logistical advances described above make the application of these tags more feasible, future work will undoubtedly focus on increasing the integration of multiple sensors and devices to provide multivariate context to the aspects of behavior, physiology, and ecology that can be garnered from these devices. Although the use of data-logging tags in conjunction with conventional telemetry is a natural combination to provide spatial context to behavior and physiology, the use of multiple data loggers or single loggers with multiple sensors can provide true integration of spatial and behavioral data to generate a wealth of new discoveries. The integration of sensors such as swim speed, depth, compass heading, and acceleration in proprietary tags has been used for years for dead reckoning: reconstructing three-dimensional dive paths for marine animals (Johnson and Tyack, 2003; Wilson and Wilson, 1988; Wilson et al., 2007). These types of analyses can provide unparalleled information about how animals are using habitats and provide high-resolution information about behaviors performed at various points along the animal's movement path (Wilson et al., 2007). Tags that can be used for this purpose are now commercially available (e.g., Muramoto et al., 2004), with multiple manufacturers developing their own models. Of more immediate value may be the use of sensors to validate information obtained from other devices. For instance, accelerometers can be used to validate movement data obtained from telemetry or depth sensors, gastric pH loggers can be used to validate putative foraging events detected using accelerometery, and camera loggers can provide context to bursts of speed or diving behaviors detected from swim speed or pressure sensors (e.g., Nakamura et al., 2011). Each new dimension measured will bring us closer to seeing the big picture of what these animals do and why.

9.6 Summary and Conclusions

The application of high-resolution data loggers to the study of elasmobranch biology is rapidly expanding and helping to address previously unanswerable questions. Although it has limitations (see Section 9.5), accelerometry represents one of the most promising new techniques for quantifying the fine-scale behavior and activities of elasmobranchs and other marine animals. Whereas certain behaviors (resting, various types of swimming, diving, tail-beat activity, and body angles) can be quantified without additional confirmation, others (e.g., various reproductive behaviors, foraging) require validation of the data through direct observation or additional sensors. For these behaviors, studies on species that can be observed in their natural habitat or in which behaviors can be observed in captive animals are especially important. Even with validation of complex behaviors, ADLs are not able to directly measure the physiological changes required to confirm activities such as successful feeding events. To quantify changes in physiology associated with feeding, various types of gastric data loggers are available. Temperature loggers can be used to quantify feeding events in endothermic sharks, whereas pH and motility data loggers can be used to measure feeding time and relative meal size in ectothermic species. Jaw movement, sound, and visual images provide additional potential to quantify feeding events, although they are more likely to be used as a qualitative measure of feeding rather than to quantify meal type or size. The use of data loggers (particularly accelerometers) to quantify the absolute or relative energetic expenditure of free-living elasmobranchs may be their most exciting application. Although it has been applied to relatively few species thus far, this technique has the potential to be transformative in our understanding of how and where sharks choose to spend their time, as well as their energetic impact on the ecosystem.

Acknowledgments

Work presented in this chapter was funded in part by a NSF EAGER grant (IOS #1010567) to NMW and ACG and a NGS CRE grant to NMW. YPP was also funded by a NGS CRE grant. We thank S. Mulvany for the digital video used in Figure 9.3, J. Gathercole for work with captive shark data, and M. Feldner for help with figures and references.

References

Aguilar Soto, N., Johnson, M.P., Madsen, P.T., Díaz, F., Domínguez, I., Brito, A., and Tyack, P. (2008). Cheetahs of the deep sea: deep foraging sprints in short finned pilot whales off Tenerife (Canary Islands). *J. Anim. Ecol.* 77(5):936–947.

Austin, D., Bowen, W.D., McMillan, J.I., and Boness, D.J. (2006). Stomach temperature telemetry reveals temporal patterns of foraging success in a free ranging marine mammal. *J. Anim. Ecol.* 75(2):408–420.

Bainbridge, R. (1958). The speed of swimming of fish as related to size and to the frequency and amplitude of the tail beat. *J. Exp. Biol.* 35(1):109–133.

Bestley, S., Patterson, T.A., Hindell, M.A., and Gunn, J.S. (2008). Feeding ecology of wild migratory tunas revealed by archival tag records of visceral warming. *J. Anim. Ecol.* 77(6):1223–1233.

Bestley, S., Patterson, T.A., Hindell, M.A., and Gunn, J.S. (2010). Predicting feeding success in a migratory predator: integrating telemetry, environment, and modeling techniques. *Ecology* 91(8):2373–2384.

Brown, J.H., Gillooly, J.F., Allen, A.P., Savage, V.M., and West, G.B. (2004). Toward a metabolic theory of ecology. *Ecology* 85(7):1771–1789.

Butler, P.J., Green, J.A., Boyd, I.L., and Speakman, J.R. (2004). Measuring metabolic rate in the field: the pros and cons of the doubly labelled water and heart rate methods. *Funct. Ecol.* 18(2):168–183.

Campana, S.E., Joyce, W., and Manning, M.J. (2009). Bycatch and discard mortality in commercially caught blue sharks *Prionace glauca* assessed using archival satellite pop-up tags. *Mar. Ecol. Prog. Ser.* 387:241–253.

Carey, F.G. and Scharold, J.V. (1990). Movements of blue sharks (*Prionace glauca*) in depth and course. *Mar. Biol.* 106(3):329–342.

Carey, F.G., Teal, J.M., and Kanwisher, J.W. (1981). The visceral temperatures of mackerel sharks (Lamnidae). *Physiol. Zool.* 54(3):334–344.

Carey, F.G., Kanwisher, J.W., Brazier, O., Gabrielson, G., Casey, J.G., and Pratt, Jr., H.L. (1982). Temperature and activities of a white shark, *Carcharodon carcharias. Copeia* (2):254–260.

Carey, F.G., Kanwisher, J.W., and Stevens, E.D. (1984). Bluefin tuna warm their viscera during digestion. *J. Exp. Biol.* 109:1–20.

Carlson, J.K. and Parsons, G.R. (1999). Seasonal differences in routine oxygen consumption rates of the bonnethead shark. *J. Fish Biol.* 55(4):876–879.

Carlson, J.K. and Parsons, G.R. (2001). The effects of hypoxia on three sympatric shark species: physiological and behavioral responses. *Environ. Biol. Fish.* 61(4):427–433.

Carlson, J.K., Goldman, K.J., and Lowe, C.G. (2004). Metabolism, energetic demand, and endothermy. In: Carrier, J.C., Musick, J.A., and Heithaus, M.R. (Eds.), *Biology of Sharks and Their Relatives.* CRC Press, Boca Raton, FL, pp. 203–224.

Carrier, J.C., Pratt, Jr., H.L., and Martin, L.K. (1994). Group reproductive behaviour in free-living nurse sharks, *Ginglymostoma cirratum. Copeia* 1994:646–656.

Cartamil, D.P., Wegner, N.C., Aalbers, S., Sepulveda, C.A., Baquero, A., and Graham, J.B. (2010). Diel movement patterns and habitat preferences of the common thresher shark (*Alopias vulpinus*) in the Southern California Bight. *Mar. Freshw. Res.* 61:596–604.

Castro, J.I. (1988). The reproductive biology of the chain dogfish, *Scyliorhinus retifer. Copeia* 1988:740–746.

Clark, T.D., Sandblom, E., Hinch, S.G., Patterson, D.A., Frappell, P.B., and Farrell, A.P. (2010). Simultaneous biologging of heart rate and acceleration, and their relationships with energy expenditure in free-swimming sockeye salmon (*Oncorhynchus nerka*). *J. Comp. Physiol. B Biochem. Syst. Environ. Physiol.* 180(5):673–684.

Clarke, A. and Johnston, N.M. (1999). Scaling of metabolic rate with body mass and temperature in teleost fish. *J. Anim. Ecol.* 68(5):893–905.

Cooke, S.J., Bunt, C.M., Schreer, J.F., and Wahl, D.H. (2001). Comparison of several techniques for mobility and activity estimates of smallmouth bass in lentic environments. *J. Fish Biol.* 58(2):573–587.

Cooke, S.J., Hinch, S.G., Wikelski, M., Andrews, R.D., Kuchel, L.J. et al. (2004a). Biotelemetry: a mechanistic approach to ecology. *Trends Ecol. Evol.* 19(6):334–343.

Cooke, S.J., Thorstad, E.B., and Hinch, S.G. (2004b). Activity and energetics of free-swimming fish: insights from electromyogram telemetry. *Fish Fish.* 5(1):21–52.

Cortés, E., Papastamatiou, Y.P., Carlson, J.K., Ferry-Graham, L., Wetherbee, B.M. et al. (2008). An overview of the feeding ecology and physiology of elasmobranch fishes. In: Cyrino, J.E.P., Bureau, D.P., and Kapoor, B.G. (Eds.), *Feeding and Digestive Functions of Fishes.* Science Publishers, Enfield, NH, pp. 393–443.

Daly-Engel, T.S., Grubbs, R.D., Bowen, B.W., and Toonen, R.J. (2007). Frequency of multiple paternity in an unexploited tropical population of sandbar sharks (*Carcharhinus plumbeus*). *Can. J. Fish. Aquat. Sci.* 64:198–204.

Daly-Engel, T.S., Grubbs, R.D., Feldheim, K.A., Bowen, B.W., and Toonen, R.J. (2010). Is multiple mating beneficial or unavoidable? Low multiple paternity and genetic diversity in the shortspine spurdog *Squalus mitsukurii. Mar. Ecol. Prog. Ser.* 403:255–267.

Dewar, H., Deffenbaugh, M., Thurmond, G., Lashkari, K., and Block, B.A. (1999). Development of an acoustic telemetry tag for monitoring electromyograms in free-swimming fish. *J. Exp. Biol.* 202(19):2693–2699.

Di Santo, V. and Bennett, W.A. (2011). Is post feeding thermotaxis advantageous in elasmobranch fishes? *J. Fish Biol.* 78(1):195–207.

DiBattista, J.D., Feldheim, K.A., Gruber, S.H., and Hendry, A.P. (2008). Are indirect genetic benefits associated with polyandry? Testing predictions in a natural population of lemon sharks. *Mol. Ecol.* 17:783–795.

Dickson, K.A., Donley, J.M., Sepulveda, C.A., and Bhoopat, L. (2002). Effects of temperature on sustained swimming performance and swimming kinematics of the chub mackerel *Scomber japonicus. J. Exp. Biol.* 205(7):969–980.

Dowd, W.W., Brill, R.W., Bushnell, P.G., and Musick, J.A. (2006). Estimating consumption rates of juvenile sandbar sharks (*Carcharhinus plumbeus*) in Chesapeake Bay, Virginia, using a bioenergetics model. *Fish. Bull.* 104(3):332–342.

DuBois, A.B. and Ogilvy, C.S. (1978). Forces on the tail surface of swimming fish: thrust, drag and acceleration in bluefish (*Pomatomus saltatrix*). *J. Exp. Biol.* 77:225–241.

DuBois, A.B., Cavagna, G.A., and Fox, R.S. (1976). Locomotion of bluefish. *J. Exp. Zool.* 195(2):223–235.

Estes, J.A., Riedman, M.L., Staedler, M.M., Tinker, M.T., and Lyon, B.E. (2003). Individual variation in prey selection by sea otters: patterns, causes and implications. *J. Anim. Ecol.* 72(1):144–155.

Fitzpatrick, R., Abrantes, K.G., Seymour, J., and Barnett, A. (In press). Activity patterns of whitetip reef sharks: do shark feeds change their behavior? *Environ. Biol. Fish.*

Gillooly, J.F., Brown, J.H., West, G.B., Savage, V.M., and Charnov, E.L. (2001). Effects of size and temperature on metabolic rate. *Science* 293(5538):2248.

Gleiss, A.C., Gruber, S.H., and Wilson, R.P. (2009a). Multi-channel data-logging: towards determination of behaviour and metabolic rate in free-swimming sharks. In: Nielsen, J.L., Arrizabalaga, H., Fragoso, N., Hobday, A., Lutcavage, M., and Sibert, J. (Eds.), *Tagging and Tracking of Marine Animals with Electronic Devices*, Springer, New York, pp. 211–228.

Gleiss, A.C., Norman, B., Liebsch, N., Francis, C., and Wilson, R.P. (2009b). A new prospect for tagging large free-swimming sharks with motion-sensitive data-loggers. *Fish. Res.* 97(1–2):11–16.

Gleiss, A.C., Dale, J.J., Holland, K.N., and Wilson, R.P. (2010a). Accelerating estimates of activity-specific metabolic rate in fish: testing the applicability of acceleration data-loggers. *J. Exp. Mar. Biol. Ecol.* 385(1–2):85–91.

Gleiss, A.C., Norman, B., and Wilson, R.P. (2010b). Moved by that sinking feeling: variable energetic costs of diving geometry underlie movement strategies in the whale shark. *Funct. Ecol.* 25(3):595–607.

Gleiss, A.C., Wilson, R.P., and Shepard, E.L.C. (2010c). Making overall dynamic body acceleration work: on the theory of acceleration as a proxy for energy expenditure. *Meth. Ecol. Evol.* 2(1):23–33.

Goldman, K.J., Anderson, S.D., Latour, R.J., and Musick, J.A. (2004). Homeothermy in adult salmon sharks, *Lamna ditropis*. *Environ. Biol. Fish.* 71(4):403–411.

Graham, J.B., Dewar, H., Lai, N.C., Lowell, W.R., and Arce, S.M. (1990). Aspects of shark swimming performance determined using a large water tunnel. *J. Exp. Biol.* 151:175–192.

Green, J.A. (2010). The heart rate method for estimating metabolic rate: review and recommendations. *Comp. Biochem. Physiol. A Mol. Integr. Physiol.* 158(3):287–304.

Gruber, S.H., Nelson, D.R., and Morrissey, J.F. (1988). Patterns of activity and space utilization of lemon sharks, *Negaprion brevirostris*, in a shallow Bahamian lagoon. *Bull. Mar. Sci.* 43:61–76.

Grundy, E., Jones, M.W., Laramee, R.S., Wilson, R.P., and Shepard, E.L.C. (2009). Visualisation of sensor data from animal movement. *Comput. Graph. Forum* 28(3):815–822.

Gunn, J., Stevens, J.D., Davis, T.L.O., and Norman, B.M. (1999). Observations on the short-term movements and behaviour of whale sharks (*Rhincodon typus*) at Ningaloo Reef, Western Australia. *Mar. Biol.* 135(3):553–559.

Gunn, J., Hartog, J., and Rough, K. (2001). The relationship between food intake and visceral warming in southern bluefin tuna (*Thunnus maccoyii*). In: Sibert, J.R. and Nielsen, J.L. (Eds.), *Electronic Tagging and Tracking in Marine Fisheries*, Vol. 1. Kluwer, Dordrecht, pp. 109–130.

Guttridge, T.L., Gruber, S.H., Krause, J., Sims, D.W., and Goldstien, S.J. (2010). Novel acoustic technology for studying free-ranging shark social behaviour by recording individuals' interactions. *PloS ONE* 5(2):e9324.

Halsey, L.G., Green, J.A., Wilson, R.P., and Frappell, P.B. (2008). Accelerometry to estimate energy expenditure during activity: best practice with data loggers. *Physiol. Biochem. Zool.* 82(4).

Halsey, L.G., Shepard, E.L.C., Quintana, F., Gomez Laich, A., Green, J.A., and Wilson, R.P. (2009). The relationship between oxygen consumption and body acceleration in a range of species. *Comp. Biochem. Physiol. A Mol. Integr. Physiol.* 152(2):197–202.

Halsey, L.G., Shepard, E.L.C., and Wilson, R.P. (2011). Assessing the development and application of the accelerometry technique for estimating energy expenditure. *Comp. Biochem. Physiol. A Mol. Integr. Physiol.* 158(3):305–314.

Hammill, M.O., Lesage, V., Lobb, G., Carter, P., and Voegeli, F. (1999). A remote release mechanism to recover time–depth recorders from marine mammals. *Mar. Mam. Sci.* 15:584–588.

Heithaus, M.R. (2004). Predator–prey interactions. In: Carrier, J.C., Musick, J.A., and Heithaus, M.R. (Eds.), *Biology of Sharks and Their Relatives*. CRC Press, Boca Raton, FL, pp. 487–523.

Heithaus, M.R., Marshall, G.J., Buhleier, B.M., and Dill, L.M. (2001). Employing Crittercam to study habitat use and behavior of large sharks. *Mar. Ecol. Prog. Ser.* 209:307–310.

Heithaus, M.R., Dill, L.M., Marshall, G.J., and Buhleier, B. (2002). Habitat use and foraging behavior of tiger sharks (*Galeocerdo cuvier*) in a seagrass ecosystem. *Mar. Biol.* 140(2):237–248.

Heithaus, M.R., Frid, A., Wirsing, A.J., and Worm, B. (2008). Predicting ecological consequences of marine top predator declines. *Trends Ecol. Evol.* 23(4):202–210.

Heithaus, M.R., Frid, A., Vaudo, J.J., Worm, B., and Wirsing, A.J. (2010). Unraveling the ecological importance of elasmobranchs. In: Carrier, J., Musick, J.A., and Heithaus, M. (Eds.), *Sharks and Their Relatives II: Biodiversity, Adaptive Physiology, and Conservation*. CRC Press, Boca Raton, FL, pp. 611–638.

Heupel, M.R., Carlson, J.K., and Simpfendorfer, C.A. (2007). Shark nursery areas: concepts, definition, characterization and assumptions. *Mar. Ecol. Prog. Ser.* 337:287–297.

Hines, A.H., Whitlatch, R.B., Thrush, S.F., Hewitt, J.E., Cummings, V.J., Dayton, P.K., and Legendre, P. (1997). Nonlinear foraging response of a large marine predator to benthic prey: eagle ray pits and bivalves in a New Zealand sandflat. *J. Exp. Mar. Biol. Ecol.* 216(1–2):191–210.

Hoffmayer, E.R. and Parsons, G.R. (2001). The physiological response to capture and handling stress in the Atlantic sharpnose shark, *Rhizoprionodon terraenovae*. *Fish Physiol. Biochem.* 25(4):277–285.

Holland, K.N., Meyer, C.G., and Dagorn, L.C. (2009). Inter-animal telemetry: results from first deployment of acoustic 'business card' tags. *Endangered Species Res.* 10:287–293.

Holmgren, S. and Holmberg, A. (2005). Control of gut motility and secretion in fasting and fed non-mammalian vertebrates. In: Starck, J.M. and Wang, T. (Eds.), *Physiological and Ecological Adaptations to Feeding in Vertebrates*. Science Publishers, Enfield, NH, pp. 325–362.

Hume, I.D. (2005). Concepts of digestive efficiency. In: Starck, J.M. and Wang, T. (Eds.), *Physiological and Ecological Adaptations to Feeding in Vertebrates*. Science Publishers, Enfield, NH, pp. 43–58.

Insley, S.J., Robson, B.W., and Burgess, W.C. (2004). Acoustic monitoring of northern fur seals in the Bering Sea. *J. Acoust. Soc. Am.* 116(4):2555.

Jacoby, D.M.P., Busawon, D.S., and Sims, D.W. (2010). Sex and social networking: the influence of male presence on social structure of female shark groups. *Behav. Ecol.* 21(4):808–818.

Jobling, M.X. (1994). *Fish Bioenergetics*. Chapman & Hall, London.

Johnson, M.P. and Tyack, P.L. (2003). A digital acoustic recording tag for measuring the response of wild marine mammals to sound. *IEEE J. Oceanic Eng.* 28:3–12.

Jorgensen, S.J., Reeb, C.A., Chapple, T.K., Anderson, S., Perle, C., Van Sommeran, S.R., Fritz-Cope, C., Brown, A.C., Klimley, A.P., and Block, B.A. (2010). Philopatry and migration of Pacific white sharks. *Proc. R. Soc. B Biol. Sci.* 277:679–688.

Kawabe, R., Naito, Y., Sato, K., Miyashita, K., and Yamashita, N. (2004). Direct measurement of the swimming speed, tailbeat, and body angle of Japanese flounder (*Paralichthys olivaceus*). *ICES J. Mar. Sci.* 61(7):1080–1087.

Killen, S.S., Atkinson, D., and Glazier, D.S. (2010). The intraspecific scaling of metabolic rate with body mass in fishes depends on lifestyle and temperature. *Ecol. Lett.* 13(2):184–193.

Klimley, A.P. (1993). Highly directional swimming by scalloped hammerhead sharks, *Sphyrna lewini*, and subsurface irradiance, temperature, bathymetry, and geomagnetic field. *Mar. Biol.* 117(1):1–22.

Klimley, A.P., Le Boeuf, B.J., Cantara, K.M., Richert, J.E., Davis, S.F., Van Sommeran, S., and Kelly, J.T. (2001). The hunting strategy of white sharks (*Carcharodon carcharias*) near a seal colony. *Mar. Biol.* 138(3):617–636.

Liebsch, N., Wilson, R.P., Bornemann, H., Adelung, D., and Plotz, J. (2007). Mouthing off about fish capture: jaw movement in pinnipeds reveals the real secrets of ingestion. *Deep Sea Res. II* 54(3–4):256–269.

Lowe, C.G. (2002). Bioenergetics of free-ranging juvenile scalloped hammerhead sharks (*Sphyrna lewini*) in Kane'ohe Bay, O'ahu, HI. *J. Exp. Mar. Biol. Ecol.* 278(2):141–156.

Lowe, C.G. and Goldman, K.J. (2001). Thermal and bioenergetics of elasmobranchs: bridging the gap. *Environ. Biol. Fish.* 60:251–266.

Lowe, C.G., Holland, K.N., and Wolcott, T.G. (1998). A new acoustic tailbeat transmitter for fishes. *Fish. Res.* 36(2–3):275–283.

Manire, C., Hueter, R., Hull, E., and Spieler, R. (2001). Serological changes associated with gill-net capture and restraint in three species of sharks. *Trans. Am. Fish. Soc.* 130:1038–1048.

Marshall, G., Bakhtiari, M., Shepard, M., Tweedy, J., Rasch, D., Abernathy, K., Joliff, B., Carrier, J.C., and Heithaus, M.R. (2007). An advanced solid-state animal-borne video and environmental data-logging device ("crittercam") for marine research. *Mar. Technol. Soc. J.* 41(2):31–38.

Matern, S.A., Cech, Jr., J.J., and Hopkins, T.E. (2000). Diel movements of bat rays, *Myliobatis californica*, in Tomales Bay, California: evidence for behavioral thermoregulation? *Environ. Biol. Fish.* 58(2):173–182.

Matich, P., Heithaus, M.R., and Layman, C.A. (2011). Contrasting patterns of individual specialization and trophic coupling in two marine apex predators. *J. Anim. Ecol.* 80(1):294–305.

McCosker, J.E. (1987). The white shark, *Carcharodon carcharias*, has a warm stomach. *Copeia* 1987(1):195–197.

Metcalfe, J.D., Fulcher, M.C., Clarke, S.R., Challiss, M.J., and Hetherington, S. (2009). An archival tag for monitoring key behaviours (feeding and spawning) in fish. In: Nielsen, J.L. (Ed.), *Tagging and Tracking of Marine Animals with Electronic Devices*, Springer, Berlin, pp. 243–254.

Meyer, C.G., Burgess, W.C., Papastamatiou, Y.P., and Holland, K.N. (2007). Use of an implanted sound recording device (Bioacoustic Probe) to document the acoustic environment of a blacktip reef shark (*Carcharhinus melanopterus*). *Aquat. Living Resour.* 20(4):291–298.

Moll, R.J., Millspaugh, J.J., Beringer, J., Sartwell, J., and He, Z. (2007). A new 'view' of ecology and conservation through animal-borne video systems. *Trends Ecol. Evol.* 22:660–668.

Motta, P.J. (2004). Prey capture behavior and feeding mechanics of elasmobranchs. In: Carrier, J.C., Musick, J.A., and Heithaus, M.R. (Eds.), *Biology of Sharks and Their Relatives*. CRC Press, Boca Raton, FL, pp. 165–202.

Motta, P.J., Maslanka, M., Hueter, R.E., Davis, R.L., de la Parra, R. et al. (2010). Feeding anatomy, filter-feeding rate, and diet of whale sharks, *Rhincodon typus*, during surface ram filter feeding off the Yucatan Peninsula, Mexico. *Zoology* 113(4):199–212.

Moyes, C.D., Fragoso, N., Musyl, M.K., and Brill, R.W. (2006). Predicting post release survival in large pelagic fish. *Trans. Am. Fish. Soc.* 135:1389–1397.

Muramoto, H., Ogawa, M., Suzuki, M., and Naito, Y. (2004). Little Leonardo digital data logger: its past, present, and future role in bio-logging science. *Mem. Natl. Inst. Polar Res. Spec. Issue* 58:196–202.

Murchie, K.J., Cooke, S.J., Danylchuk, A.J., and Suski, C.D. (2011). Estimates of field activity and metabolic rates of bonefish (*Albula vulpes*) in coastal marine habitats using acoustic tri-axial accelerometer transmitters and intermittent-flow respirometry. *J. Exp. Mar. Biol. Ecol.* 396(2):147–155.

Myrberg, A.A. (2001). The acoustical biology of elasmobranchs. *Environ. Biol. Fish.* 60(1):31–46.

Nagy, K.A., Huey, R.B., and Bennett, A.F. (1984). Field energetics and foraging mode of Kalahari lacertid lizards. *Ecology* 65(2):588–596.

Naito, Y. (2007). How can we observe the underwater feeding behavior of endotherms? *Polar Sci.* 1(2–4):101–111.

Nakamura, I., Watanabe, Y., Papastamatiou, Y.P., Sato, K., and Meyer, C.G. (2011). Yo-yo vertical movements suggest a foraging strategy for tiger sharks *Galeocerdo cuvier*. *Mar. Ecol. Prog. Ser.* 424:237–246.

Nelson, D.R. (1967). Hearing thresholds, frequency discrimination, and acoustic orientation in the lemon shark, *Negaprion brevirostris* (Poey). *Bull. Mar. Sci.* 17(3):741–768.

Nelson, D.R. (1990) Telemetry studies of sharks: a review, with applications in resource management. In: Pratt, Jr., H.L., Gruber, S.H., and Taniuchi, T. (Eds.), *Elasmobranchs as Living Resources: Advances in the Biology, Ecology, Systematics, and the Status of the Fisheries*, NOAA Tech. Rep. NMFS 90. U.S. Department of Commerce, Washington, D.C., pp. 239–256.

Nelson, D.R. and Gruber, S.H. (1963). Sharks: attraction by low-frequency sounds. *Science* 142:975–977.

Nelson, J.D. and Eckert, S.A. (2007). Foraging ecology of whale sharks (*Rhincodon typus*) within Bahia de Los Angeles, Baja California Norte, Mexico. *Fish. Res.* 84(1):47–64.

Nixon, A.J. and Gruber SH. (1988). Diel metabolic and activity patterns of the lemon shark (*Negaprion brevirostris*). *J. Exp. Zool.* 248(1):1–6.

O'Toole, A.C., Murchie, K.J., Pullen, C., Hanson, K.C., Suski, C.D., Danylchuk, A.J., and Cooke, S.J. (2010). Locomotory activity and depth distribution of adult great barracuda (*Sphyraena barracuda*) in Bahamian coastal habitats determined using acceleration and pressure biotelemetry transmitters. *Mar. Freshw. Res.* 61:1446–1456.

Papastamatiou, Y.P. and Lowe, C.G. (2004). Postprandial response of gastric pH in leopard sharks (*Triakis semifasciata*) and its use to study foraging ecology. *J. Exp. Biol.* 207(2):225–232.

Papastamatiou, Y.P. and Lowe, C.G. (2005). Variations in gastric acid secretion during periods of fasting between two species of shark. *Comp. Biochem. Physiol. A* 141(2):210–214.

Papastamatiou, Y.P., Purkis, S.J., and Holland, K.N. (2007). The response of gastric pH and motility to fasting and feeding in free swimming blacktip reef sharks, *Carcharhinus melanopterus*. *J. Exp. Mar. Biol. Ecol.* 345(2):129–140.

Papastamatiou, Y.P., Meyer, C.G., and Holland, K.N. (2008). A new acoustic pH transmitter for studying the feeding habits of free-ranging sharks. *Aquat. Living Resour.* 20(4):287–290.

Parrish, F.A., Marshall, G.J., Buhleier, B., and Antonelis, G.A. (2008). Foraging interaction between monk seals and large predatory fish in the northwestern Hawaiian Islands. *Endangered Species Res.* 4:299–308.

Payne, N.L., Gillanders, B.M., Seymour, R.S., Webber, D.M., Snelling, E.P., and Semmens, J.M. (2010). Accelerometry estimates field metabolic rate in giant Australian cuttlefish *Sepia apama* during breeding. *J. Anim. Ecol.* 80(2):422–430.

Peters, G. (1997a). A new device for monitoring gastric pH in free-ranging animals. *Am. J. Physiol. Gastrointest. Liver Physiol.* 273(3):G748–G753.

Peters, G. (1997b). A reference electrode with free-diffusion liquid junction for electrochemical measurements under changing pressure conditions. *Anal. Chem.* 69(13):2362–2366.

Peters, G. (2004). Measurement of digestive variables in free-living animals: gastric motility in penguins during foraging. *Mem. Natl. Inst. Polar Res. Spec. Issue* 58:203–209.

Portnoy, D.S., Piercy, A.N., Musick, J.A., Burgess, G.H., and Graves, J.E. (2007). Genetic polyandry and sexual conflict in the sandbar shark, *Carcharhinus plumbeus*, in the western North Atlantic and Gulf of Mexico. *Mol. Ecol.* 16:187–197.

Pratt, Jr., H.L. and Carrier, J.C. (2001). A review of elasmobranch reproductive behavior with a case study on the nurse shark, *Ginglymostoma cirratum*. *Environ. Biol. Fish.* 60:157–188.

Ropert-Coudert, Y. and Wilson, R.P. (2005). Trends and perspectives in animal-attached remote sensing. *Front. Ecol. Environ.* 3(8): 437–444.

Ropert-Coudert, Y., Beaulieu, M., Hanuise, N., and Kato, A. (2009). Diving into the world of biologging. *Endangered Species Res.* 10:21–27.

Sakamoto, K.Q., Sato, K., Ishizuka, M., Watanuki, Y., Takahashi, A., Daunt, F., and Wanless, S. (2009). Can ethograms be automatically generated using body acceleration data from free-ranging birds? *PLoS ONE* 4(4):e5379.

Sasko, D.E. (2000). The Prey Capture Behavior of the Atlantic Cownose Ray, *Rhinoptera bonasus*, master's thesis, University of South Florida, Tampa.

Sato, K., Watanuki, Y., Takahashi, A., Miller, P.J.O., Tanaka, H. et al. (2007). Stroke frequency, but not swimming speed, is related to body size in free-ranging seabirds, pinnipeds and cetaceans. *Proc. R. Soc. B Biol. Sci.* 274(1609):471–477.

Schaller, P. (2004). Observation of whitetip reef shark (*Triaenodon obesus*) parturition in captivity. *Drum Croaker* 35:13–17.

Scharold, J. and Gruber, S.H. (1991). Telemetered heart rate as a measure of metabolic rate in the lemon shark, *Negaprion brevirostris*. *Copeia* 1991:942–953.

Sciarrotta, T.C. and Nelson, D.R. (1977). Diel behavior of the blue shark, *Prionace glauca*, near Santa Catalina Island, California. *Fish. Bull.* 75(3):519–528.

Sepulveda, C.A., Kohin, S., Chan, C., Vetter, R., and Graham, J.B. (2004). Movement patterns, depth preferences, and stomach temperatures of free-swimming juvenile mako sharks, *Isurus oxyrinchus*, in the Southern California Bight. *Mar. Biol.* 145(1):191–199.

Shepard, E.L.C., Ahmed, M.Z., Southall, E.J., Witt, M.J., Metcalfe, J.D., and Sims, D.W. (2006). Diel and tidal rhythms in diving behaviour of pelagic sharks identified by signal processing of archival tagging data. *Mar. Ecol. Prog. Ser.* 328:205–213.

Shepard, E.L.C., Wilson, R.P., Quintana, F., Laich, A.G. et al. (2008). Identification of animal movement patterns using tri-axial accelerometry. *Endangered Species Res.* 10:47–60.

Shepard, E.L.C., Wilson, R.P., Halsey, L.G., Quintana, F., Laich, A.G., Gleiss, A.C., Liebsch, N, Myers, A.E., and Norman, B. (2009a). Derivation of body motion via appropriate smoothing of acceleration data. *Aquat. Biol.* 4(3):235–241.

Shepard, E.L.C., Wilson, R.P., Quintana, F., Gómez Laich, A., and Forman, D.W. (2009b). Pushed for time or saving on fuel: fine-scale energy budgets shed light on currencies in a diving bird. *Proc. R. Soc. B Biol. Sci.* 276(1670):3149.

Simpfendorfer, C.A. and Heupel, M.R. (2004). Assessing habitat use and movement. In: Carrier, J.C., Musick, J.A., and Heithaus, M.R. (Eds.), *Biology of Sharks and Their Relatives*. CRC Press, Boca Raton, FL, pp. 553–572.

Sims, D.W. (1996). The effect of body size on the standard metabolic rate of the lesser spotted dogfish. *J. Fish Biol.* 48(3):542–544.

Sims, D.W. (2010). Tracking and analysis techniques for understanding free-ranging shark movements and behaviour. In: Carrier, J., Musick, J.A., and Heithaus, M. (Eds.), *Sharks and Their Relatives II: Biodiversity, Adaptive Physiology, and Conservation*. CRC Press, Boca Raton, FL, pp. 341–392.

Sims, D.W., Davies, S.J., and Bone, Q. (1996). Gastric emptying rate and return of appetite in lesser spotted dogfish, *Scyliorhinus canicula* (Chondrichthyes: Elasmobranchii). *J. Mar. Biol. Assoc. U.K.* 76(2):479–491.

Sims, D.W., Southall, E.J., Wearmouth, V.J., Hutchinson, N., Budd, G.C., and Morritt, D. (2005). Refuging behaviour in the nursehound *Scyliorhinus stellaris* (Chondrichthyes: Elasmobranchii): preliminary evidence from acoustic telemetry. *J. Mar. Biol. Assoc. U.K.* 85(5):1137–1140.

Sims, D.W., Wearmouth, V.J., Southall, E.J., Hill, J.M., Moore, P. et al. (2006). Hunt warm, rest cool: bioenergetic strategy underlying diel vertical migration of a benthic shark. *J. Anim. Ecol.* 75(1):176–190.

Sims, D.W., Southall, E.J., Humphries, N.E., Hays, G.C., Bradshaw, C.J.A. et al. (2008). Scaling laws of marine predator search behaviour. *Nature* 451(7182):1098–1095.

Skomal, G.B. (2006). The Physiological Effects of Capture Stress on Post-Release Survivorship of Sharks, Tunas, and Marlin, doctoral dissertation, Boston University.

Skomal, G.B. and Bernal, D. (2010). Physiological responses to stress in sharks. In: Carrier, J.C., Musick, J.A., and Heithaus, M.R. (Eds.), *Sharks and Their Relatives*. CRC Press, Boca Raton, FL, pp. 459–490.

Skomal, G.B. and Chase, B.C. (2002). The physiological effects of angling on post-release survivorship in tunas, sharks, and marlin. *Am. Fish. Soc. Symp.* 30:135–138.

Springer, S. (1967). Social organization of shark populations. In: Gilbert, P.W. et al. (Eds.), *Sharks, Skates, and Rays*. The Johns Hopkins University Press, Baltimore, MD, pp. 149–174.

Sundstrom, L.F. and Gruber, S.H. (1998). Using speed-sensing transmitters to construct a bioenergetics model for sub-adult lemon sharks, *Negaprion brevirostris* (Poey), in the field. *Hydrobiologia* 372:241–247.

Sundstrom, L.F., Gruber, S.H., Clermont, S.M., Correia, J.P.S., de Marignac, J.R.C. et al. (2001). Review of elasmobranch behavioral studies using ultrasonic telemetry with special reference to the lemon shark, *Negaprion brevirostris*, around Bimini Islands, Bahamas. *Environ. Biol. Fish.* 60(1–3):225–250.

Tanaka, H., Takagi, Y., and Naito, Y. (2001). Swimming speeds and buoyancy compensation of migrating adult chum salmon *Oncorhynchus keta* revealed by speed/depth/acceleration data logger. *J. Exp. Biol.* 204:3895–3904.

Thorarensen, H., Gallaugher, P.E., and Farrell, A.P. (1996). The limitations of heart rate as a predictor of metabolic rate in fish. *J. Fish Biol.* 49(2):226–236.

Thrush, S.F., Pridmore, R.D., Hewitt, J.E., and Cummings, V.J. (1991). Impact of ray feeding disturbances on sandflat macrobenthos: do communities dominated by polychaetes or shellfish respond differently? *Mar. Ecol. Prog. Ser.* 69(3):245–252.

Uchida, S., Toda, M., and Kamei, Y. (1990). Reproduction of elasmobranchs in captivity. In: Pratt, Jr., H.L., Gruber, S.H., and Taniuchi, T. (Eds.), *Elasmobranchs as Living Resources: Advances in the Biology, Ecology, Systematics, and the Status of the Fisheries*, NOAA Tech. Rep. NMFS 90. U.S. Department of Commerce, Washington, D.C., pp. 211–237.

Voegeli, F.A., Smale, M.J., Webber, D.M., Andrade, Y., and O'Dor, R.K. (2001). Ultrasonic telemetry, tracking and automated monitoring technology for sharks. *Environ. Biol. Fish.* 60(1):267–282.

Watanabe, Y. and Sato, K. (2008). Functional dorsoventral symmetry in relation to lift-based swimming in the ocean sunfish *Mola mola*. *PLoS ONE* 3(10):1–7.

Wearmouth, V.J. and Sims, D.W. (2009). Movement and behaviour patterns of the critically endangered common skate *Dipturus batis* revealed by electronic tagging. *J. Exp. Mar. Biol. Ecol.* 380:77–87.

Webber, D.M., Boutilier, R.G., Kerr, S.R., and Smale, M.J. (2001). Caudal differential pressure as a predictor of swimming speed of cod (*Gadus morhua*). *J. Exp. Biol.* (20):3561–3570.

Weihs, D. (1973). Mechanically efficient swimming techniques for fish with negative buoyancy. *J. Mar. Res.* 31:194–209.

Wetherbee, B.M. and Cortés, E. (2004). Food consumption and feeding habits. In: Carrier, J.C., Musick, J.A., and Heithaus, M.R. (Eds.), *Biology of Sharks and Their Relatives*. CRC Press, Boca Raton, FL, pp. 223–244.

Whitney, N.M., Pratt, Jr., H.L., and Carrier, J.C. (2004). Group courtship, mating behaviour, and siphon sac function in the whitetip reef shark, *Triaenodon obesus*. *Anim. Behav.* 68(6):1435–1442.

Whitney, N.M., Papastamatiou, Y.P., Holland, K.N., and Lowe, C.G. (2007). Use of an acceleration data logger to measure diel activity patterns in captive whitetip reef sharks, *Triaenodon obesus*. *Aquat. Living Resour.* 20(4):299–305.

Whitney, N.M., Pratt, Jr., H.L., and Carrier, J.C. (2010). Identifying shark mating behaviour using three-dimensional acceleration loggers. *Endangered Species Res.* 10:71–82.

Wilga, C.D. and Motta, P.J. (2000). Durophagy in sharks: feeding mechanics of the hammerhead *Sphyrna tiburo*. *J. Exp. Biol.* 203(Pt. 18):2781–2796.

Wilson, R.P. and Wilson, M.P. (1988). Dead reckoning: a new technique for determining penguin movements at sea. *Meeresforschung/Rep. Mar. Res.* 32:155–158.

Wilson, R.P., Cooper, J., and Plötz, J. (1992). Can we determine when marine endotherms feed? A case study with seabirds. *J. Exp. Biol.* 167:267–275.

Wilson, R.P., Steinfurth, A., Ropert-Coudert, Y., Kato, A., and Kurita, M. (2002). Lip-reading in remote subjects: an attempt to quantify and separate ingestion, breathing and vocalisation in free-living animals using penguins as a model. *Mar. Biol.* 140(1):17–27.

Wilson, R.P., Scolaro, A., Quintana, F., Siebert, U., Straten, M. et al. (2004). To the bottom of the heart: cloacal movement as an index of cardiac frequency, respiration and digestive evacuation in penguins. *Mar. Biol.* 144:813–827.

Wilson, R.P., White, C.R., Quintana, F., Halsey, L.G., Liebsch, N., Martin, G.R., and Butler, P.J. (2006). Moving towards acceleration for estimates of activity-specific metabolic rate in free-living animals: the case of the cormorant. *J. Anim. Ecol.* 75:1081–1090.

Wilson, R.P., Liebsch, N., Davies, I.M., Quintana, F., Weimerskirch, H. et al. (2007). All at sea with animal tracks: methodological and analytical solutions for the resolution of movement. *Deep Sea Res. II* 54:193–210.

Wilson, R.P., Shepard, E.L.C., and Liebsch, N. (2008). Prying into the intimate details of animal lives: use of a daily diary on animals. *Endangered Species Res.* 4(1–2):123–137.

Wilson, R.P., Shepard, E.L.C., Gomez Laich, A., Frere, E., and Quintana, F. (2010). Pedalling downhill and freewheeling up; a penguin perspective on foraging. *Aquat. Biol.* 8(3):193–202.

Woo, K.J., Elliott, K.H., Davidson, M., Gaston, A.J., and Davoren, G.K. (2008). Individual specialization in diet by a generalist marine predator reflects specialization in foraging behaviour. *J. Anim. Ecol.* 77(6):1082–1091.

Wood, C.M., Kajimura, M., Bucking, C., and Walsh, P.J. (2007). Osmoregulation, ionoregulation and acid–base regulation by the gastrointestinal tract after feeding in the elasmobranch (*Squalus acanthias*). *J. Exp. Biol.* 210(Pt. 8):1335.

Wright, S., Norman, B., Liebsch, N., Wilson, R.P., and Gleiss, A.C. (submitted). Anticipating your next meal: mismatch in locomotory activity and vertical movement in whale sharks at Ningaloo Reef, Australia. *Aquat. Biol.*

Yoda, K., Naito, Y., Sato, K., Takahashi, A., Nishikawa, J., Ropert-Coudert, Y., Kurita, M., and Le Maho, Y. (2001). A new technique for monitoring the behaviour of free-ranging Adelie penguins. *J. Exp. Biol.* 204(4):685–690.

10

Reproductive Biology of Elasmobranchs

Christina L. Conrath and John A. Musick

CONTENTS

10.1 Introduction

The species that comprise class Chondrichthyes exhibit a wide diversity of reproductive strategies, but they all have internal fertilization and a relatively small number of large offspring. Although the Chondrichthyes are comprised of two sister taxa (Elasmobranchii and Holocephali), the number of extant elasmobranchs is considerably larger, and this chapter emphasizes the reproductive biology of this group. The diversity of reproductive modes found within the elasmobranchs is often reflected in reproductive anatomy and tends to be related to how the embryo receives nourishment. Historically,

elasmobranch reproductive modes were divided into oviparity, aplacental viviparity, and placental viviparity; however, the aplacental viviparous group is problematic, as it includes a diversity of unrelated reproductive strategies and this classification will not be used. Reproductive modes in this chapter are distinguished based on the maternal contribution to development. Recent research that emphasizes the importance of maternal contributions via mucoid histotroph and the continuum in reproductive strategies from yolk-sac viviparity to oviparity and mucoid histotrophy to placentotrophy and oophagy will be discussed. This research suggests that dividing species into discrete modes of reproduction while convenient may not always be realistic.

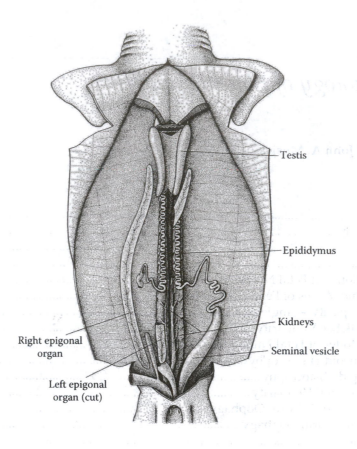

FIGURE 10.1
Male reproductive system. (Adapted from Castro, J.I., *The Sharks of North American Waters*, Texas A&M University Press, College Station, 1983. With permission.)

This chapter discusses the reproductive anatomy of elasmobranchs, the modes of reproduction, reproductive cycles, and behavior exhibited by this group. Recent developments within this field include new evidence about the pleisiomorphic condition of the elasmobranchs, examinations of multiple paternity and sexual conflict, and the recent confirmation of reproductive plasticity within some populations.

10.2 Reproductive Anatomy

10.2.1 Male

The male elasmobranch reproductive system is comprised of the testes, ductus efferens, epididymis, ductus deferens, seminal vesicle, Leydig glands, and alkaline gland (Hamlett, 1999) (Figure 10.1). The testes are paired organs suspended by a mesorchium and in some species enveloped by the epigonal organ (Wourms, 1977). Spermatogenesis occurs within the testes, and the testes appear to have a role in creating

and secreting steroid hormones (Hamlett, 1999). Three types of elasmobranch testes have been identified with differing patterns of seminiferous follicle origin and propagation: diametric, radial, and compound (Figure 10.2). Within the diametric testis follicle development occurs along the cross-sectional width of the testis, whereas in the radial testis follicle development occurs in multiple lobes of the testis from the central germinal zone of the lobe toward the circumference of the lobe. The compound testis is comprised of follicles that develop in both radial and diametric directions (Engel and Callard, 2005; Pratt, 1988). Mature sperm travels through the ductus efferens located at the anterior end of the mesorchium into the epididymis. The epididymis is a convoluted and complex tubule or mass of tubules that may function in protein secretion (Hamlett, 1999). The epididymis is continuous with the ductus deferens and the seminal vesicle. The final sperm products are formed into clumps in the ductus deferens and the seminal vesicle (Hamlett, 1999; Wourms, 1977). These sperm clumps may be simply aggregated, aggregated with sperm tails protruding (spermatozeugmata), or encapsulated (spermatophores) (Jamieson, 2005; Jones et al., 2005). Squalomorphii and Heterodontiformes

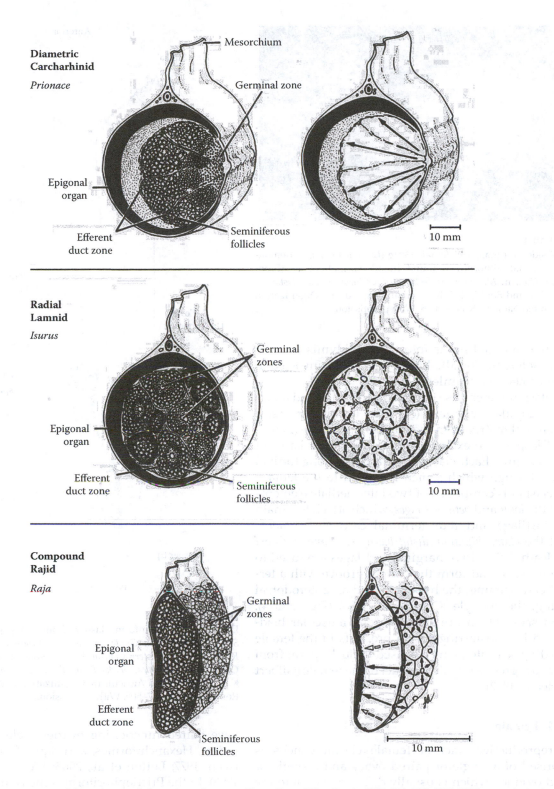

FIGURE 10.2
Diagrammatic representation of three forms of elasmobranch testes. (From Pratt, H.L., *Copeia*, 1988, 719–729, 1988. With permission.)

tend to have simple sperm aggregations, whereas the batoids have been reported to have either simple aggregations or spermatozeugmata. The Carcharhinformes form spermatozeugmata, and the Lamniformes have spermatophores (Jamieson, 2005; Tanaka et al., 1995).

The seminal vesicle in some species is expanded to store sperm prior to mating (Wourms, 1977). The reproductive system terminates in the common urogenital sinus, which vents into a common cloaca by means of a single large papilla. The Leydig glands are thought

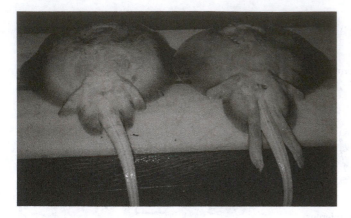

FIGURE 10.3
Ventral side of a male and female skate (*Leucoraja erinacea*) demonstrating sexual dimorphism found in all elasmobranchs. (From Conrath, C.L., in *Management Techniques for Elasmobranch Fisheries*, Musick, J.A. and Bonfil, R., Eds., Food and Agriculture Organization of the United Nations, Rome, 2005. With permission.)

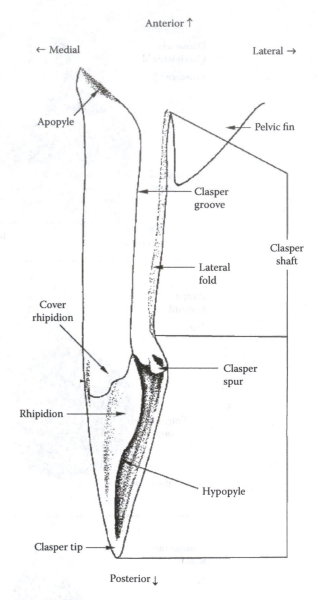

FIGURE 10.4
Diagram of a lamniform clasper (from Compagno, L.J.V., in *FAO Species Identification Guide for Fishery Purposes: The Living Marine Resources of the Western Central Atlantic*. Vol. 1. *Introduction, Molluscs, Crustaceans, Hagfishes, Sharks, Batoid Fishes, and Chimaeras*, Carpenter, K.E., Ed., Food and Agriculture Organization of the United Nations, Rome, 2003, pp. 357–505. With permission.)

to secrete seminal fluids into the epididymis and the ductus deferens. The alkaline gland may be involved in sperm protection (Hamlett, 1999).

All elasmobranchs use internal fertilization and males possess copulatory organs (claspers) to transfer sperm and seminal matrix from the male to the female (Figure 10.3). Claspers are extensions of the posterior bases of the pelvic fins. Each clasper has a dorsal longitudinal groove through which semen is passed to the female. Each clasper is comprised of two intermediate elements called the *joint* and *beta cartilages*, which attach to a main stem cartilage and four terminal cartilage elements called the *claw, rhipidion, distal basal*, and *spur* (Gilbert and Heath, 1972). Two marginal cartilages are fused to the main stem and form the clasper groove with a terminal end opening, the hypopyle, and an anterodorsal opening, the apopyle (Compagno, 1988) (Figure 10.4). Siphon sacs, which are subcutaneous muscular bladders, aid in transferring sperm products to the female by holding seawater, which is used to flush sperm from the clasper groove into the oviduct of the female (Gilbert and Heath, 1972).

10.2.2 Female

The reproductive tract of female elasmobranchs is comprised of a single or paired ovary and a single or paired oviduct, which is usually differentiated into the following structures: ostium, anterior oviduct, oviducal gland, isthmus, uterus, cervix, and urogenital sinus (Hamlett and Koob, 1999) (Figure 10.5). The ovaries and oviducts of vertebrates always originate from bilateral primordia but often become asymmetrical during development through fusion or failure of one of the pair to develop (Hoar, 1969). Groups in which most species

have paired functioning ovaries include some batoid orders, Hexanchiformes, and Squaliformes (Jones and Geen, 1977; Lutton et al., 2005; Wenbin and Shuyuan, 1993). In the Pristiophoriformes the right ovary is functional, and in the Squatiniformes there is ovarian asymmetry, with the left ovary developing alone or always larger than the right depending on the species (Capapé et al., 1990; Lutton et al., 2005; Sunye and Vooren, 1997). Only one ovary (usually the right) is developed in the Heterodontiformes, Orectolobiformes, Lamniformes, and Carcharhiniformes (Castro, 2000; Compagno,

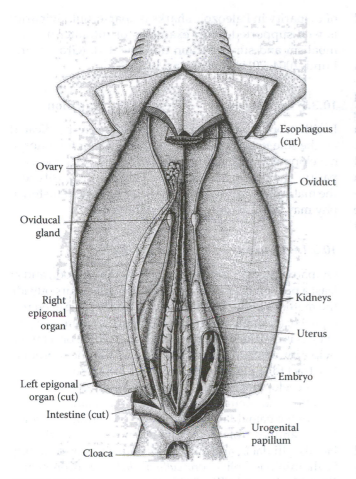

FIGURE 10.5
Female reproductive system. (Adapted from Castro, J.I., *The Sharks of North American Waters*, Texas A&M University Press, College Station, 1983. With permission.)

1988; Gilmore et al., 2005; Lutton et al., 2005; Smale and Goosen, 1999; Teshima, 1981; Tovar-Avila et al., 2007). The ovaries are suspended from the body wall by a mesentery called the *mesovarium* (Wourms, 1977). The ovaries generate germ cells, accumulate yolk, and are involved in the synthesis and secretion of hormones (Hamlett and Koob, 1999). Two types of ovary have been identified: (1) an internal form found in lamnid sharks that is hollow, with the germinal epithelium encapsulated in the epigonal organ; and (2) an external form found in other elasmobranchs in which the ovary is borne externally on the distal surface of the epigonal organ or suspended directly from the mesovarium (Pratt, 1988). Fully developed eggs move from ovarian follicles into the body cavity, where cilia act to move the eggs into the ostium. In the ostium, eggs are moved into the anterior oviduct by peristaltic and/or ciliary action (Gilbert, 1984). The ostium is a funnel-shaped opening designed to assist in ova movement to the rest of the oviduct. The anterior oviduct leads to the oviducal gland. The oviducal gland functions in egg encapsulation, sperm

storage, and fertilization in some species (Hamlett and Koob, 1999; Pratt, 1979; Wourms, 1977). Hamlett et al. (1998) described four fundamental zones of the elasmobranch oviducal gland based on the morphology of the epithelium: the proximal club zone, the papillary zone, the baffle zone, and the terminal zone. The jelly coats that surround the egg are produced within the proximal club and papillary zones, and various types of egg investments are produced within the baffle zone (Hamlett and Koob, 1999). In oviparous species, the uterus is specialized for chemical and structural modification of the egg envelope to form a horny egg capsule and may have structural modifications to enable movement of the capsule through the uterine lumen (Koob and Hamlett, 1998). In viviparous species, the uterus is highly developed for egg retention and embryo development, and its structure depends on the mode of viviparity (Figure 10.6). The uterus in most chondrichthyans is paired, except in some advanced Myliobatiformes (Fahy et al., 2007; Hamlett and Koob, 1999). Also, some deepwater species tend to have a reduced reproductive tract with only one functional uterus, as found in *Chlamydoselachus anguineus* (Tanaka et al., 1990). The cervix is a constriction between the uterus and the urogenital sinus (Hamlett and Koob, 1999).

Sperm storage occurs in the females of holocephalans and elasmobranch species in all orders except Lamniformes and appears to be widespread (Hamlett et al., 1998, 2005a; Pratt, 1993). It was first proposed when aquarium specimens of *Raja* continued to lay fertilized eggs after periods of separation from male specimens (Clark, 1922). Sperm are stored within the oviducal gland, generally within the terminal zone tubules in the lower portion of the gland, where they

FIGURE 10.6
(See color insert.) Uterine wall and embryo of *Ginglymostoma cirratum*; photograph taken with an endoscope *in utero*. (Photograph © Jeffrey C. Carrier and used with permission.)

are nourished and protected from immunological attack by the mother until they are needed for fertilization (Hamlett et al., 2005a). Sperm storage uncouples the female ovarian cycle from the male spermatogenic and mating cycles and is an adaptation of particular value to the Chondrichthyes, within which geographic sexual segregation is widespread. In addition, sperm storage eliminates the need for frequent multiple mating which can lead to injury to the female (Pratt and Carrier, 2005). This is particularly relevant in oviparous species and viviparous species with large litters where ovulation may occur over an extended period of time. The duration of sperm storage is species specific and may range from a few months to a year or possibly longer (Hamlett et al., 2005a; Walker, 2005, 2007).

10.3 Modes of Reproduction and Fetal Nutrition

Chondrichthyan reproductive strategies include several kinds of fetal nutrition that can be classified as lecithotrophic or matrophic. In lecithotrophic species, the embryos derive their nutrition entirely from yolk reserves present within the egg case. Lecithotrophic forms of reproduction include yolk sac viviparity and oviparity. In matrotrophic species, the energy reserves present in the egg are supplemented by additional maternally derived nutrients obtained during gestation. Matrotrophic forms of reproduction include mucoid histotrophy, lipid histotrophy, placental viviparity, carcharhiniform oophagy, and lamniform oophagy. Although it appears the basal mode of reproduction of elasmobranchs is some type of lecithotrophy, recently there has been some debate about which mode is truly ancestral. The traditional view of oviparity as the pleisiomorphic condition (Dulvey and Reynolds, 1997; Wourms, 1977) has been challenged; for the Neoselachii, yolk sac viviparity is strongly supported as ancestral, based on the distribution of the different reproductive modes within the most recently accepted phylogeny (Musick, 2010; Musick and Ellis, 2005). In addition, because all chondrichthyans, both recent and fossil, are characterized by the presence of claspers and internal fertilization and because all living chondrichthyans exhibit urea retention as their principal mode of osmoregulation, the plesiomorphic reproductive state for the Chondrichthyes was probably yolk sac viviparity. Griffith (1991) argued that urea retention evolved early on in vertebrates in order to detoxify ammonia produced *in utero* by developing embryos. Recent fossil evidence further supports this contention, as evidence

of oviparity in Paleozoic sharks is sparse, but viviparity is well supported, with examples of intrauterine cannibalism and superfetation documented. (Grogan and Lund, 2004, 2011; Musick and Ellis, 2005).

10.3.1 Lecithotrophic Modes of Fetal Nutrition

Lecithotrophic modes of fetal nutrition are predicated on the sole use of yolk from within the egg for embryo development and include oviparity and yolk sac viviparity. The distinction between yolk sac viviparity and the matrophic reproductive strategy of limited histotrophy may be difficult to distinguish visually.

10.3.1.1 Oviparity—Single and Retained

Oviparous species extrude eggs onto the seafloor, and at least a portion of embryonic development occurs outside the mother's body. Two types of oviparity have been recognized: (1) single or extended oviparity, where one pair of eggs is extruded at a time; and (2) multiple or retained oviparity, where eggs are retained within the reproductive tract for a period of time during which some embryonic development occurs. Single oviparity is much more common and occurs within the holocephalans, all members of the batoid family Rajidae, all Heterodontiformes, two families of Orectolobiformes (Parascylliidae, Hemiscylliidae), and in the carcharhiniform family Scyliorhinidae, plus *Proscyllium habereri* (Compagno, 1988; Musick and Ellis, 2005). Retained oviparity occurs in at least one orectolobid species, *Stegostoma fasciatum* (Goto, 2001) and five scyliorhinid species of the genus *Halaelurus* and the genus *Galeus* (Compagno, 1990, 2001). In at least one *Halaelurus* species, eggs are retained until the embryos are well developed, with a short period of time between oviposition and hatching (White, 2007). In contrast, *Galeus melastomus* appears to have an extended laying period with batches of up to eight eggs laid throughout a protracted period of the year (Capapé et al., 2008). Oviparity is absent from the Superorder Squalomorphii (Musick and Ellis, 2005).

The egg cases of oviparous species vary greatly in shape and size, but all have a leathery exterior and a morphology designed to retain them at the location of oviposition (Figure 10.7). Usually only one embryo is contained in an egg case, with the exception of at least two species of skates, *Raja binoculata* and *R. pulchra* (Rajidae), which have up to four embryos per egg case (Ebert and Winton, 2010). The egg case generally has tendrils and sticky filaments that will attach to either structure or the substrate. It is possible that the behavior of the pregnant female in determining site selection and additional egg placement also aids in egg retention within a chosen locality (Compagno, 1990). Some oviparous species use

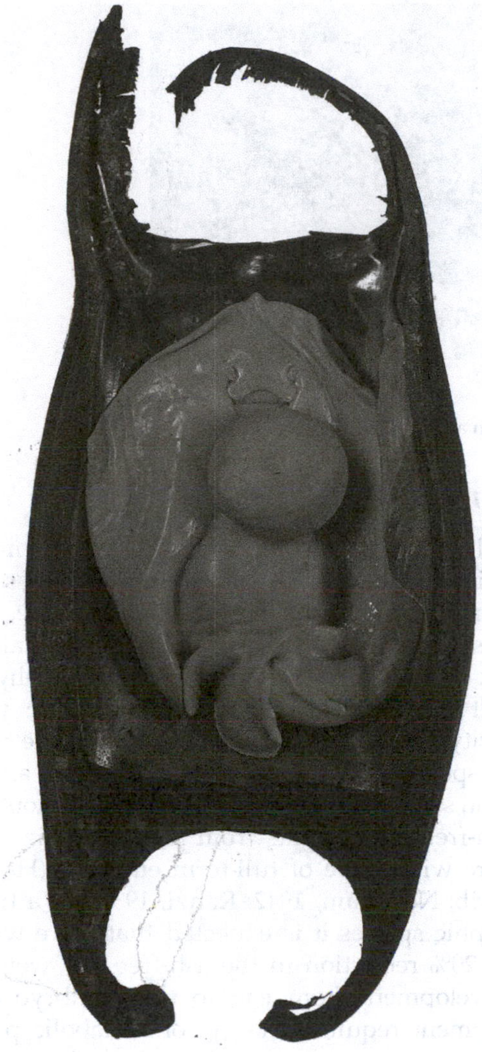

FIGURE 10.7
(See color insert.) Egg case and late-stage embryo of *Bathyraja parmifera*. (From Hoff, G.R., *J. Fish Biol.*, 74(1), 250–269, 2009. With permission.)

communal egg deposition sites (Compagno, 1984, 2001; Hoff, 2010). Development within the egg case is temperature dependent (Ellis and Shackley, 1995; Perkins, 1965), and reported incubation periods range from 2 to 3 months to over 15 months (Luer and Gilbert, 1985; Wourms, 1977). Recent studies, however, suggest that development in coldwater species may be much more protracted, with some skates in Alaskan waters, such as *Bathyraja parmifera*, having an estimated incubation time of 3.5 years (Hoff, 2008).

All oviparous species are demersal, and most are small in size (Musick and Ellis, 2005). The size of term oviparous embryos is constrained by the amount of yolk necessary to support development within the yolk sac; therefore, oviparous term embryos tend to be smaller than matrotrophic term embryos, which obtain additional nutrients from the mother (Carrier et al., 2004; Hamlett, 1997). Annual fecundity of oviparous species (approximately 20 to 100) tends to be an order of magnitude higher than viviparous species of comparable size. This form of reproduction may be advantageous as a method to increase fecundity in small elasmobranchs that have limited space within the body cavity for the care and storage of embryos (Musick and Ellis, 2005).

Oviparity is restricted to only seven families of living elasmobranchs: the Rajidae, Heterodontidae, Parascylliidae, Hemiscylliidae, Stegostomatidae, the Scyliorhinidae, and Proscylliidae; however, two of these (rajids and scyliorhinids) are among the most speciose elasmobranch families. The large number of species in these families does not mean that they are more successful than viviparous families; rather, it reflects their lack of vagility engendered by small size and benthic habit, which increases the probability of population isolation and speciation (Musick et al., 2004). The most speciose taxon among viviparous elasmobranchs is the genus *Etmopterus*, which is comprised of small and even diminutive species (Musick et al., 2004). That small size and benthic habit are conducive to more frequent speciation has been recognized and well documented in teleost fishes for 50 years (Rosenblatt, 1963).

10.3.1.2 Yolk Sac Viviparity

All viviparous elasmobranchs, even those that are matrotrophic, depend on the yolk sac as the initial source of fetal nutrition (Hamlett et al., 2005b). Yolk sac viviparity occurs when the yolk sac is the principal source of fetal nutrition throughout development to parturition. This mode of reproduction is widespread among the elasmobranchs and occurs in all living orders except the oviparous Heterodontiformes and the Lamniformes (Compagno, 1990). Lecithotrophic viviparous species tend to produce a smaller number of larger offspring than comparably sized oviparous species (Musick and Ellis, 2005). Potential advantages of larger offspring include reducing the number of predators and competitors, greater resistance to starvation, a higher tolerance of environmental conditions, increasing potential prey items, and greater swimming efficiency (Sogard, 1997; Wourms, 1977).

In yolk sac viviparous species, the egg envelope that forms is much thinner than that of oviparous species. In most species, each embryo is contained in an egg envelope, but in some species of squaliform, squatiniform, and batoid fishes several fertilized eggs are contained in a single egg membranous envelope called a *candle* (Gilbert, 1984; Meagher, 2011; Sunye and Vooren, 1997) (Figure 10.8). As development begins, yolk is stored in

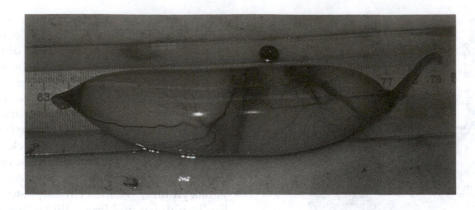

FIGURE 10.8
(See color insert.) Candle of a *Squalus acanthias*. (Photograph © Christina Conrath and used with permission.)

an external yolk sac, but during gestation yolk is usually transferred to an internal yolk sac, separate from the embryonic stomach. Gestation times vary widely within species with this mode of reproduction. Gestation times tend to be between 4 and 12 months but may be as long as 2 years for *Squalus* species (Jones and Geen, 1977; Wilson and Seki, 1994). A gestation of longer than 3.5 years has been proposed for *Chlamydoselachus anguineus* (Tanaka et al., 1990) but remains to be confirmed.

A diverse group of elasmobranchs employ yolk sac viviparity, and the length of the reproductive cycle, the structure of the reproductive tract, and the fecundity of these species vary widely. In addition to varying times of gestation, some species have a resting period of up to a year between pregnancies while other species have a nearly continuous reproductive cycle. Fecundity of species with this mode of reproduction varies from one pup in the deepwater species *Centrophorus granulosis* (Guallart and Vicent, 2001) to 300 in the whale shark, *Rhincodon typus* (Joung et al., 1996). It should be cautioned, however, that it can be difficult to visually distinguish between species that employ yolk sac viviparity and those with limited histotrophy (see below). Ranzi (1932, 1934) and Needham (1942) showed that several elasmobranch species thought to be yolk sac viviparous were actually mucoid histotrophs. We will discuss this topic in the section below.

10.3.2 Matrophic Modes of Fetal Nutrition

Matrophic modes of reproduction include mucoid histotrophy, lipid histotrophy, placental viviparity, carcharhiniform oophagy, and lamniform oophagy. In all forms of matrophic reproduction, the embryo completes development within the body of the mother and receives additional nutrition during the developmental process. Maternal contributions during development lead to the production of larger, more capable, and less vulnerable offspring.

10.3.2.1 Mucoid Histotrophy

Mucoid histotrophy is the simplest form of matrotrophic viviparity wherein developing embryos receive additional nutrients from the female in the form of mucous produced by the uterus (Hamlett et al., 2005b; Musick and Ellis, 2005). This type of matrophy can be difficult to distinguish from lecithotrophic yolk sac viviparity because there are no easily visible morphological specializations of the uterus and the amount of nutrition supplied may not be apparent without obtaining ash-free dry weights from newly fertilized ova to compare with those of full-term embryos (Hamlett et al., 2005b; Needham, 1942; Ranzi, 1934). In a truly lecithotrophic species it is expected that there will be at least a 20% reduction in the ash-free dry weight during development from egg to term embryo because development requires energy for metabolic processes as well as organic matter for anabolism (Hamlett et al., 2005b). The egg yolk must support not only the embryonic incorporation of organic matter (anabolism) but also embryonic metabolic processes (catabolism) during development. Until recently the use of at least 20% organic loss as a metric to define lecithotrophy has been verified only in two oviparous elasmobranchs (Diez and Davenport, 1987; Tullis and Peterson, 2000). However, in an important new study using calorimetry, Meagher (2011) showed that the lecithotrophic ray *Aptychotrema rostrata* lost 21% dry mass and 19% organic mass during development. These values are similar to those for the oviparous species and verify the use of ~20% as a valid metric for determining strict lecithotrophy. Ranzi (1932, 1934) noted that some species of the houndshark family Triakidae actually gained organic content, indicating that additional nutrients are obtained from the mother during development. Evidence for mucoid histotrophy may also be provided by histological examination of the uterine walls, which should exhibit high mucus secretory activity at least during early and mid-term development (Hamlett et al., 2005b).

Hamlett et al. (2005b) distinguished between incipient histotrophy and minimal histotrophy. Both are forms of mucoid histotrophy; however, in incipient histotrophy, mass gain offsets some catabolic loss but still results in an embryonic ash-free dry weight less than that of the fertilized ovum. Minimal histotrophy results in a mass gain, usually a significant mass gain (>100%). Incipient histotrophy has been documented in a small number of squaliforms (Cotton, 2010; Ranzi, 1934) and a pristiophoriform (Stevens, 2002) and has been inferred equivocally in some other species of squaliforms and batoids (Setna and Sarangdhar, 1950) and in a hexanchiform (Tanaka et al., 1990). Recent studies of an orectolobiform (Castro, 2000), a rhinobatid (Meagher, 2011), and two species of squaliforms (Braccini et al., 2007; Cotton, 2010) have failed to find incipient histotrophy. Its importance to fetal nutrition remains to be determined with regard to the extent to which it is found among elasmobranch taxa and its quantitative contribution to embryonic development. By definition, incipient histotrophy contributes 20% or less to embryonic development in those yolk sac viviparous elasmobranchs in which it occurs. Consequently, we consider such species to be principally yolk sac viviparous. Minimal histotrophy seems restricted primarily to sharks in the order Carcharhiniformes, and in most of those species it is not a minimal matrotrophic contributor. Mucoid histotrophy is the principal source of fetal nutrition in many triakids (Farrell et al., 2010; Ranzi, 1932, 1934; Walker, 2007) and perhaps contributes significantly to fetal development in the proscyliid *Eridacnis* sp. (Compagno, 1988), the ovophagous pseudotriakids *Gollum* and *Pseudotriakis* (Yano, 1992, 1993), and most species of placental sharks (Hamlett et al., 2005b,c). For that reason, we suggest dropping the term *minimal histotrophy* and replacing it simply with *mucoid histotrophy*.

Triakids offer an interesting case in that some members are placental and others are mucoid histotrophic (Farrell et al., 2010; Lopez et al., 2006; Ranzi, 1932, 1934; Walker, 2007). Lopez et al. (2006) offered evidence that the division between the two reproductive modes followed phylogenetic patterns within the Triakidae and that mucoid histotrophy represented an evolutionary reversal in the family. Walker (2007) noted that within *Mustelus*, which has both placental and histotrophic clades, term embryos from both clades are similar in mass (comparing species of similar size with similar size ova). Teshima (1981) had earlier noted that *M. manazo*, a nonplacental species, and *M. griseus*, a placental species, both had similar size term embryos. Thus, mucoid histotrophy may approach placentotrophy (discussed in Section 10.3.2.3) in efficiency of fetal nutrient transfer in some triakids. Increase in ash-free dry weight in histotrophic triakids has been reported to range from 11% in *Galeorhinus galeus* (Ranzi, 1932) to 784% in *Mustelus antarcticus* (Storrie et al., 2009).

10.3.2.2 Lipid Histotrophy

This type of development occurs in rays of the order Myliobatiformes which produce and secrete a protein- and lipid-rich histotroph from highly developed secretory structures within the uterine lining called *trophonemata* (Hamlett et al., 2005b) (Figure 10.9). Embryonic development is initially supported by yolk reserves, but as development proceeds trophonemata increase in size and secrete uterine secretions known as *histotroph* (Hamlett et al., 2005b; Wourms, 1981). In some species, trophonemata increase in length and enter the gill, spiracles, and mouths of developing embryos in the uterus (White et al., 2001). The transfer of nutrients via lipid histotrophy appears to be extremely efficient, and the myliobatiform rays are distinguished from other groups by the extremely large dry weight gain (in some species 1680 to 4900%) that occurs during embryonic development. The amount of lipid in the histotroph is significantly lower in more primitive species such as *Urobatis jamaicensis* than in the more advanced genera *Dasyatis*, *Myliobatis*, and *Rhinoptera* (Hamlett et al., 2005b). As a whole, the reproductive biology of this group is not well studied but it appears that most species have very low fecundity, generally less than ten (Garayzar et al., 1994; Jacobsen et al., 2009), with a minimum of one pup per litter found in *Rhinoptera bonasus* (Neer and Thompson, 2005).

10.3.2.3 Placental Viviparity

Embryos of species that exhibit placental viviparity develop a yolk sac placenta after the yolk within the external yolk sac is exhausted. The yolk sac forms an attachment with the uterine epithelium and the yolk stalk elongates to form a placenta (Musick and Ellis, 2005) (Figure 10.10). Placentation generally occurs within 2 to 4 months after fertilization after the embryo has reached at least 100 mm, but in at least one species, *Scoliodon laticaudis*, a placental connection is formed between the embryo and the mother in a very short period of time when the embryo is about 3 mm in length (Wourms, 1993). All placental species develop within uterine compartments. This specialized type of reproductive development is only found within five families of carcharhiniform sharks: Leptochariidae, Triakidae, Hemigaleidae, Carcharhinidae, and Sphyrnidae, of which the leptochariids are the most primitive (Lopez et al., 2006; Musick and Ellis, 2005). The elasmobranch placenta formed from the fetal yolk sac and stalk is functionally and morphologically distinct from the mammalian placenta. Fetal development in all placental sharks is initially supported by the yolk, then by mucoid histotrophy, and finally by implantation of the placenta. Fetal development in all placental sharks takes place in

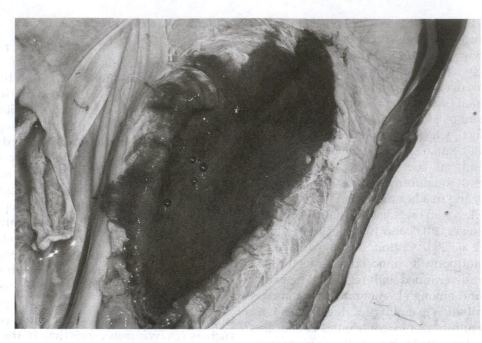

FIGURE 10.9
(See color insert.) *Gymnura altivela* uterus containing trophonemata. (Photograph © John A. Musick and used with permission.)

individual uterine compartments that form during the first weeks of pregnancy (Otake, 1990). These compartments are unique to placentotrophs but are also retained in most histotrophic *Mustelus* (Lopez et al., 2006). Uterine compartments increase the surface area of the uterus and stabilize the growing embryos after attachment (Hamlett et al., 2005c). Placentotrophy in elasmobranchs may be viewed as a highly modified form of histotrophy (Hamlett et al., 2005c). Contrary to some reports in the literature (Wourms, 1993), no elasmobranch has ever been found to have hemotrophic transfer of nutrients from mother to fetus; rather, all transfer is through uterine secretions (Hamlett et al., 2005c). Several types of placentae have been identified and defined based on shape and development, including columnar, discoidal, entire, and globular (Compagno, 1988; Hamlett et al., 2005c). The uterine epithelium at the attachment site is highly vascularized and serves to enhance uterine secretions and support embryonic respiration, osmoregulation, and waste disposal. Adjacent areas of the uterus continue to secrete mucus, and mucoid histotrophy may occur simultaneously with placentotrophy. In some species, long threadlike or lobular flaps form on the yolk stalk or placenta called *appendiculae*. These are richly vascularized and provide enhanced surface area for absorption of uterine secretions (Hamlett et al., 2005c; Wourms, 1977).

Placentotrophic species tend to have gestation times of equal to or less than one year, although *Carcharhinus obscurus* appears to have an 18-month gestation (Romine et al., 2009). Placental species that have biennial reproductive cycles have a year resting period between pregnancies. Litter size tends to be quite small in placental species; generally litters consist of fewer than 20 embryos. One exception is the pelagic blue shark, *Prionace glauca*, which appears to have 80+ pups per litter (Pratt, 1979). The size at parturition varies among species, but larger members of this group tend to have embryos that attain lengths of 60 to 90 cm (Compagno, 1984).

10.3.2.4 Carcharhiniform Oophagy

Oophagy is a form of matrophy in which embryonic development is supported through the ingestion of unfertilized eggs produced by the mother. Carcharhiniform oophagy is functionally and evolutionarily distinct from

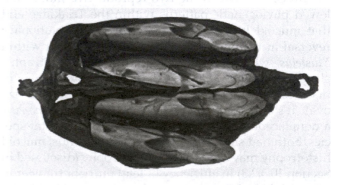

FIGURE 10.10
(See color insert.) Reproductive tract of term female *Carcharhinus acronotus* with visible placentae. (Photograph © Jose I. Castro and used with permission.)

lamniform oophagy and is considered separately in this account. In carcharhiniform oophagy, a multitude of unfertilized ova are contained within the same egg envelope as the developing embryo and are ingested by the embryo (Musick and Ellis, 2005). The deepwater carcharhiniform false catshark family, Pseudotriakidae, is the only family with this reproductive mode (Yano, 1992, 1993). This family of benthic deepwater sharks contains only five rare species, three of which have only recently been identified (Compagno et al., 2005). This reproductive mode has been documented for the two described species, *Pseudotriakis microdon* and *Gollum attenuatus*. These species are characterized by low fecundity of two to possibly four embryos per litter and may also use histotrophy during development (Yano, 1992, 1993).

10.3.2.5 Lamniform Oophagy

Lamniform oophagy is distinguished from carcharhiniform oophagy by the continued production and delivery of ova throughout a portion or most of the pregnancy. Embryos within the uterus hatch out of the egg capsule when the yolk reserves within it are nearly depleted and then consume additional eggs, which continue to be ovulated while the embryo develops (Gilmore et al., 2005). Several small ova are packed together in the oviducal gland into egg capsules before being transferred to the uterus. Ova ingested by the developing embryo are stored within a prominent yolk stomach in all but the *Alopias* genus (Gilmore et al., 2005; Musick and Ellis,

2005) (Figure 10.11). This type of oophagy occurs in all species of the Lamniformes and is limited to this group (Compagno, 1990; Gilmore, 1993, 2005). In the sandtiger shark, *Carcharias taurus*, the first embryo to hatch within each uterus consumes all the other embryos within that uterus then continues to feed on ovulated eggs (Gilmore et al., 1983; Springer, 1948). This type of oophagy is known as *adelphophagy*.

Lamniform oophagy tends to lead to a small number of large embryos. Litter sizes of six or fewer have been documented in *Alopias* (Moreno and Moron, 1992), *Lamna* (Francis and Stevens, 1999), *Pseudocarcharias* (Oliveira et al., 2010), and *Carcharias* (Gilmore et al., 1983); however, larger litter sizes of up to 25 have been documented for some populations of *Carcharodon carcharias* (Sanzo, 1912, as cited in Gilmore, 1993) and *Isurus oxyrinchus* (Mollet et al., 2000). This difference may be a result of habitat and may reflect the general pattern that pelagic species tend to have a larger number of small embryos (Gilmore, 1993). With the exception of the diminutive *Pseudocarcharias kamoharai*, the members of this order tend to have the largest embryos found within the Elasmobranchii, with embryos ranging in size from around 60 to 150 cm in length.

10.3.3 Parthenogenesis

Recently the ability of some female sharks in captivity to reproduce without mating has been documented. Genetic studies have shown that these females are

FIGURE 10.11
(See color insert.) *Lamna ditropis* embryo with prominent yolk stomach. (Photograph © Kenneth J. Goldman and used with permission.)

most likely undergoing automictic parthenogenesis (Chapman, 2007, 2008). Automictic parthenogenesis occurs when an ovum and a polar body fuse to produce a diploid offspring after meiosis has occurred. This phenomenon is thought to have occurred in captive *Sphyrna tiburo*, *Carcharhinus limbatus*, and *Chiloscyllium plagiosum*. In the first two species, only one embryo was formed and neither was documented to live for a substantial period of time, but in the last species at least two parthenogens were produced and survived at least 5 years (Chapman, 2007, 2008; Feldheim et al., 2010). The likelihood or prevalence of this phenomenon in the natural environment is unknown and unlikely to be high, but it could have implications for isolated or depleted shark populations.

10.4 Reproductive Cycles

10.4.1 Reproductive Interval

The reproductive cycle (interval) in elasmobranchs is the time between discrete ovulation periods. The reproductive cycle is comprised of three phases: (1) vitellogenesis, wherein the ova within the ovary develop; (2) gestation, the period from conception to parturition; and (3) a resting stage in some species during which the liver sequesters nutrients in preparation for vitellogenesis. Vitellogensis and parturition may occur concurrently or sequentially, and only some species undergo a resting stage (Castro, 2009). The length of the reproductive interval varies widely among elasmobranch species and ranges from several months to over 3 years. Very short development/gestation times have been reported for two *Dasyatis* rays and for the ray *Urobatis jamaicensis*, and these species may have two or three reproductive cycles in a single year (Capapé, 1993; Capapé and Zaouali, 1995; Fahy et al., 2007). Reproductive cycles of 3 years or longer have been reported for some sharks, such as *Chlamydoselachus anguineus* and *Galeorhinus galeus* (Peres and Vooren, 1991; Tanaka et al., 1990). Gestation periods also range widely between species; in the previous example, *Chlamydoselachus anguineus* is reported to have a gestation period of 3.5 years but *Galeorhinus galeus* has a much shorter gestation period of only 12 months with a resting period of one year and another year for ovarian development before ovulation occurs. The gestation period may be as short as a few months or as long as several years. Some species undergo a lengthy resting period between pregnancies, whereas others have only a period of days to weeks before ovulation occurs again. *Squalus acanthias* has a gestation period of 23 months, but ovulation and gestation cycles occur

concurrently so this species has a reproductive interval of 2 years (Jones and Geen, 1977). Similarly, *Carcharhinus limbatus* has a 2-year reproductive interval, but gestation only lasts 12 months, after which these sharks undergo a resting period before vitellogenesis and oogenesis resume (Castro, 1996).

10.4.2 Reproductive Seasonality

Wourms (1977) identified three types of elasmobranch reproductive cycles: reproducing throughout the year, having a partially defined annual cycle with one to two peaks, or having a well-defined annual or biennial cycle. Although it is not always easy to distinguish between the first two types listed above and reproductive intervals longer than 2 years have been identified, it is useful to distinguish between species with reproductive cycles that are aseasonal, partially seasonal, and seasonal. The type of reproductive interval of a species appears to be strongly related to habitat. Deepwater species of several different orders, including *Chlamydoselachus anguineus* (a hexanchiform), *Galeus melastomus* (a carcharhiniform), and *Centroscyllium fabricii* (a squaliform), tend to lack a defined reproductive season likely due to the constancy of their habitat (Costa et al., 2005; Tanaka et al., 1990; Yano, 1995). Some pelagic and tropical species also tend to have an aseasonal reproductive cycle, including *Pseudocarcharias kamoharai* and *Alopias pelagicus* (Liu et al., 1999; Oliveira et al., 2010). The second type of reproductive cycle is prevalent among oviparous species that tend to have a protracted egg-laying period, such as *Scyliorhinus canicula* and *Raja clavata* (Holden, 1975; Sumpter and Dodd, 1979). This type of cycle also occurs in tropical species such as *Rhizoprionodon porosus* (Mattos et al., 2001). The third type of reproductive cycle is prevalent among species that occur in coastal areas in subtropical regions of the world where there is a marked difference in seasonal water temperatures and food availability. Seasonal cycles often involve seasonal migrations over long distances from resting or mating areas to pupping areas (Grubbs, 2010; Musick et al., 1993). This type of cycle is prevalent in members of *Carcharhinus*, *Sphyrna*, and *Mustelus* that occur in shallow coastal waters.

10.4.3 Embryonic Diapause

Embryonic diapause is a delay in the development of fertilized or young embryos within the uterus during gestation. This phenomenon has been noted in several species of elasmobranchs and may allow embryos to be born during times of year of either optimal temperatures or prey abundance. Embryonic diapause has been described for the ray species *Trygonoptera personata* (White et al., 2002), *Trygonorrhina fasciata* (Marshall et al., 2007), *Rhinoptera bonasus* (Smith and Merriner, 1986), and *Dasyatis say* (Snelson et al., 1989) and for the shark species *Rhizoprionodon taylori* (Simpfendorfer, 1992).

10.4.4 Plasticity

Intraspecific variation in reproductive parameters has been noted for several species that have geographically separated populations. Differences in size at maturity, fecundity, and reproductive timing have been noted between populations of several species. These parameters vary for discrete populations of *Carcharhinus plumbeus* found throughout the world's oceans, including the northwest Atlantic Ocean, the Hawaiian Islands, the western Indian Ocean, and the East China Sea (Cliff et al., 1988; Joung and Chen, 1995; McAuley et al., 2007; Wass, 1973). Parsons (1993a,b) found that two populations of *Sphyrna tiburo* in two locations in Florida had differing sizes and ages at maturity and size at birth. Differences in reproductive interval appear to be more rare, but this phenomenon has also been documented. Three species of histotrophic *Mustelus* (*M. antarcticus*, *M. asterias*, and *M. manazo*) have annual cycles in warmer water populations and biannual cycles in colder water populations (Farrell et al., 2010; Walker, 2007; Yamaguchi et al., 2000).

How reproductive parameters of individuals within populations may depend on, or vary with, environmental factors is largely unknown. Litter size, embryo size, and reproductive periodicity are generally assumed to exhibit little variation within species or populations, with changes in young of the year, juvenile, or adult survivorship considered to be the most likely method of density dependence for these populations. Recent studies, however, have found that some populations may also have differing reproductive intervals. Driggers and Hoffmayer (2009) found that within the same population of *Carcharhinus isodon* there was evidence of females undergoing annual and biennial reproductive cycles and suggested that a shift between cycle length could be energetically based. Bansemer and Bennet (2009), in a study using underwater census, acoustic tracking, and photo identification to track individual *Carcharias taurus*, found that some females underwent a biennial cycle and some underwent a triennial cycle between pregnancies with either a 1- or 2-year resting phase. It is difficult to determine the reproductive periodicity of large sharks that have long reproductive intervals, and similar variation within populations may be widespread. Several teleost species have also been shown to suppress reproduction and skip a reproductive season, and it is generally assumed that this occurs when food is limiting and fish condition is not robust (Bell et al., 1992; Engelhard and Heino, 2005; Rideout and Rose, 2006).

10.5 Reproductive Behavior

10.5.1 Mating Behavior

All elasmobranchs employ internal fertilization, which requires the male to grasp the female and insert one or (rarely) both claspers into her cloaca, where the clasper gland everts and becomes anchored with cartilaginous hook and gaff elements. Few data are available on the full complexities and likely diversity of elasmobranch reproductive behavior. Pratt and Carrier (2005) summarized the published literature on elasmobranch reproductive behavior, and Table 10.1 summarizes the behaviors that were observed in these investigations. Observed precopulatory or courtship behaviors include following, cupping and flaring of fins, and biting (Pratt and Carrier, 2005). Biting may aid in maintaining position and proximity and has been proposed to act as a precopulatory release mechanism for female sharks (Kajiura et al., 2000; Tricas and LeFeuvre, 1985). Sexual dimorphism in female skin thickness and/or male dentition has been documented in several elasmobranch species, and these dichotomies are most likely related to the prevalence of biting associated within mating events. Several female avoidance behaviors have been documented as well, including shielding, arching, lying motionless, and rolling away from the male (Gordon, 1993; Klimley, 1980; Pratt and Carrier, 2005). In the natural environment, precopulatory activities and mate selection may be prolonged and involve many complex behaviors. *In situ* studies of *Ginglymostoma cirratum* have revealed complex precopulatory behaviors on the part of both males and females as well as complex social interactions between males during mating events (Carrier et al., 1994).

10.5.2 Multiple Paternity

It has been reported that *Ginglymostoma cirratum* females mate with more than one male, and genetic analysis of broods also supports multiple paternity for this species (Carrier et al., 2003; Saville et al., 2002). Multiple paternity has been identified for many other species, including *Negaprion brevirostris* (Feldheim et al., 2004), *Carcharhinus plumbeus* (Portnoy et al., 2007), and *Raja clavata* (Chevolot et al., 2007). Multiple paternity may allow for increased genetic diversity and provide a defense against genetic incompatibility, which will lead to increased fitness and increased reproductive output, particularly for viviparous females (Zeh and Zeh, 2001). It has also been suggested that multiple paternity may simply be a consequence of male competition, and the violence of shark reproduction may make multiple matings disadvantageous for female elasmobranchs (Daly-Engel, 2010; Portnoy, 2010; Portnoy et al., 2007).

TABLE 10.1

Summary of Observed Courtship and Mating Behaviors in Elasmobranch Fishes

General Behavior and Species	Descriptions/Notes	Ref.
Sharks		
Precopulatory and courtship		
Following		
Carcharhinus melanopterus	"Close follow" near female's vent possibly olfactory-mediated	Johnson and Nelson (1978)
Ginglymostoma cirratum	Male and female swim parallel and synchronously side by side	Klimley (1980)
Negaprion brevirostris	Swimming with body axes in parallel	Clark (1963)
Female avoidance		
Carcharias taurus	Female "shielding" with pelvics close to substrate	Gordon (1993)
Ginglymostoma cirratum	"Lying on back," the female rests motionless and rigid	Gordon (1993)
	Female "pivots and rolls" on her back when a male bites her pectoral fin	Klimley (1980)
Female acceptance		
Carcharias taurus	"Submissive" body, "cupping" and "flaring" of pelvic fins	Gordon (1993)
Ginglymostoma cirratum	Female arches body toward male, "cups" pelvic fins	Carrier et al. (1994)
Biting		
Heterodontus francisci	Male bites and wraps female pectoral fin body, tail, gills	Dempster and Herald (1961)
Scyliorhinus retifer	Male bites and wraps female pectoral fin body, tail, gills	Castro et al. (1988)
Scyliorhinus torazame	Male bites and wraps female pectoral fin body, tail, gills	Uchida et al. (1990)
Ginglymostoma cirratum	Male bites and holds female's pectoral fin	Klimley (1980); Carrier et al. (1994)
Carcharhinus sp.	Male bites and holds female's pectoral fin	Clark (1975)
Triaenodon obesus	Male bites and holds female's pectoral fin	Uchida et al. (1990); Tricas and LeFeuvre (1985)
Positioning and alignment		
Ginglymostoma cirratum	"Nudging" female into position with head	Klimley (1980)
Ginglymostoma cirratum	After "pectoral bite," male rolls female, then aligns for insertion	Carrier et al. (1994)
Sphyrna lewini	"Torso thrust" with "clasper flexion" possibly filling siphon sacs	Klimley (1985)
Carcharias taurus	"Crossing" or "splaying" claspers as position requires	Gordon (1993)
Group		
Ginglymostoma cirratum	Multiple males compete or cooperate for a mate; a cooperative behavior or a single male "blocking" a mating pair	Carrier et al. (1994)
Insertion and copulation	Insertion of one or more claspers into the cloaca, leading to ejaculation	Carrier et al. (1994)
Copulatory		
Male bites female while at rest		
Heterodontus francisci	Male wraps around female's body	Dempster and Herald (1961)
Scyliorhinus retifer	Male wraps around female's body	Castro et al. (1988)
Smaller shark species	Male wraps around female's body	Dempster and Herald (1961); Castro et al. (1988); Gilbert and Heath (1972); Dral (1980)
Triaenodon obesus	Heads to substrate, sharks undulate to keep tails elevated	Tricas and LeFeuvre (1985)
Ginglymostoma cirratum	"Lying parallel on substrate" less than two pectoral widths apart during bouts of "parallel swimming"	Klimley (1980)
Ginglymostoma cirratum	Heads to substrate, tails elevated or lying parallel "Copulation" sometimes in groups of many males	Carrier et al. (1994)
Heterodontus francisci	Male crosses female's body, rhythmic motion for up to 35 minutes	Dempster and Herald (1961)
Parallel swimming "in copula"		
Negaprion brevirostris	Coordinated pair swimming while copulating	Clark (1963)
Carcharodon carcharias	Possible coordinated pair swimming while copulating	Francis (1996)
Polygyny		
Ginglymostoma cirratum	Males will mate with many females over several weeks	Pratt and Carrier (2001)
Polyandry		
Ginglymostoma cirratum	Females will mate with many males over several weeks	Pratt and Carrier (2001)

TABLE 10.1 (continued)

Summary of Observed Courtship and Mating Behaviors in Elasmobranch Fishes

General Behavior and Species	Descriptions/Notes	Ref.
Postcopulatory	Pair remains together or departs rapidly	Carrier et al. (1994)
Stalking		
Carcharias taurus	Male aggression toward other species in a captive environment	Gordon (1993)
Batoids		
Precopulatory and courtship		
Following		
Aetobatus narinari	Rapid "chase," close to tail of female	Uchida et al. (1990)
Manta birostris	Rapid "chase," close to tail of female	Yano et al. (1999)
Myliobatis californica	Male ventral to female with wingbeats synchronized	Tricas (1980)
Myliobatis california	Males "follow" females	Feder et al. (1974)
Female avoidance		
Urolophus halleri	Females bury in sand to "avoid" males	Tricas et al. (1995)
Aetobatus narinari	Females raise back out of water and slap wing on surface in response to male "nipping"	Tricas (1980)
Urolophus halleri	Females spine males with caudal spine	Michaels (1993)
Female acceptance		
Raja eglanteria	"Back arching," "pectoral fin undulations" to attract males	Luer and Gilbert (1985)
Biting		
Aetobatus narinari	"Gouging," "nibbling," bites on female dorsal surface	Uchida et al. (1990)
Rhinoptera bonasus	"Gouging," "nibbling," bites on female dorsal surface	Tricas (1980)
Rhinoptera javanica	"Gouging," "nibbling," bites on female dorsal surface	Uchida et al. (1990)
Manta birostris	Male grasps pectoral fin tips (nipping)	Yano et al. (1999)
Group		
Dasyatid and Myliobatid rays	Common for multiple males to "follow" single females	Uchida et al. (1990); Tricas (1980); Feder et al. (1974)
Rhinoptera javanica	Many captive males overwhelmed a female for multiple matings	Uchida et al. (1990)
	Mortality sometimes resulted from wounds and exhaustion	
Other behaviors		
Aetobatus narinari	Males "bob" and "sway" while "following" "avoiding" females	Tricas (1980)
Copulatory		
While reposed on bottom		
Raja eglanteria	Copulate for 1 to 4 hours while at rest on bottom; male holds trailing edge of female's pectoral fin, swings tail beneath hers, and inserts one clasper	Luer and Gilbert (1985)
While swimming		
Manta birostris	Copulation near the surface, abdomen to abdomen	Yano et al. (1999)
Aetobatus narinari	Mating abdomen to abdomen in the mid-depths of the tank; insertion time was 0.5 to 1.5 minutes	Uchida et al. (1990)
Rhinoptera javanica	Starts at the surface or mid-depth, abdomen to abdomen; continues on the bottom	Uchida et al. (1990)
Rhinoptera bonasus	Starts at the surface or mid-depth, abdomen to abdomen; continues on the bottom	Uchida et al. (1990)
Polyandry		
Aetobatus narinari	A captive female mated many times in succession with three to four males in 1 hour	Uchida et al. (1990)
Rhinoptera javanica	Multiple matings common	Uchida et al. (1990)
Postcopulatory		
Manta birostris	Males remains attached to pectoral fin tip briefly	Yano et al. (1999)

Source: Pratt, Jr., H.L. and Carrier, J.C., *Environ. Biol. Fish.*, 60, 157–188, 2001. With permission.

10.6 Summary and Conclusions

Chondrichthyan reproduction is based on internal fertilization. All male chondrichthyans have intromittant organs called *claspers*. Fetal nutrition may be lecithotrophic or matrotrophic. Two forms of lecithotrophic fetal nutrition have been recognized: (1) oviparity, wherein females deposit large numbers of leathery egg capsules over several weeks, or months; and (2) yolk sac viviparity, wherein the developing embryos are retained in the uterus and nurtured by yolk from the egg through gestation. Fetal nutrition from the yolk sac is universal in the initial stages of development for all viviparous chondrichthyans, even those that are clearly matrotrophic. Phylogenetic patterns and evidence from the structure of the oviducal gland and uterus and the universal presence of urea retention suggest that yolk sac viviparity, not oviparity, is the plesiomorphic reproductive mode in the Neoselachii, if not Chondrichthyes.

Matrotrophic forms of fetal nutrition include mucoid histotrophy, lipid histotrophy, placentotrophy, carcharhiniform oophagy, and lamniform oophagy. Mucoid histotrophy may be divided into two forms: (1) incipient histotrophy, wherein the embryos are primarily yolk sac dependent and mucoid secretions contribute 20% or less of the organic material in term embryos; and (2) full mucoid histotrophy, wherein the histotrophic contribution results in an often significant gain in organic material in term embryos. Incipient histotrophy has been documented in some squaloids and batoids. Mucoid histotrophy is important in the Carcharhiniformes; it is the principal source of fetal nutrition in many if not most nonplacental triakids and probably occurs in the viviparous proscylliids and pseudotriakids. In addition, mucoid histotrophy probably occurs in all placental sharks between the time of yolk sac absorption and placentation and even during placentation in many species. Lipid histotrophy is limited to the myliobatiform stingrays and involves the secretion of a lipid-rich histotroph, which can result in term embryo organic gains of almost 5000%. Placentotrophy is restricted to the carcharhiniform families Leptochariidae, Triakidae, Hemigaleidae, Carcharhinidae, and Sphyrnidae. The shark placenta is in some ways analogous but not homologous to the mammalian placenta. Contrary to some earlier reports in the literature, no sharks have hemotrophic transfer of nutrients via the placenta, which rather functions to absorb mucoid secretions from the uterus, facilitated at the site of attachment. Thus, placentotrophy in sharks may be viewed as a highly modified form of mucoid histotrophy. Carcharhiniform oophagy is restricted to the small family Pseudotriakidae, wherein each embryo is supported by unfertilized ova enclosed within the egg envelope, a one-time maternal contribution. Pseudotriakids may also be supported by mucoid histotroph later in gestation. Lamniform oophagy is limited to lamniform sharks, wherein developing embryos are supported by a multitude of small ova produced by the mother during most of gestation.

The recent documentation of multiple paternity in several species of chondrichthyans has raised questions about the cost and benefits of multiple matings. Males and females have different reproductive strategies relative to maximizing evolutionary fitness mitigated by their relatively violent mating behavior. Males achieve maximum fitness by mating with as many females as many times as they can (sperm are energetically cheap). Conversely, females may achieve maximum fitness by mating only a sufficient number of times to ensure that all her ova are fertilized while minimizing the threat of serious injury during mating. These ideas go a long way toward explaining the evolution of sperm storage and geographic sexual segregation phenomena widespread among the living Chondrichthyes and for which evidence exists back into the Paleozoic.

Acknowledgments

Thanks are due to Jose Castro for critically reviewing this manuscript and to Will Hamlett, Eileen Grogan, and Dick Lund for stimulating discussions leading to some of the ideas espoused in this manuscript. All errors are the responsibility of the authors. This is Contribution Number VIMS 3160 from the Virginia Institute of Marine Science.

References

Bansemer, C.S. and Bennett, M.B. (2009). Reproductive periodicity, localized movements and behavioral segregation of pregnant *Carcharias taurus* at Wolf Rock, southeast Queensland, Australia. *Mar. Ecol. Prog. Ser.* 374:215–227.

Bell, J.D., Lyle, J.M., Bulman, C.M., Graham, K.J., Newton, G.M., and Smith, D.C. (1992). Spatial variation in reproduction and occurrence of non-reproductive adults, in orange roughy, *Hopostethus atlanticus* Collett (Trachichthyidae), from south-eastern Australia. *J. Fish Biol.* 40:107–122.

Braccini, J.M., Hamlett, W.C., and Gillanders, B.M. (2007). Embryo development and maternal-embryo nutritional relationships of piked spurdog (*Squalus megalops*). *Mar. Biol.* 150:727–737.

Capapé, C. (1993). New data on the reproductive biology of the thorny stingray, *Dasyatis centroura* (Pisces: Dasyatidae) from off the Tunisian coasts. *Environ. Biol. Fish.* 38:73–80.

Capapé, C. and Zaouali, J. (1995). Reproductive biology of the marbled sting ray, *Dasyatis marmorata* (Steindachner, 1892) (Pisces: Dasyatidae) in Tunisian waters (Central Mediterranean). *J. Aquaricult. Aquat. Sci.* 7:108–119.

Capapé, C., Quignard, J.P., and Mellinger, J. (1990). Reproduction and development of two angel sharks, *Squatina squatina* and *S. oculata* (Pisces, Squatinidae), off Tunisian coasts—semi-delayed vitellogenesis, lack of egg capsules, and lecithotrophy. *J. Fish Biol.* 37:347–356.

Capapé, C., Guelorget, O., Vergne, Y., and Reynaud C. (2008). Reproductive biology of the blackmouth catshark, *Galeus melastomus* (Chondrichthyes: Scyliorhinidae) off the Languedocian coast (southern France, northern Mediterranean). *J. Mar. Biol. Assoc. U.K.* 88:415–421.

Carrier, J.C., Pratt, Jr., H.L., and Martin, L.K. (1994). Group reproductive behaviors in free-living nurse sharks, *Ginglymostoma cirratum. Copeia* 1994:646–656.

Carrier, J.C., Murru, F.L., Walsh, M.T., and Pratt, H.L. (2003). Assessing reproductive potential and gestation in nurse sharks (*Ginglymostoma cirratum*) using ultrasonography and endoscopy: an example of bridging the gap between field research and captive studies. *Zool. Biol.* 22:179–187.

Carrier, J.C., Pratt, Jr., H.L., and Castro, J.I. (2004). Reproductive biology of elasmobranchs. In: Carrier, J.C., Musick, J.A., and Heithaus, M.R. (Eds.), *Biology of Sharks and Their Relatives*, CRC Press, Boca Raton, FL, pp 269–286.

Castro, J.I. (1996). Biology of the blacktip shark, *Carcharhinus limbatus*, off the southeastern United States. *Bull. Mar. Sci.* 59:508–522.

Castro, J.I. (2000). The biology of the nurse shark, *Ginglymostoma cirratum*, off the Florida east coast and the Bahama Islands. *Environ. Biol. Fish.* 58:1–22.

Castro, J.I. (2009). Observations on the reproductive cycles of some viviparous North American sharks. *Intl. J. Ichthyol.* 15:205–222.

Castro, J.I., Bubucis, P.M., and Overstrom, N.A. (1988). The reproductive biology of the chain dogfish, *Scyliorhinus rotifer. Copeia* 1988:740–746.

Chapman, D.D., Shivji, M.S., Louis, E., Sommer, J., Fletcher, H., and Prodohl, P.A. (2007). Virgin birth in a hammerhead shark. *Biol. Lett.* 3:425–427.

Chapman, D.D., Firchau, B., and Shivji, M.S. (2008). Parthenogenesis in a large-bodied requiem shark, the blacktip *Carcharhinus limbatus. J. Fish Biol.* 73:1473–1477.

Chevolot, M., Ellis, J.R., Rijnsdorp, A.D., Stam, W.T., and Olsen, J.L. (2007). Multiple paternity analysis in the thornback ray *Raja clavata*, L. *J. Hered.* 98:712–715.

Clark, E. (1963). The maintenance of sharks in captivity, with a report on their instrumental conditioning. In: Gilbert, P.W. (Ed.), *Sharks and Survival*, Heath, Boston, pp. 115–149.

Clark, E. (1975). The strangest sea. *Natl. Geogr. Mag.* 148:338–343.

Clark, R.S. (1922). Rays and skates, Part 1. *J. Mar. Biol. Assoc. U.K.* 12:577–643.

Cliff, G., Dudley, S.F.J., and Davis, B. (1988). Sharks caught in the protective gill nets off Natal, South Africa. 1. The sandbar shark, *Carcharhinus plumbeus* (Nardo). *S. Afr. J. Mar. Sci.* 7:255–265.

Compagno, L.J.V. (1984). *FAO Species Catalogue.* Vol. 4, Part 1. *Sharks of the World: An Annotated and Illustrated Catalogue of Shark Species Known to Date.* Food and Agriculture Organization of the United Nations, Rome.

Compagno, L.J.V. (1988). *Sharks of the Order Carcharhiniformes.* Princeton University Press, Princeton, NJ.

Compagno, L.J.V. (1990). Alternative life-history styles of cartilaginous fishes in time and space. *Environ. Biol. Fish.* 28:33–75.

Compagno, L.J.V. (2001). *Sharks of the World: An Annotated and Illustrated Catalogue of Shark Species Known to Date.* Vol. 2. *Bullhead, Mackerel and Carpet Sharks (Heterodontiformes, Lamniformes and Orectolobiformes).* Food and Agriculture Organization of the United Nations, Rome.

Compagno, L.J.V., Dando, M., and Fowler, S. (2005). *Sharks of the World.* Princeton University Press, Princeton, NJ.

Costa, M.E., Erzini, K., and Borges, T.C. (2005). Reproductive biology of the blackmouth catshark, *Galeus melastomus* (Chondrichthyes: Scyliorhinidae) off the south coast of Portugal. *J. Mar. Biol. Assoc. U.K.* 85:1173–1183.

Cotton, C.F. (2010). Age, Growth, and Reproductive Biology of Deep-Water Chondrichthyans, doctoral dissertation, Virginia Institute of Marine Science, College of William and Mary, Gloucester Point.

Daly-Engel, T.S., Grubbs, R.D., Feldheim, K.A., Bowen, B.W., and Toonen, R.J. (2010). Is multiple mating beneficial or unavoidable? Low multiple paternity and genetic diversity in the shortspine spurdog *Squalus mitsukurii. Mar. Ecol. Prog. Ser.* 403:255–267.

Dempster, R.P. and Herald, E.S. (1961). Notes on the hornshark, *Heterodontus francisci*, with observations on mating activities. *Occ. Pap. Calif. Acad. Sci.* 33:1–7.

Diez, J.M. and Davenport, J. (1987). Embryonic respiration in the dogfish (*Scyliorhinus canicula* L.). *J. Mar. Biol. Assoc. U.K.* 67:249–261.

Dral, A.J. (1980). Reproduction en aquarium du requin de fond tropical, *Chiloscyllium griseum* Mull. et Henle (Orectolobides). *Rev. Fr. Aquariol.* 7:99–104.

Driggers, W.B. and Hoffmayer, E.R. (2009). Variability in the reproductive cycle of finetooth sharks, *Carcharhinus isodon*, in the northern Gulf of Mexico. *Copeia* 2009:390–393.

Dulvey, N.K. and Reynolds, J.D. (1997). Evolutionary transitions among egg-laying, live-bearing, and maternal inputs in sharks and rays. *Proc. R. Soc. B Biol. Sci.* 264:1309–1315.

Ebert, D.A. and Winton, M.V. (2010). Chondrichthyans of high latitude seas. In: Carrier, J.C., Musick, J.A., and Heithaus, M.R. (Eds.), *Biodiversity, Adaptive Physiology, and Conservation.* CRC Press, Boca Raton, FL, pp. 115–158.

Ellis, J.R. and Shackley, S.E. (1995). Observations on egg-laying in the thornback ray. *J. Fish Biol.* 46:903–904.

Engel, K.B. and Callard, G.V. (2005). The testis and spermatogenesis. In: Hamlett, W.C. (Ed.), *Reproductive Biology and Phylogeny of Chondrichthyes.* Science Publishers, Enfield, NH, pp. 171–200.

Engelhard, G.H. and Heino, M. (2005). Scale analysis suggests frequent skipping of the second reproductive season in Atlantic herring. *Biol. Lett.* 1:172–175.

Fahy, D.P., Spieler, R.E., and Hamlett, W.C. (2007). Preliminary observations on the reproductive cycle and uterine fecundity of the yellow stingray, *Urobatis jamaicensis* (Elasmobranchii: Myliobatiformes: Urolophidae) in Southeast Florida, USA. *Raffles Bull. Zool.* 2007:131–139.

Farrell, E.D., Mariani, S., and Clarke, M.W. (2010). Reproductive biology of the starry smooth-hound shark *Mustelus asterias*: geographic variation and implications for sustainable exploitation. *J. Fish Biol.* 77:1505–1525.

Feder, H.M., Turner, C.H., and Limbaugh, C. (1974). Observations on fishes associated with kelp beds in Southern California. *Calif. Fish Biol.* 1160:1–144.

Feldheim, K.A., Gruber, S.H., and Ashley, M.V. (2004). Reconstruction of parental microsatellite genotypes reveals female polyandry and philopatry in the lemon shark, *Negaprion brevirostris. Evolution* 58:2332–2342.

Feldheim, K.A., Chapman, D.D., Sweet, D., Fitzpatrick, S., Prodohl, P.A., Shivji, M.S., and Snowden, B. (2010). Shark virgin birth produces multiple, viable offspring. *J. Hered.* 101:374–377.

Francis, M.P. (1996). Observations on pregnant white shark with a review of reproductive biology. In: Klimley, A.P. and Ainley, D.G. (Eds.), *Great White Sharks: The Biology of Carcharodon carcharias*, Academic Press, San Diego, CA, pp. 157–172.

Francis, M.P. and Stevens, K.D. (1999). Reproduction, embryonic development, and growth of the porbeagle shark, *Lamna nasus*, in the southwest Pacific Ocean. *Fish. Bull.* 98:41–63.

Garayzar, C.J.V., Hoffman, C.D., and Melendez, E.M. (1994). Size and reproduction of the ray, *Dasyatis longus* (Pisces, Dasyatidae) in Bahia Almejsa, Baja California Sur, Mexico. *Rev. Biol. Trop.* 42:375–377.

Gilbert, P.W. (1984). Biology and behaviour of sharks. *Endeavour* 8:179–187.

Gilbert, P.W. and Heath, G.W. (1972). The clasper–siphon sac mechanism in *Squalus acanthias* and *Mustelus canis. Comp. Biochem. Physiol.* 42A:97–119.

Gilmore, R.G. (1993). Reproductive biology of lamnoid sharks. *Environ. Biol. Fish.* 38:95–114.

Gilmore, R.G., Dodrill, J.W., and Linley, P.A. (1983). Reproduction and embryonic development of the sand tiger shark, *Odontaspis taurus* (Rafinesque). *Fish. Bull.* 81:201–225.

Gilmore, R.G., Putz, O., and Dodrill, J.W. (2005). Oophagy, intrauterine cannibalism and reproductive strategy in lamnoid sharks. In: Hamlett, W.C. (Ed.), *Reproductive Biology and Phylogeny of Chondrichthyes*. Science Publishers, Enfield, NH, pp. 435–462.

Gordon, I. (1993). Pre-copulatory behavior of captive sand tiger sharks, *Carcharias taurus. Environ. Biol. Fish.* 38:159–164.

Goto, T. (2001). Comparative anatomy, phylogeny and cladisitic classification of the order Orectolobiformes (Chondrichthyes, Elasmobranchii). *Mem. Grad. Sch. Fish. Sci. Hokkaido Univ.* 48:1–100.

Griffith, R.W. (1991). Guppies, toadfish, lungfish, coelacanths, and frogs: a scenario for the evolution of urea retention in fishes. *Environ. Biol. Fish.* 32:199–218.

Grogan, E.D. and Lund, R. (2004). Origin and relationships of early Chondrichthyes. In: Carrier, J.C., Musick, J.A., and Heithaus, M.R. (Eds.), *Biology of Sharks and Their Relatives*. CRC Press, Boca Raton, FL, pp. 3–31.

Grogan, E.D. and Lund, R. (2011). Superfoetative viviparity in a Carboniferous chondrichthyan and reproduction in early gnathostomes. *Zool. J. Linn. Soc.* 161:587–594.

Grubbs, R.D. (2010). Ontogenetic shifts in movements and habitat use. In: Carrier, J.C., Musick, J.A., and Heithaus, M.R. (Eds.), *Biodiversity, Adaptive Physiology, and Conservation*. CRC Press, Boca Raton, FL, pp. 319–350.

Guallart, J. and Vicent, J.J. (2001). Changes in composition during embryo development of the gulper shark, *Centrophorus granulosus* (Elasmobranchii, Centrophoridae): an assessment of maternal–embryonic nutritional relationships. *Environ. Biol. Fish.* 61:135–150.

Hamlett, W.C. (1997). Reproductive modes of elasmobranchs. *Shark News* 9:1–3.

Hamlett, W.C. (1999). Male reproductive system. In: Hamlett, W.C. (Ed.), *Sharks, Skates, and Rays: The Biology of Elasmobranch Fishes*. The Johns Hopkins University Press, Baltimore, MD, pp. 444–470.

Hamlett, W.C. and Koob, T.J. (1999). Female reproductive system. In: Hamlett, W.C. (Ed.), *Sharks, Skates, and Rays: The Biology of Elasmobranch Fishes*. The Johns Hopkins University Press, Baltimore, MD, pp. 398–443.

Hamlett, W.C., Knight, D.P., Koob, T.J., Jezior, M., Luong, T. et al. (1998). Survey of oviducal gland structure and function in elasmobranchs. *J. Exp. Zool.* 282:399–420.

Hamlett, W.C., Knight, D.P., Pereira, F.T.V., Steele, J., and Sever, D.M. (2005a). Oviducal glands in Chondrichthyans. In: Hamlett, W.C. (Ed.), *Reproductive Biology and Phylogeny of Chondrichthyes*. Science Publishers, Enfield, NH, pp. 301–335.

Hamlett, W.C., Kormarik, C.G., Storrie, M., Serevy, B., and Walker, T.I. (2005b). Chondrichthyan parity, lecithotrophy and matrotrophy. In: Hamlett, W.C. (Ed.), *Reproductive Biology and Phylogeny of Chondrichthyes*. Science Publishers, Enfield, NH, pp. 395–434.

Hamlett, W.C., Jones, C.J.P., and Paulesu, L.R. (2005c). Placentotrophy in sharks. In: Hamlett, W.C. (Ed.), *Reproductive Biology and Phylogeny of Chondrichthyes*. Science Publishers, Enfield, NH, pp. 463–502.

Hoar, W.S. (1969). Reproduction. In: Hoar, W.S. and Randall, D.J. (Eds.), *Fish Physiology.* Vol. III. *Reproduction and Growth: Bioluminescence, Pigments and Poisons*. Academic Press, New York, pp. 1–72.

Hoff, G.R. (2008). A nursery site of the Alaska skate (*Bathyraja parmifera*) in the eastern Bering Sea. *Fish. Bull.* 106:233–244.

Hoff, G.R. (2010). Identification of skate nursery habitat in the eastern Bering Sea. *Mar. Ecol. Prog. Ser.* 403:243–254.

Holden, M.J. (1975). The fecundity of *Raja clavata* in British waters. *J. Cons. Int. Explor. Mer* 36:110–118.

Jacobsen, I.P., Johnson, J.W., and Bennett, M.B. (2009). Diet and reproduction in the Australian butterfly ray, *Gymnura australis*, from northern and north-eastern Australia. *J. Fish Biol.* 75:2475–2489.

Jamieson, G.M. (2005). Chondrichthyan spermatozoa and phylogeny. In: Hamlett, W.C. (Ed.), *Reproductive Biology and Phylogeny of Chondrichthyes*. Science Publishers, Enfield, NH, pp. 201–237.

Johnson, R.H. and Nelson, D.R. (1978). Copulation and possible olfaction-mediated pair formation in two species of carcharhinid sharks. *Copeia* 1978:539–542.

Jones, B.C. and Geen, G.H. (1977). Reproduction and embryonic development of spiny dogfish (*Squalus acanthias*) in the Strait of Georgia, British Columbia. *J. Fish. Res. Bd. Can.* 34:1286–1292.

Jones, C.J.P., Walker, T.I., Bell, J.D., Reardon, M.B., Ambrosio, C.E., Almeida, A., and Hamlett, W.C. (2005). Male genital ducts and copulatory appendages in Chondrichthyans. In: Hamlett, W.C. (Ed.), *Reproductive Biology and Phylogeny of Chondrichthyes*. Science Publishers, Enfield, NH, pp. 361–393.

Joung, S.J. and Chen, C.T. (1995). Reproduction in the sandbar shark, *Carcharhinus plumbeus*, in the waters off Northeastern Taiwan. *Copeia* 3:659–665.

Joung, S.J., Chen, C.T., Clark, E., Uchinda, S., and Huang, W.Y.P. (1996). The whale shark, *Rhincodon typus*, is a livebearer: 300 embryos found in one 'megamamma' supreme. *Environ. Biol. Fish.* 46:219–223.

Kajiura, S.M., Sebastian, A.P., and Tricas, T.C. (2000). Dermal bite wounds as indicators of reproductive seasonality and behaviour in the Atlantic stingray, *Dasyatis sabina*. *Environ. Biol. Fish.* 58:23–31.

Klimley, A.P. (1980). Observations of courtship and copulation in the nurse shark, *Ginglymostoma cirratum*. *Copeia* 1980:878–882.

Klimley, A.P. (1985). Schooling in the large predator, *Sphyrna lewini*, a species with low risk of predation: a non-egalitarian state. *Z. Tierpsychol.* 70:297–319.

Koob, T. and Hamlett, W.C. (1998). Microscopic structure of the gravid uterus in *Raja erinacea*. *J. Exp. Zool.* 282:421–437.

Liu, K.-M., Chen, C.-T., Liao, T.-H., and Joung, S.-J. (1999). Age, growth, and reproduction of the pelagic thresher shark, *Alopias pelagicus*, in the northwestern Pacific. *Copeia* 1999:68–74.

Lopez, J.A., Ryburn, J.A., Fedrigo, O., and Naylor, G.J.P. (2006). Phylogeny of sharks of the family Triakidae (Carcharhiniformes) and its implications for the evolution of carcharhiniform placental viviparity. *Mol. Phylogenet. Evol.* 40:50–60.

Luer, C.A. and Gilbert, P.W. (1985). Mating behavior, egg deposition, incubation period and hatching in the clearnose skate, *Raja eglanteria*. *Environ. Biol. Fish.* 13:161–171.

Lutton, B.V., St. George, J., Murrin, C.R., Fileti, L.A., and Callard, I.P. (2005). The elasmobranch ovary. In: Hamlett, W.C. (Ed.), *Reproductive Biology and Phylogeny of Chondrichthyes*. Science Publishers, Enfield, NH, pp. 237–281.

Marshall, L.J., White, W.T., and Potter, I.C. (2007). Reproductive biology and diet of the southern fiddler ray, *Trygonorrhina fasciata* (Batoidea: Rhinobatidae), an important trawl bycatch species. *Mar. Freshw. Res.* 58:104–115.

Mattos, S.M.G., Broadhurst, M.K., Hazin, F.H.V., and Jones, D.M. (2001). Reproductive biology of the Caribbean sharpnose shark, *Rhizoprionodon porosus*, from northern Brazil. *Mar. Freshw. Res.* 52:745–752.

McAuley, R.B., Simpfendorfer, C.A., Hyndes, G.A., and Lenanton, R.C.J. (2007). Distribution and reproductive biology of the sandbar shark, *Carcharhinus plumbeus* (Nardo), in western Australian waters. *Mar. Freshw. Res.* 58:116–126.

Meagher, P. (2011). Reproductive Biology, Maternal–Fetal Exchange and Fishery Impact in the Viviparous Eastern Shovenose Ray *Aptychtrema rostrata* in New South Wales, Australia, doctoral dissertation, University of Sydney.

Michaels, S. (1993). *Reef Sharks and Rays of the World*. Sea Challengers, Monterey, CA, 107 pp.

Mollet, H.F., Cliff, G., Pratt, H.L., and Stevens, J.D. (2000). Reproductive biology of the female shortfin mako, *Isurus oxyrinchus* Rafinesque, 1810, with comments on the embryonic development of lamnoids. *Fish. Bull.* 98:299–318.

Moreno, J.A. and Moron, J. (1992). Reproductive biology of the bigeye thresher shark, *Alopias superciliosus* (Lowe, 1839). *Aust. J. Mar. Freshw. Res.* 43:77–86.

Musick, J.A. (2010). Chondrichthyan reproduction. In: Coles, K. (Ed.), *Reproduction and Sexuality of Marine Fishes*. University of California Press, Los Angeles, pp. 3–20.

Musick, J.A. and Ellis, J. (2005). Reproductive evolution of chondrichthyans. In: Hamlett, W.C. (Ed.), *Reproductive Biology and Phylogeny of Chondrichthyans*. Science Publishers, Enfield, NH, pp. 45–79.

Musick, J.A., Branstetter, S., and Colvocoresses, J.A. (1993). Trends in shark abundance from 1974 to 1991 for the Chesapeake Bight region of the U.S. mid-Atlantic coast. In: Branstetter, S. (Ed.), *Conservation Biology of Elasmobranchs*, NOAA Tech. Rep. NMFS 115. U.S. Department of Commerce, Washington, D.C., pp. 1–18.

Musick, J.A., Harbin, M.M., and Compagno, L.J.V. (2004). Historical zoogeography of the Selachii. In: Carrier, J., Musick, J.A., and Heithaus, M. (Eds.), *Biology of Sharks and Their Relatives*. CRC Press, Boca Raton, FL, pp. 33–78.

Needham, J. (1942). *Biochemistry and Morphogenesis*. Cambridge University Press, Cambridge, U.K.

Neer, J.A. and Thompson, B.A. (2005). Life history of the cownose ray, *Rhinoptera bonasus*, in the northern Gulf of Mexico, with comments on geographic variability in life history traits. *Environ. Biol. Fish.* 73:321–331.

Oliveira, P., Hazin, F.H.V., Carvalho, F., Rego, M., Coelho, R., Piercy, A., and Burgess, G. (2010). Reproductive biology of the crocodile shark *Pseudocarcharias kamoharai*. *J. Fish Biol.* 76:1655–1670.

Otake, T. (1990). Classification of reproductive modes in sharks with comments on female reproductive tissues and structures. In: Pratt, Jr., H.L., Gruber, S.H., and Taniuchi, T. (Eds.), *Elasmobranchs as Living Resources: Advances in the Biology, Ecology, Systematics, and Status of the Fisheries*, NOAA Tech. Rep. NMFS 90. U.S. Department of Commerce, Washington, D.C., pp. 111–130.

Parsons, G.R. (1993a). Geographic variation in reproduction between two populations of the bonnethead shark, *Sphyrna tiburo*. *Environ. Biol. Fish.* 38:25–35.

Parsons, G.R. (1993b). Age determination and growth of the bonnethead shark, *Sphyrna tiburo*: a comparison of two populations. *Mar. Biol.* 117:23–31.

Peres, M.B. and Vooren, C.M. (1991). Sexual development, reproductive cycle, and fecundity of the school shark, *Galeorhinus galeus*, off southern Brazil. *Fish. Bull.* 89:655–667.

Perkins, F.E. (1965). Incubation of fall-spawned eggs of the little skate, *Raja erinacea*. *Copeia* 1965:114–115.

Portnoy, D.S. (2010). Molecular insights into elasmobranch reproductive behavior for conservation and management. In: Carrier, J., Musick, J.A., and Heithaus, M. (Eds.), *Sharks and Their Relatives II: Biodiversity, Adaptive Physiology, and Conservation.* CRC Press, Boca Raton, FL, pp. 435–458.

Portnoy, D.S., Piercy, A.N., Musick, J.A., Burgess, G.H., and Graves, J.E. (2007). Genetic polyandry and sexual conflict in the sandbar shark, *Carcharhinus plumbeus*, in the western North Atlantic and Gulf of Mexico. *Mol. Ecol.* 16:187–197.

Pratt, Jr., H.L. (1979). Reproduction in the blue shark, *Prionace glauca. Fish. Bull.* 77:445–470.

Pratt, Jr., H.L. (1988). Elasmobranch gonad structure: A description and survey. *Copeia* 1988:719–729.

Pratt, Jr., H.L. (1993). The storage of spermatozoa in the oviducal glands of western North Atlantic sharks. *Environ. Biol. Fish.* 38:139–149.

Pratt, Jr., H.L. and Carrier, J.C. (2001). A review of elasmobranch reproductive behavior with a case study on the nurse shark, *Ginglymostoma cirratum. Environ. Biol. Fish.* 60:157–188.

Pratt, Jr., H.L. and Carrier, J.C. (2005). Elasmobranch courtship and mating behavior. In: Hamlett, W.C. (Ed.), *Reproductive Biology and Phylogeny of Chondrichthyes.* Science Publishers, Enfield, NH, pp. 129–170.

Ranzi, S. (1932). Le basi fisio-morfologische dello sviluppo embrionale dei Selaci, Parti I. *Pubblicazioni Della Stazione Zoologica di Napoli* 13:209–240.

Ranzi, S. (1934). Le basi fisio-morfologische dello sviluppo embrionale dei Selaci, Parti II and III. *Pubblicazioni Della Stazione Zoologica di Napoli* 13:331–437.

Rideout, R.M. and Rose, G.A. (2006). Suppression of reproduction in Atlantic cod *Gadus morhua. Mar. Ecol. Prog. Ser.* 320:267–277.

Romine, J.G., Musick, J.A., and Burgess, G.H. (2009). Demographic analyses of the dusky shark, *Carcharhinus obscurus*, in the Northwest Atlantic incorporating hooking mortality estimates and revised reproductive parameters. *Environ. Biol. Fish.* 84:277–289.

Rosenblatt, R. (1963). Some aspects of speciation in marine shore fishes. In: Harding, J.P. and Tebble, N. (Eds.), *Speciation in the Sea.* Systematics Association, London, pp. 117–180.

Sanzo, L. (1912). Embrione di *Carcharadon rondoletii*, M. Hle. Con particolare disposizione del sacco vitellino. *R. Comitato Talassografico Italiano, Memoria* 11:3–9.

Saville, K.J., Lindley, A.M., Maries, E.G., Carrier, J.C., and Pratt, Jr., H.L. (2002). Multiple paternity in the nurse shark, *Ginglymostoma cirratum. Environ. Biol. Fish.* 63:347–351.

Setna, P.B. and Sarangdhar, P.N. (1950). Breeding habits of Bombay elasmobranchs. *Rec. Indian Mus.* 47:107–124.

Simpfendorfer, C.A. (1992). Reproductive strategy of the Australian sharpnose shark, *Rhizoprionodon taylori* (Elasmobranchii: Carcharhinidae) from Cleveland Bay, Northern Queensland. *Aust. J. Mar. Freshw. Res.* 43:67–75.

Smale, M.J. and Goosen, A.J.J. (1999). Reproduction and feeding of spotted gully shark, *Triakis megalopterus*, off the Eastern Cape, South Africa. *Fish. Bull.* 97:987–998.

Smith, J.W. and Merriner, J.V. (1986). Observations on the reproductive biology of the cownose ray, *Rhinoptera bonasus*, in Chesapeake Bay. *Fish. Bull.* 84:871–877.

Snelson, F.F., Williams-Hooper, S.E., and Schmid, T.H. (1989). Biology of the bluntnose stingray, *Dasyatis sayi*, in Florida coastal lagoons. *Bull. Mar. Sci.* 45:15–25.

Sogard, S.M. (1997). Size-selective mortality in the juvenile stage of teleost fishes: a review. *Bull. Mar. Sci.* 60:1129–1157.

Springer, S. (1948). Oviphagous embryos of the sand shark, *Odontaspis taurus. Copeia* 1983:153–157.

Stevens, B. (2002). Uterine and Oviducal Mechanisms for Gestation in the Common Sawshark, *Pristiophorus cirratus*, bachelor of science thesis, University of Melbourne, Victoria, Australia.

Storrie, M.T., Walker, T.I., Laurenson, L.J., and Hamlett, W.C. (2009). Gestational morphogenesis of the uterine epithelium of the gummy shark (*Mustelus antarcticus*). *J. Morphol.* 270:319–336.

Sumpter, J.P. and Dodd, J.M. (1979). The annual reproductive cycle of the female small-spotted catshark, *Scyliorhinus canicula*, L., and its endocrine control. *J. Fish Biol.* 15:687–695.

Sunye, P.S. and Vooren, C.M. (1997). On cloacal gestation in angel sharks from southern Brazil. *J. Fish Biol.* 50:86–94.

Tanaka, S., Shiobara, Y., Hioki, S., Abe, H., Nishi, G., Yano, K., and Suzuki, K. (1990). The reproductive biology of the frilled shark, *Chlamydoselachus anguineus*, from Suruga Bay, Japan. *Jpn. J. Ichthyol.* 37:273–291.

Teshima, K. (1981). Studies on the reproduction of Japanese dusky smooth-hounds, *Mustelus manazo* and *M. griseus. J. Shimonoseki Univ. Fish.* 29:113–199.

Tovar-Avila, J., Walker, T.I., and Day, R.W. (2007). Reproduction of *Heterodontus portusjacksoni* in Victoria, Australia: evidence of two populations and reproductive parameters for the eastern population. *Mar. Freshw. Res.* 58:956–965.

Tricas, T.C. (1980). Courtship and mating-related behaviors in myliobatid rays. *Copeia* 1980:553–556.

Tricas, T.C. and LeFeuvre, E.M. (1985). Mating in the reef white-tip shark *Triaenodon obesus. Mar. Biol.* 84:233–237.

Tricas, T.C., Michael, S.W., and Sisneros, J.A. (1995). Electrosensory optimization to conspecific phasic signals for mating. *Neurosci. Lett.* 202:129–132.

Tullis, A. and Peterson, G. (2000). Growth and metabolism in the embryonic white-spotted bamboo shark, *Chiloscyllium plagiosum*: comparison with embryonic birds and reptiles. *Physiol. Biochem. Zool.* 73:271–282.

Uchida, S., Toda, M., and Kamei, Y. (1990). Reproduction of elasmobranchs in captivity. In: Pratt, Jr., H.L., Gruber, S.H., and Taniuchi, T. (Eds.), *Elasmobranchs as Living Resources: Advances in the Biology, Ecology, Systematics, and Status of the Fisheries*, NOAA Tech. Rep. NMFS 90. U.S. Department of Commerce, Washington, D.C., pp. 211–237.

Walker, T.I. (2005). Reproduction in fisheries science. In: Hamlett, W.C. (Ed.), *Reproductive Biology and Phylogeny of Chondrichthyans.* Science Publishers, Enfield, NH, pp. 81–128.

Walker, T.I. (2007). Spatial and temporal variation in the reproductive biology of gummy shark *Mustelus antarcticus* (Chondrichthyes: Triakidae) harvested off southern Australia. *Mar. Freshw. Res.* 58:67–97.

Wass, R.C. (1973). Size, growth, and reproduction of the sandbar shark *Carcharhinus milberti*, in Hawaii. *Pac. Sci.* 27:305–308.

Wenbin, Z. and Shuyuan, Q. (1993). Reproductive biology of the guitarfish, *Rhinobatos hynnicephalus*. *Environ. Biol. Fish.* 38:81–93.

White, W.T. (2007). Aspects of the biology of carcharhiniform sharks in Indonesian waters. *J. Mar. Biol. Assoc. U.K.* 87:1269–1276.

White, W.T., Platell, M.E., and Potter, I.C. (2001). Relationship between reproductive biology and age composition and growth in *Urolophus lobatus* (Batoidea: Urolophidae). *Mar. Biol.* 138:135–147.

White, W.T., Hall, N.G., and Potter, I.C. (2002). Reproductive biology and growth during pre- and postnatal life of *Trygonoptera personata* and *T. mucosa* (Batoidea: Urolophidae). *Mar. Biol.* 140:699–712.

Wilson, C.D. and Seki, M.P. (1994). Biology and population characteristics of *Squalus mitsukurii* from a seamount in the central North Pacific Ocean. *Fish. Bull.* 92:851–864.

Wourms, J.P. (1977). Reproduction and development in chondrichthyan fishes. *Am. Zool.* 17:379–410.

Wourms, J.P. (1981). Viviparity: the maternal–fetal relationship in fishes. *Am. Zool.* 21:473–515.

Wourms, J.P. (1993). Maximization of evolutionary trends for placental viviparity in the spadenose shark, *Scoliodon laticaudus*. *Environ. Biol. Fish.* 38:269–294.

Yamaguchi, A., Taniuchi, T., and Shimizu, M. (2000). Geographic variations in reproductive parameters of the starspotted dogfish, *Mustelus manazo*, from five localities in Japan and in Taiwan. *Environ. Biol. Fish.* 57:221–233.

Yano, K. (1992). Comments on the reproductive mode of the false cat shark, *Pseudotriakis microdon*. *Copeia* 1992:460–468.

Yano, K. (1993). Reproductive biology of the slender smoothhound, *Gollum attenuatus*, collected from New Zealand waters. *Environ. Biol. Fish.* 38:59–71.

Yano, K. (1995). Reproductive biology of the black dogfish, *Centroscyllium fabricii*, collected from waters off western Greenland. *J. Mar. Biol. Assoc. U.K.* 75:285–310.

Yano, K., Sato, F., and Takahashi, T. (1999). Observation of the mating behavior of the manta ray, *Manta birostris*, at the Ogasawara Islands. *Ichthyol. Res.* 46:289–296.

Zeh, J.A. and Zeh, D.W. (2001). Reproductive mode and the genetics of polyandry. *Anim. Behav.* 61:1051–1063.

Hormonal Regulation of Elasmobranch Physiology

James Gelsleichter and Andrew N. Evans

CONTENTS

11.1 Introduction

The field of "elasmobranch endocrinology" began at the same time as the field of vertebrate endocrinology itself, when Bayliss and Starling (1903) used extracts from the intestines of sharks and skates to demonstrate the actions of secretin, the first described vertebrate hormone. Although perhaps coincidental, a pivotal role for sharks and their relatives in the birth of this field is prophetic to some extent, given that many vertebrate hormones appear to have first appeared in the cartilaginous fishes. Because sharks and their relatives occupy such a critical position in the evolution of the vertebrate endocrine system, studies on endocrinology of these fishes contribute to a better understanding of the roles that hormones exert in all higher vertebrates. Furthermore, because hormones regulate virtually all aspects of elasmobranch physiology, knowledge concerning the function of the elasmobranch endocrine system is essential for developing a full comprehension of how these fishes develop, grow, reproduce, and survive.

Because the structure and comparative aspects of the elasmobranch endocrine system are generally well addressed in most comparative endocrinology texts (e.g., Norris, 2006), this chapter focuses on the manner in which hormones participate in the regulation of processes vital for survival of sharks and their relatives. Although a "functional approach" has been used in this updated chapter, an extensive list of references regarding the structure and comparative homologies of elasmobranch hormones in addition to their known or putative actions has been provided for the reader more concerned with these topics.

11.2 Digestion and Energy Metabolism

11.2.1 Overview

The survival of an individual elasmobranch depends on its ability to convert food items into usable nutrients through actions of the digestive system. With the exception of certain specialized adaptations such as the spiral valve intestine, both the structure and function of the elasmobranch gastrointestinal tract are generally similar to that in other vertebrate groups (see review by Holmgren and Nilsson, 1999). Following its capture and maceration by the oral cavity, food is transferred through the esophagus to a two-chambered stomach, where it is stored and partially disrupted through the actions of acid-secreting and proteolytic enzyme-secreting cells. Afterward, the acidic slurry of incompletely digested food (generally referred to as *chyme*) enters the duodenum, where it is broken down further by intestinal and pancreatic enzymes, the latter of which are transferred to the duodenum via the pancreatic duct. Bile produced by the liver and stored by the gallbladder also contributes to food digestion, particularly the hydrolysis of fat, following its transport to the duodenum via the bile duct. Bile salts emulsify dietary fat globules, a process that causes them to be dispersed as smaller droplets more prone to digestion by pancreatic enzymes. The duodenum also receives bicarbonate-rich pancreatic secretions, which neutralize chyme prior to its movement to more delicate sites of nutrient absorption in the intestine. Once this passage occurs, nutrients are assimilated, presumably through both passive and active forms of uptake. Nondigested material is transported through the rectum and discharged to the environment via the cloaca.

11.2.2 Digestive Hormones

Based on the presence and distribution of major vertebrate gut hormones in the elasmobranch gastrointestinal tract, endocrine regulation of digestion in these fishes may be very similar to that occurring in higher vertebrates (Figure 11.1). The secretion of digestive acids in the foregut may be hormonally regulated by gastrin, which is capable of stimulating this process in spiny dogfish, *Squalus acanthias* (Vigna, 1983) and has been localized in endocrine cells of the stomach, intestine, and pancreas of this and other shark species (Aldman et al., 1989; El-Salhy, 1984; Holmgren and Nilsson, 1983; Johnsen et al., 1997; Jonsson, 1991, 1995). Once chyme enters the duodenum, localized declines in pH likely trigger intestinal release of the hormone secretin, which stimulates secretion of bicarbonate-rich pancreatic juices in higher vertebrates and has been identified in the intestine of skates and sharks (Bayliss and Starling, 1903). The arrival of chyme in the elasmobranch midgut also is believed to stimulate intestinal release of cholecystokinin (CCK), the hormone primarily responsible for regulating the supply of bile and pancreatic enzymes to the duodenum in mammals. A similar role for CCK in sharks and their relatives is supported by detection of CCK-like substances in the elasmobranch intestine and pancreas (Aldman et al., 1989; El-Salhy, 1984; Hansen, 1975; Holmgren and Nilsson, 1983; Johnsen et al., 1997; Jonsson et al., 1991, 1995; Vigna, 1979), as well as evidence for CCK-binding activity (Oliver and Vigna, 1996) and CCK-like actions (Andrews and Young, 1988) in the elasmobranch gallbladder. Last, inhibition of the digestive process may be regulated by somatostatin (SS), which is known to suppress production of gastric acid via inhibition of gastrin release. Cells containing SS have been localized in several components of the elasmobranch gut, including the gastric mucosa (Conlon

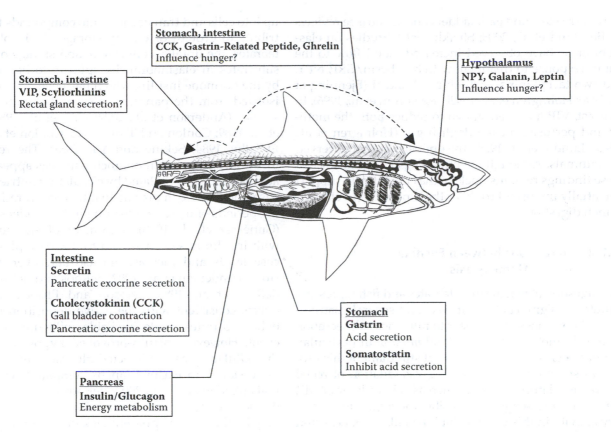

FIGURE 11.1
Proposed mechanism for the hormonal regulation of digestion and energy metabolism in elasmobranchs. Following ingestion of food, gastric acid secretion is likely stimulated by gastrin but may be inhibited by somatostatin at some point in the digestive process. Transport of chyme into the intestine causes release of secretin and cholecystokinin (CCK), which increases production and supply of pancreatic enzymes/bicarbonate secretions and hepatic bile. Release of CCK may also influence hunger and the production of other putative satiety hormones at the level of the hypothalamus. Ingestion of prey with high salt content may influence salt release by the rectal gland via the actions of vasoactive intestinal polypeptide (VIP) or scyliorhinins, but the physiological significance of these hormones remains unclear. Following absorption of energy substrates, the endocrine pancreatic hormone insulin appears to promote energy storage in the liver and other tissues. The endocrine pancreatic hormone glucagon may influence the mobilization of energy stores, but this has not been well studied. Other hormones are believed to influence gut motility or circulation (see text), but are not included in this figure.

et al., 1985; El-Salhy, 1984; Holmgren and Nilsson, 1983; King and Millar, 1979; Tagliafierro et al., 1985, 1989). Virtually all of the hormones discussed above also have been detected in nerves of the elasmobranch gut, and some (e.g., CCK, SS) have been shown to be capable of influencing gut motility in dogfish and skates (Aldman et al., 1989; Andrews and Young, 1988; Lundin et al., 1984). Therefore, they may additionally function as neurotransmitters and play important roles in regulating the passage of food through the gastrointestinal tract (Holmgren and Nilsson, 1999).

11.2.3 Hormones Involved in Gut Motility and Blood Flow

Although numerous other hormones have been detected in the gastrointestinal system of elasmobranchs, the digestive functions of few have been investigated. Four of these compounds—peptide YY (PYY), the structurally

similar compound neuropeptide Y (NPY), bombesin/gastrin-releasing peptide (GRP), and vasoactive intestinal polypeptide (VIP)—and the family of peptides known as the tachykinins are of notable interest because they are believed to exert significant actions on gut motility and circulation in higher vertebrates and have been consistently localized in the elasmobranch digestive tract (Bjenning and Holmgren, 1988; Bjenning et al., 1990, 1991, 1993; Chiba, 1998; Chiba et al., 1995; Cimini et al., 1985, 1989, 1992; Conlon et al., 1987, 1992; Dimaline et al., 1986, 1987; El-Salhy, 1984; Holmgren and Nilsson, 1983; Larsson et al., 2009; Pan et al., 1992, 1994; Shaw et al., 1990). In general, bombesin/GRP, VIP, and the tachykinins appear to promote vertebrate digestion by increasing blood flow to the gut in addition to exerting varied effects on acid or enzyme secretion and gut motility. In contrast, both NPY and PYY are believed to suppress vertebrate digestion by reducing gastrointestinal blood flow and inhibiting gastric acid secretion, pancreatic enzyme release, and gallbladder contraction (see reviews by Berglund et al., 2003; Sheikh, 1991). Studies on elasmobranchs have observed increased blood flow to the gut in response to treatment with bombesin/GRP, NPY, and two tachykinins, scyliorhinin I and II (Bjenning et al., 1990; Holmgren et al., 1992b; Kågstrom et al., 1996). In contrast, VIP has been shown to reduce both the motility and perfusion of the dogfish gut (Holmgren et al., 1992a; Lundin et al., 1984), the opposite of that observed in mammals. Although the physiological significance of these findings remains unclear, the current data support potentially important roles for these peptides in elasmobranch digestion.

11.2.4 Interactions between Feeding and Ion Homeostasis

The ingestion of marine invertebrates and fish represents a route for significant salt intake, and several gastrointestinal hormones also appear to influence salt secretion by the elasmobranch rectal gland *in vitro*. In particular, VIP has been shown to be a potent stimulant of this process in spiny dogfish by causing vasodilation of rectal gland vasculature as well as increases in cellular cAMP, the second messenger responsible for regulating secretory activity in this organ (Chipkin et al., 1988; Ecay and Valentich, 1991; Epstein et al., 1981; Forrest et al., 1983; Lehrich et al., 1998; Stoff et al., 1979). These actions are presumably mediated by hormone binding to G-protein-coupled VIP receptors, which have recently been isolated and cloned from rectal gland of spiny dogfish (Bewley et al., 2006). Although VIP is unable to elicit this response in the common dogfish, *Scyliorhinus canicula* (Thorndyke and Shuttleworth, 1985), rectal gland secretion in this species is similarly increased by treatment with scyliorhinin II (Anderson et al., 1995). In contrast, SS, GRP, and NPY inhibit rectal gland secretion, although the effects of SS and GRP appear to be mediated via inhibition of VIP-stimulated responses (Silva et al., 1985, 1990, 1993). Based on the localization of VIP, SS, and GRP in the elasmobranch rectal gland (Chipkin et al., 1988; Holmgren and Nilsson, 1983), the effects of these compounds in intact animals may be a consequence of local rather than gastrointestinal sources; however, because rectal gland secretion increases significantly following feeding activity (MacKenzie et al., 2002), hormones originating from the gut may exert at least some physiological role in regulating salt release by this organ.

11.2.5 Hormones Involved in Energy Metabolism

Following their absorption by the gastrointestinal system, the molecular products of food digestion (i.e., monosaccharides, amino acids, fatty acids, and glycerol) are directly utilized for production of energy or taken up into cells and transformed into compounds that contribute to growth or energy storage. As in other vertebrates, the uptake, conversion, and storage of energy substrates in elasmobranchs appears to be promoted by the hormone insulin, which has been detected in or isolated from the pancreas of several chondrichthyan species (Anderson et al., 2002; Bajaj et al., 1983; Berks et al., 1989; Conlon and Thim, 1986; Conlon et al., 1989; El-Salhy, 1984; Sekine and Yui, 1981). The release of insulin from the elasmobranch pancreas appears to be at least partially regulated by circulating nutrient levels based on the rise in plasma insulin concentrations during periods of increased feeding in *Scyliorhinus canicula* (Gutiérrez et al., 1988). Treatment of elasmobranchs with insulin has been shown to decrease plasma glucose levels and increase muscle and liver glycogen stores (Anderson et al., 2002; deRoos and deRoos, 1979; deRoos et al., 1985; Leibson and Plisetskaia, 1972), effects consistent with the active deposition of metabolic substrates made available following a feeding event. However, insulin-provoked hypoglycemia and the cellular uptake of injected glucose generally occur more slowly in elasmobranchs compared with mammals (Anderson et al., 2002; deRoos and deRoos, 1979; deRoos et al., 1985; Patent, 1970). These phenomena may be due to the apparent lack of insulin-dependent glucose transporters in elasmobranch tissues, the factors responsible for the rapid clearance of circulating glucose in mammals following insulin treatment (Anderson et al., 2002). Although the absence of these transporters in elasmobranch tissues remains unconfirmed, this argument is persuasive based on the relative lack and limited importance of direct sources of glucose (i.e., carbohydrates) in the protein- and fat-rich diet of sharks and their relatives. Furthermore, because insulin also promotes a reduction in circulating amino acid levels in elasmobranchs (deRoos et al., 1985), it likely plays a more important role in stimulating the cellular uptake, use, and transformation of these compounds compared with its actions on dietary glucose.

Despite the importance of hepatic lipid storage in elasmobranchs, little is known regarding the role that insulin may play in regulating this process. Nonetheless, insulin may be involved in stimulating postprandial uptake of lipids in the elasmobranch liver based on the association between increases in feeding activity, total plasma lipids, circulating insulin concentrations, and hepatosomatic index in *Scyliorhinus canicula* (Gutiérrez et al., 1988). Although insulin also promotes lipid storage in mammals by inhibiting the mobilization of stored fats, this does not appear to be the case for sharks and their relatives. Treatment of elasmobranchs with insulin has no effect on circulating levels of ketone bodies, the primary end products of hepatic lipid metabolism in these fishes. The maintenance of high endogenous levels of ketone bodies in even recently fed elasmobranchs appears to be due to their use as key fuels for aerobic metabolism, a practice that is unusual in non-starved vertebrates (Anderson et al., 2002; deRoos et al., 1985; Gutiérrez et al., 1988; Watson and Dickson, 2001). As suggested by several authors, the use of ketones as an energy source in cartilaginous fish is likely due to their limited capacity for the transport and utilization of non-esterified fatty acids compared with that in other vertebrate groups (Anderson et al., 2002; deRoos et al., 1985; Watson and Dickson, 2001).

Although it is produced by the chondrichthyan pancreas (Berks et al., 1989; Conlon and Thim, 1985; Conlon et al., 1989, 1994; El-Salhy, 1984; Faraldi et al., 1988; Gutiérrez et al., 1986; Sekine and Yui, 1981; Tagliafierro et al., 1989), the insulin antagonist glucagon does not appear to stimulate a rise in circulating glucose levels in sharks (Patent, 1970). However, because unfed elasmobranchs appear to derive energy primarily from ketone bodies (Anderson et al., 2002; deRoos et al., 1985), these observations may reflect the greater reliance of these animals on lipid stores rather than glycogen reserves during periods of undernourishment. Although the metabolism of stored lipids in fasting mammals and birds is regulated by glucagon, no studies have directly investigated if this hormone has similar actions in sharks and their relatives. Clearly, this topic should be addressed in future studies, especially due to the often sporadic feeding habits of large migratory sharks.

In addition to the pancreas, a number of other hormonal systems also appear to influence energy metabolism in sharks and their relatives. For example, thyroid hormones have been shown to alter levels of enzymes involved in amino acid and lipid metabolism in *Squalus acanthias* (Battersby et al., 1996), an action similar to that observed in higher vertebrates. Hormones involved in regulating growth and the stress response in elasmobranchs also contribute to the regulation of energy metabolism in these fishes and are discussed in later sections of this chapter. In contrast, the peptides of the caudal neurosecretory system (i.e., urotensin I and II), the urophysis, do not appear to influence carbohydrate or lipid metabolism in elasmobranchs, as they have been shown to do in certain teleosts (Conlon et al., 1994a).

11.2.6 Possible Mechanisms of Satiation

Considering the general fascination with shark feeding behavior, it is interesting to note that most of the hormones believed to regulate appetite in mammals and some nonmammalian vertebrates (see reviews by Jensen, 2001; Klok et al., 2006) also are present in elasmobranchs. This includes the anorexigenic or appetite-reducing hormones CCK and GRP, both of which appear to suppress the intake of food in mammals, birds, and teleosts following their postprandial release from the gastrointestinal system. More recently, the orexigenic or appetite-stimulating hormone ghrelin also has been isolated from the gastrointestinal tract of several elasmobranchs (Kaiya et al., 2009; Kawakoshi et al., 2007), perhaps suggesting a comparable role for this hormone in these fishes. These compounds generally regulate hunger by altering hypothalamic production of various anorexigenic and orexigenic neuropeptides; CCK-like binding activity and the presence of several appetite-regulating neuropeptides (i.e., NPY, galanin, melanin-concentrating hormone) also have been detected in the brain of several elasmobranchs (Chiba and Honma, 1992; Chiba et al., 2002; Conlon et al., 1992; McVey et al., 1996; Oliver and Vigna, 1996; Vallarino et al., 1988a, 1991). Last, evidence for encephalic expression of leptin, a hormone considered to be a major factor regulating satiety in birds and mammals, has been observed in the bonnethead shark, *Sphyrna tiburo*, and the smooth dogfish, *Mustelus canis* (Londraville, pers. comm.). Because no studies to date have investigated the effects of these or other potential "satiety hormones" on elasmobranch feeding behavior, this is a topic in need of considerable attention.

11.3 Growth

The factors that regulate elasmobranch growth are of interest, particularly to fisheries scientists who use estimates of growth rates in determining the resilience of shark and ray populations to exploitation. Unfortunately, very little is known regarding the hormonal control of growth in cartilaginous fishes; however, virtually all major hormones involved in the endocrine growth axis of higher vertebrates also are present in elasmobranchs, so the regulation of growth in sharks and their relatives is probably similar to that in mammals (see review by Le Roith et al., 2001). If this is the case, the primary factor controlling elasmobranch growth is likely to be growth hormone (GH), which has been isolated from the pituitary gland of two elasmobranchs: the blue shark, *Prionace glauca* (Hayashida and Lewis, 1978; Yamaguchi et al., 1989), and *Squalus acanthias* (Moriyama et al., 2008). Secretion of GH in vertebrates is generally regulated by stimulatory (growth hormone-releasing hormone, or GHRH) and inhibitory (SS) factors originating from the hypothalamus, both of which also have been detected in the elasmobranch brain (Conlon et al., 1985; Plesch et al., 2000). In mammals, GH promotes somatic and skeletal growth by stimulating cell proliferation and differentiation in skeletal muscle and cartilage. In addition, GH increases production of insulin-like growth factors (IGFs), highly conserved compounds (Bautista et al., 1990) that stimulate cell hypertrophy and extracellular matrix production in skeletal muscle, fat, and cartilage through autocrine and paracrine mechanisms. Evidence for anabolic actions of these growth factors in elasmobranchs has been provided by Gelsleichter and Musick (1999), who observed increased growth of skate vertebral cartilage in response to treatment with IGF-I; however, until relationships among production of GHRH, SS, GH, and IGF-I in elasmobranchs have been characterized, the regulatory scheme proposed above is largely speculative.

11.4 Stress

11.4.1 Chromaffin Tissue and Catecholamines

As the primary factors responsible for maintaining vertebrate homeostasis, hormones play key roles in the response to physiological imbalances caused by exposure to stressful stimuli (Figure 11.2). Like that in other vertebrates, the response to acute forms of stress in elasmobranchs appears to be partially regulated by the chromaffin tissue, small masses of neurosecretory cells distributed along the dorsal surface of the kidney. In response to neural signals resulting from exposure of elasmobranchs to diverse physiological stressors (e.g., hypoxia, hemorrhage, capture, handling, exercise), chromaffin cells secrete epinephrine and norepinephrine, the neurohormones known collectively as catecholamines (Butler et al., 1986; Carroll et al., 1984; Metcalfe and Butler, 1984; Opdyke et al., 1982, 1983). Catecholamines increase the supply of glucose and oxygen to the brain and muscles, preparing an organism for a "fight or flight" response. In elasmobranchs, catecholamines promote the mobilization of energy reserves, as demonstrated by the reduction in hepatic lipid stores or increase in circulating nutrient levels following treatment with these compounds (deRoos and deRoos, 1978; Grant et al., 1969; Lipshaw et al., 1972; Patent, 1970) or during stressful events (Hoffmayer and Parsons, 2001; Torres et al., 1986, 1994). Catecholamines also increase blood pressure in elasmobranchs (Opdyke et al., 1982), an action that promotes the transport of metabolic substrates to muscles and organs such as the brain and heart. In contrast, blood flow to the elasmobranch gut is reduced in response to catecholamine treatment (Holmgren et al., 1992a), a logical outcome considering that digestion is an unnecessary process during stressful periods. Catecholamines also have been shown to increase the perfusion and ventilation frequency of

elasmobranch gills, effects that stimulate the uptake of oxygen and its delivery to tissues (Butler et al., 1986; Metcalfe and Butler, 1984).

11.4.2 Hypothalamic–Pituitary–Interrenal Axis

In most vertebrates, a central component of the stress response (both acute and chronic) is the timely production of corticosteroids called *glucocorticoids* (e.g., cortisol, corticosterone) by the adrenal gland or its non-mammalian homologue, the interrenal body. In response to a stressor, the sympathetic nervous system induces the release of corticotropin-releasing factor (CRF) from the hypothalamus. CRF stimulates the production of adrenocorticotropic hormone (ACTH) in the pituitary, ultimately inducing the synthesis and secretion of glucocorticoids from adrenal/interrenal tissue. Glucocorticoids facilitate a sustained stress response by directly increasing blood glucose levels, metabolic rate, and blood pressure. It is also thought that glucocorticoids limit the stress-induced inflammatory reaction, thereby minimizing tissue damage (Bamberger et al., 1996). Additionally, elevated corticosteroid levels facilitate the suppression of physiological systems nonessential for immediate survival, such as immunity, growth, and reproduction (Mommsen et al., 1999). Inappropriate or incomplete responsiveness of the stress system results in many adverse effects, including impaired growth and development, abnormal behavior, and ultimately decreased survival (Charmandari et al., 2005).

Although CRF has yet to be characterized in elasmobranchs, they appear to possess a functional hypothalamic–pituitary–interrenal (HPI) axis based on the presence of ACTH (Amemiya et al., 2000; Denning-Kendall et al., 1982; Lowry et al., 1974; Okamoto et al., 1979; Shimamura et al., 1978; Vallarino and Ottonello, 1987) and a unique interrenal corticosteroid produced only in these fishes, 1α-hydroxycorticosterone (1α-OHB) (Idler and Truscott, 1966) (Figure 11.2). ACTH induces interrenal production of 1α-OHB *in vitro* (Armour et al., 1993b; Hazon and Henderson, 1985; Klesch and Sage, 1975; Nunez and Trant, 1999; O'Toole et al., 1990) and acts at least in part by increasing transcription of primary steroidogenic enzymes (Evans and Nunez, 2010).

Few studies have investigated the response of the elasmobranch HPI axis to stress, in large part because there are no antibodies specific to 1α-OHB. Because of this, researchers have used assays for corticosterone (B), the precursor to 1α-OHB, to quantify seasonal and stress-induced changes in serum corticosteroid concentrations in elasmobranchs (Manire et al., 2007; Rasmussen and Crow, 1993). Methods for the specific quantification of 1α-OHB using a synthetic standard have only recently been developed (Evans et al., 2010a). If corticosteroids

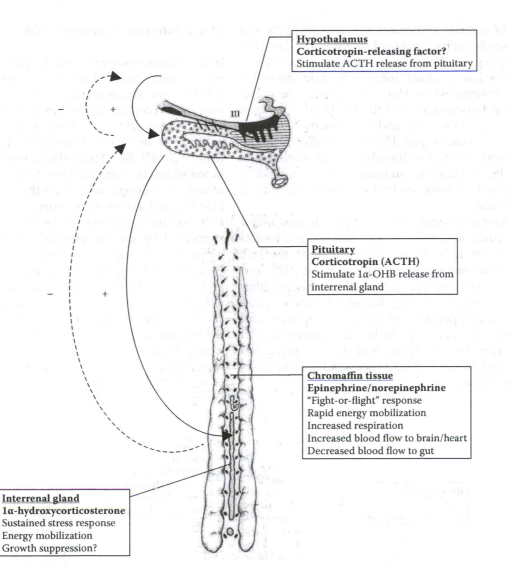

Hypothalamus
Corticotropin-releasing factor?
Stimulate ACTH release from pituitary

Pituitary
Corticotropin (ACTH)
Stimulate 1α-OHB release from
interrenal gland

Chromaffin tissue
Epinephrine/norepinephrine
"Fight-or-flight" response
Rapid energy mobilization
Increased respiration
Increased blood flow to brain/heart
Decreased blood flow to gut

Interrenal gland
1α-hydroxycorticosterone
Sustained stress response
Energy mobilization
Growth suppression?

FIGURE 11.2
Proposed mechanism for hormonal regulation of stress in elasmobranchs. Perception of stressful stimuli causes neurally mediated release of the catecholamines epinephrine and norepinephrine from chromaffin tissue, which is distributed as small, isolated pockets of neurosecretory cells along the dorsal surface of the kidney. Stress is also believed to cause release of corticotropin-releasing factor (CRF) from the hypothalamus, which promotes secretion of corticotropin (ACTH) from the pituitary gland. Release of ACTH stimulates production of the unique corticosteroid 1α-hydroxycorticosterone (1α-OHB) from the interrenal gland, which is situated along the dorsomedial surface of the kidney. Both catecholamines and 1α-OHB are believed to have effects on branchial function and cardiovascular pressure, as well as the utilization of energy substrates. 1α-OHB and ACTH are believed to influence the stress response via negative feedback on pituitary or hypothalamus function. (Adapted from Bentley, P.J., *Comparative Vertebrate Endocrinology*, Cambridge University Press, Cambridge, U.K., 1998.)

are involved in the stress response in elasmobranchs, they are likely to influence energy metabolism in these fishes as they do in other vertebrates. This premise is based on the ability of ACTH to induce hyperglycemia in sharks, perhaps by stimulating increased production of substrates (e.g., amino acids, lactate) for glucose synthesis (deRoos and deRoos, 1973, 1992). Corticosteroids also may suppress elasmobranch growth given that corticosterone is capable of inhibiting extracellular matrix production in skate vertebral cartilage *in vitro* (Gelsleichter and Musick, 1999).

11.5 Osmoregulation

11.5.1 Overview

Marine and euryhaline elasmobranchs adapt to changes in environmental salinity primarily by regulating endogenous concentrations of the organic salt urea (Evans et al., 2004). In marine elasmobranchs, urea retention stems the potential osmotic loss of water, whereas salt secretion by the rectal gland counteracts the influx of sodium and chloride across the gills and gut. In elasmobranchs capable of surviving in freshwater systems (e.g., bull sharks), reductions in the retention of urea lower blood osmolarity and diminish water gain to some extent. However, because these animals remain hyperosmotic to the environment, some uptake of water does occur and is compensated for by an increase in urine output. The maintenance of solute concentrations in freshwater-adapted elasmobranchs appears to be regulated by an increase in ion uptake in the gills, as well as a decline in the secretory activity of the rectal gland.

At least four hormonal systems appear to play major roles in regulating water and ion balance in elasmobranchs (Figure 11.3). These factors include the HPI axis, the renin–angiotensin system (RAS), VIP, and C-type natriuretic peptide (CNP). A number of other endocrine factors, including thyroid hormones, catecholamines, and peptides of the gut, urophysis, and neurohypophysis, also may influence osmoregulation and ionic regulation in sharks and their relatives, but their potential roles in these processes have not been extensively studied.

11.5.2 Interrenal Corticosteroids

In mammals, distinct adrenal steroid hormones mediate ionoregulation (mineralocorticoids, such as aldosterone) vs. the stress response (glucocorticoids, such as cortisol and corticosterone). In contrast, a single corticosteroid is thought to mediate both of these physiological systems in teleosts (cortisol) and the elasmobranch fishes (1α-OHB). Mineralocorticoids influence ion homeostasis in mammals by stimulating the retention of sodium through actions on the kidney, gut, urinary bladder, and accessory organs. A similar role for the HPI axis in regulating ion levels in elasmobranchs is supported by the presence of 1α-OHB-binding activity in the gills, kidney, and rectal gland of these fishes (Burton and Idler, 1986; Idler and Kane, 1980) and the ability of 1α-OHB to stimulate sodium transport *in vitro* (Grimm et al., 1969). Also, changes in the secretory activity of the rectal gland have been observed in interrenalectomized skates (Holt and Idler, 1975; Idler and Kane, 1976). Circulating 1α-OHB concentrations are increased in dogfish adapted to 50% seawater (Armour et al., 1993a), supporting a function for 1α-OHB in the

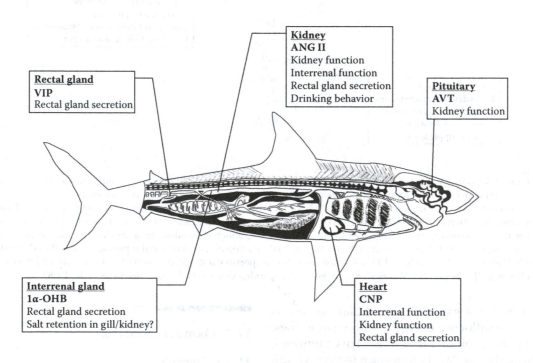

FIGURE 11.3
Proposed mechanism for hormonal regulation of osmoregulation in elasmobranchs. Water and ion balance in sharks and their relatives is modulated through four major systems: the hypothalamo–pituitary–interrenal (HPI) axis, the renal renin–angiotensin system (RAS), vasoactive intestinal polypeptide (VIP) secretion from the rectal gland, and C-type natriuretic peptide (CNP) release from the heart. Although their specific functions remain unclear, both the HPI axis and the RAS are believed to influence osmotic balance through effects on the gill, kidney, and rectal gland that promote retention of salt. The RAS also stimulates production of 1α-OHB and drinking behavior. Production of VIP in the rectal gland occurs in response to changes in blood volume and leads to increased secretion of sodium and chloride from this organ. The release of CNP from the heart also occurs in response to changes in cardiovascular pressure and increases both the production of VIP and the release of salt from the rectal gland. Arginine vasotocin (AVT) released from the pituitary gland may also contribute to the regulation of water and ion balance in elasmobranchs via actions on the kidney that promote water retention.

retention of sodium in hypoosmotically challenged cartilaginous fishes. 1α-OHB may also play a role in the osmoregulation of marine sharks and rays, especially in cases when urea homeostasis is compromised; for example, Armour et al. (1993a) observed a rise in plasma 1α-OHB concentrations in *Scyliorhinus canicula* maintained in 130% seawater and fed a low-protein diet. Because this dietary restriction limits urea biosynthesis, increased levels of 1α-OHB are believed to have been involved in stimulating the retention of sodium and chloride, an alternative osmoregulatory strategy employed by these fishes.

11.5.3 Renin-Angiotensin System

The RAS includes a series of biochemical steps that begin with the conversion of the hepatic glycoprotein angiotensinogen to angiotensin I (ANG I) via actions of renin, an enzyme secreted by the juxtaglomerular (JG) cells of the kidney. ANG I is further cleaved via angiotensin-converting enzyme (ACE), resulting in an eight-amino-acid peptide, angiotensin II (ANG II). Although subsequent enzymatic reactions produce additional peptides (ANG III and IV), ANG II is the most biologically active. In terrestrial vertebrates and bony fish, ANG II participates in the regulation of ion and water balance by stimulating corticosteroid production, drinking behavior, and changes in kidney function, actions that generally result in the uptake and retention of sodium, as well as an increase in cardiovascular pressure. Although initially thought to be absent in elasmobranchs (Bean, 1942; Nishimura et al., 1970), both the presence of the RAS (Henderson et al., 1981; Takei et al., 1993; Uva et al., 1992) and its actions on ion and water homeostasis in these fishes have been confirmed.

Isolation of the elasmobranch ANG II peptide from the banded houndshark, *Triakis scyllia*, revealed a significant amino acid substitution (proline at position 3, or Pro3) that likely induces a folded tertiary structure in contrast to the linear conformation of all other known ANG II peptides (Takei et al., 1993). Pro3 is conserved in ANG II peptides from all elasmobranch species examined to date (Evans et al., 2010a; Watanabe et al., 2009), supporting the hypothesis that the elasmobranch angiotensin receptor co-evolved to accommodate the unique structure of its ligand (Hamano et al., 1998; Nishimura, 2001). ANG II binding has been detected in a number of elasmobranch tissues, including interrenal gland, gill, rectal gland, and intestine (Tierney et al., 1997), and an elasmobranch angiotensin receptor has recently been cloned (Evans et al., 2010b). ANG II plays a significant role in the elasmobranch HPI axis, as demonstrated by the superlative number of ANG II receptors and elevated receptor mRNA levels in the interrenal gland (Evans et al., 2010b; Tierney et al., 1997), as well as the

ability of ANG II to stimulate 1α-OHB secretion *in vitro* (Anderson et al., 2001a; Armour et al., 1993b; Evans and Nunez, 2010; Nunez and Trant, 1999; O'Toole et al., 1990) and *in vivo* (Hazon and Henderson, 1984, 1985).

In addition to promoting sodium retention via effects on interrenal corticosteroidogenesis, ANG II also appears to influence electrolyte balance in elasmobranchs by reducing rates of glomerular filtration (GFR) and urine flow (UFR) in the kidney (Anderson et al., 2001a) and inhibiting salt release by the rectal gland (Anderson et al., 1995, 2001a). Both of these responses are likely to result from ANG II-stimulated reductions in blood flow through these organs (Anderson et al., 2001a). Furthermore, ANG II has been shown to increase drinking rates in elasmobranchs despite the earlier belief that this process was both unnecessary and not present in these fishes (Anderson et al., 2001b; Hazon et al., 1989). Much like the retention of salt by 1α-OHB, ingestion of seawater may enable elasmobranchs to adapt to hyperosmotic environments, a premise supported by the positive correlation between salinity and drinking rate in *Scyliorhinus canicula* (Hazon et al., 1997). A role for ANG II in adaptation to hyperosmotic environments is further supported by the demonstration that plasma ANG II increases in bull sharks upon transfer to increased salinity (Anderson et al., 2006).

11.5.4 Vasoactive Intestinal Polypeptide

As previously discussed in the section regarding gut hormones, VIP influences ion homeostasis in at least some elasmobranchs by stimulating salt secretion by epithelial cells of the rectal gland (Chipkin et al., 1988; Ecay and Valentich, 1991; Epstein et al., 1981, 1983; Lehrich et al., 1998). Although this process may partially result from gastrointestinal sources of this peptide, it is primarily regulated by the release of VIP from nerves surrounding this organ. Supporting a central role for VIP in the elasmobranch rectal gland, VIP receptor mRNA expression is highest in this tissue, followed by intestine and brain (Bewley et al., 2006). The effect of VIP on epithelial cell salt secretion is mediated by increased production of intracellular cAMP, which activates the efflux of chloride to the rectal gland lumen via chloride channels (Olson, 1999). The transport of sodium from the epithelial cells to the neighboring blood supply and, afterward, to the rectal gland lumen is largely a consequence of decreased intracellular chloride levels and the negative potential in the lumen, both of which are established by chloride secretion. VIP also appears to regulate salt secretion by analogous organs in other vertebrates, such as the salt glands of certain reptiles (Belfry and Cowan, 1995; Franklin et al., 1996; Reina and Cooper, 2000) and birds (Gerstberger, 1988; Gerstberger et al., 1988; Martin and Shuttleworth, 1994); therefore, much like its structure, the physiological functions of this compound appear to be highly conserved throughout vertebrate evolution.

11.5.5 C-Type Natriuretic Peptide

Increased production of VIP and a subsequent rise in salt secretion by the elasmobranch rectal gland are stimulated by CNP, which is thought to be the only natriuretic peptide in sharks and their relatives (Kawakoshi et al., 2001; Takei, 2000). Elasmobranch CNP is synthesized primarily in the heart and released in response to osmotic loading (Anderson et al., 2005) at concentrations far exceeding those of any natriuretic peptide in other taxa (Suzuki et al., 1994). In addition to its indirect, VIP-mediated effect on rectal gland activity, CNP also appears to stimulate salt secretion by this organ through direct actions on epithelial cells. Binding of CNP to natriuretic peptide type-B receptors (NPR-B) in the rectal gland causes an increase in intracellular levels of cyclic guanosine monophosphate (cGMP) and stimulation of protein kinase C (PKC), producing a synergistic effect on chloride transport via cellular chloride channels (Aller et al., 1999; Silva et al., 1999). CNP also is capable of causing dilation of the rectal gland vasculature, a process that results in a rise in salt release through increased perfusion of this organ (Evans and Piermarini, 2001). Although CNP binding and cGMP generation (Sakaguchi and Takei, 1998) as well as NPR-B mRNA expression (Evans et al., 2010b) have been detected in the elasmobranch gill, kidney, intestine, and interrenal gland, few studies have examined the osmoregulatory role of CNP in these tissues. CNP increases the renal clearance of urea, sodium, and chloride in *Scyliorhinus canicula* (Wells et al., 2006) and also decreases the levels of primary steroidogenic mRNAs in the interrenal gland of *Dasyatis sabina* (Evans et al., 2010a); therefore, it is likely that this peptide mediates elasmobranch ion regulation by multiple pathways.

11.5.6 Neurohypophysial Hormones

Because it is a major factor influencing ion and water balance in most vertebrates, the neurohypophyseal hormone arginine vasotocin (AVT), also known in mammals as arginine vasopressin (AVP) or antidiuretic hormone (ADH), may also play a significant role in regulating these phenomena in cartilaginous fishes. This premise is supported by the high degree of homology between AVT/AVP from elasmobranchs and other vertebrate groups, whereas the diverse nature of the other major neurohypophyseal hormone, oxytocin, may reflect a lack of conserved function (Acher et al., 1999). In terrestrial vertebrates, AVT/AVP reduces urinary water loss (i.e., diuresis) and GFR in response to rises in osmotic pressure (e.g.,

dehydration). Similar actions also have been reported in the teleost kidney (Amer and Brown, 1995), along with possible effects of this hormone on branchial ion transfer (Guibbolini et al., 1988). In elasmobranchs, AVT reduces diuresis in isolated kidney preparations from *Scyliorhinus canicula* (Wells et al., 2002, 2006), and hypothalamic AVT mRNA expression and circulating peptide concentrations increase in *Triakis scyllia* exposed to a hyperosmotic environment (Hyodo et al., 2004); therefore, AVT may be involved in regulating osmoregulation in sharks and rays by modulating urine production. Other studies have suggested that AVT may also influence urea reabsorption in elasmobranchs as it does in other vertebrates (Acher et al., 1999), but this has yet to be directly investigated.

11.6 Physiological Color Change and Bioluminescence

11.6.1 Overview

As first described by Schaeffer (1921), many elasmobranchs are capable of dramatically altering their skin color in response to the color or shade of their immediate environment (Figure 11.4). This process, which is generally termed *physiological color change*, occurs through the migration of pigment-containing organelles within specialized dermal cells known as *chromatophores*. The most abundant type of chromatophore in elasmobranch skin is the melanophore, a dermal cell that contains the brown-black pigment melanin within organelles called *melanosomes* (Figure 11.4). Elasmobranch melanophores are normally "punctuate" in appearance; that is, melanosomes are concentrated in the center of the cell, an arrangement that is associated with pallor or "lightening" of the skin. When certain sharks, skates, or rays are situated above a dark background, melanosomes become dispersed throughout the cytoplasm of the melanophore, giving the skin a darker appearance. The ability to chromatically adapt to their background benefits elasmobranchs by reducing the risk of predation, as well as enhancing opportunities for prey capture.

11.6.2 Melanocyte-Stimulating Hormone

Physiological color change in elasmobranchs is primarily regulated by α-melanocyte-stimulating hormone (α-MSH), a 13-amino-acid peptide produced in the neurointermediate lobe (NIL) of the pituitary gland (Figure 11.4). The presence of this regulatory system was established by early studies that demonstrated that removal of the NIL resulted in skin lightening in sharks and skates, which could be reversed by treatment with NIL

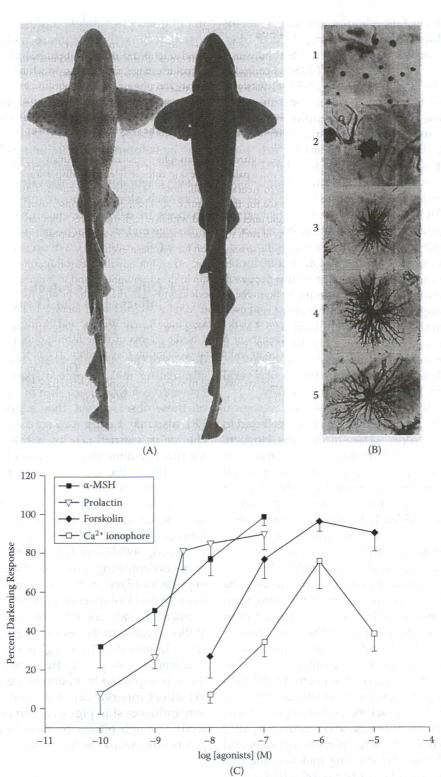

FIGURE 11.4

Regulation of skin coloration in elasmobranchs. (A) Two *Scyliorhinus canicula*, demonstrating variations in skin coloration resulting from physiological color change. (B) Microscopic appearance of dermal melanophores of *Scyliorhinus canicula*, demonstrating the dispersion of the melanin-containing organelles (melanosomes) that occurs when an elasmobranch adopts a darker skin coloration (15). (C) Dose–response curves, demonstrating darkening effects in isolated *Potamotrygon reticulates* dorsal skin pieces exposed to α-melanocyte-stimulating hormone (α-MSH), prolactin, the adenylyl cyclase activator forskolin, and calcium ionophore A23187 (which is used to increase intracellular Ca^{2+} levels and test for the role of calcium signaling in hormone-regulated processes). (Parts A and B from Bentley, P.J., *Comparative Vertebrate Endocrinology*, Cambridge University Press, Cambridge, U.K., 1998. With permission. Part C from Visconti, M.A. et al., *J. Exp. Zool.*, 284, 485–491, 1999. With permission.)

extracts or blood from dark-adapted animals (Chevins and Dodd, 1970; Lundstrom and Bard, 1932; Parker, 1936; Waring, 1936). Subsequent experiments that examined the effects of the multiple types of MSH produced by the NIL (i.e., α-, β-, and γ-MSH only; δ-MSH was not yet described) on dogfish skin coloration confirmed that the alpha form is the principal factor influencing elasmobranch color change (Sumpter et al., 1984; Wilson and Dodd, 1973a). Although the elasmobranch pituitary produces both acetylated and deacetylated forms of α-MSH (Bennett et al., 1974; Denning-Kendall et al., 1982; Eberle et al., 1978; Love and Pickering, 1974; Lowry and Chadwick, 1970), the acetylated form appears to be more active in regulating skin coloration based on its greater effect in both *in vitro* and *in vivo* melanophore bioassays (Sumpter et al., 1984). α-MSH also contributes to the regulation of morphological color change, a gradual, long-term adjustment in elasmobranch skin color associated with changes in the total amount of melanin present in an animal's epidermis (Wilson and Dodd, 1973b). Furthermore, α-MSH also may function as a neurotransmitter or neuromodulator in elasmobranchs, as its presence has been detected in the brains of several shark species (Chiba, 2001; Vallarino et al., 1988b).

The manner by which α-MSH regulates physiological color change in elasmobranchs (Figure 11.5) is believed to be similar to that first proposed for amphibians (Hogben, 1942). Under environmental conditions that favor skin pallor, the release of α-MSH appears to be suppressed by neural signals originating from the rostral lobe of the hypothalamus. This premise has been supported by experimental studies, which have observed irreversible skin darkening in sharks and skates following removal of the rostral lobe or damage to the connections between it and the NIL (Chevins and Dodd, 1970). When an elasmobranch is repositioned above a dark surface, the visual system likely receives new stimuli resulting from a reduction in the amount of light that is reflected from the background to the upper portion of the retina. Neural information associated with differential stimulation of the upper and lower retina is subsequently conveyed to the hypothalamus, which relaxes the normal inhibition of α-MSH release or promotes the production of this hormone through stimulatory factors. Following its release, circulating α-MSH binds to hormone receptors (i.e., MC1 receptors) (Mountjoy et al., 1992) in melanophores and promotes skin darkening by eliciting melanosome dispersion, presumably via actions of cytoskeletal filaments such as microtubules and/or microfilaments.

The importance of the visual system in triggering physiological color change in elasmobranchs has been validated by studies on blinded dogfish, which lack the ability to undergo this process in response to changes in background coloration (Wilson and Dodd, 1973c); however, as these animals do exhibit limited pallor when maintained in complete darkness, the presence of non-visual factors that influence melanophore function is likely. In particular, the pineal gland is believed to regulate changes in skin coloration resulting from nonvisual perception of light levels based on the lack of such responses in pinealecotomized dogfish. These findings suggest that melatonin (MT), the hormone primarily secreted by the pineal gland during the dark cycle, may be responsible for inducing skin pallor in elasmobranchs during nocturnal periods as it appears to do in other vertebrates. If so, the effect of melatonin on elasmobranch skin coloration may be mediated through changes in α-MSH release given that it is unable to influence melanosome dispersion in freshwater ray skin *in vitro* (Visconti and Castrucci, 1993).

11.6.3 Other Factors Potentially Influencing Color Change

Visconti et al. (1999) determined that prolactin (PRL), a 190- to 200-amino acid peptide produced in the par distalis of the pituitary gland, is as potent as α-MSH in stimulating melanosome dispersion in freshwater ray, *Potamotrygon reticulatus*, skin *in vitro* (Figure 11.4). Based on these observations, they suggested that circulating PRL also may function in regulating physiological color change in elasmobranchs, a role previously proposed for this hormone in amphibians (Camargo et al., 1999). Interestingly, the same researchers did not observe significant melanosome translocation in *P. reticulatus* skin in response to treatment with endothelins, catecholamines, or purines, compounds that have been shown to influence color change in other vertebrates (Visconti and Castrucci, 1993; Visconti et al., 1999). Treatment with melanin-concentrating hormone (MCH), a 17-amino-acid peptide localized in the brain and pars distalis of elasmobranchs (Vallarino et al., 1989), also had no effect on *P. reticulatus* skin color *in vitro* despite its well-described ability to cause melanosome aggregation and skin lightening in teleosts. Physiological color change in teleosts, however, differs greatly from that in elasmobranchs in that it is regulated by neural as well as hormonal signals via direct innervation of dermal melanophores. MCH may influence skin pigmentation in elasmobranchs indirectly through effects on α-MSH release, a regulatory process that would be overlooked in *in vitro* experiments.

11.6.4 Hormones Influencing Luminescence in Sharks

Certain elasmobranchs—in particular, the deepwater lantern sharks of the Family Etmopteridae—are capable of emitting a complex pattern of skin luminescence that is intrinsically produced by numerous photogenic or light-producing cells (also referred to as *photophores*)

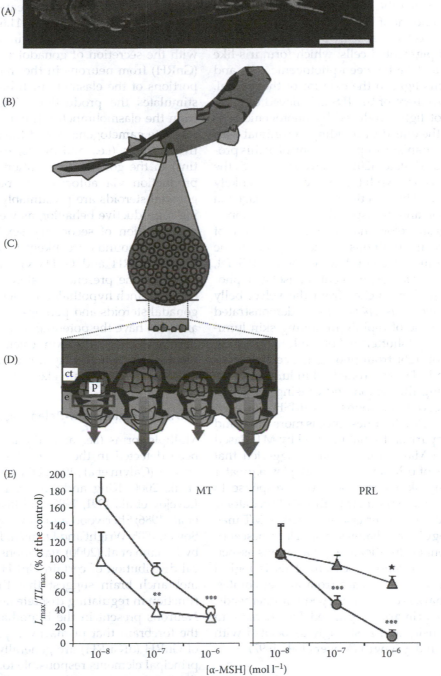

FIGURE 11.5

(See color insert.) Regulation of skin luminescence in the velvet belly lantern shark, *Etmopterus spinax*. (A) Photograph of the ventral view of luminescing *E. spinax* (scale bar: 5 cm). (From Claes, J.M. and Mallefet, J., *J. Exp. Biol.*, 213, 1852–1858, 2010. With permission.) (B) Illustration of the ventral view of *E. spinax*, demonstrating (C) a ventral skin patch. (D) Illustration of cross-section of skin from *E. spinax*, demonstrating the epidermis (e), underlying connection tissue (ct), and mechanism of light production and emission. Light is produced in photocytes (p), passes through lenses (in yellow), and is emitted to the outside (large arrows). The amount of light that is emitted is believed to be regulated by expansion or contraction of iris-like structures (ILS, red rectangle) that are part of pigmented sheath (s) cells that surround photocytes and are connected to blood sinuses. (E) Dose–response curves demonstrating inhibition of maximum intensity of light emitted (L_{max}, circles) and the time between the start of light emission to the point at which *Lmax* is reached (TL_{max}, triangles) in isolated *E. spinax* skin in response to exposure to α-melanocyte-stimulating hormone (α-MSH) after treatment with melatonin (MT) and prolactin (PRL). Graph demonstrates that α-MSH can significantly inhibit light emission that is stimulated by MT and PRL. (Parts B, C, D, and E from Claes, J.M. and Mallefet, J., *J. Exp. Biol.*, 212, 3684–3692, 2009. With permission.)

embedded in connective tissue lying between the dermis and epidermis (Claes and Mallefet, 2009, 2010a,b) (Figure 11.5). Much like the regulation of physiological color change, the amount of light that is projected to the outside from these cells is at least partially controlled via the actions of pigmented cells, which form iris-like structures (ILSs) that lie between photogenic cells and lens cells that focus light to the exterior of the animal. "Shutter-like" movement of the ILS is believed to regulate the amount of light produced by photogenic cells that is emitted to the outside, providing the animal with a physiological mechanism for precise control of this process. As Claes and Mallefet (2010a) have suggested, the pattern of luminescence exhibited by these sharks likely benefits their survival by functioning in camouflage via counterillumination and intraspecific communication.

Like physiological color change, the regulation of skin luminescence in luminous sharks appears to be largely under hormonal control (Claes and Mallefet, 2009) (Figure 11.5). *In vitro* experiments on isolated, photophore-containing skin patches from the velvet belly lantern shark, *Etmopterus spinax*, have demonstrated that α-MSH is capable of rapidly inhibiting skin luminescence by inducing "shuttering" of the ILS, occluding the transmission of light from photogenic cells. In contrast, both PRL and MT can promote skin luminescence *in vitro* by opposing this action and causing retraction of the ILS. As observed in response to α-MSH, however, the effect of PRL on skin luminescence is more rapid and short-lived in comparison to that induced by MT. Based on this, Claes and Mallefet (2009) have suggested that the quicker effects of α-MSH or PRL may play a greater role in modulating skin luminescence in response to irregularly timed visual stimuli (e.g., threat of predators), whereas the slower but longer lasting effects of MT may be important in regulating changes in skin luminescence that occur over longer durations (e.g., variations associated with vertical migrations, season). This is logical to consider given that MT release from all vertebrates, including some sharks such as scalloped hammerheads, *Sphyrna lewini*, is primarily regulated by changes in ambient light intensity and is strongly associated with seasonal changes in day length (Mayer et al., 1997).

11.7 Reproduction

11.7.1 Overview

The diversity of breeding strategies in sharks and their relatives (Conrath and Musick, 2011; see Chapter 10 of this volume) makes it imprudent to generalize concerning the hormonal control of elasmobranch reproduction.

However, it is valid to assume that the brain–pituitary–gonadal (BPG) axis is the primary endocrine system involved in regulating procreation in most, if not all, cartilaginous fishes (Figure 11.6). Environmental signals likely initiate this endocrine cascade, which begins with the secretion of gonadotropin-releasing hormone (GnRH) from neurons in the hypothalamus and other portions of the elasmobranch brain. Release of GnRH stimulates the production of gonadotropins (GTHs) from the elasmobranch pituitary gland, which in turn promote gametogenesis and the secretion of reproductive steroids (i.e., androgens, estrogens, and progestins) in the gonads. In addition to regulating gamete production via autocrine or paracrine mechanisms, gonadal steroids are presumably involved in modulating reproductive behavior, as well as the development and function of secondary sex organs. Furthermore, these compounds are likely to influence the production of GnRH and GTHs via feedback mechanisms, based on the presence of steroid binding sites in the elasmobranch hypothalamus (Jenkins et al., 1980). Last, gonadal steroids and perhaps other aspects of the BPG axis also have the potential to alter production of other hormones such as relaxin, calcitonin, and thyroid hormones, which may play accessory roles in regulating reproduction in certain elasmobranchs.

11.7.2 Gonadotropin-Releasing Hormone

Multiple forms (i.e., as many as seven) of GnRH have been detected in the brain of several chondrichthyan species (Calvin et al., 1993; D'Antonio et al., 1995; Forlano et al., 2000; King and Millar, 1980; King et al., 1992; Lovejoy et al., 1991, 1992a,b; Masini et al., 2008; Powell et al., 1986; Sherwood and Lovejoy, 1993; Sherwood and Sower, 1985; Wright and Demski, 1991). As demonstrated by Forlano et al. (2000), variations in the neuroanatomical distribution of certain GnRH subtypes in the elasmobranch brain suggest that these compounds may function in regulating discrete aspects of reproduction. Neurons present in the hypothalamus and regions of the forebrain that primarily express the dogfish form of GnRH (dfGnRH) are generally considered to be the principal elements responsible for regulating GTH production in the ventral lobe of the pituitary gland, the primary site of gonadotropic activity. This premise is particularly well supported for male Atlantic stingrays, *Dasyatis sabina*, in which changes in dfGnRH expression in certain regions of the forebrain appear to be associated with the seasonal reproductive cycle (Forlano et al., 2000). Because elasmobranchs lack a neural or vascular conduit between the hypothalamus and the ventral lobe, transport of GnRH to pituitary gonadotrophs presumably occurs via the general circulation. This suggestion appears feasible based on the presence of both GnRH

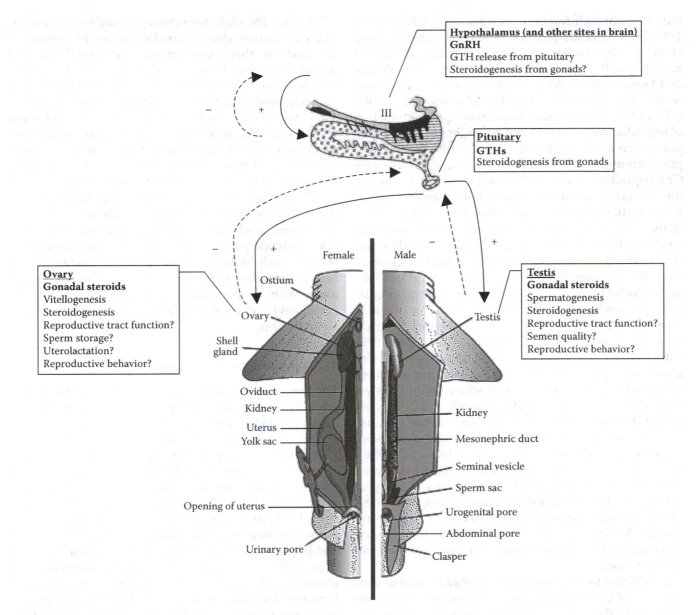

FIGURE 11.6

Proposed mechanism for the regulation of elasmobranch reproduction by the brain–pituitary–gonad (BPG) axis. Perception of environmental cues that signal seasonality are believed to trigger release of gonadotropin-releasing hormone (GnRH) from the hypothalamus. GnRH likely stimulates the production of gonadotropins (GTHs) from the ventral lobe of the pituitary gland, which in turn promote production of gonadal steroids (androgens, estrogens, progesterone) from the testis (right) and ovary (left) elasmobranchs. GnRH may also directly stimulate steroidogenesis in the gonads. Gonadal steroids regulate gonadal gametogenesis and steroidogenesis and likely influence various aspects of reproductive tract function, reproductive behavior, and other processes. Gonadal steroids and gonadotropins are believed to influence the BPG axis via negative feedback on pituitary or hypothalamus function.

and GnRH-binding proteins (GnRHBPs) in the blood of certain elasmobranchs (D'Antonio et al., 1995; King and Millar, 1980; King et al., 1992; Pierantoni et al., 1993; Powell et al., 1986; Sherwood and Lovejoy, 1993). The transport of GnRH in systemic circulation also provides a route for its direct actions on the gonads, which are likely to occur because removal of the pituitary gland is capable of only partially impairing steroidogenesis and gametogenesis in these fishes (Dobson and Dodd, 1977a,b; Sumpter et al., 1978b). More specific evidence

for a direct relationship between GnRH bioactivity and gonadal function in elasmobranchs has been provided by Jenkins et al. (1980), Fasano et al. (1989), and Callard et al. (1993), all of whom observed changes in gonadal (i.e., ovarian and testicular) steroidogenesis following administration of GnRH-like substances.

The terminal nerve (TN; also known as the nervus terminalis, or NT), a cranial nerve that connects the brain and peripheral olfactory structures, represents an additional major site of GnRH production in the elasmobranch forebrain (Chiba, 2000; Chiba et al., 1996; Demski et al., 1987; Forlano et al., 2000; Lovejoy et al., 1992b; Moeller and Meredith, 2010; Nozaki et al., 1984; Stell, 1984; White and Meredith, 1995; Wright and Demski, 1993). As a result of its direct association with chemoreceptive structures, GnRH-producing cells in the TN have been implicated in the regulation of reproductive processes and behaviors resulting from perception of olfactory cues linked with breeding (i.e., pheromones) (Demski and Northcutt, 1983). Because GnRH-positive fibers in this nerve project to sites in the forebrain generally believed to regulate ventral lobe function (Forlano et al., 2000), the putative effects of the TN on GTH or gonadal steroid production may be indirectly mediated through increased GnRH secretion in these regions. Additionally, GnRH originating from the TN may have direct effects on ventral lobe and gonadal function via transport in systemic circulation and cerebrospinal fluid (CSF), the latter of which contains increased GnRH levels following electrical stimulation of the TN (Moeller and Meredith, 1998). Although the scenarios proposed above appear both logical and plausible, concrete evidence for a change in TN activity in response to olfactory stimuli has yet to be demonstrated in sharks (Bullock and Northcutt, 1984; White and Meredith, 1995). Nonetheless, because the elasmobranch TN does appear to respond to at least some peripheral signals (White and Meredith, 1995), a function for this nerve in regulating sensory-mediated reproductive events remains a possibility. A reciprocal role for the TN in modulating the responsiveness of elasmobranch sensory organs to chemosensory and visual (Demski et al., 1987) cues through efferent pathways also has been proposed, but similarly requires confirmation.

In addition to locations in the forebrain, sizable populations of GnRH-containing neurons also have been detected in the midbrain and hindbrain of certain elasmobranchs (Forlano et al., 2000; Wright and Demski, 1991, 1993). Because fibers from these sites project to regions of the central nervous system that are involved in processing visual, electrosensory, and mechanosensory stimuli, they are alleged to function in regulating the sensitivity of the eyes, ampullae of Lorenzini, and lateral line system during the copulatory period (Forlano et al., 2000). Such actions would significantly influence reproductive success in elasmobranchs, particularly rays, which are known to use electroreception to detect potential mates (Tricas et al., 1995). Wright and Demski (1993) also proposed that GnRH fibers in the midbrain may serve to regulate movement of the claspers, given that they project to regions in the spinal cord where motor neurons for these copulatory organs are located (Liu and Demski, 1993; Wright and Demski, 1991, 1993).

Like the TN, GnRH-producing neurons in the midbrain of certain elasmobranchs project to regions of the forebrain that appear to regulate ventral lobe and gonadal function (Forlano et al., 2000). Therefore, these cells have the potential to influence steroidogenesis and gametogenesis, in addition to their purported actions on sensory perception and locomotor activity. In fact, some recent data suggest that the GnRH nucleus in the elasmobranch midbrain may be involved in conveying information regarding environmental cues that are believed to initiate cyclic activity of the BPG axis. Mandado et al. (2001) reported the presence of neural projections from the pineal organ to GnRH-immunoreactive neurons in the dogfish midbrain, signifying that these cells may alter hormone production in relation to photoreceptive stimuli. Such findings add weight to the long-held but largely unexplored premise that changes in day length, along with temperature or food availability, are the major environmental signals regulating elasmobranch reproduction.

11.7.3 Pituitary Gonadotropins

The presence of immunoreactive GTHs in the ventral lobe of the elasmobranch pituitary has been demonstrated by both immunocytochemistry (Mellinger and DuBois, 1973) and radioimmunoassay (Scanes et al., 1972). Furthermore, extracts of this organ have been shown to possess biologically active GTHs due to their ability to stimulate steroidogenesis in chondricthyan, reptilian, and avian testicular cells (Lance and Callard, 1978; Sourdaine et al., 1990; Sumpter et al., 1980), as well as follicular and luteal components of the elasmobranch ovary (Callard and Klosterman, 1988). Treatment of both male (Sumpter et al., 1978b) and female (Callard and Klosterman, 1988) elasmobranchs with ventral lobe extracts also may increase steroidogenesis *in vivo*, but these responses may vary depending on the stage of reproduction. Last, hypophysectomy or more selective removal of the ventral lobe has been shown to cause partial regression of the testis (Dobson and Dodd, 1977a,b; Dodd et al., 1960) and reduced androgen concentrations (Fasano et al., 1989; Sumpter et al., 1978a) in male elasmobranchs, as well as follicular atresia and impaired oviposition in female *Scyliorhinus canicula* (Norris, 2006). The nature of some of these responses also may depend on reproductive stage or environmental stimuli such as water temperature (Dobson and Dodd, 1977c). Although the effects of ventral lobectomy support a role for elasmobranch GTHs in regulating gonadal activity, this procedure is not capable of completely suppressing gametogenesis and steroidogenesis (Dobson and Dodd, 1977a,b; Sumpter et al., 1978b); therefore, the direct actions of GnRH on elasmobranch gonadal function may represent a vital determinant of reproductive efficiency in these fishes.

Although a gonadotropic fraction has been purified from the ventral lobe of *Scyliorhinus canicula* (Sumpter et al., 1978a,b,c), the number and biochemical structures of GTHs produced by the elasmobranch pituitary have long been unresolved. Quérat et al. (2001) demonstrated the presence of two GTHs in the dogfish ventral lobe, which are structurally similar to paired gonadotropins from both tetrapods—follicle-stimulating hormone (FSH) and luteinizing hormone (LH)—and teleosts (GTH1 and GTH2). Given the discrete actions that FSH/LH and GTH1/GTH2 exert on gonadal function in these groups, future studies should investigate the distribution of GTH receptors in the elasmobranch gonad and examine if the two elasmobranch GTHs serve to regulate dissimilar aspects of steroidogenesis and/or gametogenesis.

11.7.4 Gonadal Steroid Hormones in the Female

The ovary of female elasmobranchs produces three major gonadal steroids: 17β-estradiol (E_2), testosterone (T), and progesterone (P_4). As indicated by measurements of steroid production by *Squalus acanthias* ovarian subcomponents *in vitro*, the synthesis of these compounds appears to be a shared function of both granulosa and theca cells (Callard et al., 1993; Tsang and Callard, 1992). Granulosa cells from active ovarian follicles secrete low levels of T and E_2 in unstimulated cultures and are capable of dramatically increasing production of E_2 and P_4 in response to stimulation by ventral lobe extracts. In contrast, isolated theca cells also synthesize T and E_2 but do not appear to contribute to the production of P_4 or increase steroidogenesis in response to gonadotropic stimulation. The cooperative nature of ovarian steroidogenesis was revealed through co-incubation of these cell layers, which resulted in a more modest increase in P_4 production, as well as a significant rise in T concentrations in stimulated cultures. Based on these results, it appears likely that stimulated theca cells utilize P_4 secreted by granulosa cells to produce heightened levels of T. The nature of these findings suggests that both of these cell layers contribute to total follicular steroidogenesis in intact animals.

Production of P_4 in female elasmobranchs also is a function of ovarian corpora lutea, which form primarily from granulosa cells either prior to or after ovulation. Evidence for expression of 3β-hydroxysteroid dehydrogenase (3β-HSD), the key enzyme involved in P_4 synthesis, has been demonstrated in corpora lutea from *Squalus acanthias* (Callard et al., 1992). Also, luteal minces from both *S. acanthias* (Tsang and Callard, 1987a) and little skate, *Raja erinacea* (Fileti and Callard, 1988), have been shown to be capable of secreting substantial quantities of P_4 *in vitro*. As shown in these studies, production of P_4 increases with the maturation of the corpus luteum, but declines with its age.

As demonstrated in oviparous and viviparous species (Fasano et al., 1992; Heupel et al., 1999; Koob et al., 1986; Manire et al., 1995; Rasmussen et al., 1999; Snelson et al., 1997; Sulikowski et al., 2004, 2005; Tricas et al., 2000; Tsang and Callard, 1987b), circulating concentrations of E_2 in female elasmobranchs generally peak during the period of follicular development (Figure 11.7). Such increases are believed to reflect the common role of E_2 on synthesis of vitellogenin, the precursor to egg yolk proteins (Koob and Callard, 1999). Production of vitellogenin occurs in the liver and is stimulated by E_2 through interactions with hepatic estrogen receptors (ERs), which have been identified in at least one elasmobranch, *Raja erinacea* (Koob and Callard, 1999). Afterward, it is transported to the ovary via systemic circulation and is sequestered by oocytes through receptor-mediated endocytosis. The stimulatory effect that E_2 exerts on vitellogenin production in elasmobranchs has been demonstrated in female *Scyliorhinus canicula* (Craik, 1978), *Squalus acanthias* (Ho et al., 1980), *R. erinacea* (Perez and Callard, 1992, 1993), and *Torpedo marmorata* (Prisco et al., 2008) in response to hormone treatment. Furthermore, increased levels of circulating vitellogenin have been shown to correspond with the preovulatory rise in E_2 in female *S. canicula* (Craik, 1978, 1979) and *R. erinacea* (Perez and Callard, 1993). Male elasmobranchs also possess the ability to synthesize vitellogenin but normally do not express this protein, presumably due to lower levels of circulating E_2; however, treatment with E_2 can result in induction of vitellogenin synthesis in at least some male elasmobranchs, as demonstrated in *R. erinacea* (Perez and Callard, 1992, 1993) and *T. marmorata* (Prisco et al., 2008b). Because the presence of vitellogenin in nonmammalian male vertebrates is commonly used as a tool for detecting exposure of these animals to estrogens or certain estrogen-like pollutants (e.g., synthetic hormones, organochlorine pesticides) (see review by Denslow et al., 1999), similar use of this procedure for male elasmobranchs may be a valuable tool for evaluating the impacts of pollution on this group (Gelsleichter and Walker, 2010).

Because elevated concentrations of E_2 during the follicular stage coincide with increased growth of the oviducal gland in female *Raja erinacea* (Koob et al., 1986), *Hemiscyllium ocellatum* (Heupel et al., 1999), and winter skate, *Leucoraja ocellata* (Sulikowski et al., 2004), it is reasonable to consider that E_2 may regulate the development and functions of this organ (Figure 11.7). A similar relationship likely exists in most female elasmobranchs for which data regarding E_2 profiles are available (e.g., *Dasyatis sabina*, *Squalus acanthias*, *Raja eglanteria*, *Sphyrna. tiburo*), but presumably was not observed because changes in oviducal gland size generally were not measured. Presence of ER in the oviducal gland has been demonstrated in *Raja erinacea* (Reese and Callard, 1991),

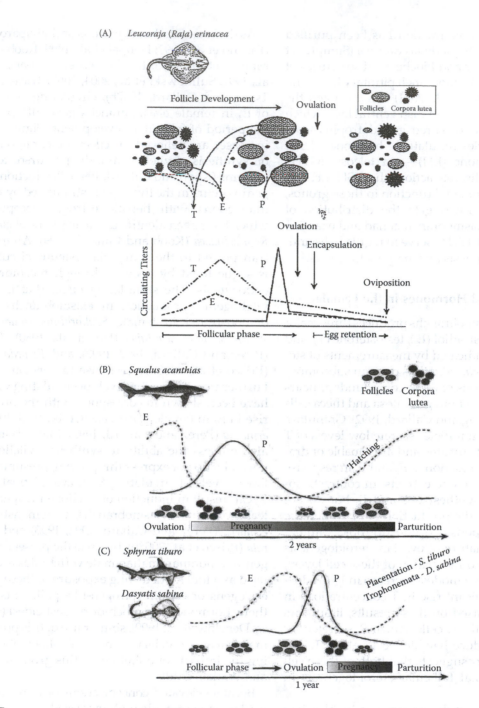

FIGURE 11.7

Gonadal steroid hormone profiles of selected female elasmobranchs in relation to the reproductive cycle. (A) Tissue source and profiles of plasma 17β-estradiol (E), progesterone (P), and testosterone (T) in the oviparous female little skate (*Leucoraja* [*Raja*] *erinacea*) during a single ovulatory cycle. The principal source of each steroid is indicated by solid lines in the upper panel; additional sources are indicated by dashed lines. Both E and T are elevated during follicular development, whereas P levels rise only for a short period just prior to ovulation. (From Maruska, K.P. and Gelsleichter, J., in *Hormones and Reproduction in Vertebrates*, Norris, D.O. and Lopez, K.H., Eds., Academic Press, San Diego, CA, 2011, pp. 209–237. With permission.) (B) Profiles of plasma E and P in the aplacental viviparous female spiny dogfish (*Squalus acanthias*). Plasma E concentrations are highest in *S. acanthias* during periods of follicular development, whereas circulating P concentrations begin to rise during ovulation and remain elevated during the first half of the 2-year gestation period. The decline that occurs in circulating P concentrations in *S. acanthias* during late pregnancy coincides with the increase in plasma E concentrations that occurs during the development of follicles for the subsequent breeding cycle. (C) Profiles of plasma E and P concentrations in the aplacental viviparous Atlantic stingray (*Dasyatis sabina*) and the placental viviparous bonnethead shark (*Sphyrna tiburo*). Plasma E concentrations are elevated in both *D. sabina* and *S. tiburo* during follicular development but drop during early pregnancy. Circulating P concentrations are elevated during early pregnancy in both species but decline in late pregnancy. E levels rise in both species during late pregnancy, when embryonic nourishment shifts from yolk dependency to matrotrophy.

and E_2 treatment has been shown to cause enlargement and increased protein secretion by this organ in female *Scyliorhinus canicula* (Dodd and Goddard, 1961). The response of the oviducal gland to hormone treatment, as well as the rise in E_2 observed in *R. eglanteria* specifically during egg capsulation (Rasmussen et al., 1999), would seem to indicate that E_2 functions to regulate this event. There is also the possibility, however, that E_2 may influence other functions of the oviducal gland such as the storage of spermatozoa, which occurs at the same time as vitellogenesis in several of these species.

In female *Sphyrna tiburo* (Manire et al., 1995), *Squalus acanthias* (Tsang and Callard, 1987b), and *Dasyatis sabina* (Snelson et al., 1997; Tricas et al., 2000), the rise in circulating E_2 concentrations beginning prior to ovulation overlaps to some extent with the passage of fertilized ova to the uterus (Figure 11.7). Because of this, it seems possible that E_2 may play a role in regulating uterine function in a manner that ensures the success of this transport process. This notion is supported by the observation that E_2 increases the compliance of the isthmus in female *S. acanthias* (Koob et al., 1983), the region of the reproductive tract lying between the oviducal gland and the uterus. Furthermore, prior studies have demonstrated a high expression of ERs in the isthmus of female *S. acanthias* (Callard et al., 2005). As discussed by Koob and Callard (1999), increased extensibility of this tissue is probably necessary for permitting the movement of delicate ova without damage to its integrity.

A secondary rise in endogenous levels of E_2 has been observed to occur in female spiny dogfish during the latter stages of gestation (Tsang and Callard, 1987b) (Figure 11.7). This species is a continuous breeder, and the change in E_2 concentrations at this time appears to regulate vitellogenesis and development of follicles for the succeeding reproductive cycle. There is also evidence, however, that E_2 may potentiate the effects of peptide hormones on uterine function in a manner that influences the maintenance of pregnancy (Koob and Callard, 1999). The actions of one of these hormones, relaxin, is discussed later in this chapter.

Post-oogenic elevations in circulating E_2 concentrations have been reported to also occur in seasonally breeding viviparous elasmobranchs during pregnancy (Figure 11.7). As follicular development for the subsequent year does not begin until after parturition in these species, increased levels of E_2 are believed to reflect a role for this hormone in the gestation process. In female *Sphyrna tiburo*, a rise in serum E_2 concentrations occurs coincident with the formation of placental connections between the gravid female and developing embryos (Manire et al., 1995). Similarly, in *Dasyatis sabina*, elevated E_2 concentrations coincide with secretion of uterine histotroph, which functions to nourish embryos between the middle and late stages of pregnancy (Gelsleichter et al., 2006; Snelson et al., 1997; Tricas et al., 2000). Because E_2 has well-characterized actions on uterine function in many vertebrates, it is reasonable to consider that it may be involved in modulating placental function in *S. tiburo* and the secretion of nutritive substances by the stingray uterus. Alternatively, Koob and Callard (1999) have postulated that E_2 may influence embryonic sustenance in these species by regulating nutrient availability in the pregnant female.

In viviparous elasmobranchs such as *Sphyrna tiburo* (Manire et al., 1995), *Squalus acanthias* (Tsang and Callard, 1987b), *Dasyatis sabina* (Snelson et al., 1997; Tricas et al., 2000), and *Torpedo marmorata* (Fasano et al., 1992), endogenous P_4 concentrations generally peak during or shortly after the ovulatory period (Figure 11.7). As detailed in Koob and Callard (1999), these changes may reflect a role for P_4 in suppressing further production of vitellogenin, which, until this point, was stimulated by the actions of E_2. Receptors for P_4 have been detected in the liver of female *Raja eglanteria* (Paolucci and Callard, 1998), and P_4 treatment is capable of blocking E_2-stimulated vitellogenesis (Perez and Callard, 1992, 1993) and overall follicular development (Koob and Callard, 1985) in this species. Furthermore, attempts to induce vitellogenin production in pregnant *Squalus acanthias* have been shown to be unsuccessful until later stages of pregnancy, when circulating P_4 concentrations decline (Ho et al., 1980). Thus, the reduction in P_4 levels in *S. acanthias* that occurs during this period (Tsang and Callard, 1987b) is believed to permit the development of follicles for the subsequent pregnancy. More recent evidence for a role for P_4 in inhibiting vitellogenin production has been observed in female *T. marmorata*, in which circulating P_4 concentrations and vitellogenin expression were found to be negatively correlated (Prisco et al., 2008b).

Because the strongest support for an inhibitory action of P_4 on elasmobranch vitellogenesis has been derived from studies on *Raja erinacea* (Koob and Callard, 1985; Paolucci and Callard, 1998; Perez and Callard, 1992, 1993), this response may also occur in oviparous elasmobranchs. This appears to be the case for female *R. erinacea*, which experience an ephemeral surge in endogenous P_4 levels just prior to ovulation (Koob et al., 1986) (Figure 11.7); however, occurrence of this process is less supported in female *Raja eglanteria*, in which the only significant rise in serum P_4 concentrations during the egg-laying period occurs at the time of oviposition (Rasmussen et al., 1999). Although the reasons for these dissimilarities remain unclear, they warrant further investigation because they are the most distinct difference in what is known regarding the reproductive endocrinology of these species. In total, these findings may reflect important roles for P_4 in both suppression of vitellogenesis and egg laying in skates. The latter of these two proposed functions is supported by the presence of P_4 receptors (PRs) in the skate reproductive tract (Callard et al., 1993), as well as the observation that P_4 treatment can cause early oviposition in *R. erinacea* (Callard and Koob, 1993). A relationship between circulating P_4 levels and the time of egg laying also has been recently observed in the smooth skate, *Malacoraja senta* (Kneebone et al., 2007).

In virtually all female elasmobranchs for which data regarding steroid hormone profiles are available, endogenous concentrations of androgens rise specifically during the period of follicular development (Koob et al., 1986; Manire et al., 1995; Rasmussen et al., 1999; Snelson et al., 1997; Sulikowski et al., 2004, 2005; Tricas et al., 2000; Tsang and Callard, 1987b) (Figure 11.7). Because elevations in T in particular overlap with the preovulatory peak in E_2, it is possible that it may partially serve as a precursor for E_2 synthesis during this stage. Alternatively, because seasonal peaks in T and dihydrotestosterone (DHT) production slightly precede those for E_2 in *Raja eglanteria* and *Dasyatis sabina*, Rasmussen et al. (1999) and Tricas et al. (2000) have suggested that androgens may play a role in modulating copulatory behavior. In addition, because elevated levels of T continue 6 months beyond the mating period in female *Sphyrna tiburo*, Manire et al. (1995) hypothesized that T may be involved in the regulation of oviducal sperm storage, which has been shown to occur at this time. Finally, based on an increase in circulating T and DHT levels specifically during late stages of the egg-laying process in *R. eglanteria* (Rasmussen et al., 1999) and draughtsboard shark, *Cephaloscyllium laticeps* (Awruch et al., 2008), it has been suggested that androgens might function in regulating oviposition in oviparous elasmobranchs; however, this is not supported by observations on *Raja erincea*, in which T levels are minimal during this same period (Koob et al., 1986). Notwithstanding this myriad of hypotheses, no published studies to date have described distribution of androgen receptors or effects of androgens in female elasmobranchs; therefore, the role of T and DHT in these animals remains largely unresolved.

11.7.5 Gonadal Steroids, Sex Differentiation, and Puberty

Although sex differentiation in elasmobranchs has been poorly studied, it appears to progress in a manner similar to that in amphibians and amniotes (for a review, see Hayes, 1998). As in all vertebrates, proper development of the gonads and secondary sex organs in sharks and their relatives appears to be sensitive to endogenous levels of steroid hormones. In embryonic *Torpedo ocellata*, for example, Chieffi (1967) observed feminization of embryonic gonads and accessory ducts following injections of

E_2, P_4, T, and deoxycorticosterone into the external yolk supply. Thiebold (1953, 1954) observed similar effects of E_2 and T in embryonic *Scyliorhinus canicula*. Although limited in number, these studies underscore the need for a clearer understanding of levels of exposure and effects of steroid hormones in developing elasmobranchs. In an effort to partially address this topic, Manire et al. (2004) examined yolk concentrations of E_2, P_4, T, and DHT in preovulatory (i.e., ovarian), ovulatory (i.e., oviducal), and postovulatory (i.e., uterine) ova in reproductively mature *Sphyrna tiburo*. The results from this study indicated that significant concentrations of E_2, P_4, and T are transferred from female *S. tiburo* to early-stage embryos via yolk. Furthermore, reductions in yolk concentrations of E_2 and T during early development suggest the active use of these steroids. Interestingly, increased levels of all three steroids during the later stages of yolk dependency in these animals may reflect the period during which embryonic steroidogenesis is initiated.

In most vertebrates, the period of sexual maturation is associated with activation of the BPG axis, which results in heightened production and release of gonadal steroids (for reviews, see Bourguignon and Plant, 2000; Okuzawa, 2002). Increased concentrations of these hormones are believed to be essential in regulating the development of the gonads and secondary sex organs, in addition to influencing activity of the hypothalamus and pituitary gland via feedback mechanisms. Changes in gonadal steroidogenesis with maturity also appear to occur in elasmobranchs, based on comparisons of circulating steroid concentrations in immature and mature elasmobranchs in field studies (Awruch et al., 2008; Gelsleichter et al., 2002; Manire et al., 1999; Rasmussen and Gruber, 1993; Sulikowski et al., 2005, 2006). More concrete evidence of these changes has been observed in serially examined captive male *Sphyrna tiburo*, which exhibit significant but stage-specific increases in serum T, DHT, E_2, and P_4 concentrations during pubertal development (Gelsleichter et al., 2002). Although increases in the concentrations of these hormones coincide with development of the testis and accessory sex organs (i.e., epididymis, seminal vesicle, clasper), their roles in such processes remain unclear. Nonetheless, because E_2 treatment has been shown to be capable of promoting maturation of the reproductive tract in immature female *Mustelus canis* (Hisaw and Abramowitz, 1939) and *Scyliorhinus canicula* (Dodd and Goddard, 1961), it seems likely that the pubertal surge in steroid concentrations in maturing elasmobranchs has a functional significance.

11.7.6 Gonadal Steroids in the Male

The presence of numerous gonadal steroids has been reported in male elasmobranchs (Callard, 1988; Manire et al., 1999), but several of these observations require confirmation via analysis of compounds produced by testicular tissues or cells cultured in the presence of radiolabeled precursors. Unlike many other male vertebrates, in which gonadal steroids are largely produced by cells that lie between testicular spermatocysts (i.e., Leydig or interstitial cells), male elasmobranchs appear to synthesize the bulk of these compounds in Sertoli cells. This notion was first established by studies that demonstrated that these cells possess both the cytological (Holstein, 1969; Pudney and Callard, 1984a) and enzymatic (Simpson and Wardle, 1967) characteristics of steroid producers and has been further validated via direct measurement of steroids secreted by isolated Sertoli cell monolayers (DuBois et al., 1989). Although Leydig-like interstitial cells with steroidogenic features generally occur in the testis of male elasmobranchs, Pudney and Callard (1984b) reported that these cells are undifferentiated in appearance and do not undergo structural changes that occur in Sertoli cells in association with spermatogenic progression (Pudney and Callard, 1984a). Because of this, the involvement of these cells in testicular steroidogenesis in male elasmobranchs has long been questioned; however, observations on the testis of male *Torpedo marmorata* have argued for the presence of true Leydig cells in this species, which exhibit both ultrastructural and enzymatic attributes of steroid-producing cells (Prisco et al., 2002a). Moreover, because these cells appear most active in regions bordering early-stage spermatocysts, they have been proposed to function in partially regulating the initial stages of spermatogenesis. Thus, it is reasonable to consider that Leydig-like cells supplement gonadal steroidogenesis in at least some male elasmobranchs. There is also some evidence for the production of gonadal steroid hormones by elasmobranch germ cells, which has been shown to occur in some other male vertebrates (Prisco et al., 2008a).

Despite the large number of gonadal steroids that have been detected in male elasmobranchs, patterns in the endogenous concentrations of only T, DHT, E_2, and P_4 have been well investigated in relation to the reproductive cycle. Associations between testicular and circulating levels of these hormones and certain breeding stages suggest that they function in regulating essential aspects of male reproduction; for example, in most male elasmobranchs that have been examined to date (Awruch et al., 2008; Garnier et al., 1999; Heupel et al., 1999; Manire and Rasmussen, 1997; Mull et al., 2008; Snelson et al., 1997; Sulikowski et al., 2004; Tricas et al., 2000), serum T or DHT concentrations significantly increase during the middle to late stages of spermatogenesis (Figure 11.8). In all these species, this period is characterized by an increase in gonadosomatic index (GSI), as well as a rise in the presence of mature spermatocysts in the testis. The increase in androgen concentrations experienced

during this period likely reflects increased production of these compounds by late-stage, postmeiotic spermatocysts, which has been demonstrated to occur in the testis of both *Squalus acanthias* (Callard et al., 1985; Cuevas et al., 1993) and *Scyliorhinus canicula* (Sourdaine and Garnier, 1993; Sourdaine et al., 1990). Although this initially suggested that androgens directly regulate the final stages of sperm maturation, it has been demonstrated that androgen receptors in the dogfish testis are primarily localized in early-stage (i.e., premeiotic and meiotic) spermatocysts (Cuevas and Callard, 1992; Engel and Callard, 2005). Therefore, T and DHT produced by Sertoli cells in mature spermatocysts more likely function to regulate the developmental advance of spermatogonia (Callard, 1992). This phenomenon appears to be made possible by the route of blood flow through the elasmobranch testis, which proceeds from more advanced to less advanced stages of spermatocyst differentiation (Cuevas et al., 1992).

Because serum androgen concentrations in male *Sphyrna tiburo* (Manire and Rasmussen, 1997), *Dasyatis sabina* (Snelson et al., 1997; Tricas et al., 2000), *Hemiscyllium ocellatum* (Heupel et al., 1999), and *Scyliorhinus canicula* (Garnier et al., 1999) are elevated during periods of increased semen transport, these compounds probably influence development and function of the gonaducts, as well as the maturation and viability of spermatozoa. Whereas such actions are likely based on the roles of these compounds in other vertebrates, no published studies have reported on the effects of androgens on these aspects of male elasmobranch reproduction. Nonetheless, multiple routes of steroid hormone transfer between the testis and urogenital system in male chondrichthyans exist. In addition to transport in the general circulation, steroid hormones appear capable of accessing putative binding sites in spermatozoa and the male reproductive tract through the occurrence of Sertoli cell cytoplasts or remnants (Prisco et al., 2002; Pudney and Callard, 1986) and steroidogenic enzyme activity (Simpson et al., 1964) in elasmobranch semen. This mechanism may represent a significant contribution to the regulation of reproductive events occurring after spermiation because the seminal fluid of some sharks contains high concentrations of certain steroids (Gottfried and Chieffi, 1967; Simpson et al., 1963).

Although increased clasper size coincides with peak T concentrations in some mature male elasmobranchs (Garnier, 1999; Heupel et al., 1999), no studies have confirmed androgen sensitivity of this organ. Even during puberty, when growth of the clasper and other sexually dimorphic skeletal elements (e.g., the cephalofoil of male *Sphyrna tiburo*) (Kajiura et al., 2005) are at their maximum, Gelsleichter et al. (2002) found no direct relationship between circulating androgen concentrations and rates of clasper elongation in serially

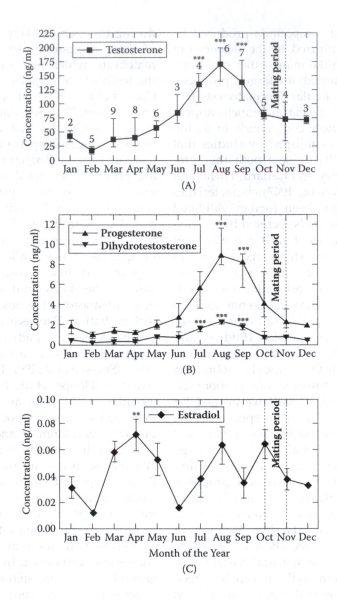

FIGURE 11.8
Gonadal steroid hormone profiles of male bonnethead sharks (*Sphyrna tiburo*) illustrating the changes that generally occur in plasma androgen (testosterone, dihydrotestosterone), 17β-estradiol (estradiol), and progesterone concentrations in seasonally reproducing male elasmobranchs in relation to the reproductive cycle. Androgens and progesterone rise steadily during spermatogenesis. Androgens may also be elevated during the mating period in some male elasmobranchs. Plasma estradiol concentrations vary irregularly in male *S. tiburo* but may be slightly elevated during early stages of spermatogenesis. (From Manire, C.A. and Rasmussen, L.E., *Gen. Comp. Endocrinol.*, 107, 414–420, 1997. With permission.)

examined captive male *S. tiburo*. Furthermore, both hypophysectomy and administration of T are incapable of altering clasper growth *in vivo* in immature male elasmobranchs (Wourms, 1977). These findings appear to argue against a function for androgens in the development and growth of the male elasmobranch copulatory organs, but the effects of androgens on skeletal growth in mammals are currently believed to be largely mediated through estrogen-regulated (i.e., obtained via aromitization of T) increases in GH and IGF-I production (Grumbach, 2000). Androgens also may influence the pubertal growth of the mammalian skeletal system

by stimulating increased production of IGF-I receptors (Phillip et al., 2001); therefore, future studies should evaluate the possible links between the BPG and growth axes in the cartilaginous fishes to fully clarify the putative roles of T and DHT on the external genitalia.

Elevated serum androgen concentrations occur during copulatory activity in male *Dasyatis sabina* (Snelson et al., 1997; Tricas et al., 2000), *Hemiscyllium ocellatum* (Heupel et al., 1999), and *Negaprion brevirostris* (Rasmussen and Gruber, 1993), suggesting that these hormones may function in modulating certain aspects of reproductive behavior. Although this topic has not been extensively studied, recent evidence supports the notion that androgens are capable of influencing mating activity of elasmobranchs through effects on sensory organ responsiveness. Sisneros and Tricas (2000) have demonstrated that the electrosensory abilities of male *D. sabina* are significantly improved during the seasonal peak in circulating androgen levels. Similarly, increased sensitivity of ampullary electroreceptors in these animals occurred following treatment with DHT. Androgen-mediated changes in electroreception specifically improve the ability of male stingrays to detect low-frequency stimuli, such as those generated by conspecifics; therefore, these changes are likely to influence both the detection of potential mates and overall reproductive success. As bioelectric information produced by the typical prey of this species is generally of a higher frequency, it is doubtful that the seasonal changes in electroreceptive ability are more associated with feeding behavior rather than reproduction.

Unlike those observed for androgens, seasonal patterns in circulating E_2 concentrations in most male elasmobranchs generally reveal little about the role of this hormone in reproduction. For example, in both male *Sphyrna tiburo* (Manire and Rasmussen, 1997) and *Scyliorhinus canicula* (Garnier et al., 1999), endogenous levels of E_2 vary irregularly during the reproductive cycle; however, in male *Dasyatis sabina* (Snelson et al., 1997; Tricas et al., 2000), serum E_2 concentrations exhibit a clear pattern of variation in which levels of this hormone rise specifically during the early to middle stages of spermatogenesis. These changes better reflect patterns in testicular E_2 production in *Squalus acanthias*, which appears to be highest in spermatocysts undergoing meiosis (Callard et al., 1985; Cuevas and Callard, 1992). Whereas evidence for peak E_2 production during meiosis suggests a role for this hormone during mid-spermatogenesis, ERs are primarily localized in regions of the testis containing premeiotic spermatocysts (Callard, 1992; Callard et al., 1985). As suggested for androgens, testicular E_2 appears to have its greatest effect on downstream germ cells undergoing earlier stages of spermatogenesis. Because treatment of these premeiotic cells with E_2 results in a dose-dependent reduction in both cell proliferation and programmed cell death, this hormone appears to regulate spermatogenic progression through developmental arrest via a negative feedback system (Betka and Callard, 1998). As this effect is largely paracrine in nature, circulating levels of E_2 may not necessarily reflect its rate of production or role in the testis in certain male elasmobranchs (e.g., *Sphyrna tiburo*).

A function for E_2 in regulating the development and actions of the reproductive tract in male vertebrates is well supported by studies that have demonstrated the presence of ER in both the epididymis and seminal vesicle of several taxa (e.g., Kwon et al., 1997; Misao et al., 1997). Prior experiments using the transgenic ER-alpha knockout mouse model have confirmed that estrogens play vital roles in maintaining virtually all aspects of genital tract function, particularly in the epididymis (Eddy et al., 1996). No published studies have investigated the presence or distribution of ER in the gonaducts of male elasmobranchs; however, because a peak in E_2 concentrations coincides with increased cell proliferation and growth in the epididymis and seminal vesicle of male *Dasyatis sabina* (Piercy et al., 2003), this is a topic that should be addressed in future studies.

Because changes in circulating levels of P_4 mirror those of T and DHT in mature male *Sphyrna tiburo*, P_4 may function as a substrate for androgen synthesis during the latter stages of spermatogenesis (Manire and Rasmusssen, 1997); however, in serially examined pubertal male *S. tiburo*, elevations in serum T concentrations precede those of P_4 by several months (Gelsleichter et al., 2002). Similarly, in male *Dasyatis sabina*, increased levels of P_4 both occur later and persist longer than the peak in serum androgen levels. Together, along with the lack of correlation between endogenous T and P_4 concentrations in *Scyliorhinus canicula* (Garnier et al., 1999), these findings suggest that P_4 functions as more than merely a precursor for other steroids in the elasmobranch testis. This notion is supported by the presence of PRs in the testis of *Squalus acanthias*, which are primarily localized in late-stage (postmeiotic) spermatocysts (Cuevas and Callard, 1992). Such observations suggest that P_4 plays a role in regulating spermiogenesis or spermiation in male elasmobranchs. As testicular P_4 synthesis is greatest in postmeiotic spermatocysts (Callard et al., 1985), these actions may be regulated through an autocrine or paracrine mechanism.

11.7.7 Other Hormones Involved in Reproduction: Corticosteroids

As demonstrated in other non-mammalian vertebrates (Romero, 2002), steroid hormones produced by the interrenal gland also may influence various aspects of elasmobranch reproduction, perhaps in relation to energy balance or stress. This is supported by several studies that have observed differences in circulating corticosteroid concentrations in sharks and rays in association with sex, maturity, or stage of reproduction (Manire et al., 2007; Rasmussen and Crow, 1993; Snelson et al., 1997). For example, Manire et al. (2007) found that plasma corticosterone concentrations covaried with reproductive stage in both male and female *Sphyrna tiburo* and *Dasyatis sabina*, although differences between species were observed. In males, plasma corticosterone concentrations increased significantly during spermatogenesis and mating in both bonnethead sharks and Atlantic rays, providing compelling evidence for its potential effects on testicular function and male behavior. In contrast, plasma corticosterone profiles differed in female *D. sabina* and *S. tiburo*, suggesting possible roles for corticosteroids during pregnancy in the former and during the stages preceding pregnancy in the latter. Because roles for adrenal steroids in regulating gonadal and gonaduct function in vertebrates are well supported in the literature (Michael et al., 2003), future studies are necessary to clarify these initial findings.

11.7.8 Other Hormones Involved in Reproduction: Relaxin

Relaxin, a 6-kDa polypeptide hormone best known for its ability to prepare the mammalian reproductive tract for successful parturition, has been detected in the ovaries of *Squalus acanthias* (Bullesbach et al., 1986), *Raja erinacea* (Bullesbach et al., 1987), and the sandtiger shark, *Carcharias taurus* (Gowan et al., 1981; Reinig et al., 1981). Koob et al. (1984) determined that relaxin and its structural homologue insulin were capable of increasing cervical cross-sectional area in late-stage, pregnant *S. acanthias*, leading to the premature loss of developing fetuses. In similar studies, treatment of female *R. erincea* with homologues of porcine relaxin resulted in increased compliance of the cervix and other portions of the reproductive tract (Callard et al., 1993). As these effects mirror the ability of relaxin to increase circumference of the mammalian birth canal (Steinetz et al., 1983), it appears likely that this hormone may participate in pupping and/or egg laying in female elasmobranchs. Relaxin also has been shown to reduce the frequency of myometrial contractions *in vitro* and *in vivo* in uterus of pregnant *S. acanthias*, suggesting that it functions in the maintenance of pregnancy prior to parturition (Sorbera and Callard, 1995). This response is also analogous to one of the roles proposed for relaxin in certain mammals (Downing and Sherwood, 1985).

Although typically considered a "female hormone" due to the effects previously discussed, relaxin is also produced by the reproductive organs (e.g., testis, prostate, seminal vesicle) of some male vertebrates and is believed to play a role in regulating male fertility (Weiss, 1989). Relaxin has been purified from the testis of male *Squalus acanthias* (Steinetz et al., 1998), and Gelsleichter et al. (2003) demonstrated that serum relaxin concentrations in male *Sphyrna tiburo* are elevated specifically during late spermatogenesis and the copulatory period. These observations tentatively suggest that relaxin regulates certain aspects of sperm production or function, a hypothesis also proposed for its role in mammals (Weiss, 1989); however, as the concentration of relaxin in semen of male *S. tiburo* is approximately 1000 times greater than that in circulation (Gelsleichter and

Steinetz, unpublished), it is also reasonable to consider that this hormone may facilitate insemination through regulating uterine contractibility in postmated females. A similar function has been proposed for a relaxin-like compound produced by the alkaline gland of skates and stingrays (i.e., "raylaxin"), a structure analogous to the mammalian prostrate gland (Bullesbach et al., 1997).

11.7.9 Other Hormones Involved in Reproduction: Thyroid Hormones

Thyroid hormones are believed to play a permissive role in regulating vertebrate reproduction, largely through interactions with the BPG axis (Karsch et al., 1995). Evidence for a similar function in elasmobranchs was first proposed in early studies that reported sexual dimorphism in this organ or its increased activity during oogenesis (for a review, see Dodd, 1975). Further support for this premise was provided by the observation that thyroidectomy is capable of impairing follicular development in female *Scyliorhinus canicula* (Lewis and Dodd, 1974); however, despite these findings, the role of thyroid hormones in elasmobranch reproduction has long been unresolved.

In one of the few published studies that have readdressed the relationship between thyroid activity and reproduction in the cartilaginous fishes, Volkoff et al. (1999) determined that both thyroid gland activity and circulating levels of triiodothyronine (T_3) were significantly elevated in female *Dasyatis sabina* during ovulation and throughout the period of gestation. These observations contradict the earlier notion that thyroid gland function in female elasmobranchs is greatest during follicular development. An unpublished study by Gash (2000) also observed an increase in thyroid gland activity and hormone production during pregnancy in two populations of female *Sphyrna tiburo*, with peak levels occurring during formation of the maternal–fetal placental connection. Because both species provide nourishment to developing embryos through energetically demanding processes, Gash (2000) hypothesized that increased production of thyroid hormones may address greater metabolic need during this period.

Gash (2000) also observed a seasonal pattern in thyroid gland activity of mature male *Sphyrna tiburo* which was characterized by increased hormone production during both the spring and fall. Although thyroid hormones may serve a function during the mating period, which occurs between September and November in the population in question, immature males exhibited a similar hormonal pattern. Because of this, Gash (2000) acknowledged the possibility that activity of the thyroid gland in male *S. tiburo* may be associated more with migratory activity, which increases dramatically during both of these periods due to changes in environmental

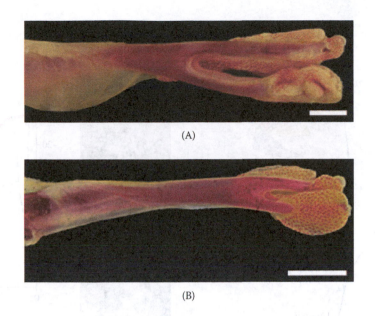

(A)

(B)

FIGURE 4.3

Cleared and stained pelvic claspers of representative specimens of *Chimaera* (A) and *Hydrolagus* (B) showing bifurcate internal skeletal morphology. Morphology of pelvic claspers, particularly the point at which the internal skeleton divides, is useful for species identification when used in combination with other morphological characters. Scale = 1 cm.

(A)   (B)   (C)

FIGURE 4.5

Egg capsules from representative species of the three families of chimaeroid fishes. (A) *Callorhinchus milii*, shown in dorsal view (left) and ventral view (right). Scale = 3 cm. (B) *Rhinochimaera atlantica*, preserved specimen, shown in dorsal view (left) and ventral view (right). Scale = 3 cm. (C) *Hydrolagus colliei* shown in dorsolateral view (left) and ventral view (right). Scale = 1 cm.

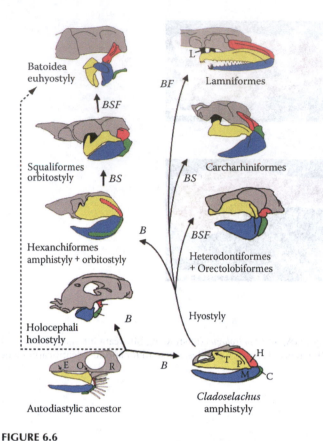

FIGURE 6.6
Morphoclinal transition predicted for the evolution of jaw suspension and feeding modes in chondrichthyans. Biting appears to be the basal feeding mode. Other feeding modes evolved with hyostyly, orbitostyly, and euhyostyly. *Abbreviations:* C, ceratohyals; E, ethmoidal articulation; H, hyomandibula; L, palatobasal articulation; M, lower jaw; O, orbital articulation; P, palatoquadrate; R, cranium; T, postorbital articulation; B, bite feeding; F, filter feeding; S, suction feeding. (From Wilga, C.D. et al., *Integr. Comp. Biol.*, 47, 55–69, 2007. With permission.)

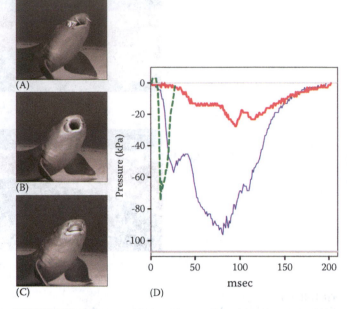

FIGURE 6.12
Suction food capture in an 85-cm TL nurse shark, *Ginglymostoma cirratum*. (A) Mandible depression, which averages 26 ms, is occurring during the expansive phase. (B) Peak gape, which occurs at 32 ms, is visible with the food entering the mouth (36 ms). (C) Upper jaw protrusion is visible as the white band inside the mouth during the compressive phase. Total bite time averages 92 ms. (D) Three representative buccal pressure profiles from different semicaptive adult *Ginglymostoma cirratum* demonstrating the variability in suction performance during feeding. Some captures are extremely rapid with large subambient pressures (green dashed line), others can approach –1 atmosphere but be more prolonged (blue solid line), whereas others may generate little subambient pressure (red dash–dot line). The lower gray line indicates –1 atmosphere pressure at the average depth of the probe (~0.5 m). (From Motta, P.J. et al., *J. Morphol.*, 269, 1041–1055, 2008. With permission.)

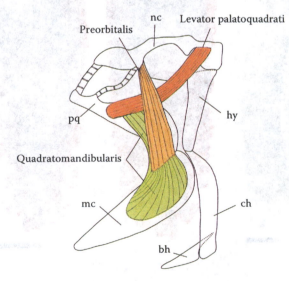

FIGURE 6.15
Diagram of the skeleton, muscles, and ligaments associated with jaw movements in the megamouth shark, *Megachasma pelagios*. *Abbreviations:* bh, basihyal; ch, ceratohyals; hy, hyomandibula; mc, Meckel's cartilage; nc, neurocranium; pq, palatoquatrate. (From Nakaya, K. et al., *J. Fish Biol.*, 73, 17–34, 2008. With permission.)

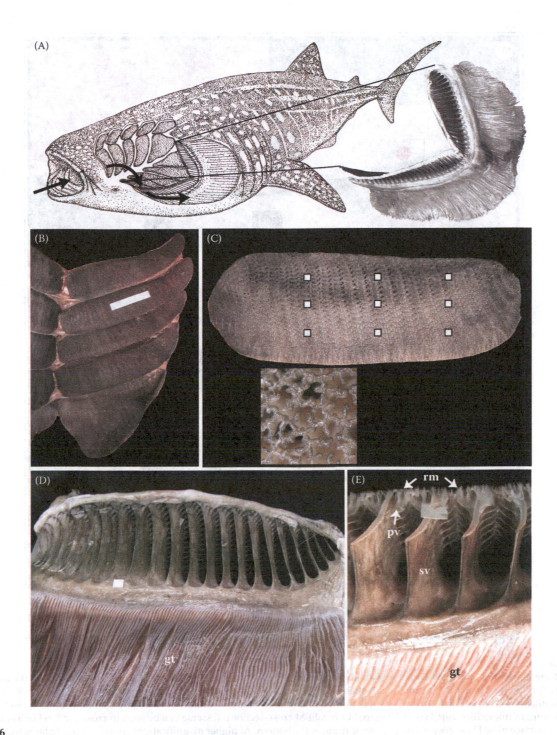

FIGURE 6.16

(A) Schematic representation of a surface ram filter feeding whale shark, *Rhincodon typus*, showing the approximate position of the filtering pads and the direction of water flow through them. Inset shows a lateral view of the vanes deep to the filtering mesh, as well as the primary gill filaments on the first branchial arch over which the water flows. (B) Gross morphology of the whale shark filtering pads. Dorsal view of the lower filtering pads of a shark of approximately 622-cm TL. The fifth most posterior lower pad at the bottom is triangular in shape, and the lateral side of the pads is to the left. The lateral raphe between the lower and upper pads is visible toward the left. All other soft tissue has been removed. White ruler is 15 cm. (C) The upper second filtering pad of a shark of approximately 593-cm TL. Because it is an upper pad, lateral is to the left and posterior toward the top. Upper pads are not as falcate on their medial margin as the lower pads. The 1-cm squares indicate areas sampled to measure mesh diameter, and the inset is a representative 1-cm square area showing the irregularly shaped holes of the reticulated mesh. (D) External view of the first upper left pad of 622-cm TL shark with lateral margin toward the left. Note that the secondary vanes direct water laterally into the parabranchial chamber and over the gill tissue (gt) before it exits the pharyngeal slit (not shown). White square is 1 cm. (E) Close-up of a section through the third left lower filtering pad of 622-cm TL shark showing the reticulated mesh (rm), primary vanes (pv), secondary vanes (sv), and gill tissue (gt). Water flow is through the mesh, between the primary and secondary vanes, and over the gill tissue. White square is 1 cm. (From Motta, P.J. et al., *Zoology*, 113, 199–212, 2010. With permission.)

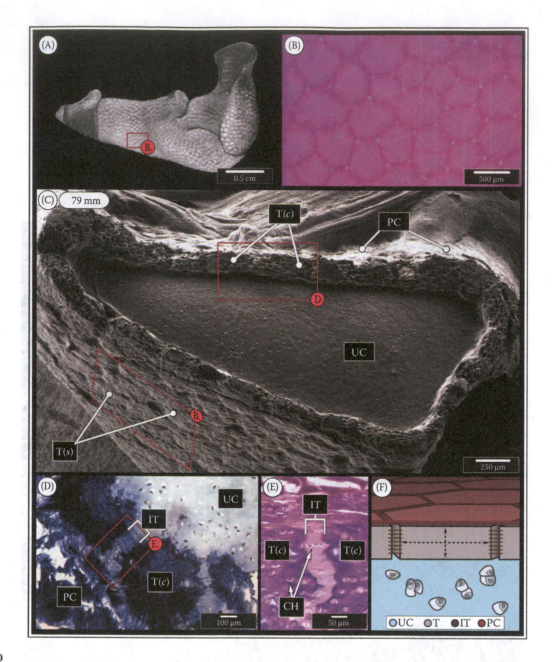

FIGURE 6.20

Anatomy of the tessellated lower jaw skeleton of an elasmobranch; lettered and inset red boxes reference other panels in the figure (e.g., the box in panel A references panel B). Elasmobranch skeletal elements are tiled superficially with abutting mineralized blocks called tesserae (T) (panel A: microCT scan, left lateral view; panel B: cleared and stained tissue), overlain by a fibrous perichondrium (PC) and surmounting a monolithic core of uncalcified cartilage (UC) (panel C: cryoSEM cross-section). Tesserae can be seen in cross-section [T(c)] at the top of the image and in surface view [T(s)], covered by perichondrium, at the bottom. At higher magnifications (panels D and E: hematoxylin and eosin stained cross-sections), the margins of tesserae are less regular, and vital chondrocytes (CH) can be seen in mineralized lacunae in tesserae and extending into the intertesseral fibrous joints (IT). The tessellated skeleton can therefore be thought of simply as unmineralized cartilage wrapped in a composite fibro-mineral bark (panel F: schematic cross-section). (From Dean, M.N. et al., *J. Anat.*, 215, 227–239, 2009. With permission.)

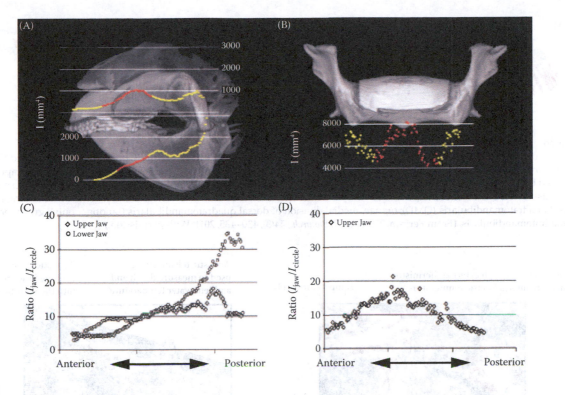

FIGURE 6.22

(A) Second moment of area of the cross-section of the upper and lower jaws of the horn shark, *Heterodontus francisci*, and (B) upper jaw of the spotted eagle ray, *Aetobatus narinari*. The *x*-axis position of each point on the graph corresponds to the position of the section through jaws in the background. Points corresponding to sections with teeth involved in crushing hard prey are in red. (C) Moment ratio (ratio of the second moment of area of the jaw cross-section to the second moment of area of a circle with the same cross-sectional area) plotted vs. position along the jaw for the upper and lower jaws of *H. francisci*. (D) Moment ratio vs. position along the jaw for the upper jaw of *A. narinari*. (Adapted from Summers, A.P. et al., *J. Morphol.*, 260, 1–12, 2004. With permission.)

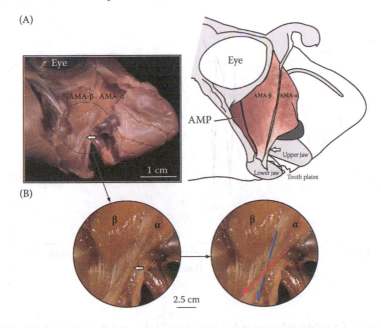

FIGURE 6.25

Right lateral view of the cranial musculature of the spotted ratfish, *Hydrolagus colliei*. (A) The schematic on the right illustrates the musculature labeled on the left. (B) The tendon (circled in panel A) has been magnified to show the twisted portion. Although all three adductors insert on the lower jaw, only the anterior adductor (AMA-α) exhibits a pronounced twist in its tendon (its approximate middle indicated by a white arrow in A and B) where the anterior face (red arrow) inserts more posteriorly than the posterior face (blue arrow). *Abbreviations:* AMA-α, anterior subdivision of the adductor mandibulae anterior; AMA-β, posterior subdivision of the adductor mandibulae anterior; AMP, adductor mandibulae posterior. (From Dean, M.N. et al., *J. Exp. Biol.* 210, 3395–3406, 2007. With permission.)

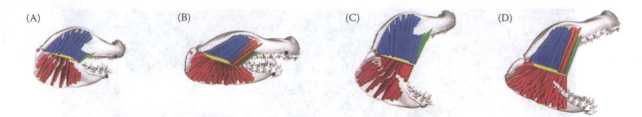

FIGURE 6.26

Arrangement of muscle fibers in finite element models (FEMs) of the jaws of (A, C) white sharks, *Carcharodon carcharias*, and (B, D) sandtiger sharks, *Carcharias taurus*, at 15° (A, B) and 55° (C, D) gape angles. Each jaw adductor muscle group inserts on the mid-lateral raphe (yellow) and is represented by a series of trusses that are used to approximate muscle forces and insertion angles of muscle fibers. In both species, the angle of muscle trusses becomes more orthogonal at 55° due to their insertion on the MLR. Truss colors correspond to the following muscle groups: blue, dorsal quadratomandibularis (QMD); orange, medial division of dorsal quadratomandibularis (sandtiger only); green, preorbitalis; red, ventral quadratomandibularis. (From Ferrara, T.L. et al., *J. Biomech.*, 44(3), 430–435, 2011. With permission.)

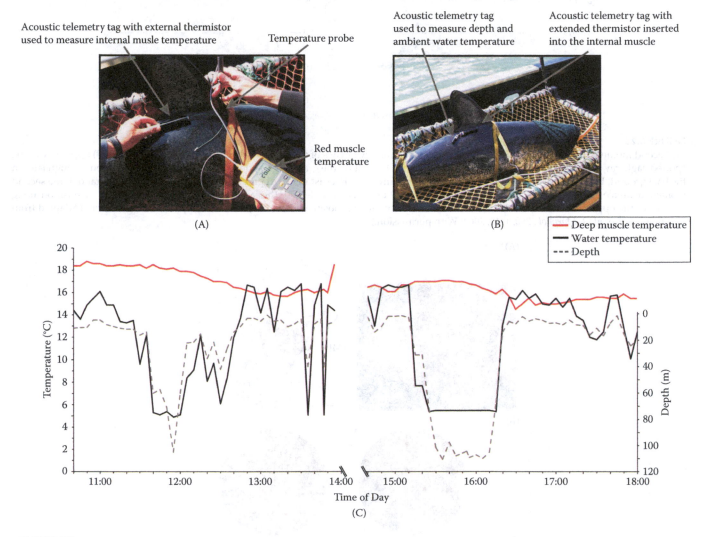

FIGURE 7.7

(A) Close-up of the dorsal fin area of a 140-kg salmon shark (*Lamna ditropis*) showing the acoustic telemetry tag with an external thermistor and a real-time reading of red muscle temperature (i.e., 26.0°C) using a temperature probe. (B) Final tag placement with the extended thermistor inserted ~15 cm into the internal muscle. (Photographs courtesy of Kenneth J. Goldman.) (C) An example of vertical movement patterns of a salmon shark (shown in part B) in the Gulf of Alaska showing 9 hours of depth, ambient temperature, and internal muscle temperature recorded during an ~15.5-hour acoustic telemetry track (no data available between 14:00 and 14:40). Notice the degree to which the internal muscle temperature remains elevated relative to that of ambient water temperature, particularly during the dive between 15:00 and 16:30. (Data from K.J. Goldman, unpublished.)

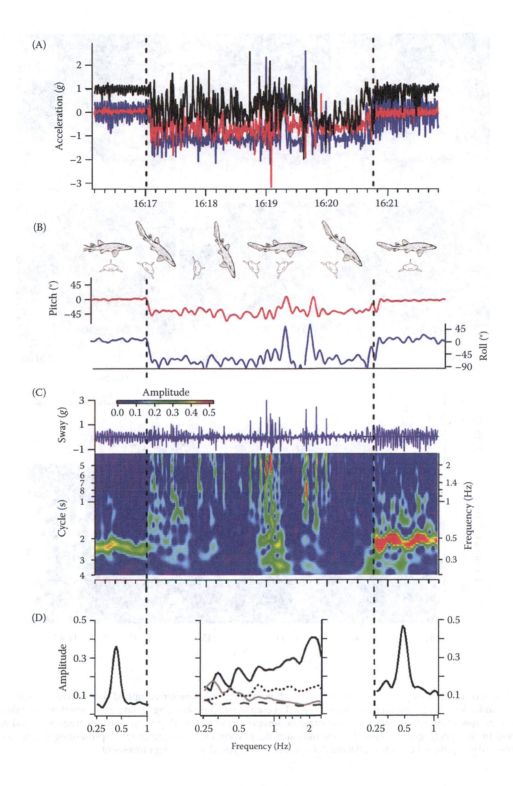

FIGURE 9.2

A mating event between a logger-equipped female nurse shark and two males as represented by (A) the raw triaxial acceleration data; (B) the static component of acceleration (converted to degrees) from the surge (pitch) and sway (roll) axes showing the female moving into a head-down pitch of as much as 67° while rolling to 90° and 45° to her left and right, respectively; (C) a spectrogram of dynamic acceleration from the swaying (lateral) axis; and (D) the behavioral spectra of the event showing typical swimming (TBF = 0.4 Hz) before the event, sporadic movements across a broad range of frequencies during the event, and faster (TBF = ~0.5 Hz) swimming with high signal amplitude after the event. Vertical dashed lines indicate the start and end of the mating event respectively. Timing and details of this event were confirmed through direct observation from the surface (i.e., researcher in a kayak). (From Whitney, N.M. et al., *Endangered Species Res.*, 10, 71–82, 2010. With permission.)

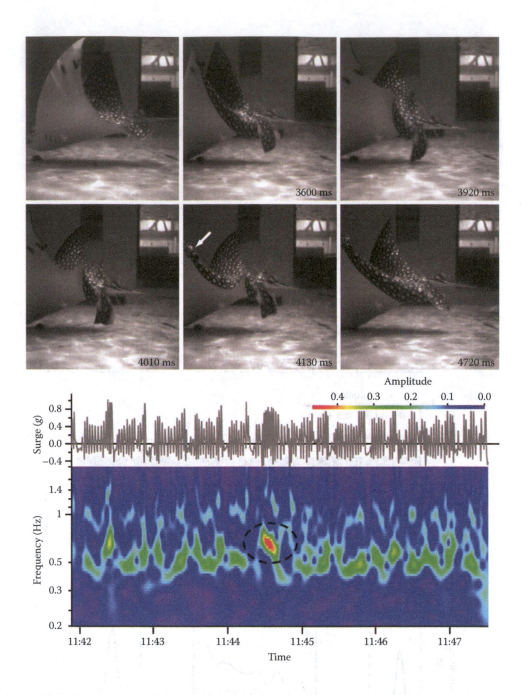

FIGURE 9.3

Frames from high-speed (125 frames per second) digital video showing an accelerometer-equipped (white arrow) spotted eagle ray feeding on clam meat in captivity and exhibiting fin undulations typical of excavation-type foraging behavior observed in the wild (Whitney, unpublished). Gray lines are raw acceleration data from the surge (anteroposterior) axis with pectoral finbeat frequency and acceleration signal amplitude depicted in the spectrograph. Dashed circle indicates the portion of the spectrograph representing the behavior in the frames above, which is easily distinguished from the adjacent data produced by typical swimming movements.

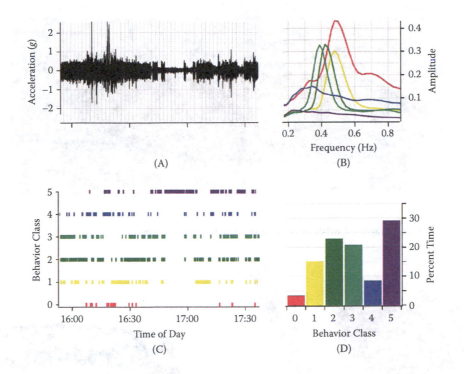

(A)

(B)

(C)

(D)

FIGURE 9.10

Nurse shark behaviors classified from acceleration data using wavelet analysis and *k*-means clustering using Ethographer software (Japanese Society of Biologging Science; Sakamoto et al., 2009). A 100-minute period of swimming and resting is shown as raw acceleration data from the sway (lateral) axis (A). Behaviors were grouped into six clusters based on their tail-beat frequency and amplitude (B) and used to generate an ethogram (C) showing the animal's behavior on a per-second basis. The percent time spent on each behavior is automatically calculated from the ethogram and presented in a frequency histogram (D).

FIGURE 10.6

Uterine wall and embryo of *Ginglymostoma cirratum*; photograph taken with an endoscope *in utero*. (Photograph © Jeffrey C. Carrier and used with permission.)

FIGURE 10.7

Egg case and late-stage embryo of *Bathyraja parmifera*. (From Hoff, G.R., *J. Fish Biol.*, 74(1), 250–269, 2009. With permission.)

FIGURE 10.8

Candle of a *Squalus acanthias*. (Photograph © Christina Conrath and used with permission.)

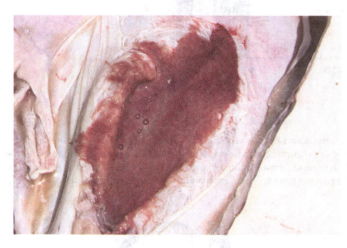

FIGURE 10.9

Gymnura altivela uterus containing trophonemata. (Photograph © John A. Musick and used with permission.)

FIGURE 10.10

Reproductive tract of term female *Carcharhinus acronotus* with visible placentae. (Photograph © Jose I. Castro and used with permission.)

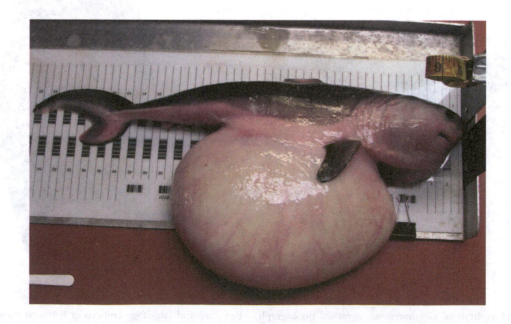

FIGURE 10.11

Lamna ditropis embryo with prominent yolk stomach. (Photograph © Kenneth J. Goldman and used with permission.)

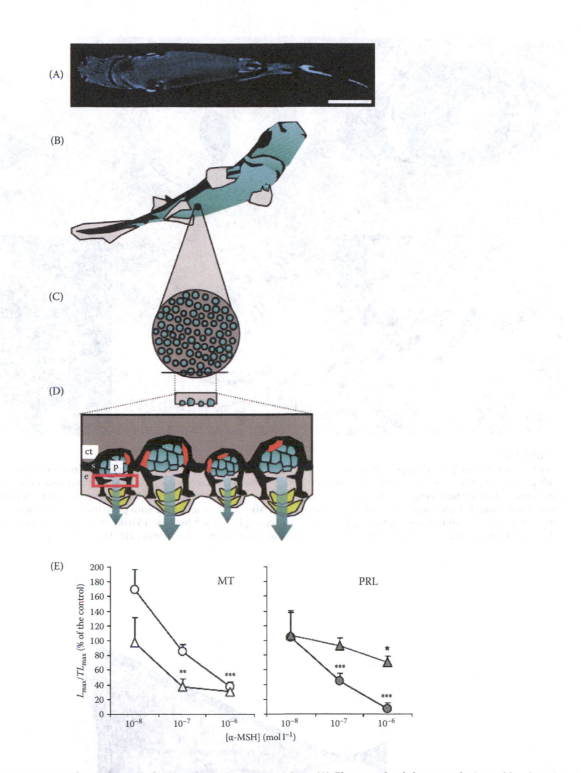

FIGURE 11.5

Regulation of skin luminescence in the velvet belly lantern shark, *Etmopterus spinax*. (A) Photograph of the ventral view of luminescing *E. spinax* (scale bar: 5 cm). (From Claes, J.M. and Mallefet, J., *J. Exp. Biol.*, 213, 1852–1858, 2010. With permission.) (B) Illustration of the ventral view of *E. spinax*, demonstrating (C) a ventral skin patch. (D) Illustration of cross-section of skin from *E. spinax*, demonstrating the epidermis (e), underlying connection tissue (ct), and mechanism of light production and emission. Light is produced in photocytes (p), passes through lenses (in yellow), and is emitted to the outside (large arrows). The amount of light that is emitted is believed to be regulated by expansion or contraction of iris-like structures (ILS, red rectangle) that are part of pigmented sheath (s) cells that surround photocytes and are connected to blood sinuses. (E) Dose–response curves demonstrating inhibition of maximum intensity of light emitted (L_{max}, circles) and the time between the start of light emission to the point at which *Lmax* is reached (TL_{max}, triangles) in isolated *E. spinax* skin in response to exposure to α-melanocyte-stimulating hormone (α-MSH) after treatment with melatonin (MT) and prolactin (PRL). Graph demonstrates that α-MSH can significantly inhibit light emission that is stimulated by MT and PRL. (Parts B, C, D, and E from Claes, J.M. and Mallefet, J., *J. Exp. Biol.*, 212, 3684–3692, 2009. With permission.)

FIGURE 12.4
Diversity of pupil shapes among elasmobranchs. (A) Circular pupil in a gulper shark, *Centrophorus* sp. (photograph by José Castro and used with permission); (B) vertical slit in the whitetip reef shark, *Triaenodon obesus* (photograph by Christian Loader and used with permission); (C) horizontal slit in the bonnethead, *Sphyrna tiburo* (photograph by D.M. McComb and S.M. Kajiura and used with permission); (D) oblique slit in the Pacific angel shark, *Squatina californica* (photograph by Alison Vitsky and used with permission); (E) crescent-shaped pupil with papillary apertures in the shovelnose guitarfish, *Rhinobatos productus* (photograph by Alison Vitsky and used with permission); and (F) the yellow stingray, *Urobatis jamaicensis* (adapted from McComb, D.M. and Kajiura, S.M., *J. Exp. Biol.*, 211, 482–490, 2008).

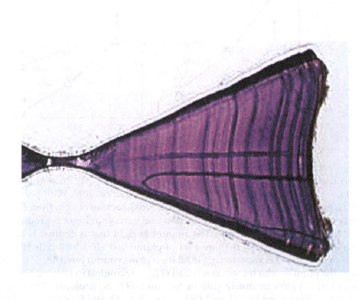

FIGURE 14.5
Vertebral section stained with hemotoxylin. (Staining by S. Tanaka; photograph courtesy of K.G. Yudin and G.M. Cailliet.)

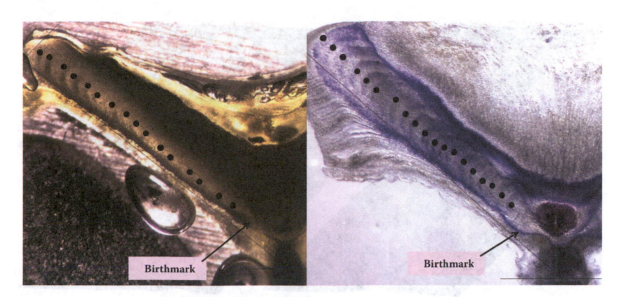

FIGURE 14.6
Comparison showing gross sectioned vertebral centrum vs. histologically prepared vertebral section of vertebrae from the same individual Commander skate, *Bathyraja lindbergi*. (Photograph courtesy of J. Maurer.)

FIGURE 17.2
A group of pink whiprays, *Himantura fai*, swimming over seagrass in Shark Bay, Western Australia. These groups will forage together over sand and seagrass, but it is not clear whether groups form to enhance foraging or for other reasons, such as reducing predation risk. (Photograph by Kirk Gastrich.)

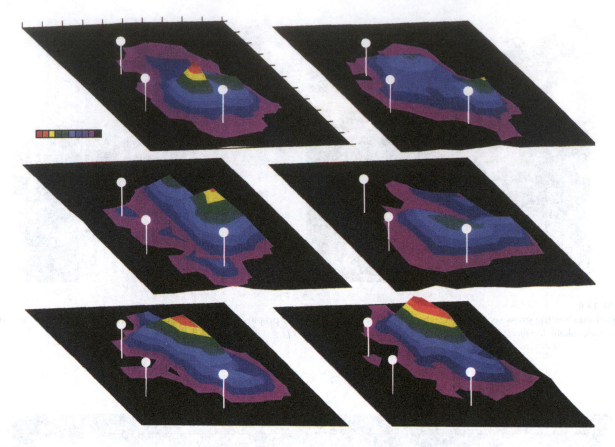

FIGURE 19.7
Three-dimensional contour maps of habitat use by two white sharks, *Carcharodon carcharias*, at Año Nuevo Island, CA. (From Klimley, A.P. et al., *Mar. Biol.*, 138, 429–446, 2001. With permission.)

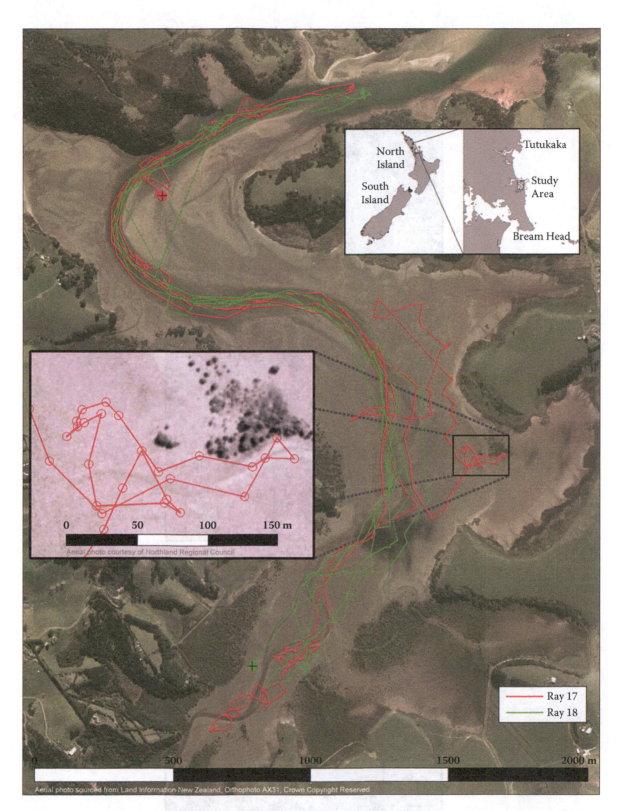

FIGURE 19.9

Movement trajectories of two individual *Myliobatis tenuicaudatus* in Taiharuru estuary, New Zealand. The entrance to the estuary from the Pacific Ocean is at the top right-hand side of the picture. Colored circles represent the locations where global positioning devices on buoys were attached to rays with monofilament tethers (4 to 7 m in length). Dashed lines connect points where fixes were >10 minutes apart (i.e., indicate areas where instantaneous fix rates were very low). Left inset: close-up of Ray 17 swimming at the edge of a stand of gray mangroves, *Avicennia marina* var. *resinifera*; open circles are locations of fixes made at a sampling interval of 150 seconds. (From Riding, T.A.C. et al., *Mar. Ecol. Prog. Ser.*, 377, 255–262, 2009. With permission.)

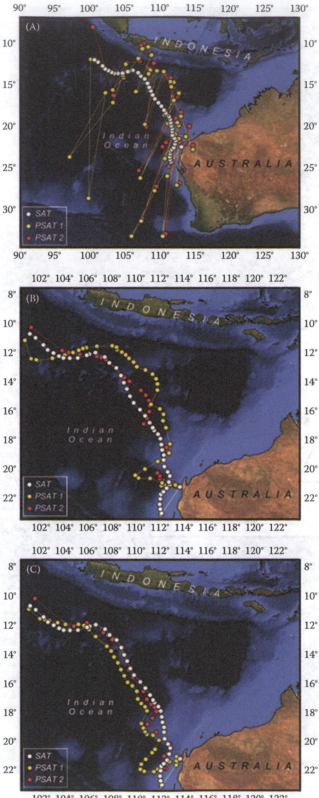

FIGURE 19.10
SAT tag track vs. location estimates from PSATs 1 and 2 derived from three levels of data processing: (A) raw light level, (B) Kalman-filtered light level, and (C) Kalman-filtered light level with SST integration. (From Wilson, S.G. et al., *Fish. Oceanogr.*, 16, 547–554, 2007. With permission.)

stimuli. Similar caution in interpreting changes in thyroid gland activity of seasonally breeding elasmobranchs is stressed, because the reproductive cycle is more than likely associated with environmental cues.

Maternally provided thyroid hormones are critically important during early stages of development in many vertebrates. Because of this, McComb et al. (2005) examined the maternal transfer of thyroid hormones via yolk to developing elasmobranchs and the use of these compounds during gestation. Results from this study have characterized both the presence and abundance of T_3 and thyroxine (T_4) in yolk from preovulatory (i.e., ovarian) and postovulatory (i.e., uterine) ova of *Sphyrna tiburo* from two populations. Interestingly, levels of thyroid hormones in yolk were lower in sharks from the population exhibiting comparatively lower rates of embryonic development, size at birth, size at maturity, and maternal investment in reproduction. Therefore, these findings may reflect a role for thyroid hormones in dictating the rate of embryonic development of sharks and their relatives.

11.7.10 Other Hormones Involved in Reproduction: Calcitonin

In addition to its ability to regulate calcium balance in mammals through the inhibition of bone demineralization, the hormone calcitonin is generally believed to play a role in regulating vertebrate reproduction (Zaidi et al., 2002). Relationships between production of gonadal steroids and calcitonin have been observed in virtually all major vertebrate groups, and evidence has been presented to link calcitonin with a number of reproductive processes, including pregnancy and lactation in mammals, follicular development in birds and teleosts, and embryonic development in all of these groups (e.g., Dacke et al., 1976; Björnsson et al., 1986; Lu et al., 1998). Although produced in parafollicular C cells of the thyroid gland in mammals, calcitonin is largely produced in a separate organ, the ultimobranchial gland, in all other jawed vertebrates (Wendelaar Bonga and Pang, 1991). In elasmobranchs, this paired organ is embedded in the musculature lying between the pharynx and pericardial cavity. A role for calcitonin in regulating reproduction in cartilaginous fishes is supported by the presence of ERs in the ultimobranchial gland of the red stingray *Dasyatis akajei* (Yamamoto et al., 1996), as well as the ability of E_2 to cause an increase in calcitonin production in the same species (Takagi et al., 1995). In contrast, a role for calcitonin in regulating calcium metabolism in elasmobranchs is less supported (Wendelaar Bonga and Pang, 1991).

Published studies have indicated that calcitonin may have important functions in viviparous elasmobranchs during the period of gestation. Nichols et al. (2003) reported a temporal pattern in serum calcitonin concentrations in female *Sphyrna tiburo*, in which peak levels of this hormone were observed during the yolk-dependent stage of pregnancy. Although immunoreactive calcitonin was not detected in any reproductive or major nonreproductive tissues other than its site of production in pregnant females, it was localized in both the duodenum and pancreas of developing embryos during the same reproductive stage. These findings suggest that calcitonin may be involved in digestion of yolk and overall fetal nutrition in this species; however, this action appears to be limited to early stages of development, as calcitonin was not detected in the gastrointestinal system of late-stage, placental embryos. The ultimobranchial gland of embryonic *S. tiburo* does not appear to be active during yolk dependency; thus, calcitonin present in the digestive tract of these animals reflects either *in situ* production or maternal transfer through the fetal egg capsule.

11.8 Conclusions and Future Directions

Although discussions concerning elasmobranch endocrinology often bemoan the amount of available data on this topic, it is clear that there is a wealth of information regarding the manner in which hormones regulate the biology of sharks and their relatives. As demonstrated in this chapter, many of these regulatory mechanisms are strikingly similar to those in advanced vertebrates, indicating that they have been highly conserved throughout evolution. Thus, in addition to providing a better understanding of how elasmobranchs function, studies on hormonal regulation in these fishes contribute significantly to that which is known regarding vertebrate endocrinology as a whole.

Although recent decades have proved to be fruitful with regard to uncovering the functions of certain hormones in elasmobranch physiology, the premise that less is known about the endocrine system in elasmobranchs than in any other vertebrate group except perhaps the jawless fish still rings true. With that in mind, future studies should address deficiencies regarding major aspects of endocrine regulation in these fishes that still remain largely unclear (e.g., the role of hormones in growth, stress, and development). In addition, as illustrated by studies regarding the roles for gonadal steroids in elasmobranch steroidogenesis, there is a critical need to continue and expand investigations on the distribution of hormone receptors in order to truly understand the roles of certain hormones. Although the preceding comments are largely a repeat of the recommendations for future study made in the previous version of this review, they remain as true today as they were then.

Acknowledgments

JG acknowledges the University of North Florida for providing the time and resources needed to prepare this chapter. AE acknowledges B. Scott Nunez for providing assistance in preparing this chapter.

References

Acher, R., Chauvet, J., Chauvet, M.T., and Rouille, Y. (1999). Unique evolution of neurohypophysial hormones in cartilaginous fishes: possible implications for urea-based osmoregulation. *J. Exp. Zool.* 284:475–484.

Aldman, G., Jonsson, A.C., Jensen, J., and Holmgren, S. (1989). Gastrin/CCK-like peptides in the spiny dogfish, *Squalus acanthias*; concentrations and actions in the gut. *Comp. Biochem. Physiol. C Toxicol. Pharmacol.* 92:103–108.

Aller, S.G., Lombardo, I.D., Bhanot, S., and Forrest, J.N. (1999). Cloning, characterization, and functional expression of a CNP receptor regulating CFTR in the shark rectal gland. *Am. J. Physiol. Cell Physiol.* 276:C442–C449.

Amemiya, Y., Takahashi, A., Suzuki, N., Sasayama, Y., and Kawauchi, H. (2000). Molecular cloning of proopiomelanocortin cDNA from an elasmobranch, the stingray, *Dasyatis akajei. Gen. Comp. Endocrinol.* 118:105–112.

Amer, S. and Brown, J.A. (1995). Glomerular actions of arginine vasotocin in the *in situ* perfused trout kidney. *Am. J. Physiol.* 269:R775–R780.

Anderson, W.G., Conlon, J.M., and Hazon, N. (1995). Characterization of the endogenous intestinal peptide that stimulates the rectal gland of *Scyliorhinus canicula. Am. J. Physiol.* 268:R1359–R1364.

Anderson, W.G., Cerra, M.C., Wells, A., Tierney, M.L., Tota, B., Takei, Y., and Hazon, N. (2001a). Angiotensin and angiotensin receptors in cartilaginous fishes. *Comp. Biochem. Physiol. A Mol. Integr. Physiol.* 128:31–40.

Anderson, W.G., Takei, Y., and Hazon, N. (2001b). The dipsogenic effect of the renin-angiotensin system in elasmobranch fish. *Gen. Comp. Endocrinol.* 124:300–307.

Anderson, W.G., Ali, M.F., Einarsdottir, I.E., Schaffer, L., Hazon, N., and Conlon, J.M. (2002). Purification, characterization, and biological activity of insulins from the spotted dogfish, *Scyliorhinus canicula*, and the hammerhead shark, *Sphyrna lewini. Gen. Comp. Endocrinol.* 126:113–122.

Anderson, W.G., Hyodo, S., Tsukada, T., Meischke, L., Pillans, R., Good, J., Takei, Y., Cramb, G., Franklin, C., and Hazon, N. (2005). Sequence, circulating levels, and expression of C-type natriuretic peptide in a euryhaline elasmobranch, *Carcharhinus leucas. Gen. Comp. Endocrinol.* 144:90–98.

Anderson, W.G., Pillans, R.D., Hyodo, S., Tsukada, T., Good, J.P., Takei, Y., Franklin, C.E., and Hazon, N. (2006). The effects of freshwater to seawater transfer on circulating levels of angiotensin II, C-type natriuretic peptide and arginine vasotocin in the euryhaline elasmobranch, *Carcharhinus leucas. Gen. Comp. Endocrinol.* 147:39–46.

Andrews, P.L. and Young, J.Z. (1988). The effect of peptides on the motility of the stomach, intestine and rectum in the skate (*Raja*). *Comp. Biochem. Physiol. C Toxicol. Pharmacol.* 89:343–348.

Armour, K.J., O'Toole, L.B., and Hazon, N. (1993a). The effect of dietary protein restriction on the secretory dynamics of 1α-hydroxycorticosterone and urea in the dogfish *Scyliorhinus canicula*: a possible role for 1α-hydroxycorticosterone in sodium retention. *J. Endocrinol.* 138:275–282.

Armour, K.J., O'Toole, L.B., and Hazon, N. (1993b). Mechanisms of ACTH- and angiotensin II-stimulated 1 alpha-hydroxycorticosterone secretion in the dogfish, *Scyliorhinus canicula. J. Mol. Endocrinol.* 10:235–244.

Awruch, C.A., Pankhurst, N.W., Frusher, S.D., and Stevens, J.D. (2008). Endocrine and morphological correlates of reproduction in the draughtboard shark *Cephaloscyllium laticeps* (Elasmobranchii: Scyliorhinidae). *J. Exp. Zool. A Ecol. Genet. Physiol.* 309:184–197.

Bagnell, C.A., Zhang, Q., Downey, B., and Ainsworth, L. (1993). Sources and biological actions of relaxin in pigs. *J. Reprod. Fertil. Suppl.* 48:127–138.

Bajaj, M., Blundell, T.L., Pitts, J.E., Wood, S.P., Tatnell, M.A., Falkmer, S., Emdin, S.O., Gowan, L.K., Crow, H., and Schwabe, C. (1983). Dogfish insulin. Primary structure, conformation and biological properties of an elasmobranchial insulin. *Eur. J. Biochem.* 135:535–542.

Bamberger, C.M., Schulte, H.M., and Chrousos, G.P. (1996). Molecular determinants of glucocorticoid receptor function and tissue sensitivity to glucocorticoids. *Endocr. Rev.* 17:245–261.

Battersby, B.J., McFarlane, W.J., and Ballantyne, J.S. (1996). Short-term effects of 3,5,3′-triiodothyronine on the intermediary metabolism of the dogfish shark *Squalus acanthias*: evidence from enzyme activities. *J. Exp. Zool.* 274:157–162.

Bautista, C.M., Mohan, S., and Baylink, D.J. (1990). Insulin-like growth factors I and II are present in the skeletal tissues of ten vertebrates. *Metabolism* 39:96–100.

Bayliss, W.M. and Starling, E.H. (1903). On the uniformity of the pancreatic mechanisms in vertebrata. *J. Physiol. (Lond.)* 29:174–180.

Bean, J.W. (1942). Specificity in the renin-hypertensinogen reaction. *Am. J. Physiol.* 136:731–742.

Belfry, C.S. and Cowan, F.B. (1995). Peptidergic and adrenergic innervation of the lachrymal gland in the euryhaline turtle, *Malaclemys terrapin. J. Exp. Zool.* 273:363–375.

Bennett, H.P., Lowry, P.J., and McMartin, C. (1974). Structural studies of alpha-melanocyte-stimulating hormone and a novel beta-melanocyte-stimulating hormone from the neurointermediate lobe of the pituitary of the dogfish *Squalus acanthias. Biochem. J.* 141:439–444.

Berglund, M.M., Hipskind, P.A., and Gehlert, D.R. (2003). Recent developments in our understanding of the physiological role of PP-fold peptide receptor subtypes. *Exp. Biol. Med. (Maywood)* 228:217–244.

Berks, B.C., Marshall, C.J., Carne, A., Galloway, S.M., and Cutfield, J.F. (1989). Isolation and structural characterization of insulin and glucagon from the holocephalan species *Callorhynchus milii* (elephantfish). *Biochem. J.* 263:261–266.

Betka, M. and Callard, G.V. (1998). Negative feedback control of the spermatogenic progression by testicular oestrogen synthesis: insights from the shark testis model. *APMIS* 106:252–258.

Bewley, M.S., Pena, J.T., Plesch, F.N., Decker, S.E., Weber, G.J., and Forrest, Jr., J.N. (2006). Shark rectal gland vasoactive intestinal peptide receptor: cloning, functional expression, and regulation of CFTR chloride channels. *Am. J. Physiol. Regul. Integr. Comp. Physiol.* 291:R1157–R1164.

Bjenning, C. and Holmgren, S. (1988). Neuropeptides in the fish gut. An immunohistochemical study of evolutionary patterns. *Histochemistry* 88:155–163.

Bjenning, C., Jonsson, A.C., and Holmgren, S. (1990). Bombesin-like immunoreactive material in the gut, and the effect of bombesin on the stomach circulatory system of an elasmobranch fish, *Squalus acanthias. Regul. Pept.* 28:57–69.

Bjenning, C., Farrell, A.P., and Holmgren, S. (1991). Bombesin-like immunoreactivity in skates and the *in vitro* effect of bombesin on coronary vessels from the longnose skate, *Raja rhina. Regul. Pept.* 35:207–219.

Bjenning, C., Hazon, N., Balasubramaniam, A., Holmgren, S., and Conlon, J.M. (1993). Distribution and activity of dogfish NPY and peptide YY in the cardiovascular system of the common dogfish. *Am. J. Physiol.* 264:R1119–R1124.

Björnsson, B.T., Haux, C., Forlin, L., and Deftos, L.J. (1986). The involvement of calcitonin in the reproductive physiology of the rainbow trout. *J. Endocrinol.* 108:17–23.

Bourguignon, J.P. and Plant, T.M. (Eds.). (2000). *The Onset of Puberty in Perspective.* Elsevier, Amsterdam.

Bullesbach, E.E., Gowan, L.K., Schwabe, C., Steinetz, B.G., O'Byrne, E., and Callard, I.P. (1986). Isolation, purification, and the sequence of relaxin from spiny dogfish (*Squalus acanthias*). *Eur. J. Biochem.* 161:335–341.

Bullesbach, E.E., Schwabe, C., and Callard, I.P. (1987). Relaxin from an oviparous species, the skate (*Raja erinacea*). *Biochem. Biophys. Res. Comm.* 143:273–280.

Bullesbach, E.E., Schwabe, C., and Lacy, E.R. (1997). Identification of a glycosylated relaxin-like molecule from the male Atlantic stingray, *Dasyatis sabina. Biochemistry* 36:10735–10741.

Bullock, T.H. and Northcutt, R.G. (1984). Nervus terminalis in dogfish (*Squalus acanthias*, Elasmobranchii) carries tonic efferent impulses. *Neurosci. Lett.* 44:155–160.

Burton, M. and Idler, D.R. (1986). The cellular location of 1 alpha-hydroxycorticosterone binding protein in skate. *Gen. Comp. Endocrinol.* 64:260–266.

Butler, P.J., Metcalfe, J.D., and Ginley, S.A. (1986). Plasma catecholamines in the lesser spotted dogfish and rainbow trout at rest and during different levels of exercise. *J. Exp. Biol.* 123:409–421.

Callard, G.V. (1988). Reproductive physiology. Part B. The male. In: Shuttleworth, T.J. (Ed.), *Physiology of Elasmobranch Fishes.* Springer-Verlag, Heidelberg, pp. 292–317.

Callard, G.V. (1992). Autocrine and paracrine role of steroids during spermatogenesis: studies in *Squalus acanthias* and *Necturus maculosus. J. Exp. Zool.* 261:132–142.

Callard, G.V. and Mak, P. (1985). Exclusive nuclear location of estrogen receptors in *Squalus* testis. *Proc. Natl. Acad. Sci. U.S.A.* 82:1336–1340.

Callard, G.V., Pudney, J.A., Mak, P., and Canick, J.A. (1985). Stage-dependent changes in steroidogenic enzymes and estrogen receptors during spermatogenesis in the testis of the dogfish, *Squalus acanthias. Endocrinology* 117:1328–1335.

Callard, I.P. and Klosterman, L. (1988). Reproductive physiology. Part A. The female. In: Shuttleworth, T.J. (Ed.), *Physiology of Elasmobranch Fishes.* Springer-Verlag, Heidelberg, pp. 279–292.

Callard, I.P. and Koob T.J. (1993). Endocrine regulation of the elasmobranch reproductive tract. *J. Exp. Zool.* 266:368–377.

Callard, I.P., Fileti, L.A., Perez, L.E., Sorbera, L.A., Giannoukos, G., Klosterman, L.L., Tsang, P., and McCracken, J.A. (1992). Role of the corpus luteum and progesterone in the evolution of vertebrate viviparity. *Am. Zool.* 32:264–275.

Callard, I.P., Fileti, L.A., and Koob, T.J. (1993). Ovarian steroid synthesis and the hormonal control of the elasmobranch reproductive tract. *Environ. Biol. Fish.* 38:175–186.

Callard, I.P., St. George, J., and Koob, T.J. (2005). Endocrine control of the female reproductive tract. In: Hamlett, W.C. (Ed.), *Reproductive Biology and Phylogeny of Chondrichthyes: Sharks, Batoids and Chimaeras.* Science Publishers, Enfield, NH, pp. 283–300.

Calvin, J.L., Slater, C.H., Bolduc, T.G., Laudano, A.P., and Sower, S.A. (1993). Multiple molecular forms of gonadotropin-releasing hormone in the brain of an elasmobranch: evidence for IR-lamprey GnRH. *Peptides* 14:725–729.

Camargo, C.R., Visconti, M.A., and Castrucci, A.M. (1999). Physiological color change in the bullfrog, *Rana catesbeiana. J. Exp. Zool.* 283:160–169.

Carroll, R.G., Opdyke, D.F., and Keller, N.E. (1984). Vascular recovery following hemorrhage in the dogfish shark *Squalus acanthias. Am. J. Physiol.* 246:R825–R828.

Charmandari, E., Tsigos, C., and Chrousos, G. (2005). Endocrinology of the stress response. *Annu. Rev. Physiol.* 67:259–284.

Chevins, P.F. and Dodd, J.M. (1970). Pituitary innervation and control of colour change in the skates *Raia naevus, R. clavata, R. montagui,* and *R. radiata. Gen. Comp. Endocrinol.* 15:232–241.

Chiba, A. (1998). Ontogeny of serotonin-immunoreactive cells in the gut epithelium of the cloudy dogfish, *Scyliorhinus torazame,* with reference to coexistence of serotonin and neuropeptide Y. *Gen. Comp. Endocrinol.* 111:290–298.

Chiba, A. (2000). Immunohistochemical cell types in the terminal nerve ganglion of the cloudy dogfish, *Scyliorhinus torazame,* with special regard to neuropeptide Y/FMRFamide-immunoreactive cells. *Neurosci. Lett.* 286:195–198.

Chiba, A. (2001). Marked distributional difference of alpha-melanocyte-stimulating hormone (alpha-MSH)-like immunoreactivity in the brain between two elasmobranchs (*Scyliorhinus torazame* and *Etmopterus brachyurus*): an immunohistochemical study. *Gen. Comp. Endocrinol.* 122:287–295.

Chiba, A. and Honma, Y. (1992). Distribution of neuropeptide Y-like immunoreactivity in the brain and hypophysis of the cloudy dogfish, *Scyliorhinus torazame. Cell Tissue Res.* 268:453–461.

Chiba, A., Honma, Y., and Oka, S. (1995). Ontogenetic development of neuropeptide Y-like-immunoreactive cells in the gastroenteropancreatic endocrine system of the dogfish. *Cell Tissue Res.* 282:33–40.

Chiba, A., Oka, S., and Honma, Y. (1996). Ontogenetic changes in neuropeptide Y-like-immunoreactivity in the terminal nerve of the chum salmon and the cloudy dogfish, with special reference to colocalization with gonadotropin-releasing hormone-immunoreactivity. *Neurosci. Lett.* 213:49–52.

Chiba, A., Oka, S., and Saitoh, E. (2002). Ontogenetic changes in neuropeptide Y-immunoreactive cerebrospinal fluid-contacting neurons in the hypothalamus of the cloudy dogfish, *Scyliorhinus torazame* (Elasmobranchii). *Neurosci. Lett.* 329:301–304.

Chieffi, G. (1967). The reproductive system of elasmobranchs: developmental and endocrinological aspects. In: Gilbert, P.W., Mathewson, R.F., and Rall, D.P. (Eds.), *Sharks, Skates, and Rays*. The Johns Hopkins University Press, Baltimore, MD, pp. 553–580.

Chipkin, S.R., Stoff, J.S., and Aronin, N. (1988). Immunohistochemical evidence for neural mediation of VIP activity in the dogfish rectal gland. *Peptides* 9:119–124.

Cimini, V., van Noorden, S., Giordano-Lanza, G., Nardini, V., McGregor, G.P., Bloom, S.R., and Polak, J.M. (1985). Neuropeptides and 5-HT immunoreactivity in the gastric nerves of the dogfish (*Scyliorhinus stellaris*). *Peptides* 6(Suppl. 3):373–377.

Cimini, V., van Noorden, S., and Polak, J.M. (1989). Co-localisation of substance P-, bombesin- and peptide histidine isoleucine (PHI)-like peptides in gut endocrine cells of the dogfish *Scyliorhinus stellaris*. *Anat. Embryol. (Berlin)* 179:605–614.

Cimini, V., van Noorden, S., and Sansone, M. (1992). Neuropeptide Y-like immunoreactivity in the dogfish gastroenteropancreatic tract: light and electron microscopical study. *Gen. Comp. Endocrinol.* 86:413–423.

Claes, J.M. and Mallefet, J. (2009). Hormonal control of luminescence from lantern shark (*Etmopterus spinax*) photophores. *J. Exp. Biol.* 212:3684–3692.

Claes, J.M. and Mallefet, J. (2010a). Functional physiology of lantern shark (*Etmopterus spinax*) luminescent pattern: differential hormonal regulation of luminous zones. *J. Exp. Biol.* 213:1852–1858.

Claes, J.M. and Mallefet, J. (2010b). The lantern shark's light switch: turning shallow water crypsis into midwater camouflage. *Biol. Lett.* 6:685–687.

Conlon, J.M. and Thim, L. (1986). Primary structure of insulin and a truncated C-peptide from an elasmobranchian fish, *Torpedo marmorata*. *Gen. Comp. Endocrinol.* 64:199–205.

Conlon, J.M., Agoston, D.V., and Thim, L. (1985). An elasmobranchian somatostatin: primary structure and tissue distribution in *Torpedo marmorata*. *Gen. Comp. Endocrinol.* 60:406–413.

Conlon, J.M., Henderson, I.W., and Thim, L. (1987). Gastrin-releasing peptide from the intestine of the elasmobranch fish, *Scyliorhinus canicula* (common dogfish). *Gen. Comp. Endocrinol.* 68:415–420.

Conlon, J.M., Goke, R., Andrews, P.C., and Thim, L. (1989). Multiple molecular forms of insulin and glucagon-like peptide from the Pacific ratfish (*Hydrolagus colliei*). *Gen. Comp. Endocrinol.* 73:136–146.

Conlon, J.M., Bjenning, C., and Hazon, N. (1992). Structural characterization of neuropeptide Y from the brain of the dogfish, *Scyliorhinus canicula*. *Peptides* 13:493–497.

Conlon, J.M., Agius, L., George, K., Alberti, M.M., and Hazon, N. (1994a). Effects of dogfish urotensin II on lipid mobilization in the fasted dogfish, *Scyliorhinus canicula*. *Gen. Comp. Endocrinol.* 93:177–180.

Conlon, J.M., Hazon, N., and Thim, L. (1994b). Primary structures of peptides derived from proglucagon isolated from the pancreas of the elasmobranch fish, *Scyliorhinus canicula*. *Peptides* 15:163–167.

Craik, J.C. (1978). The effects of oestrogen treatment on certain plasma constituents associated with vitellogenesis in the elasmobranch *Scyliorhinus canicula* L. *Gen. Comp. Endocrinol.* 35:455–464.

Craik, J.C. (1979). Simultaneous measurement of rates of vitellogenin synthesis and plasma levels of oestradiol in an elasmobranch. *Gen. Comp. Endocrinol.* 38:264–266.

Cuevas, M.E. and Callard, G. (1992). Androgen and progesterone receptors in shark (*Squalus*) testis: characteristics and stage-related distribution. *Endocrinology* 130:2173–2182.

Cuevas, M.E., Miller, W., and Callard, G. (1992). Sulfoconjugation of steroids and the vascular pathway of communication in dogfish testis. *J. Exp. Zool.* 264:119–129.

Cuevas, M.E., Collins, K., and Callard, G.V. (1993). Stage-related changes in steroid-converting enzyme activities in *Squalus* testis: synthesis of biologically active metabolites via 3 beta-hydroxysteroid dehydrogenase/isomerase and 5 alpha-reductase. *Steroids* 58:87–94.

Dacke, C.G., Furr, B.J., Boelkins, J.N., and Kenny, A.D. (1976). Sexually related changes in plasma calcitonin levels in Japanese quail. *Comp. Biochem. Physiol. A Mol. Integr. Physiol.* 55:341–344.

D'Antonio, M., Vallarino, M., Lovejoy, D.A., Vandesande, F., King, J.A., Pierantoni, R., and Peter, R.E. (1995). Nature and distribution of gonadotropin-releasing hormone (GnRH) in the brain, and GnRH and GnRH binding activity in serum of the spotted dogfish *Scyliorhinus canicula*. *Gen. Comp. Endocrinol.* 98:35–49.

Demski, L.S. and Northcutt, R.G. (1983). The terminal nerve: a new chemosensory system in vertebrates? *Science* 22:435–437.

Demski, L.S., Fields, R.D., Bullock, T.H., Schriebman, M.P., and Margolis-Nunno, H. (1987). The terminal nerve of sharks and rays: EM, immunocytochemical and electrophysiological studies. *Ann. N.Y. Acad. Sci.* 519:15–32.

Denning-Kendall, P.A., Sumpter, J.P., and Lowry, P.J. (1982). Peptides derived from pro-opiocortin in the pituitary gland of the dogfish, *Squalus acanthias*. *J. Endocrinol.* 93:381–390.

Denslow, N.D., Chow, M.C., Kroll, K.J., and Green, L. (1999). Vitellogenin as a biomarker of exposure for estrogen or estrogen mimics. *Ecotoxicology* 8:385–398.

deRoos, R. and deRoos, C.C. (1973). Elevation of plasma glucose levels by mammalian ACTH in the spiny dogfish shark (*Squalus acanthias*). *Gen. Comp. Endocrinol.* 21:403–409.

deRoos, R. and deRoos, C.C. (1978). Elevation of plasma glucose levels by catecholamines in elasmobranch fish. *Gen. Comp. Endocrinol.* 34:447–452.

deRoos, R. and deRoos, C.C. (1979). Severe insulin-induced hypoglycemia in the spiny dogfish shark (*Squalus acanthias*). *Gen. Comp. Endocrinol.* 37:186–191.

deRoos, R. and deRoos, C.C. (1992). Effects of mammalian ACTH on potential fuels and gluconeogenic substrates in the plasma of the spiny dogfish shark (*Squalus acanthias*). *Gen. Comp. Endocrinol.* 87:149–158.

deRoos, R., deRoos, C.C., Werner, C.S., and Werner, H. (1985). Plasma levels of glucose, alanine, lactate, and beta-hydroxybutyrate in the unfed spiny dogfish shark (*Squalus acanthias*) after surgery and following mammalian insulin infusion. *Gen. Comp. Endocrinol.* 58:28–43.

Dimaline, R., Thorndyke, M.C., and Young, J. (1986). Isolation and partial sequence of elasmobranch VIP. *Regul. Pept.* 14:1–10.

Dimaline, R., Young, J., Thwaites, D.T., Lee, C.M., Shuttleworth, T.J., and Thorndyke, M.C. (1987). A novel vasoactive intestinal peptide (VIP) from elasmobranch intestine has full affinity for mammalian pancreatic VIP receptors. *Biochim. Biophys. Acta* 930:97–100.

Dobson, S. and Dodd, J.M. (1977a). Endocrine control of the testis in the dogfish *Scyliorhinus canicula* L. I. Effects of partial hypophysectomy on gravimetric, hormonal and biochemical aspects of testis function. *Gen. Comp. Endocrinol.* 32:41–52.

Dobson, S. and Dodd, J.M. (1977b). Endocrine control of the testis in the dogfish *Scyliorhinus canicula* L. II. Histological and ultrastructural changes in the testis after partial hypophysectomy (ventral lobectomy). *Gen. Comp. Endocrinol.* 32:53–71.

Dobson, S. and Dodd, J.M. (1977c). The roles of temperature and photoperiod in the response of the testis of the dogfish, *Scyliorhinus canicula* L., to partial hypophysectomy (ventral lobectomy). *Gen. Comp. Endocrinol.* 32:114–115.

Dodd, J.M. (1975). The hormones of sex and reproduction and their effects in fish and lower chordates: twenty years on. *Am. Zool.* 12:325–339.

Dodd, J.M. and Goddard, C.K. (1961). Some effects of oestradiol benzoate on the reproductive ducts of the female dogfish *Scyliorhinus canciulus*. *Proc. Zool. Soc. London* 137:325–331.

Dodd, J.M., Evennett, P.J., and Goddard, C.K. (1960). Reproductive endocrinology in cyclostomes and elasmobranchs. *Symp. Zool. Soc. London* 1:77–103.

Downing, S.J. and Sherwood, O.D. (1985). The physiological role of relaxin in the pregnant rat. I. The influence of relaxin on parturition. *Endocrinology* 116:1200–1205.

DuBois, W., Mak, P., and Callard, G.V. (1989). Sertoli cell functions during spermatogenesis: the shark testis model. *Fish Physiol. Biochem.* 7:221–227.

Eberle, A., Chang, Y.-S., and Schwyzer, R. (1978). Chemical synthesis and biological activity of the dogfish (*Squalus acanthias*) alpha-melanotropins I and II and of related peptides. *Helv. Chim. Acta* 61:2360–2374.

Ecay, T.W. and Valentich, J.D. (1991). Chloride secretagogues stimulate inositol phosphate formation in shark rectal gland tubules cultured in suspension. *J. Cell. Physiol.* 146:407–416.

Eddy, E.M., Washburn, T.F., Bunch, D.O., Goulding, E.H., Gladen, B.C., Lubahn, D.B., and Korach, K.S. (1996). Targeted disruption of the estrogen receptor gene in male mice causes alteration of spermatogenesis and infertility. *Endocrinology* 137:4796–4805.

El-Salhy, M. (1984). Immunocytochemical investigation of the gastro–entero–pancreatic (GEP) neurohormonal peptides in the pancreas and gastrointestinal tract of the dogfish *Squalus acanthias*. *Histochemistry* 80:193–205.

Engel, K.B. and Callard, G.V. (2005). The testis and spermatogenesis. In: Hamlett, W.C. (Ed.), *Reproductive Biology and Phylogeny of Chondrichthyes: Sharks, Batoids and Chimaeras*. Science Publishers, Enfield, NH, pp. 171–200.

Epstein, F.H., Stoff, J.S., and Silva, P. (1981). Hormonal control of secretion in shark rectal gland. *Ann. N.Y. Acad. Sci.* 372:613–625.

Epstein, F.H., Stoff, J.S., and Silva, P. (1983). Mechanism and control of hyperosmotic NaCl-rich secretion by the rectal gland of *Squalus acanthias*. *J. Exp. Biol.* 106:25–41.

Evans, A.N. and Nunez, B.S. (2010). Regulation of mRNAs encoding the steroidogenic acute regulatory protein and cholesterol side-chain cleavage enzyme in the elasmobranch interrenal gland. *Gen. Comp. Endocrinol.* 168:121–132.

Evans, A.N., Henning, T., Gelsleichter, J., and Nunez, B.S. (2010). Molecular classification of an elasmobranch angiotensin receptor: quantification of angiotensin receptor and natriuretic peptide receptor mRNAs in saltwater and freshwater populations of the Atlantic stingray. *Comp. Biochem. Physiol. B Biochem. Mol. Biol.* 157:423–431.

Evans, A.N., Rimoldi, J.M., Gadepalli, R.S., and Nunez, B.S. (2010). Adaptation of a corticosterone ELISA to demonstrate sequence-specific effects of angiotensin II peptides and C-type natriuretic peptide on 1alpha-hydroxycorticosterone synthesis and steroidogenic mRNAs in the elasmobranch interrenal gland. *J. Steroid Biochem. Mol. Biol.* 120:149–154.

Evans, D.H. and Piermarini, P.M. (2001). Contractile properties of the elasmobranch rectal gland. *J. Exp. Biol.* 204:59–67.

Faraldi, G., Bonini, E., Farina, L., and Tagliafierro, G. (1988). Distribution and ontogeny of glucagon-like cells in the gastrointestinal tract of the cartilaginous fish *Scyliorhinus stellaris* (L.). *Acta Histochem.* 83:57–64.

Fasano, S., Pierantoni, R., Minucci, S., Di Matteo, L., D'Antonio, M., and Chieffi, G. (1989). Effects of intratesticular injections of estradiol and gonadotropin-releasing hormone (GnRHA, HOE 766) on plasma androgen levels in intact and hypophysectomized *Torpedo marmorata* and *Torpedo ocellata*. *Gen. Comp. Endocrinol.* 75:349–354.

Fasano, S., D'Antonio, M., Pierantoni, R., and Chieffi, G. (1992). Plasma and follicular tissue steroid levels in the elasmobranch fish, *Torpedo marmorata*. *Gen. Comp. Endocrinol.* 85:327–333.

Fileti, L.A. and Callard, I.P. (1988). Corpus luteum function and regulation in the skate, *Raja erinacea*. *Bull. Mt. Desert Isl. Biol. Lab.* 29:129–130.

Forlano, P.M., Maruska, K.P., Sower, S.A., King, J.A., and Tricas, T.C. (2000). Differential distribution of gonadotropin-releasing hormone-immunoreactive neurons in the stingray brain: functional and evolutionary considerations. *Gen. Comp. Endocrinol.* 118:226–248.

Forrest, Jr., J.N., Wang, F., and Beyenbach, K.W. (1983). Perfusion of isolated tubules of the shark rectal gland. Electrical characteristics and response to hormones. *J. Clin. Invest.* 72:1163–1167.

Franklin, C.E., Holmgren, S., and Taylor, G.C. (1996). A preliminary investigation of the effects of vasoactive intestinal peptide on secretion from the lingual salt glands of *Crocodylus porosus. Gen. Comp. Endocrinol.* 102:74–78.

Garnier, D.H., Sourdaine, P., and Jegou, B. (1999). Seasonal variations in sex steroids and male sexual characteristics in *Scyliorhinus canicula. Gen. Comp. Endocrinol.* 116:281–290.

Gelsleichter, J. and Musick, J.A. (1999). Effects of insulin-like growth factor-I, corticosterone, and 3,3′,5-tri-iodo-L-thyronine on glycosaminoglycan synthesis in vertebral cartilage of the clearnose skate, *Raja eglanteria. J. Exp. Zool.* 284:549–556.

Gelsleichter, J., Rasmussen, L.E.L., Manire, C.A., Tyminski, J., Chang, B., and Lombardi-Carson, L. (2002). Serum steroid concentrations and development of reproductive organs during puberty in male bonnethead sharks, *Sphyrna tiburo. Fish Physiol. Biochem.* 26:389–401.

Gelsleichter, J., Steinetz, B.G., Manire, C.A., and Ange, C. (2003). Serum relaxin concentrations and reproduction in male bonnethead sharks, *Sphyrna tiburo. Gen. Comp. Endocrinol.* 132:27–34.

Gelsleichter, J., Walsh, C.J., Szabo, N.J., and Rasmussen, L.E. (2006). Organochlorine concentrations, reproductive physiology, and immune function in unique populations of freshwater Atlantic stingrays (*Dasyatis sabina*) from Florida's St. Johns River. *Chemosphere* 63:1506–1522.

Gelsleichter, J. and Walker, C.J. (2010). Pollutant exposure and effects in sharks and their relatives. In: Carrier, J., Musick, J.A., and Heithaus, M. (Eds.), *Sharks and Their Relatives II: Biodiversity, Adaptive Physiology, and Conservation.* CRC Press, Boca Raton, FL, pp. 491–537.

Gerstberger, R. (1988). Functional vasoactive intestinal polypeptide (VIP)-system in salt glands of the Pekin duck. *Cell. Tissue Res.* 252:39–48.

Gerstberger, R., Sann, H., and Simon, E. (1988). Vasoactive intestinal peptide stimulates blood flow and secretion of avian salt glands. *Am. J. Physiol.* 255:R575–R582.

Gottfried, H. and Chieffi, G. (1967). The seminal steroids of the dogfish *Scylliorhinus stellaris. J. Endocrinol.* 37:99–100.

Gowan, L.K., Reinig, J.W., Schwabe, C., Bedarkar, S., and Blundell, T.L. (1981). On the primary and tertiary structure of relaxin from the sand tiger shark (*Odontaspis taurus*). *FEBS Lett.* 129:80–82.

Grant, Jr., W.C., Hendler, F.J., and Banks, P.M. (1969). Studies on the blood-sugar regulation in the little skate *Raja erinacea. Physiol. Zool.* 42:231–247.

Grimm, A., O'Halloran, M., and Idler, D. (1969). Stimulation of sodium transport across the isolated toad bladder by 1-alpha-hydroxycorticosterone from an elasmobranch. *J. Fish. Res. Bd. Can.* 26:1823–1835.

Grumbach, M.M. (2000). Estrogen, bone, growth and sex: a sea change in conventional wisdom. *J. Pediatr. Endocrinol. Metab.* 13(Suppl. 6):1439–1455.

Guibbolini, M.E., Henderson, I.W., Mosley, W., and Lahlou, B. (1988). Arginine vasotocin binding to isolated branchial cells of the eel: effect of salinity. *J. Mol. Endocrinol.* 1:125–130.

Gutiérrez, J., Fernandez, J., Blasco, J., Gesse, J.M., and Planas, J. (1986). Plasma glucagon levels in different species of fish. *Gen. Comp. Endocrinol.* 63:328–333.

Gutiérrez, J., Fernandez, J., and Planas, J. (1988). Seasonal variations of insulin and some metabolites in dogfish plasma, *Scyliorhinus canicula, L. Gen. Comp. Endocrinol.* 70:1–8.

Hamano, K., Tierney, M.L., Ashida, K., Takei, Y., and Hazon, N. (1998). Direct vasoconstrictor action of homologous angiotensin II on isolated arterial ring preparations in an elasmobranch fish. *J. Endocrinol.* 158:419–423.

Hansen, D. (1975). Evidence of a gastrin-like substance in *Rhinobatis productus. Comp. Biochem. Physiol. C Toxicol. Pharmacol.* 52:61–63.

Hayashida, T. and Lewis, U.J. (1978). Immunochemical and biological studies with antiserum to shark growth hormone. *Gen. Comp. Endocrinol.* 36:530–542.

Hayes, T.B. (1998). Sex determination and primary sex differentiation in amphibians: genetic and developmental mechanisms. *J. Exp. Zool.* 281:373–399.

Hazon, N. and Henderson, I.W. (1984). Secretory dynamics of 1-alpha-hydroxycorticosterone in the elasmobranch fish, *Scyliorhinus canicula. J. Endocrinol.* 103:205–211.

Hazon, N. and Henderson, I.W. (1985). Factors affecting the secretory dynamics of 1-alpha-hydroxycorticosterone in the dogfish, *Scyliorhinus canicula. Gen. Comp. Endocrinol.* 59:50–55.

Hazon, N., Balment, R.J., Perrott, M., and O'Toole, L.B. (1989). The renin–angiotensin system and vascular and dipsogenic regulation in elasmobranchs. *Gen. Comp. Endocrinol.* 74:230–236.

Hazon, N., Tierney, M.L., Anderson, W.G., MacKenzie, S., Cutler, C., and Cramb, G. (1997). Ion and water balance in elasmobranch fish. In: Hazon, N., Eddy, F.B., and Flik, G. (Eds.), *Ionic Regulation in Animals.* Springer-Verlag, Heidelberg, pp. 70–86.

Henderson, I.W., Oliver, J.A., McKeever, A., and Hazon, N. (1981). Phylogenetic aspects of the renin–angiotensin system. In: Pethes, G. and Frenyo, V.L. (Eds.), *Advances in Physiological Sciences.* Vol. 2. *Advances in Animal and Comparative Physiology.* Pergamon Press, London, pp. 355–363.

Heupel, M.R., Whittier, J.M., and Bennett, M.B. (1999). Plasma steroid hormone profiles and reproductive biology of the epaulette shark, *Hemiscyllium ocellatum. J. Exp. Zool.* 284:586–594.

Hisaw, L. and Abramowitz, A. (1939). Physiology of reproduction in the dogfishes *Mustelus canis* and *Squalus acanthias. Rep. Woods Hole Oceanogr. Inst.* 1938:22.

Ho, S.-M., Wulczyn, G., and Callard, I.P. (1980). Induction of vitellogenin synthesis in the spiny dogfish, *Squalus acanthias. Bull. Mt. Desert Isl. Biol. Lab.* 19:37–38.

Hoffmayer, E.R. and Parsons, G.R. (2001). The physiological response to capture and handling stress in the Atlantic sharpnose shark, *Rhizoprionodon terranovae. Fish Physiol. Biochem.* 25:277–285.

Hogben, L.T. (1942). Chromatic behaviour. *Proc. R. Soc. Lond. B Biol. Sci.* 131:111–136.

Holmgren, S. and Nilsson, S. (1983). Bombesin-, gastrin/CCK-, 5-hydroxytryptamine-, neurotensin-, somatostatin-, and VIP-like immunoreactivity and catecholamine fluorescence in the gut of the elasmobranch, *Squalus acanthias. Cell Tissue Res.* 234:595–618.

Holmgren, S. and Nilsson, S. (1999). Digestive system. In: Hamlett, W.C. (Ed.), *Sharks, Skates and Rays: The Biology of Elasmobranch Fish*. The Johns Hopkins University Press, Baltimore, MD, pp. 144–173.

Holmgren, S., Axelsson, M., and Farrell, A.P. (1992a). The effects of catecholamines, substance P, and vasoactive intestinal polypeptide on blood flow to the gut in the dogfish *Squalus acanthias*. *J. Exp. Biol.* 168:161–175.

Holmgren, S., Axelsson, M., and Farrell, A.P. (1992b). The effects of neuropeptide Y and bombesin on blood flow to the gut in dogfish *Squalus acanthias*. *Regul. Pept.* 40:169.

Holstein, A.F. (1969). Zur Frange der lokalen Steuerung der Spermatogenese beim Dornhai (*Squalus acanthias* L.). *Z. Zellforsch.* 93:265–281.

Holt, W.F. and Idler, D.R. (1975). Influence of the interrenal gland on the rectal gland of a skate. *Comp. Biochem. Physiol. C Toxicol. Pharmacol.* 50:111–119.

Hyodo, S., Tsukada, T., and Takei, Y. (2004). Neurohypophysial hormones of dogfish, *Triakis scyllium*: structures and salinity-dependent secretion. *Gen. Comp. Endocrinol.* 138:97–104.

Idler, D.R. and Kane, K.M. (1976). Interrenalectomy and Na-K-ATPase activity in the rectal gland of the skate *Raja ocellata*. *Gen. Comp. Endocrinol.* 28:100–102.

Idler, D.R. and Kane, K.M. (1980). Cytosol receptor glycoprotein for 1 alpha-hydroxycorticosterone in tissues of an elasmobranch fish (*Raja ocellata*). *Gen. Comp. Endocrinol.* 42:259–266.

Idler, D.R. and Truscott, B. (1966). 1α-hydroxycorticosterone from cartilaginous fish: a new adrenal steroid in blood. *J. Fish. Res. Bd. Can.* 23:615–619.

Jenkins, N., Joss, J.P., and Dodd, J.M. (1980). Biochemical and autoradiographic studies on the oestradiol-concentrating cells in the diencephalon and pituitary gland of the female dogfish (*Scyliorhinus canicula* L.). *Gen. Comp. Endocrinol.* 40:211–219.

Jensen, J. (2001). Regulatory peptides and control of food intake in non-mammalian vertebrates. *Comp. Biochem. Physiol. A Mol. Integr. Physiol.* 128:471–479.

Johnsen, A.H., Jonson, L., Rourke, I.J., and Rehfeld, J.F. (1997). Elasmobranchs express separate cholecystokinin and gastrin genes. *Proc. Natl. Acad. Sci. U.S.A.* 94:10221–10226.

Jonsson, A.C. (1991). Regulatory peptides in the pancreas of two species of elasmobranchs and in the Brockmann bodies of four teleost species. *Cell Tissue Res.* 266:163–172.

Jonsson, A.C. (1995). Endocrine cells with gastrin/cholecystokinin-like immunoreactivity in the pancreas of the spiny dogfish, *Squalus acanthias*. *Regul. Pept.* 59:67–78.

Kågstrom, J., Axelsson, M., Jensen, J., Farrell, A.P., and Holmgren, S. (1996). Vasoactivity and immunoreactivity of fish tachykinins in the vascular system of the spiny dogfish. *Am. J. Physiol.* 270:R585–R593.

Kaiya, H., Kodama, S., Ishiguro, K., Matsuda, K., Uchiyama, M., Miyazato, M., and Kangawa, K. (2009). Ghrelin-like peptide with fatty acid modification and O-glycosylation in the red stingray, *Dasyatis akajei*. *BMC Biochem.* 10:30.

Kajiura, S.M., Tyminski, J.P., Forni, J.B., and Summers, A.P. (2005). The sexually dimorphic cephalofoil of bonnethead sharks, *Sphyrna tiburo*. *Biol. Bull.* 209:1–5.

Karsch, F.J., Dahl, G.E., Hachigian, T.M., and Thrun, L.A. (1995). Involvement of thyroid hormones in seasonal reproduction. *J. Reprod. Fertil. Suppl.* 49:409–422.

Kawakoshi, A., Hyodo, S., and Takei, Y. (2001). CNP is the only natriuretic peptide in an elasmobranch fish, *Triakis scyllia*. *Zool. Sci.* 18:861–868.

Kawakoshi, A., Kaiya, H., Riley, L.G., Hirano, T., Grau, E.G., Miyazato, M., Hosoda, H., and Kangawa, K. (2007). Identification of a ghrelin-like peptide in two species of shark, *Sphyrna lewini* and *Carcharhinus melanopterus*. *Gen. Comp. Endocrinol.* 151:259–268.

King, J.A. and Millar, R.P. (1979). Phylogenetic and anatomical distribution of somatostatin in vertebrates. *Endocrinology* 105:1322–1329.

King, J.A. and Millar, R.P. (1980). Comparative aspects of luteinizing hormone-releasing hormone structure and function in vertebrate phylogeny. *Endocrinology* 106:707–717.

King, J.A., Steneveld, A.A., Millar, R.P., Fasano, S., Romano, G., Spagnuolo, A., Zanetti, L., and Pierantoni, R. (1992). Gonadotropin-releasing hormone in elasmobranch (electric ray, *Torpedo marmorata*) brain and plasma: chromatographic and immunological evidence for chicken GnRH II and novel molecular forms. *Peptides* 13:27–35.

Klesch, W. and Sage, M. (1975). The stimulation of corticosteroidogenesis in the interrenal of the elasmobranch *Dasyatis sabina* by mammalian ACTH. *Comp. Biochem. Physiol. A Mol. Integr. Physiol.* 52:145–146.

Klok, M.D., Jakobsdottir, S., and Drent, M.L. (2007). The role of leptin and ghrelin in the regulation of food intake and body weight in humans: a review. *Obes. Rev.* 8:21–34.

Kneebone, J., Ferguson, D.E., Sulikowski, J.A., and Tsang, P.C.W. (2007). Endocrinological investigation into the reproductive cycles of two sympatric skate species, *Malacoraja senta* and *Amblyraja radiata*, in the western Gulf of Maine. *Environ. Biol. Fish.* 80:257–265.

Koob, T.J. and Callard, I.P. (1985). Effect of progesterone on the ovulatory cycle of the little skate *Raja erinacea*. *Bull. Mt. Desert Isl. Biol. Lab.* 25:138–139.

Koob, T.J. and Callard, I.P. (1999). Reproductive endocrinology of female elasmobranchs: lessons from the little skate (*Raja erinacea*) and spiny dogfish (*Squalus acanthias*). *J. Exp. Zool.* 284:557–574.

Koob, T.J., Laffan, J.J., Elger, B., and Callard, I.P. (1983). Effects of estradiol on the Verschlussvorrichtung of *Squalus acanthias*. *Bull. Mt. Desert Isl. Biol. Lab.* 23:67–68.

Koob, T.J., Laffan, J.J., and Callard, I.P. (1984). Effects of relaxin and insulin on reproductive tract size and early fetal loss in *Squalus acanthias*. *Biol. Reprod.* 31:231–238.

Koob, T.J., Tsang, P., and Callard, I.P. (1986). Plasma estradiol, testosterone, and progesterone levels during the ovulatory cycle of the skate (*Raja erinacea*). *Biol. Reprod.* 35:267–275.

Kwon, S., Hess, R.A., Bunick, D., Kirby, J.D., and Bahr, J.M. (1997). Estrogen receptors are present in the epididymis of the rooster. *J. Androl.* 18:378–384.

Lance, V. and Callard, I.P. (1978). Gonadotrophic activity in pituitary extracts from a elasmobranch (*Squalus acanthias* L.). *J. Endocrinol.* 78:149–150.

Le Roith, D., Bondy, C., Yakar, S., Liu, J.-L., and Butler, A. (2001). The somatomedin hypothesis: 2001. *Endocr. Rev.* 22:53–74.

Lehrich, R.W., Aller, S.G., Webster, P., Marino, C.R., and Forrest, Jr., J.N. (1998). Vasoactive intestinal peptide, forskolin, and genistein increase apical CFTR trafficking in the rectal gland of the spiny dogfish, *Squalus acanthias*. Acute regulation of CFTR trafficking in an intact epithelium. *J. Clin. Invest.* 101:737–745.

Leibson, L.G. and Plisetskaia, E.M. (1972). Hormones and their role in regulating metabolism in cold-blooded vertebrates. *Usp. Fiziol. Nauk.* 3:26–44.

Lewis, M. and Dodd, J.M. (1974). Proceedings: thyroid function and the ovary in the spotted dogfish, *Scyliorhinus canicula. J. Endocrinol.* 63:63P.

Lipshaw, L.A., Patent, G.J., and Foa, P.P. (1972). Effects of epinephrine and norepinephrine on the hepatic lipids of the nurse shark, *Ginglymostoma cirratum. Horm. Metab. Res.* 4:34–38.

Liu, Q. and Demski, L.S. (1993). Clasper control in the round stingray, *Urolophus halleri*: lower sensorimotor pathways. *Environ. Biol. Fish.* 38:219–232.

Love, R.M. and Pickering, B.T. (1974). A beta-MSH in the pituitary gland of the spotted dogfish (*Scyliorhinus canicula*): isolation and structure. *Gen. Comp. Endocrinol.* 24:398–404.

Lovejoy, D.A., Sherwood, N.M., Fischer, W.H., Jackson, B.C., Rivier, J.E., and Lee, T. (1991). Primary structure of gonadotropin-releasing hormone from the brain of a holocephalan (ratfish: *Hydrolagus colliei*). *Gen. Comp. Endocrinol.* 82:152–161.

Lovejoy, D.A., Fischer, W.H., Ngamvongchon, S., Craig, A.G., Nahorniak, C.S., Peter, R.E., Rivier, J.E., and Sherwood, N.M. (1992a). Distinct sequence of gonadotropin-releasing hormone (GnRH) in dogfish brain provides insight into GnRH evolution. *Proc. Natl. Acad. Sci. U.S.A.* 89:6373–6377.

Lovejoy, D.A., Stell, W.K., and Sherwood, N.M. (1992b). Partial characterization of four forms of immunoreactive gonadotropin-releasing hormone in the brain and terminal nerve of the spiny dogfish (Elasmobranchii; *Squalus acanthias*). *Regul. Pept.* 37:39–48.

Lowry, P.J. and Chadwick, A. (1970). Purification and amino acid sequence of melanocyte-stimulating hormone from the dogfish *Squalus acanthias. Biochem. J.* 118:713–718.

Lowry, P.J., Bennett, H.P., and McMartin, C. (1974). The isolation and amino acid sequence of an adrenocorticotrophin from the pars distalis and a corticotrophin-like intermediate-lobe peptide from the neurointermediate lobe of the pituitary of the dogfish *Squalus acanthias. Biochem. J.* 141:427–437.

Lu, C.C., Tsai, S.C., Wang, S.W., Tsai, C.L., Lau, C.P., Shih, H.C., Chen, Y.H., Chiao, Y.C., Liaw, C., and Wang, P.S. (1998). Effects of ovarian steroid hormones and thyroxine on calcitonin secretion in pregnant rats. *Am. J. Physiol.* 274:E246–E252.

Lundin, K., Holmgren, S., and Nilsson, S. (1984). Peptidergic functions in the dogfish rectum. *Acta Physiol. Scand.* 121:46A.

Lundstrom, H.M. and Bard, P. (1932). Hypophysial control of cutaneous pigmentation in an elasmobranch fish. *Biol. Bull.* 62:1–9.

MacKenzie, S., Cutler, C.P., Hazon, N., and Cramb, G. (2002). The effects of dietary sodium loading on the activity and expression of Na, K-ATPase in the rectal gland of the European dogfish (*Scyliorhinus canicula*). *Comp. Biochem. Physiol. B Biochem. Mol. Biol.* 131:185–200.

Mandado, M., Molist, P., Anadon, R., and Yanez, J. (2001). A DiI-tracing study of the neural connections of the pineal organ in two elasmobranchs (*Scyliorhinus canicula* and *Raja montagui*) suggests a pineal projection to the midbrain GnRH-immunoreactive nucleus. *Cell Tissue Res.* 303:391–401.

Manire, C.A. and Rasmussen, L.E. (1997). Serum concentrations of steroid hormones in the mature male bonnethead shark, *Sphyrna tiburo. Gen. Comp. Endocrinol.* 107:414–420.

Manire, C.A., Rasmussen, L.E., Hess, D.L., and Hueter, R.E. (1995). Serum steroid hormones and the reproductive cycle of the female bonnethead shark, *Sphyrna tiburo. Gen. Comp. Endocrinol.* 97:366–376.

Manire, C.A., Rasmussen, L.E., and Gross, T.S. (1999). Serum steroid hormones including 11-ketotestosterone, 11-keto-androstenedione, and dihydroprogesterone in juvenile and adult bonnethead sharks, *Sphyrna tiburo. J. Exp. Zool.* 284:595–603.

Manire, C.A., Rasmussen, L.E., Gelsleichter, J., and Hess, D.L. (2004). Maternal serum and yolk hormone concentrations in the placental viviparous bonnethead shark, *Sphyrna tiburo. Gen. Comp. Endocrinol.* 136:241–247.

Manire, C.A., Rasmussen, L.E., Maruska, K.P., and Tricas, T.C. (2007). Sex, seasonal, and stress-related variations in elasmobranch corticosterone concentrations. *Comp. Biochem. Physiol. A Mol. Integr. Physiol.* 148:926–935.

Martin, S.C. and Shuttleworth, T.J. (1994). Vasoactive intestinal peptide stimulates a cAMP-mediated Cl⁻ current in avian salt gland cells. *Regul. Pept.* 52:205–214.

Maruska, K.P. and Gelsleichter, J. (2011). Hormones and reproduction in chondricthyan fishes. In: Norris, D.O. and Lopez, K.H. (Eds.), *Hormones and Reproduction in Vertebrates*. Academic Press, San Diego, CA, pp. 209–237.

Masini, M.A., Prato, P., Vacchi, M., and Uva, B.M. (2008). GnRH immunodetection in the brain of the holocephalan fish *Chimaera monstrosa* L.: correlation to oocyte maturation. *Gen. Comp. Endocrinol.* 156:559–563.

Mayer, I., Bornestaf, C., and Borg, B. (1997). Melatonin in non-mammalian vertebrates: physiological role in reproduction? *Comp. Biochem. Physiol. A Mol. Integr. Physiol.* 118:515–531.

McComb, D.M., Gelsleichter, J., Manire, C.A., Brinn, R., and Brown, C.L. (2005). Comparative thyroid hormone concentration in maternal serum and yolk of the bonnethead shark (*Sphyrna tiburo*) from two sites along the coast of Florida. *Gen. Comp. Endocrinol.* 144:167–173.

McVey, D.C., Rittschof, D., Mannon, P.J., and Vigna, S.R. (1996). Localization and characterization of neuropeptide Y/peptide YY receptors in the brain of the smooth dogfish (*Mustelis canis*). *Regul. Pept.* 61:167–173.

Mellinger, J.C.A. and Dubois, M.P. (1973). Confirmation, par l'immunofluorescence, de la fonction corticotrope du lobe rostral et de la fonction gonadotrope du lobe ventral de l'hypophyse d'un poisson cartilagineux, la torpille marbree (*Torpedo marmorata*). *C. R. Acad. Sci.* 276:1979–1981.

Metcalfe, J.D. and Butler, P.J. (1984). Changes in activity and ventilation in response to hypoxia in unrestrained, unoperated dogfish (*Scyliorhinus canicula* L.). *J. Exp. Biol.* 108:411–418.

Michael, A.E., Thurston, L.M., and Rae, M.T. (2003). Glucocorticoid metabolism and reproduction: a tale of two enzymes. *Reproduction* 126:425–441.

Misao, R., Fujimoto, J., Niwa, K., Morishita, S., Nakanishi, Y., and Tamaya, T. (1997). Immunohistochemical expressions of estrogen and progesterone receptors in human epididymis at different ages: a preliminary study. *Int. J. Fertil. Womens Med.* 42:39–42.

Moeller, J.F. and Meredith, M. (1998). Increase in gonadotropin-releasing hormone (GnRH) levels in CSF after stimulation of the nervus terminalis in Atlantic stingray, *Dasyatis sabina. Brain Res.* 806:104–107.

Moeller, J.F. and Meredith, M. (2010). Differential co-localization with choline acetyltransferase in nervus terminalis suggests functional differences for GnRH isoforms in bonnethead sharks (*Sphyrna tiburo*). *Brain Res.* 1366:44–53.

Mommsen, T., Vijayan, M., and Moon, T. (1999). Cortisol in teleosts: dynamics, mechanisms of action, and metabolic regulation. *Rev. Fish Biol. Fisher.* 9:211–268.

Moriyama, S., Oda, M., Yamazaki, T., Yamaguchi, K., Amiya, N., Takahashi, A., Amano, M., Goto, T., Nozaki, M., Meguro, H., and Kawauchi, H. (2008). Gene structure and functional characterization of growth hormone in dogfish, *Squalus acanthias. Zool. Sci.* 25:604–613.

Mountjoy, K.G., Robbins, L.S., Mortrud, M.T., and Cone, R.D. (1992). The cloning of a family of genes that encode the melanocortin receptors. *Science* 257:1248–1251.

Nichols, S., Gelsleichter, J., Manire, C.A., and Caillet, G.M. (2003). Calcitonin-like immunoreactivity in serum and tissues of the bonnethead shark, *Sphyrna tiburo. J. Exp. Zool.* 298A:150–161.

Nishimura, H. (2001). Angiotensin receptors: evolutionary overview and perspectives. *Comp. Biochem. Physiol. A Mol. Integr. Physiol.* 128:11–30.

Nishimura, H., Oguri, M., Ogawa, M., Sokabe, H., and Imai, M. (1970). Absence of renin in kidneys of elasmobranchs and cyclostomes. *Am. J. Physiol.* 218:911–915.

Norris, D.O. (2006). *Vertebrate Endocrinology*, 4th ed. Academic Press, San Diego, CA.

Nozaki, M., Tsukahara, T., and Kobayashi, H. (1984). An immunocytological study of the distribution of neuropeptides in the brain of fish. *Biomed. Res.* 4(Suppl.):135–145.

Nunez, S. and Trant, J. (1999). Regulation of interrenal gland steroidogenesis in the Atlantic stingray (*Dasyatis sabina*). *J. Exp. Zool.* 284:517–525.

Okamoto, K., Yasumura, K., Shimamura, S., Nakamura, M., Tanaka, A., and Yajima, H. (1979). Synthesis of the non-atriacontapeptide corresponding to the entire amino acid sequence of dogfish adrenocorticotropic hormone (*Squalus acantias*). *Chem. Pharm. Bull. (Tokyo)* 27:499–507.

Okuzawa, K. (2002). Puberty in teleosts. *Fish Physiol. Biochem.* 26:31–41.

Oliver, A.S. and Vigna, S.R. (1996). CCK-X receptors in the endothermic mako shark (*Isurus oxyrinchus*). *Gen. Comp. Endocrinol.* 102:61–73.

Olson, K.R. (1999). Rectal gland and volume homeostasis. In: Hamlett, W.C. (Ed.), *Sharks, Skates and Rays: The Biology of Elasmobranch Fish.* The Johns Hopkins University Press, Baltimore, MD, pp. 329–352.

Opdyke, D.F., Carroll, R.G., and Keller, N.E. (1982). Catecholamine release and blood pressure changes induced by exercise in dogfish. *Am. J. Physiol.* 242:R306–R310.

Opdyke, D.F., Bullock, J., Keller, N.E., and Holmes, K. (1983). Dual mechanism for catecholamine secretion in the dogfish shark *Squalus acanthias. Am. J. Physiol.* 244:R641–R645.

O'Toole, L.B., Armour, K.J., Decourt, C., Hazon, N., Lahlou, B., and Henderson, I.W. (1990). Secretory patterns of 1 alpha-hydroxycorticosterone in the isolated perifused interrenal gland of the dogfish, *Scyliorhinus canicula. J. Mol. Endocrinol.* 5:55–60.

Pan, J., McFerran, N.V., Shaw, C., and Halton, D.W. (1994). *Torpedo marmorata* gut PLY and the NPY/PP family: phylogenetic and structural considerations. *Biochem. Soc. Trans.* 22:6S.

Pan, J.Z., Shaw, C., Halton, D.W., Thim, L., Johnston, C.F., and Buchanan, K.D. (1992). The primary structure of peptide Y (PY) of the spiny dogfish, *Squalus acanthias*: immunocytochemical localisation and isolation from the pancreas. *Comp. Biochem. Physiol. B Biochem. Mol. Biol.* 102:1–5.

Paolucci, M. and Callard, I.P. (1998). Characterization of progesterone-binding moieties in the little skate *Raja erinacea. Gen. Comp. Endocrinol.* 109:106–118.

Parker, G.H. (1936). Color change in elasmobranchs. *Proc. Natl. Acad. Sci. U.S.A.* 22:55–60.

Patent, G.J. (1970). Comparison of some hormonal effects on carbohydrate metabolism in an elasmobranch (*Squalus acanthias*) and a holocephalan (*Hydrolagus colliei*). *Gen. Comp. Endocrinol.* 14:215–242.

Perez, L.E. and Callard, I.P. (1992). Identification of vitellogenin in the little skate (*Raja erinacea*). *Comp. Biochem. Physiol. B Biochem. Mol. Biol.* 103:699–705.

Perez, L.E. and Callard, I.P. (1993). Regulation of hepatic vitellogenin synthesis in the little skate (*Raja erinacea*): use of a homologous enzyme-linked immunosorbent assay. *J. Exp. Zool.* 266:31–39.

Phillip, M., Maor, G., Assa, S., Silbergeld, A., and Segev, Y. (2001). Testosterone stimulates growth of tibial epiphyseal growth plate and insulin-like growth factor-1 receptor abundance in hypophysectomized and castrated rats. *Endocrine* 16:1–6.

Pierantoni, R., D'Antonio, M., and Fasano, S. (1993). Morphofunctional aspects of the hypothalamus–pituitary–gonadal axis of elasmobranch fishes. *Environ. Biol. Fish.* 38:187–196.

Plesch, F., Scott, C., Calhoun, D., Aller, S., and Forrest, J. (2000). Cloning of the GHRH-like hormone/PACAP recursor polypeptide from the brain of the spiny dogfish shark, *Squalus acanthias. Bull. Mt. Desert Biol. Lab.* 39:127–129.

Powell, R.C., Millar, R.P., and King, J.A. (1986). Diverse molecular forms of gonadotropin-releasing hormone in an elasmobranch and a teleost fish. *Gen. Comp. Endocrinol.* 63:77–85.

Prisco, M., Liguoro, A., D'Onghia, B., Ricchiari, L., Andreuccetti, P., and Angelina, F. (2002). Fine structure of Leydig and Sertoli cells in the testis of immature and mature spotted ray *Torpedo marmorata*. *Mol. Reprod. Dev.* 63:192–201.

Prisco, M., Liguoro, A., Ricchiari, L., Del Giudice, G., Angelini, F., and Andreuccetti, P. (2008a). Immunolocalization of 3β-HSD and 17β-HSD in the testis of the spotted ray *Torpedo marmorata*. *Gen. Comp. Endocrinol.* 155:157–163.

Prisco, M., Valiante, S., Maddalena Di Fiore, M., Raucci, F., Del Giudice, G., Romano, M., Laforgia, V., Limatola, E., and Andreuccetti, P. (2008b). Effect of 17beta-estradiol and progesterone on vitellogenesis in the spotted ray *Torpedo marmorata* Risso 1810 (Elasmobranchii: Torpediniformes): studies on females and on estrogen-treated males. *Gen. Comp. Endocrinol.* 157:125–132.

Pudney, J. and Callard, G.V. (1984a). Development of agranular reticulum in Sertoli cells of the testis of the dogfish *Squalus acanthias* during spermatogenesis. *Anat. Rec.* 209:311–321.

Pudney, J. and Callard, G.V. (1984b). Identification of Leydig-like cells in the testis of the dogfish *Squalus acanthias*. *Anat. Rec.* 209:323–330.

Pudney, J. and Callard, G.V. (1986). Sertoli cell cytoplasts in the semen of the spiny dogfish *Squalus acanthias*. *Tissue Cell.* 18:375–382.

Quérat, B., Tonnerre-Doncarli, C., Genies, F., and Salmon, C. (2001). Duality of gonadotropins in gnathostomes. *Gen. Comp. Endocrinol.* 124:308–314.

Rasmussen, L.E. and Crow, G.L. (1993). Serum corticosterone concentrations in immature captive whitetip reef sharks, *Triaenodon obesus*. *J. Exp. Zool.* 267:283–287.

Rasmussen, L.E. and Gruber, S.H. (1993). Serum concentrations of reproductively related circulating steroid hormones in the free-ranging lemon shark, *Negaprion brevirostris*. *Environ. Biol. Fish.* 38:167–174.

Rasmussen, L.E., Hess, D.L., and Luer, C.A. (1999). Alterations in serum steroid concentrations in the clearnose skate, *Raja eglanteria*: correlations with season and reproductive status. *J. Exp. Zool.* 284:575–585.

Reese, J.C. and Callard, I.P. (1991). Characterization of a specific estrogen receptor in the oviduct of the little skate, *Raja erinacea*. *Gen. Comp. Endocrinol.* 84:170–181.

Reina, R.D. and Cooper, P.D. (2000). Control of salt gland activity in the hatchling green sea turtle, *Chelonia mydas*. *J. Comp. Physiol. B Biochem. Syst. Environ. Physiol.* 170:27–35.

Reinig, J.W., Daniel, L.N., Schwabe, C., Gowan, L.K., Steinetz, B.G., and O'Byrne, E.M. (1981). Isolation and characterization of relaxin from the sand tiger shark (*Odontaspis taurus*). *Endocrinology* 109:537–543.

Romero, L.M. (2002). Seasonal changes in plasma glucocorticoid concentrations in free-living vertebrates. *Gen. Comp. Endocrinol.* 128:1–24.

Sakaguchi, H. and Takei, Y. (1998). Characterisation of C-type natriuretic peptide receptors in the gill of dogfish *Triakis scyllia*. *J. Endocrinol.* 156:127–134.

Scanes, C.G., Dobson, S., Follett, B.K., and Dodd, J.M. (1972). Gonadotrophic activity in the pituitary gland of the dogfish (*Scyliorhinus canicula*). *J. Endocrinol.* 54:343–344.

Schaeffer, J.G. (1921). Beiträge zur Physiologie des Farbenwechsels der Fische. *Arch. Physiol.* 188:25–42.

Sekine, Y. and Yui, R. (1981). Immunohistochemical study of the pancreatic endocrine cells of the ray, *Dasyatis akajei*. *Arch. Histol. Jpn.* 44:95–101.

Shaw, C., Whittaker, V.P., and Agoston, D.V. (1990). Characterization of gastrin-releasing peptide immunoreactivity in distinct storage particles in guinea pig myenteric and *Torpedo* electromotor neurones. *Peptides* 11:69–74.

Sheikh, S.P. (1991). Neuropeptide Y and peptide YY: major modulators of gastrointestinal blood flow and function. *Am. J. Physiol.* 261:G701–G715.

Sherwood, N.M. and Lovejoy, D.A. (1993). Gonadotropin-releasing hormone in cartilaginous fishes: structure, location, and transport. *Environ. Biol. Fish.* 38:197–208.

Sherwood, N.M. and Sower, S.A. (1985). A new family member for gonadotropin-releasing hormone. *Neuropeptides* 6:205–214.

Shimamura, S., Okamoto, K., Nakamura, M., Tanaka, A., and Yajima, H. (1978). Synthesis of the nonatriacontapeptide corresponding to the entire amino acid sequence of dogfish adrenocorticotropin. *Int. J. Pept. Protein Res.* 12:170–172.

Silva, P., Stoff, J.S., Leone, D.R., and Epstein, F.H. (1985). Mode of action of somatostatin to inhibit secretion by shark rectal gland. *Am. J. Physiol.* 249:R329–R334.

Silva, P., Lear, S., Reichlin, S., and Epstein, F.H. (1990). Somatostatin mediates bombesin inhibition of chloride secretion by rectal gland. *Am. J. Physiol.* 258:R1459–R1463.

Silva, P., Epstein, F.H., Karnaky, Jr., K.J., Reichlin, S., and Forrest, Jr., J.N. (1993). Neuropeptide Y inhibits chloride secretion in the shark rectal gland. *Am. J. Physiol.* 265:R439–R446.

Silva, P., Solomon, R.J., and Epstein, F.H. (1999). Mode of activation of salt secretion by C-type natriuretic peptide in the shark rectal gland. *Am. J. Physiol.* 277:R1725–R1732.

Simpson, T.H. and Wardle, C.S. (1967). A seasonal cycle in the testis of the spurdog, *Squalus acanthias*, and the sites of 3'-hydroxysteroid dehydrogenase activity. *J. Mar. Biol. Assoc. U.K.* 47:699–708.

Simpson, T.H., Wright, R.S., and Gottfried, H. (1963). Steroids in the semen of dogfish *Squalus acanthias*. *J. Endocrinol.* 26:489–498.

Simpson, T.H., Wright, R.S., and Renfrew, J. (1964). Steroid biosynthesis in the semen of dogfish *Squalus acanthias*. *J. Endocrinol.* 31:11–20.

Sisneros, J.A. and Tricas, T.C. (2000). Androgen-induced changes in the response dynamics of ampullary electrosensory primary afferent neurons. *J. Neurosci.* 20:8586–8595.

Snelson, Jr., F.F., Rasmussen, L.E., Johnson, M.R., and Hess, D.L. (1997). Serum concentrations of steroid hormones during reproduction in the Atlantic stingray, *Dasyatis sabina*. *Gen. Comp. Endocrinol.* 108:67–79.

Song, L., Ryan, P.L., Porter, D.G., and Coomber, B.L. (2001). Effects of relaxin on matrix remodeling enzyme activity of cultured equine ovarian stromal cells. *Anim. Reprod. Sci.* 66:239–255.

Sorbera, L.A. and Callard, I.P. (1995). Myometrium of the spiny dogfish *Squalus acanthias*: peptide and steroid regulation. *Am. J. Physiol.* 269:R389–R397.

Sourdaine, P. and Garnier, D.H. (1993). Stage-dependent modulation of Sertoli cell steroid production in dogfish (*Scyliorhinus canicula*). *J. Reprod. Fertil.* 97:133–142.

Sourdaine, P., Garnier, D.H., and Jegou, B. (1990). The adult dogfish (*Scyliorhinus canicula* L.) testis: a model to study stage-dependent changes in steroid levels during spermatogenesis. *J. Endocrinol.* 127:451–460.

Steinetz, B.G., O'Byrne, E.M., Butler, M., and Hickman, L. (1983). Hormonal regulation of the connective tissue of the symphysis pubis. In: Bigazzi, M., Greenwood, F., and Gasparri, F. (Eds.), *Biology of Relaxin and Its Role in the Human*. Excerpta Medica, Amsterdam, pp. 71–92.

Steinetz, B.G., Schwabe, C., Callard, I.P., and Goldsmith, L.T. (1998). Dogfish shark (*Squalus acanthias*) testes contain a relaxin. *J. Androl.* 19:110–115.

Stell, W.K. (1984). Luteinizing hormone-releasing hormone (LHRH)- and pancreatic polypeptide (PP)-immunoreactive neurons in the terminal nerve of spiny dogfish, *Squalus acanthias*. *Anat. Rec.* 208:173A–174A.

Stoff, J.S., Rosa, R., Hallac, R., Silva, P., and Epstein, F.H. (1979). Hormonal regulation of active chloride transport in the dogfish rectal gland. *Am. J. Physiol.* 237:F138–F144.

Sulikowski, J.A., Tsang, P.C.W., and Howell, W.H. (2004). An annual cycle of steroid hormone concentrations and gonad development in the winter skate, *Leucoraja ocellata*, from the western Gulf of Maine. *Mar. Biol.* 144:845–853.

Sulikowski, J.A., Tsang, P.C.W., and Howell, W.H. (2005). Age and size at sexual maturity for the winter skate, *Leucoraja ocellata*, in the western Gulf of Maine based on morphological, histological and steroid hormone analyses. *Environ. Biol. Fish.* 72:429–441.

Sumpter, J.P., Follett, B.K., Jenkins, N., and Dodd, J.M. (1978a). Studies on the purification and properties of gonadotrophin from ventral lobes of the pituitary gland of the dogfish (*Scyliorhinus canicula* L.). *Gen. Comp. Endocrinol.* 36:264–274.

Sumpter, J.P., Jenkins, N., and Dodd, J.M. (1978b). Gonadotrophic hormone in the pituitary gland of the dogfish (*Scyliorhinus canicula* L.): distribution and physiological significance. *Gen. Comp. Endocrinol.* 36:275–285.

Sumpter, J.P., Jenkins, N., and Dodd, J.M. (1978c). Hormonal control of steroidogenesis in an elasmobranch fish (*Scyliorhinus canicula* L.): studies using a specific anti-gonadotrophic antibody (proceedings). *J. Endocrinol.* 79:28P–29P.

Sumpter, J.P., Jenkins, N., Duggan, R.T., and Dodd, J.M. (1980). The steroidogenic effects of pituitary extracts from a range of elasmobranch species on isolated testicular cells of quail and their neutralisation by a dogfish anti-gonadotropin. *Gen. Comp. Endocrinol.* 40:331.

Sumpter, J.P., Denning-Kendall, P.A., and Lowry, P.J. (1984). The involvement of melanotrophins in physiological colour change in the dogfish *Scyliorhinus canicula*. *Gen. Comp. Endocrinol.* 56:360–367.

Suzuki, R., Togashi, K., Ando, K., and Takei, Y. (1994). Distribution and molecular forms of C-type natriuretic peptide in plasma and tissue of a dogfish, *Triakis scyllia*. *Gen. Comp. Endocrinol.* 96:378–384.

Tagliafierro, G., Faraldi, G., and Pestarino, M. (1985). Interrelationships between somatostatin-like cells and other endocrine cells in the pancreas of some cartilaginous fish. *Cell. Mol. Biol.* 31:201–207.

Tagliafierro, G., Farina, L., Faraldi, G., Rossi, G.G., and Vacchi, M. (1989). Distribution of somatostatin and glucagon immunoreactive cells in the gastric mucosa of some cartilaginous fishes. *Gen. Comp. Endocrinol.* 75:1–9.

Takagi, T. et al. (1995). Plasma calcitonin levels in the stingray (cartilaginous fish) *Dasyatis akajei*. *Zool. Sci.* 10:134.

Takei, Y. (2000). Structural and functional evolution of the natriuretic peptide system in vertebrates. *Int. Rev. Cytol.* 194:1–66.

Takei, Y., Hasegawa, Y., Watanabe, T.X., Nakajima, K., and Hazon, N. (1993). A novel angiotensin I isolated from an elasmobranch fish. *J. Endocrinol.* 139:281–285.

Thiebold, J.J. (1953). Action du benzoate d'oestradiol sur la differenciation sexuelle des embryons de *Scylliorhinus canicula* C. *Compt. Rend. Soc. Biol.* 236:2174–2175.

Thiebold, J.J. (1954). Étude preliminaire de l'action des hormones sexuelles sur la morphogenese des voies genitales chez *Scylliorhinus canicula* L. *Bull. Biol. Fr. Belg.* 88:130–145.

Thorndyke, M.C. and Shuttleworth, T.J. (1985). Biochemical and physiological studies on peptides from the elasmobranch gut. *Peptides* 6(Suppl. 3):369–372.

Tierney, M., Takei, Y., and Hazon, N. (1997). The presence of angiotensin II receptors in elasmobranchs. *Gen. Comp. Endocrinol.* 105:9–17.

Torres, P., Tort, L., Planas, J., and Flos, R. (1986). Effect of confinement stress and additional zinc treatment on some blood parameters in the dogfish *Scyliorhinus canicula*. *Comp. Biochem. Physiol. C Toxicol. Pharmacol.* 83:89–92.

Tort, L., Gonzalez-Arch, F., and Balasch, J. (1994). Plasma glucose and lactate and hematological changes after handling stresses in the dogfish. *Rev. Esp. Fisiol.* 50:41–46.

Tricas, T.C., Michael, S.W., and Sisneros, J.A. (1995). Electrosensory optimization to conspecific phasic signals for mating. *Neurosci. Lett.* 202:129–132.

Tricas, T.C., Maruska, K.P., and Rasmussen, L.E. (2000). Annual cycles of steroid hormone production, gonad development, and reproductive behavior in the Atlantic stingray. *Gen. Comp. Endocrinol.* 118:209–225.

Tsang, P.C. and Callard, I.P. (1987a). Luteal progesterone production and regulation in the viviparous dogfish, *Squalus acanthias*. *J. Exp. Zool.* 241:377–382.

Tsang, P.C. and Callard, I.P. (1987b). Morphological and endocrine correlates of the reproductive cycle of the aplacental viviparous dogfish, *Squalus acanthias*. *Gen. Comp. Endocrinol.* 66:182–189.

Tsang, P.C. and Callard, I.P. (1992). Regulation of ovarian steroidogenesis *in vitro* in the viviparous shark, *Squalus acanthias*. *J. Exp. Zool.* 261:97–104.

Uva, B., Masini, M.A., Hazon, N., O'Toole, L.B., Henderson, I.W., and Ghiani, P. (1992). Renin and angiotensin converting enzyme in elasmobranchs. *Gen. Comp. Endocrinol.* 86:407–412.

Vallarino, M. and Ottonello, I. (1987). Neuronal localization of immunoreactive adrenocorticotropin-like substance in the hypothalamus of elasmobranch fishes. *Neurosci. Lett.* 80:1–6.

Vallarino, M., Danger, J.M., Fasolo, A., Pelletier, G., Saint-Pierre, S., and Vaudry, H. (1988a). Distribution and characterization of neuropeptide Y in the brain of an elasmobranch fish. *Brain Res.* 448:67–76.

Vallarino, M., Delbende, C., Jegou, S., and Vaudry, H. (1988b). Alpha-melanocyte-stimulating hormone (alpha-MSH) in the brain of the cartilagenous fish. Immunohistochemical localization and biochemical characterization. *Peptides* 9:899–907.

Vallarino, M., Andersen, A.C., Delbende, C., Ottonello, I., Eberle, A.N., and Vaudry, H. (1989). Melanin-concentrating hormone (MCH) immunoreactivity in the brain and pituitary of the dogfish *Scyliorhinus canicula*. Colocalization with alpha-melanocyte-stimulating hormone (alpha-MSH) in hypothalamic neurons. *Peptides* 10:375–382.

Vallarino, M., Feuilloley, M., Vandesande, F., and Vaudry, H. (1991). Immunohistochemical mapping of galanin-like immunoreactivity in the brain of the dogfish *Scyliorhinus canicula*. *Peptides* 12:351–357.

Vigna, S.R. (1979). Distinction between cholecystokinin-like and gastrin-like biological activities extracted from gastrointestinal tissues of some lower vertebrates. *Gen. Comp. Endocrinol.* 39:512–520.

Vigna, S.R. (1983). Evolution of endocrine regulation of gastrointestinal function in lower vertebrates. *Am. Zool.* 23:729–738.

Visconti, M.A. and Castrucci, A.M. (1993). Melanotropin receptors in the lungfish, *Lepidosiren paradoxa*, and in the cartilaginous fish, *Potamotrygon reticularis*. *Comp. Biochem. Physiol. C Toxicol. Pharmacol.* 106:523–528.

Visconti, M.A., Ramanzini, G.C., Camargo, C.R., and Castrucci, A.M. (1999). Elasmobranch color change: a short review and novel data on hormone regulation. *J. Exp. Zool.* 284:485–491.

Volkoff, H., Wourms, J.P., Amesbury, E., and Snelson, F.F. (1999). Structure of the thyroid gland, serum thyroid hormones, and the reproductive cycle of the Atlantic stingray, *Dasyatis sabina*. *J. Exp. Zool.* 284:505–516.

Waring, H. (1936). Colour change in the dogfish (*Scyliorhinus canicula*). *Proc. Liverpool Biol. Soc.* 49:17–68.

Watanabe, T., Inoue, K., and Takei, Y. (2009). Identification of angiotensinogen genes with unique and variable angiotensin sequences in chondrichthyans. *Gen. Comp. Endocrinol.* 161:115–122.

Watson, R.R. and Dickson, K.A. (2001). Enzyme activities support the use of liver lipid-derived ketone bodies as aerobic fuels in muscle tissues of active sharks. *Physiol. Biochem. Zool.* 74:273–282.

Weiss, G. (1989). Relaxin in the male. *Biol. Reprod.* 40:197–200.

Wells, A., Anderson, W.G., and Hazon, N. (2002). Development of an *in situ* perfused kidney preparation for elasmobranch fish: action of arginine vasotocin. *Am. J. Physiol. Regul. Integr. Comp. Physiol.* 282:R1636–R1642.

Wells, A., Anderson, W.G., Cains, J.E., Cooper, M.W., and Hazon, N. (2006). Effects of angiotensin II and C-type natriuretic peptide on the *in situ* perfused trunk preparation of the dogfish, *Scyliorhinus canicula*. *Gen. Comp. Endocrinol.* 145:109–115.

Wendelaar Bonga, S.E. and Pang, P.K.T. (1991). Control of calcium regulating hormones in the vertebrates: parathyroid hormone, calcitonin, prolactin, and stanniocalcin. *Int. Rev. Cytol.* 128:139–213.

White, J. and Meredith, M. (1995). Nervus terminalis ganglion of the bonnethead shark (*Sphyrna tiburo*): evidence for cholinergic and catecholaminergic influence on two cell types distinguished by peptide immunocytochemistry. *J. Comp. Neurol.* 351:385–403.

Wilson, J.F. and Dodd, J.M. (1973a). Effects of pharmacological agents on the *in vivo* release of melanophore-stimulating hormone in the dogfish, *Scyliorhinus canicula*. *Gen. Comp. Endocrinol.* 20:556–566.

Wilson, J.F. and Dodd, J.M. (1973b). The role of melanophore-stimulating hormone in melanogenesis in the dogfish, *Scyliorhinus canicula* L. *J. Endocrinol.* 58:685–686.

Wilson, J.F. and Dodd, J.M. (1973c). The role of the pineal complex and lateral eyes in the colour change response of the dogfish, *Scyliorhinus canicula* L. *J. Endocrinol.* 58:591–598.

Wourms, J.P. (1977). Reproduction and development of chondrichthyan fishes. *Am. Zool.* 17:379–410.

Wright, D.E. and Demski, L.S. (1991). Gonadotropin hormone-releasing hormone (GnRH) immunoreactivity in the mesencephalon of sharks and rays. *J. Comp. Neurol.* 307:49–56.

Wright, D.E. and Demski, L.S. (1993). Gonadotropin-releasing hormone (GnRH) pathways and reproductive control in elasmobranchs. *Environ. Biol. Fish.* 38:209–218.

Yamaguchi, K., Yasuda, A., Lewis, U.J., Yokoo, Y., and Kawauchi, H. (1989). The complete amino acid sequence of growth hormone of an elasmobranch, the blue shark (*Prionace glauca*). *Gen. Comp. Endocrinol.* 73:252–259.

Yamamoto, K., Suzuki, N., Takahashi, A., Sasayama, Y., and Kikuyama S. (1999). Estrogen receptors in the stingray (*Dasyatis akajei*) ultimobranchial gland. *Gen. Comp. Endocrinol.* 110:107–114.

Zaidi, M., Inzerillo, A.M., Moonga, B.S., Bevis, P.J., and Huang, C.L. (2002). Forty years of calcitonin: where are we now? A tribute to the work of Iain Macintyre, FRS. *Bone* 30:655–663.

12

Sensory Physiology and Behavior of Elasmobranchs

Jayne M. Gardiner, Robert E. Hueter, Karen P. Maruska, Joseph A. Sisneros,
Brandon M. Casper, David A. Mann, and Leo S. Demski

CONTENTS

12.1 Introduction

Sharks are practically legendary for their sensory capabilities, with some of this reputation deserved and some exaggerated. Accounts of sharks being able to smell or hear a single fish from miles away may be fish stories, but controlled measurements of elasmobranch sensory function have revealed that these animals possess an exquisite array of sensory systems for detecting prey and conspecifics, avoiding predators and obstacles, and orienting in the sea. This sensory array provides information to a central nervous system (CNS) that includes a relatively large brain, particularly in the rays and galeomorph sharks, whose brain-to-body weight ratios are comparable to those of birds and mammals (Northcutt, 1978).

Sensory system performance can be quantified in many ways. In the end, elasmobranch biologists wish to know, "How 'good' is elasmobranch hearing … smell … vision?" in a given behavioral or ecological context. To approach this basic question, sensory performance can be scaled in two general ways: *sensitivity*, which involves the minimum stimulus detectable by the system, and *acuity*, which is the ability of the system to discriminate stimulus characteristics, such as its location (e.g., direction of a sound or odor, resolution of a visual image) and type (e.g., frequency of sound, odorant chemical, wavelength of light). These parameters apply to all senses in one way or another and help to make comparisons across phylogenetic lines.

This chapter reviews the anatomy, physiology, and performance of elasmobranch senses within the context of sensory ecology and behavior. Special emphasis is placed on information that has come to light since publication of Hodgson and Mathewson's 1978 volume on elasmobranch senses (Hodgson and Mathewson, 1978a). Generalizations across all elasmobranch species are difficult and unwise; with about 1000 extant species and only a fraction studied for their sensory capabilities, much still remains to be discovered about the diversity of sensory system function in elasmobranchs.

12.2 Vision

My nose is sufficiently good. My eyes are large and gray; although, in fact, they are weak to a very inconvenient degree, still no defect in this regard would be suspected from their appearance.

Edgar Allan Poe ("The Spectacles," 1844)

Poe could have been writing about the eyes and nose of a shark, for prior to the 1960s the perception, both scholarly and popular, was that vision in sharks was poor compared with the other senses, especially olfaction. This perception was pervasive even though visual scientists (e.g., Walls, 1942) recognized that elasmobranch ocular anatomy was highly developed. Sensory research in the 1960s and subsequent decades began to transform our understanding of shark visual capabilities. Several comprehensive reviews can be consulted for detailed research findings on elasmobranch vision (see Gilbert, 1963; Gruber and Cohen, 1978; Hueter and Cohen, 1991). This section summarizes what is known about the visual systems of sharks, skates, and rays with an emphasis on special adaptations for elasmobranch behavior and ecology.

12.2.1 Ocular Anatomy and Optics

Elasmobranch eyes are situated laterally on the head in the case of selachians and on the dorsal surface of the head in batoids, although the more benthic sharks (e.g., orectolobids, squatinids) have more dorsally positioned eyes and the less benthic rays (e.g., myliobatids, rhinopterids, mobulids) have more laterally positioned eyes, obvious adaptations for benthic vs. pelagic habits. Eye size in elasmobranchs is generally small in relation to body size but relatively larger in juveniles (Lisney et al., 2007) and in some notable species, such as the bigeye thresher shark, *Alopias superciliosus*. In general, sharks have larger eyes than batoids, but eye size differences also correlate with habitat type, activity level, and prey type. Oceanic species have relatively larger eyes than coastal and benthic species, and more active swimmers that feed on active, mobile prey have relatively larger eyes than more sluggish species that feed on sedentary prey (Lisney and Collin, 2007). As with osteichthyan fishes (Warrant and Locket, 2004), relative eye size in mesopelagic deep-sea sharks is often large to allow for enhanced light gathering.

In all elasmobranchs, the two eyes oppose each other, which can allow for a nearly 360° visual field in at least one plane of vision (Figure 12.1). In the case of swimming sharks using a laterally sinusoidal swimming pattern, the dynamic visual field can be extended beyond 360°. Limited eye movements are observed in some species, primarily to compensate for swimming movements and to stabilize the visual field (Harris, 1965). Binocular overlap is generally small, except in the hammerhead sharks (Sphyrnidae) and some batoids, but their enhanced frontal vision comes at the expense of larger posterior blind areas (Litherland et al., 2009a; McComb and Kajiura, 2008; McComb et al., 2009) (Figure 12.1). Blind areas exist directly in front of the snout or behind the head when the animal is still. The sizes of these blind areas depend on the configuration of the head and the separation of the eyes, but typically the forward blind area extends less than one body length in front of the rostrum.

The ocular adnexa are well developed and more elaborate than in most teleosts, although the upper and lower eyelids in most elasmobranchs do not move appreciably or cover the entire eyeball (Gilbert, 1963). Benthic shark species such as orectolobids have more mobile lids, which serve to protect the eyes while burrowing. Some sharks, especially the carcharhinids and sphyrnids, possess a third eyelid, the nictitating membrane, which can be extended from the lower nasal corner of the eye to cover the exposed portion of the eye (Gilbert, 1963) (Figure 12.2). This membrane functions to protect the eye from damaging abrasion and may be extended when the shark feeds or comes into contact with an object. It does not naturally respond to bright light, although it can be conditioned to do so (Gruber and Schneiderman, 1975). Some other sharks not equipped with a nictitating membrane, including the white shark, *Carcharodon carcharias* (Tricas and McCosker, 1984), and the whale shark, *Rhincodon typus* (Hueter, pers. obs.), use the extraocular muscles to rotate the entire eye back into the orbit to protect it from abrasion during feeding and other activities.

The outer layer of the elasmobranch eye (Figure 12.3) is comprised of a thick cartilaginous sclera and a gently curving, transparent cornea, the fine structure of which includes sutural fibers that resist corneal swelling and loss of transparency in challenging chemical environments (Tolpin et al., 1969). Unlike teleosts, most elasmobranchs have a dynamic iris that can increase the size of the pupil in dim light or decrease it in bright light.

Depending on species, the shape of the pupil can be circular (e.g., most deep-sea sharks, which have less mobile pupils for the more constant, low-light conditions), vertical slit (e.g., *Carcharhinus* spp., *Negaprion brevirostris*), horizontal slit (e.g., *Sphyrna tiburo*), oblique slit (e.g., *Scyliorhinus canicula*, *Ginglymostoma cirratum*), or crescent shaped (e.g., many skates and rays) (Figure 12.4). Mobile slit pupils are typically found in active predators with periods of activity in both photopic (bright light) and scotopic (dim light) conditions, such as the lemon shark, *N. brevirostris* (Gruber, 1967). A slit pupil that can be closed down to a pinhole is thought to be the most effective way to achieve the smallest aperture under photopic conditions, because a circular pupil is mechanically constrained from closing to a complete pinhole (Walls, 1942). In skates and rays, the combination of a U-shaped crescent pupil with multiple pupillary apertures (Figure 12.4E,F) under photopic conditions provides optical benefits, including enhanced visual resolution, contrast, and focusing ability (Murphy and Howland, 1991).

The elasmobranch cornea is virtually optically absent underwater due to its similarity in refractive index to that of seawater (Hueter, 1991), leaving the crystalline lens to provide the total refractive power of the eye. Elasmobranch lenses are typically large, relatively free of optical aberration, and ellipsoidal in shape, although the spiny dogfish, *Squalus acanthias*, and clearnose skate, *Raja eglanteria*, have nearly spherical lenses (Sivak, 1978a, 1991). In the juvenile lemon shark, *Negaprion brevirostris*, the principal power (D_p) of the lens is nearly +140 diopters (D), about seven times the optical power of the human lens (Hueter, 1991).

Some elasmobranch lenses contain yellowish pigments that are enzymatically formed oxidation products of tryptophan, similar to lens pigments found in many teleosts and diurnal terrestrial animals. These pigments filter near-ultraviolet (UV) light, which helps to minimize defocus of multiple wavelengths (chromatic aberration), enhance contrast sensitivity, and reduce light scatter and glare under conditions of bright sunlight (Zigman, 1991). They may also help to protect the retina from UV damage in shallow benthic and epipelagic species. Zigman (1991) found yellow lens pigments in coastal and surface-dwelling species such as the sandbar shark (*Carcharhinus plumbeus*), the dusky shark (*Carcharhinus obscurus*), and the tiger shark (*Galeocerdo cuvier*), but interestingly not in another carcharhinid and shallow-water shark, the lemon shark (*Negaprion brevirostris*) or in the shallow-dwelling nurse shark (*Ginglymostoma cirratum*). Both lemon and nurse sharks inhabit tropical waters where UV damage to the eye could be a problem, so the ecological correlations are unclear, and there may be other factors selecting for the presence or absence of these lens filters. Nelson et al. (2003) described a related UV-filtering mechanism in the corneas of scalloped

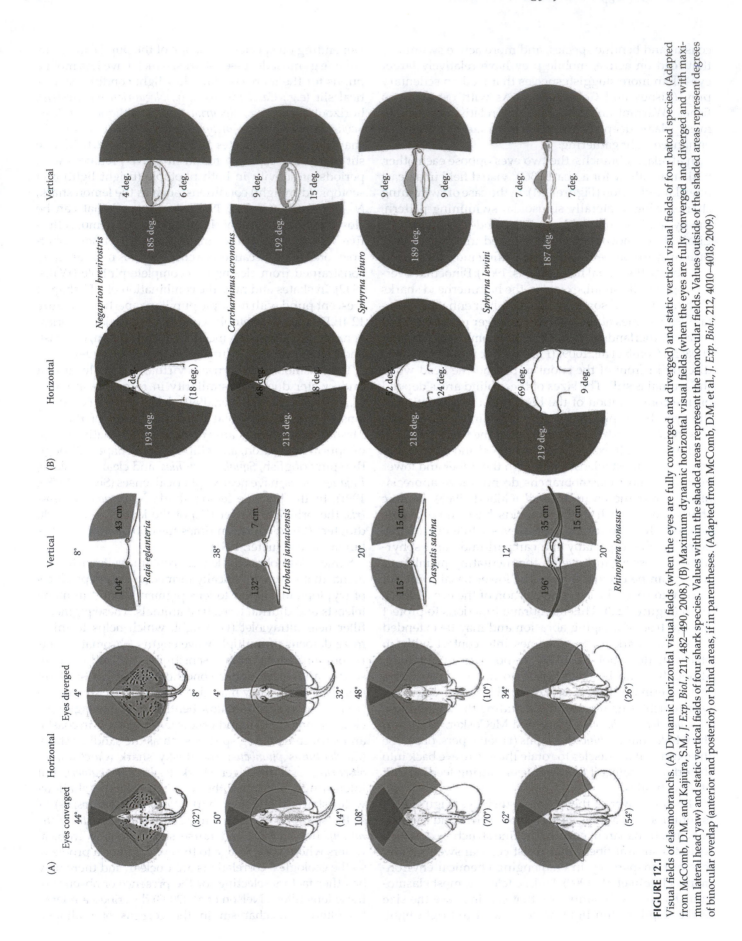

FIGURE 12.1
Visual fields of elasmobranchs. (A) Dynamic horizontal visual fields (when the eyes are fully converged and diverged) and static vertical visual fields of four batoid species. (Adapted from McComb, D.M. and Kajiura, S.M., *J. Exp. Biol.*, 211, 482–490, 2008.) (B) Maximum dynamic horizontal visual fields (when the eyes are fully converged and diverged and with maximum lateral head yaw) and static vertical visual fields of four shark species. Values within the shaded areas represent the monocular fields. Values outside of the shaded areas represent degrees of binocular overlap (anterior and posterior) or blind areas, if in parentheses. (Adapted from McComb, D.M. et al., *J. Exp. Biol.*, 212, 4010–4018, 2009.)

FIGURE 12.2
Lemon shark, *Negaprion brevirostris*, with its nictitating membrane partially retracted. (From Gruber, S.H. and Cohen, J.L., in *Sensory Biology of Sharks, Skates, and Rays*, Hodgson, E.S. and Mathewson, R.F., Eds., U.S. Office of Naval Research, Arlington, VA, 1978, pp. 11–105. Photograph by E. Fisher and used with permission.)

pseudocampanule, a papilla with ostensibly contractile function (Sivak and Gilbert, 1976). Evidence of accommodation in elasmobranchs has been inconsistent across species, and many of the species studied have appeared to be hyperopic (farsighted) in the resting state of the eye (Hueter, 1980; Hueter and Gruber, 1982; Sivak, 1978b; Spielman and Gruber, 1983). This condition is problematic in that objects at optical infinity would be out of focus and the closer an object approaches an eye, the more out of focus it becomes.

Hueter et al. (2001), however, discovered that unrestrained, free-swimming lemon sharks, *Negaprion brevirostris*, were not hyperopic and could accommodate, in contrast to previous findings for the same species under restraint (Hueter, 1980; Hueter and Gruber, 1982), suggesting that the hyperopia and absence of accommodation observed in many elasmobranchs under restraint could be an induced, unnatural artifact resulting from handling stress. Eliminating this artifact, it is possible that most elasmobranchs would be emmetropic (neither farsighted nor nearsighted) in the resting state and have accommodative ability. This complication aside, there is some indication that benthic elasmobranchs, such as the nurse shark, *Ginglymostoma cirratum*, and the bluntnose stingray, *Dasyatis say*, have greater accommodative range than more active, mobile elasmobranchs (Sivak, 1978b). This may be attributable to the stability of the visual field in sedentary species, providing advantages for a more refined focusing mechanism, but more research into the interrelationship between vision and locomotion in elasmobranchs is needed.

hammerhead sharks, *Sphyrna lewini*, in which the degree of UV protection by the cornea increased with duration of exposure to solar radiation.

Accommodation is the ability to change the refractive power of the eye to focus on objects at varying distances. Without accommodative ability, the focal plane of the eye is static, and in the absence of other optical adaptations the image of any object in front of or behind that plane will be out of focus on the retina. Elasmobranchs that accommodate do not vary lens shape as humans do, but instead change the position of the lens by moving it toward the retina (for distant targets) or away from the retina (for near targets). The lens is supported dorsally by a suspensory ligament and ventrally by the

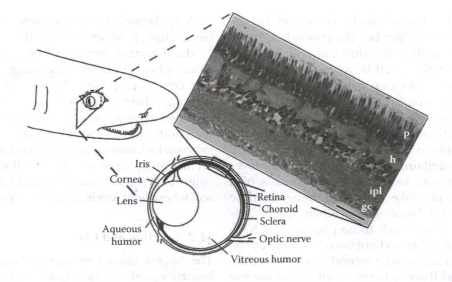

FIGURE 12.3
Cross-section through an elasmobranch eye showing ocular and retinal anatomy. (Adapted from Hueter, R.E. and Gilbert, P.W., in *Discovering Sharks*, Gruber, S.H., Ed., American Littoral Society, Highlands, NJ, 1990, pp. 48–55.) Inset: Light micrograph of the retina of the giant shovelnose ray, *Rhinobatos typus*, showing the photoreceptive layer (longer receptors are rods, shorter receptors are cones). *Abbreviations:* gc, ganglion cell layer; h, horizontal cell layer; ipl, inner plexiform layer; p, photoreceptor layer. Scale bar: 100 μm. (Adapted from Hart, N.S. et al., *J. Exp. Biol.*, 207, 4587–4594, 2004.)

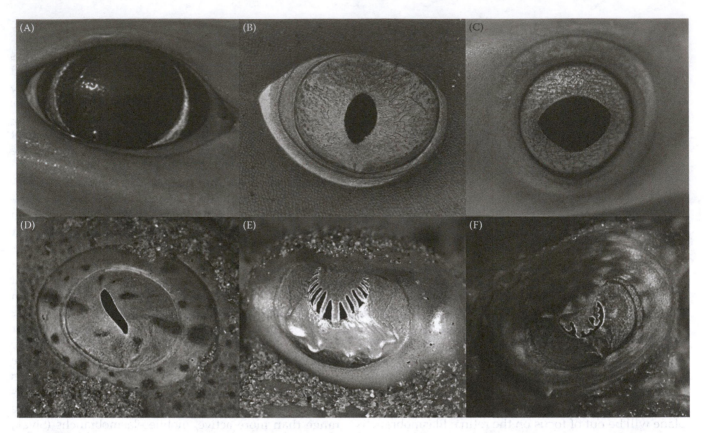

FIGURE 12.4
(See color insert.) Diversity of pupil shapes among elasmobranchs. (A) Circular pupil in a gulper shark, *Centrophorus* sp. (Photograph by José Castro and used with permission.) (B) Vertical slit in the whitetip reef shark, *Triaenodon obesus*. (Photograph by Christian Loader and used with permission.) (C) Horizontal slit in the bonnethead, *Sphyrna tiburo*. (Photograph by D.M. McComb and S.M. Kajiura and used with permission.) (D) Oblique slit in the Pacific angel shark, *Squatina californica*. (Photograph by Alison Vitsky and used with permission.) (E) Crescent-shaped pupil with papillary apertures in the shovelnose guitarfish, *Rhinobatos productus*. (Photograph by Alison Vitsky and used with permission.) (F) The yellow stingray, *Urobatis jamaicensis*. (Adapted from McComb, D.M. and Kajiura, S.M., *J. Exp. Biol.*, 211, 482–490, 2008.)

At the back of the elasmobranch eye behind the retina and in front of the sclera lies the choroid, the only vascularized tissue within the adult elasmobranch eye. The elasmobranch retina itself is not vascularized and typically contains no obvious landmarks other than the optic disk (corresponding to a small blind spot in the visual field), which contains no photoreceptors and marks the exit of retinal ganglion cell fibers via the optic nerve from the retina to the CNS. The choroid in nearly all elasmobranchs contains a specialized reflective layer known as the tapetum lucidum, which consists of a series of parallel, platelike cells containing guanine crystals (Denton and Nicol, 1964; Gilbert, 1963). This layer functions to reflect back those photons that have passed through the retina and not been absorbed by the photoreceptor layer, allowing a second chance for detection of photons and thereby boost sensitivity of the eye in dim light. The alignment of the tapetal cells provides for specular reflection; that is, photons are reflected back along the same path and are not scattered within the eye, which would blur the image.

Many elasmobranchs, furthermore, possess an occlusible tapetum, in which the reflective layer can be occluded by dark pigment granules that migrate within tapetal melanophores to block the passage of light under photopic conditions (Heath, 1991; Nicol, 1964) (Figure 12.5). Although there are exceptions, occlusible tapeta tend to be found in more surface-dwelling, arrhythmic species with both diurnal and nocturnal activity, which selects for visual adaptation to widely varying light levels. Non-occlusible tapeta in which the reflective layer is permanently exposed are found in sharks that inhabit the deep sea, where light levels are consistently dim (Nicol, 1964).

12.2.2 Retina and CNS

The largest impact on our understanding of elasmobranch visual function came with the realization that nearly all elasmobranchs have duplex retinas containing both rod and cone photoreceptors (Gruber and Cohen, 1978) (Figure 12.6), beginning with the discovery by Gruber et al. (1963) of cones in the retina of the

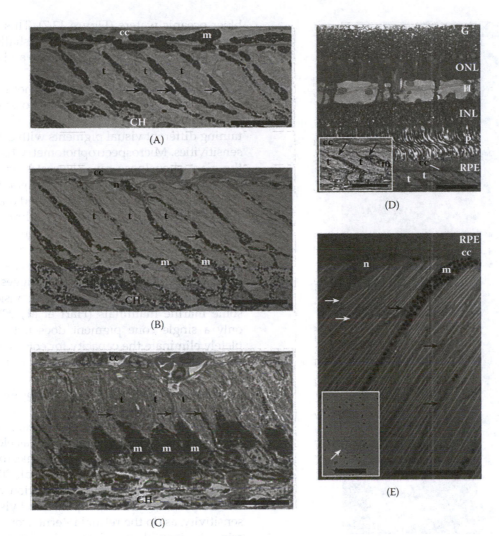

FIGURE 12.5
Morphological variation in the structure of the tapetum lucidum. (A–C) Light micrographs of the occlusible tapetum lucidum of the sandbar shark, *Carcharhinus plumbeus*, showing the occlusion of the tapetal cells by pigment migration in a light-adapted tapetum (A), a partially dark-adapted tapetum (B), and a fully dark-adapted tapetum (C). Note the dispersal of melanosomes along the melanocyte cell processes to occlude the tapetal cells in the light-adapted state and the aggregations of the melanosomes toward the choroid in the dark-adapted state. Scale bar: 20 μm. (D) Transverse section of the retina of the shortspine spurdog, *Squalus mitsukurii*. Inset highlights the tapetal cells. Scale bar: 50 μm; inset scale bar: 20 μm. (E) High-power electronmicrograph illustrating the arrangement of the reflective crystals within a *C. plumbeus* tapetal cell. Inset: light micrograph of crystal plates. Scale bars: 10 μm. Black arrows: melanocyte processes; white arrows: reflective crystals. *Abbreviations:* cc, choriocapillaris; CH, choroid; G, ganglion cell layer; H, horizontal cells; INL, inner nuclear layer; m, melanocyte containing melanosomes; n, nucleus of tapetal cell; ONL, outer nuclear layer; P, photoreceptor layer; rc, reflective crystals; RPE, retinal pigment epithelium; t, tapetal cell. (From Litherland, L. et al., *J. Exp. Biol.*, 212, 3583–3594, 2009. With permission.)

lemon shark, *Negaprion brevirostris*. Cones subserve photopic and color vision and are responsible for higher visual acuity; rods subserve scotopic vision and are involved in setting the limits of visual sensitivity in the eye. Prior to 1963, elasmobranchs were thought to possess all-rod retinas and thus were thought to have poor visual acuity and no capability for color vision. The only elasmobranchs that appear to have no cone photoreceptors are skates (*Raja* spp.), but even their rods appear to have conelike functions under certain photic conditions (Dowling and Ripps, 1991; Ripps and Dowling, 1991).

Both rods and cones contain visual pigments that absorb photons and begin the process of vision. These pigments consist of a protein called *opsin* and a chromophore prosthetic group related to either vitamin A_1 or A_2, the former type called *rhodopsins* or *chrysopsins* and the latter called *porphyropsins* (Cohen, 1991). Rhodopsins are maximally sensitive to blue–green light, chrysopsins to deep-blue light, and porphyropsins to yellow–red light. Most elasmobranchs have been found to possess rhodopsins, which provides maximum sensitivity for clearer, shallow ocean waters associated with

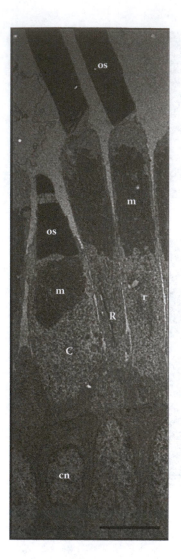

FIGURE 12.6
Photoreceptor ultrastructure in the giant shovelnose ray, *Rhinobatos typus*, showing the typical morphology of rods (R) and cones (C). *Abbreviations:* cn, cone nucleus; m, mitochondria; os, outer segment. Scale bar: 5 μm. (Adapted from Hart, N.S. et al., *J. Exp. Biol.*, 207, 4587–4594, 2004.)

epipelagic environments (Cohen, 1991). Chrysopsin has been found in deep-sea squaliform sharks such as *Centrophorus*, *Centroscymnus*, and *Deania* (Denton and Shaw, 1963), which inhabit regions where the little available light is deep blue. Porphyropsin, which is common in freshwater teleosts and is more suited for turbid, yellowish photic conditions, is rare in elasmobranchs, even freshwater species. Cohen et al. (1990), however, found a porphyropsin with maximum sensitivity (λ_{max}) of 522 nm (yellow–green) in the juvenile lemon shark, *Negaprion brevirostris*, whereas adult lemon sharks have a rhodopsin with λ_{max} = 501 nm (blue–green). In this species, the visual pigment apparently changes from a porphyropsin adapted for maximum sensitivity in inshore, shallow waters to a rhodopsin better suited for clearer,

bluer oceanic waters (Figure 12.7). This visual adaptation matches a habitat shift from shallow to oceanic waters that occurs between juvenile and adult stages of this shark (Cohen et al., 1990).

A duplex (rod–cone) retina does not necessarily provide for color vision in all cases. Color discrimination normally requires at least two types of cones, each containing different visual pigments with different spectral sensitivities. Microspectrophotometry has revealed that the giant shovelnose ray, *Rhinobatos typus* (Hart et al., 2004), the eastern shovelnose ray, *Aptychotrema rostrata* (Hart et al., 2004), and the blue-spotted maskray, *Dasyatis kuhlii* (Theiss et al., 2007), possess three different cone pigments with different spectral sensitivities, suggesting that these animals are capable of color vision. By contrast, only one cone pigment per species was found in 17 species of sharks examined, suggesting that these animals may have monochromatic vision, similar to some marine mammals (Hart et al., 2011). Possessing only a single cone pigment does not, however, completely eliminate the capacity for color vision. If the rod and cone pigments have different spectral sensitivities and the retina and brain are capable of comparing signals between them, dichromatic color vision is possible. This may be the case in the blacknose shark, *Carcharhinus acronotus*, the scalloped hammerhead, *Sphyrna lewini*, and the bonnethead, *Sphyrna tiburo*, as electroretinography has revealed two absorbance peaks (blue and green) in their photoreceptors (McComb et al., 2010).

The density and spatial distribution of photoreceptors in the retina fundamentally affect visual acuity and sensitivity, as do the retinal interneurons (bipolar, amacrine, horizontal, ganglion cells) (Figure 12.3), which transmit impulses ultimately to visual centers in the CNS. Elasmobranch retinas are rod dominated, ranging from the skates with all-rod retinas (Dowling and Ripps, 1991) to species with apparently few cones such as *Mustelus* (Sillman et al., 1996; Stell and Witkovsky, 1973) to lamnid and carcharhinid sharks with as many as one cone for every 4 to 13 rods (Gruber and Cohen, 1978; Gruber et al., 1963). Some authors have suggested a correlation between greater rod-to-cone ratios and more scotopic habits (such as nocturnal behavior) or habitats (visually murky environments or deep-sea) of elasmobranch species. That sharks, skates, and rays have rod-dominated retinas does not in itself allow us to conclude that their vision is adapted primarily for low-light conditions, sensitivity to movement, and crude visual acuity; the human retina also has many more rods than cones, and our diurnal vision and acuity are among the best in the animal kingdom.

On the other hand, the spatial topography of retinal cells can reveal much about the quality of vision in these animals. Although elasmobranchs do not have all-cone foveas, they do have retinal areas (areae) of higher cone

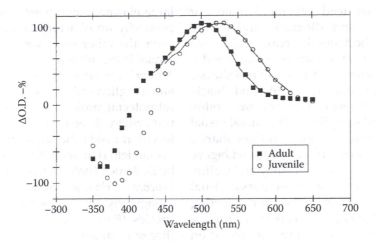

FIGURE 12.7

Normalized difference spectra for visual pigment absorption characteristics of adult vs. juvenile lemon sharks (*Negaprion brevirostris*). Peak absorption for the juvenile pigment is 522 nm, whereas the adult peak is 501 nm, demonstrating a shift in this species from a more yellow–red-sensitive porphyropsin in the juvenile to a more blue–green-sensitive rhodopsin in the adult. (From Cohen, J.L. et al., *Vision Res.*, 30, 1949–1953, 1990. With permission.)

or ganglion cell density, indicating regional specializations for higher visual acuity (Collin, 1999; Hueter, 1991). Higher cone concentrations have been found in the "central" retina of the nurse shark, *Ginglymostoma cirratum* (Hamasaki and Gruber, 1965), whitespotted bamboo shark, *Chiloscyllium plagiosum* (Yew et al., 1984), and white shark, *Carcharodon carcharias* (Gruber and Cohen, 1985). Franz (1931) was the first to report horizontal streaks of higher ganglion cell density in the small-spotted catshark, *Scyliorhinus canicula*, and smoothhound, *Mustelus mustelus*.

Retinal whole-mount techniques have been used to map the topographic distributions of retinal cells in 33 elasmobranch species representing 17 families of sharks, skates, and rays and one family of chimaera (Bozzano, 2004; Bozzano and Collin, 2000; Collin, 1988, 1999; Hueter, 1991; Lisney and Collin, 2008; Litherland and Collin, 2008; Litherland et al., 2009a; Logiudice and

Laird, 1994; Peterson and Rowe, 1980; Theiss et al., 2007). Most species have horizontal visual streaks with one or more areas of increased photoreceptor and ganglion cell density (areae centrales) (Figure 12.8A,B). The position and extent of the horizontal streak appear to vary with habitat and ecology. Benthic species such as batoids (Bozzano and Collin, 2000; Collin, 1988; Litherland and Collin, 2008; Logiudice and Laird, 1994; Theiss et al., 2007), chimaeras (Collin, 1999; Lisney and Collin, 2008), catsharks (Bozzano and Collin, 2000; Lisney and Collin, 2008), bamboo and carpet sharks (Bozzano and Collin, 2000; Lisney and Collin, 2008; Litherland and Collin, 2008), lantern sharks (Bozzano and Collin, 2000), horn sharks (Collin, 1999; Peterson and Rowe, 1980), and sleeper sharks (Bozzano, 2004) generally have dorsally located horizontal streaks, providing increased sampling of the ventral visual field. This is thought to reflect the importance of the horizon at the substrate–water

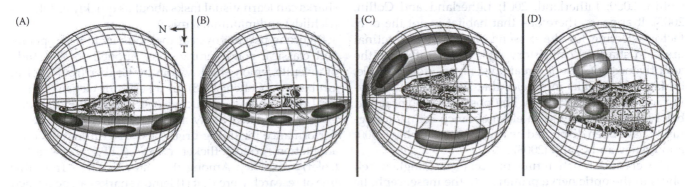

FIGURE 12.8

Diagrammatic representation of regions of the visual field subserved by regions of higher retinal cell density, depicting horizontal streaks (lightly shaded bands) with multiple areae (darkly shaded ovals) in the eastern shovelnose ray, *Aptychotrema rostrata* (A) and the epaulette shark, *Hemiscyllium ocellatum* (B); concentric retinal areae in the whitetip reef shark, *Triaenodon obesus* (C) and the ornate wobbegong, *Orectolobus ornatus* (D). *Abbreviations:* N, nasal; T, temporal. (Adapted from Litherland, L. and Collin, S.P., *Visual Neurosci.*, 25, 549–561, 2008.)

interface in animals that feed off the benthos or bury themselves in the sand (Bozzano and Collin, 2000). Centrally located horizontal streaks have been found in benthopelagic species such as lemon sharks, *Negaprion brevirostris* (Hueter, 1991), blacktip reef sharks, *Carcharhinus melanopterus* (Collin, 1999), and blackmouth dogfish, *Galeus melastomus* (Bozzano and Collin, 2000), providing increased sampling of the lateral visual field. Ventral horizontal streaks found in tiger sharks, *Galeocerdo cuvier* (Bozzano and Collin, 2000), and bigeye thresher sharks, *Alopias superciliosus* (Lisney and Collin, 2008), provide increased sampling of the dorsal visual field. This may represent an adaptation for detecting prey from below. Tiger sharks prey on birds, sea turtles, and marine mammals, which are commonly found on or near the sea surface (Lowe et al., 1996), and common thresher sharks, *Alopias vulpinus*, a sister species to the bigeye thresher, have been demonstrated to attack prey from below, using the elongated dorsal lobe of their caudal fin to stun their prey (Aalbers et al., 2010).

In contrast, concentric retinal areae (Figure 12.8C,D) are more applicable for visualizing a limited spot in the visual field or for operating in complex, three-dimensional visual environments, such as reefs. Areae have been found in phylogenetically and ecologically diverse species of sharks, including blue sharks (*Prionace glauca*), sand tiger sharks (*Carcharias taurus*), hammerheads (*Sphyrna* spp.), bull sharks (*Carcharhinus leucas*), gray reef sharks (*Carcharhinus amblyrhynchos*) (Lisney and Collin, 2008), sandbar sharks (*Carcharhinus plumbeus*) (Litherland et al., 2009a), and whitetip reef sharks (*Triaenodon obesus*) (Litherland and Collin, 2008). These species range from pelagic, open ocean environments to reef, coastal, and even riverine habitats, yet all have areae of one kind or another. Cookie-cutter sharks, *Isistius brasiliensis*, and white sharks, *Carcharodon carcharias*, are both ambush predators in open water, while ornate wobbegongs, *Orectolobus ornatus*, are benthic ambush predators, and all three have retinal areae, not streaks (Bozzano and Collin, 2000; Litherland, 2001; Litherland and Collin, 2008). It appears, therefore, that habitat is not the only factor selecting for the presence or absence of retinal areae in sharks. Locomotory style could influence the adaptiveness of visual streaks vs. areae—for example, by favoring streaks in species that are constantly moving forward (Hueter, 1991). The possible ecological and behavioral correlates with elasmobranch retinal topography have been discussed by Bozzano and Collin (2000) and Lisney and Collin (2008).

The elasmobranch retina projects via ganglion cell fibers in the optic nerve primarily to the mesencephalic optic tectum, but most species also possess at least ten other retinofugal targets in the brain, similar to the pattern in other vertebrates (Graeber and Ebbesson, 1972; Northcutt, 1979, 1991). These targets include the large elasmobranch telencephalon, once believed to be primarily an olfactory center but now known to subserve the other senses as well, particularly for multimodal integration (Bodznick, 1991). In the lemon shark, *Negaprion brevirostris*, the visual streak found in the cone and ganglion cell layers of the retina is preserved in the retinotectal projection to the surface of the optic tectum, where three times more tectal surface is dedicated to vision inside the streak than in the periphery of the visual field (Hueter, 1991). A similar result was reported by Bodznick (1991) in the optic tectum of the little skate, *Leucoraja erinacea* (formerly *Raja erinacea*). The retinal topography of this skate is unknown, but a related species (*Raja bigelowi*) has a prominent visual streak (Bozzano and Collin, 2000). Bodznick (1991) furthermore found that a spatial map of electroreceptive input, aligned with the visual map, also overrepresented the animal's sensory horizon in the tectum. These findings give tantalizing insights into the coordination of multimodal sensory function in the elasmobranch brain, but much more work needs to be done in this area.

12.2.3 Visual Performance

Controlled experiments to test visual performance in sharks began in 1959 when Clark trained adult lemon sharks, *Negaprion brevirostris*, to locate a square white target for food reward (Clark, 1959). Later, Clark (1963) trained lemon sharks to discriminate visually between a square vs. diamond and a white vs. black-and-white striped square. Parameters such as visual angle, contrast, and luminance of targets were not quantified, but the demonstration that sharks could learn certain visually mediated tasks was noteworthy at the time. Wright and Jackson (1964) and Aronson et al. (1967) added to Clark's findings with further conditioning experiments on lemon, bull (*Carcharhinus leucas*), and nurse (*Ginglymostoma cirratum*) sharks, again without quantified visual parameters but providing evidence that sharks can learn visual tasks about as quickly as teleosts (cichlids) and mammals (mice).

Rigorous psychophysical methods including operant and classical conditioning were applied to the study of juvenile lemon shark vision by Gruber (reviewed in Gruber and Cohen, 1978). In a series of elegant behavioral experiments conducted over nearly two decades, Gruber elucidated many aspects of lemon shark visual performance including brightness discrimination, dark adaptation, critical flicker fusion (CFF), and spectral (color) sensitivity. Among the many findings from this line of research were that (1) lemon sharks can be trained to discriminate the brighter of two visual targets down to a 0.3-log unit difference (as opposed to a 0.2-log unit threshold in human subjects); (2) lemon sharks slowly dark-adapt to scotopic conditions over the course of about

1 hour, eventually becoming more than 1 million times (6 log units) more sensitive to light than under photopic conditions (and more sensitive than dark-adapted human subjects); (3) a kink in the CFF vs. light intensity curve for the lemon shark demonstrates the rod–cone break characteristic of a duplex retina; and (4) a shift in the lemon shark's light-adapted vs. dark-adapted spectral sensitivity, also confirmed electrophysiologically by Cohen et al. (1977), provides further evidence of duplex visual function in this shark. This work confirmed that the lemon shark possesses superior scotopic vision in extremely dim light and also is potentially capable of color vision under photopic conditions.

The ultimate behavioral test of whether elasmobranchs use color vision in the wild to discriminate visual targets has yet to be reported. Sharks can be attracted to bright colors, including the brilliant orange of life vests—a source of concern to the U.S. Navy, which funded many shark sensory studies in the 1960s and 1970s to understand shark behavior—but it is unclear whether the animals are visually cueing on color, brightness, or contrast. Similarly, the functional visual acuity of sharks in the wild is poorly known. Hueter (1991) calculated that the juvenile lemon shark has a theoretical resolving power of 4.5′ of arc, based on the closest separation of cones in the retina and the eye's optics. This acuity is about one ninth that of the human eye, which can resolve down to about 30″ of arc, but the prediction remains to be behaviorally tested.

The importance of vision in the daily lives of elasmobranchs certainly finds support in the complexity of their anatomical and physiological visual adaptations, many of which appear to be correlated with species behavior and ecology. Field reports of sharks appearing to use vision during the final approach to prey items are common, but controlled tests are not. In a study of the Pacific angel shark, *Squatina californica*, by Fouts and Nelson (1999), chemical, mechanical, and electrical cues were eliminated to determine that visual stimuli released an ambush attack by these benthic sharks on nearby prey items. Based on their observations, the authors hypothesized that the angel shark visual system probably is specialized for anterodorsally directed vision. A study of retinal topography in this species would help to confirm this hypothesis. Gardiner and Atema (2007) demonstrated that smooth dogfish, *Mustelus canis*, can perform rheotaxis behaviors using vision when the lateral line system had been chemically ablated. Gardiner et al. (2011) also used sensory knockout techniques to determine that vision is used to line up strikes on live prey in blacktip sharks, *Carcharhinus limbatus*, and bonnetheads, *Sphyrna tiburo*. Strong (1996) tested behavioral preferences of white sharks, *Carcharodon carcharias*, approaching differently shaped visual targets. The sharks were attracted to the testing area with olfactory stimuli, but

they appeared to use vision as they approached the objects, which were ≥15-cm-diameter surface-borne targets to which the sharks appeared to orient visually from depths of ≥17 m. At that depth, a 15-cm target would subtend a visual angle of about 0.5°, or 30′ of arc, which is more than six times as large as the theoretical minimum separable angle of the juvenile lemon shark eye. This visual task should not be a problem for a white shark with a relatively large, cone-rich eye (Gruber and Cohen, 1985).

12.3 Hearing

Hearing in sharks is of great interest because sound in the ocean presents a directional signal that is capable of propagating over large distances. The explorations of the ear and hearing in elasmobranchs are also important as they reveal a basal stage in the evolution of vertebrate audition within a group of fishes that have evolved little over hundreds of millions of years. Sharks are not known to make sounds, so their hearing abilities have likely been shaped by the ambient noise (both physical and biological) in their environment. Hearing in sharks and rays has been reviewed by numerous authors (Corwin, 1981, 1989; Myrberg, 2001; Popper and Fay, 1977; Wisby et al., 1964). These reviews provide both an excellent overview of shark hearing research and a historical perspective on the scientific approaches to studying shark hearing. The purpose of this section is to describe what is known about shark hearing with an emphasis on what remains to be learned.

12.3.1 Anatomy

12.3.1.1 Inner Ear

The inner ear of sharks, skates, and rays consists of a pair of membranous labyrinths with three semicircular canals and four sensory maculae each (Maisey, 2001; Retzius 1881) (Figure 12.9). The semicircular canals are similar to those in other vertebrates and are used to sense angular acceleration. They are not known to be involved in sound perception.

The saccule, lagena, and utricle are three sensory areas that are thought to be involved in both balance and sound perception. They consist of a patch of sensory hair cells on an epithelium overlain by an otoconial mass. The otoconia, made of calcium-carbonate granules embedded in a mucopolysaccharide matrix, act as an inertial mass (Tester et al., 1972). As in fishes, these otolith organs are thought to be responsive to accelerations produced by a sound field, which accelerate the

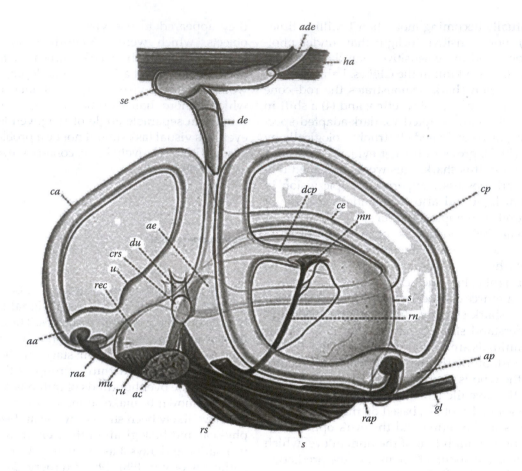

FIGURE 12.9
Anatomy of the ear of the thornback ray, *Raja clavata*. *Abbreviations:* aa, ampulla of anterior canal; ac, acoustic nerve; ade, opening of endolymphatic duct; ae, ampulla of horizontal canal; ap, ampulla of posterior canal; ca, anterior semicircular canal; ce, horizontal semicircular canal; cp, posterior semicircular canal; crs, saccular recess; dcp, posterior canal duct; de, endolymphatic duct; du, utricular duct; ha, chondrocranium; l, lagena; mn, macula neglecta; mu, utricule macula; pl, lagena macula; raa, ramus anterior ampulla; rap, ramus posterior ampulla; rec, utricular recess; rn, ramus neglectus; rs, ramus sacculus; ru, ramus utriculus; s, saccule; se, endolymphatic sac; u, utricule. (Adapted from Retzius, G., *Das Gehörorgan der Wirbelthiere*, Vol. 1, Samson and Wallin, Stockholm, 1881.)

shark and the sensory macula relative to the otoconial mass. Some elasmobranchs, such as the spiny dogfish, *Squalus acanthias*, have been found to incorporate exogenous sand grains as a way to increase the endogenous otoconial mass (Lychakov et al., 2000).

12.3.1.2 Macula Neglecta

Sharks are unique among fishes in having a tympanic connection, the fenestra ovalis, to the posterior semicircular canal that enhances audition (Howes, 1883). The fenestra ovalis is located in the base of the parietal fossa, which makes a depression in the posterior portion of the skull (Figure 12.10). The fenestrae lead to the posterior canal ducts of the semicircular canals, each of which contains a sensory macula, the macula neglecta, that is not overlain by otoconia (Tester et al., 1972). Elasmobranchs also have an endolymphatic duct that connects to the saccule and leads to a small opening on the dorsal surface of the shark (Figure 12.10).

This connection has been hypothesized to act as a site of release of displacement waves (Tester et al., 1972), as any flow induced over the fenestrae ovalis would propagate down the posterior canal duct and into the sacculus.

Because of the specialization of the posterior canal in sharks, most hearing research has focused on the macula neglecta. The macula neglecta consists of one patch of sensory hair cells in rays and two patches of sensory hair cells in carcharhinid sharks (Corwin, 1977, 1978). The macula neglecta lacks otoconia but does have a crista like other hair cells in the semicircular canals. In rays, the hair cells show a variety of orientations. In carcharhinids, the hair cells are oriented in opposite directions in each sensory patch, and the orientation patterns are positioned so that fluid flows in the posterior canal would stimulate the hair cells. Variation of the structure of the macula neglecta has been hypothesized to be linked to the foraging behavior of different elasmobranchs (Corwin, 1978). A more recent analysis of the entire auditory structure of 17 different species of

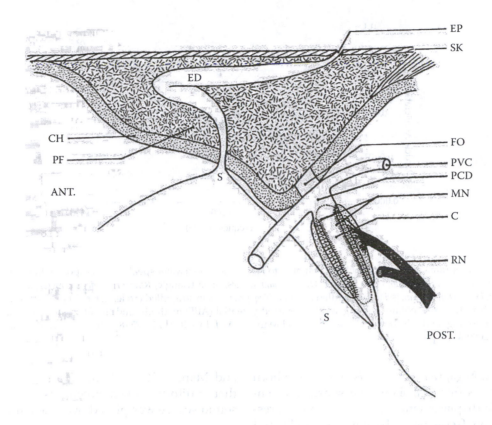

FIGURE 12.10
Cross-section of the elasmobranch ear focusing on the location of the parietal fossa and macula neglecta relative to the saccular chamber. *Abbreviations:* C, cupula; CH, chondrocranium; ED, endolymphatic duct; EP, endolymphatic pore; FO, fenestra ovalis; MN, macula neglecta; PCD, posterior canal duct; PF, parietal fossa; PVC, posterior vertical canal; RN, ramus neglectus nerve; S, saccule; SK, skin covering fossa. (From Fay, R.R. et al., *Comp. Biochem. Physiol. A Physiol.*, 47, 1235–1240, 1974. With permission.)

sharks and rays suggests that variations within the ear may be a combination of phylogeny as well as behavior and ecology (Evangelista et al., 2010); however, until the function of the macula neglecta is determined, this hypothesis will be difficult to test.

The macula neglecta in rays has been shown to add hair cells continually as the fish grows (Barber et al., 1985; Corwin, 1983). Sex differences have also been found: Females have been found to have more hair cells than males. The increase in hair cell number has been shown to increase vibrational sensitivity in neurons innervating the macula neglecta.

12.3.1.3 Central Pathways

As in other vertebrates, the ear of elasmobranchs is innervated by the VIIIth cranial (octaval) nerve. Studies of afferent connections and the physiology of the octaval nerve from individual end organs (saccule, lagena, utricle, and macula neglecta) show projections ipsilaterally to five primary octaval nuclei: magnocellular, descending, posterior, anterior, and periventricular (Barry, 1987; Corwin and Northcutt, 1982). Much work remains to be done regarding both the anatomy and neurophysiology of the CNS as it relates to audition.

12.3.2 Physiology

12.3.2.1 Audiograms

Audiograms are measures of hearing sensitivity to sounds of different frequencies. Audiograms are the most basic information that is collected about hearing systems in animals. To date, there are only six published audiograms in elasmobranchs (summarized in Figure 12.11). Given the diversity of the group, more audiograms are warranted.

The greatest issue in measuring audiograms is what component of sound is relevant to acoustic detection in sharks. Fishes without swimbladders, including all elasmobranchs, detect the *particle motion* component of sound (can be described in terms of acceleration, velocity, and displacement). Fishes with swimbladders, especially those with connections between the swimbladder and ear, such as the goldfish, also detect the *pressure* component of sound. In these fishes, the swimbladder acts as a pressure-to-displacement transducer.

One way to determine the importance of particle motion vs. pressure is to measure hearing sensitivity at different distances from a sound projector. The ratio of pressure to particle displacement changes as the distance from the sound changes. Measurements in

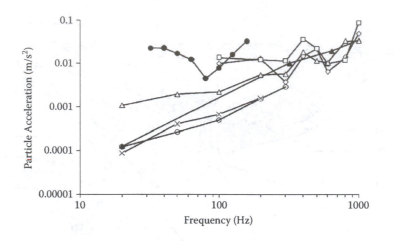

FIGURE 12.11

Particle acceleration audiograms of elasmobranchs in response to monopole (underwater speaker) and dipole (vibrating bead) sound stimuli. Open diamond, *Ginglymostoma cirratum*; open square, *Urobatis jamaicensis*; open triangle, *Rhizoprionodon terraenovae*; x, *Chiloscyllium griseum* (dipole); open circle, *Heterodontus francisci* (dipole); filled triangle, *Negaprion brevirostris*; filled circle, *Heterodontus francisci*. The open shapes and x's are elasmobranch audiograms obtained using auditory evoked potential (AEP) methods, and the filled shapes are audiograms obtained using classical conditioning methods. (From Casper, B.M. and Mann, D.A., *J. Exp. Biol.*, 210, 75–81, 2007; Casper, B.M. and Mann, D.A., *J. Fish Biol.*, 75, 2768–2776, 2009. With permission.)

the lemon shark, *Negaprion brevirostris*, and in the horn shark, *Heterodontus francisci*, show that sharks are sensitive to particle displacement rather than sound pressure, at least at low frequencies (Banner, 1967; Kelly and Nelson, 1975). It was not clear whether higher frequency thresholds (640 Hz in Banner, 1967; 100 to 160 Hz in Kelly and Nelson, 1975) in these species are dominated by either pressure or particle displacement sensitivity. This could be because of measurement errors or because the sharks are detecting some other measurement of the sound field, such as the pressure gradient. Particle motion thresholds have also been measured in the nurse shark (*Ginglymostoma cirratum*), yellow stingray (*Urobatis jamaicensis*), and the Atlantic sharpnose shark (*Rhizoprionodon terraenovae*) by measuring auditory evoked potentials elicited from the brain in response to acoustic stimuli (Casper and Mann, 2006, 2009).

Despite these issues, laboratory studies indicate that shark hearing is not as sensitive as that of some other fishes, especially those with hearing adaptations coupling a swimbladder to the inner ear. All the sharks tested show mainly low-frequency sensitivity, and there is no evidence that they are more sensitive at low frequencies than other fishes (Banner, 1967; Casper and Mann, 2006, 2009; Casper et al., 2003; Kelly and Nelson, 1975; Kritzler and Wood, 1961; Nelson, 1967).

Several papers show the importance of the macula neglecta in detecting sound or vibration (Lowenstein and Roberts, 1951). Fay et al. (1974) measured the response of the macula neglecta to vibrational stimuli applied to the parietal fossa. This showed that the parietal fossa is indeed in some way linked to hearing in the macula neglecta. Bullock and Corwin (1979) and Casper

and Mann (2007a) obtained similar results in finding that auditory evoked potentials were highest when a sound source was placed over the parietal fossa.

12.3.2.2 Pressure Sensitivity

Isolated preparations of small-spotted catshark, *Scyliorhinus canicula*, hair cells from the horizontal semicircular canals have recently been shown to respond to changes in ambient pressure (Fraser and Shelmerdine, 2002). Increased ambient pressure led to increased spike rates in response to an oscillation at 1 Hz. This result shows that sharks have a sensor that could be used to sense depth and atmospheric pressure, and studies by Heupel et al. (2003) demonstrate that blacktip sharks, *Carcharhinus limbatus*, behaviorally respond to decreases in atmospheric pressure associated with tropical storms. The physiological findings need to be pursued in other parts of the ear to determine whether responses to sound are modulated by pressure as well, and if shark hair cells can detect sound pressures directly. The ambient pressures tested were on the order of 200 dB re 1 µPa, which would be extremely loud for a sound.

12.3.3 Behavior

12.3.3.1 Attraction of Sharks with Sound

Several studies have shown that sharks can be attracted with low-frequency sounds in the field (Myrberg et al., 1969, 1972; Nelson and Gruber, 1963). In some of these tests, the received sound pressure levels were likely well below thresholds obtained from laboratory studies of shark hearing. This apparent disconnect between field and laboratory studies needs to be addressed. There are problems with each type of study. In the laboratory, sound fields are very complicated near-field stimuli that are rarely quantified. In the field, it is often difficult to know the distribution of sharks prior to playback and difficult to control for other stimuli, such as visual stimuli. The fact that sharks show a behavioral response to sound presentation should present a good system for testing hypotheses about shark hearing abilities. An implanted data logger has been used to record the acoustic environment of a free-swimming shark (Meyer et al., 2008), but it was not used for measuring behavioral responses to sounds as has been accomplished with marine mammals (Johnson and Tyack, 2003).

12.3.3.2 Other Aspects of Hearing

There is more to hearing than just detection of sound. The ability to localize a sound source is just as an important. The otolithic organs in other fishes respond directionally to sound presentations due to the polarizations of the sensory hair cells (Lu and Popper, 2001). This is likely to be the case with sharks as well. One reason why the debate over the ability of sharks to detect sound pressure has been intense is that theoretical arguments have been made that sharks must be able to detect sound pressure to resolve a 180° ambiguity about the location of a source (see Kalmijn, 1988b; van den Berg and Schuijf, 1983). The acoustic attraction experiments show that sharks have the ability to localize a sound source, and laboratory experiments show that the lemon shark can localize a sound source to about 10° (Nelson, 1967). Directional sensitivity was also measured in two species of bamboo sharks, with results suggesting that these sharks were able to detect sounds equally well from all directions (Casper and Mann, 2007b). Clearly, we need to collect more data with regard to hearing sensitivity, masking by noise, frequency discrimination, intensity discrimination, and temporal sensitivity. Regardless of the actual mechanism of sound detection, data collected on these attributes of sound will be important for understanding the acoustic world of sharks.

12.4 Mechanosenses

The ability to detect water movements at multiple scales is essential in the lives of fishes. The detection of large tidal currents provides information important for orientation and navigation, and small-scale flows can reveal the location of prey, predators, and conspecifics during social behaviors. The mechanosensory lateral line system is stimulated by differential movement between the body and surrounding water and is used by fishes to detect both dipole sources (e.g., prey) and uniform flow fields (e.g., currents). This sensory system functions to mediate behaviors such as rheotaxis (orientation to water currents), predator avoidance, hydrodynamic imaging to localize objects, prey detection, and social communication including schooling and mating (for reviews, see Bleckmann, 2008; Coombs and Montgomery, 1999). In contrast to the amount of information available on lateral line morphology, physiology, and function in bony fishes, relatively little is known about mechanosensory systems in elasmobranchs.

12.4.1 Peripheral Organization

The functional unit of all lateral line end organs is the mechanosensory neuromast, which is a group of sensory hair cells and support cells covered by a gelatinous cupula (Figure 12.12A). Ultrastructural studies have revealed that the support cells of the neuromast have apical microvilli that are taller than those observed in other vertebrate lateral line organs, but the function of this morphological difference is not yet known (Peach and Marshall, 2009). Elasmobranch fishes have several different types of mechanosensory end organs that are classified by morphology and location: superficial neuromasts (also called pit organs or free neuromasts), pored and nonpored canals, spiracular organs, and vesicles of Savi. The variety of surrounding morphological structures and spatial distribution of these sensory neuromasts determine functional parameters such as response properties, receptive field area, distance range of the system, and which component of water motion (velocity or acceleration) is encoded (Denton and Gray, 1983, 1988; Kroese and Schellart, 1992; Maruska and Tricas, 2004; Münz, 1989).

Superficial neuromasts (SNs) are distributed on the skin surface either in grooves positioned on raised papillae (skates, rays, and some sharks) or between modified placoid scales/denticles (sharks) with their cupulae directly exposed to the environment (Peach and Marshall, 2000, 2009; Tester and Nelson, 1967) (Figure 12.12B). There is considerable diversity in the morphology and position of SNs among elasmobranch taxa (e.g., SNs covered by overlapping denticles, in grooves bordered by denticles, or in grooves without associated denticles), and these morphological features may have functional implications related to water flow, filtering properties, and directionality, but this remains to be tested (Maruska, 2001; Peach, 2003; Peach and Marshall, 2000, 2009). Superficial neuromasts in the few batoids examined thus far are located in bilateral rows along the dorsal midline from the spiracle to the tip of the tail (dorsolateral neuromasts), a pair anterior to the endolymphatic pores, and a small group lateral to the eyes associated with the spiracle (spiracular neuromasts),

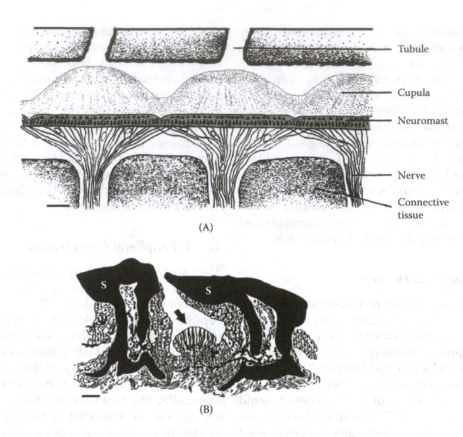

FIGURE 12.12

Morphology of the lateral line canal system and superficial neuromasts in elasmobranchs. (A) Diagrammatic longitudinal section of a pored canal from a juvenile gray reef shark, *Carcharhinus amblyrhynchos*. Innervated canal neuromasts are arranged in a nearly continuous sensory epithelium and covered by gelatinous cupulae. Pored canals are connected to the environment via tubules that terminate in openings on the skin surface. Scale bar: 150 µm. (Adapted from Tester, A.L. and Kendall, J.I., *Pac. Sci.*, 23, 1–16, 1969.) (B) Schematic transverse section of a single superficial neuromast (pit organ) in the nurse shark, *Ginglymostoma cirratum*. The sensory neuromast (arrow) is positioned between modified scales (S). Scale bar: 50 µm. Cupula is not shown. (Adapted from Budker, P., in *Traité de Zoologie. Anatomie, Systémique, Biologie*. Tome XIII. *Agnathes et Poissons*, Grassé, P.P., Ed., Masson et Cie, Paris, 1958, pp. 1033–1062.)

which may have been lost in myliobatiform rays (Ewart and Mitchell 1892; Maruska, 2001; Maruska and Tricas, 1998; Peach and Rouse, 2004) (Figure 12.13A). The number of SNs in batoids examined thus far ranges from ~25 per side in some skates (*Raja* spp.) to >100 in some rhinobatids (Maruska, 2001; Peach and Rouse, 2004).

In sharks, SNs are positioned on the dorsolateral and lateral portions of the body and caudal fin (dorsolateral neuromasts), posterior to the mouth (mandibular row), between the pectoral fins (umbilical row; disappears during ontogeny in some species), and as a pair anterior to each endolymphatic pore (Budker, 1958; Peach and Marshall, 2000; Tester and Nelson, 1967) (Figure 12.13B,C,D). However, the distribution pattern varies among taxa with one or more of the neuromast groups absent in some species. The number of SNs ranges from less than 50 per side in the horn shark (*Heterodontus* spp.) to 80 per side in the spiny dogfish, *Squalus acanthias*, to more than 600 per side in the scalloped hammerhead, *Sphyrna lewini* (Tester and Nelson, 1967, 1969) (Figure 12.13C,D). A phylogenetic analysis of the distribution

and abundance of SNs also showed that (1) the distinctive overlapping denticles covering the SNs in many sharks are a derived feature, (2) plesiomorphic elasmobranchs have SNs in open slits with widely spaced accessory denticles, (3) SN number on the ventral surface of rays has been reduced during evolution, and (4) spiracular SNs have changed position or were lost on several occasions in elasmobranch evolution (Peach and Rouse, 2004). In general, elasmobranchs with the fewest SNs include many benthic/demersal rays and sharks, while those with the most abundant SNs are pelagic sharks. Exceptions to this rule, however, such as high SN abundance in some demersal batoids and low SN abundance in some pelagic rays, indicate there is likely no straightforward relationship between SN abundance and pelagic lifestyle in elasmobranchs (Peach and Rouse, 2004).

The position of the SN sensory epithelium within grooves or between scales differs from bony fishes and may enhance water flow parallel to the cupula to provide greater directional sensitivity. Superficial neuromasts likely encode the velocity of water motion and

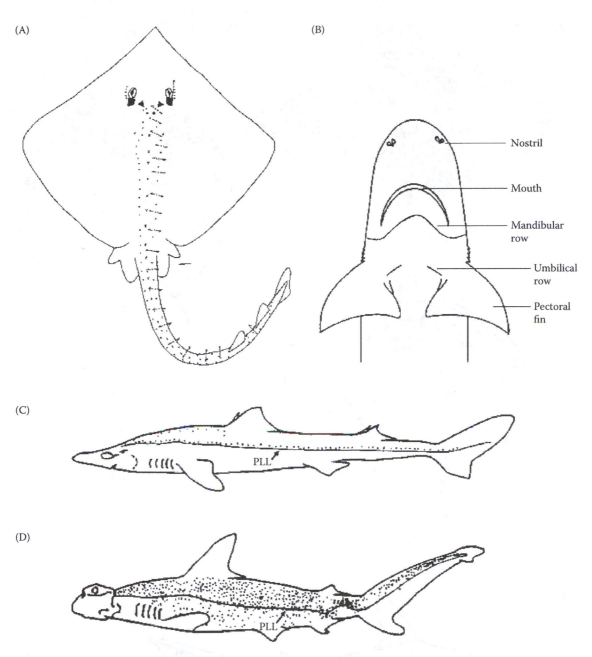

FIGURE 12.13

Distribution of superficial neuromasts (pit organs) in elasmobranchs. Each dot represents a single superficial neuromast. (A) Superficial neuromasts on the clearnose skate, *Raja eglanteria*, are located in bilateral rows along the dorsal midline to the end of the tail, a pair anterior to each endolymphatic pore (arrowheads), and a small group positioned lateral to each eye. Arrows indicate the groove orientation on every other neuromast. Scale bar: 1 cm. (Adapted from Maruska, K.P., *Environ. Biol. Fish.*, 60, 47–75, 2001.) (B) Ventral surface of the lemon shark, *Negaprion brevirostris* (67 cm total length), shows the mandibular and umbilical rows of superficial neuromasts found on many shark species. (C) Superficial neuromasts on the spiny dogfish, *Squalus acanthias* (79 cm total length), are relatively few in number and positioned along the dorsal aspect of the posterior lateral line canal (PLL). (D) Superficial neuromasts on the scalloped hammerhead, *Sphyrna lewini* (61 cm total length), are more numerous (>600 per side) and located both dorsal and ventral to the posterior lateral line canal. (Parts B, C, and D adapted from Tester, A.L. and Nelson, G.J., in *Sharks, Skates, and Rays*, Gilbert, P.W. et al., Eds., The Johns Hopkins University Press, Baltimore, MD, 1967, pp. 503–531.)

may function to detect water movements generated by predators, conspecifics, or currents similar to that demonstrated for bony fishes (Blaxter and Fuiman, 1989; Kroese and Schellart, 1992; Montgomery et al., 1997), but physiological studies on the response properties of SNs in elasmobranchs are lacking.

The most visible part of the mechanosensory system is the network of subepidermal fluid-filled canals distributed throughout the body. The main lateral line canals located on the head of elasmobranchs include the supraorbital, infraorbital, hyomandibular, and mandibular canals (Boord and Campbell, 1977; Chu and Wen,

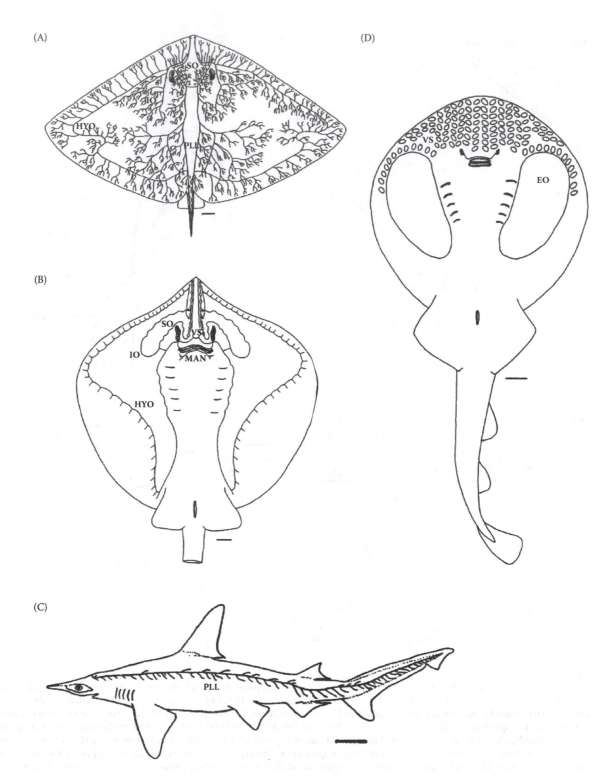

FIGURE 12.14
Distribution of lateral line canals and vesicles of Savi in elasmobranchs. The end of each line represents a pore opening on the skin surface. (A) Distribution of lateral line canals on the dorsal surface of the butterfly ray, *Gymnura micrura*. Canals are interconnected with extensive tubule branching that covers the majority of the disk surface. Scale bar: 1 cm. (B) Ventral lateral line system of the Atlantic stingray, *Dasyatis sabina*, contains pored canals along the disk margin, nonpored canals along the midline and around the mouth, and vesicles of Savi (ovals) on the rostral midline. Scale bar: 1 cm. (C) Lateral view of the posterior lateral line canal on the bonnethead shark, *Sphyrna tiburo*, which extends from the endolymphatic pores on the head to the upper lobe of the caudal fin. Scale bar: 0.5 cm. (D) Vesicles of Savi (ovals) on the ventral surface of the lesser electric ray, *Narcine brasiliensis*, are located in rows on the rostrum and along the anterior edge of the electric organ (EO). Scale bar: 1 cm. *Abbreviations:* HYO, hyomandibular canal; IO, infraorbital canal; MAN, mandibular canal; PLL, posterior lateral line canal; SO, supraorbital canal; VS, vesicles of Savi. (Adapted from Maruska, K.P., *Environ. Biol. Fish.*, 60, 47–75, 2001.)

1979; Maruska, 2001; Roberts, 1978; Tester and Kendall, 1969) (Figure 12.14). These canals show varying degrees of complex bifurcations on the head in sharks or branching patterns that extend laterally onto the pectoral fins in skates and rays (Figure 12.14A). The principal canal on the remainder of the body is the posterior lateral line canal, which extends caudally from the endolymphatic pores on the dorsal surface of the head to the tip of the tail (Figure 12.14C). These lateral line canals all contain between tens and thousands of neuromasts organized into an almost continuous sensory epithelium that results in multiple neuromasts between pores (Ewart and Mitchell, 1892; Johnson, 1917) (Figure 12.12A). This differs from bony fishes that have a single discrete neuromast positioned between adjacent pores, but the extent of this morphological organization among different canal subtypes or among species, as well as its functional significance, is still unclear.

Elasmobranchs contain two different morphological classes of lateral line canals: pored and nonpored. Pored canals are in contact with the surrounding water via neuromast-free tubules that terminate in pores on the skin surface. These canals are abundant on the dorsal head of sharks and dorsal surface of batoids, where they often form complex branching patterns that increase the mechanosensory receptive field on the disk (Chu and Wen, 1979; Jordan, 2008; Maruska, 2001) (Figure 12.14A). In general, pored canals encode water accelerations and are best positioned to detect water movements generated by prey, predators, conspecifics during social interactions or schooling, and distortions in the animal's own flow field to localize objects while swimming, as demonstrated in bony fishes (Coombs and Montgomery, 1999; Hassan, 1989; Kroese and Schellart, 1992; Montgomery et al., 1995). Neurophysiological recordings from primary afferent neurons that innervate pored canal neuromasts in the stingray, *Dasyatis sabina*, also demonstrate that, similar to bony fishes, pored canals show response properties consistent with acceleration detectors (Maruska and Tricas, 2004).

The presence of an extensive plexus of nonpored canals represents one of the most significant differences between teleost and elasmobranch lateral line systems. Nonpored canals are isolated from the environment and thus will not respond to pressure differences established across the skin surface. These canals are most common on the ventral surface of skates and rays but are also found on the head of many shark species (Chu and Wen, 1979; Jordan, 2008; Maruska, 2001; Maruska and Tricas, 1998; Wueringer and Tibbetts, 2008). In the batoids, these nonpored canals have wide diameters, are located beneath compliant skin layers, and are concentrated along the midline, around the mouth and on the rostrum (Jordan, 2008; Maruska, 2001; Maruska and Tricas, 1998) (Figure 12.14B). These morphological characteristics indicate that nonpored canals may function as tactile receptors that encode the velocity of skin movements caused by contact with prey, the substrate, or conspecifics during social interactions (Maruska, 2001; Maruska and Tricas, 2004). The number and distribution of pored vs. nonpored canals differ widely among species and may be explained by phylogeny and/or correlated with ecology and behavior. Jordan et al. (2009a), for example, showed that morphological variation in lateral line canals of several stingray species (*Urobatis halleri*, *Myliobatis californica*, *Pteroplatytrygon violacea*) was related to functional differences in detection capabilities and also corresponded well to their individual feeding ecologies.

Specialized mechanoreceptors in elasmobranchs are the spiracular organs and vesicles of Savi, both of which are isolated from the surrounding water. Spiracular organs are bilaterally associated with the first (spiracular) gill cleft and consist of a tube or pouch lined with sensory neuromasts and covered by a cupula (Barry and Bennett, 1989). This organ is found in both sharks and batoids and is stimulated by flexion of the cranial–hyomandibular joint; although its biological role is unclear, morphological and physiological studies indicate it functions as a joint proprioceptor (Barry and Bennett, 1989; Barry et al., 1988a,b). Vesicles of Savi consist of neuromasts enclosed in sub-epidermal pouches, are most abundant on the ventral surface of the rostrum, and are thus far only found in some torpedinid, narcinid, and dasyatid batoids (Barry and Bennett, 1989; Chu and Wen, 1979; Maruska, 2001; Savi 1844) (Figure 12.14B,D). Vesicular morphology differs slightly among these taxa and, although these mechanoreceptors are hypothesized to represent an obsolescent canal condition or serve as specialized touch or substrate-borne vibration receptors, their proper biological function also remains unclear (Barry and Bennett, 1989; Maruska, 2001; Nickel and Fuchs, 1974; Norris, 1932).

12.4.2 Adequate Stimulus and Processing

The necessary stimulus for the lateral line system is differential movement between the body surface and surrounding water. Because the flow amplitude of a dipole stimulus falls off rapidly with distance from the source (rate of $1/r^3$), the lateral line can only be stimulated within the inner regions of the so-called near-field (e.g., within one to two body lengths of a dipole source) (Denton and Gray, 1983; Kalmijn, 1989). Movement of the overlying cupula by viscous forces is coupled to stereocilia and kinocilia motions such that displacement of stereocilia toward the single kinocilium causes depolarization of the hair cell and an increase in the spontaneous discharge rate of the primary afferent neuron. Displacement in the opposite direction causes hyperpolarization of the hair cell and an inhibition or decrease in the spontaneous primary afferent firing rate. Thus, water motion stimuli effectively modulate

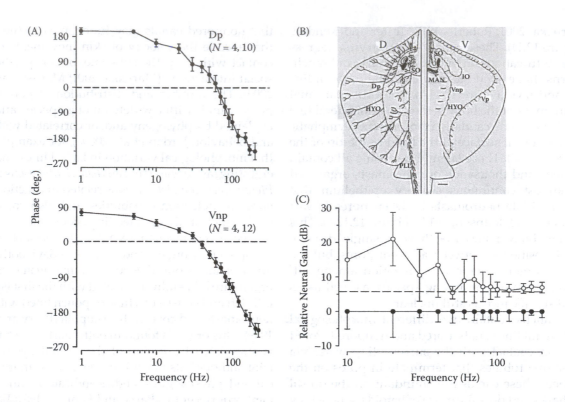

FIGURE 12.15

Response properties of primary afferent neurons that innervate canal neuromasts in the Atlantic stingray, *Dasyatis sabina*. (A) Phase plots for frequency responses of primary afferent neurons from dorsal pored (Dp) and ventral nonpored (Vnp) hyomandibular canals show a low-frequency phase lead of ~180° (acceleration-sensitive) for Dp canals and ~90° (velocity-sensitive) for Vnp canals. Phase of the peak neural response is expressed in degrees (mean ± SEM) relative to the peak displacement of a vibrating sphere. (N = number of animals, number of neurons). (B) Lateral line canal system on the dorsal (D) and ventral (V) surface of the stingray shows the distribution of Dp and ventral pored (Vp) and Vnp canals. *Abbreviations:* HYO, hyomandibular canal; IO, infraorbital canal; MAN, mandibular canal; PLL, posterior lateral line canal; SO, supraorbital canal. Scale bar: 1 cm. (C) Increase in relative neural gain to tactile stimulation (open circles) over hydrodynamic flow (closed circles) for primary afferents from Vnp canals. The average neural response is 6 to 20 dB greater (or 2 to 10 times more sensitive) to tactile stimuli compared to hydrodynamic stimuli above the canal, especially at low frequencies. (Adapted from Maruska, K.P. and Tricas, T.C., *J. Exp. Biol.*, 207, 3463–3476, 2004.)

the spontaneous primary afferent neuron discharges sent to the mechanosensory processing centers in the hindbrain. This modulation of neural activity from spatially distributed end organs throughout the body provides the animal with information about the frequency, intensity, location, and identity of the stimulus source (Bleckmann et al., 1989; Denton and Gray, 1988; Kalmijn, 1989). In general, neuromasts are sensitive to low-frequency stimuli (≤200 Hz), and neurophysiology studies indicate that the lateral line system is sensitive to velocities in the μm s⁻¹ range and accelerations in the mm s⁻² range (Bleckmann et al., 1989; Coombs and Janssen, 1990; Maruska and Tricas, 2004; Münz, 1985). Recordings from primary afferent neurons in the stingray *Dasyatis sabina* also show that pored canals exhibit response characteristics consistent with acceleration detectors (best frequencies of 20 to 30 Hz) whereas ventral nonpored canals better encode the velocity of canal fluid induced by skin movements (best frequencies of ≤10 Hz) at a 20-fold or greater sensitivity than that of the cutaneous tactile receptor system (Maruska and Tricas, 2004) (Figure 12.15).

Lateral line neuromasts are innervated by a distinct set of nerves separate from the traditional 11 to 12 cranial nerves described in most vertebrates (Northcutt, 1989a). The cephalic region of elasmobranchs is innervated by the ventral root of the anterior lateral line nerve complex and the body and tail by the posterior lateral line nerve complex (Koester, 1983). Both complexes contain efferents as well as afferent axons that enter the brain and terminate somatotopically within octavolateralis nuclei of the hindbrain (Bleckmann et al., 1987; Bodznick and Northcutt, 1980; Koester, 1983; Puzdrowski and Leonard, 1993). Ascending lateral line pathways continue to the lateral mesencephalic nucleus and tectum in the midbrain and to the thalamic and pallial nuclei in the forebrain (Bleckmann et al., 1987; Boord and Montgomery, 1989). Bleckmann et al. (1987) also demonstrated that mechanosensory receptive fields are somatotopically organized in a point-to-point rostrocaudal body map within the midbrain of the thornback ray (Figure 12.16). Further neurophysiological studies show bimodal and multimodal neurons

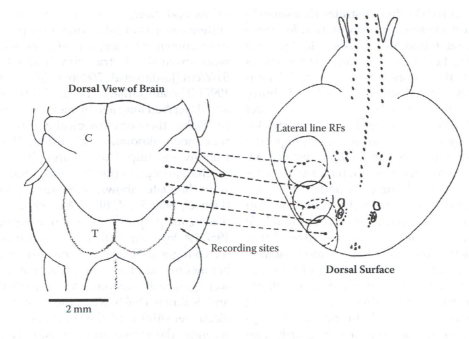

Dorsal View of Brain

Lateral line RFs

Recording sites

Dorsal Surface

2 mm

FIGURE 12.16

Mechanosensory lateral line receptive fields (RFs) on the body are somatotopically organized in a point-to-point rostrocaudal map in the midbrain of the thornback ray, *Platyrhinoidis triseriata*. Receptive fields on the anterior, mid-, and posterior body are mapped onto the contralateral rostral, mid-, and caudal dorsomedial nucleus of the midbrain. *Abbreviations:* C, cerebellum; T, tectum. (Adapted from Bleckmann, H. et al., *J. Comp. Physiol. A*, 161, 67–84, 1987.)

within midbrain and forebrain centers that respond to hydrodynamic flow as well as to auditory, visual, or electrosensory stimuli (Bleckmann and Bullock, 1989; Bleckmann et al., 1989). Thus, these processing regions can integrate information from several sensory systems to help mediate appropriate behavioral responses to complex biological stimuli.

12.4.3 Behavior and Function

Among bony fishes, the lateral line system is known to function in schooling behavior, social communication, hydrodynamic imaging, predator avoidance, rheotaxis, and prey detection; however, behavioral experiments to demonstrate these lateral-line-mediated behaviors in elasmobranch species are available only for prey detection and rheotaxis. The best-known behavioral use of the lateral line system is in prey detection. The concentration of mechanoreceptors on the cephalic region of sharks and ventral surface of batoids, as well as the low-frequency close range of the system, indicates an important role in the detection, localization, and capture of prey. Swimming and feeding movements of invertebrates and vortex trails behind swimming fish can produce water movements within the frequency and sensitivity range of the lateral line system (Montgomery et al., 1995). Montgomery and Skipworth (1997) showed that the ventral lateral line canal system of the short-tailed stingray, *Dasyatis brevicaudata*, could detect small transient water flows similar to those produced by the bivalves found in their diet. Similarly, Jordan et al. (2009a) compared behavioral responses of several ray species to water jets that mimicked signals produced by potential prey and demonstrated that a greater proportion of pored canals, high degree of canal branching, and high pore numbers corresponded with an increased behavioral response to water flow. Furthermore, based on the peripheral morphology of the lateral line system and feeding behavior of the Atlantic stingray, *Dasyatis sabina*, Maruska and Tricas (1998) hypothesized that the nonpored canals on the ventral surface of the ray function as specialized tactile receptors that encode the velocity of skin movements caused by contact with small benthic prey. Early neurophysiology experiments also demonstrated that touching the skin near the nonpored canals caused a transient stimulation of the neuromasts (Sand, 1937), and more recent recordings showed that the ventral nonpored canals in the stingray *D. sabina* are 2 to 10 times more sensitive to direct skin depression velocity than to hydrodynamic dipole stimulation near the skin, which supports the hypothesized mechanotactile function (see Figure 12.15) (Maruska and Tricas, 2004). Although prey detection is mediated by the integration of multiple sensory inputs (i.e., electroreception, olfaction, vision), the mechanosensory lateral line likely also plays an important role in feeding behavior across elasmobranch taxa.

Recent evidence in sharks demonstrates that superficial neuromasts provide sensory information for rheotaxis, similar to that found in teleosts (Montgomery et al., 1997). Resting Port Jackson sharks, *Heterodontus portusjacksoni*, with their dorsolateral superficial neuromasts (pit organs) ablated showed a reduced ability to orient upstream in a flume when compared to intact individuals (Peach, 2001). Positive rheotaxis in sharks, skates, and rays may be important for species-specific behaviors and is hypothesized to facilitate water flow over the gills, to help maintain position on the substratum, to help orient to tidal currents, and to facilitate prey detection by enabling the animal to remain within an odor plume (see Peach, 2001). A recent study in the smooth dogfish, *Mustelus canis*, also demonstrated that in addition to olfaction an intact lateral line system is required for efficient and precise tracking of odor-flavored wakes used for eddy chemotaxis (e.g., simultaneous analysis of chemical and hydrodynamic dispersal fields) (Gardiner and Atema, 2007). In the smooth dogfish, as well as other species that are primarily crepuscular or nocturnal hunters, reliance on lateral line information is likely essential.

The structure and function of the elasmobranch mechanosensory system are ripe for future study. For example, the variety of morphological specializations (e.g., nonpored canals, vesicles of Savi, neuromast morphology) found in elasmobranchs requires quantitative examinations of response properties among receptor types. Comparisons of specific mechanoreceptor distributions on the body are needed across elasmobranch taxa to test hypotheses on whether species-specific distributions have some ecological significance and represent specializations driven by evolutionary selective pressures or are possibly explained by phylogeny. The ability of the lateral line system to separate signal from noise is also critical, and future studies should examine the behavioral and physiological strategies used by elasmobranch fishes to enhance signal detection in a noisy environment (Montgomery et al., 2009). Finally, direct behavioral studies are sorely needed to clarify the many putative functions of the mechanosensory system in elasmobranch fishes, other than prey detection and rheotaxis, such as schooling, object localization, predator avoidance, and social communication.

12.5 Electrosenses

All elasmobranch fishes possess an elaborate ampullary electroreceptor system that is exquisitely sensitive to low-frequency electric stimuli (see review by Bodznick and Boord, 1986; see also Montgomery, 1984; New, 1990;

Tricas and New, 1998). The ampullary electroreceptor system consists of subdermal groups of electroreceptive units known as the *ampullae of Lorenzini*, which can detect weak extrinsic electric stimuli at intensities less than 5 nV/cm (Jordan et al., 2009b; Kajiura, 2003; Kalmijn, 1982, 1997). The ampullae of Lorenzini were first recognized and described long ago by Stenonis (1664) and Lorenzini (1678), but their physiological and behavioral functions remained unknown for almost another three centuries. Initially, the ampullae of Lorenzini were thought to be mechanoreceptors (Dotterweich, 1932; Parker, 1909), but were then later shown to be also temperature sensitive (Hensel, 1955; Sand, 1937). A mechanoreceptive function was again proposed later (Loewenstein, 1960; Murray, 1957, 1960b) along with a proposed function as detectors for changes in salinity (Loewenstein and Ishiko, 1962) before current ideas about their use in electroreception were generally accepted. Murray (1960a) and Dijkgraaf and Kalmijn (1962) were the first to demonstrate the electrosensitivity of the ampullae of Lorenzini. More recently, the temperature sensitivity of ampullae was reconfirmed by Brown (2003; but for a complete review of this topic, see Brown, 2010), who demonstrated that the extracellular gel from the ampullae develops significant voltages in response to very small temperature gradients. Thus, temperature can be translated into electrical information by elasmobranchs without the need of cold-sensitive ion channels as used by mammals (Reid and Flonta, 2001; Viana et al., 2002). The extremely sensitive ampullary electroreceptor system of elasmobranchs is now known to mediate orientation to local inanimate electric fields (Kalmijn, 1974, 1982; Pals et al., 1982a), is hypothesized to function in geomagnetic navigation (Kalmijn, 1974, 1988a, 2000; Paulin, 1995), and is known to be important for the detection of the bioelectric fields produced by prey (Blonder and Alevizon, 1988; Jordan et al., 2009b; Kajiura, 2003; Kajiura and Fitzgerald, 2009; Kalmijn, 1971, 1982; Tricas, 1982), potential predators (Sisneros et al., 1998), and conspecifics during social interactions (Tricas et al., 1995).

12.5.1 Anatomy

12.5.1.1 Ampullae of Lorenzini

Single ampullae of Lorenzini consist of a small chamber (the ampulla) and a subdermal canal about 1 mm wide that projects to the surface of the skin (Figure 12.17A) (Waltman, 1966). Small bulbous pouches known as alveoli form the ampulla chamber. Within each alveolus, hundreds of sensory hair-cell receptors and pyramidal support cells line the alveoli wall with only the apical surface of the sensory receptors and support cells exposed to the internal lumen of the ampulla chamber. Tight junctions unite the support cells and sensory

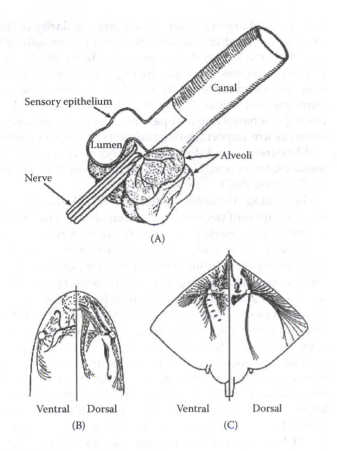

FIGURE 12.17
Ampullary electroreceptor organ of elasmobranchs. (A) The ampulla of Lorenzini consists of a small ampulla chamber composed of multiple alveoli that share a common lumen and a subdermal ampullary canal that projects to a pore on the surface of the skin. The sensory epithelium forms a high-resistance ampulla wall composed of a single layer of sensory receptor cells and support cells. The basal surface of the sensory receptor cells is innervated by primary afferents of the VIIIth cranial nerve. (Adapted from Waltman, B., *Acta Physiol. Scand.*, 66(Suppl. 264), 1–60, 1966.) (B) Diagrammatic representation of the horizontal distribution of the subdermal ampullary clusters and their radial canals that terminate at surface pores on the ventral and dorsal surfaces of the small-spotted catshark, *Scyliorhinus canicula*. (Adapted from Dijkgraaf, S. and Kalmijn, A.J., *Z. Vergl. Physiol.*, 47, 438–456, 1963.) (C) Horizontal distribution of the ampullae of Lorenzini in the skate, *Raja clavata*. (Adapted from Murray, R.W., *J. Exp. Biol.*, 37, 417–424, 1960.)

receptors to create a high-resistance electrical barrier between the basal and apical surfaces of the sensory epithelium, which form the ampulla wall (Sejnowski and Yodlowski, 1982; Waltman, 1966). The basal surface of the sensory receptor cell is innervated by 5 to 12 primary afferents of the VIIIth cranial nerve with no efferents present (Kantner et al., 1962). The wall of the canal consists of a double layer of connective tissue fibers and squamous epithelial cells that are tightly joined together to form a high electrical resistance (6 MΩ-cm) between the outer and inner surface of the canal wall. In contrast, the canal and ampulla are filled with a low-resistance uniform hydrogel (25 to 31 Ω-cm) composed of sulfated

glycoprotein molecules with an ionic composition similar to that of seawater (Brown et al., 2005; Doyle, 1963; Waltman, 1966). The shark hydrogel has a lower admittance than seawater or synthetic (collagen) hydrogels, and it promotes a charge-induced voltage gradient along the interior length of the canal rather than acting as previously thought as a core conductor providing direct electrical contact to the external seawater environment (Brown et al., 2004).

In marine elasmobranchs, many individual ampullae are grouped into discrete, bilateral cephalic clusters from which project the subdermal canals that radiate in many directions and terminate at individual skin pores on the head of sharks (Figure 12.17B) and the head and pectoral fins of skates and rays (Figure 12.17C). The ampullary clusters, which usually vary in number (three to six per side of animal) and location depending on species, are innervated by different branches of the anterior lateral line nerve (VIII) (Norris, 1929). The special arrangement of the contiguously grouped ampullae within the cluster creates a common internal potential near the basal region of the sensory receptors within each cluster. The sensory receptor cells within individual ampullae detect potential differences between the animal's common internal potential at the ampullary cluster and seawater at the surface pore of the canal which projects to the internal lumen of the ampulla (Bennett, 1971). In effect, electroreceptors measure the voltage drop of the electric field gradient along the length of the ampullary canal. Thus, ampullae with long canals sample across a greater distance within a uniform field, provide a larger potential difference for the sensory receptors, and thus have a greater sensitivity than do ampullae with short canals (Broun et al., 1979; Sisneros and Tricas, 2000). The morphological arrangement of the ampullary canals and clusters permits detection of both small local fields produced by small prey organisms and also the uniform electric fields of inanimate origins for possible use in orientation and navigation (Kalmijn, 1974; Tricas, 2001).

In contrast to marine species, freshwater elasmobranchs have a very different morphology and organization of the ampullary electroreceptors that are thought to reflect sensory adaptations to the highly resistive environment of freshwater (Kalmijn, 1974, 1982, 1988a; Raschi and Mackanos, 1989). One such adaptation is a thicker epidermis that functions to increase transcutaneous electrical resistance. In addition, the size of the ampullary electroreceptors in freshwater elasmobranchs is greatly reduced, thus the ampullae are referred to as *microampullae* or *miniampullae*. Furthermore, the ampullary electroreceptors are distributed individually, rather than in clusters, over the head and pectoral fins and have very short subdermal canals (~0.3 to 2.1 mm long) that extend to the surface pores on the skin.

12.5.1.2 Central Pathways

The ampullae of Lorenzini are innervated by primary afferent neurons that convey sensory information to the brain via the dorsal root projections of the anterior lateral line (VIII). The electrosensory primary afferents from ipsilateral ampullae terminate in a somatotopic order within the central zone of the dorsal octavolateralis nucleus (DON), the first-order hindbrain electrosensory nucleus (Bodznick and Northcutt, 1980; Bodznick and Schmidt, 1984; Koester, 1983). The large electrosensory multipolar principal cells in the DON known as ascending efferent neurons (AENs) receive afferent input from the dorsal granular ridge and both the peripheral and central zones of the DON. AENs ascend to the midbrain via a lateral line lemniscus and terminate in somatotopic order in a part of the contralateral midbrain known as the lateral mesencephalic nucleus (LMN) and in deep layers of the tectum (Bodznick and Boord, 1986). The LMN is one of the three elasmobranch midbrain nuclei that compose the lateral mesencephalic nuclear complex (Boord and Northcutt, 1982), which is a midbrain region considered to be homologous to the torus semicircularis in electrosensory teleost fishes (Northcutt, 1978; Platt et al., 1974). Electrosensory information processed in the LMN is sent to the posterior lateral nucleus of the thalamus, where it is then relayed to the medial pallium of the forebrain (Bodznick and Northcutt, 1984; Bullock, 1979; Schweitzer and Lowe, 1984). Some electrosensory information is also conveyed to the cerebellum (Fiebig, 1988; Tong and Bullock, 1982).

12.5.2 Physiology

12.5.2.1 Peripheral Physiology

Electrosensory primary afferent neurons that innervate the ampullae of Lorenzini exhibit a regular pattern of discharge activity in the absence of electrical stimulation. The average resting or "spontaneous" discharge rates of electrosensory afferents in batoid elasmobranchs range from 8.6 impulses/s at 7°C in the little skate, *Leucoraja erinacea* (New, 1990), to 18.0 impulses/s at 16 to 18°C in the thornback guitarfish, *Platyrhinoidis triseriata* (Montgomery, 1984), 34.2 impulses/s at 18°C in the round stingray, *Urolophus halleri* (Tricas and New, 1998), 44.9 impulses/s at 20°C in the clearnose skate, *Raja eglanteria* (Sisneros et al., 1998), and 52.1 impulses/s at 21 to 23°C in the Atlantic stingray, *Dasyatis sabina* (Sisneros and Tricas, 2002b). These differences in resting discharge rates among batoids are most likely due to the influence of temperature, which in the case of higher temperatures can decrease the thresholds required for membrane depolarization of the sensory receptors and spike initiation of the electrosensory primary afferents (Carpenter, 1981; Montgomery and MacDonald, 1990).

Resting discharge rates and discharge regularity of the electrosensory afferents are influenced by the animal's age. Both the rate and discharge regularity of electrosensory afferents increase during development from neonates to adults in both *R. eglanteria* and *D. sabina* (Sisneros and Tricas, 2002b; Sisneros et al., 1998). The resting discharge rate and pattern of the electrosensory afferents are important determinants of the sensitivity and low-frequency information encoding of the electric sense (Ratnam and Nelson, 2000; Sisneros and Tricas, 2002b; Stein, 1967).

The resting discharge patterns of the electrosensory primary afferent neurons in all elasmobranch fishes are modulated by extrinsic electric fields as a function of stimulus polarity and intensity. Presentation of a cathodal (negative) stimulus at the ampullary pore increases the neural discharge activity of electrosensory afferents, whereas an anodal (positive) stimulus decreases discharge activity (Murray, 1962, 1965). Stimulation of the electroreceptors with a sinusoidal electric field modulates the neural discharges of electrosensory afferents as a linear function of the stimulus intensity over the dynamic range of the peripheral electrosensory system, which is from 20 nV/cm to 25 μV/cm (Montgomery, 1984; Murray, 1965; Tricas and New, 1998). Electrosensory afferents are most responsive to electric fields oriented parallel to the vector between the ampullary canal opening on the skin surface and the respective ampulla. Within the intensity range of natural biologically relevant electric fields, electroreceptors are broadly tuned to low-frequency electric stimuli and respond maximally to sinusoidal stimuli from approximately 0.1 to 15 Hz (Andrianov et al., 1984; Montgomery, 1984; New, 1990; Peters and Evers, 1985; Sisneros and Tricas, 2000; Sisneros et al., 1998; Tricas and New, 1998; Tricas et al., 1995). Sensitivity (gain) of the electrosensory afferents to a sinusoidal uniform electric field is 0.9 spikes/s per μV/cm for the little skate, *Leucoraja erinacea* (Montgomery and Bodznick, 1993), 4 spikes/s per μV/cm for the thornback guitarfish, *Platyrhinoidis triseriata* (Montgomery, 1984), 7.4 spikes/s per μV/cm average for the Atlantic stingray, *Dasyatis sabina* (Sisneros and Tricas, 2000, 2002b), 17.7 spikes/s per μV/cm average for the clearnose skate, *Raja eglanteria* (Sisneros et al., 1998), and 24 spikes/s per μV/cm average for the round stingray, *Urolophus halleri* (Tricas and New, 1998).

12.5.2.2 Central Physiology

Although neurophysiological studies of the elasmobranch central electrosensory system have been limited, several features of electrosensory processing in the hindbrain and midbrain, and to a lesser extent in the thalamus and forebrain, have been well characterized. The principal cells of the DON known as AENs exhibit lower resting discharge rates and are more phasic in response than primary afferent neurons found in the peripheral electrosensory system (Bodznick and Schmidt, 1984; New, 1990). Average resting discharge rates of AENs range from 0 to 5 spikes/s in the little skate, *Leucoraja erinacea* (Bodznick and Schmidt, 1984; New, 1990) to 10 spikes/s in the thornback guitarfish, *Platyrhinoidis triseriata* (Montgomery, 1984). However, AENs are similar to electrosensory primary afferents in that they are excited by cathodal stimuli and inhibited by anodal stimuli (New, 1990). Sensitivity to sinusoidal uniform electric fields is higher for second-order AENs than the primary afferent neurons. The sensitivity of AENs ranges from 2.2 spikes/s per μV/cm for *L. erinacea* (Conley and Bodznick, 1994) to 32 spikes/s per μV/cm for *P. triseriata* (Montgomery, 1984). The increased gain of AENs is most likely due to the convergent input of multiple electrosensory primary afferents onto AENs, which have excitatory receptive fields that comprise two to five adjacent ampullary electroreceptor pores (Bodznick and Schmidt, 1984). AENs are also similar to electrosensory primary afferents in their frequency response, with a maximum response in the range 0.5 to 10 Hz, followed by a sharp cutoff frequency between 10 and 15 Hz (Andrianov et al., 1984; Montgomery, 1984; New, 1990; Tricas and New, 1998).

One important function of the second-order AENs is to filter out unwanted noise or reafference created by the animal's own movements, which could interfere with the detection of biologically relevant signals (Montgomery and Bodznick, 1994). Electrosensory AENs show a greatly reduced response to sensory reafference that is essentially similar or common mode across all electrosensory primary afferents. An adaptive filter model was proposed by Montgomery and Bodznick (1994) to account for the ability of electrosensory AENs to suppress common mode reafference. The suppression of common mode signals by AENs is mediated by the balanced excitatory and inhibitory components of their spatial receptive fields (Bodznick and Montgomery, 1992; Bodznick et al., 1992, 1999; Montgomery and Bodznick, 1993).

The response properties of the central electrosensory system have also been studied in the midbrain of elasmobranchs. The midbrain electrosensory neurons of *Platyrhinoidis triseriata* are usually "silent" and exhibit no resting discharge activity (Schweitzer, 1986). Midbrain unit thresholds range from less than 0.3 μV/cm, the lowest intensity tested in this study, to 5 μV/cm in *P. triseriata* (Schweitzer, 1986), to even lower thresholds of 0.015 μV/cm measured with evoked potentials in the blacktip reef shark, *Carcharhinus melanopterus* (Bullock, 1979). Midbrain neurons respond maximally to frequency stimuli from 0.2 Hz (lowest frequency tested) to 4 Hz in *P. triseriata*, 10 to 15 Hz in the freshwater stingray,

Potamotrygon sp., and at higher frequencies from 20 to 30 Hz in the blacktip reef shark, *C. melanopterus* (Bullock, 1979; Schweitzer, 1986). Such discrepancies in frequency sensitivity may be due to differences in methodology or to variation among species. Electrosensory neurons in the LMN of the midbrain may have small, well-defined minimum excitatory receptive fields that include 2 to 20 ampullary pores in *P. triseriata* (Schweitzer, 1986) and 4 to 8 ampullary pores in the thorny skate, *Raja radiata* (Andrianov et al., 1984). Electroreceptive fields are somatotopically mapped in the midbrain such that the anterior, middle, and posterior body surfaces are represented in the rostral, middle, and caudal levels of the contralateral midbrain. Like electrosensory primary afferents and AENs, the electrosensory midbrain neurons are also sensitive to the orientation of uniform electric fields with maximal response corresponding to the vector parallel to the length of the ampullary canal.

Neurophysiological recordings of electrosensory processing areas in the thalamus and forebrain have been limited at best. Multiunit and evoked potential recordings have localized electrosensory activity in the lateral posterior nucleus of the thalamus in *Leucoraja erinacea* (Bodznick and Northcutt, 1984) and in *Platyrhinoidis triseriata* (Schweitzer, 1983). Bodznick and Northcutt (1984) also recorded electrosensory evoked potentials and multiple-unit activity throughout the central one third of the skate forebrain in a pallial area that corresponds to the medial pallium.

12.5.3 Behavior

12.5.3.1 Prey and Predator Detection

The first demonstrated use of the elasmobranch electric sense was for the detection of the bioelectric fields produced by prey organisms (Kalmijn, 1971). In laboratory behavioral experiments, Kalmijn (1971) demonstrated that both the small-spotted catshark, *Scyliorhinus canicula*, and the thornback ray, *Raja clavata*, executed well-aimed feeding responses to small, visually inconspicuous buried flounder (Figure 12.18A) and to flounder buried in a seawater agar-screened chamber that permitted the emission of the prey's bioelectric field but not its odor (Figure 12.18B). When the agar-screened prey was covered by a thin plastic film that insulated the prey electrically, the flounder remained undetected (Figure 12.18C). Feeding responses indistinguishable from those mediated by natural prey were observed again directed toward dipole electrodes that simulated bioelectric prey fields when buried under the sand or agar (Figure 12.18D). In later field experiments, Kalmijn (1982) also demonstrated that free-ranging sharks such as the smooth dogfish, *Mustelus canis*, and the blue shark, *Prionace glauca*, were attracted to an area by odor but preferentially attacked an active

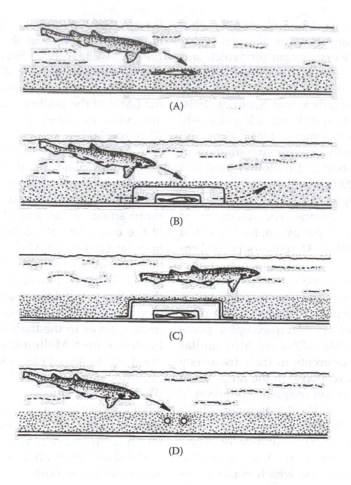

FIGURE 12.18

Use of the elasmobranch electric sense for the detection of electric fields produced by prey organisms. Behavioral responses of the small-spotted catshark, *Scyliorhinus canicula*, to a small flounder buried in the sand (A), a flounder buried in a seawater agar-screened chamber permeable to bioelectric fields (B), a flounder in an agar chamber covered by a plastic film that insulates the prey electrically (C), and electrodes simulating the bioelectric fields produced by a flounder (D). Solid arrows indicate path of attack by the catshark; broken arrows indicate flow of seawater. (Adapted from Kalmijn, A.J., *J. Exp. Biol.*, 55, 371–383, 1971.)

dipole source that simulated the prey's bioelectric field rather than the odor source of the prey. In addition, Tricas (1982) showed that the swell shark, *Cephaloscyllium ventriosum*, uses its electric sense to capture prey during nocturnal predation on small reef fish. Subsequently, other elasmobranch species were shown to demonstrate well-aimed feeding responses at electrically simulated prey; these elasmobranch species include the Atlantic stingray (*Dasyatis sabina*) (Blonder and Alevizon, 1988), sandbar shark (*Carcharhinus plumbeus*), scalloped hammerhead (*Sphyrna lewini*), and neonate bonnethead (*Sphyrna tiburo*) (Kajiura, 2003; Kajiura and Fitzgerald, 2009; Kajiura and Holland, 2002), as well as more recently three batoid species: round stingray (*Urobatis halleri*), pelagic stingray (*Pteroplatytrygon violacea*), and bat ray (*Myliobatis californica*) (Jordan et al., 2009b).

McGowan and Kajiura (2009) recently showed that the euryhaline Atlantic stingray, *Dasyatis sabina*, responded similarly to prey-simulating stimuli when tested across a broad range of salinities from freshwater (0 ppt) to full-strength seawater (35 ppt), but there was a reduction in the electrosensitivity and detection range of stingrays in freshwater environments that is most likely due to the water's electrical resistivity and the physiological function of the stingray's ampullary canals. Other work by Kajiura and Holland (2002) demonstrated that the "hammer" head morphology of sphyrnid sharks does not appear to confer a greater electroreceptive sensitivity to prey-simulating dipole electric fields than the "standard" head shark morphology, but it may provide a greater lateral search area to increase the probability of prey encounter and enhance maneuverability for prey capture.

Another important function of the elasmobranch electric sense is for use in predator detection and avoidance. Work on the clearnose skate, *Raja eglanteria*, demonstrates that the electric sense of egg-encapsulated embryonic skates is well suited to detect potential egg predators (Sisneros et al., 1998), which include other elasmobranchs, teleost fishes, marine mammals, and

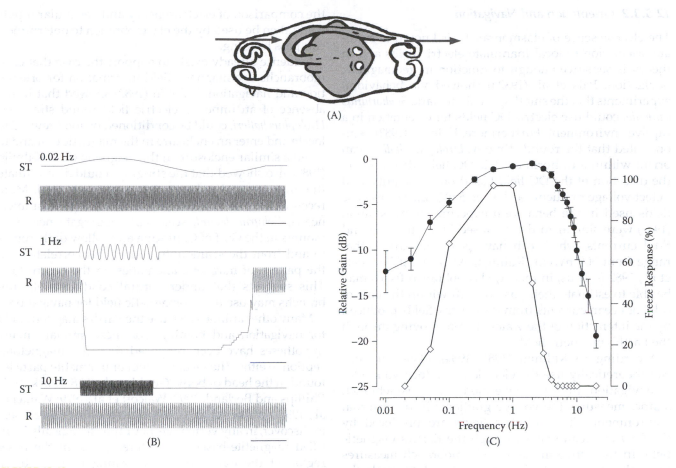

FIGURE 12.19

Behavioral response of embryonic clearnose skates, *Raja eglanteria*, to weak electric stimuli. (A) Ventilation behavior of embryonic skates. Diagram depicts a late-term embryonic skate circulating seawater within the egg case by undulating its tail in one corner of the egg near ventilation pores found in the horn of the egg case. The tail-beating action of the skate draws fresh seawater through pores on the opposite end of the case and creates a localized vortex near the exit pore by the tail. Arrow indicates flow of seawater. (B) Behavioral responses of skate embryos to sinusoidal uniform electric fields at stimulus (ST) frequencies of 0.02, 1, and 10 Hz. Stimuli were applied at an intensity of 0.56 μV cm^{-1} across the longitudinal axis of the skate. The response (R) is expressed as a change in the peak-to-peak (PTP) tail displacement of the skate within the egg case. Prestimulus tail displacement for each record was 10 mm PTP. At 1 Hz, note the large tail displacement that occurs during coiling of the tail around the body after the onset of the electrical ST and a period of no tail movement during and after stimulation. Time bars: 5 s. (C) Freeze response of embryonic skates to weak electric stimuli. Behavioral responses (open diamonds) are shown as a percentage of total ST presentation to 0.02 to 20 Hz. Note that the peak frequency sensitivity of electrosensory primary afferent neurons (solid dots) for embryonic skates is at 1 to 2 Hz and is aligned with the freeze response peak of 0.5 to 1 Hz. (Adapted from Sisneros, J.A. et al., *J. Comp. Physiol. A*, 183, 87–99, 1998.)

molluscan gastropods (for a review, see Cox and Koob, 1993). Late-term embryonic skates circulate seawater within the egg case by undulating their tail in one corner of the egg near ventilation pores found in the horn of the egg case (Figure 12.19A). This action draws fresh seawater through pores on the opposite end of the egg case and creates a localized vortex near the exit pore by the tail, which can provide potential predators with olfactory, electrosensory, and mechanosensory cues needed for the detection and localization of the egg-encapsulated embryo. The peak frequency sensitivity of the peripheral electrosensory system in embryonic clearnose skates matches the frequency of phasic electric stimuli produced by large fish predators during

ventilatory activity (0.5 to 2 Hz) and also corresponds to the same frequency of phasic electric stimuli that interrupts the respiratory movements of skate embryos and elicits an antipredator freeze behavior (Figure 12.19B,C) (Sisneros et al., 1998). This freeze response exhibited by embryonic skates stops the ventilatory streaming of seawater from the egg case and decreases the likelihood of sensory detection by predators. Phasic electric stimuli of 0.1 to 1 Hz are also known to interrupt the ventilatory activity of newly posthatched catsharks, *Scyliorhinus canicula* (Peters and Evers, 1985) and thus may represent an adaptive response in skates and other elasmobranchs to enhance survival during their early life history.

12.5.3.2 Orientation and Navigation

The electric sense of elasmobranchs is known to mediate orientation to local inanimate electric fields and in theory is sensitive enough to function in geomagnetic navigation. Pals et al. (1982a) showed via behavioral experiments that the small-spotted catshark, *Scyliorhinus canicula*, could use electric DC fields for orientation in a captive environment. Furthermore, Kalmijn (1982) demonstrated that the round stingray, *Urolophus halleri*, can orient within a uniform electric DC field, discriminate the direction of the DC field based on its polarity, and detect voltage gradients as low as 5 nV/cm. The electric fields used in the behavioral experiments by Kalmijn (1982) were similar to those caused by both ocean and tidal currents, which can have peak amplitudes that range from 500 nV/cm (Kalmijn, 1984) to 8 μV/m (Pals et al., 1982b). Thus, in theory, elasmobranch fishes may be able to estimate their passive drift within the flow of tidal or ocean currents from the electric fields produced by the interaction of the water current moving through the Earth's magnetic field.

According to Kalmijn (1981, 1984), elasmobranchs can theoretically use the electric sense for two modes of navigation. In the passive mode, the elasmobranch simply measures the voltage gradients in the external environment. These electric fields are produced by the flow of ocean water through the Earth's magnetic field. In the active mode, the elasmobranch measures the voltage gradients that are induced through the animal's body due to its own swimming movements through the geomagnetic field (Figure 12.20). A different hypothesis of active electronavigation proposed by Paulin (1995) maintains that directional information is acquired from the modulation of electrosensory inputs caused by head turning during swimming movements. Sufficient electrosensory information is obtained during head turns to allow the elasmobranch to extract directional cues from electroreceptor voltages induced in the animal as it swims in different directions. Thus,

the comparison of electrosensory and vestibular inputs could then be used by the elasmobranch to determine a compass heading.

Evidence already exists to support the case that elasmobranchs use magnetic field information for orientation and navigation. Kalmijn (1982) showed that in the absence of an imposed electric field round stingrays, *Urolophus halleri*, could be conditioned by food reward to locate and enter an enclosure in the magnetic east and to avoid a similar enclosure in the magnetic west. Kalmijn (1982) also showed that the stingrays could discriminate the direction and polarity of the magnetic field. More recently, Klimley (1993) showed that scalloped hammerheads, *Sphyrna lewini*, seasonally aggregate near seamounts in the Gulf of California and follow daily routes to and from the seamounts, routes that correlate with the pattern of magnetic anomalies on the ocean floor. This suggests that under natural conditions elasmobranchs may use the geomagnetic field for navigation.

Many other animals also use the Earth's magnetic field for navigation and homing. For these animals, many hypotheses have been proposed that link magnetoreception to either the visual system or magnetite particles found in the head or body (Gould et al., 1978; Leask, 1977; Phillips and Borland, 1992; Walcott et al., 1979; Walker et al., 1997). Walker et al. (1997) were the first researchers to discover, in any vertebrate, neurophysiologically identified magnetite-based magnetoreceptors, in the nasal region of the long-distance migrating rainbow trout, *Oncorhynchus mykiss*. Based on their behavioral, anatomical, and neurophysiological experiments, Walker et al. (1997) provided the best evidence to date of a structure and function for a magnetite-based vertebrate magnetic sense. The identification of the key components of the magnetic sense in the rainbow trout will no doubt lead to new perspectives in the study of long-distance orientation and navigation in a variety of vertebrate groups.

12.5.3.3 Conspecific Detection

Work on non-electric stingrays demonstrates that the elasmobranch electric sense is used for conspecific detection and localization during social and reproductive behaviors (Sisneros and Tricas, 2002a; Tricas et al., 1995). Male and female round stingrays, *Urolophus halleri*, use the electric sense to detect and locate the bioelectric fields of buried conspecifics during the mating season (Figure 12.21A). Stingrays produce a standing DC bioelectric field that is partially modulated by the ventilatory movements of the mouth, spiracles, and gill slits (Figure 12.21B) (Kalmijn, 1984; Tricas et al., 1995). Male rays use the electric sense to detect and locate females for mating, and females use their electric sense to locate and join other buried, less receptive females for refuge (Sisneros and Tricas, 2002a; Tricas et al., 1995). The round

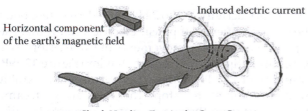

Induced electric current

Horizontal component
of the earth's magnetic field

Shark Heading East in the Open Ocean

FIGURE 12.20
Use of the elasmobranch electric sense in the active mode of navigation. Diagram depicts the induction of electric current induced in the head and body of the animal as the shark swims through the horizontal component of the Earth's geomagnetic field. (Adapted from Kalmijn, A.J., in *Sensory Biology of Aquatic Animals*, Atema, J. et al., Eds., Springer-Verlag, New York, 1988, pp. 151–186.)

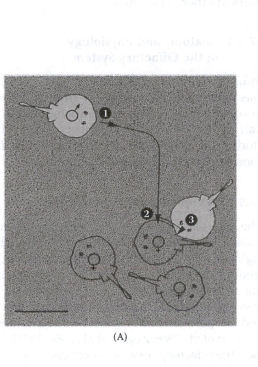

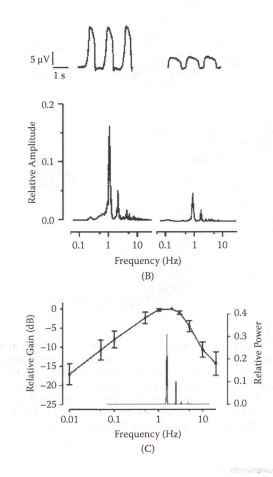

FIGURE 12.21

Detection of conspecific mates, bioelectric stimuli, and the frequency response of the peripheral electrosensory system in the round stingray, *Urolophus halleri*. (A) Orientation response by a male round stingray to cryptically buried conspecific females during the mating season. Males localize, orient toward, and inspect buried females in the sandy substrate. Search path of the male ray (1) changes abruptly after the detection of the female's bioelectric field. Males inspect buried females near the margins of her body disk (2) and pelvic fins (3). Active courtship and copulation begin after the male excavates the buried female and grasps the female's body disk with his mouth. Scale bar: 25 cm. (B) Bioelectric potentials recorded from a female stingray on the ventral surface near the gill slits (top, left record) and dorsal surface above the spiracle (top, right record). Recorded potentials are similar for both male (not shown) and female rays. Scales apply to both top records. Bottom graphs are Fourier transforms that show strong frequency components near 1 to 2 Hz that result from ventilatory movements. (C) Match between the peak frequency sensitivity of electrosensory primary afferent neurons and the frequency spectrum of the modulated bioelectric waveforms produced by round stingrays. The response dynamics of the electrosensory primary afferents in *U. halleri* show greatest frequency sensitivity at approximately 1 to 2 Hz with a 3-dB drop at approximately 0.5 and 4 Hz. Data are plotted as the relative gain of mean discharge peak (±1 SD). (Adapted from Tricas, T.C. et al., *Neurosci. Lett.*, 202, 29–131, 1995.)

stingray's peak frequency sensitivity of the peripheral electrosensory system matches the modulated frequency components of the bioelectric fields produced by conspecific stingrays (Figure 12.21C). Thus, the stingray's electric sense is "tuned" to social bioelectric stimuli and is used in a sex-dependent context for conspecific localization during the mating season. In addition to the detection of conspecific bioelectric fields, the electric sense is also used by skates to detect the weak electric organ discharges (EODs) produced by conspecifics during social and reproductive behaviors (New, 1994; Sisneros et al., 1998). All marine skates of the family Rajidae produce intermittently pulsed, weak electric discharges from spindle-shaped electric organs found bilaterally in the tail (Figure 12.22). The EODs of skates are relatively low in amplitude and species specific in duration, and they are thought to serve an important communication function during social and reproductive interactions (Bratton and Ayers, 1987; Mikhailenko, 1971; Mortenson and Whitaker, 1973). Peak frequency sensitivity of the peripheral electrosensory system in the clearnose skate, *Raja eglanteria*, matches the pulse rate of EODs produced by conspecific skates during social and mating behaviors (Sisneros et al., 1998). A similar match between peak frequency sensitivity of the peripheral electrosensory and EOD pulse rate also occurs in the little skate, *Leucoraja erinacea* (Bratton and Ayers, 1987; New, 1990). Thus, the match between the electrosensory-encoding and EOD properties in these skates likely facilitates electric communication during social and reproductive behaviors.

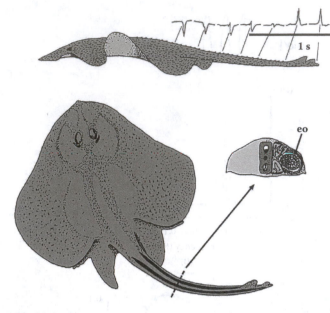

FIGURE 12.22
Diagram of the little skate, *Leucoraja erinacea*, showing the position of the electric organ (eo; black) in the tail and the corresponding monophasic, head-negative electric organ discharge waveform recorded 1 cm from the skin in the tail regions indicated. Note that the cross-section of the tail shows the position of the electric organ and lateral displacement of muscle bundles around the electric organ. (Adapted from Bratton, B.O. and Ayers, J.L., *Environ. Biol. Fish.*, 20, 241–254, 1987.)

12.6 Olfaction and Other Chemical Senses

Elasmobranchs, sharks in particular, are renowned for their olfactory capabilities. Often described as "swimming noses," sharks are the subject of several pervasive myths, such as possessing the ability to detect a single drop of blood in an Olympic-sized swimming pool. These popular perceptions have been fueled by anecdotal observations and early experimental studies that identified olfaction as an important, if not the primary, means by which sharks find food (Parker, 1909, 1914; Parker and Sheldon, 1913; Sheldon, 1909, 1911). In addition, shark olfaction has been thought to be important due to the relatively large size of their olfactory structures, compared to those of other vertebrates (reviewed in Northcutt, 1978). Interest in preventing shark attacks on military personnel in World War II sparked a second generation of investigations on shark feeding and its olfactory control. This work continued into the mid-1970s (Hodgson and Mathewson, 1978a). More recent studies on olfaction in elasmobranchs have detailed aspects of the anatomy and physiology of olfactory systems, identified mechanisms of olfactory control of feeding, and suggested that female sex pheromones attract males and that predators may be detected by smell. Limited information on gustation

and the common chemical sense, or chemesthesis, in elasmobranchs suggests similarities to their counterparts in other vertebrates.

12.6.1 Anatomy and Physiology of the Olfactory System

Information on the anatomical pathways for smell in elasmobranchs derives mostly from considerable work in comparative vertebrate neuroanatomy in the second half of the 20th century (Smeets, 1998). Physiological studies on elasmobranch olfaction, while limited, are consistent with the anatomical and behavioral data.

12.6.1.1 Peripheral Organ and Epithelium

The two elasmobranch olfactory organs are ellipsoid saclike structures, situated in laterally placed cartilaginous capsules on the ventral aspect of the head, in front of the mouth. They are open to the environment via nostrils (nares), which are typically divided by skin-covered flaps into a more lateral incurrent nostril (naris) and a more medial excurrent nostril (Tester, 1963a; Theisen et al., 1986; Zeiske et al., 1986, 1987). In most species, the olfactory organs are entirely separate from the mouth, but in a few species they are in close association with the mouth or even connected to it via a deep groove, called the *nasoral groove*, which extends posteriorly from the excurrent naris, forming a virtual tube between the naris and the mouth (e.g., Orectolobidae, Heterodontidae) (Bell, 1993; Tester, 1963a). The external nasal morphology varies greatly among species, though some broad trends have been found based on lifestyle. Benthic and sedentary species tend to have large nasal openings, while benthopelagic and faster-swimming species tend to have smaller, slit-like openings or large nasal flaps (Schluessel et al., 2008). An anterior depression or groove may be present, helping to channel water into the incurrent opening, and the excurrent opening may be associated with a shallow posterior depression (Tester, 1963a; Zeiske et al., 1986, 1987). In hammerhead sharks (Sphyrinidae), these prenarial grooves are particularly well developed (Gilbert, 1967). In addition to the deep, narrow (prenarial) grooves, which extend along the anterior edge of each side of the head, linking to the incurrent nares (major nasal grooves), a second set of smaller grooves (minor nasal grooves) run parallel and anterior to each incurrent nostril on the dorsal side of the head, further assisting with channeling water into the incurrent naris (Abel et al., 2010) (Figure 12.23).

The olfactory sac is nearly completely filled by an olfactory rosette consisting of two rows of stacked wing-shaped plates, called *lamellae*, which originate from a central ridge (raphe) and attach to the wall of the olfactory cavity (Kajiura et al., 2005; Meredith and Kajiura,

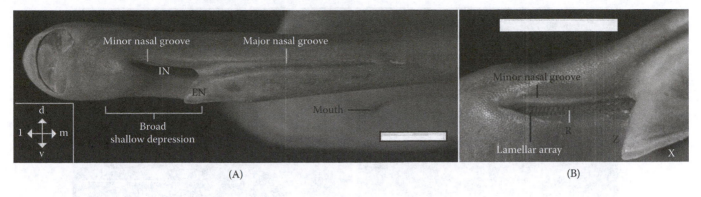

FIGURE 12.23
Nasal grooves in the golden hammerhead, *Sphyrna tudes*. (A) anterior and (B) anteroventral views of the right nasal region, with the lamellae visible through the incurrent nostril. *Abbreviations:* d, dorsal; l, lateral; m, medial; v, ventral; EN, excurrent nostril region; IN, incurrent nostril; R, raphe; X and Z, ventral and lateral edges of the excurrent nostril, respectively. Scale bars: 1 cm. (Adapted from Abel, R.L. et al., *Comp. Biochem. Physiol. A Comp. Physiol.*, 155(4), 464–475, 2010.)

2010; Tester, 1963a; Theisen et al., 1986; Zeiske et al., 1986, 1987) (Figure 12.24A,B). The lamellae are largest in the middle, decreasing in size toward both the medial and lateral ends (Meredith and Kajiura, 2010; Theisen et al., 198, 2009). Each lamella is covered with secondary folds (secondary lamellae), which greatly increase the surface area of the olfactory epithelium. The olfactory epithelium is divided into sensory and nonsensory areas. The nonsensory, squamous epithelium is composed of cells that bear microvilli only and numerous goblet cells (Schluessel et al., 2008; Theisen et al., 1986, 2009; Zeiske et al., 1986, 1987). It is generally found on the margins of the lamellae, although in some species it extends along the ridges of the secondary folds, and in other species a patchy, irregular distribution of sensory and nonsensory areas is found (Schluessel et al., 2008; Theiss et al., 2009) (Figure 12.24C,D). The much larger, centrally located sensory epithelium is composed of pseudostratified, columnar epithelium. It contains receptor cells, supporting cells (which bear numerous cilia), and basal cells, along with occasional goblet cells. It is similar to that found in olfactory systems of most vertebrates, with the major exception that the elasmobranch bipolar receptor cells are not ciliated but rather have a dendritic knob (olfactory knob) from which extends a tuft of microvilli (Reese and Brightman, 1970; Schluessel et al., 2008; Theisen et al., 1986, 2009; Zeiske et al., 1986, 1987) (Figure 12.24E). Similar microvillous receptors have been found along with the "typical" ciliated type in certain bony fishes. Cell surface lectin-binding patterns also differentiate the elasmobranch microvillous receptors (small-spotted catshark, *Scyliorhinus canicula*) from the ciliated receptors of amphibians, rodents, and some bony fishes (Francheschini and Ciani, 1993). Studies on the clearnose skate, *Raja eglanteria*, identify two types of nonciliated olfactory receptor neurons (Takami et al., 1994). Type 1 is typical of those found in the other fishes (as above);

the type 2 cell, so far unique to elasmobranchs, is distinguished from the type 1 by its thicker dendritic knob and microvilli that are shorter, thicker, and more regularly arranged. The functional meaning of the morphological differences in receptor types has yet to be determined.

The olfactory morphology of numerous elasmobranch species has been examined. Olfactory rosette size, lamellar number, and sensory surface area vary by species (Kajiura et al., 2005; Meredith and Kajiura, 2010; Schluessel et al., 2008; Theiss et al., 2009); these differences can be correlated with habitat type but not phylogeny or prey type (Schluessel et al., 2008). Benthopelagic sharks and rays possess higher numbers of lamellae, larger olfactory surface areas, and larger rosettes than benthic species (Meredith and Kajiura, 2010; Schluessel et al., 2008). The ontogeny of the olfactory system has been examined in only a handful of species, but it appears to be well developed at birth, undergoing only minor changes as the animal grows. The morphology of the nares and olfactory rosettes and the ultrastructure of the epithelium of juveniles closely resemble those of adults (Schluessel et al., 2010). The olfactory bulbs undergo growth, increasing with body size, although not proportionally (Schluessel et al., 2010), such that the olfactory bulbs represent a larger proportion of the brain volume in adults as compared with juveniles (Lisney et al., 2007). The olfactory rosettes undergo similar growth; whereas lamellar surface area increases with body size, lamellar number does not, except in the spotted eagle ray, *Aetobatus narinari* (Meredith and Kajiura, 2010; Schluessel et al., 2010).

Interspecific differences in olfactory morphological data have often been used to assess olfactory capability, with increased sensitivity inferred from increased size (Kajiura et al., 2005; Lisney et al., 2007; Schluessel et al., 2008, 2010; Theisen et al., 1986, 2009; Zeiske et al., 1986, 1987), but electrophysiological data refute this. The

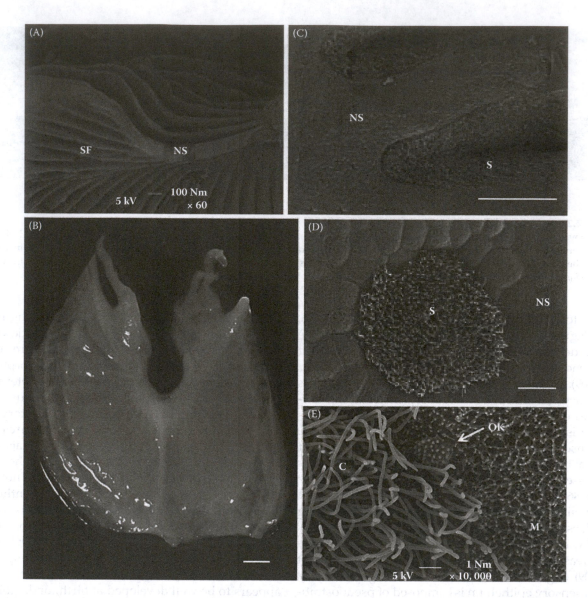

FIGURE 12.24

The olfactory rosette. (A) Low-power SEM image of stacks of lamellae in the bull shark, *Carcharhinus leucas*. (B) Whole lamella from the western wobbegong, *Orectolobus hutchinsi* (scale bar: 1 mm). (C) High-power SEM showing extension of the nonsensory epithelium along the secondary folds in the spotted eagle ray, *Aetobatus narinari* (scale bar: 100 µm). (D) High-power SEM showing division between the sensory (ciliated region) and nonsensory (nonciliated region with microvilli) in the blue-spotted maskray, *Dasyatis kuhlii* (scale bar: 100 µm). (E) High-power SEM showing an olfactory knob present on the lamellae of *O. hutchinsi*. *Abbreviations:* C, cilia; M, microvilli; NS, non-sensory epithelium; OK, olfactory knob; S, sensory epithelium; SF, secondary folds. (Parts A, C, and D adapted from Schluessel, V. et al., *J. Morphol.*, 269, 1365–1386, 2008. Photographs in Parts B and E by Susan M. Theiss and used with permission.)

underwater electroolfactogram (EOG) is a tool for recording the extracellular DC field potentials or analog of the summed electrical activity of the olfactory epithelium in response to chemical stimulation (Silver et al., 1976). EOG responses have been studied in eight elasmobranchs: the nurse shark, *Ginglymostoma cirratum* (Hodgson and Mathewson, 1978b); the Atlantic stingray, *Dasyatis sabina* (Meredith and Kajiura, 2010; Silver, 1979; Silver et al., 1976); the lemon shark, *Negaprion brevirostris* (Meredith and Kajiura, 2010; Zeiske et al., 1986); the thornback ray, *Raja clavata* (Nikonov et al., 1990); the scalloped hammerhead,

Sphyrna lewini (Tricas et al., 2009); the clearnose skate, *Raja eglanteria*; the yellow stingray, *Urobatis jamaicensis*; and the bonnethead, *Sphyrna tiburo* (Meredith and Kajiura, 2010). Several amino acids, known to be effective stimuli for evoking EOGs in bony fishes and behavioral responses in both bony fishes and elasmobranchs, were tested in these species, and extracts of squid muscle were also used in the lemon shark study (Zeiske et al., 1986). The thresholds for individual amino acids varied by species, but, in general, neutral amino acids are more stimulatory, while valine, proline, and isoleucine

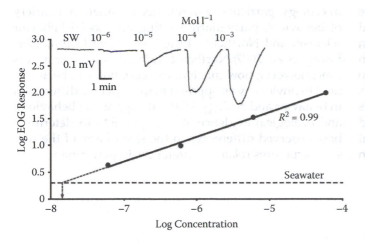

FIGURE 12.25

Representative electroolfactogram (EOG) concentration–response curve for a bonnethead, *Sphyrna tiburo*. The magnitude of the log EOG response (a percentage of the 10^{-3} mol/L alanine standard) is linearly related to the log amino acid stimulus concentration (10^x mol/L). The horizontal dashed black line indicates the averaged response to the seawater (SW) control. The olfactory threshold is calculated as the point where the regression line for the best-fit line of the response intersects the averaged response to the SW control. The inset shows representative EOG responses to the SW control and to increasing log concentrations of L-alanine. Based on absorbance calculations of diluted dye, all stimuli were diluted to 6% of their injected concentration at the entrance to the incurrent naris. The estimated diluted stimulus concentrations are plotted at arrival to the olfactory organ. (From Meredith, T.M. and Kajiura, S.M., *J. Exp. Biol.*, 213, 3449–3456, 2010. With permission.)

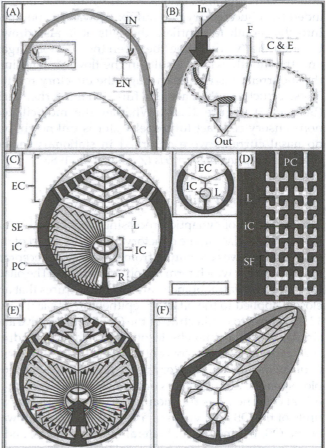

FIGURE 12.26

Schematics of the functional morphology of the nasal region of an active elasmobranch. (A) Ventral surface of the head of a lemon shark, *Negaprion brevirostris* (based on Zeiske et al., 1986). The dotted line is the approximate location of the olfactory chamber. (B) Boxed region in part A. Lines labeled C, E, and F indicate positions of the front face of sections in panels (C), (E), and (F), respectively. (C) Sagittal section through the olfactory chamber, toward the medial end of the chamber (based on Zeiske et al., 1986), with secondary lamellae shown on the left lamella only. Scale bar: 5 mm. Inset: Outlines of incurrent and excurrent channels created by lamellae and the roof of the olfactory chamber. (D) Transverse section through two lamellae, toward the side wall of the olfactory chamber, showing the convoluted nature of the interlamellar channel. (E) Flow through the olfactory chamber, same view as part C. (F) Cut-away view to one side of the olfactory chamber, showing principal flow (arrowed line) through incurrent and excurrent channels (interlamellar gaps and secondary lamellae omitted for clarity). Gray arrows, incurrent flow; white arrows, excurrent flow; dark arrows, flow in interlamellar channels. *Abbreviations:* EC, excurrent channel; EN, excurrent nostril; iC, interlamellar channel; IC, incurrent channel; IN, incurrent nostril; L, lamella; PC, peripheral channel; R, raphe; SF, secondary fold. (From Abel, R.L. et al., *Comp. Biochem. Physiol. A Comp. Physiol.*, 155(4), 464–475, 2010. With permission.)

(also neutral, but with branched side-chains or secondary amine groups) are the least stimulatory. These results are similar for elasmobranchs and teleost fishes (Hara, 1994; Meredith and Kajiura, 2010). The EOG magnitude increased exponentially with the log of the stimulus concentration (Figure 12.25), and calculated thresholds ranged between 10^{-6} and 10^{-11} M. These levels are similar to those reported for bony fishes (teleosts) (Hara, 1994), as well as the levels of free amino acids in seawater (Kuznetsova et al., 2004; Pocklington, 1971). Despite differences in lamellar number and surface area, olfactory thresholds do not differ significantly among elasmobranch species. Because behavioral evidence is lacking, the functional significance of interspecific differences in olfactory morphology and physiology are unknown.

The dynamics of nasal water circulation (nasal ventilation) have been analyzed in a series of detailed studies on several sharks. Briefly, water enters the incurrent nostril, passes along the incurrent channel, and is drawn through the interlamellar channels, out into peripheral channels on the outer edges of the lamellae, and then into the excurrent channel; it then passes back out to the environment via the excurrent nostril (Abel et al., 2010; Theisen et al., 1986; Zeiske et al., 1986, 1987) (Figure 12.26). In actively swimming elasmobranchs, this water

flow is likely generated by differences in pressure between the incurrent and excurrent nostrils that are primarily caused by the forward motion of the animal (Theisen et al., 1986; Zeiske et al., 1986, 1987). In benthic and more sedentary species, nasal ventilation may be aided by a buccopharygeal pump: As water is pumped into the mouth to ventilate the gills, it is also drawn through the "virtual tube" between the olfactory organ and the mouth, which results in the flow of water into the incurrent nostril and through the olfactory rosette. These structures thus act as a functional internal naris (Bell, 1993) (Figure 12.27). Whether the multiciliated non-sensory cells act to propel water is unknown, but no nasal currents were observed in stationary lemon sharks, *Negaprion brevirostris* (Zeiske et al., 1986).

12.6.1.2 Olfactory Bulb

The first level of synaptic processing of olfactory information takes place in the olfactory bulb (OB), a part of the brain that receives the output from the olfactory receptors via their axons, which form the olfactory nerve. The olfactory bulbs of elasmobranchs are large structures that are closely applied to the olfactory epithelium or sac (Figure 12.28). The cytoarchitecture of the OB is conservative and similar in elasmobranchs to other vertebrates (Andres, 1970; Smeets, 1998). Its concentric layers (from superficial to deep) include the olfactory nerve fibers; a layer of complex synaptic arrangements or glomeruli; a layer of large mitral cells, neurons functioning as the chief integrative units of the OB and, via their axons, the output pathway of the OB, the medial and lateral olfactory tracts; and a layer containing many small local circuit neurons, the granular cells. The olfactory tracts or peduncles travel to the cerebral hemispheres or telencephalon proper to make contact with secondary olfactory areas.

Only fairly recently has information on the ultrastructure and electrophysiology of the OB of elasmobranchs become available. Studies on the topography of inputs and synaptic organization of the OB of bonnetheads, *Sphyrna tiburo* (Dryer and Graziadei, 1993, 1994a, 1996) and electrophysiology of the OB of the small-spotted catshark, *Scyliorhinus canicula* (Bruckmoser and Dieringer, 1973), and the little skate, *Leucoraja erinacea* (Cinelli and Salzberg, 1990), have greatly advanced the understanding of the structure in elasmobranchs and permit some useful comparisons to the OB of other, better studied "model" species. Unlike other vertebrates, the OB of elasmobranchs is compartmentalized in a series of swellings or independent sub-bulbs, each exclusively receiving input from the adjacent olfactory epithelium. The mitral cells in fishes (teleosts and elasmobranchs) lack the basal dendrites characteristic of mitral cells of tetrapods, a finding that suggests differences in information processing, especially lateral inhibition (for details, see Andres, 1970; Dryer and Graziadei, 1993, 1994a, 1996).

Species differences in the size of the OB relative to total brain mass or volume have been calculated in several elasmobranch species and used to suggest differences

in ecology, particularly in reliance on smell in a variety of behaviors, particularly feeding and social behavior (Demski and Northcutt, 1996; Lisney and Collin, 2006; Lisney et al., 2007; Northcutt, 1978). It is unclear at this time, however, how much of the variation can be attributed to phylogeny as opposed to interspecific differences in behavior and ecology. Without supporting behavioral and ecological evidence, it is impossible to determine how observed differences in the size of any of the sensory structures relate to differences in performance.

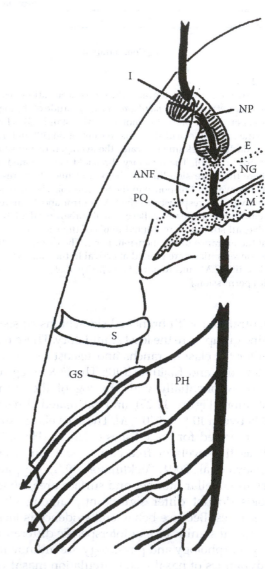

FIGURE 12.27

Proposed path of water drawn through the olfactory chamber by the buccopharyngeal pump in a sedentary elasmobranch (ventral view; anterior is up). Water (arrows) enters through the incurrent opening (I), flows through the nasal pouch (olfactory chamber; NP), and exits the excurrent opening (E) into the nasoral groove (NG), which is covered by the anteromedial nasal flap (ANF). Water continues through the nasoral groove across the palatoquadrate (PQ) and into the mouth (M). Finally, water exits the pharynx (PH) through the gill slits (GS). The first two complete gills slits and spiracle (S) are shown. (From Bell, M.A., *Copeia*, 1993, 144–158, 1993. With permission.)

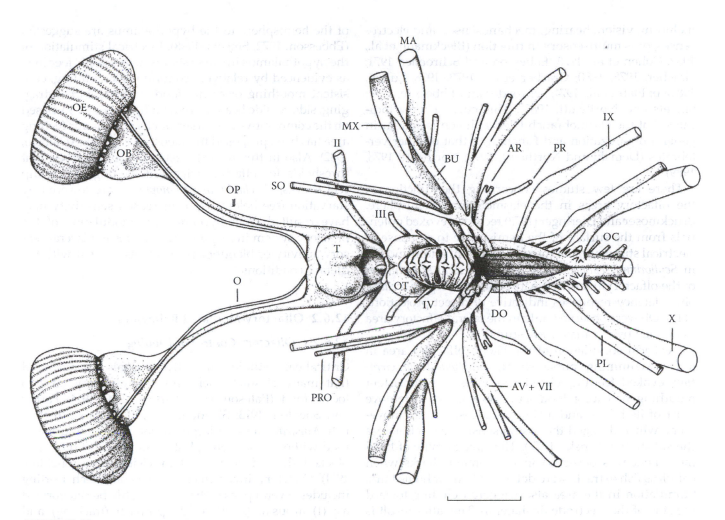

FIGURE 12.28

Dorsal view of the brain and olfactory system of the white shark, *Carcharodon carcharias*. The large partially divided olfactory bulb (OB) is closely applied to peripheral olfactory sac or epithelium (OE). Receptor cells in the epithelium project axons into the olfactory bulb (as the olfactory nerve) to make connections in complex synaptic arrangements. The mitral cells of the olfactory bulb distribute their axons to the secondary olfactory areas of the telencephalic hemisphere (T) via the elongated olfactory tracts or peduncles (OP). The terminal nerve or cranial nerve zero (O), which also extends from the olfactory epithelium to the hemisphere, may have chemosensory-related function(s) (see Demski and Schwanzel-Fukuda, 1987). *Abbreviations:* AR, anterior ramus of the octaval nerve; AV, anteroventral lateral-line nerve; BU, buccal ramus of the anterodorsal lateral line nerve; DO, dorsal octavolateralis nucleus; MA, mandibular ramus of the trigeminal nerve; MX, maxillary ramus of the trigeminal nerve; OC, occipital nerves; OT, optic tectum; PL, posterior lateral line nerve; PR, posterior ramus of the octaval nerve; PRO, profundal nerve; SO, superficial ophthalmic ramus of the anterodorsal lateral line nerve; II, optic nerve; III, oculomotor nerve; IV, trochlear nerve; VII, facial nerve; IX, glossopharyngeal nerve; X, vagus nerve. Scale bar: 3 cm. (From Demski, L.S. and Northcutt, R.G., in *Great White Sharks: The Biology of Carcharodon carcharias*, Klimley A.P. and Ainley, D.G., Eds., Academic Press, San Diego, CA, 1996, pp. 121–130. With permission.)

12.6.1.3 Higher Level Systems

Projections from the OB to the telencephalic hemisphere have been mapped using contemporary neuro-anatomical techniques in a variety of species (Dryer and Graziadei, 1994b; Ebbesson, 1972, 1980; Ebbesson and Heimer, 1970; Ebbesson and Northcutt, 1976; Northcutt, 1978; Smeets, 1983, 1998; Smeets et al., 1983). The results are in general agreement that the primary olfactory tract projection is to the lateral region of the ipsilateral hemisphere. Less well-developed contralateral projections are reported in some species but not others. Spatial mapping of the projection of the medial and lateral olfactory tracts has been documented in the bonnethead, *Sphyrna tiburo* (Dryer and Graziadei, 1994b).

The findings refute earlier claims (see Aronson, 1963) that the entire hemisphere was dominated by the olfactory inputs and consequently that the enlarged hemispheres of sharks and rays could be attributed to their highly developed sense of smell. Other neuroanatomical, physiological, and behavioral studies have demonstrated that, other than the modest area of olfactory tract projection, most of the remainder of the hemisphere either receives specific inputs from other senses, including vision, hearing, mechanosenses, and electro-senses, or is multisensory in function (Bleckmann et al., 1987; Cohen et al., 1973; Ebbesson and Schroeder, 1971; Graeber, 1978, 1980; Graeber et al., 1973, 1978; Luiten, 1981a,b; Platt et al., 1974; Schroeder and Ebbesson, 1974; Smeets and Northcutt, 1987). This current view indi-cates that the elasmobranch telencephalon is similar in general organization and function to that of other ver-tebrates (Demski and Northcutt, 1996; Northcutt, 1978, 1989).

There are few studies concerning the function of the olfactory areas in the elasmobranch hemisphere. Bruckmoser and Dieringer (1973) recorded evoked poten-tials from the surface of the hemisphere in response to electrical stimulation of the olfactory epithelium and OB in *Scyliorhinus canicula* and from electrical stimulation of the olfactory tracts in the torpedo ray, *Torpedo ocellata*. Short latency responses indicative of direct projections of the OB were observed only in the lateral olfactory area as defined by the anatomical studies.

Electrical stimulation of the lateral olfactory area in a free-swimming nurse shark, *Ginglymostoma cirra-tum*, evoked feeding-related responses of inconsistent mouthing or eating food (cut fish soaked to remove most of its juices) and a slow side-to-side head move-ment, which dragged the rostral sensory barbels across the substrate (Demski, 1977). The specific type of head movement was observed in unoperated sharks when colorless fish extracts were delivered to their home tank. Stimulation in the area also triggered circling toward the side of the electrode (ipsilateral). The latter result is consistent with Parker's (1914) observation that sharks with a unilateral occlusion of the nostril circle toward the side of the open nostril. Thus, the physiological and behavioral studies available are consistent with the ana-tomical projections and suggest that the olfactory area of the lateral hemisphere is involved in the arousal of feeding by olfactory stimulation.

Bruckmoser and Dieringer (1973) recorded potentials of longer latency (20 to 800 ms), including regular EEG-synchronous afterpotentials in other areas of the hemi-spheres. This secondary activity was more labile than the primary responses and differed in the two species. It is most likely indicative of areas involved in higher level processing of the olfactory information or regions for multisensory or sensorimotor integration.

It should be noted that in bony fishes the OBs project to the hypothalamus of the diencephalon (Bass, 1981; Finger, 1975; Murakami et al., 1983; Prasada Rao and Finger, 1984), an area from which feeding activity has been evoked by electrical stimulation (Demski, 1983) and potentials triggered by olfactory tract stimulation (Demski, 1981). Although a direct olfactory bulb projec-tion to the hypothalamus has not been reported for elas-mobranchs, projections from the lateral olfactory area

of the hemisphere to the hypothalamus are suggested (Ebbesson, 1972; Smeets, 1998). Electrical stimulation of the hypothalamus in nurse sharks has evoked "feeding" as evidenced by relatively continuous swimming, con-sistent mouthing or eating food, and the barbel-drag-ging, side-to-side head movement (Demski, 1977). Based on the comparative data, a similar hypothalamic feeding area has been proposed for teleosts and sharks (Demski, 1982). Also in this regard, Tester (1963b) observed that thresholds for olfactory-triggered feeding in blacktip reef sharks, *Carcharhinus melanopterus*, are lowered by starvation (see below). Such increased sensitivity may have resulted from hypothalamic modulation of the olfactory system in response to changes in visceral sen-sory activity or bloodborne factors associated with the dietary conditions.

12.6.2 Olfactory-Mediated Behaviors

12.6.2.1 Olfactory Control of Feeding

Critical early studies on captive animals demonstrated that many elasmobranchs rely on olfactory cues to locate food (Bateson 1890; Parker, 1909, 1914; Parker and Sheldon, 1913; Sheldon, 1909, 1911). Smooth dog-fish, *Mustelus canis*, in large outdoor pens could locate food without visual cues, but animals with both nares blocked showed no interest in visible prey (Sheldon, 1911). Olfactory involvement in elasmobranch feeding includes several phases that can roughly be categorized as: (1) arousal, (2) directed approach (tracking) and attack, and (3) continued search, if the prey or bait is not located or is lost. These components vary depending on circumstance and species. Arousal is often indicated by a sudden change from normal swimming (cruising) behavior, such as sudden tight circling by, for example, bonnetheads, *Sphyrna tiburo* (Johnsen and Teeter, 1985); by a sharp turn; or by a sudden drop or spiral to the bottom, such as for smooth dogfish, *M. canis* (Parker, 1914), or blacktip reef sharks, *Carcharhinus melanop-terus* (Tester, 1963a,b). Elasmobranchs were previously thought to accomplish this initial odor orientation by performing bilateral comparisons between the two nares, turning toward the highest concentration, termed *tropotaxis* (Hodgson and Mathewson, 1971; Johnsen and Teeter, 1985). This notion dates back to the early stud-ies of Parker (1914). Control (unblocked) smooth dog-fish, *M. canis*, located food using an equal frequency of turns to either side; blocking one naris resulted in a predominance of turning behavior to the unblocked side. In the aquatic environment, however, water flow is inevitable, be it from currents, the tail beats of prey, or the tail beats of the predator (self-generated noise). Flowing water causes turbulent mixing, resulting in an odor plume that is highly chaotic and intermittent, with a high degree of variance in concentration (reviewed in Webster, 2007). A spatial concentration gradient can only be obtained by averaging over several minutes, far slower than the tracking speed of most animals, including elasmobranchs. Using animals fitted with headstages driven by computer-synchronized pumps, Gardiner and Atema (2010) demonstrated that, for instantaneous bilateral comparisons, *M. canis* responds to differences in the timing of arrival of odor at the two nares, not concentration. Even when the animals receive a weak odor pulse ahead of a strong one, the animals turn toward the naris that first receives an odor cue. This likely aids the animals in initially orienting to odor patches, steering them into the plume. Further work is needed to determine if concentration information is used over time, by comparing concentrations detected across several subsequent odor patch encounters (i.e., through klinotaxis) during odor tracking.

During tracking, many elasmobranchs approach odors from downstream, including white sharks, *Carcharodon carcharias* (Strong et al., 1992, 1996); gray reef sharks, *Carcharhinus amblyrhynchos*; blacktip reef sharks, *Carcharhinus melanopterus*; whitetip reef sharks, *Triaenodon obesus* (Hobson, 1963); and smooth dogfish, *Mustelus canis* (Gardiner and Atema, 2007). Tight circles and figure-eight patterns are common (Gardiner and Atema, 2007; Parker, 1914; Tester, 1963a,b), and animals cover a greater area in the presence of food odors than when these odors are absent (searching behavior, such as in nurse-hounds, *Scyliorhinus stellaris*, and smoothhounds, *Mustelus mustelus*) (Kleerekoper, 1978, 1982). Hodgson and Mathewson suggested two different tracking tactics based on their work with lemon sharks, *Negaprion brevirostris*, and nurse sharks, *Ginglymostoma cirratum*, in large outdoor pens (Hodgson and Mathewson, 1971; Mathewson and Hodgson, 1972). When presented with an attractive odor stimulus, such as glutamic acid and trimethylamine oxide (TMAO), lemon sharks swam upstream into the strongest current, regardless of where the odor source was actually located. In contrast, nurse sharks began moving up the odor corridor and were always able to localize the source. The authors concluded that in lemon sharks the reaction to an odor stimulus is dominated by a rheotactic (orientation to the mean current) bias or release mechanism, a behavior referred to as *odor-stimulated rheotaxis*, whereas in nurse sharks a chemical stimulus triggers true concentration gradient searching (sequential comparisons of concentrations at different points), termed *klinotaxis* (Hodgson and Mathewson, 1971; Mathewson and Hodgson, 1972). Kleerekoper et al. (1975), however, found that nurse sharks in stagnant water could not locate the source of odor release. Gardiner and Atema (2007) demonstrated that the smooth dogfish, *M. canis*, requires information from the lateral line system to locate the source of turbulent food odors (squid rinse). This species can navigate upstream through an odor field to the general area of a turbulent odor source using either vision (visual flow field) or the lateral line system (hydrodynamic flow field), performing odor-stimulated rheotaxis, but the lateral line is necessary to precisely locate the source of coincident odor and flow (i.e., the source). This suggests that these animals are tracking the fine-scale structure of the plume—a turbulent wake flavored with food or prey odor, shed by a moving prey item in still water, or a still piece of food in flowing water (termed *eddy chemotaxis*) (Atema, 1996). In the event that the target is not located, continued search can involve repeated bouts of swimming back downstream and then retracing the plume (e.g., smooth dogfish, *M. canis*) (Gardiner and Atema, 2007) or continuous circling, sometimes for hours (e.g., white sharks, *C. carcharias*) (Strong et al., 1996, 1992).

Most studies of feeding behavior have used live prey (Sheldon, 1911), pieces of bait (Hobson, 1963; Parker, 1914), or food rinses or extracts (Gardiner and Atema, 2007, 2010; Johnsen and Teeter, 1985; Kleerekoper et al., 1975; Tester, 1963a). Tester (1963b) recorded responses of several shark species to a variety of extracts of fish and invertebrates as well as human urine, blood, and sweat. Essentially all food substance extracts were "attractive." Regarding responses to human materials, sharks demonstrated "attraction" to blood, "sensing" but otherwise indifference to urine, and, although highly variable, "repulsion" to sweat. Sharks were "attracted" to introduction of water from containers with prey fish that were not stressed but the sharks soon adapted to the stimuli; in contrast, the sharks showed concerted "hunting reactions" to the test water when the prey fish were "frightened and excited by threatening them with a stick" (Tester, 1963b) and could accurately pinpoint a source of water flowing from tanks of stressed fish (Hobson, 1963), suggesting that sharks can use odors to discriminate between stressed and unstressed prey fish. Hodgson and Mathewson (Hodgson and Mathewson, 1971; Mathewson and Hodgson, 1972) successfully elicited feeding behavior from nurse sharks, *Ginglymostoma cirratum*, and lemon sharks, *Negaprion brevirostris*, using a mixture of chemical attractants (glutamic acid and TMAO) at concentrations of $0.1\ M$ and released into the water in the pen. Thus, the actual concentration of chemicals at the olfactory epithelium is unknown in behavioral experiments. Meanwhile, electrophysiological studies (see Sections 12.6.1.1 and 12.6.1.2) have examined brain and olfactory receptor responses to precisely measured concentrations of single amino acids. Only one study has matched behavior and electrophysiology. Hodgson et al. (1967) performed experiments on free-swimming lemon sharks in which EEG responses were correlated with changes in swimming behavior when the animals were exposed to 10^{-4} M glycine, betaine, trimethylamine, and TMAO. The stimuli were released into flowing water in the test tank, however, so the exact concentration at the olfactory receptors remains unknown. There is, therefore, a disconnect between electrophysiological and behavioral studies, an area that certainly warrants further investigation.

12.6.2.2 Sex Pheromones in Mating

The evidence for use of olfactory cues in social–sexual behavior of elasmobranchs is indirect; nevertheless, it is consistent across several groups of sharks and batoids. The most compelling suggestion of olfactory sex attraction was reported by Johnson and Nelson (1978), who recounted an incident of "close following" behavior of blacktip reef sharks, *Carcharhinus melanopterus*, at Rangiroa Atoll in French Polynesia. One shark tracked down another, which was initially out of its view, and then followed it closely with its snout directed toward the leader's vent. The latter swam close to the substrate in an atypical slow, sinuous manner with its head inclined downward and its tail uplifted. The authors concluded that only an olfactory cue could have guided the second shark to the position of the other. Although sex was not determined in this incidence, other observations indicated that unusual swimming and following behaviors appeared to be sex specific to the females and males, respectively.

There are scattered observations of males of other elasmobranch species following closely behind females, usually with their nose directed to the female's vent, sometimes pushing on it. This has been reported for the bonnethead, *Sphyrna tiburo* (Myrberg and Gruber, 1974); nurse shark, *Ginglymostoma cirratum* (Carrier et al., 1994; Klimley, 1980); spotted eagle ray, *Aetobatus narinari* (Tricas, 1980); clearnose skate, *Raja eglanteria* (Luer and Gilbert, 1985); and sand tiger shark, *Carcharias taurus* (Gordon, 1993) (see also review by Demski, 1991). Other indications of the sex-related nature of the encounters include the presence of scars on the females or swelling of the pelvic fins and cloacal area suggestive of recent mating, male attempts to mount the female, and in captive female sand tiger sharks "cupping and flaring" of the pelvic fins in response to the close presence of the male. Thus, although there are no direct experimental findings to document female sex-attraction pheromones, behavioral observations in natural and captive environments strongly suggest their existence.

12.6.2.3 Olfaction and Predator Avoidance

Lemon sharks, *Negaprion brevirostris*, and American crocodiles, *Crocodylus acutus*, overlap in their distributions, and where such is the case the crocodiles may prey on the sharks. Rasmussen and Schmidt (1992) demonstrated that water samples taken from ponds holding *C. acutus* and delivered to the nares of juvenile lemon sharks consistently aroused the sharks from a state of tonic mobility (induced by inversion and restraint), an established bioassay for chemical awareness. Water from ponds containing alligators, *Alligator mississippiensis*, which have no substantial natural contact with lemon sharks, had no such effect. The authors identified three organic compounds produced by the crocodiles (2-ethyl-3-methyl maleimide, 2-ethyl-3-methyl succinimide, and 2-ethylidene-3-methyl succinimide) that accounted for the positive results. Synthetic versions of the chemicals were also effective. The results strongly suggest that lemon sharks and perhaps other elasmobranchs use olfactory cues to avoid potential predators.

12.6.3 Gustation

Anatomical studies in elasmobranchs have identified receptors that closely resemble taste organs in other vertebrates. A few behavioral observations suggest gustation is important for the acceptance of food in sharks (Sheldon, 1909; see also review by Tester, 1963a). Cook and Neal (1921) mapped the distribution of taste buds in the oral–pharyngeal cavity of the spiny dogfish, *Squalus acanthias*. While located over the entire region, the receptor organs appear most numerous on the roof of the cavity. In microscopic section, the taste buds are characterized as small papillae covered with a multilayer epithelium that has a central cluster of elongate sensory receptor cells. Nerve fibers are associated with the base of the receptors (Figure 12.29A). Older descriptive anatomical studies of several sharks indicate that the taste organs are supplied by branches of the facial (VII), glossopharyngeal (IX), and vagus (X) nerves (Aronson, 1963; Daniel, 1928; Herrick, 1924; Norris and Hughes, 1920), as is the case with other vertebrates (reviewed in Northcutt, 2004).

Whitear and Moate (1994a) carried out a detailed ultrastructural analysis of the taste buds of the small-spotted catshark, *Scyliorhinus canicula*. The apical regions of the receptors with their protruding microvilli form pores, which are clearly visible in their scanning electron micrographs. Nerve fibers were associated with the receptors as well as possible free nerve endings. Part of a taste bud was reconstructed from serial transmission electron micrographs. In general, the organization of the peripheral gustatory system of sharks appears comparable with that of other vertebrates. Unfortunately, detailed physiological and behavioral studies are not available to further support this observation. It seems reasonable to assume that the gustatory apparatus in sharks functions primarily in the final determination of food vs. nonfood.

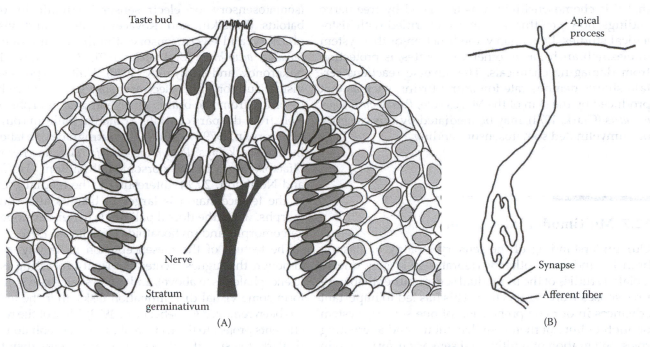

FIGURE 12.29
Line drawings of (A) an elasmobranch taste bud (Cook and Neal, 1921; Whitear and Moate, 1994a) and (B) solitary chemosensory cell (Whitear and Moate, 1994b). (From Gardiner, J.M., *Multisensory Integration in Shark Feeding Behavior*, University of South Florida, Tampa, 2011. With permission.)

12.6.4 Solitary Chemosensory Cells

Solitary chemosensory cells (SCCs) are found in a number of lower vertebrate taxa. These spindle-shaped, epidermal sensory cells are found protruding between the squamous cells of the superficial layer of the epidermis, with a single apical process that bears one or a few microvilli (fish, amphibians) or many microvilli (oligovillous cells in lampreys), and are innervated by spinal or cranial (VII, facial) nerves (reviewed in Kotrschal, 1995) (Figure 12.29B). Their structure resembles that of taste buds, suggesting a chemosensory function, verified through electrophysiological experiments on teleost fish (Peters et al., 1987; Silver and Finger, 1984) and lampreys (Baatrup and Doving, 1985) which demonstrated that they are sensitive to skin washes and bile from other fish, but not amino acids. It has been hypothesized that in rocklings SCCs allow for bulk water sampling, mainly for detecting the presence of predators upstream (Kotrschal et al., 1996), while in sea robins they may be used to find food (Silver and Finger, 1984). SCCs have been examined in only a handful of species; thus, their biological function remains poorly understood, particularly in elasmobranchs, and to date no term for the sense that they mediate has been developed. In elasmobranchs, SCCs have only been confirmed in one species, the thornback ray, *Raja clavata* (Whitear and Moate, 1994b), where they are found in the oral cavity. It has recently been suggested, however, that they may be present on the dorsolateral surface of the skin, near the pit organs, in Port Jackson sharks, *Heterodontus portusjacksoni*, and in whitetip reef sharks, *Triaenodon obesus* (Peach, 2005). Further work is needed to determine the distribution and function of SCCs in elasmobranchs.

12.6.5 Common Chemical Sense

The common chemical sense, the ability to detect irritating substances, is considered separate from olfaction and gustation. Free nerve endings, which in fish occur in the oral and nasal cavities, as well as all over the skin, serve as receptors (Tester, 1963a). Studies in other vertebrates indicate that the nerves involved in such reactions are part of the somatosensory system and appear to represent a subset of temperature- and pain-sensitive fibers, including spinal nerves and cranial nerves V (trigeminal), VII (facial), IX (glossopharyngeal), and vagus (X). The sense conveyed by these chemosensitive components has been renamed *chemesthesis* to reflect this relationship (Bryant and Silver, 2000).

Studies in *Mustelus canis* demonstrated that sharks respond behaviorally to injections of certain chemicals (irritants) into the nostrils, even with the olfactory tracts severed. In these cases, detection was through components of the maxillary branch of the trigeminal nerve (Sheldon, 1909). The animals reacted similarly to applications on the body surface. The latter responses were triggered via spinal nerves. Sheldon (1909) considered that this chemosensitivity was mediated by free nerve endings; however, this has not been verified with histological studies. Presumably, the function of this system in elasmobranchs, as in other vertebrates, is protection from damaging chemicals. The adverse reactions certain sharks demonstrate to natural toxins, such as that produced by the skin of the Moses sole, *Pardachirus marmoratus* (Clark, 1974), may be mediated by this category of unmyelinated somatosensory ending.

12.7 Multimodal Integration

Our understanding of the sensory biology of elasmobranchs and most other vertebrates is largely due to isolated studies of the individual senses rather than multiple senses working together. This has led to important advances in our comprehension of one sensory system or another but not their complementary and alternating roles. Integration of multimodal sensory information in the elasmobranch CNS ultimately leads to a behavioral response at the level of the whole animal. How sharks, skates, and rays integrate the complex input of environmental information through their various senses to form an adaptive response is among the most interesting questions in elasmobranch sensory biology.

12.7.1 Multimodal Integration in the Brain

Early studies (reviewed in Aronson, 1963) concluded there was little multisensory integration in the elasmobranch brain, and those conclusions influenced the naming of the brain regions; for example, the tectum of the mesencephalon was called the *optic tectum*, as it was presumed to be dominated by vision, and the telencephalon was called the *olfactory lobe*, as it was presumed to be dominated by olfactory inputs (Ariëns Kappers et al., 1936). However, electrophysiology has revealed areas of the telencephalon that show responses to multiple sensory stimuli. The pallium, or roof, of the telencephalon can be divided into lateral, medial, and dorsal portions. The lateral pallium has been found to be dominated by olfaction (Hoffman and Northcutt, 2008; Smeets, 1983), while the dorsal (or general) pallium and medial pallium appear to be multisensory. The dorsal pallium is the site of recordings in response to visual (optic nerve) and trigeminal nerve stimuli, which may represent cutaneous mechanoreceptors or electroreceptors, in the nurse shark, *Ginglymostoma cirratum* (Cohen et al., 1973; Ebbesson, 1980). The medial (or hippocampal) pallium has been found to respond to visual, electrosensory, and lateral line stimuli in the little skate, *Leucoraja erinacea* (Bodznick and Northcutt, 1984), visual and cutaneous

(somatosensory or electrosensory) stimuli in other batoids (Veselkin and Kovacevic, 1973), and visual, olfactory, and electrosensory stimuli in spiny dogfish, *Squalus acanthias* (Bodznick, 1991; Nikaronov, 1983; Nikaronov and Lukyanov, 1980). Units responsive to visual, auditory, and electrosensory stimuli have been recorded from the brains of several galeomorphs, possibly from the pars centralis or medial pallium (Bullock and Corwin, 1979). Additionally, retrograde dye labeling in thornback rays, *Platyrhinoidis triseriata*, has revealed olfactory areas in the dorsomedial pallium (Hoffman and Northcutt, 2008). Interestingly, the medial portion of the telencephalon is larger in batoids and squalomorphs, while the dorsal pallium is better developed in galeomorphs and myliobatoids (Northcutt, 1978).

The tectum of the mesencephalon is heavily visual, although the highest center of visual processing is the telencephalon (see above), and nurse sharks can still perform some visual discrimination tasks after the tectum has been removed (Graeber et al., 1973). Most of the retinal efferents project to the tectum of the mesencephalon, particularly the superficial tectal laminae, where they form a topographic map (reviewed in Bodznick, 1991; Hueter, 1991). The deeper layers, however, are multimodal. The electrosensory and mechanosensory medullar nuclei project to a nucleus in the roof of the midbrain, called the *lateral mesencephalic nucleus* (Boord and Northcutt, 1982, 1988). Recordings from the tectum have been made in response to electrosensory, common cutaneous, and auditory stimuli in several species of rays and sharks (Platt et al., 1974), and single multimodal (visual, electrosensory, tactile/lateral line) neurons have been found in the tectum of the little skate, *Leucoraja erinacea* (Bodznick, 1991). Hoffman and Northcutt's (2008) retrograde dye labeling study on thornback rays, *Platyrhinoidis triseriata*, suggested that olfactory, electrosensory, and mechanosensory (lateral line) information converges in the lateral mesencephalic nucleus. Although this has yet to be confirmed with electrophysiology, all of these senses are important for locating prey buried in the substrate.

12.7.2 Multimodal Integration in Behavior

A biological target (prey item, predator, or potential mate) might simultaneously emit several signals: odor; a hydrodynamic disturbance (sound), such as from gill movements or tail beats (reviewed in Bleckmann, 1994); or a weak electrical field (Kalmijn, 1972) (summarized in Figure 12.30). Based on the threshold of the elasmobranch electrosensory system (reviewed above under Section 12.5) for electric fields produced by aquatic animals (Kalmijn, 1972), the limit of detection for most bioelectric stimuli translates to a distance of less than a meter from the source. The detection limits of the visual system of most aquatic animals are not well known but

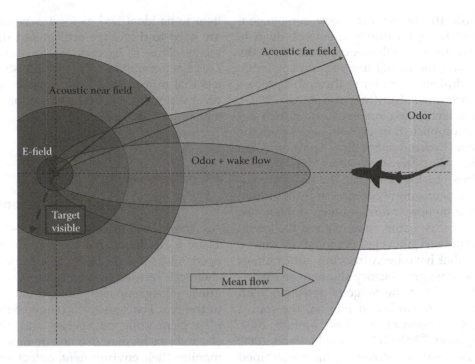

FIGURE 12.30

Summary of the hypothetical stimulus fields emitted by a biological target (small dark-gray circle) in an unbounded, laminar flow environment. In the natural world, any number of environmental, physical, or biological variables could attenuate any of these sensory inputs to the elasmobranch. In very clear, well-lit waters, the visual stimulus could range much farther than depicted, and the acoustic regime is frequency dependent, such that low-frequency sounds will extend over a greater range, possibly even as far as olfaction, and the near-/far-field boundary will be found at a greater distance from the source. (From Gardiner, J.M., *Multisensory Integration in Shark Feeding Behavior*, University of South Florida, Tampa, 2011. With permission.)

depend on the amount of available light, the amount of scatter (Duntley, 1963; Mazur and Beauchamp, 2003), and the background contrast in intensity, polarization, and pattern of reflected light (Johnsen, 2005; Johnsen and Sosik, 2004). In clear, well-lit waters the visual detection range rarely exceeds tens of meters. Sound can be divided into the near acoustic field, primarily particle motion detected in teleosts by the otoliths as particle acceleration (Kalmijn, 1988b; Schellart and Popper, 1992), and the far field, primarily pressure transmitted in teleosts by the swimbladder to the inner ear (Popper and Fay, 1999). For a dipole sound source, the near field dominates at a distance from the source of less than one sixth of the wavelength of the sound ($\lambda/2\pi$); for a sound of 100 Hz, a frequency in the hearing range of many fishes, this translates to approximately 2.5 m (Kalmijn, 1988b). The maximum range of detection of the lateral line has been shown to be one to two body lengths from the source (Coombs et al., 2001).

Odor, on the other hand, may be carried a great distance from the source by the mean flow. In flowing water, odors are dispersed by two mechanisms: advection and turbulent mixing (reviewed in Webster, 2007). Advection refers to the transportation of a filament or patch of odor by the mean or bulk flow. Turbulent flow generates swirling packets, referred to as *eddies*, that

break up into a series of successively smaller eddies through a process known as the *Kolmogorov cascade* (reviewed in Weissburg, 2000). The hydrodynamic motion of these eddies can be detected by the lateral line system. Intermolecular viscous forces dissipate the energy until they reach the smallest eddy size that still contains turbulent energy, known as the *Kolmogorov length scale*, on the order of millimeters. Beyond this scale, in the odor far field, only very patchy odor information is available, carried by the bulk flow (Figure 12.30). Thus, locating a biological target involves: (1) initially detecting and orienting to a patchy odor field, (2) tracking the odor plume, and (3) localizing and orienting to the target. In the case of food, this progression culminates with striking at and capturing the prey.

The sequence in which each of the sensory modalities comes into play depends on a multitude of factors, however, and there is no single sensory hierarchy that operates under all circumstances for all elasmobranch species. How animals use sensory information depends not only on what sensory stimuli are available, as determined by the animal's proximity to the prey, the physics of the stimulus fields (Figure 12.30), and the thresholds of detection for each species, but also on which stimulus or stimuli the animal chooses to focus upon when information from multiple senses is available simultaneously.

For a given task, the senses may have complementary or alternating roles; for example, smooth dogfish, *Mustelus canis*, requires simultaneous input from the olfactory system and the lateral line to precisely locate the source of a turbulent odor plume, through a process known as *eddy chemotaxis* (Gardiner and Atema, 2007). On the other hand, navigating large-scale flow, such as a current, can be accomplished using either cues from the lateral line system (hydrodynamic flow field) or vision (visual flow field) (Gardiner and Atema, 2007). As an animal approaches a biological target, the sensory environment becomes increasingly complex and the animal might either integrate new information encountered in an additive fashion (e.g., using olfaction and the lateral line for eddy chemotaxis, above) or demonstrate *sensory switching*. Sharks that have been tracking odor plumes switch their focus from an olfactory signal to an electrical signal once it is within the range of detection, with a sudden sharp turn toward, and bite on, the source of the electric field (Jordan et al., 2009b; Kajiura, 2003; Kajiura and Fitzgerald, 2009; Kajiura and Holland, 2002; Kalmijn, 1982). Few studies, however, have examined more than one or two senses at a time.

Recently, Gardiner (2011) conducted a study on three species of sharks, examining their ability to capture live, tethered prey items after selective blocks of the visual, olfactory, lateral line, and electrosensory systems. Nurse sharks, *Ginglymostoma cirratum*, rely heavily on olfaction to feed. This species is capable of orienting to the prey using non-olfactory cues but will not ingest the food unless an attractive odor is present. This suggests that this species relies entirely on olfactory cues to confirm the identity of a target as food. Bonnetheads, *Sphyrna tiburo*, generally rely on an attractive odor to initiate tracking behaviors but will strike at prey based on visual cues. The electrosensory system, however, is essential for a successful strike. Animals with the electrosensory system blocked are capable of precisely lining up a strike, based on visual cues, but the jaws do not begin to move without the appropriate electrical cues, and thus the prey is not ingested. Blinded animals display an inability to line up a strike, suggesting that orientation to the prey is visually mediated. Blacktip sharks, *Carcharhinus limbatus*, also rely on olfactory cues to perform tracking behaviors. They demonstrate sensory switching, focusing on visual cues to line up a rapid, ram strike. If the olfactory system is blocked, they will strike visually if their swimming motions bring them within visual range of the prey. They can successfully orient to prey using non-visual cues, but the process is much slower and typically involves a number of misses before a successful capture. If olfaction and vision are simultaneously blocked, feeding behaviors cease altogether. This suggests that appropriate olfactory or visual cues are essential in these species for an item to be identified as food, whereas non-visual cues are used to direct the strikes and time the jaw movements. This is in contrast to short-tailed stingrays, *Dasyatis brevicaudata*, which will strike at weak water jets that mimic the hydrodynamic signature of buried bivalve prey in the absence of odor cues (Montgomery and Skipworth, 1997).

12.8 Summary and Conclusions

Are sharks and their relatives sensory marvels or not? There is no doubt that the combination of well-developed visual, acoustical, mechanical, electrical, and chemical sensing systems in elasmobranchs distinguishes the group and makes them well adapted for life in the sea. The sensory ecology of these fishes is complex. Depending on species and ambient conditions, elasmobranchs may use one or more of their senses to monitor their environment, detect and locate prey and mates, avoid predators, and find their way in the ocean.

Our understanding of these sensory processes progressed rapidly in the latter half of the 1900s. A lull in this research toward the end of the century has been replaced with a renewed interest in the field, which has been gaining increasing momentum over the last decade. Exciting new technologies have opened the door for fine-scale investigations into the behavior and ecology of these animals, both in captivity and in the wild (see Chapter 9). How sharks, skates, and rays integrate complex multimodal environmental information through their various senses and which cues they choose to focus on to form an adaptive response are among the most interesting questions left in elasmobranch sensory biology. Investigations into multimodal integration have begun, but this remains a ripe area for further research.

Acknowledgments

The authors wish to thank Jelle Atema, Samuel Gruber, Joel Cohen, William Tavolga, and Timothy Tricas for many useful discussions and collaborations in elasmobranch sensory biology. The work of JMG and REH has been supported by funds from the University of South Florida, Mote Marine Laboratory, and the National Science Foundation. LSD's work on elasmobranchs has been supported by grants from the National Science Foundation and the Florsheim endowment to the New College Foundation.

References

Aalbers, S.A., Bernal, D., and Sepulveda, C.A. (2010). The functional role of the caudal fin in the feeding ecology of the common thresher shark *Alopias vulpinus*. *J. Fish Biol.* 76:1863–1868.

Abel, R.L., Maclaine, J.S., Cotton, R. et al. (2010). Functional morphology of the nasal region of a hammerhead shark. *Comp. Biochem. Physiol. A Comp. Physiol.* 155(4):464–475.

Andres, K.H. (1970). Anatomy and ultrastructure of the olfactory bulb in fish, amphibia, reptiles, birds and mammals. In: Wolstenhome, G.E.W. and Knight, J. (Eds.), *Taste and Smell in Vertebrates*. Churchill, London, pp. 177–196.

Andrianov, G.N., Broun, G.R., Il'inskii, O.B., and Muraveiko, V.M. (1984). Frequency characteristics of skate electroreceptive central neurons responding to electric and magnetic stimulation. *Neurophysiology* 16:365–376.

Ariëns Kappers, C.U., Huber, G.C., and Crosby, E.C. (1936). *The Comparative Anatomy of the Nervous System of Vertebrates, Including Man*. Hafner Publishing, New York.

Aronson, L.R. (1963). The central nervous system of sharks and bony fishes with special reference to sensory and integrative mechanisms. In: Gilbert, P.W. (Ed.), *Sharks and Survival*. Heath, Boston, pp. 165–241.

Aronson, L.R., Aronson, F.R., and Clark, E. (1967). Instrumental conditioning and light-dark discrimination in young nurse sharks. *Bull. Mar. Sci.* 17:249–256.

Atema, J. (1996). Eddy chemotaxis and odor landscapes: exploration of nature with animal sensors. *Biol. Bull.* 191:129–138.

Baatrup, E. and Doving, K.B. (1985). Physiological studies on solitary receptors of the oral disk papillae in adult brook lamprey, *Lampetra planeri* (Bloch). *Chem. Senses* 10:559–566.

Banner, A. (1967). Evidence of sensitivity to acoustic displacements in the lemon shark, *Negaprion brevirostris* (Poey). In: Cahn, P. (Ed.), *Lateral Line Detectors*. Indiana University Press, Bloomington, pp. 265–273.

Barber, V.C., Yake, Y.I., Clark, V.F., and Pungur, J. (1985). Quantitative analyses of sex and size differences in the macula neglecta and ramus neglectus in the inner ear of the skate, *Raja ocellata*. *Cell Tissue Res.* 241:597–605.

Barry, M.A. (1987). Afferent and efferent connections of the primary octaval nuclei in the clearnose skate, *Raja eglanteria*. *J. Comp. Neurol.* 266:457–477.

Barry, M.A. and Bennett, M.V.L. (1989). Specialized lateral line receptor systems in elasmobranchs: the spiracular organs and vesicles of Savi. In: Coombs, S., Görner, P., and Münz, H. (Eds.), *The Mechanosensory Lateral Line: Neurobiology and Evolution*. Springer-Verlag, New York, pp. 591–606.

Barry, M.A., Hall, D.H., and Bennett, M.V.L. (1988a). The elasmobranch spiracular organ I. Morphological studies. *J. Comp. Physiol. A Neuroethol. Sens. Neural Behav. Physiol.* 163:85–92.

Barry, M.A., Hall, D.H., and Bennett, M.V.L. (1988b). The elasmobranch spiracular organ II. Physiological studies. *J. Comp. Physiol. A Neuroethol. Sens. Neural Behav. Physiol.* 163:93–98.

Bass, A. (1981). Olfactory bulb efferents in the channel catfish, *Ictalurus punctatus*. *J. Morphol.* 169:91–111.

Bateson, W. (1890). The sense-organs and perceptions of fishes; with some remarks on the supply of bait. *J. Mar. Biol. Assoc. U.K.* 1:225–256.

Bell, M.A. (1993). Convergent evolution of nasal structure in sedentary elasmobranchs. *Copeia* 1993:144–158.

Bennett, M.V.L. (1971). Electroreception. In: Hoar, W.S. and Randall, D.J. (Eds.), *Fish Physiology*, Vol. 5. Academic Press, New York, pp. 493–574.

Blaxter, J.H.S. and Fuiman, L.A. (1989). Function of the free neuromasts of marine teleost larvae. In: Coombs, S. et al. (Eds.), *The Mechanosensory Lateral Line: Neurobiology and Evolution*. Springer-Verlag, New York, pp. 481–499.

Bleckmann, H. (1994). *Reception of Hydrodynamic Stimuli in Aquatic and Semi-Aquatic Environments*. Gustav Fischer Verlag, New York.

Bleckmann, H. (2008). Peripheral and central processing of lateral line information. *J. Comp. Physiol. A Neuroethol. Sens. Neural Behav. Physiol.* 194:145–148.

Bleckmann, H. and Bullock, T.H. (1989). Central nervous physiology of the lateral line, with special reference to cartilaginous fishes. In: Coombs, S., Görner, P., and Münz, H. (Eds.), *The Mechanosensory Lateral Line: Neurobiology and Evolution*. Springer-Verlag, New York, pp. 387–408.

Bleckmann, H., Bullock, T.H., and Jorgensen, J.M. (1987). The lateral line mechanoreceptive mesencephalic, diencephalic, and telencephalic regions in the thornback ray, *Platyrhinoidis triseriata* (Elasmobranchii). *J. Comp. Physiol. A Neuroethol. Sens. Neural Behav. Physiol.* 161:67–84.

Bleckmann, H., Weiss, O., and Bullock, T.H. (1989). Physiology of lateral line mechanoreceptive regions in the elasmobranch brain. *J. Comp. Physiol. A Neuroethol. Sens. Neural Behav. Physiol.* 164:459–474.

Blonder, B.I. and Alevizon, W.S. (1988). Prey discrimination and electroreception in the stingray *Dasyatis sabina*. *Copeia* 1988:33–36.

Bodznick, D. (1991). Elasmobranch vision: multimodal integration in the brain. *J. Exp. Zool.* 5(Suppl.):108–116.

Bodznick, D. and Boord, R.L. (1986). Electroreception in Chondrichthyes. In: Bullock, T.H. and Heiligenberg, W. (Eds.), *Electroreception*. John Wiley & Sons, New York, pp. 225–256.

Bodznick, D. and Montgomery, J.C. (1992). Suppression of ventilatory reference in the elasmobranch electrosensory system: medullary neuron receptive fields support a common mode rejection mechanism. *J. Exp. Biol.* 171:127–137.

Bodznick, D. and Northcutt, R.G. (1980). Segregation of electro- and mechanoreceptive inputs to the elasmobranch medulla. *Brain Res.* 195:313–321.

Bodznick, D. and Northcutt, R.G. (1984). An electrosensory area in the telencephalon of the little skate, *Raja erinacea*. *Brain Res.* 298:117–124.

Bodznick, D. and Schmidt, A.W. (1984). Somatotopy within the medullary electrosensory nucleus of the skate, *Raja erinacea*. *J. Comp. Neurol.* 225:581–590.

Bodznick, D., Montgomery, J.C., and Bradley, D.J. (1992). Suppression of common mode signals within the electrosensory system of the little skate, *Raja erinacea*. *J. Exp. Biol.* 171:107–125.

Bodznick, D., Montgomery, J.C., and Cary, M. (1999). Adaptive mechanism in the elasmobranch hindbrain. *J. Exp. Biol.* 202:1357–1364.

Boord, R.L. and Campbell, C.B.G. (1977). Structural and functional organization of the lateral line system of sharks. *Am. Zool.* 17:431–441.

Boord, R.L. and Montgomery, J.C. (1989). Central mechanosensory lateral line centers and pathways among the elasmobranchs. In: Coombs, S., Görner, P., and Münz, H. (Eds.), *The Mechanosensory Lateral Line: Neurobiology and Evolution.* Springer-Verlag, New York, pp. 323–340.

Boord, R.L. and Northcutt, R.G. (1982). Ascending lateral line pathways to the midbrain of the clearnose skate, *Raja eglanteria. J. Comp. Neurol.* 207:274–282.

Boord, R.L. and Northcutt, R.G. (1988). Medullary and mesencephalic pathways and connections of lateral line neurons in the spiny dogfish *Squalus acanthias. Brain Behav. Evol.* 32:76–88.

Bozzano, A. (2004). Retinal specialisations in the dogfish *Centroscymnus coelolepis* from the Mediterranean deepsea. *Sci. Mar. (Barc.)* 68(Suppl. 3):185–195.

Bozzano, A. and Collin, S.P. (2000). Retinal ganglion cell topography in elasmobranchs. *Brain Behav. Evol.* 55:191–208.

Bratton, B.O. and Ayers, J.L. (1987). Observations on the electric discharge of two skate species (Chondrichthyes: Rajidae) and its relationship to behavior. *Environ. Biol. Fish.* 20:241–254.

Broun, G.R., Il'inskii, O.B., and Krylov, B.V. (1979). Responses of the ampullae of Lorenzini in a uniform electric field. *Neurophysiology* 11:118–124.

Brown, B.R. (2003). Sensing temperature with ion channels. *Nature* 421:495.

Brown, B.R. (2010). Temperature response in electrosensors and thermal voltages in electrolytes. *J. Biol. Phys.* 36:121–134.

Brown, B.R., Hughes, M.E., and Russo, C. (2004). Thermoelectricity in natural and synthetic hydrogels. *Phys. Rev. E* 70:319171–319177.

Brown, B.R., Hughes, M.E., and Russo, C. (2005). Infrastructure in the electric sense: admittance data from shark hydrogels. *J. Comp. Physiol. A Neuroethol. Sens. Neural Behav. Physiol.* 191:115–123.

Bruckmoser, P. and Dieringer, N. (1973). Evoked potentials in the primary and secondary olfactory projection areas of the forebrain in Elasmobranchia. *J. Comp. Physiol.* 87:65–74.

Bryant, B. and Silver, W.L. (2000). Chemesthesis: the common chemical sense. In: Finger, T.E., Silver, W.L., and Restrepo, D. (Eds.), *The Neurobiology of Taste and Smell*, 2nd ed. Wiley-Liss, New York, pp. 73–98.

Budker, P. (1958). Les organes sensoriels cutanés des sélaciens. In: Grassé, P.P. (Ed.), *Traité de Zoologie. Anatomie, Systémique, Biologie.* Tome XIII. *Agnathes et Poissons.* Masson et Cie, Paris, pp. 1033–1062.

Bullock, T.H. (1979). Processing of ampullary input in the brain: comparisons of sensitivity and evoked responses among siluroids and elasmobranchs. *J. Physiol. (Paris)* 75:315–317.

Bullock, T.H. and Corwin, J.T. (1979). Acoustic evoked activity in the brain in sharks. *J. Comp. Physiol.* 129:223–234.

Carpenter, D.O. (1981). Ionic and metabolic bases of neuronal thermosensitivity. *Fed. Proc.* 40:2808–2813.

Carrier, J.C., Pratt, Jr., H.L., and Martin, L.K. (1994). Group reproductive behaviors in free-living nurse sharks, *Ginglymostoma cirratum. Copeia* 1994:646–656.

Casper, B.M. and Mann, D.A. (2006). Evoked potential audiograms of the nurse shark (*Ginglymostoma cirratum*) and the yellow stingray (*Urobatis jamaicensis*). *Environ. Biol. Fish.* 76:101–108.

Casper, B.M. and Mann, D.A. (2007a). Dipole hearing measurements in elasmobranch fishes. *J. Exp. Biol.* 210:75–81.

Casper, B.M. and Mann, D.A. (2007b). The directional hearing abilities of two species of bamboo sharks. *J. Exp. Biol.* 210:505–511.

Casper, B.M. and Mann, D.A. (2009). Field hearing measurements of the Atlantic sharpnose shark *Rhizoprionodon terraenovae. J. Fish Biol.* 75:2768–2776.

Casper, B.M., Lobel, P.S., and Yan, H.Y. (2003). The hearing sensitivity of the little skate, *Raja erinacea*: a comparison of two methods. *Environ. Biol. Fish.* 68:371–379.

Chu, Y.T. and Wen, W.C. (1979). A study of the lateral-line canal system and that of Lorenzini ampullae and tubules of elasmobranchiate fishes of China. In: *Monograph of Fishes of China*, Vol. 2. Science and Technology Press, Shanghai, pp. 117–126.

Cinelli, A.R. and Salzberg, B.M. (1990). Multiple optical recording of transmembrane voltage (MSORTV), single-unit recordings, and evoked potentials from the olfactory bulb of skate (*Raja erinacea*). *J. Neurophysiol.* 64:1767–1790.

Clark, E. (1959). Instrumental conditioning of sharks. *Science* 130:217–218.

Clark, E. (1963). Maintenance of sharks in captivity with a report on their instrumental conditioning. In: Gilbert, P.W. (Ed.), *Sharks and Survival.* Heath, Boston, pp. 115–149.

Clark, E. (1974). The red sea's sharkproof fish. *Natl. Geog.* 146:719–727.

Cohen, D.H., Duff, T.A., and Ebbesson, S.O.E. (1973). Electrophysiological identification of a visual area in the shark telencephalon. *Science* 182:492–494.

Cohen, J.L. (1991). Adaptations for scotopic vision in the lemon shark (*Negaprion brevirostris*). *J. Exp. Zool.* 5(Suppl.):76–84.

Cohen, J.L., Gruber, S.H., and Hamasaki, D.I. (1977). Spectral sensitivity and Purkinje shift in the retina of the lemon shark, *Negaprion brevirostris* (Poey). *Vision Res.* 17:787–792.

Cohen, J.L., Hueter, R.E., and Organisciak, D.T. (1990). The presence of a porphyropsin-based visual pigment in the juvenile lemon shark (*Negaprion brevirostris*). *Vision Res.* 30:1949–1953.

Collin, S.P. (1988). The retina of the shovel-nosed ray, *Rhinobatos batillum* (Rhinobatidae): morphology and quantitative analysis of the ganglion, amacrine and bipolar cell populations. *Exp. Biol.* 47:195–207.

Collin, S.P. (1999). Behavioural ecology and retinal cell topography. In: Archer, S.N., Djamgoz, M.B.A., Loew, E.R., Partridge, J.C., and Vallerga, S. (Eds.), *Adaptive Mechanisms in the Ecology of Vision.* Kluwer, Dordrecht, pp. 509–535.

Conley, R.A. and Bodznick, D. (1994). The cerebellar dorsal granular ridge in an elasmobranch has proprioceptive and electroreceptive representations and projects homotopically to the medullary electrosensory nucleus. *J. Comp. Physiol. A Neuroethol. Sens. Neural Behav. Physiol.* 174:707–721.

Cook, M.H. and Neal, H.V. (1921). Are taste buds of elasmobranchs endodermal in origin? *J. Comp. Neurol.* 33:45–63.

Coombs, S., Braun, C.B., and Donovan, B. (2001). The orienting response of Lake Michigan mottled sculpin is mediated by canal neuromasts. *J. Exp. Biol.* 204(2):337–348.

Coombs, S. and Janssen, J. (1990). Behavioral and neurophysiological assessment of lateral line sensitivity in the mottled sculpin, *Cottus bairdi. J. Comp. Physiol. A Neuroethol. Sens. Neural Behav. Physiol.* 167:557–567.

Coombs, S. and Montgomery, J.C. (1999). The enigmatic lateral line system. In: Fay, R.R. and Popper, A.N. (Eds.), *Comparative Hearing: Fish and Amphibians.* Springer-Verlag, New York, pp. 319–362.

Corwin, J.T. (1977). Morphology of the macula neglecta in sharks of the genus *Carcharhinus. J. Morphol.* 152:341–362.

Corwin, J.T. (1978). The relation of inner ear structure to the feeding behavior in sharks and rays. In: Johari, O.M. (Ed.), *Scanning Electron Microscopy,* Vol. 2. SEM, Chicago, pp. 1105–1112.

Corwin, J.T. (1981). Audition in elasmobranchs. In: Tavolga, W.N., Popper, A.N., and Fay, R.R. (Eds.), *Hearing and Sound Communication in Fishes.* Springer-Verlag, New York, pp. 81–105.

Corwin, J.T. (1983). Postembryonic growth of the macula neglecta auditory detector in the ray, *Raja clavata*: continual increases in hair cell number, neural convergence, and physiological sensitivity. *J. Comp. Neurol.* 217:345–356.

Corwin, J.T. (1989). Functional anatomy of the auditory system in sharks and rays. *J. Exp. Zool.* 2(Suppl.):62–74.

Corwin, J.T. and Northcutt, R.G. (1982). Auditory centers in the elasmobranch brain stem: deoxyglucose autoradiography and evoked potential recording. *Brain Res.* 236:261–273.

Cox, D.L. and Koob, T.J. (1993). Predation on elasmobranch eggs. *Environ. Biol. Fish.* 38:117–125.

Daniel, J.F. (1928). *The Elasmobranch Fishes.* University of California Press, Berkeley.

Demski, L.S. (1977). Electrical stimulation of the shark brain. *Am. Zool.* 17:487–500.

Demski, L.S. (1981). Hypothalamic mechanisms of feeding in fishes. In: Laming, P.J. (Ed.), *Brain Mechanisms of Behaviour in Lower Vertebrates.* Cambridge University Press, Cambridge, U.K., pp. 225–237.

Demski, L.S. (1982). A hypothalamic feeding area in the brains of sharks and teleosts. *Fla. Sci.* 45:34–40.

Demski, L.S. (1983). Behavioral effects of electrical stimulation of the brain. In: Davis, R.E. and Northcutt, R.G. (Eds.), *Fish Neurobiology,* Vol. 2. University of Michigan Press, Ann Arbor, pp. 317–359.

Demski, L.S. (1991). Elasmobranch reproductive behavior: implications for captive breeding. *J. Aquaricult. Aquat. Sci.* 5:84–95.

Demski, L.S. and Northcutt, R.G. (1996). The brain and cranial nerves of the white shark: an evolutionary perspective. In: Klimley, A.P. and Ainley, D.G. (Eds.), *Great White Sharks: The Biology of Carcharodon carcharias.* Academic Press, San Diego, pp. 121–130.

Denton, E.J. and Gray, J.A.B. (1983). Mechanical factors in the excitation of clupeid lateral lines. *Proc. R. Soc. Lond. B Biol. Sci.* 218(1210):1–26.

Denton, E.J. and Gray, J.A.B. (1988). Mechanical factors in the excitation of the lateral lines of fishes. In: Atema, J. et al. (Eds.), *Sensory Biology of Aquatic Animals.* Springer-Verlag, New York, pp. 595–617.

Denton, E.J. and Nicol, J.A.C. (1964). The chorioidal tapeta of some cartilaginous fishes (Chondrichthyes). *J. Mar. Biol. Assoc. U.K.* 44:219–258.

Denton, E.J. and Shaw, T.I. (1963). The visual pigments of some deep-sea elasmobranchs. *J. Mar. Biol. Assoc. U.K.* 43:65–70.

Dijkgraaf, S. and Kalmijn, A.J. (1962). Verhaltensversuche zur Funktion der Lorenzinischen Ampullen. *Naturwissenschaften* 49:400.

Dotterweich, H. (1932). Baud und Funktion der Lorenzinischen Ampullen. *Zool. Jahrb. Abt.* 50:347–418.

Dowling, J.E. and Ripps, H. (1991). On the duplex nature of the skate retina. *J. Exp. Zool.* 5(Suppl.):55–65.

Doyle, J. (1963). The acid mucopolysaccharides in the glands of Lorenzini of elasmobranch fish. *Biochem. J.* 88:7–8.

Dryer, L. and Graziadei, P.P.C. (1993). A pilot study on morphological compartmentalization and heterogeneity in the elasmobranch olfactory bulb. *Anat. Embryol. (Berl.)* 188:41–51.

Dryer, L. and Graziadei, P.P.C. (1994a). Mitral cell dendrites: a comparative approach. *Anat. Embryol. (Berl.)* 189:91–106.

Dryer, L. and Graziadei, P.P.C. (1994b). Projections of the olfactory bulb in an elasmobranch fish, *Sphyrna tiburo*: segregation of inputs to the telencephalon. *Anat. Embryol. (Berl.)* 190:563–572.

Dryer, L. and Graziadei, P.P.C. (1996). Synaptology of the olfactory bulb of an elasmobranch fish, *Sphyrna tiburo. Anat. Embryol. (Berl.)* 193:101–114.

Duntley, S.Q. (1963). Light in the sea. *J. Opt. Soc. Am.* 53:214–233.

Ebbesson, S.O.E. (1972). New insights into the organization of the shark brain. *Comp. Biochem. Physiol. A Comp. Physiol.* 42:121–129.

Ebbesson, S.O.E. (1980). On the organization of the telencephalon in elasmobranchs. In: Ebbesson, S.O.E. (Ed.), *Comparative Neurology of the Telencephalon.* Plenum Press, New York, pp. 1–16.

Ebbesson, S.O.E. and Heimer, L. (1970). Projections of the olfactory tract fibers in the nurse shark (*Ginglymostoma cirratum*). *Brain Res.* 17:47–55.

Ebbesson, S.O.E. and Northcutt, R.G. (1976). Neurology of anamniotic vertebrates. In: Masterton, R.B., Bitterman, M.E., Campbell, C.B.G., and Hotton, N. (Eds.), *Evolution of Brain and Behavior in Vertebrates.* Erlbaum, Hillsdale, NJ, pp. 115–146.

Ebbesson, S.O.E. and Schroeder, D.M. (1971). Connections of the nurse shark's telencephalon. *Science* 173:254–256.

Evangelista, C., Mills, M., Siebeck, U.E., and Collin, S.P. (2010). A comparison of the external morphology of the membranous inner ear in elasmobranchs. *J. Morphol.* 271:483–495.

Ewart, J.C. and Mitchell, H.C. (1892). On the lateral sense organs of elasmobranchs. II. The sensory canals of the common skate (*Raja batis*). *Trans. R. Soc. Edinburgh* 37:87–105.

Fay, R.R., Kendall, J.I., Popper, A.N., and Tester, A.L. (1974). Vibration detection by the macula neglecta of sharks. *Comp. Biochem. Physiol. A Physiol.* 47:1235–1240.

Fiebig, E. (1988). Connections of the corpus cerebelli in the thornback guitarfish, *Platyrhinoidis triseriata* (Elasmobranchii): a study with WGA-HRP and extracellular granule cell recording. *J. Comp. Neurol.* 268:567–583.

Finger, T.E. (1975). Distribution of the olfactory tracts in the bullhead catfish, *Ictalurus nebulosus*. *J. Comp. Neurol.* 161:125–142.

Fouts, W.R. and Nelson, D.R. (1999). Prey capture by the Pacific angel shark, *Squatina californica*: visually mediated strikes and ambush-site characteristics. *Copeia* 1999:304–312.

Francheschini, V. and Ciani, F. (1993). Lectin binding to the olfactory system in a shark, *Scyliorhinus canicula*. *Fol. Histochem. Cytobiol.* 31:133–137.

Franz, V. (1931). Die Akkommodation des Selachierauges und seine Abblendungsapparate, nebst Befunden an der Retina. *Zool. Jahrb. Abt. Allg. Zool. Physiol. Tiere* 49:323–462.

Fraser, P.J. and Shelmerdine, R.L. (2002). Dogfish hair cells sense hydrostatic pressure. *Nature* 415:495–496.

Gardiner, J.M. (2011). *Multisensory Integration in Shark Feeding Behavior*. University of South Florida, Tampa.

Gardiner, J.M. and Atema, J. (2007). Sharks need the lateral line to locate odor sources: rheotaxis and eddy chemotaxis. *J. Exp. Biol.* 210:1925–1934.

Gardiner, J.M. and Atema, J. (2010). The function of bilateral timing differences in olfactory orientation of sharks. *Curr. Biol.* 20:1187–1191.

Gilbert, C.R. (1967). A revision of the hammerhead sharks (Family Sphyrnidae). *Proc. U.S. Natl. Mus.* 119:1–88.

Gilbert, P.W. (1963). The visual apparatus of sharks. In: Gilbert, P.W. (Ed.), *Sharks and Survival*. Heath, Boston, pp. 283–326.

Gordon, I. (1993). Pre-copulatory behaviour of captive sandtiger sharks, *Carcharias taurus*. *Environ. Biol. Fish.* 38:159–164.

Gould, J.L., Kirschvink, J.L., and Deffeyes, K.D. (1978). Bees have magnetic remanence. *Science* 201:1026–1028.

Graeber, R.C. (1978). Behavioral studies correlated with central nervous system integration of vision in sharks. In: Hodgson, E.S. and Mathewson, R.F. (Eds.), *Sensory Biology of Sharks, Skates, and Rays*. U.S. Office of Naval Research, Arlington, VA, pp. 195–225.

Graeber, R.C. (1980). Telencephalic function in elasmobranchs, a behavioral perspective. In: Ebbesson, S.O.E. (Ed.), *Comparative Neurology of the Telencephalon*. Plenum Press, New York, pp. 17–39.

Graeber, R.C. and Ebbesson, S.O.E. (1972). Retinal projections in the lemon shark (*Negaprion brevirostris*). *Brain Behav. Evol.* 5:461–477.

Graeber, R.C., Ebbesson, S.O.E., and Jane, J.A. (1973). Visual discrimination in sharks without optic tectum. *Science* 180:413–415.

Graeber, R.C., Schroeder, D.M., Jane, J.A., and Ebbesson, S.O.E. (1978). Visual discrimination following partial telencephalic ablations in nurse sharks (*Ginglymostoma cirratum*). *J. Comp. Neurol.* 180:325–344.

Gruber, S.H. (1967). A behavioral measurement of dark adaptation in the lemon shark, *Negaprion brevirostris*. In: Gilbert, P.W., Mathewson, R.F., and Rall, D.P. (Eds.), *Sharks, Skates, and Rays*. The Johns Hopkins University Press, Baltimore, MD, pp. 479–490.

Gruber, S.H. and Cohen, J.L. (1978). Visual system of the elasmobranchs: state of the art 1960–1975. In: Hodgson, E.S. and Mathewson, R.F. (Eds.), *Sensory Biology of Sharks, Skates, and Rays*. U.S. Office of Naval Research, Arlington, VA, pp. 11–105.

Gruber, S.H. and Cohen, J.L. (1985). Visual system of the white shark, *Carcharodon carcharias*, with emphasis on retinal structure. *Mem. S. Calif. Acad. Sci.* 9:61–72.

Gruber, S.H. and Schneiderman, N. (1975). Classical conditioning of the nictitating membrane response of the lemon shark (*Negaprion brevirostris*). *Behav. Res. Meth. Instr.* 7:430–434.

Gruber, S.H., Hamasaki, D.I., and Bridges, C.D.B. (1963). Cones in the retina of the lemon shark (*Negaprion brevirostris*). *Vision Res.* 3:397–399.

Hamasaki, D.I. and Gruber, S.H. (1965). The photoreceptors of the nurse shark, *Ginglymostoma cirratum*, and the sting ray, *Dasyatis sayi*. *Bull. Mar. Sci.* 15:1051–1059.

Hara, T.J. (1994). The diversity of chemical stimulation in fish olfaction and gustation. *Rev. Fish Biol. Fish.* 4:1–35.

Harris, A.J. (1965). Eye movements of the dogfish *Squalus acanthias* L. *J. Exp. Biol.* 43:107–130.

Hart, N.S., Lisney, T.J., Marshall, N.J., and Collin, S.P. (2004). Multiple cone visual pigments and the potential for trichromatic colour vision in two species of elasmobranch. *J. Exp. Biol.* 207:4587–4594.

Hart, N.S., Theiss, S.M., Harahush, B.K., and Collin, S.P. (2011). Microspectrophotometric evidence for cone monochromacy in sharks. *Naturwissenschaften* 98:193–201.

Hassan, E.S. (1989). Hydrodynamic imaging of the surroundings by the lateral line of the blind cave fish, *Anoptichthys jordani*. In: Coombs, S., Görner, P., and Münz, H. (Eds.), *The Mechanosensory Lateral Line: Neurobiology and Evolution*. Springer-Verlag, New York, pp. 217–227.

Heath, A.R. (1991). The ocular tapetum lucidum: a model system for interdisciplinary studies in elasmobranch biology. *J. Exp. Zool.* 5(Suppl.):41–45.

Hensel, H. (1955). Quantitative Beziehungen zwischen Temperaturreiz und Aktionspotentialen der Lorenzinischen Ampullen. *Z. Vergl. Physiol.* 37:509–526.

Herrick, C.J. (1924). *Neurological Foundations of Animal Behavior*. Henry Holt and Company; reprint edition 1965 by Hafner, New York.

Heupel, M.R., Simpendorfer, C.A., and Hueter, R.E. (2003). Running before the storm: sharks respond to falling barometric pressure associated with Tropical Storm Gabrielle. *J. Fish Biol.* 63:1357–1363.

Hobson, E.S. (1963). Feeding behavior in three species of sharks. *Pac. Sci.* 17:171–194.

Hodgson, E.S. and Mathewson, R.F. (1971). Chemosensory orientation in sharks. *Ann. N.Y. Acad. Sci.* 188:175–182.

Hodgson, E.S. and Mathewson, R.F. (Eds.). (1978a). *Sensory Biology of Sharks, Skates, and Rays.* U.S. Office of Naval Research, Arlington, VA.

Hodgson, E.S. and Mathewson, R.F. (1978b). Electrophysiological studies of chemoreception in elasmobranchs. In: Hodgson, E.S. and Mathewson, R.F. (Eds.), *Sensory Biology of Sharks, Skates, and Rays.* Office of Naval Research, Arlington, VA, pp. 227–267.

Hodgson, E.S., Mathewson, R.F., and Gilbert, P.W. (1967). Electroencephalographic studies of chemoreception in sharks. In: Gilbert, P.W., Mathewson, R.F., and Rall, D.P. (Eds.), *Sharks, Skates, and Rays.* The Johns Hopkins Press, Baltimore, MD, pp. 491–501.

Hoffman, M.H. and Northcutt, R.G. (2008). Organization of major telencephalic pathways in an elasmobranch, the thornback ray *Platyrhinoidis triseriata. Brain Behav. Evol.* 72:307–325.

Howes, G.B. (1883). The presence of a tympanum in the genus *Raja. J. Anat. Physiol.* 17:188–191.

Hueter, R.E. (1980). Physiological Optics of the Eye of the Juvenile Lemon Shark (*Negaprion brevirostris*), master's thesis, University of Miami, Coral Gables, FL.

Hueter, R.E. (1991). Adaptations for spatial vision in sharks. *J. Exp. Zool.* 5(Suppl.):130–141.

Hueter, R.E. and Cohen, D.H. (1991). Vision in elasmobranchs: a comparative and ecological perspective. *J. Exp. Zool.* 5(Suppl.):1–182.

Hueter, R.E. and Gilbert, P.W. (1990). The sensory world of sharks. In: Gruber, S.H. (Ed.), *Discovering Sharks.* American Littoral Society, Highlands, NJ.

Hueter, R.E. and Gruber, S.H. (1982). Recent advances in studies of the visual system of the juvenile lemon shark (*Negaprion brevirostris*). *Fla. Sci.* 45:11–25.

Hueter, R.E., Murphy, C.J., Howland, M. et al. (2001). Refractive state and accommodation in the eyes of free-swimming versus restrained juvenile lemon sharks (*Negaprion brevirostris*). *Vision Res.* 41:1885–1889.

Johnsen, P.B. and Teeter, J.H. (1985). Behavioral responses of bonnethead sharks (*Sphyrna tiburo*) to controlled olfactory stimulation. *Mar. Behav. Physiol.* 11:283–291.

Johnsen, S. (2005). Visual ecology on the high seas. *Mar. Ecol. Prog. Ser.* 287:281–285.

Johnsen, S. and Sosik, H.M. (2004). Shedding light on light in the sea. *Oceanus* 43:24–28.

Johnson, M.P. and Tyack, P.L. (2003). A digital acoustic recording tag for measuring the response of wild marine mammals to sound. *IEEE J. Ocean. Eng.* 28:3–12.

Johnson, R.H. and Nelson, D.R. (1978). Copulation and possible olfaction-mediated pair formation in two species of carcharhinid sharks. *Copeia* 1978:539–542.

Johnson, S.E. (1917). Structure and development of the sense organs of the lateral canal system of selachians (*Mustelus canis* and *Squalus acanthias*). *J. Comp. Neurol.* 28(1):1–74.

Jordan, L.K. (2008). Comparative morphology of stingray lateral line canal and electrosensory systems. *J. Morphol.* 269:1325–1339.

Jordan, L.K., Kajiura, S.M., and Gordon, M.S. (2009a). Functional consequences of structural differences in stingray sensory systems. Part I: mechanosensory lateral line canals. *J. Exp. Biol.* 212:3037–3043.

Jordan, L.K., Kajiura, S.M., and Gordon, M.S. (2009b). Functional consequences of structural differences in stingray sensory systems. Part II. Electrosensory system. *J. Exp. Biol.* 212:3044–3050.

Kajiura, S.M. (2003). Electroreception in neonatal bonnethead sharks, *Sphyrna tiburo. Mar. Biol.* 143(3):603–611.

Kajiura, S.M. and Fitzgerald, T.P. (2009). Response of juvenile scalloped hammerhead sharks to electric stimuli. *Zoology* 112:241–250.

Kajiura, S.M. and Holland, K.N. (2002). Electroreception in juvenile scalloped hammerhead and sandbar sharks. *J. Exp. Biol.* 205:3609–3621.

Kajiura, S.M., Forni, J.B., and Summers, A.P. (2005). Olfactory morphology of carcharhinid and sphyrnid sharks: does the cephalofoil confer a sensory advantage? *J. Morphol.* 264(3):253–263.

Kalmijn, A.J. (1971). The electric sense of sharks and rays. *J. Exp. Biol.* 55(2):371–383.

Kalmijn, A.J. (1972). Bioelectric fields in seawater and the function of the ampullae of Lorenzini in elasmobranch fishes. *Scripps Inst. Oceanogr. Ref. Ser.* 72–83:1–21.

Kalmijn, A.J. (1974). The detection of electric fields from inanimate and animate sources other than electric organs. In: Albe-Fessard, D.G. (Ed.), *Handbook of Sensory Physiology,* Vol. III. Springer-Verlag, New York, pp. 147–200.

Kalmijn, A.J. (1981). Biophysics of geomagnetic field detection. *IEEE Trans. Magn.* 17:1113–1124.

Kalmijn, A.J. (1982). Electric and magnetic field detection in elasmobranch fishes. *Science* 218:916–918.

Kalmijn, A.J. (1984). Theory of electromagnetic orientation: a further analysis. In: Bolis, L., Keynes, R.D., and Madrell, S.H.P. (Eds.), *Comparative Physiology of Sensory Systems.* Cambridge University Press, Cambridge, U.K.

Kalmijn, A.J. (1988a). Detection of weak electric fields. In: Atema, J., Fay, R.R., Popper, A.N., and Tavolga, W.N. (Eds.), *Sensory Biology of Aquatic Animals.* Springer-Verlag, New York, pp. 151–186.

Kalmijn, A.J. (1988b). Hydrodynamic and acoustic field detection. In: Atema, J., Fay, R.R., Popper, A.N., and Tavolga, W.N. (Eds.), *Sensory Biology of Aquatic Animals.* Springer-Verlag, New York, pp. 83–130.

Kalmijn, A.J. (1989). Functional evolution of lateral line and inner ear sensory systems. In: Coombs, S., Görner, P., and Münz, H. (Eds.), *The Mechanosensory Lateral Line: Neurobiology and Evolution.* Springer-Verlag, New York, pp. 187–215.

Kalmijn, A.J. (1997). Electric and near-field acoustic detection, a comparative study. *Acta Physiol. Scand.* 161(Suppl. 638):25–38.

Kalmijn, A.J. (2000). Detection and processing of electromagnetic and near-field acoustic signals in elasmobranch fishes. *Phil. Trans. R. Soc. Lond. Ser. B Biol. Sci.* 355:1135–1141.

Kantner, M., Konig, W.F., and Reinbach, W. (1962). Baud und Innervation der Lorenzinischen Ampullen und deren Bedeutung als niederes Sinnesorgan. *Z. Zellforsch.* 57:124–135.

Kelly, J.C. and Nelson, D.R. (1975). Hearing thresholds of the horn shark, *Heterodontus francisci*. *J. Acoust. Soc. Am.* 58:905–909.

Kleerekoper, H. (1978). Chemoreception and its interaction with flow and light perception in the locomotion and orientation of some elasmobranchs. In: Hodgson, E.S. and Mathewson, R.F. (Eds.), *Sensory Biology of Sharks, Skates, and Rays*. U.S. Office of Naval Research, Arlington, VA, pp. 269–329.

Kleerekoper, H. (1982). The role of olfaction in the orientation of fishes. In: Hara, T.J. (Ed.), *Chemoreception in Fishes: Developments in Aquaculture and Fisheries Science*. Elsevier, Amsterdam, pp. 201–225.

Kleerekoper, H., Gruber, D., and Matis, J. (1975). Accuracy of localization of a chemical stimulus in flowing and stagnant water by the nurse shark, *Ginglymostoma cirratum*. *J. Comp. Physiol. A Neuroethol. Sens. Neural Behav. Physiol.* 98(3):257–275.

Klimley, A.P. (1980). Observations of courtship and copulation in the nurse shark, *Ginglymostoma cirratum*. *Copeia* 1980:878–882.

Klimley, A.P. (1993). Highly directional swimming by scalloped hammerhead sharks, *Sphyrna lewini*, and subsurface irradiance, temperature, bathymetry, and geomagnetic field. *Mar. Biol.* 117:1–22.

Koester, D.M. (1983). Central projections of the octavolateralis nerves of the clearnose skate, *Raja eglanteria*. *J. Comp. Neurol.* 221:199–215.

Kotrschal, K. (1995). Ecomorphology of solitary chemosensory cell systems in fish: a review. *Environ. Biol. Fish.* 1995:143–155.

Kotrschal, K., Peters, R.C., and Doving, K.B. (1996). Chemosensory and tactile nerve responses from the anterior dorsal fin of a rockling, *Gaidropsarus vulgaris* (Gadidae, Teleostei). *Primary Sensory Neuron* 1:297–308.

Kritzler, H. and Wood, L. (1961). Provisional audiogram for the shark, *Carcharhinus leucas*. *Science* 133:1480–1482.

Kroese, A.B. and Schellart, N.A.M. (1992). Velocity- and acceleration-sensitive units in the trunk lateral line of the trout. *J. Neurophysiol.* 68:2212–2221.

Kuznetsova, M., Lee, C., Aller, J., and Frew, N. (2004). Enrichment of amino acids in the sea surface microlayer at coastal and open ocean sites in the North Atlantic Ocean. *Limnol. Oceanogr.* 49:1605–1619.

Leask, M.J.M. (1977). A physicochemical mechanism for magnetic field detection by migratory birds and homing pigeons. *Nature* 359:142–144.

Lisney, T.J., Bennett, M.B., and Collin, S.P. (2007). Volumetric analysis of sensory brain areas indicates ontogenetic shifts in the relative importance of sensory systems in elasmobranchs. *Raffles Bull. Zool.* 14(Suppl.):7–15.

Lisney, T.J. and Collin, S.P. (2006). Brain morphology in large pelagic fishes: a comparison between sharks and teleosts. *J. Fish Biol.* 68:532–554.

Lisney, T.J. and Collin, S.P. (2008). Retinal ganglion cell distribution and resolving power in elasmobranchs. *Brain Behav. Evol.* 72:59–77.

Litherland, L. (2001). Retinal Topography in Elasmobranchs: Interspecific and Ontogenetic Variations, honors thesis, University of Queensland, Brisbane, Australia.

Litherland, L. and Collin, S.P. (2008). Comparative visual function in elasmobranchs: spatial arrangement and ecological correlates of photoreceptor and ganglion cell distributions. *Visual Neurosci.* 25:549–561.

Litherland, L., Collin, S.P., and Fritsches, K.A. (2009a). Eye growth in sharks: ecological implications for changes in retinal topography and visual resolution. *Visual Neurosci.* 26:397–409.

Litherland, L., Collin, S.P., and Fritsches, K.A. (2009b). Visual optics and ecomorphology of the growing shark eye: a comparison between deep and shallow water species. *J. Exp. Biol.* 212:3583–3594.

Loewenstein, W.R. (1960). Mechanisms of nerve impulse initiation in a pressure receptor (*Lorenzinian ampulla*). *Nature* 188:1034–1035.

Loewenstein, W.R. and Ishiko, N. (1962). Sodium chloride sensitivity and electrochemical effects in a *Lorenzinian ampulla*. *Nature* 194:292–294.

Logiudice, F.T. and Laird, R.J. (1994). Morphology and density distribution of cone photoreceptor in the retina of the Atlantic stingray, *Dasyatis sabina*. *J. Morphol.* 221:277–289.

Lorenzini, S. (1678). *Osservazioni intorno alle Torpedini, fatte da Stefano Lorenzini, Fiorentino*. In Firenze per l'Onofri, Florence, Italy.

Lowe, C.G., Wetherbee, B.M., Crow, G.L., and Tester, A.L. (1996). Ontogenetic dietary shifts and feeding behavior of the tiger shark, *Galeocerdo cuvier*, in Hawaiian waters. *Environ. Biol. Fish.* 47:203–211.

Lowenstein, O. and Roberts, T.D.M. (1951). The localization and analysis of the responses to vibration from the isolated elasmobranch labyrinth: a contribution to the problem of the evolution of hearing in vertebrates. *J. Physiol. (Lond.)* 114:471–489.

Lu, Z. and Popper, A.N. (2001). Neural response directionality correlates with hair cell orientation in a teleost fish. *J. Comp. Physiol. A Neuroethol. Sens. Neural Behav. Physiol.* 187:453–465.

Luer, C.A. and Gilbert, P.W. (1985). Mating behavior, egg deposition, incubation period, and hatching in the clearnose skate, *Raja eglanteria*. *Environ. Biol. Fish.* 13:161–171.

Luiten, P.G.M. (1981a). Two visual pathways to the telencephalon in the nurse shark (*Ginglymostoma cirratum*). 1. Retinal projections. *J. Comp. Neurol.* 196:531–538.

Luiten, P.G.M. (1981b). Two visual pathways to the telencephalon in the nurse shark (*Ginglymostoma cirratum*). 2. Ascending thalamo–telencephalic connections. *J. Comp. Neurol.* 196:539–548.

Lychakov, D.V., Boyadzhieva-Mikhailova, A., Christov, I., and Evdokimov, I.I. (2000). Otolithic apparatus in Black Sea elasmobranchs. *Fish. Res.* 46:27–38.

Maisey, J.G. (2001). Remarks on the inner ear of elasmobranchs and its interpretation from skeletal labyrinth morphology. *J. Morphol.* 250:236–264.

Maruska, K.P. (2001). Morphology of the mechanosensory lateral line system in elasmobranch fishes: ecological and behavioral considerations. *Environ. Biol. Fish.* 60:47–75.

Maruska, K.P. and Tricas, T.C. (1998). Morphology of the mechanosensory lateral line system in the Atlantic stingray, *Dasyatis sabina*: the mechanotactile hypothesis. *J. Morphol.* 238(1):1–22.

Maruska, K.P. and Tricas, T.C. (2004). Test of the mechanotactile hypothesis: neuromast morphology and response dynamics of mechanosensory lateral line primary afferents in the stingray. *J. Exp. Biol.* 207:3463–3476.

Mathewson, R.F. and Hodgson, E.S. (1972). Klinotaxis and rheotaxis in orientation of sharks toward chemical stimuli. *Comp. Biochem. Physiol. A Comp. Physiol.* 42(1):79–84.

Mazur, M.M. and Beauchamp, D.A. (2003). A comparison of visual prey detection among species of piscivorous salmonids: effects of light and low turbidities. *Environ. Biol. Fish.* 67:397–405.

McComb, D.M. and Kajiura, S.M. (2008). Visual fields of four batoid species: a comparative study. *J. Exp. Biol.* 211:482–490.

McComb, D.M., Tricas, T.C., and Kajiura, S.M. (2009). Enhanced visual fields in hammerhead sharks. *J. Exp. Biol.* 212:4010–4018.

McComb, D.M., Frank, T.M., Hueter, R.E., and Kajiura, S.M. (2010). Temporal resolution and spectral sensitivity of the visual system of three coastal shark species from different light environments. *Physiol. Biochem. Zool.* 83:299–307.

McGowan, D.W. and Kajiura, S.M. (2009). Electroreception in the euryhaline stingray, *Dasyatis sabina. J. Exp. Biol.* 212:1544–1552.

Meredith, T.M. and Kajiura, S.M. (2010). Olfactory morphology and physiology of elasmobranchs. *J. Exp. Biol.* 213:3449–3456.

Meyer, C.G., Burgess, W.C., Papastamatiou, Y.P., and Holland, K.N. (2008). Use of an implanted sound recording device (Bioacoustic Probe) to document the acoustic environment of a blacktip reef shark (*Carcharhinus melanopterus*). *Aquat. Living Resour.* 20:291–298.

Mikhailenko, N.A. (1971). Biological significance and dynamics of electrical discharges in weak electric fishes of the Black Sea (in Russian). *Zool. Zh.* 50:1347–1352.

Montgomery, J.C. (1984). Frequency response characteristics of primary and secondary neurons in the electrosensory system of the thornback ray. *Comp. Biochem. Physiol. A Comp. Physiol.* 79:189–195.

Montgomery, J.C. and Bodznick, D. (1993). Hindbrain circuitry mediating common mode suppression of ventilatory reafference in the electrosensory system of the little skate, *Raja erinacea. J. Exp. Biol.* 183:203–215.

Montgomery, J.C., Baker, C.F., and Carton, A.G. (1997). The lateral line can mediate rheotaxis in fish. *Nature* 389:960–963.

Montgomery, J.C. and Bodznick, D. (1994). An adaptive filter that cancels self-induced noise in the electrosensory and lateral line mechanosensory systems of fish. *Neurosci. Lett.* 174:145–148.

Montgomery, J.C., Coombs, S., and Halstead, M. (1995). Biology of the mechanosensory lateral line in fishes. *Rev. Fish Biol. Fish.* 5:399–416.

Montgomery, J.C. and MacDonald, J.A. (1990). Effects of temperature on the nervous system: implications for behavioral performance. *Am. J. Physiol. Regul. Integr. Comp. Physiol.* 259:191–196.

Montgomery, J.C. and Skipworth, E. (1997). Detection of weak water jets by the short-tailed stingray *Dasyatis brevicaudata* (Pisces: Dasyatidae). *Copeia* 1997:881–883.

Montgomery, J.C., Windsor, S., and Bassett, D. (2009). Behavior and physiology of mechanoreception: separating signal and noise. *Int. Zool.* 4:3–12.

Mortenson, J. and Whitaker, R.H. (1973). Electric discharges in free swimming female winter skates (*Raja ocellata*). *Am. Zool.* 13:1266.

Münz, H. (1985). Unit activity in the peripheral lateral line system of the cichlid fish *Sarotherodon niloticus* L. *J. Comp. Physiol. A Neuroethol. Sens. Neural Behav. Physiol.* 157:555–568.

Münz, H. (1989). Functional organization of the lateral line periphery. In: Coombs, S., Görner, P., and Münz, H. (Eds.), *The Mechanosensory Lateral Line: Neurobiology and Evolution*. Springer-Verlag, New York, pp. 285–297.

Murakami, T., Morita, Y., and Ito, H. (1983). Extrinsic and intrinsic fiber connections of the telencephalon in a teleost, *Sebastiscus marmoratus. J. Comp. Neurol.* 216(2):115–131.

Murphy, C.J. and Howland, H.C. (1991). The functional significance of crescent-shaped pupils and multiple pupillary apertures. *J. Exp. Zool.* 5(Suppl.):22–28.

Murray, R.W. (1957). Evidence for a mechanoreceptive function of the ampullae of Lorenzini. *Nature* 179:106–107.

Murray, R.W. (1960a). Electrical sensitivity of the ampullae of Lorenzini. *Nature* 187:957.

Murray, R.W. (1960b). The response of ampullae of Lorenzini of elasmobranchs to mechanical stimulation. *J. Exp. Biol.* 37:417–424.

Murray, R.W. (1962). The response of the ampullae of Lorenzini in elasmobranchs to electrical stimulation. *J. Exp. Biol.* 39:119–128.

Murray, R.W. (1965). Receptor mechanisms in the ampullae of Lorenzini of elasmobranch fishes. *Cold Spring Harb. Symp. Quant. Biol.* 30:235–262.

Myrberg, Jr., A.A. (2001). The acoustical biology of elasmobranchs. *Environ. Biol. Fish.* 60:31–45.

Myrberg, Jr., A.A. and Gruber, S.G. (1974). The behavior of the bonnethead shark, *Sphyrna tiburo. Copeia* 1974:358–374.

Myrberg, Jr., A.A., Banner, A., and Richard, J.D. (1969). Shark attraction using a video-acoustic system. *Mar. Biol.* 2:264–276.

Myrberg, Jr., A.A., Ha, S.J., Walewski, S., and Banbury, J.C. (1972). Effectiveness of acoustic signals in attracting epipelagic sharks to an underwater sound source. *Bull. Mar. Sci.* 22:926–949.

Nelson, D.R. (1967). Hearing thresholds, frequency discrimination, and acoustic orientation in the lemon shark, *Negaprion brevirostris* (Poey). *Bull. Mar. Sci.* 17:741–767.

Nelson, D.R. and Gruber, S.H. (1963). Sharks: attraction by low-frequency sounds. *Science* 142:975–977.

Nelson, P.A., Kajiura, S.M., and Losey, G.S. (2003). Exposure to solar radiation may increase ocular UV-filtering in the juvenile scalloped hammerhead shark, *Sphyrna lewini. Mar. Biol.* 142:53–56.

New, J.G. (1990). Medullary electrosensory processing in the little skate. I. Response characteristics of neurons in the dorsal octavolateralis nucleus. *J. Comp. Physiol. A Neuroethol. Sens. Neural Behav. Physiol.* 167:285–294.

New, J.G. (1994). Electric organ discharge and electrosensory reafference in skates. *Biol. Bull.* 187:64–75.

Nickel, E. and Fuchs, S. (1974). Organization and ultrastructure of mechanoreceptors (Savi vesicles) in the elasmobranch *Torpedo*. *J. Neurocytol.* 3:161–177.

Nicol, J.A.C. (1964). Reflectivity of the chorioidal tapeta of selachians. *J. Fish. Res. Bd. Can.* 21:1089–1100.

Nikaronov, S.I. (1983). Electrophysiological analysis of convergent relations between the olfactory and visual analyzers in the forebrain of *Squalus acanthias*. *J. Evol. Biochem. Physiol.* 18:268–273.

Nikaronov, S.I. and Lukyanov, A.S. (1980). Electrophysiological investigations of visual afferentation pathways in the forebrain of *Squalus acanthias*. *J. Evol. Biochem. Physiol.* 16:132–139.

Nikonov, A.A., Illyin, Y.N., Zherelove, O.M., and Fesenko, E.E. (1990). Odour thresholds of the black sea skate (*Raja clavata*). Electrophysiological study. *Comp. Biochem. Physiol. A Comp. Physiol.* 95:325–238.

Norris, H.W. (1929). The distribution and innervation of the ampullae of Lorenzini of the dogfish, *Squalus acanthias*: some comparisons with conditions in other plagiostomes and corrections of prevalent errors. *J. Comp. Neurol.* 47:449–465.

Norris, H.W. (1932). The laterosensory system of *Torpedo marmorata*, innervation and morphology. *J. Comp. Neurol.* 56:129–178.

Norris, H.W. and Hughes, S.P. (1920). The cranial, occipital, and anterior spinal nerves of the dogfish, *Squalus acanthias*. *J. Comp. Neurol.* 21:293–402.

Northcutt, R.G. (1978). Brain organization in the cartilaginous fishes. In: Hodgson, E.S. and Mathewson, R.F. (Eds.), *Sensory Biology of Sharks, Skates, and Rays*. U.S. Office of Naval Research, Arlington, VA, pp. 117–193.

Northcutt, R.G. (1979). Retinofugal pathways in fetal and adult spiny dogfish, *Squalus acanthias*. *Brain Res.* 162:219–230.

Northcutt, R.G. (1989a). The phylogenetic distribution and innervation of craniate mechanoreceptive lateral lines. In: Coombs, S., Görner, P., and Münz, H. (Eds.), *The Mechanosensory Lateral Line: Neurobiology and Evolution*. Springer-Verlag, New York, pp. 17–78.

Northcutt, R.G. (1989b). Brain variation and phylogenetic trends in elasmobranch fishes. *J. Exp. Zool.* 2(Suppl.):83–100.

Northcutt, R.G. (1991). Visual pathways in elasmobranchs: organization and phylogenetic implications. *J. Exp. Zool.* 5(Suppl.):97–107.

Northcutt, R.G. (2004). Taste buds: development and evolution. *Brain Behav. Evol.* 64:198–206.

Pals, N., Peters, R.C., and Schoenhage, A.A.C. (1982a). Local geo-electric fields at the bottom of the sea and their relevance for electrosensitive fish. *Neth. J. Zool.* 32:479–494.

Pals, N., Valentijn, P., and Verwey, D. (1982b). Orientation reactions of the dogfish, *Scyliorhinus canicula*, to local electric fields. *Neth. J. Zool.* 32:495–512.

Parker, G.H. (1909). The influence of eyes and ears and other allied sense organs on the movement of *Mustelus canis*. *Bull. U.S. Bur. Fish.* 29:43–58.

Parker, G.H. (1914). The directive influence of the sense of smell in the dogfish. *Bull. U.S. Bur. Fish.* 33:61–68.

Parker, G.H. and Sheldon, R.E. (1913). The sense of smell in fishes. *Bull. U.S. Bur. Fish.* 32:33–46.

Paulin, M.G. (1995). Electroreception and the compass sense of sharks. *J. Theor. Biol.* 174:325–339.

Peach, M.B. (2001). The dorso-lateral pit organs of the Port Jackson shark contribute sensory information for rheotaxis. *J. Fish Biol.* 59:696–704.

Peach, M.B. (2003). Inter- and intraspecific variation in the distribution and number of pit organs (free neuromasts) of sharks and rays. *J. Morphol.* 256:89–102.

Peach, M.B. (2005). New microvillous cells with possible sensory function on the skin of sharks. *Mar. Freshw. Behav. Physiol.* 38:275–279.

Peach, M.B. and Marshall, N.J. (2000). The pit organs of elasmobranchs: a review. *Phil. Trans. R. Soc. Lond. B Biol. Sci.* 355(1401):1131–1134.

Peach, M.B. and Marshall, N.J. (2009). The comparative morphology of pit organs in elasmobranchs. *J. Morphol.* 270:688–701.

Peach, M.B. and Rouse, G.W. (2004). Phylogenetic trends in the abundance and distribution of pit organs of elasmobranchs. *Acta Zool.* 85(4):233–244.

Peters, R.C. and Evers, H.P. (1985). Frequency selectivity in the ampullary system of an elasmobranch fish (*Scyliorhinus canicula*). *J. Exp. Biol.* 118:99–109.

Peters, R.C., van Steenderen, G.W., and Kotrschal, K. (1987). A chemoreceptive function for the anterior dorsal fin in rocklings (*Gaidropsarus* and *Ciliata*: Teleostei: Gadidae): electrophysiological evidence. *J. Mar. Biol. Assoc. U.K.* 67:819–823.

Peterson, E.H. and Rowe, M.H. (1980). Different regional specializations of neurons in the ganglion cell layer and inner plexiform layer of the California horned shark, *Heterodontus francisci*. *Brain Res.* 201:195–201.

Phillips, J.B. and Borland, S.C. (1992). Behavioral evidence for use of a light-dependent magnetoreception mechanism in a vertebrate. *Nature* 359:142–144.

Platt, C.J., Bullock, T.H., Czéh, G., Kovacevic, N., Konjevic, D.J., and Gojkovic, M. (1974). Comparison of electroreceptor, mechanoreceptor, and optic evoked potentials in the brain of some rays and sharks. *J. Comp. Physiol. A Neuroethol. Sens. Neural Behav. Physiol.* 95:323–355.

Pocklington, R. (1971). Physical sciences: free amino-acids dissolved in North Atlantic Ocean waters. *Nature* 230:374–375.

Popper, A.N. and Fay, R.R. (1977). Structure and function of the elasmobranch auditory system. *Am. Zool.* 17:443–452.

Popper, A.N. and Fay, R.R. (1999). The auditory periphery in fishes. In: Fay, R.R. and Popper, A.N. (Eds.), *Comparative Hearing: Fish and Amphibians*. Springer-Verlag, New York, pp. 43–100.

Prasada Rao, P.D. and Finger, T.E. (1984). Asymmetry of the olfactory system in the winter flounder, *Pseudopleuronectes americanus*. *J. Comp. Neurol.* 225:492–510.

Puzdrowski, R.L. and Leonard, R.B. (1993). The octavolateral systems in the stingray, *Dasyatis sabina*. I. Primary projections of the octaval and lateral line nerves. *J. Comp. Neurol.* 332:21–37.

Raschi, W. and Mackanos, L.A. (1989). The structure of the ampullae of Lorenzini in *Dasyatis garouaensis* and its implications on the evolution of freshwater electroreceptive systems. *J. Exp. Zool.* 2:101–111.

Rasmussen, L.E.L. and Schmidt, M.J. (1992). Are sharks chemically aware of crocodiles? In: Doty, R.L. and Müller-Schwarze, D. (Eds.), *Chemical Signals in Vertebrates*, Vol. IV. Plenum Press, New York, pp. 335–342.

Ratnam, R. and Nelson, M.E. (2000). Nonrenewal statistics of electrosensory afferent spike trains: implications for the detection of weak sensory signals. *J. Neurosci.* 20:6672–6683.

Reese, T.S. and Brightman, W.M. (1970). Olfactory surface and central olfactory connections in some vertebrates. In: Wolstenhome, G.E.W. and Knight, J. (Eds.), *Taste and Smell in Vertebrates*. Churchill, London, pp. 115–149.

Reid, G. and Flonta, M.L. (2001). Physiology: cold current in thermoreceptive neurons. *Nature* 413:480.

Retzius, G. (1881). *Das Gehörorgan der Wirbelthiere*, Vol. 1. Samson and Wallin, Stockholm.

Ripps, H. and Dowling, J.E. (1991). Structural features and adaptive properties of photoreceptors in the skate retina. *J. Exp. Zool.* 5(Suppl.):46–54.

Roberts, B.L. (1978). Mechanoreceptors and the behavior of elasmobranch fishes with special reference to the acoustico-lateralis system. In: Hodgson, E.S. and Mathewson, R.F. (Eds.), *Sensory Biology of Sharks, Skates, and Rays*. U.S. Office of Naval Research, Arlington, VA, pp. 331–390.

Sand, A. (1937). The mechanism of the lateral sense organs of fishes. *Proc. R. Soc. Lond. B Biol. Sci.* 123:472–495.

Savi, P. (1844). Études anatomiques sur le système nerveux et sur l'organe électrique de la torpille. In: Matteucci, C. (Ed.), *Traité des Phénomènes Électrophysiologiques des Animaux*. Chez Fortin-Masson et Cie, Paris, pp. 272–348.

Schellart, N.A.M. and Popper, A.N. (1992). Functional aspects of the evolution of the auditory system of actinopterygian fish. In: Webster, D.B., Fay, R.R., and Popper, A.N. (Eds.), *The Evolutionary Biology of Hearing*. Springer-Verlag, New York, pp. 295–322.

Schluessel, V., Bennett, M.B., Bleckmann, H., Blomberg, S., and Collin, S.P. (2008). Morphometric and ultrastructural comparison of the olfactory system of elasmobranchs: the significance of structure–function relationships based on phylogeny and ecology. *J. Morphol.* 269:1365–1386.

Schluessel, V., Bennett, M.B., Bleckmann, H., and Collin, S.P. (2010). The role of olfaction throughout juvenile development: functional adaptations in elasmobranchs. *J. Morphol.* 271:451–461.

Schroeder, D.M. and Ebbesson, S.O.E. (1974). Nonolfactory telencephalic afferents in the nurse shark (*Ginglymostoma cirratum*). *Brain Behav. Evol.* 9:121–155.

Schweitzer, J. (1983). The physiological and anatomical localization of two electroreceptive diencephalic nuclei in the thornback ray, *Platyrhinoidis triseriata*. *J. Comp. Physiol. A Neuroethol. Sens. Neural Behav. Physiol.* 153:331–341.

Schweitzer, J. (1986). Functional organization of the electroreceptive midbrain in an elasmobranch (*Platyrhinoidis triseriata*): a single unit study. *J. Comp. Physiol. A Neuroethol. Sens. Neural Behav. Physiol.* 158:43–48.

Schweitzer, J. and Lowe, D.A. (1984). Mesencephalic and diencephalic cobalt-lysine injections in an elasmobranch: evidence for two parallel electrosensory pathways. *Neurosci. Lett.* 44:317–322.

Sejnowski, T.J. and Yodlowski, M.L. (1982). A freeze fracture study of the skate electroreceptors. *J. Neurocytol.* 11:897–912.

Sheldon, R.E. (1909). The reactions of the dogfish to chemical stimuli. *J. Comp. Neurol. Psychol.* 19:273–311.

Sheldon, R.E. (1911). The sense of smell in selachians. *J. Exp. Zool.* 10:51–62.

Sillman, A.J., Letsinger, G.A., Patel, S., Loew, E.R., and Klimley, A.P. (1996). Visual pigments and photoreceptors in two species of shark, *Triakis semifasciata* and *Mustelus henlei*. *J. Exp. Zool.* 276:1–10.

Silver, W. and Finger, T.E. (1984). Electrophysiological examination of a non-olfactory, non-gustatory chemosense in the sea robin. *J. Comp. Physiol. A Neuroethol. Sens. Neural Behav. Physiol.* 154:167–174.

Silver, W.L. (1979). Olfactory responses from a marine elasmobranch, the Atlantic stingray, *Dasyatis sabina*. *Mar. Behav. Physiol.* 6:297–305.

Silver, W.L, Caprio, J., Blackwell, J.F., and Tucker, D. (1976). The underwater electro-olfactogram: a tool for the study of the sense of smell of marine fishes. *Experientia* 32:1216–1217.

Sisneros, J.A. and Tricas, T.C. (2000). Androgen-induced changes in the response dynamics of ampullary electrosensory primary afferent neurons. *J. Neurosci.* 20:8586–8595.

Sisneros, J.A. and Tricas, T.C. (2002a). Neuroethology and life history adaptations of the elasmobranch electric sense. *J. Physiol. (Paris)* 96:379–389.

Sisneros, J.A. and Tricas, T.C. (2002b). Ontogenetic changes in the response properties of the peripheral electrosensory system in the Atlantic stingray (*Dasyatis sabina*). *Brain Behav. Evol.* 59:130–140.

Sisneros, J.A., Tricas, T.C., and Luer, C.A. (1998). Response properties and biological function of the skate electrosensory system during ontogeny. *J. Comp. Physiol. A Neuroethol. Sens. Neural Behav. Physiol.* 183:87–99.

Sivak, J.G. (1978a). Optical characteristics of the eye of the spiny dogfish (*Squalus acanthias*). *Rev. Can. Biol.* 37:209–217.

Sivak, J.G. (1978b). Refraction and accommodation of the elasmobranch eye. In: Hodgson, E.S. and Mathewson, R.F. (Eds.), *Sensory Biology of Sharks, Skates, and Rays*. U.S. Office of Naval Research, Arlington, VA, pp. 107–116.

Sivak, J.G. (1991). Elasmobranch visual optics. *J. Exp. Zool.* 5(Suppl.):13–21.

Sivak, J.G. and Gilbert, P.W. (1976). Refractive and histological study of accommodation in two species of sharks (*Ginglymostoma cirratum* and *Carcharhinus milberti*). *Can. J. Zool.* 54:1811–1817.

Smeets, W.J.A.J. (1983). The secondary olfactory connections in two chondrichthians, the shark *Scyliorhinus canicula* and the ray *Raja clavata*. *J. Comp. Neurol.* 218:334–344.

Smeets, W.J.A.J. (1998). Cartilaginous fishes. In: Nieuwenhuys, R., ten Donkelaar, H.J., and Nicholson, C. (Eds.), *The Central Nervous System of Vertebrates*, Vol. 1. Springer, Berlin, pp. 551–654.

Smeets, W.J.A.J. and Northcutt, R.G. (1987). At least one thalamotelencephalic pathway in cartilaginous fishes projects to the medium pallium. *Neurosci. Lett.* 78:277–282.

Smeets, W.J.A.J., Nieuwenhuys, R., and Roberts, B.L. (1983). *The Central Nervous System of Cartilaginous Fishes: Structure and Functional Correlations*. Springer, Berlin.

Spielman, S.L. and Gruber, S.H. (1983). Development of a contact lens for refracting aquatic animals. *Ophthal. Physiol. Opt.* 3:255–260.

Stein, R.B. (1967). The information capacity of nerve cells using a frequency code. *Biophys. J.* 7:797–826.

Stell, W.K. and Witkovsky, P. (1973). Retinal structure in the smooth dogfish, *Mustelus canis*: light microscopy of photoreceptor and horizontal cells. *J. Comp. Neurol.* 148:33–46.

Stenonis, N. (1664). *De Musculis et Glandulis Observationum Specimen cum Epistolis duabus Anatomicis.* Amsterdam.

Strong, Jr., W.R. (1996). Shape discrimination and visual predatory tactics in white sharks. In: Klimley, A.P. and Ainley, D.G. (Eds.), *Great White Sharks: The Biology of Carcharodon carcharias.* Academic Press, San Diego, CA, pp. 229–240.

Strong, Jr., W.R., Murphy, R.C., Bruce, B.D., and Nelson, D.R. (1992). Movements and associated observations of bait-attracted white sharks, *Carcharodon carcharias*: a preliminary report. *Aust. J. Mar. Freshw. Res.* 43:13–20.

Strong, Jr., W.R., Bruce, B.D., Nelson, D.R., and Murphy, R.C. (1996). Population dynamics of white sharks in Spencer Gulf, South Australia. In: Klimley, A.P. and Ainley, D.G. (Eds.), *Great White Sharks: The Biology of Carcharodon carcharias.* Academic Press, San Diego, pp. 401–414.

Takami, S., Luer, C.A., and Graziadei, P.P.C. (1994). Microscopic structure of the olfactory organ of the clearnose skate, *Raja eglanteria. Anat. Embryol. (Berl.)* 190:211–230.

Tester, A.L. (1963a). Olfaction, gustation, and the common chemical sense in sharks. In: Gilbert, P.W. (Ed.), *Sharks and Survival.* D.C. Heath, Boston, pp. 255–282.

Tester, A.L. (1963b). The role of olfaction in shark predation. *Pac. Sci.* 17:145–170.

Tester, A.L. and Kendall, J.I. (1969). Morphology of the lateralis canal system in shark genus *Carcharhinus. Pac. Sci.* 23:1–16.

Tester, A.L. and Nelson, G.J. (1967). Free neuromasts (pit organs) in sharks. In: Gilbert, P.W., Mathewson, R.F., and Rall, D.P. (Eds.), *Sharks, Skates, and Rays.* The Johns Hopkins Press, Baltimore, MD, pp. 503–531.

Tester, A.L., Kendall, J.I., and Milisen, W.B. (1972). Morphology of the ear of the shark genus *Carcharhinus*, with particular reference to the macula neglecta. *Pac. Sci.* 26:264–274.

Theisen, B., Zeiske, E., and Breucker, H. (1986). Functional morphology of the olfactory organs in the spiny dogfish (*Squalus acanthias* L.) and the small-spotted catshark (*Scyliorhinus canicula* L.). *Acta Zool. (Stockh.)* 67:73–86.

Theiss, S.M., Lisney, T.J., Collin, S.P., and Hart, N.S. (2007). Colour vision and visual ecology of the blue-spotted maskray, *Dasyatis kuhlii* Müller & Henle, 1814. *J. Comp. Physiol. A Neuroethol. Sens. Neural Behav. Physiol.* 193:67–79.

Theiss, S.M., Hart, N.S., and Collin, S.P. (2009). Morphological indicators of olfactory capability in wobbegong sharks (Orectolobidae, Elasmobranchii). *Brain Behav. Evol.* 73:91–101.

Tolpin, W., Klyce, D., and Dohlman, C.H. (1969). Swelling properties of dogfish cornea. *Exp. Eye Res.* 8:429–437.

Tong, S.L. and Bullock, T.H. (1982). The sensory functions of the cerebellum of the thornback ray, *Platyrhinoidis triseriata. J. Comp. Physiol. A Neuroethol. Sens. Neural Behav. Physiol.* 148:399–410.

Tricas, T.C. (1980). Courtship and mating-related behaviors in myliobatid rays. *Copeia* 1980:553–556.

Tricas, T.C. (1982). Bioelectric-mediated predation by swell sharks, *Cephaloscyllium ventriosum. Copeia* 1982:948–952.

Tricas, T.C. (2001). The neuroecology of the elasmobranch electrosensory world: why peripheral morphology shapes behavior. *Environ. Biol. Fish.* 60:77–92.

Tricas, T.C. and McCosker, J.E. (1984). Predatory behavior of the white shark (*Carcharodon carcharias*), with notes on its biology. *Proc. Calif. Acad. Sci.* 43:221–238.

Tricas, T.C. and New, J.G. (1998). Sensitivity and response dynamics of elasmobranch electrosensory primary afferent neurons to near threshold fields. *J. Comp. Physiol. A Neuroethol. Sens. Neural Behav. Physiol.* 182:89–101.

Tricas, T.C., Michael, S.W., and Sisnero, J.A. (1995). Electrosensory optimization to conspecific phasic signals for mating. *Neurosci. Lett.* 202:129–132.

Tricas, T.C., Kajiura, S.M., and Summers, A.P. (2009). Response of the hammerhead shark olfactory epithelium to amino acid stimuli. *J. Comp. Physiol. A Neuroethol. Sens. Neural Behav. Physiol.* 195(10):947–954.

van den Berg, A.V. and Schuijf, A. (1983). Discrimination of sounds based on the phase difference between particle motion and acoustic pressure in the shark *Chiloscyllium griseum. Proc. R. Soc. Lond. B Biol. Sci.* 218:127–134.

Veselkin, N.P. and Kovacevic, N. (1973). Nonolfactory afferent projections of the telencephalon of elasmobranchii. *J. Evol. Biochem. Physiol.* 9:512–518.

Viana, F., de la Pena, E., and Belmonte, C. (2002). Specificity of cold thermotransduction is determined by differential ionic channel expression. *Nat. Neurosci.* 5:254–260.

Walcott, C., Gould, J.L., and Kirschvink, J.L. (1979). Pigeons have magnets. *Science* 205:1027–1029.

Walker, M.M., Diebel, C.C., Haugh, C.V., Pankhurst, P.M., Montgomery, J.C., and Green, C.R. (1997). Structure and function of the vertebrate magnetic sense. *Nature* 390:371–376.

Walls, G.L. (1942). *The Vertebrate Eye and Its Adaptive Radiation.* Cranbrook Institute of Science, Bloomfield Hills, MI; reprint edition 1967 by Hafner, New York.

Waltman, B. (1966). Electrical properties and fine structure of the ampullary canals of Lorenzini. *Acta Physiol. Scand.* 66(Suppl. 264):1–60.

Warrant, E.J. and Locket, N.A. (2004). Vision in the deep sea. *Biol. Rev. Camb. Philos. Soc.* 79:671–712.

Webster, D.R. (2007). Structure of turbulent chemical plumes. In: Woodfin, R.L. (Rd.), *Trace Chemical Sensing of Explosives.* John Wiley & Sons, New York, pp. 109–129.

Weissburg, M.J. (2000). The fluid dynamical context of chemosensory behavior. *Biol. Bull.* 198:188–202.

Whitear, M. and Moate, R.M. (1994a). Microanatomy of the taste buds in the dogfish, *Scyliorhinus canicula. J. Submicrosc. Cytol. Pathol.* 26:357–367.

Whitear, M. and Moate, R.M. (1994b). Chemosensory cells in the oral epithelium of *Raja clavata* (Chondrichtyes). *J. Zool.* 232:295–312.

Wisby, W.J., Richard, J.D., Nelson, D.R., and Gruber, S.H. (1964). Sound perception in elasmobranchs. In: Tavolga, W.N. (Ed.), *Marine Bio-Acoustics.* Pergamon Press, New York, pp. 255–268.

Wright, T. and Jackson, R. (1964). Instrumental conditioning of young sharks. *Copeia* 1964:409–412.

Wueringer, B.E. and Tibbetts, I.R. (2008). Comparison of the lateral line and ampullary systems of two species of shovelnose ray. *Rev. Fish Biol. Fish.* 18:47–64.

Yew, D.T., Chan, Y.W., Lee, M., and Lam, S. (1984). A biophysical, morphological and morphometrical survey of the eye of the small shark (*Hemiscyllium plagiosum*). *Anat. Anz. Jena* 155:355–363.

Zeiske, E., Caprio, J., and Gruber, S.H. (1986). Morphological and electrophysiological studies on the olfactory organ of the lemon shark *Negaprio brevirostris* (Poey). In: Uyeno, T., Arai, R., Taniuchi, T., and Matsuura, K. (Eds.), *Indo-Pacific Fish Biology*. Ichthyological Society of Japan, Tokyo, pp. 381–391.

Zeiske, E., Theisen, B., and Gruber, S.H. (1987). Functional morphology of the olfactory organ of two carcharhinid shark species. *Can. J. Zool.* 65:2406–2412.

Zigman, S. (1991). Comparative biochemistry and biophysics of elasmobranch lenses. *J. Exp. Zool.* 5(Suppl.):29–40.

13

Recent Advances in Elasmobranch Immunology

Carl A. Luer, Catherine J. Walsh, and Ashby B. Bodine

CONTENTS

13.1 Introduction

The first edition of *Biology of Sharks and Their Relatives* included a chapter describing the immune system of sharks, skates, and rays (Luer et al., 2004) in terms of the cellular components and tissue sites involved, an overview of various nonspecific and specific immune responses identified in elasmobranch fishes and how they compare to immune responses in higher vertebrates, the ontogeny of immune tissues and cells, and examples of how experimental approaches using elasmobranch models are advancing our knowledge of comparative immunology. For clarity and continuity with the previous volume, this chapter will begin with a brief overview of elasmobranch immune cells and their sites of origin. The rest of the chapter is devoted to sections describing recent advances in areas of antigen receptor molecules, immune system genes and their transcripts, and applications for elasmobranch immune cell-derived factors as immunomodulators of mammalian cells.

13.2 Elasmobranch Immune Cells and Their Sites of Origin

13.2.1 Leukocytes

The white blood cells, or leukocytes, characteristic of peripheral blood in higher vertebrates are also found in elasmobranch blood (Fänge, 1987; Hyder et al., 1983). These include lymphocytes, granulocytes, and monocytes (Table 13.1) and are typically responsible for vertebrate immune functions. In addition to fully differentiated cell

TABLE 13.1

Relative Abundance of Leukocyte Cell Types in Peripheral Blood of Representative Elasmobranch Species

Cell Type	Black Tip (*Carcharhinus limbatus*) (*n* = 15)	Nurse Shark (*Ginglymostoma cirratum*) (*n* = 13)	Atlantic Stingray (*Dasyatis sabina*) (*n* = 10)	Clearnose Skate (*Raja eglanteria*) (*n* = 7)
Lymphocyte	72.5 ± 2.3	73.3 ± 2.2	69.1 ± 2.6	77.4 ± 3.5
	(47–86)	(62–87)	(56–83)	(59–83)
Granulocyte	24.8 ± 2.0	24.5 ± 2.2	29.3 ± 2.7	21.4 ± 3.6
	(12–46)	(11–38)	(17–42)	(15–40)
Monocyte	2.8 ± 0.4	2.2 ± 0.5	1.6 ± 0.4	1.3 ± 0.3
	(0–4)	(0–4)	(0–4)	(0–3)

Note: Values are the mean % ± SEM from a minimum of 1000 leukocytes counted; the range is given in parentheses.

types, elasmobranch blood includes leukocytes at varying stages of mitosis as well as cells in "immature" stages (Hyder et al., 1983; Luer et al., 2004). Although the presence of differentiating cells complicates classification of cells into recognizable categories, many morphological similarities with higher vertebrate cells exist. Even so, attempts to correlate function with specific cell types in elasmobranchs are, in many cases, inconclusive.

The most common leukocyte in elasmobranch peripheral blood is the lymphocyte. Morphologically, elasmobranch lymphocytes are similar to lymphocytes from other vertebrates (Figure 13.1) and occur in varying sizes reflecting their degree of maturation (Luer et al., 2004). The majority of circulating lymphocytes are small (mature) or medium (maturing), but large (immature) lymphocytes are present as well.

The two principal subsets of lymphocytes in vertebrates are B lymphocytes (bursa- or bone-marrow-derived lymphocytes, or B cells) and T lymphocytes (thymus-derived lymphocytes, or T cells), representing the earliest phylogenetic appearance for both of these immune cell types. B lymphocytes and T lymphocytes are morphologically indistinguishable, but their existence in elasmobranch fishes has been established by the presence of immunoglobulins (Igs) (see Section 13.3.1), the identification of genes coding for T-cell antigen receptors (see Section 13.4.1) and major histocompatibility gene complexes (see Section 13.4.2), and the expression of genes associated with B-lymphocyte and T-lymphocyte function in higher vertebrates (see Section 13.5.1).

Granulocytes have been described in several species of elasmobranchs but have been inconsistently identified and classified, probably as a result of great variability in size, shape, and staining properties of these cells (Luer et al., 2004; Rowley et al., 1988; Walsh and Luer, 2004). Not all granulocytes found in elasmobranch blood have a clear mammalian counterpart, and attempts to

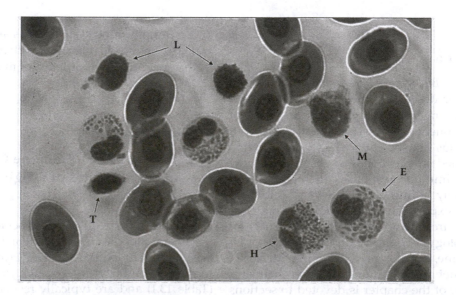

FIGURE 13.1

Peripheral blood smear from a blacktip shark, *Carcharhinus limbatus*, showing lymphocytes (L), heterophilic granulocytes (H), eosinophilic granulocytes (E), a monocyte (M), and a thrombocyte (T). (Stain, Wright–Giemsa; original magnification, 1000×.)

classify these cells have complicated the issue. As mentioned previously, immature cells are common in elasmobranch blood (Ellis, 1977), and cells with different morphologies or staining properties can be mistakenly considered to be different end cells when they actually may be different developmental stages from the same lineage (Hine and Wain, 1987).

The most common granulocyte in elasmobranch blood is referred to in non-mammalian hematology as the *heterophil* (analogous to the mammalian neutrophil) (Luer et al., 2004; Walsh and Luer, 2004). Heterophils have cytoplasmic granules of varying shapes, sizes, and staining intensity, all of which can vary among species as well as with maturity of the cell (Figure 13.1). Although heterophils are the predominant granulocyte, their numbers vary widely among elasmobranch species, ranging from 20 to 50% of the total leukocytes. Eosinophilic granulocytes are also present in elasmobranch peripheral blood, although typically in much fewer numbers than heterophils. This type of granulocyte is referred to as an *eosinophil* and characteristically contains intensely staining granules. Eosinophils usually account for only 2 to 3% of total leukocytes, but they can range from nonexistent to more than 10% of the total leukocyte count. A third type of granulocyte is the *basophil*, which, as in higher vertebrates, is uncommon and accounts for less than 1% of the total leukocytes in elasmobranch peripheral blood.

Morphologically, elasmobranch monocytes and macrophages resemble those of higher vertebrates. Monocytes are large, agranular cells with abundant cytoplasm and account for less than 3% of the leukocytes in elasmobranch peripheral blood (Luer et al., 2004; Walsh and Luer, 2004). They are typically larger than lymphocytes and are often irregular in shape due to pseudopodial processes. The nucleus is eccentric in location and has a characteristic kidney shape, often appearing to be bilobed or indented (Figure 13.1).

Among higher vertebrates, the term *monocyte* typically refers to an immature, circulating cell, and the term *macrophage* describes a mature cell type found in tissues. In fish, however, a distinction is not often made, with this cell type being referred to as the *monocyte/macrophage* (Secombes, 1996). Hyder et al. (1983) suggested that, in the nurse shark, differentiation of immature monocyte-like cells to fully differentiated macrophage-like cells takes place in the circulation, complicating the distinction between these cell types in the peripheral blood.

Because of the relative prevalence of thrombocytes, this non-leukocyte cell type is worthy of inclusion in this discussion. Although not routinely included as part of a differential cell count, thrombocytes can account for as much as 20% of the nonerythroid cells in the peripheral circulation (Walsh, unpublished). In peripheral blood smears, thrombocytes can assume a variety of shapes, including spindle-shaped (Figure 13.1), elliptical, or round, probably varying with the stage of maturity or degree of reactivity.

Although their role in blood clotting has not been experimentally demonstrated in elasmobranchs, thrombocytes are thought to play a role in coagulation comparable to platelets in mammals (Ellis, 1977; Stokes and Firkin, 1971). Unlike platelets, however, elasmobranch thrombocytes may have an immune function, based on observations that they can accumulate dyes and engulf latex beads and yeast cells (Stokes and Firkin, 1971; Walsh and Luer, 1998).

13.2.2 Lymphoid and Lymphomyeloid Tissues

Tissue sites that provide the environments for immune cell production in elasmobranch fishes consist of sites that are common to other vertebrate immune systems as well as some that are unique to sharks, skates, and rays. Thymus and spleen, both vital to immune cell production in higher vertebrates, have their earliest phylogenetic appearance in the cartilaginous fishes. In the absence of bone marrow and lymph nodes, however, alternative tissue sites, often referred to as *bone marrow equivalents*, have evolved to serve remarkably similar functions.

The spleen is easily recognized among elasmobranch visceral organs by its rich dark red to purplish color. Histologically, the elasmobranch spleen is composed of regions of red and white pulp, giving it a structural organization that is surprisingly similar in appearance to that of higher vertebrates (Figure 13.2A). The scattered regions of white pulp are dense accumulations of small lymphocytes with asymmetrically placed central arteries. Areas of white pulp are surrounded by less dense areas of red pulp containing venous sinuses. Instead of being filled with lymph as in mammals, these sinuses are filled primarily with erythrocytes and to a lesser extent with lymphocytes (Andrew and Hickman, 1974). Whereas the presence of mature, immature, and dividing cells in splenic imprints confirms this tissue as a site for lymphocyte production, granulocytes may also be produced in the spleen (Figure 13.2B).

The thymus is a paired organ situated dorsomedial to both gill regions (Luer et al., 1995). As in higher vertebrates, the elasmobranch thymus is organized into distinct lobules (Figure 13.3A), each lobule consisting of an outer cortex and an inner medulla. Tissue imprints reveal that the cortex and medulla contain thymocytes at varying stages of maturation (Figure 13.3B).

The most conspicuous of the bone marrow equivalent tissues are the epigonal and Leydig organs, which represent immune tissues that are unique to elasmobranch fishes. The epigonal organ continues caudally

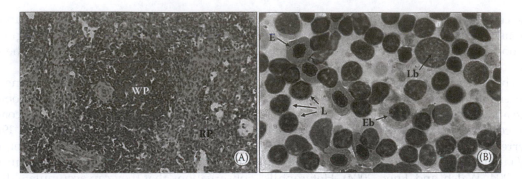

FIGURE 13.2

(A) Paraffin-embedded 10-μm section of spleen from a clearnose skate, *Raja eglanteria*, showing characteristic red pulp (RP) composed of venous sinuses filled with red blood cells and white pulp (WP) composed of dense accumulations of leukocytes. (Stain: hematoxylin and eosin; original magnification, 40×.) (B) Tissue imprint of spleen from an Atlantic stingray, *Dasyatis sabina*, showing lymphocytes (L), lymphoblasts (Lb), erythrocytes (E), and erythroblasts (Eb). (Stain, Wright–Giemsa; original magnification, 1000×.)

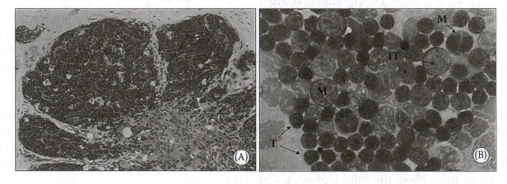

FIGURE 13.3

(A) Paraffin-embedded 10-μm section of thymus from a juvenile nurse shark, *Ginglymostoma cirratum*, showing characteristic lobular architecture composed of cortical regions of tightly packed thymocytes and medullary regions of less densely populated thymocytes. (Stain, hematoxylin and eosin; original magnification, 100×.) (B) Tissue imprint of thymus from a juvenile nurse shark, *Ginglymostoma cirratum*, showing small, darkly staining mature thymocytes (T), large immature thymocytes (IT) of varying sizes, and thymocytes in the process of mitosis (M). (Stain, methylene blue; original magnification, 1000×.)

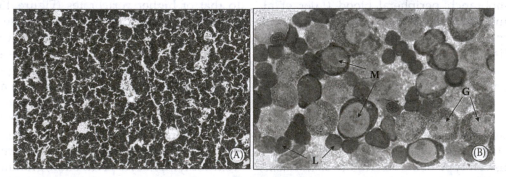

FIGURE 13.4

(A) Paraffin-embedded 10-μm section of epigonal organ from a nurse shark, *Ginglymostoma cirratum*, showing leukocyte-filled sinuses reminiscent of mammalian bone marrow. (Stain, hematoxylin and eosin; original magnification, 100×.) (B) Tissue imprint of epigonal organ from a bonnethead shark, *Sphyrna tiburo*, showing the presence of granulocytes (G), myeloblasts (M), and lymphocytes, L. (Stain, Wright–Giemsa; original magnification, 1,000×.)

from the posterior margin of the gonads in all shark and batoid species. Histologically, the epigonal is composed of sinuses reminiscent of mammalian bone marrow (Figure 13.4A), except for the absence of adipose (fat) cells. Tissue imprints demonstrate that epigonal sinuses are filled with leukocytes at various stages of

maturation. Most of the cells are granulocytes, with lymphocytes present to a significant but lesser degree (Figure 13.4B).

Unlike the epigonal organ, the Leydig organ is not ubiquitous among elasmobranch species. Anecdotal observations support the notion that species possessing

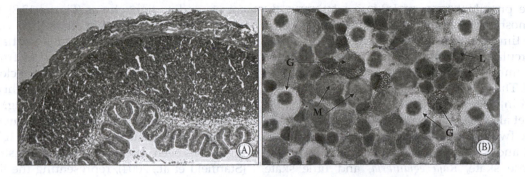

FIGURE 13.5

(A) Paraffin-embedded 10-µm transverse section across the esophagus of a clearnose skate, *Raja eglanteria*, showing the bilobed Leydig organ situated between the epithelium and submucosal layer on both dorsal and ventral sides of the esophagus. The Leydig organ consists of leukocyte-filled sinuses much like the epigonal organ. (Stain, hematoxylin and eosin; original magnification, 40×.) (B) Tissue imprint of Leydig organ from an Atlantic guitarfish, *Rhinobatos lentiginosus*, showing the presence of granulocytes (G), myeloblasts (M), and lymphocytes (L). Granulocytes with darkly staining granules, as well as granulocytes with neutrally staining granules, are visible in this species. (Stain, Wright–Giemsa; original magnification, 1000×.)

Leydig organs tend to have smaller epigonal organs, fueling speculation that Leydig tissue may compensate for the lack of lymphomyeloid tissue when epigonal tissue is limited. When present, Leydig organs can be visualized as whitish masses beneath the epithelium on both dorsal and ventral sides of the esophagus. Leydig organ histology is virtually identical to that of the epigonal organ, composed of sinuses that again are reminiscent of mammalian bone marrow (Figure 13.5A). Tissue imprints are also similar to those of epigonal tissue, indicating leukocytes at various stages of maturation (Figure 13.5B). Again, cells are primarily granulocytes, although lymphocytes are also present. In addition to the well-defined, encapsulated lymphomyeloid tissues described previously, pockets or aggregations of leukocytes can be found in various locations ranging from the intestinal mucosa to the meninges of the brain (Chiba et al., 1988; Zapata et al., 1996) and occasionally in the rectal gland (Luer and Walsh, unpublished). Intestinal aggregations known as *gut-associated lymphoid tissue*, or GALT, can often be substantial (Tomonaga et al., 1986) but appear to be sites where immune cells accumulate rather than sites of immune cell production (Hart et al., 1988).

13.3 Antigen Receptor Molecules

It is widely viewed that members of Subclass Elasmobranchii represent the earliest phylogenetic group of jawed vertebrates to possess all the components necessary for an adaptive immune system (Flajnik and Rumfelt, 2000; Litman et al., 1999). These components include immunoglobulin (Ig) molecules, T-cell receptors (TCRs), major histocompatibility complex (MHC) products, and recombination activator genes (RAGs).

13.3.1 Immunoglobulins

The first true immunoglobulin to be identified in elasmobranchs was immunoglobulin M (IgM) (Marchalonis and Edelman, 1965, 1966), isolated initially from the smooth dogfish, *Mustelus canis*, and confirmed soon after in lemon sharks, *Negaprion brevirostris* (Clem and Small, 1967) and nurse sharks, *Ginglymostoma cirratum* (Clem et al., 1967). As in mammals, elasmobranch IgM exists as a high-molecular-weight 19S pentamer of 7S monomeric subunits, each consisting of two heavy and two light chains with constant and variable regions covalently linked together by disulfide bonds (Figure 13.6).

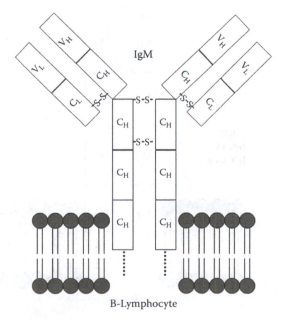

FIGURE 13.6

Schematic representation of the immunoglobulin M (IgM) molecule, composed of two heavy chains (H) and two light chains (L), each with variable (V) and constant (C) regions and interchain disulfide bonds.

The more prevalent form of IgM circulating in the serum of most elasmobranchs is the monomeric form and at one time was thought to be the only immunoglobulin circulating in elasmobranch blood. Additional monomeric immunoglobulins unrelated to IgM are now recognized. The first to be identified, termed IgX or IgR, were found in skates (Kobayashi and Tomonaga, 1988; Kobayashi et al., 1984) and primitive sharks (Kobayashi et al., 1992). Two forms of IgX, a short form (Anderson et al., 1994) and a long form (Anderson et al., 1999), in the clearnose skate, *Raja eglanteria*, and little skate, *Leucoraja* (formerly *Raja*) *erinacea* (Harding et al., 1990b), are orthologs of two other monomeric immunoglobulins: IgW from sandbar sharks, *Carcharhinus plumbeus* (Bernstein et al., 1996a), and IgNARC (Ig new antigen receptor) from cartilaginous fish (Greenberg et al., 1996) (Figure 13.7). Another monomeric form of immunoglobulin, new antigen receptor (NAR), has been identified from nurse shark (Greenberg et al., 1995). Like the other immunoglobulins, IgNAR is composed of heavy chains; however, unlike the others, there is no dimerization with

corresponding light chains (Figure 13.8). In the absence of covalent linkage to light chains, the variable regions are relatively unrestricted and potentially more flexible. Interestingly, peptide sequences from the IgNAR heavy-chain variable regions are more closely related to variable regions of T-cell receptor or light chains than to heavy-chain variable regions of either IgM or IgW.

In a significant advancement in elasmobranch immunology, the crystal structure of the IgNAR variable region has been determined at 1.45-angstrom resolution (Stanfield et al., 2004), representing the first non-mammalian antibody structure to be deciphered. Although details of the crystallographic analyses are beyond the scope of this chapter, the noteworthy observations are in the structure of the IgNAR variable domain (V domain), where one of the three conventional complementarity-determining regions (CDRs) has been deleted. Hypervariable regions CDR1 and CDR3 are present, but CDR2 has been replaced with a much shorter strand, termed HV2. Deletion of the CDR2 region is ultimately responsible for the characteristically small size of the

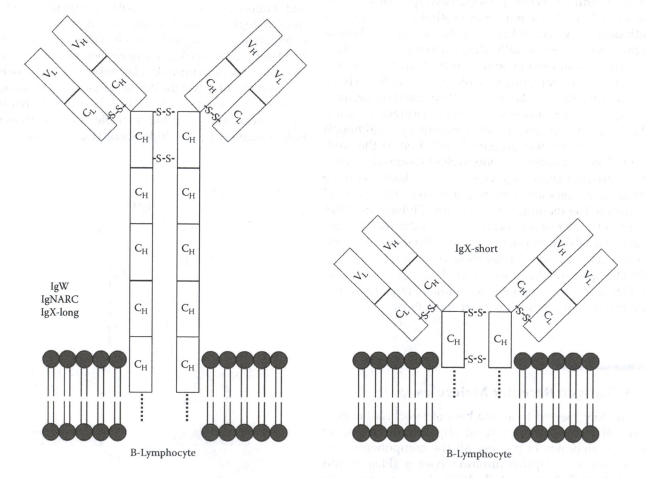

FIGURE 13.7
Schematic representation of related immunoglobulin isotypes identified in various elasmobranch fishes. IgW, IgNARC, IgX-long, and IgX-short are orthologs identified from sandbar shark, *Carcharhinus plumbeus*; nurse shark, *Ginglymostoma cirratum*; little skate, *Leucoraja* (formerly *Raja*) *erinacea*; and clearnose skate, *Raja eglanteria*, respectively.

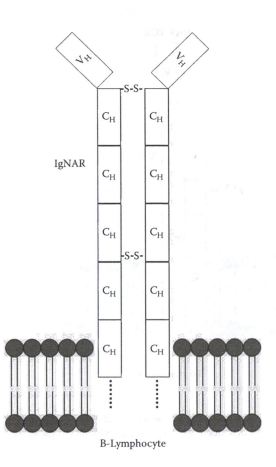

FIGURE 13.8

Schematic representation of an atypical immunoglobulin isotype, IgNAR, identified in nurse shark, *Ginglymostoma cirratum*, and composed of heavy-chain dimers with single variable regions and no associated light chains. Absence of light chains allows for flexible V domains.

IgNAR V region. Compared with human and murine CDR3 regions, IgNAR CDR3 is longer and contains unusual disulfide bonds that effectively cause the CDR3 loop to shield that portion of the heavy chain where light chains would typically bind (Stanfield et al., 2007). Dimeric heavy-chain immunoglobulins are also found in camels and llamas (Hamers-Casterman et al., 1993).

13.3.2 T-Cell Receptors

Whereas B-cell antigen receptors are found in a variety of unique immunoglobulin isotypes, T-cell receptors tend to be more evolutionarily conserved and closely resemble their higher vertebrate counterparts in their inferred structures. T-cell antigen receptors are found as heterodimers of either α and β chains or γ and δ chains (Figure 13.9). As in all other vertebrates examined, diversity regions appear to be absent in α and γ chains but present in β and δ chains. In humans, T cells with the γ/δ heterodimer are much more prevalent than those expressing α/β; greater than 95% of human T

lymphocytes are α/β cells. A major functional difference between α/β and γ/δ T cells is that, unlike α/β receptors, γ/δ receptors do not require MHC for the recognition of proteins and non-protein antigens (Carding and Egan, 2002; Chien et al., 1996). T cells expresssing the α/β heterodimer are present in elasmobranch fishes, but there is also significant expression of γ/δ lymphocytes in clearnose skates, *Raja eglanteria* (Miracle et al., 2001) and nurse sharks, *Ginglymostoma cirratum* (Criscitiello et al., 2010). The prominence of γ/δ T cells suggests that they may play a major role in the elasmobranch immune system.

Recently, sandbar shark (*Carcharhinus plumbeus*) TCR γ-chain cDNA has been successfully cloned (Chen et al., 2009), with a Basic Local Alignment Search Tool (BLAST) database search confirming clear identity to known TCR γ chains. Not surprisingly, the highest identity (50%) was to skate γ chain, while the identity to human γ chain was 23%. This value was considered too low to suggest specific relatedness to higher vertebrate chains but does support significant evolutionary divergence of TCR γ chains. Also of considerable interest is the recent identification of a unique TCR antigen receptor chain with features of both Ig and TCR chains (Criscitiello et al., 2006). This chain contains two variable regions, each encoded by separate VDJ segments (see discussion in Section 13.4.1), linked to a membrane-anchored δ-chain C domain (Figure 13.9). The distal of the two V domains is called NAR-TcRV and is closely related to the V domains of IgNAR, with amino-acid alignments establishing both of these domains as clearly distinct from the conventional Ig and TCR variable regions.

13.4 Immune System Genes

13.4.1 Immunoglobulin and TCR Genes

In most cases, the identification of genes coding for various elasmobranch immune function molecules has preceded identification of the transcript. IgM genes were the first to be isolated, including genes coding for IgM heavy and light chains from sharks and skates (Anderson et al., 1995; Harding et al., 1990a; Hinds and Litman, 1986; Hohman et al., 1992, 1993; Kokubu et al., 1988; Shamblott and Litman, 1989), IgX heavy chains from skates (Anderson et al., 1994, 1999; Harding et al., 1990b), and IgNAR (Greenberg et al., 1995), IgW (Bernstein et al., 1996a), and IgNARC (Greenberg et al., 1996) heavy chains from sharks.

Closer examination of how immune function genes in elasmobranch fishes are organized has revealed unique insights into the evolutionary origins and diversification of immunoglobulins and T-cell receptors. Unlike

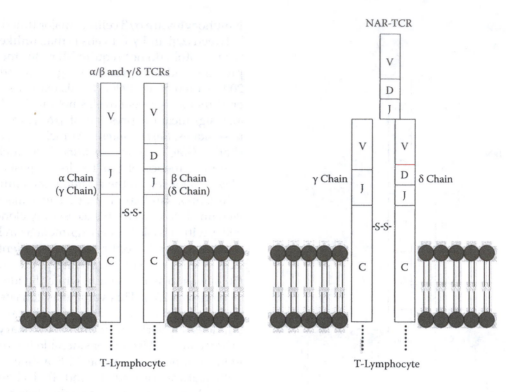

FIGURE 13.9
Schematic representation of the T-cell receptor (TCR) molecules. The receptor can be found as a heterodimer of either alpha and beta chains or gamma and delta chains. Also depicted is the inferred structure of the recently described NAR–TCR chain, with its unique arrangement of two V domains on a membrane-anchored TCR δ C domain.

mammalian immunoglobulin genes, in which gene segments coding for V (variable), D (diverse), J (joining), and C (constant) regions of the antibody molecule are located in separate clusters on the same chromosome, elasmobranch immunoglobulin genes are arranged in more than 100 clusters distributed on several different chromosomes, with each cluster containing one V segment, two D segments, and one J segment (linked to a C segment) (Kokubu et al., 1987). Instead of generating antibody diversity by recombination of one gene segment from each cluster as in mammals, elasmobranchs utilize junctional diversity (related to the additional D segment) and somatic mutation to create diversity (Hinds-Frey et al., 1993). In addition, the identification of a significant number of completely or partially prejoined gene segments suggests a form of inherited diversity.

Although phylogenetically "primitive" features appear to be maintained in Ig gene organization, the opposite is true for TCR genes, which closely resemble their higher vertebrate counterparts with respect to overall inferred structure (Rast and Litman, 1994) as well as diversity (Hawke et al., 1996). In the clearnose skate, *Raja eglanteria*, all four TCR genes (α, β, γ, and δ) have been identified and are similar to the four mammalian gene types in comparisons of both V and C region sequences, junctional characteristics, absence

of D regions in TCR α and γ genes, and presence of D regions in TCR β and γ genes (Rast et al., 1997). In contrast to the cluster arrangement of Ig genes, TCR genes are organized in the classic translocon arrangement found in higher vertebrates. That the organization and diversity of TCRs have changed little over the course of vertebrate phylogeny suggests that these four TCR types were likely present in the common ancestor of the living jawed vertebrates.

13.4.2 Major Histocompatibility Complex

Major histocompatibility complex class I and class II molecules are fundamental components of the adaptive immune system, and, as the most phylogenetically primitive animal possessing components of an adaptive immune system, cartilaginous fish represent important models for study of MHC. In contrast to teleost fish, class I and class II genes are closely linked in sharks (Ohta et al., 2000). Genes in the transporter associated with antigen processing family, or TAP genes, play important roles in MHC class I antigen processing/presentation through transport of antigenic peptides (Ohara et al., 2008). TAP genes are members of the ABC transporter family and play crucial roles in the processing and presentation of MHC class I restricted antigens (Ritz and Selig, 2001).

TAP consists of two subunits, TAP1 and TAP2, and is the only ABC transporter with a unique function in the immune system. Genes belonging to this family, TAP1 and TAP2, have been identified in horn sharks (Ohta et al., 1999, 2000). LMP7 and LMP2, two subunits of proteasomes encoded in MHC, play a role in generating endogenous peptides for presentation by class I molecules. Homologs of LMP7 and LMP2, as well as a proteasome gene unique to cartilaginous fish, called LMP7-like, have been identified in nurse shark (Kandil et al., 1996; Ohta et al., 2002). Proteasome genes LMP2 and LMP7, shark-specific LMP7-like, and the TAP1/TAP2 genes are linked to classical class I and class II genes in nurse shark MHC (Ohta et al., 2002). In all nonmammalian species examined to date, MHC class I is closely linked to proteasome/transporter genes, observations that suggest this MHC organization is primitive.

13.4.3 Transcription Factors

In addition to the major components responsible for adaptive immunity (i.e., Ig, TCR, and MHC), genes required for lymphopoiesis and Ig gene rearrangement/diversification, such as PU.1, Ikaros, RAG1, RAG2, and TdT, are also present in elasmobranch fishes (Anderson et al., 2001; Bernstein et al., 1996b; Haire et al., 2000; Rumfelt et al., 2001; Schluter and Marchalonis, 2003; Zapata et al., 1996). Ikaros, Aiolos, Helios, and Eos were identified in clearnose skate (Haire et al., 2000), the only species other than mouse in which all four Ikaros family members have been identified. Expression patterns of Ikaros and Helios were consistent with expression of their orthologs in corresponding tissues of higher vertebrate species. In addition, prominent expression of Ikaros is demonstrated in the Leydig organ and in epigonal tissue. Ikaros is expressed in thymus and spleen of both higher vertebrates and skate. Ikaros multigene family members are critical determinants in development of B and T lymphocytes as well as NK, and dendritic APC lineages also associated with immune function. Anderson et al. (2004) investigated structure and regulation of homologs of specific transcription factors that regulate mammalian T- and B-cell development in the clearnose skate, with skate orthologs of mammalian GATA-3, GATA-1, EBF-1, Pax-5, Runx2, and Runx3 characterized. GATA-3, Pax-5, Runx3, EBF-1, SpiC, and most members of the Ikaros family were shown through ontogeny to be coregulated with TCR or Ig expression and coexpressed with each other in combinations that generally correspond to known mouse T- and B-cell patterns, observations that support conservation of function. Most of the transcription factors involved in lymphocyte development in humans appear to function in similar roles in lymphocyte development in elasmobranch species (Anderson et al., 2004).

13.5 Gene Expression by Elasmobranch Immune Cells

13.5.1 Expression of Ig and TCR

Depending on the species studied and the stage of development or maturation, patterns of immunoglobulin and T-cell receptor expression are providing interesting glimpses into the roles of the primary and secondary lymphoid tissues of elasmobranch fishes. In perhaps the earliest study of the ontogenetic appearance of Ig-expressing cells, seven lymphomyeloid tissue sites in embryonic and post-hatch small-spotted catshark, *Scyliorhinus canicula*, were screened for Ig-positive cells (Lloyd-Evans, 1993). At 2 months of development, Ig-positive cells were detected in the liver, with interstitial kidney cells positive at 3 months, demonstrating the importance of these tissues in the early differentiation of Ig-producing cells. At 4 months, Ig expression appeared in thymus, spleen, and Leydig organ; epigonal and GALT tissue were the last to possess Ig-positive cells, at 6 months of the approximately 10-month development. With the exception of kidney and thymus, the remaining lymphomyeloid tissues in neonates (monitored up to 5 months post-hatch) retained expression of Ig. Because only Ig-positive cells were detected in these studies, the results could only reflect the presence of potential B lymphocytes.

With the identification in the clearnose skate, *Raja eglanteria*, of genes for IgM, IgX, and all four T-cell receptor antigens, the presence of potential B and T cells has been identified in a variety of tissues from embryonic and adult clearnose skates (Table 13.2). In characterizing the expression of TCR and Ig as a function of embryonic age, Miracle et al. (2001) found that all four classes of TCR (α, β, γ, and δ) were expressed in the skate thymus at the end of the second trimester of embryonic development (8 weeks into the 12-week developmental period). At this stage of embryogenesis, TCR gene expression was restricted to the thymus. Later in development and in hatchlings and adults, TCR gene expression also occurred in peripheral sites, suggesting that T lymphocytes originate in the elasmobranch thymus as they do in the thymus of higher vertebrates.

In contrast to expression of TCR genes, tissue expression of Ig genes during ontogeny was far more complex. With Ig expression in the spleen, Leydig, and epigonal organs of 8-week embryos, B-cell development occurred at multiple sites in the developing embryo, with relative abundance of IgX greater than IgM in most tissues. During embryonic development, several Ig genes, including IgX and IgM heavy chains and light chains I and II, were also expressed in the embryonic skate thymus, again with peak expression

TABLE 13.2

Expression of IgM and IgX Heavy-Chain Genes and TCR Antigen Genes in Tissues from Embryo and Adult Clearnose Skate (*Raja eglanteria*)

Tissue	Immunoglobulin	TCR	Major Expression	Minor Expression
8-Week Embryo				
Spleen	IgM, IgX	—	B cells	—
Thymus	IgX	$\alpha, \beta, \gamma, \delta$	T cells	B cells
Intestine	—	—	—	—
Liver	IgX	—	—	B cells
Leydig	IgX	—	—	B cells
Gonad	IgX	—	—	B cells
Adult				
Spleen	IgM, IgX	$\alpha, \beta, \gamma, \delta$	T and B cells	—
Thymus	—	$\alpha, \beta, \gamma, \delta$	T cells	—
Intestine	IgM, IgX	β, δ	—	T and B cells
Liver	—	—	—	—
Leydig	IgM, IgX	—	B cells	—
Epigonal	IgM, IgX	—	—	B cells

occurring 8 weeks into development. The expression of Ig genes in the developing thymus suggests that this tissue may serve as an early site for B-cell development during embryogenesis, as thymus in mature skates does not exhibit B-cell gene expression (Miracle et al., 2001). In agreement with the earlier findings of Lloyd-Evans et al. (1993), significant lymphoid gene expression was observed in embryonic skate liver, but not in adult liver.

Expression of two important genes with major roles in generation of immune receptor repertoire, RAG1 and TdT, was also found in embryonic skate thymus. RAG1 is an integral component in the segmental rearrangement of Ig and TCR genes; TdT functions in junctional diversification of both Ig and TCR. Expression of these genes in the embryonic skate thymus, along with expression of Ig and TCR genes, suggests that rearrangement and junctional diversification are occurring at this stage of development (Miracle et al., 2001).

More recently, differences in neonatal and adult expression and/or secretion of B-cell antigens have been characterized in nurse sharks, *Ginglymostoma cirratum* (Rumfelt et al., 2001, 2002, 2004). Low amounts of IgM are detected in neonatal nurse shark serum. It is not until about one month of age that levels approaching those of adult serum begin to appear. In place of the adult IgM, neonatal serum contains a novel IgM, termed IgM$_{1gj}$, expressed and secreted by cells in the neonatal spleen and epigonal organ. By 5 months of age, serum IgM reaches adult levels, but IgM$_{1gj}$ is no longer detectable. IgNAR, only slightly detectable by 3 months, does not approach adult levels until 5 months of age.

13.5.2 Cell Surface Markers

Cell surface markers have not been characterized in the elasmobranch immune system or in most non-mammalian species. CD83, an adhesion molecule for cell-mediated immunity, is a cell surface membrane glycoprotein (~45 kDa) whose surface expression is primarily restricted to dendritic cells (DCs), and it is a member of the Ig superfamily. Surface expression of CD83 is the standard lineage marker for activated or differentiated DCs, which play critical roles in antigen presentation and T-cell regulation. In mammalian cells, CD83 is believed to play a role in the induction of immune responses as expression of CD83 on DCs is also accompanied by upregulation of costimulatory molecules (CD80 and CD86). CD83 transcription is largely controlled by SP1 and NF-κB elements within the CD83 promoter, which is in agreement with the upregulation of CD83 upon infection or TNF-α, IL-1β, and mitogen activation. CD83 has been isolated and characterized from nurse shark (Ohta et al., 2004) and found to be largely expressed within immunologically important tissues, results suggesting that the role of CD83 has been conserved over 450 million years of vertebrate evolution.

13.5.3 Experimentally Induced Changes in Gene Expression

In the immune system, immunotoxic mechanisms serve as a first line of defense to protect cells and tissues from xenobiotic exposure. At the interface of divergent immune systems, elasmobranch fishes provide an excellent animal model to investigate the evolution

TABLE 13.3

Upregulated Genes in Bonnethead Shark (*Sphyrna tiburo*) Peripheral Blood Leukocytes Exposed to Red Tide Toxin (500 ng/mL PbTx-2) for 18 hr

Gene Transcript	Functional Category	E Value	Homologous Species
Sulfiredoxin-1	Oxidative stress	1.80E-38	*Homo sapiens*
XPA-binding protein 2 (HCNP protein)	Excision repair	4.39E-20	*Homo sapiens*
ATP synthase beta chain, mitochondrial precursor	ATP synthesis	5.75E-50	*Bos taurus*
ATP-binding cassette subfamily B member 9 precursor	Transporter function Antigen processing	3.72E-55	*Mus musculus*
Nuclear receptor coactivator 6	Transcription	2.49E-36	*Mus musculus*

and function of immunotoxic mechanisms, functions that have not yet been documented in immune cells at the phylogenetic level of elasmobranchs. Preliminary studies to assess the impact of naturally occurring toxins in the shark habitat have been conducted to begin to understand immunotoxic defense mechanisms in these animals. Using methods similar to those reported for loggerhead sea turtles (Walsh et al., 2010), bonnethead shark, *Sphyrna tiburo*, peripheral blood leukocytes (PBLs) were exposed *in vitro* to red tide toxins (brevetoxins, PbTx); the effects on immune cell gene expression were evaluated using suppression subtractive hybridization (SSH).

Peripheral blood leukocytes isolated from whole blood of healthy captive bonnethead sharks (*n* = 4) were divided into two cultures: (1) 500 ng/mL PbTx-2 (~558 n*M*), or (2) ethanol (0.035% v/v) as vehicle control. Cultures were incubated (25°C, 5% CO_2) for 24 hr, RNA was isolated, and suppression subtractive hybridization was conducted (EcoArray, Inc.; Gainesville, FL). Up- and downregulated genes affected by PbTx-2 exposure are shown in Tables 13.3 and 13.4. The tables list several genes sorted into broad functional categories. Many sequences were classified as "no hit," meaning that no closely matched sequences in publicly available databases could be identified. An expectation value of E-4 or lower was used for analysis. Gene functions were assigned based on the Gene Ontology database (http://amigo.geneontology.org/).

Identifiable genes that were upregulated in shark PBL after exposure to PbTx-2 included sulfiredoxin-1, XPA-binding protein-2, ATP synthase beta chain, ATP-binding cassette subfamily B member 9 precursor, and nuclear receptor coactivator 6 (Table 13.3). According to the AmiGO Gene Ontology website (http://amigo.geneontology.org/), sulfiredoxin-1 in humans contributes to oxidative stress resistance by reducing cysteine-sulfinic acid formed under exposure to oxidants in the peroxiredoxins and may act as a thioltransferase. It is located predominantly in the cytoplasm and is widely expressed in highest levels in kidney, lung, spleen, and thymus. XAB2 is a novel component involved in

transcription-coupled repair and transcription (Nakatsu et al., 2000). Other upregulated genes include ATP synthase beta chain and ATP-binding cassette subfamily B member 9 precursor (ATP-binding cassette transporter 9; TAP-like protein), proteins that are involved in xenobiotic response. Sequences corresponding to nuclear receptor coactivator 6 were also upregulated. The protein encoded by nuclear receptor coactivator 6 is a transcriptional coactivator that can interact with nuclear hormone receptors to enhance transcriptional activator functions and may also act as a general coactivator as it has been shown to interact with some basal transcription factors, histone acetyltransferases, and methyltransferases. Nuclear receptor coactivator 6 has emerged as an important coactivator not only for nuclear receptors, but also for a number of other well-known transcription factors such as c-Fos, c-Jun, CREB, NF-κB, ATF-2, heat shock factors, E2F-1, SRF, Rb, p53, and Stat2 (reviewed in Mahajan and Samuels, 2008). With regard to xenobiotic metabolism, nuclear receptor coactivator 6 has recently been shown to differentiate regulation of expression of genes in the cytochrome P-450 superfamily (Surapureddi et al., 2011).

An even greater number of identifiable downregulated genes were observed following the exposure of bonnethead shark PBLs to red tide toxin (Table 13.4). Downregulated genes include hemoglobin subunits alpha and beta, heat shock protein 30C, brain protein I3, CDH1-D, nipped-B-like protein, zinc finger protein HRX (ALL-1), cytochrome *c* oxidase subunit 3, probable ubiquitin carboxyl-terminal hydrolase CYLD, proteasome subunit beta type 6 precursor, and T-complex protein 1 subunit alpha (TCP-1α). Hemoglobins alpha and beta are involved in oxygen transport. Heat shock protein 30C belongs to the small heat shock protein (HSP20) family and is involved in stress response. Brain protein I3 participates in TNF-α-induced cell death. CDH1-D plays a role in controlling cell division. Nipped-B-like protein is a transcription factor that also plays a role in cell division. Zinc finger protein HRX (ALL-1) plays a role in development and hematopoiesis. Cytochrome *c* oxidase subunit 3 is involved in mitochondrial respiration.

TABLE 13.4

Downregulated Genes in Bonnethead Shark (*Sphyrna tiburo*) Peripheral Blood Leukocytes Exposed to
Red Tide Toxin (500 ng/mL PbTx-2) for 18 hr

Gene Transcript	Functional Category	E Value	Homologous Species
Hemoglobin beta chain	Oxygen transport	1.97E-18	*Staphylococcus aureus*
Hemoglobin alpha chain	Oxygen transport	6.13E-49	*Mustelus griseus*
Heat shock protein 30C	Cell stress	9.84E-11	*Squalus acanthias*
Brain protein I3 (pRGR2)	Apoptosis	7.22E-05	*Homo sapiens*
CDH1-D	Mitosis	3.64E-16	*Gallus gallus*
Nipped-B-like protein (Delangin) (SCC2 homolog)	Mitosis	2.40E-18	*Homo sapiens*
Zinc finger protein HRX (ALL-1) Trithorax-like protein	Transcription	1.58E-59	*Homo sapiens*
Cytochrome *c* oxidase subunit 3	Oxidation/reduction	3.67E-57	*Scyliorhinus canicula*
Probable ubiquitin carboxyl-terminal hydrolase CYLD	Signal transduction	3.72E-13	*Mus musculus*
Proteasome subunit beta type 6 precursor	Antigen processing	4.96E-14	*Homo sapiens*
T-complex protein 1 subunit alpha (TCP-1-alpha)	Protein modification	1.15E-68	*Paleosuchus palpebrosus*

Probable ubiquitin carboxy-terminal hydrolase CYLD likely functions as an ubiquitin–protein hydrolase involved in processing of ubiquitin precursors and ubiquitinated proteins and thus may play an important regulatory role at the level of protein turnover by preventing degradation of proteins through removal of conjugated ubiquitin. This protein is also an essential component of the TGF-β/BMP signaling cascade. Proteasome subunit beta type 6 precursor is likely a component of a multicatalytic proteinase complex and may catalyze basal processing of intracellular antigens. TCP-1 functions as a molecular chaperone and is part of a complex that folds various proteins, including actin and tubulin.

A similar experiment conducted in nurse shark PBLs resulted in upregulation of genes in response to treatment with PbTx-2 that were related to T cells, including T-cell receptor homolog, MHC class II, and butyrophilin, a gene that belongs to a family of proteins that play a prominent role in regulation of T-cell responses (Arnett et al., 2009). Because of the role of TAP1 gene in antigen presentation and potential upregulation in shark PBLs exposed to brevetoxin, real-time polymerase chain reaction (PCR) was conducted to quantify expression of this functionally important gene in brevetoxin-treated cells. PBLs from nurse sharks were used for these experiments because the sequence generated from the SSH corresponded with a nurse shark sequence (Nucleotide Accession Number: AF363579). Effects of PbTx-2 on an ATP-binding cassette transporter gene following *in vitro* exposure of nurse shark PBL were measured using primers designed to nurse shark TAP1 sequence. At 125 ng/mL PbTx-2 for 24 hr, expression of TAP1 was significantly upregulated ($n = 4$; $P = 0.04$; fold-change = 1.79) (Figure 13.10).

To summarize these observations, genes in shark immune cells are impacted by exposure to brevetoxin, a naturally occurring biotoxin in the shark habitat. The identification of gene sequences in response to red tide toxin exposure was limited by the number of publicly available sequences for these or related species that are

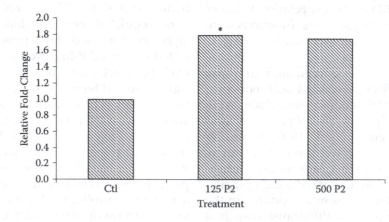

FIGURE 13.10

Quantitative polymerase chain reaction (Q-PCR) of TAP1 gene in nurse shark, *Ginglymostoma cirratum*, peripheral blood leukocytes (PBLs) following *in vitro* exposure to PbTx-2 (P2) at 125 or 500 ng/mL for 24 hr. $N = 4$ for each treatment. *Significantly different ($P = 0.04$) from control.

currently referenced in available databases. A large number of unidentified genes were affected by breve-toxin exposure, and it is highly possible that this pool of unidentified genes contains several genetic markers of brevetoxin exposure that currently cannot be identified. Available genes, however, suggest important impacts on oxidative stress, transcription, antigen processing, and protein modification in elasmobranch immune cells.

13.6 Immune Factors as Modulators of Apoptosis

13.6.1 Apoptotic Pathway

Caspases are cysteinyl aspartate proteinases, cysteine proteases that cleave substrates after an aspartic acid residue. Several enzymes that belong to the caspase family of enzymes exist within cells, and 11 caspases have been described in human cells (Kumar, 2007). Caspases exist normally in cells as inactive precursors, or procaspases. When they receive upstream apoptotic signals, procaspase enzymes undergo proteolytic processing to generate active or cleaved forms of the enzymes. The apoptotic cascade can be initiated through two major pathways involving (1) release of cytochrome c from mitochondria, or (2) activation of death receptors in response to ligand binding. With triggering of either pathway, caspases are activated and lead to pro-grammed cell death, or apoptosis. Caspases that are activated through recruitment to signaling complexes are known as initiator caspases due to their role in link-ing cell signaling with apoptosis. The main initiator cas-pases are caspases-2, -8, -9, and -10. Key effector caspases include caspase-3, -6, and -7. Caspase-9 is thought to be a key caspase in the intrinsic mitochondrial pathway and caspase-8 a key initiator of death receptor-mediated apoptosis (Kumar, 2007). In the intrinsic pathway, cell death signals lead to cytochrome c release from mito-chondria, which binds to apoptosome that recruits and activates caspase-9. Apoptosome-bound caspase-9 cleaves and activates caspase-3. Caspase-8 is an essen-tial component of the extrinsic cell death pathway initi-ated by TNF family members. In response to activation of death receptors by TNF family members, caspase-8 is recruited to death-inducing signaling complex (DISC) through binding to the adaptor protein, FADD, which leads to caspase-8 activation and cell death. Caspase-3 is activated following cleavage by caspase-8 or -9; it is the main downstream effector caspase and cleaves most of the substrates within apoptotic cells. Caspase-3 and caspase-7 function in amplifying mitochondrial caspase activation signaling. The cascade of enzymes is a very complex system and ultimately leads to programmed cell death, or apoptosis, within the cell upon receipt of appropriate signals. A simplified diagram of these path-ways is shown in Figure 13.11.

13.6.2 Induction of Apoptosis by Shark Immune-Cell-Derived Factors

Unique lymphomyeloid sites in elasmobranch fish, such as the epigonal organ, have potential as a source of novel immune regulators. Preliminary results with short-term cultures of the epigonal organ have supported this hypothesis. Potent cytotoxic or growth inhibitory activity has been demonstrated in conditioned media generated from cultures of elasmobranch epigonal cells, referred to as *epigonal conditioned medium* (ECM) (Walsh et al., 2004, 2006). In Walsh et al. (2006), cytotoxic activ-ity in ECM was demonstrated against nine different tumor cell lines, with preferential growth inhibition of malignant cells observed in assays using ECM against two different malignant/nonmalignant cell line pairs. Preliminary results indicating that cytotoxic activity proceeds through apoptotic mechanisms (Walsh et al., 2004) are described here.

Studies to investigate caspase activation in an ECM-treated T-cell leukemia cell line (Jurkat clone E6-1; ATCC, Manassas, VA) were conducted using western blotting as well as enzyme activity assays. For both types of assays, Jurkat cells were exposed to different concen-trations (0, 1, and 2 mg/mL) of ECM protein for 24 hr and then lysed. For western blotting, proteins in Jurkat cell lysates were separated by molecular size using SDS-PAGE (polyacrylamide gel electrophoresis), transferred to a nitrocellulose membrane, and incubated overnight at 4°C with primary antibodies for procaspase-9, acti-vated caspase-9, procaspase-3, and activated caspase-3 (Cell Signaling Technology; Danvers, MA) at a dilution of 1:1000. A secondary antibody, horseradish peroxidase (HRP)-conjugated anti-rabbit IgG, was used for detec-tion of antibody binding through enhanced chemilu-minescence (ECL). Antibody binding was documented and relative band densities were compared using a ChemiDoc gel documentation system (BioRad).

Caspase enzyme results from western blotting exper-iments are summarized in Table 13.5. A decrease in procaspase-9 binding and a corresponding increase in cleaved (active) caspase-9 were observed in Jurkat cells treated with ECM for 24 hr compared with untreated cells. At 1 mg ECM protein/mL, the amount of inactive procaspase-9 was approximately 49% of control while active caspase-9 increased by approximately 183% compared to control. At 2 mg ECM protein/mL, the amount of inactive procaspase-9 decreased to approxi-mately 35% of control while active capsase-9 increased by approximately 203% compared to control. In Jurkat

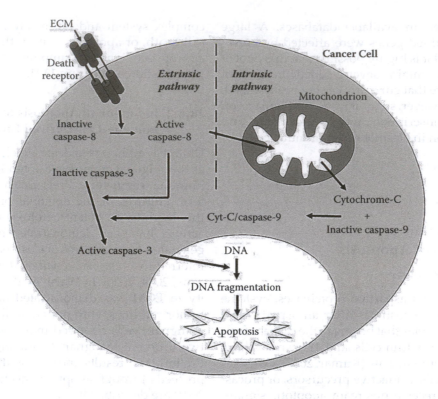

FIGURE 13.11
Simplified schematic diagram of a cancer cell demonstrating how ECM might bind to death receptors on the cell surface, triggering the activation of enzymes in the TNF-related apoptosis-inducing ligand (TRAIL) pathway, leading to nuclear damage and programmed cell death (apoptosis).

cells not exposed to ECM (control), there was no measurable conversion of the inactive to the active form of caspase-9. These results are consistent with activation of caspase-9 in Jurkat cells following treatment with ECM.

Although expected, a decrease in procaspase-3 binding was not observed in Jurkat cells treated with ECM for 24 hr compared with untreated cells (Table 13.5). An increase in the amount of cleaved caspase-3 present in cell lysates treated with ECM, however, was observed. At 1 mg ECM protein per mL, the amount of cleaved caspase-3 was approximately 250% of control; at 2 mg ECM protein per mL, the amount was approximately 267% of control. In Jurkat cells not exposed to ECM (control), there was no visible cleaved caspase-3 present. These results suggest activation of caspase-3 in Jurkat cells following treatment with ECM.

Measurement of caspase enzyme activity for caspases -3, -8, and -9 are summarized in Table 13.6. Activity of caspase-8 toward a commercially available substrate (*p*-nitroaniline; Sigma Chemical Company) increased four- to fivefold in Jurkat cells treated with ECM at 1 mg/mL and approximately twofold in Jurkat cells treated with ECM at 2 mg/mL. Caspase-9 assays to measure enzymatic activity against a substrate (LEHD conjugated to *p*-nitroaniline; R&D Systems) demonstrated that Jurkat cells treated with ECM possess a greater than 2.5-fold increase in caspase-9 activity compared to untreated cells. At 2 mg/mL ECM protein caspase-9 activity was 2.73-fold greater, and at 2 mg/mL ECM protein caspase-9 activity was 2.83-fold greater than control (untreated) Jurkat cells. Caspase-3 activity was measured with a fluorimetric assay kit (Invitrogen; Carlsbad, CA)

TABLE 13.5

Activation of Precursor to Active Forms of Caspase Enzymes in Lysates of Jurkat Cells Treated for 24 hr with Epigonal Conditioned Medium (ECM) Visualized Using Western Blots

Treatment	Procaspase-3	Caspase-3	Procaspase-9	Caspase-9
1 mg ECM/mL	110.56 ± 9.65	250.17 ± 28.94	49 ± 12.54	183 ± 37.35
2 mg ECM/mL	96.95 ± 8.60	266.75 ± 47.53	35 ± 7.21	203 ± 45.96

Note: Numbers represent percent change in band densities between treated and untreated cells. Each value represents three trials of the experiment.

TABLE 13.6

Fold-Increase in Caspase Activity in Lysates from Jurkat Cells Treated for 24 hr with Epigonal Conditioned Medium (ECM) Compared to Activity in Lysates from Untreated Control Jurkat Cells

Treatment	Caspase-3	Caspase-8	Caspase-9
1 mg ECM/mL	4.21 ± 1.05	4.87 ± 0.87	2.73 ± 0.41
2 mg ECM/mL	4.85 ± 1.21	1.63 ± 0.24	2.83 ± 0.59

Note: Each value represents four trials of the experiment.

and was found to increase in Jurkat cells treated with ECM for 24 hr. In response to 1 mg/mL ECM protein, caspase-3 activity increased approximately fourfold, and, in response to 2 mg/mL ECM protein, this activity approached fivefold greater than control (untreated) Jurkat cells.

Based on western blotting results and cellular enzyme activity assays, shark immune-cell-derived compounds are capable of inducing key apoptotic pathway enzymes (i.e., caspase-3, caspase-8, and caspase-9) in a T-cell leukemia cell line (Jurkat) and correlate with significant increases in activity of respective caspase enzymes in ECM-treated tumor cell targets.

13.7 Summary and Conclusions

The first decade of the 21st century has seen significant progress in contributing to both basic and applied aspects of elasmobranch immunology. As presented in this chapter, advances in knowledge range from pertinent findings in the evolution of vertebrate immune function to the realization that elasmobranch immune regulatory factors may someday contribute to the development of novel drug therapies. Not only has progress been achieved in developing a clearer picture of the organization of immune system genes, but detailed structural aspects of immune system gene transcripts are also beginning to be revealed. For the first time, the crystal structure of an elasmobranch antibody has been determined. It will not be surprising to see crystal structures of other elasmobranch antibodies in the decade to come. Also, progress has been made in recognizing the importance of the γ/δ receptor. Not only might γ/δ T cells play a more significant functional role in elasmobranch fishes than once thought, but there is also considerable speculation that the primordial antigen receptor was a primitive γ/δ molecule. In addition, important transcription factors have been identified in elasmobranch immune systems, and cell surface markers are beginning to be characterized, leading to the potential to further understanding of the evolution of adaptive immunity and antigen presentation. Finally, with the realization that immune regulatory factors originating from unique elasmobranch lymphomyeloid tissues, such as the epigonal organ, are capable of inducing apoptosis in human tumor cell lines, the possibility that elasmobranch immunology could some day impact human health is indeed exciting.

Acknowledgments

The authors gratefully acknowledge the use of the facilities at Mote Marine Laboratory (Sarasota, FL) and Clemson University (Clemson, SC). The authors wish to thank Stephanie Leggett and Theresa Cantu, for technical assistance in the laboratory, and Jack Morris, for maintenance of experimental tanks and experimental animals. All photographs and drawings appearing in figures are original and were prepared by CAL and CJW. Portions of the research described in this chapter were funded by grants to CAL and CJW from the Henry L. and Grace Doherty Charitable Foundation, the Polly Loomis Endowment, the National Institutes of Health (NIH Award No. 1R21CA100523-01A2), the University of South Florida High Tech Corridor, and the Mote Extended Red Tide Monitoring and Research Program Continuation of FWC Agreement No. 06125. U.S. Patents 6,908,627 and 7,309,501 have been issued to CJW and CAL for ECM-related technology.

References

Anderson, M., Amemiya, C., Luer, C., Litman, R., Rast, J., Niimura, Y., and Litman, G. (1994). Complete genomic sequence and patterns of transcription of a member of an unusual family of closely related, chromosomally dispersed immunoglobulin gene clusters in *Raja*. *Int. Immunol.* 6:1661–1670.

Anderson, M., Shamblott, M.J., Litman, R.T., and Litman, G.W. (1995). The generation of immunoglobulin light chain diversity in *Raja erinacea* is not associated with somatic rearrangement, an exception to a central paradigm of B cell immunity. *J. Exp. Med.* 181:109–119.

Anderson, M., Strong, S.J., Litman, R.T., Luer, C.A., Amemiya, C.T., Rast, J.P., and Litman, G.W. (1999). A long form of the skate IgX gene exhibits a striking resemblance to the new shark IgW and IgNARC genes. *Immunogenetics* 49:56–67.

Anderson, M.K., Sun, X., Miracle, A.L., Litman, G.W., and Rothenberg, E.V. (2001). Evolution of hematopoiesis: three members of the PU.1 transcription factor family in a cartilaginous fish, *Raja eglanteria*. *Proc. Natl. Acad. Sci. U.S.A.* 98:553–558.

Anderson, M.K., Pant, R., Miracle, A.L., Sun, X., Luer, C.A., Walsh, C.J., Telfer, J.C., Litman, G.W., and Rothenberg, E.V. (2004). Evolutionary origins of lymphocytes: ensembles of T-cell and B-cell transcriptional regulators in a cartilaginous fish. *J. Immunol.* 172:5851–5860.

Andrew, W. and Hickman, C.P. (1974). *Histology of the Vertebrates: A Comparative Text.* Mosby, St. Louis, MO, pp. 133–165.

Arnett, H.A., Escobar, S.S., and Viney, J.L. (2009). Regulation of costimulation in the era of butyrophilins. *Cytokine* 46(3):370–375.

Bernstein, R.M., Schluter, S.F., Shen, S., and Marchalonis, J.J. (1996a). A new high molecular weight immunoglobulin class from the carcharhine shark: implications for the properties of the primordial immunoglobulin. *Proc. Natl. Acad. Sci. U.S.A.* 93:3289–3293.

Bernstein, R.M., Schluter, S.F., Bernstein, H., and Marchalonis, J.J. (1996b). Primordial emergence of the recombination activating gene 1 (RAG1): sequence of the complete shark gene indicates homology to microbial integrases. *Proc. Natl. Acad. Sci. U.S.A.* 93:9454–9459.

Carding, S.R. and Egan, P.J. (2002). Gammadelta T cells: functional plasticity and heterogeneity. *Nat. Rev. Immunol.* 2:336–345.

Chen, H., Kshirsagar, S., Jensen, I., Lau, K., Covarrubias, R., Schluter, S.F., and Marchalonis, J.J. (2009). Characterization of arrangement and expression of the T cell receptor γ locus in the sandbar shark. *Proc. Natl. Acad. Sci. U.S.A.* 106:8591–8596.

Chiba, A., Torroba, M., Honma, Y., and Zapata, A.G. (1988). Occurrence of lymphohaemopoietic tissue in the meninges of the stingray *Dasyatis akajei* (Elasmobranchii, chondrichthyes). *Am. J. Anat.* 183:268–276.

Chien, Y.H., Jores, R., and Crowley, M.P. (1996). Recognition by gamma/delta T cells. *Annu. Rev. Immunol.* 14:511–532.

Clem, L.W. and Small, Jr., P.A. (1967). Phylogeny of immunoglobulin structure and function. I. Immunoglobulins of the lemon shark. *J. Exp. Med.* 125:893–920.

Clem, L.W., DeBoutaud, F., and Sigel, M.M. (1967). Phylogeny of immunoglobulin structure and function. II. Immunoglobulins of the nurse shark. *J. Immunol.* 99:1226–1235.

Criscitiello, M.F., Saltis, M., and Flajnik, M.F. (2006). An evolutionarily mobile antigen receptor variable region gene: doubly rearranging NAR-TcR genes in sharks. *Proc. Natl. Acad. Sci. U.S.A.* 103:5036–5041.

Criscitiello, M.F., Ohta, Y., Saltis, M., McKinney, E.C., and Flajnik, M.F. (2010). Evolutionarily conserved TCR binding sites, identification of T cells in primary lymphoid tissues, and surprising *trans*-rearrangements in nurse shark. *J. Immunol.* 184:6950–6960.

Ellis, A.E. (1977). The leucocytes of fish: a review. *J. Fish Biol.* 11:453–491.

Fänge, R. (1987). Lymphomyeloid system and blood cell morphology in elasmobranchs. *Arch. Biol.* 98:187–208.

Flajnik, M.F. and Dooley, H. (2009). The generation and selection of single-domain, V region libraries from nurse sharks. *Meth. Mol. Biol.* 562:71–82.

Flajnik, M.F. and Rumfelt, L.L. (2000). The immune system of cartilaginous fish. *Curr. Top. Microbiol. Immunol.* 248:249–270.

Greenberg, A.S., Avila, D., Hughes, M., Hughes, A., McKinney, E.C., and Flajnik, M.F. (1995). A new antigen receptor gene family that undergoes rearrangement and extensive somatic diversification in sharks. *Nature* 374:168–173.

Greenberg, A.S., Hughes, A.L., Guo, J., Avila, D., McKinney, E.C., and Flajnik, M.F. (1996). A novel 'chimeric' antibody class in cartilaginous fish: IgM may not be the primordial immunoglobulin. *Eur. J. Immunol.* 26:1123–1129.

Haire, R.N., Miracle, A.L., Rast, J.P., and Litman, G.W. (2000). Members of the Ikaros gene family are present in early representative vertebrates. *J. Immunol.* 165:306–312.

Hamers-Casterman, C., Atarhouch, T., Muyldermans, S., Robinson, G., Hamers, C., Bajyana Songa, E., Bendahamen, H., and Hamers, R. (1993). Naturally occurring antibodies devoid of light chains. *Nature* 363:446–448.

Harding, F.A., Cohen, N., and Litman, G.W. (1990a). Immunoglobulin heavy chain gene organization and complexity in the skate, *Raja erinacea. Nucleic Acids Res.* 18:1015–1020.

Harding, F.A., Amemiya, C.T., Litman, R.T., Cohen, N., and Litman, G.W. (1990b). Two distinct immunoglobulin heavy chain isotypes in a primitive, cartilaginous fish, *Raja erinacea. Nucleic Acids Res.* 18:6369–6376.

Hart, S., Wrathmell, A.B., Harris, J.E., and Grayson, T.H. (1988). Gut immunology in fish: a review. *Dev. Comp. Immunol.* 12:453–480.

Hawke, N.A., Rast, J.P., and Litman, G.W. (1996). Extensive diversity of transcribed TCR-β in a phylogenetically primitive vertebrate. *J. Immunol.* 156:2458–2464.

Hinds, K.R. and Litman, G.W. (1986). Major reorganization of immunoglobulin VH segmental elements during vertebrate evolution. *Nature* 320:546–549.

Hinds-Frey, K.R., Nishikata, H., Litman, R.T., and Litman, G.W. (1993). Somatic variation precedes extensive diversification of germline sequences and combinatorial joining in the evolution of immunoglobulin heavy chain diversity. *J. Exp. Med.* 178:825–834.

Hine, P.M. and Wain, J.M. (1987). The enzyme cytochemistry and composition of elasmobranch granulocytes. *J. Fish Biol.* 30:465–476.

Hohman, V.S., Schluter, S.F., and Marchalonis, J.J. (1992). Complete sequence of a cDNA clone specifying sandbar shark immunoglobulin light chain: gene organization and implications for the evolution of light chains. *Proc. Natl. Acad. Sci. U.S.A.* 89:276–280.

Hohman, V.S., Schuchman, D.B., Schluter, S.F., and Marchalonis, J.J. (1993). Genomic clone for the sandbar shark lambda light chain: generation of diversity in the absence of gene rearrangement. *Proc. Natl. Acad. Sci. U.S.A.* 90:9882–9886.

Hyder, S.L., Cayer, M.L., and Pettey, C.L. (1983). Cell types in peripheral blood of the nurse shark: an approach to structure and function. *Tissue Cell* 15:437–455.

Kandil, E., Namikawa, C., Nonaka, M., Greenberg, A.S., Flajnik, M.F., Ishibashi, T., and Kasahara, M. (1996). Isolation of low molecular mass polypeptide complementary DNA clones from primitive vertebrates. Implications for the origin of MHC class I-restricted antigen presentation. *J. Immunol.* 156:4245–4253.

Kobayashi, K. and Tomonaga, S. (1988). The second immunoglobulin class is commonly present in cartilaginous fish belonging to the order Rajiformes. *Mol. Immunol.* 25:115–120.

Kobayashi, K., Tomonaga, S., and Kajii, T. (1984). A second class of immunoglobulin other than IgM present in the serum of a cartilaginous fish, the skate, *Raja kenojei*: isolation and characterization. *Mol. Immunol.* 21:397–404.

Kobayashi, K., Tomonaga, S., and Tanaka, S. (1992). Identification of a second immunoglobulin in the most primitive shark, the frill shark, *Chlamydoselachus anguineus. Dev. Comp. Immunol.* 16:295–299.

Kokubu, F., Hinds, K., Litman, R., Shamblott, M.J., and Litman, G.W. (1987). Extensive families of constant region genes in a phylogenetically primitive vertebrate indicate an additional level of immunoglobulin complexity. *Proc. Natl. Acad. Sci. U.S.A.* 84:5868–5872.

Kokubu, F., Hinds, K., Litman, R., Shamblott, M.J., and Litman, G.W. (1988). Complete structure and organization of immunoglobulin heavy chain constant region genes in a phylogenetically primitive vertebrate. *EMBO J.* 7:1979–1988.

Kumar, S. (2007). Caspase function in programmed cell death. *Cell Death Different.* 14:32–43.

Litman, G.W., Anderson, M.K., and Rast, J.P. (1999). Evolution of antigen binding receptors. *Annu. Rev. Immunol.* 17:109–147.

Lloyd-Evans, P. (1993). Development of the lymphomyeloid system in the dogfish, *Scyliorhinus canicula. Dev. Comp. Immunol.* 17:501–514.

Luer, C.A., Walsh, C.J., Bodine, A.B., Wyffels, J.T., and Scott, T.R. (1995). The elasmobranch thymus: anatomical, histological, and preliminary functional characterization. *J. Exp. Zool.* 273:342–354.

Luer, C.A., Walsh, C.J., and Bodine, A.B. (2004). The immune system of sharks, skates, and rays. In: Carrier, J.C., Musick, J.A., and Heithaus, M.R. (Eds.), *Biology of Sharks and Their Relatives.* CRC Press, Boca Raton, FL, pp. 369–395.

Mahajan, M.A. and Samuels, H.H. (2008). Nuclear receptor coactivator/coregulator NCoA6(NRC) is a pleiotropic coregulator involved in transcription, cell survival, growth, and development. *Nucl. Recept. Signal.* 6:e002.

Marchalonis, J.J. and Edelman, G.M. (1965). Phylogenetic origins of antibody structure. I. Multichain structure of immunoglobulins in the smooth dogfish (*Mustelus canis*). *J. Exp. Med.* 122:601–618.

Marchalonis, J.J. and Edelman, G.M. (1966). Polypeptide chains of immunoglobulins of the smooth dogfish (*Mustelus canis*). *Science* 154:1567–1568.

Miracle, A.L., Anderson, M.K., Litman, R.T., Walsh, C.J., Luer, C.A., Rothenberg, E.V., and Litman, G.W. (2001). Complex expression patterns of lymphocyte-specific genes during the development of cartilaginous fish implicate unique lymphoid tissues in generating an immune repertoire. *Int. Immunol.* 13:567–580.

Nakatsu, Y., Asahina, H., Citterio, E., Rademakers, S., Vermeulen, W. et al. (2000). XAB2, a novel tetratricopeptide repeat protein involved in transcription-coupled DNA repair and transcription. *J. Biol. Chem.* 275:34931–34937.

Ohara, T., Ohashi-Kobayashi, A., and Maeda, M. (2008). Biochemical characterization of transporter associated with antigen processing (TAP)-like (ABCB9) expressed in insect cells. *Biol. Pharm. Bull.* 31:1–5.

Ohta, Y., Haliniewski, D.E., Hansen, J., and Flajnik, M.F. (1999). Isolation of transporter associated with antigen processing genes, *TAP1* and *TAP2*, from the horned shark, *Heterodontus francisci. Immunogenetics* 49:981–986.

Ohta, Y., Okamura, K., McKinney, E.C., Bartl, S., Hashimoto, K., and Flajnik, M.F. (2000). Primitive synteny of vertebrate major histocompatibility complex class I and class II genes. *Proc. Natl. Acad. Sci. U.S.A.* 97:4712–4717.

Ohta, Y., McKinney, E.C., Criscitiello, M.F., and Flajnik, M.F. (2002). Proteasome, transporter associated with antigen processing, and class I genes in the nurse shark, *Ginglymostoma cirratum*: evidence for a stable class I region and MHC haplotype lineages. *J. Immunol.* 168:771–781.

Ohta, Y., Landis, E., Boulay, T., Phillips, R.B., Collet, B., Secombes, C.J., Flajnik, M.F., and Hansen, J.D. (2004). Homologs of CD83 from elasmobranch and teleost fish. *J. Immunol.* 173:4553–4560.

Rast, J.P. and Litman, G.W. (1994). T cell receptor gene homologs are present in the most primitive jawed vertebrates. *Proc. Natl. Acad. Sci. U.S.A.* 91:9248–9252.

Rast, J.P., Anderson, M.K., Strong, S.J., Luer, C., Litman, R.T., and Litman, G.W. (1997). α, β, γ, and δ T cell antigen receptor genes arose early in vertebrate phylogeny. *Immunity* 6:1–11.

Ritz, U. and Seliger, B. (2001). The transporter associated with antigen processing (TAP): structural integrity, expression, function, and its clinical relevance. *Mol. Med.* 7:149–158.

Rowley, T.C., Hunt, T.C., Page, M., and Mainwaring, G. (1988). Fish. In: Rawley, A.F. and Ratcliffe, N.A. (Eds.), *Vertebrate Blood Cells.* Cambridge University Press, Cambridge, U.K., pp. 19–127.

Rumfelt, L.L., Avila, D., Diaz, M., Bartl, S., McKinney, E.C., and Flajnik, M.F. (2001). A shark antibody heavy chain encoded by a nonsomatically rearranged VDJ is preferentially expressed in early development and is convergent with mammalian IgG. *Proc. Natl. Acad. Sci. U.S.A.* 98:1775–1780.

Rumfelt, L.L., McKinney, E.C., Taylor, E., and Flajnik, M.F. (2002). The development of primary and secondary lymphoid tissues in the nurse shark, *Ginglymostoma cirratum*: B-cell zones precede dendritic cell immigration and T-cell zone formation during ontogeny of the spleen. *Scand. J. Immunol.* 56:130–148.

Rumfelt, L.L., Lohr, R.L., Dooley, H., and Flajnik, M.F. (2004). Diversity and repertoire of IgW and IgM VH families in the newborn nurse shark. *BMC Immunol.* 5:8.

Schluter, S.F. and Marchalonis, J.J. (2003). Cloning of shark RAG2 and characterization of the RAG1/RAG2 gene locus. *FASEB J.* 17:470–472.

Secombes, C.J. (1996). The nonspecific immune system: cellular defenses. In: Iwama, G. and Nakanishi, T. (Eds.), *The Fish Immune System.* Academic Press, San Diego, CA, pp. 63–103.

Shamblott, M.J. and Litman, G.W. (1989). Complete nucleotide sequence of primitive vertebrate immunoglobulin light chain genes. *Proc. Natl. Acad. Sci. U.S.A.* 86:4684–4688.

Stanfield, R.L., Dooley, H., Flajnik, M.F., and Wilson, I.A. (2004). Crystal structure of a shark single-domain antibody V region in complex with lysozyme. *Science* 305:1770–1773.

Stanfield, R.L., Dooley, H., Verdino, P., Flajnik, M.F., and Wilson, I.A. (2007). Maturation of shark single-domain (IgNAR) antibodies: evidence for induced-fit binding. *J. Mol. Biol.* 367:358–372.

Stokes, E.E. and Firkin, B.G. (1971). Studies of the peripheral blood of the Port Jackson shark (*Heterodontus portusjacksoni*) with particular reference to the thrombocytes. *Br. J. Haematol.* 20:427–435.

Surapureddi, S., Rana, R., and Goldstein, J.A. (2011). NCOA6 differentially regulates the expression of the CYP2C9 and CYP3A4 genes. *Pharmacol. Res.* 63(5):405–413.

Tomonaga, S., Kobayashi, K., Hagiwara, K., Yamaguchi, K., and Awaya, K. (1986). Gut-associated lymphoid tissue in the elasmobranchs. *Zool. Sci.* 3:453–458.

Walsh, C.J. and Luer, C.A. (1998). Comparative phagocytic and pinocytic activities of leucocytes from peripheral blood and lymphomyeloid tissues of the nurse shark (*Ginglymostoma cirratum* Bonaterre) and the clearnose skate (*Raja eglanteria* Bosc). *Fish Shellfish Immunol.* 8:197–215.

Walsh, C.J. and Luer, C.A. (2004). Elasmobranch hematology: identification of cell types and practical applications. In: Smith, M., Warmolts, D., Thoney, D., and Hueter, R. (Eds.), *Elasmobranch Husbandry Manual: Captive Care of Sharks, Rays and Their Relatives*. Ohio Biological Survey, Columbus, OH, pp. 307–323.

Walsh, C.J., Luer, C.A., Noyes, D.R., Smith, C.A., Gasparetto, M., and Bhalla, K.N. (2004). Characterization of shark immune cell factor (*Sphyrna tiburo* epigonal factor, STEF) that inhibits tumor cell growth by inhibiting S-phase and inducing apoptosis via the TRAIL pathway. *FASEB J.* 18(4):A60.

Walsh, C.J., Luer, C.A., Bodine, A.B., Smith, C.A., Cox, H.L., Noyes, D.R., and Gasparetto, M. (2006). Elasmobranch immune cells as a source of novel tumor cell inhibitors: implications for public health. *Integr. Comp. Biol.* 46:1072–1081.

Walsh, C.J., Leggett, S.R., Carter, B.J., and Colle, C. (2010). Effects of brevetoxin exposure on the immune system of loggerhead sea turtles. *Aquat. Toxicol.* 97:293–303.

Zapata, A.G., Torroba, M., Sacedon, R., Varas, A., and Vicente, A. (1996). Structure of the lymphoid organs of elasmobranchs. *J. Exp. Zool.* 275:125–143.

Section III

Ecology and Life History

14

Assessing the Age and Growth of Chondrichthyan Fishes

Kenneth J. Goldman, Gregor M. Cailliet, Allen H. Andrews, and Lisa J. Natanson

CONTENTS

14.1 Introduction

The ability to perform age determinations based on the examination of hard anatomical parts is of fundamental importance in fisheries research. Precise and accurate age information is the key to obtaining quality estimates of growth and other vital rates such as natural mortality and longevity and is essential for successful fisheries management. The effect of inaccurate age determinations on our understanding of population dynamics can lead to serious errors in stock assessment, often resulting in overexploitation (Beamish et al., 2006;

Cailliet and Andrews, 2008; Campana, 2001; Heppell et al., 2005; Hoenig and Gruber, 1990; Hoff and Musick, 1990; Longhurst, 2002; Musick, 1999; Officer et al., 1996). Fish age and growth are also critical correlates with which to evaluate many other biological (and pathological) processes, such as productivity, yield per recruit, prey availability, habitat suitability, and even feeding kinematics (Campana, 2001; DeVries and Frie, 1996; Robinson and Motta, 2002). Whereas age and growth are usually used together in phraseology, it is important to remember that each term has its own distinct meaning, which was eloquently stated by DeVries and Frie (1996):

Age refers to some quantitative description of the length of time that an organism has lived, whereas growth is the change in body or body part size between two points in time, and growth rate is a measure of change in some metric of fish size as a function of time.

It is important to understand the ages, growth characteristics, maturation processes, and longevity of fishes to assess their current population status and to predict how their populations will change in time (Cailliet et al., 1986a; Ricker, 1975). Fishery biologists have used age, length, and weight data as important tools for their age-based population models. Especially important are details about growth and mortality rates, age at maturity, and life span (Beamish et al., 2006; Cailliet and Andrews, 2008; Cortés, 1997; Heppell et al., 2005; Longhurst, 2002; Ricker, 1975). Over the past several decades, it has become obvious that many fisheries for chondrichthyan fishes are not sustainable at current exploitation rates or at all. As early as 1974, Holden suggested that these fishes had life histories, including late age at maturity, few offspring, and lengthy gestation periods, that made them vulnerable to overexploitation. Since then, fishing mortality on elasmobranchs, both as directed and as non-target catch in fisheries has increased (Baum et al., 2003; Bonfil, 1994; Casey and Myers, 1998; Stevens et al., 2000), and discards at sea are either underestimated or unknown (Camhi, 1999). These facts make the study of their life histories, including age, growth, and reproduction, even more important.

In the first review of elasmobranch ageing by Cailliet et al. (1986a), the age verification studies were relatively few. They included some statistical analyses, direct measurements of growth; marking anatomical features, such as vertebrae or spines, with oxytetracycline (OTC) and then describing their location over time in both laboratory and field studies; and several relatively new chemical studies of calcified structures (Welden et al., 1987). In a second review, Cailliet (1990) updated progress made and showed that many additional studies had derived age estimates based on opaque and translucent band pairs in calcified structures and that more studies were attempting to verify the periodicity with which these band pairs were deposited. Sufficient information to validate banding patterns in chondrichthyan hard parts was only available for six species at that time. In some cases, poor calcification and only partially verified band patterns prevented a full understanding of growth patterns. One species, the Pacific angel shark, *Squatina californica*, did not deposit any predictable growth bands in their vertebral centra (Natanson and Cailliet, 1990).

In the most recent review of age determination and validation studies in chondrichthyan fishes by Cailliet and Goldman (2004; see also Cailliet et al., 2006), which updated progress since the Cailliet (1990) review, 115 publications on at least 91 species of chondrichthyans had been produced using some form of age verification or validation, and approximately 68 were new to the list. Roughly 70% of the studies reviewed used vertebrae, either whole or sectioned, some of which were stained in one way or another; however, dorsal spines, jaws, and neural arches were also used. Other techniques, not necessarily involving calcified structures, were also employed to calculate growth coefficients or annual increments of growth. These include captive growth, field tag–recapture, and embryonic growth methodologies. In many cases, combinations of techniques were used. Cailliet and Goldman (2004) showed that precision analyses were beginning to become more common, as 21 studies calculated some form of reader precision estimates, such as average percent error (APE), Chang's (1982) APE value (D), percent agreement (PA), or coefficient of variation (CV); however, they felt it was still not a high proportion of the studies they reviewed and that greater efforts should be undertaken by researchers to conduct these types of analyses. There had also been an increase in the use of both verification and validation methodologies. The most common method employed was some form of marginal increment analysis or ratio and centrum edge analysis. Even though they are not very robust methods, some authors retained the use of size-frequency modal analysis and back-calculation techniques. Tag–recapture and laboratory growth studies were also used to provide growth estimates, and 18 studies used OTC to attempt age validation. It was apparent in many studies, and advocated for by Cailliet and Goldman (2004), that using combinations of verification and validation approaches is most likely to produce statistical results that provide biological meaning.

With such thorough recent reviews (Cailliet and Goldman, 2004; Cailliet et al., 2006), the purpose of this chapter is not to further update and summarize results from age and growth studies, but to instead provide a concise description of the processes, methodologies, and statistical analyses that can be used to quantify, verify, and validate age estimates in chondrichthyan fishes, with particular attention to emerging technologies being applied to both age determination and validation studies (e.g., histological processing and bomb radiocarbon dating, respectively). Finally, we briefly touch on the implications of growth rate, age at maturity, longevity, and the demographic traits of chondrichthyan fishes relative to their management and conservation.

14.2 Methodology

The age determination process consists of the following steps: collection of hard part samples, preparation of the hard part for age determination, examination (age reading), assessment of the validity and reliability of the resulting data, and interpretation (modeling growth). This section briefly discusses the hard parts that have been used to age chondrichthyan fishes and how to collect and prepare them for age determination. The examination of hard parts and assessment of validity and reliability of age estimates and modeling growth are discussed later in this chapter.

14.2.1 Structures

14.2.1.1 *Vertebrae*

Whole vertebral centra, as well as transverse and sagittally (i.e., longitudinally) sectioned centra, have been used for ageing elasmobranchs (Figure 14.1). Transverse sectioning will prevent bands on opposing halves from obscuring each other when illuminated from below. However, determining the age of older animals can still be problematic as bands become more tightly grouped at the outer edge of vertebrae and may be inadvertently grouped and counted together if transverse sections or whole centra are used for ageing, thereby causing underestimates of age (Cailliet, 1990; Cailliet et al., 1983a, 1986a). As such, sagittally sectioned vertebrae should be used for ageing unless it can be unequivocally demonstrated that identical ages can repeatedly be obtained from a given species using whole centra (Campana, 2001; Goldman, 2005). Because vertebral centra vary in shape from species to species, along the

column and sometimes within an individual centum, a variety of investigatory sagittal cuts should be made (e.g., top to bottom, side to side) in order to reduce the potential for introducing error in centrum radius measurements, which could result in poor centrum radius to body length correlations, and to identify which type of sagittal cut provides intermedialia to assist with identifying annuli.

14.2.1.2 *Spines*

Dorsal fin spines (Figure 14.2) have been another useful hard part for ageing some elasmobranchs, most notably dogfish sharks (Family Squalidae) (Ketchen, 1975; McFarlane and Beamish, 1987a; Nammack et al., 1985; Tribuzio, 2010; Tribuzio et al., 2010). As dorsal fin spines become increasingly popular as an ageing structure, we recommend that Clarke and Irvine's (2006) guide to spine ageing terminology be used. Spines from the second dorsal fin are preferred for ageing, as the tips of first dorsal fin spines tend to be more worn down, leading to an underestimation of age. Correction factors can be calculated to estimate ages of individuals with worn spines (Ketchen, 1975; Sullivan, 1977). Additionally, a cautious approach should be used, as spines may have different growth internally and externally (Cotton, 2010; Irvine, 2006a,b).

Spines can be read whole (without further preparation) by wet-sanding the enamel and pigment off the surface and polishing the spine or from the exposed surface resulting from a longitudinal cut (Ketchen, 1975; McFarlane and Beamish, 1987a). Cross-sectioned dorsal fin spines have also proved useful in assessing ages in some squaloids and chimaeras (Calis et al., 2005; Clarke et al., 2002a,b; Freer and Griffiths, 1993; Sullivan, 1977).

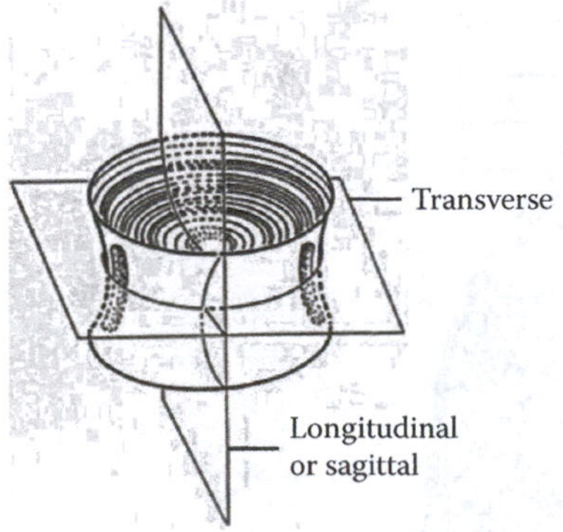

FIGURE 14.1
The two sectioning planes that can be used on vertebral centra. (Courtesy of G.M. Cailliet, Moss Landing Marine Laboratories, California State University.)

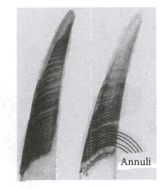

FIGURE 14.2
Spiny dogfish, *Squalus acanthias*, second dorsal fin spines showing annuli. First spine was aged at 42 years; second spine was aged at 46 years. (Courtesy of G.A. McFarlane, Pacific Biological Station, Nanaimo, British Columbia, Canada.)

14.2.1.3 Neural Arches

Calcium deposits have been documented in the neural arches of elasmobranch fishes (Cailliet, 1990; Peignoux-Deville et al., 1982), but they had not been used for ageing. In 2002, McFarlane and colleagues introduced the first attempt to use this structure for ageing elasmobranchs by silver nitrate staining the neural arches of sixgill sharks, *Hexanchus griseus*. The results from that preliminary study indicate that neural arches may provide another ageing structure for elasmobranch species with poorly calcified vertebral centra, but the method has not been followed up on or validated (Figure 14.3).

14.2.1.4 Caudal Thorns and Other Structures

Novel approaches to ageing various elasmobranchs continue to arise, and researchers may want to begin collecting additional hard parts from specimens in the field to be experimented with in the laboratory. Gallagher and Nolan (1999), for example, used caudal thorns (Figure 14.4) along with vertebral centra to determine age in four bathyrajid species, demonstrating high precision in ages between the two parts. Gallagher et al. (2005a,b, 2006) further elaborated on the structure and growth processes in caudal thorns. Comparing counts in more than one hard part is a

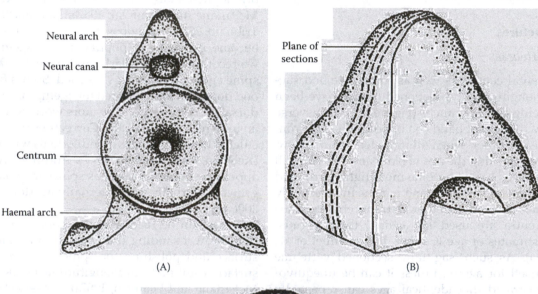

(A)   (B)

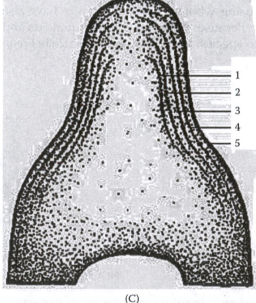

(C)

FIGURE 14.3
Neural arch diagrams from sixgill sharks, *Hexanchus griseus*. (A) The whole centra, (B) the planes at which sectioning took place, and (C) the resulting banding pattern after silver nitrate staining. (From McFarlane, G.A. et al., *Fish. Bull.* 100, 861–864, 2002. With permission.)

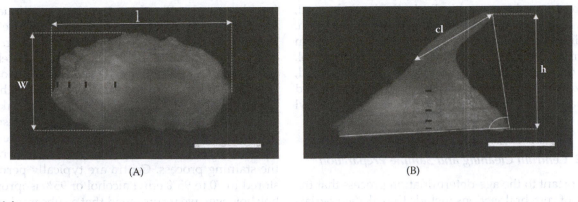

(A)   (B)

FIGURE 14.4
Caudal thorn of a 2-year-old *Raja clavata*, 297-mm TL male, in (A) superior and (B) lateral view. Measurements are BP length (l), width (w), and height (h) and crown length (cl) and angle. The black marks correspond to hyaline bands. Band-counting criteria, from the crown to the edge, were applied as follows: The first band corresponds to the proto-thorn margin, the second and third are birthmarks from the first and second years, and the fourth mark is the beginning of the third year, but because the sampling date was before the birth date this band was not considered. (Scale bar: 1 mm). (From Serra-Pereira et al., *ICES J. Mar. Sci.*, 65, 1701–1709, 2008. With permission.)

common age verification technique used in teleost ageing studies; however, it is not frequently conducted on cartilaginous fishes because of the lack of multiple hard parts available for comparison. Thorns can be a reliable hard part for ageing, and they allow more accurate age assessment than vertebrae do (Serra-Pereira et al., 2005, 2008). Some studies found similar results with vertebrae and caudal thorns, but still chose to use vertebrae for age assessment purposes (e.g., Matta and Gunderson, 2007). Several recent studies (Ainsley, 2009; Davis et al., 2007; Maurer, 2009; Moura et al., 2007; Perez et al., 2011) indicate that vertebral counts have provided more consistent and reliable age assessment, whereas counts from caudal thorns have provided more variable and consistently fewer annuli for the same individual. Where appropriate, however, these structures have the potential to greatly aid our understanding of the life histories of several species of skate and ray; this is a novel approach for chondrichthyans, as it represents the first possible nonlethal age assessment method for this group of fishes. Tanaka (1990) experimented with growth bands in the upper jaw of one specimen of the wobbegong, *Orectolobus japonicus*, kept in captivity and found evidence for growth bands there; however, his search for nonvertebral cartilaginous tissues in this species and the swell shark, *Cephaloscyllium umbratile*, were not productive.

14.2.2 Sampling and Processing Specimens

14.2.2.1 Taking Samples

The location in the vertebral column from which samples are taken for ageing can have a statistically significant effect on increment counts (Natanson and Cailliet, 1990; Natanson et al., 2008; Officer et al., 1996); therefore,

it is important to use the larger, more anterior (thoracic) centra for age studies because smaller centra from the caudal region may lack some bands (Cailliet et al., 1983b). This emphasizes the importance of standardizing the vertebral sampling region for all ageing studies, allowing for precise, valid comparisons among individuals within a population and for more accurate comparisons between populations. If possible, entire vertebral columns should be examined before starting an ageing project. By examining a sample of vertebrae from the various regions along the column (for example, every fifth centrum from head to tail), the ager can determine if the count changes along the column. This is important when the ager is not the collector. In many cases, centra are taken from the easiest possible area; if all centra provide the same count, then these vertebrae can be interchanged for counts but not for measurements. Additionally, a large change in count along the vertebral column may indicate that this is not the appropriate technique for ageing the species; examples of such species include the angel shark (Natanson and Cailliet, 1990) and the basking shark (Natanson et al., 2008). All ageing techniques require centra free of tissue; however, we recommend that neural arches be left on several centra based on their potential for use in ageing if vertebral centra or spines show no banding pattern (McFarlane et al., 2002). The position of the neural and hemal arches can also indicate the approximate position of the centrum along the column and can be used to determine if the measurements are comparable. This may be especially useful if samples are not collected by the ager or are from an area of the vertebral column that differs from the standard location being sampled. Dorsal fin spines should be removed by cutting horizontally just above the notochord to ensure that the spine base and stem are intact.

Vertebral samples are typically individually bagged, labeled, and stored frozen until ready for preparation. If freezing is not an option, vertebrae can be fixed in 10% formalin for 24 hr and then preserved in alcohol. Second dorsal fin spines are typically bagged, labeled, and frozen until returned to the laboratory or are placed immediately in 70 to 95% ethyl alcohol or 95% isopropyl alcohol.

14.2.2.2 Centrum Cleaning and Sample Preparation

It is important to the age-determination process that the majority of vertebral sections include the calcified radials of the intermedialia, but this is not always easy. The radials of the intermedialia of carcharhinid sharks, for example, are relatively hard, robust, and numerous, making centra nearly solid; in contrast, the radials of the intermedialia in lamnoid sharks are less numerous, softer, and quite fragile. Large interstitial spaces between radials can prevent intermedialia from being present in a sectioned centrum. Conducting several preliminary "test cuts" should reveal the best location to make a sagittal cut that will include intermedialia. When the best location has been found, all cuts must be consistent (i.e., made in the same location on each centrum) to minimize error in centrum measurements, which are critically important for centrum edge analyses and back-calculations. In the experience of the authors, the best cut to obtain the radials of the intermedialia has most frequently been a side-to-side cut from the vertebral centrum vs. a top-to-bottom one. Additionally, this cut provides symmetrical sides (relative to the focus), which provides four corpus calcarea to use for ageing; for example, as the girth greatly expands with age in the large lamnids, the vertebra also become wider in response. Along with this growth, the vertebra widens at the bottom, thus a top-to-bottom cut results in a wide "V" from the focus to the bottom of the section and a thin "V" from the focus to the top of the section. This nonsymmetrical type of sagittal section does not allow for measurements from each half of the sample to be compared, thereby relegating the ager to choosing either the top or bottom part of the section and limiting analyses to two corpus calcareum arms from which to age the sample.

The following provides a synopsis of methods for cleaning vertebral centrum; however, parts or all of the recommendations below may or may not be necessary depending on species and condition of the samples at the time of preparation. Vertebral samples need to be thawed if frozen or washed if preserved in alcohol, cleaned of excess tissue, and separated into individual centra. Tissue-removal techniques vary with species. For many, soaking the centrum in distilled water for 5 min followed by air-drying allows the connective tissue to be peeled away. Soaking in bleach may be required

for other species. Bleaching time is proportional to centrum size and ranges from 5 to 30 min. After bleaching, the centrum is rinsed thoroughly in water. Another simple and effective method is to soak vertebral sections in a 5% sodium hypochlorite solution. Soak times can range from 5 min to 1 hr depending on the size of the vertebrae and should be followed by soaking centra in distilled water for 30 to 45 min (Johnson, 1979; Schwartz, 1983). This method also assists in removal of the vertebral fascia between centra and does not affect the staining process. Centra are typically permanently stored in 70 to 95% ethyl alcohol or 95% isopropyl alcohol; however, we recommend that a subsample of centra be permanently stored in a freezer in case it is needed for staining and because long-term exposure to alcohol may reduce the resolution of the banding pattern (Allen and Wintner, 2002; Wintner et al., 2002).

Vertebrae can be analyzed whole or sectioned, but sectioning is typically ideal. Vertebral sectioning is often done with a low-speed, diamond-blade saw; however, a wide variety of saws are available that can be used for this purpose. Each centrum should be sagittally sectioned immediately adjacent to the center of its focus (so the center of the focus is at the edge of the cut) and then cut again approximately 1.5 mm off-center. Accuracy and precision in these cuts (i.e., always including the center point of the focus) will reduce centrum measurement error among individuals. A double-blade saw can be used to eliminate the problem of cutting a small section off of half of a vertebral centrum (spacing between blades should be no less than 0.6 mm to allow for some sanding or polishing). Large vertebrae can be handheld for cutting, whereas imbedding small vertebrae in resin (thermoplastic cement) and then cutting may prove easier. If a rotary saw is not used, small vertebrae can be sanded in half, mounted, sanded thin, and polished. A grinder may be used to section large vertebrae, which can then be mounted, sanded thin, and polished. If necessary, sections can be cut with small handsaws and even scalpels when working with very small centra, or half of the centrum can be worn away with aluminum-oxide wheel points and fine sandpaper attachments for the same tool (Cailliet et al., 1983a,b). Large vertebrae may be handheld or secured in a vise and cut with a small circular saw attachment on a jeweler's drill or even ground in half with a grinder.

If working with vertebrae with small numbers of radials (e.g., lamniform), pressing the sagittally cut (bow-tie-shaped) sections between two pieces of Plexiglas® and placing weight on the top sheet during drying will prevent warping, which can effect increment and centrum radius measurements. Sectioned vertebrae should be air-dried for 12 to 24 hr (under a ventilation hood, if possible), and then mounted onto microscope slides. The focus side of the vertebral section must consistently be placed face down on the slide when mounting in order to avoid adding to centrum measurement error that will lead to subsequent analysis error. Any typical slide-mounting medium will suffice for attaching vertebral sections. After mounting the sections to slides, they should be sanded with wet fine-grit sandpaper in a series (grades 320, 400, and finally 600 for polishing) to approximately 0.3 to 0.5 mm and air-dried. Alternatively, sections can be stored in 70% EtOH and subsequently viewed under a microscope submerged in a small amount of water or EtOH. It may be prudent to attempt both wet and dry reads on vertebral samples as one way may provide easier band interpretation for a given species. A binocular dissecting microscope with transmitted light is generally used for identification of growth rings and image analysis.

14.2.2.3 Centrum Staining

Numerous techniques have been used in attempts to enhance the visibility of growth bands in elasmobranch vertebral centra. Many are simply stained (Figure 14.5), but the list of techniques includes alcohol immersion (Richards et al., 1963), xylene impregnation (Daiber, 1960), histology (Casey et al., 1985; Ishiyama, 1951; Natanson, 1992; Natanson and Cailliet, 1990; Natanson and Kohler, 1996; Natanson et al., 1995, 2007; Skomal and Natanson, 2003), x-radiography (Aasen, 1963; Cailliet et al., 1983a,b; Martin and Cailliet, 1988; Natanson and Cailliet, 1990), x-ray spectrometry (Jones and Green, 1977), cedarwood oil (Cailliet et al., 1983a; Neer and Cailliet, 2001), alizarin red (Cailliet et al., 1983a; Goosen and Smale, 1997; Gruber and Stout, 1983; LaMarca, 1966), silver nitrate (Cailliet et al., 1983a,b; Schwartz, 1983; Stevens, 1975), crystal violet (Anislado-Tolentino and Robinson-Mendoza, 2001; Carlson et al., 2003; Johnson, 1979; Schwartz, 1983), graphite microtopography (Neer

and Cailliet, 2001; Parsons, 1983), a combination of cobalt nitrate and ammonium sulfide (Hoenig and Brown, 1988), and the use of copper-, lead-, and iron-based salts (Gelsleichter et al., 1998a). Many of these studies used multiple techniques on a number of species for comparison, particularly Schwartz (1983) and Cailliet et al. (1983a). These studies show that the success of each technique is often species specific and that slight modifications in technique may enhance the results.

In addition to their effectiveness, the various techniques mentioned vary in their simplicity, cost, and technological requirements. Histological processes have proved useful but require specialized equipment and a number of chemicals and are relatively time consuming. The resulting staining process is long lasting, with no color change in vertebral sections after 15 years (Casey et al., 1985). X-radiography has proved useful in many studies but has the obvious necessity of an appropriate x-ray machine and film-processing capabilities. Although x-ray spectrometry may hold promise (Cailliet et al., 1983a, 1986b; Casselman, 1983; Jones and Green, 1977), it is also time consuming and expensive. Simpler, less expensive, and more time-efficient staining techniques, such as crystal violet, silver nitrate, cedarwood oil, graphite microtopography, and alizarin red, should be used prior to considering other methods. Although these techniques have been tried, many have not yet been thoroughly evaluated; for example, the cobalt nitrate and ammonium sulfide stain suggested by Hoenig and Brown (1988) is easy to use and time efficient and has provided quality results for two species, but it has not been extensively applied. A microradiographic method using injected fluorochrome dyes to aid in resolving individual hypermineralized increments was applied to captive gummy sharks, *Mustelus antarcticus*, with success (Officer et al., 1997), but this method has not been extensively applied or thoroughly evaluated. The possibility that this method may also have application as a validation technique needs to be investigated.

14.2.2.4 Histology

Histological processing typically produces finer detail and improved clarity of band patterns compared to what is obtained by gross sectioning. In many cases, enough detail is seen on gross-sectioned vertebra, and the time and expense of histology are not warranted. In some instances, however, gross sectioning of vertebral centra does not produce clear band patterns, and it is necessary to explore other means of elucidating band pairs.

Histological processing of vertebral centra of elasmobranchs was used as far back as 1951 for various skate species (Ishiyama, 1951). Since that time the method has

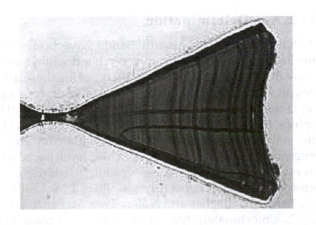

FIGURE 14.5
(See color insert.) Vertebral section stained with hemotoxylin. (Staining by S. Tanaka; photograph courtesy of K.G. Yudin and G.M. Cailliet.)

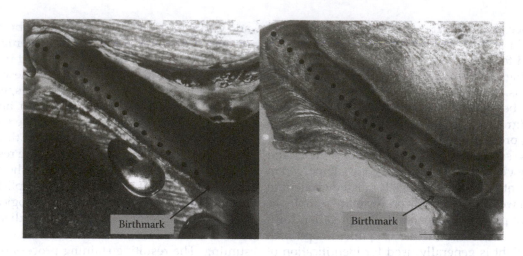

FIGURE 14.6
(See color insert.) Comparison showing gross sectioned vertebral centrum vs. histologically prepared vertebral section of vertebrae from the same individual Commander skate, *Bathyraja lindbergi*. (Photograph courtesy of J. Maurer.)

been used sporadically (Casey et al., 1985; Natanson, 1993; Natanson and Kohler, 1996; Natanson et al., 1995), due in part to few laboratories having the equipment and training and the expense in terms of both funds and time. The current method for processing vertebra for histology follows methods developed by Casey et al. (1985) for the sandbar shark and has evolved slightly as more species are being aged. The method for skates was described in detail in Natanson et al. (2007).

The band pairs of various skate species and many carcharhinid species have been more easily interpreted using histology. Recently, with an increase in the number of ageing studies on skate species and the difficulty surrounding the band pair clarity, histology has become an increasingly important tool for assessing age (Ainsley, 2009; Maurer, 2009; Natanson et al., 2007). Maurer (2009) found that band pair clarity was more distinct and more uniform using the histologically prepared sections vs. those prepared with gross sectioning (Figure 14.6). Additionally, band-pair counts using the histologically prepared samples showed less reader bias than those with gross sectioning. This typically led to older ages being assessed due to band elucidation and provided better repeatable age assessments, thus leading to higher reader precision.

An overview of the process follows: Vertebra should be cleaned of muscle but do not have to be soaked or otherwise scraped before processing. Vertebra need to be preserved in 70% ethanol (EtOH), which can be done when they are whole or after the initial sectioning. In general, one vertebra is sectioned along the lateral plane and around the focus of the vertebra using a rough saw to 3- to 5-mm thickness (larger vertebra may need to be cut thicker). Cut sections are then stored in 70% EtOH. Sections are decalcified with RDO (Dupage Kinetics Laboratories; Plainfield, IL), a rapid bone decalcifier,

though other products can be used and embedded in paraffin. When the sections are in paraffin "blocks," they are sectioned to approximately 80 to 100 μm using a sledge microtome. The final sections used for age assessment are those that are cut directly through the focus. These are placed in xylene and must be stained within 24 hr or less. After staining, sections are mounted on glass slides using an aqueous mounting media and a coverslip. Chemicals, standard embedding, and staining times are modified for use with the vertebra (Natanson et al., 2007). Processing times may have to be adjusted based on the size of the sections and the strengths of the solutions (i.e., older stain may take longer). Once final sections are mounted, they can be examined under a binocular microscope or photographed for image analysis.

14.3 Age Determination

Although concentric growth bands have been documented in the vertebral centra of chondrichthyans for more than 90 years (Ridewood, 1921), ageing these fishes has proved a slow and difficult process. Counts of opaque and translucent banding patterns in vertebrae, dorsal fin spines, caudal thorns, and neural arches have provided the only means of obtaining information on growth rates in these fishes, as they lack the hard parts, such as otoliths, scales, and bones, typically used in age and growth studies of teleost fishes (Cailliet, 1990; Cailliet et al., 1986a,b; Gallagher and Nolan, 1999; McFarlane et al., 2002). Unfortunately, the vertebral centra of many elasmobranch species (such as numerous deepwater species) are too poorly calcified to provide information on age, most species have no dorsal spines, and there may

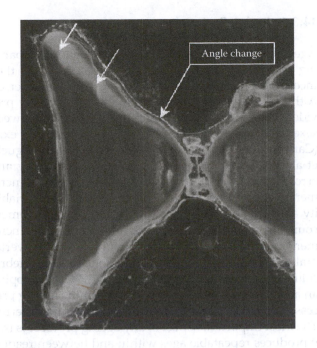

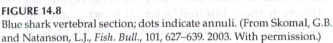

FIGURE 14.7
Sagittal section of a vertebral centrum from a 2-year-old smooth dogfish, *Mustelus canis*, showing the distinct notching pattern (white arrows) that accompanied the distinct banding pattern. (Courtesy of C. Conrath, NOAA/NMFS, Kodiak, Alaska.)

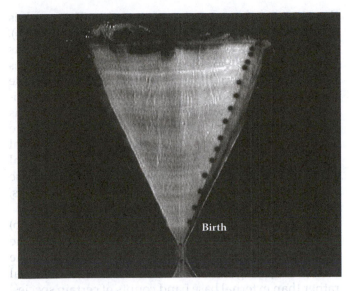

FIGURE 14.8
Blue shark vertebral section; dots indicate annuli. (From Skomal, G.B. and Natanson, L.J., *Fish. Bull.*, 101, 627–639. 2003. With permission.)

be no tangible relationship between observed banding patterns and growth (Cailliet, 1990; Cailliet et al., 1986b; McFarlane et al., 2002; Natanson and Cailliet, 1990). These circumstances continue to cause difficulties in making age estimates for many species.

The most commonly distinguishable banding pattern in sectioned centra when viewed microscopically is one of wide bands separated by distinct narrow bands (Figure 14.7). The terms *opaque* and *translucent* are commonly used to describe these bands, and they tend to occur in summer and winter, respectively; however, the opacity and translucency of these bands vary considerably with species, light source, and methodology (Cailliet, 1990; Cailliet et al., 1986a; Goldman, 2002; Wintner et al., 2002). It should not be assumed that the opaque and translucent nature of vertebral bands in different species will be similar; however, the pattern of wide/narrow banding tends to be consistent. An annulus is usually defined as the winter band. The difference in appearance between summer (wide) and winter (narrow) growth bands provides the basis for age determinations. In many species, this so-called winter band actually forms in the spring (Sminkey and Musick, 1995a).

In elasmobranch vertebral sections, each pair of wide/narrow bands extends across one arm of the corpus calcareum, across the intermedialia, and across the opposing corpus calcareum arm and is considered to represent an annual growth cycle; the narrow bands, hereafter referred to as *rings* or *annuli*, are what are

counted (Figure 14.8). In many skate species processed using histological methods, the intermedialia does not show bands; in these cases, criteria must be adjusted to accommodate the species. It must be noted that counting these rings at this point in the process carries with it the assumption that each one represents a year's growth; however, the validity of this assumption must be tested. (The term *annulus* is defined as a ringlike figure, part, structure, or marking, but annuli must be shown to be annual in their deposition.) The age determination process for spines is virtually identical to that for vertebrae; however, Ketchen's (1975) method for calculating age from worn spines should be considered instead of discarding the spines. This method uses an age to spine-base-diameter regression for unworn spines to allow an estimation of age for individuals with worn spines.

In recent studies on deep-sea elasmobranch age and growth using spines, it was found that spines exhibited different growth internally and externally. Cotton (2010) found discrepancies between counts of internal and external base bands on fin spines of several species. In most of the centrophorids, the internal bands outnumbered those on the enamel cap or on the external base of the spine. In centrophorids, the thickness of the dentine layers was not greatly reduced near the base as in somniosid and etmopterid spines. In those species, this reduction in thickness of the dentine layers near the spine base allowed for better visualization of growth bands, whereas the thicker base of the centrophorid fin spine occluded growth bands in this region of the spine.

Also, for two etmopterids and one somniosid included in his study, Cotton (2010) found that more bands were visible on the external base of the spine than in a transverse section of the spine. This result is in agreement with the findings of Irvine et al. (2006a,b), who reported a discrepancy in the external band counts (base bands) of *Centroselachus crepidater* and *Etmopterus baxteri*, compared with those formed internally, visible in a transverse section. The spine grows longitudinally at a much faster rate than it does centrifugally, as the spine is much longer than it is wide. The distance between growth bands deposited along the external (longitudinal) surface is greater than between internal bands in the dentine layer, thus making external base bands more easily discernable than internal bands, especially late in life when growth slows markedly.

The type of discrepancy found in Irvine et al. (2006a,b) and Cotton (2010) could lead to an underestimate of the shark's age and an overestimation in the growth rate constant (k) if age estimates are derived from internal rather than external base band counts of certain species. Although these internal/external band count discrepancies may be unique to these species (or genera), it is important for future fin spine ageing studies to examine the possibility of such a discrepancy in the species being investigated.

Centrum banding patterns in vertebral centra may be related to physiological changes induced by changes in environmental parameters such as temperature and photoperiod (Branstetter, 1987; Cailliet et al., 1986a). Some species, however, such as the little skate, *Leucoraja erinacea* (Natanson, 1993), and the Pacific angel shark, *Squatina californica*, do not reflect such relationships (Cailliet et al., 1992; Natanson and Cailliet, 1990; Natanson et al., 2008). Vertebral growth is inevitably linked to food intake, and a lack of food for short periods of time can cause subtle bands to appear in vertebral centra of some species (Gelsleichter et al., 1995; J. Gelsleichter, pers comm.; K.J. Goldman, pers. obs.). Considerable variability exists in the amount and pattern of calcification within and among taxonomic groups of elasmobranch fishes, and much of the variation observed in several species has not yet been explained (Branstetter, 1990; Branstetter and Musick, 1994; Wintner and Cliff, 1999). These factors make it inherently risky to assume that the vertebral banding pattern of one species is representative of another species or under all conditions, necessitating a species-specific approach.

Transmitted light is the most commonly used method of illuminating sectioned centra, but we strongly recommend comparing transmitted light with reflected light, translucent or other filtered light, and ultraviolet (UV) illumination, even if staining or tetracycline injection has not been conducted. Altering the intensity of each type of light and making finite adjustments to the optical focus of the microscope can often provide visual enhancement of the banding pattern.

14.3.1 Ageing Protocols

Age and growth studies require interpretation of banding patterns in the hard parts of fishes. As such, they incorporate several sources of variability and error. Although the individuals used in an ageing study provide a source of natural variability, variability between sexes and among geographic locations may also exist (Carlson and Parsons, 1997; Parsons, 1993; Yamaguchi et al., 1999). Other potential sources of variability and error include the method used to count growth increments, effects of within- and between-reader variability and bias, effects of staining, variation in increment counts from different hard parts, and variation in increment counts from within the same region of the vertebral column and from different regions of the vertebral column (Campana, 2001; Officer et al., 1996). Developing an ageing protocol brings consistency in the ageing process, leading to better precision and minimizing error. The most important aspect of any ageing protocol is that it produces repeatable ages within and between readers (i.e., precision). Ageing protocols have two key components: (1) determining which marks on vertebral centra or spines will be counted, and (2) checking for reader agreement and precision and testing for bias within and between readers after age determinations are completed. A standard part of every ageing protocol, whenever possible, should be to have two readers independently age all centra two times in blind, randomized trials without knowledge of each specimen's length or disc width.

One of the more common problems in age determination occurs as a result of deviations in typical growth patterns observed in vertebral centra, which can lead to inaccurate counts. These deviations can result from false checks or split bands occurring within the corpus calcareum, the intermedialia, or both, and the vertebral intermedialia of many species possess a great deal of background noise. As such, it is important that these accessory bands be recognized as anomalies when assigning an age to a specimen. Checks tend to be discontinuous, weak or diffuse, and inconsistent with the general growth pattern of true annuli. Developing some familiarity with the typical look of the banding pattern in a given species' centra to aid in distinguishing checks from annuli is recommended. If the ageing study is an ongoing one, regular review of reference collections (i.e., a subsample of previously agreed upon, verified, or validated samples that readers can use to reacquaint themselves with the age assessment protocol for a given species) and comparing summaries of age–length data from one season to the next can also help maintain accuracy and precision and reduce bias in age determinations (Campana, 2001; Officer et al., 1996). In addition, because the intermedialia of the centrum in many species is not very robust, it may warp in a concave manner during the drying process. When this occurs, the rings near the outer edge of the intermedialia bunch up and become indistinguishable. The rings on the corpus calcareum also become more tightly grouped at the outer edge, particularly in larger or older animals; however, they have a tendency to remain distinguishable due to the stronger (more robust) nature of the structure. For these reasons, the corpus calcareum should always be used as the primary counting and measuring surface, with the distinct rings in the intermedialia and any additional features (see below) used as confirmation of a ring or annulus. Ageing from digital photographs and image analysis has become more common in recent years. Sections can be photographed under a microscope and examined in a much larger capacity on a screen. Advantages include the ability of several readers to interact when counts are being compared. Additionally, sections such as the difficult-to-read edge can be enhanced and enlarged with computer software. Measurements can be taken more easily, as a cursor is simply placed on the band in question and the measurement calculated. It is often helpful to have the actual section available to compare, as sometimes the photographs are not as distinct as the actual section.

Additional difficulties in ageing elasmobranch fishes can include determining the birthmark and first growth ring. Birthmarks are usually represented by an angle change along the centrum face of whole vertebrae or along the intermedialia–corpus calcareum interface with an associated ring on the corpus calcareum in sectioned centra, but this feature may not be distinct in either. The birthmark usually can be found on the whole centrum surface (i.e., the outside wall of the corpus calcareum), but the variability in this mark is such that it may appear distinctly only within the sagittally cut section. Additionally, pre-birth rings have been reported in some species (Branstetter and Musick, 1994; Casey et al., 1985; Goldman, 2002; Goldman et al., 2006; Nagasawa, 1998). Once the angle change is located, pre-birth rings can easily be distinguished from the first growth ring. The first growth ring may consist of minimal growth around the focus of a vertebra, can be faint relative to other annuli (Campana, 2001), and can also differ in its opacity or translucency (Allen and Wintner, 2002; Wintner and Dudley, 2000). Being able to consistently locate a birthmark and (particularly) the first annulus are obviously of critical importance to accurate age assessment. Knowledge of the pupping (or hatching) time of a given species can help in determining if the first annulus is expected to be very small (first winter is soon after birth) or large (first winter is a considerable time after birth).

The vertebral centra of some species may also possess features that can assist in ageing specimens. For example, sagittally cut vertebral sections of some species reveal distinct notches along either the inside or outside edge of the corpus calcareum at each ring, providing an additional ageing feature (Goldman and Musick, 2006; Goldman et al., 2006). This can be particularly useful in ageing vertebral sections where the cut has excluded the radials of the intermedialia and in distinguishing growth checks from annuli. If examination of vertebral centra reveals no discernable banding patterns or reveals rings that are difficult to interpret, centra (either whole or sectioned) can be stained or histologically processed to attempt enhancement of growth bands for enumeration.

14.3.2 Precision and Bias

Precise and accurate age estimation is a critical component of any ageing study. It is important to keep in mind that the consistent reproducibility of age estimates from vertebral centra will achieve high precision but these age estimates may not be accurate (i.e., reflect the true or absolute age), and precision should never be used as a substitute for accuracy. Accurate age determination requires validation of absolute age, not just the frequency of increment formation in vertebral centra or spines (Beamish and McFarlane, 1983; Cailliet, 1990; Campana, 2001).

Two readers independently ageing all centra two times in blind, randomized trials without knowledge of each specimen's length or disc width allows two calculations of between-reader agreement and precision and helps prevent reader bias that can be caused by predetermination of age based on knowledge of length (i.e., prevent subjectivity). It also allows for within-reader comparisons, which may be critical if only one reader is assessing ages and no between-reader comparisons are possible. When there is a disagreement between readers, a final age determination should be made by the two readers viewing the ageing structure together, as a single age is needed from each specimen for input into growth models. If no consensus can be reached, the sample should be eliminated from the study.

The most commonly used methods for evaluating precision among age determinations have been the average percent error (APE) technique of Beamish and Fournier (1981) and the modification of their method by Chang (1982). Hoenig et al. (1995) and Evans and Hoenig (1998), however, demonstrated that there can be differences in precision that these methods obscure because the APE assumes that the variability among observations of individual fish can be averaged over all age groups, and this variability can be expressed in relative terms. Also, APE does not result in values that are independent of the age estimates. APE indices do not test for systematic differences, do not distinguish all sources of variability (such as differences in precision with age), and do not take experimental design into account (i.e., number of times each sample was read in each study) (Hoenig et al., 1995). Within a given ageing study, however, APE indices may serve as good relative indicators of precision within and between readers provided that each reader ages each vertebra the same number of times. Even this, though, appears to tell us only which reader was less variable, not which was better or if either was biased, which is more critical to discern in ageing data. Comparing precision between studies would seem to hold importance only if the study species is the same, but caution should be used if samples are from different geographic areas or if samples were prepared using different methods.

Goldman (2005) provided a simple and accurate approach to estimating precision: (1) calculate the percent reader agreement (PA), which is equal to the (No. agreed/No. read) × 100, within and between readers for all samples; (2) calculate the percent agreement plus or minus one year (PA ± 1 year) within and between readers for all samples; (3) calculate the percent agreement within and between readers, with individuals divided into appropriate length or disk-width groups (e.g., 5- to 10-cm increments) as an estimate of precision (this should be done with sexes separate and together); and (4) test for bias using one or more of the methods discussed below. The criticism of percent agreement as a measure of precision has been that it varies widely

among species and ages within a species (Beamish and Fournier, 1981; Campana, 2001). Precision estimates of percent agreement varying among species is not a valid concern, as there is no purpose in comparing PA estimates between studies or species. We are not aware of any literature where this has been done. A more valid concern about percent agreement is the variation among ages within a species, because the ages used to obtain percent agreement are typically only assessed and not validated. Age could be used if, and only if, validation of absolute age for all available age classes had been achieved. There is, however, validity in using percent agreement with individuals grouped by length as a test of precision because it does not rely on ages (which have been estimated), but rather on lengths, which are empirical values (Cailliet and Goldman, 2004; Goldman, 2005; Goldman and Musick, 2006; Goldman et al., 2006).

Several methods can be used to compare counts (ages) by multiple readers, such as regression analysis of the first reader counts vs. the second reader counts, a paired *t*-test of the two readers' counts, and a Wilcoxon matched-pairs signed-ranks test (DeVries and Frie, 1996). Campana et al. (1995) stated the importance of a separate measure for bias, and even that bias should be tested for prior to running any tests for precision. They suggest an age-bias plot (Figure 14.9), which graphs one reader vs. the other and is interpreted by referencing the results to the equivalence line of the two readers (45° line through the origin). Similarly, Hoenig et al. (1995) and Evans and Hoenig (1998) stated that comparisons of precision are only of interest if there is no evidence

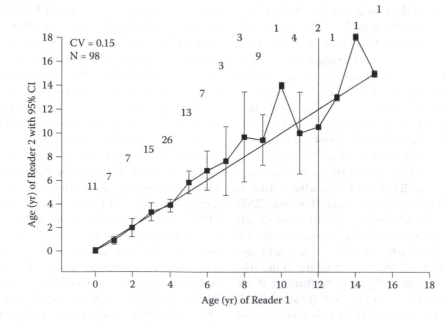

FIGURE 14.9
Age-bias curve. (From Skomal, G.B. and Natanson, L.J., *Fish. Bull.*, 101, 627–639. 2003. With permission.)

of systematic disagreement among readers or methods; they suggested testing for systematic differences between readers using chi-square tests of symmetry, such as Bowker's (1948), McNemar's (1947), or their Evans–Hoenig test, to determine whether differences between and within readers were systematic (biased) or due to random error. This is of particular importance if initial percent agreement and precision estimates are low. We recommend these tests of symmetry for testing for bias regardless of precision because they place all age values in contingency tables and test the hypothesis that values in a given table are symmetrical about the main diagonal, and because they can be set up to test among all individual age classes or groups of age classes. The test statistic (the chi-square variable) will tend to be large if a systematic difference exists between the two readers.

14.3.3 Back-Calculation

Back-calculation is a method for describing the growth history of each individual sampled, and numerous variations in methodology exist (for a thorough review, see Francis, 1990; for a description and application to elasmobranchs, see Goldman, 2005). Back-calculations estimate lengths at previous ages for each individual and should be used if sample sizes are small and if samples have not been obtained from each month. Back-calculation formulas that follow a hard part or body proportion hypothesis are recommended (Campana, 1990; Francis, 1990; Ricker, 1992). The proportional relationship between animal length or disk width and the radius of the vertebral centrum among different length animals within a population is used as a basis for empirical relationships regarding population and individual growth, as is the distance from the focus to each annulus within a given centrum. Centrum radius (CR) and distance to each ring should be measured as a straight line from the central focus to the outer margin of the corpus calcareum to the finest scale possible. Lengths or disk widths should then be plotted against CR to determine the proportional relationship between somatic and vertebral growth, which will assist in determining the most appropriate back-calculation method.

Providing biological and statistical reasoning behind the choice of a back-calculation method is extremely important for obtaining accurate life-history parameter estimates from a growth function (e.g., Gompertz) when using back-calculated data. Although one method may prove to be more statistically appropriate for back-calculation, researchers should conduct several methods for comparison to available sample length-at-age data to verify that statistical significance equates to biological accuracy. Biological accuracy can be determined by plotting the sample mean length-at-age data against the difference between mean back-calculated length-at-age estimates and the sample mean length-at-age data to see which method provides results that most accurately reflect sample data (Goldman and Musick, 2006; Goldman et al., 2006). Although the most commonly used back-calculation method has been the Dahl–Lea direct proportions method (Carlander, 1969), linear and quadratic modified Dahl–Lea methods (Francis, 1990) and the Frazer–Lee birth-modified back-calculation method (Campana, 1990; Ricker, 1992) should be conducted, where appropriate, and compared to sample length-at-age data (Goldman and Musick, 2006; Goldman et al., 2006).

14.4 Verification and Validation

Cailliet (1990) stated that the process of evaluating growth zone deposition in fishes can be categorized as *verification* or *validation*. Verification is defined as "confirming an age estimate by comparison with other indeterminate methods," and validation as "proving the accuracy of age estimates by comparison with a determinate method." These definitions are used throughout this discussion.

Estimates of age, growth rate, and longevity in chondrichthyans assume that the growth rings are an accurate indicator of age. Although this is probably true for most species, few studies on elasmobranch growth have validated the temporal periodicity of band deposition in vertebral centra, and even fewer have validated the absolute age (Cailliet, 1990; Cailliet et al., 1986a; Campana, 2001).

Validation can be achieved via several methods, such as chemically tagging wild fish, conducting mark–recapture studies of known-age individuals, and bomb radiocarbon dating (the latter two can also be used to validate absolute age). A combination of using known-aged individuals, tag and recapture, and chemical marking is probably the most robust method for achieving complete validation (Beamish and McFarlane, 1983; Cailliet, 1990; Campana, 2001; Natanson et al., 2002). Although this is a rather daunting task to accomplish with most elasmobranch species, the current necessity to obtain age–growth data for fisheries management purposes dictates that it be attempted. The most frequently applied method used with elasmobranchs has been chemical marking of wild fish, even though recaptures can be difficult to obtain for many species. Because validation has proved difficult in elasmobranchs, verification methods such as centrum edge analysis and relative marginal increment analysis are frequently employed.

Obtaining the absolute age of individual fish (complete validation) is the ultimate goal of every ageing study, yet it is the frequency of growth ring formation for which validation is typically attempted. The distinction between validating absolute age and validating the periodicity of growth-ring formation is important (Beamish and McFarlane, 1983; Cailliet, 1990; Campana, 2001). Validation of the frequency of growth-ring formation must prove that the mark being considered an annulus forms once a year (Beamish and McFarlane, 1983); however, it is the consistency of the marks in "number per year" that really matters, be it one or more than one. Two or more marks (rings) may make up an annulus if, and only if, consistent multiple marks per year can be proved. Strictly speaking, validation of absolute age is only complete when it has been done for all age classes available, with validation of the first growth ring being the critical component for obtaining absolute ages (Beamish and McFarlane, 1983; Cailliet, 1990; Campana, 2001).

In the following sections, both verification and validation are discussed. It is important to remember that some techniques, especially if used in conjunction with others, can be verification and/or validation.

14.4.1 Size Mode Analysis

This technique monitors the progression of discrete length modes of fish over time. Although commonly considered a basic approach to studying age composition and even growth, its use as a growth tool is primarily verification; that is, if the size modes seen in data from a presumed random sample of all sizes of fish in a population appear to coincide with the mean or median sizes in an age class (as determined by ageing studies or other means), then this lends support to the contention that these age classes are real. Kusher et al. (1992), for example, used this method to show that young leopard sharks, *Triakis semifasciata*, in Elkhorn Slough, CA, followed growth patterns that would have been predicted by the von Bertalanffy growth function determined by size at age patterns from vertebral sections. Similarly, Natanson et al. (2002) determined growth rates for age 0 and 1 porbeagle sharks, *Lamna nasus*, by monitoring the progression of those two discrete length modes across months within a year.

14.4.2 Tag–Recapture

In addition to size mode analysis, tag–recapture data are often used to produce growth curves. This usually involves capturing, measuring, weighing, and tagging specimens in the field and then releasing them. Through recaptures obtained from either dedicated surveys or recreational or commercial fishers, tagged specimens

provide information on growth (length or weight) over a distinct period of time. This has been done in many studies; for example, in the Pacific angel shark, *Squatina californica*, the von Bertalanffy growth functions are based on size at capture and recapture but not oxytetracycline (OTC) (Cailliet et al., 1992). A significant amount of literature exists on the procedures of estimating growth parameters from tag–recapture data (Cailliet et al., 1992; Gulland and Holt, 1959; Fabens, 1965). The method developed by Gulland and Holt (1959) is fairly straightforward; however, efforts should be made to use several methods, such as GROTAG (Francis, 1988; Natanson et al., 2002) when analyzing growth increment data.

14.4.3 Marking, Field Tag–Recapture, and Laboratory Studies

Validation of absolute age is extremely difficult to achieve with elasmobranch fishes; hence, the (few) studies that have attempted validation in these fishes have focused on validating the temporal periodicity of ring (growth increment) formation. The oxytetracycline validation method is a standard among fisheries biologists for marking free-swimming individuals (Cailliet, 1990; Campana, 2001; DeVries and Frie, 1996; Smith et al., 2003) to test the assumption of annual periodicity of growth rings. OTC, a general antibiotic that can be purchased through veterinary catalogs, binds to calcium and is subsequently deposited at sites of active calcification. It is typically injected intramuscularly at a dose of 25 mg kg^{-1} body weight (Gelsleichter et al., 1998b; Tanaka, 1990), and an external identification tag is simultaneously attached to each injected animal. OTC produces highly visible marks in vertebral centra and dorsal fin spines of recaptured sharks when viewed under ultraviolet light (Beamish and McFarlane, 1985; Branstetter, 1987; Brown and Gruber, 1988; Gelsleichter et al., 1998b; Goldman, 2002; Goldman et al., 2006; Gruber and Stout, 1983; Holden and Vince, 1973; Kusher et al., 1992; McFarlane and Beamish, 1987a,b; Natanson, 1993; Natanson and Cailliet, 1990; Natanson et al., 2002; Simpfendorfer et al., 2002; Skomal and Natanson, 2003; Smith, 1984; Tanaka, 1990; Wintner and Cliff, 1999).

The combination of body growth information and a discrete mark in the calcified structure permits direct comparison of time at liberty with growth band deposition, such that the number of rings deposited in the vertebra or spine since the OTC injection can be counted and related to the time at liberty. Although there may be problems associated with using captive growth as a surrogate to growth in the wild and with recapturing animals that have been at large for a sufficiently long period of time, this method has been used on a number of species in the laboratory and field (Cailliet, 1990;

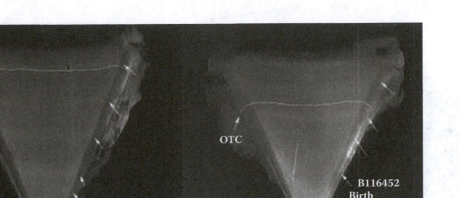

FIGURE 14.10
Sagittally cut vertebral section from tagged and recaptured OTC-injected blue sharks. (From Skomal, G.B. and Natanson, L.J., *Fish. Bull.*, 101, 627–639. 2003. With permission.)

Cailliet et al., 1986a; Goldman et al., 2006). The use of OTC in a laboratory setting was an instrumental component in showing that vertebral banding was not annual but related to somatic growth in Pacific angel sharks, *Squatina californica* (Natanson and Cailliet, 1990), and in showing that temperature had no effect on annual band deposition in mature little skates, *Raja erinacea* (Natanson, 1993).

Nevertheless, this technique, when successful, has proved to be invaluable at validating growth characteristics of chondrichthyans. The best recent examples are tag–recaptures, some with OTC, of the blue shark, *Prionace glauca*, by Skomal and Natanson (2003) (Figure 14.10) and the 20-year tag return of a leopard shark, *Triakis semifasciata*, reported by Smith et al. (2003) (Figure 14.11). In both cases, it was possible to define birth years and to identify individual growth characteristics for individual years from zones on sections of the vertebrae, relative to the OTC mark from the original release.

Several other chemical markers such as fluorescein and calcein have been used to validate growth ring periodicity in teleost otoliths, but very few studies have evaluated these in elasmobranchs (Gelsleichter et al., 1997; Officer et al., 1997). Gelsleichter et al. (1997) found that doses of 25 mg kg^{-1} body weight (typical dose for teleosts) induced physiological stress and mortality in the nurse shark, *Ginglymostoma cirratum*, but doses of 5 to 10 mg kg^{-1} body weight produced suitable marks without causing physiological trauma or death. Based on this evaluation, any alternative chemical markers tested should consider that doses for teleosts might be too high for elasmobranchs. Calcein, however, has been successfully used in the field to validate ages in sandbar sharks, *Carcharhinus plumbeus*, in Western Australia (McAuley et al., 2006).

14.4.4 Centrum Edge and Relative Marginal Increment Analysis

Centrum edge analysis compares the opacity and translucency (width and density) of the centrum edge over time in many different individuals to discern seasonal changes in growth. The centrum edge is categorized as opaque or translucent, and the bandwidth is measured or graded, then compared to season or time of year (Kusher et al., 1992; Wintner and Dudley, 2000; Wintner et al., 2002). A more detailed centrum edge analysis can be conducted by analyzing the levels of calcium and phosphorus at the centrum edge using x-ray or electron microprobe spectrometry (Cailliet and Radtke, 1987; Cailliet et al., 1986a). This technique has only been applied in a single study on recaptured nurse sharks that had been injected with tetracycline (Carrier and Radtke, 1988, as cited in Cailliet, 1990).

Relative marginal increment (RMI) analysis, sometimes referred to as marginal increment ratio (MIR) analysis, is a useful, direct technique with which to assess seasonal band and ring deposition (Figure 14.12). The margin, or growth area of a centrum from the last growth ring to the centrum edge, is divided by the width of the last (previously) fully formed annulus (Branstetter and Musick, 1994; Conrath et al., 2002; Goldman, 2002; Natanson et al., 1995; Smith et al., 2007; Wintner et al., 2002). Resulting RMI values are then plotted against month of capture to determine temporal periodicity of band formation. Age 0 animals cannot be used in this analysis because they have no fully formed increments.

Recently, ecologists have employed stable isotope composition to trace the early life histories of fishes, including analyses of habitats and environments occupied, as well as biochronologies (Campana and Thorrold, 2001).

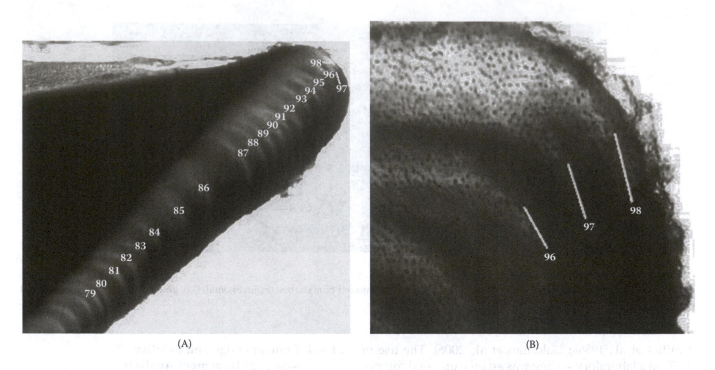

(A)   (B)

FIGURE 14.11
Sagittally cut vertebral section from tagged and recaptured OTC-injected leopard shark after 20 years at large. (A) 20 years of annual ring formation from 1978 to 1998, and (B) the last 3 years at the edge of the section. (From Smith, S.E. et al., *Fish. Bull.*, 101, 194–198, 2003. With permission.)

This approach has not been used to study either process in sharks and rays, but it is certainly a field open to providing useful, additional information on the ecology of chondrichthyans.

14.4.5 Captive Rearing

The operation, often seen in public aquaria but also in research laboratories, of keeping chondrichthyans alive in captivity can produce some useful growth information. Van Dykhuizen and Mollet (1992) analyzed growth of captive sevengill sharks, *Notorynchus cepedianus*, and provided the first estimates of growth for this species, which has poorly calcified vertebrae that cannot easily be analyzed for growth characteristics (Cailliet et al., 1983a). Similar studies have been done for such open-water species as the pelagic stingray, *Dasyatis violacea* (Mollet et al., 2002). Laboratory growth has also been used as a way of determining the periodicity of growth zone formation in the vertebral centra of several species, such as the Atlantic sharpnose shark, *Rhizoprionodon terraenovae* (Branstetter, 1987), and the wobbegong, *Orectolobus japonicus* (Tanaka, 1990).

The public aquarium trade is beginning to emphasize research as part of its husbandry practices, especially on topics such as the relationship between food intake and growth in chondrichthyan fishes. In a recent review on age and growth of captive sharks, Mohan et al. (2004)

pointed out how carefully taken morphometric data on captive sharks can result in useful information on their age and growth. In addition, such growth information can provide data on their life histories that could not be obtained from animals in the wild. Unfortunately, there is also the caveat that captivity in itself can influence growth in a way that does not reflect what might occur in the wild.

One of the biggest problems with captive growth is accurately measuring individual specimens without harming them for display (Mohan et al., 2004); however, recent handling techniques and the advent of remote measuring techniques have made this less of a problem. Mohan et al. (2004) provided very useful weight–length relationships for 17 species of sharks. Their study also provides a detailed life history summary of ten species: eight kept in captivity, ranging from lamnoids and charharinids to the angel shark, and the other two not kept in captivity (*Carcharodon carcharias* and *Isurus oxyrinchus*). In addition, von Bertalanffy growth curves are presented as a result of this summary of life histories for 16 species.

14.4.6 Bomb Radiocarbon Dating

Bomb radiocarbon dating is a technique that has evolved as a unique application in the age validation of fishes (Kalish, 1995). The approach relies on a preserved record of the rapid increase in radiocarbon

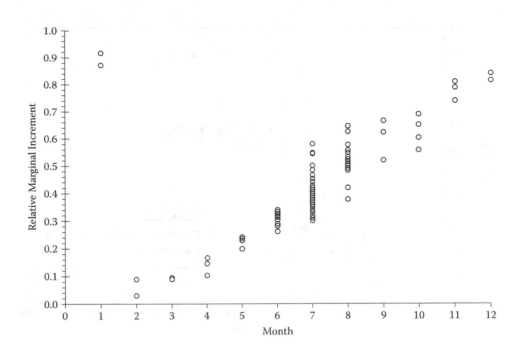

FIGURE 14.12
Results of relative marginal increment analysis showing that annulus formation in salmon sharks occurs between January and March. (From Goldman, K.J. and Musick, J.A., *Fish. Bull.*, 104, 278–292, 2006. With permission.)

(^{14}C) that occurred in the world's oceans as a result of atmospheric testing of thermonuclear devices in the 1950s and 1960s (Broecker and Peng, 1982). The uptake of bomb-produced ^{14}C, reported as Δ^{14}C in reference to an established prenuclear age ^{14}C record (Stuiver and Pollach, 1977), was virtually synchronous in the mixed layer of mid-latitude oceans. This was first recorded from marine carbonates in hermatypic corals (Druffel and Linick, 1978). Application to fishes began with an innovative comparison of Δ^{14}C values recorded in otolith carbonate relative to regional Δ^{14}C records from hermatypic corals (Kalish, 1993). The temporal specificity of the measured levels provided an independent determination of age for corroboration of age estimates from growth zone counting in otoliths (Campana, 2001; Kalish, 2001a). Bomb radiocarbon dating has since been applied successfully to numerous teleost fishes in otolith applications (e.g., Andrews et al., 2007; Ewing et al., 2007; Neilson and Campana, 2008), has expanded to applications with other marine organisms (e.g., Frantz et al., 2005; Kilada et al., 2009; Roark et al., 2006; Stewart et al., 2006), and was recently applied to elasmobranchs (Ardizzone et al., 2006; Campana et al., 2002; Kneebone et al., 2008). This section outlines the approach and several nuances that must be considered in the application of this technique.

Samples selected for an application of bomb radiocarbon dating must have birth dates or times of formation that range from a time prior to significant changes in Δ^{14}C in the marine environment from atmospheric

nuclear testing (pre-1957) to the post-bomb period (after approximately 1967). There are typically two approaches to the design of the extracted sample series: (1) a number of individuals, each with an estimated age; or (2) within an individual, with a series of samples ranging from the oldest to youngest portions of the growth structure. In either case, the approach is the same in terms of age; the approach uses the rise in radiocarbon as a time-specific marker for age validation, and it is the agreement or disagreement of the sample series relative to the Δ^{14}C reference that provides a measure of age estimate accuracy. Hence, the utility of this approach for determining age or lifespan is dependent on the difference between the collection year and its correlations with the rise in ^{14}C. In cases where there was considerable uncertainty in age estimates, bomb radiocarbon dating was used to assign an age that was independent of any age estimation procedures related to growth zone counting (Andrews et al., 2005; Campana, 1997; Piner and Wischniowski, 2004).

The first application of bomb radiocarbon dating to validate ages in long-lived sharks was with porbeagle shark, *Lamna nasus*, with preliminary results reported for shortfin mako shark, *Isurus oxyrinchus* (Campana et al., 2002). Unlike otoliths, in which the uptake of ^{14}C has been mostly synchronous with the marine environment, the vertebrae of porbeagle sharks provided evidence for a phase lag of approximately 3 years in the timing of the rise in Δ^{14}C. This was attributed to a trophic-level delay or depth-related dilution of carbon sources to the formation of vertebrae. Use of known-age individuals ruled

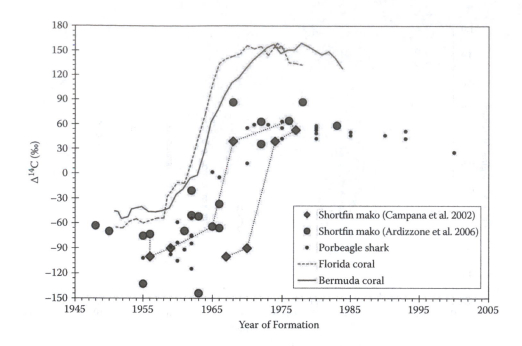

FIGURE 14.13
Radiocarbon data from two studies on shortfin mako shark revealing support for one band pair per year based on an alignment with the porbeagle shark $\Delta^{14}C$ reference record. This plot provides both the underestimated age scenario (connected diamonds on right) from Campana et al. (2002) and the age-validated scenario supplemented with additional data points (connected circles and diamonds on left) from Ardizzone et al. (2006) for the same shark. Plotted as an upper limit for the rapid rise in $\Delta^{14}C$ are two hermatypic coral records from the region (Druffel, 1989; Druffel and Linick, 1978).

out other possible violations of the system requirements that would obfuscate the technique (e.g., reworking of vertebral carbon through the life of the fish); hence, age validation was achieved for porbeagle shark, and the reported preliminary results for shortfin mako shark were well supported. In a follow-up study to the mako shark work by Campana et al. (2002), age and growth were further described and validated using bomb radiocarbon dating (Ardizzone et al., 2006). The hypothesis of two growth bands per year through the life of mako shark was not supported by the reference bomb radiocarbon time series (Ardizzone et al., 2006; Pratt and Casey, 1983) (Figure 14.13). Age estimates approaching 30 years were consistent with the measured $\Delta^{14}C$ values, as well as other age validation studies (Natanson et al., 2002).

A factor in bomb radiocarbon dating that was initially described by Campana et al. (2002) and has become an even greater consideration is the variability in uptake of ^{14}C through time. Further study of the mako shark provided additional evidence that ^{14}C-depleted sources of carbon were making their way into the diet of these sharks and ultimately being measured in the vertebrate (Ardizzone et al., 2006). This represents a potential complication for assigning ages because the time specificity of bomb radiocarbon dating becomes less certain. In general, age validation for sharks must be initially qualified with an unknown level of age estimate uncertainty

due to this factor; however, in most cases there is empirical evidence to support the temporal correlation and to conclude that the age estimate uncertainty is a few years.

Perhaps the most evident case of how this complication can affect the results of bomb radiocarbon dating was with a study performed on white sharks, *Carcharodon carcharias*, from the eastern North Pacific Ocean (Kerr et al., 2006). In the study, age could not be validated because of highly variable $\Delta^{14}C$ values for any given estimated year of formation (Figure 14.14). Initially, the approach was to consider the age estimation procedure as invalid and that age is actually greater because of the unexpectedly attenuated values assigned to years of formation well after the rise in $\Delta^{14}C$. However, measurements from vertebral edge material (constrained in time by the collection date) and juvenile white sharks (age less in question) prevented what might have been a logical shift of those data to older ages. Hence, a phase lag was observed relative to a regional carbonate reference chronology of approximately 10 to 15 years—considerably greater than the 3-year phase lag observed by Campana et al. (2002) for porbeagle sharks in the North Atlantic Ocean. The cause of the phase-lagged $\Delta^{14}C$ measurements was attributed to some combination of dietary carbon sources (e.g., Tricas and McCosker, 1984), large-scale ontogenetic movement patterns on and off shore (e.g., Boustany et al., 2002), and precipitous $\Delta^{14}C$ gradients with depth in the

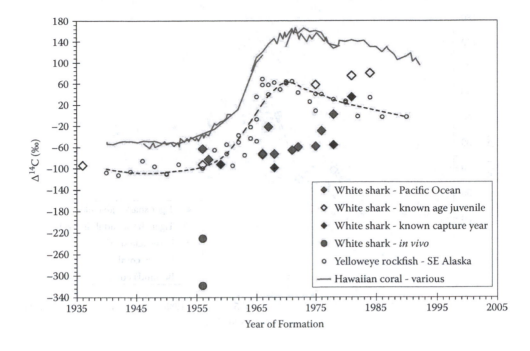

FIGURE 14.14

Radiocarbon data from a study on white shark vertebrae from the northeastern Pacific Ocean plotted with regional Δ14C reference records (Kerr et al., 2006). Juvenile and known capture year samples provided constraints on interpretations of age, and it was clear that there were problems with uptake of carbon sources depleted of 14C. Although age could not be validated, information can be gleaned from these data in terms of feeding behavior, with two of the lowest Δ14C values recorded from sharks taken from the *in vivo* portion of the adult vertebra. The yelloweye rockfish otolith record (Kerr et al., 2004) and a series of hermatypic coral records (Druffel, 1987; Druffel et al., 2001; Roark et al., 2006) were plotted as the nearest regional Δ14C references for comparative purposes.

region (Broecker and Peng, 1982). The potential effect of depleted 14C sources on vertebral composition was exemplified by two of the lowest Δ14C measurements made to date in sharks; two samples extracted from the apex of the corpus calcareum in an adult white shark were depleted by over 100‰ and 200‰ relative to pre-bomb levels for the northeastern Pacific Ocean (Figure 14.14). This portion of the vertebrae would have been formed *in vivo* and provides an indication that the pregnant mother of this shark was feeding in deep water or on dietary sources from deep water where Δ14C levels can be very depleted. This finding provides further evidence of a deep-water life history and that feeding is taking place on the observed deep forays between Hawaii and the mainland (Jorgensen et al., 2010).

Consistent with the unexpected phase lag documented for some species was a result from early work on school sharks, *Galeorhinus galeus*, off Australia (Kalish and Johnston, 2001) that also revealed a puzzling Δ14C phase lag of approximately 10 years. The study provided empirical evidence from δ13C measurements in support of metabolic uptake of 14C and long-term stability of the vertebral collagen. Based on this conclusion, the phase lag was attributed to underestimation of age. In support of this notion was a single recaptured school shark that was at liberty for 35.4 years and known to be 36 years of age (Kalish and Johnston, 2001). This kind of discovery

was more recently described for New Zealand porbeagle sharks where bomb radiocarbon dating did not support ages exceeding, 20 years; it was concluded that growth band width decreased to an irresolvable level and that age was underestimated by both band counting and bomb radiocarbon dating (Francis et al., 2007).

Complexities tied to ontogenetic changes in feeding were also observed with a bomb radiocarbon dating study of tiger shark, *Galeocerdo cuvier*, in the northwestern Atlantic Ocean (Kneebone et al., 2008). In addition to validating age estimates up to 20 years for this species, information in terms of carbon sources was apparent in the measured levels of Δ14C. The interesting finding with this study in terms of 14C uptake was the differences in the correlation of Δ14C values between juveniles and an adult shark (Figure 14.15). Juvenile sharks were in agreement with a hermatypic coral record from Florida, indicating there was no phase lag in terms of the timing of the Δ14C signal for the early growth of vertebrae. In contrast, the older adult, one that lived through the period of bomb testing to nearly the end of the marine Δ14C record, was mostly in phase with the porbeagle shark record as an adult and deviated to match the coral record in what would have been the juvenile portion of the adult vertebrae. These findings can be logically attributed to tiger shark juveniles feeding on short-lived and near-surface dietary sources and adults shifting to

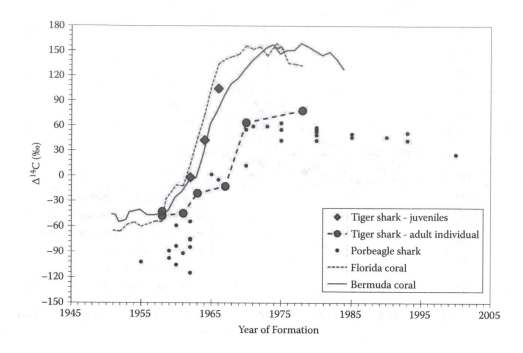

FIGURE 14.15

Radiocarbon data from a study on tiger shark vertebrae from the northwestern Atlantic Ocean plotted with regional $\Delta^{14}C$ reference records (Kneebone et al., 2008). Inference of these data is the importance of understanding ontogenetic changes in dietary sources of carbon and its affect on $\Delta^{14}C$ stored in vertebrae through time. Juvenile tiger shark samples provided a record that was in synch with hermatypic coral $\Delta^{14}C$ records, an indication of the most rapid rise in $\Delta^{14}C$ for the region (Druffel, 1998; Druffel and Linick, 1978). In contrast, an adult tiger shark shifted away from food sources in synch with corals to a more depleted record, similar to the porbeagle shark $\Delta^{14}C$ record (Campana et al., 2002).

older dietary sources, represented as a phase lag that can be attributed to trophic level changes (Kneebone et al., 2008). A similar scenario was measured for great hammerheads, *Sphyrna mokarran*, where there was close agreement with a coral $\Delta^{14}C$ record, yet some years of formation provided an indication there was an attenuation of the $\Delta^{14}C$ signal (Passerotti et al., 2010). These studies highlight the necessity for understanding: (1) ontogenetic changes in feeding, and (2) the ^{14}C content of dietary sources in an application of bomb radiocarbon dating to sharks.

Bomb radiocarbon dating of elasmobranchs was also applied to the vertebrae of skates and dorsal fin spines of Atlantic and Pacific spiny dogfish: *Squalus acanthias* in the north Atlantic and *Squalus suckleyi* in the north Pacific (see Ebert et al., 2010). An application to three species of skate from off the eastern coast of Canada provided confirmation of age, as well as some degree of age estimate calibration (McPhie and Campana, 2009). An offset of a measured value in the thorny skate, *Amblyraja radiata*, led to the conclusion of underestimated age. Based on knowledge of its depth distribution and trophic level, phase-lagged $\Delta^{14}C$ and depleted ^{14}C sources could not explain the apparent offset of approximately 5 years; hence, the porbeagle shark reference series was used to substantiate an increase in the growth-zone-counting age of 23 years to at least 28 years. In an innovative application

of bomb radiocarbon dating, age was validated for two populations of spiny dogfish (North Atlantic and North Pacific) using dorsal spine enamel (Campana et al., 2006). Growth bands from known-age fish corroborated most of the age determinations with an agreement, on average, of estimated age with expected $\Delta^{14}C$ levels. In this case, the expected $\Delta^{14}C$ levels correlated well with carbonate-based $\Delta^{14}C$ references and no phase lag was measured (Campana et al., 2006). Once again, the importance of understanding the depth distribution and feeding behavior of the species in question was an important factor in securing the temporal specificity of the rise in $\Delta^{14}C$. Although there was some limitation to the degree of age validation because of a lack of known age spine material from the North Pacific for the pre-nuclear period, agreement was good relative to other regional carbonate records (Kerr et al., 2004; Piner and Wischniowski, 2004). The North Atlantic portion of the study provided a more comprehensive record, and age estimates were consistent with both carbonate and known age spine material. The study provided conclusive evidence that: (1) age estimation procedures were accurate, with perhaps a minor amount of imprecision; (2) spiny dogfish in the north Pacific Ocean (*Squalus suckleyi*) possess a slower growth rate and live longer than *S. acanthias* in the north Atlantic; and (3) longevity is confirmed to exceed 45 years in the northwestern Pacific Ocean (maximum estimated age was over 100 years, but bomb radiocarbon dating is limited to the difference in collection date and the mid-1950s).

Bomb radiocarbon dating is a proven and reliable technique in age validation of elasmobranchs, with an imperative admonition in terms of defining regional specificity and dietary carbon uptake. While it is apparent from numerous carbonate $\Delta^{14}C$ records throughout the marine environment that the timing of the first rise in ^{14}C is consistent in much of the world's oceans, the magnitude of the rise and the manner in which the signal declines vary considerably (Druffel, 2002; Druffel and Linick, 1978). Proper age validation using $\Delta^{14}C$ values from skeletal structures requires confidence in the comparison of measured values with a reference series. The greatest confidence can be gained by having known age material from the species being studied that covers the period of nuclear testing (i.e., Campana et al., 2002; Piner and Wischniowski, 2004); however, this is often not possible due to a lack of sample availability. The second best application is to resort to other regional records with the necessary assumptions on the uptake of ^{14}C (Kalish, 1993). Bomb radiocarbon dating of elasmobranchs has added another facet to the technique with greater consideration for dietary carbon sources. It has been documented that bomb radiocarbon dating of otoliths from deep-sea or southern ocean fishes can be confounded by uptake of depleted ^{14}C from the water column—for example, in the Patagonian toothfish, *Dissostichus eleginoides* (Kalish et al., 2001b), and orange roughy, *Hoplostethus atlanticus* (Andrews et al., 2009). However, the addition of metabolic pathways and a potential trophic-level phase lag makes elasmobranch age validation more complicated and potentially unsuccessful (Kerr et al., 2006). Despite the potential complications, proper application with full consideration for the factors described herein holds promise as a unique tool for determining and validating the age and growth of elasmobranchs.

14.5 Growth Models

A number of models and variations of models exist for estimating growth parameters in fishes, of which the von Bertalanffy and Gompertz growth models are the most commonly applied (for thorough reviews, see Haddon, 2001; Ricker, 1979; Summerfelt and Hall, 1987). The von Bertalanffy growth function (VBGF) is most often used to describe fish growth through ontogeny, and the Gompertz curve is often used to describe larval and early life growth of fishes and growth in many invertebrates (Ricker, 1979; Zweifel and Lasker, 1976). These functions, though, are flexible in application; weight can be used in place of length in the VBGF, and length may

be substituted for weight in the Gompertz model. Many statistical packages include modules (i.e., functions) that can be used to calculate the best-fitting growth parameters for the available length-at-age or weight-at-age data pairs. A nonlinear least-squares regression algorithm (e.g., "nls" in S-Plus) (Mathsoft, Inc., 2000), a maximum-likelihood function, or the PROC NLIN function in SAS (SAS Institute, Inc., 1999) can be used to fit the von Bertalanffy and Gompertz curves to the data, and programs such as PC-YIELD (Punt and Hughes, 1989) can calculate a wide range of growth models for comparison (Wintner et al., 2002). Additionally, FISHPARM (Prager et al., 1987), a fishery-based statistics program, is simple to use and provides quality statistical results for the two models presented here. Both models can also be fit to data on a spreadsheet via a nonlinear regression using the "solver" function in Microsoft Excel.

The von Bertalanffy growth function has been widely used since its introduction into fisheries by Beverton and Holt (1957). Although the VBGF has received much criticism over the years (e.g., Roff, 1982), it is the most widely used growth function in fisheries biology today (Haddon, 2001). It maintains its attractiveness, in part, because its approach to modeling growth is based on the biological premise that the size of an organism at any moment depends on the resultant of two opposing forces: anabolism and catabolism. Additionally, it is convenient to use and allows for much easier comparison between populations, but several alternative forms of the model can be fit to the age–length data. Haddon (2001) presented a variety of growth models, including generalized models, as possible alternatives to VBGF.

Small sample size, particularly of the smallest or largest individual sizes (and hence age classes) can cause poor parameter estimates using the VBGF (Cailliet and Tanaka, 1990; Francis and Francis, 1992). In lieu of using t_0 due to its lack of biological meaning, it is suggested that an estimate of length at birth (L_0) is a more robust approach (Cailliet et al., 2006; Carlson et al., 2003; Goosen and Smale, 1997). This method was first introduced by Fabens (1965) as an alternative equation to the VBGF. Although only a few studies have used the equation published by Fabens (1965) to estimate growth parameters in elasmobranchs, it has provided growth parameter estimates for some species that appear to be more realistic when the sample size was small (Goosen and Smale, 1997). Also, growth estimates were very similar to those obtained using VBGF when sample size was adequate (Carlson et al., 2003). This appears to be an excellent alternative to the von Bertalanffy model and should be applied where appropriate for comparison to other models.

From a mathematical point of view, it does not matter whether L_0 or t_0 is used as the third parameter; for mathematical manipulations, it may even be advantageous to use t_0 as the third parameter (Cailliet et al., 2006). From a biological point of view, however, and in particular for elasmobranchs, the size at birth (L_0) is often well defined and has biological meaning and significance, whereas t_0 has no biological meaning. Holden (1974) originally proposed t_0 to be an estimate of gestation time, which unreasonably implies that embryonic growth is governed by the same growth parameters as postnatal or posthatch growth. The use of L_0 as the third parameter allows an easy evaluation of the growth curve. If L_0 is needed for comparison with published data, it can easily be calculated from reported L_∞, k, and t_0 using the formula $L_0 = L_\infty[1 - \exp(k \times t_0)]$. This would indicate whether previously reported results were reasonable, because L_0 is often the best-known parameter.

The Gompertz growth function is an S-shaped model function similar to the logistic function; for use of the logistic function and several alternatives to the Gompertz function, see Ricker (1975, 1979). The estimated instantaneous growth rate in the Gompertz function is proportional to the difference between the logarithms of the asymptotic disk width or length and the actual disk width or length (Ricker, 1975, 1979). This growth function has been used most often for skates and stingrays (Mollet et al., 2002) and may be better suited to elasmobranchs that hatch from eggs, but it may also be the most appropriate model for some shark species (Wintner et al., 2002). This model may offer a better option when the volume of an organism greatly expands with age, such as myliobatiform rays, where considerable thickness is added to the animal over time but not so much disk width or length (Smith et al., 2007). The body mass may be distributed differently than would be readily detectable by length measurement and by the von Bertalanffy model. Additionally, captive growth rates (particularly when starting with young, small animals) may be better estimated by this function, as newly captured specimens may not grow in their typical fashion due to physiological stress or a reduction in feeding that often accompanies that stress, which may cause growth rates to slow (Mollet et al., 2002).

14.6 Implication of Growth, Longevity, and Demography to Fisheries Management

As stated in the introduction, a better understanding of age and growth processes and associated life history and demographic vital rates will lead to better estimates of the potential for chondrichthyan populations to grow or remain stable, particularly in response to additional sources of mortality from fisheries, and to rebuild overexploited populations and stocks (Beamish et al., 2006; Cailliet and Andrews, 2008; Heppell et al., 2005; Longhurst, 2002; Restrepo et al., 1998; also see Chapter 15 of this volume). In particular, parameters such as age

at first reproduction and longevity of these organisms are necessary to foster effective management strategies. Organisms that have high intrinsic rates of population growth often also have early ages at first reproduction and low longevities, resulting in population turnover times that may be able to respond to fishing mortality better than populations with low intrinsic rates of population growth, late age at first reproduction, and high longevity. One of the largest remaining gaps in modeling population and stock dynamics of elasmobranchs is a paucity of age and growth data, including intraspecific variation in growth rates. As new techniques continue to be employed and refined, the next several years should see a substantial increase in our understanding of age and growth processes and patterns in elasmobranchs.

The demographic consequences of these age and growth studies are, therefore, very important and have stimulated numerous authors to apply the results of these studies to their demographic analyses and stock assessments. This has been done with specific species, such as the leopard shark, *Triakis semifasciata* (Au and Smith, 1997), and the Atlantic sharpnose shark, *Rhizoprionodon terraenovae* (Cortés, 1994). It also has been done with different populations of the same species, such as the bonnethead shark, *Sphyrna tiburo*, by Cortés and Parsons (1996) and Carlson and Parsons (1997), and sandbar sharks, *Carcharhinus plumbeus*, before and after fishery depletion (Sminkey and Musick, 1995a,b). It also has been more broadly applied to many species of sharks using matrix and life-table demographic approaches (Cortés, 1997, 2000, 2002; Frisk et al., 2001; Goldman, 2002) and a related approach, called the *intrinsic rebound potential*, by Smith et al. (1998). This approach was also recently used for deep-sea chondrichthyans (Kyne and Simpfendorfer, 2007; Simpfendorfer and Kyne, 2009). Deep-sea fishes, especially chondrichthyans, appear to be quite susceptible to overfishing. Thus, solid estimates of their age, growth, longevity, and age-specific reproductive output are essential. Providing precise and accurate life-history parameters is essential to the ability of demographic and stock assessment models to accurately assess population status and provide for sustainable fisheries. Population dynamics, demography, and stock assessment are reviewed and elaborated on in detail in Chapter 15 of this book.

Acknowledgments

We thank all the authors who published papers in the past two decades for doing such a good job of starting to fill the gaps in knowledge about the life histories of chondrichthyan fishes. We thank Susan Smith, Jasmine Maurer, Bárbara Serra-Pereira, Katherine Yudin, Sho

Tanaka, Gregory Skomal, and their coauthors for letting us use the figures illustrating bomb radiocarbon dating, tag–recapture age validation, and histology/section comparison, and Gordon (Sandy) McFarlane, and Christina Conrath for other images. Finally, we thank Jack Musick and Michael Heithaus for their reviews and comments on the manuscript, and Chip Cotton and Sarah Irvine for their contribution to the section on spines as ageing tools. This is contribution PP-268 of the Commercial Fisheries Division of the Alaska Department of Fish and Game.

References

Aasen, O. (1963). Length and growth of the porbeagle (*Lamna nasus* Bonnaterre) in the North West Atlantic. *Fiskeridir. Skr. Ser. Havunders.* 13(6):20–37.

Ainsley, S.M. (2009). Age, Growth and Reproduction of the Bering Skate, *Bathyraja interrupta* (Gill and Townsend, 1897), from the eastern Bering Sea and Gulf of Alaska, master's thesis, California State University, Monterey Bay.

Allen, B.R. and Wintner, S.P. (2002). Age and growth of the spinner shark *Carcharhinus brevipinna* (Muller and Henle, 1839) off the KwaZulu-Natal coast, South Africa. *S. Afr. J. Mar. Sci.* 24:1–8.

Andrews, A.H., Burton, E.J., Kerr, L.A., Cailliet, G.M., Coale, K.H., Lundstrom, C.C., and Brown, T.A. (2005). Bomb radiocarbon and lead–radium disequilibria in otoliths of bocaccio rockfish (*Sebastes paucispinis*): a determination of age and longevity for a difficult-to-age fish. *Mar. Freshw. Res.* 56:517–528.

Andrews, A.H., Kerr, L.A., Cailliet, G.M., Brown, T.A., Lundstrom, C.C., and Stanley, R.D. (2007). Age validation of canary rockfish (*Sebastes pinniger*) using two independent otolith techniques: lead–radium and bomb radiocarbon dating. *Mar. Freshw. Res.* 58:531–541.

Andrews, A.H., Tracey, D.M., and Dunn, M.R. (2009). Lead–radium dating of orange roughy (*Hoplostethus altanticus*): validation of a centenarian life span. *Can. J. Fish. Aquat. Sci.* 66:1130–1140.

Anislado-Tolentino, V. and Robinson-Mendoza, C. (2001). Age and growth for the scalloped hammerhead shark, *Sphyrna lewini* (Griffith and Smith, 1834), along the central Pacific coast of Mexico. *Cien. Mar.* 27:501–520.

Ardizzone, D.A., Cailliet, G.M., Natanson, L.J., Andrews, A.H., Kerr, L.A., and Brown, T.A. (2006). Application of bomb radiocarbon chronologies to shortfin mako (*Isurus oxyrinchus*) age validation. In: Carlson, J.K. and Goldman, K.J. (Eds.), *Age and Growth of Chondrichthyan Fishes: New Methods, Techniques and Analysis*. Springer, Dordrecht, pp. 355–366.

Au, D.W. and Smith, S.E. (1997). A demographic method with population density compensation for estimating productivity and yield per recruit of the leopard shark (*Triakis semifasciata*). *Can. J. Fish. Aquat. Sci.* 54:415–420.

Baum, J.K., Myers, R.A., Kehler, D.G., Worm, B., Harley, S.J., and Doherty, P.A. (2003). Collapse and conservation of shark populations in the northwest Atlantic. *Science* 299:389–392.

Beamish, R.J. and Fournier, D.A. (1981). A method for comparing the precision of a set of age determinations. *Can. J. Fish. Aquat. Sci.* 38:982–983.

Beamish, R.J. and McFarlane, G. (1983). The forgotten requirement for validation in fisheries biology. *Trans. Am. Fish. Soc.* 112:735–743.

Beamish, R.J. and McFarlane, G. (1985). Annulus development on the second dorsal spines of the spiny dogfish (*Squalus acanthias*) and its validity for age determinations. *Can. J. Fish. Aquat. Sci.* 42:1799–1805.

Beamish, R.J., McFarlane, G.A., and Benson, A. (2006). Longevity overfishing. *Prog. Oceanogr.* 68:289–302.

Beverton, R.J.H. and Holt, S.J. (1957). *On the Dynamics of Exploited Fish Populations*, Fisheries Investigations Series II, Vol. 19. Ministry of Agriculture, Fisheries, and Food, London.

Bonfil, R. (1994). Overview of world elasmobranch fisheries. *FOA Fish. Tech. Pap.* 341:119.

Boustany, A.M., Davis, S.F., Pyle, P., Anderson, S.D., Le Boef, B.J., and Block, B.A. (2002). Satellite tagging: expanded niche for white sharks. *Nature* 415:35–36.

Bowker, A.H. (1948). A test for symmetry in contingency tables. *J. Am. Stat. Assoc.* 43:572–574.

Branstetter, S. (1987). Age and growth validation of newborn sharks held in laboratory aquaria, with comments on the life history of the Atlantic sharpnose shark, *Rhizoprionodon terraenovae*. *Copeia* 1987:291–300.

Branstetter, S. (1990). Early life-history implications of selected carcharhinoid and lamnoid sharks of the northwest Atlantic. In: Pratt, Jr., H.L., Gruber, S.H., and Taniuchi, T. (Eds.), *Elasmobranchs as Living Resources: Advances in the Biology, Ecology, Systematics, and Status of the Fisheries*, NOAA Tech. Rep. NMFS 90. U.S. Department of Commerce, Washington, D.C., pp. 17–28.

Branstetter, S. and Musick, J.A. (1994). Age and growth estimates for the sand tiger in the Northwestern Atlantic Ocean. *Trans. Am. Fish. Soc.* 123:242–254.

Broecker, W.S. and Peng, T.H. (1982). *Tracers in the Sea*. Lamont-Doherty Geological Observatory, Palisades, New York, 690 pp.

Brown, C.A. and Gruber, S.H. (1988). Age assessment of the lemon shark, *Negaprion brevirostris*, using tetracycline validated vertebral centra. *Copeia* 1988:747–753.

Cailliet, G.M. (1990). Elasmobranch age determination and verification: an updated review. In: Pratt, Jr., H.L., Gruber, S.H., and Taniuchi, T. (Eds.), *Elasmobranchs as Living Resources: Advances in the Biology, Ecology, Systematics, and Status of the Fisheries*, NOAA Tech. Rep. NMFS 90. U.S. Department of Commerce, Washington, D.C., pp. 157–165.

Cailliet, G.M. and Andrews, A.H. (2008). Age-validated longevity of fishes: its importance for sustainable fisheries. In: Tsukamoto, K., Kawamura, T., Takeuchi, T., Beard, Jr., T.D., and Kaiser, M.J. (Eds.), *Fisheries for Global Welfare and Environment: Memorial Book of the 5th World Fisheries Congress*. TERRAPUB, Tokyo, Japan, pp. 103–120.

Cailliet, G.M. and Goldman, K.J. (2004). Age determination and validation in chondrichthyan fishes. In: Carrier, J., Musick, J.A., and Heithaus, M. (Eds.), *Biology of Sharks and Their Relatives*. CRC Press, Boca Raton, FL, pp. 399–447.

Cailliet, G.M. and Radtke, R.L. (1987). A progress report on the electron microprobe analysis technique for age determination and verification in elasmobranchs. In: Summerfelt, R.C. and Hall, G.E. (Eds.), *The Age and Growth of Fish*, Iowa State University Press, Ames, pp. 359–369.

Cailliet, G.M. and Tanaka, S. (1990). Recommendations for research needed to better understand the age and growth of elasmobranchs. In: Pratt, Jr., H.L., Gruber, S.H., and Taniuchi, T. (Eds.), *Elasmobranchs as Living Resources: Advances in the Biology, Ecology, Systematics, and Status of the Fisheries*, NOAA Tech. Rep. NMFS 90. U.S. Department of Commerce, Washington, D.C., pp. 505–507.

Cailliet, G.M., Martin, L.K., Harvey, J.T., Kusher, D., and Welden, B.A. (1983a). Preliminary studies on the age and growth of blue (*Prionace glauca*), common thresher (*Alopias vulpinus*), and shortfin mako (*Isurus oxyrinchus*) sharks from California waters. In: Prince, E.D. and Pulos, L.M. (Eds.), *Proceedings of the International Workshop on Age Determination of Oceanic Pelagic Fishes: Tunas, Billfishes, Sharks*, NOAA Tech. Rep. NMFS 8. U.S. Department of Commerce, Washington, D.C., pp. 179–188.

Cailliet, G.M., Martin, L.K., Kusher, D., Wolf, P., and Welden, B.A. (1983b). Techniques for enhancing vertebral bands in age estimation of California elasmobranchs. In: Prince, E.D. and Pulos, L.M. (Eds.), *Proceedings of the International Workshop on Age Determination of Oceanic Pelagic Fishes: Tunas, Billfishes, Sharks*, NOAA Tech. Rep. NMFS 8. U.S. Department of Commerce, Washington, D.C., pp. 157–165.

Cailliet, G.M., Radtke, R.L., and Welden, B.A. (1986a). Elasmobranch age determination and verification: a review. In: Uyeno, T., Arai, R., Taniuchi, T., and Matsuura, K. (Eds.), *Indo-Pacific Fish Biology: Proceedings of the Second International Conference on Indo-Pacific Fishes*. Ichthyological Society of Japan, Tokyo, pp. 345–359.

Cailliet, G.M., Love, M.S., and Ebeling, A.W. (1986b). *Fishes: A Field and Laboratory Manual on Their Structure, Identification, and Natural History*. Wadsworth, Belmont, CA.

Cailliet, G.M., Mollet, H.F., Pittenger, G.G., Bedford, D., and Natanson, L.J. (1992). Growth and demography of the Pacific angel shark (*Squatina californica*), based on tag returns off California. *Aust. J. Mar. Freshwater Res.* 43:1313–1330.

Cailliet, G.M., Smith, W.D., Mollet, H.F., and Goldman, K.J. (2006). Chondrichthyan growth studies: an updated review, stressing terminology, sample size sufficiency, validation, and curve fitting. In: Carlson, J.K. and Goldman, K.J. (Eds.), *Age and Growth of Chondrichthyan Fishes: New Methods, Techniques and Analysis*. Springer, Dordrecht, pp. 211–228.

Calis, E., Jackson, E.H., Nolan, C.P., and Jeal, F. (2005). Preliminary age and growth estimates of the rabbitfish, *Chimaera monstrosa*, with implications for future resource management. *J. Northw. Atl. Fish. Sci.* 35:15–26.

Camhi, M. (1999). *Sharks on the Line*. II. *An Analysis of Pacific State Shark Fisheries*. National Audubon Society, Washington, D.C., 114 pp.

Campana, S.E. (1990). How reliable are growth back-calculations based on otoliths? *Can. J. Fish. Aquat. Sci.* 47:2219–2227.

Campana, S.E. (1997). Use of radiocarbon from nuclear fallout as a dated marker in the otoliths of haddock *Melanogrammus aeglefinus. Mar. Ecol. Prog. Ser.* 150:49–56.

Campana, S.E. (2001). Accuracy, precision and quality control in age determination, including a review of the use and abuse of age validation methods. *J. Fish Biol.* 59:197–242.

Campana, S.E. and Thorrold, S.R. (2001). Otoliths, increments, and elements: keys to a comprehensive understanding of fish populations? *Can. J. Fish. Aquat. Sci.* 58:30–38.

Campana, S.E., Annand, M.C., and McMillan, J.I. (1995). Graphical and statistical methods for determining the consistency of age determinations. *Trans. Am. Fish. Soc.* 124:131–138.

Campana, S.E., Natanson, L.J., and Myklevoll, S. (2002). Bomb dating and age determination of large pelagic sharks. *Can. J. Fish. Aquat. Sci.* 59:450–455.

Campana, S.E., Jones, C., McFarlane, G.A., and Myklevoll, S. (2006). Bomb dating and age validation using the spines of spiny dogfish (*Squalus acanthias*). In: Carlson, J.K. and Goldman, K.J. (Eds.), *Age and Growth of Chondrichthyan Fishes: New Methods, Techniques and Analysis*. Springer, Dordrecht, pp. 327–336.

Carlander, K.D. (1969). *Handbook of Freshwater Fishery Biology*, Vol. 1. Iowa University Press, Ames.

Carlson, J.K. and Parsons, G.R. (1997). Age and growth of the bonnethead shark, *Sphyrna tiburo*, from northwest Florida, with comments on clinal variation. *Environ. Biol. Fish.* 50:331–341.

Carlson, J.K., Cortés, E., and Bethea, D.M. (2003). Life history and population dynamics of the finetooth shark (*Carcharhinus isodon*) in the northeastern Gulf of Mexico. *Fish. Bull.* 101:281–292.

Carrier, J.C. and Radtke, R. (1988). Preliminary evaluation of age and growth in juvenile nurse sharks (*Ginglymostoma cirratum*) using visual and electron microprobe assessment of tetracycline-labeled vertebral centra, unpublished paper available from author, Biology Department, Albion College, Albion, MI 49224.

Casey, J.G., Pratt, Jr., H.L., and Stillwell, C. (1985). Age and growth of the sandbar shark (*Carcharhinus plumbeus*) from the western North Atlantic. *Can. J. Fish. Aquat. Sci.* 42(5):963–975.

Casey, J.M. and Myers, R.A. (1998). Near extinction of a large, widely distributed fish. *Science* 281:690–692.

Casselman, J.M. (1983). Age and growth assessments of fish from their calcified structures & techniques and tools. In: Prince, E.D. and Pulos, L.M. (Eds.), *Proceedings of the International Workshop on Age Determination of Oceanic Pelagic Fishes: Tunas, Billfishes, Sharks*, NOAA Tech. Rep. NMFS 8. U.S. Department of Commerce, Washington, D.C., pp. 1–17.

Chang, W.Y.B. (1982). A statistical method for evaluating the reproducibility of age determination. *Can. J. Fish. Aquat. Sci.* 39:1208–1210.

Clarke, M.W. and Irvine, S.B. (2006). Terminology for the age-ing of chondrichthyan fish using dorsal-fin spines. In: Carlson, J.K. and Goldman, K.J. (Eds.), *Age and Growth of Chondrichthyan Fishes: New Methods, Techniques and Analysis*. Springer, Dordrecht, pp. 273–277.

Clarke, M.W., Connolly, P.L., and Bracken, J.J. (2002a). Catch, discarding, age estimation, growth and maturity of the squalid shark *Deania calceus* west and north of Ireland. *Fish. Res.* 56:139–153.

Clarke, M.W., Connolly, P.L., and Bracken, J.J. (2002b). Age estimation of the exploited deepwater shark *Centrophorus squamosus* from the continental slopes of the Rockall Trough and Porcupine Bank. *J. Fish Biol.* 60:501–514.

Conrath, C.L., Gelsleichter, J., and Musick, J.A. (2002). Age and growth of the smooth dogfish (*Mustelus canis*) in the northwest Atlantic. *Fish. Bull.* 100:674–682.

Cortés, E. (1994). Demographic analysis of the Atlantic sharpnose shark, *Rhizoprionodon terraenovae*, in the Gulf of Mexico. *Fish. Bull.* 93:57–66.

Cortés, E. (1997). Demographic analysis as an aid in shark stock assessment and management. *Fish. Res.* 39:199–208.

Cortés, E. (2000). Life history patterns and correlations in sharks. *Rev. Fish. Sci.* 8(4):299–344.

Cortés, E. (2002). Incorporating uncertainty into demographic modeling: application to shark populations. *Conserv. Biol.* 16(4):1048–1062.

Cortés, E. and Parsons, G.R. (1996). Comparative demography of two populations of the bonnethead shark (*Sphyrna tiburo*). *Can. J. Fish. Aquat. Sci.* 53:709–718.

Cotton, C. (2010). Age, Growth, and Reproductive Biology of Deep-Water Chondrichthyans, doctoral dissertation, College of William and Mary, Virginia Institute of Marine Science, Williamsburg.

Daiber, F.C. (1960). A technique for age determination in the skate *Raja eglantaria*. *Copeia* 1960:258–260.

Davis, C.D., Cailliet, G.M., and Ebert, D.A. (2007). Age and growth of the roughtail skate, *Bathyraja trachura* (Gilbert, 1892). *Environ. Biol. Fish.* 80(2–3):325–336.

DeVries, D.R. and Frie, R.V. (1996). Determination of age and growth. In: Murphy, B.R. and Willis, D.W. (Eds.), *Fisheries Techniques*, 2nd ed. American Fisheries Society, Bethesda, MD, pp. 483–512.

Druffel, E.R.M. (1987). Bomb radiocarbon in the Pacific: annual and seasonal timescale variations. *J. Mar. Res.* 45:667–698.

Druffel, E.R.M. (1989). Decadal time scale variability of ven-tilation in the North Atlantic: high-precision measure-ments of bomb radiocarbon in banded corals. *J. Geophys. Res.* 94:3271–3285.

Druffel, E.R.M. (2002). Radiocarbon in corals: records of the car-bon cycle, surface circulation and climate. *Oceanography* 15:122–127.

Druffel, E.R.M. and Linick, T.W. (1978). Radiocarbon in annual coral rings of Florida. *Geophys. Res. Lett.* 5:913–916.

Druffel, E.R.M., Griffin, S., and Guilderson, T.P. (2001). Changes of subtropical north Pacific radiocarbon and correlation with climate variability. *Radiocarbon* 43(1):15–25.

Ebert, D., White, W.T., Goldman, K.J., Compagno, L.J.V., Daly-Engel, T.S., and Ward, R.D. (2010). Resurrection and redescription of *Squalus suckleyi* (Girard, 1854) from the North Pacific, with comments on the *Squalus acanthias* subgroup (Squaliformes: Squalidae). *Zootaxa* 2612:22–40.

Evans, G.T. and Hoenig, J.M. (1998). Testing and viewing sym-metry in contingency tables, with application to readers of fish ages. *Biometrics* 54:620–629.

Ewing, G.P., Lyle, J.M., Murphy, R.J., Kalish, J.M., and Ziegler, P.E. (2007). Validation of age and growth in a long-lived temperate reef fish using otolith structure, oxytetracy-cline and bomb radiocarbon methods. *Mar. Freshw. Res.* 58:944–955.

Fabens, A.J. (1965). Properties and fitting of the von Bertalanffy growth curve. *Growth* 29:265–289.

Francis, M.P. and Francis, R.I.C.C. (1992). Growth rate esti-mates for New Zealand rig (*Mustelus lenticulatus*). *Aust. J. Mar. Freshw. Res.* 43:1157–1176.

Francis, M.P., Campana, S.E., and Jones, C.M. (2007). Age under-estimation in New Zealand porbeagle sharks (*Lamna nasus*): is there an upper limit to ages that can be deter-mined from shark vertebrae? *Mar. Freshw. Res.* 58:10–23.

Francis, R.I.C.C. (1988). Maximum likelihood estimation of growth and growth variability from tagging data. *N.Z. J. Mar. Freshw. Res.* 22:43–51.

Francis, R.I.C.C. (1990). Back-calculation of fish length: a criti-cal review. *J. Fish Biol.* 36:883–902.

Frantz, B.R., Foster, M.S., and Riosmensa-Rodriguez, R. (2005). *Clathromorphum nereostratum* (Corralinales, Rhodophyta): the oldest alga? *J. Phycol.* 41:770–773.

Freer, D.W.L. and Griffiths, C.L. (1993). Estimation of age and growth in the St. Joseph *Callorhinchus capensis* (Dumeril). *S. Afr. J. Mar. Sci.* 13:75–82.

Frisk, M.G., Miller, T.J., and Fogarty, M.J. (2001). Estimation and analysis of biological parameters in elasmobranch fishes: a comparative life history study. *Can. J. Fish. Aquat. Sci.* 58:969–981.

Gallagher, M. and Nolan, C.P. (1999). A novel method for the estimation of age and growth in rajids using caudal thorns. *Can. J. Fish. Aquat. Sci.* 56:1590–1599.

Gallagher, M.J., Nolan, C.P., and Jeal, F. (2005a). Age, growth and maturity of the commercial ray species from the Irish Sea. *J. Northw. Atl. Fish. Sci.* 35:47–66.

Gallagher, M.J., Nolan, C.P., and Jeal, F. (2005b). The structure and growth processes of caudal thorns. *J. Northw. Atl. Fish. Sci.* 35:125–129.

Gallagher, M.J., Green, M.J., and Nolan, C.P. (2006). The poten-tial use of caudal thorns as a non-invasive ageing struc-ture in the thorny skate (*Amblyraja radiata* Donovan, 1808). In: Carlson, J.K. and Goldman, K.J. (Eds.), *Age and Growth of Chondrichthyan Fishes: New Methods, Techniques and Analysis*. Springer, Dordrecht, pp. 265–272.

Gelsleichter, J., Musick, J.A., and Van Veld, P. (1995). Proteoglycans from vertebral cartilage of the clearnose skate, *Raja eglantaria*: inhibition of hydroxyapetite forma-tion. *Fish. Physiol. Biochem.* 14:247–251.

Gelsleichter, J., Cortés, E., Manire, C.A., Hueter, R.E., and Musick, J.A. (1997). Use of calcein as a fluorescent marker for elasmobranch vertebral cartilage. *Trans. Am. Fish. Soc.* 126:862–865.

Gelsleichter, J., Cortés, E., Manire, C.A., Hueter, R.E., and Musick, J.A. (1998a). Evaluation of toxicity of oxytetracycline on growth of captive nurse sharks, *Ginglymostoma cirratum. Fish. Bull.* 96:624–627.

Gelsleichter, J., Piercy, A., and Musick, J.A. (1998b). Evaluation of copper, iron, and lead substitution techniques in elasmobranch age determination. *J. Fish Biol.* 53:465–470.

Goldman, K.J. (2002). Aspects of Age, Growth, Demographics and Thermal Biology of Two Lamniform Shark Species, doctoral dissertation, College of William and Mary, School of Marine Science, Virginia Institute of Marine Science, Williamsburg, 220 pp.

Goldman, K.J. (2005). Age and growth of elasmobranch fishes. In: Musick, J.A. and Bonfil, R. (Eds.), *Management Techniques for Elasmobranch Fisheries*, FAO Fisheries Tech. Paper 474. Food and Agriculture Organization, Rome, pp. 97–132.

Goldman, K.J. and Musick, J.A. (2006). Growth and maturity of salmon sharks in the eastern and western North Pacific, and comments on back-calculation methods. *Fish. Bull.* 104:278–292.

Goldman, K.J., Branstetter, S., and Musick, J.A. (2006). A re-examination of the age and growth of sand tiger sharks, *Carcharias taurus*, in the western North Atlantic. In: Carlson, J.K. and Goldman, K.J. (Eds.), *Age and Growth of Chondrichthyan Fishes: New Methods, Techniques and Analysis*. Springer, Dordrecht, pp. 241–252.

Goosen, A.J.J. and Smale, M.J. (1997). A preliminary study of age and growth of the smoothhound shark *Mustelus mustelus* (Triakidae). *S. Afr. J. Mar. Sci.* 18:85–92.

Gruber, S.H. and Stout, R.G. (1983). Biological materials for the study of age and growth in a tropical marine elasmobranch, the lemon shark, *Negaprion brevirostris* (Poey). In: Prince, E.D. and Pulos, L.M. (Eds.), *Proceedings of the International Workshop on Age Determination of Oceanic Pelagic Fishes: Tunas, Billfishes, Sharks*, NOAA Tech. Rep. NMFS 8. U.S. Department of Commerce, Washington, D.C., pp. 193–205.

Gulland, J.A. and Holt, S.J. (1959). Estimation of growth parameters for data at unequal time intervals. *J. Cons. Int. Explor. Mer* 25:47–49.

Haddon, M. (2001). *Modelling and Quantitative Measures in Fisheries*. Chapman & Hall/CRC Press, Boca Raton, FL, pp. 197–246.

Heppell, S.S., Heppell, S.A., Read, A.J., and Crowder, L.B. (2005). Effects of fishing on long-lived marine organisms. In: Norse, E.A. and Crowder, L.B. (Eds.), *Marine Conservation Biology: The Science of Maintaining the Sea's Biodiversity*. Marine Conservation Biology Institute, Island Press, Washington, D.C., pp. 211–231.

Hoenig, J.M. and Brown, C.A. (1988). A simple technique for staining growth bands in elasmobranch vertebrae. *Bull. Mar. Sci.* 42(2):334–337.

Hoenig, J.M. and Gruber, S.H. (1990). Life-history patterns in elasmobranchs: implications for fisheries management. In: Pratt, Jr., H.L., Gruber, S.H., and Taniuchi, T. (Eds.), *Elasmobranchs as Living Resources: Advances in the Biology, Ecology, Systematics, and Status of the Fisheries*, NOAA Tech. Rep. NMFS 90. U.S. Department of Commerce, Washington, D.C., pp. 1–16.

Hoenig, J.M., Morgan, M.J., and Brown, C.A. (1995). Analyzing differences between two age determination methods by tests of symmetry. *Can. J. Fish. Aquat. Sci.* 52:364–368.

Hoff, T.B. and Musick, J.A. (1990). Western North Atlantic shark-fishery management problems and informational requirements. In: Pratt, Jr., H.L., Gruber, S.H., and Taniuchi, T. (Eds.), *Elasmobranchs as Living Resources: Advances in the Biology, Ecology, Systematics, and Status of the Fisheries*, NOAA Tech. Rep. NMFS 90. U.S. Department of Commerce, Washington, D.C., pp. 455–472.

Holden, M.J. (1974). Problems in the rational exploitation of elasmobranch populations and some suggested solutions. In: Jones, F.R. (Ed.), *Sea Fisheries Research*. Halstead Press/John Wiley & Sons, New York, pp. 117–138.

Holden, M.J. and Vince, M.R. (1973). Age validation studies on the centra of *Raja clavata* using tetracycline. *J. Cons. Int. Explor. Mer.* 35:13–17.

Ishiyama, R. (1951). Studies on the rays and skates belonging to the family *Rajidae*, found in Japan and adjacent regions. 2. On the age determination of Japanese black-skate *Raja fusca* Garman (preliminary report). *Bull. Jpn. Soc. Sci. Fish.* 16(12):112–118.

Irvine, S.B., Stevens, J.D., and Laurenson, L.J.B. (2006a). Surface bands on deepwater squalid dorsal-fin spines: an alternative method for ageing *Centroselachus crepidater. Can. J. Fish. Aquat. Sci.* 63:617–627.

Irvine, S.B., Stevens, J.D., and Laurenson, L.J.B. (2006b). Comparing external and internal dorsal-spine bands to interpret the age and growth of the giant lantern shark, *Etmopterus baxteri* (Squaliformes: Etmopteridae). In: Carlson, J.K. and Goldman, K.J. (Eds.), *Age and Growth of Chondrichthyan Fishes: New Methods, Techniques and Analysis*. Springer, Dordrecht, pp. 253–264.

Johnson, A.G. (1979). A simple method for staining the centra of teleosts vertebrae. *Northeast. Gulf Sci.* 3:113–115.

Jones, B.C. and Green, G.H. (1977). Age determination of an elasmobranch (*Squalus acanthias*) by x-ray spectrometry. *J. Fish. Res. Bd. Can.* 34:44–48.

Jorgensen, S.J., Reeb, C.A., Chapple, T.K., Anderson, S., Perle, C. et al. (2010). Philopatry and migration of Pacific white sharks. *Proc. R. Soc. B Biol. Sci.* 277:679–688.

Kalish, J.M. (1993). Pre and post bomb radiocarbon in fish otoliths. *Earth Planet. Sci. Lett.* 114:549–554.

Kalish, J.M. (1995). Radiocarbon and fish biology. In: Secor, D.H., Dean, J.M., and Campana, S.E. (Eds.), *Recent Developments in Fish Otolith Research*. University of South Carolina Press, Columbia, pp. 637–653.

Kalish, J.M. and Johnston, J. (2001). Determination of school shark age based on analysis of radiocarbon in vertebral collagen. In: Kalish, J.M. (Ed.), *Use of the Bomb Radiocarbon Chronometer to Validate Fish Age: Final Report*, FDRC Project 93/109. Fisheries Research and Development Corporation, Canberra, Australia, pp. 116–122.

Kalish, J.M., Nydal, R., Nedreaas, K.H., Burr, G.S., and Eine, G.L. (2001a). A time history of pre- and post-bomb radiocarbon in the Barents Sea derived from Arcto-Norwegian cod otoliths. *Radiocarbon* 43:843–855.

Kalish, J.M., Timmiss, T., Pritchard, J., Johnstone, J., and Dunhamel, G. (2001b). Validation and direct estimation of age and growth of Patagonian toothfish *Dissostichus eleginoides* based on otoliths. In: Kalish, J.M. (Ed.), *Use of the Bomb Radiocarbon Chronometer to Validate Fish Age: Final Report*, FDRC Project 93/109. Fisheries Research and Development Corporation, Canberra, Australia, pp. 164–182.

Kerr, L.A., Andrews, A.H., Frantz, B.R., Coale, K.H., Brown T.A., and Cailliet, G.M. (2004). Radiocarbon in otoliths of yelloweye rockfish (*Sebastes ruberrimus*): a reference time series for the coastal waters of southeast Alaska. *Can. J. Fish. Aquat. Sci.* 61:443–451.

Kerr, L.A., Andrews, A.H., Cailliet, G.M., Brown, T.A., and Coale, K.H. (2006). Investigations of $\Delta^{14}C$, $\delta^{15}N$, and $\delta^{13}C$ in vertebrae of white shark (*Carcharodon carcharias*) from the eastern Pacific Ocean. In: Carlson, J.K. and Goldman, K.J. (Eds.), *Age and Growth of Chondrichthyan Fishes: New Methods, Techniques and Analysis*. Springer, Dordrecht, pp. 337–353.

Ketchen, K.S. (1975). Age and growth of dogfish (*Squalus acanthias*) in British Columbia waters. *J. Fish. Res. Bd. Can.* 32:13–59.

Kilada, R., Campana, S.E., and Roddick, D. (2009). Validated age, growth and mortality estimates of the ocean quahog (*Arctica islandica*) in the western Atlantic. *ICES J. Mar. Sci.* 64:31–38.

Kneebone, J., Natanson, L.J., Andrews, A.H., and Howell, H. (2008). Using bomb radiocarbon analyses to validate age and growth estimates for the tiger shark, *Galeocerdo cuvier*, in the western North Atlantic. *Mar. Biol.* 154:423–434.

Kusher, D.I., Smith, S.E., and Cailliet, G.M. (1992). Validated age and growth of the leopard shark, *Triakis semifasciata*, from central California. *Environ. Biol. Fish.* 35:187–203.

Kyne, P.M. and Simpfendorfer, C.A. (2007). *A Collation and Summarization of Available Data on Deepwater Chondrichthyans: Biodiversity, Life History and Fisheries*, prepared by the IUCN Shark Specialist Group for the Marine Conservation Biology Institute, Bellevue, WA (http://www.flmnh.ufl.edu/fish/organizations/ssg/deepchondreport.pdf).

LaMarca, M.J. (1966). A simple technique for demonstrating calcified annuli in the vertebrae of large elasmobranchs. *Copeia* 1966:351–352.

Longhurst, A. (2002). Murphy's law revisited: longevity as a factor in recruitment to fish populations. *Fish. Res.* 56:125–131.

Martin, L.K. and Cailliet, G.M. (1988). Age and growth of the bat ray, *Myliobatis californica*, off central California. *Copeia* 1988:762–773.

Mathsoft, Inc. (2000). *S-Plus 2000 Professional Release 1*. Mathsoft, Seattle, WA.

Matta, M.E. and Gunderson, D.R. (2007). Age, growth, maturity, and mortality of the Alaska skate, *Bathyraja parmifera*, in the eastern Bering Sea, *Environ. Biol. Fish.* 80(2–3):309–323.

Maurer, J.R. (2009). Life History of Two Bering Sea Slop Skates: *Bathyraja lindbergi* and *B. maculata*, master's thesis, Moss Landing Marine Laboratories and California State University, Monterey Bay.

McAuley, R.B., Simpfendorfer, C.A., Hyndes, G.A., Allison, R.R., Chidlow, J.A., Newman, S.J., and Lenanton, R.C.J. (2006). Validated age and growth of the sandbar shark, *Carcharhinus plumbeus* (Nardo 1827) in the waters off Western Australia. In: Carlson, J.K. and Goldman, K.J. (Eds.), *Age and Growth of Chondrichthyan Fishes: New Methods, Techniques and Analysis*. Springer, Dordrecht, pp. 385–400.

McFarlane, G.A. and Beamish, R.J. (1987a). Validation of the dorsal spine method of age determination for spiny dogfish. In: Summerfelt, R.C. and Hall, G.E. (Eds.), *The Age and Growth of Fish*, Iowa State University Press, Ames, pp. 287–300.

McFarlane, G.A. and Beamish, R.J. (1987b). Selection of dosages of oxytetracycline for age validation studies. *Can. J. Fish. Aquat. Sci.* 44:905–909.

McFarlane, G.A., King, J.R., and Saunders, M.W. (2002). Preliminary study on the use of neural arches in the age determination of bluntnose sixgill sharks (*Hexanchus griseus*). *Fish. Bull.* 100:861–864.

McNemar, Q. (1947). Note on the sampling error of the difference between correlated proportions or percentages. *Psychometrika* 12:153–157.

McPhie, R.P. and Campana, S.E. (2009). Bomb dating and age determination of skates (family Rajidae) off the eastern coast of Canada. *ICES J. Mar. Sci.* 66:546–560.

Mohan, P.J., Clar, S.T., and Schmid, T.H. (2004). Age and growth of captive sharks. In: Smith, M., Warmolts, D., Thoney, D., and Hueter, R. (Eds.), *Elasmobranch Husbandry Manual: Captive Care of Sharks, Rays and Their Relatives*. Ohio Biological Survey, Columbus, OH, pp. 201–226.

Mollet, H.F., Ezcurra, J.M., and O'Sullivan, J.B. (2002). Captive biology of the pelagic stingray, *Dasyatis violacea* (Bonaparte, 1830). *Mar. Freshw. Res.* 53:531–541.

Moura, T., Figueiredo, I., Farias, I., Serra-Pereira, B., Coelho, R., Erzini, K., Neves, A., and Serrano Gordo, L. (2007). The use of caudal thorns for ageing *Raja undulata* from the Portuguese continental shelf, with comments on its reproductive cycle. *Mar. Freshw. Res.* 58:983–992.

Musick, J.A. (1999). Ecology and conservation of long-lived marine animals. In: Musick, J.A. (Ed.), *Life in the Slow Lane: Ecology and Conservation of Long-Lived Marine Animals*. American Fisheries Society, Bethesda, MD, pp. 1–10.

Nagasawa, K. (1998). Predation by salmon sharks (*Lamna ditropis*) on Pacific salmon (*Oncorhynchus* spp.) in the North Pacific Ocean. *Bull. N. Pac. Anadromous Fish Comm.* 1: 419–433.

Nakano, H. (1994). Age, reproduction and migration of blue shark in the North Pacific Ocean. *Bull. Nat. Res. Inst. Far Seas Fish.* 31:141–256.

Nammack, M.F., Musick, J.A., and Colvocoresses, J.A. (1985). Life history of spiny dogfish off the northeastern United States. *Trans. Am. Fish. Soc.* 114:367–376.

Natanson, L.J. (1993). Effect of temperature on band deposition in the little skate, *Raja erinacea*. *Copeia* 1993:199–206.

Natanson, L.J. and Cailliet, G.M. (1990). Vertebral growth zone deposition in Pacific angel sharks. *Copeia* 1990:1133–1145.

Natanson, L.J. and Kohler, N.E. (1996). A preliminary estimate of age and growth of the dusky shark *Carcharhinus obscurus* from the south-west Indian Ocean, with comparisons to the western North Atlantic population. *S. Afr. J. Mar. Sci.* 17:217–224.

Natanson, L.J., Casey, J.G., and Kohler, N.E. (1995). Age and growth estimates for the dusky sharks, *Carcharhinus obscurus*, in the western North Atlantic Ocean. *Fish. Bull.* 93:116–126.

Natanson, L.J., Mello, J.J., and Campana, S.E. (2002). Validated age and growth of the porbeagle shark (*Lamna nasus*) in the western North Atlantic Ocean. *Fish. Bull.* 100:266–278.

Natanson, L.J., Kohler, N.E., Ardizzone, D., Cailliet, G.M., Wintner, S.P., and Mollet, H.F. (2006). Validated age and growth estimates for the shortfin mako, *Isurus oxyrinchus*, in the North Atlantic Ocean. In: Carlson, J.K. and Goldman, K.J. (Eds.), *Age and Growth of Chondrichthyan Fishes: New Methods, Techniques and Analysis*. Springer, Dordrecht, pp. 367–383.

Natanson, L.J., Sulikowski, J.A., Kneebone, J.R., and Tsang, P.C. (2007). Age and growth estimates for the smooth skate, *Malacoraja senta*, in the Gulf of Maine. *Environ. Biol. Fish.* 80:293–308.

Natanson, L.J., Wintner, S.P., Johansson, F., Piercy, A., Campbell, P. et al. (2008). Ontogenetic vertebral growth patterns in the basking shark *Cetorhinus maximus*. *Mar. Ecol. Prog. Ser.* 361:267–278.

Neer, J.A. and Cailliet, G.M. (2001). Aspects of the life history of the Pacific electric ray, *Torpedo californica* (Ayers). *Copeia* 2001:842–847.

Neilson, J.D. and Campana, S.E. (2008). A validated description of age and growth of western Atlantic bluefin tuna (*Thunnis thynnus*). *Can. J. Fish. Aquat. Sci.* 65:1523–1527.

Officer, R.A., Gason, A.S., Walker, T.I., and Clement, J.G. (1996). Sources of variation in counts of growth increments in vertebrae from gummy shark, *Mustelus antarcticus*, and school shark, *Galeorhinus galeus*: implications for age determination. *Can. J. Fish. Aquat. Sci.* 53:1765–1777.

Officer, R.A., Day, R.W., Clement, J.G., and Brown, L.P. (1997). Captive gummy sharks, *Mustelus antarcticus*, for hypermineralized bands in their vertebrae during winter. *Can. J. Fish. Aquat. Sci.* 54:2677–2683.

Parsons, G.R. (1983). An examination of the vertebral rings of the Atlantic sharpnose shark, *Rhizoprionodon terraenovae*. *Northeast. Gulf Sci.* 6:63–66.

Parsons, G.R. (1993). Age determination and growth of the bonnethead shark *Sphyrna tiburo*: a comparison of two populations. *Mar. Biol.* 117:23–31.

Passerotti, M.S., Carlson, J.K., Piercy, A.N., and Campana, S.E. (2010). Age validation of great hammerhead shark (*Sphyrna mokarran*), determined by bomb radiocarbon analysis. *Fish. Bull.* 108:346–351.

Peignoux-Deville, J., Lallier, F., and Vidal, B. (1982). Evidence for the presence of osseous tissue in dogfish vertebrae. *Cell Tissue Res.* 222:605–614.

Perez, C.R., Cailliet, G.M., and Ebert, D.A. (2011). Age and growth of the sandpaper skate, *Bathyraja kincaidii*, using vertebral centra, with an investigation of caudal thorns. *J. Mar. Biol. Assoc. U.K.* 91(6):1149–1156.

Piner, K.R. and Wischniowski, S.G. (2004). Pacific halibut chronology of bomb radiocarbon in otoliths from 1941 to 1981 and a validation of ageing methods. *J. Fish Biol.* 64:1060–1071.

Prager, M.H., Saila, S.B., and Recksick, C.W. (1987). *FISHPARM: A Microcomputer Program for Parameter Estimation of Nonlinear Models in Fishery Science*, Tech. Rep. 87-10. Department of Oceanography, Old Dominion University, Norfolk, VA.

Pratt, H.L. and Casey, J.G. (1983). Age and growth of the shortfin mako, *Isurus oxyrinchus*, using four methods. *Can. J. Fish. Aquat. Sci.* 40:1944–1957.

Punt, A.E. and Hughes, G.S. (1989). *PC-YIELD II User's Guide*, Benguela Ecology Programme Report 18. Foundation for Research and Development, University of Natal, Pietermaritzburg, South Africa, 55 pp.

Restrepo, V.R., Thompson, G.G., Mace, P.M., Gabriel, W.L., Low, L.L. et al. (1998). *Technical Guidance on the use of Precautionary Approaches to Implementing National Standard 1 of the Magnuson–Stevens Fishery Conservation and Management Act*, NOAA Tech. Memo. NMFS-F/SPO. U.S. Department of Commerce, Washington, D.C., 54 pp.

Richards, S.W., Merriman, D., and Calhoun, L.H. (1963). Studies of the marine resources of southern New England. IX. The biology of the little skate, *Raja erinacea* Mitchell. *Bull. Bingham Oceanogr. Collect. Yale Univ.* 18(3):5–67.

Ricker, W.E. (1975). Computation and interpretation of biological statistics of fish populations. *Bull. Fish. Res. Board Can.* 191:1–382.

Ricker, W.E. (1979). *Growth Rates and Models*. Academic Press, San Diego, CA, pp. 677–743.

Ricker, W.E. (1992). Back-calculation of fish lengths based on proportionality between scale and length increments. *Can. J. Fish. Aquat. Sci.* 49:1018–1026.

Ridewood, W.G. (1921). On the calcification of the vertebral centra in sharks and rays. *Phil. Trans. R. Soc. B Biol.* 210:311–407.

Roark, E.B., Guilderson, T.P., Dunbar, R.B., and Ingram, B.L. (2006). Radiocarbon based ages and growth rates: Hawaiian deep sea corals. *Mar. Ecol. Prog. Ser.* 327:1–14.

Robinson, M.P. and Motta, P.J. (2002). Patterns of growth and the effects of scale on the feeding kinematics of the nurse shark (*Ginglymostoma cirratum*). *J. Zool. (Lond.)* 256(4):449–462.

Roff, D.A. (1982). *The Evolution of Life Histories: Theory and Analysis*. Chapman & Hall, New York.

SAS Institute, Inc. (1999). *SAS/ETS User's Guide, Version 8.0.* Statistical Analysis Systems Institute, Inc., Cary, NC.

Schwartz, F.J. (1983). Shark ageing methods and age estimates of scalloped hammerhead, *Sphyrna lewini*, and dusky, *Carcharhinus obscurus*, sharks based on vertebral ring counts. In: Prince, E.D. and Pulos, L.M. (Eds.), *Proceedings of the International Workshop on Age Determination of Oceanic Pelagic Fishes: Tunas, Billfishes, Sharks*, NOAA Tech. Rep. NMFS 8. U.S. Department of Commerce, Washington, D.C., pp. 167–174.

Serra-Pereira, B., Figueiredo, I., Bordalo-Machado, P., Farias, I., Moura, T., and Gordo, L.S. (2005). Age and growth of *Raja clavata* Linnaeus, 1758: evaluation of ageing precision using different types of caudal denticles. *ICES CM Documents* 2005 17:1–10.

Serra-Pereira, B., Figueiredo, I., Farias, I., Moura, T., and Gordo, L.S. (2008). Description of dermal denticles from the caudal region of *Raja clavata* and their use for the estimation of age and growth. *ICES J. Mar. Sci.* 65:1701–1709.

Simpfendorfer, C.A. and Kyne, P.M. (2009). Limited potential to recover from overfishing raises concerns for deep-sea sharks, rays and chimaeras. *Environ. Conserv.* 36:97–103.

Simpfendorfer, C.A., McAuley, R.B., Chidlow, J., and Unsworth, P. (2002). Validated age and growth of the dusky shark, *Carcharhinus obscurus*, from Western Australia waters. *Mar. Freshw. Res.* 53:567–573.

Skomal, G.B. and Natanson, L.J. (2003). Age and growth of the blue shark, *Prionace glauca*, in the North Atlantic Ocean. *Fish. Bull.* 101:627–639.

Sminkey, T.R. and Musick, J.A. (1995a). Age and growth of the sandbar shark, *Carcharhinus plumbeus*, before and after population depletion. *Copeia* 1995:871–883.

Sminkey, T.R. and Musick, J.A. (1995b). Demographic analysis of the sandbar shark, *Carcharhinus plumbeus*, in the western North Atlantic. *Fish. Bull.* 94:341–347.

Smith, S.E. (1984). Timing of vertebral band deposition in tetracycline injected leopard sharks. *Trans. Am. Fish. Soc.* 113:308–313.

Smith, S.E., Au, D.W., and Show, C. (1998). Intrinsic rebound potentials of 26 species of Pacific sharks. *Mar. Freshw. Res.* 49:663–678.

Smith, S.E., Mitchell, R.A., and Fuller, D. (2003). Age, validation of a leopard shark (*Triakis semifasciata*) recaptured after 20 years. *Fish. Bull.* 101:194–198.

Smith, W.D., Cailliet, G.M., and Melendez, E.M. (2007). Maturity and growth characteristics of a commercially exploited stingray, *Dasyatis dipterura*. *Mar. Freshw. Res.* 58:54–66.

Stevens, J.D. (1975). Vertebral rings as a means of age determination in the blue shark (*Prionace glauca*). *J. Mar. Biol. Assoc. U.K.* 55:657–665.

Stevens, J.D., Bonfil, R., Dulvey, N.K., and Walker, P.A. (2000). The effects of fish on sharks, rays, and chimaeras (chondrichthyans), and the implications for marine ecosystems. *ICES J. Mar. Sci.* 57:476–494.

Stewart, R.E.A., Campana, S.E., Jones, C.M., and Stewart, B.E. (2006). Bomb radiocarbon dating calculated beluga (*Delphinapterus leucas*) age estimates. *Can. J. Zool.* 84:1840–1852.

Stuiver, M. and Polach, H.A. (1977). Discussion: reporting of ^{14}C data. *Radiocarbon* 19:355–363.

Sullivan, K.J. (1977). Age and growth of the elephant fish *Callorhinchus milii* (Elasmobranchii: Callorhynchidae). *N.Z. J. Mar. Freshw. Res.* 11(4):745–753.

Summerfelt, R.C. and Hall, G.E. (1987). *Age and Growth of Fish.* Iowa State University Press, Ames.

Tanaka, S. (1990). Age and growth studies on the calcified structures of newborn sharks in laboratory aquaria using tetracycline. In: Pratt, Jr., H.L., Gruber, S.H., and Taniuchi, T. (Eds.), *Elasmobranchs as Living Resources: Advances in the Biology, Ecology, Systematics, and Status of the Fisheries*, NOAA Tech. Rep. NMFS 90. U.S. Department of Commerce, Washington, D.C., pp. 189–202.

Tribuzio, C.A. (2010). Life History, Demography, and Ecology of the Spiny Dogfish (*Squalus acanthias*) in the Gulf of Alaska, doctoral dissertation, University of Alaska Fairbanks, 183 pp.

Tribuzio, C.A., Kruse, G.H., and Fujioka, J.T. (2010). Age and growth of spiny dogfish (*Squalus acanthias*) in the Gulf of Alaska: analysis of alternative growth models. *Fish. Bull.* 108:119–135.

Tricas, T.C. and McCosker, J.E. (1984). Predatory behavior of the white shark (*Carcharodon carcharias*), with notes on its biology. *Proc. Calif. Acad. Sci.* 43:221–238.

Van Dykhuizen, G. and Mollet, H.F. (1992). Growth, age estimation and feeding of captive sevengill sharks, *Notorynchus cepedianus*, at the Monterey Bay Aquarium. *Aust. J. Mar. Freshw. Res.* 43:297–318.

Welden, B.A., Cailliet, G.M., and Flegal, A.R. (1987). Comparison of radiometric with vertebral band age estimates in four California elasmobranchs. In: Summerfelt, R.C. and Hall, G.E. (Eds.), *The Age and Growth of Fish*, Iowa State University Press, Ames, pp. 301–315.

Wintner, S.P. and Cliff, G. (1999). Age and growth determination of the white shark, *Carcharodon carcharias*, from the east coast of South Africa. *Fish. Bull.* 97:153–169.

Wintner, S.P. and Dudley, S.F.J. (2000). Age and growth estimates for the tiger shark, *Galeocerdo cuvier*, from the east coast of South Africa. *Mar. Freshw. Res.* 51:43–53.

Wintner, S.P., Dudley, S.F.J., Kistnasamy, N., and Everett, B. (2002). Age and growth estimates for the Zambezi shark, *Carcharhinus leucas*, from the east coast of South Africa. *Mar. Freshw. Res.* 53:557–566.

Yamaguchi, A., Huang, S.Y., Chen, C.T., and Taniuchi, T. (1999). Age and growth of the starspotted smooth-hound, *Mustelus manazo* (Chondrichthyes: Triakidae) in the waters of north-eastern Taiwan. In: Séret, B. and Sire, J.Y. (Eds.), *Proceedings of the 5th Indo-Pacific Fishes Conference, Noumea*, Society of French Ichthyologists, Paris, pp. 505–513.

Zweifel, J.R. and Lasker, R. (1976) Prehatch and posthatch growth of fishes: a general model. *Fish. Bull.* 74:609–621.

15

Population Dynamics, Demography, and Stock Assessment

Enric Cortés, Elizabeth N. Brooks, and Todd Gedamke

CONTENTS

15.1 Introduction

There is mounting evidence of declines in elasmobranch populations worldwide mainly as a result of overharvesting (Baum et al., 2003; Campana et al., 1999, 2001; Cortés et al., 2002, 2006; Dulvy et al., 2010; Myers et al., 2007; Simpfendorfer, 2000), although habitat loss and degradation are also of increasing concern (Dulvy and Forrest, 2010; Field et al., 2009; García et al., 2008; Simpfendorfer, 2007). The severity of several of these reported declines, however, has been questioned on the grounds that the data and methods of analysis may not have captured the real changes in relative abundance exhibited by the

populations under study (Baum and Myers, 2004; Baum et al., 2003, 2005; Burgess et al., 2005a,b). Within the past three decades, the links between life histories and the risk of overexploitation or even extinction of elasmobranch populations have been increasingly studied, mainly through demographic (life table or matrix population) models and formal stock assessment models.

Surprisingly our knowledge of life history traits of many species is still precarious, and we are only beginning to gain insight into the life history patterns shared by some species, the relationships among life history traits, and the links between those life history traits and population dynamics (Compagno, 1990; Cortés, 2000; Dulvy and Forrest, 2010; Frisk, 2010; Frisk et al., 2001).

The main goal of this chapter is to review our knowledge of the population dynamics of elasmobranch populations. In doing so we will necessarily introduce the frameworks used to incorporate our knowledge of the biology of each species into population models. The first step is to present an overview of methodological issues relevant to the study of demography and dynamics of unexploited and exploited elasmobranch populations, which is critical to understanding the data requirements, limitations, and advantages of different population modeling approaches. After setting the methodological background, we review the complementary approaches used to model elasmobranch populations and arrange the individual studies in a summary table. We conclude with a synthesis of the review and recommendations for future work. We refer the reader particularly interested in life history strategies to reviews by Compagno (1990), Cortés (2000, 2004), and Stevens (2010) for sharks; Frisk (2010) and Frisk et al. (2001) for batoids; and Dulvy and Forrest (2010) for chondrichthyans in general.

15.2 Population Dynamics

Populations are made up of individuals with a life cycle consisting of a series of sequential and recognizable states of development that can be described by age, stage, or size (cohorts). Population dynamics attempts to describe changes in the cohort-specific abundance of a population in space and with time as a result of multiple sources of variability. In general terms, the sources of variability governing population dynamics are both ecological and genetic processes (Cortés, 1999). The cohort-specific abundance of individuals over time and space is determined by three basic vital rates—birth, growth, and death—and the demographic processes of emigration and immigration, which are subject to genetic, demographic, environmental, sampling, and human-induced stochasticity. The effect that these sources of variability have on vital rates and demographic processes ultimately determines the fate of the population. Ideally, a population dynamics model should thus capture the interaction of vital rates and demographic processes with all sources of variability to provide knowledge on population abundance in time and space.

15.2.1 Methodological Background

The reality for elasmobranch population modeling is quite different, however. Our knowledge of vital rates and demographic processes is still fragmentary for most species, and even less information is available on the spatial distribution of populations, stock–recruitment

dynamics, and the effect of most sources of stochasticity on elasmobranch populations. Despite this state of affairs, considerable progress has been made in the recent past in the fields of demographic analysis and population modeling of elasmobranchs. Two main approaches with separate philosophies and purposes have emerged. Life tables and population matrix models have been developed to gain a basic understanding of the population ecology of some species while assessing their vulnerability to fishing and to address conservation issues by producing population metrics that can be used to generate mostly qualitative management measures. In contrast, stock assessment models traditionally used in fisheries research have been applied to several stocks to produce estimates of population status that can be used for implementing quantitative management measures. Table 15.1 summarizes elasmobranch population models to the best of our knowledge arranged into several groups according to the following factors: (1) whether the model was cohort structured or considered lumped biomass only; (2) whether the model was static or dynamic (i.e., whether or not it explicitly incorporated time in the equations describing the population dynamics); (3) whether the cohort structure of the population was classified as age, stage, or size; (4) whether the model dealt with uncertainty or not (deterministic vs. stochastic); and (5) whether the model was linear or nonlinear (i.e., with density independence or density dependence, respectively) (Chaloupka and Musick, 1997). Table 15.1 also includes the modeling approach, species, geographic location, purpose of the study, and citation. Publications based only on a few of the methods discussed in Section 15.3.3 are included in this table. The distinction between static and dynamic population models is arbitrary because in a strict sense only models that incorporate temporal variation in demographic rates and allow for feedback mechanisms such as potential density-dependent responses reflect the dynamics of a population (Chaloupka and Musick, 1997). In studies of elasmobranch populations, the year, age, and cohort effects are often confounded because a year-specific state–space vector (Getz and Haight, 1989) of absolute abundance is not available, thus the transient or time-dependent behavior of the population is being modeled in relative, rather than absolute, terms. Before describing the various population modeling approaches, it is convenient to define some terms and describe the limitations of sampling design in relation to the data requirements of the different methods.

15.2.1.1 Demographic Unit or Stock

One of the main assumptions of a population dynamics model is that the stock, population, or demographic unit under study can be distinguished in time and

TABLE 15.1

Summary of Elasmobranch Demography and Population Status Evaluation Studies

Structure	Time	Cohort Type	Mode	Shape	Model Type(s)	Species	Area	Aim	Refs.
Biomass	Dyn	—	Det	NL	Schaefer	*Squalus acanthias*	NEA	Sa/Ma	Aasen (1964)
Biomass	Dyn	—	Det	NL	Schaefer	Large sharks	NWA	Sa/Ma	Otto et al. (1977)
Biomass	Dyn	—	Det	NL	Fox, Pella-Tomlinson	Pelagic sharks	NWA	Sa/Ma	Anderson (1980)
Biomass	Dyn	—	Det	NL	Fox	*Dalatias licha*	Azores	Sa/Ma	Silva (1983, 1987)
Biomass	Dyn	—	Stoch	NL	Schaefer, Fox	—	—	Sa/Ma	Bonfil (1996)
Biomass	Dyn	—	Det	NL	Schaefer, Fox, Pella-Tomlinson	Rajid assemblage	Falkland Islands	Sa/Ma	Agnew et al. (2000)
Biomass	Dyn	—	Stoch	NL	Schaefer (Bayesian)	*Carcharhinus plumbeus, C. limbatus*	NWA	Sa/Ma	McAllister et al. (2001)
Biomass	Dyn	—	Stoch	NL	Schaefer (Bayesian)	Small coastal sharks	NWA	Sa/Ma	Cortés (2002b)
Biomass	Dyn	—	Stoch	NL	Schaefer (Bayesian)	Large coastal sharks	NWA	Sa/Ma	Cortés et al. (2002)
Biomass	Dyn	—	Stoch	NL	Schaefer (Bayesian)	*Prionace glauca, Isurus oxyrinchus*	NWA	Sa/Ma	Babcock and Cortés (2005)
Biomass	Dyn	—	Stoch	NL	Schaefer (Bayesian)	*Squalus acanthias*	NEA	Sa/Ma	Hammond and Ellis (2005)
Biomass	Dyn	—	Stoch	NL	Schaefer (Bayesian)	*Prionace glauca*	NEP	Sa/Ma	Kleiber et al. (2009)
Biomass	Dyn	—	Stoch	NL	Schaefer, Fox (maximum likelihood)	*Sphyrna lewini*	NWA	Sa/Ma	Hayes et al. (2009)
Biomass	Dyn	—	Stoch	NL	Schaefer (Bayesian, hierarchical and non-hierarchical)	Hammerhead shark complex (three species)	NWA	Sa/Ma	Jiao et al. (2009)
Cohort	Static	Age	Det	Linear	Life table	*Carcharhinus plumbeus*	NWA	Da/Ma	Hoff (1990)
Cohort	Static	Age	Det	Linear	Life table	*Triakis semifasciata*	California	Da/Ma	Cailliet (1992)
Cohort	Static	Age	Det	Linear	Life table	*Squatina californica*	California	Da/Ma	Cailliet et al. (1992)
Cohort	Static	Age	Det	Linear	Life table	*Rhizoprionodon terraenovae*	NWA	Da/Ma	Cortés (1995)
Cohort	Static	Age	Det	Linear	Life table	*Sphyrna tiburo*	EGM	Da	Cortés and Parsons (1996)
Cohort	Static	Age	Det	Linear	Life table	*Carcharhinus plumbeus*	NWA	Da/Ma	Sminkey and Musick (1996)
Cohort	Static	Age	Det	Linear	Life table	5 carcharhinid, 1 sphyrnid shark	NWA	Da/Ma	Cortés (1998)
Cohort	Static	Age	Det	Linear	Life table	*Rhizoprionodon terraenovae*	SEGM	Da/Ma	Márquez and Castillo (1998)
Cohort	Static	Age	Det	Linear	Life table	*Sphyrna tiburo*	SEGM	Da/Ma	Márquez et al. (1998)
Cohort	Static	Age	Det	Linear	Life table	*Sphyrna lewini*	NWP	Da/Ma	Liu and Chen (1999)
Cohort	Static	Age	Det	Linear	Life table	*Rhizoprionodon taylori*	Northern Australia	Da/Ma	Simpfendorfer (1999a)
Cohort	Static	Age	Det	Linear	Life table	*Carcharhinus obscurus*	Southwest Australia	Da/Ma	Simpfendorfer (1999b)
Cohort	Static	Age	Det	Linear	Life table	*Pristis pectinata, P. perotteti*	WA	Da/Ma	Simpfendorfer (2000)
Cohort	Static	Age	Det	Linear	Life table	*Torpedo californica*	California	Da	Neer and Cailliet (2001)
Cohort	Static	Age	Det	Linear	Life table	*Lamna nasus*	NWA	Da/Ma	Campana et al. (2002)
Cohort	Static	Age	Stoch	Linear	Life table	*Carcharhinus falciformis*	NWA	Da	Beerkircher et al. (2003)
Cohort	Static	Age	Det	Linear	Life table	18 shark species	Multiple locations	Da	Chen and Yuan (2006)

(continued)

TABLE 15.1 (continued)

Summary of Elasmobranch Demography and Population Status Evaluation Studies

Structure	Time	Cohort Type	Mode	Shape	Model Type(s)	Species	Area	Aim	Refs.
Cohort	Static	Age	Stoch	Linear	Life table	*Carcharhinus obscurus, C. plumbeus*	EI	Da/Ma	McAuley et al. (2007)
Cohort	Static	Age	Det	Linear	Life table	21 species of pelagic elasmobranch	Multiple locations	Da/Ma	Dulvy et al. (2008)
Cohort	Static	Age	Det	Linear	Modified Euler–Lotka equation	31 shark species, 1 ray species	Multiple locations	Da/Ma	Smith et al. (1998, 2008), Au et al. (2008)
Cohort	Static	Age	Det	Linear	Modified "dual" Euler–Lotka equation	*Galeorhinus galeus, Mustelus antarcticus*	Southern Australia	Da	Xiao and Walker (2000)
Cohort	Static	Age	Det	Linear	Leslie matrix	*Negaprion brevirostris*	NWA	Da/Ma	Hoenig and Gruber (1990)
Cohort	Static	Age	Det	NL	Leslie matrix	*Squalus acanthias*	NWA	Sa/Ma	Silva (1993)
Cohort	Static	Age	Det	Linear	Leslie matrix	*Raja batis, R. clavata, R. montagui, R. naevus, R. radiata*	North Sea	Da/Ma	Walker and Hislop (1998)
Cohort	Static	Age	Det	Linear	Leslie matrix	*Triakis semifasciata, Squatina californica*	California	Da/Ma	Heppell et al. (1999)
Cohort	Static	Age	Stoch	Linear	Leslie matrix	*Leucoraja erinacea, L. ocellata*	NWA	Sa/Ma	Frisk et al. (2002)
Cohort	Static	Age	Det	Linear	Leslie matrix	*Carcharodon carcharias, Alopias pelagicus, Carcharias taurus, Dasyatis violacea*	Multiple locations	Da	Mollet and Cailliet (2002)
Cohort	Static	Age	Det	Linear	Leslie matrix	*Carcharias taurus*	SWP	Da/Ma	Otway et al. (2004)
Cohort	Static	Age	Stoch	Linear	Leslie matrix	*Squalus acanthias, Rhizoprionodon taylori*	NEP and Northern Australia	Da/Ma	Gallucci et al. (2006)
Cohort	Static	Age	Stoch	Linear	Leslie matrix	*Prionace glauca*	NA	Da/Ma	Aires-da-Silva and Gallucci (2007)
Cohort	Static	Age	Det	NL	Leslie matrix	*Dipturus laevis, Negaprion brevirostris*	NWA	Da/Ma	Gedamke et al. (2007)
Cohort	Static	Age	Stoch	Linear	Leslie matrix, life table	*Carcharhinus plumbeus, C. limbatus*	NWA	Input to Sa	McAllister et al. (2001)
Cohort	Static	Age	Stoch	Linear	Leslie matrix, life table	41 shark species	Multiple locations	Da/Ma	Cortés (2002a)
Cohort	Static	Age	Stoch	Linear	Leslie matrix, life table	Small coastal sharks	NWA	Input to Sa	Cortés (2002b)
Cohort	Static	Age	Stoch	Linear	Leslie matrix, life table	8 pelagic shark species	Multiple locations	Da/Ma	Cortés (2008)
Cohort	Dyn	Stage	Stoch	Linear	Lefkovitch matrix	*Carcharhinus plumbeus*	NWA	Da/Ma	Cortés (1999)
Cohort	Static	Stage	Det	Linear	Lefkovitch matrix	*Carcharhinus plumbeus*	NWA	Da/Ma	Brewster-Geisz and Miller (2000)
Cohort	Static	Stage	Det	Linear	Lefkovitch matrix	*Dipturus laevis*	NWA	Da/Ma	Frisk et al. (2002)
Cohort	Static	Stage	Det	Linear	Lefkovitch matrix	*Carcharodon carcharias, Alopias pelagicus, Carcharias taurus, Dasyatis violacea*	Multiple locations	Da	Mollet and Cailliet (2002)
Cohort	Static	Stage	Det	Linear	Lefkovitch matrix	*Carcharias taurus*	SWP	Da/Ma	Otway et al. (2004)
Cohort	Static	Stage	Det	Linear	Lefkovitch matrix	55 species of elasmobranch	Multiple locations	Da/Ma	Frisk et al. (2005)

Cohort	Static	Age	Det	Linear	Yield per recruit, Cohort analysis	*Galeorhinus galeus*	Australia	Sa/Ma	Grant et al. (1979)
Cohort	Static	Age	Det	Linear	Yield per recruit	*Leucoraja erinacea*	NWA	Ma	Waring (1984)
Cohort	Static	Age	Det	Linear	Yield per recruit, VPA	*Triakis semifasciata*	California	Sa/Ma	Smith and Abramson (1990)
Cohort	Static	Age	Det	Linear	Recruitment-adjusted yield per recruit	*Triakis semifasciata*	California	Sa/Ma	Au and Smith (1997)
Cohort	Static	Age	Det	Linear	Yield per recruit	*Carcharhinus plumbeus*	NWA	Sa/Ma	Cortés (1998)
Cohort	Static	Age	Det	Linear	Yield per recruit	*Lamna nasus*	NWA	Sa/Ma	Campana et al. (1999, 2001, 2002)
Cohort	Static	Age	Stoch	NL	Spawning per recruit and VPA	*Alopias pelagicus*	NWP	Sa/Ma	Liu et al. (2006)
Cohort	Dyn	Age	Det	NL	Dynamic pool	*Mustelus antarcticus*	Southern Australia	Sa/Ma	Walker (1992, 1994a,b)
Cohort	Dyn	Age	Stoch	NL	Age-structured (Bayesian)	*Galeorhinus galeus*	Southern Australia	Sa/Ma	Punt and Walker (1998)
Cohort	Static	Age	Det	Linear	Age-structured	*Squalus acanthias*	NWA	Sa/Ma	Rago et al. (1998)
Cohort	Static	Age	Det	Linear	Age-structured	12 species of deepwater dogsharks (Squaliformes)	Australia	Sa/Ma	Forrest and Walters (2009)
Cohort	Dyn	Age	Stoch	NL	Spatially explicit, age-structured (Bayesian)	*Galeorhinus galeus*	Southern Australia	Sa/Ma	Punt et al. (2000)
Cohort	Dyn	Age	Stoch	NL	Age-structured (Maximum likelihood)	*Furgaleus macki*	Southwest Australia	Sa/Ma	Simpfendorfer et al. (2000)
Cohort	Dyn	Age	Stoch	NL	Spatially explicit, stage-based, age-structured (Bayesian)	*Carcharhinus limbatus*	NWA	Sa/Ma	Apostolaki et al. (2002a)
Cohort	Dyn	Age	Stoch	NL	Age-structured (Bayesian)	Large coastal sharks	NWA	Sa/Ma	Apostolaki et al. (2002b)
Cohort	Dyn	Age	Stoch	NL	Age-structured production (Bayesian and maximum likelihood)	*Carcharhinus plumbeus, C. limbatus*	NWA	Sa/Ma	Brooks et al. (2002); Cortés et al. (2002)
Cohort	Dyn	Age	Stoch	NL	Length-based, age-structured (Bayesian)	*Lamna nasus*	NWA	Sa/Ma	Harley (2002), Gibson and Campana (2005), Campana et al. (2010)
Cohort	Dyn	Age	Stoch	NL	Age-structured (Bayesian)	*Rhizoprionodon terraenovae*	NWA	Sa/Ma	Simpfendorfer and Burgess (2002)
Cohort	Dyn	Age	Det	NL	Age-structured	*Galeorhinus galeus, Mustelus antarcticus*	SEI	Sa/Ma	Prince (2005)
Cohort	Dyn	Age	Stoch	NL	Length- and age-based, age-structured (Bayesian)	*Mustelus antarcticus*	Southern Australia	Sa/Ma	Pribac et al. (2005)
Cohort	Dyn	Age	Stoch	NL	Catch-free, age-structured production (Bayesian and maximum likelihood)	*Carcharhinus obscurus*	NWA	Sa/Ma	Cortés et al. (2006)

(continued)

TABLE 15.1 (continued)

Summary of Elasmobranch Demography and Population Status Evaluation Studies

Structure	Time	Cohort Type	Mode	Shape	Model Type(s)	Species	Area	Aim	Refs.
Cohort	Dyn	Age	Stoch	NL	Age-structured (Bayesian)	*Raja rhina*	NEP	Sa/Ma	Gertseva (2009)
Cohort	Dyn	Age	Stoch	NL	Length-based, age-structured (Bayesian)	*Prionace glauca*	NEP	Sa/Ma	Kleiber et al. (2009)
Cohort	Dyn	Age	Stoch	NL	Age-structured (Bayesian)	*Leucoraja ocellata*	NWA	Sa/Ma	Frisk et al. (2010)
Delay difference	Dyn	Age	Det	NL	Deriso–Schnute	*Galeorhinus galeus*	Southern Australia	Sa/Ma	Walker (1995)
Delay difference	Dyn	Age	Stoch	NL	Deriso–Schnute	—	—	Da/Ma	Bonfil (1996)
Delay difference	Dyn	Age	Stoch	NL	Lagged recruitment, survival and growth (Bayesian)	Small coastal sharks	NWA	Sa/Ma	Cortés (2002b)
Delay difference	Dyn	Age	Stoch	NL	Lagged recruitment, survival and growth (Bayesian)	Large coastal sharks	NWA	Sa/Ma	Cortés et al. (2002)
—	—	—	Det	—	Tag–recapture (Modified Petersen estimate)	*Carcharodon carcharias*	South Africa	Sa/Ma	Cliff et al. (1996)
—	—	—	Det	—	Tag–recapture (Jolly-Seber method)	*Carcharodon carcharias*	South Australia	Sa/Ma	Strong et al. (1996)
—	—	—	Det	—	Tag–recapture (Leslie's depletion estimator)	*Negaprion brevirostris*	Bahamas	Sa/Ma	Gruber et al. (2001)
—	—	—	Det	—	Tag–recapture	*Ginglymostoma cirratum*	SWA	Sa/Ma	Castro and Rosa (2005)
Cohort	Dyn	Age	Det	NL	Tag–recapture (maximum likelihood)	*Prionace glauca*	NWA	Sa/Ma	Aires-da-Silva et al. (2009)

Note: Studies are listed according to the type of structure and in chronological order. *Abbreviations:* Da, demographic analysis; Det, deterministic; Dyn, dynamic; EGM, Eastern Gulf of Mexico; EI, Eastern Indian; Ma, management advice; NA, North Atlantic; NEA, Northeastern Atlantic; NEP, Northeastern Pacific; NL, Nonlinear; NWA, Northwestern Atlantic; NWP, Northwestern Pacific; Sa, stock assessment; SEGM, Southeastern Gulf of Mexico; SEI, Southeastern Indian; Stoch, stochastic; SWP, Southwestern Pacific; WA, Western Atlantic.

space from other similar units. Furthermore, most models implicitly assume a closed population and that the model equations therefore adequately explain gains and losses to population abundance. Although movement, migratory patterns, and genetic stock identification of elasmobranchs are beginning to be better understood (Heist, 2004; Musick et al., 2004; Shivji, 2010; Stevens, 2010), identifying discrete demographic units or stocks still remains a major challenge in the study of elasmobranch populations. Many shark species, for example, are widely distributed and highly migratory, posing an especially difficult problem because individuals from potentially different stocks are likely to co-occur in some areas or habitats. In some other cases, as with the spiny dogfish, *Squalus acanthias*, and school shark, *Galeorhinus galeus*, genetically separate stocks have been identified and little mixing is believed to occur (Walker, 1998). Ideally, demographic and population modeling of elasmobranchs should focus on genetically distinct stocks. In practice, the transboundary nature of many populations or stocks poses a practical problem for management, which is generally restricted geographically because of jurisdictional issues.

15.2.1.2 Population Sampling Design

Vital rates and demographic processes are affected by three separate, yet often confounded, time effects (Chaloupka and Musick, 1997). Indeed, vital rates may vary from year to year due to external factors, may differ among cohorts due to genetic factors, and most likely vary by age. A realistic population dynamics model thus needs to uncouple the effects of year, age, and cohort factors; however, it is not always possible to separate these time effects because of shortcomings in the modeling framework or, more often, owing to sampling limitations and lack of comprehensive empirical estimates to parameterize complex models. This is the case with elasmobranch population modeling studies, which usually rely on only one set of estimates of demographic rates that are often not age specific. These models thus do not consider year effects, let alone cohort effects, and only characterize one state of nature of the more complex population dynamics that occur in a dynamic environment.

At present we simply do not know how these confounding time effects may bias estimates of population parameters for elasmobranchs. Given the life histories of elasmobranchs, it is reasonable to assume that year factors will not have the pronounced effect they can have on other fishes because vital rates of elasmobranchs are believed to be less sensitive to environmental influences and therefore more stable and predictable (Stevens, 1999). It is unknown how genetic influences, expressed through cohort factors, affect vital rates of elasmobranchs. In terms of age factors, we know from

life history theory that natural mortality, for example, varies with age (Roff, 1992). In sharks, it is believed that intraspecific mortality generally remains fairly low and stable once individuals attain a certain size, but that juvenile mortality decreases from birth to adulthood as individuals grow and predation risk decreases (Cortés and Parsons, 1996).

There are few direct estimates of instantaneous natural mortality rate (M) or instantaneous total mortality rate (Z) for elasmobranchs based on mark–recapture techniques or catch curves. Direct estimates of natural mortality were obtained only in the mark–depletion experiments conducted for age-0 (Manire and Gruber, 1993) and juvenile (Gruber et al., 2001) lemon sharks, *Negaprion brevirostris*. Estimates of natural mortality derived from Z were obtained in mark–recapture studies for school shark, *Galeorhinus galeus* (Grant et al., 1979); little skate, *Raja erinacea* (Waring, 1984); juvenile blacktip sharks, *Carcharhinus limbatus* (Heupel and Simpfendorfer, 2002); and lesser spotted dogfish, *Scyliorhinus canicula* (Rodríguez-Cabello and Sánchez, 2005). Mortality rates were also obtained from length-converted catch curves for bonnethead, *Sphyrna tiburo* (Cortés and Parsons, 1996); rays, *Raja clavata* and *Raja radiata* (Walker and Hislop, 1998); porbeagle, *Lamna nasus* (Campana et al., 2001, 2002); and juvenile tiger sharks, *Galeocerdo cuvier* (Driggers et al., 2008). Total mortality rates have been estimated by Campana et al. (2002) for porbeagle shark through Paloheimo Zs (Ricker, 1975) and by Gedamke et al. (2008) for the barndoor skate, *Dipturus laevis*, through an analysis of mean lengths and catch per unit effort (CPUE) of recruits and adults.

The majority of population modeling studies for elasmobranchs has relied on indirect estimates of mortality obtained through methods based on predictive equations of life history traits. Most of these methods make use of parameters estimated from the von Bertalanffy growth (VBG) function, including those of Pauly (1980), Hoenig (1983), Chen and Watanabe (1989), and Jensen (1996) (for reviews of these methods, see Cortés, 1998, 1999; Roff, 1992; Simpfendorfer, 1999a, 2005). These equations do not yield age-specific estimates of natural mortality, except for the Chen and Watanabe (1989) method. In contrast, methods proposed by Peterson and Wroblewski (1984) and Lorenzen (1996, 2000) allow estimation of size-specific natural mortality, which can then be transformed into age-specific estimates through the VBG function. These age-specific methods show a monotonic decrease in mortality with age. The use of U-shaped curves (Walker, 1998) has also been advocated to account for the fact that individuals must die off in their terminal year of life. A modified U-shape curve, the so-called "bathtub" curve (Chen and Watanabe, 1989; Siegfried, 2006) is probably more adequate for elasmobranch fishes because the initial decrease in natural mortality (*M*) at young ages is followed by a flatter pro-file, and *M* only increases sharply toward the oldest ages, possibly due to senescence.

Back-transformation of lengths into ages through the VBG function has been the usual method for estimat-ing age-specific life history traits in elasmobranchs, because determining age of individuals is much more difficult than simply measuring their lengths. Thus, few studies have determined age at maturity or age-specific reproductive parameters directly (Gedamke et al., 2005; Harry et al., 2010; McAuley et al., 2007; Sulikowski et al., 2009). Use of ages at maturity or age-specific fecun-dity estimates derived from length can result in biased estimates of population metrics because this procedure does not account for variability in age at length, and *vice versa*. Many elasmobranch population models in the past have described maturity as a knife-edge pro-cess in which it is assumed that 100% of females reach maturity at the same size (age). This assumption is a direct consequence of reproductive studies that do not attempt to fit an ogive (logistic function) to describe the proportion of mature females at size or age in a popula-tion. Additionally, pup production per female is often assumed to be a constant value. Another potential bias introduced when estimating age–length and length–maturity relationships is length-selective fishing mor-tality, which is rarely considered (Walker, 1998, 2005a, 2007). Also, using the proportion of mature females at age in population models can lead to overestimation of reproductive output. When possible, the proportion of females in maternal state should be used instead to account for the time it takes females to mate, become gravid, and gestate before contributing offspring to the population. An additional confounding factor is that females of several species are known to store sperm, which would further delay offspring production.

15.2.1.3 Exponential vs. Logistic Population Growth

As mentioned earlier, two separate approaches with dif-ferent philosophies have been used to gain an under-standing of the productivity and risk of overexploitation of elasmobranch fishes. The primary difference between the two approaches is whether density-dependent compensation is modeled and the resulting trade-offs between data requirements, realism of assumptions, and complexity of interpretation. Density-independent models, which are based on exponential growth, have limited data requirements and assume that the popu-lation growth rate is constant at all population sizes (Figure 15.1). This implies that there are no limiting factors—such as food resources or space—to popula-tion growth rate and, in turn, no limits on population size (i.e., no carrying capacity of the environment). Traditional fisheries models used in assessments of the status of exploited stocks are based on density-depen-dent or logistic growth, wherein positive population growth rates are only attainable if there is some level of exploitation that reduces population size from its maxi-mum (carrying capacity, *K*). The per-capita growth rate (*r*) increases linearly from a state of equilibrium (*r* = 0) at *K*, where there is maximum density dependence, to a maximum at very low population size, where there is no density-dependent compensation in survival or reproductive rates (Figures 15.1 and 15.2). The maxi-mum growth rate that occurs at the lowest population sizes is also known as the *intrinsic rate of increase* ($r_{intrinsic}$) and can only be observed, or estimated from modeling a very depleted population.

Comprehensive empirical estimates of vital rates in relationship to population size are not available for any elasmobranch population to fully parameterize the more complex density-dependent logistic models.

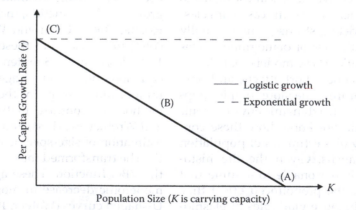

FIGURE 15.1
Per-capita growth rate (*r*) for populations exposed to exponential or logistic growth. The exponential growth model results in constant (den-sity-independent) growth, whereas the logistic growth model results in a linear (density-dependent) increase with decreasing population size. (A) Zero per-capita growth rate at carrying capacity; (B) per-capita growth rate at 0.5*K*; and (C) maximum per-capita growth rate at very low population size. See Figure 15.2 for the corresponding position of (A), (B), and (C) in a plot of population size vs. time.

As a result, two types of analyses based on exponential growth—life tables and matrix population models—have typically been used in conservation contexts to characterize the potential productivity of species and assess the risk of overexploitation or even extinction (Dulvy et al., 2003, 2004, 2005). These approaches are appealing for use with elasmobranchs because they only require a schedule of reproduction and survival and can be used to make interspecific comparisons of productivity and gain insights on the importance of different life stages and response to fishing pressure (Smith et al., 1998). However, the density-independent structure requires that significant consideration be given to model parameterization, interpretation of results, and any attempts to quantify sustainable levels of exploitation. To illustrate the point, consider two different parameterizations of the same model. In the first case, empirical estimates of survival rates from a tagging study are available; in the second, survival rates from life-history invariant methods are used. Although empirical estimates from a tagging study are preferable and most estimates will be derived from a population undergoing some level of fishing mortality, it is unlikely that enough individuals would be available at the lowest population sizes for a tagging study to be successful. Therefore, maximum potential survival rates will not be reported, and the predicted rate of population growth will not reflect the intrinsic or maximum population growth rate. In turn, survival rates from life-history invariant methods reflect general patterns or equilibrium formulations, neither of which will reflect survival rates at the lowest population sizes, which would tend to underestimate the intrinsic rate of increase. Alternatively, if the assumed life-history invariant relationship (or its assumed coefficients) does not accurately characterize the species to which

it is applied, this would also bias the estimate of intrinsic rate of increase, and the direction of bias would be related to the mismatch in the assumed relationship.

Because empirical or life-history invariant estimates of maximum survival and reproductive rates are rare, at best, density-independent life table or matrix analyses are generally parameterized to predict population growth rates that underestimate shark productivity (Gedamke et al., 2007; Walker, 1998), thus potentially overestimating the risk of overexploitation or extinction. The solution lies in recognizing that, for the overall elasmobranch life history strategy, it is generally believed that the density-dependent compensation occurs on survival during the first few years of life and that these rates are likely to be the most pliable and directly tied to stock size (Cortés, 2002a, 2004; Hoenig and Gruber, 1990; Keeny and Simpfendorfer, 2009). Gruber et al. (2001) documented this process in the first-year survival of lemon sharks in a Bimini nursery lagoon where survival rates ranged from approximately 40% at larger population sizes to approximately 70% at lower population sizes. The scope of density-dependent compensation and resulting range in early life stage survival will be species and population specific, but, given that no other empirical estimates of how vital rates respond to population size in an elasmobranch are available, the relatively wide range of first-year survival in lemon sharks should be considered when parameterizing and interpreting models for other species. To minimize potential bias, Cortés (2007) suggested using the maximum age-specific estimate of survival from multiple life-history invariant methods as a proxy to simulate a density-dependent response. Gedamke et al. (2007, 2009) discussed the challenges of parameterizing and interpreting predicted rates of population growth from exponentially based life tables and matrix analyses in detail

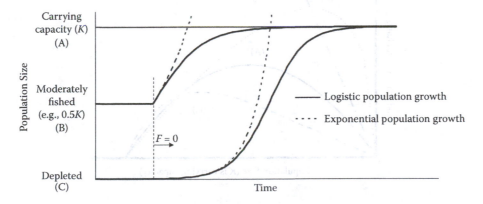

FIGURE 15.2
Steady-state and logistic growth of populations recovering from three levels of exploitation and its relationship to model predictions of exponential growth. Labels (A), (B), and (C) correspond to those in Figure 15.1. When exploitation ceases ($F = 0$ to the right of the vertical dashed line), a severely depleted population (C) will take longer to recover than a moderately exploited population (B). An unexploited population (A) will continue in a steady state (equilibrium). Note that the logistic and exponential models give similar results (overlap) in the initial phases after exploitation ends, but increasingly diverge as recovery time elapses.

and developed an approach that uses auxiliary information in conjunction with exponential growth models to obtain estimates of the intrinsic rate of increase. The approach circumvents the fundamental problem of providing life history parameters from a depleted population (i.e., maximum rates) by using an index of abundance to extrapolate rates of population growth toward a zero population size. Two illustrations of the method were presented that use different types of information. For the lemon shark, estimates of first-year survival corresponding to different population sizes were available so model predictions of population growth were used to estimate the maximum population growth rate ($r_{intrinsic}$) that would occur at a zero population size. In contrast, for the barndoor skate, the density-dependent response in juvenile survival was unknown. Using survey indices and observed rates of population change, a matrix model was used to solve for the unknown parameters, and an extrapolation provided an estimate of $r_{intrinsic}$.

15.2.1.3.1 Position of the Inflection Point of Population Growth Curves

Fowler (1981, 1988) found that the relative position of the inflection point of population growth curves or its corresponding maximum in production curves (the proportion of K at which maximum net production occurs) varies along a continuum across animal taxa. Whereas in the traditional parabolic curve the maximum net change in abundance occurs at 0.5K, very productive species such as insects or many fish populations reach their maximum net change at lower population sizes and, conversely, low-productivity species such as many large mammals reach this maximum net change near carrying capacity (Figure 15.3). A shift to the right of 0.5K in the net change in abundance results from delayed density dependence (when negative effects of density dependence do not reduce growth until population size is near K) (Skalski et al., 2005). This spectrum of possibilities of where the maximum net change in abundance is reached roughly corresponds to the r–K continuum of life histories (Fowler, 1981). Using intrinsic rates per generation obtained from life tables and population matrix models, Cortés (2008) estimated the position of the inflection point of population growth curves for a suite of pelagic sharks using an equation derived by Fowler (1988):

$$R = 0.633 - 0.187\ln(rT)$$

where R is the position of the inflection point, r is the intrinsic rate of increase, and T is generation time. Cortés (2008) found that, with the exception of blue shark, *Prionace glauca*, for which R was close to the traditional 0.5K, the seven other pelagic species had inflection points >0.5K, with some being close to K. The implications of these results are twofold: First, depleted stocks will take longer to recover and reach the point of maximum surplus production (maximum sustainable yield, or MSY) because the production curve is no longer parabolic. Second, as can be seen in Figure 15.3, the magnitude of the MSY is smaller than for more productive stocks.

15.2.1.3.2 Allee Effects (Depensation)

The Allee effect describes a situation that can occur at very low population sizes where population growth rates not only do not reach a maximum as prescribed in traditional fishery theory but may even become negative

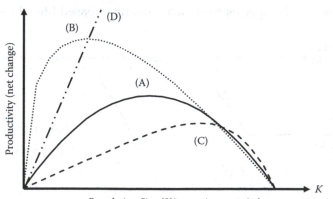

FIGURE 15.3

Productivity (net change or production) as a function of population size for populations exhibiting different types of growth. (A) Productivity under the traditional logistic growth model; (B) productivity for insects and other "r-selected" taxa where net change is maximized at low population sizes; (C) productivity pattern for "K-selected" taxa such as large mammals that exhibit delayed density dependence (i.e., productivity does not decrease until the population reaches a size near carrying capacity); and (D) productivity under exponential growth. Many species of elasmobranchs, especially large sharks, would have a productivity curve of type (C). (Adapted from Skalski, J.R. et al., *Ecol. Model.*, 212, 528–535, 2005.)

(hence depensation). This is generally attributed to the inability of individuals to find mates and is of particular relevance in conservation contexts where extinction risks are of interest (e.g., in population viability analyses). The importance of the Allee effect to population growth is that it produces an unstable equilibrium at low population sizes, thereby increasing extinction risk (Dulvy et al., 2003; Skalski et al., 2005). In mathematical terms, dN/dt has a positive slope when it crosses the N (abundance)-axis in a plot of N against dN/dt at very low population size, whereas the slope of dN/dt is negative when it crosses the N-axis at carrying capacity, which is a stable equilibrium (Pastor, 2008). It must be pointed out, however, that simple models of homogeneous populations may not apply to natural populations, which are not truly homogeneous but rather age and size structured. This is particularly important in the context of predator–prey interactions where recent work suggests that predator population size may increase as its own mortality rate increases in both stable and unstable systems (Abrams and Quince, 2005), a phenomenon known as the *hydra* effect (Abrams and Matsuda, 2005). This effect is caused by a saturating functional response of the predator to prey density or when the prey exhibit an Allee effect, whereby the prey's per capita growth rate increases with prey density at low population sizes (Abrams and Quince, 2005).

15.2.1.4 Stock–Recruitment Curves

Knowledge of the relationship between stock size and recruitment is central to the understanding of the population dynamics of marine organisms. For elasmobranchs, it is generally assumed that recruitment is directly related to spawning (pupping) stock size due to their overall life history strategy and relatively small number of offspring (Holden, 1977); however, until recently no empirical data on this relationship had been published for any species of elasmobranch. Walker (1994a) had first produced some indirect support for a Beverton–Holt type of stock–recruitment curve. By assuming that a density-dependent response was elicited through natural mortality of pre-recruit ages, he found that the number of gummy shark recruits off southeastern Australia predicted by an age-structured model remained relatively constant over a fairly wide range of high stock biomass levels. More recently, Gedamke et al. (2009) found a strong stock–recruitment relationship for the barndoor skate using groundfish survey CPUE data from Georges Bank in the northeastern United States. There was weak evidence supporting a Ricker stock-recruit curve, but due to a limited number of data points at larger stock sizes the authors carried both the Ricker and Beverton–Holt curves through their analysis. In this study, the primary parameter of interest was the slope at the origin, which can be converted to an estimate of the maximum annual reproductive rate (see Section 15.3.3.4), and also an estimate of the intrinsic rate of population growth (Myers et al., 1999). The results of this stock–recruit analysis were consistent with those of a Leslie matrix analysis and suggest that the barndoor skate is capable of growing at an instantaneous rate in excess of 35% at low population sizes.

In the past decade, several stock assessments of elasmobranchs have used a reparameterized version of the Beverton–Holt stock–recruitment curve, which includes the term *steepness*. Analogous to the slope at the origin of the original formulations, steepness is defined simply as the recruitment relative to virgin levels when spawning stock has been reduced to 20% of its virgin size. A steepness of 0.2 indicates that recruitment is directly proportional to spawning stock, and 1 is the theoretical maximum (Hilborn and Mangel, 1997). Simpfendorfer et al. (2000) constrained steepness between 0.205 and a maximum given by recruitment at virgin biomass and unexploited egg production in an age-structured model for whiskery shark, *Furgaleus macki*, off southwestern Australia. Harley (2002) estimated steepness values ranging from 0.25 to 0.67 for porbeagle through a relationship between steepness and maximum reproductive rate proposed by Myers et al. (1999) (see Section 15.3.3.4). Apostolaki et al. (2002a) estimated pup survival at low densities, a function of steepness and pup production and recruitment under virgin conditions, in an age-structured model application to blacktip shark. Brooks et al. (2002) and Brooks and Cortés (2006) also estimated steepness, parameterized similarly to Apostolaki et al. (2002a) in age-structured production models for sandbar sharks, *Carcharhinus plumbeus*, and blacktip sharks. Cortés (2002b) and Cortés et al. (2002) assigned uniform prior distributions for steepness ranging from 0.2 to 0.9 in Bayesian lagged recruitment, survival, and growth models for small and large coastal sharks, respectively.

15.2.1.5 Compensatory Responses to Exploitation

Density-dependent responses as a result of increased exploitation are thought to be expressed as increases in age-0 survivorship (Gedamke et al., 2007, 2009), increased survivorship for ages prior to selection into the fishery (Walker, 1994a), increased growth rates leading to an earlier age or size at maturity (Carlson and Baremore, 2003; Cassoff et al., 2007; Silva, 1993; Sosebee, 2005; Taylor and Gallucci, 2009) and increased fecundity (Taylor and Gallucci, 2009). Irrespective of methodological issues that may have affected results in some studies, the growing weight of evidence points towards increased survival of immature ages (see also Section 15.2.1.3) along with accelerated growth and the induced earlier attainment of maturity as the main compensatory mechanisms in response to exploitation. These results are supported by predictions from demographic modeling (see Section 15.3.2) that identify the juvenile (immature) stage and age at maturity as those traits having the largest influence on population growth rates. Increased reproductive output expressed as an increase in litter size has little supporting evidence, except for recent work by Taylor and Gallucci (2009) on spiny dogfish. Increased reproductive periodicity (e.g., switching from a biennial to an annual reproductive cycle) is another possible compensatory mechanism, but it currently lacks supporting evidence. Taylor and Gallucci (2009) further cautioned that it was very difficult to tease apart the effects of density dependence and climate change in trying to explain the abundance of long-lived spiny dogfish in the Northeast Pacific Ocean.

15.3 Modeling Approaches

15.3.1 Biomass Dynamic Models

Biomass dynamic models, also known as (surplus) production models, have been and still are fairly widely used in the assessment of teleost stocks. Use of these models in assessment of elasmobranch stocks, however, has been criticized because of violation of the underlying assumptions, notably the presupposition that r responds immediately to changes in stock density and that it is independent of the age structure of the stock (Holden, 1977; Walker, 1998). In general, production models trade biological realism for mathematical simplicity, combining growth, recruitment, and mortality into one single "surplus production" term; however, they are useful in situations where only catch and effort data on the stock are available and for practical stock assessments because they are easy to implement and provide management parameters, such as maximum sustainable yield (MSY) and virgin biomass (Meyer and Millar, 1999a).

Walker (1998) cited some of the early assessment work on elasmobranchs (Aasen, 1964; Anderson, 1980; Holden, 1974; Otto et al., 1977; Silva, 1983, 1987), which was based on application of production models and therefore was thought to produce questionable results, but the lack of quality data for many species of elasmobranchs and the need for management benchmarks prompted the resurgence of this method in recent times. Bonfil (1996) used simulation to compare the performance of several dynamic production models and a delay difference model in estimating assessment and management parameters of elasmobranchs, concluding that only the Schaefer (1954) model gave acceptable results. Agnew et

al. (2000) used what they called a constant recruitment model, a Schaefer production model, a Fox (1970) model, and a Pella–Tomlinson (1969) model to assess the multispecies skate and ray fishery off the Falkland Islands. They were able to demonstrate that there are two distinct rajid communities off the islands, with different sustainable yields, and that species composition was affected by fishing, such that smaller and earlier maturing species took over larger and slower maturing species. More sophisticated applications of surplus production models have been used for assessment of large coastal (Cortés et al., 2002, 2006; McAllister et al., 2001) and small coastal (Cortés, 2002b; NMFS, 2007) sharks off the east coast of the United States. More recently, Hayes et al. (2009) used Schaefer and Fox production curves to assess the status of scalloped hammerhead, *Sphyrna lewini*, sharks, also off the eastern United States, and found that the Fox model provided the best fit to catch and index data.

The biomass dynamic models used in the last decade have characterized uncertainty through the use of either Bayesian inference or classical frequentist methods. Typically, in stock assessment work two stochastic components must be taken into consideration (Hilborn and Mangel, 1997): (1) natural variability affecting the annual change in population biomass (process error) and (2) uncertainty in the observed indices of relative abundance owing to sampling and measurement error (observation error). Bayesian surplus production models have been used by a number of researchers to assess the status of shark populations. The Bayesian surplus production (BSP) model (McAllister and Babcock, 2006; McAllister et al., 2001), a Schaefer production model that uses the Sampling Importance Resampling (SIR) algorithm for numerical integration, has now been used in numerous assessments of shark stocks in the Atlantic Ocean (Cortés, 2002b; Cortés et al., 2002, 2006; McAllister et al., 2001, 2008; to cite a few). The BSP considers observation error only, which is integrated along with q (catchability coefficient) from the joint posterior distribution using the analytical approach described by Walters and Ludwig (1994).

Both process and observation errors can be incorporated when using a dynamic state–space modeling framework of time series (Meyer and Millar, 1999a). This approach relates observed states (CPUE observations) to unobserved states (biomasses) through a stochastic model. State–space models allow for stochasticity in population dynamics because they treat the annual biomasses as unknown states, which are a function of previous states, other unknown model parameters, and explanatory variables (e.g., catch). The observed states are in turn linked to the biomasses in a way that includes observation error by specifying the distribution of each observed CPUE index given the biomass of the stock in that year. A Bayesian approach to state–space modeling has only been applied fairly recently to fisheries (Meyer and Millar, 1999a). One advantage of using a Bayesian approach is that it allows fitting nonlinear and highly parameterized models that are more likely to capture the complex dynamics of natural populations. Meyer and Millar (1999a,b) advocated the use of the Gibbs sampler, a special Markov chain Monte Carlo (MCMC) method, to compute posterior distributions in nonlinear state–space models. This Bayesian nonlinear state–space surplus production model has been adapted and applied in several assessments of Atlantic shark stocks (Cortés et al., 2002, 2006). Additionally, Jiao et al. (2009) compared hierarchical and nonhierarchical Bayesian production models applied to a complex of three hammerhead species (*Sphyrna lewini*, *Sphyrna mokarran*, and *Sphyrna zygaena*) to address the problem of assessing fish complexes for which there are no species-specific data. They found that the fit of the Bayesian hierarchical models was better than that of the traditional Bayesian models, possibly due to the addition of multilevel prior distributions, among which was a multilevel prior of *r* intended to capture the variability of intrinsic rates of increase across species and populations of the hammerhead shark complex.

15.3.2 Cohort-Structured Models

15.3.2.1 Age-Structured Demographic Models

Demographic studies of elasmobranchs are typically based on deterministic, density-independent population growth theory, whereby populations grow at an exponential rate *r* and converge to a stable age distribution (see Section 15.2.1.3). Indeed, most of the age-structured life tables and matrix population models reviewed here assumed time-invariant (stationary with respect to time) and density-independent demographic rates; that is, the estimates of demographic rates were generally collected from a single point in time and thus provide only a snapshot of the population.

The majority of demographic analyses of elasmobranch populations are deterministic (1) life tables based on a discrete implementation of the Euler–Lotka equation (Euler, 1760; Lotka, 1907) or (2) age-based Leslie (also known as Bernardelli–Leslie–Lewis) (Manly, 1990) matrix population models. The basic Euler–Lotka equation may be written as

$$\sum_{x=\alpha}^{\omega} l_x m_x e^{-rx} = 1$$

where α is age at first breeding, ω is maximum age, l_x is cumulative survival from age 0 to *x*, and m_x is age-specific fecundity (the number of offspring produced per breeding female of age *x*). Because all demographic

models applied to elasmobranch populations have only considered females, the number of offspring in the term m_x is divided by two to account for a 1:1 male-to-female ratio generally seen in elasmobranch litters. The Leslie matrix formulation is more amenable to extending the model to the two sexes or multiple populations (Caswell, 2001).

Hoff (1990) and Cailliet (1992) pioneered the use of life tables for elasmobranch fishes, as did Hoenig and Gruber (1990) for Leslie matrices (Table 15.1), with the aim of producing basic population statistics, measuring the sensitivity of *r* to variation in some demographic rates, and assessing the vulnerability of each population to fishing. The latter is generally accomplished by adding a constant instantaneous fishing mortality (*F*) term to *M* starting at a given age and thereafter and then recalculating *r* while still assuming fixed demographic rates with time and exponential population growth. This approach is straightforward but has the limitation identified in Section 15.2.1.3. Nevertheless, it has become a common framework for evaluating the effect of harvesting on population growth of elasmobranchs, having been used for leopard shark, *Triakis semifasciata* (Cailliet, 1992); Pacific angel shark, *Squatina californica* (Cailliet et al., 1992); Atlantic sharpnose shark, *Rhizoprionodon terraenovae* (Cortés, 1995, 1998; Márquez and Castillo, 1998); sandbar shark (Cortés, 1998, 1999; McAuley et al., 2007; Sminkey and Musick, 1996); blacktip shark (Cortés, 1998); bonnethead shark (Márquez et al., 1998); dusky shark, *Carcharhinus obscurus* (Cortés, 1998; McAuley et al., 2007; Simpfendorfer, 1999b); lemon shark (Cortés, 2008); Australian sharpnose shark, *Rhizoprionodon taylori* (Simpfendorfer, 1999a); scalloped hammerhead (Liu and Chen, 1999); Atlantic sawfishes, *Pristis* spp. (Simpfendorfer, 2000); Pacific electric ray, *Torpedo californica* (Neer and Cailliet, 2001); porbeagle (Campana et al., 2002); and blue shark (Aires-da-Silva and Gallucci, 2007).

Several modifications of the Euler–Lotka equation have been proposed, all of which can be solved iteratively for *r*. Myers et al. (1997) proposed the expression:

$$(e^r)^\alpha - p_s(e^r)^{\alpha-1} - \tilde{\alpha} = 0$$

where α is age at maturity; p_s is probability of adult survival from natural mortality only (e^{-M}); and $\tilde{\alpha} = m_0 l_\alpha$, where m_0 is the number of pups produced per spawner and l_α is the cumulative survival from age 0 to age at maturity. This expression, or a modification thereof (Myers and Mertz, 1998), has been used to calculate extinction risk in marine fishes (Dulvy et al., 2004; Myers and Worm, 2005). Another modification of the Euler–Lotka equation was derived earlier by Eberhardt et al. (1982) for mammals:

$$e^{r\alpha} - S(e^r)^{\alpha-1} - FS_{0,\alpha}\left(1 - \left(\frac{S}{e^r}\right)^{\omega-\alpha+1}\right) = 0$$

where $S = p_s$, $F = m_0$ (constant fecundity rate), $S_{0,\alpha} = l_\alpha$ (cumulative survival from age 0 to age at first reproduction vs. age at maturity in Myers et al.'s equation), and ω is maximum life expectancy. Because longevity is often unknown or uncertain, Skalski et al. (2008) proposed a further modification of the above equation that explicitly allows ω to go to infinity and thus does not require estimates of longevity:

$$0 = S(e^r)^{\alpha-1} + FS_{0,\alpha} - e^{r\alpha}$$

This equation is the same as that proposed by Myers et al. (1997) and has been used to calculate population growth rates of terrestrial vertebrates (Skalski et al., 2008) and humpback whales (*Megaptera novaeangliae*) (Zerbini et al., 2010).

Au and Smith (1997) introduced a demographic technique applied to leopard shark that combines the traditional Euler–Lotka equation with concepts of density dependence from standard fisheries models. The premise of this method is that the growth potential of each species can be approximated for a given level of exploitation, which then becomes its potential population growth rate after harvest is removed, or its "rebound" potential. The density-dependent compensation is assumed to be manifested in pre-adult survival as a result of increased mortality in the adult ages. Starting from the Euler–Lotka equation, if one replaces l_x with $l_\alpha e^{-M(x-\alpha)}$, where $l_\alpha = S_{0\alpha}$ (from previous notations), and m_x with b (average fecundity), completing the summation term yields:

$$e^{-(M+r)} + l_\alpha b e^{-r\alpha}\left(1 - e^{-(M+r)(\omega-\alpha+1)}\right) = 1$$

Pre-adult survival $l_\alpha = l_{\alpha/Z}$ that makes increased mortality $Z \; (= M + F)$ sustainable ($r = 0$) is calculated from the above equation by setting $M = Z$ and $r = 0$. If F is then removed ($Z = M$), the population under survival $l_{\alpha/Z}$ will rebound at a productivity rate of r_Z, which is found by solving the above equation iteratively (Au and Smith, 1997; Au et al., 2008). Rebound potentials have been calculated for a suite of elasmobranch species (Au et al., 2008; Smith et al., 1998, 2008) and were found to be strongly affected by age at maturity. Au et al. (2008) further derived total mortalities (Z) producing MSY for a suite of shark and large pelagic teleost species by linking stock–recruitment and abundance-per-recruit relationships through the Euler–Lotka equation and found $Z = 1.5M$ for sharks and $Z = 2.0M$ for large pelagic teleosts. Xiao and Walker (2000) developed

another modification of the Lotka equation that allowed calculation of the rate of increase with time and the rate of decrease with age and applied it to gummy sharks, *Mustelus antarcticus*, and school sharks. They concluded that the rate of increase with time is a function of the reproductive and total mortality schedules, but that the rate of decrease with age is a function of the reproductive schedules only.

The matrix form analog of the Euler–Lotka equation, the Leslie matrix, has been used in a number of studies of elasmobranch populations. Walker and Hislop (1998) compared the demography of four *Raja* species. Heppell et al. (1999) compared the demography of several long-lived marine vertebrates, including the leopard and Pacific angel sharks. Mollet and Cailliet (2002) modeled the demography of the pelagic stingray, *Dasyatis violacea*; pelagic thresher, *Alopias pelagicus*; white shark, *Carcharodon carcharias*; and sandtiger, *Carcharias taurus*. Frisk et al. (2002) compared the demography of two *Leucoraja* species. Otway et al. (2004) used both deterministic age- and stage-based matrix models to estimate quasi-extinction time (defined as the time it would take population size to be <50 individuals) for a *C. taurus* population off the east coast of Australia and found it would take anywhere from 6 to 324 years depending on model structure and assumptions. Elasticities were calculated in all these studies, leading to the almost unanimous conclusion that juvenile survival (i.e., survival of all pre-adult or immature ages) was the vital rate that had the largest effect on population growth rate.

Uncertainty in estimates of demographic rates has been incorporated into various forms of demographic analysis of elasmobranchs using Monte Carlo simulation. Cortés (2002a) used Monte Carlo simulation applied to age-structured life tables and Leslie matrices to reflect uncertainty in estimates of demographic rates and to calculate population statistics and elasticities in a comparative analysis of 41 shark populations. He also used correlation analysis to identify the demographic rates that explained most of the variance in population growth rates. He reported that the populations examined fell along a continuum of life history characteristics that could be linked to elasticity patterns. Early-maturing, short-lived, and fecund sharks that generally had high values of λ and short generation times were at the fast end of the spectrum, whereas late-maturing, long-lived, and less fecund sharks that had low values of λ and long generation times were placed at the slow end of the spectrum. "Fast" sharks tended to have comparable adult and juvenile survival elasticities, whereas "slow" sharks had high juvenile survival elasticity and low age-0 survival (or fertility) elasticity. Ratios of adult survival to fertility elasticities and juvenile survival to fertility elasticities suggested that many of the 41 populations considered were biologically incapable of withstanding even moderate levels of exploitation. Although elasticity analysis suggested that changes in juvenile survival would have the greatest effect on λ, correlation analysis indicated that variation in juvenile survival, age at maturity, and reproduction accounted for most of the variance in λ. Combined results from the application of elasticity and correlation analyses in tandem led Cortés (2002a) to recommend that research, conservation, and management efforts be focused on those demographic traits.

Beerkircher et al. (2003) highlighted the importance of considering gear selectivity in demographic models incorporating harvesting in a study on silky shark, *Carcharhinus falciformis*, off the eastern coast of the United States. Aires-da-Silva and Gallucci (2007) used Monte Carlo simulation of vital rates to generate population growth rates for blue sharks through Leslie matrices and explored the effect that different levels of exploitation would have on *r*, concluding that annual harvest rates of *F* < 0.2 on juveniles and adults could be considered an acceptable risk. McAuley et al. (2007) incorporated uncertainty in the calculation of several population metrics obtained through life tables through a hybrid method that included directly resampling empirical biological data for most vital rates and from a uniform distribution bounded by the minimum and maximum estimates obtained from seven life history methods for natural mortality. These authors also used empirically estimated age-specific rates of fishing mortality to evaluate the impacts on the demography of sandbar and dusky sharks off the west coast of Australia.

Monte Carlo simulation of demographic rates has also been used to generate statistical distributions of the intrinsic rate of increase for use as informative prior distributions (priors) in Bayesian stock assessments or in ecological risk assessments (ERAs) (see Section 15.3.3.3). Both McAllister et al. (2001) and Cortés (2002b) used a variety of statistical distributions to describe vital rates of sandbar and blacktip sharks and four species of small coastal shark, respectively, in the western North Atlantic Ocean, producing probability density functions for *r* that were subsequently used in Bayesian stock assessments of these species. Braccini et al. (2006), Cortés et al. (2010), and Tovar-Avila et al. (2010) used Monte Carlo simulation of vital rates to produce estimates of *r* that were subsequently used in ERAs of Australian piked spurdog, *Squalus megalops*; Atlantic pelagic sharks; and Port Jackson shark, *Heterodontus portusjacksoni*, respectively.

15.3.2.2 Stage-Structured Demographic Models

Stage-structured analogs of the age-based Leslie matrix models, referred to as Lefkovitch or Usher models (for details, see Getz and Haight, 1989; Manly, 1990), have also been applied in analyses of some elasmobranch populations. Brewster-Geisz and Miller (2000) used this approach in combination with stage-based matrix elasticity analysis to examine management implications for the sandbar shark. They concluded that of the five stages they considered (neonate, juvenile, subadult, pregnant adult, and resting adult) juveniles and subadults affected λ the most. Frisk et al. (2002) also applied a stage-based matrix model and elasticity analysis to the barndoor skate but found that adult survival contributed the most to λ. Mollet and Cailliet (2002) applied life tables and age- and stage-based matrix models to the pelagic stingray, sandtiger, pelagic thresher, and white shark to demonstrate the effect of various methodological issues on population statistics. When using stage-based models, they found that if stage duration was fixed population growth rates were identical to those obtained with the other methods, but net reproductive rates and generation times differed. Frisk et al. (2005) used three-stage matrix models to calculate population metrics and elasticities for a suite of 55 elasmobranch species and found that juvenile and adult mortality, as well as age at maturity, were the mechanisms through which compensatory responses to harvesting would more likely be elicited.

There has been debate about the adequacy of age- vs. stage-based matrix models for modeling the demography and generating management advice for elasmobranch fishes (Cortés, 2007). One of the topics deserving further research is the effect of the number of stages used on elasticity patterns (Miller et al., 2003; Mollet and Cailliet, 2002, 2003), but until more work is available it seems sensible that the choice of model be dictated by the data available (Miller et al., 2003) or the feasibility to obtain them. For example, estimating mortality rates of juveniles of a certain size may often be easier than estimating the corresponding age-specific mortality rate of those juveniles (Cortés, 2007). It is also important to note that elasticities of age- and stage-based models are not directly comparable unless the number of age classes (i.e., stage duration) included in a given stage is known to allow comparison with the sum of the corresponding ages in an age-based model.

Uncertainty in estimates of demographic rates has also been incorporated into stage-based models through Monte Carlo simulation. Cortés (1999) first used a stage-based matrix population model to incorporate uncertainty in size-specific estimates of fecundity and survivorship for sandbar shark but fixed the values of age at maturity and maximum age. He added a constant exploitation vector separately to each of the six stages identified and considered three fixed-quota harvesting strategies to simulate the effect of fishing on population abundance 20 years into the future. The model was dynamic in that it included a vector of stage-specific abundance that was updated at each time step (year), and the transition matrix varied yearly as a result of different values being drawn randomly from the distributions describing fecundity and survivorship. This author found that removal of large juveniles resulted in the greatest population declines, whereas removal of age-0 individuals at low values of fishing ($F = 0.1$) could be sustainable. These results were in agreement with findings from a deterministic stage-structured matrix population model by Brewster-Geisz and Miller (2000), who found that population growth rates of sandbar sharks were most sensitive to variations in the juvenile and subadult stages.

15.3.2.3 Yield-per-Recruit Models

Yield-per-recruit (YPR) models are a form of age-structured analysis that takes account of age-specific weight and survival but does not include fecundity rates and assumes constant and density-independent recruitment. As originally devised by Beverton and Holt (1957), the main application of this model in elasmobranchs has been to determine the fishing mortality rate (F) that maximizes the yield per recruit when considering different ages of entry into the fishery (age at first capture). It is often applied in combination with methods that analyze tag–recapture or length–frequency information to estimate mortality, which is then used in the YPR model.

Most researchers who have used YPR analysis to model elasmobranch populations have concluded that the predicted maximum YPR is likely not to be sustainable. Grant et al. (1979) first applied this methodology to the school shark in Australia after estimating natural and fishing mortality rates through cohort analysis (Pope, 1972) and found that to achieve the maximum YPR the fishery should be expanded, but they cautioned that such action could reduce the breeding stock. Waring (1984) used catch curves to estimate Z, which he then used in a YPR analysis of little skate off the northeastern United States, also concluding that the value of F that maximized yield per recruit could result in overexploitation given the low fecundity of little skate. Smith and Abramson (1990) used YPR analysis in combination with backward virtual population analysis (VPA) to estimate population replacement of leopard sharks off California and concluded that imposition of a 100-cm total length size limit would allow the stock to be maintained while providing a yield per recruit close to the predicted maximum. Au and Smith (1997) used their modified demographic method described earlier to adjust the estimates of YPR obtained by Smith and Abramson (1990) for the effects of reduction in recruitment as a result of fishing. Their results showed that the leopard shark is much easier to overfish than originally thought when the adjustment for reduced recruitment is introduced. Cortés (1998) used estimates of M and Z

from life table analysis in a YPR analysis of the sandbar shark in the Northwest Atlantic Ocean and estimated that the maximum YPR when using the value of F that results in MSY would be attained at an age of 22 years. He also concluded that sustainable YPR values for this population could be reached only with ages of entry into the fishery of 15+ years and at low F values. Campana et al. (1999, 2001, 2002) used F estimates from Petersen analysis of tag–recaptures (Ricker, 1975), Paloheimo Zs (Paloheimo, 1961), and M from catch curves in a YPR analysis of the porbeagle in the Northwest Atlantic Ocean, concluding that the fishing mortality that would result in MSY is very low for this stock. Finally, Liu et al. (2006) used spawning-per-recruit (SPR) analysis in combination with a VPA to estimate abundance for the pelagic thresher shark, _Alopias pelagicus_, in eastern Taiwan waters and concluded that the stock was slightly overfished based on the SPR being below a biological reference point of 35% SPR.

15.3.2.4 Age-Structured Stock Assessment Models

Models described under this section incorporate time explicitly in the equations describing the population dynamics, and many include nonlinear terms to account for density dependence in the three main components: growth, recruitment, and mortality. While the structure of age-based dynamic models is biologically more realistic than that of biomass dynamic models, for example, it comes at the price of having to fix or estimate values for an increased number of parameters. Age-structured models are thus more sophisticated, but also more assumption laden (Chaloupka and Musick, 1997). Some of the major assumptions of a typical fully age-structured model as applied to elasmobranch fishes are that (1) growth is described adequately by a VBG or similar function; (2) catch-at-age can be obtained by back-transforming catch-at-length through the growth function in the absence of an age–length key or similar method, but even if an age–length key is available, it is still year and cohort invariant; (3) age at maturity (or age at first breeding) and lifespan are fixed, year- and cohort-invariant parameters; (4) recruitment is constant from year to year (although this can be modified in nonlinear models); (5) all members of a cohort become vulnerable to the fishing gear at the same age and size; (6) natural mortality is time invariant (also modifiable in some nonlinear models); and (7) removals are adequately described by a Baranov-type catch equation (Quinn and Deriso, 1999).

Before addressing age-structured population models, we will discuss delay difference models, a class of models that bridge the gap between the simple, but biologically unrealistic production models and the more complex age-structured population models (Quinn and Deriso, 1999). Unlike production models, delay difference models consider the age-specific structure of the population, including the lag that exists between spawning and recruitment, and consider separately growth, recruitment, and natural mortality processes. Unlike fully age-structured models, no age data are required for fitting delay difference models because the age-specific equations are collapsed into a single equation for the entire population (Meyer and Millar, 1999b). Walker (1995) applied a Deriso–Schnute delay difference model (Quinn and Deriso, 1999) to the school shark off southern Australia using a Beverton–Holt (1957) curve to describe the stock–recruitment relationship. The model estimated the catchability coefficient q and the stock–recruitment parameters through maximum likelihood estimation techniques, but assumed knife-edge selectivity and did not fully utilize all available information on reproduction. Cortés (2002b) and Cortés et al. (2002) also applied a simplified version of the delay difference model developed by Meyer and Millar (1999b) to small and large coastal sharks, respectively, in the Northwest Atlantic Ocean using the Gibbs sampler for numerical integration. The lagged recruitment, survival, and growth model (Hilborn and Mangel, 1997) is an approximation of the Deriso (1980) delay difference model that describes annual changes in biomass through a parameter combining natural mortality and growth, incorporates a lag phase to account for the time elapsed between reproduction and recruitment to the fishery, and describes the stock–recruitment relationship through a Beverton–Holt curve. The model assumes that fish reach sexual maturity and recruit to the fishery at the same age, although some alternative models that alleviate this assumption have been developed (Mangel, 1992).

Wood et al. (1979) developed the first dynamic pool (or age-structured) model (Quinn and Deriso, 1999) to describe the population dynamics of spiny dogfish off western Canada. Their model simulated alternative hypotheses about density-dependent regulation of mortality, reproduction, and growth, leading them to conclude that adult natural mortality was the compensatory mechanism regulating stock abundance in this species. Walker (1992) applied an age-structured simulation model to gummy shark off southern Australia that was sex specific, included terms to account for selectivity of the fishing gear, and assumed that density-dependent regulation operated through pre-recruit natural mortality. He subsequently refined the model for gummy shark with updated data and the ability to estimate some parameters, such as catchability and natural mortality (Walker, 1994a), and replaced the assumption of constant natural mortality for sharks recruited to the fishery with an asymmetric U-shaped function that varied with age (Walker, 1994b). Silva (1993) developed an analogous approach using a Leslie nonlinear model for

spiny dogfish in the Northwest Atlantic Ocean, which incorporated density-dependent terms for growth, fecundity, and recruitment. He concluded that the observed increase in abundance of spiny dogfish in the 1980s was due at least in part to an increase in juvenile growth rate during the early 1970s compared to an earlier period.

Punt and Walker (1998) and Simpfendorfer et al. (2000) developed age- and sex-structured population dynamics models for school and whiskery shark, respectively, off southern Australia and used probabilistic risk analysis to predict stock status under several harvesting strategies. Both studies accounted for the effect of selectivity of different gears. Punt and Walker (1998) used a Bayesian statistical framework in which they incorporated an observation error component in the catch rate series and a process error term to account for recruitment variability under virgin conditions, both of which were assumed normally distributed. These authors incorporated two forms of assumed density dependence: (1) in pup production, which the model related to the number of breeding females and their fecundity, and (2) in natural mortality, which they described with a U-shaped curve consisting of a decreasing exponential function for ages 0 to 2, a constant value for adults, and values increasing toward an asymptote for old ages (30+ years). Simpfendorfer et al. (2000) used a likelihood approach, fixed the value of the process error term based on Punt and Walker (1998), estimated the observation error, assumed that the stock–recruitment relationship was described by a Beverton–Holt curve, and fixed the value of natural mortality. Punt et al. (2000) later refined their model to consider explicitly the spatial structure of multiple stocks of school shark obtained from extensive tagging studies. They identified two sources of uncertainty in their study: uncertainty in the model structural assumptions and statistical uncertainty in the variability of parameter estimates.

Apostolaki et al. (2002a), Harley (2002), Brooks et al. (2002), Cortés et al. (2002), and Brooks (2006) presented detailed models with the ability to incorporate fleet-disaggregated, fully explicit age- and sex-structured (Apostolaki et al., 2002a; Harley, 2002) population dynamics based on Bayesian inference for parameter estimation. The model of Apostolaki et al. (2002a), applied to blacktip shark in the Northwest Atlantic Ocean as an example, used a Beverton–Holt stock–recruitment curve in its baseline application, but they also investigated the effects of considering a generalized hockey stick model (Barrowman and Myers, 2000) and a Ricker (1954) function. Apostolaki et al. (2002a) reported the somewhat surprising finding that stock depletion was essentially unaffected by the form of the stock–recruitment curve. Their model also allowed for incorporation of separate spatial areas, considered observation uncertainty only, and used the SIR algorithm for numerical integration. Harley (2002) used a statistical catch-at-length approach applied to the porbeagle shark in the Northwest Atlantic Ocean. The model assumed that the stock–recruitment relationship could be described by a Beverton–Holt curve, allowed for interannual recruitment variability, considered observation error, and allowed for incorporation of mark–recapture data. Gibson and Campana (2005) and Campana et al. (2010) later used an adaptation of the model by Harley (2002) to provide further status evaluations for the porbeagle shark. The application to blacktip and sandbar sharks from the Northwest Atlantic Ocean by Brooks et al. (2002), Cortés et al. (2002), and Brooks (2006) was based on a model developed by Porch (2003). The model is a state–space implementation of an age-structured production model, a step up in complexity with respect to a production model, which can incorporate age-specific vectors for fecundity, maturity, and fleet-specific gear selectivity while considering both observation error and process error for several parameters. The model assumed that the stock–recruitment relationship is described by a Beverton–Holt curve and allowed specification of either maximum likelihood or Bayesian techniques for parameter estimation.

Even more detailed models have recently been applied in some shark stock assessments. Kleiber et al. (2009) used an integrated length-based, age- and spatially structured model to test several structural assumptions in a stock assessment of blue shark in the North Pacific Ocean. This model, known as MULTIFAN-CL (Fournier et al., 1998), has been used extensively in stock assessments of several large pelagic teleosts, such as tunas. Frisk et al. (2010) investigated the dramatic increase in abundance of winter skate, *Leucoraja ocellata*, on Georges Bank in the 1980s using a statistical catch-at-age model that incorporated length, migration, and recruitment process errors. They found that the steep increase in abundance in the Georges Bank region not only was due to increases in recruitment but also likely involved immigration of adults into the population from the Scotian Shelf. This influx of adults coincided with an equally drastic decline in abundance and condition for the species in the Scotian shelf, suggesting that these two events were related (Frisk et al., 2008). A model developed by Methot (2000) has now been applied to a few elasmobranch species. Stock synthesis (SS) is an integrated statistical catch-at-age model, widely used for fish stock assessments in the United States and throughout the world; it is very versatile and allows incorporation of many types of fishery and biological data. Gertseva (2009) used SSII (Methot, 2007) to assess stock status of the longnose skate, *Raja rhina*, in the Northeast Pacific Ocean. Aires-da-Silva et al. (2010) used SSIII in a preliminary stock assessment of the silky shark, *Carcharhinus falciformis*, in the Eastern Pacific Ocean.

Although highly sophisticated models are now available, data-poor situations are the norm, rather than the exception, in elasmobranch population assessments. We will next cover a suite of approaches that have been, or can be, used in situations with varying degrees of data availability.

15.3.3 Models for Data-Poor Situations

It is somewhat ironic that a group of species with very little known biological information is also one of the most vulnerable to exploitation and therefore directly in need of population assessment and management advice. Despite clear progress in assessment of elasmobranch populations in the past two decades, it has also become apparent that uncertainty in our knowledge of the biology of many species and the degree and nature of their exploitation is rather pervasive. We must clarify that this uncertainty is almost always epistemic (due to lack of knowledge), rather than based on natural variability. As a result, data-poor (few or limited data), but also poor-data (data of low quality) situations are common. It would thus be very useful to have a set of models or approaches that are only as complex as needed to make full use of the available data. A number of such methods have been developed in the last decade that best use whatever information is available to derive important quantities to provide managers with stock status estimates. Next we present several methods that can be considered applicable in relatively data-poor situations and that allow assessment of the status and vulnerability of elasmobranch populations and provision of reasonable management advice.

15.3.3.1 Rapid Assessment Techniques

One of the simplest approaches is based on the premise that fishing pressure proportionally removes larger and older fish from the population and that increases (or decreases) in mortality rates are reflected by decreases (or increases) in mean length. These approaches generally have minimal data requirements and are therefore appealing for use in many elasmobranchs; however, the trade-off for model simplicity is a set of relatively stringent assumptions, which can be difficult to meet in a long-lived species. The widely used Beverton–Holt (Berverton and Holt, 1956, 1957) formulation, for example, only requires the von Bertalanffy growth parameters, a size at full vulnerability, and mean length of fully vulnerable animals, but assumes that growth, recruitment, and mortality have been in equilibrium for a time period equal to at least the maximum age of the species. Ehrhardt and Ault (1992) developed a variant of this approach that relaxes these assumptions by excluding the oldest animals from the analysis, therefore reducing the required time period of constant (i.e., equilibrium) conditions. The Gedamke and Hoenig (2006) non-equilibrium formulation allows for trends to be inspected through a time-series analysis of mean length data and provides the ability to estimate multiple mortality rates and the years in which mortality changed.

Even with these advances, the application of length-based approaches to relatively long-lived elasmobranchs should be done cautiously. Model assumptions should be carefully considered prior to application and during the evaluation of available data, the interpretation of results, and in producing management advice. To illustrate this point, consider that constant recruitment is a potentially problematic standard assumption of length-based approaches, including the Ehrhardt and Ault (1992) and Gedamke and Hoenig (2006) variants noted above. Although elasmobranchs might be expected to have relatively low annual variability in recruitment rates owing to their life history strategy, the assumed strong stock–recruitment relationship would result in trends in recruitment, and assumption violations, for declining or recovering populations. The analyst must ensure that the available data and chosen model can separate, for example, decreases in minimum size that result from increased exploitation from decreases in minimum size that are caused by an increasing trend in recruitment or very abundant year class. In both cases, the mean length in the population would be expected to decline, but the implications from each are completely opposite: Increasing recruitment rates or a strong year class are good events from the population perspective, while increased exploitation that drives down the population mean length would be a negative event for the population. In any situation where there is evidence of changing stock size, recruitment dynamics should be taken into account. In practice, and in response to observed increases in survey indices, Gedamke et al. (2008) chose to recast the Gedamke and Hoenig (2006) non-equilibrium formulation in a discrete form for application to the barndoor skate. The discrete version allowed a vector of recruitment, which was generated from survey data, to be included in the analysis.

Froese (2004) proposed three simple indicators to deal with overfishing of fish populations, which he described as "let them spawn," "let them grow," and "let the megaspawners live." The first indicator is measured as the percentage of mature specimens in the catch, and the target would be to let 100% of fish spawn at least once before capture. This seems like the most sensible approach for elasmobranch fishes because results from elasticity analysis (see Section 15.3.2.1) routinely show, especially for sharks, that the juvenile stage exerts the greatest influence on population growth due to the late attainment of maturity in many species. Other work by Au et al. (2008) using the modified demographic model

presented in Section 15.3.2.1 linking stock–recruitment and abundance-per-recruit relationships also proposed protection of the first few reproductive ages (young adults). This is further supported by the prediction from demographic models that the reproductive value distribution (the number of offspring that remain to be born to a female of a given age) peaks shortly after maturity. Based on matrix and yield-per-recruit analyses, Gallucci et al. (2006) also concluded that protection of juveniles and the associated preservation of reproductive potential (the sum of the reproductive values of all individuals in a population) represented the preferred management strategy for exploited shark populations. Protection of the first few breeding ages can be achieved through imposition of minimum size limits, as is the case in some elasmobranch fisheries worldwide.

The second indicator is measured as the percentage of fish caught at optimum length, defined as "the length where the number of fish in a given unfished year class multiplied by their mean individual weight is maximum and where thus the maximum yield and revenue can be obtained," and the target would be to catch 100% of fish within a predetermined range around that length. Froese (2004) further added that optimum length is generally slightly larger than length at first maturity, thus this does not seem like a sensible strategy for elasmobranch fishes given that this optimum size could include young breeding females. However, a possible management strategy could be to allow retention of only male specimens at optimum length.

The third indicator is measured as the percentage of old, large fish in the catch, and the target is 0%. Although there is support for this strategy for many teleost stocks based on the fact that larger individuals generally have exponentially greater fecundity and that the larvae they produce may have higher survival than those of younger individuals (Birkeland and Dayton, 2005), in elasmobranch fishes the relationship between maternal size and fecundity is not so strong. Cortés (2000) reported that larger female sharks in a population tend to produce more offspring, but the relationship is often not statistically significant; in other cases, larger females may produce the same number of possibly larger offspring, and yet in other cases there is evidence of a trade-off between the number and size of offspring. Additionally, as mentioned earlier, the reproductive value distribution tends to peak shortly after reaching maturity, decreasing thereafter. Thus, it appears that the contribution of older females to population growth is not so important in elasmobranch fishes and current evidence does not support strategies that protect the oldest females in a population only, such as maximum size limits.

Another method that could be applied to elasmobranch populations in data-poor situations is the depletion-corrected average catch proposed by MacCall

(2009). In this method, the catch over an extended period of time is divided into a sustainable and an unsustainable component termed *windfall*, which is associated with a one-time reduction in stock biomass and whose size is expressed as being equivalent to a number of years of sustainable yield in the form of a *windfall ratio*. Data required for calculation of the depletion-corrected average catch include the sum of catches and respective number of years, the relative depletion in biomass during that period, natural mortality (M), and an assumed ratio of F_{MSY} to M, which in the case of elasmobranch fishes could be set to 0.5 (see Section 15.3.2.1).

15.3.3.2 Statistical Analysis of Relative Abundance Trends

Temporal changes in catch per unit effort (CPUE) have been used as indicators of stock status without having to fit a population dynamics model when only catch–effort data are available and can thus be considered to be data-poor situations to some extent. The implicit, fundamental assumption in this type of analysis is that CPUE reflects true changes in the relative abundance of the stock. Time-series trends of CPUE can come from scientific surveys and thus be fishery independent or they can be fishery dependent. Especially in the case of fishery-dependent indices, there are factors that can affect the observed indices and are unrelated to true abundance. Standardization of CPUE time series through statistical techniques is undertaken in an attempt to remove the effect of those factors. The reader is referred to Maunder and Punt (2004) for a review of some of the statistical techniques used to standardize CPUE time series. Interestingly, several of the most highly publicized reports of drastic declines in elasmobranch abundance have used this type of approach (Baum and Myers, 2004; Baum et al., 2003, 2005; Casey and Myers, 1998), and a large number of studies have now examined changes in relative abundance of elasmobranchs based on this method (e.g., Aires-da-Silva et al., 2008; Carlson et al., 2007; Cortés et al., 2007; Dudley and Simpfendorfer, 2006; Ferretti et al., 2008; Graham et al., 2001, 2010; Myers et al., 2007; O'Connell et al., 2007; Robbins et al., 2006; Ward and Myers, 2005). Standardized CPUE time series are also often used as one of the inputs to more formal stock assessments. It is important, however, to examine those derived trends because they often show interannual fluctuations that seem incompatible with the biology of the species under study (Cortés et al., 2007).

When this type of model is used to determine stock status, it requires a reference point proxy on the same scale as the index being considered. These proxies may be a time-series median, a smoothed average during a period where exploitation was considered sustainable or where the population was believed to be stable, or some

other heuristic proxy. When a proxy is agreed on, stock status is determined by comparing the current index value (or a recent average) to the proxy reference point. This approach was recently taken for a complex of skates assessed in the Northwest Atlantic Ocean. For all skates assessed, except the barndoor skate, a proxy reference point for B_{MSY} (the stock biomass associated with the maximum sustainable yield) was estimated as the 75th percentile of the species-specific mean biomass index from a bottom trawl survey; for the barndoor skate, B_{MSY} was estimated as the average 1963 to 1966 biomass index (Northeast Data Poor Stocks Working Group, 2009). With this type of approach, it should be recognized that if the full-time series for a survey is used to derive the reference point, then future assessment updates would likely estimate new reference points as additional observations become available from the survey.

A more analytical framework for interpreting abundance trends is found in the AIM (An Index Method) model (NOAA Fisheries Toolbox, 2011), which relates survey trends to fishery removals. The AIM model estimates a relative fishing mortality rate from a ratio of catch to a smoothed index of abundance. The second calculated quantity is the replacement ratio, which is obtained by taking the abundance index values divided by a moving average of the abundance index. The idea behind the replacement ratio is that values greater than one indicate that the population increased and values less than one suggest negative population growth. A regression of the natural logarithm of the replacement ratio against the natural logarithm of relative F can be solved for the relative F value that produces ln(replacement ratio) = 0 (i.e., stable population growth). The F producing stable growth can be considered as an F reference point, against which the relative F time series can be compared to evaluate overfishing. Implicit in this approach is that the catch and abundance indices have the same selectivity. This method fundamentally assumes linear (density-independent) population growth, and the user should consider the same caveats as given for age- and stage-based matrix models (see Sections 15.3.2.1 and 15.3.2.2). Furthermore, there is no age structure; thus, biological parameters that have strong age trends or long time lags in population dynamics due to late, protracted maturation and generation time are ignored.

15.3.3.3 Ecological Risk Assessments

Ecological risk assessments (ERAs), also known as productivity and susceptibility analyses (PSAs), were originally developed to assess the vulnerability of stocks of species caught as bycatch in the Australian prawn fishery (Milton, 2001; Stobutzki et al., 2001a,b). Although they appeared only about a decade ago, they have now been used rather extensively to assess vulnerability to fishing of elasmobranch fishes and other marine taxa. Ecological risk assessments are in fact a family of models that can range from purely qualitative analyses in their simplest form to more quantitative analyses, depending on data availability (Hobday et al., 2007; Walker, 2005b). Most PSAs have been semiquantitative approaches where the vulnerability of a stock to fishing is expressed as a function of its productivity, or capacity to recover after it has been depleted, and its susceptibility, or propensity to capture and mortality from fishing (Stobutzki et al., 2001a). Each of these two components, productivity and susceptibility, is in turn defined by a number of attributes that are given a score on a predetermined scale. Scores are then typically averaged for each index and displayed graphically on an *x,y* plot (PSA plot). Additionally, vulnerability can be computed as the Euclidean distance of the productivity and susceptibility scores on the PSA plot. Applications to elasmobranch fishes have ranged from semiquantitative PSAs (Griffiths et al., 2006; Patrick et al., 2010; Rosenberg et al., 2007; Stobutzki et al., 2002) to different degrees of quantitative analyses where the productivity component was estimated directly as *r* in stochastic demographic models (Braccini et al., 2006; Cortés et al., 2010; Simpfendorfer et al., 2008; Tovar-Avila et al., 2010; Zhou and Griffiths, 2008). The main advantages of PSAs can be summarized as: (1) being a practical tool to evaluate the vulnerability of a stock to becoming overfished based on its biological characteristics and susceptibility to the fishery or fisheries exploiting it, (2) they can be used to help management bodies identify which stocks are more vulnerable to overfishing so they can monitor and adjust their management measures to protect the viability of these stocks, and (3) they can also be used to prioritize research efforts for species that are very susceptible but for which biological information is too sparse.

An extension of the ERA approach that also included a climate change vulnerability assessment framework was recently developed by Chin et al. (2010) to assess the vulnerability of sharks and rays on Australia's Great Barrier Reef to climate change. These authors concluded that freshwater–estuarine and reef-associated sharks and rays are most vulnerable to climate change and predicted that changes in temperature, freshwater input, and ocean circulation will have the most pervasive effects on these species.

15.3.3.4 Analytical Reference Points

Reference points for evaluating stock status (overfished and overfishing conditions) are an important component of assessment model estimates, but assembling the data required for most assessment models can be especially challenging for data-poor stocks. Methodology to analytically calculate reference points without an assessment model was first introduced in Brooks et al. (2006) and Brooks and Powers (2007), where it was demonstrated that reference points corresponding to maximum excess recruitment (MER) (Goodyear, 1980) could be derived simply from biological parameters and an assumption about the form of the stock recruit function. Brooks et al. (2010) re-derived those analytical solutions to calculate the spawning potential ratio (SPR) at MER, then demonstrated how stock status could be determined given auxiliary information and illustrated the method for 11 shark stocks.

Before presenting the derivation for SPR_{MER}, we first define SPR as a measure of the extent to which an individual's potential lifetime reproductive output has been reduced by fishing. SPR ranges from 0 to 1, with a value of 1 corresponding to unexploited conditions ($F = 0$). For any given fishing mortality (F) and selectivity at age (s_a), the value for SPR_F is

$$SPR_F = \frac{\sum_{a=1}^{A} E_a \mu_a \prod_{j=1}^{a-1} \exp\left(-M_j - Fs_j\right)}{\sum_{a=1}^{A} E_a \mu_a \prod_{j=1}^{a-1} \exp(-M_j)} = \frac{\varphi_F}{\varphi_0}$$

where E_a is egg (or pup) production at age, μ_a is maturity at age, and M_j is natural mortality at age (Goodyear, 1993). The term in the denominator, φ_0, is the number of spawners produced by a recruit over its lifetime in the absence of fishing and is equivalent to the net reproductive rate calculated in matrix models or life tables (referred to as R_0, *sensu* Leslie, 1945).

By definition, SPR calculations are deterministic and density independent because the calculations start at the age of recruitment ($a = 1$, in the example above), and maturity, fecundity, and post-recruit survival is assumed to be time invariant. Density-dependent considerations are accounted for in the stock–recruit curve, where first-year survival ranges from its minimum at unexploited stock sizes (where density is greatest) to a maximum survival rate as stock size approaches 0 (density is minimal). This property is referred to as *compensation*, because as stock size is reduced first-year survival increases by an amount that compensates for the reduction in reproductive output. Goodyear (1977) defined the term *compensation ratio* to describe the factor by which survival (or another vital rate) would have to increase in order for an exploited population to persist. Goodyear's compensation ratio is simply the inverse of SPR_F for any given fishing mortality rate, F (Brooks et al., 2010; Goodyear, 1977). As SPR_F ranges from 0 to 1, then $1/SPR_F$ would range from 1 to an infinitely large number to reflect the degree to which a vital rate would need to increase in order to sustain the given amount of F.

There is a level of SPR_F corresponding to a point of stock reduction where surplus production (that which exceeds replacement) is maximized. The F that achieves this reduction is denoted F_{MSY} if the resulting yield in weight is maximized and F_{MER} if the resulting yield in number is maximized. Brooks et al. (2010) found that F_{MSY} and F_{MER} were very similar, given likely ranges of elasmobranch vital rates. There is an analytical solution for SPR_{MER}, and it can be expressed solely as a function of φ_0 and the slope at the origin of the stock recruit curve (Brooks and Powers, 2007; Brooks et al., 2006, 2010). Myers et al. (1997, 1999) defined the product of these two parameters, φ_0 and slope at the origin, to be $\hat{\alpha}$, the maximum lifetime reproductive rate. Given this definition, the analytical solutions for SPR_{MER} for the Beverton–Holt and Ricker stock recruit functions are (Brooks et al., 2006, 2010):

(Beverton–Holt)

$$SPR_{MER} = \frac{1}{\sqrt{S_0 \varphi_0}} = \frac{1}{\sqrt{\hat{\alpha}}}$$

(Ricker, first-order approximation)

$$SPR_{MER} = \frac{\exp\left(1 - \sqrt{1/S_0 \varphi_0}\right)}{S_0 \varphi_0} = \frac{\exp\left(1 - \sqrt{1/\hat{\alpha}}\right)}{\hat{\alpha}}$$

For teleosts, the slope at the origin incorporates first-year survival and a scalar to convert spawning biomass (kg or other weight unit) into numbers of recruits. In the case of elasmobranchs, the units on the x-axis are the number of pups produced and the units on the y-axis are the number of pups that survive to recruitment age (typically age 1); therefore, the slope at the origin is simply first-year survival (pup survival, S_0). Pup survival could be estimated from tagging studies on nursery grounds, for example (see Section 15.2.1.2) (Heupel and Simpfendorfer, 2002), but it must be recognized that these estimates would likely be biased low if observations are made from populations that are not very depleted.

Typical SPR values thought to be sustainable for teleosts range from 0.2 to 0.4 (Clark, 1991, 1993, 2002; Mace, 1994; Mace and Sissenwine, 1993). Of the 11 Atlantic shark stocks for which Brooks et al. (2010) calculated SPR_{MER} values, 10 were ≥0.54, indicating that reproductive potential could not be driven as low as might be recommended for teleosts. Only one stock, the blue shark, had a low SPR value, suggesting that it could sustain more exploitation than the others, as supported by multiple demographic analyses (Cortés, 2002a, 2008; Smith et al., 1998, 2008).

As discussed in Section 15.3.3.3, PSAs provide qualitative rankings of species vulnerabilities to exploitation. The methodology described in this section provides a quantitative basis for directly calculating a metric (SPR_{MER}) to rank vulnerabilities. The SPR metric should be expected to give similar qualitative results to PSA analyses, however, because most PSA studies use the same vital rate information and differ only in consideration of the interaction with susceptibility to fishing gear. An advantage of calculating SPR_{MER} is that it provides a reference point for evaluating the overfishing status of a stock. If an estimate of current fishing mortality can be obtained, then that F can be compared to the F_{MER} (or F_{MSY}) that produces SPR_{MER}. If the current F is greater than F_{MER}, then overfishing is occurring and management is required to reduce fishing mortality.

Although SPR_{MER} provides a means for evaluating overfishing, an analytical solution for evaluating the overfished criterion was also presented in Brooks et al. (2006, 2010). Defining S as "spawning" biomass (or "pupping stock" biomass, or number of pups produced), then an analytical solution for depletion of S at MER is calculated as

(Beverton–Holt)

$$\frac{S_{MER}}{S_0} = \frac{\sqrt{\hat{\alpha}} - 1}{\hat{\alpha} - 1}$$

(Ricker, first-order approximation)

$$\frac{S_{MER}}{S_0} = \frac{1 - \sqrt{1/\hat{\alpha}}}{\ln(\hat{\alpha})}$$

As was true for the SPR_{MER} solutions, the MER level of spawner depletion also depends only on vital rates and pup survival, the product of which is captured in $\hat{\alpha}$. In order to determine if a stock is overfished, one would need to estimate current depletion ($D_{t=current}$, current abundance divided by unexploited abundance) and compare it to MER depletion. Although MER depletion is typically the management target, spawning biomass is often allowed to drop below MER by some scalar, p, before management action is required. Thus, a stock is only considered overfished if $D_{t=current}$ is below pS_{MER}/S_0. For some stocks, p is allowed to be as large as 0.5, meaning that a stock can decline to half of the level that would produce maximum excess recruitment before it would be considered overfished. In other applications, it is recommended that p be the greater of 0.5 or $(1 - M)$, where M is adult natural mortality (Restrepo et al., 1998).

Goodyear (2002) suggested that an index of abundance (I), scaled by the index value at unexploited levels ($I_{unfished}$), would provide a way to evaluate the overfished condition. Goodyear (2002) approached depletion from a surplus production perspective, where B_{MSY} is 0.5 of B_0, but the analytical solution for MER depletion in Brooks et al. (2010) allows stock-specific calculation of the fraction of S_0 that is sustainable. Given a management decision about the depletion threshold proportion, p, a stock would be considered to be overfished if the following condition is true:

$$\frac{D_{t=current}}{S_{MER}/S_0} = \frac{I_{t=current}/I_{unfished}}{S_{MER}/S_0} < p$$

An appealing aspect of this method is that no fishery information (catch, CPUE) is needed. This is desirable because catch histories are often short or incomplete, may not be disaggregated to the species level, and may not include discards. For many of the same reasons, fisheries-dependent CPUE series may not be reflective of changes in abundance (see Section 15.3.3.2). Due to the highly migratory nature of elamobranchs and the small number of sampling opportunities, encounter rates in fisheries-independent surveys can also be low, making it difficult to derive abundance indices. Although only vital rates are necessary to derive these analytical reference points, an estimate of current F is necessary to evaluate overfishing, and an estimate of current biomass or a time series of relative abundance is needed to evaluate the overfished criterion. Although catch is not needed to estimate these reference points, the lack of knowledge about fishery removals makes it challenging to estimate magnitude for scaling sustainable catch levels. Although this method has to be further tested, initial results are encouraging. Brooks et al. (2010) compared results for overfished status from stock assessments with predictions from the analytical method and found total agreement for the nine stocks of sharks for which an assessment estimate was available.

15.3.3.5 Catch-Free Assessment Model

A method known as the catch-free model was developed by Porch et al. (2006). The original application was to goliath grouper, *Epinephelus itajara*, for which three indices of abundance were available, but no reliable catch history. This age-structured production model expresses all population dynamics on a relative scale (relative to unexploited levels) to account for the fact that catch is not included in the model. Model inputs include the usual age-specific vital rates, indices of abundance, and specification of a form for the stock recruit curve, which is parameterized in terms of maximum lifetime reproductive rate ($\hat{\alpha}$). The model estimates relative biomass trends, fishing mortality rates, predicted values for indices, and MSY-based reference points (abundance-related values are expressed relative to the unexploited level). This model has been used to conduct assessments of several shark stocks, including shortfin mako, *Isurus oxyrinchus*, and blue shark (Anon., 2005; Brooks, 2005),

dusky shark (Cortés et al., 2006), and sandbar shark (Brooks, 2006). In each of these applications, catch histories were poor but estimates of vital rates and some indices of abundance were available. The ability to estimate stock status and to characterize sustainable depletion and target F levels provided much-needed management advice, but because all input and output are on a relative scale, no estimates of sustainable catch could be determined. In the case of dusky shark, the stock was already a prohibited species and still estimated to be overfished, so no catch was permitted anyway. For sandbar shark, the catch-free model was a sensitivity run to evaluate how influential catches in the base model were for determining stock status; as conclusions from the catch-free model supported the base model (which included catch), then the catch advice from the base model was adopted. Finally, for shortfin mako and blue shark, stock status of blue shark was uncertain, but general consensus was that the stock was not likely to be overfished, thus catch advice was that current levels might be sustainable. In the case of shortfin mako, status was highly uncertain due to questions about the vital rates and assumptions about gestation period.

The parameterization of stock recruit functions in terms of $\hat{\alpha}$ made it possible to calculate that parameter directly from the best estimates of vital rates and pup survival to determine whether there were any issues with mathematical boundaries. Just as steepness is bound between 0.2 and 1, it is easy to show that $\hat{\alpha}$ has a lower bound at 1.0 but no upper bound. In applying the catch-free model to sandbar shark, Brooks (2006) determined that the best estimates for vital rates and pup survival led to a value for $\hat{\alpha}$ that was less than one. By decomposing $\hat{\alpha}$ into pup survival and φ_0, Brooks and Cortés (2006) addressed this boundary problem by sequentially varying each parameter to determine how much it needed to be increased to produce a plausible value of $\hat{\alpha}$. It is recommended that $\hat{\alpha}$ be routinely calculated directly from assumed values for vital rates (or Bayesian prior means) before any modeling to check that the lower bound condition is satisfied and will not interfere with model convergence.

15.4 Applications to Management

Ultimately all modeling approaches are intended to provide advice for conservation and management of elasmobranch resources. One such piece of management advice that can be considered qualitative and can be obtained from more or less data-poor methods is the life stage that can be sustainably exploited. As we saw in Section 15.3.3.1, there are at least three related lines of evidence in support of minimum sizes that protect individuals until a few years after they first breed: elasticity analyses, reproductive output and potential, and stock–recruitment considerations. We also discussed that setting an optimum length for fisheries that was a bit larger than length at first maturity did not seem like a viable strategy for elasmobranch fishes, unless it is applied perhaps to male individuals only. The reality is that very few models of elasmobranch fishes have considered separate sexes (e.g., Rago and Sosebee, 2009) due primarily to the lack of sex-specific fishery data. It was also noted that available evidence from demographic and other models on the contribution of older females to population growth does not presently support strategies that protect the oldest females in a population, such as maximum size limits. However, there is also the view that harvesting a certain size class or few age classes of juveniles may be advantageous based on the premise that natural mortality at young ages is high and one would only be replacing natural with fishing mortality, whereas protecting older females that have already been through the "gauntlet" and are exposed to low levels of M is preferable because they can immediately contribute to the population (Walker, 1998). This hypothesis is not necessarily in contradiction with a minimum size limit that would protect a sufficiently large number of females, but it would be very difficult to verify if indeed F replaces M, rather than adding to it. The level and duration of exploitation of any particular size or age group is obviously important, too. Cortés (1999), for example, found that removal of age-0 sandbar sharks at low values of F (= 0.1) was sustainable but only projected the stage-based model used one generation time. Simpfendorfer (1999b) found that removal of up to 65% of age-0 dusky sharks in a static life table framework could be sustainable. More recently, McAuley et al. (2007) conducted a stochastic demographic analysis of dusky and sandbar shark and noted that the population was more susceptible to harvesting of older sharks than previously believed. For sandbar sharks, they reported that fishing mortality rates on adults ages 18 to 24 had increased by 172% in 2 consecutive years, resulting in large declines in estimated productivity. Because the median age at maturity reported in their study was 17 years (mean longevity was 33 years), the 18- to 24-year-old age group corresponds to first breeders and young adults, the class of adults that we argue should receive more protection. Prince (2005) used an age-structured model to make population projections in support of the gauntlet fishery hypothesis, concluding that concentrating the fishery on a few year classes of pups, juveniles, and sub-adults was a robust management strategy for Australian school and gummy shark provided adult stages were protected. However, he only explored scenarios that harvested ages 2+ or

ages 4 to 6, but not adult age groups exclusively, thus crucially limiting the conclusions that can be drawn from his study. That the level and duration of exploitation are important is best illustrated by a hypothetical example depicting an extreme—and obviously undesirable—case: If 100% of any immature age class of a given species is harvested on a continuous basis, the population will crash in about one lifespan; in contrast, completely harvesting all mature ages except the first breeding class will still produce offspring and the population will persist.

Size limits are obviously not the only management measure for elasmobranch conservation, which can include many strategies. It is not our intent to conduct a review of such strategies here (for a more in-depth review, see Walker, 2005b), but we will briefly discuss some further measures, including quotas and spatio-temporal closures. Apostolaki et al. (2006) compared deterministic and probabilistic model projections to evaluate the consequences of multiple fishery management actions for the sandbar shark in the Northwest Atlantic Ocean and found that size limits were likely less effective in achieving stock recovery than catch quotas and that predictions about the relative effectiveness of alternative management actions may differ between deterministic and probabilistic models. A strategy combining catch quotas with a minimum size limit of 110 cm total length (~age 5), for example, was deemed the best for stock recovery according to deterministic calculations, but the worst according to the probabilistic analysis, under which catch quotas alone performed the best. They also found that a strategy based on protection of adults performed better than one based on protecting large juveniles and sub-adults. This finding does not conflict with the "let them breed" strategy, as the latter includes both immature and several age classes of mature fish; however, the authors concluded that the most risk-averse management approach was to allow exploitation of all age groups rather than allow targeting of specific groups without reducing fishing mortality. Apostolaki et al. (2006) also concluded that establishment of a marine protected area (MPA) may not be appropriate to manage a highly migratory species such as the sandbar shark because it only spends part of the year in the MPA. This is in contrast to conclusions from Baum et al. (2003), who reported that MPAs could protect shark populations occurring therein but could have negative effects on species living outside the MPAs because of redistribution of fishing effort; however, Baum et al. (2003) did not account for migration into and out of the MPAs. Kinney and Simpfendorfer (2009) concluded that nursery areas alone are not sufficient for recovery of shark stocks because they do not protect segments of the population, such as older juveniles and mature individuals, that occur outside those areas and are important to population persistence as we have previously discussed. Clearly, more comprehensive simulation modeling is still required to identify the relative merits of alternative management measures on shark populations.

15.5 Summary and Conclusion

To gain a good understanding of elasmobranch population dynamics, we should invest in obtaining empirical estimates of vital rates and demographic processes. Uncritical use of some measures of productivity alone to assess vulnerability to exploitation is potentially dangerous because these measures are correlated with population size. This is problematic because calculation of productivity measures requires extensive biological data, while assessment of absolute population abundance in elasmobranchs is particularly difficult. Given the chronic paucity of information for many species and the increasing evidence of their high vulnerability to exploitation and other anthropogenic sources, we advocate a pragmatic approach that includes the use of data-poor methods that still allow us to assess to varying degrees the status of populations and provide reasonable management advice for the conservation of this group. We also advocate further simulation testing of these data-poor methods to identify potential pitfalls and to prescribe robust techniques for avoiding any bias in management advice.

Despite significant development of population models of elasmobranchs for conservation and stock assessment purposes in the recent past, we reiterate that empirical research is still limited. Highly sophisticated age-structured population dynamics models describe reality better by incorporating a large number of parameters, but their greater realism is also their pitfall in that they require many parameter estimates. There may be greater predictive return from investing in increased data quantity and quality rather than model sophistication.

In all, much still remains to be done in the field of elasmobranch population modeling. In addition to validation of ages for the majority of species, very little is known of crucial vital rates such as mortality or of the relationship between parental stock and recruitment. Implicitly related to the latter is also an understanding of the density-dependent mechanisms that control the size of elasmobranch populations. Little is still known of the temporal and spatial structure of populations, despite the increased number of mark–recapture programs and telemetry studies in existence. Satellite telemetry is providing invaluable insight into our knowledge of the detailed behavior of individuals, but we are still not in a position to transfer that information to the population level and ultimately understand the main mechanisms that control population abundance of elasmobranchs.

Acknowledgments

In reviewing materials used in the chapter we may have inadvertently omitted literature that we were not aware of and for that we apologize.

References

Aasen, O. (1964). The exploitation of the spiny dogfish (*Squalus acanthias* L.) in European waters. *Fiskeridir. Skr. Ser. Havunders.* 13:5–16.

Abrams, P.A. and Matsuda, H. (2005). The effect of adaptive change in the prey on the dynamics of an exploited predator population. *Can. J. Fish. Aquat. Sci.* 62:358–366.

Abrams, P.A. and Quince, C. (2005). The impact of mortality on predator population size and stability in systems with stage-structured prey. *Theor. Popul. Biol.* 68:253–266.

Agnew, D.J., Nolan, C.P., Beddington, J.R., and Baranowski, R. (2000). Approaches to the assessment and management of multispecies skate and ray fisheries using the Falkland Islands fishery as an example. *Can. J. Fish. Aquat. Sci.* 57:429–440.

Aires-da-Silva, A.M. and Gallucci, V.F. (2007). Demographic and risk analyses applied to management and conservation of the blue shark (*Prionace glauca*) in the North Atlantic Ocean. *Mar. Freshw. Res.* 58:570–580.

Aires-da-Silva, A.M., Hoey, J.J., and Gallucci, V.F. (2008). A historical index of abundance for the blue shark (*Prionace glauca*) in the western North Atlantic. *Fish. Res.* 58:41–52.

Aires-da-Silva, A.M., Maunder, M.N., Gallucci, V.F., Kohler, N.E., and Hoey, J.J. (2009). A spatially structured tagging model to estimate movement and fishing mortality rates for the blue shark *Prionace glauca* in the North Atlantic Ocean. *Mar. Freshw. Res.* 60:1029–1043.

Aires-da-Silva, A.M., Maunder, M.N., and Lennert-Cody, C. (2010). Early steps in the construction of a stock assessment for the silky shark, *Carcharhinus falciformis*, in the eastern Pacific Ocean. In: *Proceedings of Inter-American Tropical Tuna Commission Technical Meeting on Sharks*, August 30, La Jolla, CA.

Anderson, E.D. (1980). *MSY Estimate of Pelagic Sharks in the Western North Atlantic*, Woods Hole Laboratory Reference Document 80-18. U.S. Department of Commerce, Washington, D.C.

Anon. (2005). Report of the 2004 inter-sessional meeting of the ICCAT sub-committee on by-catches: shark stock assessment. *Col. Vol. Sci. Pap. ICCAT* 58(3):799–890.

Apostolaki, P., Babcock, E.A., and McAllister, M.K. (2006). Contrasting deterministic and probabilistic ranking of catch quotas and spatially and size-regulated fisheries management. *Can. J. Fish. Aquat. Sci.* 63: 1777–1792.

Apostolaki, P., McAllister, M.K., Babcock, E.A., and Bonfil, R. (2002a). Use of a generalized stage-based, age, and sex-structured model for shark stock assessment. *Col. Vol. Sci. Pap. ICCAT* 54:1182–1198.

Apostolaki, P., Babcock, E.A., Bonfil, R., and McAllister, M.K. (2002b). *Assessment of Large Coastal Sharks Using a Two-Area, Fleet-Disaggregated, Age-Structured Model,* 2002 Shark Evaluation Workshop Document SB-02-1. National Oceanographic and Atmospheric Administration, Panama City, FL.

Au, D.W. and Smith, S.E. (1997). A demographic method with population density compensation for estimating productivity and yield per recruit. *Can. J. Fish. Aquat. Sci.* 54:415–420.

Au, D.W., Smith, S.E., and Show, C. (2008). Shark productivity and reproductive protection, and a comparison with teleosts. In: Camhi, M.D., Pikitch, E.K., and Babcock, E.A. (Eds.), *Sharks of the Open Ocean.* Blackwell Publishing, Oxford, pp. 298–308.

Babcock, E.A. and Cortés, E. (2005). Surplus production model applied to the data for blue and mako sharks available at the 2001 ICCAT Bycatch Working Group and other published data. *Col. Vol. Sci. Pap. ICCAT* 58:1044–1053.

Barrowman, N.J. and Myers, R.A. (2000). Still more spawner-recruitment curves: the hockey stick and its generalizations. *Can. J. Fish. Aquat. Sci.* 57:665–676.

Baum, J.K. and Myers, R.A. (2004). Shifting baselines and the decline of pelagic sharks in the Gulf of Mexico. *Ecol. Lett.* 7:135–145.

Baum, J.K., Kehler, D.G., and Myers, R.A. (2005). Robust estimates of decline for pelagic shark populations in the northwest Atlantic and Gulf of Mexico. *Fisheries* 30(10):27–29.

Baum, J.K., Myers, R.A., Kehler, D.G., Worm, B., Harley, S.J., and Doherty, P.A. (2003). Collapse and conservation of shark populations in the northwest Atlantic. *Science* 299:389–392.

Beerkircher, L., Shivji, M., and Cortés, E. (2003). A Monte Carlo demographic analysis of the silky shark (*Carcharhinus falciformis*): implications of gear selectivity. *Fish. Bull.* 101:168–174.

Beverton, R.J.H. and Holt, S.J. (1956). A review of methods for estimating mortality rates in fish populations, with special reference to sources of bias in catch sampling. *Rapp. P.-v. Réun. Cons. Int. Explor. Mer* 140:67–83.

Beverton, R.J.H. and Holt, S.J. (1957). *On the Dynamics of Exploited Fish Populations.* Chapman & Hall, New York.

Birkeland, C. and Dayton, P.K. (2005). The importance in fisheries management of keeping the big ones. *Trends Ecol. Evol.* 20:356–358.

Bonfil, R. (1996). Elasmobranch Fisheries: Status, Assessment and Management, doctoral dissertation, University of British Columbia, Vancouver, Canada.

Braccini, J.M., Gillanders, B.M., and Walker, T.I. (2006). Hierarchical approach to the assessment of fishing effects on non-target chondrichthyans: case study of *Squalus megalops* in southeastern Australia. *Can. J. Fish. Aquat. Sci.* 63:2456–2466.

Brewster-Geisz, K. and Miller, T.J. (2000). Management of the sandbar shark, *Carcharhinus plumbeus*: implications of a stage-based model. *Fish. Bull.* 98:236–249.

Brooks, E.N. (2005). Re-visiting benchmark estimates from the catch-free model application to blue shark and shortfin mako shark. *Col. Vol. Sci. Pap. ICCAT* 58:1200–1203.

Brooks, E.N. (2006). *A State-Space, Age-Structured Production Model for Sandbar Shark,* Southeast Data Assessment Review 11-03. Southeast Fisheries Science Center, Miami, FL.

Brooks, E.N. and Cortés, E. (2006). *Issues Related to Biological Inputs to Blacktip and Sandbar Shark Assessments,* Southeast Data Assessment Review 11-10. Southeast Fisheries Science Center, Miami, FL.

Brooks, E.N. and Powers, J.E. (2007). Generalized compensation in stock–recruit functions: properties and implications for management. *ICES J. Mar. Sci.* 64:413–424.

Brooks, E.N., Cortés, E., and Porch, C. (2002). *An Age-Structured Production Model (ASPM) for Application to Large Coastal Sharks,* Sustainable Fisheries Division Contribution SFD-01/02-166. NOAA Fisheries, Miami, FL.

Brooks, E.N., Powers, J.E., and Cortés, E. (2006). Analytic benchmarks for age-structured models: application to data-poor fisheries, paper presented at the Sixth William R. and Lenore Mote International Symposium in Fisheries Ecology: Life History in Fisheries Ecology and Management, November 13–16, Sarasota, FL.

Brooks, E.N., Powers, J.E., and Cortés, E. (2010). Analytical reference points for age-structured models: application to data-poor fisheries. *ICES J. Mar. Sci.* 67:165–175.

Burgess, G., Beerkircher, L., Cailliet, G.M., Carlson, J.K., Cortés, E., Goldman, K.J., Grubbs, D., Musick, J.A., Musyl, M.K., and Simpfendorfer, C.A. (2005a). Is the collapse of shark populations in the northwest Atlantic and Gulf of Mexico real? *Fisheries* 30(10):19–26.

Burgess, G., Beerkircher, L., Cailliet, G.M., Carlson, J.K., Cortés, E., Goldman, K.J., Grubbs, D., Musick, J.A., Musyl, M.K., and Simpfendorfer, C.A. (2005b). Reply to 'Robust estimates of decline for pelagic shark populations in the northwest Atlantic and Gulf of Mexico.' *Fisheries* 30(10):30–31.

Cailliet, G.M. (1992). Demography of the Central California population of the leopard shark (*Triakis semifasciata*). *Aust. J. Mar. Freshw. Res.* 43:183–193.

Cailliet, G.M., Mollet, H.F., Pittinger, G.G., Bedford, D., and Natanson, L.J. (1992). Growth and demography of the Pacific angel shark (*Squatina californica*), based upon tag returns off California. *Aust. J. Mar. Freshw. Res.* 43:1313–1330.

Campana, S., Marks, L., Joyce, W., and Harley, S. (2001). *Analytical Assessment of the Porbeagle Shark (Lamna nasus) Population in the Northwest Atlantic, with Estimates of Long-Term Sustainable Yield,* CSAS Res. Doc. 2001/067. Canadian Science Advisory Secretariat, Fisheries and Oceans Canada, Ottawa, Ontario.

Campana, S., Gibson, A.J.F., Fowler, M., Dorey, A., and Joyce, W. (2010). Population dynamics of porbeagle in the northwest Atlantic, with an assessment of status to 2009 and projections for recovery. *Col. Vol. Sci. Pap. ICCAT* 65:2109–2182.

Campana, S., Marks, L., Joyce, W., Hurley, P., Showell, M., and Kulka, D. (1999). *An Analytical Assessment of the Porbeagle Shark (Lamna nasus) Population in the Northwest Atlantic*, CSAS Res. Doc. 99/158. Canadian Science Advisory Secretariat, Fisheries and Oceans Canada, Ottawa.

Campana, S., Joyce, W., Marks, L., Natanson, L.J., Kohler, N.E., Jensen, C.F., Mello, J.J., Pratt, Jr., H.L., and Myklevoll, S. (2002). Population dynamics of the porbeagle in the northwest Atlantic Ocean. *N. Am. J. Fish. Manage.* 22:106–121.

Carlson, J.K. and Baremore, I.E. (2003). Changes in biological parameters of Atlantic sharpnose shark *Rhizoprionodon terraenovae* in the Gulf of Mexico: evidence for density-dependent growth and maturity? *Mar. Freshw. Res.* 54:227–234.

Carlson, J.K., Osborne, J., and Schmidt, T.W. (2007). Monitoring the recovery of smalltooth sawfish, *Pristis pectinata*, using standardized relative indices of abundance. *Biol. Conserv.* 136:195–202.

Casey, J.M. and Myers, R.A. (1998). Near extinction of a large, widely distributed fish. *Science* 281:690–692.

Cassoff, R., Campana, S.E., and Myklevoll, S. (2007). Changes in baseline growth and maturation parameters of northwest Atlantic porbeagle, *Lamna nasus*, following heavy exploitation. *Can. J. Fish. Aquat. Sci.* 64:19–29.

Castro, A.L.F. and Rosa, R.S. (2005). Use of natural marks on population estimates of the nurse shark, *Ginglymostoma cirratum*, at Atol das Rocas Biological Reserve, Brazil. *Environ. Biol. Fish.* 72:213–221.

Caswell, H. (2001). *Matrix Population Models: Construction, Analysis, and Interpretation*, 2nd ed. Sinauer Associates, Sunderland, MA.

Chaloupka, M.Y. and Musick, J.A. (1997). Age, growth, and population dynamics. In: Lutz, P.L. and Musick, J.A. (Eds.), *Biology of Sea Turtles*. CRC Press, Boca Raton, FL, pp. 233–276.

Chen, S.B. and Watanabe, S. (1989). Age dependence of natural mortality coefficient in fish population dynamics. *Nip. Suisan Gak.* 55:205–208.

Chen, P. and Yuan, W. (2006). Demographic analysis based on the growth parameter of sharks. *Fish. Res.* 78:374–379.

Chin, A., Kyne, P.M., Walker, T.I., and McAuley, R. (2010). An integrated risk assessment for climate change: analysing the vulnerability of sharks and rays on Australia's Great Barrier Reef. *Global Change Biol.* 16:1936–1953.

Clark, W.G. (1991). Groundfish exploitation rates based on life history parameters. *Can. J. Fish. Aquat. Sci.* 48:734–750.

Clark, W.G. (1993). The effect of recruitment variability on the choice of a target level of spawning biomass per recruit. In: Kruse, G., Marasco, R.J., Pautzke, C., and Quinn, T.J. (Eds.), *Proceedings of the International Symposium on Management Strategies for Exploited Fish Populations*, Alaska Sea Grant College Program Report 93-02. University of Alaska, Fairbanks, pp. 233–246.

Clark, W.G. (2002). $F_{35\%}$ revisited ten years later. *N. Am. J. Fish. Manage.* 22:251–257.

Cliff, G., van der Elst, R.P., Govender, A., Witthuhn, T.K., and Bullen, E.M. (2006). First estimates of mortality and population size of white sharks on the South African coast. In: Klimley, A.P. and Ainley, D.G. (Eds.), *Great White Sharks: The Biology of Carcharodon carcharias*. Academic Press, San Diego, CA, pp. 393–400.

Compagno, L.J.V. (1990). Alternative life history styles of cartilaginous fishes in time and space. *Environ. Biol. Fish.* 28:33–75.

Cortés, E. (1995). Demographic analysis of the Atlantic sharpnose, *Rhizoprionodon terraenovae*, in the Gulf of Mexico. *Fish. Bull.* 93:57–66.

Cortés, E. (1998). Demographic analysis as an aid in shark stock assessment and management. *Fish. Res.* 39:199–208.

Cortés, E. (1999). A stochastic stage-based population model of the sandbar shark in the western North Atlantic. In: Musick, J.A. (Ed.), *Life in the Slow Lane: Ecology and Conservation of Long-Lived Marine Animals*. American Fisheries Society, Bethesda, MD, pp. 115–136.

Cortés, E. (2000). Life history patterns and correlations in sharks. *Rev. Fish. Sci.* 8:299–344.

Cortés, E. (2002a). Incorporating uncertainty into demographic modeling: application to shark populations and their conservation. *Conserv. Biol.* 16:1048–1062.

Cortés, E. (2002b). *Stock Assessment of Small Coastal Sharks in the U.S. Atlantic and Gulf of Mexico*, Sustainable Fisheries Division Contribution SFD-01/02-152. National Oceanographic and Atmospheric Administration, Panama City, FL.

Cortés, E. (2004). Life history patterns, demography, and population dynamics. In: Carrier, J., Musick, J.A., and Heithaus, M. (Eds.), *Biology of Sharks and Their Relatives*. CRC Press, Boca Raton, FL, pp. 449–470.

Cortés, E. (2007). Chondrichthyan demographic modelling: an essay on its use, abuse and future. *Mar. Freshw. Res.* 58:4–6.

Cortés, E. (2008). Comparative life history and demography of pelagic sharks. In: Camhi, M.D., Pikitch, E.K., and Babcock, E.A. (Eds.), *Sharks of the Open Ocean*. Blackwell Publishing, Oxford, pp. 309–322.

Cortés, E. and Parsons, G.R. (1996). Comparative demography of two populations of the bonnethead shark (*Sphyrna tiburo*). *Can. J. Fish. Aquat. Sci.* 53:709–718.

Cortés, E., Brooks, E.N., and Scott, G. (2002). *Stock Assessment of Large Coastal Sharks in the U.S. Atlantic and Gulf of Mexico*, Sustainable Fisheries Division Contribution SFD-02/03-177. National Oceanographic and Atmospheric Administration, Panama City, FL.

Cortés, E., Brown, C.A., and Beerkircher, L. (2007). Relative abundance of pelagic sharks in the western North Atlantic Ocean, including the Gulf of Mexico and Caribbean Sea. *Gulf Caribb. Res.* 19: 37–52.

Cortés, E., Brooks, E.N., Apostolaki, P., and Brown, C.A. (2006). *Stock Assessment of Dusky Shark in the U.S. Atlantic and Gulf of Mexico*, Sustainable Fisheries Division Contribution SFD-2006-014. National Oceanographic and Atmospheric Administration, Miami, FL, 155 pp.

Cortés, E., Arocha, F., Beerkircher, L., Carvalho, F., Domingo, A., Heupel, M., Holtzhausen, H., Neves, M., Ribera, M., and Simpfendorfer, C.A. (2010). Ecological risk assessment of pelagic sharks caught in Atlantic pelagic longline fisheries. *Aquat. Living Resour.* 23:25–34.

Deriso, R.B. (1980). Harvesting strategies and parameter estimation for an age-structured model. *Can. J. Fish. Aquat. Sci.* 37:268–282.

Driggers, W.B., Ingram, G.W., Grace, M.A., Gledhill, C.T., Henwood, T.A., Horton, C.N., and Jones, C.M. (2008). Pupping areas and mortality rates of young tiger sharks *Galeocerdo cuvier* in the western North Atlantic Ocean. *Aquat. Biol.* 2:161–170.

Dudley, S.F.J. and Simpfendorfer, C.A. (2006). Population status of 14 shark species caught in the protective gillnets off KwaZulu-Natal beaches, South Africa, 1978–2006. *Mar. Freshw. Res.* 57:225–240.

Dulvy, N.K. and Reynolds, J.D. (2002). Predicting extinction vulnerability in skates. *Conserv. Biol.* 16:440–450.

Dulvy, N.K. and Forrest, R.E. (2010). Life histories, population dynamics, and extinction risks in chondrichthyans. In: Carrier, J., Musick, J.A., and Heithaus, M. (Eds.), *Sharks and Their Relatives II: Biodiversity, Adaptive Physiology, and Conservation.* CRC Press, Boca Raton, FL, pp. 639–679.

Dulvy, N.K., Sadovy, Y., and Reynolds, J.D. (2003). Extinction vulnerability in marine populations. *Fish Fish.* 4:25–64.

Dulvy, N.K., Jennings, S., Goodwin, N.B., Grant, A., and Reynolds, J.D. (2005). Comparison of threat and exploitation status in Northeast Atlantic marine populations. *J. Appl. Ecol.* 42:883–891.

Dulvy, N.K., Ellis, R.J., Goodwin, N.B., Grant, A., Reynolds, J.D., and Jennings, S. (2004). Methods of assessing extinction risk in marine fishes. *Fish Fish.* 5:255–276.

Dulvy, N.K., Baum, J.K., Clarke, S., Compagno, L.J.V., and Cortés, E. et al. (2008). You can swim but you can't hide: the global status and conservation of oceanic pelagic sharks and rays. *Aquat. Conserv.* 18:459–482.

Eberhardt, L.L., Majorowicz, A.K., and Wilcox, J.A. (1982). Apparent rates of increase for two feral horse herds. *J. Wildl. Manage.* 46:367–374.

Ehrhardt, N.M. and Ault, J.S. (1992). Analysis of two length-based mortality models applied to bounded catch length frequencies. *Trans. Am. Fish. Soc.* 121:115–122.

Euler, L. (1760). Recherches générales sur la mortalité et la multiplication du genre humain. *Mem. Acad. R. Sci. Belles Lett. (Belg.)* 16:144–164.

Ferretti, F., Myers, R.A., Serena, F., and Lotze, H.K. (2008). Loss of large predatory sharks from the Mediterranean Sea. *Conserv. Biol.* 22:952–964.

Field, I.C., Meekan, M.G., Buckworth, R.C., and Bradshaw, C.J.A. (2009). Susceptibility of sharks, rays and chimaeras to global extinction. *Adv. Mar. Biol.* 56:275–363.

Forrest, R.E. and Walters, C.J. (2009). Estimating thresholds to optimal harvest rate for long-lived, low-fecundity sharks accounting for selectivity and density dependence in recruitment. *Can. J. Fish. Aquat. Sci.* 66:2062–2080.

Fournier, D.A., Hampton, J., and Sibert, J.R. (1998). MULTIFAN-CL: a length-based, age-structured model for fisheries stock assessment, with application to South Pacific albacore, *Thunnus alalunga. Can. J. Fish. Aquat. Sci.* 55:2105–2116.

Fowler, C.W. (1981). Comparative population dynamics in large mammals. In: Fowler, C.W. and Smith, T. (Eds.), *Dynamics of Large Mammal Populations.* John Wiley & Sons, New York, pp. 437–455.

Fowler, C.W. (1988). Population dynamics as related to rate of increase per generation. *Evol. Ecol.* 2:197–204.

Fox, W.W. (1970). An exponential surplus-yield model for optimizing exploited fish populations. *Trans. Am. Fish. Soc.* 99:80–88.

Frisk, M.G. (2010). Life history strategies of batoids. In: Carrier, J., Musick, J.A., and Heithaus, M. (Eds.), *Sharks and Their Relatives II: Biodiversity, Adaptive Physiology, and Conservation.* CRC Press, Boca Raton, FL, pp. 283–316.

Frisk, M.G., Miller, T.J., and Fogarty, M.J. (2001). Estimation and analysis of biological parameters in elasmobranch fishes: a comparative life history study. *Can. J. Fish. Aquat. Sci.* 58:969–981.

Frisk, M.G., Miller, T.J., and Fogarty, M.J. (2002). The population dynamics of little skate *Leucoraja erinacea*, winter skate *Leucoraja ocellata*, and barndoor skate *Dipturus laevis*: predicting exploitation limits using matrix analyses. *ICES J. Mar. Sci.* 59:576–586.

Frisk, M.G., Miller, T.J., and Dulvy, N.K. (2005). Life histories and vulnerability to exploitation of elasmobranchs: inferences from elasticity, perturbation and phylogenetic analyses. *J. Northw. Atl. Fish. Sci.* 35:27–45.

Frisk, M.G., Miller, T.J., Martell, S.J.D., and Sosebee, K. (2008). New hypothesis helps explain elasmobranch 'outburst' on Georges Bank in the 1980s. *Ecol. Appl.* 18:234–245.

Frisk, M.G., Martell, S.J.D., Miller, T.J., and Sosebee, K. (2010). Exploring the population dynamics of winter skate (*Leucoraja ocellata*) in the Georges Bank region using a statistical catch-at-age model incorporating length, migration, and recruitment process errors. *Can. J. Fish. Aquat. Sci.* 67:774–792.

Froese, R. (2004). Keep it simple: three indicators to deal with overfishing. *Fish Fish.* 5:86–91.

Gallucci, V.F., Taylor, I.G., and Erzini, K. (2006). Conservation and management of exploited shark populations based on reproductive value. *Can. J. Fish. Aquat. Sci.* 63: 931–942.

García, V.B., Lucifora, L.O., and Myers, R.A. (2008). The importance of habitat and life history to extinction risk in sharks, skates, rays and chimaeras. *Proc. R. Soc. B Biol. Sci.* 275:83–89.

Gedamke, T. and Hoenig, J.M. (2006). Estimating mortality from mean length data in non-equilibrium situations, with application to the assessment of goosefish (*Lophius americanus*). *Trans. Am. Fish. Soc.* 135:476–487.

Gedamke, T., DuPaul, W.D., and Musick, J.A. (2005). Observations on the life history of the barndoor skate, *Dipturus laevis*, on Georges Bank (western North Atlantic). *J. Northw. Atl. Fish. Sci.* 35:67–78.

Gedamke, T., Hoenig, J.M., DuPaul, W.D., and Musick, J.A. (2008). Total mortality rates of the barndoor skate, *Dipturus laevis*, in the northeast United States, 1963–2005. *Fish. Res.* 89:17–25.

Gedamke, T., Hoenig, J.M., DuPaul, W.D., and Musick, J.A. (2009). Stock recruitment dynamics and the maximum population rate of population growth of the barndoor skate. *N. Am. J. Fish. Manage.* 26:512–529.

Gedamke, T., Hoenig, J.M., DuPaul, W.D., Musick, J.A., and Gruber, S.H. (2007). Using demographic models to determine intrinsic rate of increase and sustainable fishing for elasmobranchs: pitfalls, advances and applications. *N. Am. J. Fish. Manage.* 27:605–618.

Gertseva, V.V. (2009). The population dynamics of the long-nose skate, *Raja rhina*, in the northeast Pacific Ocean. *Fish. Res.* 95:146–153.

Getz, W.N. and Haight, R.G. (1989). *Population Harvesting*. Princeton University Press, Princeton, NJ.

Gibson, A.J.F. and Campana, S.E. (2005). *Status and Recovery Potential of Porbeagle Shark in the Northwest Atlantic*, CSAS Res. Doc. 2005/053. Canadian Science Advisory Secretariat, Fisheries and Oceans Canada, Ottawa.

Goodyear, C.P. (1977). Assessing the impact of power plant mortality on the compensatory reserve of fish populations. In: van Winkle, W. (Ed.), *Proceedings of the Conference on Assessing the Effects of Power Plant Induced Mortality on Fish Populations*, May 3–6, Gatlinburg, Tennessee. Pergamon Press, New York, pp. 186–195.

Goodyear, C.P. (1980). Compensation in fish populations. In: Hocutt, C.H. and Stauffer, J.R. (Eds.), *Biological Monitoring of Fish*. Lexington Books, Lexington, MA, pp. 253–280.

Goodyear, C.P. (1993). Spawning stock biomass per recruit in fisheries management: foundation and current use. In: Smith, S.J., Hunt, J.J., and Rivard, D. (Eds.), *Risk Evaluation and Biological Reference Points for Fisheries Management*, Canadian Special Publication in Fisheries and Aquatic Sciences. National Research Council Canada, Ottawa, Ontario, pp. 67–81.

Goodyear, C.P. (2002). Biological reference points without models. *Col. Vol. Sci. Pap. ICCAT* 55:633–648.

Graham, K.J., Andrew, N.L., and Hodgson, K.E. (2001). Changes in relative abundance of sharks and rays on Australian South East Fishery trawl grounds after twenty years of fishing. *Mar. Freshw. Res.* 52:549–561.

Graham, N.A.J., Spalding, M.D., and Sheppard, C.R.C. (2010). Reef shark declines in remote atolls highlight the need for multi-faceted conservation action. *Aquat. Conserv.* 20:543–548.

Grant, C.J., Sandland, R.L., and Olsen, A.M. (1979). Estimation of growth, mortality and yield per recruit of the Australian school shark, *Galeorhinus galeus* (Macleay), from tag recoveries. *Aust. J. Mar. Freshw. Res.* 30:625–637.

Griffiths, S.P., Brewer, D.T., Heales, D.S., Milton, D.A., and Stobutzki, I.C. (2006). Validating ecological risk assessments for fisheries: assessing the impacts of turtle excluder devices on elasmobranch bycatch populations in an Australian trawl fishery. *Mar. Freshw. Res.* 57:395–401.

Gruber, S.H., de Marignac, J.R.C., and Hoenig, J.M. (2001). Survival of juvenile lemon sharks at Bimini, Bahamas, estimated by mark-depletion experiments. *Trans. Am. Fish. Soc.* 130:376–384.

Hammond, T.R. and Ellis, J.R. (2005). Bayesian assessment of Northeast Atlantic spurdog using a stock production model, with prior for intrinsic population growth rate set by demographic methods. *J. Northw. Atl. Fish. Sci.* 35:299–308.

Harley, S.J. (2002). Statistical catch-at-length model for porbeagle shark (*Lamna nasus*) in the northwest Atlantic. *Col. Vol. Sci. Pap. ICCAT* 54:1314–1332.

Harry, A.V., Simpfendorfer, C.A., and Tobin, A.J. (2010). Improving age, growth, and maturity estimates for aseasonally reproducing chondrichthyans. *Fish. Res.* 106:393–403.

Hayes, C.G., Jiao, Y., and Cortés, E. (2009). Stock assessment of scalloped hammerheads in the western North Atlantic Ocean and Gulf of Mexico. *N. Am. J. Fish. Manage.* 29:1406–1417.

Heist, E.J. (2004). Genetics of sharks, skates, and rays. In: Carrier, J., Musick, J.A., and Heithaus, M. (Eds.), *Biology of Sharks and Their Relatives*. CRC Press, Boca Raton, FL, pp. 471–485.

Heppell, S.S., Crowder, L.B., and Menzel, T.R. (1999). Life table analysis of long-lived marine species, with implications for conservation and management. In: Musick, J.A. (Ed.), *Life in the Slow Lane: Ecology and Conservation of Long-Lived Marine Animals*. American Fisheries Society, Bethesda, MD, pp. 137–148.

Heupel, M.R. and Simpfendorfer, C.A. (2002). Estimation of mortality of juvenile blacktip sharks, *Carcharhinus limbatus*, within a nursery area using telemetry data. *Can. J. Fish. Aquat. Sci.* 59:624–632.

Hilborn, R. and Mangel, M. (1997). *The Ecological Detective*. Princeton University Press, Princeton, NJ.

Hobday, A.J., Smith, A., Webb, H., Daley, R., Wayte, S. et al. (2007). *Ecological Risk Assessment for the Effects of Fishing: Methodology*, Report R04/1072. Australian Fisheries Management Authority, Canberra (www.afma.gov.au/environment/eco_based/eras/docs/methodology.pdf).

Hoenig, J.M. (1983). Empirical use of longevity data to estimate mortality rates. *Fish. Bull.* 82:898–903.

Hoenig, J.M. and Gruber, S.H. (1990). Life history patterns in the elasmobranchs: implications for fisheries management. In: Pratt, Jr., H.L., Gruber, S.H., and Taniuchi, T. (Eds.), *Elasmobranchs as Living Resources: Advances in the Biology, Ecology, Systematics, and Status of the Fisheries*, NOAA Tech. Rep. NMFS 90. U.S. Department of Commerce, Washington, D.C., pp. 1–16.

Hoff, T.B. (1990). Conservation and Management of the Western North Atlantic Shark Resource Based on the Life History Strategy Limitations of the Sandbar Shark, doctoral dissertation, University of Delaware, Newark.

Holden, M.J. (1974). Problems in the rational exploitation of elasmobranch populations and some suggested solutions. In: Harden-Jones, F.R. (Ed.), *Sea Fisheries Research*. Halsted Press, New York, pp. 117–137.

Holden, M.J. (1977). Elasmobranchs. In: Gulland, J.A. (Ed.), *Fish Population Dynamics*. John Wiley & Sons, New York, pp. 187–214.

Jensen, A.L. (1996). Beverton and Holt life history invariants result from optimal trade-off of reproduction and survival. *Can. J. Fish. Aquat. Sci.* 53:820–822.

Jiao, Y., Hayes, C., and Cortés, E. (2009). Hierarchical Bayesian approach for population dynamics modelling of fish complexes without species-specific data. *ICES J. Mar. Sci.* 66: 367–377.

Kinney, M.J. and Simpfendorfer, C.A. (2009). Reassessing the value of nursery areas to shark conservation and management. *Cons. Lett.* 2:53–60.

Kleiber, P., Clarke, S., Bigelow, K., Nakano, H., McAllister, M., and Takeuchi, Y. (2009). *North Pacific Blue Shark Stock Assessment*, NOAA Tech. Memo. NMFS-PIFSC-17. U.S. Department of Commerce, Washington, D.C.

Leslie, P.H. (1945). On the use of matrices in certain population mathematics. *Biometrika* 33:213–245.

Liu, K.M. and Chen, C.T. (1999). Demographic analysis of the scalloped hammerhead, *Sphyrna lewini*, in the northwestern Pacific. *Fish. Sci.* 65:218–223.

Liu, K.M., Chang, Y.T., Ni, I.H., and Jin, C.B. (2006). Spawning per recruit analysis of the pelagic thresher shark, *Alopias pelagicus*, in the eastern Taiwan waters. *Fish. Res.* 82:56–64.

Lorenzen, K. (1996). The relationship between body weight and natural mortality in juvenile and adult fish: a comparison of natural ecosystems and aquaculture. *J. Fish Biol.* 49:627–647.

Lorenzen, K. (2000). Allometry of natural mortality as a basis for assessing optimal release size in fish-stocking programmes. *Can. J. Fish. Aquat. Sci.* 57:2374–2381.

Lotka, A.J. (1907). Studies on the mode of growth of material aggregates. *Am. J. Sci.* 24:199–216.

MacCall, A.D. (2009). Depletion-corrected average catch: a simple formula for estimating sustainable yields in data-poor situations. *ICES J. Mar. Sci.* 66:2267–2271.

Mace, P.M. (1994). Relationships between common biological reference points used as thresholds and targets of fisheries management strategies. *Can. J. Fish. Aquat. Sci.* 51:110–122.

Mace, P.M. and Sissenwine, M.P. (1993). How much spawning per recruit is enough? In: Smith, S.J., Hunt, J.J., and Rivard, D. (Eds.), *Risk Evaluation and Biological Reference Points for Fisheries Management*, Canadian Special Publication in Fisheries and Aquatic Sciences. National Research Council Canada, Ottawa, Ontario, pp. 101–118.

Mangel, M. (1992). Comparative analyses of the effects of high seas driftnets on the northern right whale dolphin *Lissodelphus borealis*. *Ecol. Appl.* 3:221–229.

Manire, C.A. and Gruber, S.H. (1993). A preliminary estimate of natural mortality of age-0 lemon sharks, *Negaprion brevirostris*. In: Branstetter, S. (Ed.), *Conservation Biology of Elasmobranchs*, NOAA Tech. Rep. NMFS 115. U.S. Department of Commerce, Washington, D.C., pp. 65–71.

Manly, B.J.F. (1990). *Stage-Structured Populations: Sampling, Analysis, and Simulation*. Chapman & Hall, London.

Márquez, J.F. and Castillo, J.L. (1998). Fishery biology and demography of the Atlantic sharpnose shark, *Rhizoprionodon terraenovae*, in the southern Gulf of Mexico. *Fish. Res.* 39:183–198.

Márquez, J.F., Castillo, J.L., and Rodríguez de la Cruz, M.C. (1998). Demography of the bonnethead shark, *Sphyrna tiburo* (Linnaeus, 1758), in the southeastern Gulf of Mexico. *Cienc. Mar.* 24:13–34.

Maunder, M.N. and Punt, A.E. (2004). Standardizing catch and effort data: a review of recent approaches. *Fish. Res.* 70:141–159.

McAllister, M.K. and Babcock, E.A. (2006). *Bayesian Surplus Production Model with the Sampling Importance Resampling Algorithm (BSP): A User's Guide* (www.iccat.int/en/AssessCatalog.htm).

McAllister, M.K., Pikitch, E.K., and Babcock, E.A. (2001). Using demographic methods to construct Bayesian priors for the intrinsic rate of increase in the Schaefer model and implications for stock rebuilding. *Can. J. Fish. Aquat. Sci.* 58:1871–1890.

McAllister, M.K., Pikitch, E.K., and Babcock, E.A. (2008). Why are Bayesian methods useful for the stock assessment of sharks? In: Camhi, M.D., Pikitch, E.K., and Babcock, E.A. (Eds.), *Sharks of the Open Ocean*. Blackwell Publishing, Oxford, pp. 351–368.

McAuley, R.B., Simpfendorfer, C.A., and Hall, N.G. (2007). A method for evaluating the impacts of fishing mortality and stochastic influences on the demography of two long-lived shark stocks. *ICES J. Mar. Sci.* 64:1710–1722.

McAuley, R.B., Simpfendorfer, C.A., Hyndes, G.A., Allison, R.R., Chidlow, J.A. et al. (2006). Validated age and growth of the sandbar shark *Carcharhinus plumbeus* (Nardo 1827) in the waters off western Australia. *Environ. Biol. Fish.* 77:385–400.

Methot, R.D. (2000). *Technical Description of the Stock Synthesis Assessment Program*, NOAA Tech. Memo. NMFS-NWFSC-43. U.S. Department of Commerce, Washington, D.C.

Methot, R.D. (2007). *User Manual for the Integrated Analysis Program Stock Synthesis 2 (SS2), Version 2.00a*. U.S. Department of Commerce, Washington, D.C.

Meyer, R. and Millar, R.B. (1999a). BUGS in Bayesian stock assessments. *Can. J. Fish. Aquat. Sci.* 56:1078–1086.

Meyer, R. and Millar, R.B. (1999b). Bayesian stock assessment using a state–space implementation of the delay difference model. *Can. J. Fish. Aquat. Sci.* 56:37–52.

Miller, T.J., Frisk, M.G., and Fogarty, M.J. (2003). Comment on Mollet and Cailliet. (2002): confronting models with data. *Mar. Freshw. Res.* 54:737–738.

Milton, D.A. (2001). Assessing the susceptibility to fishing of populations of rare trawl bycatch: sea snakes caught by Australia's Northern Prawn Fishery. *Biol. Cons.* 101:281–290.

Mollet, H.F. and Cailliet, G.M. (2002). Comparative population demography of elasmobranchs using life history tables, Leslie matrices and stage-based matrix models. *Mar. Freshw. Res.* 53:503–516.

Mollet, H.F. and Cailliet, G.M. (2003). Reply to comments by Miller et al. (2003) on Mollet and Cailliet (2002): confronting models with data. *Mar. Freshw. Res.* 54:739–744.

Musick, J.A., Harbin, M.H., and Compagno, L.J.V. (2004). Historical zoogeography of the selachii. In: Carrier, J., Musick, J.A., and Heithaus, M. (Eds.), *Biology of Sharks and Their Relatives*. CRC Press, Boca Raton, FL, pp. 33–78.

Myers, R.A. and Mertz, G. (1998). The limits of exploitation: a precautionary approach. *Ecol. Appl.* 8(Suppl.):165–169.

Myers, R.A. and Worm, B. (2005). Extinction, survival or recovery of large predatory fishes. *Phil. Trans. R. Soc. B. Biol. Sci.* 360:13–20.

Myers, R.A., Mertz, G., and Fowlow, P.S. (1997). Maximum population growth rates and recovery times for Atlantic cod, *Gadus morhua*. *Fish. Bull.* 95:762–772.

Myers, R.A., Bowen, K.G., and Barrowman, N.J. (1999). Maximum reproductive rate of fish at low population sizes. *Can. J. Fish. Aquat. Sci.* 56:2404–2419.

Myers, R.A., Baum, J.K., Shepherd, T.D., Powers, S.D., and Peterson, C.H. (2007). Cascading effects of the loss of apex predatory sharks from a coastal ocean. *Science* 315:1846–1850.

Neer, J.A. and Cailliet, G.M. (2001). Aspects of the life history of the Pacific electric ray, *Torpedo californica* (Ayres). *Copeia* 2001:842–847.

NMFS. (2007). *Southeast Data, Assessment and Review (SEDAR) 13: Stock Assessment Report—Small Coastal Sharks*. National Marine Fisheries Service, Silver Spring, MD.

NOAA Fisheries Toolbox. (2011). *An Index Method, Version 2.2.0* (http://nft.nefsc.noaa.gov).

Northeast Data Poor Stocks Working Group. (2009). *The Northeast Data Poor Stocks Working Group Report, December 8–12, 2008 Meeting. Part A. Skate Species Complex, Deep Sea Red Crab, Atlantic Wolffish, Scup, and Black Sea Bass.* NEFSC Ref. Doc. 09-02. U.S. Department of Commerce, Washington, D.C. Available from: National Marine Fisheries Service, 166 Water Street, Woods Hole, MA 02543-1026, or online at http://www.nefsc.noaa.gov/nefsc/publications/.

O'Connell, M.T., Shepherd, T.D., O'Connell, A.M.U., and Myers, R.A. (2007). Long-term declines in two apex predators, bull sharks (*Carcharhinus leucas*) and alligator gar (*Atractosteus spatula*), in Lake Pontchartrain, an oligohaline estuary in southeastern Louisiana. *Estuar. Coasts* 30:567–574.

Otto, R.S., Zuboy, J.R., and Sakagawa, G.T. (1977). *Status of Northwest Atlantic Billfish and Shark Stocks*, Report of the La Jolla Working Group, March 28–April 8.

Otway, N.M., Bradshaw, C.J.A., and Harcourt, R.G. (2004). Estimating the rate of quasi-extinction of the Australian grey nurse shark (*Carcharias taurus*) population using deterministic age-and stage-classified models. *Biol. Cons.* 119:341–350.

Paloheimo, J.E. (1961). Studies on estimation of mortalities. I. Comparison of a method described by Beverton and Holt and a new linear formula. *J. Fish. Res. Bd. Can.* 18:645–662.

Pastor, J. (2008). *Mathematical Ecology of Populations and Ecosystems*. Wiley-Blackwell, Chichester, U.K.

Patrick, W.S., Spencer, P., Link, J., Ormseth, O., Cope, J. et al. (2010). Using productivity and susceptibility indices to assess the vulnerability of United States fish stocks to overfishing. *Fish. Bull.* 108:305–322.

Pauly, D. (1980). On the interrelationship between natural mortality, growth parameters, and mean environmental temperature in 175 fish stocks. *J. Cons. Int. Explor. Mer* 39:175–192.

Pella, J.J. and Tomlinson, P.K. (1969). A generalized stock production model. *Inter-Am. Trop. Tuna Comm. Bull.* 13:419–496.

Peterson, I. and Wroblewski, J.S. (1984). Mortality rates of fishes in the pelagic ecosystem. *Can. J. Fish. Aquat. Sci.* 41:1117–1120.

Pope, J.G. (1972). An investigation of the accuracy of virtual population analysis using cohort analysis. *Res. Bull. Int. Comm. Northw. Atl. Fish.* 9:65–74.

Porch, C.E. (2003). A preliminary assessment of Atlantic white marlin (*Tetrapturus albidus*) using a state-space implementation of an age-structured production model. *Col. Vol. Sci. Pap. ICCAT* 55:559–527.

Porch, C.E., Ecklund, A.M., and Scott, G.P. (2006). A catch-free stock assessment model with application to goliath grouper (*Epinephelus itajara*) off southern Florida. *Fish. Bull.* 104:89–101.

Pribac, F., Punt, A.E., Taylor, B.L., and Walker, T.I. (2005). Using length, age and tagging data in a stock assessment of a length selective fishery for gummy shark (*Mustelus antarcticus*). *J. Northw. Atl. Fish. Sci.* 35:267–290.

Prince, J.D. (2005). Gauntlet fisheries for elasmobranchs: the secret of sustainable shark fisheries. *J. Northw. Atl. Fish. Sci.* 35:407–416.

Punt, A.E. and Walker, T.I. (1998). Stock assessment and risk analysis for the school shark *Galeorhinus galeus* (Linnaeus) off southern Australia. *Mar. Freshw. Res.* 49:719–731.

Punt, A.E., Pribac, F., Walker, T.I., Taylor, B.L., and Prince, J.D. (2000). Stock assessment of school shark *Galeorhinus galeus*, based on a spatially explicit population dynamics model. *Mar. Freshw. Res.* 51:205–220.

Quinn, T.J. and Deriso, R.B. (1999). *Quantitative Fish Dynamics*. Oxford University Press, New York.

Rago, P.J. and Sosebee, K.A. (2009). The agony of recovery: scientific challenges of spiny dogfish recovery programs. In: Gallucci, V.F., McFarlane, G.A., and Bargmann, G.G. (Eds.), *Biology and Management of Dogfish Sharks*. American Fisheries Society, Bethesda, MD, pp. 343–372.

Rago, P.J., Sosebee, K.A., Brodziak, J.K.T., Murawski, S.A., and Anderson, E.D. (1998). Implications of recent increases in catches on the dynamics of northwest Atlantic spiny dogfish (*Squalus acanthias*). *Fish. Res.* 39:165–181.

Restrepo, V.R., Thompson, G.G., Mace, P.M., Gabriel, W.L., Low, L.L. et al. (1998). *Technical Guidance on the Use of Precautionary Approaches to Implementing National Standard 1 of the Magnuson–Stevens Fishery Conservation and Management Act*, NOAA Tech. Memo. NMFS-F/SPO. U.S. Department of Commerce, Washington, D.C.

Ricker, W.E. (1954). Stock and recruitment. *J. Fish. Res. Bd. Can.* 11:559–623.

Ricker, W.E. (1975). Computation and interpretation of biological statistics of fish populations. *Bull. Fish. Res. Bd. Can.* 191:1–382.

Robbins, W.D., Hisano, M., Connolly, S.R., and Choat, J.H. (2006). Ongoing collapse of coral-reef shark populations. *Curr. Biol.* 16:2314–2319.

Rodríguez-Cabello, C. and Sánchez, F. (2005). Mortality estimates of *Scyliorhinus canicula* in the Cantabrian Sea using tag recapture data. *J. Fish Biol.* 66:1116–1126.

Roff, D.A. (1992). *The Evolution of Life Histories: Theory and Analysis*. Chapman & Hall, New York.

Rosenberg, A., Agnew, D., Babcock, E., Cooper, A., Mogensen, C et al. (2007). *Setting Annual Catch Limits for U.S. Fisheries: An Expert Working Group Report*. MRAG Americas, Washington, D.C.

Schaefer, M.B. (1954). Some aspects of the dynamics of populations important to the management of commercial marine fisheries. *Inter-Am. Trop. Tuna Comm. Bull.* 2:247–285.

Shivji, M.S. (2010). DNA forensic applications in shark management and conservation. In: Carrier, J., Musick, J.A., and Heithaus, M. (Eds.), *Sharks and Their Relatives II: Biodiversity, Adaptive Physiology, and Conservation*. CRC Press, Boca Raton, FL, pp. 593–610.

Siegfried, K.I. (2006). Fishery Management in Data-Limited Situations: Applications to Stock Assessment, Marine Reserve Design and Fish Bycatch Policy, doctoral dissertation, University of California, Santa Cruz.

Silva, H.M. (1983). *Preliminary Studies of the Exploited Stock of Kitefin Shark Scymnorhinus licha (Bonnaterre, 1788) in the Azores*, ICES Council Meeting Papers No. ICES CM 1983/G:18. International Council for the Exploration of the Sea, Copenhagen, Denmark.

Silva, H.M. (1987). *An Assessment of the Azorean Stock of Kitefin Shark Scymnorhinus licha (Bonnaterre, 1788) in the Azores*, ICES Council Meeting Papers No. ICES CM 1987/G:66. International Council for the Exploration of the Sea, Copenhagen, Denmark.

Silva, H.M. (1993). *A Density-Dependent Leslie Matrix-Based Population Model of Spiny Dogfish, Squalus acanthias, in the Northwest Atlantic*, ICES Council Meeting Papers No. ICES CM 1993/G:54. International Council for the Exploration of the Sea, Copenhagen, Denmark.

Simpfendorfer, C.A. (1999a). Mortality estimates and demographic analysis for the Australian sharpnose shark, *Rhizoprionodon taylori*, from northern Australia. *Fish. Bull.* 97:978–986.

Simpfendorfer, C.A. (1999b). Demographic analysis of the dusky shark fishery in southwestern Australia. In: Musick, J.A. (Ed.), *Life in the Slow Lane: Ecology and Conservation of Long-Lived Marine Animals*. American Fisheries Society, Bethesda, MD, pp. 149–160.

Simpfendorfer, C.A. (2000). Predicting population recovery rates for endangered western Atlantic sawfishes using demographic analysis. *Environ. Biol. Fish.* 58:371–377.

Simpfendorfer, C.A. (2005). Demographic models, life tables, matrix models and rebound potential. In: Musick, J.A. and Bonfil, R. (Eds.), *Elasmobranch Fisheries Management Techniques*. Asia-Pacific Economic Cooperation Secretariat, Singapore, pp. 187–203.

Simpfendorfer, C.A. (2007). The importance of mangroves as nursery habitat for smalltooth sawfish (*Pristis pectinata*) in South Florida. *Bull. Mar. Sci.* 80:933–934.

Simpfendorfer, C.A. and Burgess, G.H. (2002). *Assessment of the Status of the Atlantic Sharpnose Shark (Rhizoprionodon terraenovae) Using an Age-Structured Population Model*, NAFO SCR Doc. 02/116. Northwest Atlantic Fisheries Organization, Dartmouth, Nova Scotia.

Simpfendorfer, C.A., Donohue, K., and Hall, N.G. (2000). Stock assessment and risk analysis for the whiskery shark (*Furgaleus macki* (Whitley)) in south-western Australia. *Fish. Res.* 47:1–17.

Simpfendorfer, C.A., Cortés, E., Heupel, M., Brooks, E., Babcock, E. et al. (2008). *An Integrated Approach to Determining the Risk of Over-Exploitation for Data-Poor Pelagic Atlantic Sharks: An Expert Working Group Report*, Lenfest Ocean Program, Washington, D.C.

Skalski, J.R., Ryding, K.E., and Millspaugh, J.J. (2005). *Wildlife Demography: Analysis of Sex, Age, and Count Data*. Elsevier, Burlington, MA.

Skalski, J.R., Millspaugh, J.J., and Ryding, K.E. (2008). Effects of asymptotic and maximum age estimates on calculated rates of population change. *Ecol. Model.* 212:528–535.

Sminkey, T.R. and Musick, J.A. (1996). Demographic analysis of the sandbar shark, *Carcharhinus plumbeus*, in the western North Atlantic. *Fish. Bull.* 94:341–347.

Smith, S.E. and Abramson, N.J. (1990). Leopard shark *Triakis semifasciata* distribution, mortality rate, yield and stock replenishment estimates based on a tagging study in San Francisco Bay. *Fish. Bull.* 88:371–381.

Smith, S.E., Au, D.W., and Show, C. (1998). Intrinsic rebound potentials of 26 species of Pacific sharks. *Mar. Freshw. Res.* 49:663–678.

Smith, S.E., Au, D.W., and Show, C. (2008). Intrinsic rates of increase in pelagic elasmobranchs. In: Camhi, M.D., Pikitch, E.K., and Babcock, E.A. (Eds.), *Sharks of the Open Ocean*. Blackwell Publishing, Oxford, pp. 288–297.

Sosebee, K.A. (2005). Are density-dependent effects on elasmobranch maturity possible? *J. Northw. Atl. Fish. Sci.* 35:115–124.

Stevens, J.D. (1999). Variable resilience to fishing pressure in two sharks: the significance of different ecological and life history parameters. In: Musick, J.A. (Ed.), *Life in the Slow Lane: Ecology and Conservation of Long-Lived Marine Animals*. American Fisheries Society, Bethesda, MD, pp. 11–15.

Stevens, J.D. (2010). Epipelagic oceanic elasmobranchs. In: Carrier, J., Musick, J.A., and Heithaus, M. (Eds.), *Sharks and Their Relatives II: Biodiversity, Adaptive Physiology, and Conservation*. CRC Press, Boca Raton, FL, pp. 3–35.

Stobutzki, I.C., Miller, M.J., and Brewer, D.T. (2001a). Sustainability of fishery bycatch: a process for assessing highly diverse and numerous bycatch. *Environ. Cons.* 28:167–181.

Stobutzki, I.C., Miller, M.J., Jones, P., and Salini, J.P. (2001b). Bycatch diversity and variation in a tropical Australian penaeid fishery: the implications for monitoring. *Fish. Res.* 53:283–301.

Stobutzki, I.C., Miller, M.J., Heales, D.S., and Brewer, D.T. (2002). Sustainability of elasmobranchs caught as bycatch in a tropical prawn (shrimp) trawl fishery. *Fish. Bull.* 100:800–821.

Strong, Jr., W.R., Nelson, D.R., Bruce, B.D., and Murphy, R.D. (2006). Population dynamics of white sharks in Spencer Gulf, South Australia. In: Klimley, A.P. and Ainley, D.G. (Eds.), *Great White Sharks: The Biology of Carcharodon carcharias*. Academic Press, San Diego, CA, pp. 401–414.

Sulikowski, J.A., Cicia, A.M., Kneebone, J.R., Natanson, L.J., and Tsang, P.C.W. (2009). Age and size at maturity of the smooth skate *Malacoraja senta* from the western Gulf of Maine. *J. Fish Biol.* 75:2832–2838.

Taylor, I.G. and Gallucci, V.F. (2009). Unconfounding the effects of climate and density dependence using 60 years of data on spiny dogfish (*Squalus acanthias*). *Can. J. Fish. Aquat. Sci.* 66:351–366.

Tovar Ávila, J., Day, R.W., and Walker, T.I. (2010). Using rapid assessment and demographic methods to evaluate the effects of fishing on *Heterodontus portusjacksoni* off far eastern Victoria, Australia. *J. Fish Biol.* 77:1564–1578.

Walker, P.A. and Hislop, J.R.G. (1998). Sensitive skates or resilient rays? Spatial and temporal shifts in ray species composition in the central and north-western North Sea between 1930 and the present day. *ICES J. Mar. Sci.* 55:392–402.

Walker, T.I. (1992). A fishery simulation model for sharks applied to the gummy shark, *Mustelus antarcticus* Günther, from southern Australian waters. *Aust. J. Mar. Freshw. Res.* 43:195–212.

Walker, T.I. (1994a). Fishery model of gummy shark, *Mustelus antarcticus*, for Bass Strait. In: Bishop, I. (Ed.), *Proceedings of Resource Technology '94 New Opportunities Best Practice*. Centre for Geographic Information Systems and Modelling, University of Melbourne, Australia, pp. 422–438.

Walker, T.I. (1994b). Stock assessments of the gummy shark, *Mustelus antarcticus* Günther, in Bass Strait and off South Australia. In: Hancock, D.A. (Ed.), *Population Dynamics for Fisheries Management*. Australian Government Printing Service, Canberra, pp. 173–187.

Walker, T.I. (1995). *Stock Assessment of the School Shark, Galeorhinus galeus (Linnaeus), off Southern Australia by Applying a Delay-Difference Model*, report to Southern Shark Fishery Assessment Group Workshop, February 27–March 3. Victorian Fisheries Research Institute, Department of Conservation and Natural Resources, Queenscliff, Victoria, Australia.

Walker, T.I. (1998). Can shark resources be harvested sustainably? A question revisited with a review of shark fisheries. *Mar. Freshw. Res.* 49:553–572.

Walker, T.I. (2005a). Reproduction in fisheries science. In: Hamlett, W.C. (Ed.), *Reproductive Biology and Phylogeny of Chondrichthyes: Sharks, Batoids, and Chimaeras*. Science Publishers, Enfield, NH, pp. 81–127.

Walker, T.I. (2005b). Management measures. In: Musick, J.A. and Bonfil, R. (Eds.), *Elasmobranch Fisheries Management Techniques*. Asia-Pacific Economic Cooperation Secretariat, Singapore, pp. 285–321.

Walker, T.I. (2007). Spatial and temporal variation in the reproductive biology of gummy shark *Mustelus antarcticus* (Chondrichthyes: Triakidae) harvested off southern Australia. *Mar. Freshw. Res.* 58:1–3.

Walters, C.J. and Ludwig, D. (1994). Calculation of Bayes posterior probability distributions for key population parameters: a simplified approach. *Can. J. Fish. Aquat. Sci.* 51:713–722.

Ward P. and Myers, R.A. (2005). Shifts in open-ocean fish communities coinciding with the commencement of commercial fishing. *Ecology* 86:835–847.

Waring, G.T. (1984). Age, growth, and mortality of the little skate off the northeast coast of the United States. *Trans. Am. Fish. Soc.* 113:314–321.

Wood, C.C., Ketchen, K.S., and Beamish, R.J. (1979). Population dynamics of spiny dogfish (*Squalus acanthias*) in British Columbia waters. *J. Fish. Res. Bd. Can.* 36:647–656.

Xiao, Y. and Walker, T.I. (2000). Demographic analysis of gummy shark (*Mustelus antarcticus*) and school shark (*Galeorhinus galeus*) off southern Australia by applying a generalized Lotka equation and its dual equation. *Can. J. Fish. Aquat. Sci.* 57:214–222.

Zerbini, A.N., Clapham, P.J., and Wade, P.R. (2010). Assessing plausible rates of population growth in humpback whales from life-history data. *Mar. Biol.* 157:1225–1236.

Zhou, S. and Griffiths, S.P. (2008). Sustainability assessment for fishing effects (SAFE): a new quantitative ecological risk assessment method and its application to elasmobranch bycatch in an Australian trawl fishery. *Fish. Res.* 91:56–68.

16

Genetics of Sharks, Skates, and Rays

Edward J. Heist

CONTENTS

16.1 Elasmobranch Cytogenetics

The genetic code of animals including elasmobranchs is compartmentalized into two cellular organelles: the nucleus and the mitochondrion. The vast majority of DNA is found in the nucleus, where it is packaged into discrete chromosomes (Futuyma, 1998). Chromosomes segregate during meiosis in germ cells, a process that ultimately leads to the formation of haploid gametes (sperm and egg). Thus, nuclear DNA exhibits biparental inheritance; each diploid parent contributes a haploid chromosome complement to form a new diploid offspring. Mitochondrial (mt) DNA is a haploid code that typically exhibits maternal inheritance in vertebrates (Futuyma, 1998) including, presumably, elasmobranchs. Every cell contains numerous mitochondria, each with multiple copies of the mitochondrial genome. When an egg is fertilized, only the mtDNA derived from the female parent is retained in the developing embryo; thus, all mtDNA in an elasmobranch is derived from a small number of copies present in the ovum.

16.1.1 Genome Sizes

The size of a nuclear genome, measured in picograms of DNA per haploid nucleus, is directly proportional to the number of base pairs in the genetic code of the organism. Shark genomes measured so far range from 3 to 34 pg (Stingo and Rocco, 2001) compared to 3.4 pg found in the human genome. With the exception of dipnoans (lungfishes) and urodeles (salamanders and allies), elasmobranchs possess the largest vertebrate genomes. Genome size varies widely among elasmobranch species, and the size of the genome does not seem to exhibit an evolutionary trend among primitive and derived forms (Schwartz and Maddock, 2002; Stingo and Rocco, 2001). DNA sequencing technology is rapidly improving. Although the first nearly complete human genome sequence took 13 years and cost $437 million when it was published in 2007 (U.S. Department of Energy, 2011), modern next-generation sequencing technology is orders of magnitudes less expensive. Whole-genome sequencing projects in Chondrichthyes are currently underway for the chimaerid elephant fish (*Callorhinchus milii*) and the little skate (*Leucoraja erinacea*), which has one of the smallest elasmobranch genomes (NCBI, 2011). We will likely see additional sequenced chondrichthyan genomes in the near future.

Complete mtDNA sequences in elasmobranchs have been published for small-spotted catshark, *Scyliorhinus canicula* (Delarbre et al., 1998); starspotted smoothhound, *Mustelus manazo* (Cao et al., 1998); spiny dogfish, *Squalus acanthias* (Rasmussen and Arnason, 1999b); thorny skate, *Amblyraja radiata* (Rasmussen and Arnason, 1999a); ocellate spot skate, *Okamejei kenojei* (Kim et al., 2005); and horn shark, *Heterodontus francisci* (Arnason et al., 2001). Unpublished mtDNA genomes of goblin shark, *Mitsukurina owstoni*, whitespotted bamboo shark, *Chiloscyllium plagiosum*, and deepwater stingray, *Plesiobatis daviesi*, are available through the GenBank database (NCBI, 2011). Inoue et al. (2010) recently published a phylogeny of holocephalans based on eight complete mtDNA genomes. The sizes of the elasmobranch mitochondrial genomes are similar to those of other vertebrates, ranging from 16,707 base pairs (bp) to 16,783 bp. Gene order and arrangement of RNAs and noncoding regions are identical to those of mammals and bony fishes but differ slightly from those of sea lamprey (Lee and Kocher, 1995). Consistent with other vertebrates, elasmobranch mtDNA contains 13 uninterrupted protein-coding genes (12 of which are found on the "heavy" strand), 22 tRNAs, 2 rRNAs, and a noncoding control region or D-loop approximately 1000 to 1100 bp in length. Interestingly, two families of holocephalans have among the largest mtDNA genomes due to the insertion of a large noncoding region (Inoue et al., 2010).

16.1.2 Chromosome Complements

Among vertebrates, fish have the least studied chromosome complements, and the chromosomes of cartilaginous fishes are not as well studied as those of bony fishes (Solari, 1994). Stingo and Rocco (2001) reported that of approximately 1100 species of Selachii (elasmobranchs and holocephalans) karyotypes have been described for only 63 species. The limited data present indicate that relative to other vertebrates, elasmobranchs possess large genomes comprising a large number of chromosomes, some of which are very small in size. Chromosome counts in elasmobranchs range from 28 to 106 chromosomes in a full diploid complement (Stingo and Rocco, 2001), and some elasmobranch chromosomes are so small that they are near the limit of resolution of the light microscope (Maddock and Schwartz, 1996). Thus, discrepancies among authors in the chromosome counts for particular species can arise through differences in the ability to resolve the presence of tiny "microchromosomes." Stingo and Rocco (2001) surmised that poyploidy played an important role in the evolution of elasmobranchs and that

the evolutionary trend from primitive (e.g., Hexanchidae) to more derived (e.g., Carcharhinidae) forms was a reduction in the number of telocentric chromosomes with fusion into a smaller number of metacentric chromosomes accompanied by a loss of microchromosomes. This trend is apparent both among superorders and within superorders; for example, galeomorphs tend to have fewer and larger chromosomes than squalomorphs, but within the galeomorphs the primitive horn sharks (Heterodontidae) have a greater number of shorter chromosomes than do more derived requiem sharks (Carcharhinidae) and hammerheads (Sphyrnidae). Similar trends are seen within the batoids with a reduction in the number of chromosomes and a loss of microchromosomes in the Myliobatiformes relative to the Rajiformes, and within the Myliobatiformes there is a further reduction in chromosome number in Myliobatidae relative to Dasyatidae (Rocco et al., 2007). There is also an evolutionary trend toward a reduction in the quantity of AT-rich DNA, which is presumably associated with repetitive noncoding regions in more advanced forms (Stingo and Rocco, 2001; Stingo et al., 1989).

16.1.3 Sex Determination

In many gonochoristic species, the separate sexes have morphologically distinguishable chromosome complements that can be used to infer the genetic mechanism of sex determination. Often one sex possesses a matched set of chromosomes (i.e., it is homogametic), whereas the other has one single chromosome or one pair of unmatched chromosomes (heterogametic). In mammals, males are heterogametic (XY), whereas female birds are the heterogametic (WZ) sex. Fishes exhibit XY, WZ, and at least six other genetic sex-determining systems as well as several varieties of hermaphroditism (Tave, 1993). Fishes commonly lack dimorphic sex chromosomes, and inferences about sex-determining systems, which can vary among genera, species, and even strains within species (Sandra and Norma, 2010), are based on other evidence (e.g., from the sex of parthenogenic offspring or gene linkage studies) (Charlesworth and Mank, 2010). To date, there has been very little investigation into the sex-determining mechanisms in elasmobranchs. Based on chromosome morphologies, Maddock and Schwartz (1996) determined that two species of guitarfish (*Rhinobatus*) exhibited XY sex determination and found evidence of male heterogamy in white shark, *Carcharodon carcharias*, Atlantic sharpnose shark, *Rhizoprionodon terraenovae*, blacknose shark, *Carcharhinus acronotus*, and blacktip shark, *C. limbatus*, as well as evidence of female heterogamy in southern stingray, *Dasyatis americana*. They suggested that male heterogamy is the predominate sex-determining mechanism in elasmobranchs. Chapman et al. (2007) noted that a viable female parthenogenic bonnethead shark, *Sphyrna tiburo*, was consistent with XY (as opposed to WZ) because parthenogenisis in a WZ system (like birds) would have produced only viable ZZ males and nonviable WW females; however, WW female (and also ZZ male) fish are often viable (Charlesworth and Mank, 2010). Nevertheless, all four confirmed parthenogenic sharks recorded to date are female, which is consistent with, if not proving, XY sex determination in these species. Castro (1996) described a hermaphroditic blacktip shark and stated that hermaphroditism in elasmobranchs is very rare.

16.1.4 Parthenogenesis and Genetic Oddities

Parthenogenesis (production of offspring by females without genetic contributions from males) has been reported in all major vertebrate taxa except mammals (Feldheim et al., 2010b). Parthenogenesis was first confirmed in a bonnethead shark born in captivity from a female parent that was captured as a juvenile and was never housed with a male conspecific (Chapman et al., 2007). The mother and offspring were genotyped at four microsatellite loci, and the offspring was homozygous for only alleles found in the mother, indicating no evidence of paternal contribution. Chapman et al. (2007) concluded that the offspring was produced via automictic parthenogenesis, a process in which an ovum fuses with the second polar body to produce a parthenogenic zygote that has reduced heterozygosity. Similar results were found for a single embryo in blacktip shark (Chapman et al., 2008) and in two viable whitespotted bamboo shark that survived for more than 5 years (Feldheim et al., 2010b). Whether parthenogenesis occurs in any appreciable frequency in wild populations is unknown. Albinism has been reported in at least nine species of elasmobranchs (Smale and Heemstra, 1997, and references cited therein). To date, no cases of hybridization in elasmobranchs have been documented, although the morphological similarity of many sympatric species might make recognition of hybrids in the field problematic. The only record of triploidy in an elasmobranch is for a nurse shark, *Ginglymostoma cirratum*, by Kendall et al. (1994).

16.2 Population Genetics, Stock Structure, and Forensics

16.2.1 Molecular Markers

Since the development of isozyme electrophoresis in the 1960s, molecular markers have increasingly been used to partition genetic variation among species and to define the presence of multiple discrete units (stocks) within

species (for a historical review, see Utter, 1991). During the 1980s and 1990s, analysis of mtDNA restriction fragment length polymorphisms (RFLP) was popular, fueled by the compact size and therefore manageability of the mitochondrial genome coupled with the ability to isolate mtDNA away from the much larger and more complex nuclear genome (Avise, 1994). The revolution of polymerase chain reaction (PCR), which began in the late 1980s and has continued to this day, has provided access to specific segments of DNA. As more is learned about the nuclear genomes of organisms, techniques that explore nuclear DNA, including analysis of highly polymorphic major histocompatibility complex (MHC) genes and DNA microsatellites, have provided the resolution to go beyond the species and population level and to examine genetic traits at the level of the family and the individual.

16.2.1.1 Isozymes and Allozymes

Isozymes are enzymes with similar catalytic properties that differ in the rate of migration in an electric field and can thus be resolved as discrete zones of activity on an electrophoretic medium (e.g., starch gel or cellulose acetate plate) (Murphy et al., 1996). Whereas some isozymes are the result of products at different gene loci, allozymes are a subset of isozymes that possess allelic variation at a single locus. During the 1970s, studies of allozymes in *Drosophila* and other organisms demonstrated that natural populations contained far more genetic variation than was previously assumed. These studies ultimately led to development of the neutral theory of molecular evolution, which stated, briefly, that the majority of genetic variation found at the molecular level is selectively neutral and thus is subject to such random forces as genetic drift (Futuyma, 1998).

In the first published study of allozymes in elasmobranchs, Smith (1986) reported that variation in allozymes (as indicated by mean heterozygosity and the percentage of loci that are polymorphic) is low in sharks. In that study, mean heterozygosity ranged from 0.001 in spotted estuary smoothhound, *Mustelus lenticulatus*, to 0.037 in blue shark, *Prionace glauca*. In a review paper, Ward et al. (1994) reported a mean heterozygosity of 0.064 for marine fishes. Low levels of allozyme heterozygosity were subsequently reported in gummy shark, *M. antarcticus* (mean heterozygosity = 0.006), by MacDonald (1988) and in sandbar shark (mean heterozygosity = 0.005) by Heist et al. (1995). Larger amounts of intraspecific variation were observed in two species of *Carcharhinus* (*C. tilstoni* and *C. sorrah*) (mean heterozygosity = 0.037 and 0.035, respectively) by Lavery and Shaklee (1989) and in Pacific angel sharks (*Squatina*) (mean heterozygosity = 0.056) by Gaida (1997).

The amount of genetic variation present within a species is a function of the mutation rate, the long-term effective population size of the organism, and natural selection. With some markers (e.g., allozymes), the amount of variation detected also depends upon the resolution obtained by different research protocols (e.g., number of buffer systems employed, separatory media employed, experience and skills of researchers). MacDonald (1988), for example, detected variation in only 1 of 32 presumed allozyme loci with a mean heterozygosity of 0.006 in gummy shark from the waters of southern Australia. Gardner and Ward (1998) found variation in 7 of 28 presumed loci with a mean heterozygosity of 0.099 in the same species from many of the same locations. Although the latter study scored polymorphism at four loci not surveyed by MacDonald, they also detected variation at two loci MacDonald scored as monomorphic and suggested that their use of additional buffer systems afforded them greater resolution.

16.2.1.2 Mitochondrial DNA

After allozymes, the next type of molecular marker to be widely used for determining stock structure of fishes was mtDNA. The reasons for the initial use of mtDNA rather than nuclear DNA have to do with the compact size of mtDNA relative to nuclear DNA (see above) coupled with the ability to isolate mtDNA from nuclear DNA. Although mitochondrial genes tend to evolve more rapidly than nuclear genes (Brown, 1979), mtDNA evolves more slowly in sharks than in mammals (Martin, 1995; Martin et al., 1992). Because mtDNA is haploid and maternally inherited, isolated populations will drift to different haplotype frequencies faster and achieve approximately twice the level of differentiation relative to nuclear markers. The first studies of mtDNA employed whole-molecule analysis of RFLPs. As "universal" PCR primers for mtDNA genes were developed (Kocher et al., 1989), smaller fragments of PCR-amplified mtDNA were analyzed using either RFLP or direct sequencing. Early studies employed RFLP analysis of whole-molecule mtDNA (Heist et al., 1995, 1996a,b), but as DNA sequencing technology improved and became more affordable most recent studies are based on direct sequences of particular mtDNA regions, typically the noncoding control region or sodium dehydrogenase subunit 4 (ND4) gene.

Mitochondrial DNA also forms the basis for DNA barcoding, a procedure in which an animal specimen (or part of a specimen) can be identified to species typically using the DNA sequence of a particular segment of the mitochondrial cytochrome oxidase I (COI) gene (Hebert et al., 2003). DNA barcode data are maintained and updated via the Barcoding of Life Database

(BOLD) at www.barcodinglife.org (Ratnasingham and Hebert, 2007). Because there is generally less COI variation within species than between species, over 98% of marine fishes and 93% of freshwater fishes can be accurately identified based solely on COI sequences (Ward et al., 2009). Cartilaginous fishes are better represented than bony fishes in terms of the percentage of taxa catalogued, and the tiger shark, *Galeocerdo cuvier*, is currently the most extensively barcoded fish (Ward et al., 2009). In a study of 210 species of elasmobranchs, 99% could be correctly identified using DNA barcodes (Ward et al., 2008). Attempts have been made to use COI sequence divergence as the benchmark for determining whether allopatric populations are conspecific or heterospecific based on either absolute levels of divergence (Lefebure et al., 2006) or the ratio of intrapopulation variation and interpopulation variation (Hebert et al., 2004). However, in cases where species recently diverged or where there has been hybridization, DNA barcodes will fail to resolve species (Hickerson et al., 2006). Sharing of COI sequences among species of *Urolophus*, *Carcharhinus*, and *Pristiophorus* may be due to either hybridization or misidentification, highlighting the need to have DNA barcode data supported by voucher specimens (Ward et al., 2009). DNA barcodes can be the first step for identifying cryptic species but need to be confirmed with nuclear DNA markers and traditional taxonomic methods. DNA barcodes have been used to confirm species distinctiveness in river sharks (*Glyphis*) (Wynen et al., 2009) and sharpnose sharks (*Rhizoprionodon*) (Mendonca et al., 2011) and can also distinguish among three species of "blacktip" sharks (*C. amblyrhynchoides*, *C. limbatus, and C. tilstoni*) in Australian waters that are difficult to discriminate morphologically (Boomer et al., 2010). DNA barcodes are also useful for forensic analysis of elasmobranch fins (Holmes et al., 2009).

16.2.1.3 Nuclear DNA

The nuclear genomes of vertebrates are far larger and more complex than the mitochondrial genomes. Most segments of nuclear DNA evolve very slowly and thus exhibit very little intra- and interspecific variation. Because of the combination of size, complexity, and low variation in the nuclear genome, most studies of population genetics and systematics in elasmobranchs have utilized mitochondrial data; however, as more is learned about the makeup of vertebrate nuclear genomes and as nuclear entities with higher levels of variation are characterized, more studies are employing nuclear data. Types of nuclear markers that have been employed to study elasmobranchs include ribosomal internal transcribed spacers (ITS), microsatellites, and MHC genes.

Among the most conserved nuclear genes in vertebrates are the 5.8S, 18S, and 28S ribosomal RNA (rRNA) genes, which are found as multiple copies of a single long transcript of all three conserved genes separated by more polymorphic ITS segments. Because rRNA gene sequences are so highly conserved, PCR primers developed in one species have very broad taxonomic utility and can be used to amplify the more variable ITS regions. Using a combination of conserved PCR primers located in the 5.8S and 28S rRNA genes and species-specific primers in the ITS2 region, Pank et al. (2001) and Shivji et al. (2002) produced forensic tools for the identification of shark species (see Section 16.2.3).

Microsatellites are short repetitive segments of DNA (e.g., $(GT)_n$ and $(GA)_m$, where n refers to the number of repeats of the core motif) that are highly variable for the number of repeats and hence the size of PCR fragments that are produced by primers flanking the specific repeat (Ashley and Dow, 1994; O'Connell and Wright, 1997; Wright and Bentzen, 1994). Microsatellites are among the most polymorphic markers yet developed, with many loci possessing more than 20 alleles and heterozygosities exceeding 95%. Microsatellites are useful for studies of population genetics; however, they tend to underestimate genetic divergence among populations because of the large amount of variation and high rate of homoplasy (Balloux et al., 2000). To date, polymorphic microsatellite loci have been developed in sandbar shark (Heist and Gold, 1999b; Portnoy et al., 2006); white shark (Pardini et al., 2000); lemon shark, *Negaprion brevirostris* (Feldheim et al., 2001a,b); shortfin mako shark (Schrey and Heist, 2002); nurse shark (Heist et al., 2003); blacktip shark (Keeney and Heist, 2003); bonnethead shark (Chapman et al., 2004); spiny dogfish (McCauley et al., 2004; Veríssimo et al., 2010); thornback ray (Chevolot et al., 2005); spot-tail shark, *Carcharhinus sorrah* (Ovenden et al., 2006); Australian blacktip shark, *Carcharhinus tilstoni* (Ovenden et al., 2006); sand tiger shark, *Carcharias taurus* (Feldheim et al., 2007); whale shark, *Rhincodon typus* (Schmidt et al., 2009); sixgill shark, *Hexanchus griseus* (Larson et al., 2009); longheaded eagle ray, *Aetobatus flagellum* (Yagishita and Yamaguchi, 2009); scalloped hammerhead, *Sphyrna lewini* (Nance et al., 2009); little skate (El Nagar et al., 2010); Australian gummy shark, *Mustelus antarcticus* (Boomer and Stow, 2010); longnose velvet dogfish, *Centroselachus crepidater* (Helyar et al., 2011); blue shark, *Prionace glauca* (Fitzpatrick et al., 2011); Portuguese dogfish, *Centroscymnus coelolepis* (Veríssimo et al., 2011); smalltooth sawfish, *Pristis pectinata* (Feldheim et al., 2010a); spotted eagle ray, *Aetobatus narinari* (Sellas et al., 2011); and tope, *Galeorhinus galeus* (Chabot and Nigenda, 2011). Development of microsatellite loci can be a difficult and time-consuming process, but when a set of primers has been developed in one species they often retain utility in closely related species. Of the five polymorphic microsatellite loci developed in shortfin mako by Schrey and Heist (2002), all were polymorphic in porbeagle, *Lamna nasus*, and salmon shark, *Lamna ditropis*, and two were polymorphic in white shark and common thresher, *Alopias vulpinus*.

With more than 100 alleles in some species, MHC genes are the most highly polymorphic markers known in vertebrates (Potts and Wakeland, 1990). MHC genes have received considerable study in elasmobranchs due to the presumed basal location of elasmobranchs in the lineage that includes bony fishes and tetrapods coupled with the lack of MHC genes in jawless fishes (Bartl, 1998; Flajnik et al., 1999; Ohta et al., 2011). Thus, elasmobranchs are an important group for studying the evolution of immunity in vertebrates. Variation at MHC loci is typically scored via amplification of a particular locus using primers designed in conserved regions flanking the highly polymorphic antigen-binding cleft. This is followed by digesting with a restriction enzyme that cuts the products of both alleles into a population of DNA fragments, the sizes of which are resolved via gel electrophoresis. By comparing the patterns produced by parents and offspring, the DNA restriction fragments associated with individual alleles can be resolved. The high allelic diversity makes each MHC locus potentially more powerful than a single microsatellite locus for studies of relatedness and paternity (see Section 16.3). The first documentation of multiple paternity in elasmobranchs was an unexpected outcome of a study of gene linkage of MHC loci in a litter of nurse sharks (Ohta et al., 2000); however, given the limited number of loci available and the difficulty in resolving individual alleles, microsatellites are ultimately the more powerful marker for many applications. Variations at MHC loci appear to be maintained by balancing selection (Edwards and Hedrick, 1998); thus, population genetics models that assume neutrality may not be suitable for analysis of MHC data.

Single nucleotide polymorphisms (SNPs) (Morin et al., 2004) have many advantages over allozymes, mtDNA, and microsatellites, including reliable scoring, ease of automation, and transferability of assays across laboratories (Smith et al., 2005). SNPs have levels of variation comparable to allozymes (typically two alleles per locus), and because they typically occur with a frequency of one every few hundred base pairs an almost unlimited number of loci is available. Scoring of SNPs from a large number of unlinked loci has the potential to provide a more detailed understanding of the evolutionary history of populations than any other marker currently available (Brumfield et al., 2003). Although SNPs are becoming the preferred marker for many applications, they have not yet been widely applied to elasmobranchs (Portnoy, 2010).

16.2.2 Measuring Stock Structure with Molecules

Since the development of allozyme electrophoresis in the 1960s, molecular markers have been increasingly used to determine stock structure in fishes including elasmobranchs (Utter, 1991; Ward, 2000). When a species is divided into multiple reproductively isolated populations, the evolutionary forces of mutation and genetic drift cause frequencies of neutral alleles to change such that over time significant differences in gene frequencies develop. These disruptive forces are countered by migration, which has a tendency to homogenize allele frequencies throughout the range of the species. When equilibrium has been achieved between the disruptive forces of mutation and drift and the homogenizing force of migration, the magnitude of the variance in allele frequencies among geographic units, as determined by various estimators of Wright's (1969) F_{ST}, is indicative of the reproductive isolation, and hence stock structure, of the units involved. If we can assume that the rate at which new mutations spread by migration is large relative to the rate at which new mutations arise in isolated populations, mutation can be effectively ignored and F_{ST} is a function of migration and drift (Ward and Grewe, 1994). Generally, F_{ST} values of less than 0.05 indicate little genetic differentiation, whereas F_{ST} values greater than 0.15 indicate great genetic differentiation (Hartl and Clark, 1997). These guidelines, however, are based on markers with low levels of variation (e.g., allozymes); where markers with high variation (e.g., microsatellites) are employed, the maximum value that F_{ST} can obtain is equal to the sum of the squares of the allele frequencies (i.e., the expected homozygosity under Hardy–Weinberg equilibrium) within populations, which for microsatellites may be 0.1 or less (Hedrick, 1999). Methods exist for rescaling F_{ST} values based on the maximum F_{ST} value that could be obtained given the level of intrapopulation variation (Hedrick, 2005), but these are not routinely applied, and magnitudes of raw F_{ST} values based on microsatellite studies should not be directly compared. Because elasmobranchs tend to have long generation lengths, species whose distributions have been altered by the geologically recent Pleistocene glaciations may not yet have reached equilibrium between migration and drift (Hauser, 2009).

Although it may seem counterintuitive, the magnitude of F_{ST} among locations is determined not by the rate of migration among regions but by the absolute number of migrants, abbreviated by $N_e m$, which stands for the product of the effective population size (N_e) and the migration rate (m). The reason for this relationship is that the rate at which populations diverge due to genetic drift is inversely proportional to population size (N_e); therefore, smaller populations require a larger migration rate to arrive at the same F_{ST} as larger populations with a smaller migration rate.

Under the island model of migration, which assumes that multiple same-sized populations exist with an equal rate of exchange among all populations, the relationship between F_{ST} and $N_e m$ is

$$F_{ST} \approx \frac{1}{4N_e m + 1} \quad (16.1)$$

This relationship has been widely used (and abused) (Neigel, 2002) in estimating the degree of reproductive isolation among fishery stocks. Given the unrealistic assumptions that accompany this model (e.g., large number of populations with constant equal migration among the populations, equilibrium between migration and drift), this equation should really be considered an approximation rather than an absolute measure of migration. The above relationship holds only for nuclear genes, which are diploid and biparentally inherited. For the haploid maternally inherited mitochondrial DNA the relationship is

$$F_{ST} \approx \frac{1}{2N_e m_f + 1} \quad (16.2)$$

where $N_e m_f$ refers to the number of female migrants. Estimates of gene flow based on nuclear markers indicate movement by both (or either) sex and will tend to indicate the pattern of gene flow caused by the most dispersive sex. Conversely, mitochondrial DNA reflects only the movements of females; in situations where females are philopatric and males are more likely to roam, there can be large discrepancies in the estimates of gene flow and stock structure based on nuclear and mitochondrial markers (Pardini et al., 2001). Using more detailed models, it is also possible to estimate N_e and migration rate directly from genetic data (reviewed in Portnoy, 2010).

Mitochondrial markers have some decided advantages over nuclear markers in estimating stock structure. In species with equal levels of male- and female-mediated gene flow, the magnitude of F_{ST} for mitochondrial markers is larger than that of nuclear markers (Figure 16.1); thus, there can be greater statistical power for detecting nonzero F_{ST} values (see below). Furthermore, it is easier to unambiguously interpret the magnitude of genetic divergence among mtDNA haplotypes based on the number of nucleotide substitutions between haplotypes than it is for many kinds of nuclear DNA data (e.g., allozymes and microsatellites), where two alleles that appear very different may differ by only a single mutation.

Some significant difficulties are encountered in the use of gene frequencies to estimate stock structure in highly motile species (such as many elasmobranchs)

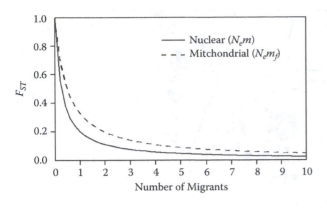

FIGURE 16.1
Relationship between F_{ST} and $N_e m$ for nuclear loci and $N_e m_f$ for mitochondrial loci.

that live in environments with few barriers to migrations (i.e., the seas) (Waples, 1998). Figure 16.2 demonstrates the relationship between F_{ST} and $N_e m$ (or $N_e m_f$) for nuclear and mitochondrial markers. At low levels of gene flow (i.e., less than one individual per generation), the relationship between a measured estimate of F_{ST} and an inferred level of gene flow is robust because F_{ST} values are large, and modest errors in the measurement of F_{ST} result in only small changes in the estimate of $N_e m$. However, at greater levels of gene flow, F_{ST} approaches zero, and small errors in the measurement of F_{ST} produce large errors in the estimate of $N_e m$. Furthermore, it is very difficult to interpret the meaning of a small but statistically significant (nonzero) F_{ST}. Small factors in the sampling regimen (e.g., collection of related individuals within samples, variation in gene frequencies among sampling years) coupled with statistically powerful tests of genetic homogeneity can produce significant F_{ST} values that are not representative of the long-term genetic structure of populations.

The inability to resolve stock structure in the presence of moderate amounts of gene flow in the marine environment may lead to improper management and conservation practices. Studies that employ small sample sizes or markers with low levels of variation may result in a failure to reject the null hypothesis of a single stock (type II error) when in fact multiple stocks do exist. Managing multiple, largely independent stocks as a single stock may be more injurious to the resource than managing for multiple stocks. For this reason, Dizon et al. (1995) recommended evaluating the consequences of type I and type II errors and perhaps lowering the rejection criterion (alpha) to balance the risks associated with both types of error. For such a strategy to be effective, calculations of statistical power associated with hypothesis testing should be employed (e.g., Schrey and Heist, 2003; Veríssimo et al., 2010).

16.2.3 Elasmobranch Population Structure

Patterns emerging from numerous studies of genetic variation in elasmobranchs indicate little genetic heterogeneity in pelagic species and within the contiguous ranges of vagile species and more genetic heterogeneity in sedentary species and those with multiple discrete populations. Generally, coastal sharks exhibit little divergence in nuclear and mitochondrial gene frequencies where they are continuously distributed along continental margins—for example, sandbar sharks in the western North Atlantic (Heist et al., 1995); sharpnose sharks from the United States and Mexico (Heist et al., 1996b) and Brazil (Mendonca et al., 2009); narrownose sharks, *Mustelus schmitti*, from Uruguay (Pereyra et al., 2010); dusky sharks, *Carcharhinus obscurus*, and scalloped hammerhead sharks in Australia and Indonesia (Ovenden et al., 2009); and lemon sharks from the Bahamas to Brazil (Feldheim et al., 2001b). Exceptions include the study of zebra sharks, *Stegostoma fasciatum* (Dudgeon et al., 2009), and leopard sharks (Lewallen et al., 2007), which are both relatively slow-moving benthic species that exhibit mtDNA haplotype heterogeneity across continuous habitat. Perhaps the smallest scale over which stock structure was indicated in elasmobranchs is recorded in the study of Gaida (1997) of Pacific angel sharks. The mean F_{ST} value of 0.085 among California's Channel Islands separated by a distance of less than 100 km was attributed to the tendency of angel sharks to remain in less than 100-m water depth and the deep (greater than 500 m) channel between islands. Similarly, Plank et al. (2010) found negligible heterogeneity on round stingray, *Urolophus halleri*, along the California mainland but significant structure between the mainland and Catalina Island, which is separated from the mainland by 42 km of deep water.

Coastal species that are genetically homogeneous along continuous coastlines are typically heterogeneous across basins. Examples include sandbar shark (Portnoy et al., 2010), scalloped hammerhead (Duncan et al., 2006), and tope (Chabot and Allen, 2009). Keeney et al. (2004) found significant differences in mtDNA haplotype frequencies among blacktip shark nursery areas from South Carolina, Florida, Texas, and the Yucatan (see Figure 16.2); however, nursery areas separated by tens of kilometers in Florida did not exhibit significant differences. Thornback ray, *Raja clavata*, populations in British waters exhibit only slight nuclear and mitochondrial differences (Chevolot et al., 2006a), but there is considerable heterogeneity among populations from the Mediterranean, European continental shelf, and the Azores (Chevolot et al., 2006b). Conversely, Chevolot et al. (2007b) found no significant heterogeneity in mtDNA haplotype frequencies of thorny skate from the North Sea, Iceland, and Newfoundland. Feldheim et al. (2001b)

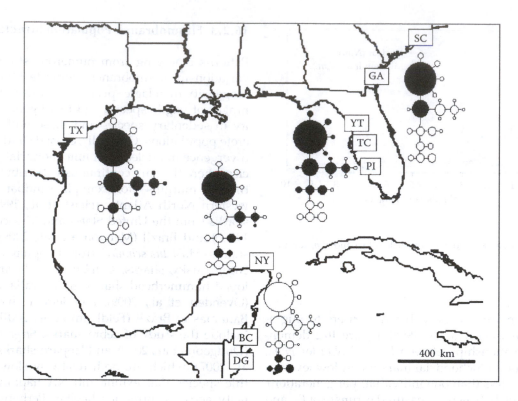

FIGURE 16.2

Heterogeneity in blacktip shark (*Carcharhinus limbatus*) mtDNA from the western Atlantic and Gulf of Mexico. Sample sites are South Carolina (SC); Georgia (GA); Pine Island Sound, Florida (PI); Terra Ceia Bay, Florida (TC); Yankeetown, Florida (YT); Texas (TX); northern Yucatan (NY); Belize City (BC); and Dangriga, Belize (DG). Each circle represents a unique sequence (haplotype) with the sizes of the circles proportional to the number of individuals observed and the lines between circles representing the inferred evolutionary relationships among haplotypes. Haplotypes recovered in each region are darkened. All haplotypes are connected to nearest haplotypes by one mutation except the five haplotypes at the bottom of each network which were found only in northern Yucatan and Belize and are separated from the rest of the network by two mutations. Although there were no significant differences among the three nursery areas in Florida, or between SC and GA, comparisons among the five regions as shown were all significant ($P < 0.001$), with the common haplotypes in the north replaced by different haplotypes in the south. (From Keeney, D.B. et al., *Mol. Ecol.*, 14, 1911–1923, 2005. With permission.)

found only slight differences in microsatellite allele frequencies in lemon sharks between the Bahamas and Brazil, indicating a significant amount of gene flow in this philopatric species, but Schultz et al. (2008) found significant differences across ocean basins. Sand tiger sharks, *Carcharias taurus*, exhibit mtDNA patterns consistent with distinct populations between ocean basins, although there was some sharing of haplotypes between eastern and western Australia, and surprisingly there was sharing of haplotypes between Brazil and South Africa (Ahonen et al., 2009).

Pelagic species sometimes exhibit small but significant differences among ocean basins—for example, whale shark (Castro et al., 2007) and shortfin mako (Heist et al., 1996a; Schrey and Heist, 2003). In contrast, Hoelzel et al. (2006) found no heterogeneity in a worldwide study of basking sharks, *Cetorhinus maximus*, but samples sizes were small and further work would help clarify population structure. In the first published study of population genetics in a deep-sea squaloid, Veríssimo et al. (2011) found no heterogeneity among Portuguese dogfish sampled from Ireland to South Africa.

16.2.4 Forensic Identification and Cryptic Species

Many elasmobranch genera include multiple species that are morphologically similar to one another. This has caused considerable confusion in species identification for management and scientific purposes; thus, new species of elasmobranchs are continually being described as subtle differences in morphology and genetic differences are detected within nominal species. The tools of molecular genetics, including allozymes and DNA sequencing, have provided aids for identifying specimens (reviewed in Shivji, 2010) and for identifying the presence of cryptic species (Lavery and Shaklee, 1991; Quattro et al., 2006).

In diploid organisms, allozymes exhibit disomic inheritance, meaning that each individual inherits two alleles, one from each parent. When a locus exhibits multiple alleles in a large outbreeding population, the mixture of homozygotes and heterozygotes typically conform to Hardy–Weinberg equilibrium (i.e., the frequency of heterozygotes is equal to twice the product of the individual allele frequencies). A deficit of heterozygotes indicates that two or more groups within the population are reproductively isolated. It is easier to identify the presence of cryptic species when two similar species occur in sympatry by using molecular genetics to identify the presence of reproductive isolation. It is more difficult to decide whether allopatric forms that are morphologically similar constitute discrete species, although molecular genetics can provide estimates of the genetic divergence that can be used in concert with morphological data to assign species status. Solé-Cava et al. (1983) confirmed that two morphotypes of Brazilian angel shark (*Squatina*) were distinct species based on fixed allelic differences at two esterase loci. Later, Solé-Cava and Levy (1987) identified a third, less common species. Similarly, Gardner and Ward (2002) examined mtDNA RFLPs, allozymes, and morphology in 550 specimens of *Mustelus* from Australia and concluded that in addition to the two named species (*M. lenticulatus* and *M. antarcticus*) two additional species (now referred to as *M. ravidus* and *M. stevensi*) were present. In a study that included all known species of thresher sharks (*Alopias*), Eitner (1995) suggested the presence of an unrecognized species based on fixed allelic differences among individuals in a group of specimens identified as *A. superciliosus* by fishers in Baja California. Unfortunately, the specimens were not retained so no morphological comparison could be made, nor are there any published DNA sequence data that would support the presence of an undescribed thresher shark; thus, the status of this "unrecognized species" remains unsettled. Using a combination of nuclear and mitochondrial markers, Quattro et al. (2006) determined that a cryptic species of scalloped hammerhead occurs in the western North Atlantic. Other examples of cryptic elasmobranchs discovered or confirmed with genetic markers include two species of ornate wobbegong (*Orectolobus*) (Corrigan et al., 2008), two genetically differentiated and spatially segregated forms of common skate (*Dipturus*) (Griffiths et al., 2010), and an additional species of Antarctic skate (*Bathyraja*) (Smith et al., 2008). Undoubtedly, additional species of elasmobranchs will continue to be identified and molecular genetics will play a large role in providing evidence for species discrimination; however, given the propensity of misidentification of elasmobranch specimens, accurate collection of morphological data and retention of voucher specimens are crucial.

Although reproductive isolation in sympatry can clearly identify the presence of distinct species, it is more difficult to determine how much divergence is needed among allopatric populations in order to recognize distinct species. Independent studies of spiny dogfish (Hauser, 2009; Veríssimo et al., 2010) concluded that the populations from the North Pacific were distinct enough from those in the Atlantic and South Pacific to warrant separate species recognition. Similarly,

Chabot and Allen (2009) concluded that worldwide populations of tope comprised at least two species and likely more. Shovelnose guitarfish, *Rhinobatos productus*, from the Gulf of California and the Pacific coast exhibit distinct mtDNA profiles and may be different species (Sandoval-Castillo et al., 2004). Richards et al. (2009) found that mtDNA and nuclear ribosomal ITS sequences between populations of spotted eagle ray, *Aetobatus narinari*, were greater than those between other pairs of batoid taxa and recommended that either three species (Atlantic, Eastern Pacific, and Western/Central Pacific) be recognized or, alternatively, two taxa be recognized, with Eastern Pacific and Atlantic populations recognized as subspecies. They noted that this designation was consistent with differences in parasite species found in each population. Similar segregation of mtDNA haplotypes was found in an independent study by Schluessel et al. (2010).

Blacktip sharks (*Carcharhinus limbatus* and closely related species) are especially in need of revision. Lavery and Shaklee (1991) showed that two color morphs of blacktip shark in the waters of northern Australia were heterospecific. Although it had previously been assumed that all blacktip sharks in northern Australia were the widely distributed *C. limbatus*, Lavery and Shaklee (1991) concluded that *C. limbatus* was rare in northern Australia and that the common species was actually *C. tilstoni*, as suggested from a previous morphological study (Stevens and Wiley, 1986). Conversely, Ovenden et al. (2010) concluded that both *C. limbatus* and *C. tilstoni* occurred in similar frequencies in Northern Australia and that a third species (*C. amblyrhinchoides*) is present as well, based on diagnostic mtDNA sequences. Keeney and Heist (2006) determined that *C. limbatus* in Australia was genetically more similar to *C. tilstoni* than it was to *C. limbatus* in the Atlantic; thus, not only are Atlantic and Australian *C. limbatus* distinct species but they are not even each other's closest relative. Ovenden et al. (2010) amplified the results of Keeney and Heist (2006) by demonstrating that both *C. tilstoni* and *C. amblyrhinchoides* from Northern Australia were genetically more similar to Australian *C. limbatus* than Australian *C. limbatus* was to Atlantic *C. limbatus*. The three species of "blacktip sharks" in Australia (*C. limbatus*, *C. tistoni*, and *C. amblyrhynchoides*) are morphologically and genetically similar (Ovenden et al., 2010; Ward et al., 2008), and additional genetic analyses will likely be needed to resolve this "species complex." As more data on the relationships among elasmobranch species and populations become available, the number of recognized taxa is likely to increase considerably, albeit not without controversy.

DNA sequences are proving useful for forensically identifying fins, carcasses, and other shark parts (reviewed in Shivji, 2010). Because heads and fins are typically removed at sea by commercial fishers, it is very difficult to identify whether prohibited species are present in the landed catch. Smith and Benson (2001) used isoelectric focusing, a protein-based technique, to demonstrate that 40% of the shark filets labeled as *Mustelus lenticulatus* in New Zealand were actually other species and in some cases prohibited species. In the United States, landings of dusky shark, *Carcharhinus obscurus*, are currently prohibited, while landings of very similar species (e.g., sandbar sharks, bignose sharks, *C. altimus*) are allowed. Heist and Gold (1999a) provided a diagnostic means of discriminating among the most commonly utilized species of *Carcharhinus* in a U.S. Atlantic large coastal shark fishery through the use of PCR RFLP. A segment of the mitochondrial cytochrome *b* gene was amplified and digested with a panel of seven restriction enzymes, and a unique restriction profile was generated for each species. Pank et al. (2001) described an innovative approach for distinguishing among sandbar and dusky sharks without the added time and expense of digesting with restriction enzymes (Figure 16.3). By using a species-specific primer that matches the DNA sequence of only one species between two "universal" primers that amplify a larger "positive control" fragment in many species, a species-specific PCR fragment profile is produced. Shivji et al. (2002) expanded this approach to show that six species of sharks—shortfin mako; longfin mako, *Isurus paucus*; porbeagle; dusky; silky, *Carcharhinus falciformis*; and blue—could be distinguished from each other and from all but one other species likely to be encountered in North Atlantic fisheries (dusky sharks could not be distinguished from oceanic whitetip, *C. longimanus*). The above methods were developed to distinguish among a limited suite of taxa represented in a reference species

database and will either be uninformative or potentially mis-informative for species missing from the database. For truly unknown specimens, sequencing a segment of mtDNA (e.g., COI; see discussion of DNA barcodes above) and then comparing the sequences to published sequences on the GenBank database (i.e., the phylogenetic approach of Baker and Palumbi, 1994) is an effective means of identification (Ward et al., 2008; Wong et al., 2009). However, methods based on mtDNA may not be effective in identifying hybrids, which would possess the mtDNA of the female parent. Where species are divided into multiple discrete genetic populations, it may be possible to identify the population of origin as well as the species (Shivji, 2010).

16.3 Molecular Ecology

The development of highly polymorphic molecular markers has fostered a revolution in studies of ecology and evolutionary biology. The high resolution provided by these markers allows for assessments of relatedness among individuals to determine familial relationships within populations to determine—for example, how many fathers sired a litter of offspring, whether related individuals associate or cooperate, or which adults in a population are successful breeders. Results of various studies have either confirmed or contradicted data from field observations of mating behavior concerning fidelity to mates and reproductive success of dominant individuals.

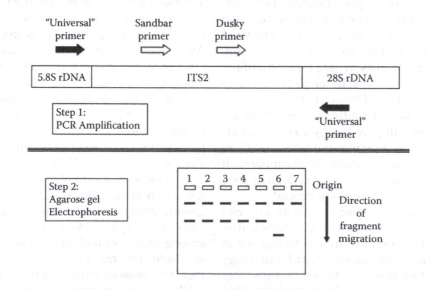

FIGURE 16.3

Protocol employed by Pank et al. (2001) for producing species-specific fragment profiles to distinguish between sandbar shark (*Carcharhinus plumbeus*) and dusky shark (*C. obscurus*). Results in lanes 1 to 5 indicate that the unknown tissue came from sandbar shark, lane 6 is a dusky shark, and lane 7 is tissue from another species.

16.3.1 Philopatry and Sex-Biased Dispersal

Philopatry, or the tendency of an animal to return to or stay in a particular location, has been confirmed in several elasmobranchs based on tracking and telemetry data (Hueter et al., 2005). Strong reproductive philopatry has been demonstrated based on tagging and field observations in lemon (DiBattista et al., 2008b; Feldheim et al., 2002), blacktip (Heupel and Hueter, 2001), and nurse (Pratt and Carrier, 2001) sharks. Heupel and Hueter (2001) found that after juvenile blacktip sharks from southwest Florida left their nursery area for the winter they faithfully returned each of the following two summers. Juvenile lemon sharks from Bimini, Bahamas, remain in their nursery areas for 2 to 3 years and then are likely to remain in the general vicinity for several years; the fraction of sharks that were locally born decreases as they age (Chapman et al., 2009). Family reconstructions based on microsatellite genotyping of young lemon sharks in Marquesas Key, Florida, and Bimini, Bahamas, indicate that female lemon sharks generally return to the same nursery areas every 2 years, but some females return on more irregular schedules (DiBattista et al., 2008b; Feldheim et al., 2004). In the same studies, it was uncommon to detect the siring of offspring by the same male over multiple years, indicating that males are more likely to stray among nursery areas. Offspring produced by the same male and female parents in different years were likely attributed to sperm storage in females and not pair bonding (DiBattista et al., 2008b).

Differences between maternally inherited mitochondrial markers and nuclear encoded loci have been used to demonstrate sex-biased dispersal and female philopatry in a variety of marine taxa including marine mammals (Gladden et al., 1999; Lyrholm et al., 1999; Palumbi and Baker, 1994) and sea turtles (Bowen and Karl, 1997; Karl et al., 1992). Several species of sea turtles exhibit significant mtDNA differences among nesting beaches, indicating strong female natal philopatry, accompanied by much lower levels of divergence in nuclear markers, indicating considerable male-mediated gene flow (Bowen and Karl, 1997). Humpback whale, *Megaptera novaeangliae* (Palumbi and Baker, 1994); beluga whale, *Delphinapterus leucas* (Gladden et al., 1999); and sperm whale, *Physeter macrocephalus* (Lyrholm et al., 1999), populations also exhibit higher levels of genetic structure in mitochondrial than nuclear markers, indicating female natal philopatry and male dispersal.

A number of studies have cited differences in genetic heterogeneity between mtDNA and microsatellites to infer female natal philopatry (Hueter et al., 2005). Pardini et al. (2001) found that, in white shark, mtDNA haplotypes from South Africa, Australia, and New Zealand clustered into two highly divergent clades. One clade was found only in 48 of 49 individuals surveyed

in Australia and New Zealand, whereas the other clade was found in 39 individuals from South Africa and in 1 of the 49 individuals surveyed in the Australia/New Zealand sample. The high degree of mtDNA divergence among white sharks between Australia/New Zealand and South Africa detected by Pardini et al. (2001) was accompanied by data from five microsatellite loci that exhibited no significant difference in allele frequency. The most obvious explanation for this discrepancy is that female white sharks either do not travel far or return to natal nursery areas; thus, genetic drift operating on maternally inherited mtDNA has resulted in significant structure. Males do occasionally move great distances and in doing so homogenize allele frequencies at nuclear microsatellite loci. These conclusions were corroborated by the observation of a telemetered female white shark that was tracked from South Africa to western Australia and then was observed back in South Africa (Bonfil et al., 2005). Similarly the moderate levels of mtDNA divergence in shortfin mako reported by Heist et al. (1996a) are accompanied by a lack of divergence at microsatellite loci (Schrey and Heist, 2003). Large discrepancies in the magnitudes of heterogeneity for mtDNA and microsatellites have been reported for blacktip shark (Keeney et al., 2005), shortfin mako shark (Schrey and Heist, 2003), thornback ray (Chevolot et al., 2006b), lemon shark (Schultz et al., 2008), sandbar shark (Portnoy et al., 2010), and bull shark (Karl et al., 2011).

Differences between estimates of gene flow based on nuclear and mitochondrial markers do not necessarily imply sex-specific dispersal because the differences in the rate of genetic drift, the high mutation rate, and high degree of homoplasy in microsatellite data can produce very different values for F_{ST} under several scenarios (Buonaccorsi et al., 2001). F_{ST} values are expected to be lower for markers with a large number of alleles (Hedrick, 1999); however, given the magnitude of the differences in the studies referenced above, it appears that female fidelity accompanied by male dispersal is emerging as a common pattern of elasmobranch population structure (for a detailed discussion, see Karl et al., 2011). Perhaps the similarities in reproductive biology among viviparous sharks, sea turtles, and whales, notably internal fertilization followed by parturition or egg laying in nursery areas temporally and spatially removed from mating areas, are responsible for the similar patterns in genetic structure. Conclusions based on nuclear and mitochondrial data do not always disagree. Gardner and Ward (1998) found concordant differences in mtDNA and allozyme frequencies in gummy sharks from Australia, consistent with the segregation of both males and females into multiple stocks, and Veríssimo et al. (2010) found concordant results for microsatellites and mtDNA in spiny dogfish.

16.3.2 Parentage and Multiple Paternity

The extremely high levels of genetic variation provided by DNA microsatellite and MHC loci make it possible to determine whether a single clutch of offspring was sired by one or more fathers. In a study of linkage relationships among MHC loci in sharks, Ohta et al. (2000) observed that at least four fathers must have sired a litter of 17 nurse sharks. Saville et al. (2002) independently discovered that in another clutch of 32 nurse shark pups the number of MHC genotypes required a minimum of 4 fathers. The high resolution of multiple microsatellite loci has the potential to provide even greater power for determining the number of sires, and with a sufficiently large number of microsatellite loci it will be possible to assign all of the pups to full-sib (same mother and father) and half-sib (same mother, different father) groups. Heist et al. (2011) used 12 microsatellite loci to show that the litters examined by Ohta et al. (2000) and Saville et al. (2002) had seven sires each. Microsatellites have also been used to demonstrate philopatry and multiple paternity in lemon shark (DiBattista et al., 2008b; Feldheim et al., 2001a, 2002); bonnethead (Chapman et al., 2004); sandbar shark, *Carcharhinus plumbeus* (Daly-Engel et al., 2006; Portnoy et al., 2007); bignose shark, *C. altimus* (Daly-Engel et al., 2006); thornback ray (Chevolot et al., 2007a); shortspine spurdog (Daly-Engel et al., 2010); and spiny dogfish (Veríssimo et al., 2011). Polyandry appears to be ubiquitous in some species (nurse shark, thornback ray, and lemon shark) and uncommon in others (bonnethead, shortspine spurdog, and spiny dogfish) and may even vary among populations of the same species. Daly-Engel et al. (2007), for example, found polyandry in 8 of 20 (40%) sandbar sharks from Hawaii, and Portnoy et al. (2007) detected polyandry in 17 of 20 (85%) sandbar sharks from the Atlantic.

There has been considerable discussion as to the evolutionary role of polyandry in elasmobranchs. Because there is no paternal care after insemination, it seems unlikely that direct benefits are accrued by the female through the involvement of additional males. Under some scenarios, polyandry may increase genetic variation indirectly through an increase in the effective population size (i.e., if it allows the contribution of males that would otherwise not mate) (Chapman et al., 2004), but it can also decrease effective population size by allowing dominant males to have disproportionately large impacts (Karl, 2008). Also, because elasmobranchs are iteroparous and do not pair-bond, they are all polyandrous over their lifetimes. A recent study by DiBattista et al. (2008a) found that offspring from multiply-sired litters had no greater survival to year 2 than did offspring from single-sired litters and inferred that there was no evidence of indirect genetic benefits to polyandry in lemon shark. The most commonly cited explanations for polyandry is convenience polyandry, which posits that mating in sharks is often a violent process that results in physical injury to females (Pratt and Carrier, 2001, 2005), so female sharks may perceive matings with multiple males as less costly than avoiding additional unnecessary matings (Daly-Engel et al., 2010).

Just as multiple microsatellite loci can be used to determine relatedness of pups within a litter, they are very powerful for matching offspring with parents and even inferring relationships among siblings. Feldheim et al. (2002) were able to identify the female parent of 89 subadult lemon sharks in Bimini Lagoon and the male parent of an additional 15 lemon sharks by matching the sharing of alleles at nine highly polymorphic microsatellite loci. The presence of sharks of different age classes with the same maternal parent indicated that female lemon sharks return to Bimini in alternate years to pup, a confirmation of previous observational data. DiBattista et al. (2008b) used microsatellites to reconstruct sib groups in lemon sharks from Marquesas Key, Florida, and similarly infer philopatry and periodicity of female reproduction.

16.4 Summary

The following points summarize the state of our understanding of elasmobranch genetics and serve to identify critical areas for future research:

- Elasmobranchs possess large nuclear genomes comprised of a large number of small chromosomes. There are evolutionary trends in some lineages to reduction in chromosome number and increase in chromosome size. Elasmobranch mitochondrial genomes are similar to those of fish and mammals. We already have numerous complete mtDNA genomes in elasmobranchs, and given advances in genome sequencing technologies we are likely to see multiple elasombranch nuclear genomes published in the near future.

- Elasmobranchs apparently exhibit both XY and WZ sex-determining systems and perhaps other mechanisms, as well. Little is known about sex determination in elasmobranchs.

- Elasmobranchs tend to have relatively low levels of allozyme variation and a reduced rate of mtDNA evolution compared to mammals.

- Because many elasmobranchs live in environments with few barriers, migration is expected to homogenize gene frequencies across vast distances. High levels of gene flow make detection of stock structure a challenge; nevertheless, there is evidence of significant stock structure among ocean basins for even pelagic species and within the contiguous distributions of some relatively sedentary species.

- Molecular markers are useful tools for forensic identification of elasmobranch tissues and for detection and confirmation of previously unrecognized species. The number of recognized species is expanding and will continue to do so based in part on genetic data.

- Several species of elasmobranchs show evidence of sex-biased dispersal, with males moving more than females.

- Highly polymorphic microsatellite markers indicate the presence of multiple paternity in elasmobranchs and can be used to determine paternity and relatedness among individuals.

References

Ahonen, H., Harcourt, R.G., and Stow, A.J. (2009). Nuclear and mitochondrial DNA reveals isolation of imperilled grey nurse shark populations (*Carcharias taurus*). *Mol. Ecol.* 18(21):4409–4421.

Arnason, U., Gullberg, A., and Janke, A. (2001). Molecular phylogenetics of gnathostomous (jawed) fishes: old bones, new cartilage. *Zool. Scr.* 30(4):249–255.

Ashley, M.V. and Dow, B.D. (1994). The use of microsatellite analysis in population biology: background, methods and potential applications. In: Schierwater, B., Streit, B., and DeSalle, R. (Eds.), *Molecular Ecology and Evolution: Approaches and Applications*. Birkhauser Verlag, Basel, Switzerland, pp. 185–201.

Avise, J.C. (1994). *Molecular Markers, Natural History and Evolution*. Chapman & Hall, New York.

Baker, C.S. and Palumbi, S.R. (1994). Which whales are hunted: a molecular genetic approach to monitoring whaling. *Science* 265(5178):1538–1539.

Balloux, F., Brunner, H., Lugon-Moulin, N., Hausser, J., and Goudet, J. (2000). Microsatellites can be misleading: an empirical and simulation study. *Evolution* 54(4):1414–1422.

Bartl, S. (1998). What sharks can tell us about the evolution of MHC genes. *Immunol. Rev.* 166:317–331.

Bonfil, R., Meyer, M., Scholl, M.C., Johnson, R., O'Brien, S., Oosthuizen, H., Swanson, S., Kotze, D., and Paterson, M. (2005). Transoceanic migration, spatial dynamics, and population linkages of white sharks. *Science* 310(5745):100–103.

Boomer, J. and Stow, A. (2010). Rapid isolation of the first set of polymorphic microsatellite loci from the Australian gummy shark, *Mustelus antarcticus*, and their utility across divergent shark taxa. *Cons. Genet. Resour.* 2(0):393–395.

Boomer, J.J., Peddemors, V., and Stow, A.J. (2010). Genetic data show that *Carcharhinus tilstoni* is not confined to the tropics, highlighting the importance of a multi-faceted approach to species identification. *J. Fish Biol.* 77(5):1165–1172.

Bowen, B.W. and Karl, S.A. (1997). Population genetics, phylogeography, and molecular evolution. In: Lutz, P.L. and Musick, J.A. (Eds.), *The Biology of Sea Turtles*. CRC Press, Boca Raton, FL, pp. 29–50.

Brown, W.M. (1979). Rapid evolution of animal mitochondrial DNA. *Proc. Natl. Acad. Sci. U.S.A.* 78(4):1967–1197.

Brumfield, R.T., Beerli, P., Nickerson, D.A., and Edwards, S.V. (2003). The utility of single nucleotide polymorphisms in inferences of population history. *Trends Ecol. Evol.* 18(5):249–256.

Cao, Y., Waddell, P.J., Okada, N., and Hasegawa, M. (1998). The complete mitochondrial DNA sequence of the shark *Mustelus manazo*: evaluating rooting contradictions to living bony vertebrates. *Mol. Biol. Evol.* 15(12):1637–1646.

Castro, A.L.F., Stewart, B.S., Wilson, S.G., Hueter, R.E., Meekan, M.G., Motta, P.J., Bowen, B.W., and Karl, S.A. (2007). Population genetic structure of Earth's largest fish, the whale shark (*Rhincodon typus*). *Mol. Ecol.* 16(24):5183–5192.

Castro, J.I. (1996). Biology of the blacktip shark, *Carcharhinus limbatus*, off the southeastern United States. *Bull. Mar. Sci.* 59(3):508–522.

Chabot, C.L. and Allen, L.G. (2009). Global population structure of the tope (*Galeorhinus galeus*) inferred by mitochondrial control region sequence data. *Mol. Ecol.* 18(3):545–552.

Chabot, C.L. and Nigenda, S. (2011). Characterization of 13 microsatellite loci for the tope shark, *Galeorhinus galeus*, discovered with next-generation sequencing and their utility for eastern Pacific smooth-hound sharks (*Mustelus*). *Cons. Genet. Resour.* 3(3):553–555.

Chapman, D.D., Prodohl, P.A., Gelsleichter, J., Manire, C.A., and Shivji, M.S. (2004). Predominance of genetic monogamy by females in a hammerhead shark, *Sphyrna tiburo*: implications for shark conservation. *Mol. Ecol.* 13(7):1965–1974.

Chapman, D.D., Shivji, M.S., Louis, E., Sommer, J., Fletcher, H., and Prodohl, P.A. (2007). Virgin birth in a hammerhead shark. *Biol. Lett.* 3(4):425–427.

Chapman, D.D., Firchau, B., and Shivji, M.S. (2008). Parthenogenesis in a large-bodied requiem shark, the blacktip *Carcharhinus limbatus*. *J. Fish Biol.* 73(6):1473–1477.

Chapman, D.D., Babcock, E.A., Gruber, S.H., Dibattista, J.D., Franks, B.R. et al. (2009). Long-term natal site-fidelity by immature lemon sharks (*Negaprion brevirostris*) at a subtropical island. *Mol. Ecol.* 18(16):3500–3507.

Charlesworth, D. and Mank, J.E. (2010). The birds and the bees and the flowers and the trees: lessons from genetic mapping of sex determination in plants and animals. *Genetics* 186(1):9–31.

Chevolot, M., Reusch, T.B.H., Boele-Bos, S., Stam, W.T., and Olsen, J.L. (2005). Characterization and isolation of DNA microsatellite primers in *Raja clavata* L. (thornback ray, Rajidae). *Mol. Ecol. Notes* 5(2):427–429.

Chevolot, M., Ellis, J.R., Hoarau, G., Rijnsdorp, A.D., Stain, W.T., and Olsen, J.L. (2006a). Population structure of the thornback ray (*Raja clavata* L.) in British waters. *J. Sea Res.* 56(4):305–316.

Chevolot, M., Hoarau, G., Rijnsdorp, A.D., Stam, W.T., and Olsen, J.L. (2006b). Phylogeography and population structure of thornback rays (*Raja clavata* L., Rajidae). *Mol. Ecol.* 15(12):3693–3705.

Chevolot, M., Ellis, J.R., Rijnsdorp, A.D., Stam, W.T., and Olsen, J.L. (2007a). Multiple paternity analysis in the thornback ray *Raja clavata* L. *J. Hered.* 98(7):712–715.

Chevolot, M., Wolfs, P.H.J., Palsson, J., Rijnsdorp, A.D., Stam, W.T., and Olsen, J.L. (2007b). Population structure and historical demography of the thorny skate (*Amblyraja radiata*, Rajidae) in the North Atlantic. *Mar. Biol.* 151(4):1275–1286.

Corrigan, S., Huveneers, C., Schwartz, T.S., Harcourt, R.G., and Beheregaray, L.B. (2008). Genetic and reproductive evidence for two species of ornate wobbegong shark *Orectolobus* spp. on the Australian east coast. *J. Fish Biol.* 73(7):1662–1675.

Daly-Engel, T.S., Grubbs, R.D., Holland, K.M., Toonen, R.J., and Bowen, B.W. (2006). Assessment of multiple paternity in single litters from three species of carcharhinid sharks in Hawaii. *Environ. Biol. Fish.* 76:419–424.

Daly-Engel, T.S., Grubbs, R.D., Bowen, B.W., and Toonen, R.J. (2007). Frequency of multiple paternity in an unexploited tropical population of sandbar sharks (*Carcharhinus plumbeus*). *Can. J. Fish. Aquat. Sci.* 64(2):198.

Daly-Engel, T.S., Grubbs, R.D., Feldheim, K.A., Bowen, B.W., and Toonen, R.J. (2010). Is multiple mating beneficial or unavoidable? Low multiple paternity and genetic diversity in the shortspine spurdog *Squalus mitsukurii*. *Mar. Ecol. Prog. Ser.* 403:255–267.

Delarbre, C., Spruyt, N., Delmarre, C., Gallut, C., Barriel, V., Janvier, P., Laudet, V., and Gachelin, G. (1998). The complete nucleotide sequence of the mitochondrial DNA of the dogfish, *Scyliorhinus canicula*. *Genetics* 150(1):331–344.

DiBattista, J.D., Feldheim, K.A., Gruber, S.H., and Hendry, A.P. (2008a). Are indirect genetic benefits associated with polyandry? Testing predictions in a natural population of lemon sharks. *Mol. Ecol.* 17(3):783.

DiBattista, J.D., Feldheim, K.A., Thibert-Plante, X., Gruber, S.H., and Hendry, A.P. (2008b). A genetic assessment of polyandry and breeding-site fidelity in lemon sharks. *Mol. Ecol.* 17(14):3337–3351.

Dizon, A.E., Taylor, B.L., and O'Corry-Crowe, G.M. (1995). Why statistical power is necessary to link analyses of molecular variation to decisions about population structure. In: Neilson, J.L. and Powers, D.A. (Eds.), *Evolution and the Aquatic Ecosystem*. American Fisheries Society, Bethesda, MD, pp. 288–294.

Dudgeon, C.L., Broderick, D., and Ovenden, J.R. (2009). IUCN classification zones concord with, but underestimate, the population genetic structure of the zebra shark *Stegostoma fasciatum* in the Indo-West Pacific. *Mol. Ecol.* 18(2):248–261.

Duncan, K.M., Martin, A.P., Bowen, B.W., and De Couet, H.G. (2006). Global phylogeography of the scalloped hammerhead shark (*Sphyrna lewini*). *Mol. Ecol.* 15(8):2239–2251.

Edwards, S.V. and Hedrick, P.W. (1998). Evolution and ecology of MHC molecules: from genomics to sexual selection. *Trends Ecol. Evol.* 13(8):305–311.

Eitner, B.J. (1995). Systematics of the genus *Alopias* (Lamniformes: Alopiidae) with evidence for the existence of an unrecognized species. *Copeia* (3):562–571.

El Nagar, A., McHugh, M., Rapp, T., Sims, D.W., and Genner, M.J. (2010). Characterisation of polymorphic microsatellite markers for skates (Elasmobranchii: Rajidae) from expressed sequence tags. *Cons. Genet.* 11(3):1203–1206.

Feldheim, K.A., Gruber, S.H., and Ashley, M.V. (2001a). Multiple paternity of a lemon shark litter (Chondrichthyes: Carcharhinidae). *Copeia* (3):781–786.

Feldheim, K.A., Gruber, S.H., and Ashley, M.V. (2001b). Population genetic structure of the lemon shark (*Negaprion brevirostris*) in the western Atlantic: DNA microsatellite variation. *Mol. Ecol.* 10(2):295–303.

Feldheim, K.A., Gruber, S.H., and Ashley, M.V. (2002). The breeding biology of lemon sharks at a tropical nursery lagoon. *Proc. R. Soc. B Biol. Sci.* 269(1501):1655–1661.

Feldheim, K.A., Gruber, S.H., and Ashley, M.V. (2004). Reconstruction of parental microsatellite genotypes reveals female polyandry and philopatry in the lemon shark, *Negaprion brevirostris*. *Evolution* 58(10):2332–2342.

Feldheim, K.A., Stow, A.J., Ahonen, H., Chapman, D.D., Shivji, M., Peddemors, V., and Wintner, S. (2007). Polymorphic microsatellite markers for studies of the conservation and reproductive genetics of imperilled sand tiger sharks (*Carcharias taurus*). *Mol. Ecol. Notes* 7(6):1366–1368.

Feldheim, K.A., Chapman, D.D., Simpfendorfer, C., Richards, V., Shivji, M. et al. (2010a). Genetic tools to support the conservation of the endangered smalltooth sawfish, *Pristis pectinata*. *Cons. Genet. Resour.* 2(1):105–113.

Feldheim, K.A., Chapman, D.D., Sweet, D., Fitzpatrick, S., Prodohl, P.A., Shivji, M.S., and Snowden, B. (2010b). Shark virgin birth produces multiple, viable offspring. *J. Hered.* 101(3):374–377.

Fitzpatrick, S., Shivji, M., Chapman, D., and Prodöhl, P. (2011). Development and characterization of 10 polymorphic microsatellite loci for the blue shark, *Prionace glauca*, and their cross shark-species amplification. *Cons. Genet. Resour.* 3:523–527.

Flajnik, M.F., Ohta, Y., Namikawa-Yamada, C., and Nonaka, M. (1999). Insight into the primordial MHC from studies in ectothermic vertebrates. *Immunol. Rev.* 167:59–67.

Futuyma, D.J. (1998). *Evolutionary Biology*, 3rd ed. Sinauer Associates, Sunderland, MA.

Gaida, I.H. (1997). Population structure of the Pacific angel shark, *Squatina californica* (Squatiniformes: Squatinidae), around the California Channel Islands. *Copeia* (4):738–744.

Gardner, M.G. and Ward, R.D. (1998). Population structure of the Australian gummy shark (*Mustelus antarcticus* Gunther) inferred from allozymes, mitochondrial DNA and vertebrae counts. *Mar. Freshw. Res.* 49(7):733–745.

Gardner, M.G. and Ward, R.D. (2002). Taxonomic affinities within Australian and New Zealand *Mustelus* sharks (Chondrichthyes: Triakidae) inferred from allozymes, mitochondrial DNA and precaudal vertebrae counts. *Copeia* (2):356–363.

Griffiths, A.M., Sims, D.W., Cotterell, S.P., El Nagar, A., Ellis, J.R. et al. (2010). Molecular markers reveal spatially segregated cryptic species in a critically endangered fish, the common skate (*Dipturus batis*). *Proc. R. Soc. B Biol. Sci.* 277(1687):1497–1503.

Hartl, D.L. and Clark, A.G. (1997). *Principles of Population Genetics*. Sinauer Associates, Sunderland, MA.

Hauser, L. (2009). The molecular ecology of dogfish sharks. In: Gallucci, V.F., Hauser, L., and Franks, J. (Eds.), *Biology and Management of Dogfish Sharks*. American Fisheries Society, Bethesda, MD, pp. 169–180.

Hebert, P.D.N., Ratnasingham, S., and deWaard, J.R. (2003). Barcoding animal life: cytochrome *c* oxidase subunit 1 divergences among closely related species. *Proc. R. Soc. B Biol. Sci.* 270:S96–S99.

Hebert, P.D.N., Stoeckle, M.Y., Zemlak, T.S., and Francis, C.M. (2004). Identification of birds through DNA barcodes. *PLoS Biol.* 2(10):1657–1663.

Hedrick, P.W. (1999). Perspective: highly variable loci and their interpretation in evolution and conservation. *Evolution* 53(2):313–318.

Hedrick, P.W. (2005). A standardized genetic differentiation measure. *Evolution* 59(8):1633–1638.

Heist, E.J. and Gold, J.R. (1999a). Genetic identification of sharks in the U.S. Atlantic large coastal shark fishery. *Fish. Bull.* 97(1):53–61.

Heist, E.J. and Gold, J.R. (1999b). Microsatellite DNA variation in sandbar sharks (*Carcharhinus plumbeus*) from the Gulf of Mexico and mid-Atlantic Bight. *Copeia* 5(1):182–186.

Heist, E.J., Graves, J.E., and Musick, J.A. (1995). Population genetics of the sandbar shark (*Carcharhinus plumbeus*) in the Gulf of Mexico and Mid-Atlantic Bight. *Copeia* 18(3):555–562.

Heist, E.J., Musick, J.A., and Graves, J.E. (1996a). Genetic population structure of the shortfin mako (*Isurus oxyrinchus*) inferred from restriction fragment length polymorphism analysis of mitochondrial DNA. *Can. J. Fish. Aquat. Sci.* 53(3):583–588.

Heist, E.J., Musick, J.A., and Graves, J.E. (1996b). Mitochondrial DNA diversity and divergence among sharpnose sharks, *Rhizoprionodon terraenovae*, from the Gulf of Mexico and Mid-Atlantic Bight. *Fish. Bull.* 94(4):664–668.

Heist, E.J., Jenkot, J.L., Keeney, D.B., Lane, R.L., Moyer, G.R., Reading, B.J., and Smith, N.L. (2003). Isolation and characterization of polymorphic microsatellite loci in nurse shark (*Ginglymostoma cirratum*). *Mol. Ecol. Notes* 3:59–61.

Heist, E.J., Carrier, J.C., Pratt, Jr., H.L., and Pratt, T.C. (2011). Exact enumeration of sires in the polyandrous nurse shark (*Ginglymostoma cirratum*). *Copeia* 2011:539–544.

Helyar, S., Coscia, I., Sala-Bozano, M., and Mariani, S. (2011). New microsatellite loci for the longnose velvet dogfish *Centroselachus crepidater* (Squaliformes: Somniosidae) and other deep sea sharks. *Cons. Genet. Resour.* 3(1):173–176.

Heupel, M.R. and Hueter, R.E. (2001). Use of an automated acoustic telemetry system to passively track juvenile blacktip shark movements. In: Silbert, J.R. and Nielsen, J.L. (Eds.), *Electronic Tagging and Tracking in Marine Fisheries*. Kluwer, Dordrecht, pp. 217–236.

Hickerson, M.J., Meyer, C.P., and Moritz, C. (2006). DNA barcoding will often fail to discover new animal species over broad parameter space. *Syst. Biol.* 55(5):729–739.

Hoelzel, A.R., Shivji, M.S., Magnussen, J., and Francis, M.P. (2006). Low worldwide genetic diversity in the basking shark (*Cetorhinus maximus*). *Biol. Lett.* 2(4):639–642.

Holmes, B.H., Steinke, D., and Ward, R.D. (2009). Identification of shark and ray fins using DNA barcoding. *Fish. Res.* 95(2–3):280–288.

Hueter, R.E., Heupel, M.R., Heist, E.J., and Keeney, D.B. (2005). The implications of philopatry in sharks for the management of shark fisheries. *J. Northw. Atl. Fish. Sci.* 35:239–247.

Inoue, J.G., Miya, M., Lam, K., Tay, B.H., Danks, J.A., Bell, J., Walker, T.I., and Venkatesh, B. (2010). Evolutionary origin and phylogeny of the modern holocephalans (Chondrichthyes: Chimaeriformes): a mitogenomic perspective. *Mol. Biol. Evol.* 27(11):2576–2586.

Karl, S.A. (2008). The effect of multiple paternity on the genetically effective size of a population. *Mol. Ecol.* 17(18):3973–3977.

Karl, S.A., Bowen, B.W., and Avise, J.C. (1992). Global population genetic-structure and male-mediated gene flow in the green turtle (*Chelonia-Mydas*): RFLP analyses of anonymous nuclear loci. *Genetics* 131:163–173.

Karl, S.A., Castro, A.L.F., Lopez, J.A., Charvet, P., and Burgess, G.H. (2011). Phylogeography and conservation of the bull shark (*Carcharhinus leucas*) inferred from mitochondrial and microsatellite DNA. *Cons. Genet.* 12(2):371–382.

Keeney, D.B. and Heist, E.J. (2003). Characterization of microsatellite loci isolated from the blacktip shark and their utility in requiem and hammerhead sharks. *Mol. Ecol. Notes* 3:501–504.

Keeney, D.B. and Heist, E.J. (2006). Worldwide phylogeography of the blacktip shark (*Carcharhinus limbatus*) inferred from mitochondrial DNA reveals isolation of western Atlantic populations coupled with recent Pacific dispersal. *Mol. Ecol.* 15(12):3669–3679.

Keeney, D.B., Heupel, M.R., Hueter, R.E., and Heist, E.J. (2004). Genetic heterogeneity among blacktip shark, *Carcharhinus limbatus*, continental nurseries along the U.S. Atlantic and Gulf of Mexico. *Mar. Biol.* 3:1039–1046.

Keeney, D.B., Heupel, M.R., Hueter, R.E., and Heist, E.J. (2005). The genetic structure of blacktip shark (*Carcharhinus limbatus*) nurseries in the western Atlantic, Gulf of Mexico, and Caribbean Sea inferred from control region sequences and microsatellites. *Mol. Ecol.* 14:1911–1923.

Kendall, C., Valentino, S., Bodine, A.B., and Luer, C.A. (1994). Triploidy in a nurse shark, *Ginglymostoma cirratum*. *Copeia* 17(3):825–827.

Kim, I.C., Jung, S.O., Lee, Y.M., Lee, C.J., Park, J.K., and Lee, J.S. (2005). The complete mitochondrial genome of the rayfish *Raja porosa* (Chondrichthyes, Rajidae). *DNA Seq.* 16(3):187–194.

Kocher, T.D., Thomas, W.K., Meyer, A., Edwards, S.V., Paabo, A., Villablanca, F.X., and Wilson, A.C. (1989). Dynamics of mitichondrial DNA evolution in animals: amplification and sequencing with conserved primers. *Proc. Natl. Acad. Sci. U.S.A.* 86:6196–6200.

Larson, S., Tinnemore, D., and Amemiya, C. (2009). Microsatellite loci within sixgill sharks, *Hexanchus griseus*. *Mol. Ecol. Resour.* 9(3):978–981.

Lavery, S. and Shaklee, J.B. (1989). Population genetics of two tropical sharks *Carcharhinus tilstoni* and *C. sorah*, in Northern Australia. *Aust. J. Mar. Freshw. Res.* 40:541–557.

Lavery, S. and Shaklee, J.B. (1991). Genetic evidence for separation of two sharks, *Carcharhinus limbatus* and *C. tilstoni*, from Northern Australia. *Mar. Biol.* 108:1–4.

Lee, W.J. and Kocher, T.D. (1995). Complete sequence of a sea lamprey (*Petromyzon marinus*) mitochondrial genome: early establishment of the vertebrate genome organization. *Genetics* 139(2):873–887.

Lefebure, T., Douady, C.J., Gouy, M., and Gibert, J. (2006). Relationship between morphological taxonomy and molecular divergence within Crustacea: proposal of a molecular threshold to help species delimitation. *Mol. Phylogenet. Evol.* 40(2):435–447.

Lewallen, E.A., Anderson, T.W., and Bohonak, A.J. (2007). Genetic structure of leopard shark (*Triakis semifasciata*) populations in California waters. *Mar. Biol.* 152(3):599–609.

MacDonald, C.M. (1988). Genetic variation, breeding structure and taxonomic status of the gummy shark *Mustelus antarcticus* in southern Australian waters. *Aust. J. Mar. Freshw. Res.* 39:641–648.

Maddock, M.B. and Schwartz, F.J. (1996). Elasmobranch cytogenetics: methods and sex chromosomes. *Bull. Mar. Sci.* 58(1):147–155.

Martin, A.P. (1995). Metabolic rate and directional nucleotide substitution in animal mitochondrial DNA. *Mol. Biol. Evol.* 12(6):1124–1131.

Martin, A.P, Naylor, G.J.P., and Palumbi, S.R. (1992). Rates of mitochondrial DNA evolution in sharks are slow compared with mammals. *Nature* 357:153–155.

McCauley, L., Goecker, C., Parker, P., Rudolph, T., Goetz, F., and Gerlach, G. (2004). Characterization and isolation of DNA microsatellite primers in the spiny dogfish (*Squalus acanthias*). *Mol. Ecol. Notes* 4(3):494–496.

Mendonca, F.F., Hashimoto, D.T., Porto-Foresti, F., Oliveira, C., Gadig, O.B.F., and Foresti, F. (2009). Identification of the shark species *Rhizoprionodon lalandii* and *R. porosus* (Elasmobranchii, Carcharhinidae) by multiplex PCR and PCR-RFLP techniques. *Mol. Ecol. Resour.* 9(3):771–773.

Mendonca, F.F., Oliveira, C., Burgess, G., Coelho, R., Piercy, A., Gadig, O.B.F., and Foresti, F. (2011). Species delimitation in sharpnose sharks (genus *Rhizoprionodon*) in the western Atlantic Ocean using mitochondrial DNA. *Cons. Genet.* 12(1):193–200.

Morin, P.A., Luikart, G., and Wayne, R.K. (2004). SNPs in ecology, evolution and conservation. *Trends Ecol. Evol.* 19(4):208–216.

Murphy, R.W., Sites, J.M.J., Buth, D.G., and Haufler, C.H. (1996). Proteins: isozyme electrophoresis. In: Hillis, D.M., Moritz, C., and Mable, B.K. (Eds.), *Molecular Systematics*. Sinauer Associates, Sunderland, MA, pp. 51–120.

Nance, H.A., Daly-Engel, T.S., and Marko, P.B. (2009). New microsatellite loci for the endangered scalloped hammerhead shark, *Sphyrna lewini*. *Mol. Ecol. Resour.* 9(3):955–957.

NCBI. (2011). *Genome: Information by Genome Sequence*. National Center for Biotechnology Information, Bethesda, MD (http://www.ncbi.nlm.nih.gov/sites/genome).

Neigel, J.E. (2002). Is F_{ST} obsolete? *Cons. Genet.* 3(2):167–173.

O'Connell, M. and Wright, J.M. (1997). Microsatellite DNA in fishes. *Rev. Fish Biol. Fish.* 7(3):331–363.

Ohta, Y., Okamura, K., McKinney, E.C., Bartl, S., Hashimoto, K., and Flajnik, M.F. (2000). Primitive synteny of vertebrate major histocompatibility complex class I and class II genes. *Proc. Natl. Acad. Sci. U.S.A.* 97(9):4712–4717.

Ohta, Y., Shiina, T., Lohr, R.L., Hosomichi, K., Pollin, T.I. et al. (2011). Primordial linkage of β_2-microglobulin to the MHC. *J. Immunol.* 186(6):3563–3571.

Ovenden, J.R., Street, R., and Broderick, D. (2006). New microsatellite loci for Carcharhinid sharks (*Carcharhinus tilstoni* and *C. sorrah*) and their cross-amplification in other shark species. *Mol. Ecol. Notes* 6(2):415–418.

Ovenden, J.R., Kashiwagi, T., Broderick, D., Giles, J., and Salini, J. (2009). The extent of population genetic subdivision differs among four co-distributed shark species in the Indo-Australian archipelago. *BMC Evol. Biol.* 9:40.

Ovenden, J.R., Morgan, J.A.T., Kashiwagi, T., Broderick, D., and Salini, J. (2010). Towards better management of Australia's shark fishery: genetic analyses reveal unexpected ratios of cryptic blacktip species *Carcharhinus tilstoni* and *C. limbatus*. *Mar. Freshw. Res.* 61(2):253–262.

Pank, M., Stanhope, M., Natanson, L., Kohler, N., and Shivji, M. (2001). Rapid and simultaneous identification of body parts from the morphologically similar sharks *Carcharhinus obscurus* and *Carcharhinus plumbeus* (Carcharhinidae) using multiplex PCR. *Mar. Biotechnol.* 3(3):231–240.

Pardini, A.T., Jones, C.S., Scholl, M.C., and Noble, L.R. (2000). Isolation and characterization of dinucleotide microsatellite loci in the Great White Shark, *Carcharodon carcharias*. *Mol. Ecol.* 9(8):1176–1178.

Pardini, A.T., Jones, C.S., Noble, L.R., Kreiser, B., Malcolm, H. et al. (2001). Sex-biased dispersal of great white sharks: in some respects, these sharks behave more like whales and dolphins than other fish. *Nature* 412(6843):139–140.

Pereyra, S., Garcia, G., Miller, P., Oviedo, S., and Domingo, A. (2010). Low genetic diversity and population structure of the narrownose shark (*Mustelus schmitti*). *Fish. Res.* 106(3):468–473.

Plank, S.M., Lowe, C.G., Feldheim, K.A., Wilson, R.R., and Brusslan, J.A. (2010). Population genetic structure of the round stingray *Urobatis halleri* (Elasmobranchii: Rajiformes) in southern California and the Gulf of California. *J. Fish Biol.* 77(2):329–340.

Portnoy, D. (2010). Molecular insights into elasmobranch reproductive behavior for conservation and management. In: Carrier, J., Musick, J.A., and Heithaus, M. (Eds.), *Sharks and Their Relatives II: Biodiversity, Adaptive Physiology, and Conservation*. CRC Press, Boca Raton, FL, pp. 435–457.

Portnoy, D.S., McDowell, J.R., Thompson, K., Musick, J.A., and Graves, J.E. (2006). Isolation and characterization of five dinucleotide microsatellite loci in the sandbar shark, *Carcharhinus plumbeus*. *Mol. Ecol. Notes* 6(2):431.

Portnoy, D.S., Piercy, A.N., Musick, J.A., Burgess, G.H., and Graves, J.E. (2007). Genetic polyandry and sexual conflict in the sandbar shark, *Carcharhinus plumbeus*, in the western North Atlantic and Gulf of Mexico. *Mol. Ecol.* 16(1):187.

Portnoy, D.S., McDowell, J.R., Heist, E.J., Musick, J.A., and Graves, J.E. (2010). World phylogeography and male-mediated gene flow in the sandbar shark, *Carcharhinus plumbeus*. *Mol. Ecol.* 19(10):1994–2010.

Potts, W.K. and Wakeland, E.K. (1990). Evolution of diversity at the major histocompatibility complex. *Trends Ecol. Evol.* 5(6):181–187.

Pratt, Jr., H.L. and Carrier, J.C. (2001). A review of elasmobranch reproductive behavior with a case study on the nurse shark, *Ginglymostoma cirratum*. *Environ. Biol. Fish.* 60(1–3):157–188.

Pratt, Jr., H.L. and Carrier, J.C. (2005). Elasmobranch courtship and mating behavior. In: Hamlett, W.C. (Ed.), *Reproductive Biology and Phylogeny of Chodrichthyes*. Science Publishers, Queensland Australia, pp. 129–169.

Quattro, J.M., Stoner, D.S., Driggers, W.B., Anderson, C.A., Priede, K.A. et al. (2006). Genetic evidence of cryptic speciation within hammerhead sharks (genus *Sphyrna*). *Mar. Biol.* 148(5):1143–1155.

Rasmussen, A.S. and Arnason, U. (1999a). Molecular studies suggest that cartilaginous fishes have a terminal position in the piscine tree. *Proc. Natl. Acad. Sci. U.S.A.* 96(5):2177–2182.

Rasmussen, A.S. and Arnason, U. (1999b). Phylogenetic studies of complete mitochondrial DNA molecules place cartilaginous fishes within the tree of bony fishes. *J. Mol. Evol.* 48(1):118–123.

Ratnasingham, S. and Hebert, P.D.N. (2007). BOLD: The Barcode of Life data system (www.barcodinglife.org). *Mol. Ecol. Notes* 7(3):355–364.

Richards, V.P., Henning, M., Witzell, W., and Shivji, M.S. (2009). Species delineation and evolutionary history of the globally distributed spotted eagle ray (*Aetobatus narinari*). *J. Hered.* 100(3):273–283.

Rocco, L., Liguori, I., Costagliola, D., Morescalchi, M.A., Tinti, F., and Stingo, V. (2007). Molecular and karyological aspects of Batoidea (Chondrichthyes, Elasmobranchi) phylogeny. *Gene* 389(1):80–86.

Sandoval-Castillo, J., Rocha-Olivares, A., Villavicencio-Garayzar, C., and Balart, E. (2004). Cryptic isolation of Gulf of California shovelnose guitarfish evidenced by mitochondrial DNA. *Mar. Biol.* 145(5):983–988.

Sandra, G.E. and Norma, M.M. (2010). Sexual determination and differentiation in teleost fish. *Rev. Fish Biol. Fish.* 20(1):101–121.

Saville, K.J., Lindley, A.M., Maries, E.G., Carrier, J.C., and Pratt, H.L. (2002). Multiple paternity in the nurse shark, *Ginglymostoma cirratum*. *Environ. Biol. Fish.* 63(3):347–351.

Schluessel, V., Broderick, D., Collin, S.P., and Ovenden, J.R. (2010). Evidence for extensive population structure in the white-spotted eagle ray within the Indo-Pacific inferred from mitochondrial gene sequences. *J. Zool.* 281(1):46–55.

Schmidt, J.V., Schmidt, C.L., Ozer, F., Ernst, R.E., Feldheim, K.A., Ashley, M.V., and Levine, M. (2009). Low genetic differentiation across three major ocean populations of the whale shark, *Rhincodon typus*. *PLoS One* 4(4):e4988.

Schrey, A.W. and Heist, E.J. (2002). Microsatellite markers for the shortfin mako and cross-species amplification in Lamniformes. *Cons. Genet.* 3(4):459–461.

Schrey, A.W. and Heist, E.J. (2003). Microsatellite analysis of population structure in the shortfin mako (*Isurus oxyrinchus*). *Can. J. Fish. Aquat. Sci.* 60:670–675.

Schultz, J.K., Feldheim, K.A., Gruber, S.H., Ashley, M.V., McGovern, T.M., and Bowen, B.W. (2008). Global phylogeography and seascape genetics of the lemon sharks (genus *Negaprion*). *Mol. Ecol.* 17(24):5336–5348.

Schwartz, F.J. and Maddock, M.B. (2002). Cytogenetics of the elasmobranchs: genome evolution and phylogenetic implications. *Mar. Freshw. Res.* 53(2):491–502.

Sellas, A., Bassos-Hull, K., Hueter, R., and Feldheim, K. (2011). Isolation and characterization of polymorphic microsatellite markers from the spotted eagle ray (*Aetobatus narinari*). *Cons. Genet. Resour.* 3:609–611.

Shivji, M. (2010). DNA forensic applications in shark management and conservation. In: Carrier, J., Musick, J.A., and Heithaus, M. (Eds.), *Sharks and Their Relatives II: Biodiversity, Adaptive Physiology, and Conservation*. CRC Press, Boca Raton, FL, pp. 593–610.

Shivji, M., Clarke, S., Pank, M., Natanson, L., Kohler, N., and Stanhope, M. (2002). Genetic identification of pelagic shark body parts for conservation and trade monitoring. *Cons. Biol.* 16(4):1036–1047.

Smale, M.J. and Heemstra, P.C. (1997). First record of albinism in the great white shark, *Carcharodon carcharias* (Linnaeus, 1758). *S. Afr. J. Sci.* 93:243–245.

Smith, C.T., Templin, W.D., Seeb, J.E., and Seeb, U.W. (2005). Single nucleotide polymorphisms provide rapid and accurate estimates of the proportions of U.S. and Canadian Chinook salmon caught in Yukon River fisheries. *North Am. J. Fish. Manage.* 25(3):944–953.

Smith, P.J. (1986). Low genetic variation in sharks (Chondrichthyes). *Copeia* 1986(1):202–207.

Smith, P.J. and Benson, P.G. (2001). Biochemical identification of shark fins and fillets from the coastal fisheries in New Zealand. *Fish. Bull.* 99(2):351–355.

Smith, P.J., Steinke, D., McVeagh, S.M., Stewart, A.L., Struthers, C.D., and Roberts, C.D. (2008). Molecular analysis of Southern Ocean skates (*Bathyraja*) reveals a new species of Antarctic skate. *J. Fish Biol.* 73(5):1170–1182.

Solari, A.J. (1994). *Sex Chromosomes and Sex Determination in Vertebrates*. CRC Press, Boca Raton, FL.

Sole-Cava, A.M. and Levy, J.A. (1987). Biochemical evidence for a third species of angel shark off the east coast of South America. *Biochem. Syst. Ecol.* 15(1):139–144.

Sole-Cava, A.M., Voreen, C.M., and Levy, J.A. (1983). Isozymic differentiation of two sibling species of *Squatina* (Chondrichthyes) in South Brazil. *Comp. Biochem. Physiol.* 75B(2):355–358.

Stevens, J.D. and Wiley, P.D. (1986). Biology of two commercially important carcharhinid sharks from Northern Australia. *Aust. J. Mar. Freshw. Res.* 37:671–688.

Stingo, V. and Rocco, L. (2001). Selachian cytogenetics: a review. *Genetica* 111(1–3):329–347.

Stingo, V., Capriglione, T., Rocco, L., Improta, R., and Morescalchi, A. (1989). Genome size and A-T rich DNA in selachians. *Genetica* 79(3):197–205.

Tave, D. (1993). *Genetics for Fish Hatchery Managers*, 2nd ed. Van Nostrand Reinhold, New York.

U.S. Department of Energy. (2011). *Human Genome Project Information* (http://www.ornl.gov/sci/techresources/Human_Genome/home.shtml).

Utter, F. (1991). Biochemical genetics and fishery management: an historical perspective. *J. Fish Biol.* 39(Suppl. A):1–20.

Veríssimo, A., McDowell, J.R., and Graves, J.E. (2010). Global population structure of the spiny dogfish *Squalus acanthias*, a temperate shark with an antitropical distribution. *Mol. Ecol.* 19(8):1651–1662.

Veríssimo, A., Grubbs, D., McDowell, J., Musick, J., and Portnoy, D. (2011a). Frequency of multiple paternity in the spiny dogfish *Squalus acanthias* in the western North Atlantic. *J. Hered.* 102(1):88–93.

Veríssimo, A., Moura, T., McDowell, J., Graves, J., Gordo, L., and Hoelzel, R. (2011b). Isolation and characterization of ten nuclear microsatellite loci for the Portuguese dogfish *Centroscymnus coelolepis*. *Cons. Genet. Resour.* 3(2):299–301.

Waples, R.S. (1998). Separating the wheat from the chaff: patterns of genetic differentiation in high gene flow species. *J. Hered.* 89(5):438–450.

Ward, R.D. (2000). Genetics in fisheries management. *Hydrobiologia* 420:191–201.

Ward, R.D. and Grewe, P.M. (1994). Appraisal of molecular genetic techniques in fisheries. *Rev. Fish Biol. Fish.* 4(3):300–325.

Ward, R.D., Elliott, N.G., Grewe, P.M., and Smolenski, A.J. (1994). Allozyme and mitochondrial DNA variation in yellowfin tuna (*Thunnus albacares*) from the Pacific Ocean. *Mar. Biol.* 118(4):531–539.

Ward, R.D., Holmes, B.H., White, W.T., and Last, P.R. (2008). DNA barcoding Australasian chondrichthyans: results and potential uses in conservation. *Mar. Freshw. Res.* 59(1):57–71.

Ward, R.D., Hanner, R., and Hebert, P.D.N. (2009). The campaign to DNA barcode all fishes, FISH-BOL. *J. Fish Biol.* 74(2):329–356.

Wong, E.H.K., Shivji, M.S., and Hanner, R.H. (2009). Identifying sharks with DNA barcodes: assessing the utility of a nucleotide diagnostic approach. *Mol. Ecol. Resour.* 9:243–256.

Wright, J.M. and Bentzen, P. (1994). Microsatellites: genetic markers for the future. *Rev. Fish Biol. Fish.* 4(3):384–388.

Wright, S. (1969). *Evolution and the Genetics of Populations*. Vol. 2. *The Theory of Gene Frequencies*. University of Chicago Press, Chicago.

Wynen, L., Larson, H., Thorburn, D., Peverell, S., Morgan, D., Field, I., and Gibb, K. (2009). Mitochondrial DNA supports the identification of two endangered river sharks (*Glyphis glyphis* and *Glyphis garricki*) across northern Australia. *Mar. Freshw. Res.* 60(6):554–562.

Yagishita, N. and Yamaguchi, A. (2009). Isolation and characterization of eight microsatellite loci from the long-headed eagle ray, *Aetobatus flagellum* (Elasmobranchii, Myliobatidae). *Mol. Ecol. Resour.* 9(3):1034–1036.

17

Predator–Prey Interactions

Michael R. Heithaus and Jeremy J. Vaudo

CONTENTS

17.1 Introduction

Predator–prey interactions play a central role in the behavior, ecology, and population biology of most taxa and are critical in community dynamics. For elasmobranchs, most studies focus only on their role as a predator. This is an important oversight, because most species are both predator and prey, at least for periods of their life. In this chapter, we place predator–prey interactions in the rich theoretical framework that has developed over the last several decades, from both a behavioral and a trophic perspective (but see Chapter 8 of this volume for a detailed consideration of elasmobranch diets). Rather than compiling an exhaustive list of predator–prey interactions, we develop a framework for these interactions and highlight relevant elasmobranch examples. Considerable progress has been made in our understanding of predator–prey interactions, including insights into the behavioral intricacies of these interactions, since the first edition of *Biology of Sharks and Their Relatives*; however, there remains much work to do, especially in light of changes in elasmobranch population sizes (for a review, see Feretti et al., 2010) and a growing realization that elasmobranchs may play important roles as both apex and mesopredators in some situations (Feretti et al., 2010; Heithaus et al., 2008a, 2010). We hope that this chapter will stimulate further studies that will help answer unresolved issues in the behavior and ecology of elasmobranchs.

In this chapter, we consider predator–prey interactions at a variety of levels. Initially, we investigate the behavioral strategies and tactics used by elasmobranchs to capture food and to avoid becoming another predator's meal. Then, we consider factors driving variation in foraging patterns and predation avoidance within elasmobranch individuals and populations. Finally, we consider how predator–prey interactions might influence both elasmobranch populations and those of their prey. We do not investigate the cascading effects of elasmobranch predator–prey interactions in detail here. Although we have retained much from the original version of this chapter, we have expanded our theoretical framework in accordance with advances in the field and focused on recent studies of elasmobranchs. There also are considerable changes to the 2004 version of this chapter in sections on the consequences of elasmobranch predator–prey interactions to reflect changes in the wider ecological understanding of such interactions. We have, however, reduced portions of this section and refer readers to our recent review in *Sharks and Their Relatives II* (Heithaus et al., 2010), as well as treatments in Heithaus et al. (2008a) and Ferretti et al. (2010).

Before investigating the behavioral ecology of predator–prey interactions, it is important to distinguish between a strategy and a tactic. A *strategy* is a genetically based decision rule, or set of rules, that results in the use of particular tactics. Animals use *tactics* (which include behaviors) to pursue a strategy (Gross, 1996). Tactics may be fixed or flexible and may depend on the condition of the individual or environmental conditions (including predators and prey); for example, a juvenile shark's strategy may be to use the tactic that will optimize energy intake and survival probability. The shark may pursue this strategy by switching between habitat use tactics that place the shark in dangerous but productive areas for some time periods and low-risk and low-food habitats during others.

17.2 Elasmobranchs as Prey

17.2.1 Predators of Elasmobranchs

Although we generally do not think of sharks as prey, they are often included in the diets of other species, as are skates and rays. Large teleosts have been found with small sharks and batoids in their stomachs, and odontocete cetaceans (for a review, see Heithaus, 2001a) and some pinnipeds (Allen and Huveneers, 2005) are also occasional elasmobranch predators. Although there are no published accounts, popular accounts and reactions of juvenile lemon shark pups to chemical stimuli suggest that crocodiles (Rasmussen and Schmidt, 1992) may be important predators of small sharks in at least the Indo-Pacific and the Americas. Even the largest sharks, such as whale sharks, *Rhincodon typus*, are not entirely free of predation pressures and are often observed with scars from predation attempts, presumably from large sharks or killer whales, *Orcinus orca* (Fitzpatrick et al., 2006; Speed et al., 2008).

A number of dolphins and small-toothed whales may opportunistically feed on small sharks and rays, but killer whales are the only cetaceans that will regularly take elasmobranchs or take large species and size classes of elasmobranchs. Killer whales have been noted taking several species of carcharhinid sharks and species as large as a 3- to 4-m great white shark, *Carcharodon carcharias*, and even larger basking sharks, *Cetorhinus maximus*, and whale sharks (Fertl et al., 1996; Pyle et al., 1999). In New Zealand, elasmobranchs may be an important component of killer whale diets, with individuals often observed preying upon eagle rays, *Myliobatis tenuicaudatus*; short-tailed stingrays, *Dasyatis brevicaudata*; and occasionally medium-sized sharks (Visser, 1999, 2005; Visser et al., 2000). In the northeast Pacific, "offshore" killer whales consumed at least 16 Pacific sleeper sharks, *Somniosus pacificus*, during two research encounters and have been found with other shark species in their stomachs (Ford et al., 2011). Extreme tooth wear relative to other forms of killer whales (i.e., "transients" and "residents") in the region suggests that sharks may be an important component of "offshore" killer whale diets (Ford et al., 2011).

In most cases, large sharks are the most important predators of other sharks and batoids. Many species of large sharks include sharks or rays in their diets and some, such as white; bull, *Carcharhinus leucas*; great hammerhead, *Sphyrna mokarran*; and broadnose seven-gill, *Notorynchus cepedianus*, sharks, regularly consume other elasmobranchs (Barnett et al., 2010a; Braccini, 2008; Cliff, 1995; Cliff and Dudley, 1991; Cliff et al., 1989; Ebert, 2002; Tricas and McCosker, 1984). Cannibalism has been recorded in a number of species of large sharks, and in some areas the most important predators of pups and juveniles may be adult conspecifics, including scalloped hammerhead sharks, *Sphyrna lewini* (Clarke, 1971), and bull sharks (Snelson et al., 1984). In general, even within species that consume other elasmobranchs, it is only the larger size classes that are a predation threat to other elasmobranchs; for example, copper sharks, *Carcharhinus brachyurus*, prey on chondrichthyans after reaching 200 cm total length (TL), but sharks of all sizes are able to cut prey into pieces, suggesting that gape limitation does not produce this pattern (Lucifora et al., 2009a). Similar to copper sharks, the proportion of elasmobranchs in the diets of broadnose sevengill; sixgill, *Hexanchus griseus*; tiger, *Galeocerdo cuvier*; and sand tiger, *Carcharias taurus*, sharks increases with increasing body size (Braccini, 2008; Ebert, 1994, 2002; Lowe et al., 1996, Lucifora et al., 2009b). In general, quantitative data on the relative predation rates of large sharks on smaller elasmobranchs are lacking; however, DNA analysis of unidentifiable prey types found in the stomachs of broadnose sevengill sharks revealed that the proportion of gummy sharks, *Mustelus antarcticus*, in their diets was considerably higher than that estimated from visual inspection of stomach contents (Barnett et al., 2010a). With increasing interest in the potential role of elasmobranchs that are both predators and prey ("mesopredators") in mediating top-down effects in ecosystems (for reviews, see Heithaus et al., 2008a, 2010; Feretti et al., 2010), a greater understanding of predation rates on these species is a high priority. It is important to note, however, that high predation rates on elasmobranch mesopredators are not necessary for their potential predators to have large impacts on their populations or modifying the spatiotemporal pattern or intensity of their impacts on their own prey (e.g., Creel and Christianson, 2008; Heithaus et al., 2008a, 2010) (see Section 17.5.2).

17.2.2 Avoiding Predators

Elasmobranchs can use many tactics to avoid being killed by predators. These range from immediate responses to a threat, such as flight or defense, to longer term tactics, such as habitat use and group formation (Figure 17.1).

17.2.2.1 Habitat Use

By selecting habitats where predators are relatively rare or absent, prey can greatly reduce their probability of being killed. Predation risk faced in a habitat, however, is not determined by predator abundance alone (e.g., Heithaus et al., 2009a). Indeed, predation risk in a habitat is determined by the probability of encountering a predator, as well as the probability that the prey is killed in an encounter situation (sometimes called *intrinsic habitat risk*) (e.g., Heithaus, 2001b; Hugie and Dill, 1994). Intrinsic habitat risk can be influenced by the presence of cover (habitat complexity), substrate color, light level, water depth, and water turbidity (Gotceitas and Colgan, 1989; Hugie and Dill, 1994; Lima and Dill, 1990; Miner and Stein, 1996; Werner and Hall, 1988), and in some situations intrinsic habitat risk may be the primary determinant of habitat use by prey species (e.g., Heithaus, 2001b; Hugie and Dill, 1994) (see Section 17.3.5.1). Indeed, some species will accept higher encounter rates with predators when these high-encounter habitats result in greater probabilities of escape during encounters with predators (through lower probabilities of predator attack or higher probabilities of escaping an attack) (Heithaus and Dill, 2006; Heithaus et al., 2009b; Wirsing et al., 2007a, 2010). In general, it is now widely recognized that the direction and nature of predator impacts on prey habitat use decisions are context dependent and can vary with prey escape behavior, landscape features, predator hunting mode, and all possible interactions among these three factors (Heithaus et al., 2009a; Schmitz, 2008; Schmitz et al., 2004; Wirsing et al., 2010). The strength of antipredator responses also can be influenced by the condition of prey. Individuals that are closer to starvation are more willing to take risks (e.g., forage in profitable but dangerous habitats) than those that are in better condition (e.g., Clark, 1994; Heithaus et al., 2007a; Sinclair and Arcese, 1995).

Habitat use decisions that reduce the risk of predation likely are extremely important to many juvenile elasmobranchs, which because of their small size, are more vulnerable to predation (e.g., Branstetter, 1990; Grubbs and Musick, 2007; Heithaus, 2007; Heithaus et al., 2009b; Heupel et al., 2007; Musick and Colvocoresses, 1988; Musick et al., 1993). Indeed, the use of shallow-water nursery habitats by juvenile sharks of many species is hypothesized to be driven largely by predator

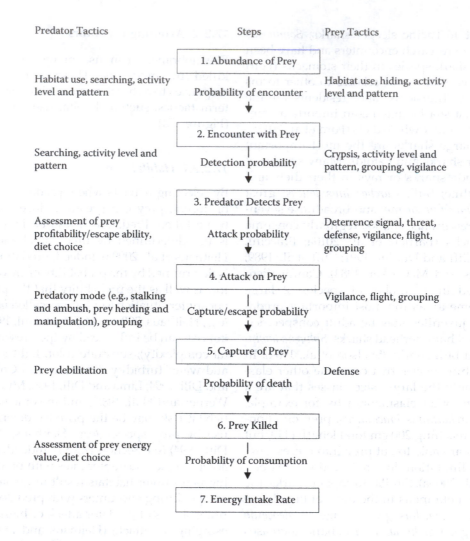

Predator Tactics — Steps — Prey Tactics

Habitat use, searching, activity level and pattern

1. Abundance of Prey

Probability of encounter

Habitat use, hiding, activity level and pattern

2. Encounter with Prey

Searching, activity level and pattern

Detection probability

Crypsis, activity level and pattern, grouping, vigilance

3. Predator Detects Prey

Assessment of prey profitability/escape ability, diet choice

Attack probability

Deterrence signal, threat, defense, vigilance, flight, grouping

4. Attack on Prey

Predatory mode (e.g., stalking and ambush, prey herding and manipulation), grouping

Capture/escape probability

Vigilance, flight, grouping

5. Capture of Prey

Prey debilitation

Probability of death

Defense

6. Prey Killed

Assessment of prey energy value, diet choice

Probability of consumption

7. Energy Intake Rate

FIGURE 17.1

Steps in a predator–prey interaction leading to energy intake for the predator. Many tactics may be used by both predator and prey throughout the predatory process, and the energy intake of the predator (foraging success) and survival probability of prey will depend on the tactics of the other. (Adapted from Lima, S.L. and Dill, L.M., *Can. J. Zool.*, 68, 619–640, 1990; Sih, A. and Christensen, B., *Anim. Behav.*, 61, 379–390, 2001; Heithaus, M.R. et al., *Mar. Biol.*, 140, 229–236, 2002.)

avoidance (e.g., Castro, 1987, 1993; Duncan and Holland, 2006; Heithaus, 2007; Morrissey and Gruber, 1993a; Simpfendorfer and Milward, 1993; Springer, 1967; Wetherbee et al., 2007). Morrissey and Gruber (1993a) found that juvenile lemon sharks, *Negaprion brevirostris*, selected warm, shallow waters with sand or rock bottoms, as opposed to slightly deeper seagrass habitats. They suggested that this habitat choice was driven largely by the need to avoid larger sharks. Similarly, juvenile blacktip shark, *Carcharhinus limbatus*, habitat use within their nursery may be driven largely by the need to avoid predators (Heupel and Heuter, 2002), and the presence of juvenile bull sharks in some low-salinity nurseries appears to be primarily to reduce the risk of predation rather than for access to resources (e.g., Heithaus et al., 2009b; Matich et al., 2011; Simpfendorfer et al., 2005). In addition, juvenile scalloped hammerheads in Hawaii remain in coastal nurseries despite

limited food resources in order to avoid predators (Duncan and Holland, 2006) and select the most turbid waters of their nursery area during the day, potentially as a refuge from predators (Clarke, 1971). Large sharks, the primary predator of small sharks, tend to be more abundant in deeper waters (e.g., Morrissey and Gruber, 1993a, but see Heithaus et al., 2002a). By selecting shallow waters, juveniles probably reduce the probability of encounter with predators as well as increase their likelihood of detecting a predator (fewer directions for a predator approach) and avoiding an attack (waters too shallow for effective attacks by predators). It is possible, however, that other demands (e.g., energy intake) may also be an important factor in juvenile shark habitat use (see Section 17.3.5.1), especially within nurseries.

Even outside of traditional nursery areas, elasmobranchs, including adults of some species, appear to select predator-free habitats, at least during some activities. In Shark Bay, Western Australia, several species of rays, including reticulate whiprays, *Himantura uarnak*, and giant shovelnose rays, *Glaucostegus typus*, rest in extremely shallow waters with sand bottoms that are free of predatory hammerhead (*Sphyrna* spp.) and tiger sharks (Vaudo and Heithaus, 2009). These rays do not appear to forage heavily within these shallow habitats and appear to leave them to forage (Vaudo and Heithaus, 2011); instead, they spend their time in these habitats resting despite potential metabolic costs (Vaudo, 2011). Also, use of shallow habitats increases during low tides, when predators have the least access to these areas. Together, these findings suggest an antipredator function for using these shallow habitats. Although most individual rays are juveniles and these habitats may provide similar benefits as some nursery areas (i.e., reduced predation risk), these habitats may not meet all the criteria outlined in the recently proposed definition of elasmobranch nursery areas (Heupel et al., 2007).

17.2.2.2 Activity Levels and Patterns

By reducing their activity level (i.e., movement speed or duration), prey can often reduce their probability of being attacked by predators (Anholt and Werner, 1995; Gerritsen and Strickler, 1997; Taylor, 1984; Werner and Anholt, 1993). This can operate through reduced encounter rates with predators as well as reduced probability of being detected by a predator (Lima, 1998a,b). It is currently unknown if elasmobranchs use this antipredator tactic. Choosing an appropriate time of day to be active can greatly influence predation risk, and changes in light level can cause prey to modify their activities because of increased susceptibility to predators (Lima and Dill, 1990). Nocturnal foragers, for example, may reduce activity on nights with bright moonlight, and diurnal foragers reduce activity during crepuscular periods because predators enjoy detection advantages at these times (Lima and Dill, 1990). Many species of elasmobranchs show distinct differences in movement patterns and rates between diurnal and nocturnal periods (see Section 17.3.2.3). These patterns have generally been interpreted as preferred foraging times, but it is also possible that the diel patterns of activity level are in response to predation risk. Hawaiian stingrays, *Dasyatis lata*, exhibit low activity rates during the day, which may be a mechanism for avoiding predators (Cartamil et al., 2003).

17.2.2.3 Hiding and Crypsis

Hiding behavior and cryptic coloration can reduce the probability of being killed by a predator, and examples of both are found in elasmobranchs. Skates and rays are the most obvious; they bury themselves in the substrate, thereby reducing the probability of being detected by a predator. Some small sharks, skates, and rays hide in structures such as reefs and rocks where predators are unlikely to detect and capture them; for example, ornate wobbegong sharks, *Orectolobus ornatus*, select daytime resting positions with high complexity and crevice volumes. Sharks do not select areas based on prey availability and are likely avoiding potential predators (Carraro and Gladstone, 2006). Swell sharks, *Cephaloscyllium ventriosum*, clustering underneath rocks during day also appears to be an antipredator tactic (Dewit, 1977; Nelson and Johnson, 1970).

Cryptic coloration is found in a number of species. Benthic species may have dorsal surfaces that are the color of the substrate, and the complex body color patterns of reef-dwelling sharks probably help them blend into their surroundings. Sometimes, coloration changes through ontogeny; for example, nurse sharks, *Ginglymostoma cirratum*, are born with black spots that help them blend into their reef and sponge habitat. As body size increases, they slowly develop the brown dorsal coloration common in adults (Castro, 2000). Finally, some species of elasmobranchs are able to change color to mimic those of their surroundings (physiological color change; see Chapter 11), which would help them avoid detection by predators.

Some small-bodied deepwater sharks may even use bioluminescence as a means to reduce predation risk. Velvet belly lantern sharks, *Etmopterus spinax*, emit light from photophores on their ventral surface that closely matches the light in their environment, allowing for effective camouflage by counterillumination (Claes et al., 2010). Interestingly, these sharks do not appear to modify the intensity of their illumination (although the possibility cannot be fully discounted yet), suggesting that they migrate vertically in a manner that keeps them within a constant light level at 500 nm (Claes et al., 2010). Counterillumination likely serves an antipredator function in these small-bodied sharks, but it may also enhance their foraging efficiency by camouflaging them from prey they approach from above (Claes et al., 2010).

17.2.2.4 Group Formation

Group formation can reduce the risk of being killed by a predator, and predation has been suggested as the selective force leading to sociality in many taxa (Bertram, 1978; Pulliam and Caraco, 1984). Many species of small sharks and rays occur in groups, but there have been few studies of the factors leading to group formation and the dynamics of group living in elasmobranchs. There are four major ways in which groups can reduce the risk of predation. First, they may increase the probability that predators are detected. Second, fleeing groups may confuse predators. Third, groups may provide active defense against predators. Rays forming groups (see below) might plausibly benefit in this way if a predator approaches close enough for a group member to use its spine to deter an attack. A fourth benefit of group living is dilution, whereby increasing the number of individuals in a group reduces an individual's probability of being captured and killed by a predator. Dilution is likely a benefit to elasmobranch groups. Scalloped hammerhead pups in nurseries of Kaneohe Bay, Oahu, Hawaii, form aggregations during the day, which may provide protection from their predators in the bay (Clarke, 1971; Holland et al., 1993). Simpfendorfer and Milward (1993) suggested that multispecies nurseries may provide additional benefits for pups in the form of reduced predation risk, probably due to dilution. It is important to note that the antipredator benefits of grouping are often not mutually exclusive and sometimes inseparable; for example, it is not possible to separate dilution and detection benefits when group members do not share information about predatory attacks perfectly (Bednekoff and Lima, 1998).

Cowtail stingrays, *Pastinachus atrus*, resting in shallow waters of Shark Bay, Western Australia, show facultative group formation, which is apparently driven by predation risk (Semeniuk and Dill, 2005, 2006). When visibility is good, rays generally rest alone but form groups under conditions that would decrease their ability to detect predators, such as low light levels and high turbidity (Semeniuk and Dill, 2005). Consistent with the antipredator benefit of grouping, ray groups exhibited greater flight initiation distances (FIDs) (see Section 17.2.2.7) in response to the approach of a model predator than solitary rays, suggesting that groups detected predators sooner. Further, cowtail stingrays appear to preferentially rest with reticulate whiprays, which respond earlier to mock predators than conspecifics, thereby providing cowtail stingrays earlier warning of an approaching predator (Semeniuk and Dill, 2006). The longer tails of reticulate whiprays presumably allow for earlier mechanical detection of predators. Indeed, rays preferentially settled with model rays that had abnormally long tails over those with normal and shortened tails (Semeniuk and Dill, 2006). Rays in these groups also tended to adopt specific geometries to increase detection area (i.e., a rosette formation with individuals' heads oriented toward each other and tails extended outward) and displayed coordinated escape responses from which they may also benefit from predator confusion and dilution (Semeniuk and Dill, 2005). Group formation in good visibility conditions, however, may actually increase predation risk. Although not fully resolved, groups show slower escape velocities and escape impedance (rays may have to move toward a predator before initiating flight), which may reduce the benefits of grouping if predators can be detected at sufficient distance by solitary rays in good visibility conditions (Semeniuk and Dill, 2005).

17.2.2.5 Vigilance

Vigilance, where prey cease other activities and watch for predators, is a common behavior used to reduce predation risk. Optimal vigilance level depends on both the risk of attack and the size of the group an animal is in, with higher levels occurring when predation risk is high and when group size is low (Brown et al., 1999; Lima, 1995; Lima and Dill, 1990). We are not aware of any studies of vigilance in elasmobranchs, and it is difficult to determine what behaviors may be considered vigilance; however, species that can remain stationary may stop searching for food to scan for predators. Vigilance may overlap substantially with searching behavior and may be difficult to operationally isolate in elasmobranchs.

17.2.2.6 Deterrence and Defense

Pursuit-deterrence signals, where prey signal to predators that they have been observed, can result in a reduced probability of predatory attack when signals are honest (Caro, 1995). There are currently no studies of such signaling in elasmobranchs; however, they are likely to be effective. Some elasmobranch predators are unlikely to initiate attacks on potential prey that are vigilant (e.g., tiger sharks) (Heithaus et al., 2002a). Thus, prey behaviors that signal their readiness to flee are likely to reduce attack probability.

Behavioral threat displays may also serve to thwart potential predators. Threat displays and subsequent attacks on divers and submersibles by gray reef sharks, *Carcharhinus amblyrhynchos*, may have an antipredator function. Nelson et al. (1986) approached gray reef sharks with submersibles and elicited threat displays and attacks. They suggested that the behavior served an antipredator function rather than to defend a territory or resources, because (1) both solitary and grouped sharks attacked, (2) attacks were elicited primarily when sharks were pursued, (3) sharks fled after attacks rather than continue an attack as would be expected if trying to drive off an intruder or competitor, (4) cornering a shark against a barrier increased the likelihood of attack, and (5) there were no threat displays directed at, or attacks observed on, conspecifics.

A number of defensive morphological characters have evolved in elasmobranchs. These include the spines found on the tail of many ray species, which can be used to inflict a painful wound on a potential attacker, as well as spines found anterior to the dorsal fins of horn sharks (Heterodontiformes) and many squaliform sharks, which may result in them being ejected from a predator's mouth after being engulfed. The ability of swell sharks to inflate their bodies when attacked, although not capable of harming a predator, could be considered a form of defense. Also, although larger piscivorous torpedo rays may use their electric organ discharges (EODs) primarily for foraging and secondarily for predator deterrence (Belbenoit, 1986), EODs in smaller electric rays are more likely to be used primarily for defense. In the lesser electric ray, *Narcine brasiliensis*, for example, EODs are not used in foraging and are probably primarily a predator defense that also may be used for intraspecific communication (Macesic and Kajiura, 2009).

17.2.2.7 Flight

Fleeing is an obvious reaction to the immediate threat of a predator (see Chapter 5 for a brief discussion of the biomechanics of escape response behavior) and involves decisions of escape direction, at what distance from the predator to initiate flight (flight initiation distance, or FID), and escape velocity. Because flight is costly (in terms of both energetic expenditure and lost foraging or mating opportunities), animals should not necessarily flee as soon as a predator is detected; instead, there should be an optimal flight response (Dill, 1990; Ydenberg and Dill, 1986). The cost of flight, the speed and angle of approach of the predator (i.e., the loom rate), and distance to safety (which is often influenced by habitat characteristics) influence escape responses (e.g., Bonenfant and Kramer, 1994; Dill, 1990; Lima and Dill, 1990; Ydenberg and Dill, 1986). FIDs will be shorter (predators allowed to approach closer) when costs of flight are high, the loom rate is lower, and habitat characteristics favor easier escape (e.g., distance to cover is shorter). Often, distance or time to safety is equivalent to the distance to a physical refuge, but in marine systems time to safety might be the amount of time it takes a prey animal to reach a critical velocity or maneuvering ability such that they are inaccessible to a predator (Heithaus et al., 2002b). Dill (1990) proposed that animals would maintain a consistent "margin of safety" by selecting a combination of escape velocity and FID; however, the margin of safety may increase as distance from cover increases. Indeed, some species appear to flee at well below their maximum speed. This choice of escape velocity would allow prey to respond to acceleration by the predator, and having such flexibility may be relatively more advantageous as distance from safety increases (Bonenfant and Kramer, 1994). For cryptic species, the decision to flee will also be based on the probability that a predator will detect and capture the prey if it flees relative to the probability of detection and capture if the prey remains motionless and cryptic; cryptic individuals show shorter FIDs than those that are not as well camouflaged (Lima and Dill, 1990).

There are few studies of FIDs in elasmobranchs, but solitary cowtail stingrays in Shark Bay, Western Australia, initiate flight when approached by a model predator at a significantly greater distance when water

visibility conditions are relatively good (Semeniuk and Dill, 2005). This can be interpreted in several ways. First, it is possible that rays are able to detect the model predator at a greater distance in clear conditions and initiate flight as soon as a predator is detected. Alternatively, rays detected models at a similar distance in good and poor visibility conditions but waited to initiate flight in poor-visibility conditions because remaining motionless during these conditions had a lower probability of being detected than if flight had been initiated.

17.2.3 Apparent Competition

Apparent competition occurs when two (or more) species share a common predator and high productivity by one prey species supports predators at a population level sufficient to eliminate another prey species (Holt, 1977, 1984). Apparent competition may also be manifested through behavioral mechanisms where predator abundance in a habitat is driven by one prey species and leads to habitat abandonment by another species (Dill et al., 2003; Heithaus and Dill, 2002a). We are not aware of apparent competition among elasmobranchs, but tiger sharks regulate apparent competition between dugongs, *Dugong dugon*, and Indo-Pacific bottlenose dolphins, *Tursiops aduncus* (Heithaus and Dill, 2002a), and sharks evidently mediate apparent competition between gray seals, *Halichoerus grypus*, and harbor seals, *Phoca vitulina*, on Sable Island, Canada (Bowen et al., 2003) (see Section 17.5.1).

17.3 Elasmobranchs as Predators

Elasmobranchs feed on an amazing array of species from plankton and benthic invertebrates to marine mammals and other large vertebrates. Species also vary from batch-feeding filter feeders and scavengers to active predators and from opportunists with catholic diets to highly specialized feeders. Although a review of elasmobranch diets is beyond the scope of this chapter (see Chapter 8), in this section we review optimal diet theory and the predatory tactics of elasmobranchs.

17.3.1 Diets of Elasmobranchs and Optimal Diet Theory

Diet composition is the result of behavioral decisions associated with locating and capturing or rejecting potential prey items (Stephens and Krebs, 1986). Optimal diet theory (ODT) has a long history but has not been widely applied to studies of elasmobranchs. Most studies have been limited to descriptions of prey items found in a particular population or life history stage, but recent work has begun to explore relationships between dietary occurrence and the abundance of particular prey items in the environment (see below). Although there is likely to be some consistency in the diets of a particular species throughout its range, there is certain to be a high degree of flexibility as ecological conditions vary. Bonnethead sharks, *Sphyrna tiburo*, in Florida; tiger sharks in Western Australia; mako sharks, *Isurus oxyrinchus*, in the northwest Atlantic; leopard sharks, *Triakis semifasciata*, in Tomales Bay, California; and scalloped hammerhead pups within a Hawaiian nursery ground show geographic variation in their diets (Clarke, 1971; Cortés et al., 1996; Simpfendorfer et al., 2001a,b; Stillwell and Kohler, 1982; Webber and Cech, 1998).

The most basic form of optimal diet theory describes when a prey item should be accepted or rejected (Stephens and Krebs, 1986). If prey items vary in their net energy gain (energy content minus the energy expended in capture and handling), handling time, and encounter rate, then it is possible to make simple predictions about which prey should be eaten. In this simple situation, predators are predicted to rank prey items in order of their profitability; they either always consume or always reject a particular prey item (Stephens and Krebs, 1986). It is important to note that the predictions based on this theory do not specifically state whether a predator should diversify or narrow its diet choice as more prey types become available to the predator (e.g., seasonal changes in prey availability, ontogenetic changes in foraging abilities). Instead, the nature of prey selection will be based on the relative profitability of each potential prey item. The general theory is upheld in some situations (e.g., Werner and Hall, 1974), but animals often do not conform to the prediction of always accepting or rejecting particular prey items (e.g., Hart and Ison, 1991; see also Clark and Mangel, 2000). This partial preference, where individuals sometimes accept prey items of lower profitability, can be explained by differences in the state of the individual (e.g., body condition, energetic reserves, gut fullness); for example, three-spined stickleback, *Gasterosteus aculeatus*, feeding on larger invertebrates (*Asellus*) sometimes accept prey items that are not predicted to be taken (Hart and Ison, 1991). When guts are relatively empty, large prey items with very long handling times are accepted, but as gut fullness increases the acceptance of these relatively unprofitable prey items decreases. Although traditional diet theory fails to explain such results, dynamic state variable models that include gut fullness and prey catchability show a good match with data (Hart and Gill, 1992a). Predictions from these types of models suggest that individuals in good condition or with relatively full stomachs should be the most selective for high-quality prey items. Dynamic state variable

models (see Clark and Mangel, 2000) allow predictions to be made about prey choice when both ecological and internal conditions are variable and are likely to benefit studies of elasmobranch diets.

The diets of elasmobranchs are almost certainly a result of the decision processes described above, and there is evidence for selection of relatively profitable prey items. Southern stingrays, *Dasyatis americana*, preferentially ingest only large size classes of lancelets (Stokes and Holland, 1992), and mako sharks appear to selectively feed on large bluefish, *Pomatomus saltatrix*, to maximize energy intake (Stillwell and Kohler, 1982). School sharks, *Galeorhinus galeus*, in Anegada Bay, Argentina, avoided abundant pelagic teleosts as prey (Lucifora et al., 2006), although pelagic teleosts make up the dominant prey in other areas (Morato et al., 2003), and Atlantic angel sharks, *Squatina dumeril*, show a preference for squid, scorpionfishes, hakes, and croakers (Baremore et al., 2008). A predator selecting an optimal diet may not stop the decision-making process with what type of prey to capture but may also decide what portions of a prey item to consume to maximize energy intake rate. White sharks; blue sharks, *Prionace glauca*; and Greenland sharks, *Somniosus microcephalus*, seem to selectively feed on the blubber layer of marine mammal carcasses, thus maximizing their energy intake through ingestion of only high-quality food (Beck and Mansfield, 1969; Carey et al., 1982; Klimley, 1994; Klimley et al., 1996a; Long and Jones, 1996). Rays foraging on thin-shelled bivalves usually consume the entire shell, but when feeding on thick-shelled bivalves they consume the body of the animal and largely avoid ingesting the shell (e.g., cownose rays, *Rhinoptera bonasus*) (Smith and Merriner, 1985). In addition to not consuming shells of prey, cownose rays can eject inedible portions of prey, such as crustacean exoskeletons and the pens of squid (Sasko et al., 2006).

Recent reviews of ODT have shown that, although it is very good at predicting the diet of foragers consuming immobile prey, predictions are often not supported for species consuming mobile prey (Lima et al., 2003; Sih and Christensen, 2001). The reason for failure of ODT in the mobile prey scenario is likely to lie in two factors. First, diet preferences of a predator (i.e., prey are consumed disproportionately relative to their abundance in the environment) can be the result of both active predator choice (unequal attack probabilities in encounter situations with prey; attack probability) (see Figure 17.1) and antipredator tactics that result in differences among prey in their probability of detection or probability of escape (see Figure 17.1). ODT only considers active predator choice and does not account for antipredator behavior (Lima et al., 2003; Sih and Christensen, 2001). Second, most studies of ODT do not include information on all aspects of prey profitability, which would include probability of capture, handling time, and probability of prey escape after capture (Sih and Christensen, 2001). Indeed, a recent review of predators across 16 vertebrate communities showed that, in general, predators tend to favor larger bodied, smaller brained prey and prey in smaller groups (Shultz and Finlayson, 2010); therefore, predators might not show preferential attacks on prey taxa with the highest energy content (Lima et al., 2003; Sih and Christensen, 2001). If energy-rich prey are rarely encountered or usually escape, predators may show no prey preferences at all, and diets may be more reflective of variation among prey in escape abilities. At Seal Island, South Africa, for example, young fur seals, *Arctocephalus pusillus*, engage in movements that put them at higher risk than adults, which would provide a more energy-rich meal, and as a consequence are the primary target of white sharks in this system (Laroche et al., 2008). Therefore, to truly understand the diet selection process of predators it is important to understand the antipredatory tactics of their prey. This aspect of predator–prey interactions has largely been ignored in studies of the diets of elasmobranchs that consume highly mobile prey.

17.3.2 Finding Prey

17.3.2.1 Habitat Use

A predator can increase its probability of encountering and capturing prey through optimal selection of a foraging habitat. In the absence of other individuals, this may entail selecting the habitat with the highest prey density; however, selecting the optimal patch to maximize energy intake does not necessarily equate to selecting the habitat with the highest density (i.e., abundance) of prey. Prey in highly structured environments (e.g., reefs or seagrass beds) may escape easily and thus become unavailable (e.g., Gotceitas and Colgan, 1989). Also, when other factors such as the presence of competitors or a third trophic level (the prey's food) are considered, then there may not be a good relationship between the abundance of prey and the distribution of predators even if energy intake is the only factor determining the distribution of the predator. Finally, if a predator is at risk of predation itself, then its distribution may be unrelated to prey distributions (see Section 17.3.5).

If a predator is pursuing a strategy to maximize net energetic gains, it should select the habitat with the highest energetic return rate (energy gained per unit time minus energetic costs of being in the habitat and capturing and consuming prey). Some elasmobranchs appear to conform to this prediction. Basking sharks are found in highest abundance in warm-water oceanic fronts that concentrate their planktonic prey, and they appear to select higher density plankton patches with abundant large copepods (Sims and Merrett, 1997; Sims and Quayle, 1998). Thus, these sharks appear

to preferentially select habitat patches based on their potential energetic intake rate. Bat rays, *Myliobatis californica*, in Tomales Bay, California, appear to adopt foraging tactics to pursue a strategy that maximizes energy intake rate. Rays move into shallow waters of the inner bay from 2:50 to 14:50, then return to the cooler waters of the outer bay (Matern et al., 2000). The infaunal prey of the rays may be buried 0.5 to 1.0 m deep and require substantial energy to excavate; thus, a high metabolic rate during foraging would be beneficial (Matern et al., 2000). Over the temperature range observed in the bay, bat rays have an extremely high Q_{10} (the increase in metabolic rate with a 10°C change in temperature) of 6.8 (Hopkins and Chech, 1994). Moving into the foraging areas during the day likely increases the foraging efficiency of the rays (Matern et al., 2000). Moving into cooler waters to rest would reduce metabolic rate (less energy wasted compared to warm water), which can save a considerable amount of energy (Matern et al., 2000; also see Chapter 7 of this volume). Similarly, captive Atlantic stingrays, *Dasyatis sabina*, prefer to rest in cooler waters after feeding (Wallman and Bennett, 2006). Also, small-spotted catsharks, *Scyliorhinus canicula*, in Lough Hyne, Ireland, prefer deeper, cooler waters, only moving into shallower, warmer waters to feed; models suggest that adopting this behavior can reduce daily energy costs by ~4% (Sims et al., 2006a). Experimental work has also confirmed that in Atlantic stingrays the decreased evacuation rates, which result from exposure to cooler water, more than offset concomitant decreases in absorption rates resulting in an overall increase in total absorption (Di Santo and Bennett, 2011).

In the absence of other factors, foragers should begin using a patch only when the energy intake rate available in that habitat is above, or equal to, the energy intake rate available in other areas within the environment. In some cases, this results in threshold foraging responses. For example, basking sharks will only forage on zooplankton concentrations above 0.48 to 0.70 g/m³, which is close to the theoretically predicted threshold at which sharks would be foraging at a net energetic loss (Sims, 1999). Similarly, whale sharks in Bahía de Los Angeles only foraged in zooplankton densities above ~10 × 10³ individuals m⁻³ and altered foraging behaviors based on zooplankton densities and composition (Nelson and Eckert, 2007). Threshold foraging responses have also been observed in rays. Cownose rays will consume all bay scallops, *Argopecten irradians concentricus*, in some high-density patches but not forage in areas with low bay scallop density (Peterson et al., 2001). Similarly, infaunal bivalves, *Macomona lilliana*, have a refuge in low densities because eagle rays show a threshold foraging response at about 44 *Macomona* per 0.25 m² (Hines et al., 1997). Furthermore, the rays forage on a relatively large spatial scale and respond to prey patches on a scale of

75 to 100 m, so rays did not forage on small but dense aggregations of bivalves (Hines et al., 1997). It is unclear how eagle rays avoid unprofitable prey patches, but they may make use of their electoreception abilities or be able to detect water jets from bivalves (Montgomery and Skipworth, 1997).

Many studies suggest that habitat use by elasmobranchs is influenced by the distribution of their prey but often are constrained by the difficulty of collecting rigorous measurements of predator habitat preference (i.e., whether habitat use patterns differ from random expectations) or prey densities or availability. Morrissey and Gruber (1993a) suggested that juvenile lemon sharks may show a preference for sand and rock bottoms over seagrass areas because prey have a refuge in seagrass and thus are less available. The distribution of large white sharks along the California coast may be influenced by the distribution of their pinniped prey (Klimley, 1985), and individuals show site fidelity to particular seal colonies across years (e.g., Goldman and Anderson, 1999; Klimley and Anderson, 1996; Strong et al., 1992, 1996). In southern Australia white sharks may select habitats where dolphin densities are high (Bruce, 1992), and at Seal Island white shark presence and attacks on fur seals are centered around the south side of the island, where fur seals are most concentrated (Laroche et al., 2008; Martin et al., 2009). Pacific angel sharks, *Squatina californica*, also appear to select ambush sites in response to prey availability (Fouts and Nelson, 1999), and sandbar sharks, *Carcharhinus plumbeus*, show some degree of site fidelity to ocean fish farms in Hawaii (Papastamatiou et al., 2010a). In contrast, although shark abundance was correlated with the abundance of potential teleost prey over large spatial scales in Florida Bay, there was no relationship between shark abundance and prey abundance at smaller spatial scales (Torres et al., 2006). Such mismatches at local scales can be due to antipredator behavior of prey, imperfect information of small-scale prey abundance, or sharks being captured in areas and during times where and when they are not foraging. Mismatches between predator abundance and prey availability also can occur during times when prey become unavailable; for example, tidal mudflats and sandflats are often inaccessible, except at high tide, and leopard sharks move into these muddy littoral zones on incoming tides, presumably to feed on worms and clam siphons (Ackerman et al., 2000).

Predators that forage on sessile prey and are in low densities may be able to make decisions in the manner described above, but often a predator's decision on which habitat to select will be influenced by antipredator decisions and behaviors of their prey, by foraging decisions made by other predators, or by both. As a predator spends more time in a particular habitat or more predators accumulate in a given habitat, prey should become more vigilant or leave a particular habitat, and prey availability will decline (for a discussion of the "ecology of fear," see Brown et al., 1999). Therefore, there may be selection for a predator to switch among habitats as prey increase their investment in antipredator behavior. In such circumstances, a tactic that involves covering an extremely large area may be optimal, as is often observed in large sharks that consume highly mobile prey not tied to a specific haul-out site. In Hawaii, tiger sharks make large moves between different areas of what appears to be a large home range (Holland et al., 1999; Meyer et al., 2009, 2010). In Shark Bay, Western Australia, tiger sharks appear to move within extremely large home ranges, and, although they preferentially select habitat and microhabitat types that are rich in prey, they do not remain within a particular habitat patch for extended periods (Heithaus, 2001c; Heithaus et al., 2002a, 2006). Even within nursery areas, where they have relatively small home ranges, elasmobranchs may use a habitat-shifting tactic. Juvenile lemon sharks shift locations within their home ranges from day to day, which may allow prey availability to rebound (presumably through habitat-use shifts in prey or changes in prey vigilance) (Morrissey and Gruber, 1993b), but may also be in response to temporarily reduced prey availability in the area.

Often the intake rate of an individual predator will be influenced by the number of other predators in a habitat (i.e., frequency-dependent energy intake). A vast array of models has been developed to describe such situations. The most basic is the ideal free distribution (IFD) (Fretwell and Lucas, 1969). This model assumes that (1) prey distribution is fixed, (2) animals forage to maximize energy intake rate, (3) resources are split evenly among predators, and (4) energy intake of each predator is reduced with the addition of another predator. Under this model, predators are expected to be distributed across habitats proportional to available food resources. This results in the basic prediction that intake rates of individuals in patches with high and low availability are identical due to the higher density of foragers in more productive patches. Although there is empirical support for this model in some non-elasmobranch situations (e.g., Tregenza, 1995), many factors may cause deviations from this distribution, including differences in competitive ability among foragers (see Tregenza, 1995) and predation risk (see Section 17.3.5.1).

Studies of elasmobranch foraging are inconclusive. If habitat patches of all prey densities are considered, eagle rays foraging on benthic invertebrates do not conform to IFD predictions but may conform to the predictions of the IFD at prey densities above a threshold level (Hines et al., 1997). Juvenile blacktip sharks within a coastal nursery in Tampa Bay, Florida, did not appear to distribute themselves across the nursery area in relation to prey density (Heupel and Hueter, 2002), and neither do bull sharks in the Shark River in Florida (Heithaus et al., 2009b); however, a mismatch between prey distribution and predator distribution does not necessarily indicate that factors other than food are important in habitat use decisions during foraging (see below). Finally, the IFD would not predict complete depletion of prey patches, as is observed in cownose rays foraging on bay scallops (Peterson et al., 2001). This result may be due to either of the following: (1) the energy gained by completely depleting bay scallops is greater than the energy gained by leaving early and searching for new prey patches (i.e., high search costs relative to rates of energetic return when prey are at low densities); or (2) although all bay scallops were removed, rays were still harvesting other resources in the patch (e.g., infaunal invertebrates), but bay scallop profitability was higher than other prey so they were consumed first, before the profitability of the habitat had dropped below that available in other habitats.

Antipredator decisions by prey are another important factor in habitat use decisions made by predators. This is especially likely for elasmobranchs that consume mobile prey, since these prey species show a diverse array of antipredator behaviors and can adaptively modify their behaviors in response to predation risk (for reviews, see Brown and Kotler, 2004; Lima, 1998a; Lima and Dill, 1990). When decisions made by predators are influenced by decisions made by prey, and *vice versa*, game theory can provide insights into optimal decisions of both players. Game-theoretic models of habitat use by predators and prey in systems with three trophic levels—predators (e.g., a shark), their prey (e.g., teleost fish), and food of the prey (e.g., benthic invertebrates)—have revealed counterintuitive results (Heithaus, 2001b; Hugie and Dill, 1994; Sih, 1998). In these situations, predators are actually predicted to be distributed across habitats in relation to the food of their prey (i.e., resources one trophic level removed) while prey are distributed across habitats relative to intrinsic habitat risk. This can result in predator distributions that do not relate well to those of their prey; therefore, even if the fitness of a predator is determined solely by energy intake rate, field observations may not find a match between predator distribution and prey density.

Tiger shark habitat use in Shark Bay, Western Australia, appears to be influenced primarily by the distribution of their prey species. Sharks showed a strong preference for shallow seagrass habitats (<4 m) over surrounding deeper waters covered by sand (6 to 15 m). These seagrass habitats have higher densities of potential prey, including dugongs, sea snakes, sea turtles, and birds (Heithaus et al., 2002a). Further, at finer scales, sharks prefer the edges of the shallow seagrass habitats, which have the highest abundance of prey, although prey vulnerability for most prey species is higher within the interiors of seagrass habitats (Heithaus et al., 2006). These results qualitatively match theoretical predictions of several foraging models including the IFD, a predator–prey game, or optimal foraging when frequency-dependent selection does not occur (Heithaus et al., 2002a, 2006).

Vertical migration can be considered a type of habitat use (Hugie and Dill, 1994; Iwasa, 1982) and has been observed in a variety of sharks, including cookie-cutter sharks, *Isistius* spp. (Jones, 1971); school sharks (West and Stevens, 2001); bigeye thresher sharks, *Alopias superciliosus* (Nakano et al., 2003; Weng and Block, 2004); juvenile white sharks (Weng et al., 2007a); and common thresher sharks, *Alopias vulpinus* (Cartamil et al., 2010, 2011). In many cases, such movements may mirror those of prey; for example, deep scattering layer organisms undergo daily vertical migrations from deeper waters during the day to shallower depths at night. Megamouth sharks, *Megachasma pelagios*, migrate with this layer, effectively keeping within the prey band (Nelson et al., 1997). These sharks appeared to migrate according to light level, following the 0.4-lux isolum (Nelson et al., 1997), but this is probably a proximate response that allows the sharks to maintain contact with prey patches. In fact, basking sharks display reverse diel vertical migrations (deeper during the night and shallower during the day) in habitats where their planktonic prey do the same, suggesting that factors other than light levels may also influence vertical migrations (Sims et al., 2005). In some species, habitat may also play an important role in vertical migrations. Sixgill sharks off Bermuda showed no obvious diurnal movements (Carey and Clark, 1995), but in Puget Sound, Washington, they were consistently found at greater depths during the day but maintained an association with the bottom (i.e., they moved along the slope of the bottom, not vertically through the water column) (Andrews et al., 2009). Habitat-related differences in diel vertical behavior has also been noted in the Greenland shark, with individuals in shallow areas of the St. Lawrence Estuary, Canada, displaying diel vertical movements, while an individual in a deeper portion of the estuary (>300 m) did not (Stokesbury et al., 2005). Temperature differences may also drive vertical migrations. By spending time resting in cooler deeper waters during the day and foraging in warmer shallower waters at night, small-spotted catsharks may be able to reduce their daily energy use (Sims et al., 2006a) (see Section 17.3.2.1).

Migration, where individuals move extremely large distances seasonally, can be considered an extreme case of habitat selection with preferred habitats widely separated. Seasonal migrations are common in elasmobranchs, which in many cases may be driven solely by temperature (climate) (Musick et al., 1993) because of conflicting demands of reproduction and foraging (see Section 17.3.5.2), or may be primarily food driven. Whale sharks respond to seasonal changes in resources and seasonally congregate in areas with abundant prey. Whale shark densities peak at Ningaloo Reef, Western Australia, during coral spawning (Gunn et al., 1999; Wilson et al., 2001); off the coast of Belize during spawning aggregations of snappers, *Lutjanus cyanopterus* and *L. jocu* (Heyman et al., 2001); and off the Yucatan Peninsula during plankton blooms (Hueter et al., 2008; Motta et al., 2010). Tiger sharks appear to congregate at islands offshore of the Western Australian coast when food abundance (fishing industry discards) peaks (Simpfendorfer et al., 2001b), and seasonal changes in the abundance of large size classes of tiger sharks in Shark Bay, Western Australia, coincide with changes in the abundance of dugongs, a high-quality prey item (Wirsing et al., 2007b).

In the Pacific, some tiger sharks travel thousands of kilometers before returning to French Frigate Shoals to feed on seasonally abundant fledgling albatrosses (Meyer et al., 2010). Movements of blue sharks from inshore to offshore waters at Catalina Island, California, may be due to movements of their prey, with sharks moving inshore in winter to feed on spawning schools of market squid, *Loligo opalescens* (Sciarrotta and Nelson, 1977; Tricas, 1979). White sharks in the eastern Pacific also make large-scale migrations, moving from coastal waters off California and Mexico to an offshore area approximately halfway between Baja California and Hawaii, which has been referred to as the white shark "café" and the "shared offshore foraging area" (Domeier and Nasby-Lucas, 2008; Jorgensen et al., 2010; Nasby-Lucas et al., 2009; Weng et al., 2007b). Whether this area is used primarily for foraging and whether white shark movements are driven primarily by foraging considerations, however, remain largely untested. White sharks in this area, however, do engage in oscillatory dives to depths of approximately 400 to 500 m, a practice that has been associated with foraging in pelagic sharks (e.g., Sepulveda et al., 2004), and ascent and decent rates do not match predictions of using oscillatory dives for energy conservation (Nasby-Lucas et al., 2009).

In the summer, salmon sharks, *Lamna ditropis*, aggregate at migration routes and near spawning sites of Pacific salmon in Prince William Sound, Alaska, and then disperse in autumn (Hulbert et al., 2005). Some sharks remain in Prince William Sound, while others travel thousands of kilometers to the southeast. The movements of sharks, including "focal foraging," "foraging dispersals," and "direct migration," appear to be driven by spatial and temporal variation in prey availability and quality, competition, reproductive considerations, and energetic trade-offs (Hulbert et al., 2005). Overall, more detailed studies of elasmobranch prey are needed to fully understand elasmobranch migrations.

Also, the influence of spatial scale on predator–prey behavioral interactions has been largely overlooked in studies of elasmobranchs with several notable exceptions. Given the high mobility of both elasmobranchs and their prey in many situations, this is likely to continue to be a challenging but growing and exciting avenue of research.

17.3.2.2 Search Patterns

The probability of prey encounter and capture can be further increased within a habitat by adopting an optimal searching tactic. In some cases, this may include remaining in one location, as in the case of ambush predators (see Section 17.3.3.1), or it may involve roving movements within a single location; for example, tiger sharks in Shark Bay, Western Australia, do not remain in one high-productivity habitat patch for an extended period but instead move among productive shallow seagrass patches (Heithaus et al., 2002a). This kind of pattern may also occur at larger scales. Tiger sharks in Hawaii were observed to display wide-ranging, aperiodic movements between locations and spent limited time at a given location; Meyer et al. (2009) suggested that this type of movement may be the result of sharks quickly losing the element of surprise at a location. White sharks have several searching tactics within and among seal colonies. Sharks make relatively directional travels between islands that might have pinniped prey, then more restricted movements close to the shore of islands that have high pinniped prey densities (Goldman and Anderson, 1999; Klimley et al., 2001; Strong et al., 1992, 1996). Goldman and Anderson (1999) suggested that larger white sharks may have more restricted patrolling areas than smaller sharks, but their sample size was insufficient to conclusively test this hypothesis. Similar observations have been made for white sharks in South Africa (Martin et al., 2009) and tiger sharks in Hawaii (Meyer et al., 2009).

Much recent work has focused on elucidating optimal search tactics in oceanic habitats where prey availability and predictability could vary substantially, but there are no obvious habitat differences that could be used to reliably locate prey. Not surprisingly, the optimal searching behavior is expected to change with variations in the distribution, abundance, and predictability or resources. In general, when prey are abundant, predators are predicted to adopt more random movements (Brownian motion), whereas in situations where prey are less abundant or unpredictable different movement tactics are optimal. Lévy flights are a type of random movement, typified by many short moves with less frequent long-distance displacements (Sims et al., 2008). Such movement can maximize search efficiency, and minimize energetic costs, when predators are confronted with an uncertain prey environment (Reynolds and Rhodes, 2009). Recent studies suggest that sharks, including predatory and planktivorous species, use Lévy flights that match the distribution of prey in oceanic environments and may switch between these movement tactics and Brownian motion as prey availability changes (Humphries et al., 2010; Sims et al., 2008). In the northeast Atlantic, for example, a blue shark, tracked across shelf (productive) and deep (less productive) waters, displayed dive patterns consistent with Brownian movements when in shelf waters and dive patterns consistent with Lévy flights in open ocean habitats (Humphries et al., 2010).

Vertical movements within a habitat may also be a form of searching behavior. Many species of sharks make regular movements up and down through the water column. The reason for these movements is still uncertain, and there are probably different factors that apply in certain situations, especially considering the prevalence of this behavior in space and time compared to the feeding periodicity of most sharks. Oscillatory swimming appears to be a general rule in pelagic species (e.g., Carey and Scharold, 1990; Cartamil et al., 2010, 2011; Holts and Bedford, 1993; Klimley et al., 2002). Oscillatory swimming in pelagic habitats may be used for (1) conserving energy, (2) thermoregulation, (3) obtaining olfactory information for foraging or navigation, or (4) detecting magnetic gradients for navigation (for a discussion of these mechanisms, see Klimley et al., 2002). In some species, this behavior appears to be associated directly with foraging. Whale sharks at Ningaloo Reef, Western Australia, make regular oscillations between the surface and the bottom and are probably searching for food throughout the water column (Gunn et al., 1999). Tiger sharks in Shark Bay, Western Australia, exhibit oscillatory swimming even when in shallow waters (3 to 10 m) (Heithaus et al., 2002a). This behavior appears to be a foraging tactic that aids in visual detection of air-breathing and benthic prey (Heithaus et al., 2002a) but may also aid in olfactory detection. The presence of continuous tail beats during descent suggests that this behavior is not simply for energy conservation (Heithaus et al., 2002a; Nakamura et al., 2011). Carey and Scharold (1990) suggested that blue shark dives were for both behavioral thermoregulation and foraging, with oscillations aiding in both visual and olfactory detection of prey. Vertical movements in blue sharks, however, may vary considerably within and among individuals (Carey and Scharold, 1990; Quieroz et al., 2010; Stevens et al., 2010). Finally, white sharks making movements above and below the thermocline may increase olfactory detection of whale carcasses (Carey et al., 1982).

17.3.2.3 Activity Levels and Patterns

For roving predators, increased activity levels can result in increased access to food through increased encounter rates with prey. Elasmobranchs can also increase the probability of both encountering and capturing prey by selecting an appropriate time of day to feed. Diel patterns of foraging have often been inferred by increases in the rate of movement (ROM) as measured by a tracking boat (Sundström et al., 2001) or swimming speed; however, these results must be viewed with caution. First, ROM does not necessarily reflect the swimming speeds of the elasmobranch because it does not account for the path of animals between fixes or currents (Sundström et al., 2001). Also, although increased rates of movement and swimming speeds may be indicative of foraging behavior, they may also represent transit movements between areas of a home range or between foraging sites. During summer, blue sharks patrol inshore waters during the night, then move in a relatively straight line to offshore waters during the day (Sciarrotta and Nelson, 1977). Juvenile lemon sharks also migrate between daytime and nighttime activity centers (Sundström et al., 2001), and animals in tidally influenced systems may ride the tides between areas (Carlisle and Starr, 2010; Medved and Marshall, 1983). Furthermore, species feeding on prey with extensive handling times are likely to show lower rates of movement when foraging than when traveling. Also, there is not always agreement between ROM data and feeding data obtained from stomach contents analysis. Although juvenile lemon sharks, for example, appear to become more active at night (Morrissey and Gruber, 1993a), there are no diel patterns in feeding (Cortés and Gruber, 1990); thus, increased movement rates should not be automatically interpreted as indicative of foraging activity. Also, the interpretation of ROM and swimming speeds should be done with caution unless combined with other technology to assess feeding behavior, such as ingested temperature transmitters (Goldman, 1997; Goldman et al., 2004; Klimley et al., 2001; Sepulveda et al., 2004), animal-borne video devices (Heithaus et al., 2001, 2002a), or gastric motility and pH loggers (Papastamatiou et al., 2007a,b) (see Chapter 9).

Elasmobranchs may have a large sensory advantage over their prey during crepuscular and nocturnal periods (see Chapter 12), and many studies have found evidence for increased foraging activity during these times. Scalloped hammerhead sharks in the Gulf of California spend their days in large groups swimming around seamounts and no feeding occurs. At dusk, these sharks disperse from the seamount into pelagic waters to feed (Klimley and Nelson, 1981, 1984; Klimley et al., 1988), apparently navigating along geomagnetic fields (Klimley, 1993). They appear to use the same seamounts repeatedly, probably as a reference point for their foraging excursions (Klimley et al., 1988), which would conform to a central-place foraging system. However, the sharks do not return to the seamount every day, which may be dictated largely by water temperature fluctuations (Klimley and Butler, 1988). Many gray reef sharks form daytime aggregations that occupy relatively limited areas then disperse at night, presumably to forage (Economakis and Lobel, 1998; McKibben and Nelson, 1986; Nelson and Johnson, 1980). Scalloped hammerhead pups have been observed forming groups during the day and making larger excursions at night (Holland et al., 1993) when most foraging occurs (Bush, 2003); however, many prey species of the pups are diurnal species that hide at night, and it is unclear when and how they capture these prey (Clarke, 1971). Several species of rays have been shown to become more active at night; for example, Hawaiian stingrays have higher ROM and cover larger activity spaces at night (Cartamil et al., 2003), and the highest ROM of round stingrays, *Urobatis halleri*, also occurred during the night (Vaudo and Lowe, 2006). Similarly, Pacific electric rays, *Torpedo californica*, are buried during the day but appear to actively seek prey at night (Bray and Hixon, 1978; Lowe et al., 1994) and attack potential prey items that are presented to them significantly faster at night (Lowe et al., 1994).

For visual predators, diurnal foraging may be the most common tactic. White sharks arrived at bait stations off southern Australia primarily during daylight hours, and sharks foraging on pinnipeds are probably diurnal because of their reliance on vision during prey detection and capture (Strong, 1996); however, white sharks at Año Nuevo, California, patrolled near seal haul-outs during both diurnal and nocturnal periods with equal frequency (Klimley et al., 2001). Juvenile mako sharks tend to make deeper dives during the day, and in many cases successfully capture prey during these dives (Sepulveda et al., 2004). Given their foraging tactic of using their long upper caudal lobe to stun prey (Aalbers et al., 2010), common thresher sharks are also likely to be highly dependent on vision during foraging and, similarly, make deeper vertical movements during the day (Cartamil et al., 2010, 2011). Catch rates of tiger sharks were substantially higher during daylight hours in Shark Bay, Western Australia (Heithaus, 2001c), despite previous suggestions that tiger sharks were nocturnal (Randall, 1992). Given the flexibility observed in foraging tactics of many elasmobranchs, it is likely that there is geographic variation in diel patterns of foraging and movement, so generalizations from one or a few studies should be viewed with caution. Cueing in on the bioluminescence produced by prey during the night, or at depth, has been suggested for a number of species, including Pacific angel sharks (Fouts and Nelson, 1999)

and blue sharks (Tricas, 1979). Also, based on its diet and visual capabilities, which are tuned to bioluminescent light, the deep-water blackmouth dogfish, *Galeus melastomus*, likely uses bioluminescence to detect and capture prey, whereas the sympatric small-spotted catshark does not (Bozzano et al., 2001).

Some species do not show obvious diel patterns. Juvenile lemon sharks did not show a diel pattern in the number of burst-speed events that may be associated with prey capture or possibly predator avoidance (Sundström et al., 2001), and there are no diel patterns in the swimming behavior of whale sharks (Gunn et al., 1999) or activity spaces of cownose rays (Collins et al., 2007b). And, in some cases, individuals from the same population may show different diel patterns, as was observed in juvenile bull sharks in the Caloosahatchee River, Florida (Ortega et al., 2009).

17.3.3 Capturing and Consuming Prey

17.3.3.1 Stalking and Ambush

Stalking and ambush are two tactics elasmobranchs may use to capture swift or large prey. Stalking is the process by which a predator attempts to reduce the distance between itself and its intended prey without being seen (stealth) before a rapid chase over the final distance. Ambush predators conceal themselves and allow the prey to approach before a rapid attack. For both stalking and ambush tactics a close, undetected, approach to prey is critical to success.

Sevengill sharks have been observed swimming slowly near prey and then suddenly making a speed burst to capture small leopard sharks that had not reacted to the predator (Ebert, 1991). Other ambush tactics include selecting turbid and low-light waters and slowly gliding up from below prey so they are not detected (Ebert, 1991). In Shark Bay, Western Australia, tiger sharks may use oscillatory swimming as a stalking tactic; they are able to closely approach benthic prey items when descending from above (Heithaus et al., 2002a). Oscillatory swimming patterns have also been observed in Pacific sleeper sharks and broadnose sevengill sharks and may allow these species to ambush the fast moving prey found in their diets (Barnett et al., 2010b; Hulbert et al., 2006). When swimming along the bottom or lower in the water column, sharks would have a better chance of surprising prey at the surface. White sharks also use a stalking approach, staying close to the bottom and then rushing to the surface to capture unsuspecting prey silhouetted above (Martin et al., 2005; Strong, 1996; Tricas and McCosker, 1984). The use of this tactic has also been proposed for sevengill sharks (Ebert, 1991), tiger sharks (Heithaus et al., 2002a), and blue sharks (Carey and Scharold, 1990).

Staying close to the bottom, or below prey, is an effective stalking tactic for several reasons. First, approaching from below gives a predator a detection advantage over prey where prey have limited visibility into the water due to light attenuation and predators have better visual detection capabilities of silhouettes (Strong, 1996). An approach from below surface prey also is advantageous by limiting escape routes of prey (Strong, 1996). Finally, by staying close to the bottom, an elasmobranch may be camouflaged against the substrate or darker, deeper waters (Goldman and Anderson, 1999; Heithaus et al., 2002a; Klimley, 1994). One interesting aspect of white shark attacks is their tendency to attack floating objects that are not prey. Hunting success of stalking predators such as white and tiger sharks relies on a close approach and attack before being detected because of these species' limited speed and maneuverability relative to their prey (Heithaus et al., 2002a; Strong, 1996). The costs of waiting to gather more information about prey identity (in terms of probability of being detected) likely outweigh the costs of attacking nonprey items at the surface.

Ambush predation is widespread in elasmobranchs. Crypsis, including hiding as well as body coloration, can enhance the efficiency of ambush predators. Some elasmobranchs lie in wait buried in the sand or concealed in coral or rock caves and crevices. Pyjama sharks, *Poroderma africanum*, have been observed ambushing squid by concealing themselves among squid eggs (Smale et al., 1995). When squid had habituated to the shark's presence and returned to the site, the shark would attack the squid as they approached the seafloor to lay eggs. Diamond rays, *Gymnura natalensis*, ambush spawning squid by burying themselves in the sand near egg beds (Smale et al., 2001), and torpedo rays ambush their prey by jumping from the bottom (Belbenoit, 1986; Bray and Hixon, 1978; Lowe et al., 1994; Michaelson et al., 1979). Pacific angel sharks lie in body pits to ambush prey and appear to use the same ambush sites repeatedly, occasionally making longer movements that may be to select new ambush sites (Fouts and Nelson, 1999). Ambush sites were primarily adjacent to rock–sand interfaces or patch reefs, which may serve as refugia for prey species and thus may maximize encounter rates with prey. Freshwater rays, *Potamotrygon* spp., may also use an ambush tactic where they charge into shallow waters to attack concentrated prey (Garrone-Neto and Sazima, 2009). Finally, Pacific sleeper sharks appear to rely on cryptic coloration, the cover of darkness, and slow vertical oscillations to avoid detection by prey (Hulbert et al., 2006).

17.3.3.2 Prey Herding and Manipulation

Although some swift species such as mako and salmon sharks may be capable of swimming down a meal, others rely on manipulating prey behavior to make capture more energetically efficient. Some species of elasmobranchs are thought to lure prey. Megamouth sharks might be able to attract prey toward the mouth with luminescent tissue along the upper jaw (Compagno, 1990; Diamond, 1985; also see Chapter 6 of this volume). Cookie-cutter sharks, *Isistius brasiliensis*, may be squid mimics (Jones, 1971) and lure a host close enough to attack. Widder (1998) expanded this hypothesis further, suggesting that the "collar" of nonluminescent tissue on the underside of the cookie-cutter would stand out from the general luminescence of deep-sea waters, which would mimic the search image of pelagic predators approaching from below. Myberg (1991) suggested that the white tips on the fins of oceanic whitetip sharks, *Carcharhinus longimanus*, function to lure fast-swimming prey close enough to be successfully captured. Given the generally good water visibility in which oceanic whitetips are found, potential prey might be able to distinguish a predator from a great distance, which would make lures less advantageous. Although untested, these luring hypotheses are worthy of experimental tests. A first step would be to determine if prey species (such as tuna and mackerel for whitetips) will respond to shark models in the ways predicted by these hypotheses.

Some elasmobranchs may take advantage of antipredatory behavior of prey in order to increase feeding efficiency. Blacktip reef sharks, *Carcharhinus melanopterus*, have been observed in groups chasing small teleosts up onto the shoreline, then beaching themselves to feed on the fish (Wetherbee et al., 1990). Sharks in pelagic environments feed on baitfish herded into tight schools. For example, silky sharks, *Carcharhinus falciformis*, will feed from schools herded together by bottlenose dolphins, *Tursiops truncatus* (Acevedo-Gutiérrez, 2002).

17.3.3.3 Prey Debilitation

Rather than attempting to capture prey immediately, some species debilitate it before they consume it. Thresher sharks use the greatly exaggerated upper lobe of their caudal fin to hit small fish, killing or stunning them (Aalbers et al., 2010). A great hammerhead shark was observed using its head to hit a fleeing stingray into the bottom, then pinned the ray to the bottom with its head so it could debilitate it with bites to both of the ray's pectoral fins (Strong et al., 1990). Similarly, a great hammerhead was observed to immobilize a spotted eagle ray, *Aetobatus narinari*, by initially biting the ray's pectoral fins (Chapman and Gruber, 2002). The largetooth sawfish, *Pristis microdon*, uses its rostrum to pin prey to the substrate (Wueringer et al., 2009) and is known to stun prey by hitting them with swipes of their saw (Breder, 1952). White sharks may also use a prey debilitation tactic that involves the shark making a first bite followed by a release of the prey until it bleeds to death

(Tricas and McCosker, 1984). This "bite and spit" tactic has been interpreted as sharks reducing their probability of being injured by their prey (McCosker, 1985; Tricas and McCosker, 1984), but it is still unclear whether this tactic is widely used (Klimley, 1994; Klimley et al., 1996a; Martin et al., 2005). Variation in the size and species of pinniped prey taken may partially explain variation in support for this hypothesis across studies. Electric rays may either swim over their prey when actively searching or jump from the bottom when they are concealed and envelop their prey within their disk. They then discharge an electric current that debilitates the prey, after which they can orient the prey and ingest it (e.g., Belbenoit, 1986; Bray and Hixon, 1978; Lowe et al., 1994). The electric ray, *Torpedo ocellata*, is able to emit electric discharges immediately after birth but the voltage increases substantially in the first 3 weeks (Michaelson et al., 1979).

17.3.3.4 Benthic Foraging

Benthic foraging is found in a diverse array of elasmobranchs. Some merely capture prey along the bottom, but others dig or excavate the bottom to capture infaunal prey. Epaulette sharks, *Hemiscyllium ocellatum*, have been observed burying their bodies up to the first gill slit to capture prey (Heupel and Bennett, 1998), and some rays are able to forage on infauna buried deep in the substrate (e.g., Matern et al., 2000; Smith and Merriner, 1985), although it is unclear to what degree they are able to discriminate prey before excavating them (Tillet et al., 2008). When foraging on infaunal prey, rays produce pits that leave a record of their foraging activity (Gregory et al, 1979; Howard et al., 1977). Freshwater stingrays (Potamotrygonidae) use both this excavation tactic as well as a "picking" tactic, where they remove items on the surface of the substrate or substrates that extend up into the water column (Garrone-Neto and Sazima, 2009), and some species will even flip over rocks (~1 kg) in search of prey (Rosa et al., 2010). See Chapter 6 for more details on benthic foraging tactics in sharks and rays.

17.3.3.5 Batch Feeding

Whereas most species of elasmobranchs are raptorial predators, consuming a single prey item at a time, a few species have evolved into batch feeders (see Chapter 6 for detailed descriptions of filter-feeding mechanisms in elasmobranchs). These include the whale, basking, and megamouth sharks and manta rays. There are several tactics used in batch feeding. The most common is ram-ventilation filter feeding, where the predator swims through prey patches with its mouth open, straining prey from the water. This tactic is observed in whale sharks, basking sharks, and manta rays. Whale sharks sometimes also use a gulping tactic (Heyman et

al., 2001; Motta et al., 2010; Nelson and Eckert, 2007), and megamouth sharks appear to use engulfment feeding similar to that seen in balaenopterid whales (Nakaya et al., 2008).

For years, it was thought that basking sharks hibernated over winter because of observations of sharks without gill rakers (Francis and Duffy, 2002). Satellite tracking work, however, has shown that the winter disappearance of basking sharks is associated with seasonal migrations (Sims et al., 2003; Skomal et al., 2009). Recent work also suggests that basking sharks use search tactics structured across multiple scales to find and exploit prey patches (Sims et al., 2006b) and that sharks are able to track prey vertically in the water column (Sims et al., 2005). Basking sharks are selective filter feeders (Sims and Quayle, 1998), but they forage at swimming speeds slower than predicted by optimal filter-feeding models (Sims, 2000). The mismatch between theoretical predictions and observed swimming speed is likely due to higher drag incurred by basking sharks than the small teleost fishes for which model predictions were developed (Sims, 2000).

17.3.3.6 Ectoparasitism

Although we typically think of parasites as small organisms living in or on the bodies of another organism (see Chapter 18), there is an elasmobranch feeding tactic that is more similar to parasitism than it is to predation. This tactic, sometimes called *ectoparasitism* (Heithaus, 2001a; Heithaus and Dill, 2002b; Long and Jones, 1996), involves a shark gouging a mouthful of tissue from a "host." Cookie-cutter sharks are ectoparasites of many large teleosts and marine mammals and even other sharks (Diamond, 1985; Heithaus, 2001a; Hiruki et al., 1993; Jones, 1971; Papastamatiou et al., 2010b). A cookie-cutter shark is able to create a vacuum with its tongue and fleshy lips, then spin around using its teeth to remove a plug of flesh (Jones, 1971; Shirai and Nakaya, 1992). It has been suggested that cookie-cutter sharks may make use of bioluminescence to lure fast-moving species to parasitize (Widder, 1998). Portuguese dogfish, *Centroscymmus coelolepis*, may also parasitize marine mammals in this manner (e.g., Ebert et al., 1992), and the dentition, tongue, and fleshy lips of the kitefin shark, *Dalatias licha*, suggest that this species may also use an ectoparasitic tactic (Clark and Kristof, 1990), although differences in jaw morphology make it unlikely that the kitefin shark is able to feed in the same way as the cookie-cutter shark (Shirai and Nakaya, 1992). Small carcharhinid sharks may also use an ectoparasitic strategy. Indo-Pacific bottlenose dolphins in Shark Bay, Western Australia, have been observed with scars inflicted from small shallow-water sharks (Heithaus, 2001d); however, the generality of this foraging tactic in such species is unknown.

17.3.3.7 Scavenging

Scavenging, at least opportunistically, is probably one of the most common feeding tactics of elasmobranchs (e.g., Compagno, 1984a,b; Smith and Merriner, 1985), and for some species or age classes it is probably the primary feeding tactic; for example, large great white sharks are thought to adopt a diet composed largely of cetacean carcasses in the Atlantic Ocean (Carey et al., 1982; Pratt et al., 1982). Some observations suggest that white sharks may defend carcasses from both conspecifics and other species (e.g., Long and Jones, 1996; McCosker, 1985; Pratt et al., 1982), but white sharks have been observed scavenging a whale carcass concurrently with tiger sharks (Dudley et al., 2000). Although most observations of white sharks scavenging cetacean carcasses have been for large individuals, recent observations of young of the year and juvenile white sharks scavenging a whale carcass (Dicken, 2008) suggest that all size classes may adopt this tactic opportunistically. The diets of Pacific sleeper sharks may also feature large amounts of carrion. Sleeper sharks are often observed at whale-falls (Lundsten et al., 2010; Smith and Baco, 2003), and 70% of identifiable marine mammals from shark diets were scavenged (Sigler et al., 2006).

17.3.3.8 Group Foraging

Group foraging may increase an elasmobranch's ability to gather resources. Groups may form for reasons other than foraging, such as reproduction (see Chapter 10) or

reducing predation risk (see Section 17.2.2.4), forcing individuals to forage in close proximity to each other and resulting in intraspecific competition and reduced energy intake rates (see Section 17.3.4). Cownose rays forage in groups (Collins et al., 2007a; Smith and Merriner, 1985), but it is unclear why these groups form. Similarly, pink whiprays, *Himantura fai*, are often observed swimming in groups (Vaudo and Heithaus, 2009), and the rays in these groups will forage together over both sandflats and seagrass beds (M. Heithaus, pers. obs.) (Figure 17.2). In most elasmobranchs, groups that form likely are a result of mutual attraction to prey resources. Groups of both shortnose spiny dogfish, *Squalus megalops*, and smoothhounds, *Mustelus mustelus*, have been recorded at spawning aggregations of chokka squid, *Loligo vulgaris reynaudii*, but little foraging was observed, and it is possible that groups formed for nonforaging reasons as well (Smale et al., 2001). Sevengill sharks will aggregate in large groups around potential food (Ebert, 1991). Finally, white sharks at Año Nuevo, California, appeared to hunt in relatively close proximity. Although movements are most consistent with an individual foraging tactic, sharks probably remained in relatively close proximity to take advantage of kills by other individuals (Klimley et al., 2001).

Some groups may increase the *per capita* intake of the individuals in the group through increased detection of prey or increased probability of prey capture and death. Cooperative foraging has not been conclusively shown in elasmobranchs but has been suggested for several species (Motta and Wilga, 2001); for example,

FIGURE 17.2

(See color insert.) A group of pink whiprays, *Himantura fai*, swimming over seagrass in Shark Bay, Western Australia. These groups will forage together over sand and seagrass, but it is not clear whether groups form to enhance foraging or for other reasons, such as reducing predation risk. (Photograph by Kirk Gastrich.)

sevengill sharks appear to cooperatively forage on seals, cetaceans, and large rays. These groups surround a potential prey item and then slowly circle until they all converge for the kill (Ebert, 1991). Cooperation has been defined as "an outcome that—despite individual costs— is 'good' in some appropriate sense for the members of the group, and whose achievement requires collective action" (Mesterson-Gibbons and Dugatkin, 1992), or, more recently, as "all interactions or series of interactions that, as a rule (or 'on average'), result in net gain for all participants" (Noë, 2006). Sevengill sharks appear to conform to this definition because the outcome is "good" for the group in that larger prey is taken than would normally be available. The cost, or whether there is a specific interaction that enhances energy intake above what would occur with simple attraction to the same resource, is somewhat unclear; however, there may be lost foraging opportunities on other prey while large prey are subdued (Heithaus and Dill, 2002b), or interactions are required to successfully capture large prey. Further studies would be necessary to determine if this is, indeed, an example of cooperative foraging or merely mutual attraction to resources.

Shark groups are often found with groups of tuna and dolphins in pelagic waters (Au, 1991; Leatherwood, 1977). In the eastern tropical Pacific sharks are very common around tuna schools that have congregated around logs (Au, 1991). In the Gulf of Mexico, many pods of dolphins are also followed by sharks (Leatherwood, 1977), and oceanic whitetip sharks are found in association with short-finned pilot whales, *Globicephala macrorhynchus*, off Hawaii (M. Heithaus, pers. obs.). In the first two cases, sharks may attack tuna or dolphins or they may be following to take advantage of prey detection capabilities of dolphins and tuna. Because of the body size differences between pilot whales and the smaller whitetip sharks that follow them, shark predation on whales is extremely unlikely and the sharks may follow whales for foraging reasons.

17.3.4 Competition

Competition is an almost ubiquitous aspect of elasmobranch life. This competition may be intra- or interspecific and may take the form of exploitative or interference competition. In exploitative competition, the consumption of a prey item by one individual removes it from possible consumption by another. Interference competition may take several forms, where individuals actively exclude others from prey resources (contest competition) or merely get in the way of other foragers, reducing foraging efficiency. Extreme cases of interference competition include food stealing (kleptoparasitism) and the killing of competitors (intraguild predation) (see Section 17.3.5.3).

Few studies have explicitly examined competition in elasmobranchs, although several studies have examined dietary overlap. High values of dietary overlap among elasmobranchs have been found in the North Sea (Ellis et al., 1996), Apalachicola Bay, Florida (Bethea et al., 2004), the Hawaiian Islands (Papastamatiou et al., 2006), and among sharks and between sharks and dolphins off the coast of South Africa (Heithaus, 2001a). High levels of dietary overlap are often used to infer competition, although other factors must be considered. The use of broad taxonomic categories can lead to overestimations of dietary overlap; for example, broad dietary categories suggested moderate dietary overlap among dusky, *Carcharhinus obscurus*; whiskery, *Furgaleus macki*; and gummy sharks in Western Australia, but a more detailed examination of prey revealed that each species had different feeding habitats (Simpfendorfer et al., 2001a). Partitioning habitat may also lessen competition. Juvenile carcharhinids in Apalachicola Bay, which had high values of dietary overlap, showed low overlap of habitat (Bethea et al., 2004). Similarly, Papastamatiou et al. (2006) found an inverse relationship between abundance of sandbar and gray reef sharks along a habitat gradient. But, even high dietary and habitat overlap do not necessarily imply competition because resources may not be limiting. In the sandflats of Shark Bay, Western Australia, a diverse batoid community shows considerable overlap both in habitat use (Vaudo and Heithaus, 2009) and diets (Vaudo and Heithaus, 2011). The lack of spatial or resource partitioning among species could be due to the presence of intact top predator (tiger shark and hammerhead shark) populations in Shark Bay that keep batoid densities at levels below those where competition would be predicted to lead to resource partitioning. Prey pulses can also homogenize diets. Such an example appears to take place during the sardine run off southern Africa. Blacktip, dusky, and copper sharks in this area have high dietary overlap (Heithaus, 2001a). All of these species are present during the sardine run, and sardines dominant the diets of these sharks (Cliff and Dudley, 1992; Dudley and Cliff, 1993; Dudley et al., 2005). Even white sharks make use of this seasonal abundant resource (Cliff et al., 1989). The high proportion of sardines in the diets of these sharks and dolphins drives many of the high diet overlap values in this system (Heithaus, 2001a). Because this resource is super abundant it is possible that there is little impact of apparently high dietary overlap in these species (Heithaus, 2001a).

Qualitative comparisons suggest that competition is common. Gray reef sharks engage in competition for bait, but there is no intraspecific aggression (Nelson and Johnson, 1980). Silky sharks have been observed competing with bottlenose dolphins over schooling fish off the Pacific coast of Costa Rica (Acevedo-Gutiérrez,

2002). As the number of sharks increased, the intake rate of dolphins appeared to decrease. Also, large beaked skates, *Dipturus chilensis*, compete with southern sea lions, *Otaria flavescens* (Koen Alonso et al., 2001). Intraspecific competition may explain the low foraging success of white sharks on fur seals in the area where the highest number of attacks take place (Martin et al., 2005). In some situations, competition may be relatively weak; for example, there does not appear to be interference competition among eagle rays foraging on bivalves even at the highest foraging densities, but this may be due to generally low densities of rays (Hines et al., 1997).

The implications of competition within and among elasmobranch species and between elasmobranchs and other taxa are largely unknown; however, competition appears to be important in determining the species composition and abundance of several elasmobranch communities (see Section 17.4). We know little about how elasmobranchs respond to competition behaviorally or over short time scales, but it is probably important in determining habitat use patterns and structuring inter- and intraspecific interactions. Theoretical extensions of the IFD, supported by empirical tests, suggest that if individuals differ in their competitive ability (i.e., the division of resources is not equal among individuals), then the distribution of animals may deviate substantially from that of their prey, with final distributions being dictated partially by individuals' abilities to monopolize resources (Tregenza, 1995).

Territoriality is one way that competition can be manifested, with individuals defending food resources. Currently, there is no evidence for territoriality in elasmobranchs (e.g., Klimley et al., 2001; Nelson et al., 1986), which may be due to the indefensibility of most food resources; however, size-based dominance hierarchies may exist in many shark species. Large sevengill sharks displace smaller conspecifics from baited situations (Ebert, 1991), and large white sharks may displace smaller ones from whale carcasses (e.g., Long and Jones, 1996; Pratt et al., 1982).

Kleptoparasitism, or food stealing, is an extreme case of interference competition that appears to be widespread in sharks. Sevengill sharks that have captured relatively large-bodied prey (e.g., smaller sharks) and not consumed it quickly will have some of their prey taken by conspecifics (Ebert, 1991). White sharks will compete over pinniped prey, and the shark that makes a kill may be driven away from it (e.g., Klimley et al., 1996b, 2001; Martin et al., 2005). Klimley et al. (1996b) argued that tail-slapping behavior and, in some cases, breaches observed near kills are displays directed at conspecifics that are competing for the carcass. Sharks may also kleptoparasitize other species. Hawaiian monk seals, *Monachus schauinslandi*, foraging on banks are often followed by carcharhinid sharks that attempt to capture prey flushed out by seals and may even attempt to steal food captured by seals (Parrish et al., 2008).

17.3.5 Foraging Trade-Offs

17.3.5.1 Foraging–Safety Trade-Offs

All organisms face trade-offs, which are inevitable because time and resources are limited and demands are often in conflict. One trade-off faced by many taxa is that between foraging and avoiding predators (for reviews, see Brown and Kotler, 2004; Lima, 1998a; Lima and Dill, 1990). This is because the most energetically productive habitats are often the most dangerous, and behaviors that increase foraging efficiency (e.g., increased activity levels) often increase the risk of being killed by a predator (Lima, 1998a,b; Lima and Dill, 1990). Because most species of elasmobranchs are both predators and prey, an energy intake–predation risk trade-off is certainly important for many species, but studies specifically addressing this tradeoff generally are lacking for elasmobranchs.

One way to test for the existence of food–risk trade-offs is to measure the giving up density (GUD) of foraging individuals. The GUD is the density of food remaining in a patch at the time an individual, or group of individuals, ceases foraging and abandons the patch (Brown, 1988, 1992b, 1999; Brown and Kotler, 2004). GUDs should be greater in habitats with higher risk than those with lower risk because the marginal gain of continued foraging in a high-risk habitat does not outweigh the benefits of continued foraging at low food densities; however, the exact GUD in a patch will also be influenced by food availability in other patches because time spent foraging represents lost foraging opportunities in other patches (Brown, 1988, 1992b, 1999; Brown and Kotler, 2004). This prediction has been supported in a number of species foraging on immobile prey, such as granivorous rodents, squirrels, ungulates (Berger-Tal and Kotler, 2010; Brown and Kotler, 2004), and small carnivores (Mukherjee et al., 2009). There are currently no published studies of GUDs in elasmobranchs. Although measuring GUDs is likely to be very difficult for species that consume mobile prey, studying this parameter may be useful in studies of benthic foragers. In marine communities where GUDs are not appropriate, such as most elasmobranchs, adaptations of the IFD can help test hypotheses about foraging–risk trade-offs (Heithaus et al., 2007b). Foragers are predicted to be distributed across habitats proportional to food supply when risk is low or non-existent, but their abundances relative to their food (e.g., the proportion of foragers in a habitat divided by the portion of food resources in a habitat) should increase in safer habitats and decrease

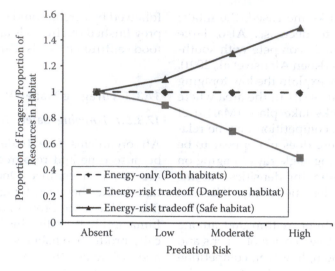

FIGURE 17.3
Conceptual model of how the ideal free distribution (IFD) can be used to estimate responses to predation risk in a two-habitat system (for details, see Heithaus et al., 2007b). The density of foragers across habitats corrected for food supply is expected to be close to 1.0 under an IFD when foragers do not respond to predators (dashed line). As predation risk increases, deviations from base IFD predictions can provide insights into perceived risk. In dangerous habitats, there are fewer foragers than expected based on food supply (gray line), whereas foragers are more abundant than predicted by food alone in safe habitats (solid black line). The shape and slope of the relationship between risk and investment in antipredator behavior are largely unexplored.

in dangerous habitats as the overall risk of predation increases. By monitoring forager distributions relative to that of their food as predation risk changes, it is possible to examine the effects of predation risk on forager habitat use (e.g., Heithaus and Dill, 2006; Heithaus et al., 2007b, 2009a; Wirsing et al., 2007a,c) (Figure 17.3).

It is possible to make predictions about how predation risk should modify habitat use decisions of foraging animals. First, there are various strategies that elasmobranchs might reasonably use to trade off risk with foraging (for reviews, see Brown, 1992a; Brown and Kotler, 2004). First, it is possible that animals must attain only some minimum energy (maintenance costs) to survive and any additional energy does not result in fitness increases. In this situation, they maximize their safety value as long as they meet this minimum energy requirement. When reproduction is relatively infrequent, individuals must survive for extended periods to realize increased reproductive output. In this situation, animals may try to maximize the product of safety (probability of survival) and the number of surviving descendants (which may be approximated by energy intake is some cases) (Brown, 1992a). Some studies suggest that animals may adopt a strategy of maximizing energy intake over their lifetimes by minimizing the risk of predation in a habitat (μ) divided by the energy intake in that habitat (f) (μ/f rule) (Gilliam and Fraser, 1987). There is empirical support for this prediction (Gilliam and Fraser, 1987); it is likely to be a good predictor of the behavior of juvenile animals that are faced primarily with the challenges of growth, and survival

and reproductive decisions are unlikely to cause deviations from such optimal habitat use. The μ/f rule, as well as incorporating predation risk into IFD-based models (e.g., Heithaus et al., 2007b; Moody et al., 1996), suggests that animals should forage in relatively less productive habitats if they are safer; however, some individuals may select higher risk habitats to take advantage of greater growth options there (e.g., Abrahams and Dill, 1989; Gilliam and Fraser, 1987; Lima, 1998a; Lima and Dill, 1990). Often, individuals found in high-risk, high-reward habitats are those that are in low condition (i.e., risk predation to reduce the risk of starvation or other state-dependent fitness loss) (e.g., Clark, 1994; Heithaus et al., 2007a; Heithaus et al., 2008a; Kotler et al., 2010; Sinclair and Arcese, 1995) or are age–sex classes that realize the greatest fitness benefits of enhanced energy intake (e.g., Corti and Shackleton, 2002; Cresswell, 1994).

Most models of food–risk trade-offs assume that predators are behaviorally inert and cannot modify their distributions in accordance with decisions made by their prey, which has been a major oversight (Lima, 2002). When both predators and prey can move freely, a predator–prey game ensues, and game theoretical modeling can help predict optimal behaviors of both predators and prey (Dugatkin and Reeve, 1998). When predator–prey games are considered, habitat selection by the middle predator (e.g., small sharks or rays) may not be driven by the amount of food in a habitat; instead, they are predicted to distribute themselves across habitats proportional to the intrinsic habitat risk (e.g., Hugie and Dill, 1994; Sih, 1998). If, however, top predators have diverse diets (i.e., alternative prey to the prey species of interest), then prey habitat selection may more closely approximate a situation where predators are behaviorally fixed (Heithaus, 2001b).

A trade-off between food availability and predation risk may be important in habitat use decisions of juvenile sharks at multiple spatial scales, including the use of nursery areas and movement patterns within these nurseries. Reducing predation risk seems to be of primary importance at both of these spatial scales (see Section 17.2.2.1); for example, a study of juvenile blacktip shark movements within a Florida nursery failed to find a link between shark movements and prey abundance (Heupel and Hueter, 2002). With current techniques, however, it is difficult to measure the relative importance of food and safety in determining habitat use of elasmobranchs because of their relatively low energetic requirements and the possibility that some species confine feeding activity to short time periods followed by long periods without foraging (see Chapter 8). Juvenile bull sharks in a coastal estuary of Florida are captured primarily in low-salinity areas but have stable isotopic signatures that are indicative of foraging in marine food webs, which would require movements of at least 10 km downstream (Matich et al., 2011). This pattern likely is due to juvenile bull sharks spending most of their time in low-salinity waters that are low risk and have lower food abundance but making occasional trips into the high predation risk habitats at the mouth of the estuary where food resources appear to be more abundant (Heithaus et al., 2009a; Matich et al., 2011). Similarly, several species of rays in Shark Bay, Western Australia, spend considerable time in shallow sand habitats (Vaudo and Heithaus, 2009), despite limited prey availability there (Vaudo, 2011) and diets suggestive of foraging in deeper, more dangerous, seagrass habitats (Vaudo and Heithaus, 2011). In general, individuals are likely to modify their habitat use decisions depending on which behavioral state they are in (e.g., Heithaus and Dill, 2002a), and most studies of elasmobranch habitat use employ techniques that rely on determining average habitat use of individuals over relatively long time intervals and are not able to assess an animal's behavior at any particular time (see Chapter, 19). For these reasons, our knowledge of how elasmobranchs might trade off food and safety is still limited, and studies that can determine habitat use during different behavioral states (e.g., foraging vs. non-foraging) will be of great value (for a discussion of advances in biologging technology and applications to studies of foraging ecology, see Sims, 2010; Chapter 9 of this volume).

The activity level of a forager is also subject to a trade-off between energy intake and risk of death. This occurs because increased activity rates are generally associated with both increased food intake and increased probability of predation. Theoretical models suggest that the optimal activity level depends on the relationship between activity rate and feeding rate and predation risk, the density of foragers relative to their prey, and the state of the individual (e.g., Walters and Juanes, 1993; Werner and Anholt, 1993). Game theoretical models of activity levels of predators and prey reveal results strikingly similar to habitat use games. Predator activity level is predicted to parallel that of the availability of *their prey's food* while prey should maintain constant activity levels (Brown et al., 2001).

Foraging under predation risk may lead to changes in diet selectivity, but the nature of this change depends on how prey items influence the risk of death (Houtman and Dill, 1998). In general, animals will increase their acceptance of food items associated with lower predation risk (e.g., items with relatively low handling times), which may lead to increased diet selectivity (if profitable prey are safer), reduced selectivity (if less profitable prey are safer), or no change in selectivity (if prey are of similar risk) (Houtman and Dill, 1998). A behavioral optimization model based on empirical data suggests that two species of pinnipeds reduce their use of energetically profitable prey when faced with the risk of predation from sleeper sharks in the depth strata where these prey are found (Frid et al., 2007, 2009). Although we are unaware of examples of such behavior of elasmobranch mesopredators, it is likely a common phenomenon.

Vigilance and other antipredator behaviors, such as hiding, are often mutually exclusive with foraging. Optimal vigilance levels will vary with risk because individuals that overinvest in vigilance are likely to realize lower fitness than those that do not because of reduced energy intake, whereas those that underinvest in vigilance are more likely to be killed by predators (Brown et al., 1999). Vigilance levels may also be influenced by temporal variation in the risk of predation (i.e., pulses of high and low risk). Counterintuitively, vigilance during high-risk periods will be *lower* when the proportion of time spent at high risk is greater (Ferrari et al., 2009; Hamilton and Heithaus, 2001; Lima and Bednekoff, 1999). This is because an animal often at high risk cannot afford to invest heavily in antipredator behavior during periods of high risk without seriously compromising energy intake. Such a system may occur for elasmobranchs in shallow waters because periods of higher water may allow increased access to predators.

Finally, food–risk trade-offs likely influence elasmobranch group sizes. Generally, predation risk selects for larger groups while increased foraging competition tends to select for smaller groups, and observed group sizes often reflect a balance of these conflicting selective pressures (Bertram, 1978; Lima and Dill, 1990). There currently are no data on factors influencing the size of elasmobranch groups.

17.3.5.2 Foraging–Reproduction Trade-Offs

Sometimes animals must trade off gathering energy efficiently or growing with securing mates or investing in reproduction, and it is possible that mating systems are actually the result of tactical decisions in response to ecological conditions rather than simple fixed strategies (e.g., Lott, 1984; Siems and Sikes, 1998). Dental sexual dimorphism appears to be one example. Although differences in male and female dentition are observed in a number of species, there are seasonal changes in this dimorphism within Atlantic stingrays, *Dasyatis sabina*. Female dentition is stable year-round, but males possess recurved cuspidiform dentition during the breeding season and female-like molariform dentition outside of the breeding season (Kajiura and Tricas, 1996). The molariform dentition is likely the most efficient for feeding but is not well suited for reproduction by males, which must use their teeth to grasp females during reproduction (Kajiura and Tricas, 1996); therefore, male changes in dentition represent a trade-off between foraging and reproductive success. Further studies are required to determine whether seasonal dental sexual dimorphism is common in elasmobranchs with well-defined mating seasons.

Reproductive effort may be reduced when predation risk is high, and the reproductive habitat use and reproductive tactics used may be influenced by risk (Lima and Dill, 1990). Sex-specific habitat use may be the result of foraging–reproduction trade-offs. Klimley (1987) suggested that female scalloped hammerhead sharks move offshore and adopt a pelagic diet at a smaller size than males, which results in higher growth rates for females. Cailliet and Goldman (see Chapter 14), however, suggest that growth rates for male and females are intrinsically different for many species. This raises the question of why juvenile males would not make the shift to a pelagic lifestyle at a smaller size to take advantage of higher growth rates. Klimley (1987) suggested that small individuals shifting to pelagic habitats incur a higher risk of predation. Females, however, are willing to accept this higher risk because large body size is important to reproductive success. Males likely do not take the risks because the benefits of increased growth are not high enough. Other species of sharks show similar sex differences in habitat use and growth rates (Klimley, 1987), raising the possibility that foraging–reproduction trade-offs are common in elasmobranchs; for example, sex differences in foraging habitat use and activity patterns of small-spotted catsharks (see Section 17.3.6.2) probably represent females making a trade-off between foraging and male harassment (Sims et al., 2001). Laboratory experiments showed that females were less responsive to prey stimuli in mixed-sex groups than single-sex groups, further

supporting the conclusion that females trade off foraging with reducing harassment from males (Kimber et al., 2009).

The need to deliver pups in nursery areas with low predation risk (Castro, 1987) may cause female elasmobranchs to abandon more productive foraging areas to migrate to nursery areas. Early work on sharks suggested another foraging–reproduction trade-off: that females might fast when they enter nursery areas in order to protect their young (e.g., Olsen, 1984; Springer, 1960). This hypothesis has yet to be verified, and there is mounting evidence that this is not the case (Wetherbee et al., 1990).

17.3.5.3 Intraguild Predation

Intraguild predation (IGP), where competitors are also predator and prey (Polis et al., 1989), creates special trade-offs for the intraguild prey. Intraguild predation may be symmetrical, where both species eat each other, or asymmetrical, where only one species eats its competitor. It may also be age structured, where only certain age–sex classes are engaged in IGP. Cannibalism is an example of asymmetrical age-structured IGP within a species. IGP appears to be common among sharks and also between sharks and other taxa such as cetaceans; for example, IGP occurs among sharks and between sharks and dolphins off South Africa (Heithaus, 2001a, and references therein). Intraguild predation can have dramatic consequences for the coexistence and spatial distribution of intraguild predators and prey (e.g., Heithaus, 2001b; Holt and Polis, 1997), and it is possible that IGP between killer whales and white sharks was responsible for the displacement of white sharks from the Farallon Islands during a season when killer whales were present (Heithaus, 2001a; Pyle et al., 1999). Ecosystem models suggest that the presence and nature of intraguild predation (sometimes labeled as "omnivory" in these papers) could be important in community stability and dynamics and that large sharks in at least some ecosystems are disproportionately involved in intraguild interactions (Bascompte et al., 2005; Kitchell et al., 2002; Kondoh, 2008). Therefore, further studies of the prevalence and nature of intraguild predation as well as behavioral interactions among intraguild predators and prey should be a focus of future research.

17.3.6 Variation in Feeding Strategies and Tactics

With their large ranges, large-scale seasonal movements, and sometimes diverse diets, it is not surprising that elasmobranchs show considerable variation in foraging tactics. This variation may occur within an individual, when individuals vary their foraging tactics in response to internal or external changes in conditions;

for example, seasonal changes in prey abundance or variation in an individual's condition may result in changes in foraging strategies and tactics. Changes may also occur over longer time scales; ontogenetic shifts in diets, foraging strategies, and tactics are common (see Chapter 8 for a review of ontogenetic shifts in diets). Consistent differences among individuals of a species or a population may also occur. Although some of these differences can be attributed to sex or size differences in foraging or regional variation within a species, there is increasing recognition that individuals within an age–sex class foraging in the same area can exhibit consistent differences in foraging patterns (individual specialization) (Bolnick et al., 2003). All of these sources of variation in foraging strategies and tactics can play an important role in population, community, and ecosystem dynamics and deserve increased attention. Indeed, many of these sources of variation likely operate simultaneously; for example, juvenile blue sharks in the eastern Atlantic Ocean display variation in apparent foraging behavior both within and among individuals. In coastal waters, shark behaviors were relatively regular in thermally stratified waters where sharks preferred surface waters but were more variable in well-mixed waters. In offshore waters, however, sharks used either surface-oriented or depth-oriented vertical movements, with individuals often switching between both, possibly in response to changes in prey type or distribution (Quieroz et al., 2010a).

17.3.6.1 Intra-Individual Variation

Variation in foraging behavior within individuals is commonplace in elasmobranchs. This variation may occur on multiple temporal scales, from changing foraging tactics or movements based on diel patterns or short-term variation in prey availability or prey types. White sharks, for example, vary their feeding tactics with pinniped prey type; they attack sea lions with greater force than elephant seals, *Mirounga angustirostris*, which may be due to sea lions' better escape abilities and probability of wounding a shark (Klimley et al., 1996a). It also appears that white sharks do not use a bite and spit tactic with elephant seals whereas they may with sea lions (Klimley et al., 1996a). Whale sharks appear to alter their feeding tactics based on plankton densities using active feeding (sharks swimming with upper jaw breaking the water surface and featuring frequent head turns and changes of direction) at extremely high zooplankton densities (~87 × 10³ m⁻³), vertical feeding (sharks maintain a vertical orientation just below the surface and gulp water) at high zooplankton densities (~19 × 10³ m⁻³), and passive feeding (swimming with mouth open) at low zooplankton densities (~6 × 10³ m⁻³) (Nelson and Eckert, 2007). In addition to displaying periods of high

activity with limited vertical and horizontal movement which may be associated with benthic foraging, common skates, *Dipturus batis*, also exhibit oscillatory vertical movements and diel vertical migrations similar to behaviors associated with foraging in pelagic predators (Wearmouth and Sims, 2009). Pyjama sharks modify their daily hunting rhythm to take advantage of diurnally spawning squid (Smale et al., 1995), and a number of elasmobranchs aggregate at these spawning sites to take advantage of this seasonally abundant prey (Smale et al., 2001). Torpedo rays appear to change their foraging tactics from a sit-in-wait ambush predator during the day to actively searching for prey at night (Bray and Hixon, 1978; Lowe et al., 1994), and broadnose sevengill sharks switch from oscillatory swimming patterns, likely associated with foraging, at night to less active opportunistic foraging tactics during the day (Barnett et al., 2010b). In other cases, individuals may shift among foraging tactics based on seasonal changes in prey or trade-offs (e.g., foraging–reproduction trade-offs). While in the white shark "café," white sharks engage in vertical oscillations thought to be related to foraging like other pelagic predators (Weng et al., 2007b). This tactic is very different from those observed at coastal pinniped rookeries.

Although not yet studied explicitly in elasmobranchs, variation in body condition may cause differences in diet selection, foraging tactics, and risk-taking behavior (e.g., Bouskila et al., 1998; Houston et al., 1993). Hungry animals tend to spend less time in refuges and engage in more risk-prone behaviors such as foraging in high-productivity and high-risk habitats (e.g., Heithaus et al., 2007a; Houston et al., 1993; Lima, 1998b). One possible elasmobranch example is southern stingrays, *Dasyatis americana*, visiting provisioning sites in the Grand Cayman Islands. Rays at the provisioning site exhibit lower body condition than those at other locations (Semeniuk and Rothley, 2008; Semeniuk et al., 2009). This may be because of suboptimal diets at the provisioning site (Semeniuk et al., 2007). Although the unnatural diet provided at the provisioning site is likely to affect ray health, the provisioning site represents a risky habitat, as witnessed by the large proportion of injured rays (85%) and rays with conspecific bite marks (100%) (Semeniuk and Rothley, 2008), yet also features high energetic rewards (i.e., provisioning). Although not tested, condition-dependent foraging by rays is consistent with observations from the provisioning site and may also contribute to the number of poor-condition rays found at the provisioning site.

17.3.6.1.1 Ontogenetic Variation

Ontogenetic shifts in diets are well documented for many species of elasmobranchs (see Chapter 8) and drive much of the intra-individual variation in foraging tactics that is observed. Young white sharks are more agile than larger ones and thus are able to capture fast-swimming teleost prey, whereas large sharks must rely on stealth to capture large mammalian prey (Tricas and McCosker, 1984). In cownose rays, the shift from non-burying and shallow-burying bivalves to deep-burrowing ones (Smith and Merriner, 1985) would result in a shift in foraging tactics from collecting benthic prey to excavation.

Ontogenetic changes in foraging tactics and habitat use can also result from changes in sensitivity to predation risk, with more susceptible juveniles selecting safe habitats and shifting into more productive but dangerous areas as their susceptibility to predators decreases (e.g., Bouskila et al., 1998; Werner and Hall, 1988). Juvenile elasmobranchs inhabiting nursery areas face this decision of when to shift from safe nurseries with relatively high intraspecific competition into more productive areas that have more predators (Heithaus, 2007). In general, juveniles should delay switching habitats as predation risk outside nurseries increases (Bouskila et al., 1998); however, if many juveniles synchronize their departure, risk can be reduced through dilution, and switching may occur sooner (Bouskila et al., 1998).

17.3.6.2 Inter-Individual Variation

Ontogenetic shifts in diets lead to apparent inter-individual variation in foraging tactics in populations, although other factors also contribute to this variation. Sex differences in foraging tactics have been observed in several different species of elasmobranchs. Male small-spotted catsharks in Lough Hyne off southwest Ireland rest in deep waters during the day, then move into shallow waters to feed on crustacean prey throughout crepuscular and nocturnal hours. Female catsharks, however, refuge in shallow-water caves during the day and through some nights, only emerging at night to forage in deeper waters (Sims et al., 2001). In experimental settings, females also show reduced interest in prey stimuli when in the presence of males, suggesting female avoidance of males (Kimber et al., 2009). Other sex differences in foraging tactics may be due to variation in selective pressures between males and females (see Section 17.3.5.2). Regional variation in diets or the behavior of prey also can lead to individual variation in diets and foraging tactics.

17.3.6.2.1 Individual Specialization

Increasingly, it is being recognized that there can be consistent differences in foraging behavior among individuals of the same age–sex class and in the same basic habitats (Bolnick et al., 2003). This individual specialization results in individuals consuming, or using, a subset of the resources of a population as a whole and has

important implications for evolutionary, ecological, and population dynamics (e.g., Baird et al., 1992; Bolnick et al., 2003). In some cases, populations that appear to be generalists at the population level are actually made up of groups of specialists (e.g., Bolnick et al., 2003; Matich et al., 2011; Quevedo et al., 2009). Individual specialization in elasmobranchs has not been well studied, but recent work using stable isotopes suggests that long-term specialization may occur within some populations. There appear to be long-term inter-individual differences in trophic interactions of several species of rays captured on sandflats in Western Australia (Vaudo and Heithaus, 2011), while there appears to be relatively little temporal stability in trophic interactions of individual tiger sharks (Matich et al., 2011). Juvenile bull sharks in a coastal nursery in Florida, however, exhibit wide variation among individuals in trophic interactions that are maintained over long time periods (Matich et al., 2011). Although these individuals are captured in low-salinity areas typified by freshwater and estuarine food webs, some individuals forage in marine food webs located over 15 km away. The differences in foraging behaviors appear to be driven by some individuals accepting higher predation risk at the mouth of the estuary, where prey are more abundant, while others remain in low-risk but lower productivity low-salinity waters. Several other examples suggest that individual specialization in foraging tactics (which may not be reflected in isotopic data used in the above examples) may be more widespread. Tiger sharks in Shark Bay, Western Australia, show individual variation in short-term habitat preference (Heithaus et al., 2002a), which may represent differences in foraging tactics. Whether these are long-term differences in foraging tactics remains to be determined. Similarly, tiger sharks tagged at French Frigate Shoals in the Northwest Hawaiian Islands showed variable movement patterns. Some individuals remained at French Frigate Shoals year-round, others moved along the Hawaiian chain, and one traveled across open water to the North Pacific transition zone chlorophyll front before returning in later years (Meyer et al., 2010).

17.4 Regulation of Elasmobranch Populations

The above sections largely investigated the behavioral mechanisms that elasmobranchs use to capture prey and to avoid predators. The end results of these interactions can have profound consequences on the equilibrium population sizes of both predators and prey; however, few studies have identified density dependence in elasmobranchs. Juvenile lemon sharks in North Sound, Bimini, Bahamas, show density-dependent survival, with a carrying capacity of about 30 pups in the sound (Gruber et al., 2001); it is not known what sets this level. In Kaneohe Bay, Oahu, Hawaii, food limitation may set carrying capacity. Growth and diet data from juvenile scalloped hammerhead sharks suggest that sharks are taking in less energy than their maintenance ration (Bush and Holland, 2002), and the low level of juvenile retention in the nursery is largely because of starvation (Duncan and Holland, 2006). Walker (1998) argues that density-dependent natural mortality occurs in young age classes of gummy sharks presumably largely due to predation. Although density-dependent survival may occur in the youngest age classes within nursery areas, population sizes and population growth rates may be determined largely by survival in older juvenile stages after they have left nursery areas (Kinney and Simpfendorfer, 2009) and may be more vulnerable to predators.

Competition, food availability, and predation may play important roles in regulating population sizes of some elasmobranchs. Fishing pressure has decreased populations of large-bodied skates (with some species disappearing) in the Irish Sea, but small-bodied species have increased in abundance and biomass, resulting in stable aggregate catch trends (Dulvy et al., 2000). Previous studies have shown significant dietary overlaps among species (Ellis et al., 1996), leading to the suggestion that small species have increased due to competitive release (Dulvy et al., 2000). Similarly, Walker and Heesen (1996) suggested that competitive release or increased food availability from fishery discards led to increases in starry ray, *Raja radiata*, populations in the North Sea. Similar arguments had been made for the increase of small and medium-sized elasmobranchs on George's Bank. Population demographics from George's Bank and population surveys of neighboring areas, however, suggest that increases are likely to be the result of movements between connected populations rather than actual population growth (Frisk et al., 2008). Small shark populations off South Africa may have been regulated to some degree by competition with, or predation by, large sharks; reductions in catches of large sharks by recreational fishers were linked with increases in catches of small sharks and other smaller elasmobranchs (Ferretti et al., 2010; van der Elst, 1979). Similar increases in smaller bodied elasmobranchs appear to have occurred in other areas of the world when large shark populations have declined (for a review, see Feretti et al., 2010). Increases in elasmobranch mesopredators, in response to relaxed predation pressure, are unlikely to be universal, and possible examples have been hotly debated; for example, Myers et al. (2007) suggested that cownose ray populations along the Atlantic coast of North America increased in response to relaxed predation pressure from large sharks. Other possible mechanisms that could cause the apparent increase in cownose rays are changes in the timing of migration, population redistribution, and range expansion. Further, some of the trophic links between large sharks and cownose rays have been questioned, as has the ability of cownose rays to reproduce at a rate required for the observed increase (Heithaus et al., 2010).

Both competition and predation may be important factors influencing population sizes of animals, but the effects of these two are often inseparable (e.g., Sih et al., 1985; Walters, 2000; Walters and Juanes, 1993; Werner and Anholt, 1996). There is empirical support for the predation-sensitive food (PSF) hypothesis, which states that food and risk both act to limit populations for species that are both predators and prey (e.g., McNamara and Houston, 1987; Sinclair and Arcese, 1995). This arises through animals taking larger risks as food becomes limited and risk-taking individuals are killed. Antipredator behavior that limits foraging can also cause prey populations to be limited by a combination of both food and predators (Lima, 1998b; Walters, 2000; Walters and Juanes, 1993) or may stabilize otherwise oscillatory predator–prey dynamics (Lima, 1998b). Spatial and activity level components of antipredator behavior can influence population dynamics of both predators and prey in several ways (for a detailed description, see Walters, 2000). Prey restricting their movements to areas that are relatively safe from predators (e.g., small shark nursery areas) results in a limited foraging arena (Walters and Juanes, 1993; Walters and Martell, 2004) that is generally much smaller than the range of the prey's food. Because of this restriction in foraging area, it may appear that food is the limiting factor for populations, even though larger population size would be possible if predators were not present (Walters and Juanes, 1993). The restriction in prey distribution may actually allow coexistence of prey species with similar diets because neither species exploits the full range of the prey species (Walters, 2000). Antipredator behavior by the prey will also influence predator populations as energy flow rates will be restricted relative to situations that ignore prey behavior and give the appearance of bottom-up control of predator populations (Walters, 2000). One important insight from these dynamics is the importance of the spatial scale of sampling of prey food, as prey surveys at too large a scale may miss the importance of intraspecific competition within restricted foraging arenas. Also, this view of population regulation challenges the traditional view that increasing predation risk acts to lower intraspecific competition because prey are kept well below the carrying capacity set by food resources. Instead, reduced activity levels or restricted foraging areas may increase intraspecific competition within these areas or during safe times and therefore increase the limiting effects of food (Walters and Juanes, 1993). Castro (1987) suggested that many shark populations may be limited by nursery area availability. This situation would fit the foraging arena scenario (Walters, 2000; Walters and Juanes, 1993), where populations would be limited by the presence of predators and antipredator behaviors of juvenile sharks and intraspecific competition among these sharks within the foraging arena.

17.5 Role of Elasmobranchs in Marine Ecosystems

As top predators, elasmobranchs are generally thought of as critical components of marine ecosystems, perhaps regulating prey populations and even community structure; however, detailed analyses are relatively few and sometimes controversial. Several recent reviews (Feretti et al., 2010; Heithaus et al., 2008a, 2010) have provided insights into the ecological role of elasmobranchs and the mechanisms through which they might influence prey populations, community dynamics, and ecosystem processes. Here, we briefly summarize studies found in these reviews and recent advances in our understanding of elasmobranch ecological roles with a focus on impacts on potential prey rather than on wider communities (the focus of previous reviews). For more information on the ecological importance of elasmobranchs, specifically, including the potential context dependence of these effects, see Heithaus et al. (2010). Further details on the potential role of large sharks in modifying community structure are provided in Feretti et al. (2010).

17.5.1 Mechanisms of Top-Down Impacts

Since the first edition of *Biology of Sharks and Their Relatives* there has been a revision of the terminology pertaining to the two classes of predator impacts on their prey (Creel et al., 2008; Heithaus et al., 2008a). The effects of direct predation (or "consumptive effects") had previously been referred to as "density-mediated interactions" and "density-mediated indirect interactions." The latter two terms should be avoided because prey densities can be modified by processes other than predators killing prey—for example, by modifications of prey behavior that lead to reduced reproductive rates and nonlethal impacts on reproduction. Although the majority of ecological literature has been focused on the effects of direct predation, or assumed direct predation as the primary mechanism through which predators impact prey (Peckarsky et al., 2008), it is becoming increasingly apparent that predators may have a profound influence on prey through nonconsumptive interactions, which may be either direct risk effects or indirect: trait-mediated indirect interactions (TMIIs)

or behaviorally mediated indirect interactions (BMII). In the past decade, the importance of risk effects and TMIIs has begun to gain general acceptance, especially in the terrestrial literature, and it is now clear that in some cases they may equal or exceed the impacts of direct predation (Dill et al., 2003; Heithaus et al., 2008a,b; Lima, 1998; Peacor and Werner, 2000, 2001; Preisser et al., 2005; Schmitz et al., 1997, 2004; Werner and Peacor, 2003; Wirsing et al., 2008a). Although somewhat counterintuitive, such results occur because direct mortality usually removes a limited number of individuals from a population, which may result in decreased intraspecific feeding or reproductive competition. This, in turn, can result in increased reproduction or growth among remaining prey individuals (compensatory reproduction or growth) with an end result of no reduction in population size. In contrast, antipredator behaviors, which may include leaving high-risk but high-productivity habitats or reduced foraging rates, are generally performed by all (or most) individuals in a population and can result in lower access to food and a resulting reduction in the population's reproductive potential. Antipredator behaviors can also reverse competitive asymmetries between prey species (e.g., Lima, 1998b; Relyea, 2000) and allow coexistence of competitor species (Lima, 1998b).

17.5.1.1 Consumptive Effects

Many studies of elasmobranch feeding comment that elasmobranchs, especially sharks, are responsible for regulating prey populations through direct predation, and this claim is often made simply because numerous prey individuals are killed. However, because we do not know where density dependence operates in these prey species, it is currently not possible to evaluate these hypotheses. Indeed, even high rates of predation on a species may not affect equilibrium population sizes if density dependence operates at a life history stage different from that where most predation occurs (e.g., Piraino et al., 2002). Somewhat surprisingly, recent studies suggest that predators may actually *increase* equilibrium population sizes of their prey (Abrams, 2009). Such "hydra effects" can occur if predators preferentially feed on larger prey individuals, allowing for larger numbers of smaller individuals in the population. They also can occur if increased mortality decreases population fluctuations in prey species so a higher average prey population is achieved and if density-independent mortality and the density-dependent processes that counteract it are temporally separated (Abrams, 2009).

Direct predation by sharks has played an important role in the population decline of harbor seals on Sable Island (Bowen et al., 2003; Stobo and Lucas, 2000). Based on carcasses washing ashore with shark bites that were obviously not scavenged, shark attacks have been steadily increasing and sharks now regularly kill all age–sex classes. Predation, especially on adult females, influenced the substantial decline in pup production between 1980 and 1997. In fact, observed shark-inflicted mortality from 1994 to 1996 accounted for around 50% of the decline in pup production from 1995 to 1997 (Bowen et al., 2003; Lucas and Stobo, 2000). It is still unclear why there has been an increase in apparent shark-inflicted mortality (Lucas and Stobo, 2000), but it is likely that shark abundance has increased in response to the substantial population increases in gray seals on Sable Island (Bowen et al., 2003). Although gray seals also are killed by sharks (Bowen et al., 2003; Brodie and Beck, 1983), their populations have not been affected by shark predation because predation is extremely low relative to pup production, and adults likely face much lower risk than harbor seals (Bowen et al., 2003).

Predatory attacks by sharks are not always successful and often leave injured individuals in prey populations. Some studies have attempted to estimate the effects of shark predation on prey populations by measuring the rate of scarring or injury in the population. Such methods are fraught with biases (Heithaus, 2001a); for example, differences in wounding rates may reflect the probability of escape after capture rather than differences in attack and death rate. It is difficult, therefore, to make comparisons among populations or species that either face different sizes of predators or differ substantially in body size, antipredator behavior, or escape abilities. Despite these biases, such studies may provide some useful insights into the importance of shark predation on the populations and behavior of their prey. The rate of scars and wounds from white shark attacks found on pinnipeds along the California coast has led some investigators to suggest top-down control of pinniped populations (e.g., McCosker, 1985); however, further work is needed to verify this hypothesis. Nonetheless, even nonlethal white shark attacks have substantial reproductive consequences for female elephant seals. At Año Nuevo, California, only 8 of 11 adult females with fresh bites successfully weaned their pups, and the successful seals had the least severe injuries (LeBoeuf et al., 1982). Furthermore, none of the injured females was observed copulating before returning to sea, resulting in a probable loss of 2 years of reproduction. A similar result was found at the Farallon Islands (Ainley et al., 1981).

Large shark injury rates on Hawaiian monk seals in the northwest Hawaiian Islands are relatively low, with generally fewer than 3.5% injured annually (Bertilsson-Friedman, 2006), but there are large differences in age–sex classes attacked throughout the chain. At French Frigate Shoals, the largest subpopulation, pups are attacked more frequently than expected compared to other age classes. In contrast, attacks on pups are infrequent at Laysan and Lisianski Islands. Instead, juveniles are attacked more frequently than expected based on their relative abundance (Bertilsson-Friedman, 2006). These differences may be due to variation in the physical habitats and accessibility for large sharks. At French Frigate, it appears that Galapagos sharks, *Carcharhinus galapegensis*, may be responsible for a large number of attacks on pups (18 to 30% of the annual cohort between 2000 and 2003), before or near the time of weaning, with most attacks observed in the very shallow waters of a small sand island where the density of pups is quite high (Antonelis et al., 2006; Bertilsson-Friedman, 2006). No such attacks have been recorded elsewhere in the archipelago. At Lisianski and Laysan Islands, the opportunities for such predation attempts appear to be quite low, and attacks are made in different habitats and by other species of large shark such as tiger sharks (Bertilsson-Friedman, 2006). The population consequences of shark predation are unknown.

Large sharks are the primary predators of adult sea turtles while they are at sea (for a review, see Heithaus et al., 2008b). Data on predation rates are lacking, but scarring rates indicate substantial variation in predation rates among species, including those that are sympatric. In Shark Bay, Western Australia, loggerhead turtles, *Caretta caretta*, are injured much more often than sympatric green turtles, *Chelonia mydas*, which likely is due to the greater maneuverability and speed of green turtles and, therefore, higher probabilities of escape in an encounter situation (Heithaus et al., 2002b, 2005). In contrast, in Eastern Australia, shark-inflicted injuries on green and loggerhead turtles are virtually absent (Limpus et al., 1994a,b), suggesting relatively low predation rates (Heithaus et al., 2008b). Ridley sea turtles, *Lepidochelys kempi*, also tend to have low rates of shark-inflicted scarring (Shaver, 1988; Witzell, 2007), but this may be due to their small body sizes and a higher probability of being killed or eaten whole when attacked (Heithaus et al., 2008b).

Off Natal, South Africa, between 10 and 19% of bottlenose dolphins exhibit bite scars and an estimated 2.2% of the population is killed annually by sharks (Cockcroft et al., 1989); off Fernando de Noronha Archipelago, Brazil, 55 scars from large shark bites were recorded from 418 photographic records of spinner dolphins, *Stenella longirostris* (Silva et al., 2007). In other areas, dolphins with smaller body sizes facing predation risk from the same shark species have much higher rates of wounds and scars; thus, sharks probably kill a higher proportion of these populations each year (Heithaus, 2001d). In Moreton Bay, Queensland, 36.6% of dolphins bear wounds (Corkeron et al., 1987) and 74.2% of dolphins in Shark Bay, Western Australia, have been attacked at least once in their lives, with at least 10% of the population being attacked unsuccessfully each year (Heithaus,

2001d). The shark-inflicted mortality rates of dolphins in these locations and, thus, the effects of direct predation are unknown.

Studies of wounding have shown that prey age–sex classes may be affected differentially by shark predators. Male dolphins in several locations have higher rates of scarring or multiple scarring than do females (Heithaus, 2001d). In Shark Bay, male loggerhead turtles have significantly higher rates of major shark-inflicted injuries (58%) than females (12%), whereas there are no sex differences in wounding rates of sympatric green turtles (Heithaus et al., 2002b). It is likely that intraspecific variation in wounding is the result of different attack rates, possibly due to sex differences in risk taking (Heithaus et al., 2002b), rather than variation in escape ability (Wirsing et al., 2008b).

Wounds from cookie-cutter sharks have been found on a diverse array of species (e.g., Heithaus, 2001a; Hiruki et al., 1993; Papastamatiou et al., 2010b). Almost every adult spinner dolphin observed off Hawaii and Fernando de Noronha Archipelago, Brazil, shows signs of attacks from these sharks (Norris and Dohl, 1980; Silva et al., 2007). Nearly 90% of swordfish examined at the Hawaii Fish Auction had scars from cookie-cutter shark attacks, and over 60% of those individuals had multiple scars (Papastamatiou et al., 2010b). The implications of these attacks for their prey species are unknown, and, although they are certainly less detrimental than predatory attacks, they may have fitness consequences, because energy must be used for recuperation that could have been invested in growth or reproduction.

Understanding the role of sharks in regulating prey populations can be very difficult because of the mobility of both predators and prey. Benthic foraging rays, however, offer an opportunity for experimental studies, and rays can have a large impact on their prey. Exclusion experiments have shown that ray predation and disturbance of sediments can have a negative effect on abundance and number of invertebrate species in soft-bottom communities (Thrush et al., 1994; VanBlaricom, 1982). Cownose rays have been observed to completely remove bay scallops from the most productive habitat patches in the Cape Lookout lagoonal system in North Carolina, causing a population sink (Myers et al., 2007; Peterson et al., 2001). Rays removed scallops before reproduction occurred, resulting in the individuals in these habitats having no contribution to future generations; however, bay scallops are an annual species, and individuals remaining in habitats with low initial densities produce enough offspring to maintain population levels (Peterson et al., 2001). Thus, rays do not appear to regulate equilibrium population sizes but are an important factor in population dynamics of their prey and have been implicated as a potential factor in the closure of North Carolina's scallop fishery (Myers et al., 2007).

Such impacts are likely to be highly context dependent; for example, eagle rays on New Zealand sandflats have been estimated to consume only 1.6% of their main bivalve prey population (Hines et al., 1997).

In some situations, there appears to be little effect of elasmobranchs on populations of their prey. Mako sharks consume between 4 and 14% of the available bluefish biomass between Cape Hatteras, North Carolina, and Georges Bank (Stilwell and Kohler, 1982). Although the bluefish is a very important prey item of the sharks, there does not appear to be a significant impact on bluefish populations (Stillwell and Kohler, 1982). Similarly, ecosystem models suggest that elasmobranchs may not have strong top-down effects on prey populations through direct predation. In a model of the northern Gulf of Mexico, small sharks were not predicted to have large impacts on the dynamics of prey populations (Carlson, 2007) and elasmobranch mesopredators were not predicted to have strong impacts on prey in the Gulf of Tortugas, Colombia (Navia et al., 2010).

Our understanding of indirect effects of direct predation by elasmobranchs is even less developed, but such interactions may be important in community dynamics (for reviews, see Feretti et al., 2010; Heithaus et al., 2010) and commercial operations. Oyster growers in Humboldt Bay, California, have tried to reduce populations of the bat ray because of its supposed role in destroying oysters; however, this may have negative consequences for oyster farms because bat rays do not appear to regularly feed on oysters and instead are major predators on the primary oyster predator, red rock crabs, *Cancer productus* (Gray et al., 1997). Thus, a reduction in ray populations may actually result in increased losses of oysters. Future empirical research on the impacts of elasmobrach predation on prey populations and the contexts in which it is more or less important in prey population dynamics is of great importance to management and our general understanding of marine communities. A full understanding of elasmobranchs as predators, however, must include further appreciation of risk effects on prey.

17.5.1.2 Risk Effects

For some prey species, the probability of being killed by an elasmobranch predator is quite low (e.g., dolphins) (Heithaus, 2001a; Simpfendorfer et al., 2001b), but this does not mean that these prey are unlikely to be influenced by the risk of predation from elasmobranchs. Especially in long-lived species with slow reproductive rates, even a low risk of predation can lead to extreme antipredator behaviors because longevity can be a major determinant of fitness (Heithaus et al., 2008a; Lima, 1998b). A growing number of studies demonstrate that behavioral shifts in the face of shark predation risk are widespread, at least in long-lived prey taxa, including marine mammals, marine reptiles, and marine birds (Heithaus et al., 2008a). Cape fur seals appear to respond to the risk of white shark predation in several ways. When traveling near rookeries in groups, fur seals behave as a "selfish herd," with individual fur seals attempting to swim in the center of groups where predation risk is lower than at the periphery (DeVos and O'Riain, 2010). Also, adult, but not juvenile, fur seals shift their behavior when moving to and from haul-outs to reduce risk, which suggests that some period of learning is necessary to adopt optimal antipredator behavior (Laroche et al., 2008). In the northeastern Pacific, both harbor seals and Steller sea lions *Eumetopias jubatus* (Frid et al., 2007, 2009), appear to shift their diving behavior to reduce risk of predation from Pacific sleeper sharks at depth. These shifts could have important implications for predation rates of seals on different prey species.

Multiple prey species of tiger sharks in Shark Bay, Western Australia, make behavioral adjustments to enhance safety, even at the expense of foraging opportunities. Indo-Pacific bottlenose dolphins in Shark Bay are rarely found in the stomach contents of tiger sharks (Heithaus, 2001c; Simpfendorfer et al., 2001b), but dolphin habitat use is greatly influenced by the presence of tiger sharks. When sharks are absent from the bay in winter months, foraging bottlenose dolphins distribute themselves between shallow seagrass habitats and deeper waters proportional to the food available in each as predicted by the IFD (Heithaus and Dill, 2002a). When tiger sharks move into the bay in warmer months, the sharks prefer shallow seagrass habitats, which contain high densities of prey and are also the most productive for dolphins. This results in dolphins largely avoiding the productive shallow habitats and instead foraging in the lower productivity but safer deep habitats (Heithaus and Dill, 2002a). Dolphins foraging within shallow habitats when sharks are present shift microhabitats in order to enhance safety, selecting safer edge microhabitats to a greater degree than interior ones; when sharks are absent, dolphins are distributed across shallow microhabitats according to an IFD (Heithaus and Dill, 2006). Dugongs display similar habitat (Wirsing et al., 2007c) and microhabitat (Wirsing et al., 2007a) shifts between periods of high and low shark abundance. In addition, dugongs manage risk by shifting foraging tactics as the risk of shark predation varies. When sharks are scarce, dugongs excavate seagrass rhizomes, which are higher quality forage than seagrass leaves, creating large clouds of sediment in the water that could mask the approach of tiger sharks (Wirsing et al., 2007d). As shark abundance increases, dugongs largely abandon this tactic, instead choosing to crop seagrass leaves. Finally, those dugongs that do use the excavation tactic when predation risk is high modify their diving behavior to

make more and shorter dives, whereas those that are cropping do not modify their diving behavior seasonally (Wirsing et al., 2011). Changes in diving behavior in response to predation risk also have been found for pied cormorants, *Phalacrocorax varius*, but only in the most dangerous habitats (Dunphy-Daly et al., 2010). Cormorants also shift habitats and microhabitats to reduce the risk of tiger shark predation (Heithaus, 2005; Heithaus et al., 2009b). Two species of sea snakes also make predation-sensitive changes in their foraging locations. Olive-headed sea snakes, *Disteria major*, shift microhabitats on offshore banks to reduce tiger shark predation risk seasonally (Wirsing and Heithaus, 2009), and bar-bellied sea snakes, *Hydrophis elegans*, only forage in dangerous nearshore sand habitats, where prey is abundant, when tides are low and access by tiger sharks is reduced (Kerford et al., 2008). Finally, green turtles shift their use of shallow banks in a condition-dependent manner. Turtles in good condition are found close to the edges of banks, where their escape ability is maximized but the forage quality is lower. Turtles in poor condition are found near the middle of banks, where the risk of being killed by a tiger shark is higher but seagrass quality is greater (Heithaus et al., 2007a). As predation risk decreases, turtles in good condition shift further toward the middle of the bank. Theoretically, the habitat changes described above could reduce equilibrium population size of these species through reduced access to food; therefore, it is possible that tiger sharks are important in determining population sizes of their prey through behavioral effects. For some species, such as dolphins and dugongs, predation rates are probably very low and impacts of tiger sharks are likely primarily though risk effects. For sea birds and sea snakes, which are found more often in tiger shark stomach contents (Heithaus, 2001a; Simpfendorfer et al., 2001b), overall impacts of tiger sharks likely are driven by a combination of direct predation and risk effects.

Behaviorally mediated indirect interactions may create, enhance, ameliorate, or even reverse the sign (i.e., a species actually has a positive effect on its competitor) of interactions between species; thus, understanding the dynamics of BMIIs is important in understanding community dynamics and conservation biology (Dill et al., 2003; Heithaus et al., 2008a). In Shark Bay, Western Australia, tiger sharks are an important *transmitter* of a BMII between their primary prey (dugongs) and less common prey (e.g., dolphins). In this interaction, the seasonal occurrence and habitat use of dugongs results in tiger sharks selecting shallow seagrass habitats during warm months and being largely absent during winter months. This causes dolphins to switch from using high-productivity shallow waters for foraging in the winter to the less-productive deeper waters in the summer (Heithaus and Dill, 2002a). Tiger sharks may also initiate important BMIIs that are transmitted to seagrasses. By shifting the foraging locations and tactics of herbivores such as dugongs and green turtles, tiger sharks could indirectly modify the seagrass community composition and nutrient content (Heithaus et al., 2007a, 2008a). Experimental studies are still needed to verify the existence of this BMII. The growing number of examples of behavioral shifts in response to shark predation, however, suggest that BMIIs involving elasmobranchs are common in marine communities and are likely an important feature of community dynamics (for further discussion, see Heithaus et al., 2008a, 2010).

17.6 Summary

Our understanding of elasmobranch predator–prey interactions has grown considerably since the first edition of *Biology of Sharks and Their Relatives*, but there remains much to be done to gain a more general understanding of the tactics of both predators and prey and how these interactions shape elasmobranch populations and the communities of which they are a part. Of particular importance are studies that specifically address the magnitude and importance of elasmobranch predation and predation risk on prey, including other elasmobranchs. As was the case when the first edition of *Biology of Sharks and Their Relatives* was published, there is a need to continue incorporating game theoretical ideas into studies of shark foraging and antipredator behavior and further field studies that simultaneously study both elasmobranchs and their predators and prey. These studies should not be limited to coarse-scale surveys of predator and prey distribution but should endeavor to understand underlying mechanisms and tactics that cause these distributions. It is hoped that such studies will allow us to gain a functional understanding of elasmobranch behavior and give us the ability to make predictions about how changes in ecological conditions will affect them and how changes to their populations are likely to influence marine communities.

Acknowledgments

During the preparation of this chapter JJV was supported by a Dissertation Year Fellowship at Florida International University. Financial support also was provided by National Science Foundation. This is Contribution No. 49 of the Shark Bay Ecosystem Research Project (www.sberp.org).

References

Aalbers, S.A., Bernal, D., and Sepulveda, C.A. (2010). The functional role of the caudal fin in the feeding ecology of the common thresher shark *Alopias vulpinus*. *J. Fish Biol.* 76:1863–1868.

Abrahams, M.V. and Dill, L.M. (1989). A determination of the energetic equivalence of the risk of predation. *Ecology* 70:999–1007.

Abrams, P.A. (2009). When does greater mortality increase population size? The long history and diverse mechanisms underlying the hydra effect. *Ecol. Lett.* 12:462–474.

Acevedo-Gutiérrez, A. (2002). Interactions between marine predators: dolphin food intake is related to the number of sharks. *Mar. Ecol. Prog. Ser.* 240:267–271.

Ackerman, J.T., Kondratieff, M.C., Matern, S.A., and Cech, Jr., J.J. (2000). Tidal influences on spatial dynamics of leopard sharks, *Triakis semifasciata*, in Tomales Bay, California. *Environ. Biol. Fish.* 58:33–43.

Ainley, D, G., Strong, C.S., Huber, H.R., Lewis, T.J., and Morrell, S.H. (1981). Predation by white sharks on pinnipeds at the Farallon Islands. *Fish. Bull.* 78:941–945.

Allen, S. and Huveneers, C. (2005). First record of an Australian fur seal (*Arctocephalus pusillus doriferus*) feeding on a wobbegong shark (*Orectolobus ornatus*). *Proc. Linn. Soc. N.S.W.* 126:95–97.

Andrews, K.S., Williams, G.D., Farrer, D., Tolimiero, N., Harvey, C.J., Bargmann, G., and Levin, P.S. (2009). Diel activity patterns of sixgill sharks, *Hexanchus griseus*: the ups and downs of an apex predator. *Anim. Behav.* 78:525–536.

Anholt, B.R. and Werner, E.E. (1995). Interaction between food availability and predation mortality mediated by adaptive behavior. *Ecology* 76:2230–2234.

Antonelis, G.A., Baker, J.D., Johanos, T.C., Braun, R.C., and Harting, A.L. (2006). Hawaiian monk seal (*Monachus schauinslandi*): status and conservation status. *Atoll Res. Bull.* 543:75–101.

Au, D.W. (1991). Polyspecific nature of tuna schools: shark, dolphin, and seabird associates. *Fish. Bull.* 89:343–354.

Baird, R.W., Abrams, P.A., and Dill, L.M. (1992). Possible indirect interactions between transient and resident killer whales: implications for the evolution of foraging specializations in the genus *Orcinus*. *Oecologia* 89:125–132.

Baremore, I.E., Murie, D.J., and Carlson, J.K. (2008). Prey selection by the Atlantic angel shark *Squatina dumeril* in the northeastern Gulf of Mexico. *Bull. Mar. Sci.* 82:297–313.

Barnett, A., Redd, K.S., Frusher, S.D., Stevens, J.D., and Semmens, J.M. (2010a). Non-lethal method to obtain stomach samples from a large marine predator and the use of DNA analysis to improve dietary information. *J. Exp. Mar. Biol. Ecol.* 393:188–192.

Barnett, A., Abrantes, K., Stevens, J.D., Bruce, B.D., and Semmens, J.M. (2010b). Fine-scale movements of the broadnose sevengill shark and its main prey, the gummy shark. *PLoS One* 5:e15464.

Bascompte, J.C., Melián, C.J., and Sala, E. (2005). Interaction strength combinations and the overfishing of a marine food web. *Proc. Natl. Acad. Sci. U.S.A.* 102:5443–5447.

Beck, B. and Mansfield, A.W. (1969). Observations on the Greenland shark, *Somniosus microcephalus*, in northern Baffin Island. *J. Fish. Res. Bd. Can.* 26:143–145.

Bednekoff, P.A. and Lima, S.L. (1998). Re-examining safety in numbers: interactions between risk dilution and collective detection depend upon predator targeting behaviour. *Proc. R. Soc. B Biol. Sci.* 265:2021–2026.

Belbenoit, P. (1986). Fine analysis of predatory and defensive motor events in *Torpedo marmorata* (Pisces). *J. Exp. Biol.* 121:197–226.

Berger-Tal, O. and Kotler, B.P. (2010). State of emergency: behavior of gerbils is affected by the hunger state of their predators. *Ecology* 91:593–600.

Bertilsson-Friedman, P. (2002). Distribution and frequencies of shark-inflicted injuries to the endangered Hawaiian monk seal (*Monachus schauinslandi*) *J. Zool. (Lond.)* 268:361–368.

Bertram, B.C.R. (1978). Living in groups: predators and prey. In: Krebs, J.R. and Davies, N.B. (Eds.), *Behavioural Ecology: An Evolutionary Approach*. Blackwell Press, Oxford, pp. 64–96.

Bethea, D.M., Buckel, J.A., and Carlson, J.K. (2004). Foraging ecology of the early life stages of four sympatric shark species. *Mar. Ecol. Prog. Ser.* 268:245–264.

Bolnick, D.I., Svanbäck, R., Fordyce, J.A., Yang, L.H., David, J.M., Hulsey, C.D., and Forister, M.L. (2003). The ecology of individuals: incidence and implications of individual specialization. *Am. Nat.* 161:1–28.

Bonenfant, M. and Kramer, D.L. (1994). The influence of distance to burrow on flight initiation distance in the woodchuck, *Marmota monax*. *Behav. Ecol.* 7:299–303.

Bouskila, A., Robinson, M.E., Roitberg, B.D., and Tenhumberg, B. (1998). Life-history decisions under predation risk: importance of a game perspective. *Evol. Ecol.* 12:701–715.

Bowen, W.D., Ellis, S.L., Iverson, S.J., and Boness, D.J. (2003). Maternal and newborn life-history traits during periods of contrasting population trends: implications for explaining the decline of harbour seals, *Phoca vitulina*, on Sable Island. *J. Zool. (Lond.)* 261:155–163.

Bozzano, A., Murgia, R., Vallerga, S., Hirano, J., and Archer, S. (2001). The photoreceptor system in the retinae of two dogfishes, *Scyliorhinus canicula* and *Galeus melastomus*: possible relationship with depth distribution and predatory lifestyle. *J. Fish Biol.* 59:1258–1278.

Braccini, J.M. (2008). Feeding ecology of two high-order predators from south-eastern Australia: the coastal broadnose and the deepwater sharpnose sevengill sharks. *Mar. Ecol. Prog. Ser.* 371:273–284.

Branstetter, S. (1990). Early life-history implications of selected carcharinoid and lamnoid sharks of the northwest Atlantic. In: Pratt, Jr., H.L., Gruber, S.H., and Taniuchi, T. (Eds.), *Elasmobranchs as Living Resources: Advances in the Biology, Ecology, Systematics, and Status of the Fisheries*, NOAA Tech. Rep. NMFS 90. U.S. Department of Commerce, Washington, D.C., pp. 17–28.

Bray, R.N. and Hixon, M.A. (1978). Night shocker: predatory behavior of the Pacific electric ray (*Torpedo californica*). *Science* 200:333–334.

Breder, C.M. (1952). On the utility of the saw of the sawfish. *Copeia* 1952:90–91.

Brodie, P. and Beck, B. (1983). Predation by sharks on the grey seal (*Halichoerus grypus*) in eastern Canada. *Can. J. Fish. Aquat. Sci.* 40:267–271.

Brown, J.S. (1988). Patch use as an indicator of habitat preference, predation risk, and competition. *Behav. Ecol. Sociobiol.* 22:37–47.

Brown, J.S. (1992a). Patch use under predation risk. I. Models and prediction. *Ann. Zool. Fenn.* 29:301–309.

Brown, J.S. (1992b). Patch use under predation risk. II. A test with fox squirrels, *Sciurus niger*. *Ann. Zool. Fenn.* 29:311–318.

Brown, J.S. (1999). Vigilance, patch use and habitat selection: foraging under predation risk. *Evol. Ecol. Res.* 1:49–71.

Brown, J.S. and Kotler, B. (2004). Hazardous duty pay and the foraging cost of predation. *Ecol. Lett.* 10:999–1014.

Brown, J.S., Laundré, J.W., and Gurung, M. (1999). The ecology of fear: optimal foraging, game theory, and trophic interactions. *J. Mamm.* 80:385–399.

Brown, J.S., Kotler, B.P., and Bouskila, A. (2001). Ecology of fear: foraging games between predators and prey with pulsed resources. *Ann. Zool. Fenn.* 38:55–70.

Bruce, B.D. (1992). Preliminary observations on the biology of the white shark, *Carcharodon carcharias*, in South Australian waters. *Aust. J. Mar. Freshw. Res.* 43:1–11.

Bush, A. (2003). Diet and diel feeding periodicity of juvenile scalloped hammerhead sharks, *Sphyrna lewini*, in Kaneohe Bay, Oahu, Hawaii. *Environ. Biol. Fish.* 67:1–11.

Bush, A. and Holland, K. (2002). Food limitation in a nursery area: estimates of daily ration in juvenile scalloped hammerhead, *Sphyrna lewini* (Griffith and Smith, 1834) in Kaneohe Bay, Oahu, Hawaii. *J. Exp. Mar. Biol. Ecol.* 278:156–178.

Carey, F.G. and Clark, E. (1995). Depth telemetry from the sixgill shark, *Hexanchus griseus*, at Bermuda. *Environ. Biol. Fish.* 42:7–14.

Carey, F.G. and Scharold, J.V. (1990). Movements of blue sharks (*Prionace glauca*) in depth and course. *Mar. Biol.* 106:329–342.

Carey, F.G., Gabrielson, G., Kanwisher, J.W., and Brazier, O. (1982). The white shark, *Carcharodon carcharias*, is warmbodied. *Copeia* 1982:254–260.

Carlisle, A.B. and Starr, R.M. (2010). Tidal movements of female leopard sharks (*Triakis semifasciata*) in Elkhorn Slough, California. *Environ. Biol. Fish.* 89:31–45.

Carlson, J.K. (2007). Modeling the role of sharks in the trophic dynamics of Apalachicola Bay, Florida. *Am. Fish. Soc. Symp.* 50:281–300.

Caro, T.M. (1995). Pursuit-deterrence revisited. *Trends Ecol. Evol.* 10:500–503.

Carraro, R. and Gladstone, W. (2006). Habitat preferences and site fidelity of ornate wobbegong shark (*Orectolobus ornatus*) on rocky reefs of New South Wales. *Pac. Sci.* 60:207–223.

Cartamil, D.P., Vaudo, J.J., Lowe, C, G., Wetherbee, B.M., and Holland KN. (2003). Diel movement patterns of the Hawaiian stingray *Dasyatis lata*: implications for ecological interactions between sympatric elasmobranch species. *Mar. Biol.* 142:841–847.

Cartamil, D.P., Wegner, N.C., Aalbers, S., Sepulveda, C.A., Baquero, A., and Graham, J.B. (2010). Diel movement patterns and habitat preferences of the common thresher shark (*Alopias vulpinus*) in the Southern California Bight. *Mar. Freshw. Res.* 61:596–604.

Cartamil, D.P., Sepulveda, C.A., Wegner, N.C., Aalbers, S.A., Baquero, A., and Graham, J.B. (2011). Archival tagging of subadult and adult common thresher sharks (*Alopias vulpinus*) off the coast of southern California. *Mar. Biol.* 158:935–944.

Castro, J.I. (1987). The position of sharks in marine biological communities: an overview. In: Cook, S. (Ed.), *Sharks, An Inquiry into Biology, Behavior, Fisheries, and Use*. Oregon State University Extension Service, Corvallis, pp. 11–17.

Castro, J.I. (1993). The shark nursery area of Bulls Bay, South Carolina, with a review of the shark nurseries of the southeastern coast of the United States. *Environ. Biol. Fish.* 38:37–48.

Castro, J.I. (2000). The biology of the nurse shark, *Ginglymostoma cirratum*, off the Florida east coast and the Bahamas Islands. *Environ. Biol. Fish.* 58:1–22.

Chapman, D.D. and Gruber, S.H. (2002). A further observation of the prey-handling behavior of the great hammerhead shark, *Sphyrna mokarran*: predation upon the spotted eagle ray, *Aetobatus narinari*. *Bull. Mar. Sci.* 70:947–952.

Claes, J.M., Aksnes, D.L., and Mallefet, J. (2010). Phantom hunter of the fjords: camouflage by counterillumination in a shark (*Etmopterus spinax*). *J. Exp. Mar. Biol. Ecol.* 388:28–32.

Clark, C.W. (1994). Antipredator behaviour and the asset-protection principle. *Behav. Ecol.* 5:159–170.

Clark, C.W. and Mangel, M. (2000). *Dynamic State Variable Models in Ecology: Methods and Application*. Oxford University Press, Oxford.

Clark, E. and Kristof, E. (1990). Deep-sea elasmobranchs observed from submersibles off Bermuda, Grand Cayman, and Freeport, Bahamas. In: Pratt, Jr., H.L., Gruber, S.H., and Taniuchi, T. (Eds.), *Elasmobranchs as Living Resources: Advances in the Biology, Ecology, Systematics, and Status of the Fisheries*, NOAA Tech. Rep. NMFS 90. U.S. Department of Commerce, Washington, D.C., pp. 269–284.

Clarke, T.A. (1971). The ecology of the scalloped hammerhead shark, *Sphyrna lewini*, in Hawaii. *Pac. Sci.* 25:133–144.

Cliff, G. (1995). Sharks caught in the protective gill nets off Kwazulu-Natal, South Africa. 8. The great hammerhead shark *Sphyrna mokarran* (Rüppell). *S. Afr. J. Mar. Sci.* 15:105–114.

Cliff, G. and Dudley, S.F.J. (1991). Sharks caught in the protective gill nets off Natal, South Africa. 4. The bull shark *Carcharhinus leucas* Valenciennes., *S. Afr. J. Mar. Sci.* 10:253–270.

Cliff, G. and Dudley, S.F.J. (1992). Sharks caught in the protective gill nets off Natal, South Africa. 6. The copper shark *Carcharhinus brachyurus* (Gunther). *S. Afr. J. Mar. Sci.* 12:663–674.

Cliff, G., Dudley, S.F.J., and Davis, B. (1989). Sharks caught in the protective gill nets off Natal, South Africa. 2. The great white shark *Carcharodon carcharias* (Linnaeus). *S. Afr. J. Mar. Sci.* 8:131–144.

Cockcroft, V.G., Cliff, G., and Ross, G.J.B. (1989). Shark predation on Indian Ocean bottlenose dolphins *Tursiops truncatus* off Natal, South Africa. *S. Afr. J. Zool.* 24:305–310.

Collins, A.B., Heupel, M.R., Hueter, R.E., and Motta, P.J. (2007a). Hard prey specialists or opportunistic generalists? An examination of the diet of the cownose ray, *Rhinoptera bonasus*. *Mar. Freshw. Res.* 58:135–144.

Collins, A.B., Heupel, M.R., and Motta, P.J. (2007b). Residence and movement patterns of cownose rays *Rhinoptera bonasus* within a south-west Florida estuary. *J. Fish Biol.* 71:1159–1178.

Compagno, L.J.V. (1984a). *FAO Species Catalogue*. Vol. 4. *Sharks of the World: An Annotated and Illustrated Catalogue of Shark Species Known to Date. Part 1. Hexanchiformes to Lamniformes*. United Nations Food and Agriculture Organization, Rome, pp. 1–250.

Compagno, L.J.V. (1984b). *FAO Species Catalogue*. Vol. 4. *Sharks of the World: An Annotated and Illustrated Catalogue of Shark Species Known to Date. Part 2. Carchariniformes*. United Nations Food and Agriculture Organization, Rome, pp. 250–655.

Compagno, L.J.V. (1990). Relationships of the megamouth shark, *Megachasma pelagios* (Lamniformes: Megachasmidae), with comments on its feeding habits. In: Pratt, Jr., H.L., Gruber, S.H., and Taniuchi, T. (Eds.), *Elasmobranchs as Living Resources: Advances in the Biology, Ecology, Systematics, and Status of the Fisheries*, NOAA Tech. Rep. NMFS 90. U.S. Department of Commerce, Washington, D.C., pp. 357–379.

Corkeron, P.J., Morris, R.J., and Bryden, M.M. (1987). Interactions between bottlenose dolphins and sharks in Moreton Bay, Queensland. *Aquat. Mamm.* 13:109–113.

Cortés, E. and Gruber, S.H. (1990). Diet, feeding habits and estimation of daily ration of young lemon sharks, *Negaprion brevirostris* (Poey). *Copeia* 1990:204–218.

Cortés, E., Manire, C.A., and Hueter, R.E. (1996). Diet, feeding habits, and diel feeding chronology of the bonnethead shark, *Sphyrna tiburo*, in southwest Florida. *Bull. Mar. Sci.* 58:353–367.

Corti, P. and Shackleton, D.M. (2002). Relationship between predation-risk factors and sexual segregation in Dall's sheep (*Ovis dalli dalli*). *Can. J. Zool.* 80:2108–2117.

Creel, S. and Christianson, D. (2008). Relationships between direct predation and risk effects. *Trends Ecol. Evol.* 23:194–201.

Cresswell, W. (1994). Age-dependent choice of redshank (*Tringa tetanus*) feeding location: profitability or risk? *J. Anim. Ecol.* 63:589–600.

DeVos, A. and O'Riain, M.J. (2010). Sharks shape the geometry of a selfish herd: experimental evidence from seal decoys. *Biol. Lett.* 6:48–50.

Dewitt, L.A. (1977). Notes on shark species in nearshore waters near Point Conception, California. *Cal. Fish Game* 63:273–274.

Di Santo, V. and Bennett, W.A. (2011). Is post-feeding thermotaxis advantageous in elasmobranch fishes? *J. Fish Biol.* 78:195–207.

Diamond, J.M. (1985). Filter-feeding on a grand scale. *Nature* 316:679–680.

Dicken, M.L. (2008). First observations of young of the year and juvenile great white sharks (*Carcharodon carcharias*) scavenging from a whale carcass. *Mar. Freshw. Res.* 59:596–602.

Dill, L.M. (1990). Distance-to-cover and the escape decisions of an Africa cichlid fish, *Melanochromis chipokae*. *Environ. Biol. Fish.* 27:147–152.

Dill, L.M., Heithaus, M.R., and Walters, C.J. (2003). Behaviorally-mediated indirect species interactions in marine communities and their importance to conservation and management. *Ecology* 84:1151–1157.

Domeier, M.L. and Nasby-Lucas, N. (2008). Migration patterns of white sharks *Carcharodon carcharias* tagged at Guadalupe Island, Mexico, and identification of an eastern Pacific shared offshore foraging area. *Mar. Ecol. Prog. Ser.* 370:221–237.

Dudley, S.F.J. and Cliff, G. (1993). Sharks caught in the protective gill nets off Natal, South Africa. 7. The blacktip shark *Carcharhinus limbatus* (Valenciennes). *S. Afr. J. Mar. Sci.* 13:237–254.

Dudley, S.F.J., Anderson-Reade, M.D., Thompson, G.S., and McMullen, P.B. (2000). Concurrent scavenging off a whale carcass by great white sharks, *Carcharodon carcharias*, and tiger sharks, *Galeocerdo cuvier*. *Fish. Bull.* 98:646–649.

Dudley, S.F.J., Cliff, G., Zungu, M.P., and Smale, M.J. (2005). Sharks caught in the protective gill nets off Kwazulu-Natal, South Africa. 10. The dusky shark *Carcharhinus obscurus* (Lesueur 1818). *S. Afr. J. Mar. Sci.* 27:107–127.

Dugatkin, L.A. and Reeve, H.K. (1998). *Game Theory and Animal Behavior*. Oxford University Press, Oxford.

Dulvy, N.K., Metcalfe, J.D., Glanville, J., Pawson, M.G., and Reynolds, J.D. (2000). Fishery stability, local extinctions, and shifts in community structure. *Conserv. Biol.* 14:283–293.

Duncan, K.M. and Holland, K.N. (2006). Habitat use, growth rates and dispersal patterns of juvenile scalloped hammerhead sharks *Sphyrna lewini* in a nursery habitat. *Mar. Ecol. Prog. Ser.* 312:211–221.

Dunphy-Daly, M.M., Heithaus, M.R., Wirsing, A.J., Mardon, J.S.F., and Burkholder, D.A. (2010). Predation risk influences the diving behavior of a marine mesopredator. *Open Ecol. J.* 3:8–15.

Ebert, D.A. (1991). Observations on the predatory behavior of the sevengill shark *Notorynchus cepedianus*. *S. Afr. J. Mar. Sci.* 11:455–465.

Ebert, D.A. (1994). Diet of the sixgill shark *Hexanchus griseus* off southern Africa. *S. Afr. J. Mar. Sci.* 14:213–218.

Ebert, D.A. (2002). Ontogenetic changes in the diet of the sevengill shark (*Notorynchus cepedianus*). *Mar. Freshw. Res.* 53:517–523.

Ebert, D.A., Compagno, L.J.V., and Cowley, P.D. (1992). A preliminary investigation of the feeding ecology of squaloid sharks off the west coast of southern Africa. *S. Afr. J. Mar. Sci.* 12:601–609.

Economakis, A.E. and Lobel, P.S. (1998). Aggregation behavior of grey reef sharks, *Carcharhinus amblryhynchos*, at Johnson Atoll, Central Pacific Ocean. *Environ. Biol. Fish.* 51:129–139.

Ellis, J.R., Pawson, M.G., and Shackley, S.E. (1996). The comparative feeding ecology of six species of shark and four species of ray (Elasmobranchii) in the north-east Atlantic. *J. Mar. Biol. Assoc. U.K.* 76:89–106.

Ferrari, M.C.O., Sih, A., and Chivers, D.P. (2009). The paradox of risk allocation: a review and prospectus. *Anim. Behav.* 78:579–585.

Ferretti, F., Worm, B., Britten, G.L., Heithaus, M.R., and Lotze, H.K. (2010). Patterns and ecosystem consequences of shark declines in the ocean. *Ecol. Lett.* 13:1055–1071.

Fertl, D., Acevedo-Guiterrez, A., and Darby, F.L. (1996). A report of killer whales (*Orcinus orca*) feeding on a carcharhinid shark in Costa Rica. *Mar. Mamm. Sci.* 12:606–611.

Fitzpatrick, B., Meekan, M., and Richards, A. (2006). Shark attacks on a whale shark (*Rhincodon typus*) at Ningaloo Reef, Western Australia. *Bull. Mar. Sci.* 78:397–402.

Ford, J.K.B., Ellis, G.M., Matkin, C.O., Wetklo, M.H., Barrett-Lennard, L.G., and Withler, R.E. (2011). Shark predation and tooth wear in a population of northeastern Pacific killer whales. *Aquat. Biol.* 11:213–224.

Fouts, W.R. and Nelson, D.R. (1999). Prey capture by the Pacific angel shark, *Squatina californica*: visually mediated strikes and ambush-site characteristics. *Copeia* 1999:304–312.

Francis, M.P. and Duffy., C. (2002). Distribution, seasonal abundance and bycatch of basking sharks (*Cetorhinus maximus*) in New Zealand, with observations on their winter habitat. *Mar. Biol.* 140:831–842.

Fretwell, S.D. and Lucas, H.L. (1969). On territorial behavior and other factors influencing habitat distribution in birds. *Acta Biotheor.* 19:16–36.

Frid, A., Dill, L.M., Thorne, R.E., and Blundell GM. (2007). Inferring prey perception of relative danger in large-scale marine systems. *Evol. Ecol. Res.* 9:635–649.

Frid, A., Burns, J., Baker, G.G., and Thorne, R.E. (2009). Predicting synergistic effects of resources and predators on foraging decisions by juvenile Steller sea lions. *Oecologia* 158:775–786.

Frisk, M.G., Miller, T.J., Martell, S.J.D., and Sosebee, K. (2008). New hypothesis helps explain elasmobranch 'outburst' on George's Bank in the 1980s. *Ecol. Appl.* 18:234–245.

Garrone-Neto, D. and Sazima, I. (2009). Stirring, charging, and picking: hunting tactics of potamotrygonid rays in the upper Parana River. *Neotrop. Ichth.* 7:113–116.

Gerritsen, J. and Strickler, J.R. (1977). Encounter probabilities and community structure in zooplankton: a mathematical model. *J. Fish. Res. Bd. Can.* 34:73–82.

Gilliam, J.F. and Fraser, D.F. (1987). Habitat selection under predation hazard: test of a model with foraging minnows. *Ecology* 68:1856–1862.

Goldman, K.J. (1997). Regulation of body temperature in the white shark, *Carcharodon carcharias*. *J. Comp. Physiol. B Biochem. Syst. Environ. Physiol.* 167:423–429.

Goldman, K.J. and Anderson, S.D. (1999). Space utilization and swimming depth of white sharks, *Carcharodon carcharias*, at the South Farallon Islands, central California. *Environ. Biol. Fish.* 56:351–364.

Goldman, K.J., Anderson, S.D., Latour, R.J., and Musick, J.A. (2004). Homeothermy in adult salmon sharks, *Lamna ditropis*. *Environ. Biol. Fish.* 71:403–411.

Gotceitas, V. and Colgan, P. (1989). Predator foraging success and habitat complexity: quantitative test of the threshold hypothesis. *Oecologia* 80:158–166.

Gray, A.E., Mulligan, T.J., and Hannah, R.W. (1997). Food habits, occurrence, and population structure of the bat ray, *Myliobatis californica*, in Humboldt Bay, California. *Environ. Biol. Fish.* 49:227–238.

Gregory, M.R., Balance, P.F., Gibson, G.W., and Ayling, A.M. (1979). On how some rays (Elasmobranchia) excavate feeding depressions by jetting water. *J. Sed. Petrol.* 49:1125–1130.

Gross, M.R. (1996). Alternative reproductive strategies and tactics: diversity within sexes. *Trends Ecol. Evol.* 11:92–98.

Grubbs, R.D. and Musick, J.A. (2007). Spatial delineation of summer nursery areas for juvenile sandbar sharks, *Carcharhinus plumbeus*, in the Mid-Atlantic Bight. *Am. Fish. Soc. Symp.* 50:63–86.

Gruber, S.H., de Marignac, J.R.C., and Hoenig, J.M. (2001). Survival of juvenile lemon sharks at Bimini, Bahamas, estimated by mark-depletion experiments. *Trans. Am. Fish. Soc.* 130:376–384.

Gunn, J.S., Stevens, J.D., Davis, T.L.O., and Norman, B.M. (1999). Observations on the short-term movements and behaviour of whale sharks (*Rhincodon typus*) at Ningaloo Reef, Western Australia. *Mar. Biol.* 135:553–559.

Hamilton, I.M. and Heithaus, M.R. (2001). The effects of temporal variation in predation risk on anti-predator behaviour: an empirical test using marine snails. *Proc. R. Soc. B Biol. Sci.* 268:2585–2588.

Hart, P.J.B. and Gill, A.B. (1992). Choosing prey size: a comparison of static and dynamic foraging models for predicting prey choice by fish. *Mar. Behav. Phys.* 22:93–106.

Hart, P.J.B. and Ison, S. (1991). The influence of prey size and abundance and individual phenotype on prey choice by three-spined stickleback, *Gasterosteus aculeatus*, L. *J. Fish Biol.* 38:359–372.

Heithaus, M.R. (2001a). Predator–prey and competitive interactions between sharks (order Selachii) and dolphins (suborder Odontoceti): a review. *J. Zool. (Lond.)* 253:53–68.

Heithaus, M.R. (2001b). Habitat selection by predators and prey in communities with asymmetrical intraguild predation. *Oikos* 92:542–554.

Heithaus, M.R. (2001c). The biology of tiger sharks (*Galeocerdo cuvier*) in Shark Bay, Western Australia: sex ratio, size distribution, diet, and seasonal changes in catch rates. *Environ. Biol. Fish.* 61:25–36.

Heithaus, M.R. (2001d). Shark attacks on bottlenose dolphins (*Tursiops aduncus*) in Shark Bay, Western Australia: attack rate, bite scar frequencies, and attack seasonality. *Mar. Mamm. Sci.* 17:526–539.

Heithaus, M.R. (2005). Habitat use and group size of pied cormorants (*Phalacrocorax varius*) in a seagrass ecosystem: possible effects of food abundance and predation risk. *Mar. Biol.* 147:27–35.

Heithaus, M.R. (2007). Nursery areas as essential shark habitats: a theoretical perspective. *Am. Fish. Soc. Symp.* 50:3–13.

Heithaus, M.R. and Dill, L.M. (2002a). Food availability and tiger shark predation risk influence bottlenose dolphin habitat use. *Ecology* 83:480–491.

Heithaus, M.R. and Dill, L.M. (2002b). Feeding strategies and tactics. In: Perrin, W.F., Würsig, B., and Thewissen, J.G.M. (Eds.), *Encyclopedia of Marine Mammals*. Academic Press, New York, pp. 412–422.

Heithaus, M.R. and Dill, L.M. (2006). Does tiger shark predation risk influence foraging habitat use by bottlenose dolphins at multiple spatial scales? *Oikos* 114:257–264.

Heithaus, M.R., Marshall, G.J., Buhleier, B.M., and Dill LM. (2001). Employing Crittercam to study the behavior and habitat use of large sharks. *Mar. Ecol. Prog. Ser.* 209:307–310.

Heithaus, M.R., Dill, L.M., Marshall, G.J., and Buhleier, B. (2002a). Habitat use and foraging behavior of tiger sharks (*Galeocerdo cuvier*) in a seagrass ecosystem. *Mar. Biol.* 140:237–248.

Heithaus, M.R., Frid, A., and Dill, L.M. (2002b). Shark-inflicted injury frequencies, escape ability, and habitat use of green and loggerhead turtles. *Mar. Biol.* 140:229–236.

Heithaus, M.R., Frid, A., Wirsing, A.J., Bejder, L., and Dill, L.M. (2005). Biology of sea turtles under risk from tiger sharks at a foraging ground. *Mar. Ecol. Prog. Ser.* 288:285–294.

Heithaus, M.R., Hamilton, I.M., Wirsing, A.J., and Dill, L.M. (2006). Validation of a randomization procedure to assess animal habitat preferences: microhabitat use of tiger sharks in a seagrass ecosystem. *J. Anim. Ecol.* 75:666–676.

Heithaus, M.R., Frid, A., Wirsing, A.J., Dill, L.M., Fourqurean, J.W., Burkholder, D., Thomson, J., and Bejder, L. (2007a). State-dependent risk-taking by green turtles mediates top-down effects of tiger sharks. *J. Anim. Ecol.* 76:837–844.

Heithaus, M.R., Wirsing, A.J., Frid, A., and Dill, L.M. (2007b). Species interactions and marine conservation: lessons from an undisturbed ecosystem. *Israel J. Ecol. Evol.* 53:355–370.

Heithaus, M.R., Frid, A., Wirsing, A.J., and Worm, B. (2008a). Predicting ecological consequences of marine top predator declines. *Trends Ecol. Evol.* 23:202–210.

Heithaus, M.R., Wirsing, A.J., Thomson, J.A., and Burkholder, D. (2008b). A review of lethal and non-lethal effects of predators on adult marine turtles. *J. Exp. Mar. Biol. Ecol.* 356:43–51.

Heithaus, M.R., Delius, B.K., Wirsing, A.J., and Dunphy-Daly, M.M. (2009a). Physical factors influencing the distribution of a top predator in a subtropical oligotrophic estuary. *Limnol. Oceanogr.* 54:472–482.

Heithaus, M.R., Wirsing, A.J., Burkholder, D., Thomson, J.A., and Dill, L.M. (2009b). Towards a predictive framework for predator risk effects: the interaction of landscape features and prey escape tactics. *J. Anim. Ecol.* 78:556–562.

Heithaus, M.R., Frid, A., Vaudo, J.J., Worm, B., and Wirsing, A.J. (2010). Unraveling the ecological importance of elasmobranchs. In: Carrier, J., Musick, J.A., and Heithaus, M. (Eds.), *Sharks and Their Relatives II: Biodiversity, Adaptive Physiology, and Conservation*. CRC Press, Boca Raton, FL, pp. 611–637.

Heupel, M.R. and Bennett, M.B. (1998). Observations on the diet and feeding habits of the epaulette shark, *Hemiscyllium ocellatum* (Bonnaterre), on Heron Island Reef, Great Barrier Reef, Australia. *Mar. Freshw. Res.* 49:753–756.

Heupel, M.R. and Hueter, R.E. (2002). The importance of prey density in relation to the movement patterns of juvenile sharks within a coastal nursery area. *Mar. Freshw. Res.* 53:543–550.

Heupel, M.R., Carlson, J.K., and Simpfendorfer, C.A. (2007). Shark nursery areas: concepts, definition, characterization and assumptions. *Mar. Ecol. Prog. Ser.* 337:287–297.

Heyman, W.D., Graham, R.T., Kjerfve, B., and Johannes, R.E. (2001). Whale sharks *Rhincodon typus* aggregate to feed on fish spawn in Belize. *Mar. Ecol. Prog. Ser.* 215:275–282.

Hines, A.H., Whitlatch, R.B., Thrush, S.F., Hewitt, J.E., Cummings, V.J., Dayton, P.K., and Legendre, P. (1997). Nonlinear foraging response of a large marine predator to benthic prey: eagle ray pits and bivalves in a New Zealand sandflat. *J. Exp. Mar. Biol. Ecol.* 216:191–201.

Hiruki, L.M., Gilmartin, W.G., Becker, B.L., and Stirling, I. (1993). Wounding in Hawaiian monk seals (*Monachus schauinslandi*). *Can. J. Zool.* 71:458–468.

Holland, K.N., Wetherbee, B.M., Peterson, J.D., and Lowe, C.G. (1993). Movements and distribution of hammerhead shark pups on their natal grounds. *Copeia* 1993:495–502.

Holland, K.N., Wetherbee, B.M., Lowe, C.G., and Meyer, C.G. (1999). Movements of tiger sharks (*Galeocerdo cuvier*) in coastal Hawaiian waters. *Mar. Biol.* 134:665–673.

Holt, R.D. (1977). Predation, apparent competition and the structure of prey communities. *Theor. Pop. Biol.* 12:197–229.

Holt, R.D. (1984). Spatial heterogeneity, indirect interactions, and the coexistence of prey species. *Am. Nat.* 124:377–406.

Holt, R.D. and Polis, G.A. (1997). A theoretical framework for intraguild predation. *Am. Nat.* 149:745–764.

Holts, D.B. and Bedford, D.W. (1993). Horizontal and vertical movements of shortfin mako shark, *Isurus oxyrinchus*, in the southern California Bight. *Aust J. Mar. Freshw. Res.* 44:901–909.

Hopkins, T.E. and Cech, Jr., J.J. (1994). Effect of temperature on oxygen consumption of the bat ray, *Myliobatis californica* (Chondrichthyes, Myliobatidae). *Copeia* 1994:529–532.

Houston, A.I., McNamara, J.M., and Hutchinson, J.M.C. (1993). General results concerning the trade-off between gaining energy and avoiding predation. *Phil. Trans. R. Soc. Lond. B* 341:375–397.

Houtman, R. and Dill, L.M. (1998). The influence of predation risk on diet selectivity: a theoretical analysis. *Evol. Ecol.* 12:251–262.

Howard, J.D., Mayou, T.V., and Heard, R.W. (1977). Biogenic sedimentary structures formed by rays. *J. Sediment. Petrol.* 47:339–346.

Hugie, D.M. and Dill, L.M. (1994). Fish and game: a game theoretic approach to habitat selection by predators and prey. *J. Fish Biol.* 45(Suppl. A):151–169.

Hulbert, L.B., Aires-Da-Silva, A.M., Gallucci, V.F., and Rice, J.S. (2005). Seasonal foraging movements and migratory patterns of female *Lamna ditropis* tagged in Prince William Sound. *J. Fish Biol.* 67:490–509.

Hulbert, L.B., Sigler, M.F., and Lunsford, C.R. (2006). Depth and movement behaviour of the Pacific sleeper shark in the north-east Pacific Ocean. *J. Fish Biol.* 69:406–425.

Humphries, N.E., Queiroz, N., Dyer, J.R.M., Pade, N.G., Musyl, M.K., Schaefer, K.M., Fuller, D.W., Brunnschweiler, J.M., Doyle, T.K., Houghton, J.D.R., Hays, G.C., Jones, C.S., Noble, L.R., Wearmouth, V.J., Southall, E.J., and Sims, D.W. (2010). Environmental context explains Lévy and Brownian movement patterns of marine predators. *Nature* 465:1066–1069.

Iwasa, Y. (1982). Vertical migration of zooplankton: a game between predator and prey. *Am. Nat.* 120:171–180.

Jones, E.C. (1971). *Isistius brasiliensis*, a squaloid shark, the probable cause of crater wounds on fishes and cetaceans. *Fish. Bull.* 69:791–798.

Jorgensen, S.J., Reeb, C.A., Chapple, T.K., Anderson, S., Perle, C., Van Sommeran, S.R., Fritz-Cope, C., Brown, A.C., Klimley, A.P., and Block, B.A. (2010). Philopatry and migration of Pacific white sharks. *Proc. R. Soc. B Biol. Sci.* 277:679–688.

Kajiura, S.M. and Tricas, T.C. (1996). Seasonal dynamics of dental sexual dimorphism in the Atlantic stingray (*Dasyatis sabina*). *J. Exp. Biol.* 199:2297–2306.

Kerford, M.R., Wirsing, A.J., Heithaus, M.R., and Dill, L.M. (2008). Danger on the rise: diurnal tidal state mediates an exchange of food for safety by the bar-bellied sea snake *Hydrophis elegans*. *Mar. Ecol. Prog. Ser.* 358:289–294.

Kimber, J.A., Sims, D.W., Bellamy, P.H., and Gill, A.B. (2009). Male–female interactions affect foraging behaviour within groups of small-spotted catshark, *Scyliorhinus canicula*. *Anim. Behav.* 77:1435–1440.

Kinney, M.J. and Simpfendorfer, C.A. (2009). Reassessing the value of nursery areas to shark conservation and management. *Conserv. Lett.* 2:53–60.

Kitchell, J.F., Essington, T.E., Boggs, C.H., Schindler, D.E., and Walters, C.J. (2002). The role of sharks and longline fisheries in a pelagic ecosystem of the central Pacific. *Ecosystems* 5:2002–2016.

Klimley, A.P. (1985). The areal distribution and autoecology of the white shark, *Carcharodon carcharias*, off the west coast of North America. *Mem. S. Calif. Acad. Sci.* 9:15–40.

Klimley, A.P. (1987). The determinants of sexual selection in the scalloped hammerhead shark, *Sphyrna lewini*. *Environ. Biol. Fish.* 18:27–40.

Klimley, A.P. (1993). Highly directional swimming by scalloped hammerhead sharks, *Sphyrna lewini*, and subsurface irradiance, temperature, bathymetry, and geomagnetic field. *Mar. Biol.* 117:1–22.

Klimley, A.P. (1994). The predatory behavior of the white shark. *Am. Sci.* 52:122–133.

Klimley, A.P. and Anderson, S.D. (1996). Residency patterns of white sharks at the South Farallon Islands, California. In: Klimley, A.P. and Ainley, D.G. (Eds.), *Great White Sharks: The Biology of Carcharodon carcharias*. Academic Press, New York, pp. 365–374.

Klimley, A.P. and Butler, S.B. (1988). Immigration and emigration of a pelagic fish assemblage to seamounts in the Gulf of California related to water mass movements using satellite imagery. *Mar. Ecol. Prog. Ser.* 49:11–20.

Klimley, A.P. and Nelson, D.R. (1981). Schooling of the scalloped hammerhead shark, *Sphyrna lewini*, in the Gulf of California. *Fish. Bull.* 79:356–360.

Klimley, A.P. and Nelson, D.R. (1984). Diel movement patterns of the scalloped hammerhead shark (*Sphyrna lewini*) in relation to El Bajo Espiritu Santo: a refuging central-position social system. *Behav. Ecol. Sociobiol.* 15:45–54.

Klimley, A.P., Butler, S.B., Nelson, D.R., and Stull, A.T. (1988). Diel movements of scalloped hammerhead sharks, *Sphyrna lewini* Griffith and Smith, to and from a seamount in the Gulf of California. *J. Fish Biol.* 33:751–761.

Klimley, A.P., Anderson, S.D., Pyle, P., and Henderson, R.P. (1992). Spatiotemporal patterns of white shark (*Carcharodon carcharias*) predation at the South Farallon Islands, California. *Copeia* 1992:680–690.

Klimley, A.P., Pyle, P., and Anderson, S.D. (1996a). The behavior of white sharks and their pinniped prey during predatory attacks. In: Klimley, A.P. and Ainley, D.G. (Eds.), *Great White Sharks: The Biology of Carcharodon carcharias*. Academic Press, New York, pp. 175–191.

Klimley, A.P., Pyle, P., and Anderson, S.D. (1996b). Tail slap and breach: agonistic displays among white sharks? In: Klimley, A.P. and Ainley, D.G. (Eds.), *Great White Sharks: The Biology of Carcharodon carcharias*. Academic Press, New York, pp. 241–255.

Klimley, A.P., Le Boeuf, B.J., Cantara, K.M., Richert, J.E., Davis, S.F., Van Sommeran, S., and Kelly, J.T. (2001). The hunting strategy of white sharks (*Carcharodon carcharias*) near a seal colony. *Mar. Biol.* 138:617–636.

Klimley, A.P., Beavers, S.C., Curtis, T.H., and Jorgensen, S.J. (2002). Movements and swimming behavior of three species of sharks in La Jolla Canyon, California. *Environ. Biol. Fish.* 63:117–135.

Koen Alonso, M., Crespo, E.A., García, N.A., Pedraza, S.N., Mariotti, P.A., Berón Vera, B., and Mora, N.J. (2001). Food habits of *Dipturus chilensis* (Pisces: Rajidae) off Patagonia, Argentina. *ICES J. Mar. Sci.* 58:288–297.

Kondoh, M. (2008). Building trophic modules into a persistent food web. *Proc. Natl. Acad. Sci. U.S.A.* 105:16631–16635.

Kotler, B.P., Brown, J., Mukherjee, S., Berger-Tal, O., and Bouskila, A. (2010). Moonlight avoidance in gerbils reveals a sophisticated interplay among time allocation, vigilance and state-dependent foraging. *Proc. R. Soc. B Biol. Sci.* 277:1469–1474.

Laroche, R.K., Kock, A.A., Dill, L.M., and Oosthuizen, W.H. (2008). Running the gauntlet: a predator–prey game between sharks and two age classes of seals. *Anim. Behav.* 76:1901–1917.

Leatherwood, S. (1977). Some preliminary impressions of the numbers and social behavior of free-swimming bottle-nosed dolphin calves (*Tursiops truncatus*) in the northeastern Gulf of Mexico. In: Ridgway, S.H. and Benirschke, K. (Eds.), *Breeding Dolphins: Present Status, Suggestions for the Future*, Natl. Tech. Info. Serv. PB-0273 673. U.S. Marine Mammal Commission, Washington, D.C., pp. 143–167.

LeBoeuf, B.J., Reidman, M., and Keyes, R.S. (1982). White shark predation on pinnipeds in California coastal waters. *Fish. Bull.* 80:891–895.

Lima, S.L. (1995). Back to the basics of anti-predatory vigilance: the group size effect. *Anim. Behav.* 49:11–20.

Lima, S.L. (1998a). Stress and decision making under the risk of predation: recent developments from behavioral, reproductive, and ecological perspectives. *Adv. Stud. Behav.* 27:215–290.

Lima, S.L. (1998b). Nonlethal effects in the ecology of predator–prey interactions. *BioScience* 48:25–34.

Lima, S.L. (2002). Putting predators back into behavioral predator–prey interactions. *Trends Ecol. Evol.* 17:70–75.

Lima, S.L. and Bednekoff, P.A. (1999). Temporal variation in danger drives antipredator behavior: the predation risk allocation hypothesis. *Am. Nat.* 153:649–659.

Lima, S.L. and Dill, L.M. (1990). Behavioral decisions made under the risk of predation: a review and prospectus. *Can. J. Zool.* 68:619–640.

Lima, S.L., Mitchell, W.A., and Roth, T.C. (2003). Predators feeding on behaviourally responsive prey: some implications for classical models of optimal diet choice. *Evol. Ecol. Res.* 5:1083–1102.

Limpus, C.J., Couper, P.J., and Read, M.A. (1994a). The green turtle, *Chelonia mydas*, in Queensland: population structure in a warm temperate feeding area. *Mem. Queensl. Mus.* 35:139–154.

Limpus, C.J., Couper, P.J., and Read, M.A. (1994b). The loggerhead turtle, *Caretta caretta*, in Queensland: population structure in a warm temperate feeding area. *Mem. Queensl. Mus.* 37:195–204.

Long, D.J. and Jones, R.E. (1996). White shark predation and scavenging on cetaceans in the eastern north Pacific Ocean. In: Klimley, A.P. and Ainley, D.G. (Eds.), *Great White Sharks: The Biology of Carcharodon carcharias*. Academic Press, New York, pp. 293–307.

Lott, D.F. (1984). Intraspecific variation in the social system of wild vertebrates. *Behaviour* 88:266–325.

Lowe, C.G., Bray, R.N., and Nelson, D.R. (1994). Feeding and associated electrical behavior of the Pacific electric ray *Torpedo californica* in the field. *Mar. Biol.* 120:161–169.

Lowe, C.G., Wetherbee, B.M., Crow, G.L., and Tester, A.L. (1996). Ontogenetic dietary shifts and feeding behavior of the tiger shark, *Galeocerdo cuvier*, in Hawaiian waters. *Environ. Biol. Fish.* 47:203–211.

Lucas, Z. and Stobo, W.T. (2000). Shark-inflicted mortality on a population of harbour seals (*Phoca vitulina*) at Sable Island, Nova Scotia. *J. Zool. (Lond.)* 252:405–414.

Lucifora, L.O., García, V.B., Menni, R.C., and Escalante, A.H. (2006). Food habits, selectivity, and foraging modes of the school shark *Galeorhinus galeus*. *Mar. Ecol. Prog. Ser.* 315:259–270.

Lucifora, L.O., García, V.B., Menni, R.C., Escalante, A.H., and Hozbor, N.M. (2009a). Effects of body size, age and maturity stage on diet in a large shark: ecological and applied implications. *Ecol. Res.* 24:109–118.

Lucifora, L.O., García, V.B., and Escalante, A.H. (2009b). How can the feeding habits of the sand tiger shark influence the success of conservation programs? *Anim. Cons.* 12:291–301.

Lundsten, L., Schlining, K.L., Frasier, K., Johnson, S.B., Kuhnz, L.A., Harvey, J.B.J., Clague, G., and Vrijenhoek, R.C. (2010). Time-series analysis of six whale-fall communities in Monterey Canyon, California, USA. *Deep Sea Res. I* 57:1573–1584.

Macesic, L.J. and Kajiura, S.M. (2009). Electric organ morphology and function in the lesser electric ray, *Narcine brasiliensis*. *Zoology* 112:442–450.

Martin, R.A., Hammerschlag, N., Collier, R.S., and Fallows, C. (2005). Predatory behaviour of white sharks (*Carcharodon carcharias*) at Seal Island, South Africa. *J. Mar. Biol. Assoc. U.K.* 85:1121–1135.

Martin, R.A., Rossmo, D.K., and Hammerschlag, N. (2009). Hunting patterns and geographic profiling of white shark predation. *J. Zool.* 279:111–118.

Matern, S.A., Cech, Jr., J.J., and Hopkins, T.E. (2000). Diel movements of bat rays, *Myliobatis californica*, in Tomales Bay, California: evidence for behavioral thermoregulation? *Environ. Biol. Fish.* 58:173–182.

Matich, P., Heithaus, M.R., and Layman, C.R. (2011). Contrasting patterns of individual specialization and trophic coupling in two marine apex predators. *J. Anim. Ecol.* 80:294–305.

McCosker, J.E. (1985). White shark attack behavior: observation of and speculations about predator–prey strategies. *Mem. S. Calif. Acad. Sci.* 9:123–135.

McKibben, J.N. and Nelson, D.R. (1986). Patterns of movement and grouping of gray reef sharks, *Carcharhinus amblyrhynchos*, at Enewetak, Marshall Islands. *Bull. Mar. Sci.* 38:89–110.

McNamara, J.M. and Houston, A.I. (1987). Starvation and predation as factors limiting population size. *Ecology* 68:1515–1519.

Medved, R.J. and Marshall, J.A. (1983). Short-term movements of young sandbar sharks, *Carcharhinus plumbeus* (Pisces, Carcharhinidae). *Bull. Mar. Sci.* 33:87–93.

Mesterson-Gibbons, M. and Dugatkin, L.A. (1992). Cooperation among unrelated individuals: evolutionary factors. *Q. Rev. Biol.* 67:267–281.

Meyer, C.G., Clark, T.B., Papastamatiou, Y.P., Whitney, N.M., and Holland, K.N. (2009). Long-term movement patterns of tiger sharks *Galeocerdo cuvier* in Hawaii. *Mar. Ecol. Prog. Ser.* 381:223–235.

Meyer, C.G., Papastamatiou, Y.P., and Holland, K.N. (2010). A multiple instrument approach to quantifying the movement patterns and habitat use of tiger (*Galeocerdo cuvier*) and Galapagos sharks (*Carcharhinus galapagensis*) at French Frigate Shoals, Hawaii. *Mar. Biol.* 157:1857–1868.

Michaelson, D.M., Sternberg, D., and Fishelson, L. (1979). Observations on feeding, growth, and electric discharge of newborn *Torpedo ocellata* (Chondrichthyes, Batoidei). *J. Fish Biol.* 15:159–163.

Miner, J.G. and Stein, R.A. (1996). Detection of predators and habitat choice by small bluegills: effects of turbidity and alternative prey. *Trans. Am. Fish. Soc.* 125:97–103.

Montgomery, J. and Skipworth, E. (1997). Detection of weak water jets by the short-tailed stingray *Dasyatis brevicaudata*. *Copeia* 1997:881–883.

Moody, A.L., Houston, A.I., and McNamara, J.M. (1996). Ideal free distributions under predation risk. *Behav. Ecol. Sociobiol.* 38:131–143.

Morato, T., Solà, E., Grós, M.P., and Menezes, G. (2003). Diets of thornback ray (*Raja clavata*) and tope shark (*Galeorhinus galeus*) in the boron longline fishery of the Azores, northeastern Atlantic. *Fish. Bull.* 101:590–602.

Morrissey, J.F. and Gruber, S.H. (1993a). Habitat selection by juvenile lemon sharks, *Negaprion brevirostris*. *Environ. Biol. Fish.* 38:311–319.

Morrissey, J.F. and Gruber, S.H. (1993b). Home range of juvenile lemon sharks, *Negaprion brevirostris*. *Copeia* 1993:425–434.

Motta, P.J. and Wilga, C.D. (2001). Advances in the study of feeding behaviors, mechanisms, and mechanics of sharks. *Environ. Biol. Fish.* 60:131–156.

Motta, P.J., Maslanka, M., Hueter, R.E., Davis, R.L., de la Parra, R., Mulvany, S.L., Habegger, M.L., Strother, J.A., Mara, K.R., Gardiner, J.M., Tyminski, J.P., and Zeigler, L.D. (2010). Feeding anatomy, filter-feeding rate, and diet of whale sharks *Rhincodon typus* during surface ram filter feeding off the Yucatan Penninsula, Mexico. *Zoology* 113:199–212.

Mukherjee, S., Zelcer, M., and Kotler, B.P. (2009). Patch use in time and space for a meso-predator in a risky world. *Oecologia* 159:661–668.

Musick, J.A. and Colvocoresses, J.A. (1988). Seasonal recruitment of subtropical sharks in Chesapeake Bight, U.S.A. In: Yanez-Arancibia, A. and Pauly, D. (Eds.), *IOC/FAO Workshop on Recruitment in Tropical Coastal Demersal Communities*, April 21–25, 1986, Ciudad del Carmen, Mexico. Intergovernmental Oceanographic Commission, Paris, pp. 301–311.

Musick, J.A., Branstetter, S., and Colvocoresses, J.A. (1993). Trends in shark abundance from 1974 to 1991 for the Chesapeake Bight region of the U.S. Mid-Atlantic coast. In: Branstetter, S. (Ed.), *Conservation Biology of Sharks*, NOAA Tech. Rep. NMFS 115. U.S. Department of Commerce, Washington, D.C., pp. 1–18.

Myberg, Jr., A.A. (1991). Distinctive markings of sharks: ethological considerations of visual function. *J. Exp. Zool.* 5(Suppl.):156–166.

Myers, R.A., Baum, J.K., Shepherd, T.D., Powers, S.P., and Peterson, C.H. (2007). Cascading effects of the loss of apex predatory sharks from a coastal ocean. *Science* 315:1846–1850.

Nakamura, I., Watanabe, Y.K., Papastamatiou, Y.P., Sato, K., and Meyer, C.G. (2011). Vertical movements of free-ranging tiger sharks *Galeocerdo cuvier*: locomotory cost saving or foraging? *Mar. Ecol. Prog. Ser.* 424:237–246.

Nakano, H., Matsunaga, H., Okamoto, H., and Okazaki, M. (2003). Acoustic tracking of bigeye thresher shark *Alopias superciliosus* in the eastern Pacific Ocean. *Mar. Ecol. Prog. Ser.* 265:255–261.

Nakaya, K., Matsumoto, R., and Suda, K. (2008). Feeding strategy of the megamouth shark *Megachasma pelagios* (Lamniformes: Megachasmidae). *J. Fish Biol.* 73:17–34.

Nasby-Lucas, N., Dewar, H., Lam, C.H., Goldman, K.J., and Domeier, M.L. (2009). White shark offshore habitat: a behavioral and environmental characterization of the eastern Pacific shared offshore foraging area. *PLoS One* 4:e8163.

Navia, A.F., Cortés, E., and Mejía-Falla, P.A. (2010). Topological analysis of the ecological importance of elasmobranch fishes: a food web study on the Gulf of Tortugas, Columbia. *Ecol. Model.* 221:2918–2926.

Nelson, D.R. and Johnson, R.H. (1970). Diel activity rhythms in the nocturnal, bottom-dwelling sharks, *Heterodontus francisci* and *Cephaloscyllium ventriosum*. *Copeia* 1970:732–739.

Nelson, D.R. and Johnson, R.H. (1980). Behavior of the reef sharks of Rangiroa, French Polynesia. *Natl. Geogr. Soc. Res. Rep.* 12:479–499.

Nelson, D.R., Johnson, R.H., McKibben, J.N., and Pittenger, G.G. (1986). Agonistic attacks on divers and submersibles by gray reef sharks, *Carcharhinus amblyrhynchos*: antipredatory or competitive? *Bull. Mar. Sci.* 28:68–88.

Nelson, D.R., McKibben, J.N., Strong, Jr., W.R., Lowe, C.G., Sisneros, J.A., Schroeder, D.M., and Lavenberg, R.J. (1997). An acoustic tracking of a megamouth shark, *Megachasma pelagios*: a crepuscular vertical migrator. *Environ. Biol. Fish.* 49:389–399.

Nelson, J.D. and Eckert, S.A. (2007). Foraging ecology of whale sharks (*Rhincodon typus*) within Bahia de Los Angeles, Baja California Norte, Mexico. *Fish. Res.* 84:47–64.

Noë, R. (2006). Cooperation experiments: coordination through communication versus acting apart together. *Anim. Behav.* 71:1–18.

Norris, K.S. and Dohl, T.P. (1980). Behavior of the Hawaiian spinner dolphin, *Stenella longirostris*. *Fish. Bull.* 77:821–849.

Olsen, A.M. (1984). *Synopsis of Biological Data on the School Shark, Galeorhinus australis* (Maleay 1881). United Nations Food and Agriculture Organization, Rome.

Ortega, L.A., Heupel, M.R., Van Beynen, P., and Motta, P.J. (2009). Movement patterns and water quality preferences of juvenile bull sharks (*Carcharhinus leucas*) in a Florida estuary. *Environ. Biol. Fish.* 84:361–373.

Paine, R.T. (1966). Food web complexity and species diversity. *Am. Nat.* 100:65–75.

Papastamatiou, Y.P., Wetherbee, B.M., Lowe, C.G., and Crow, G.L. (2006). Distribution and diet of four species of carcharhinid shark in the Hawaiian Islands: evidence for resource partitioning and competitive exclusion. *Mar. Ecol. Prog. Ser.* 320:239–251.

Papastamatiou, Y.P., Meyer, C.G., and Holland, K.N. (2007a). A new acoustic pH transmitter for studying the feeding habits of free-ranging sharks. *Aquat. Living Resour.* 20:287–290.

Papastamatiou, Y.P., Purkis, S.J., and Holland, K.N. (2007b). The response of gastric pH and motility to fasting and feeding in free swimming blacktip reef sharks, *Carcharhinus melanopterus*. *J. Exp. Mar. Biol. Ecol.* 345:129–140.

Papastamatiou, Y.P., Itano, D.G., Dale, J.J., Meyer, C.G., and Holland, K.N. (2010a). Site fidelity and movements of sharks associated with ocean-farming cages in Hawaii. *Mar. Freshw. Res.* 61:1366–1375.

Papastamatiou, Y.P., Wetherbee, B.M., O'Sullivan, J., Goodmanlowe, G.D., and Lowe, C.G. (2010b). Foraging ecology of cookiecutter sharks (*Isistius brasiliensis*) on pelagic fishes in Hawaii, inferred from prey bite wounds. *Environ. Biol. Fish.* 88:361–368.

Parrish, F.A., Marshall, G.J., Buhleier, B., and Antonelis, G.A. (2008). Foraging interaction between monk seals and large predatory fish in the Northwestern Hawaiian Islands. *Endang. Spec. Res.* 4:299–308.

Peacor, S.D. and Werner, E.E. (2000). Predator effects on an assemblage of consumers through induced changes in consumer foraging behavior. *Ecology* 81:1998–2010.

Peacor, S.D. and Werner, E.E. (2001). The contribution of trait-mediated indirect effects to the net effects of a predator. *Proc. Natl. Acad. Sci. U.S.A.* 98:3904–3908.

Peckarsky, B.L., Abrams, P.A., Bolnick, D.I., Dill, L.M., Grabowski, J.H., Luttbeg, B., Orrock, J.L., Peacor, S.D., Preisser, E.L., Schmitz, O.J., and Trussell, G.C. (2008). Revisiting the classics: considering nonconsumptive effects in textbook examples of predator–prey interactions. *Ecology* 89:2416–2425.

Peterson, C.H., Fodrie, F.J., Summerson, H.C., and Powers, S.P. (2001). Site-specific and density-dependent extinction of prey by schooling rays: generation of a population sink in top-quality habitat for bay scallops. *Oecologia* 129:349–356.

Piraino, S., Fanelli, G., and Boero, F. (2002). Variability of species' roles in marine communities: change of paradigms for conservation priorities. *Mar. Biol.* 140:1067–1074.

Polis, G.A., Meyers, C.A., and Holt, R.D. (1989). The ecology and evolution of intraguild predation: potential competitors that eat each other. *Annu. Rev. Ecol. Syst.* 20:297–330.

Power, M.E., Matthews, W.J., and Stewart, A.J. (1985). Grazing minnows, piscivorous bass, and stream algae: dynamics of a strong interaction. *Ecology* 66:1448–1456.

Pratt, Jr., H.L., Jr., Casey, J.G., and Conklin, R.B. (1982). Observations on large white sharks, *Carcharodon carcharias*, off Long Island, New York. *Fish. Bull.* 80:153–156.

Preisser, E.L., Bolnick, D.I., and Benard, M.F. (2005). Scared to death? The effects of intimidation and consumption in predator–prey interactions. *Ecology* 86:501–509.

Pulliam, H.R. and Caraco, T. (1984). Living in groups: is there an optimal group size? In: Krebs, J.R. and Davies, N.B. (Eds.), *Behavioural Ecology: An Evolutionary Approach*, 2nd ed. Blackwell Scientific, Oxford, pp. 122–147.

Pyle, P., Schramm, M.J., Keiper, C., and Anderson, S.D. (1999). Predation on a white shark (*Carcharodon carcharias*) by a killer whale (*Orcinus orca*) and a possible case of competitive displacement. *Mar. Mamm. Sci.* 15:563–568.

Queiroz, N., Humphries, N.E., Noble, L.R., Santos, A.M., and Sims, D.W. (2010). Short-term movements and diving behavior of satellite-tracked blue sharks *Prionace glauca* in the northeastern Atlantic Ocean. *Mar. Ecol. Prog. Ser.* 406:265–279.

Quevedo, M., Svanbäck, R., and Eklöv, P. (2009). Intrapopulation niche partitioning in a generalist predator limits food web connectivity. *Ecology* 90:2263–2274.

Randall, J.E. (1977). Contribution to the biology of the whitetip reef shark (*Triaenodon obesus*). *Pac. Sci.* 31:143–164.

Randall, J.E. (1992). Review of the biology of the tiger shark (*Galeocerdo cuvier*). *Aust. J. Mar. Freshw. Res.* 43:21–31.

Rasmussen, L.E.L. and Schmidt, M.J. (1992). Are sharks chemically aware of crocodiles? In: Dity, R.L. and Müller-Schwarze, D. (Eds.), *Chemical Signals in Vertebrates*, Vol. IV. Plenum Press, New York, pp. 335–342.

Relyea, R.A. (2000). Trait-mediated indirect effects in larval anurans: reversing competition with the threat of predation. *Ecology* 81:2278–2289.

Reynolds, A.M. and Rhodes, C.J. (2009). The Levy flight paradigm: random search patterns and mechanisms. *Ecology* 90:877–887.

Rosa, R.S., Charvet-Almeida, P., and Quijada, C.C.D. (2010). Biology of the South American Potamotrygonid stingrays. In: Carrier, J., Musick, J.A., and Heithaus, M. (Eds.), *Sharks and Their Relatives II: Biodiversity, Adaptive Physiology, and Conservation*. CRC Press, Boca Raton, FL, pp. 241–281.

Sasko, D.E., Dean, M.N., Motta, P.J., and Hueter, R.E. (2006). Prey capture behavior and kinematics of the Atlantic cownose ray, *Rhinoptera bonasus*. *Zoology* 109:171–181.

Schmitz, O.J. (2008). Effects of predator hunting mode on grassland ecosystem function. *Science* 319:952–954.

Schmitz, O.J., Beckerman, A.P., and O'Brien, K.M. (1997). Behaviorally mediated trophic cascades: effects of predation risk on food web interactions. *Ecology* 78:1388–1399.

Schmitz, O.J., Krivan, V., and Ovadia, O. (2004). Trophic cascades: the primacy of trait-mediated indirect interactions. *Ecol. Lett.* 7:153–163.

Sciarrotta, T.C. and Nelson, D.R. (1977). Diel behavior of the blue shark, *Prionace glauca*, near Santa Catalina Island, California. *Fish. Bull.* 75:519–528.

Semeniuk, C.A.D. and Dill, L.M. (2005). Cost/benefit analysis of group and solitary resting in the cowtail stingray, *Pastinachus sephen*. *Behav. Ecol.* 16:417–426.

Semeniuk, C.A.D. and Dill, L.M. (2006). Anti-predator benefits of mixed-species groups of cowtail stingrays (*Pastinachus sephen*) and whiprays (*Himantura uarnak*) at rest. *Ethology* 112:33–43.

Semeniuk, C.A.D. and Rothley, K.D. (2008). Costs of group-living for a normally solitary forager: effects of provisioning tourism on southern stingrays *Dasyatis americana*. *Mar. Ecol. Prog. Ser.* 357:271–282.

Semeniuk, C.A.D., Speers-Roesch, B., and Rothley, K.D. (2007). Using fatty-acid profile analysis as an ecologic indicator in the management of tourist impacts on marine wildlife: a case of stingray feeding in the Caribbean. *Environ. Manage.* 40:665–677.

Semeniuk, C.A.D., Bourgeon, S., Smith, S.L., and Rothley, K.D. (2009). Hematological differences between stingrays at tourist and non-visited sites suggest physiological costs of wildlife tourism. *Biol. Conserv.* 142:1818–1829.

Sepulveda, C.A., Kohin, S., Chan, C., Vetter, R., and Graham, J.B. (2004). Movement patterns, depth preferences, and stomach temperatures of free swimming juvenile mako sharks, *Isurus oxyrinchus*, in the Southern California Bight. *Mar. Biol.* 145:191–199.

Shaver, D.J. (1998). Sea turtle strandings along the Texas coast, 1980–94. In: Zimmerman, R. (Ed.), *Characteristics and Causes of Texas Marine Strandings*, NOAA Tech. Rep. NMFS 143. U.S. Department of Commerce, Washington, D.C., pp. 57–72.

Shirai, S. and Nakaya, K. (1992). Functional morphology of feeding apparatus of the cookie-cutter shark, *Isistius brasiliensis* (Elasmobranchii, Dalatiinae). *Zool. Sci.* 9:811–821.

Shultz, S. and Finlayson, L.V. (2010). Large body and small brain and group sizes are associated with predator preferences for mammalian prey. *Behav. Ecol.* 21:1073–1079.

Siems, D.P. and Sikes, R.S. (1998). Trade-offs between growth and reproduction in response to temporal variation in food supply. *Environ. Biol. Fish.* 53:319–329.

Sigler, M.F., Hulbert, L.B., Lunsford, C.R., Thompson, N.H., Burek, K., O'Corry-Crowe, G., and Hirons, A.C. (2006). Diet of Pacific sleeper shark, a potential Steller sea lion predator, in the north-east Pacific Ocean. *J. Fish Biol.* 69:392–405.

Sih, A. (1998). Game theory and predator–prey response races. In: Dugatkin, L.A. and Reeve, H.K. (Eds.), *Game Theory and Animal Behavior*. Oxford University Press, Oxford, pp. 221–238

Sih, A. and Christensen, B. (2001). Optimal diet theory: when does it work, and when and why does it fail? *Anim. Behav.* 61:379–390.

Sih, A., Crowley, P., McPeek, M., Petranka, J., and Strohmeier, K. (1985). Predation, competition, and prey communities: a review of field experiments. *Annu. Rev. Ecol. Syst.* 16:269–311.

Silva, Jr., J.M., Silva, F.J.D.L., Sazima, C., and Sazima, I. (2007). Trophic relationships of the spinner dolphin at Fernando de Noronha Archipeligo, SW Atlantic. *Sci. Mar.* 71:505–511.

Simpfendorfer, C.A. and Milward, N.E. (1993). Utilisation of a tropical bay as a nursery area by sharks of the families Carcharhinidae and Sphyrnidae. *Environ. Biol. Fish.* 37:337–345.

Simpfendorfer, C.A., Goodreid, A., and McAuley, R.B. (2001a). Diet of three commercially important shark species from Western Australian waters. *Mar. Freshw. Res.* 52:975–985.

Simpfendorfer, C.A., Goodreid, A.B., and McAuley, R.B. (2001b). Size, sex, and geographic variation in the diet of the tiger shark, *Galeocerdo cuvier*, from Western Australian waters. *Environ. Biol. Fish.* 61:37–46.

Simpfendorfer, C.A., Freitas, G.G., Wiley, T.R., and Heupel, M.R. (2005). Distribution and habitat partitioning of immature bull sharks (*Carcharhinus leucas*) in a southwest Florida estuary. *Estuaries* 28:78–85.

Sims, D.W. (1999). Threshold foraging behavior of basking sharks on zooplankton: life on an energetic knife-edge? *Proc. R. Soc. B Biol. Sci.* 266:1437–1443.

Sims, D.W. (2000). Filter-feeding and cruising swimming speeds of basking sharks compared with optimal models: they filter-feed slower than predicted for their size. *J. Exp. Mar. Biol. Ecol.* 249:65–76.

Sims, D.W. (2010). Tracking and analysis techniques for understanding free-ranging shark movements and behavior. In: Carrier, J., Musick, J.A., and Heithaus, M. (Eds.), *Sharks and Their Relatives II: Biodiversity, Adaptive Physiology, and Conservation*. CRC Press, Boca Raton, FL, pp. 351–392.

Sims, D.W. and Merrett, D.A. (1997). Determination of zooplankton characteristics in the presence of surface feeding basking sharks *Cetorhinus maximus*. *Mar. Ecol. Prog. Ser.* 158:297–302.

Sims, D.W. and Quayle, V.A. (1998). Selective foraging behaviour of basking sharks on zooplankton in a small-scale front. *Nature* 393:460–464.

Sims, D.W., Nash, J.P., and Morritt, D. (2001). Movements and activity of male and female dogfish in a tidal sea lough: alternative behavioral strategies and apparent sexual selection. *Mar. Biol.* 139:1165–1175.

Sims, D.W., Southall, E.J., Richardson, A.J., Reid, P.C., and Metcalfe, J.D. (2003). Seasonal movements and behaviour of basking sharks from archival tagging: no evidence of winter hibernation. *Mar. Ecol. Prog. Ser.* 248:187–196.

Sims, D.W., Southall, E.J., Tarling, G.A., and Metcalfe, J.D. (2005). Habitat-specific normal and reverse diel vertical migration in the plankton-feeding basking shark. *J. Anim. Ecol.* 74:755–761.

Sims, D.W., Wearmouth, V.J., Southall, E.J., Hill, J.M., Moore, P., Rawlinson, K., Hutchinson, N., Budd, G.C., Righton, D., Metcalfe, J.D., Nash, J.P., and Morritt, D. (2006a). Hunt warm, rest cool: bioenergetic strategy underlying diel vertical migration of a benthic shark. *J. Anim. Ecol.* 75:176–190.

Sims, D.W., Witt, M.J., Richardson, A.J., Southall, E.J., and Metcalfe, J.D. (2006b). Encounter success of free-ranging marine predator movements across a dynamic prey landscape. *Proc. R. Soc. B Biol. Sci.* 273:1195–1201.

Sims, D.W., Southall, E.J., Humphries, N.E., Hays, G.C., Bradshaw, C.J.A., Pitchford, J.W., James, A., Ahmed, M.Z., Brierley, S., Hindell, M.A., Morritt, D., Musyl, M.K., Righton, D., Shepard, E.L.C., Wearmouth, V.J., Wilson, R.P., Witt, M.J., and Metcalfe, J.D. (2008). Scaling laws of marine predator search behavior. *Nature* 451:1098.

Sinclair, A.R.E. and Arcese, P. (1995). Population consequences of predation-sensitive foraging: the Serengeti wildebeest. *Ecology* 76:882–891.

Skomal, G.B., Zeeman, S.I., Chisholm, J.H., Summers, E.L., Walsh, H.J., McMahon, K.W., and Thorrold, S.R. (2009). Transequatorial migrations by basking sharks in the western Atlantic Ocean. *Curr. Biol.* 19:1019–1022.

Smale, M.J., Sauer, W.H.H., and Hanlon, R.T. (1995). Attempted ambush predation on spawning squids, *Loligo vulgaris reynaudii*, by benthic pyjama sharks, *Poroderma africanum*, off South Africa. *J. Mar. Biol. Assoc. U.K.* 75:739–742.

Smale, M.J., Sauer, W.H.H., and Roberts, M.J. (2001). Behavioural interactions of predators and spawning chokka squid of South Africa: towards quantification. *Mar. Biol.* 139:1095–1105.

Smith, C.R. and Baco, A.R. (2003). Ecology of whale falls at the deep-sea floor. *Oceanogr. Mar. Biol.* 41:311–354.

Smith, J.W. and Merriner, J.V. (1985). Food habits and feeding behavior of the cownose ray, *Rhinoptera bonasus*, in lower Chesapeake Bay. *Estuaries* 8:305–310.

Snelson, Jr., F.F., Mulligan, T.J., and Williams, S.E. (1984). Food habits, occurrence, and population structure of the bull shark, *Carcharhinus leucas*, in Florida coastal lagoons. *Bull. Mar. Sci.* 34:71–80.

Speed, C.W., Meekan, M.,G., Rowat, D., Pierce, S.J., Marshall, A.D., and Bradshaw, C.J.A. (2008). Scarring patterns and relative mortality rates of Indian Ocean whale sharks. *J. Fish Biol.* 72:1488–1503.

Springer, S. (1960). Natural history of the sandbar shark, *Eulamia milberti*. *Fish. Bull.* 61:1–38.

Springer, S. (1967). Social organization of shark populations. In: Gilbert, R.F., Matheson, R.F., and Rall, D.P. (Eds.), *Sharks, Skates, and Rays*. The Johns Hopkins University Press, Baltimore, MD, pp. 149–174.

Stephens, D.W. and Krebs, J.R. (1986). *Foraging Theory*. Princeton University Press, Princeton, NJ.

Stevens, J.D., Bradford, R.W., and West, G.J. (2010). Satellite tagging of blue sharks (*Prionace glauca*) and other pelagic sharks off eastern Australia: depth behaviour, temperature experience and movements. *Mar. Biol.* 157:575–591.

Stillwell, C.E. and Kohler, N.E. (1982). Food, feeding habits, and estimates of daily ration of the shortfin mako (*Isurus oxyrinchus*) in the northwest Atlantic. *Can. J. Fish. Aquat. Sci.* 39:407–414.

Stokes, M.D. and Holland, N.D. (1992). Southern stingray (*Dasyatis americana*) feeding on lancelets (*Branchiostoma floridae*). *J. Fish Biol.* 41:1043–1044.

Stokesbury, MJ.W., Harvey-Clark, C., Gallant, J., Block, B.A., and Myers, R.A. (2005). Movement and environmental preferences of Greenland sharks (*Somniosus microcephalus*) electronically tagged in the St. Lawrence Estuary, Canada. *Mar. Biol.* 148:159–165.

Strong, Jr., W.R. (1996). Shape discrimination and visual predatory tactics in white sharks, In: Klimley, A.P. and Ainley, D.G. (Eds.), *Great White Sharks: The Biology of Carcharodon carcharias*. Academic Press, New York, pp. 229–240.

Strong, Jr., W.R., Snelson, F.F., and Gruber, S.H. (1990). Hammerhead shark predation on stingrays: an observation of prey handling by *Spyrna mokarran*. *Copeia* 1990:836–840.

Strong, Jr., W.R., Murphy, R.D., Bruce, B.D., and Nelson, D.R. (1992). Movements and associated observations of bait-attracted white sharks, *Carcharodon carcharias*: a preliminary report. *Aust. J. Mar. Freshw. Res.* 43:13–20.

Strong, Jr., W.R., Bruce, B.D., Nelson, D.R., and Murphy, R.D. (1996). Population dynamics of white sharks. In: Klimley, A.P. and Ainley, D.G. (Eds.), *Great White Sharks: The Biology of Carcharodon carcharias*. Academic Press, New York, pp. 401–414.

Sundström, L.F., Gruber, S.H., Clermont, S.M., Correia, J.P.S., de Marignac, J.R.C., Morrissey, J.F., Lowrance, C.R., Thomassen, L., and Oliveira, M.T. (2001). Review of elasmobranch behavioral studies using ultrasonic telemetry with special reference to the lemon shark, *Negaprion brevirostris*, around Bimini Islands, Bahamas. *Environ. Biol. Fish.* 60:225–250.

Taylor, R.J. (1984). *Predation*. Chapman & Hall, London.

Thrush, S.F., Pridmore, R.D., Hewitt, J.E., and Cummings, V.J. (1994). The importance of predators on a sandflat: interplay between seasonal changes in prey densities and predator effects. *Mar. Ecol. Prog. Ser.* 107:211–222.

Tillet, B.J., Tibbets, I.R., and Whitehead, D.L. (2008). Foraging behaviour and prey discrimination in the bluespotted maskray *Dastatis kuhlii*. *J. Fish Biol.* 73:1554–1561.

Torres, L.G., Heithaus, M.R., and Delius, B. (2006). Influence of teleost abundance on the distribution and abundance of sharks in Florida Bay, USA. *Hydrobiologia* 569:449–455.

Tregenza, T. (1995). Building on the ideal free distribution. *Adv. Ecol. Res.* 26:253–307.

Tricas, T.C. (1979). Relationships of the blue shark, *Prionace glauca*, and its prey species near Santa Catalina Island, California. *Fish. Bull.* 77:175–182.

Tricas, T.C. and McCosker, J.E. (1984). Predatory behavior of the white shark (*Carcharodon carcharias*), with notes on its biology. *Proc. Calif. Acad. Sci.* 43:221–238.

van der Elst, R.P. (1979). A proliferation of small sharks in the shore-based Natal sports fishery. *Environ. Biol. Fish.* 4:349–362.

VanBlaricom, G.R. (1982). Experimental analyses of structural regulation in a marine sand community exposed to oceanic swell. *Ecol. Monogr.* 52:283–305.

Vaudo, J.J. (2011). Habitat Use and Foraging Ecology of a Batoid Community in Shark Bay, Western Australia, doctoral dissertation, Florida International University, Miami.

Vaudo, J.J. and Heithaus, M.R. (2009). Spatiotemporal variability in a sandflat elasmobranch fauna in Shark Bay, Australia. *Mar. Biol.* 156:2579–2590.

Vaudo, J.J. and Heithaus, M.R. (2011). Dietary niche overlap in a nearshore elasmobranch mesopredator community. *Mar. Ecol. Prog. Ser.* 425:247–260.

Vaudo, J.J. and Lowe, C.G. (2006). Movement patterns of the round stingray *Urobatis halleri* (Cooper) near a thermal outfall. *J. Fish Biol.* 68:1756–1766.

Visser, I.N. (1999). Benthic foraging on stingrays by killer whales (*Orcinus orca*) in New Zealand waters. *Mar. Mamm. Sci.* 15:220–227.

Visser, I.N. (2005). First observations of feeding on thresher (*Alopias vulpinus*) and hammerhead (*Sphyrna zygaena*) sharks by killer whales (*Orcinus orca*) specializing on elasmobranch prey *Aquat. Mamm.* 31:83–88.

Visser, I.N., Berghan, J., van Meurs, R., and Fertl, D. (2000). Killer whale (*Orcinus orca*) predation on a shortfin mako shark (*Isurus oxyrinchus*) in New Zealand waters. *Aquat. Mamm.* 26:229–231.

Walker, P.A. and Heesen, H.J.L. (1996). Long-term changes in ray populations in the North Sea. *ICES J. Mar. Sci.* 53:1085–1093.

Walker, T.I. (1998). Can shark resources be harvested sustainably? A question revisited with a review of shark fisheries. *Mar. Freshw. Res.* 49:553–572.

Wallman, H.L. and Bennett, W.A. (2006). Effects of parturition and feeding on thermal preference of Atlantic stingray, *Dasyatis sabina* (Lesueur). *Environ. Biol. Fish.* 75:259–267.

Walters, C. (2000). Natural selection for predation avoidance tactics: implications for marine population and community dynamics. *Mar. Ecol. Prog. Ser.* 208:309–313.

Walters, C. and Juanes, F. (1993). Recruitment limitation as a consequence of natural selection for use of restricted feeding habitats and predation risk taking by juvenile fishes. *Can. J. Fish. Aquat. Sci.* 50:2058–2070.

Walters, C. and Martel, S.J.D. (2004). *Fisheries Management and Ecology*. Princeton University Press, Princeton, NJ.

Wearmouth, V.J. and Sims, D.W. (2009). Movement and behaviour patterns of the critically endangered common skate *Dipturus batis* revealed by electronic tagging. *J. Exp. Mar. Biol. Ecol.* 380:77–87.

Webber, J.D. and Cech, Jr., J.J. (1998). Nondestructive diet analysis of the leopard shark from two sites in Tomales Bay, California. *Calif. Fish Game* 84:18–24.

Weng, K.C. and Block, B.A. (2004). Diel vertical migration of the bigeye thresher shark (*Alopias superciliosus*), a species possessing orbital retia mirabilia. *Fish. Bull.* 102:221–229.

Weng, K.C., O'Sullivan, J.B., Lowe, C.G., Winkler, C.E., Dewar, H., and Block, B.A. (2007a). Movements, behavior and habitat preferences of juvenile white sharks *Carcharodon carcharias* in the eastern Pacific. *Mar. Ecol. Prog. Ser.* 338:211–224.

Weng, K.C., Boustany, A.M., Pyle, P., Anderson, S.D., Brown, A., and Block, B.A. (2007b). Migration and habitat of white sharks (*Carcharodon carcharias*) in the eastern Pacific Ocean. *Mar. Biol.* 152:877–894.

Werner, E.E. and Anholt, B.R. (1993). Ecological consequences of the trade-off between growth and mortality rates mediated by foraging activity. *Am. Nat.* 142:242–272.

Werner, E.E. and Anholt, B.R. (1996). Predator-induced behavioral indirect effects: consequences to competitive interactions in anuran larvae. *Ecology* 77:157–169.

Werner, E.E. and Hall, D.J. (1974). Optimal foraging and size selection of prey by bluegill sunfish (*Lepomis macrochirus*). *Ecology* 55:1042–1052.

Werner, E.E. and Hall, D.J. (1988). Ontogenetic habitat shifts in bluegill: the foraging rate–predation risk trade-off. *Ecology* 69:1352–1366.

West, G.J. and Stevens, J.D. (2001). Archival tagging of school shark, *Galeorhinus galeus*, in Australia: initial results. *Environ. Biol. Fish.* 60:283–298.

Wetherbee, B.M., Gruber, S.H., and Cortés, E. (1990). Diet, feeding habits, digestion, and consumption in sharks with special reference to the lemon shark, *Negaprion brevirostris*. In: Pratt, Jr., H.L., Gruber, S.H., and Taniuchi, T. (Eds.), *Elasmobranchs as Living Resources: Advances in the Biology, Ecology, Systematics, and Status of the Fisheries*, NOAA Tech. Rep. NMFS 90. U.S. Department of Commerce, Washington, D.C., pp. 29–47.

Wetherbee, B.M., Gruber, S.H., and Rosa, R.S. (2007). Movement patterns of juvenile lemon sharks *Negaprion brevirostris* within Atol das Rocas, Brazil: a nursery area characterized by tidal extremes. *Mar. Ecol. Prog. Ser.* 343:283–293.

Widder, E.A. (1998). A predatory use of counterillumination by the squaloid shark, *Isistius brasiliensis*. *Environ. Biol. Fish.* 53:267–273.

Wilson, S.G., Taylor, J.G., and Pearce, A.F. (2001). The seasonal aggregation of whale sharks at Ningaloo Reef, Western Australia: currents, migrations and the El Niño/Southern Oscillation. *Environ. Biol. Fish.* 61:1–11.

Wirsing, A.J. and Heithaus, M.R. (2009). Olive-headed sea snakes *Disteria major* shift seagrass microhabitats to avoid shark predation. *Mar. Ecol. Prog. Ser.* 387:287–293.

Wirsing, A.J., Heithaus, M.R., and Dill, L.M. (2007a). Living on the edge: dugongs prefer to forage in microhabitats that allow for escape from rather than avoidance of predators. *Anim. Behav.* 74:93–101.

Wirsing, A.J., Heithaus, M.R., and Dill, L.M. (2007b). Can measures of prey availability improve our ability to predict the abundance of large marine predators? *Oecologia* 153:563–568.

Wirsing, A.J., Heithaus, M.R., and Dill, L.M. (2007c). Fear factor: do dugongs (*Dugong dugon*) trade food for safety from tiger sharks (*Galeocerdo cuvier*)? *Oecologia* 153:1031–1040.

Wirsing, A.J., Heithaus, M.R., and Dill, L.M. (2007d). Can you dig it? Use of excavation, a risky foraging tactic, by dugongs is sensitive to predation danger. *Anim. Behav.* 74:1085–1091.

Wirsing, A.J., Heithaus, M.R., Frid, A., and Dill, L.M. (2008a). Seascapes of fear: evaluating sublethal predator effects experienced and generated by marine mammals. *Mar. Mamm. Sci.* 24:1–15.

Wirsing, A.J., Abernethy, R., and Heithaus, M.R. (2008b). Speed and maneuverability of adult loggerhead turtles, *Caretta caretta*, under simulated predatory attack: do the sexes differ? *J. Herp.* 42:411–413.

Wirsing, A.J., Cameron, K.E., and Heithaus, M.R. (2010). Spatial responses to predators vary with prey escape mode. *Anim. Behav.* 79:531–537.

Wirsing, A.J., Heithaus, M.R., and Dill, L.M. (2011). Predator-induced modifications to diving behavior vary with foraging mode. *Oikos* 120:1005–1012.

Witzell, W.N. (2007). Kemp's Ridley (*Lepidochelys kempi*) shell damage. *Mar. Turt. Newsl.* 115:16–17.

Wueringer, B.E., Squire, Jr., L., and Collin, S.P. (2009). The biology of extinct and extant sawfish (Batoidea: Sclerorhynchidae and Pristidae). *Rev. Fish Biol. Fish.* 19:445–464.

Ydenberg, R.C. and Dill, L.M. (1986). The economics of fleeing from predators. *Adv. Stud. Behav.* 16:229–249.

18

An Updated Look at Elasmobranchs as Hosts of Metazoan Parasites

Janine N. Caira, Claire J. Healy, and Kirsten Jensen

CONTENTS

18.1 Introduction

The body of an elasmobranch offers a diversity of sites that can be, and often are, occupied by other animals. Indeed, essentially no organ system of elasmobranchs has escaped the attention of one or more groups of parasites (Figure 18.1). That is not to say that all sites of the body of an elasmobranch are equally parasitized. Certain organs and organ systems, such as the skin, digestive system, and gills, for example, tend to host particularly diverse faunas of parasites. Caira (1990), Cheung (1993), and Benz and Bullard (2004) have all treated the parasites of these hosts to various degrees within the last two or so decades. Although all three of these works provided overviews of the taxonomic diversity of the metazoan parasites, each emphasized a different aspect of the host–parasite relationship. Caira (1990) focused on the life cycles and utility of metazoan parasites as indicators of elasmobranch biology, Cheung (1993) included an extensive list of many of the parasite species reported from elasmobranchs, and Benz and Bullard (2004) concentrated on the pathology caused by, and treatment of, parasites, with emphasis on those harmful, or at least potentially harmful, to captive chondrichthyans. In this chapter, we have taken a slightly different approach and have treated the various parasites based on the sites they occupy within their hosts. We have concentrated on the parasitic metazoans, or multicellular parasites, because we did not feel we could do justice to the diversity of protistan (i.e., unicellular) taxa parasitizing elasmobranchs at this time. Readers interested in the protists are directed to Cheung (1993) for a list of the approximately 30 species reported from elasmobranchs. Also omitted from discussion here are the vertebrate associates of elasmobranchs such as hagfish, lampreys, eels, etc. (e.g., Caira et al., 1997a). For comparative purposes, the metazoans of holocephalans are briefly treated in the final section of this chapter.

Our knowledge of metazoan parasites of elasmobranchs has advanced since the production of the first edition of this book in the following ways. Most conspicuously, our tally of the total number of metazoan parasites described from elasmobranchs has risen from ~1475 to ~1917 species. This increase is in large part a result of substantial attention having been paid to the description of new cestodes. Collaborative survey work in Borneo and New Caledonia has been the source of many new taxa, particularly of cestodes, but also of a diversity of novel monogeneans and copepods. Although no additional sites within the body of elasmobranchs have been discovered to be home to parasites (a finding that is unsurprising given the vast repertoire of sites already known to house metazoan parasites in 2004), several taxa can be added to the faunas of some of the known sites. Bullard et al. (2006), for example, reported a digenean from the kidney of the shark *Carcharhinus limbatus*. Kitamura et al. (2010) reported a monogenean from the uterus of the shark *Squalus mitsukurii*. Andrade et al. (2008) reported third-stage juveniles of a nematode from the biliary ducts of the skate *Sympterygia acuta*. The classification of the cestodes has undergone some major revisions, which include reorganization of the order Trypanorhyncha (Olson et al., 2010; Palm, 2004), resurrection of the orders Cathetocephalidea and Litobothriidea (Caira et al., 2005), and establishment of the new order Rhinebothriidea (Healy et al., 2009). In this revision, the phylum Myxozoa, which includes approximately 34 species known to be associated with elasmobranchs, and which clearly represents a multicellular (rather than unicellular) taxon of parasites (Kent et al., 2001; Siddall et al., 1995), has been added to the selection of phyla treated here. As an aid to distinguishing sites occupied by larval or juvenile metazoans from those occupied by adults, we have indicated sites occupied by larval or juvenile forms by adding gray shading to the appropriate metazoan taxon icons in Figure 18.1. We have also expanded the section on General Observations. Finally, the section addressing the metazoan parasites of holocephalans is also new to this edition.

18.2 Metazoan Parasites of Elasmobranchs

The invertebrate metazoans parasitizing elasmobranchs belong to seven phyla. In ascending order of their diversity in elasmobranchs these are Mollusca, Acanthocephala, Annelida, Myxozoa, Nematoda, Arthropoda, and Platyhelminthes. To date, only one species of mollusc and approximately 12 species of acanthocephalans have been reported from elasmobranchs. The annelids, myxozoans, and nematodes of elasmobranchs are somewhat more diverse. Recent counts suggest that approximately 23 species of annelids (all leeches; E.M. Burreson, pers. comm.), 34 species of myxozoans, and perhaps as many as 83 species of nematodes are known to associate with elasmobranchs. However, by far the greatest diversity of elasmobranch parasites is found among the arthropods and platyhelminths. Each of these phyla includes several major subgroups that are worthy of individual consideration. The arthropod taxa parasitizing elasmobranchs, again in ascending order of their diversity in elasmobranchs, are mites, barnacles (i.e., Cirripedia), ostracods, amphipods, branchiurans, isopods, and copepods. Significant differences in diversity exist among these groups in elasmobranchs; for example, whereas there is a single record of a mite from an elasmobranch (Benz and Bullard, 2004) and perhaps 3 species of ostracods,

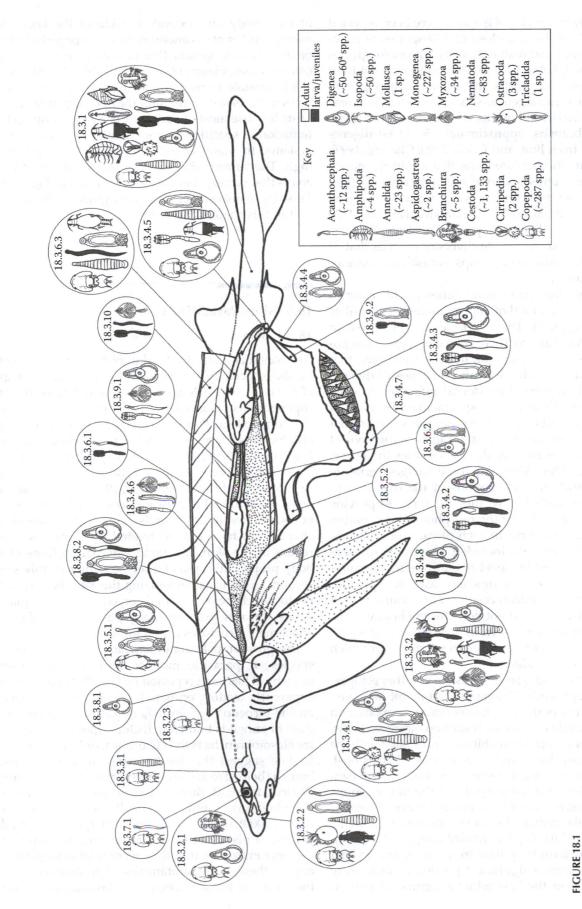

FIGURE 18.1

Overview of sites occupied by metazoan parasites of elasmobranchs, indicating text section of this chapter treating each site and approximate number of species of each parasite group found in elasmobranchs. [a]Upper end of range includes accidental infections.

approximately 50 species of isopods have been reported from elasmobranchs and close to 300 species of copepods are encountered with regularity. The major groups of parasitic platyhelminths, in ascending order of their diversity in elasmobranchs, are triclads, aspidogastreans, digeneans, monogeneans, and cestodes. Whereas only a single triclad and two aspidogastreans are known from elasmobranchs, approximately 50 to 60 digeneans (updated from Bray and Cribb, 2003; Cheung, 1993), depending on whether infections that are likely merely accidental are counted (60 total species) or not (50 total species), 226 monogeneans (updated from Whittington and Chisholm, 2003), and, we estimate, well over 1133 cestode species are known to parasitize elasmobranchs. In fact, elasmobranch cestode diversity exceeds that of all of the other metazoan groups parasitizing elasmobranchs combined.

Collectively, the metazoan parasites of elasmobranchs represent a total of approximately 119 families in these seven phyla. These families are presented in Table 18.1 according to their relavent higher categories of classification. It should, however, be noted that included in this number are nine families of digeneans that likely represent accidental infections in elasmobranchs. To illustrate the spectacular diversity of morphologies exhibited by these parasites, we present scanning electron micrographs or light micrographs of a representative of most of the 119 families in Figures 18.2 through 18.86. Although these images certainly do not substitute for descriptions of the distinguishing features of each family, they do serve to provide readers with some idea of the morphological variation found among major taxa in each phylum of parasites. In several cases, we have included images of representatives of taxa below the level of family, either because family-level taxonomy is unstable (e.g., tetraphyllidean and lecanicephalidean cestodes) or because we felt family-level diversity did not do sufficient justice to the morphological variation seen in a group (e.g., leeches). Table 18.1 also serves to indicate the families for which illustrations are provided.

It is common to categorize parasites as either ectoparasitic or endoparasitic. Ectoparasites inhabit any exterior site or orifice of their host. Leeches, arthropods, and molluscs typically occupy such sites on elasmobranchs. Endoparasites generally inhabit sites associated with the cavities, organs, ducts, and musculature of their host. With a few exceptions, the acanthocephalans, myxozoans, nematodes, and the majority of the major groups of platyhelminths (except most monogeneans) are endoparasitic in elasmobranchs. Some authors (e.g., Benz, 1993; Kabata, 1979) recognize a third category, the mesoparasites. This term is applied to those organisms that normally live with a significant portion of their body embedded within the host while a significant portion of their body also extends outside of the host. Most mesoparasites of elasmobranchs are copepods, but one of the barnacle genera that parastizes elasmobranchs (*Anelasma*; see Figure 18.26) also exhibits this lifestyle.

It is feasible to treat the metazoan parasites of elasmobranchs based on the sites they occupy in or on their hosts because most of the major parasite groups exhibit remarkable specificity for particular organs or organ systems. In many cases, this specificity is extremely high. The various sites and the major groups of metazoans that occupy them are summarized in Figure 18.1. This figure also serves as a quick guide to the sections within this chapter that follow.

18.3 Sites Parasitized

18.3.1 Skin

Parasites that attach to the skin of elasmobranchs often exhibit specificity for the skin on a particular region of the body. This site specificity is most marked in the copepods and monogeneans; for example, females of the copepod *Echthrogaleus coleopterus* are highly specific for the surfaces of the pelvic fins of their blue shark hosts (Benz, 1986), and the monogenean *Acanthocotyle greeni* is found only on the ventral surfaces of *Raja* species (MacDonald and Llewellyn, 1980). Unfortunately, a detailed treatment of parasites associated with the skin of each region of the body is beyond the scope of this chapter. We have instead treated the parasites of any region of the skin, including external surfaces of the body proper, fins, and claspers, together in this single section. Organisms parasitizing this site share the ability to attach below, around, or on top of the placoid scales of elasmobranchs. Many possess appendages or attachment structures useful for this purpose.

The prosobranch snail *Cancellaria cooperi* is the only species of parasitic mollusc known from elasmobranchs. It has been reported by O'Sullivan et al. (1987), apparently feeding on blood, on the dorsal surface of Pacific torpedo rays, *Torpedo californica*. The only subclass of annelids known to include species that parasitize elasmobranchs is the Hirudinea, or leeches. Records to date suggest that leeches associated with elasmobranchs belong to at least three subfamilies of the family Piscicolidae (E.M. Burreson, pers. comm.): Piscicolinae, Pontobdellinae, and Platybdellinae (Burreson and Kearn, 2000); however, Curran et al. (pers. comm.) discovered a piscicolid leech on the external surfaces of *Zapteryx exasperata* that they were unable to place into any of these three subfamilies. Approximately 21 of the 23 or so known species of elasmobranch leeches have been reported from the skin of their hosts (E.M. Burreson, pers. comm.). Several of these species are also known from other sites (see below), but the skin appears to be the preferred region of attachment for most leeches.

Although the skin is an unusual site in which to encounter evidence of nematodes given that most nematode species are endoparasitic, at least two species of spirurids in the family Philometridae and two species of trichinelloids in the family Trichosomoididae have been found in the skin. Juveniles of a philometrid similar to *Phlyctainophora lamnae* were reported by Ruyck and Chabaud (1960) from tumors at the bases of the fins of *Mustelus mustelus*. Adamson et al. (1987) reported adults of *Phlyctainophora squali* from lesions in the skin of several species of sharks. The trichosomoidid genus *Huffmanela* is particularly interesting because, to date, its species are known only from their darkly pigmented eggs, which are deposited by the female nematode in meandering trails around the bases of the placoid scales of the head (see Figure 18.18) and fins of a diversity of carcharhinid shark species. The first species from elasmobranchs (*Huffmanela carcharhini*) was described by MacCallum (1925) from *Carcharhinus plumbeus* (as *C. commersoni*). A detailed discussion of this genus is provided by Moravec (2001), who noted that it also includes seven species that parasitize teleosts, six of which are also known only from eggs. A second species from elasmobranchs, *Huffmanela lata*, was recently described by Justine (2005) on the skin near the gills of *Carcharhinus amblyrhynchos* in New Caledonia. We have also encountered eggs of an unidentified member of this genus on the ventral surface of *Taeniura lymma* in Borneo.

A diversity of minor arthropod groups parasitize the skin of elasmobranchs. These are barnacles, amphipods, branchiurans, and isopods. The mesoparasitic barnacle *Anelasma squalicola* has been found parasitizing several species of squaliform sharks, in which they are associated with, for example, the dorsal spines and the pectoral and pelvic fins of their hosts (e.g., Kabata, 1970; Long and Waggoner, 1993; Yano and Musick, 2000). An interesting, indirect association apparently exists between a second barnacle species, *Conchoderma virgatum*, and certain copepods found on the body surfaces of elasmobranchs. This species has been reported attached to members of two different families of copepods parasitizing the skin of large, pelagic sharks (e.g., Benz, 1984; Williams, 1978). Although most amphipods are free living, there exist a number of species that are normally found associated with hosts. Some of these—for example, the lysianassid *Opisa tridentata* and the lafystiid *Lafystius morhuanus*—have been reported from the skin of sharks and batoids; however, these species have also been reported from a diversity of teleosts (Bousfield, 1987). A few amphipod species, such as the trischizostomatid *Trischizostoma*

raschi, are known only from the body surfaces of certain squaliform sharks (Bousfield, 1987). The branchiurans, or fish lice, number approximately 150 species (Kabata, 1988) and are primarily parasitic on fishes. Only a small number of branchiurans are known from elasmobranchs. The majority of these records are of species of *Argulus* (family Argulidae) from the dorsal surfaces of either dasyatid (e.g., Cressey, 1976) or potamotrygonid (e.g., Ross, 1999) stingrays. Marques (2000) reported finding members of a second genus, *Dolops*, on freshwater stingrays in South America.

Of the thousands of species of isopods known worldwide, only about 500 are known to associate with fishes (Bunkley-Williams and Williams, 1998). Approximately 50 of these associate with elasmobranchs (e.g., Moreira and Sadowsky, 1978), and only a very small subset of these attach to the skin. Bunkley-Williams and Williams (1998) provided a useful overview of the isopods infecting fish and elasmobranchs. Of the five families of isopods infecting elasmobranchs, members of three have been found on the skin. Gnathiids are unique among these isopods in that it is the praniza larva, rather than the adult, that is parasitic. These relatively small isopods feed on the blood of their hosts, becoming more conspicuous as they feed and their body swells with host blood. Knowledge of the taxonomy and host specificity of gnathiids is limited by the fact that pranizae cannot be identified to species because the taxonomy of the family is currently based on the morphology of adult males. Although much more commonly associated with the gills and branchial chamber, pranizae of gnathiids have been reported from the skin (Heupel and Bennett, 1999). Representatives of two additional families of isopods occupy sites on the external surfaces of elasmobranchs. The cymothoid *Nerocila acuminata*, for example, occurs on the skin of several species of sharks and batoids (Brusca, 1981); several species of Aegidae have also been reported from the skin of elasmobranchs (Moreira and Sadowsky, 1978).

The copepods are not only the most diverse arthropod group parasitizing elasmobranchs, but they are also the most diverse group of ectoparasites of elasmobranchs. In 1993, Cheung listed 221 species of copepods from elasmobranchs. We estimate this number is now approaching 287 species. The copepods of elasmobranchs belong to two of the eight known copepod orders (Poecilostomatoida and Siphonostomatoida). Elasmobranchs host members of 4 families of poecilostomatoid and 12 families of siphonostomatoid copepods. Species in 8 of these 16 families have been reported from the skin. These are the Taeniacanthidae (e.g., Braswell et al., 2002) in the Poecilostomatoidea and the siphonostomatoid families Caligidae (e.g., Bere, 1936), Dissonidae (e.g., Deets and Dojiri, 1990), Euryphoridae (e.g., Lewis, 1966), Kroyeriidae (e.g., Cheung, 1993),

TABLE 18.1

Classification of Families of Metazoan Invertebrates Parasitic on Elasmobranchs

Phylum Mollusca[a]	Class Enoplea	Subphylum Neodermata
Class Gastropoda	Order Trichinellida	Class Trematoda
Subclass Prosobranchia	F. Capillariidae (Figure 18.17)	Subclass Aspidogastrea
F.[b] Cancellariidae (Figure 18.2)	F. Trichosomoididae (Figure 18.18)	F. Multicalycidae (Figure 18.40)
Phylum Acanthocephala	Phylum Arthropoda	F. Stichocotylidae
Class Palaeacanthocephala	Class Arachnida	Subclass Digenea
Order "Echinorhynchida"[c]	Order Acari	F. Acanthocolpidae (accid.)
F. Arhythmacanthidae	Class "Crustacea"	F. Azygiidae (Figure 18.41)
F. Cavisomidae (Figure 18.3)	Subclass Malacostraca	F. Bucephalidae (accid.)
F. Echinorhynchidae (accid.)	Order Amphipoda	F. Campulidae (devel.)
F. Illiosentidae (Figure 18.4)	F. Lafystiidae (Figure 18.19)	F. Derogenidae (Figure 18.42)
F. Rhadinorhynchidae (accid.)	F. Lysianassidae	F. Didymozoidae (devel.)
Order Polymorphida	F. Trischizostomatidae	F. Faustulidae (accid.)
F. Polymorphidae (accid.)	Order Isopoda	F. Gorgoderidae (Figure 18.43)
Phylum Annelida	Suborder Flabellifera	F. Hemiuridae (accid.)
Class Clitellata	F. Aegidae (Figure 18.20)	F. Hirudinellidae (accid.)
Subclass Hirudinea	F. Cirolanidae (Figure 18.21)	F. Lecithasteridae (accid.)
Order Rhynchobdellida	F. Corallanidae (Figure 18.22)	F. Lepocreadiidae (accid.)
F. Piscicolidae	F. Cymothoidae	F. Opecoelidae (accid.)
SubF.[d] Piscicolinae (Figure 18.5)	Suborder Gnathiidea	F. Ptychogonimidae
SubF. Platybdellinae	F. Gnathiidae (Figure 18.23)	F. Sanguinicolidae
SubF. Pontobdellinae (Figure 18.6)	Subclass Maxillopoda	F. Syncoeliidae (Figure 18.44)
Unassigned (Figure 18.7)	Superorder Ostracoda	F. Tandanicolidae (accid.)
Phylum Myxozoa	F. Cypridinidae (Figure 18.24)	F. Zoogonidae (Figure 18.45)
Order Bivalvulida	Superorder Branchiura	Class Monogenea
F. Ceratomyxidae (Figure 18.8)	F. Argulidae (Figure 18.25)	Subclass Monopisthocotylea
F. Chloromyxidae (Figure 18.9a,b)	Superorder Cirripedia	F. Acanthocotylidae (Figure 18.46)
F. Myxidiidae (Figure 18.9c)	F. Anelasmatidae (Figure 18.26)	F. Amphibdellidae (Figure 18.47)
F. Myxobolidae	F. Conchodermidae	F. Capsalidae (Figure 18.48)
F. Myxosomatidae	Superorder Copepoda	F. Loimoidae (Figure 18.49)
F. Sphaerosporidae	Order Siphonostomatoida	F. Microbothriidae (Figure 18.50)
Order Multivalvulida	F. Caligidae (Figure 18.27)	F. Monocotylidae (Figure 18.51)
F. Kudoidae (Figure 18.10)	F. Cecropidae (Figure 18.28)	F. Udonellidae (Figure 18.52)
F. Trilosporidae	F. Dichelesthiidae (Figure 18.29)	Subclass Polyopisthocotylea
Phylum Nematoda	F. Dissonidae (Figure 18.30)	F. Hexabothriidae (Figure 18.53)
Class Rhabditea	F. Eudactylinidae (Figure 18.31)	Class Cestoda
Order Ascaridida	F. Euryphoridae (Figure 18.32)	Order Cathetocephalidea
F. Acanthocheilidae (Figure 18.11)	F. Kroyeriidae (Figure 18.33a,b)	F. Cathetocephalidae (Figure 18.54)
F. Anisakidae (Figure 18.12)	F. Lernaeopodidae (Figure 18.34)	Order Litobothriidea
F. Ascaridae (Figure 18.13)	F. Pandaridae (Figure 18.35)	F. Litobothriidae (Figure 18.55)
Order Spirurida	F. Pennellidae (Figure 18.36)	Order Diphyllidea
F. Cucullanidae (Figure 18.14)	F. Sphyriidae (Figure 18.37)	F. Ditrachybothriidae (Figure 18.56)
F. Cystidicolidae	F. Trebiidae	F. Echinobothriidae (Figure 18.57)
F. Gnathostomatidae (Figure 18.15)	Order Poecilostomatoida	Order Rhinebothriidea
F. Philometridae	F. Chondracanthidae	F. Echeneibothriidae (Figure 18.58)
F. Physalopteridae (Figure 18.16)	F. Ergasilidae (Figure 18.38)	F. Rhinebothriidae (Figure 18.59)
F. Rhabdochonidae	F. Philichthyidae	Order Lecanicephalidea
Order Dracunculoidea	F. Taeniacanthidae	F. Anteroporidae (Figure 18.60)
F. Guyanemidae	Phylum Platyhelminthes	F. Lecanicephalidae (Figure 18.61)
F. Micropleuridae	Order Tricladida	F. Polypocephalidae (Figure 18.62)
Unassigned	F. Procerodidae (Figure 18.39)	F. Tetragonocephalidae (Figure 18.63)
		Miscellaneous (Figs. 18.64–18.66)

TABLE 18.1 (continued)

Classification of Families of Metazoan Invertebrates Parasitic on Elasmobranchs

Class Cestoda (cont.)	SuperF. Lacistorhynchoidea	Order "Tetraphyllidea"
Order Trypanorhyncha	F. "Lacistorhynchidae" (Figure 18.72)	F. Dioecotaeniidae (Figure 18.79)
Suborder Trypanobatoida	F. Pterobothriidae	F. Disculicepitidae (Figure 18.80)
SuperF.[e] Tentacularioidea	SuperF. Otobothrioidea	F. Onchobothriidae (Figure 18.81)
SuperF. Eutetrarhynchoidea	F. Otobothriidae (Figure 18.73)	F. Phyllobothriidae
F. "Eutetrarhynchidae" (Figure 18.68)	F. Paranybeliniidae	SubF. Phyllobothriinae (Figure 18.82)
F. Mixodigmatidae (Figure 18.69)	F. Pseudotobothriidae (Figure 18.74)	SubF. Thysanocephalinae (Figure 18.83)
F. Progrillotiidae	SuperF. Gymnorhynchoidea	SubF. Triloculariinae (Figure 18.84)
F. "Rhinoptericolidae" (Figure 18.70)	F. Aporhynchidae (Figure 18.75)	F. Prosobothriidae (Figure 18.85)
F. Shirleyrhynchidae (Figure 18.71)	F. "Gilquiniidae" (Figure 18.76)	F. Serendipeidae (Figure 18.86)
Suborder Trypanoselachoida	F. "Gymnorhynchidae" (Figure 18.77)	
	F. Rhopalothylacidae	
	F. Sphyriocephalidae (Figure 18.78)	

Note: accid. = likely represent accidental infections; devel. = relationship may be developing into an obligate one.

[a] Phyla are in order of their increasing diversity in elasmobranchs.
[b] Family.
[c] Quotes indicate non-monophyletic taxa.
[d] Subfamily.
[e] Superfamily.

Lernaeopodidae (e.g., Pearse, 1953), Pandaridae (e.g., Lewis, 1966), Trebiidae (e.g., Pearse, 1953), and possibly the Pennellidae (but see Benz and Bullard, 2004). The pennellids differ from the eight other skin-dwelling copepod families in that they exhibit a mesoparasitic, rather than ectoparasitic, lifestyle.

Among the ectoparasites of elasmobranchs, the platyhelminth class Monogenea is second in diversity only to the copepods. In 1993, Cheung estimated that 150 monogenean species in eight families were known to parasitize elasmobranchs; however, monogenean taxonomy is a relatively active area of investigation. In 2003, Whittington and Chisholm estimated that monogeneans known from chondrichthyans numbered 201 species, approximately only 4% of these having been reported from chimaeras; thus, their data suggest that 193 species of monogeneans were known from elasmobranchs at that time. Given the work that has been conducted since the first edition of this book, we estimate that 227 monogenean species are now known from elasmobranchs. These species represent both subclasses (Monopisthocotylea and Polyopisthocotylea) and eight

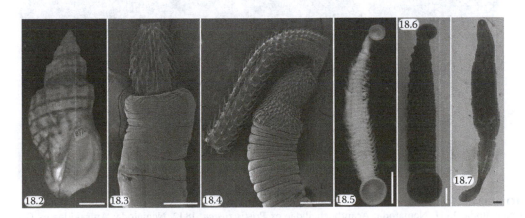

FIGURES 18.2 TO 18.7

Micrographs of Mollusca, Acanthocephala, and Annelida. 18.2. Mollusca: Cancellariidae: *Cancellaria cooperi* (USNM No. 877074). 18.3. Acanthocephala: Cavisomidae: *Megapriapus* sp. ex *Potamotrygon* sp. 18.4. Acanthocephala: Illiosentidae: *Tegorhynchus* sp. ex *Rhinoptera bonasus*. 18.5. Annelida: Piscicolidae: Piscicolinae: *Branchellion* sp. ex *Zearaja nasuta*. 18.6. Annelida: Piscicolidae: Pontobdellinae: *Stibarobdella* sp. ex *Carcharhinus plumbeus*. 18.7. Annelida: Piscicolidae: unidentified subfamily ex *Zapteryx exasperata*. Scale bars: Figure 18.2, 1 cm; Figures 18.3 and 18.4, 200 µm; Figure 18.5, 6 mm; Figure 18.6, 5.5 mm; Figure 18.7, 165 µm.

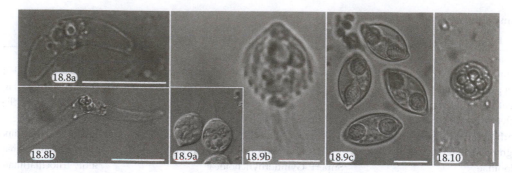

FIGURES 18.8 TO 18.10

Micrographs of spores of Myxozoa. 18.8a. Ceratomyxidae: *Ceratomyxa* sp. ex *Negaprion acutidens*. 18.8b. Ceratomyxidae: *Ceratomyxa* sp. ex *Carcharhinus melanopterus*. 18.9a. Chloromyxidae: *Chloromyxum leydigi* ex *Centroscymnus coelolepis*. 18.9b. Chloromyxidae: *Chloromyxum* sp. ex *Squalus acanthias*. 18.9c. Myxidiidae: *Myxidium* sp. ex *Squalus acanthias*. 18.10. Kudoidae: *Kudoa carcharhini* ex *Carcharhinus cautus*. Scale bars: Figures 18.8a,b, 20 µm; Figure 18.9b, 5 µm; Figures 18.9c and 18.10, 10 µm.

families of the class Monogenea. Members of seven of these families parasitize the skin of elasmobranchs. Among them, the Acanthocotylidae (e.g., Kearn, 1963), Capsalidae (e.g., Whittington and Kearn, 2009a,b), and Microbothriidae (e.g., Kearn et al., 2010) are primarily parasites of the skin. Although they are much more commonly found on the gills, loimoids have also been reported from this site (Benz and Bullard, 2004). As a family, the monocotylids are by far the least site specific of the monogenean groups; although individual species and genera are often very site specific, collectively they occupy a wide diversity of sites on the elasmobranch body. At present, the only monocotylid genus reported from the skin is *Dendromonocotyle* (e.g., Chisholm and Whittington, 2009; Kearn, 1979). Species in several of the above skin-dwelling families of monogeneans have developed an interesting mode of camouflage, whereby they sequester pigment, which appears to be derived from host skin, in their digestive tract. This pigment renders them almost invisible against the pigmented dorsal surfaces of their elasmobranch hosts (Kearn, 1979). Bullard et al. (2000) reported the postoncomiracidial

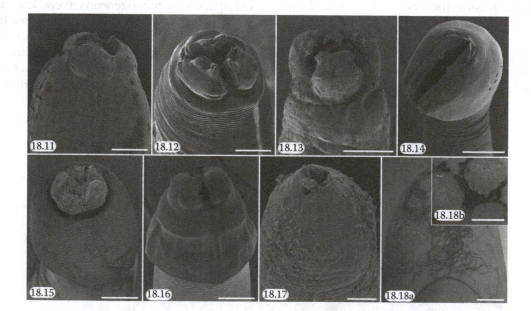

FIGURES 18.11 TO 18.18

Micrographs of Nematoda. 18.11. Nematoda: Acanthocheilidae ex *Pristis zijsron*. 18.12. Nematoda: Anisakidae ex *Galeocerdo cuvier*. 18.13. Nematoda: Ascaridae ex *Potamotrygon* sp. 18.14. Nematoda: Cucullanidae: *Cucullanus* sp. ex *Heterodontus franscisi*. 18.15. Nematoda: Gnathostomatidae: *Echinocephalus* sp. ex *Himantura granulata*. 18.16. Nematoda: Physalopteridae: *Paraleptus* sp. ex *Hemiscyllium ocellatum*. 18.17. Nematoda: Capillariidae: *Piscicapillaria* sp. ex *Rhina ancylostoma*. 18.18. Nematoda: Trichosomoididae: *Huffmanela* sp. (a) Egg trail on ventral surface of head of *Carcharhinus plumbeus*. (b) Enlarged view of characteristic pigmented, bipolar eggs around bases of placoid scales of *C. sorrah*. Scale bars: Figures 18.11, 18.13, and 18.16, 50 µm; Figures 18.12, 18.14, and 18.15, 100 µm; Figure 18.17, 2 µm; Figure 18.18a, 1 cm; Figure 18.18b, 280 µm.

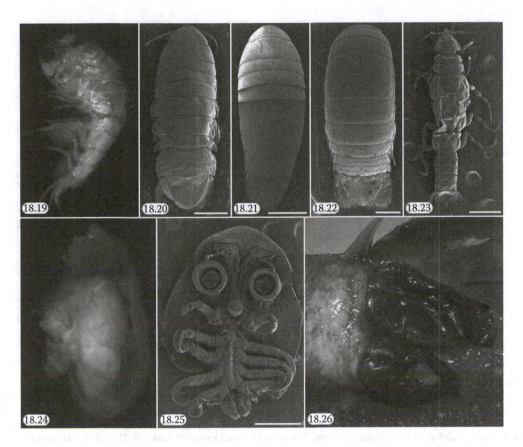

FIGURES 18.19 TO 18.26
Micrographs of Arthropoda I. 18.19. Amphipoda: Lafystiidae: *Opisa tridentata* (USNM No. 127598). 18.20. Isopoda: Aegidae: *Rocinela* sp. ex *Dasyatis* sp. 18.21. Isopoda: Cirolanidae ex *Paragaleus pectoralis*. 18.22. Isopoda: Corallanidae ex *Chiloscyllium punctatum*. 18.23. Isopoda: Gnathiidae: praniza larva of *Gnathia* sp. ex *Centrolophus niger*. 18.24. Ostracoda: Cypridinidae: *Sheina orri* (USNM No. 112675) ex *Hemiscyllium ocellatum*. 18.25. Branchiura: Argulidae: *Argulus* sp. ex *Potamotrygon magdalenae*. 18.26. Cirripedia: Anelasmatidae: *Anelasma* sp. ex *Etmopterus baxteri*. Scale bars: Figures 18.20 and 18.22, 1 mm; Figure 18.21, 2 mm; Figures 18.23 and 18.25, 500 μm.

(i.e., juvenile) stages of what is now considered a member of the family Capsalidae on the skin of *Carcharhinus limbatus*. As the final "skin-dwelling" family of monogeneans, the Udonellidae is unusual in that it includes species that attach to copepods (primarily caligids) parasitizing the body of elasmobranchs and thus can be considered as at least indirect inhabitants of this site (e.g., Price, 1938a). It appears that udonellids are particular about the site they inhabit on their copepod host (Causey, 1961), but it is possible that these monogeneans occasionally feed directly on the fish hosting their copepod host (Kearn, 1998).

Species in the remaining platyhelminth groups rarely occupy the skin, but a few exceptions exist. For example, the triclad *Micropharyx parasitica* (family Procerodidae) has been reported with some regularity from skates in the Atlantic Ocean (Beverley-Burton, 1984). This species is found on the dorsal surfaces of its host where, according to Ball and Khan (1976), it feeds on host epidermal tissue. Thus, unlike the majority of the other non-neodermatan platyhelminths, this species parasitizes vertebrates, rather than invertebrates. Cestodes and

digeneans are also found, albeit rarely, on the skin of elasmobranchs. Plerocerci (i.e., larvae) of the trypanorhynch cestode family Lacistorhynchidae were reported by Guiart (1935) encysted in the skin of several species of sharks. *Paronatrema mantae*, a digenean belonging to the family Syncoeliidae, was reported from the skin of *Manta birostris* by Manter (1940).

18.3.2 Sensory Systems

18.3.2.1 Eyes

Arthropods are the primary associates of the eyes of elasmobranchs. Benz and his coworkers (Benz et al., 1998, 2002; Borucinska et al., 1998) have done much to document the interesting association between lernaeopodid copepods of the genus *Ommatokoita* and the eyes of their shark hosts. Their work suggests that these copepods cause severe corneal displasia, resulting in at least partial blindness in their squaliform hosts. Several members of a second family of copepods, the Caligidae, have also been reported exclusively from the surface

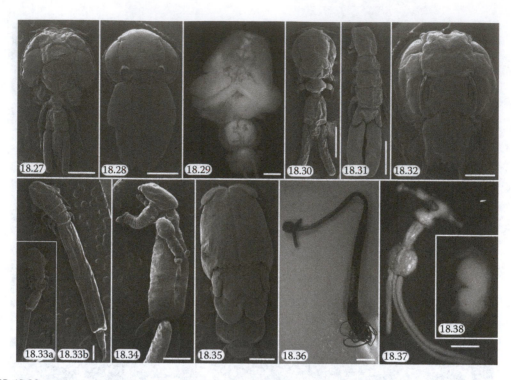

FIGURES 18.27 TO 18.38

Micrographs of Arthropoda II. 18.27. Copepoda: Caligidae ex *Himantura lobistoma*. 18.28. Copepoda: Cecropidae ex *Prionace glauca*. 18.29. Copepoda: Dichelesthiidae ex *Prionace glauca*. 18.30. Copepoda: Dissonidae ex *Chiloscyllium punctatum*. 18.31. Copepoda: Eudactylinidae ex *Himantura pastinacoides*. 18.32. Copepoda: Euryphoridae ex *Sphyrna lewini*. 18.33. Copepoda: Kroyeriidae ex *Prionace glauca*: (a) male, (b) female. 18.34. Copepoda: Lernaeopodidae ex *Galeorhinus australis*, female with small parasitic male. 18.35. Copepoda: Pandaridae: *Pandarus* sp. ex *Squalus acanthias*. 18.36. Copepoda: Pennellidae: *Pennella filosa* (USNM No. 92174) ex Actinopterygii: Perciformes: Istiophoridae: *Makaira nigricans*. 18.37. Copepoda: Sphyriidae: *Norkus cladocephalus* (USNM No. 229971) ex *Rhinobatos productus*. 18.38. Copepoda: Ergasilidae: *Ergasilus myctarothes* (USNM No. 42255) ex *Sphyrna zygaena*. Scale bars: Figures 18.27, 18.29, 18.34, and 18.35, 1 mm; Figures 18.28 and 18.32, 2 mm; Figures 18.30 and 18.33a,b, 500 μm; Figure 18.31, 200 μm; Figure 18.36, 1 cm; Figure 18.37, 3 mm.

of the eyes of a diversity of pelagic sharks off Western Australia (Newbound and Knott, 1999; Tang and Newbound, 2004). Although branchiurans have been observed on the eyes of elasmobranchs on occasion, there is no evidence to suggest that these vagile arthropods are doing anything more than traversing this site. Russo (1975) reported finding the piscicoline leech *Branchellion lobata* on the eyes of spiny dogfish. There exists a record of an adult didymozoid digenean from the back of the eye of the shark *Carcharhinus longimanus* (Pozdnyakov, 1989).

18.3.2.2 Olfactory Bulbs

The olfactory bulbs (also known as olfactory sacs, olfactory capsules, or nasal fossae) and, in particular, the lamellae of these organs are the sites of attachment of a diversity of metazoans, including arthropods such as ostracods, isopods, and copepods, as well as nematodes, leeches, and members of the platyhelminth class Monogenea. Benz (1993) suggested that the conspicuous overlap between the fauna of the olfactory bulbs and that of the gills, at least at higher taxonomic levels, is perhaps not surprising, given the remarkable similarity between the morphology and configuration of the lamellae of the olfactory bulbs and those of the gills. He hypothesized that the olfactory bulbs represent modified branchial chambers that were originally derived from gills.

Ostracods, isopods, leeches, and nematodes are, at best, only occasional associates of the olfactory bulbs of elasmobranchs; for example, although the ostracod *Vargula parasitica* has been reported from the olfactory bulbs of the smooth hammerhead, *Sphyrna zygaena*, in the West Indies (Williams and Bunkley-Williams, 1996; Wilson, 1913), this species more commonly occupies the gills of elasmobranchs (e.g., Williams and Bunkley-Williams, 1996). Praniza larvae of the isopod family Gnathiidae have been reported from the olfactory bulbs of a diversity of sharks and batoids (e.g., Smit and Basson, 2002). On occasion, leeches of the subfamily Piscicolinae have been found on the olfactory bulbs (e.g., Sawyer et al., 1975) or oronasal grooves (Llewellyn and Knight-Jones, 1984) of elasmobranchs. In such cases, however, this site appears to be one of many on which these species are found. The anisakid nematode *Terranova brevicapitata* has been reported from the olfactory bulbs of tiger and dusky sharks (Cheung, 1993).

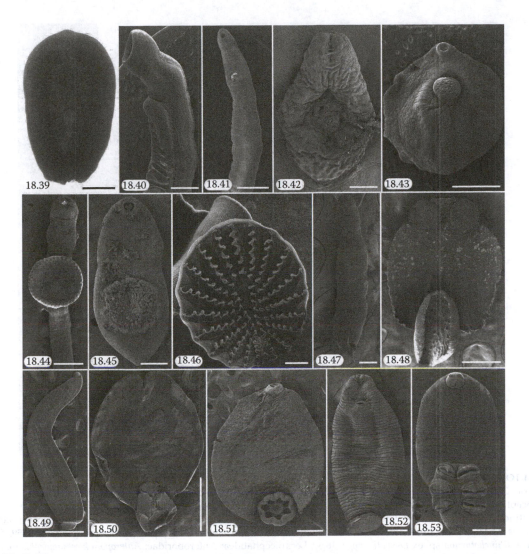

FIGURES 18.39 TO 18.53

Micrographs of non-cestode Platyhelminthes. 18.39. Tricladida: Procerodidae: *Micropharynx parasitica* (HWML No. 1904) ex *Dipturus laevis*. 18.40. Aspidogastrea: Multicalycidae: *Multicalyx cristata* ex *Dasyatis* sp. 18.41. Digenea: Azygiidae: *Otodistomum* sp. ex *Raja rhina*. 18.42. Digenea: Derogenidae: *Thometrema overstreeti* ex *Potamotrygon magdalenae*. 18.43. Digenea: Gorgoderidae: *Anaporrhutum* sp. ex *Rhinoptera* sp. 18.44. Digenea: Syncoeliidae: *Syncoelium vermilionensis* ex *Mobula japanica*. 18.45. Digenea: Zoogonidae: *Diphterostomum* sp. ex *Leptocharias smithii*. 18.46. Monogenea: Acanthocotylidae: posterior attachment structure (haptor) of *Acanthocotyle* sp. ex *Raja* sp. 18.47. Monogenea: Amphibdellidae: *Amphidelloides* sp. ex *Narcine tasmaniensis*. 18.48. Monogenea: Capsalidae ex *Dasyatis akajei*. 18.49. Monogenea: Loimoidae: *Loimopapillosum* sp. ex *Eusphyra blochii*. 18.50. Monogenea: Microbothriidae: *Dermopthirius penneri* ex *Carcharhinus limbatus*. 18.51. Monogenea: Monocotylidae: *Calicotyle* sp. ex *Rhizoprionodon terraenovae*. 18.52. Monogenea: Udonellidae: *Udonella* sp. ex caligid copepod on *Urogymnus asperrimus*. 18.53. Monogenea: Hexabothriidae: *Erpocotyle* sp. ex *Bathyraja magellanica*. Scale bars: Figure 18.39, 1.6 mm; Figures 18.40 and 18.49, 1 mm; Figures 18.41 and 18.43, 2 mm; Figure 18.42, 50 μm; Figures 18.44, 18.48, 18.50, 18.51, and 18.53, 500 μm; Figures 18.45, 18.46, 18.47, and 18.52, 100 μm.

The olfactory bulbs are parasitized by representatives of at least eight families of copepods. These include, for example, chondracanthids such as *Acanthochondrites annulatus* (e.g., Kabata, 1970), ergasilids (e.g., Wilson, 1913), all species of the kroyeriid genus *Kroyerina* (Benz, 1993) and species in several other kroyeriid genera (e.g., Rokicki and Bychawska, 1991), all species of the eudactylinid genus *Eudactylinella* (e.g., Benz, 1993), and several species of pandarids (e.g., Lewis, 1966), lernaeopodids (Benz, 1991), and at least one dissonid (e.g., Boxshall et al., 2008). The family Sphyriidae is represented on this site by the unusual mesoparasite *Thamnocephalus*

cerebrionoxius which was described by Diebakate et al. (1997) from the olfactory bulbs of the shark *Leptocharias smithii* in Senegal. The relationship between this species and its host is complicated by the fact that it simultaneously parasitizes the brain of its host; whereas the posterior region of its body extends from the olfactory bulbs, the anterior regions of the body attach to the olfactory lobes of the brain (see below).

Members of the monogenean family Monocotylidae inhabit the olfactory bulbs of sharks (e.g., Justine, 2009; Kearn and Green, 1983) and batoids (e.g., de Buron and Euzet, 2005; Whittington, 1990) with some regularity.

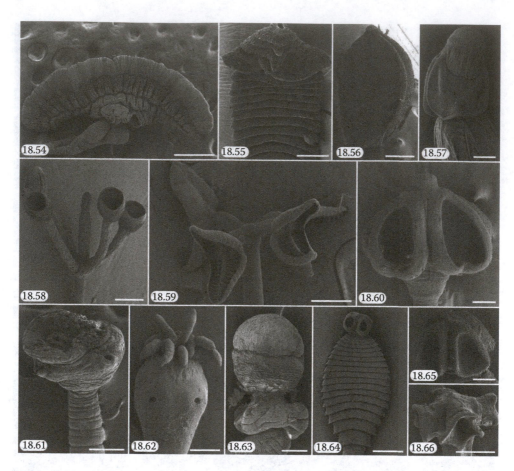

FIGURES 18.54 TO 18.66

Micrographs of Cathetocephalidea, Litobothriidea, Diphyllidea, Rhinebothriidea, and Lecanicephalidea (Cestoda: Platyhelminthes). 18.54. Cathetocephalidae: *Cathetocephalus* sp. ex *Carcharhinus leucas*. 18.55. Litobothriidae: *Litobothrium daileyi* ex *Alopias pelagicus*. 18.56. Diphyllidea: Ditrachybothriidae: *Ditrachybothridium macrocephalum* ex *Leucoraja fullonica*. 18.57. Diphyllidea: Echinobothriidae: *Echinobothrium* sp. ex *Pastinachus atrus*. 18.58. Rhinebothriidea: Echeneibothriidae: *Pseudanthobothrium* sp. ex *Leucoraja erinacea*. 18.59. Rhinebothriidea: Rhinebothriidae: *Rhinebothrium* sp. ex *Dasyatis longa*. 18.60. Lecanicephalidea: Anteroporidae: *Anteropora klosmamorphis* ex *Narcine maculata*. 18.61. Lecanicephalidea: Lecanicephalidae: *Lecanicephalum coangustatum* ex *Dasyatis centroura*. 18.62. Lecanicephalidea: Polypocephalidae: *Polypocephalus* sp. ex *Rhinoptera* cf. *steindachneri* (*sensu* Naylor et al., in press). 18.63. Lecanicephalidea: Tetragonocephalidae: *Tylocephalum* sp. ex *Rhina ancylostoma*. 18.64. Lecanicephalidea: *Hornellobothrium* sp. ex *Aetobatus ocellatus*. 18.65. Lecanicephalidea: *Aberrapex* sp. ex *Aetomylaeus vespertilio*. 18.66. Lecanicephalidea: *Quadcuspibothrium francisi* ex *Mobula japanica*. Scale bars: Figure 18.54, 500 µm; Figures 18.56, 18.58, and 18.59, 200 µm; Figures 18.55, 18.57, 18.60, 18.62, 18.63, and 18.65, 50 µm; Figures 18.61, 18.64, and 18.66, 100 µm.

In fact, the olfactory bulbs appear to represent the primary site of attachment for many species and even genera of monocotylids (Kearn, 1998), particularly the merizocotylines. Several species in the monogenean family Acanthocotylidae have also been reported from the olfactory bulbs of elasmobranchs, as have at least some species of microbothriids (Price, 1963). It should be noted that Price's records of both of these families from this site are particularly unusual and should be verified.

18.3.2.3 *Acousticolateralis System*

In their diagnosis of the philichthyid copepod genus *Colobomatus*, Deboutteville and Nunes (1952, p. 599) noted that species "vivant dans les canaux muqueux de la tête de Téléostéens (rarement des Sélaciens)." Indeed, *Colobomatus lamnae* was described by Hesse (1873) from this site on the porbeagle, *Lamna nasus* (as *L. cornubica*). This site represents one of the most poorly known regions of the elasmobranch body because it is so infrequently examined for parasites. We suspect that efforts spent examining the pores and ducts of this system in a diversity of elasmobranchs might yield additional members of this copepod family.

18.3.3 Respiratory System

18.3.3.1 *Spiracles*

Although an uncommon site of attachment for parasites, the spiracles of elasmobranchs have been reported to host copepods of the families Lernaeopodidae,

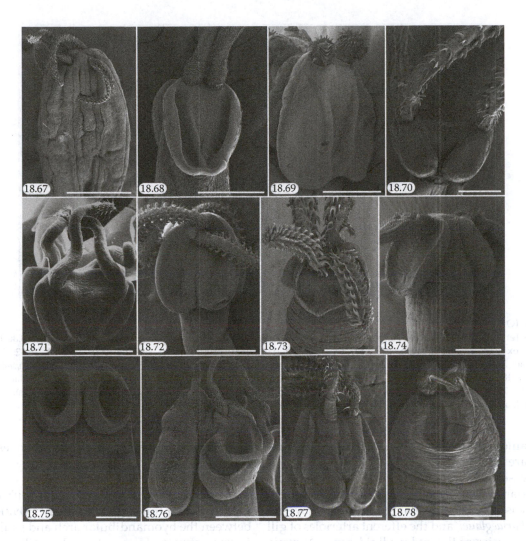

FIGURES 18.67 TO 18.78

Micrographs of Trypanorhyncha (Cestoda: Platyhelminthes). 18.67. Tentaculariidae: *Tentacularia* sp. ex *Prionace glauca*. 18.68. Eutetrarhynchidae: *Eutetrarhynchus lineatus* ex *Ginglymostoma cirratum*. 18.69. Mixodigmatidae: *Mixodigma leptaleum* ex *Megachasma pelagios*. 18.70. Rhinoptericolidae: *Rhinoptericola* sp. ex *Rhinoptera bonasus*. 18.71. Shirleyrhynchidae: *Cetorhinicola acanthocapax* ex *Cetorhinus maximus*. 18.72. Lacistorhynchidae: *Grillotia similis* ex *Ginglymostoma cirratum*. 18.73. Otobothriidae: *Otobothrium* sp. ex *Negaprion acutidens*. 18.74. Pseudotobothriidae: *Pseudotobothrium arii* ex *Lamiopsis tephrodes*. 18.75. Aporhynchidae: *Aporhynchus menezesi* ex *Etmopterus spinax*. 18.76. Gilquiniidae: *Gilquinia squali* ex *Squalus suckleyi*. 18.77. Gymnorhynchidae: *Gymnorhynchus isuri* ex *Isurus oxyrhinchus*. 18.78. Sphyriocephalidae: *Hepatoxylon trichuri* ex *Prionace glauca*. Scale bars: Figures 18.67 and 18.77, 500 μm; Figures 18.68, 18.69, 18.70, 18.73, 18.74, 18.75, and 18.76, 200 μm; Figure 18.71, 400 μm; Figure 18.72, 1 mm; Figure 18.78, 2 mm.

Taeniacanthidae, and Caligidae. For example, Kabata (1979) noted that the lernaeopodid *Pseudocharopinus bicaudatus* is commonly found within the spiracles of *Squalus acanthias* in British waters. Braswell et al. (2002) found a small percentage of the individuals of the taeniacanthid *Taeniacanthodes dojirii* from the electric ray, *Narcine entemedor*, attached in the vicinity of the spiracles. Caligid copepods have been reported from the spiracles of several species of elasmobranchs in the Gulf of Mexico (e.g., Bere, 1936). On rare occasions, piscicolid leeches of the subfamily Piscicolinae have been found to occur in, among other sites, the spiracular valves of elasmobranchs (Russo, 1975).

18.3.3.2 Gills and Branchial Chamber

The gills and branchial chamber of elasmobranchs offer a protected, oxygen-rich, compact living space for parasites (Kearn, 1998). The gill filaments are a rich source of blood for blood-feeding parasite taxa. In combination with their accessibility, these factors may explain the very high diversity of metazoan parasites found in this region of the elasmobranch body (Kearn, 1998). Although we treat the gills and branchial chamber together, in nature parasites often occupy much more specific sites within the gills and branchial chamber. This fact was nicely illustrated by Benz (1986), who

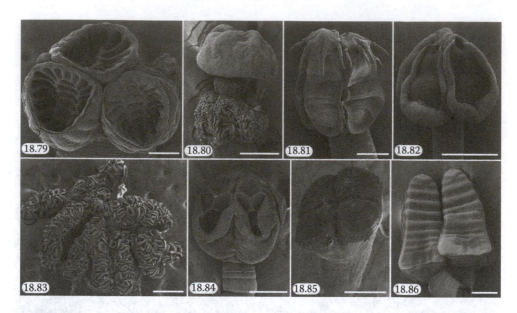

FIGURES 18.79 TO 18.86

Micrographs of Tetraphyllidea (Cestoda: Platyhelminthes). 18.79. Dioecotaeniidae: *Dioecotaenia* sp. ex *Rhinoptera bonasus*. 18.80. Disculicepitidae: *Disculiceps* sp. ex *Carcharhinus brevipinna*. 18.81. Onchobothriidae: *Acanthobothrium* sp. ex *Aetobatus laticeps*. 18.82. Phyllobothriidae: Phyllobothriinae: *Paraorygmatobothrium* sp. ex *Carcharhinus brevipinna*. 18.83. Phyllobothriidae: Thysanocephalinae: *Thysanocephalum* sp. ex *Galeocerdo cuvier*. 18.84. Phyllobothriidae: Triloculariinae: *Zyxibothrium kamienae* ex *Malacoraja senta*. 18.85. Prosobothriidae: *Prosobothrium* sp. ex *Prionace glauca*. 18.86. Serendipeidae: *Duplicibothrium minutum* ex *Rhinoptera* cf. *steindachneri* (*sensu* Naylor et al., in press). Scale bars: Figures 18.79, 18.80, 18.84, and 18.85, 200 μm; Figures 18.81, 18.82, and 18.86, 100 μm; Figure 18.83, 500 μm.

noted that, among blue shark copepods, the interbranchial septa are inhabited by the pandarid *Phyllothyreus cornutus*, the secondary lamellae of the gill filaments by the pandarid *Gangliopus pyriformis*, the excurrent water channels between gill filaments by the kroyeriid *Kroyeria carchariaeglauci*, and the efferent arterioles of gill filaments by species of the eudactylinid genus *Nemesis*. Space limitations prevent us from presenting more specific data on attachment sites for the parasites of gills and branchial chamber here.

Somewhat surprisingly, leeches are only rarely found associated with the gills of elasmobranchs. In most cases, species that have been reported from this site (e.g., the pontobdelline *Stibarobdella tasmanica* and the piscicoline *Branchellion ravenelli*) have also been reported from a diversity of other sites, most notably the skin (E.M. Burreson, pers. comm.), suggesting that the gills are not a preferred site of attachment for these annelids. Although it does not appear to be a common site of infection for myxozoans, in his list of hosts and sites occupied by myxozoans, Kudo (1920) indicated that the myxobolid *Myxobolus mülleri* has been reported from the branchiae of *Squalus*.

At least four families of nematodes include species that have been found associated with the gills and branchial chamber of elasmobranchs. Such occurrences, however, are relatively uncommon. The distinctive eggs of the trichosomoidid nematode *Huffmanela carcharhini*, discussed in more detail in the section on skin above,

have been observed in the mucosa covering the connective tissue of the gill arches of certain carcharhinid sharks (Moravec, 2001). The distinctive vesicular female of the philometrid nematode *Phlyctainophora lamnae* was reported from this region by Steiner (1921), specifically between the hyomandibular arch and skull of the shark *Lamna nasus* (as *L. cornubica*). The gills were included among the sites from which Adamson et al. (1987) collected adults of the philometrid *Phlyctainophora squali* and from which Aragort et al. (2002) collected adults of the guyanemid dracunculoid *Histodytes microocellatus*. Microfilariae of a species of dracunculoid belonging to a fourth nematode group, but one that has not yet been assigned to family, were found in gill squashes from a spotted eagle ray by Adamson et al. (pers. comm.). Cheung (1993) noted that the anisakid *Contraceacum plagiostomum* has been found on the gills of the basking shark and thorny skate.

Copepods are among the most commonly encountered parasites of the gills and branchial chamber of elasmobranchs. Of the 16 families of copepods parasitizing elasmobranchs, 11 include species that parasitize these sites. The gills and branchial chamber are the primary site of attachment for nine families of siphonostomes. All eight genera of Eudactylinidae that parasitize elasmobranchs include species that inhabit the gills; six of these eight genera are restricted to the gills (Benz, 1993). Two of the three genera of kroyeriids include species that are found on the gill lamellae or, in the case of the unusual mesoparasite *Kroyeria caseyi*, interbranchial septa of their elasmobranch hosts (Benz and Deets, 1986). Benz (1993) considered the gill filaments and branchial chamber to be among the primary sites of attachment of caligid copepods. Although not the primary site of attachment, the gills and branchial chamber are included among the sites occupied by the following additional siphonostome families: Cecropidae (e.g., Benz and Deets, 1988), Dichelesthiidae (e.g., Benz, 1993), Dissonidae (e.g., Benz, 1993), Pandaridae (e.g., Lewis, 1966), Lernaeopodidae (e.g., Wilson, 1913), and a number of species of the mesoparasitic family Sphyriidae (e.g., Kabata, 1979). In addition, members of the poecilostome copepod families Taeniacanthidae (e.g., Wilson, 1913) and Chondracanthidae (Cheung, 1993) also occur on this site.

Three additional arthropod groups have been reported from the gills of elasmobranchs, albeit infrequently. This is the primary site of attachment of cypridinid ostracods when they have been found associated with elasmobranchs (e.g., Bennett et al., 1997; Williams and Bunkley-Williams, 1996). Records of members of the branchiuran genus *Argulus* from the branchial cavity of dasyatid stingrays exist, but this does not appear to be a primary site of attachment for these arthropods (Cressey, 1976). Members of five families of isopods have been reported from the gills of elasmobranchs. Among these, the praniza larvae of the Gnathiidae are most commonly encountered in this site (e.g., Newbound and Knott, 1999; Ota and Hirose, 2009). These arthropods are known to cause injury to the epithelium and are often associated with inflammation and severe tissue hypertrophy, particularly in heavy infections (e.g., Honma et al., 1991). Delaney (1984) found corallanid isopods attached to the gills of *Aetobatus laticeps* (as *A. narinari*). At least four genera of aegid isopods have been reported from the gills (Moreira and Sadowsky, 1978). In addition, the cymothid isopod *Lironeca ovalis* has been reported from the gills of sawfish (Cheung, 1993). The gills were also among the sites occupied by the cirolanid isopod *Cirolana borealis* (Bird, 1981); however, Bunkley-Williams and Williams (1998) considered this to represent an accidental location.

Although the vast majority of the approximately 6000 species of platyhelminths belonging to the subclass Digenea live as endoparasites (Cribb et al., 2001), there are some exceptions, and many of these can be found among the digeneans of elasmobranchs. In fact, as noted by Bray and Cribb (2003), the digeneans of elasmobranchs tend to be found in a diversity of relatively unusual sites, many of which are more typical of their ectoparasitic relatives. Members of the family Syncoeliidae, for example, often associate with the gills of their hosts (Curran and Overstreet, 2000). Didymozoids, such as *Tricharrhen okenii*, have also been reported from the gill arches and

branchial chamber of elasmobranchs (Yamaguti, 1971). Bray and Cribb (2003) noted that a lepocreadiid digenean has also been reported from the gills of a porbeagle shark, but this was likely the result of an accidental infection.

On rare occasions, larval cestodes have been found associated with the gills of elasmobranchs; in all cases, these were larvae of trypanorhynchs. For example, Dollfus (1960) reported plerocerci of a species of the tentaculariid genus *Nybelinia* from the gills of *Mustelus canis*. More recently, Palm (2004) reported plerocerci of the lacistorhynchid *Grillotia scolecinus* and also of an unidentified eutetrarhynchid from the gills of a squaliform shark and a catshark, respectively. He also reported plerocercoids of the sphyriocephalid *Hepatoxylon trichuri* from the gills of mako sharks.

Representatives of at least seven of the nine families of monogeneans have been reported from elasmobranch gills. In elasmobranchs, hexabothriids are known only from the gills, where they exhibit feeding habits that are relatively unusual for monogeneans in that they feed exclusively on blood (Kearn, 1998). Approximately 53 species of hexabothriids have been reported from elasmobranchs. With only a few exceptions, the Amphibdellidae have been reported from the gills of batoids (e.g., Llewellyn, 1960). The loimoids are found primarily on the gills of their elasmobranch hosts (e.g., Cheung, 1993). The gills represent one of several sites occupied by capsalids (e.g., Cheung, 1993) and monocotylids (Chisholm et al., 1997). In addition, acanthocotylids (Bonham and Guberlet, 1938) and microbothriids (Price, 1963) have been reported from this site; however, the latter reports are so rare that they should be considered suspect until verified. Nearly half of the monogeneans described from elasmobranchs since the publication of the first edition of this book have been monocotylids found parasitizing the gills of their batoid hosts (e.g., Chisholm and Whittington, 2005; Domingues and Marques, 2007; Vaughan and Chisholm, 2010).

18.3.4 Digestive System

Parasite site specificity within the digestive system is marked, reflecting the fact that, although the digestive tract represents a single continuous tube beginning with the mouth and ending with the rectum and cloaca, it contains a diversity of physically and physiologically distinct environments where the needs of parasites are concerned.

18.3.4.1 Buccal Cavity and Esophagus

At least one species in each of the three subfamilies of piscicolid leeches has been recorded from the buccal cavity or esophagus of elasmobranchs. The pontobdelline

Stibarobdella macrothela (Sawyer et al., 1975) and the piscicoline *Branchellion lobata* (Llewellyn and Knight-Jones, 1984) have been reported from, among other sites, the buccal cavity. The platybdelline leech *Pterobdella amara* (= *Rhopalobdella japonica*), however, may be specific to the buccal cavity; this species is known only from the buccal cavity of several myliobatiform species, where it has been found dangling between the upper lip and tooth plates (Burreson, 2006; Burreson and Kearn, 2000). The unusual leech discovered by Curran et al. (pers. comm.) from *Zapteryx exasperata* in the Gulf of California, Mexico, was found in the buccal cavity in addition to the external body surfaces of this host. As noted above, to date these authors have been unable to assign this piscicolid species to a subfamily.

Copepods are less commonly encountered in the buccal cavity than they are on other external body sites of elasmobranchs, but a few species occupy this site with some regularity. The only species of dichelesthiid that parasitizes elasmobranchs, *Anthosoma crassum*, has been reported from the buccal cavity of large pelagic sharks (e.g., Lewis, 1966). Species of the eudactylinid genus *Carniforssorus* are known from the oral chamber of elasmobranchs (Benz, 1993). Bere (1936) reported a diversity of caligids from the mouth of several batoids and pandarids from the mouth of several sharks in the Gulf of Mexico. Euryphorids, lernaeopodids, and sphyriids have also been reported from the buccal cavity of elasmobranchs (Cheung, 1993). In addition, the praniza larvae of gnathiid isopods (Heupel and Bennett, 1999) and adult cirolanids (Moreira and Sadowsky, 1978) have been reported from the buccal cavity of elasmobranchs.

Digeneans, monogeneans, nematodes, and barnacles have also occasionally been found parasitizing this site. Digeneans of the families Syncoeliidae (e.g., Curran and Overstreet, 2000) and Ptychogonimidae (e.g., Cheung, 1993) are known to occur in the buccal cavity of sharks, as are the monogenean monocotylid *Tritestis ijime* (e.g., Price, 1938b) and microbothriid *Dermopristis paradoxus* (Kearn, et al., 2010). The buccal cavity was among the sites in which Adamson et al. (1987) found adults of the philometrid nematode *Phlyctainophora squali* in several species of sharks off California. The cystidicolid nematode *Parascophorus galeata* was reported from the esophagus of *Sphyrna tiburo* in North Carolina (Cheung, 1993). The mouth was among the sites reported to be infected with the barnacle *Anelasma squalicola* by Yano and Musick (2000).

18.3.4.2 Stomach

Despite the potentially inhospitable nature of the stomach as an environment, it is home to a number of platyhelminth, nematode, and, to a lesser extent, acanthocephalan and arthropod species. Most of the species that inhabit the stomach do so to the exclusion of other sites in the elasmobranch host. In some cases, this site specificity extends to higher taxonomic categories as well.

The presence of acanthocephalans in elasmobranchs is unusual. Adults of most of the 1200 or so species in this phylum are much more commonly encountered in teleosts, birds, turtles, and mammals (Crompton and Nickol, 1985). As a consequence, records of acanthocephalans from elasmobranchs are generally considered to represent accidental infections (Knoff et al., 2001). Williams et al. (1970) presented a possible explanation for this absence, suggesting that acanthocephalans may be unable to tolerate the high levels of urea found in elasmobranchs. Five of the 12 species of acanthocephalans identified from elasmobranchs have been found in the stomach. Two species belong to the polymorphid genus *Corynosoma* (Knoff et al., 2001), two to the rhadinorhynchid genus *Serrasentis* (Bilqees and Khan, 2005; Yamaguti, 1963), and one to the arhythmacanthid genus *Acanthocephalides* (Di Cave et al., 2003). In combination, these records include cysticanth stages, juveniles, and adult worms.

At least 13 species in four families of nematodes parasitize the stomach of elasmobranchs as larvae or adults. These include members of the Anisakidae (Olsen, 1952; Rajya et al., 2007), Ascaridae (McVicar, 1977), and, most commonly, the Acanthocheilidae (Diaz, 1972) and Physalopteridae (Moravec and Nagasawa, 2000). Many of the nematodes found in the stomach appear to be restricted to this site, or at least have not typically been reported from other sites in their hosts.

Although it is not clear if they represent parasites or food, members of three families of isopods have been reported from the stomach of elasmobranchs. These include the cirolanid *Cirolana borealis*, the aegid *Aega psora*, and the cymothoid *Lironeca raynaudi* (Moreira and Sadowsky, 1978).

In their treatment of the digeneans of elasmobranchs, Bray and Cribb (2003) suggested that members of as many as 18 of the 148 families of platyhelminths in the subclass Digenea have been reported parasitizing elasmobranchs; however, they considered only 7 of these families (Azygiidae, Derogenidae, Gorgoderidae, Ptychogonimidae, Sanguinicolidae, Syncoeliidae, and Zoogonidae) to represent long-term associations with elasmobranchs. Given the nature of reported associations, they considered 2 families (Campulidae and Didymozoidae) to represent developing longer term associations with elasmobranchs. The remaining 9 families (indicated by "accid." in Table 18.1) were considered likely to represent accidental infections in elasmobranchs given the frequency with which they, or their congeners, are found parasitizing teleosts in combination with the frequency with which records of these 9 families come from the digestive system of elasmobranchs. Nonetheless, among the 18 families of digeneans found in elasmobranchs (as accidental or normal infections), species in 6 have been reported from the stomach with some regularity. One of the most commonly encountered groups is the Azygiidae. In fact, large, muscular *Otodistomum* species are conspicuous inhabitants of the stomachs of a diversity of sharks and batoids (Gibson and Bray, 1977); elasmobranchs appear to be the primary hosts of these taxa. The more delicate-bodied zoogonids (e.g., Bray et al., 1995) and ptychogonimids (e.g., Bray and Gibson, 1986) have also been reported from the stomach of elasmobranchs; in fact, ptychogonimids are known only from the stomachs of elasmobranchs. Many of the zoogonid species also occur in teleosts (Bray and Gibson, 1986). Bucephalids, opecoelids, and derogenids have also been reported from the stomachs of their hosts, the latter family with some regularity (Bray and Cribb, 2003).

By far the majority of members of the platyhelminth class Cestoda reported from the stomach of elasmo-branchs are trypanorhynchs. Site data provided by Bates (1990), Cheung (1993), and Palm (2004) suggest that at least 6 of the 16 families of trypanorhynchs are able to live in the stomach as either adults or larvae. Adults of the tentaculariid genera *Heteronybelinia*, *Nybelinia*, and *Tentacularia* are routinely found attached in the pyloric stomach of a diversity of sharks and some batoids (e.g., Dollfus, 1942; Guiart, 1935; Palm, 2004), as are adult sphyriocephalids (e.g., *Hepatoxylon trichuri*) (Williams, 1960). Adults of eutetrarhynchid genera such as *Eutetrarhynchus*, *Parachristianella*, *Prochristianella*, and *Tetrarhynchobothrium*, and also of the gilquiniid genus *Gilquinia*, are occasionally found in the stomachs of their elasmobranch hosts. Larval stages of a diversity of trypanorhynch families are known, often from cyst or capsule-like structures in the wall of the stomach of elasmo-branchs. These include plerocercoids of *Hepatoxylon*, *Heteronybelinia*, *Nybelinia*, and *Sphyriocephalus*, as well as plerocerci of otobothriid genera such as *Otobothrium*, lacistorhynchids such as *Grillotia*, and pterobothriids such as *Pterobothrium* (Cheung, 1993; Dollfus, 1942; Guiart, 1935; Klimpel et al., 2001; Palm, 2004). The report of the diphyllidean cestode *Echinobothrium benedeni* from the stomach of skates in the Mediterranean Sea (Cheung, 1993) requires confirmation.

18.3.4.3 Spiral Intestine

The spiral intestine is home to acanthocephalans, digeneans, nematodes, cestodes, and infrequently monogeneans. Five families of acanthocephalans have been reported from the spiral intestine of elasmobranchs. These families are the Cavisomidae (e.g., Golvan et al., 1964), Echinorhynchidae (e.g., Arai, 1989), Illiosentidae (e.g., Buckner et al., 1978), Rhadinorhynchidae, and Polymorphidae (e.g., Knoff et al., 2001). In most of these families, however, records exist for a total of only one or two species. Like the acanthocephalans found in the stomach, most records of acanthocephalans from the spiral intestine are thought to represent accidental, rather than normal infections, because in most cases these taxa are also known to parasitize other species of vertebrates. The cavisomid genus *Megapriapus* may, however, represent an interesting exception, for it is known only from potamotrygonid stingrays (Golvan et al., 1964). The work of Marques (2000) suggests that this genus may be even more widely distributed among potamotrygonids than initially thought.

The spiral intestine of elasmobranchs hosts a greater diversity of nematodes than any other site within the elasmobranch body. Approximately 53 of the 83 or so species in 10 of the 14 families of nematodes known to parasitize elasmobranchs have been reported from this site. These records include members of both classes of nematodes, the Rhabditea and Enoplea. The spiral intestine is the primary site of attachment of the Gnathostomatidae (e.g., Deardorff and Ko, 1983) and the Capillariidae (Moravec, 2001). The following families also occur in this organ with some regularity: ascarids and anisakids (McVicar, 1977), acanthocheilids (Razi Jalali et al., 2008; Williams et al., 1970), physalopterids (Moravec et al., 2002), cystidicolids (Campana-Rouget, 1955), philometrids (Adamson et al., 1987), and cucullanids (Johnston and Mawson, 1943). In addition, Moravec et al. (1998) reported the dracunculoid *Mexiconema cichlasomae* from the spiral intestine of *Ginglymostoma cirratum*; however, these authors considered this an accidental infection because, although rare in elasmo-branchs, this nematode commonly parasitizes teleosts. Most of the above nematodes occur in the spiral intestine as adults. It is uncommon to encounter more than a single species of nematode in the spiral intestine, and, in general, infections consist of a small number of individuals (i.e., infections are of low intensity). On occasion, however, we have encountered remarkably large concentrations of individuals of some of the larger nematodes (e.g., gnathostomatids) in the spiral intestines of, for example, *Aetomylaeus* cf. *nichofii* 2 (*sensu* Naylor et al., in press) in Australia and *Heterodontus mexicanus* in Mexico.

In stark contrast to the incredible diversity of digeneans found parasitizing the intestinal tract of teleosts, only a handful of species in just 8 of the 148 known families of digeneans parasitize the spiral intestine of elasmobranchs (Bray and Cribb, 2003; Cribb et al., 2001). In the cases of most families, records from the spiral intestine are rare. With the exception of the robustly muscular azygiids such as *Otodistomum veliporum* (see Gibson and Bray, 1977), most of these species are small and thus easy to overlook. Brooks (1979) described the derogenid

Thometrema overstreeti (as *Paravitellotrema overstreeti*) from the spiral intestine of the freshwater stingray *Potamotrygon magdalenae*. Although this species also parasitizes teleosts, it appears to be a regular component of the intestinal fauna of these rays. Reports of bucephalids such as *Prosorhynchus squamatus* and *Bucephalopsis arcuatum* (e.g., Cheung, 1993) and zoogonids such as *Diptherostomum betencourti* (e.g., Bray and Gibson, 1986) from the spiral intestine also exist. Bray and Cribb (2003) also cited reports of didymozoids, hemiurids, opecoelids, syncoeliids, and other derogenids from the spiral intestine of elasmobranchs, but many of these are likely accidental infections (e.g., McVicar, 1977). On rare occasions, monocotylid monogeneans have been reported parasitizing this organ (e.g., Chisholm et al., 1997).

The spiral intestine is by far the most heavily parasitized internal organ of elasmobranchs. This is because it is the primary site occupied by adult cestodes, which unquestionably comprise the most speciose group of elasmobranch parasites. It is rare to encounter an elasmobranch in nature that does not host at least one species of cestode in its spiral intestine. The exceptions appear to be squaliform and some scyliorhinid sharks in which both the prevalence and intensities of infections are generally relatively low. In our experience, it is fairly routine for elasmobranch species, particularly batoids, to each host 10 or more species of spiral intestine cestodes. Recent work has expanded the number of cestode orders from 14 (e.g., Khalil et al., 1994), of which 4 parasitize elasmobranchs as adults, to 17, of which 7 parasitize elasmobranchs as adults. This increase is, in part, the result of recognition of the Cathetocephalidea and Litobothriidea as independent orders rather than families of Tetraphyllidea (Caira et al., 2005). It is also the result of a dismantling of the Tetraphyllidea that was begun by Healy et al. (2009) with the erection of the Rhinebothriidea as an order separate from the Tetraphyllidea. The Cathetocephalidea, Diphyllidea, Lecanicephalidea, Litobothriidea, and Rhinebothriidea are exclusive to elasmobranchs. With the exception of the Chimaerocestidae, which parasitizes holocephalans (Williams and Bray, 1984), this is also true of the Tetraphyllidea. Whereas the majority of trypanorhynchs parasitize elasmobranchs, some species are known from holocephalans (see below). Collectively, these seven cestode orders include 33 families and over 1133 species. Representatives of all but one (Chimaerocestidae) have been reported from the spiral intestine of elasmobranchs and, in fact, the spiral intestine is the only site occupied by the majority of the species in these families. These orders differ conspicuously from one another in diversity. The Cathetocephalidea and Litobothriidea are the least speciose of the orders, with five and eight species, respectively. In the most recent monographic treatment of the Diphyllidea, Tyler (2006) recognized two families,

the Ditrachybothriidae and the Echinobothriidae, each consisting of a single genus. In total, approximately 40 valid species in the order are currently recognized (Kuchta and Caira, 2010). The newly established order Rhinebothriidea, which now houses the stalk-bearing taxa formerly of the Tetraphyllidea, includes approximately 13 genera and at least 90 species (Healy, 2006; Healy et al., 2009), all of which parasitize batoids; at this point, we recognize two families in this new order, the Rhinebothriidae and the Echeneibothriidae. The Lecanicephalidea are considered now to house over 100 species and 17 valid genera (Cielocha and Jensen, 2011; Jensen, 2005; Jensen et al., 2011; Koch et al., in press). Euzet (1994a) recognized four families of lecanicephalideans: Polypocephalidae, Anteroporidae, Tetragonocephalidae, and Lecanicephalidae. There are, however, a number of forms that do not conform to the diagnoses of any of the families recognized by Euzet (1994a), and, as noted by Jensen (2005), the family-level classification is sorely in need of revision. We have included electron micrographs of some of these taxa, including *Hornellobothrium* (Figure 18.64), *Aberrapex* (Figure 18.65), and *Quadcuspibothrium* (Figure 18.66), to make their presence known to readers. Although most lecanicephalideans have been reported from batoids, records exist of members of this order parasitizing sharks (e.g., Caira et al., 1997b; Jensen, 2005). Trypanorhynch cestodes are easy to recognize by their possession of four tentacles bearing hooks. The taxonomy of the over 280 described species is based largely on the shape and arrangement of these tentacular hooks. A total of 76 genera are currently recognized. In the most recent monographic treatment of the order, Palm (2004) revised the classification scheme of Campbell and Beveridge (1994) and recognized 15 families in 5 superfamilies. This classification scheme was most recently explored in a molecular phylogenetic framework by Palm et al. (2009) and Olson et al. (2010), with the latter authors recognizing the Shirleyrhynchidae as an additional valid family. Trypanorhynchs in general parasitize both sharks and batoids; however, many families appear to be restricted to one or the other of these clades (Campbell and Beveridge, 1994). In fact, Olson et al. (2010) formally recognized the Trypanobatoida, primarily for the batoid-associated taxa, and Typanoselachoida, primarily for the shark-associated taxa. Although most trypanorhynchs parasitize the spiral intestine, they are also known to occupy other sites (e.g., gall bladder, kidneys; see below). In fact, almost all cestode species that have been found as adults in sites of the elasmobranch body other than the spiral intestine belong to this order. Unlike the other orders of cestodes, both larval and adult trypanorhynchs are found in the spiral intestine of elasmobranchs.

The tetraphyllideans are the most diverse of the seven cestode orders parasitizing elasmobranchs. The order includes approximately 65 genera and well over 650 species. Euzet (1994b) recognized eight families, seven of which are found in elasmobranchs and five of which (Dioecotaeniidae, Disculicepitidae, Onchobothriidae, Phyllobothriidae, and Prosobothriidae) remain in the order following the recent ordinal level reorganization by Healy et al. (2009) along with the Serendipeidae of Brooks and Barriga (1995) as modified by Brooks and Evenhuis (1995). The Tetraphyllidea, however, likely remains polyphyletic even in its more restricted concept. Collectively, the species in this order exhibit morphological diversity of their attachment structure, or scolex, that is unparalleled in any other cestode group. Most systematically problematic among tetraphyllidean families is the Phyllobothriidae, which includes a suite of potentially unrelated groups (Caira et al., 1999, 2001). The recent monograph of Ruhnke (2011) treating this family provides useful guidance with respect to the identity and membership of this difficult taxon. We have included illustrations of the subfamilies of phyllobothriids recognized by Euzet (1994b), with the inclusion of the Serendipeidae (Figure 18.79 through Figure 18.86) as a reminder of the diversity represented by what is currently considered to be a single family. With the exception of *Acanthobothrium*, genera of tetraphyllideans are generally restricted in distribution to either sharks or batoids.

Although we have not attempted to present detailed data here, the spiral intestine is a complex environment that consists of a diverse suite of microhabitats (Williams et al., 1970). Many cestode species, including trypanorhynchs (e.g., Caira and Gavarrino, 1990) and tetraphyllideans (e.g., Cislo and Caira, 1993; Curran and Caira, 1995; Williams et al., 1970), exhibit some degree of site specificity for particular regions within this organ.

18.3.4.4 Rectum

Platyhelminths appear to be the sole occupants of the elasmobranch rectum, and the diversity of platyhelminths parasitizing this site is limited. Members of the digenean families Gorgoderidae (e.g., Cheung, 1993), Syncoeliidae (e.g., Curran and Overstreet, 2000), and Zoogonidae (e.g., Bray and Gibson, 1986) have, on occasion, been found inhabiting the rectum of sharks. In addition, records of monogeneans of the family Monocotylidae from the rectum of elasmobranchs are not uncommon (e.g., Bullard and Overstreet, 2000; Chisholm et al., 1997).

18.3.4.5 Cloaca

The cloaca is home to a number of platyhelminths and copepods and, on occasion, leeches and isopods. The cloacal region of some batoids is the site of attachment of certain members of the monogenean family Monocotylidae (e.g., Bullard and Overstreet, 2000; Chisholm et al., 1997) and also adults of at least two species of the eutetrarhynchid trypanorhynch cestode genus *Mobulocestus* (Campbell and Beveridge, 2006). Digeneans of the family Syncoeliidae have been reported from the cloaca of several large pelagic shark species (e.g., Curran and Overstreet, 2000). Species in three families of copepods have been found in the cloaca of elasmobranchs. These include, for example, a species of chondracanthid in the genus *Acanthochondrites*, the caligid *Caligus rabidus*, and several species of lernaeopodids belonging to the genus *Lernaeopoda* (Cheung, 1993). The cloaca is among the numerous sites of the body parasitized by the piscicoline leech *Branchellion lobata* (Cheung, 1993) and also by the undescribed leech that Curran et al. (pers. comm.) found on *Zapteryx exasperata*. Hale (1940) reported the aegid isopod *Aega antillensis* from the cloaca of a tiger shark in Australia. Praniza larvae of gnathiids have also been reported from this site (Benz and Bullard, 2004).

18.3.4.6 Gallbladder and Bile Ducts

In elasmobranchs, the gallbladder and bile ducts are the primary sites of infection of members of the platyhelminth subclass Aspidogastrea. This taxon currently consists of four families (Rohde, 2002), two of which include species that parasitize elasmobranchs. In the Multicalycidae, *Multicalyx cristata* has been reported from the gallbladder of sharks and batoids (e.g., Thoney and Burreson, 1986). The only known species of Stichocotylidae, *Stichocotyle nephropis*, was found in the bile ducts of batoids (e.g., MacKenzie, 1963). By far the majority of myxozoans of elasmobranchs have been reported from the gallbladders of their hosts. This is largely because this site is home to the two most speciose families of myxozoans (Ceratomyxidae and Chloromyxidae) that parasitize elasmobranchs (Cheung, 1993; Kudo, 1920); however, this organ is also home to a few representatives of the families Myxidiidae (Kpatcha et al., 1996) and Myxosomatidae (Lom and Dykova, 1995). We were surprised to discover that the gallbladder of *Mobula japonica* in the Gulf of California, Mexico, was routinely occupied by adults of a relatively large trypanorhynch cestode belonging to the family Eutetrarhynchidae of the genus *Fellicocestus*, which was described by Campbell and Beveridge (2006).

18.3.4.7 Pancreatic Duct

Nematodes of the family Rhabdochonidae have been reported on several occasions from the pancreatic ducts of sharks and batoids. McVicar and Gibson (1975) reported adults of *Pancreatonema torriensis* from the pancreatic ducts of *Leucoraja naevus* (as *Raja*) and Moravec et al. (2001) reported a new species of rhabdochonid nematode from the pancreatic duct of the *Squalus acanthias*, in the Western Atlantic Ocean off coastal Massachusetts.

18.3.4.8 Liver

The majority of our current knowledge about the parasites of the elasmobranch liver comes from samples taken from the outer surfaces of this organ. Little is known about the organisms that may inhabit the parenchyma for these inner tissues are rarely examined. Plerocercoids of trypanorhynchs of the sphyriocephalid genus *Hepatoxylon* are commonly found attached to the exterior surfaces of the liver in some of the larger species of pelagic sharks (e.g., Waterman and Sin, 1991). Adams et al. (1998) provided an interesting account of a single gravid campulid from the liver of *Alopias vulpinus*, a shark they justifiably considered to be an atypical host for this digenean species. In the first edition of this book, we predicted that careful examination of the internal regions of the liver was likely to reveal the presence of nematodes in at least some elasmobranch species. Indeed, recently Andrade et al. (2008) reported third-stage juveniles of the anisakid nematode genus *Terranova* from the biliary ducts of *Sympterygia acuta*. We suspect that future attention to the internal structures of this organ is likely to yield additional members of this phylum.

18.3.5 Circulatory System

18.3.5.1 Heart and Vasculature

Representatives of several different groups of invertebrate metazoans have been reported, some fairly infrequently, from the heart and vasculature of elasmobranchs. Adult specimens of the dracunculoid nematode *Lockenloia sanguinis* were reported by Adamson and Caira (1991) from the heart of a nurse shark *Ginglymostoma cirratum* in the Florida Keys. Adamson et al. (pers. comm.) found microfilariae (i.e., juvenile stages) of a nematode that appears to be related to *Lockenloia* in the gill vasculature of a spotted eagle ray, *Aetobatus narinari*. Aragort et al. (2002) reported a guyanemid dracunculoid from the heart, among many other sites, of *Raja microocellata*.

The heart and vasculature are the primary sites occupied by digeneans of the family Sanguinicolidae (= Aporocotylidae). At least five genera of sanguinicolids are known from sharks and batoids. *Selachohemecus olsoni* was reported by Short (1954) from the heart of *Rhizoprionodon terraenovae* (as *Scoliodon*), and Bullard et al. (2006) described a second species of *Selachohemecus* from the heart of *Carcharhinus limbatus*. *Hyperandrotrema cetorhini* was described from the heart and blood vessels

of basking sharks by Maillard and Ktari (1978). Madhavi and Hanumantha Rao (1970) described *Orchispirium heterovitellatum* from the mesenteric vessels of *Dasyatis imbricatus*. Most recently, Bullard and Jensen (2008) described *Orchispirium heterovitellatum* and *Myliobaticola richardheardi* from the hearts of *Himantura imbricata* and *Dasyatis sabina*, respectively.

Even less frequently encountered in the elasmobranch heart are mites, isopods, and monogeneans. Benz and Bullard (2004) reported what appeared to be either a deutonymph or adult mite in the lumen of the heart of a nurse shark in Florida Bay. Bird (1981) described an interesting "outbreak" of the isopod *Cirolana borealis* in the Cape Canaveral, Florida, shark fishery from 1977 to 1978. During that period of time, this cirolanid isopod was observed to cause extensive pathology in the heart of several carcharhiniform shark species. Although it was found in a number of other sites, it showed a distinct preference for the heart. Llewellyn (1960) described the unusual occurrence of the amphibdellid monogenean *Amphibdella flavolineata* from the heart of electric rays. Euzet and Combes (1998) provided an interesting account of the sites occupied by individuals of this monogenean species in different stages of maturity as they move toward the heart to mate.

18.3.5.2 Spleen

The spleen was one of the sites from which Aragort et al. (2002) reported finding the guyanemid nematode *Histodytes microocellatus* in the skate *Raja microocellata*. To our knowledge, this is currently the only record of a metazoan parasite from the spleen of an elasmobranch.

18.3.6 Reproductive System

18.3.6.1 Gonads

Limited work has been done examining the gonads of elasmobranchs for parasites. We know of only three reports of metazoans from these organs; two of these are of nematodes, and one is of a cestode. Rosa-Molinar et al. (1983) described juvenile philometrid nematodes in granulomas associated with the ovaries of blacktip sharks. In addition, Aragort et al. (2002) found specimens of the guyanemid nematode *Histodytes microocellatus* in the gonads of *Raja microocellata*. This was only one of a number of sites parasitized by this nematode. Tandon (1972) reported plerocerci of the pterobothriid trypanorhynch *Pterobothrium* sp. from the ovary of *Himantura uarnak* (as *Dasyatis*). Although there appear to be no published records of parasites of the male gonads, we recently encountered unidentified larval trypanorhynchs from the testes of a species of *Himantura* in Borneo.

18.3.6.2 Oviducts

The oviducts are among the sites reported occupied by several species of monogeneans of the family Monocotylidae parasitizing elasmobranchs (e.g., Woolcock, 1936). The digenean syncoeliid *Paranotrema vaginacola* has also been reported from this site in a species of *Squalus* from Papua New Guinea (Dollfus, 1937).

18.3.6.3 Uterus

A surprising array of parasites, specifically leeches, copepods, isopods, monogeneans, and nematodes, has been reported from the elasmobranch uterus. All but the members of the latter group are somewhat unexpected inhabitants of the uterus because they are typically considered to be ectoparasitic. Moser and Anderson (1977) found a species of piscicoline leech inhabiting the external surfaces of embryos of the Pacific angel shark, *Squatina californica, in utero*. Similarly, Nagasawa et al. (1998) found copepods of the family Trebiidae associated with the external surfaces of embryos in the uteri of several species of angel sharks. In addition, the cirolanid isopod *Cirolana borealis* was reported from the uterus, among other sites, by Bird (1981), and we recently encountered multiple individuals of an unidentified corollanid isopod in the uterus of a specimen of the Atlantic weasel shark, *Paragaleus pectoralis*, taken off Senegal. Most recently, Kitamura et al. (2010) reported the monogenean *Calicotyle japonica* from, among other sites, the uterus of sharks identified as *Squalus mitsukurii*. Benz et al. (1987) reported juvenile philometrid nematodes in the uterus of a specimen of *Carcharhinus plumbeus* from the northeastern coast of the United States. In general, reports of metazoans from the uterus are uncommon. We suspect that the presence of at least some of these taxa in the uterus may be the result of accidental forays into this organ facilitated by the fact that the uterine pores open directly into the cloaca.

18.3.7 Nervous System

18.3.7.1 Brain

Like the liver, the elasmobranch brain and the other elements of the nervous system have been poorly sampled for parasites. Those records that do exist, however, are particularly interesting. Adamson et al. (pers. comm.) found adult individuals of a dracunculoid nematode in the brain of a spotted eagle ray. These worms were found wrapped around the optic nerves and surrounded much of the brain. As noted above, the unusual mesoparasitic sphyriid copepod described by Diebakate et al. (1997) is appropriately considered a parasite of the brain. While the posterior-most portions of its body extend from the capsule of the olfactory bulb into the external environment, the more anterior regions of the body of this animal penetrate the olfactory lobes of the brain of its shark host.

18.3.8 Body Cavities

18.3.8.1 Pericardial Cavity

Two groups of platyhelminths (Digenea and Monogenea) have been reported from the pericardial cavity of elasmobranchs. Several species of digeneans of the gorgoderid genus *Anaporrhutum* parasitize the pericardial chamber of batoids. This phenomenon was first described by Ofenheim (1900), who reported *A. albidum* from the pericardial chamber of a spotted eagle ray from the Pacific Ocean. Curran et al. (2003) reported finding one to three individuals of an *Anaporrhutum* species in the pericardial cavity of members of five different genera of batoids in the Gulf of California, Mexico. There exists one record of a monogenean from the pericardial cavity of elasmobranchs: Bullard and Overstreet (2000) reported monocotylids that may have come from the pericardial cavity of batoids in Mexico; however, this record requires verification before the pericardial cavity is considered a valid site occupied by these platyhelminths.

18.3.8.2 Peritoneal Cavity

The inhabitants of the peritoneal cavity include species in each of the three major groups of parasitic platyhelminths, the Monogenea, Cestoda, and Digenea. Although most monogeneans are ectoparasitic, several genera of monocotylids such as *Dictyocotyle* (e.g., Kearn, 1970) and *Calicotyle* (e.g., Bullard and Overstreet, 2000) have been reported from this site. The peritoneal cavity is the site most commonly parasitized by larvae of trypanorhynch cestodes such as the hepatoxylid *Hepatoxylon trichuri* (e.g., Waterman and Sin, 1991). In most cases, these larvae are generally not lying free in the body cavity itself; rather, they are attached to either the serosa surrounding organs such as the liver or the various mesenteries of the body cavity. Plerocerci (i.e., larvae) of the trypanorhynch family Grillotiidae have been reported from the peritoneal cavity of elasmobranchs with some regularity (e.g., Dollfus, 1942; Williams, 1960). Plerocerci of the tetraphyllidean cestode *Phyllobothrium radioductum* have been reported from the body cavity of skates (Cheung, 1993). Pappas (1970) reported finding adults of the trypanorhynch *Lacistorhynchus tenuis* in the body cavity of *Triakis semifasciata*; however, given the typical sites occupied by adult trypanorhynch cestodes, this report requires confirmation. Representatives of at least four genera of gorgoderid digeneans have been found in the peritoneal cavity of a diversity of sharks and batoids, suggesting that this may be a common site for these platyhelminths (e.g., Curran et al., 2003, 2009; Markell, 1953). In addition, on several occasions, species belonging to the azygiid genus *Otodistomum* have been reported from the peritoneal cavity of elasmobranchs (e.g., Gibson and Bray, 1977).

Given the typical life cycle of most monogeneans and digeneans, the occurrence of adult members of these taxa in the peritoneal cavity of elasmobranchs leads to speculation about the portals of entry and exit that might be utilized by inhabitants of this site. Gibson and Bray (1977) suggested that the abdominal pores, which open directly into the cloaca, would provide an appropriate portal both into and out of this site, making life in the peritoneal cavity possible.

The peritoneal cavity is only rarely parasitized by nematodes. Moravec and Little (1988) reported two species of micropleurid nematodes of the genus *Granulinema* that are likely to have come from the peritoneal cavity of bull sharks in Louisiana. We have occasionally encountered what we believe to be juvenile ascarid nematodes in the body cavities of a diversity of elasmobranchs in the Gulf of California.

18.3.9 Excretory System

18.3.9.1 Kidneys

The kidneys were one of the sites from which the guyanemid nematode *Histodytes microocellatus* was collected from *Raja microocellata* by Aragort et al. (2002). The kidney was also the site from which Arthur and Lom (1985) reported the sphaerosporid myxozoan *Sphaerospora araii*. More recently, the eutetrarhynchid cestode genus *Mobulocestus* was described from the nephridial system of *Mobula thurstoni* in the Gulf of California, Mexico, by Campbell and Beveridge (2006). The kidney was also among the sites from which Bullard et al. (2006) reported the sanguinicolid digenean *Selachohemecus benzi*.

18.3.9.2 Rectal Gland

The lumen of the rectal gland can be parasitized by monogeneans. Kearn (1987), for example, reported adults of monogenean monocotylid *Calicotyle kroyeri* from the lumen of the rectal gland of batoids and, more recently, Kitamura et al. (2010) reported *Calicotyle japonica* from the rectal gland of sharks identified as *Squalus mitsukurii*. Plerocerci (i.e., larvae) of the trypanorhynch family Lacistorhynchidae have also been reported from the external surface of the rectal gland (e.g., Pappas, 1970).

18.3.10 Body Musculature

Reports of metazoans from the musculature of the body are limited. Guiart (1935) found plerocerci (i.e., larvae) of a grillotiid trypanorhynch in the musculature of the shark *Pseudotriakis microdon*. On the few occasions that we examined musculature of elasmobranchs, such as *Rhinobatos* in the Gulf of California, Mexico, unidentified juvenile nematodes were found. Representatives of two families of myxozoans have been reported from the skeletal musculature of elasmobranchs. Stoffregen and Anderson (1990) reported an unidentified species of the trilosporid myxozoan *Unicapsula* from the skeletal muscles of *Carcharhinus melanopterus*. More recently, Gleeson et al. (2010) conducted an extensive survey for multivalvulid myxozoans, examining 31 species in three orders and nine families of elasmobranchs in Australia. Not only did their work result in the discovery of two new species of kudoid myxozoans, but it also revealed that 27 of the 31 elasmobranch species examined hosted myxozoans. Clearly, this site represents another poorly studied region of the elasmobranch body and effort spent examining these tissues is likely to yield additional parasite data.

18.4 General Observations

18.4.1 Site Specificity

The sites parasitized by the major groups of metazoan parasites are relatively predictable; however, each major group includes both families that exhibit fidelity for a particular site or suite of sites as well as families that exhibit more relaxed site specificity. For example, although adults of most cestode families parasitize the spiral intestine, cestodes found inhabiting other sites are generally tentaculariids or sphyriocephalids, or occasionally eutetrarhynchids and gilquiniids. Among monogeneans, most species typically inhabit the skin or gills of their hosts; the species most often found in sites other than the skin or gills are monocotylids. Most copepods inhabit the skin, gills, or olfactory bulbs of their hosts; those found in other sites are generally sphyriids. Although some digenean groups parasitize elements of the digestive system, others, such as the gorgoderids and syncoeliids, occupy sites unusual for digeneans such as the pericardial and peritoneal cavities. Nematodes, as a phylum, occupy the greatest diversity of sites within elasmobranchs, but these sites include those occupied by larvae or adults. Although many occur in the digestive system, Philometridae, Guyanemidae, Micropleuridae, and Trichosomoididae are the groups most commonly found in sites outside of that system. By far the majority of myxozoans have been reported from the gallbladder, but recent work by Gleeson et al. (2010) suggests that the diversity in skeletal muscle may be found to rival that of the gallbladder; this more difficult to examine site remains to be explored in more detail.

Overall, the skin is home to a greater assortment of metazoan higher taxa than any other region of the elasmobranch body. In fact, 5 of the 7 phyla and 12 of the 16 major lower taxa of metazoans considered here (see Figure 18.1) have been reported from the skin of the fins, claspers, head, torso, and/or tail of elasmobranchs. The spiral intestine, however, hosts the greatest number of species of metazoan parasites; this is in large part because this is the primary site occupied by cestodes.

It is important to recognize that routine fieldwork generally does not involve total necropsy of individual elasmobranchs; as a consequence, our knowledge of the parasites of some of the more difficult to examine organ systems is likely very deficient. Although they are unlikely ever to rank with the skin and digestive system in terms of the diversity of metazoans they host, sites such as the brain, circulatory system, musculature, liver, and gonads are likely to yield a greater diversity of metazoans than is currently recognized, should it become more routine to include these sites in necropsies.

18.4.2 Diversity

Recent intensive efforts to discover and describe elasmobranchs globally have increased the total number of known elasmobranch species to over 1200 (Naylor et al., in press). Yet, to date, the number of metazoan parasite species reported from sharks and batoids globally stands at only approximately 1917. Given that, in instances in which total necropsies have been conducted, multiple species of metazoans are routinely found parasitizing a single host species, it is likely that this number is a gross underestimate of the worldwide metazoan parasite fauna of elasmobranchs. Two factors conspicuously contribute to this situation. First, as noted above, studies aimed at conducting complete necropsies of elasmobranchs are rare. It is much more common for investigators to target a particular site or organ system of an elasmobranch based on their taxonomic expertise, often to the exclusion of all other regions of the elasmobranch body. Researchers interested in copepods, for example, generally focus their necropsy efforts on the gills, olfactory bulbs, and outer body surfaces, whereas researchers interested in cestodes target the spiral intestine. As a consequence, the picture of the total parasite fauna of most elasmobranch species that have been investigated remains incomplete. Second, many species of elasmobranchs have never been examined for parasites. As noted by Caira and Jensen (2001), deeper water elasmobranch taxa, such as the Scyliorhinidae, Dalatiidae, and Rajidae, are especially poorly sampled, as are some of the shallower water taxa,

such as the Urolophidae, Narcinidae, and Rhinobatidae. The description of hundreds of new species of elasmobranchs over the last few decades, although exciting, has exacerbated this situation. Particularly problematic are the cases in which recognized elasmobranch species (e.g., *Aetobatus narinari*), which were once thought to exhibit relatively broad geographical ranges, have since been determined to consist of multiple species, each more locally distributed (White et al., 2010). In such instances, all existing parasite records must be closely scrutinized and host taxa must be sampled (or resampled) on a finer geographic scale. Overall, hundreds of species of elasmobranchs have yet to be examined for parasites, and the complete parasite faunas of the majority of elasmobranchs await description.

Based on the composition of the ~1917 described species of metazoan parasites infecting elasmobranchs, the relative species diversity of these seven phyla is as follows: Platyhelminthes, 73.7%; Arthropoda, 18.3%; Nematoda, 4.3%; Myxozoa, 1.8%; Annelida, 1.2%; Acanthocephala, 0.6%; and Mollusca, 0.05%. With respect to constituent groups within these phyla, cestodes are by far the most diverse group of metazoan parasites of elasmobranchs, accounting for 59.1%, or a little over half of the described species. Copepods and monogeneans are the next most diverse groups, accounting for 15.0% and 11.8% of known species, respectively. The remaining groups are much less speciose; isopods and digeneans each account for only 2.6% of described species. In combination, molluscs, triclads, barnacles, aspidogastreans, ostracods, and amphipods account for only approximately 0.7% of species. There is no question, however, that as parasite survey work continues globally, new taxa will continue to be discovered.

18.5 Metazoan Parasites of Holocephalans

Given the close affinities between holocephalans and elasmobranchs, it is not surprising that there is substantial overlap between these two subclasses of chondrichthyans with respect to the major groups of metazoan parasites they host. Also not unexpected, given the low diversity of holocephalans relative to elasmobranchs, is that the overall diversity hosted by holocephalans is much lower than that of the elasmobranchs. In fact, whereas the ~1200 species of elasmobranchs host ~1917 species in 119 families of seven phyla of metazoans (see Table 18.1), the 50 or so species of holocephalans host a total of only 53 species of metazoans representing 25 families of five phyla (see Table 18.2). The ratio of known metazoan parasite species per species of holocephalan is, however, similar to that per species of elasmobranch (0.94 vs. 1.6, respectively).

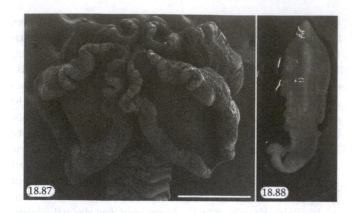

FIGURES 18.87 TO 18.88
Micrographs of selected metazoans parasitic in holocephalans but not elasmobranchs. 18.87. Tetraphyllidea: Chimaerocestidae: *Chimaerocestus* sp. ex *Rhinochimaera pacifica*. 18.88. Gyrocotylidea: Gyrocotylidae: *Gyrocotyle* sp. Scale bar: Figure 18.87, 500 μm.

Table 18.2 provides an overview of the major groups of metazoan parasites hosted by holocephalans. To date, no molluscs or acanthocephalans have been reported from holocephalans. With respect to the remaining five phyla, 20 of the families that parasitize holocephalans also parasitize elasmobranchs. Elasmobranchs host 95 families that are not found in holocephalans, and holocephalans host 5 families, all platyhelminths, that are not represented in elasmobranchs. These are the aspidogastrean family Rugogastridae, which is known only from the rectal gland of holocephalans (e.g., Amato and Pereira, 1995; Rhode et al., 1992); the monogenean family Chimaericolidae, which is known only from the gills of holocephalans (e.g., Beverley-Burton et al., 1991; Llewellyn and Simmons, 1984); the digenean family Philodistomidae (e.g., Olson et al., 1970); and the cestode families Chimaerocestidae (Williams and Bray, 1984) and Gyrocotylidae (Allison and Coakley, 1973; Karlsbakk et. al., 2002), which are each known only from the spiral intestine of holocephalans. With the exception of the Philodistomidae, which includes species that parasitize teleosts, these families are restricted to holocephalans. Moreover, the Gyrocotylidea represents an order unique to holocephalans.

Again, given the relatively lower diversity of metazoans hosted by holocephalans, it is not unexpected that the sites occupied by metazoans in or on holocephalans represent only a subset of those occupied in or on elasmobranchs. The sites parasitized, and their representative metazoans, are as follows. The skin is home to chondracanthid (Dienske, 1968), caligid (Kabata, 1968), and lernaeopodid (Karlsbakk et al., 2002) copepods, as well as to at least one aegid isopod and piscicolid leech (Dienske, 1968). A lernaeopodid copepod has been reported from the cornea (Karlsbakk et al., 2002). The gills are home to a diversity of monogeneans, including

monocotylids (e.g., Dienske, 1968), chimaericolids (e.g. Beverley-Burton et al., 1991), and hexabothriids (e.g., Kitamura et al., 2006), as well as lernaeopodid copepods (e.g., Kabata, 1988) and at least one kroyeriid copepod (e.g., Castro and Baeza, 1984). In total, the digestive system of holocephalans hosts the greatest diversity of metazoans. Several digeneans have been reported from the esophagus and stomach; these include an acanthocolpid (Machida and Kuramochi, 1994), a derogenid (Karlsbakk et al., 2002), an encysted azygiid (Dienske, 1968), and possibly a hemiurid (Hogans and Hurlbut, 1984). Perhaps as many as 10 gyrocotylidean cestode species have been reported from the spiral intestine (e.g., Allison and Coakley, 1973; Hogans and Hurlbut, 1984). Also reported from that organ were a chimaerocestid cestode (Williams and Bray, 1984), plerocerci of a lacistorhynchid and possibly a eutetrarhynchid trypanorhynch cestode (Palm, 2004), and an anisakid nematode (Hogans and Hurlbut, 1984). Monocotylid monogeneans were reported from the rectum and cloaca of two different holocephalans by Rhode et al. (1992) and Karlsbakk et al. (2002). The gallbladder of holocephalans is parasitized by aspidogastreans (Rhode, 1998) and myxozoans (Jameson, 1931). Sanguinicolid digeneans have been reported from the heart (Karlsbakk et al., 2002) and dorsal aorta (Dienske, 1968). Dienske (1968) reported finding juvenile anisakid nematodes encysted in the ovary wall. Schell (1972) reported encysted azygiid digeneans in the peritoneal cavity. Although a somewhat unusual site, the rectal gland is home to some monocotylid monogeneans (Rhode et al., 1992) and also to species of the aspidogastrean family Rugogasteridae (Amato and Pereira, 1995; Rhode et al., 1992).

To date, the olfactory bulbs, acousticolateralis system, spiracles, buccal cavity, pancreas, liver, spleen, oviducts, uterus, brain, pericardial cavity, kidneys, and body musculature are all sites from which, to our knowledge, metazoans have not been reported from holocephalans. We predict that close scrutiny of these sites is very likely to yield metazoans, as is the examination of sites already known to host metazoans in additional specimens and species of holocephalans.

Acknowledgments

We are very much indebted to Eugene Burreson for his input regarding leeches of elasmobranchs and the sites within elasmobranchs they occupy. Ian Whittington and Leslie Chisholm provided thoughtful comments on an earlier version of this manuscript. They also supplied the image of the acanthocotylid (Figure 18.46) as well

TABLE 18.2

Classification of Families of Metazoan Invertebrates Parasitic in Holocephalans

Phylum Annelida[a]	Subclass Maxillopoda	Class Monogenea
Class Clitellata	Superorder Copepoda	Subclass Monopisthocotylea
Subclass Hirudinea	Order Siphonostomatoida	F. Monocotylidae (Figure 18.51)
Order Rhynchobdellida	F. Caligidae (Figure 18.27)	Subclass Polyopisthocotylea
F.[b] Piscicolidae	F. Kroyeriidae (Figure 18.33)	F. Chimaericolidae*
SubF.[c] Piscicolinae (Figure 18.5)[d]	F. Lernaeopodidae (Figure 18.34)	F. Hexabothriidae (Figure 18.53)
Phylum Myxozoa	Phylum Platyhelminthes	Class Cestoda
F. Ceratomyxidae (Figure 18.8)	Subphylum Neodermata	Order Trypanorhyncha
Phylum Nematoda	Class Trematoda	Suborder Trypanobatoida
Class Rhabditea	Subclass Aspidogastrea	SuperF.[f] Eutetrarhynchoidea
Order Ascaridida	F. Multicalycidae (Figure 18.40)	F. Eutetrarhynchidae (Figure 18.68)
F. Anisakidae (Figure 18.12)	F. Rugogastridae*	Suborder Trypanoselachoida
Order Spirurida	Subclass Digenea	SuperF. Lacistorhynchoidea
F. Gnathostomatidae (Figure 18.15)	F. Acanthocolpidae	F. Lacistorhynchidae (Figure 18.72)
Phylum Arthropoda	F. Azygiidae (Figure 18.41)	Order "Tetraphyllidea"
Class "Crustacea"[e]	F. Derogenidae (Figure 18.42)	F. Chimaerocestidae* (Figure 18.87)
Subclass Malacostraca	F. Leptocreadiidae	Order Gyrocotylidea*
Order Isopoda	F. Opecoelidae	F. Gyrocotylidae* (Figure 18.88)
Suborder Flabellifera	F. Philodistomidae*	Order Poecilostomatoida
F. Aegidae (Figure 18.20)	F. Sanguinicolidae	F. Chondracanthidae

[a] Phyla are in order of their increasing diversity in elasmobranchs.
[b] Family.
[c] Subfamily.
[d] Figures cited illustrate taxon indicated but not necessarily specimens taken from holocephalans.
[e] Quotes = non-monophyletic taxon.
[f] Superfamily.
* Not found in elasmobranchs.

as specimens of amphibdellids and loimoids (Figures 18.47 and 18.49). Frantisek Moravec was of enormous assistance with the classification of the nematodes of elasmobranchs; in addition, he assisted with the identification of nematode specimens for illustration of the families and provided copies of much of the nematode literature. Stephen Feist graciously provided input on our compilation of myxozoan records from elasmobranchs. The photos of the myxozoans in Figures 18.8a, 18.8b, 18.9b, 18.9c, and 18.10 were kindly provided by R.J. Gleeson and R.D. Adlard of the Queensland Museum and the University of Queensland, Brisbane, Australia. Figure 18.9a was generously provided by Ivan Fiala. Burt Williams provided useful information on the isopods of elasmobranchs. Many of the images presented here were taken from specimens provided by other individuals: Clinton Duffy provided the piscicoline leech (Figure 18.5) and the gnathiid isopod (Figure 18.23), as well as the cercropid (Figure 18.28), dichelesthiid (Figure 18.29), kroyeriid (Figure 18.33), lernaeopodid (Figure 18.34), and pandarid (Figure 18.35) copepods. Stephen Curran provided the cavisomid acanthocephalan (Figure 18.3), the aspidogastrid (Figure 18.40), and the syncoeliid digenean (Figure 18.44). Gaines Tyler provided

the illiosentid acanthocephalan (Figure 18.4), Stephen Bullard provided the microbothriid monogenean (Figure 18.50), Burt Williams supplied a corallanid isopod (Figure 18.22), and Patricia Charvet-Almeida provided the ascarid nematode (Figure 18.13). The azygiid (Figure 18.41) and gilquiniid (Figure 18.76) specimens came from the collection of Nathan Riser; the hexabothriid monogenean (Figure 18.53) and the sphyriocephalid (Figure 18.78) cestode came from the collection of Ronald Campbell. The gymnorhynchid cestode (Figure 18.77) came from the collection of Murray Dailey. The figure of the pseudotobothriid cestode (Figure 18.74) was provided by Harry Palm. Clinton Duffy provided the image of the shirleyrhynchid (Figure 18.71). We appreciate the timely assistance and access to material provided by the Invertebrate Zoology Section of the Smithsonian Institution (USNM) and the Harold W. Manter Parasitology Laboratory (HWML), Lincoln, Nebraska. Maria Pickering and Florian Reyda provided very helpful comments on earlier versions of the manuscript. This work was supported in part by NSF grants PEET No. DEB 0118882; BS&I Nos. DEB 0103640, 0542846, and 0542941; and PBI Nos. DEB 0818696 and DEB 0818823.

References

Adams, A.M., Hoberg, E.P., McAlpine, D.F., and Clayden, S.L. (1998). Occurrence and morphological comparisons of *Campula oblonga* (Digenea: Campulidae), including a report from an atypical host, the thresher shark, *Alopias vulpinus*. *J. Parasitol.* 84:435–438.

Adamson, M.L. and Caira, J.N. (1991). *Lockenloia sanguinis* n. gen., n. sp. (Nematoda: Dracunculoidea) from the heart of a nurse shark, *Ginglymostoma cirratum*, in Florida. *J. Parasitol.* 77:663–665.

Adamson, M.L., Deets, G.B., and Benz, G.W. (1987). Description of male and redescription of female *Phlyctainophora squali* Mudry and Dailey, 1969 (Nematoda; Dracunculoidea) from elasmobranchs. *Can. J. Zool.* 65:3006–3010.

Allison, F.R. and Coakley, A. (1973). The two species of *Gyrocotyle* in the elephant fish, *Callorhynchus milii* (Borg). *J. R. Soc. N.Z.* 3:381–392.

Amato, J.F.R. and Pereira, Jr., J. (1995). A new species of *Rugogaster* (Aspidobothrea: Rugogastridae) parasite of the elephant fish, *Callorhinchus callorhynchi* [*callorhyncus*] (Callorhinchidae), from the estuary of the La Plata River, coasts of Uruguay and Argentina. *Rev. Bras. Parasitol. Vet.* 4:1–7.

Andrade, M.C.D., Galindez, E.J., Estecondo, S., and Tanzola, R.D. (2008). Hepatic infection by *Terranova* sp. (Nematoda, Anisakidae) larvae in *Sympterygia acuta* (Chondrichthyes, Rajidae) *acuta*. *Bull. Eur. Assoc. Fish Pathol.* 28:144–147.

Aragort, W.S., Alvarez, F., Iglesias, R., Leiro, J., and Sanmartin, M.L. (2002). *Histodytes microocellatus* gen. et sp. nov. (Dracunculoidea: Guyanemidae), a parasite of *Raja microoocellata* on the European Atlantic coast (north-western Spain). *Parasitol. Res.* 88:932–940.

Arai, J.P. (1989). Guide to the parasites of fishes of Canada. Part III. Acanthocephala and Cnidaria. *Can. Spec. Publ. Fish Aquat. Sci.* 107:1–95.

Arthur, J.R. and Lom, J. (1985). *Sphaerospora araii* n. sp. (Myxosporea: Sphaerosporidae) from the kidney of a longnose skate (*Raja rhina* Jordan and Gilbert) from the Pacific Ocean off Canada. *Can. J. Zool.* 63:2902–2906.

Ball, I.R. and Khan, R.A. (1976). On *Micropharynx parasitica* Jägerskiöld, a marine planarian ectoparasitic on the thorny skate, *Raja radiata* Donovan, from the North Atlantic Ocean. *J. Fish Biol.* 8:419–426.

Bates, R.M. (1990). *A Checklist of the Trypanorhyncha (Platyhelminthes: Cestoda) of the World. (1935–1985)*, Zoological Series 1. National Museum of Wales, Cardiff.

Bennett, M.B., Heupel, M.R., Bennett, S.M., and Parker, A.R. (1997). *Sheina orri* (Myodocopa: Cypridinidae) an ostracod parasitic on the gills of the epaulette shark, *Hemiscyllium ocellatum* (Bonnaterre, 1788) (Elasmobranchii: Hemiscyllidae). *Int. J. Parasitol.* 27:275–281.

Benz, G.W. (1984). Association of the pedunculate barnacle, *Conchoderma virgatum* (Spengler, 1790), with pandarid copepods (Siphonostomatoida: Pandaridae). *Can. J. Zool.* 62:741–742.

Benz, G.W. (1986). Distributions of siphonostomatoid copepods parasitic upon large pelagic sharks in the western North Atlantic. *Syllogeus* 58:211–219.

Benz, G.W. (1991). Description of some larval stages and augmented description of adult stages of *Albionella etmopteri* (Copepoda: Lernaeopodidae), a parasite of deep-water lanternsharks (*Etmopterus*: Squalidae). *J. Parasitol.* 77:666–674.

Benz, G.W. (1993). Evolutionary Biology of Siphonostomatoida (Copepoda) Parasitic on Vertebrates, doctoral dissertation, University of British Columbia, Vancouver.

Benz, G.W. and Bullard, S.A. (2004). Metazoan parasites and associates of chondrichthyans with emphasis on taxa harmful to captive hosts. In: Smith, M., Warmolts, D., Thoney, D., and Hueter, R. (Eds.), *Elasmobranch Husbandry Manual: Captive Care of Sharks, Rays and Their Relatives*. Ohio Biological Survey, Columbus, pp. 325–416.

Benz, G.W. and Deets, G.B. (1986). *Kroyeria caseyi* sp. nov. (Kroyeriidae: Siphonostomatoida), a parasitic copepod infesting gills of night sharks (*Carcharhinus signatus* (Poey, 1868)) in the western North Atlantic. *Can. J. Zool.* 64:2492–2498.

Benz, G.W. and Deets, G.B. (1988). Fifty-one years later: an update on *Entepherus*, with a phylogenetic analysis of Cecropidae Dana, 1849 (Copepoda: Siphonostomatoida). *Can. J. Zool.* 66:856–865.

Benz, G.W., Pratt, Jr., H.L., and Adamson, M.L. (1987). Larval philometrid nematodes (Philometridae) from the uterus of a sandbar shark, *Carcharhinus plumbeus*. *Proc. Helminthol. Soc. Wash.* 54:154–155.

Benz, G.W., Lucas, Z., and Lowry, L.F. (1998). New host and ocean records for the copepod *Ommatokoita elongata* (Siphonostomatoida: Lernaeopodidae), a parasite of the eyes of sleeper sharks. *J. Parasitol.* 84:1271–1274.

Benz, G.W., Borucinska, J.D., Lowry, L.F., and Whiteley, H.E. (2002). Ocular lesions associated with attachment of the copepod *Ommatokoita elongata* (Lernaeopodidae: Siphonostomatoida) to corneas of Pacific sleeper sharks, *Somniosus pacificus*, captured off Alaska in Prince William Sound. *J. Parasitol.* 88:474–481.

Bere, R. (1936). Parasitic copepods from Gulf of Mexico fish. *Am. Midl. Nat.* 17:577–625.

Beverley-Burton, M. (1984). Guide to the parasites of fishes of Canada. Part I. Monogenea and Turbellaria. *Can. Spec. Publ. Fish. Aquat. Sci.* 74:1–209.

Beverley-Burton, M, Chisholm, L.A., and Last, P. (1991). Two new species of *Chimaericola* Brinkmann (Monogenea: Chimaericolidae) from *Hydrolagus* spp. (Chimaeriformes: Chimaeridae) in the Pacific. *Syst. Parasitol.* 18:59–66.

Bilqees, F.M. and Khan, A. (2005). Two new helminth parasites from Pakistan, with redescription of the acanthocephalan *Centrorhynchus fasciatum* (Westrumb, 1821). *Pak. J. Zool.* 37:257–263.

Bird, P.M. (1981). The occurrence of *Cirolana borealis* (Isopoda) in the hearts of sharks from Atlantic coastal waters of Florida. *Fish. Bull.* 79:376–383.

Bonham, K. and Guberlet, J.E. (1938). Ectoparasitic trematodes of Puget Sound fishes. *Am. Midl. Nat.* 20:590–602.

Borucinska, J.D., Benz, G.W., and Whiteley, H.E. (1998). Ocular lesions associated with attachment of the parasitic copepod *Ommatokoita elongata* (Grant) to corneas of Greenland sharks, *Somniosus microcephalus* (Bloch & Schneider). *J. Fish Dis.* 21:415–422.

Bousfield, E.L. (1987). Amphipod parasites of fishes of Canada. *Can. Bull. Fish. Aquat. Sci.* 217:1–37.

Braswell, J.S., Benz, G.W., and Deets, G.B. (2002). *Taeniacanthodes dojirii* n. sp. (Copepoda: Poecilostomatoida: Taeniacanthidae) from Cortez electric rays (*Narcine entemedor*: Torpediformes: Narcinidae) captured in the Gulf of California, and a phylogenetic analysis of and key to the species of *Taeniacanthodes*. *J. Parasitol.* 88:28–35.

Bray, R.A. and Cribb, T.H. (2003). The digeneans of elasmobranchs: distribution and evolutionary significance. In: Combes, C. and Jourdane, J. (Eds.), *Taxonomy, Ecology and Evolution of Metazoan Parasites (Livre hommage à Louis Euzet)*, Vol. I. Presses Universitaire de Perpignan, Perpignan, pp. 67–96.

Bray, R.A. and Gibson, D.I. (1986). The Zoogonidae (Digenea) of fishes from the north-east Atlantic. *Bull. Br. Mus. Nat. Hist. (Zool.)* 51:127–206.

Bray, R.A., Brockerhoff, A., and Cribb, T.H. (1995). *Melogonimus rhodanometra* n. g. n. sp. (Digenea, Ptychogonimidae) from the elasmobranch *Rhina ancylostoma* Bloch and Schneider (Rhinobatidae) from the Southeastern Coastal waters of Queensland, Australia. *Syst. Parasitol.* 30:11–18.

Brooks, D.R. (1979). *Paravitellotrema overstreeti* sp. n. (Digenea: Hemiuridae) from the Colombian freshwater stingray *Potamotrygon magdalenae* Dumeril. *Proc. Helminthol. Soc. Wash.* 46:52–54.

Brooks, D.R. and Barriga, R. (1995). *Serendip deborahae* n. gen. and n. sp. (Eucestoda: Tetraphyllidea: Serendipidae n. fam.) in *Rhinoptera steindachneri* Evermann and Jenkins, 1891 (Chondrichthyes: Myliobatidae) from southeastern Ecuador. *J. Parasitol.* 81:80–84.

Brooks, D.R. and Evenhuis, N.L. (1995). Serendipidae Evenhuis, 1994 (Insecta: Diptera) and Serendipidae Brooks and Barriga, 1995 (Platyhelminthes: Eucestoda): proposed removal of homonymy. *J. Parasitol.* 81:762.

Brusca, R.C. (1981). A monograph on the Isopoda Cymothoidae (Crustacea) of the eastern Pacific. *Zool. J. Linn. Soc.* 73:117–199.

Buckner, R.L., Overstreet, R.M., and Heard, R.W. (1978). Intermediate hosts for *Tegorhynchus furcatus* and *Dollfusentis chandleri* (Acanthocephala). *Proc. Helminthol. Soc. Wash.* 45:195–201.

Bullard, S.A. and Jensen, K. (2008). Blood flukes (Digenea: Aporocotylidae) of stingrays (Myliobatiformes: Dasyatidae): *Orchispirium heterovitellatum* from *Himantura imbricata* in the Bay of Bengal and a new genus and species of Aporocotylidae from *Dasyatis sabina* in the northern Gulf of Mexico. *J. Parasitol.* 94:1311–1321.

Bullard, S.A. and Overstreet, R.M. (2000). *Calicotyle californiensis* n. sp. and *Calicotyle urobati* n. sp. (Monogenea: Calicotylinae) from elasmobranchs in the Gulf of California. *J. Parasitol.* 86:939–944.

Bullard, S.A., Benz, G.W., and Braswell, J.S. (2000). *Dioncus postonchomiracidia* (Monogenea: Dioncidae) from the skin of blacktip sharks, *Carcharhinus limbatus* (Carcharhinidae). *J. Parasitol.* 86:245–250.

Bullard, S.A., Overstreet, R.M., and Carlson, J.K. (2006). *Selachohemecus benzi* n. sp. (Digenea: Sanguinicolidae) from the black-tip shark *Carcharhinus limbatus* (Carcharhinidae) in the northern Gulf of Mexico. *Syst. Parasitol.* 63:143–154.

Bunkley-Williams, L. and Williams, Jr., E.H. (1998). Isopods associated with fishes: a synopsis and corrections. *J. Parasitol.* 84:893–896.

Burreson, E.M. (2006). A redescription of the fish leech *Pterobdella amara* (=*Rhopalobdella japonica*) (Hirudinida: Piscicolidae) based on specimens from the type locality in India and from Australia. *J. Parasitol.* 92:677–681.

Burreson, E.M. and Kearn, G.C. (2000). *Rhopalobdella japonica* n. gen., n. sp. (Hirudinea, Piscicolidae) from *Dasyatis akajei* (Chondrichthyes: Dasyatidae) in the Northwestern Pacific. *J. Parasitol.* 86:696–699.

Caira, J.N. (1990). Metazoan parasites as indicators of elasmobranch biology. In: Pratt, Jr., H.L., Gruber, S.H., and Taniuchi, T. (Eds.), *Elasmobranchs as Living Resources: Advances in the Biology, Ecology, Systematics, and Status of the Fisheries*, NOAA Tech. Rep. NMFS 90. U.S. Department of Commerce, Washington, D.C., pp. 71–96.

Caira, J.N. and Gavarrino, M.M. (1990). *Grillotia similis* (Linton, 1908) comb. n. (Cestoda: Trypanorhyncha) from nurse sharks in the Florida Keys. *J. Helminthol. Soc. Wash.* 57:15–20.

Caira, J.N. and Jensen K. (2001). An investigation of the coevolutionary relationships between onchobothriid tapeworms and their elasmobranch hosts. *Int. J. Parasitol.* 31:959–974.

Caira, J.N., Benz, G.W., Borucinska, J., and Kohler, N.E. (1997a). Pugnose eels, *Simenchelys parasiticus* (Synaphobranchidae), from the heart of a shortfin mako, *Isurus oxyrinchus* (Lamnidae). *Environ. Biol. Fish.* 49:139–144.

Caira, J.N., Jensen, K., Yamane, Y., Isobe, A., and Nagasawa, K. (1997b). On the tapeworms of *Megachasma pelagios*: description of a new genus and species of lecanicephalidean and additional information on the trypanorhynch *Mixodigma leptaleum*. In: Yano, K., Morrissey, J.F., Yabumoto, Y., and Naayam, K. (Eds.), *Biology of the Megamouth Shark*. Tokai University Press, Tokyo, pp. 181–191.

Caira, J.N., Jensen, K., and Healy, C.J. (1999). On the phylogenetic relationships among the tetraphyllidean, lecanicephalidean and diphyllidean tapeworm genera. *Syst. Parasitol.* 42:77–151.

Caira, J.N., Jensen, K., and Healy, C.J. (2001). Interrelationships among tetraphyllidean and lecanicephalidean cestodes. In: Littlewood, D.T.J. and Bray, R.A. (Eds.), *Interrelationships of the Platyhelminthes*. Taylor & Francis, London, pp. 135–158.

Caira, J.N., Mega, J., and Ruhnke, T.R. (2005). An unusual blood sequestering tapeworm (*Sanguilevator yearsleyi* n. gen., n. sp.) from Borneo with description of *Cathetocephalus resendezi* n. sp. from Mexico and molecular support for the recognition of the order Cathetocephalidea (Platyhelminthes: Eucestoda). *Int. J. Parasitol.* 35:1135–1152.

Campana-Rouget, Y. (1955). Parasites de poissons de mer ouest-africains récoltés per J. Cadenat. IV. Nematodes (1re note). Parasites de sélaciens. *Bull. Inst. Franç Afr. Sci. Noire* 14(A3):818–839.

Campbell, R.A. and Beveridge, I. (1994). Order Trypanorhyncha. In: Khalil, L.F., Jones, A., and Bray, R.A. (Eds.), *Keys to Cestode Parasites of Vertebrates*. CAB International, Wallingford, U.K., pp. 51–148.

Campbell, R.A. and Beveridge, I. (2006). Three new genera and seven new species of trypanorhynch cestodes (Family Eutetrarhynchidae) from manta rays, *Mobula* spp. (Mobulidae) from the Gulf of California, Mexico. *Folia Parasitol.* 53:255–275.

Castro, R. and Baeza, H. (1984). *Lepeophtheirus frecuens* new species and new record of *Kroyerina meridionalis* Ramirez, 1975 and new hosts record for *Lepeophtheirus chilensis* Wilson, 1905 (Copepoda: Siphonostomatoida) parasitic on fishes of Chile, South America. *Bull. Mar. Sci.* 34:197–206.

Causey, D. (1961). The site of *Udonella califorum* (Trematoda) upon parasitic copepod hosts. *Am. Midl. Nat.* 66:314–318.

Cheung, P. (1993). Parasitic diseases of elasmobranchs. In: Stoskopf, M.D. (Ed.), *Fish Medicine*. Saunders, Philadelphia, PA, pp. 782–807.

Chisholm, L.A. and Whittington, I. (2005). *Decacotyle cairae* n. sp. (Monogenea: Monocotlidae) from the gills of *Pastinachus* sp. (Elasmobranchii: Dasyatidae) from the South China Sea off Sarawak, Borneo, Malaysia. *Syst. Parasitol.* 61:79–84.

Chisholm, L.A. and Whittington, I. (2009). *Dendromonocotyle urogymni* sp. nov. (Monogenea, Monocotylidae) from *Urogymnus asperrimus* (Elasmobranchii, Dasyatidae) off eastern Australia. *Acta Parasitol.* 54:113–118.

Chisholm, L.A., Hansknecht, T.J., Whittington, I.D., and Overstreet, R.M. (1997). A revision of the Calicotylinae Monticelli, 1903 (Monogenea: Monocotylidae). *Syst. Parasitol.* 38:159–183.

Cielocha, J.J. and Jensen, K. (2011). A revision of *Hexacanalis* Perrenoud, 1931 (Cestoda: Lecanicephalidea) and description of *H. folifer* n. sp. from the zonetail butterfly ray, *Gymnura zonura* (Bleeker) (Rajiformes: Gymnuridae). *Syst. Parasitol.* 79(1):1–16.

Cislo, P.R. and Caira, J.N. (1993). The parasite assemblage in the spiral intestine of the shark *Mustelus canis*. *J. Parasitol.* 79:886–899.

Cressey, R.F. (1976). *The genus* Argulus *(Crustacea: Branchiura) of the United States*, Water Pollution Control Research Series 18050 ELDO2/72. U.S. Environmental Protection Agency, Washington, D.C.

Cribb, T.H., Bray, R.A., Littlewood, D.T.J., Pichelin, S.P., and Herniou, E.A. (2001). The Digenea. In: Littlewood, D.T.J. and Bray, R.A. (Eds.), *Interrelationships of the Platyhelminthes*. Taylor & Francis, London, pp. 168–185.

Crompton, D.W.T. and Nickol, B.B. (1985). *Biology of the Acanthocephala*. Cambridge University Press, Cambridge, U.K.

Curran, S.S. and Caira, J.N. (1995). Attachment site specificity and the tapeworm assemblage in the spiral intestine of the blue shark (*Prionace glauca*). *J. Parasitol.* 81:149–157.

Curran, S.S. and Overstreet, R.M. (2000). *Syncoelium vermilionensis* sp. n. (Hemiuroidea: Syncoeliidae) and new records for members of Azygiidae, Ptychogonimidae, and Syncoeliidae parasitizing elasmobranchs in the Gulf of California. In: Salgado-Maldonado, G., Garcia Aldrete, A.N., and Vidal-Martinez, V.M. (Eds.), *Metazoan Parasites in the Neotropics: A Systematic and Ecological Perspective*. Instituto de Biologia, UNAM, Mexico, pp. 117–133.

Curran, S.S., Blend, C.K., and Overstreet, R.M. (2003). *Anaporrhutum euzeti* sp. n. (Gorgoderidae: Anaporrhutinae) from rays in the Gulf of California, Mexico. In: Combes, C. and Jourdane, J. (Eds.), *Taxonomy, Ecology and Evolution of Metazoan Parasites (Livre hommage à Louis Euzet)*, Vol. I. Presses Universitaire de Perpignan, Perpignan, pp. 225–234.

Curran, S.S., Blend, C.K., and Overstreet, R.M. (2009). *Nagmia rodmani* n. sp., *Nagmia cisloi* n. sp., and *Probolitrema richiardii* (López, 1888) (Gorgoderidae: Anaporrhutinae) from elasmobranchs in the Gulf of California, Mexico. *Comp. Parasitol.* 76:6–18.

De Buron, I. and Euzet, L. (2005). A new species of *Thaumatocotyle* (Monogenea: Monocotylidae) from *Dasyatis sabina* (Myliobatiformes: Dasyatidae) off the coast of South Carolina. *J. Parasitol.* 91:791–793.

Deardorff, T.L. and Ko, R.C. (1983). *Echinocephalus overstreeti* sp. n. (Nematoda: Gnathostomatidae) in the stingray, *Taeniura melanopilos* Bleeker, from the Marquesas Islands, with comments on *E. sinensis* Ko, 1975. *Proc. Helminthol. Soc. Wash.* 50:285–293.

Deboutteville, D.C. and Nunes, L.P. (1952). Copépodes Philichthyidae nouveaux, parasites de poissons Européens. *Ann. Parasitol.* 27:598–609.

Deets, G.B. and Dojiri, M. (1990). *Dissonus pastinum* n. sp. (Siphonostomatoida: Dissonidae) a copepod parasitic on a horn shark from Japan. *Beaufortia* 41:49–54.

Delaney, P.M. (1984). Isopods of the genus *Excorallana* Stebbing, 1904, in the Gulf of California, Mexico, with descriptions of two new species and a key to the known species (Crustacea, Isopoda, Corallanidae). *Bull. Mar. Sci.* 34:1–20.

Di Cave, D., Orecchia, P., Ortis, M., and Paggi, L. (2003). Metazoan parasites from some elasmobranchs of Thyrrenian Sea (Metazoi parassiti di alcuni elasmobranchi del Mare Tirreno.). *Biol. Mar. Mediterr.* 10:249–252.

Diaz, J.P. (1972). Cycle évolutif d'*Acanthoceilus quadridentatus* Molin, 1988 (Nematoda). *Vie Milieu* 22:289–304.

Diebakate, C., Raibaut, A., and Kabata, Z. (1997). *Thamnocephalus cerebrinoxius* n. g., n. sp. (Copepoda: Sphyriidae), a parasite in the nasal capsules of *Leptocharias smithii* (Müller and Henle, 1839). (Pisces: Leptochariidae) off the coast of Senegal. *Syst. Parasitol.* 38:231–235.

Dienske, H. (1968). A survey of the metazoan parasites of the rabbit-fish, *Chimaera monstrosa* L. (Holocephali). *Neth. J. Sea Res.* 4:32–58.

Dollfus, R.P. (1937). Les trématodes digénea des selaciens (Plagiostomes). Catalogue par hotes. Distribution géographique. *Ann. Parasitol. Hum. Comp.* 15:259–281.

Dollfus, R.P. (1942). Etudes critiques sur les tetrarhynques du Museum de Paris. *Arch. Mus. Natl. Hist. Nat. (Paris)* 19:7–466.

Dollfus, R.P. (1960). Sur une collection de tetrarhynques home-acanthes de la famille des Tentaculariidae, recoltes principalement dans las region de Dakar. *Bull. Inst. Franç Afr. Sci. Noire* 22:788–852.

Domingues, M.V. and Marques, F.P.L. (2007). Revision of *Potamotrygonocotyle* Mayes, Brooks and Thorson, 1981 (Platyhelminthes: Monogenoidea: Monocotylidae), with descriptions of four new species from the gills of the freshwater stingrays *Potamotrygon* spp. (Rajiformes: Potamotrygonidae) from the La Plata river basin. *Syst. Parasitol.* 67:157–174.

Euzet, L. (1994a). Order Lecanicephalidea. In: Khalil, L.F., Jones, A., and Bray, R.A. (Eds.), *Keys to Cestode Parasites of Vertebrates*. CAB International, Wallingford, U.K., pp. 195–204.

Euzet, L. (1994b). Order Tetraphyllidea. In: Khalil, L.F., Jones, A., and Bray, R.A. (Eds.), *Keys to Cestode Parasites of Vertebrates*. CAB International, Wallingford, U.K., pp. 149–194.

Euzet, L. and Combes, C. (1998). The selection of habitats among the Monogenea. *Int. J. Parasitol.* 28:1645–1652.

Euzet, L. and Williams, H.H. (1960). A re-description of the trematode *Calicotyle stossichi* Braun, 1899, with an account of *Calicotyle palombi* sp. nov. *Parasitology* 50:21–30.

Gibson, D.I. and Bray, R.A. (1977). The Azygiidae, Hirudinellidae, Ptychogonimidae, Sclerodistomidae and Syncoeliidae of fishes form the north-east Atlantic. *Bull. Br. Mus. Nat. Hist. (Zool.)* 32:167–245.

Gleeson, R.J., Bennett, M.B., and Adlard, R.D. (2010). First taxonomic description of multivalvulidan parasites from elasmobranchs: *Kudoa hemiscylli* n. sp. and *Kudo carcharhini* n. sp. (Myxosporea: Multivalvulidae). *Parasitology* 137:1885–1898.

Golvan, Y.I., Garcia-Rodrigo, A., and Díaz-Ungría, C. (1964). *Megapriapus ungriai* (Garcia-Rodrigo, 1960) n. gen. (Palaeacanthocephala) parasite d'une Pastenague d'eau douce du Vénézuéla (*Pomatotrygon hystrix*). *Ann. Parasitol. Hum. Comp.* 39:53–59.

Guiart, J. (1935). Cestodes parasites provenant des campagnes scientifiques du Prince Albert 1er de Monaco (1886–1913). *Résultats des Campagnes Scientifiques Accomplies sur son Yacht par Albert 1er Monaco publies sous sa Direction avec le Concours de M. Jules Richard.* 91:1–115.

Hale, H.M. (1940). Report on the cymothoid Isopoda obtained by the F.I.S. 'Endeavour' on the coasts of Queensland, New South Wales, Victoria, Tasmania, and South Australia. *Trans. R. Soc. S. Aust.* 64:288–304.

Healy, C.J. (2006). A revision of selected Tetraphyllidea (Cestoda): *Caulobothrium, Rhabdotobothrium, Rhinebothrium, Scalithrium* and *Spongiobothrium*, doctoral dissertation, University of Connecticut, Storrs.

Healy, C.J., Caira, J.N., Jensen, K., Webster, B.L., and Littlewood, D.T.J. (2009). Proposal for a new tapeworm order, Rhinebothriidea. *Int. J. Parasitol.* 39:497–511.

Hesse, M. (1873). Mémoire sur des Crustacés rares ou nouveaux des côtes de France. *Ann. Sci. Nat.* 19:3–29.

Heupel, M.R. and Bennett, M.B. (1999). The occurrence, distribution and pathology associated with gnathiid isopod larvae infecting the epaulette shark, *Hemiscyllium ocellatum. Int. J. Parasitol.* 29:321–330.

Hogans, W.E. and Hurlbut, T.R. (1984). Parasites of the knifenose chimaera, *Rhinochimaera atlantica*, from the northwest Atlantic Ocean. *Can. Field Nat.* 98:365.

Honma, Y., Tsunaki, S., and Chiba, A. (1991). Histological studies on the juvenile gnathiid (Isopoda, Crustacea) parasitic on the branchial chamber wall of the stingray, *Dasyatis akajei*, in the Sea of Japan. *Rep. Sado Mar. Biol. Stat. Niigata Univ.* 21:37–47.

Jameson, A.P. (1931). Notes on Californian Myxosporidia. *J. Parasitol.* 18:59–69.

Jensen, K. (2005). A monograph of the order Lecanicephalidea (Platyhelminthes: Cestoda). *Bull. Univ. Neb. State Mus.* 18:1–249.

Jensen, K., Nikolov, P., and Caira, J.N. (2011). A new genus and two new species of Anteroporidae (Cestoda: Lecanicephalidea) from the darkspotted numbfish, *Narcine maculata* (Torpediniformes: Narcinidae), off Malaysian Borneo. *Folia Parasitol.* 58:95–107.

Johnston, T.H. and Mawson, P.M. (1943). Some nematodes from Australian elasmobranchs. *Trans. R. Soc. S. Aust.* 67:187–190.

Justine, J.-L. (2005). *Huffmanela lata* n. sp. (Nematoda: Trichosomoididae: Huffmanelinae) from the shark *Carcharhinus amblyrhynchos* (Elasmobranchii: Carcharhinidae) off New Caledonia. *Syst. Parasitol.* 61:181–184.

Justine, J.-L. (2009). A new species of *Triloculotrema* Kearn, 1993 (Monogenea: Monocotylidae) from a deep-sea shark, the blacktailed spurdog *Squalus melanurus* (Squaliformes: Squalidae), off New Caledonia. *Syst. Parasitol.* 74:59–63.

Kabata, Z. (1968). Some Chondracanthidae (Copepoda) from fishes of British Columbia. *J. Fish. Res. Bd. Can.* 25:321–345.

Kabata, Z. (1970). Crustacea as enemies of fishes. In: Snieszko, S.F. and Axelrod, H.R. (Eds.), *Diseases of Fishes*, Book 1. Tropical Fish Hobbiest Publications, Jersey City, NJ, pp. 157–166.

Kabata, Z. (1979). *Parasitic Copepoda of British Fishes*. Ray Society, London.

Kabata, Z. (1988). Copepoda and Branchiura. In: Margolis, L. and Kabata, Z. (Eds.), *Guide to the Parasites of Fishes of Canada. Part II. Crustacea*. Department of Fisheries and Oceans, Ottawa, pp. 3–127.

Karlsbakk, E., Aspholm, P.E., Berg, V., Hareide, N.R., and Berland, B. (2002). Some parasites of the small-eyed rabbitfish, *Hydrolagus affinis* (Capello, 1867) (Holocephali), caught in deep waters off SW Greenland. Sarsia: *North Atl. Mar. Sci.* 87:179–184.

Kearn, G.C. (1963). Feeding in some monogenean skin parasites: *Entobdella soleae* on *Solea solea* and *Acanthocotyle* sp. on *Raja clavata. J. Mar. Biol. Assoc. U.K.* 42:93–104.

Kearn, G.C. (1970). The oncomiracidia of the monocotylid monogenans *Dictocotyle coeliaca* and *Calicotyle kroyeri. Parasitology* 61:153–160.

Kearn, G.C. (1979). Studies on gut pigment in skin-parasitic monogeneans, with special reference to the monocotylid *Dendromonocotyle kuhlii. Int. J. Parasitol.* 9:545–552.

Kearn, G.C. (1987). The site of development of the monogenan *Calicotyle kroyeri*, a parasite of rays. *J. Mar. Biol. Assoc. U.K.* 67:77–87.

Kearn, G.C. (1998). *Parasitism and the Platyhelminthes.* Chapman & Hall, London.

Kearn, G.C. and Green, J.E. (1983). *Squalotrema llewellyni* gen. nov., sp. nov. a monocotylid monogenean from the nasal fossae of the spur-dog, *Squalus acanthias*, at Plymouth. *J. Mar. Biol. Assoc. U.K.* 63:17–25.

Kearn, G.C., Whittington, I., and Evans-Gowing, R. (2010). A new genus and new species of microbothriid monogenean with a functionally enigmatic reproductive system on the skin and mouth lining of the largetooth sawfish, *Pristis microdon* in Australia. *Acta Parasitol.* 55:115–122.

Kent, M.L., Andree, K.B., Bartholomew, J.L., El-Matbouli, M., Desser, S.S. et al. (2001). Recent advances in our knowledge of the Myxozoa. *J. Euk. Microbiol.* 48:395–413.

Khalil, L.F., Jones, A., and Bray, R.A. (Eds.). (1994). *Keys to Cestode Parasites of Vertebrates.* CAB International Wallingford, U.K.

Kitamura, A., Ogawa, K., Taniuchi, T., and Hirose, H. (2006). Two new species of hexabothriid monogeneans from the ginzame *Chimaera phantasma* and shortspine spurdog *Squalus mitsukurii. Syst. Parasitol.* 65:151–159.

Kitamura, A., Ogawa, K., Shimizu, T., Kurachima, A., Mano, N., Taniuchi, T., and Hirose, H. (2010). A new species of *Calicotyle* Diesing, 1850 (Monogenea: Monocotylidae) from the shortspine spurdog *Squalus mitsukurii* Jordan & Snyder and the synonym of *Gymnocalicotyle* Nybelin, 1941 with this genus. *Syst. Parasitol.* 75:117–124.

Klimpel, A., Seehagen, A., Palm, H.W., and Rosenthal, H. (2001). *Deep-Water Metazoan Fish Parasites of the World.* Lagos-Verlag, Berlin.

Knoff, M., Clemente, S.C., Pinto, R.M., and Gomes, D.C. (2001). Digenea and Acanthocephala of elasmobranch fishes from the southern coast of Brazil. *Mem. Inst. Oswaldo Cruz* 96:1095–1101.

Koch, K., Jensen, K., and Caira, J.N. (In press). Three new genera and six new species of lecanicephalideans (Cestoda) from eagle rays of the genus *Aetomylaeus* (Myliobatiformes: Myliobatidae) from Northern Australia and Borneo. *J. Parasitol.*

Kpatcha, T.K., Diebakate, C., and Toguebaye, B.S. (1996). Myxosporidia (Myxozoa, Myxosporea) of the genera *Sphaeromyxa* Thelohan, 1892, *Myxidium* Butschli, 1882, *Zschokkella* Auerbach, 1910, *Bipteria* Kovaljova, Zubtchenko & Krasin, 1983, and *Leptotheca* Thelohan, 1895, parasites of fish from the coast of Senegal (West Africa). *J. Afr. Zool.* 110:309–317.

Kuchta, R. and Caira, J.N. (2010). Three new species of *Echinobothrium* (Cestoda: Diphyllidea) from Indo-Pacific stingrays of the genus *Pastinachus* (Rajiformes: Dasyatidae). *Folia Parasitol.* 57:185–196.

Kudo, R. (1920). Studies on Myxosporidia: a synopsis of genera and species of Myxosporidia. *Ill. Biol. Monogr.* 5:1–165.

Lewis, A.G. (1966). Copepod crustaceans parasitic on elasmobranch fishes of the Hawaiian islands. *Proc. U.S. Natl. Mus.* 118:57–154.

Llewellyn, J. (1960). Amphibdellid (Monogenean) parasites of electric rays (Torpedinidae). *J. Mar. Biol. Assoc. U.K.* 39:561–589.

Llewellyn, L.C. and Knight-Jones, W. (1984). A new genus and species of marine leech from British coastal waters. *J. Mar. Biol. Assoc. U.K.* 64:919–934.

Llewellyn, J. and Simmons, J.E. (1984). The attachment of the monogenean parasite *Callorhynchicola multitesticulatus* to the gills of its holocephalan host *Callorhynchus milii. Int. J. Parasitol.* 14:191–196.

Lom, J. and Dyková, I. (1995). Myxosporea (Phylum Myxozoa). In: Woo, P.T.K. (Ed.), *Fish Diseases and Disorders*, Vol. 1. CAB International, Wallingford, U.K., pp. 97–148.

Long, D.J. and Waggoner, B.M. (1993). The ectoparasitic barnacle *Anelasma* (Cirripedia, Thoracica, Lepadomorpha) on the shark *Centroscyllium nigrum* (Chondrichthyes, Squalidae) from the Pacific sub-Antarctic. *Syst. Parasitol.* 26:133–136.

MacCallum, G.A. (1925). Eggs of a new species of nematoid worm from a shark. *Proc. U.S. Natl. Mus.* 67:1–2.

MacDonald, S. and Llewellyn, J. (1980). Reproduction in *Acanthocotyle greeni* n. sp. (Monogenea) from the skin of *Raia* spp. at Plymouth. *J. Mar. Biol. Assoc. U.K.* 60:81–88.

Machida, M. and Kuramochi, T. (1994). Two new species of trematodes (Gorgoderidae and Lepocreadiidae) from deep-sea fishes of Suruga Bay, Japan. *Bull. Natl. Sci. Mus. Ser. A (Zool.)* 20:149–153.

MacKenzie, K. (1964). *Stichocotyle nephropis* Cunningham, 1887 (Trematoda) in Scottish waters. *Ann. Mag. Nat. Hist.* 6:505–506.

Madhavi, R. and Hanumantha Rao, K. (1970). *Orchispirium heterovitellatum* gen. et sp. n. (Trematoda: Sanguinicolidae) from the ray fish, *Dasyatis imbricatus* Day, from Bay of Bengal. *J. Parasitol.* 56:41–43.

Maillard, C. and Ktari, M.-H. (1978). *Hyperandrotrema cetorhini* n. g., n. sp. (Trematoda: Sanguinicolidae) parasite du systeme circulatoire de *C. maximus* (Sel.). *Ann. Parasitol. Hum. Comp.* 53:359–365.

Manter, H.W. (1940). Digenetic trematodes of fishes from the Galapagos Islands and the neighboring Pacific. *Rep. Allan Hancock Pacif. Exped.* 2:325–497.

Markell, E.K. (1953). *Nagmia floridensis* n. sp., an anaporrhutine trematode form the coelom of the sting ray *Amphotistius sabinus. J. Parasitol.* 39:45–51.

Marques, F.P.L. (2000). Evolution of Neotropical Freshwater Stingrays and Their Parasites: Taking into Account Space and Time, doctoral dissertation, University of Toronto.

McVicar, A.H. (1977). Intestinal helminth parasites of the ray *Raja naevus* in British waters. *J. Helminthol.* 51:11–21.

McVicar, A.H. and Gibson, D.I. (1975). *Pancreatonema torriensis* gen. nov., sp. nov. (Nematoda: Rhabdochonidae) from the pancreatic duct of *Raja naevus. Int. J. Parasitol.* 5:529–535.

Moravec, F. (2001). *Trichinelloid Nematodes Parasitic in Cold-Blooded Vertebrates.* Academy of Sciences of the Czech Republic, Ceské Budejovice.

Moravec, F. and Little, M.D. (1988). *Granulinema* gen. n. a new dracunculoid genus with two new species (*G. carcharhini* sp. n. and *G. simile* sp. n.) from the bull shark, *Carcharhinus leucas* (Valenciennes), from Louisiana, U.S.A. *Folia Parasitol.* 35:113–120.

Moravec, F. and Nagasawa, K. (2000). Two remarkable nematodes from sharks in Japan. *J. Nat. Hist.* 34:1–13.

Moravec, F., Jimenez-Garcia, M.I., and Salgado Maldonado, G. (1998). New observations on *Mexiconema cichlasomae* (Nematoda: Dracunculoidea) from fishes in Mexico. *Parasite* 5:289–293.

Moravec, F., Borucinska, J.D., and Frasca, S. (2001). *Pancreatonema americanum* sp. nov. (Nematoda, Rhabdochonidae) from the pancreatic duct of the dogfish shark, *Squalus acanthias,* from the coast of Massachusetts, U.S.A. *Acta Parasitol.* 46:293–298.

Moravec, F., Van As, J.G., and Dyková, I. (2002). *Proleptus obtusus* Dujardin, 1845 (Nematoda: Physalopteridae) from the puffadder shyshark *Haploblepharus edwardsii* (Scyliiorhinidae) from off South Africa. *Syst. Parasitol.* 53:169–173.

Moreira, P.S. and Sadowsky, V. (1978). An annotated bibliography of parasitic Isopoda (Crustacea) of Chondrichthyes. *Bolm Inst. Oceanogr. (São Paulo)* 27:95–152.

Moser, M. and Anderson, S. (1977). An intrauterine leech infection: *Branchellion lobata* Moore, 1952 (Piscicolidae) in the Pacific angel shark (*Squatina californica*) from California. *Can. J. Zool.* 55:759–760.

Nagasawa, K., Tanaka, S., and Benz, G.W. (1998). *Trebius shiinoi* n. sp. (Trebiidae: Siphonostomatoida: Copepoda) from uteri and embryos of the Japanese angelshark (*Squatina japonica*) and the clouded angelshark (*Squatina nebulosa*), and redescription of *T. longicaudatus. J. Parasitol.* 84:1218–1230.

Naylor, G.J.P., Caira, J.N., Jensen, K., Rosana, K.A.M., White, W.T., and Last, P.R. (In press). A DNA sequence based approach to the identification of shark and ray species and its implications for global elasmobranch diversity and parasitology. *Bull. Am. Mus. Nat. Hist.*

Newbound, D.R. and Knott, B. (1999). Parasitic copepods from pelagic sharks in Western Australia. *Bull. Mar. Sci.* 65:715–724.

Ofenheim, E. von. (1900). Über eine neue Distomidengattung. *Z. Naturwiss.* 73:145–186.

Olsen, L.W. (1952). Some nematodes parasitic in marine fishes. *Pub. Inst. Mar. Sci. Univ. Tex.* 11:173–215.

Olson, R.E, Hanson, A.W., and Pratt, I. (1970). *Plectognathotrema hydrolagi* sp. n. (Trematoda: Cephaloporidae) from ratfish (*Hydrolagus colliei*). *J. Parasitol.* 56:724–726.

Olson, P.D., Caira, J.N., Jensen, K., Overstreet, R.M., Palm, H.W., and Beveridge, I. (2010). Evolution of the trypanorhynch tapeworms: parasite phylogeny supports independent lineages of sharks and rays. *Int. J. Parasitol.* 40:223–242.

O'Sullivan, J.B., McConnaughey, R.R., and Huber, M.E. (1987). A blood-sucking snail: the Cooper's nutmeg, *Cancellaria cooperi* Gabb, parasitizes the California electric ray, *Torpedo californica* Ayres. *Biol. Bull.* 172:362–366.

Ota, Y. and Hirose, E. (2009). Description of *Gnathia maculosa* and a new record of *Gnathia trimaculata* (Crustacea, Isopoda, Gnathiidae), ectoparasites of elasmobranchs from Okinawan coastal waters. *Zootaxa* 2114:50–60.

Palm, H.W. (2004). *The Trypanorhyncha Diesing, 1863.* PKSPL-IPB Press, Bogor.

Palm, H.W., Waeschenbach, A., Olson, P.D., and Littlewood, D.T.J. (2009). Molecular phylogeny and evolution of the Trypanorhyncha Diesing 1863 (Platyhelminthes: Cestoda). *Mol. Phylogenet. Evol.* 52:351–367.

Pappas, P.W. (1970). The trypanorhynchid cestodes from Humboldt Bay and Pacific Ocean sharks. *J. Parasitol.* 56:1034.

Pearse, A.S. (1953). Parasitic crustaceans from Alligator Harbor Florida. *Q. J. Fla. Acad. Sci.* 15:187–243.

Pozdnyakov, S.E. (1989). On the finding of didymozoids in pelagic sharks of the Pacific [in Russian]. *Parazitologiya* 23:529–532.

Price, E.W. (1938a). North American monogenetic trematodes. II. The families Monocotylidae, Microbothriidae, Acanthocotylidae, and Udonellidae (Capsaloidea). *J. Wash. Acad. Sci.* 28:183–198.

Price, E.W. (1938b). North American monogenetic trematodes. II. The families Monocotylidae, Microbothriidae, Acanthocotylidae, and Udonellidae (Capsaloidea). *J. Wash. Acad. Sci.* 28:109–126.

Price, E.W. (1963). A new genus and species of monogenetic trematode from a shark, with a review of the family Microbothriidae Price, 1936. *Proc. Helminthol. Soc. Wash.* 30:213–218.

Rajya Lakshmi, I. and Sreeramulu, K. (2007). *Hysterothylacium ganeshi* n. sp (Nematoda: Anisakidae) from the intestine of shark, *Sphyrna blochii* (Cuvier). *Geobios (Jodhpur)* 34:29–32.

Razi Jalali, M.H., Mazaheri, Y., and Peyghan, R. (2008). *Acanthocheilus rotundatus* (Nematoda: Acanthocheilidae) from the intestine of shark (*Carcharhinus macloti*) in Persian Gulf, Iran. *Iran. J. Vet. Res.* 9:178–180.

Rohde, K. (1998). Scanning electron-microscopic studies of *Multicalyx elegans* (Olsson, 1869) (Aspidogastrea, Multicalycidae). *Acta Parasitol.* 43:11–19.

Rohde, K. (2002). Subclass Aspidogastrea Faust & Tang, 1936. In: Gibson, D.I., Jones, A., and Bray, R.A. (Eds.), *Keys to the Trematoda*, Vol. 1. CAB International/Natural History Museum, London, pp. 5–14.

Rohde, K., Heap, M., Hayward, C.J., and Graham, K.J. (1992). *Calicotyle australiensis* n. sp. and *Calicotyle* sp. (Monogenea, Monopisthocotylea) from the rectum and rectal glands, and *Rugogaster hydrolagi* Schell, 1973 (Trematoda, Aspidogastrea) from the rectal glands of holocephalans off the coast of southeastern Australia. *Syst. Parasitol.* 21:69–79.

Rokicki, J. and Bychawska, D. (1991). Parasitic copepods of Carcharhinidae and Sphyridae [sic] (Elasmobranchia) from the Atlantic Ocean. *J. Nat. Hist.* 25:1439–1448.

Rosa-Molinar, E., Williams, C.S., and Lichtenfelds, J.R. (1983). Larval nematodes (Philometridae) in granulomas in ovaries of black-tip sharks, *Carcharhinus limbatus* (Valenciennes). *J. Wildl. Dis.* 19:275–277.

Ross, R. (1999). *Aqualog Special: Freshwater Stingrays from South America.* Aqualog Verlag, Rodgau, Germany.

Ruhnke, T.R. (2011). A monograph on the Phyllobothriidae (Platyhelminthes: Cestoda). *Bull. Univ. Neb. State Mus.* 25:1–208.

Russo, R.A. (1975). Notes on the external parasites of California inshore sharks. *Calif. Fish Game* 61:228–232.

Ruyck, R. and Chabaud, A.G. (1960). Un cas de parasitisme attribuable a des larves de *Phlyctainophora lamnae* Steiner chez un sélacien, et cycle évolutif probable de ce nematode. *Vie Milieu* 11:386–389.

Sawyer, R.T., Lawler, A.R., and Overstreet, R.M. (1975). Marine leeches of the eastern United States and the Gulf of Mexico with a key to the species. *J. Nat. Hist.* 9:633–667.

Schell, S.C. (1972). *Otodistomum hydrolagi* sp. n. (Trematoda: Azygiidae) from the coelom of the ratfish, *Hydrolagus colliei* (Lay and Bennett, 1839). *J. Parasitol.* 58:885–886.

Short, R.B. (1954). A new blood fluke, *Selachohemecus olsoni*, n. g., n. sp. (Aporocotylidae) from the sharp-nosed shark, *Scoliodon terra-novae. Proc. Helminthol. Soc. Wash.* 21:78–82.

Siddall, M.E., Martin, D.S., Bridge, D., Desser, S.S., and Cone, D.K. (1995). The demise of a phylum of protists: phylogeny of the Myxozoa and other parasitic cnidaria. *J. Parasitol.* 81:961–967.

Smit, N.J. and Basson, L. (2002). *Gnathia pantherina* sp. n. (Crustacea: Isopoda: Gnathiidae), a temporary ectoparasite of some elasmobranch species from southern Africa. *Folia Parasitol.* 49:137–151.

Steiner, G. (1921). *Phlyctainophora lamnae* n. g., n. sp., eine neue parasitische Nematodenform aus *Lamna cornubica* (Heringshai). *Centralbl. Bakt. Jena Abt. 1* 86:91–595.

Stoffregen, D.A. and Anderson, W.I. (1990). A myxosporidian parasite in the skeletal muscle of a blacktip reef shark, *Carcharhinus melanopterus* (Quoy and Gaimard, 1824). *J. Fish Dis.* 13:549–552.

Tandon, R.S. (1972). Some observations on the plerocercus larva of a trypanorhynchid cestode obtained from the ovary of the ray *Dasyatis uarnak. Proc. Natl. Acad. Sci. India* 42:431–435.

Tang, D. and Newbound, D.R. (2004). A new species of copepod (Siphonostomatoida: Caligidae) parasitic on the tiger shark *Galeocerdo cuvier* (Péron & Lesueur) from Western Australian waters. *Syst. Parasitol.* 58:69–80.

Thoney, D.A. and Burreson, E.M. (1986). Ecological aspects of *Multicalyx cristata* (Aspidocotylea) infections in northwest Atlantic elasmobranchs. *Proc. Helminthol. Soc. Wash.* 53:162–165.

Tyler, G.A. (2006). A monograph on the Diphyllidea (Platyhelminthes: Cestoda). *Bull. Univ. Neb. State Mus.* 20:1–142.

Vaughan, D.B. and Chisholm, L.A. (2010). A new species of *Neoheterocotyle* Hargis, 1955 (Monogenea: Monocotylidae) from the gills of *Rhinobatos annulatus* Müller & Henle (Rhinobatidae) off the southern tip of Africa. *Syst. Parasitol.* 77:205–213.

Waterman, P.B. and Sin, F.Y.T. (1991). Occurrence of the marine tapeworms, *Hepatoxylon trichiuri* and *Hepatoxylon megacephalum*, in fishes from Kaikoura, New Zealand. *N.Z. Nat. Sci.* 18:71–73.

White, W.T., Last, P.R., Naylor, G.J.P., Jensen, K., and Caira, J.N. (2010). Clarification of *Aetobatus ocellatus* (Kuhl, 1823) as a valid species, and a comparison with *Aetobatus narinari* (Euphrasen, 1790) (Rajiformes: Myliobatidae). In: Last, P.R., White, W.T., and Pogonoski, J.J. (Eds.), *Descriptions of New Sharks and Rays from Borneo*, Research Paper 032. CSIRO Marine and Atmospheric Research, Hobart, Tasmania, pp. 141–164.

Whittington, I.D. (1990). *Empruthotrema kearni* n. sp. and observations on *Thaumatocotyle pseudodasybatis* Hargis, 1955 (Monogenea: Monocotylidae) from the nasal fossae of *Aetobatus narinari* (Batiformes: Myliobatidae) from Moreton Bay, Queensland. *Syst. Parasitol.* 15:23–31.

Whittington, I.D. and Chisholm, L. (2003). Diversity of Monogenea from Chondrichthyes: do monogeneans fear sharks? In: Combes, C. and Jourdane, J. (Eds.), *Taxonomy, Ecology and Evolution of Metazoan Parasites (Livre hommage à Louis Euzet)*, Vol. II. Presses Universitaire de Perpignan, Perpignan, pp. 339–363.

Whittington, I.D. and Kearn, G.C. (2009a). Two new species of entobdelline skin parasites (Monogenea, Capsalidae) from the blotched fantail ray, *Taeniura meyeni*, in the Pacific Ocean, with comments on spermatophores and the male copulatory apparatus. *Acta Parasitol.* 54:12–21.

Whittington, I.D. and Kearn, G.C. (2009b). Two new species of *Neoentobdella* (Monogenea, Capsalidae, Entobdellinae) from the skin of Australian stingrays (Dasyatidae). *Folia Parasitol.* 56:29–35.

Williams, Jr., E.H. (1978). *Conchoderma virgatum* (Spengler) (Cirripedia, Thoracica) in association with *Dimemoura latifolia* (Steenstrup & Lutken) (Copepoda, Caligidae), a parasite of the shortfin mako, *Isurus oxyrhynchus* Rafinesque (Pisces, Chondrichthyes). *Crustaceana* 34:109–110.

Williams, Jr., E.H., and Bunkley-Williams, L. (1996). *Parasites of Offshore Big Game Fishes of Puerto Rico and the Western Atlantic*. Puerto Rico Department of Natural and Environmental Resources, San Juan, and University Puerto Rico, Mayaguez.

Williams, H.H. (1960). A list of parasitic worms, including 22 new records from marine fishes caught off the British Isles. *Ann. Mag. Nat. Hist.* 2:705–715.

Williams, H.H. and Bray, R.A. (1984). *Chimaerocestos prudhoei* gen. et sp. nov. representing a new family of tetraphyllideans and the first record of strobilate tapeworms from a holocephalan. *Parasitology* 88:105–116.

Williams, H.H., McVicar, A.V., and Ralph, R. (1970). The alimentary canal of fishes as an environment for helminth parasites. *Symp. Br. Soc. Parasitol.* 8:43–77.

Wilson, C.B. (1913). Crustacean parasites of West Indian fishes and land crabs, with descriptions of new genera and species. *Proc. U.S. Natl. Mus.* 44:189–277.

Woolcock, B. (1936). Monogenetic trematodes from some Australian fishes. *Parasitology* 28:79–91.

Yamaguti, S. (1963). *Systema Helminthum*. Vol. 5. *Acanthocephala*. Interscience Publishers, New York.

Yamaguti, S. (1971). *Synopsis of the Digenetic Trematodes of Vertebrates*, Vol. I. Keigaku, Tokyo.

Yano, K. and Musick, J.A. (2000). The effect of the mesoparasitic barnacle *Anelasma* on the development of reproductive organs of deep-sea squaloid sharks, *Centroscyllium* and *Etmopterus. Environ. Biol. Fish.* 59:329–339.

19

Assessing Habitat Use and Movement

Colin A. Simpfendorfer and Michelle R. Heupel

CONTENTS

19.1 Introduction

Sharks occur in all of the world's oceans and in waters that include deep-sea, oceanic, neritic, and estuarine habitats. In addition, a few specialized species also occur in rivers and lakes connected to the ocean. The occurrence of sharks within these broad regions is well understood for most species; for example, the gummy shark, *Mustelus antarcticus*, is known to occur in the neritic waters of southern Australia, and the salmon shark, *Lamna ditropis*, is known to inhabit the boreal waters of the north Pacific. (Chapter 2 provides a detailed consideration of the zoogeography of the sharks, skates, and rays.) A shark will not occur in all of the habitats within its range; instead, it is more likely to have specific habitats in which it spends most of its time. It is this detailed analysis of the habitats that a species uses that is provided here.

It is intuitive that information on the habitat use of sharks would be important for management and conservation. Olsen (1954) determined that newborn school sharks, *Galeorhinus galeus*, occur in protected coastal bays around Tasmania and proposed that these areas be protected (Figure 19.1). Despite some early recognition of the importance of habitat use information it was only in the 1990s that resource managers and researchers began focusing research on *essential fish habitat, critical habitat*, and *marine protected areas*.

There is a limited, but growing, literature on the habitat use in sharks, skates, and rays. By far the most widely investigated topic is that of nursery areas (i.e., habitat use patterns of juvenile sharks). A variety of studies have defined nursery areas and identified their importance (e.g., Bass, 1978; Branstetter, 1990; Castro, 1993; Clarke, 1971; Duncan and Holland, 2006, McCandless et al., 2007; Morrissey and Gruber, 1993a; Simpfendorfer

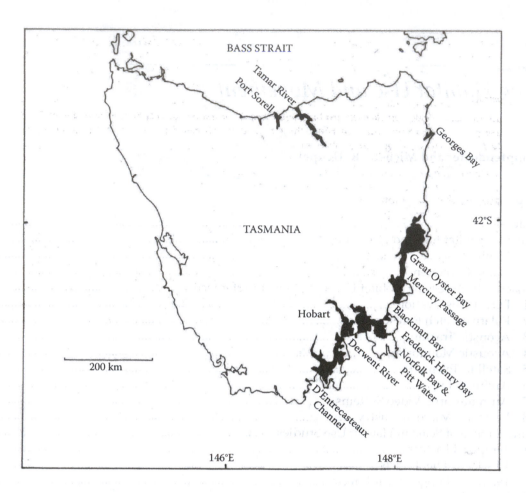

FIGURE 19.1
Nursery areas (black areas) of the school shark, *Galeorhinus galeus*, in Tasmania that are protected by state law. (From Williams, H. and Schaap, A.H., *Aust. J. Mar. Freshwater Res.*, 43, 237–250, 1992. With permission.)

and Milward, 1993; Springer, 1967), and much ongoing research seeks to better understand nursery area use and importance. Heupel et al. (2007) reviewed the shark nursery paradigm and provided a new description of shark nursery areas, including guidelines for defining these critical habitats. This paper recommended three criteria for defining a shark nursery: (1) sharks are more commonly encountered in the area than in other areas (i.e., density in the area is greater than the mean density over all areas); (2) sharks have a tendency to remain or return for extended periods—weeks or months (i.e., site fidelity is greater than the mean site fidelity for all areas); and (3) the area or habitat is repeatedly used across years, whereas others are not. This new definition of shark nursery areas has helped refine use of this term and description of these habitats. Following this redefinition of nursery areas, Knip et al. (2010) reviewed use of inshore regions by coastal sharks to further define this topic. Here, the theoretical coastal shark population proposed by Springer (1967) was reconsidered and supplemented with an additional theoretical

model for small coastal sharks that do not use discrete nursery areas and that spend their entire life in shallow coastal waters.

Only a limited number of studies have directly addressed questions of habitat preference. Morrissey and Gruber (1993b) described the habitat selection of juvenile lemon sharks, *Negaprion brevirostris*, using acoustic tracking; Heithaus et al. (2002, 2006) described the habitat selection of tiger sharks, *Galeocerdo cuvier*, using acoustic tracking and animal-borne video; Carraro and Gladstone (2006) described the habitat preferences of the ornate wobbegong, *Orectolobus ornatus*, using diver surveys; and Simpfendorfer et al. (2010) examined the habitat selection of juvenile smalltooth sawfish, *Pristis pectinata*. We return to these examples later in the chapter.

In this chapter, we consider *habitat use* to be the observed pattern of the habitats in which an individual or species occurs. This term has been used synonymously with *habitat selection* and *habitat utilization* in the ecological literature. When studying habitat use of a species, a researcher aims to identify the species' *habitat preferences* and the factors underlying these preferences. This chapter briefly considers information requirements for measuring and testing habitat use; describes the different methods by which habitat use in the sharks, skates, and rays has been investigated; and gives examples of the results from some of these studies. The final section of the chapter discusses the importance of scale in habitat use studies and examines some of the mechanisms that drive habitat preferences.

19.2 Measuring Habitat Use and Habitat Preference

There are many approaches to describing habitat use and quantifying habitat preferences. Habitat use is most commonly determined by overlaying movement or location information on habitat maps. Modern geographic information systems (GIS) have made this process relatively straightforward. Habitat preference, however, is a matter of determining if one habitat type is used more frequently than another, relative to the abundance of each habitat.

19.2.1 Habitat Use

One of the important concepts in describing habitat use is an individual's home range. The definition of home range has been refined over time. Burt (1943) originally defined home range as the area around the established home that is traversed by an animal in its normal activities of food gathering, mating, and caring for its young. Many authors have felt that Burt's original definition was too general and did not apply to animals that do not care for their young or maintain specific home or nest sites. Cooper (1978) pointed out that the home range of an animal should not be treated as an inclusive area because an animal may use a small portion of the area intensively, other areas moderately, and some areas not at all. This type of observation led several authors to define home range as the smallest subregion of an area that accounts for a specific portion (often 95%) of the space an animal uses (e.g., Anderson, 1982; Jennrich and Turner, 1969; Worton, 1987). This type of mathematical approach to defining the home range is often referred to as a *utilization distribution*. The most widely used form of the utilization distribution is the kernel distribution (Worton, 1987). Several studies of shark movements have defined home range patterns and described habitat use patterns (e.g., Dawson and Starr, 2009; Heupel et al., 2004, 2006; Holland et al., 1993; McKibben and Nelson,

1986; Morrissey and Gruber, 1993a; Papastamatiou et al., 2010; Yeiser et al., 2008). These shark studies have typically included all areas an individual uses and do not provide detailed information on whether specific habitats are selected preferentially. Recently, Kei et al. (2010) raised the question of whether the emergence of technologies such as satellite telemetry that produce large amounts of data means that the traditional measures of home range should be replaced by newer, more computationally intensive approaches.

19.2.2 Habitat Preference

The problem of determining if an individual or species shows a habitat preference can be broken down into two parts. The first is a test to determine if habitats are used in proportion to their availability; that is, are there habitat preferences? This is most often achieved using a chi-squared goodness-of-fit test. If there are differences, then the second part of the problem is to identify which habitats are preferred and which are avoided. Various indices are available. Krebs (1999) described a number of these, including the simple forage ratio, the rank preference, and more complex indices such as Manly's index. In one study of habitat selection in sharks, Morrissey and Gruber (1993b) used a simple chi-squared goodness-of-fit test to compare habitat use to habitat availability and then Strauss' index of selection (Strauss, 1979) to investigate which habitats were preferred.

The simple comparison of use to availability can often lead to difficulties, especially with wide-ranging mobile species such as sharks. The question is often how much habitat should be considered available. In enclosed systems, such as the lagoon studied by Morrissey and Gruber (1993b), the area is well defined; however, in more open areas the limits are less clear and depend much more on the temporal and spatial scales that are being considered. To account for this it is better to assume that not all habitats are equally available and instead to generate randomized tracks of animals and measure the expected proportions of habitats used. These randomized habitat use patterns can then be compared to the observed pattern of habitat use using chi-squared tests. Heithaus et al. (2002) used two methods of generating randomized habitat use patterns for tiger sharks: correlated random walk and track randomization. These methods produced expected habitat use patterns that differed from those based simply on habitat availability. These approaches were then extended using a randomization of entire samples (Heithaus et al., 2006) to investigate smaller-scale patterns of habitat preference. A similar approach was used by Simpfendorfer et al. (2010) when examining the habitat preferences of smalltooth sawfish.

19.3 Approaches to Assessing Habitat Use and Habitat Preferences

To study the habitat use and selection of a shark species requires that the movements of individuals, and the habitats in which they occur, be determined over sufficiently long time periods to obtain meaningful data. Over time, a wide variety of approaches to this problem have been taken in elasmobranch species. Below we examine these approaches, provide a brief example from the literature, and discuss the advantages and disadvantages of these approaches. Finally, we summarize the constraints of each approach in a table to allow for easy comparison (Table 19.1).

19.3.1 Direct Observation

The simplest method of examining habitat use by sharks is to directly observe individuals and record the habitats that they use over time. This technique is effective only in areas where the water clarity is sufficient to enable direct observation. This approach was used by Economakis and Lobel (1998), who studied the daily aggregations of gray reef sharks, *Carcharhinus amblyrhynchos*, at Johnston

Atoll. In this study, the numbers of sharks present at a small island were counted to provide information on how often aggregations occurred and how many animals were present. The authors combined these sightings data with water temperature, tidal cycle, and habitat descriptions to determine why this habitat use pattern occurred. More recently, Carraro and Gladstone (2006) used diver surveys to determine habitat preferences of the ornate wobbegong. They determined that this species preferred complex habitats such as sponge garden and boulder habitats over less complex habitats and those with marine plants and that the pattern did not change between sexes (Figure 19.2) or size classes. Direct observations do not have to involve underwater surveys; where conditions are suitable, surface-based techniques can be used. Vaudo and Heithaus (2009), for example, used boat-driven transects to survey the elasmobranchs of sand flats to demonstrate that batoids dominated the fauna and that diversity was much greater during warmer months.

There are a number of disadvantages to direct observation in shark studies. First, observations can normally only be made during the day, leaving any nocturnal changes in habitat use undetected. Second, the act of observing may change the behavior of individuals,

TABLE 19.1

Constraints of Various Approaches to Investigating Habitat Use in Sharks, Skates, and Rays

| Approach | Constraints | | | | | | |
	Size of Animals	Accuracy of Positions	Temporal Coverage	Geographic Coverage	Equipment Costs	Other	Best Use
Direct observation	Any	±10 m	Short, daytime	Limited	Low	Requires good water clarity; observer effects	Coral reef species
Relative catch rates	Any	N/A	Any	Any	Low	Biased by habitat-specific catch ability or movement rates	Commercially fished species
Acoustic tracking	Any	±50 m	Short (days)	Any	Moderate	Only one animal tracked at a time; chasing effects	Detailed short-term studies of habitat use
Acoustic monitoring	Any	±225 m (omnidirectional)	Any	Moderate	High	Only effective if animals stay within range of receivers	Long-term studies in defined environments
		±1 m (triangulating)		Small	High		
Satellite tracking	>1.5 m	±250 m to 10 km	Any	Global	High	Animal must surface to give location	Large species that surface regularly
Archival tags	Any	±0.5° (best)	Any	Global	High	Must recover animal or use pop-up satellite tags	Wide-ranging species
Animal-borne video systems	>1.5 m	±10 m (if tracked acoustically)	Short (hours)	Limited	High	Size of equipment	Large species in clear water
Vertebral microchemistry	Any	Broad	Long (years)	Global	Moderate	Requires vertebrae; precision is limited; demonstrates broad patterns	Movements between habitats with different chemistries (e.g., freshwater to seawater)

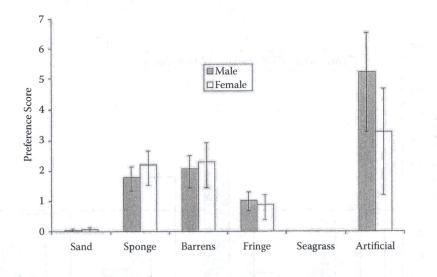

FIGURE 19.2
Habitat preferences of male and female ornate wobbegongs. Values shown are preference scores and their 95% confidence intervals. Habitat preference scores with lower confidence limit > 1 indicate a significant preference; confidence intervals that range from <1 to >1 indicate no significant preference; and an upper confidence limit < 1 indicates significant avoidance. (From Carraro, R. and Gladstone, W., *Pac. Sci.*, 60, 207–223, 2006. With permission.)

depending on the method (a disadvantage that must be addressed in any study of habitat use). It has been observed that the presence of divers can cause a response in some shark species (Johnson and Nelson, 1973; Nelson et al., 1986). Finally, it is not usually possible to identify individual animals, so information about individual habitat use is not available.

Medved and Marshall (1983) took the direct observation method one step further and overcame water clarity problems. Working with small sandbar sharks, *Carcharhinus plumbeus*, in Chincoteague Bay (an area with poor water clarity), they attached small Styrofoam floats to their dorsal fins and followed their movements. In this way, they were able to describe sandbar shark movements and habitat use, and they identified tidal flow as an important controlling factor.

19.3.2 Relative Catch Rates

Rather than directly observe sharks to determine their habitat use, it is also possible to use relative catch rates in different habitats to draw conclusions about habitat use. With this approach, a sampling gear is set in all available habitats and the catch rates between them compared using selectivity or preference index values. Michel (2002) used this approach to examine the habitat use of four species of sharks in the Ten Thousand Islands region of Florida. Gillnets were set in three habitats (gulf edge, transition, and backwater), and a preference index was used to show that blacktip sharks, *Carcharhinus limbatus*, preferred gulf edge habitats; bull sharks, *C. leucas*, preferred backwater habitats; and lemon sharks, *Negaprion brevirostris*, avoided gulf edge

habitats (Figure 19.3). Similarly, Simpfendorfer et al. (2005) compared catch rates of different sizes classes of bull sharks, *C. leucas*, to determine how they partition their use of habitat. The youngest age class occurred in a riverine area, and as they grew older they moved into coastal lagoons and finally into offshore areas. Wiley and Simpfendorfer (2007) also used catch-rate data with Ivlev's electivity index to examine the habitat use patterns of several species of sharks under varying levels of three environmental parameters (salinity, temperature, and depth) in the Everglades National Park. They found that these environmental parameters led to differences in habitat use between species.

Catch-rate comparisons are good for investigating population-level habitat use patterns; however, they have several major drawbacks. First, they are unable to resolve detailed individual movements that can help explain why the habitat use patterns occur. Second, habitat-specific movement rates can affect catch rates, which can be misidentified as habitat preference. Finally, sampling gear may be more effective in specific habitats and so also bias catch rates; for example, Heithaus et al. (2007a) demonstrated that the catches of some species were affected by the type of bait used, while others were not, demonstrating that that different catch compositions can occur depending on the bait used,

19.3.3 Acoustic Tracking

The most widely used approach in studies of shark habitat use is acoustic telemetry. With this technique, an acoustic tag that generates a series of "pings" is attached to an individual. The acoustic signal is then

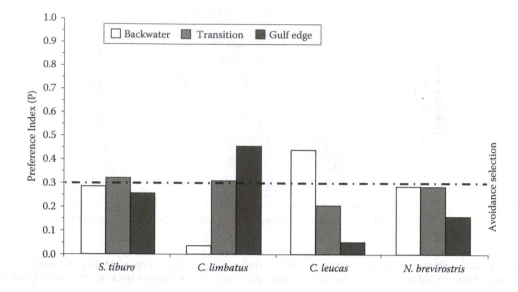

FIGURE 19.3

Habitat selection of four species of sharks in the Ten Thousand Islands region of Florida based on catch rates in gillnet and longlines. Bars above the dashed line indicate habitat preference, bars below the line indicate habitat avoidance. (From Michel, M., Environmental Factors Affecting the Distribution and Abundance of Sharks in the Mangrove Estuary of the Ten Thousand Islands, Florida, U.S.A., doctoral dissertation, University of Basel, Switzerland, 2002.)

located using a receiver and hydrophone, and the movement of the shark is followed and its position regularly recorded. The movement data can then be overlaid on habitat information to determine habitat use. We make a distinction between acoustic tracking (where an individual is followed and locations determined) and acoustic monitoring (where data-logging acoustic receivers are used to gather data remotely). Acoustic monitoring is covered in the next section.

Many good examples of acoustic tracking studies in sharks are available (e.g., Cartamil et al., 2003, 2010a,b; Holland et al., 1993; Morrissey and Gruber, 1993a; Sciarotta and Nelson, 1977; Sepulveda et al., 2004); however, few have directly addressed the issue of habitat use. Probably the best example is a series of studies conducted on lemon sharks at Bimini, Bahamas, by Gruber and his students. Gruber et al. (1988) provided an initial glimpse into the behavior and habitat use of this species, and laid the groundwork for more detailed study. Sharks were captured in the lagoon at Bimini, fitted with acoustic tags, and tracked for up to 101 hours at a time, with some individuals being reacquired and retracked for as long as 8 days. The results showed evidence of site attachment to various regions within the lagoon, as well as differences in activity space between night and day. Following this study, Morrissey and Gruber (1993a,b) used acoustic tracking over extended periods with 38 individuals, often reacquiring individuals many times, to generate a long time series of locations. The authors used these location data and data on the habitat

within the study site to demonstrate that juvenile lemon sharks preferred shallower, warmer waters with rocky or sandy substrates (Figure 19.4) and that sharks did not show any preference based on salinity (Morrissey and Gruber, 1993b). Similar research in northern California has revealed that temperate species also select habitat based on conditions; for example, bat rays and leopard sharks have been shown to select habitat to assist behavioral thermoregulation (Hight and Lowe, 2007; Matern et al., 2000).

Acoustic tracking is a method that can provide detailed spatial data over a relatively large area, depending on the range of the tracking vessel. As such, it is one of the most widely used methods for investigating habitat use patterns in sharks, skates, and rays to date; however, it does have several disadvantages in habitat use studies. First, individual sharks can only be tracked for short periods (usually less than 48 hours) because of the human resources required. This limits the technique to investigations of short-term habitat use and temporal shifts in habitat use (e.g., diurnal changes). Second, only one individual can normally be tracked at a time; thus, population-level changes in habitat use are difficult to identify. Third, the need to follow a shark, normally with the use of a boat, can possibly result in changes in behavior. This leads to the concern that the researcher in some way is chasing the shark and so not observing natural behavior. The need to capture and handle the shark adds to concerns that normal behavior is not observed following release.

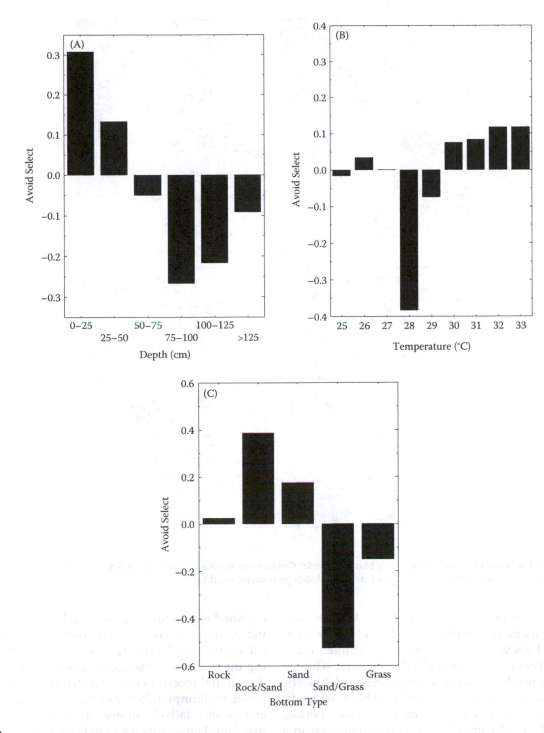

FIGURE 19.4
Habitat selection of juvenile lemon sharks, *Negaprion brevirostris*, in the lagoon at North Bimini, Bahamas, based on (A) water depth, (B) temperature, and (C) bottom type. Bars above the line indicate selection, bars below the line indicate avoidance. (From Morrissey, J.F. and Gruber, S.H., *Environ. Biol. Fish.*, 38, 311–319, 1993. With permission.)

19.3.4 Acoustic Monitoring

The results of acoustic tracking studies have been very important in defining short-term movement and habitat use patterns of sharks, but understanding longer term patterns of individuals and population-level factors is also important. Neither of these issues can be adequately tackled using acoustic tracking because of the resources required to continuously follow an individual for long periods (e.g., >1 week) or following more than one individual at a time. The development of underwater data-logging acoustic receivers opened up

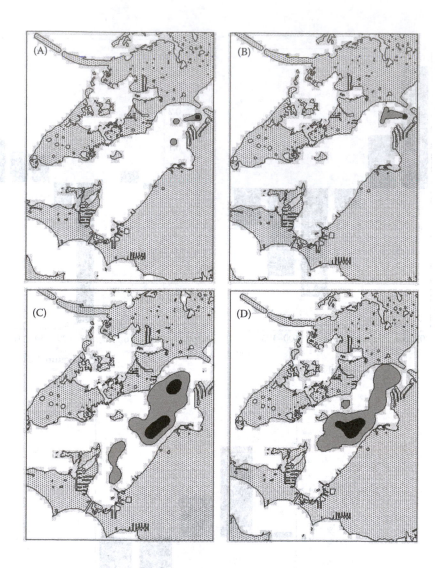

FIGURE 19.5
Increasing monthly home ranges of two juvenile blacktip sharks, *Carcharhinus limbatus*, in Terra Ceia Bay, FL, from June (A, B) to October (C, D). Home ranges were calculated using 50% (black areas) and 95% (gray areas) fixed kernels.

new possibilities in long-term population-level studies. Early equipment was large and relatively expensive, and researchers were able to use only a limited number of receivers to cover small high-use areas. Klimley et al. (1988) used two acoustic monitors to define the movements and habitat use of scalloped hammerheads, *Sphyrna lewini*, at a seamount over a 10-day period. They found that during the day sharks remained in a group at the seamount, but at night they dispersed into the surrounding area.

As technology progressed, receivers became smaller and more affordable, providing the opportunity to cover much larger areas for longer periods (Heupel et al., 2006b; Voegeli et al., 2001). Research on young blacktip sharks, *Carcharhinus limbatus*, in Terra Ceia Bay, FL (Heupel and Hueter, 2001, 2002), revealed that an array of acoustic receivers could be used to continuously monitor the movements of a population of sharks in a confined region for long periods. In this study, up to 40 individuals per year were monitored within the study site for up to 167 days. The use of an algorithm for taking the presence–absence data provided by the omnidirectional receivers and converting them to averaged positions (Simpfendorfer et al., 2002) enabled the generation of relatively fine-scale movement and habitat use data. The results provided detailed long-term data on movements and habitat use (Figure 19.5), how they vary over time (Figure 19.6), and the synchronicity in habitat use changes across the population. This study was one of the first of many to employ acoustic monitoring technology to define the long-term movements of sharks. Acoustic monitoring has been used to look at the use of complex habitats such as sandbar sharks in marsh habitats (Conrath and Musick, 2010) and blacktip reef sharks in coral reef habitats (Papastamatiou et al., 2009).

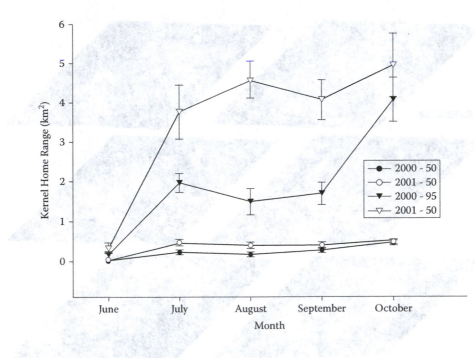

FIGURE 19.6

Increases in monthly home range size (50 and 95% kernels) of juvenile blacktip sharks, *Carcharhinus limbatus*, in Terra Ceia Bay, FL, in 2 years. (From Heupel, M.R. et al., *Environ. Biol. Fish.*, 71, 135–142, 2004. With permission.)

The data-logging receivers used by Heupel and Hueter (2001) were small, inexpensive units that provide presence–absence data. An alternative type of data-logging receiver is one that provides triangulation of acoustic signals to give submeter accuracy in position (Voegeli et al., 2001). This type of equipment is much more expensive per unit and can be used to cover only a small area; however, for animals that have small activity spaces, or that a researcher wishes to study in detail in a certain area, this equipment can provide very good results. Klimley et al. (2001) used this type of system to study white sharks, *Carcharodon carcharias*, at Año Nuevo Island off California and showed that sharks concentrated their movements to specific areas (Figure 19.7). These high-use areas were close to the islands in areas where seals haul out and provided the white sharks with the best opportunity for locating potential prey. Recent developments in the deployment and analysis of data from smaller and less expensive acoustic monitors now allows for the estimation of positional accuracy to a few meters in large arrays. Such an approach has been used to examine the habitat use patterns of the shovelnose guitarfish, *Rhinobatos productus*, in a restored estuary (Farrugia et al., 2011).

Acoustic monitoring is beginning to provide a greater understanding of habitat use in sharks, especially on longer temporal scales. As an emerging field, there still remains a significant amount of technical and analytical development to be undertaken. As with all approaches, however, it does have disadvantages. The largest drawback is that only sharks within the array of data-logging receivers can be studied. If an individual leaves the range of the array, then no data can be gathered. The introduction of large-scale networks of receivers has helped improve the monitoring of long-range migratory paths of sharks. The advent of triangulation approaches has also improved the resolution of positioning data to provide more accurate position locations. These advances will increase the scope, capability, and utility of acoustic monitoring approaches; however, users of this technology must also deal with other issues, such as variations in tag detections due to tidal depth variation, wind, bionoise, current, turbidty, and boat wakes (Heupel et al., 2006b; Payne et al., 2010; Simpfendorfer et al., 2008, Vaudo and Lowe, 2006). This means that the data collected must be carefully interpreted to understand how they represent animal movement. Despite all of these advances, multiple other study approaches are available that may provide more robust data to address some of the questions surrounding shark movement patterns.

19.3.5 Satellite Telemetry

The biggest limitation of acoustic monitoring systems is that sharks must remain in the receiver array to be studied. Once they leave this area, habitat use data can no longer be collected. One technique that can address the issue of large spatial coverage is satellite telemetry. For this method, a tag that transmits a signal to the NOAA/

FIGURE 19.7
(See color insert.) Three-dimensional contour maps of habitat use by two white sharks, *Carcharodon carcharias*, at Año Nuevo Island, CA. (From Klimley, A.P. et al., *Mar. Biol.*, 138, 429–446, 2001. With permission.)

CNES Argos system is attached to a shark. Every time the shark comes to the surface, the tag transmits and the Argos system estimates the shark's location. As the Argos system uses polar-orbiting satellites, no matter where in the world's oceans the shark is, if it is at the surface when a satellite is overhead then its location can be determined. This system is ideal for wide-ranging pelagic species of sharks that regularly come to the surface.

Satellite telemetry is becoming more commonly used for sharks and their relatives. Eckert and Stewart (2001) utilized this approach to study whale sharks, *Rhincodon typus*, in the Sea of Cortez. They attached satellite transmitters to sharks using a towed float on a long tether, so the shark did not have to fully surface for the tag to be able to transmit. Using this technique, they tagged 15 animals and tracked them from 1 to 1144 days and over distances of thousands of kilometers, including one animal that moved to the western north Pacific. Use of this technique for species that surface regularly has become quite common. Bruce et al. (2006) reported on the movements of white sharks off the Australian coast, providing documentation of the extensive movements of this species and how they remain in certain habitats for relatively long periods (Figure 19.8). The addition of Fastloc™ technology to satellite tags has improved the accuracy of locations by removing the need for positions to be estimated by Argos satellites using Doppler shift. Recoverable data-logging global positioning system

(GPS) units that have been deployed on rays (Riding et al., 2009) demonstrate the high quality of data that can be achieved if tags can be recovered (Figure 19.9).

19.3.6 Archival Tags

Archival tags—tags that store data on light level (for estimation of geographic position), depth, and temperature—overcome the problems of collecting long-term data on animals that rarely, if ever, come to the surface. These tags were originally developed for use on tuna and other pelagic teleosts but have become relatively popular for use on sharks. The use of light levels to estimate location (a process known as *light-based geolocation*) relies on the ability to accurately estimate sunrise and sunset times, relative to Greenwich Mean Time (longitude) and day length (latitude). The accuracy of location estimates using light-based geolocation is low (Musyl et al., 2001; Welch and Eveson, 1999, 2001), so this technique is only useful in habitat use studies with broad spatial scales (e.g., a species that migrates long distances) or in situations where location is of secondary importance (e.g., pelagic species where habitat use can be best defined using depth and temperature). Recent analytical developments that use other data (e.g., water temperature, maximum water depth) and a Kalman filter have improved the accuracy of geolocation estimates (Wilson et al., 2007) (Figure 19.10). These improvements have

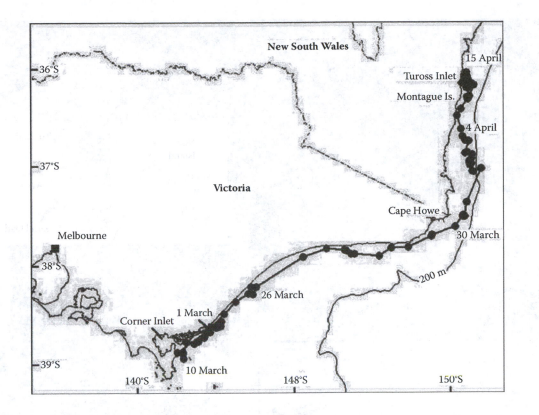

FIGURE 19.8
Track taken by a 1.8-m female white shark tagged off southeast Australia. Circles indicate satellite fixes, and lines represent the estimated track. (From Bruce, B.D. et al., *Mar. Biol.*, 150, 161–172, 2006. With permission.)

enhanced the ability of researchers to accurately track pelagic species, but their application to coastal species provides only limited improvement (Carlson et al., 2010).

West and Stevens (2001) used archival tags to study school shark, *Galeorhinus galeus*, movements in southern Australia. As a heavily fished species, the stock assessment process required information on habitat use—specifically, how often and for how long did individuals enter pelagic habitats as opposed to neritic habitats where they were fished? In the study, 30 individuals were released, and at the time of publication 9 had been recaptured. The depth data stored by the tags showed that school sharks used pelagic habitats for variable periods that lasted as long as several months (Figure 19.11). This behavior of switching between pelagic and neritic habitats is unusual in sharks. This study provided a good understanding of this phenomenon so the stock assessment process could take into account periods when school sharks were not susceptible to particular fishing gears.

A large drawback of traditional archival tags is the need to recapture tagged animals to retrieve the data. This restricts work to heavily exploited species that have high rates of recapture. Manufacturers have developed archival tags that can be programmed to detach from an animal at a specific time, float to the surface, and transmit data via the Argos system (pop-up tags). These tags eliminate the need to recapture animals, opening the way for work on species that are not heavily fished. These tags have been available for about 10 years and have become a widely used tool in the study of movement and habitat use of a range of large shark species. They are commonly used with pelagic species and have provided new insights into the movement of these species. Weng et al. (2008) used pop-up tags to described the migration of salmon sharks from boreal to mid-temperate latitudes through a number of different open ocean ecosystems. Similarly, Jorgensen et al. (2010) used them to described the movement and habitat use of white sharks, including examining levels of philopatry to mid-ocean areas. Although these tags are more suited to pelagic environments, they have also been applied to coastal species. Carlson et al. (2010) applied them on bull sharks and found that positional accuracy was low but that the depth and temperature data provided useful information on habitat use patterns of this widely distributed species.

The disadvantages of pop-up tags are that they are relatively large (restricting them to use on larger species), that a relatively small amount of information can be downloaded via satellite from a relatively small platform, and that they are high cost. Some of these disadvantages have been overcome as the technology has developed, and reductions in size mean that they can be used for smaller species (Sulikowski et al., 2010).

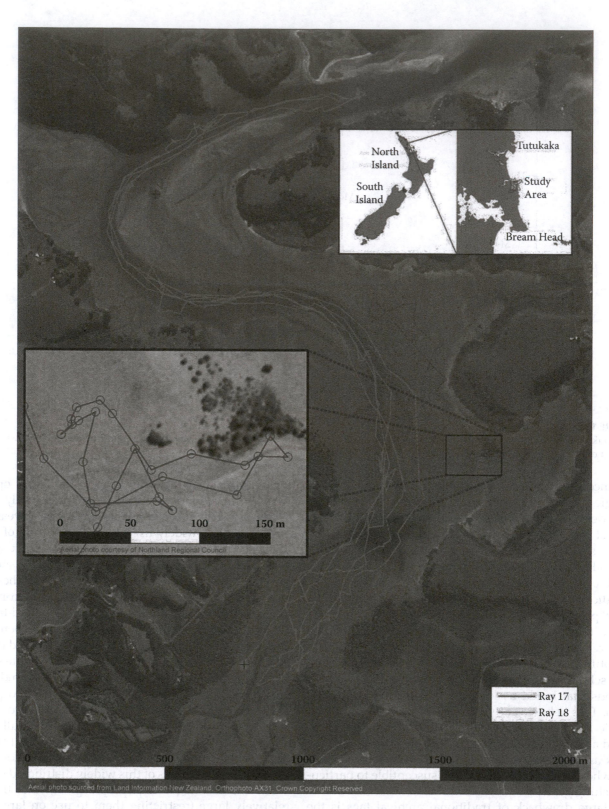

FIGURE 19.9

(See color insert.) Movement trajectories of two individual *Myliobatis tenuicaudatus* in Taiharuru estuary, New Zealand. The entrance to the estuary from the Pacific Ocean is at the top right-hand side of the picture. Colored circles represent the locations where global positioning devices on buoys were attached to rays with monofilament tethers (4 to 7 m in length). Dashed lines connect points where fixes were >10 minutes apart (i.e., indicate areas where instantaneous fix rates were very low). Left inset: close-up of Ray 17 swimming at the edge of a stand of gray mangroves, *Avicennia marina* var. *resinifera*; open circles are locations of fixes made at a sampling interval of 150 seconds. (From Riding, T.A.C. et al., *Mar. Ecol. Prog. Ser.*, 377, 255–262, 2009. With permission.)

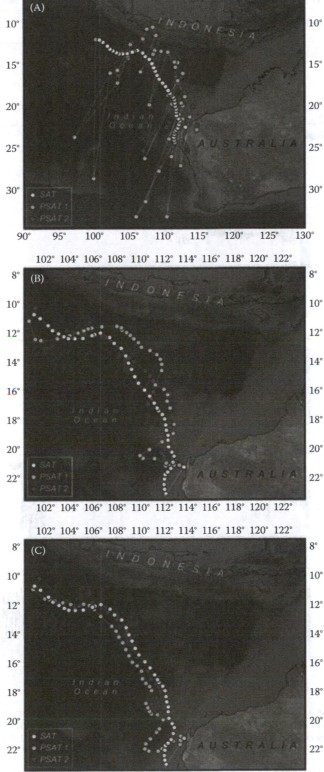

FIGURE 19.10

(See color insert.) SAT tag track vs. location estimates from PSATs 1 and 2 derived from three levels of data processing: (A) raw light level, (B) Kalman-filtered light level, and (C) Kalman-filtered light level with SST integration. (From Wilson, S.G. et al., *Fish. Oceanogr.*, 16, 547–554, 2007. With permission.)

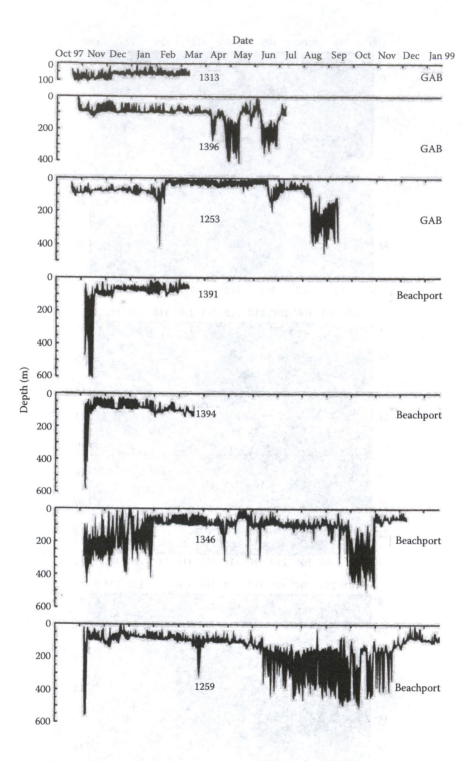

FIGURE 19.11

Depth profiles of seven school sharks, *Galeorhinus galeus*, from archival tags. Sharks were released in the Great Australian Bight (GAB) or off Beechport, South Australia. Sections of the depth profiles that show large variation between 150 and 500 m indicate periods of pelagic habitat use; the remaining sections indicate neritic habitat use. (From West, G.J. and Stevens, J.D., *Environ. Biol. Fish.*, 60, 283–298, 2001. With permission.)

19.3.7 Animal-Borne Video Systems

An additional research tool available for studying habitat use in sharks is animal-borne video and environmental data systems (AVEDs). Although most published reports on the use of video to study sharks have used National Geographic's Crittercams (Heithaus et al., 2001), advances in video and battery technology have paved the way for other video (or still-frame)-based tags. AVEDs allow the habitats in which a shark has

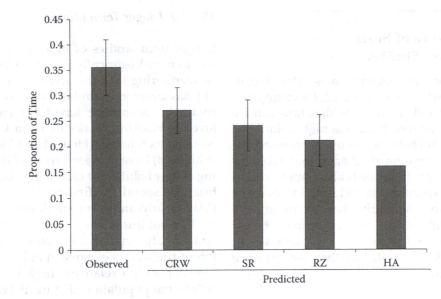

FIGURE 19.12
Habitat selection between shallow seagrass and deep sandy habitats in Shark Bay, Australia, by tiger sharks, *Galeocerdo cuvier*. Tiger sharks used shallow habitats significantly more than suggested by correlated random walk (CRW) models, sample randomization (SR), randomization (RZ), and habitat availability (HA). Error bars indicate 95% confidence intervals. (From Heithaus, M.R. et al., *Mar. Biol.*, 140, 237–248, 2002. With permission.)

swum to be directly observed when the unit has been retrieved and the videotape or digital files are viewed. In addition, telemetry data (depth, water temperature, and swimming speed) can be collected, providing a broad range of information on the shark's behavior. In studies using other telemetry methods, there is a level of error in the assignment of habitat type due to uncertainty in position information. Such errors are reduced by AVEDs because the habitats can be more accurately identified from the video. Heithaus et al. (2002) used this approach (combined with acoustic tracking) to investigate the habitat use of tiger sharks, *Galeocerdo cuvier*, in a seagrass ecosystem in Shark Bay, Western Australia. They fitted Crittercams to 37 individuals and recorded a total of 75 hours of video. They compared the habitat use data generated to track randomization, correlated random walk models, and habitat availability to determine habitat preference. They found tiger sharks preferred shallow seagrass habitats (Figure 19.12) where their prey was most abundant. This technology has also been applied to define post-release behavior of gray reef sharks to define how they respond to capture and release by fishermen (Skomal et al., 2007). Directed application of this approach can provide novel results for a variety of research questions.

Early animal-borne video systems were large pieces of equipment that could only be used with relatively large sharks (1.5 m total length and larger); however, continued development has greatly reduced the unit size. Other disadvantages of this approach are that they can only be used in areas with relatively high water clarity, are expensive to produce, and require a high degree of technical skill to use. As visual systems, they are also best used during the day, although some results can be obtained at night (M.R. Heithaus, pers. comm.). In the past several years, advances in video compression technology have greatly increased the amount of video that can be stored—from 3 to 6 hours to as long as 90 hours—with relatively little increase in system size.

19.3.8 Vertebral Microchemistry

Determining the microchemical composition of the vertebrae and how it changes over time can be used to study the habitat use of marine animals. Although this technique is in its infancy for sharks and their relatives, it has become a popular tool in teleost research. The most common method used to measure the microchemical composition of the vertebrae is inductively coupled plasma mass spectroscopy (ICPMS), although electron microprobes have also been used. One common use of this technique is to examine how the ratio of strontium and calcium changes over time, as this is an indicator of the salinity of the environment in which an individual lives. Peverell (2008) used this technique to investigate the change in habitat use of freshwater sawfish, demonstrating that this species uses freshwater areas less as they grow. Research currently under way will help demonstrate the applicability of this technique for sharks and their relatives.

19.4 The Importance of Scale in Habitat Use Studies

How habitat use data are interpreted is often dependent on the scale at which it is collected; for example, if a study collects data only during the day, how can we understand how habitat use changes at night? Similarly, we can only understand how habitat use changes as animals grow by collecting data at all ages, preferably for the same individuals. Spatial scale is also important. If a study is limited to a specific area and individuals move in and out of the area regularly, then an incomplete understanding of habitat use will be gained. Below we consider examples of temporal and spatial scales in shark habitat use to demonstrate how they can influence interpretation of results.

19.4.1 Temporal Factors

Two levels of temporal scale are considered. The first is diel effects. Many studies of shark movements from which habitat use patterns can be inferred are based on short-term acoustic tracking. In some situations, these tracks do not even last 24 hours, so diel changes in habitat use may not be fully resolved. Second, we consider longer term changes in habitat use that occur as sharks grow. Little is known of how habitat use changes for individual sharks over longer time frames; however, we know that habitat use must change over time (Grubbs, 2010), and it is important to understand how and why these changes occur.

19.4.1.1 Diel Effects

Diel changes in behavior and habitat use have been commonly observed in sharks; thus, it is important for researchers to collect data both during the day and at night to provide a full understanding of habitat use. Holland et al. (1993) demonstrated this for juvenile scalloped hammerheads in Hawaii using acoustic tracking. They found that activity space was larger at night than during the day, and that the center of activity shifted between day and night (Figure 19.13). The scalloped hammerheads used a small core daytime area but ranged more widely at night as they hunted around patch reefs. If nighttime tracking was not carried out, then the importance of dispersed feeding in patch reef areas may never have been determined. There are many other examples of diel changes in shark behavior and habitat use (e.g., Carey and Scharold, 1990; Cartamil et al., 2003, 2010a; Gruber et al., 1988; Klimley et al., 1988; Nelson and Johnson, 1970) showing that this is an important factor in this type of study.

19.4.1.2 Longer Term Effects

Longer term studies of movement and habitat use in sharks have historically been rare, but the use of acoustic monitoring systems, archival tags, or satellite tracking has overcome limitations associated with studies routinely examining long-term patterns. Studies of juvenile blacktip sharks in Terra Ceia Bay, FL, using acoustic monitoring (Heupel and Hueter, 2001; Heupel et al., 2004) have provided some of the best understanding of how habitat use can change over time. The young blacktips spend the first couple of months after birth (May to July) in a very small area at the north end of the bay, but this is followed by periods when they alter and rapidly expand their home range (Figure 19.14). Interestingly, this expansion of home range was found to occur with a relatively high level of synchronicity within the population (Figure 19.14). This pattern was repeated across years when naïve individuals were monitored each year. The persistence of this behavior suggests that consistent, expansive movement patterns may be innate rather than a response to environmental cues or based on learning. Evidence for presumably innate behavior can also be observed in the response of blacktip sharks to the presence of a tropical storm system, an experience these young individuals would not have had previously and cannot have learned from. Juvenile blacktips left the safety of a shallow nursery habitat as a storm system approached (Heupel et al., 2003). This response could not have been determined without the benefit of continuous long-term monitoring. Studies of other coastal shark species such as bull sharks (Heupel et al., 2010) and pigeye sharks (Knip et al., 2011a) are also revealing long-term patterns of habitat use and movement that are helping define how these species use space over extended periods (i.e., years).

Ontogenetic changes in habitat use in elasmobranchs have been reviewed in more detail by Grubbs (2010). With continuing work on sharks occupying nursery areas this will continue to be one aspect of habitat use that is relatively well understood. The changes that occur as individuals grow do so because of a range of factors, including changes in their food requirements, their improving ability to avoid predators (due to their size but also learned behaviors), and competition from conspecifics and others. These types of changes have been documented for species such as bull sharks (Simpfendorfer et al., 2005), lemon sharks (Sundström, 2001), sandbar sharks (Grubbs, 2010), and smalltooth sawfish (Simpfendorfer et al., 2010). Understanding these patterns of habitat use change will help develop effective conservation management strategies for not only species but also the habitats on which they rely (Simpfendorfer et al., 2010).

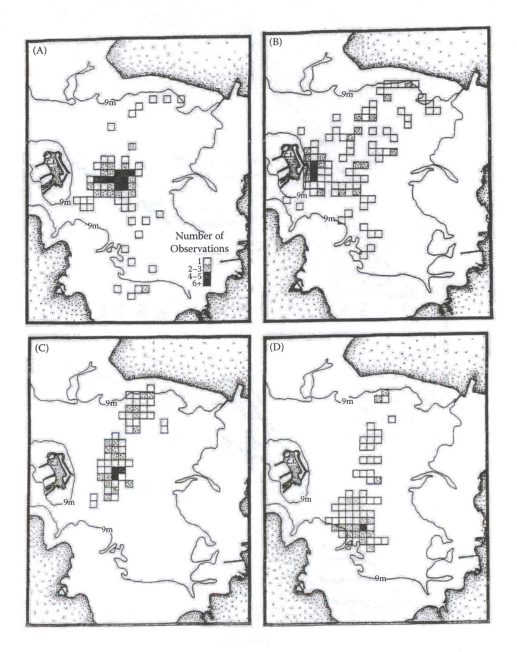

FIGURE 19.13
Diurnal habitat use of two juvenile scalloped hammerheads, *Sphyrna lewini*, in Kanahoe Bay, HI. Daytime habitat use was smaller and more concentrated (A, C), whereas nighttime habitat use was larger and more dispersed (B, D). (From Holland, K.N. et al., *Copeia*, 1993, 495–502, 1993. With permission.)

19.4.2 Spatial Factors

The spatial effects of animal movement patterns are critical to accurately defining habitat use. This is particularly true when examining large, highly migratory pelagic species. Early studies of habitat use by pelagic sharks involved acoustic tracking of individuals to define their daily movements. In these studies, horizontal and vertical movements of individuals were examined to define habitat use within the open ocean (Carey and Scharold, 1990; Holts and Bedford, 1993). These studies characterized the short-term movements of two common pelagic species: mako shark, *Isurus oxyrinchus* (Holts and Bedford, 1993), and blue shark, *Prionace glauca* (Carey and Scharold, 1990). Information was obtained regarding the use of the water column for thermal regulation and prey capture. More advanced spatial examination of habitat use was possible when archival and satellite tags were developed. These technologies have provided information on broad-scale habitat use by wide-ranging species such as white sharks, *Carcharodon carcharias* (e.g., Bonfil et al., 2005; Boustany et al., 2002;

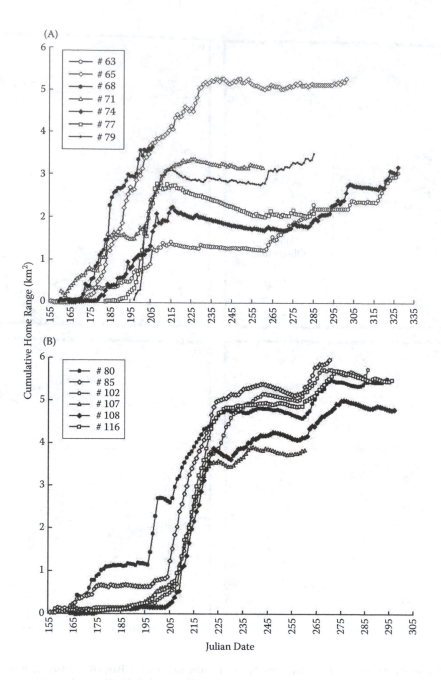

FIGURE 19.14
Increase in cumulative daily home range size for juvenile blacktip sharks, *Carcharhinus limbatus*, in Terra Ceia Bay, FL, from birth through the first summer of life; 2 years of data are presented: (A) 2000 and (B) 2001. Time is represented by days since January 1 of each year (Julian date). (From Heupel, M.R. et al., *Environ. Biol. Fish.*, 71, 135–142, 2004. With permission.)

Bruce et al., 2006; Jorgensen et al., 2010); whale sharks, *Rhincodon typus* (e.g., Eckert and Stewart, 2001; Wilson et al., 2007); school sharks (West and Stevens, 2001); and tiger sharks (Heithaus et al., 2007b; Myers et al., 2010). These studies showed that these species can at times undertake transoceanic scale movements and that to fully understand habitat use a technique that enables information to be gathered on a broad spatial scale is required.

19.5 Factors Influencing Habitat Selection by Sharks

Understanding why sharks display the habitat use patterns that they do is a much more difficult task than simply describing them. A number of factors must be taken into account and may work at different levels for different species or in different locations for the same species. Johnson (1980) recognized that habitat selection is a hierarchical process, with different factors acting at different scales, including geographic range, home range, and use of habitats within the home range. Both physical and biotic factors may shape habitat use at all spatial scales. Physical factors include temperature, salinity, depth, dissolved oxygen level, and bottom sediment characteristics. Biotic factors include benthic vegetation (e.g., seagrasses or mangroves), prey distribution and availability, predator distribution, social organization, and reproductive activity.

19.5.1 Physical Factors

Habitat use is often bound by the physical parameters of the environment and the tolerance levels of the study species. Temperature is an important physical factor affecting shark habitat use on a broad scale. Few species can survive in the full range of temperatures that occur in the world's oceans, so there are physical limits to the habitats that are available to a species. Within a species it is also common to see seasonal changes in distribution due to migrations. Although many factors may help drive these migrations, the inability of a species to tolerate seasonal changes in temperature is an important factor in many cases; for example, Heupel (2007) and Grubbs (2007) observed that the migration of coastal sharks is dependent on water temperature and day length. These results indicate the importance of these cues in the daily and seasonal movement patterns of shark populations and has potential implications for populations affected by climate change.

Physical factors also act at finer spatial scales. Salinity, for example, may be an important factor for coastal species that enter estuaries. Some species, such as the bull shark, *Carcharhinus leucas*, show a preference for lower salinity areas, particularly as juveniles (Compagno, 1984). Thus, they may actively seek these areas for their nurseries. Heupel and Simpfendorfer (2008) described the use by juvenile bull sharks of the Calooshatchee River, where individuals chose to remain in salinities ranging from 7 to 17 psu. Bull sharks were observed to move either up- or downriver to remain within this salinity regime. Knip et al. (2011b) also observed a response to freshwater flow by juvenile pigeye sharks, *Carcharhinus amboinensis*, in a nearshore environment. Here, young sharks were observed to move offshore during peak periods of the annual wet season. At the end of the wet season, individuals moved back to the nearshore habitats they appear to prefer. A species can also respond to different factors in different areas; for example, Heithaus et al. (2009) demonstrated that in the Shark River dissolved oxygen was the dominant

physical factor affecting bull shark movements, a contrast to the results found in the Calooshatchee River described above.

Tidal flow is also an important factor for species that use shallow nearshore habitats. Habitat use in leopard sharks, *Triakis semifasciata*, in coastal bays in California has been shown to be directly influenced by tidal flow in two separate studies (Ackerman et al., 2000; Carlisle and Starr, 2010). Tidal flow can act in several different ways to affect habitat use. Decreasing depth can force animals to move to other habitats as shallow areas are exposed at low tide. Alternatively, changing physical parameters within the water column can provide constraints on the type and amount of habitat available for use by a species. Temperature is also an important physical factor at fine spatial scales. In a study of bat rays, *Myliobatis californica*, Matern et al. (2000) hypothesized that in Tomales Bay this species selected areas that enabled it to behaviorally thermoregulate. Hight and Lowe (2007) also described behavioral thermoregulation in female leopard sharks and suggested that this behavior played a role in the digestion of food and gestation of young. These examples indicate the importance of physical factors in the movement ecology of sharks.

19.5.2 Biotic Factors

Intuitively, biotic factors should play an important role in habitat selection by sharks. The needs to feed, avoid predators, and reproduce are important features of a shark's life (see Chapter 17). Despite this, the importance of these needs in habitat use studies has rarely been considered for sharks. Detailed behavioral studies by Heithaus et al. (2002, 2006, 2007b) on tiger sharks in Shark Bay, Western Australia, are probably the best example of research involving prey distribution. In this study, information on shark movements and habitat use was obtained by acoustic tracking, animal-borne video systems, and prey distribution surveys. Tiger sharks showed a preference for shallow seagrass areas where their main prey (fish, turtles, sea snakes, and birds) were more commonly found. In a study of juvenile blacktip sharks in Terra Ceia Bay, FL, Heupel and Hueter (2002) found that there was no correlation between prey availability (small fish density) and blacktip occurrence. On the basis of these data, they concluded that blacktip shark habitat selection was not based on food availability and suggested that the risk of predation by larger sharks was more likely to be driving habitat selection. Further research is needed to explore the importance of predation risk in relation to habitat use, especially in smaller sharks (see Chapter 17).

19.6 The Future of Habitat Use Studies on Sharks, Skates, and Rays

The importance of habitat use studies has been increasing over the past decade. The recognition of the need to define and understand essential fish habitats for commercially fished species and critical habitats for endangered species has provided the impetus for much of this work. This increase in research, combined with the development of technologies (e.g., satellite tags, acoustic monitoring systems, archival tags, animal-borne video systems) that are making it possible for researchers to answer relevant habitat use and preference questions, has made this type of work much more accessible. As such, this type of research should continue to grow in importance. To date, however, habitat use studies have been largely qualitative. Only a handful of studies have provided quantitative evidence of habitat selection in sharks (e.g., Carraro and Gladstone, 2006; Heithaus et al., 2002; Morrissey and Gruber, 1993b; Simpfendorfer et al., 2010). Into the future, there is a need for researchers to improve the design and implementation of studies that specifically address questions of habitat use and preference and ensure that despite the broad range of tools available they choose the most appropriate one.

References

Ackerman, J.T., Kondratieff, M.C., Matern, S.A. et al. (2000). Tidal influence on spatial dynamics of leopard sharks, *Triakis semifasciata*, in Tomales Bay, California. *Environ. Biol. Fish.* 58:33–43.

Anderson, D.J. (1982). The home range: a new nonparametric estimation technique. *Ecology* 63:103–112.

Bass, A.J. (1978). Problems in studies of sharks in the southwest Indian Ocean. In Hodgson, E.S. and Mathewson, R.F. (Eds.), *Sensory Biology of Sharks, Skates and Rays*. Office of Naval Research, Department of the Navy, Arlington, VA, pp. 545–594.

Bonfil, R., Meyer, M., Scholl, M.C. et al. (2005). Transoceanic migration, spatial dynamics, and population linkages of white sharks. *Science* 310:100–103.

Boustany, A.M., Davis, S.F., Pyle P. et al. (2002). Expanded niche for white sharks. *Nature* 415:35–36.

Branstetter, S. (1990). Early life history implications of selected carcharhinoid and lamnoid sharks of the northwest Atlantic. In: Pratt, Jr., H.L., Gruber, S.H., and Taniuchi, T. (Eds.), *Elasmobranchs as Living Resources: Advances in the Biology, Ecology, Systematics, and Status of the Fisheries*, NOAA Tech. Rep. NMFS 90. U.S. Department of Commerce, Washington, D.C., pp. 17–28.

Bruce, B.D., Stevens, J.D., and Malcolm, H. (2006). Movements and swimming behaviour of white sharks (*Carcharodon carcharias*) in Australian waters. *Mar. Biol.* 150:161–172.

Burt, W.H. (1943). Territoriality and home range concepts as applied to mammals. *J. Mammal.* 24:346–352.

Carey, F.G. and Scharold, J.V. (1990). Movements of blue sharks (*Prionace glauca*) in depth and course. *Mar. Biol.* 106:329–342.

Carlisle, A.B. and Starr, R.M. (2010). Tidal movements of female leopard sharks (*Triakis semifasciata*) in Elkhorn Slough, California. *Environ. Biol. Fish.* 89:31–45.

Carlson, J.K., Ribera, M.M., Conrath, C.L. et al. (2010). Habitat use and movement patterns of bull sharks *Carcharhinus leucas* determined using pop-up satellite archival tags. *J. Fish Biol.* 77:661–675.

Carraro, R. and Gladstone, W. (2006). Habitat preferences and site fidelity of the ornate wobbegong shark (*Orectolobus ornatus*) on rocky reefs of New South Wales. *Pac. Sci.* 60:207–223.

Cartamil, D.P., Vaudo, J.J., and Lowe, C.G. (2003). Diel movement patterns of the Hawaiian stingray, *Dasyatis lata*: implications for ecological interactions between sympatric elasmobranch species. *Mar. Biol.* 142:841–847.

Cartamil, D.P., Wegner, N.C., and Aalbers, S. (2010a). Diel movement patterns and habitat preferences of the common thresher shark (*Alopias vulpinus*) in the Southern California Bight. *Mar. Freshw. Res.* 61:596–604.

Cartamil, D., Wegner, N.C., Kacev, D. et al. (2010b). Movement patterns and nursery habitat of juvenile thresher sharks *Alopias vulpinus* in the Southern California Bight. *Mar. Ecol. Prog. Ser.* 404:249–258.

Castro, J.I. (1993). The shark nursery of Bulls Bay, South Carolina, with a review of the shark nurseries of the southeastern coast of the United States. *Environ. Biol. Fish.* 38:37–48.

Clarke, T.A. (1971). The ecology of the scalloped hammerhead shark, *Sphyrna lewini*, in Hawaii. *Pac. Sci.* 25:133–144.

Compagno, L.J.V. (1984). *FAO Species Catalogue.* Vol. 4. *Sharks of the World: An Annotated and Illustrated Catalogue of Shark Species Known to Date.* Part 2. Carcharhiniformes. United Nations Food and Agriculture Organization, Rome, pp. 251–655.

Conrath, C.L. and Musick, J.A. (2010). Residency, space use and movement patterns of juvenile sandbar sharks (*Carcharhinus plumbeus*) within a Virginia summer nursery area. *Mar. Freshw. Res.* 61:223–235.

Cooper, Jr., W.E. (1978). Home range criteria based on temporal stability of area occupation. *J. Theor. Biol.* 73:687–695.

Dawson, C.L. and Starr, R.M. (2009). Movements of subadult prickly sharks *Echinorhinus cookei* in the Monterey Canyon. *Mar. Ecol. Prog. Ser.* 386:253–262.

Duncan, K.M. and Holland, K.N. (2006). Habitat use, growth rates and dispersal patterns of juvenile scalloped hammerhead sharks *Sphyrna lewini* in a nursery habitat. *Mar. Ecol. Prog. Ser.* 312:211–221.

Eckert, S.A. and Stewart, B.S. (2001). Telemetry and satellite tracking of whale sharks, *Rhincodon typus*, in the Sea of Cortez, Mexico and the north Pacific Ocean. *Environ. Biol. Fish.* 60:299–308.

Economakis, A.E. and Lobel, P.S. (1998). Aggregation behavior of the grey reef shark, *Carcharhinus amblyrhynchos*, at Johnson Atoll, Central Pacific Ocean. *Environ. Biol. Fish.* 51:129–139.

Farrugia, T.J., Espinoza, M., and Lowe, C.G. (2011). Abundance, habitat use and movement patterns of the shovelnose guitarfish (*Rhinobatos productus*) in a restored southern California estuary. *Mar. Freshw. Res.* 62(6):648–657.

Grubbs, R.D. (2010). Ontogenetic shifts in movements and habitat use. In: Carrier, J., Musick, J.A., and Heithaus, M. (Eds.), *Sharks and Their Relatives II: Biodiversity, Adaptive Physiology, and Conservation*. CRC Press, Boca Raton, FL, pp. 319–350.

Grubbs, R.D., Musick, J.A., Conrath, C.L., and Romine, J.G. (2007). Long-term movements, migration, and temporal delineation of a summer nursery for juvenile sandbar sharks in the Chesapeake Bay region. In: McCandless, C.T., Kohler, N.E., and Pratt, Jr., H.L. (Eds.), *Shark Nursery Grounds of the Gulf of Mexico and the East Coast Waters of the United States*. American Fisheries Society, Bethesda, MD, pp. 87–108.

Gruber, S.H., Nelson, D.R., and Morrissey, J.F. (1988). Patterns of activity and space utilization of lemon sharks, *Negaprion brevirostris*, in a shallow Bahamian lagoon. *Bull. Mar. Sci.* 43:61–76.

Heithaus, M.R., Marshall, G.J., Buheler, B.M. et al. (2001). Employing Crittercam to study habitat use and behavior of large sharks. *Mar. Ecol. Prog. Ser.* 209:307–310.

Heithaus, M.R., Dill, L.M., Marshall, G.J. et al. (2002). Habitat use and foraging behavior of tiger sharks (*Galeocerdo cuvier*) in a seagrass ecosystem. *Mar. Biol.* 140:237–248.

Heithaus, M.R., Hamilton, I.M., Wirsing, A.J. et al. (2006). Validation of a randomization procedure to assess animal habitat preferences: microhabitat use of tiger sharks in a seagrass ecosystem. *J. Anim. Ecol.* 75:666–676.

Heithaus, M.R., Burkholder, D., Hueter, R.E. et al. (2007a). Spatial and temporal variation in shark communities of the lower Florida Keys and evidence for historical population declines. *Can. J. Fish. Aquat. Sci.* 64:1302–1313.

Heithaus, M.R., Wirsing, A.J., Dill, L.M. et al. (2007b). Long-term movements of tiger sharks satellite-tagged in Shark Bay, Western Australia. *Mar. Biol.* 151:1455–1461.

Heithaus, M.R., Delius, B.K., Wirsing, A.J. et al. (2009). Physical factors influencing the distribution of a top predator in a subtropical oligotrophic estuary. *Limnol. Oceanogr.* 54:472–482.

Heupel, M.R. (2007). Exiting Terra Ceia Bay: an examination of cues stimulating migration from a summer nursery area. In: McCandless, C.T., Kohler, N.E., and Pratt, Jr., H.L. (Eds.), *Shark Nursery Grounds of the Gulf of Mexico and the East Coast Waters of the United States*. American Fisheries Society, Bethesda, MD, pp. 265–280.

Heupel, M.R. and Hueter, R.E. (2001). Use of an automated acoustic telemetry system to passively track juvenile blacktip shark movements. In: Sibert, J.R. and Nielsen, J.L. (Eds.), *Electronic Tagging and Tracking in Marine Fisheries*. Kluwer, Dordrecht, pp. 217–236.

Heupel, M.R. and Hueter, R.E. (2002). The importance of prey density in relation to the movement patterns of juvenile sharks within a coastal nursery area. *Mar. Freshw. Res.* 53:543–550.

Heupel, M.R. and Simpfendorfer, C.A. (2008). Movement and distribution of young bull sharks *Carcharhinus leucas* in a variable estuarine environment. *Aquat. Biol.* 1:277–289.

Heupel, M.R., Simpfendorfer, C.A., and Hueter, R.E. (2003). Running before the storm: blacktip sharks respond to falling barometric pressure associated with Tropical Storm Gabrielle. *J. Fish Biol.* 63:1357–1363.

Heupel, M.R., Simpfendorfer, C.A., Hueter, and R.E. (2004). Estimation of shark home ranges using passive monitoring techniques. *Environ. Biol. Fish.* 71:135–142.

Heupel, M.R., Simpfendorfer, C.A., Collins, A.B. et al. (2006). Residency and movement patterns of bonnethead sharks, *Sphyrna tiburo*, in a large Florida estuary. *Environ. Biol. Fish.* 76:47–67.

Heupel, M.R., Semmens, J.M., and Hobday, A.J. (2006b). Automated acoustic tracking of aquatic animals: scales, design and deployment of listening station arrays. *Mar. Freshw. Res.* 57:1–13.

Heupel, M.R., Carlson, J.K., and Simpfendorfer, C.A. (2007). Shark nursery areas: concepts, definitions, characterization and assumptions. *Mar. Ecol. Prog. Ser.* 337:287–297.

Heupel, M.R., Yeiser, B.G., Collins, A.B., Ortega, L., and Simpfendorfer, C.A. (2010). Long-term presence and movement patterns of juvenile bull sharks, *Carcharhinus leucas*, in an estuarine river system. *Mar. Freshw. Res.* 61:1–10.

Hight, B.V. and Lowe, C.G. (2007). Elevated body temperatures of adult female leopard sharks, *Triakis semifasciata*, while aggregating in shallow nearshore embayments: evidence for behavioral thermoregulation? *J. Exp. Mar. Biol. Ecol.* 352:114–128.

Holland, K.N., Wetherbee, B.M., Peterson, J.D. et al. (1993). Movements and distribution of hammerhead shark pups on their natal grounds. *Copeia* 1993:495–502.

Holts, D.B. and Bedford, D.W. (1993). Horizontal and vertical movements of the shortfin mako shark, *Isurus oxyrinchus*, in the Southern California Bight. *Aust. J. Mar. Freshw. Res.* 44:901–909.

Jennrich, R.I. and Turner, F.B. (1969). Measurement of non-circular home range. *J. Theor. Biol.* 22:227–237.

Johnson, D.H. (1980). The comparison of usage and availability measurements for evaluating resource preference. *Ecology* 61:65–71.

Johnson, R.H. and Nelson, D.R. (1973). Agonistic display in the gray reef shark, *Carcharhinus menisorrah*, and its relationship to attacks on man. *Copeia* 1973:76–84.

Jorgensen, S.J., Reeb, C.A., Chapple, T.K. et al. (2010). Philopatry and migration of Pacific white sharks. *Proc. Roy. Soc. B Biol. Sci.* 277(1682):679–688.

Kei, J.G., Matthiopoulos, J., Fieberg, J. et al. (2010). The home-range concept: are traditional estimators still relevant with modern telemetry technology? *Proc. Roy. Soc. B Biol. Sci.* 365(1550):2221–2231.

Klimley, A.P., Butler, S.B., Nelson, D.R. et al. (1988). Diel movements of the scalloped hammerhead shark, *Sphyrna lewini* Griffith and Smith, to and from a seamount in the Gulf of California. *J. Fish Biol.* 33:751–761.

Klimley, A.P., Le Boeuf, B.J., Cantara, K.M. et al. (2001). Radio-acoustic positioning as a tool for studying site-specific behavior of the white shark and other large marine species. *Mar. Biol.* 138:429–446.

Knip, D.M., Heupel, M.R., and Simpfendorfer, C.A. (2010). Sharks in nearshore environments: theories, perceptions, definitions, and consequences. *Mar. Ecol. Prog. Ser.* 402:1–11.

Knip, D.M., Heupel, M.R., Simpfendorfer, C.A. et al. (2011a). Ontogenetic shifts in movement and habitat use of juvenile pigeye sharks *Carcharhinus amboinensis* in a tropical nearshore region. *Mar. Ecol. Prog. Ser.* 425:233–246.

Knip, D.M., Heupel, M.R., Simpfendorfer, C.A. et al. (2011b). Wet-season effects on the distribution of juvenile pigeye sharks, *Carcharhinus amboinensis*, in nearshore waters. *Mar. Freshw. Res.* 62(6):658–667.

Krebs, C.J. (1999). *Ecological Methodology*, 2nd ed. Addison-Wesley/Longman, Menlo Park, CA.

Matern, S.A., Cech, J.J., and Hopkins, T.E. (2000). Diel movements of bat rays, *Myliobatis californica*, in Tomales Bay, California: evidence for behavioral thermoregulation? *Environ. Biol. Fish.* 58:173–193.

McCandless, C.T., Kohler, N.E., and Pratt, Jr., H.L. (Eds.). (2007). *Shark Nursery Grounds of the Gulf of Mexico and the East Coast Waters of the United States*. American Fisheries Society, Bethesda, MD.

McKibben, J.N. and Nelson, D.R. (1986). Patterns of movement and grouping of gray reef sharks, *Carcharhinus amblyrhynchos*, at Enewetak, Marshall Islands. *Bull. Mar. Sci.* 38:89–110.

Medved, R.J. and Marshall, J.A. (1983). Short-term movements of young sandbar sharks, *Carcharhinus plumbeus* (Pisces, Carcharhinidae). *Bull. Mar. Sci.* 33:87–93.

Michel, M. (2002). Environmental Factors Affecting the Distribution and Abundance of Sharks in the Mangrove Estuary of the Ten Thousand Islands, Florida, U.S.A., doctoral dissertation, University of Basel, Switzerland.

Morrissey, J.F. and Gruber, S.H. (1993a). Home range of juvenile lemon sharks. *Copeia* 1993:425–434.

Morrissey, J.F. and Gruber, S.H. (1993b). Habitat selection by juvenile lemon sharks, *Negaprion brevirostris*. *Environ. Biol. Fish.* 38:311–319.

Musyl, M.K., Brill, R.W., Curran, D.S. et al. (2001). Ability of archival tags to provide estimates of geographical position based on light intensity. In Sibert, J.R. and Nielsen, J.L. (Eds.), *Electronic Tagging and Tracking in Marine Fisheries*, Kluwer, Dordrecht, pp 343–367.

Nelson, D.R. and Johnson, R.H. (1970). Diel activity rhythms in the nocturnal, bottom-dwelling sharks, *Heterodontus francisci* and *Cephaloscyllium ventriosum*. *Copeia* 1970:732–739.

Nelson, D.R., Johnson, J.N., McKibben, J.N. et al. (1986). Agonistic attacks on divers and submersibles by gray reef sharks, *Carcharhinus amblyrhynchos*: antipredatory or competitive? *Bull. Mar. Sci.* 38:68–88.

Olsen, A.M. (1954). The biology, migration and growth rate of the school shark *Galeorhinus australis* (Macleay) (Carcharhinidae) in south-eastern Australian waters. *Aust. J. Mar. Freshw. Res.* 5:353–410.

Papastamatiou, Y.P., Friedlander, A.M., Caselle, J.C. et al. (2010). Long-term movement patterns and trophic ecology of blacktip reef sharks (*Carcharhinus melanopterus*) at Palmyra Atoll. *J. Exp. Mar. Biol. Ecol.* 386:94–102.

Payne, N.L., Gillanders, B.M., Webber, D.M. et al. (2010). Interpreting diel activity patterns from acoustic telemetry: the need for controls. *Mar. Ecol. Prog. Ser.* 410:295–301.

Peverell, S.C. (2008). Sawfish (Pristidae) of the Gulf of Carpentaria, master's thesis, James Cook University, Townsville, 146 pp.

Riding, T.A.C., Dennis, T.E., Stewart, C.L. et al. (2009). Tracking fish using 'buoy-based' GPS telemetry. *Mar. Ecol. Prog. Ser.* 377:255–262.

Sciarotta, T.C. and Nelson, D.R. (1977). Diel behavior of the blue shark, *Prionace glauca*, near Santa Catalina Island, California. *U.S. Fish. Bull.* 75:519–528.

Sepulveda, C.A., Kohin, S., Chan, C. et al. (2004). Movement patterns, depth preferences, and stomach temperatures of free-swimming juvenile mako sharks, *Isurus oxyrinchus*, in the Southern California Bight. *Mar. Biol.* 145:191–199.

Simpfendorfer, C.A. and Milward, N.E. (1993). Utilisation of a tropical bay as a nursery area by sharks of the families Carcharhinidae and Sphyrnidae. *Environ. Biol. Fish.* 37:337–345.

Simpfendorfer, C.A., Heupel, M.R., and Hueter, R.E. (2002). Estimation of short-term centers of activity from an array of omnidirectional hydrophones, and its use in studying animal movements. *Can. J. Fish. Aquat. Sci.* 59:23–32.

Simpfendorfer, C.A., Freitas, G.G., and Wiley, T.R. et al. (2005). Distribution and habitat partitioning of immature bull sharks (*Carcharhinus leucas*) in a southwest Florida estuary. *Estuaries* 28:78–85.

Simpfendorfer, C.A., Wiley, T.R., and Yeiser, B.G. (2010). Improving conservation planning for an endangered sawfish using data from acoustic telemetry. *Biol. Cons.* 143:1460–1469.

Skomal, G., Lobel, P.S., and Marshall, G. (2007). The use of animal-borne imaging to assess post-release behavior as it relates to capture stress in grey reef sharks, *Carcharhinus amblyrhynchos*. *Mar. Tech. Soc. J.* 41:44–48.

Springer, S. (1967). Social organisation of shark populations. In Gilbert, P.W., Mathewson, R.F., and Rall, D.P. (Eds.), *Sharks, Skates, and Rays*. The Johns Hopkins University Press, Baltimore, MD, pp. 149–174.

Strauss, R.E. (1979). Reliability for estimates for Ivlev's electivity index, the forage ratio, and a proposed linear index of food selection. *Trans. Am. Fish. Soc.* 108:344–352.

Sulikowsk, J.A., Galuardi, B., Bubley, W. et al. (2010). Use of satellite tags to reveal the movements of spiny dogfish *Squalus acanthias* in the western North Atlantic Ocean. *Mar. Ecol. Prog. Ser.* 418:249–254.

Sundström, L.F., Gruber, S.H., Clermont, S.M. et al. (2001). Review of elasmobranch behavioral studies using ultrasonic telemetry with special reference to the lemon shark, *Negaprion brevirostris*, around Bimini Islands, Bahamas. *Environ. Biol. Fish.* 60:225–250.

Vaudo, J.J. and Heithaus, M.R. (2009). Spatiotemporal variability in a sandflat elasmobranch fauna in Shark Bay, Australia. *Mar. Biol.* 156:2579–2590.

Vaudo, J.J. and Lowe, C.G. (2006). Movement patterns of the round stingray *Urobatis halleri* (Cooper) near a thermal outfall. *J. Fish Biol.* 68:1756–1766.

Voegeli, F.A., Smale, M.J., Webber, D.M. et al. (2001). Sonic telemetry, tracking and automated monitoring technology. *Environ. Biol. Fish.* 60:267–281.

Welch, D.W. and Eveson, J.P. (1999). An assessment of light-based geoposition estimates from archival tags. *Can. J. Fish. Aquat. Sci.* 56:1317–1327.

Welch, D.W. and Eveson, J.P. (2001). Recent progress in estimating geoposition using daylight. In: Sibert, J.R. and Nielsen, J.L. (Eds.), *Electronic Tagging and Tracking in Marine Fisheries*. Kluwer, Dordrecht, pp. 369–383.

Weng, K.C., Foley, D.G., Ganong, J.E. et al. (2008). Migration of an upper trophic level predator, the salmon shark *Lamna ditropis*, between distant ecoregions. *Mar. Ecol. Prog. Ser.* 372:253–264.

West, G.J. and Stevens, J.D. (2001). Archival tagging of school sharks, *Galeorhinus galeus*, in Australia: initial results. *Environ. Biol. Fish.* 60:283–298.

Wiley, T.R. and Simpfendorfer, C.A. (2007). The ecology of elasmobranchs occurring in the Everglades National Park, Florida: implications for conservation and management. *Bull. Mar. Sci.* 80:171–189.

Williams, H. and Schaap, A.H. (1992). Preliminary results of a study into the incidental mortality of sharks in gillnets in two Tasmanian shark nursery areas. *Aust. J. Mar. Freshw. Res.* 43:237–250.

Wilson, S.G., Stewart, B.S., Polovina, J.J. et al. (2007). Accuracy and precision of archival tag data: a multiple-tagging study conducted on a whale shark (*Rhincodon typus*) in the Indian Ocean. *Fish. Oceanogr.* 16:547–554.

Worton, B.J. (1987). A review of models of home range for animal movement. *Ecol. Model.* 38:277–298.

Yeiser, B.G., Heupel, M.R., and Simpfendorfer, C.A. (2008). Residence and movement patterns of bull (*Carcharhinus leucas*) and lemon (*Negaprion brevirostris*) sharks within a coastal estuary. *Mar. Freshw. Res.* 9:489–501.

Index